D0908467

PHYSICAL GEOGRAPHY

SCIENCE AND SYSTEMS OF THE HUMAN ENVIRONMENT

FIFTH CANADIAN EDITION

PHYSICAL GEOGRAPHY

SCIENCE AND SYSTEMS OF THE HUMAN ENVIRONMENT

FIFTH CANADIAN EDITION

ALAN STRAHLER
BOSTON UNIVERSITY

O.W. ARCHIBOLD
UNIVERSITY OF SASKATCHEWAN

John Wiley & Sons Canada, Ltd.

Vice President and Publisher: Veronica Visentin
Vice President, Publishing Services: Karen Bryan
Acquisitions Editor: Rodney Burke
Marketing Manager: Patty Maher
Editorial Manager: Karen Staudinger
Development Editors: Leanne Rancourt, Gail Brown
Media Editor: Channade Fenandoe
Editorial Assistant: Laura Hwee
Creative Director, Publishing Services: Ian Koo
Production Manager: Tegan Wallace
Designer: Pat Loi
Photo Researchers: Julie Pratt, Christina Beamish
Cover Image: Kenneth Ginn/Getty Images
Printing & Binding: Quad/Graphics Ltd.

This book was typeset in 10/12 Times New Roman by Prepare, Inc. and printed and bound by Quad/Graphics. The cover was printed by Quad/Graphics.

This book is printed on paper made from an environmentally responsible mill that recognizes and uses fibre certified under all credible forest certification standards, including the Forest Stewardship Council (FSC), Sustainable Forestry Initiative® (SFI®), Canadian Standards Association (CSA), Programme for the Endorsement of Forest Certification (PEFC), American Tree Farm System (ATFS), and Master Logger (ML) standards.

Library and Archives Canada Cataloguing in Publication

Strahler, Alan H.
Physical geography: science and systems of the human environment/Alan Strahler, O.W. Archibold.—5th Canadian ed.

ISBN 978-0-470-67885-5

1. Physical geography—Textbooks. I. Archibold, O. W. (Oliver William) II. Title.

GB54.5.S77 2011 910'.02 C2010-905809-7

Printed and bound in the United States of America.

3 4 5 QG 15

John Wiley & Sons Canada, Ltd.
6045 Freemont Blvd.
Mississauga, Ontario L5R 4J3
Visit our website at: www.wiley.ca

PREFACE

Welcome to *Physical Geography: Science and Systems of the Human Environment*, Fifth Canadian Edition. Although physical geography is a truly global science, Canada presents an environment with many unique geographic characteristics. Our goal is to provide a balance between local content without sacrificing the treatment of global processes. This edition makes extensive use of illustrations and examples from across Canada to provide a greater focus on local content that will be of particular relevance to both students and physical geography instructors in Canadian colleges and universities. Using reviewer feedback as our guide, we have further accentuated the coverage of Canadian geographical features to provide a strong pedagogical framework to stimulate interest in the global processes studied in physical geography.

HALLMARK FEATURES OF THIS TEXT

SYSTEMS-BASED APPROACH

In keeping with the instructional goals of previous editions of *Physical Geography*, the primary focus of this edition is to provide a view of physical geography as a key discipline in understanding the Earth's diverse environments and how they are modified by global change. This has been done by taking a systems-based approach, particularly flow systems of energy and matter as a unifying theme. In this way we provide a paradigm for understanding the underlying scientific principles and common elements of the many processes that constitute the science of physical geography.

END-OF-CHAPTER MATERIALS

Each chapter concludes with a number of aids to facilitate student learning:

- The *Chapter Summary* is worded like a scientific abstract, succinctly covering all the major concepts of the chapter.
- The list of *Key Terms* indicates the most important concepts to study to understand the chapter material.
- The *Review Questions* are designed as oral or written exercises that require description or explanation of important ideas and concepts.
- *Visualizing Exercises* use sketching or graphing as a way of motivating students to visualize key concepts.
- *Essay Questions* require more synthesis or the reorganization of knowledge in a new context.
- Some chapters have *Working It Out* features that include quantitative material, including problems. This material has been separated so that instructors have flexibility in how they want to use it. All *Working It Out* features include practice problems with answers in the back of the book.
- Some chapters have appendices that provide an in-depth, student-friendly examination of a concept discussed in the chapter.

PEDAGOGICAL EFFECTIVENESS

Today, science must be taught in an open, accessible manner to reach students, and written material needs to communicate directly and simply. Moreover, today's students are accustomed to visual learning and to an interactive methodology. The text and illustrations of *Physical Geography* have been devised to be accessible and inviting and to provide clear descriptions of scientific principles. Every map and line drawing is carefully designed and styled to make sure that its message is clear, obvious, and direct. The photos have been selected not only to provide fine illustrations of the relevant concept, but also to demonstrate it effectively and strikingly. In summary, we continue to strive to produce an illustrated text of which we are very proud.

NEW TO THIS EDITION

The fifth Canadian edition of *Physical Geography* includes more Canadian examples and contains new and updated content in virtually every chapter:

- **Chapters 4 and 10:** New appendices include detailed information on the latest Intergovernmental Panel on Climate Change (IPCC) report. Appendix 4.1 examines the changes in global climate that occurred during the twentieth century, while Appendix 10.1 discusses projected future impacts of climate change.
- **Chapter 7:** Includes updated information on tornado and hurricane development in Canada, with mention of Hurricane Igor, which hit Newfoundland in 2010.
- **Chapters 9 and 10:** Include a revised set of climographs and a new series of simplified moisture budget graphs, many of which feature Canadian locations (when appropriate).
- **Chapter 11:** New information on the Alberta tar sands has been added, including a figure showing where the oil lies and how it is extracted.

- **Chapter 13:** Includes coverage of the 2010 Haiti earthquake, with maps, photos, and a discussion of plate tectonics that make Haiti vulnerable to earthquakes; new maps showing fault lines that lead to the frequent Atlantic Ocean and California earthquakes are also included.
- **Chapter 14:** Rock structure, which is now included in this chapter, provides a link between tectonic processes and surficial landform development.
- **Chapter 15:** A new appendix discusses the state of the Great Lakes, including how they are monitored and what classification system is used to determine the condition of the lakes.
- **Chapter 18:** Information on permafrost is now included with the review of glacial landscapes and the Late-Cenozoic Ice Age.
- **Chapter 22:** Includes a new section on ecozones, focusing on Canada's terrestrial ecozones.

STREAMLINED PRESENTATION

The other principal aim of *Physical Geography* is to make the text easier to use for one-semester courses. Increasingly in Canada introductory physical geography courses are taught in one semester. To better reflect this, the content has been written to make the subject matter more manageable in a shorter teaching timeframe. Great care, however, was taken to uphold the reputation of previous editions of *Physical Geography* as a text that helps instructors teach the rigour of the discipline in a pedagogically stimulating environment without oversimplification.

In addition, the fifth Canadian edition has been redesigned to make the book more visually integrative. The number of boxes has been streamlined to minimize interruptions the flow of the text. The boxes that remain include:

- *Eye on the Environment* features examine the human impact on global systems and climate change.
- *Geographer's Tools* boxes showcase some of the tools geographers use to study the world around us. Many of these focus on how satellite technology is helping geographers learn more about the Earth.

FLY BY COORDINATES

Open source software programs such as Google Earth and World Wind by NASA provide the user access to virtual globes. To visit selected locations mentioned in the text, the map coordinates provided serve as a tool to navigate these virtual globes.

ADDITIONAL RESOURCES

The accompanying website (www.wiley.com/canada/strahler) provides access to a full range of resources for the instructor and student. Instructors will find PowerPoint presentations, an image gallery, clicker questions, and a test bank. Students will find pre- and post-lecture quizzes to help them study the material.

GEODISCOVERIES MEDIA LIBRARY

In addition, both students and instructors have access to the GeoDiscoveries Media Library. This easy-to-use website helps reinforce and illustrate key concepts from the text through the use of animations, videos, and interactive exercises. Students can use the resources for tutorials as well as self-quizzing to complement the textbook and enhance understanding of physical geography. Easy integration of this content into course management systems and homework assignments gives instructors the opportunity to incorporate multimedia with their syllabi and with more traditional reading and writing assignments. Resources include:

- **Animations:** Key diagrams and drawings from our rich signature art program have been animated to provide a virtual experience of difficult concepts. These animations have proven influential to the understanding of this content for visual learners.
- **Videos:** Brief video clips provide real-world examples of geographic features and processes, and put these examples into context with the concepts covered in the text.
- **Simulations:** Computer-based models of geographic processes allow students to manipulate data and variables to explore and interact with virtual environments.
- **Interactive Exercises:** Learning activities and games built off our presentation material give students an opportunity to test their understanding of key concepts and explore additional visual resources.

ACKNOWLEDGEMENTS

Reviewer feedback played a key role in the revision strategy of the fifth Canadian edition of *Physical Geography*. Instructors from across the country were asked to provide detailed feedback on all chapters. This call was answered by many, and the final result is the text you have here. We therefore take this opportunity to thank the following instructors who willingly gave their time to help improve this text:

- Michael Bovis, University of British Columbia
- Bill Buhay, University of Winnipeg
- Dante Canil, University of Victoria
- Krystopher Chutko, Nipissing University
- Darryl Dagesse, Brock University
- Randy Dirskowsky, Laurentian University

- Sarah Finkelstein, University of Toronto
- Jonathan Hughes, University of the Fraser Valley
- John Iacozza, University of Manitoba
- Terri Lacourse, University of Victoria
- Steven Marsh, University of the Fraser Valley
- Yvonne Martin, University of Calgary
- William Osei, Algoma University College
- Todd Randall, Lakehead University
- Anne Marie Ryan, Dalhousie University
- Maria Strack, University of Calgary
- Michele Wiens, Simon Fraser University
- Hannah Wilson, Vancouver Island University

It is with particular pleasure that we thank the staff at John Wiley & Sons Canada, Ltd. for their careful work and encouragement in the preparation and production of the fifth Canadian edition of *Physical Geography.* They include acquisitions editor, Rodney Burke; developmental editors Leanne Rancourt and Gail Brown; photo researchers, Julie Pratt and Christina Beamish; copy editor, Janice Dyer; proofreader, Ruth Wilson; and indexer, Belle Wong. Design, editing, and production thanks go to Karen Bryan, Pat Loi, and Prepare, Inc. The authors also thank Keith Bigelow for drafting many of the new illustrations.

Alan Strahler
Boston, Massachusetts

O.W. Archibold
Saskatoon, Saskatchewan

ABOUT THE AUTHORS

Alan Strahler received his B.A. degree in 1964 and his Ph.D. degree in 1969 from The Johns Hopkins University, Department of Geography and Environmental Engineering. He has held academic positions at the University of Virginia, the University of California at Santa Barbara, and Hunter College of the City University of New York, and has been Professor of Geography at Boston University since 1988. With the late Arthur Strahler, he is a co-author of eight textbook titles with eleven revised editions on physical geography and environmental science. He has published over 250 articles in the refereed scientific literature, largely on the theory of remote sensing of vegetation, and has also contributed to the fields of plant geography, forest ecology, and quantitative methods. His research has been supported by over $6 million in grant and contract funds, primarily from NASA. In 1993, he was awarded the Association of American Geographers/Remote Sensing Specialty Group Medal for Outstanding Contributions to Remote Sensing. In 2000, he received the honorary degree Doctorem Scientiarum Honoris Causa (D.S.H.C.) from the Université Catholique de Louvain, Belgium, for his academic accomplishments in teaching and research. In 2004, he was honoured by election to the rank of Fellow in the American Association for the Advancement of Science.

Bill Archibold received his B.A. degree in Geography and Biology from the University of Keele in 1968. He completed an M.Sc. in Natural Resource Survey at the University of Sussex in 1969, and received a Ph.D. degree from Simon Fraser University in 1975. He has held a teaching position in the Department of Geography at the University of Saskatchewan since 1972. His main teaching areas have been in physical geography, including biogeography, meteorology, and field techniques. His research has focused on a variety of ecological and environmental topics, such as the effects of forest fires and logging in the boreal forest, the effects of air pollution on plant communities, and various studies in microclimates. In addition, he has carried out extensive studies on the ecology of wild rice in Saskatchewan, and is currently investigating the feasibility of green-roof technology in a prairie climate. He has published numerous articles in scientific journals and is the author of *Ecology of World Vegetation,* a comprehensive textbook that describes Earth's major terrestrial, aquatic, and marine biomes. He is currently working on another textbook with John Wiley & Sons Canada, Ltd. called *Introduction to Weather and Climate.*

BRIEF CONTENTS

CONTENTS

PHYSICAL GEOGRAPHY

SCIENCE AND SYSTEMS OF THE HUMAN ENVIRONMENT

PART 1 INTRODUCTION

Chapters In Part 1

Thin layers of fog are dispersing in the morning sun to reveal the glaciated peaks of the Canadian Rockies and the conifer forests that cover the lower slopes.

Geography is concerned with the physical and human processes that differentiate places on Earth and make them unique. In this way, geography provides a fundamental understanding of the spatial connections among human activities as they relate to the Earth's physical landscape. ● The ability to model and predict human and natural spatial phenomena makes geography a vital discipline in today's world. Human impact on the environment increases inexorably as the world's population grows and makes increasing demands on Earth's natural resources. To live in harmony with the environment, the world's inhabitants will have to make difficult choices in the future, and understanding geography can help with those decisions. ● Part 1 provides a general introduction to the discipline and explains the role physical geography plays in understanding global change. We also present important ideas about systems of energy and matter flow in physical geography; these are referred to in many of the later chapters.

CHAPTER 1

INTRODUCING PHYSICAL GEOGRAPHY

Vancouver, British Columbia enjoys a spectacular setting on Canada's west coast, flanked by the Pacific Coast and Vancouver Island Ranges.

THE DISCIPLINE OF GEOGRAPHY

Geography is the study of the evolving character and organization of the Earth's surface. It investigates how, why, and where human and natural activities occur, and how these activities are interconnected. Differentiating the Earth's surface into unique **places** is covered under the general theme of **regional geography**. For example, what makes Vancouver, British Columbia, unique? Is it the city's spectacular setting where the Coast Mountains meet the Pacific Ocean? The marine west-coast climate with its mild and rainy winters and blue summer skies? Its position as a seaport gateway to Asia? In fact, these are only some of the attributes that contribute to making Vancouver the unique place that it is.

Although each place is unique, the physical, economic, and social processes that operate there may be quite similar to those found in other regions. Thus, geographers are concerned with discovering, understanding, and modelling the processes that differentiate places on the Earth's surface. Why do pineapples grow in Hawaii and not in Toronto? Why is petroleum abundant in Alberta, but not in Quebec? The answers can be found in studying the physical environment. That pineapples sell for less in Honolulu than in Inuvik reflects a simple principle of **economic geography**—prices include transportation costs, and when goods travel a longer distance they are usually more expensive. Social processes that affect selling price might include consumer demand and taxation. Discovering such principles and extending them to model and predict spatial phenomena is the domain of **systematic geography**. Thus, geographers study both the characteristics that define a place and the connections between places.

PHYSICAL GEOGRAPHY

As a field of study, systematic geography is often divided into two broad areas—**human geography**, which deals with social, economic, and behavioural processes that differentiate places, and **physical geography**, which covers the atmospheric, terrestrial, and marine environments on local, regional, and global scales. It focuses primarily on factors and processes that are part of the human environment, and provides an explanation for many common natural phenomena. The main fields of physical and human

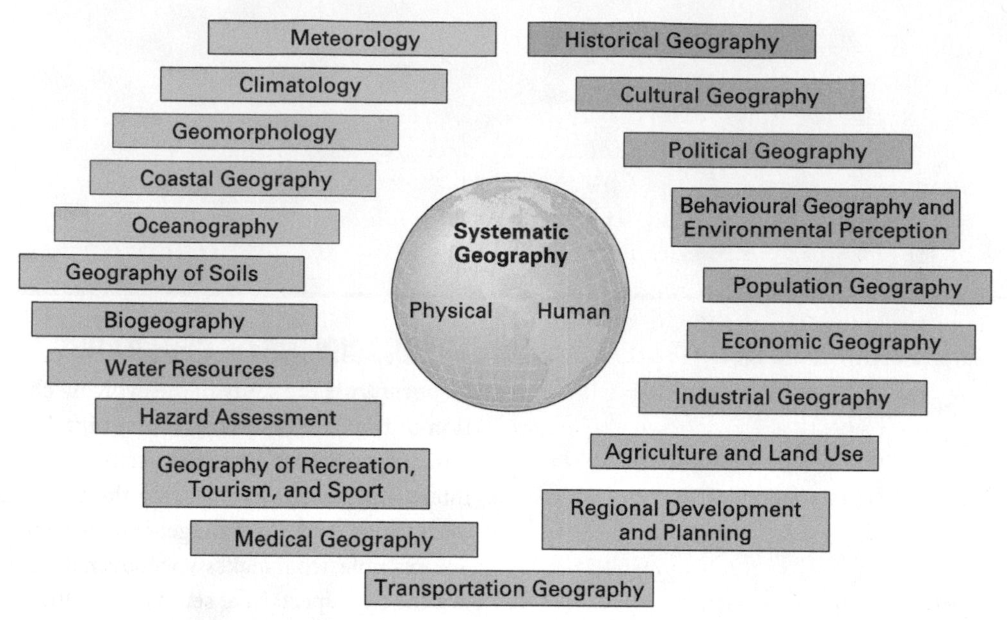

1.1 **FIELDS OF SYSTEMATIC GEOGRAPHY**

geography are shown in Figure 1.1. Aspects of physical geography, listed on the upper left side of the diagram, form the basis of this book.

Meteorology and **climatology** are both concerned with the atmosphere. Meteorology deals primarily with the processes that cause short-term fluctuations in those properties of the atmosphere that form the basis of daily weather reports. Climatology describes the results of these processes in terms of their variability in space and time. In a general sense, climate represents the average variability in weather at different places around the world. Chapters 3 to 10 discuss the processes that control weather and climate. Climatology is also concerned with climate change, both past and future. This aspect of climatology relies heavily on computer-based global climate models (GCMs) to predict how human activities might affect global climates.

Geomorphology is the science of Earth surface processes and landforms (Figure 1.2). The combined influence of human and natural factors is constantly altering the Earth's surface. The work of gravity in the collapse and movement of Earth materials, in concert with flowing water, blowing wind, breaking waves, and moving ice, acts to weaken, transport, and deposit rock and sediment. This sculpts details onto a surface that is continually renewed through volcanic activity and crustal movements. Chapters 11 to 18 describe these geomorphic processes and the basic geology of the rock materials involved. Modern geomorphology also focuses on predictive landform models. For example, in **oceanography** and **coastal geography**, computer models might be generated to predict the rate and impact of coastal erosion in relation to sea level changes that are likely to occur as a result of global climate change.

Geography of soils (Chapter 19) includes the study of the distribution of soil types and properties and the processes of soil formation. Soil formation is related to geomorphic processes of rock breakup and weathering, as well as various biological processes associated with plants and organisms living in the soil. Since both geomorphic and biologic processes are influenced by temperature and availability of moisture, broad-scale soil patterns are invariably related to climate.

1.2 **McFARLANE FALLS, SASKATCHEWAN** Water is an inexorable geomorphic agent, and over time the waterfall and bedrock remnants will be gradually eroded and the rapids will disappear.

Biogeography is the study of the distributions of organisms and the processes that produce these spatial patterns. At a local scale, the distributions of plants and animals reflect the suitability of the habitat that supports them and the functional relationships between species (Chapters 20 to 21). In this application, biogeography is closely aligned with *ecology*. Over broader scales and time periods, the migration, evolution, and extinction of plants and animals are key processes that determine their spatial distribution patterns (Chapter 21). In light of the increasing human impact on the environment, **biodiversity**—the assessment of the richness of life and life forms on Earth—is a biogeographic topic of considerable importance (Figure 1.3). The present global-scale distribution of life forms in the Earth's major biomes provides the context for biodiversity (Chapter 22).

In addition to these fields of physical geography, **water resources** and **hazards assessment** are important subfields in applied physical geography. Water resources couples the basic study of the location, distribution, and movement of water, for example in river systems, with the study of water quality and human use. This field involves many aspects of human geography, including regional development and planning, political geography, agriculture, and land use. Chapters 15 and 16 review water resources in the context of wells, dams, and water quality.

1.3 **DECIDUOUS FOREST, ONTARIO** The diversity of the forest becomes more apparent as the trees take on their autumn hues in response to shortening days and cooler temperatures caused by the revolution of Earth around the Sun.

Hazards assessment also blends physical and human geography. What are the risks of living next to a river and how do people perceive those risks? What is the government's role in protecting citizens from floods or assisting them in recovery from flood damages? Answering questions like these requires knowledge of not only how physical systems work, but also how humans interact with their environment, both as individuals and as societies. An understanding of physical processes, such as floods, earthquakes, and landslides, provides the background for assessing the impact of natural hazards.

Most of the remaining fields of human geography shown in Figure 1.1 have links to physical geography. For example, climatic and biogeographic factors may determine the spread of disease-carrying mosquitoes (medical geography). Mountain barriers may isolate populations and increase the cost of transporting goods from one place to another (cultural geography, transportation geography, and economic geography). Unique landforms and landscapes may generate many tourist activities (geography of recreation, tourism, and sport).

TOOLS IN GEOGRAPHY

Because geographers deal with spatial phenomena, they commonly use maps to display spatial information. A **map** shows point, line, or area data—that is, locations, connections, and regions (Figure 1.4a). The scale of a map links the actual ground distance between places with the represented distance on the map. The art and science of map-making is called **cartography**. Chapter 2 provides more information on this important subject.

Maps are very useful for storing information, but they have limitations—not the least of which is the fact that because traditional printed maps take a long time to produce, they can never really be up-to-date. Think, for example, of all the new construction that has occurred in your community, and compare the current arrangement of streets and neighbourhoods to a map that was produced a few years ago. The development of new cartographic techniques has greatly reduced problems of this sort. In the past two decades, advances in data collection, storage, analysis, and display have led to the development of **geographic information systems (GIS)**. These spatial databases rely on computers for analysis, manipulation, and display of spatial data (Figure 1.4b). Chapter 2 provides more information on GIS.

Another important technique for acquiring spatial information is **remote sensing**, in which aircraft or spacecraft provide images of the Earth's surface (Figure 1.4c). Depending on the scale of the remotely sensed image, the information obtained can range from fine local detail—such as the arrangement of trees in a woodland—to a large area picture—for example, the "greenness" of vegetation for an entire continent. Remote sensing is discussed further in Chapter 3.

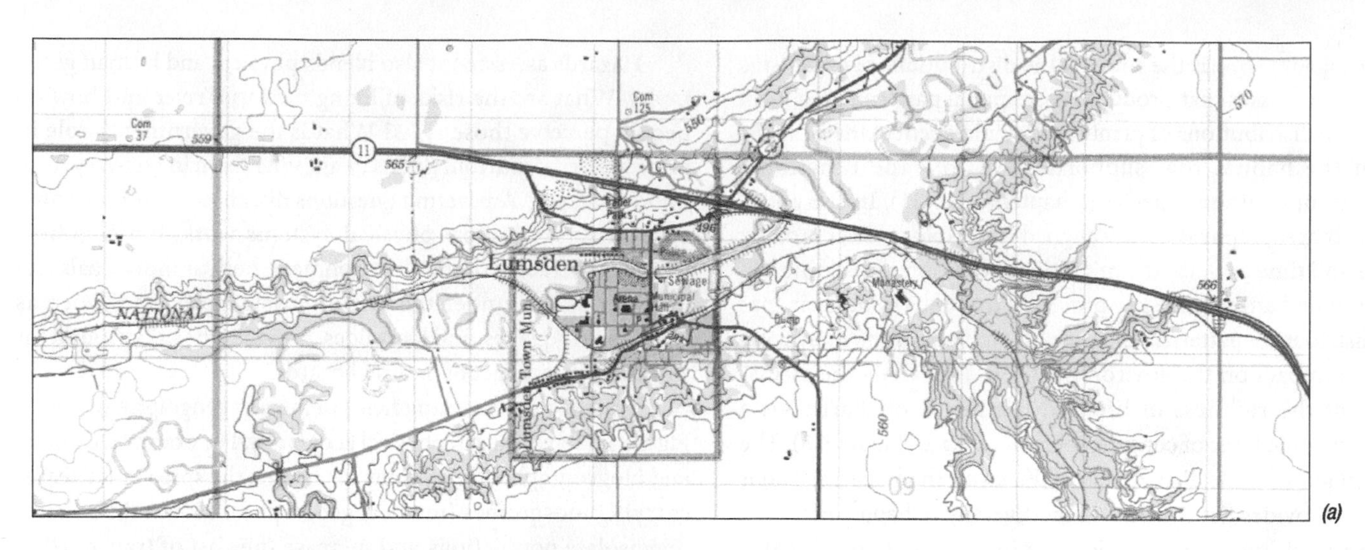

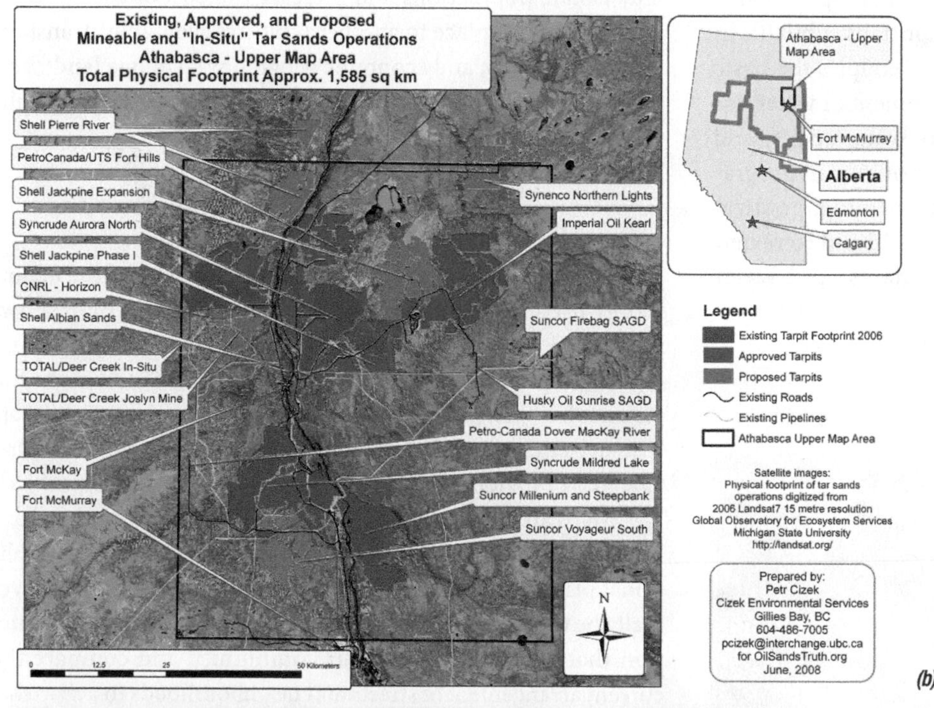

(c)

1.4 TOOLS OF PHYSICAL GEOGRAPHY (a) **Maps** Maps display geographic information in printed form. (b) **Geographic Information Systems (GIS)** Computer programs that store and manipulate geographic data are essential to modern applications of geography. (c) **Remote Sensing** Remote sensing includes observing the Earth from the perspective of an aircraft or spacecraft. This scene shows the bright blue and green swirls of phytoplankton blooms throughout Lake Erie and Lake Ontario and less extensively the other Great Lakes, captured by the Moderate Resolution Imaging Spectroradiometer (MODIS) aboard NASA's Aqua (EOS PM) satellite.

Tools in geography also include **mathematical modelling** and **statistics**. Using mathematics and computers to model geographic processes is a powerful approach to understanding both natural and human phenomena. Statistics provide methods to analyze geographic data to assess differences, trends, and patterns.

PHYSICAL GEOGRAPHY, THE ENVIRONMENT, AND GLOBAL CLIMATE CHANGE

Physical geography is concerned with the natural world around us. Because natural processes are constantly active, the Earth's environments are always changing. Sometimes the changes are slow and subtle, as when rivers slowly cut into their bedrock channels. At other times, the changes are rapid and dramatic, for example, during a landslide (Figure 1.5).

Environmental change is now produced not only by the natural processes that have acted on the planet for millions of years, but also by human activity. The human race has populated the planet so completely that few places remain free of some form of human impact (Figure 1.6). Physical geography is the key to understanding the effect this has on the environment and is discussed in various sections of this book.

GLOBAL CLIMATE CHANGE

Over the past decade, many scientists have come to the opinion that human activity has begun to change the Earth's climate. How has this happened? Popular opinion points to the greenhouse effect, a natural phenomenon that is necessary for life on Earth. Gases such as carbon dioxide (CO_2), methane (CH_4), and nitrous oxide (N_2O), which are released by human activities, reduce the radiation of heat energy to space. This induced intensification of the greenhouse effect is widely implicated in global climate change. Many of the potential impacts of **climate change**, such as increased frequency of extreme weather events and the expected rise in sea level, are discussed in later chapters.

1.6 **ATHABASCA TAR SANDS** The Athabasca oil sands in northern Alberta consist of oil-rich bitumen mixed with sand and clays. The near-surface deposits, which are about 50 m thick, occur at a depth of about 75 m in an area of about 3,400 km^2.

Fly By: 57° 01′ N; 111° 38′ W

1.5 **FRANK SLIDE** The town of Frank lies at the foot of Turtle Mountain in southern Alberta. Turtle Mountain comprises limestone thrust over weaker sedimentary rocks, including coal. In the early morning hours of April 29, 1903, more than 80 million tons (30 million m^3) of rock collapsed from the north face and buried part of the town.

One way to reduce human impact on the greenhouse effect is to slow the release of CO_2 from fossil fuel burning. Because modern civilization depends on the energy of fossil fuels to carry out almost every task, reducing consumption to stabilize CO_2 concentration in the atmosphere is not easy. However, some natural processes reduce atmospheric CO_2. Plants withdraw CO_2 from the atmosphere by absorbing it during photosynthesis. In addition, CO_2 is soluble in sea water. These two important pathways, by which carbon flows from the atmosphere to lands and oceans, are part of the **carbon cycle**.

BIODIVERSITY

There is a growing awareness that the diversity of Earth's plants and animals is an immensely valuable resource. One important reason for preserving this **biodiversity** is that, over time, species have evolved natural biochemical defences against diseases and predators. Some of these have proven useful in agriculture and medicine, where they have been used to develop such things as natural pesticides and cancer drugs.

Another important reason for maintaining biodiversity is that complex ecosystems with many species tend to be more stable and respond better to environmental change. If human activities inadvertently reduce biodiversity significantly, there is more risk of unexpected and irreversible degradation of natural environments. Chapters 21 and 22 discuss the processes that create and maintain biodiversity in Earth's many natural habitats.

POLLUTION

Unchecked human activity can degrade environmental quality. In addition to releasing CO_2, fuel consumption can release gases that are hazardous to health, especially when they react with each other to form toxic compounds—such as ozone and nitric acid—in photochemical smog. Water pollution from fertilizer runoff, industrial accidents, toxic industrial waste, and acid mine drainage can severely degrade water quality. Marine ecosystems are especially vulnerable to pollution, with many organisms threatened annually by chemicals and physical hazards such as non-biodegradable plastic waste. This pollution affects not only river ecosystems, but also human and other animal populations that depend on them as sources of water and food (Figure 1.7). Chapters 4 and 15 discuss the causes and effects of **environmental pollution** in the context of air and water.

EXTREME EVENTS

Catastrophic events, such as floods, fires, hurricanes, and earthquakes, have great and long-lasting impacts on human and natural systems. Computer models predict that extreme weather events will become more severe and frequent as Earth responds to climate changes. Droughts, and consequent wildfires and crop failures, will occur more often, as will extreme rains and floods. There have been many examples of extreme weather events in the last decade, such as Hurricane Katrina in 2005, numerous forest fires that have occurred in British Columbia and other parts of the world, and Hurricane Igor, which brought severe storms to the Maritimes in 2010.

Although the human impact on natural systems is implicated in these **extreme weather events**, a lot of questions remain unanswered. Some extreme events, such as earthquakes and volcanic eruptions, are produced by forces deep within the Earth that are not affected by human activity. But as the human population continues to expand and rely increasingly on complex technological infrastructures, the impact of such events will become more damaging and disruptive. The chapters that follow discuss many types of extreme events originating within the Earth, including tsunamis and landslides. Catastrophic events arising from other causes, such as the severe weather associated with hurricanes and tornadoes, are also described.

ORGANIZING INFORMATION IN PHYSICAL GEOGRAPHY

In all sciences, recurring principles and ideas are used to organize the wealth of accumulated knowledge. In physical geography, one approach is to focus on the Earth's **realms**—the major components of the planet, each with its own unique properties. The atmosphere, lithosphere, hydrosphere, and biosphere are the four great Earth realms that form the basis of physical geography (Figure 1.8).

The **atmosphere** is a gaseous layer that surrounds the Earth. It receives heat and moisture from the surface, redistributes them globally, and returns some heat and all the moisture to the

1.7 BP OIL SPILL In April 2010, an explosion on the Deepwater Horizon drilling rig in the Gulf of Mexico resulted in the largest accidental marine oil spill in history. The leak was not capped until July 2010, after releasing approximately 4.9 million barrels of oil into the Gulf over a period of 87 days.

1.8 The scenery of the Rocky Mountains provides fine examples of the union of Earth's four great realms.

surface. The atmosphere also supplies vital elements—carbon, hydrogen, and oxygen—that are needed to sustain life.

The outermost solid layer of the Earth is the **lithosphere**, the surface of which is sculpted into landforms, such as mountains, hills, and plains. Over most of the continents, the surface of the lithosphere is a shallow layer of soil in which nutrients are available to organisms.

The water in the world's oceans is the largest component of the **hydrosphere**, which also includes lakes, rivers, and glacier ice, as well as the water present in the atmosphere as gaseous vapour, liquid droplets, and solid ice crystals. In the lithosphere, water is found in soils and in ground water reservoirs.

The **biosphere** encompasses all of the Earth's living organisms. Life on Earth depends on the atmosphere's gases, the hydrosphere's water, and the lithosphere's nutrients, and so the biosphere interacts with all three of the other great realms. Most of the biosphere is contained in the **life layer**, which includes the land, lakes, and rivers of the continents and the upper 100 m or so of the oceans and seas.

SCALES IN PHYSICAL GEOGRAPHY

Natural processes act over a wide range of scales both in time and space. It may take millions of years for Earth forces to produce a massive chain of mountains like the Rockies; the fury of a hurricane may last for only a few days, while an earthquake can cause widespread damage in a few minutes.

On a **global scale**, Earth is a nearly spherical planet. Its shape affects how land and water surfaces absorb the Sun's energy. Unequal solar heating produces currents of air and water that are incorporated into the global atmospheric and oceanic circulation system. The resulting weather and climate patterns can be readily distinguished on a **continental scale**. Observations on a **regional scale** might be used to distinguish vegetation patterns. More detailed assessments of species distributions could be made on a **local scale**, where differences between, say, a steep slope and an adjacent valley floor become apparent. At the finest level are **individual-scale** landscape features, such as a grassy sand dune in a desert (Figure 1.9).

SYSTEMS IN PHYSICAL GEOGRAPHY

The processes that interact within the four realms to shape the life layer and differentiate global environments are varied and complex. Many of these processes are interconnected and are more readily understood by adopting a **systems approach**, which

1.9 Different scales of time and space contribute to the landscape features in Death Valley, California.

emphasizes how and where matter and energy flow in natural systems. A system is a set or collection of things that are somehow related or organized. An example is the solar system—a collection of planets that revolve around the Sun. A forest ecosystem is an ecological system comprising various trees, shrubs and herbs, together with the animals that live there and the soil on which it grows.

Physical geography uses specific types of systems. For example, **flow systems** describe how matter and energy move from one location to another over time. Understanding flow systems in physical geography is important because it helps explain how things are connected. For example, the global shortwave energy flow system describes how solar energy moves from the Sun to Earth and its atmosphere. The growth of plants in a forest ecosystem depends on the flow of energy, water, and essential elements, such as carbon.

Flow systems have a **structure** of interconnected **pathways**. The reflection of solar radiation from the top of a cloud is a pathway that turns the flow of solar energy back toward space. Similarly, in forests the nutrients stored in plant material are returned to the soil through decomposition. The various parts of a system, such as the pathways, their connections, and the types of matter and/or energy that flow within it represent the system *components*.

Flow systems need a **power source**. In *natural flow systems*, sources of power include shortwave energy flowing from the Sun to Earth, and the outward flow of heat from the Earth's interior. The term *energy* was first used in 1807, but the idea is much older, going back to Sir Isaac Newton's laws of motion, published in 1687. In 1695, Gottfried Leibnitz proposed the existence of a "vital force." Today this is termed **kinetic energy**—the energy of mass (or "matter") in motion, and is expressed mathematically as:

$$\mathbf{1/2}\ \boldsymbol{mv^2}$$

where m = mass
v = velocity

Energy is defined as the ability to do work, which in this sense is equal to the change in kinetic energy in a body. Energy "moves" or "travels" with (or within) forms of matter. For example, the gasoline engine in a car transfers the energy of fuel combustion into forward motion. The entire moving object—the car—then acquires kinetic energy. Kinetic energy is one of the forms of energy that comes under the general heading of **mechanical energy**. Should the car crash head-on into a massive power pole, it will demonstrate its ability to do work upon its own body, upon its passengers, and upon the pole. The energy released in the collision will increase in direct proportion to the mass of the car, and it will also increase with the square of the car's speed, since kinetic energy is equal to $1/2\ mv^2$.

Now that the car is motionless, it no longer has kinetic energy. Energy cannot simply be destroyed and disappear, but it can be transformed. At the point of impact, the car body would have become heated for an instant, as kinetic energy of motion was turned into **heat energy**. In this form of energy, the atoms and molecules within a solid (or a liquid or a gas) are in rapid motion. The hotter the substance, the faster the molecules vibrate. This form of kinetic energy is termed *sensible heat* because it can be sensed by touch, as well as be measured by a thermometer. Sensible heat is important in the study of temperature changes in air, water, and soil.

Radiant energy is the form of energy that is emitted from the Sun, and is the Earth's principal form of energy. In this form, energy travels through space and easily penetrates transparent substances, such as air, water, and glass.

The force of gravity is a form of energy that arises when one object is attracted to another. Imagine rolling a heavy object, such as a boulder, to the top of a cliff, then nudging it over the edge and watching it drop to the ground below. Moving the boulder to the top of the hill requires the expenditure of energy to overcome the ever-present downward pull of gravity. Thus the boulder is endowed with a certain amount of **potential energy**, also referred to as **stored energy**. With each metre of vertical distance the boulder falls, it converts a quantity of that potential energy into kinetic energy. (We can disregard the small energy loss due to friction with the surrounding air.) Upon impact, the kinetic energy is converted to heat energy, which quickly dissipates.

Potential energy, present because of gravity, must always be evaluated in terms of a given reference level, or *base level*. The standard base level is the average level of the ocean—"mean sea level." In nature, any kind of matter, on the land or in the atmosphere, holds potential energy with respect to sea level. Examples are raindrops and hailstones formed in a storm cloud, or the mineral particles of ash and cinder spewed from an erupting volcano. As these substances fall toward the Earth, their potential energy is converted to kinetic energy. Potential energy can also be converted to kinetic energy as substances descend along sloping and winding pathways. The flow of water in a river and ice in a glacier are examples.

Chemical energy, which is absorbed or released by matter when chemical reactions take place, is also important. Chemical reactions involve the coming together of atoms to form simple molecules, the recombining of simple molecules into new and more complex compounds, and the reverse changes back into simpler forms. Most of the time, chemical energy is stored energy held within the molecule. An important example is photosynthesis, in which green plants absorb radiant energy from the Sun to produce complex organic carbohydrates. Chemical energy is also stored for long periods of geologic time in the hydrocarbon compounds that are used as fuels, such as coal, oil, and natural gas. As these fuels are burned, their molecules combine with atmospheric oxygen, releasing heat energy as well as many simpler molecules, such as CO_2 and water.

Open and Closed Flow Systems

Flow systems, such as a river drainage basin, have **inputs** and **outputs**. A river drainage basin consists of water in a set of connected stream channels (Figure 1.10). In this system, the stream channels—pathways—are connected in a structure—the channel network—that organizes the flow of water and sediment from the land to the ocean. The input is in the form of rain or snow. The energy source comes from gravity, and the output of water and sediment might give rise to a delta through deposition (see Chapter 16).

A river system is an **open flow system** because there are inputs and outputs of energy and matter. Another type of matter flow system has no inputs or outputs. Instead, the system's flowing materials move endlessly in a series of interconnected paths or loops in what is termed a **closed flow system**. This type of system is also known as a **material cycle**. Examples include the global carbon cycle and the hydrologic cycle. Most natural closed flow systems have complicated structures consisting of many interconnected looping pathways. They also require a flow of energy. Whereas a material cycle may be closed, the energy flow systems that sustain them will always be open.

1.10 RIVER CHANNELS AS FLOW SYSTEM PATHWAYS The North Saskatchewan River provides an example of a natural flow system. Several tributary streams flow into the North Saskatchewan River as it flows northeast through the Rocky Mountains west of Red Deer, Alberta.

The **hydrologic cycle**, in which water circulates between the biosphere, atmosphere, lithosphere, and hydrosphere, is an example of a closed flow system in physical geography (Figure 1.11). It is maintained by condensation, precipitation, infiltration, runoff, and evapotranspiration. These processes occur simultaneously and continuously on a global scale. The loops in the hydrologic cycle are flow paths of water in gaseous, liquid, and solid forms—gaseous form as moist air currents in the atmosphere, liquid form as water flowing in rivers, and solid form as slowly advancing glaciers. The loops are interconnected and ultimately maintained by the Sun's energy flow system.

Whether the flow of matter is open or closed depends partly on the boundary of the flow system. In Figure 1.12a, the system boundary has been drawn around a single river network. Water enters the network when it runs off the land into the river system. Water exits from the network at some predefined point along its course. Defined in this way, the system is an open flow system in which water enters and leaves a specified stretch of river.

When the system boundary is redrawn to include the entire Earth and its atmosphere (Figure 1.12b), a new pathway must be added—the return flow of water from the oceans to the atmosphere by evaporation. There is no input or output because water does not leave the Earth's atmosphere or enter it from space. Thus, the system, which now represents the hydrologic cycle, is closed. Indeed, on a global scale, any matter flow system is closed, since with few exceptions (for example, meteorites), nothing enters or leaves the planet and its atmosphere.

However, energy flow systems are always open. Earth absorbs some portion of the radiant energy it receives from the Sun, so there is always an energy input. However, the Earth is warmer than the depths of space and thus emits radiant energy. A portion of that energy is ultimately radiated to space, so there is always an output energy flow, even when the system boundary encloses the Earth and its atmosphere.

Feedback and Equilibrium in Flow Systems

Two other important concepts associated with flow systems are **feedback** and **equilibrium**. Feedback occurs when the flow in one pathway acts either to reduce or increase the flow in another pathway. Negative feedback helps to dampen external changes imposed on the system. For example, as water evaporates from a soil, the soil particles hold the remaining water more tightly, and so the rate of evaporation slows. Positive feedback occurs when some force or factor creates a situation that increases the effect. This is the case when water falling on a slope starts to erode small

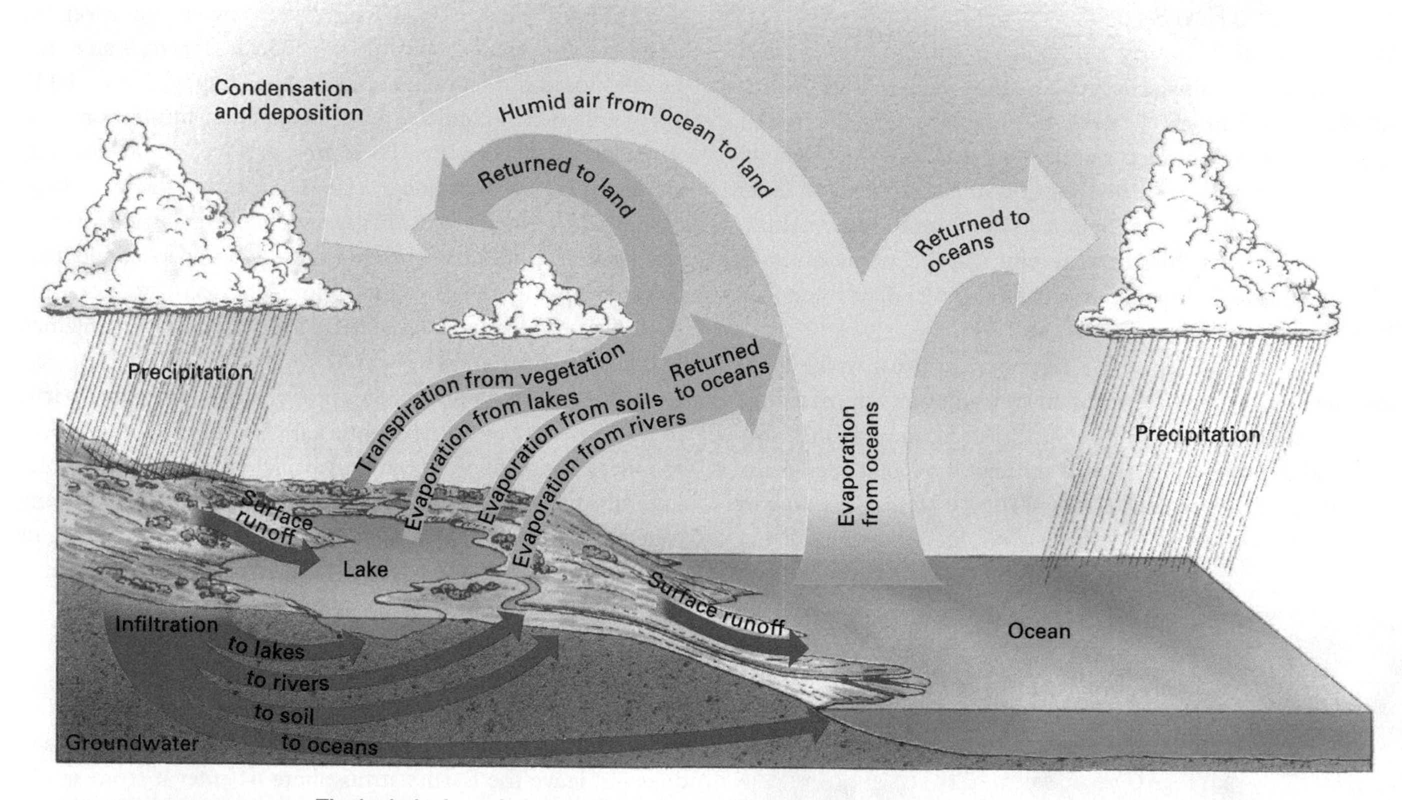

1.11 THE HYDROLOGIC CYCLE The hydrologic cycle traces the various paths of water from oceans, through the atmosphere, to land, and its return to oceans.

channels or rills. With each rainstorm, the rills capture more water and so become enlarged, directing more and more water into the channel, leading to the creation of large gullies (Figure 1.13).

Equilibrium refers to a steady condition in which the flow rates in a system's various pathways do not change significantly. The Quill Lakes of Saskatchewan are an example of an equilibrium system. Because the Quill Lakes lie in a region of internal drainage, they have no stream outlets. Water enters the lake complex from streams and groundwater discharge, but leaves by evaporation. The size and depth of the lake complex varies with inflow and evaporation rates, and during times of drought there is a marked increase in the salinity of the water.

During wetter periods, the input to the Quill Lakes from streams and groundwater increases, then the water level rises and the area of the lake expands. Because of the increased surface area, evaporation is greater. Eventually, the level rises to the point where the higher rate of evaporation equals the increased rate of inflow. That is, the lake level reaches an equilibrium. The combination of input, surface area, and evaporation is a negative feedback. Systems in equilibrium are normally stabilized by negative feedback loops.

The effect of clouds on the global climate system illustrates the role of negative and positive feedback. Low clouds reflect sunlight back to space much more efficiently than dark land or ocean surfaces. They provide an energy flow pathway in which a portion of the solar energy flow is redirected away from Earth toward space. This pathway tends to cool the surface, and so it acts as a negative feedback. High clouds are different, however. They tend to absorb the outgoing flow of heat from the Earth to space and redirect it back toward the surface. Thus, high clouds provide a positive feedback that warms the surface (Figure 1.14).

A small increase in the Earth's surface temperature will evaporate more water from the oceans and moist land surfaces, and more clouds will form. Global climate models predict that this will increase high cloud cover. Thus, the effect will be a positive feedback that is expected to make the surface even warmer.

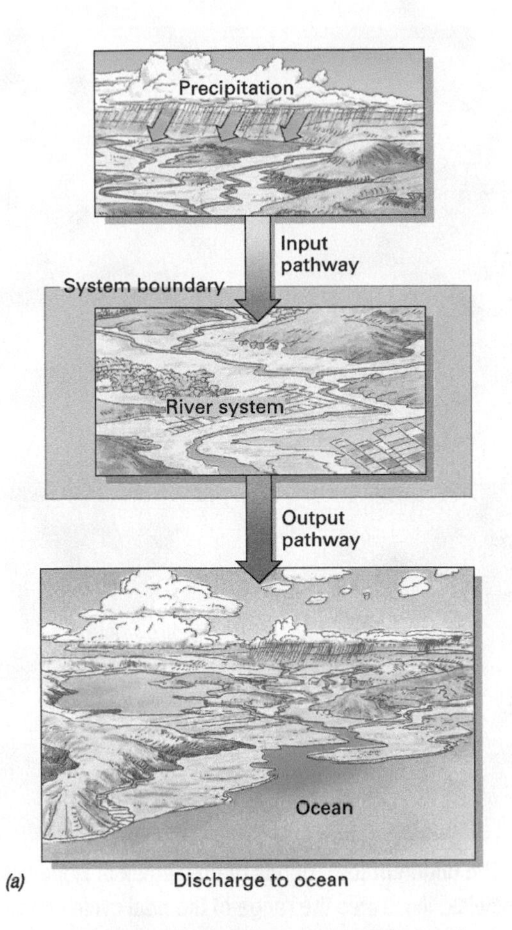

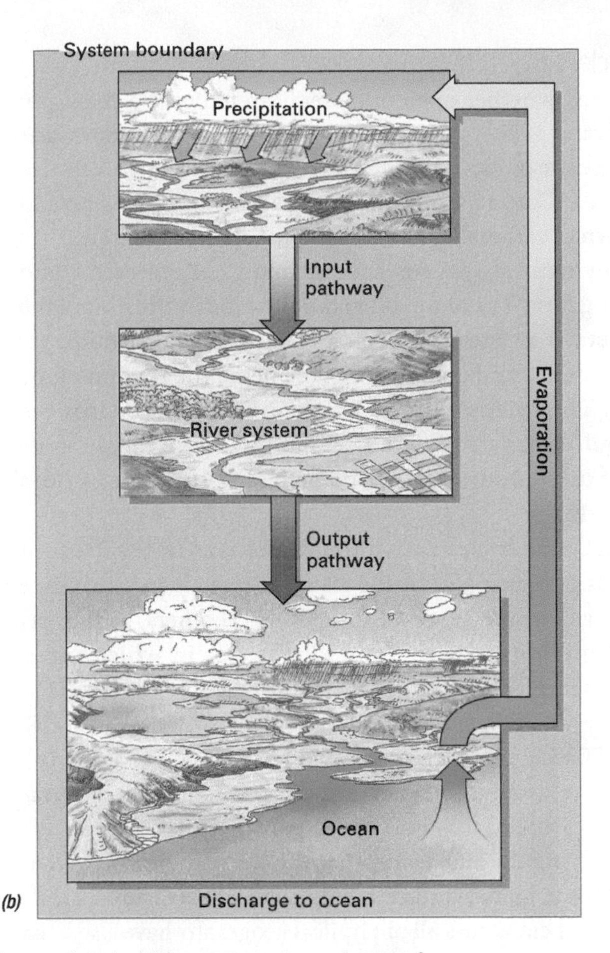

1.12 **A RIVER SYSTEM AS A FLOW SYSTEM** The river is an open flow system in (a) because water enters and leaves at specific points in the network. When the boundary is moved to enclose the entire Earth and its atmosphere (b), the system becomes a closed, global flow system—the hydrologic cycle.

1.13 Rilling and gullying, as seen here at Drumheller, Alberta, is a positive feedback process that accelerates as more water is captured by the developing channel system.

1.14 Clouds provide both positive and negative feedback controls on global temperature.

Time Cycles

Any system, whether open or closed, can undergo a change in the flow rates of energy or matter within its pathways; flow rates may accelerate or slow down. These changes in activity can also be reversed. Consequently, a rate can alternately speed up and slow down at various intervals in what is called a **time cycle**.

Many natural systems have a rhythm of increasing and decreasing flow. The annual revolution of the Earth around the Sun generates a time cycle of energy flow in many natural systems, the most fundamental outcome being the rhythm of the seasons. The rotation of the Earth on its axis produces the night-and-day cycle of darkness and light. The moon, in its monthly orbit around the Earth, generates the cycle of tides (Figure 1.15).

Other time cycles with durations of tens to hundreds of thousands of years describe the alternate growth and shrinkage of the great ice sheets (see Chapter 18). Still others, with durations of millions of years, describe cycles in which super continents form, break apart, and form anew (see Chapter 12).

1.15 The undercut form of this marine stack at Hopewell Rocks, New Brunswick, illustrates the range of the tidal cycle on the Bay of Fundy.

A Look Ahead

This chapter has presented an introduction to physical geography and to some of the tools and approaches that geographers use in studying the landscape. It has also noted some key environmental and global climate change topics. Several broad-scale concepts that cut across all of physical geography have also been described. These include the great Earth *realms* that are the basis of physical geography. Of particular importance are the *scales*, from local to global, that characterize Earth processes; the flow *systems* that power the natural processes of physical geography; and the *cycles* of processes that repeat in time and space. The chapters that follow explore these concepts further.

CHAPTER SUMMARY

- Geography is the study of the evolving character and organization of the Earth's surface. Regional geography is concerned with how the Earth's surface is differentiated into unique places, while systematic geography investigates the processes that differentiate places in time and space.
- Within systematic geography, human geography deals with social, economic, and behavioural processes that differentiate places, and physical geography examines the natural processes occurring at the Earth's surface. Meteorology and climatology are concerned with the variability in space and time of the state of the atmosphere. Geomorphology focuses on Earth surface processes and landforms. Oceanography and marine geography combine the study of geomorphic processes that shape shores and coastlines with their application to coastal development and marine resource use. Geography of soils includes the study of the distribution of soil types, their properties, and the processes that form them. Biogeography is the study of the distributions of organisms on varying spatial and temporal scales, as well as the processes that have given rise to these distribution patterns. Water resources and hazards assessment are applied fields that blend both physical and human geography.
- Important tools for studying the fields of physical geography include maps, geographic information systems, remote sensing, mathematical modelling, and statistics.
- Physical geography is concerned with the natural world, which is constantly changing. The effect of human activities on natural environments is also of interest, especially in the way it is implicated in global climate change. Maintaining global biodiversity is important both for ecosystem stability and for preserving a potential resource of bioactive compounds for human benefit. Unchecked human activity can degrade environmental quality and create environmental pollution. Extreme events—storms and droughts, for

example—may be more frequent with global climate change caused by human activity.

- Realms, scales, systems, and cycles are four overarching themes that appear in physical geography. The four great Earth realms are the atmosphere, hydrosphere, lithosphere, and biosphere. The life layer is the shallow surface layer where lands and oceans meet the atmosphere and where most forms of life are found. The systems of interaction between the realms can be examined on different scales—global, continental, regional, local, and individual.
- A systems approach to physical geography helps in understanding interconnections between natural processes by considering them as flow systems. A distinction can be made between matter flow systems and energy flow systems. Flow systems are composed of pathways of energy and/or matter that are interconnected in a structure. All flow systems have a power source.
- Open flow systems have inputs and outputs, while closed flow systems do not. Closed matter flow systems are also called material cycles in which materials move in an endless series of interconnected pathways. The hydrologic cycle and the carbon cycle are examples of material cycles in physical geography. Although *matter flow systems* may be open or closed, *energy flow systems* are always open.
- Feedback in a flow system occurs when the flow in one pathway affects the flow in another pathway. *Positive feedback* enhances or increases the flow within a pathway, while negative feedback reduces it. A system with *negative feedback* loops or pathways tends to be self-stabilizing and moves toward an equilibrium—a steady state in which flow rates in system pathways remain about the same.
- Systems may undergo periodic, repetitive changes in flow rates that constitute time cycles. Time cycles in physical geography range in length from hours to millions of years.

KEY TERMS

atmosphere
biodiversity
biogeography
biosphere
carbon cycle
cartography
chemical energy
climate change
climatology
closed flow system
coastal geography
continental scale
economic geography
environmental pollution
equilibrium
extreme weather events
feedback
flow systems
geographic information systems (GIS)
geography
geography of soils
geomorphology
global scale
hazards assessment
heat energy
human geography
hydrologic cycle
hydrosphere
individual scale
inputs
kinetic energy
life layer
lithosphere
local scale
map
material cycle
mathematical modelling
mechanical energy
meteorology
oceanography
open flow system
outputs
pathways
physical geography
place
potential energy
power source
radiant energy
realms
regional geography
regional scale
remote sensing
statistics
stored energy
structure
systematic geography
systems approach
time cycle
water resources

REVIEW QUESTIONS

1. What is geography? Regional geography? Systematic geography?
2. Identify and define three important fields of science within physical geography.
3. Identify and describe three tools that geographers use to acquire and display spatial data.
4. Why is the loss of biodiversity a concern of biogeographers and ecologists?
5. Give examples of extreme events and discuss how they might be influenced by human activity.
6. Name and describe each of the four great physical realms of Earth. What is the life layer?
7. Provide two examples of processes or systems that operate on each of the following scales: global, continental, regional, local, and individual.
8. What is a flow system? Provide an example. Distinguish between two types of flow systems.
9. What distinguishes an open flow system from a closed flow system?
10. What is a cycle (material cycle)? Why are current research efforts focused on the carbon cycle?
11. Describe the concepts of feedback and equilibrium as applied to systems. Provide an example drawn from a natural system.
12. What is a time cycle as applied to a system? Give an example of a time cycle evident in natural systems.

CHAPTER 2 THE EARTH AS A ROTATING PLANET

January on the Prairies.
A CN freight train heading west near Saskatoon, Saskatchewan.

The systematic study of physical geography starts with a discussion of the motions of the Earth in relation to the sun, and the implications for global location, world time zones, and the changing seasons.

THE SHAPE OF THE EARTH

Pictures taken from space show that the Earth is almost spherical. Technically, the Earth assumes the shape of an *oblate ellipsoid* because it bulges slightly at the equator and flattens at the poles, due to the centrifugal force of its rotation. As a result, the Earth's equatorial diameter is 12,756 km compared with its polar diameter of 12,714 km. Geodetic surveys carry out precise measurements of the Earth's shape; these are important for the accuracy of modern satellite navigation systems used in aircraft, ships, and vehicles. Satellite navigation systems determine the vehicles' exact location using the Global Positioning System (GPS). GPS also has applications in cartography and global time-keeping, and is discussed in more detail later in this chapter.

EARTH ROTATION

Earth rotation refers to the counter-clockwise turning of the Earth on its axis, an imaginary straight line passing through the centre of the planet and joining the North and South Poles. The Earth completes one rotation with respect to the sun every day (Figure 2.1). Each rotation defines the solar day and is the basis of conventional time systems. The axis of rotation also serves as a reference in setting up the system of latitude and longitude used for global positioning.

ENVIRONMENTAL EFFECTS OF EARTH ROTATION

One effect of rotation is that it imposes a daily, or diurnal, rhythm on how Earth receives solar energy. This in turn influences air temperature, air humidity, and air motion. A second effect of the Earth's rotation is that it causes the winds to turn consistently from the expected direction imposed by the pressure gradients that govern their movement. In the northern hemisphere, winds are deflected to the right; in the southern hemisphere deflection is to the left. This important phenomenon is termed the *Coriolis*

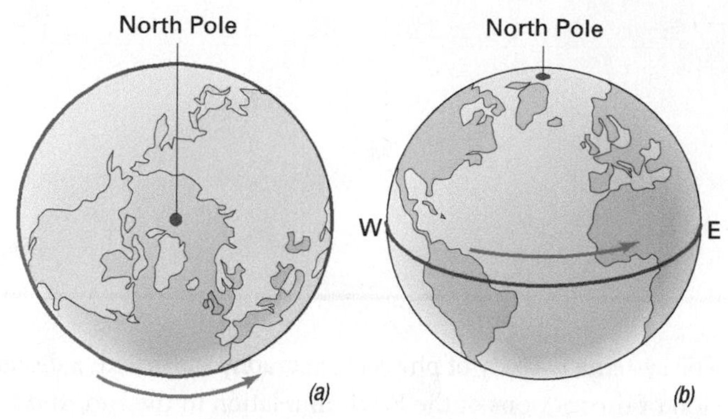

2.1 **DIRECTION OF EARTH ROTATION** The direction of rotation of the Earth can be thought of as (a) counter-clockwise at the North Pole, or (b) eastward (from left to right) at the equator.

effect (see Chapter 5). Additionally, Earth's rotation is linked to ocean currents and the diurnal rise and fall of the ocean tides (see Chapter 17).

THE GEOGRAPHIC GRID

PARALLELS AND MERIDIANS

The **geographic grid** is a spherical coordinate system used to determine the locations of features on the Earth's surface. It is constructed from two sets of intersecting circles (Figure 2.2). One set of circles, termed **parallels**, is arranged perpendicular to the axis of rotation. These intersect the second set of circles, known as **meridians**, at right angles.

The **equator** lies midway between the North Pole and South Pole, and is the Earth's longest parallel. The other parallels run parallel to the equator and are positioned according to the angular distance, measured at the centre of the Earth, between the equator and the surface. The 49th parallel, which forms much of the border between Canada and the United States, is a well-known example. Meridians run from the North Pole to the South Pole and represent half of an imaginary circle drawn around the surface of the Earth. Meridians and parallels define geographic directions. Meridians run in a true north-south direction; parallels are true east-west lines.

Although Earth is not truly spherical, for all but the most precise navigation this discrepancy has little practical significance. It is therefore customary to consider the Earth as a perfect sphere. Two types of circles—great circles and small circles—can be drawn on the surface of a sphere (Figure 2.3). **Great circles** are constructed so that the plane of intersection with the surface of the sphere passes through the centre of the globe. This effectively divides the globe into two equal halves. With **small circles** the plane of intersection passes through the surface of the sphere, but not its centre. Thus, meridians are actually halves of great circles, while all parallels, *except* the equator, are small circles. Great circles can be aligned in any direction on the sphere and can be drawn to pass through any two points on its surface. Note that the arc of the great circle between the two points represents the shortest distance between them. This is of great importance

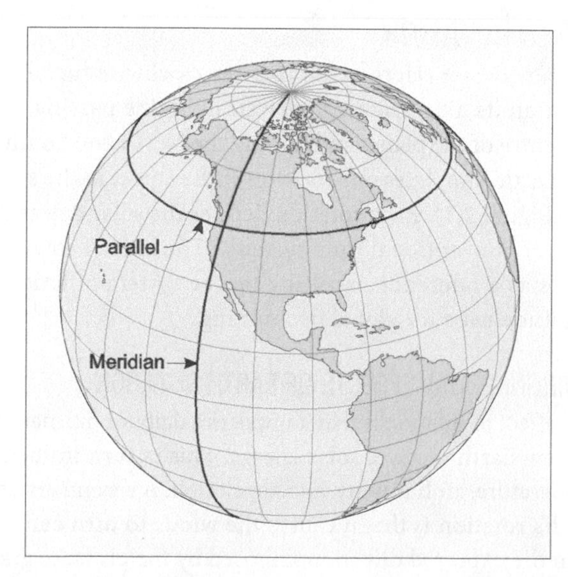

2.2 **PARALLELS AND MERIDIANS** Parallels of latitude divide the globe crosswise into a series of circles arranged perpendicular to the polar axis. Meridians of longitude divide the globe from pole to pole.

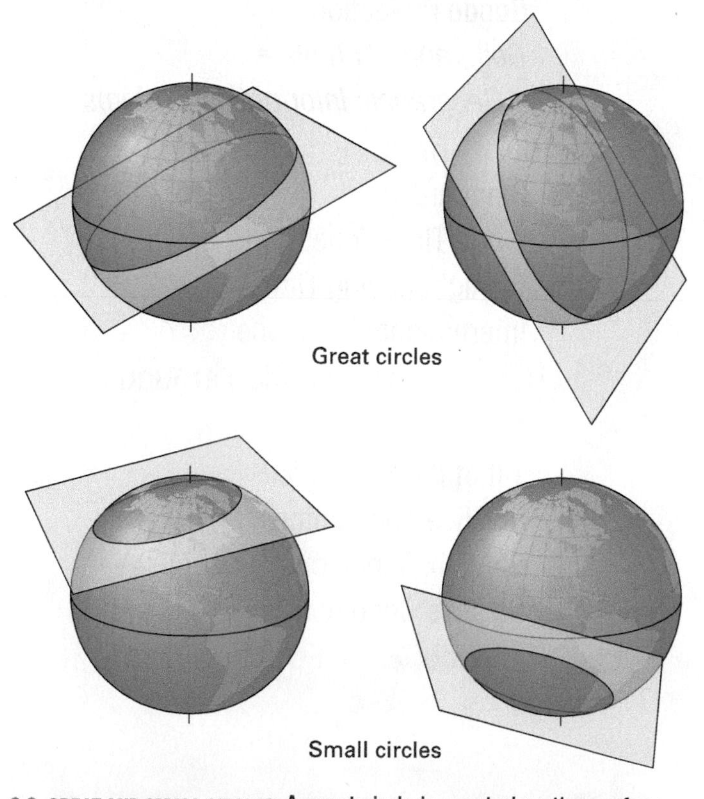

2.3 **GREAT AND SMALL CIRCLES** A great circle is created on the surface of a sphere when the plane of intersection with the surface also passes through the centre of the sphere irrespective of its inclination. Small circles are created when a plane passes through the sphere but does not intersect the centre point.

in global travel where air traffic routes and sea routes optimize travel time and transportation costs by following great circle routes wherever possible. This can be seen in the arcs representing Air Canada's long-distance routes in Figure 2.4.

LATITUDE AND LONGITUDE

The location of each point on Earth is precisely defined by the intersection of a specific meridian and parallel. Each location is represented conventionally by latitude and longitude, which represent lengths of arcs measured along the appropriate meridian and parallel (Figure 2.5). Because the equator is unique in that it is the only parallel that is a great circle, it is given the value of 0° latitude. Parallels of **latitude** measure the angular distance north or south of the equator. In the northern hemisphere, latitude ranges from 0° to 90° N at the North Pole; in the southern hemisphere, latitude increases to 90° S at the South Pole.

By international convention the meridian running through the Royal Observatory in Greenwich, England, is used as the prime meridian of the world. It is commonly referred to as the Greenwich meridian and has the value 0° longitude. Meridians of **longitude** measure angular distance east or west of the prime meridian. Longitude ranges from 0° to 180° E or 180° W.

For greater precision, degrees of latitude and longitude can be subdivided into *minutes* and *seconds*. A minute is 1/60 of a degree, and a second is 1/60 of a minute, or 1/3,600 of a degree. Latitude and longitude can also be equated with kilometres and other standard units of distance. One degree of latitude, which measures distance in a north-south direction, is equal to about 111 km; one minute of latitude represents about 1,850 m and one second of latitude about 31 m. Thus, Igloolik (69° 33′ N; 81° 18′ W) in Nunavut is located approximately 2,947 km due north of London, Ontario (43° 00′ N; 81° 20′ W). However, the distance associated with one degree of longitude varies, because meridians converge toward the poles. Thus, the length of one degree of longitude varies from approximately 111 km at the equator to 0 km at the poles (see Table 2.1). For example, Nanaimo, British Columbia (49° 10′ N; 123° 56′ W) is located about 57½ degrees west of Sept Isles, Quebec (50° 12′ N; 66° 23′ W); this is more or less the same difference in longitude that separates Vanderhoof, British Columbia (54° 01′ N; 124° 01′ W) and Schefferville, Quebec (54° 52′ N; 67° 01′ W). However, because of their respective latitudes, the actual distance between each pair of locations is quite different—approximately 3,750 km in the case of Nanaimo and Sept Isles, and 4,100 km for Vanderhoof and Schefferville. The *Working It Out • Distances from Latitude and Longitude* feature in the end-of-chapter material shows how to convert latitude and longitude coordinates to equivalent ground distances.

Traditionally, the latitude of a point on the Earth's surface was calculated from the position of the stars, and longitude was determined with the aid of an accurate clock set to Greenwich Mean Time. These procedures have now been superseded by the **Global Positioning System (GPS)**, which can provide location information to an accuracy of about 10 m horizontally and 15 m vertically. The system uses 24 satellites that orbit the Earth at an altitude of 20,200 km every 12 hours, continuously broadcasting their position and a precise time signal. The satellite constellation was completed in 1994 and is maintained by a U.S. Air Force ground station in Colorado.

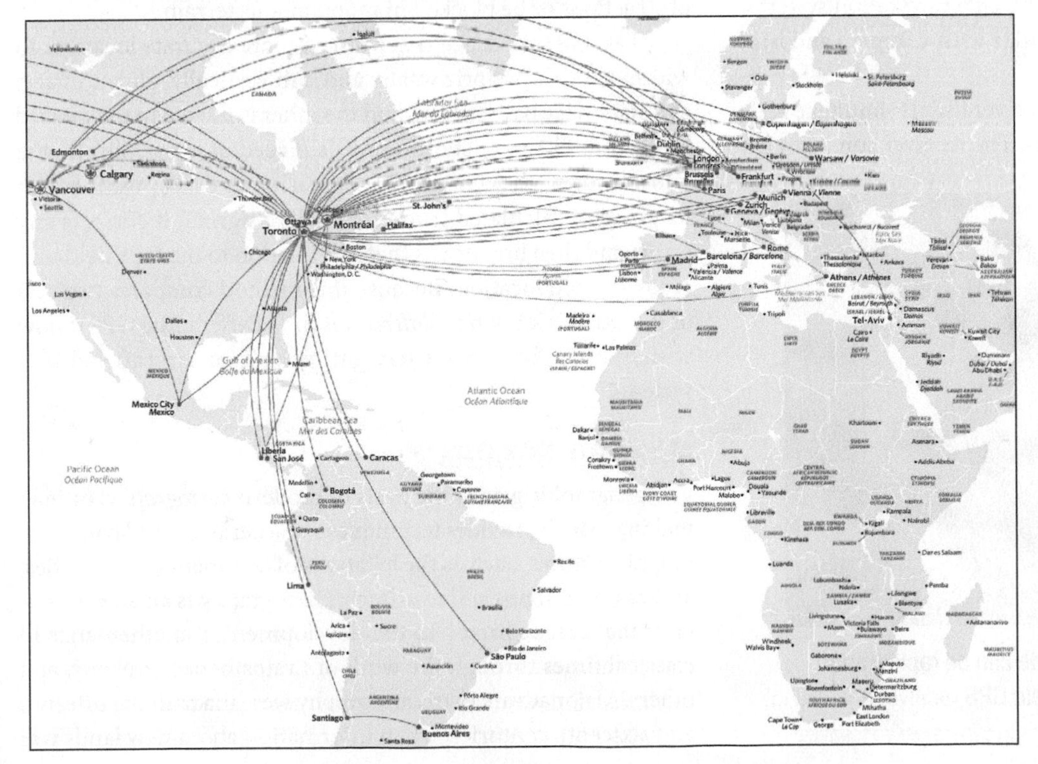

2.4 AIR CANADA ROUTE MAP Long-distance travel times and transportation costs are reduced by following great circle routes because they represent the shortest distance between two locations. Note, this is a stylized map and the routes only approximate the actual courses followed by the aircraft.

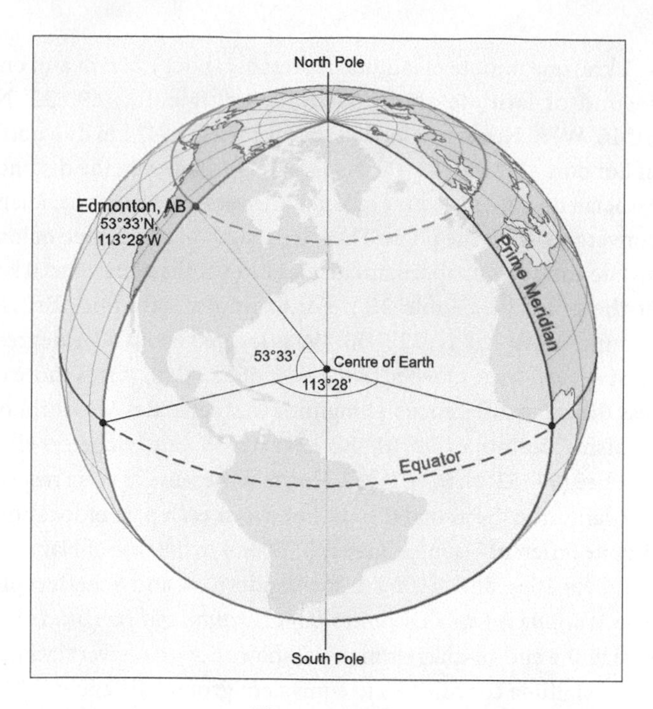

2.5 LATITUDE AND LONGITUDE The geographic coordinate system of latitude and longitude gives locations of points on Earth's surface as angular distances measured from the centre of the Earth with respect to specified planes of reference. Latitude represents the angular distance north or south of the equator. Longitude is the angular distance east or west of the prime meridian. Thus, Edmonton lies 53° 33′ north of the equator (latitude) and 113° 28′ west of the prime meridian (longitude).

Many vehicles are now equipped with GPS navigation systems, and small hand-held units are popular with campers and hikers (Figure 2.6).

To determine location, a GPS receiver listens simultaneously to signals from four or more satellites. The receiver compares the time of readings transmitted by each satellite with the receiver's own clock to determine how long it took for each signal to reach the receiver. Since the radio signal travels at a known rate, the receiver can convert the signal travel time into the distance between the receiver and the satellite. The receiver calculates its position on the ground by combining the distance to each satellite with their orbital positions at the time of the transmissions.

2.6 GPS RECEIVERS Latitude and longitude can be rapidly and accurately determined using a hand-held GPS receiver or a unit mounted in a vehicle.

TABLE 2.1 EQUIVALENT DISTANCE IN KILOMETRES OF 1 DEGREE OF LONGITUDE AT VARIOUS LATITUDES

LATITUDE	LENGTH OF 1 DEGREE OF LONGITUDE
0	111.32 km
15	107.55 km
30	96.45 km
45	78.85 km
60	55.80 km
75	28.90 km
90	0 km

Several types of errors can affect the accuracy of a GPS location. First, unpredictable events, such as solar particle showers, cause small perturbations in the orbits of the satellites. Another source of error arises from small variations in the atomic clock that each satellite carries. A larger source of error, however, is the effect of the atmosphere on the satellite signal. The layer of charged particles at the outer edge of the atmosphere (ionosphere) and water vapour slow the radio waves. Since these conditions change continuously, the speed of the radio waves varies in an unpredictable way. Another problem is that the radio waves may bounce off local obstructions or be blocked in mountainous terrain.

Two GPS units can determine a more accurate location, to within about 1 m horizontally and 2 m vertically. One unit acts as a stationary base station, and the other is moved to the desired locations. The base station unit is placed at a position that is known with high accuracy. By comparing its exact position with its position calculated from the satellite signals, it corrects any errors and then broadcasts that information to the GPS field unit at the desired location. Because this method compares two sets of signals, it is known as *differential GPS*. Differential GPS is now widely used for coastal navigation and in aircraft landing systems.

MAP PROJECTIONS

The geographic grid is the basis of modern *cartography*, or map making, which provides techniques for accurately displaying geographic features, such as the locations of continents, rivers, cities, and roads on maps and in atlases. Cartography is an ancient science that can be linked to the development of mathematics in classical times through the work of Eratosthenes, Ptolemy, and others. Major advances in cartography were made in the fifteenth and sixteenth centuries, when information about new lands was quickly accumulating following the voyages of discovery undertaken by people such as Columbus and Magellan. Mercator's world map, an indispensable navigational aid that is still used today, dates from this period; it was first published in 1569. Later voyages, such as those of Cook and Vancouver in the eighteenth century, also supplied important details of hitherto uncharted territories. The maps of the Pacific Northwest coast produced by Vancouver are remarkably accurate. It is therefore hard to imagine how he failed to locate the two biggest rivers in the region—the Columbia and the Fraser.

Because the Earth's shape is nearly spherical, it is impossible to represent it on a flat sheet of paper without cutting, stretching, or otherwise distorting the curved surface in some way. The main properties affected are direction, distance, shape, and area. A map preserves direction when the compass bearing between two points is correct, irrespective of where those points are located on the map. Similarly, distance and shape are preserved when the outlines of features and their respective areas as portrayed on a map are uniformly proportional to the real world locations they represent.

There are various methods to change the actual geographic grid of curved parallels and meridians into a flat coordinate system. This transformation is achieved mathematically using various **map projections**. Although there are many different map projections, no single projection can simultaneously preserve all of the properties that can be affected. The purpose of the map, therefore, determines projection to use; three common ones are the polar projection, Mercator projection, and Goode projection. Appendix 2.1 • *Focus on Maps*, at the end of this chapter, provides more information about how maps are made and how they display information.

POLAR PROJECTION

The **polar projection** (Figure 2.7) can be centred on either the North or South Pole. Meridians are straight lines radiating outward from the pole, and parallels are nested circles centred on the pole. The space between adjacent parallels increases outward from the centre. The equator usually forms the boundary of the map. Because the parallels and meridians intersect at right angles, this projection shows the true shapes of small areas, such as islands and lakes. However, because the scale increases away from the centre, shapes look disproportionately larger toward the edge of the map. Polar projections can be constructed in several different ways. The example shown in Figure 2.7 is known as a *gnomonic projection*. The gnomonic projection has the special property that any two points connected by a straight line is a great circle and represents the shortest distance between those points. The gnomonic projection is therefore useful for navigation and often is used in conjunction with the Mercator projection.

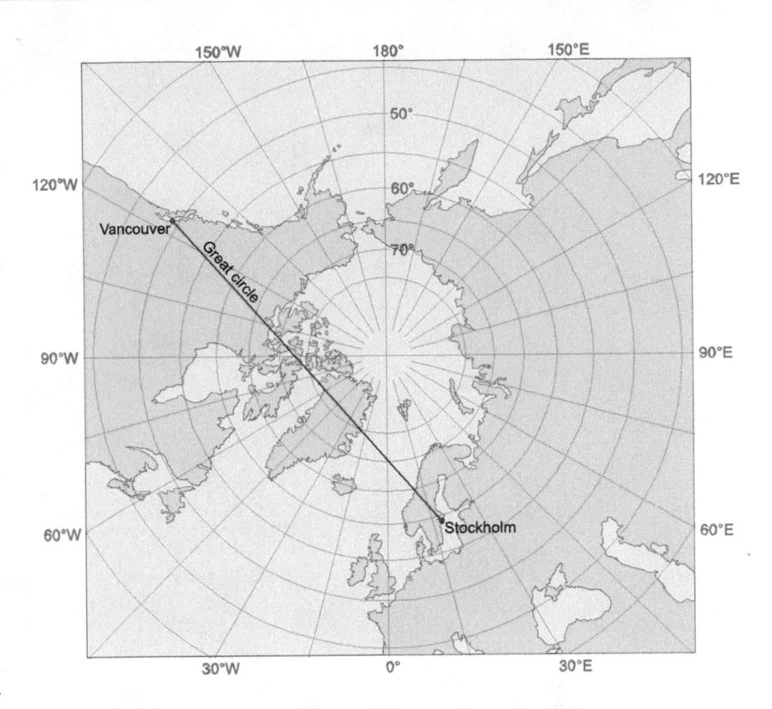

2.7 A POLAR PROJECTION This particular projection is a polar *gnomonic projection* centred on the North Pole. All meridians are straight lines radiating from the centre point, and all parallels are concentric circles. The scale increases toward the periphery, as indicated by the wider spacing between adjacent parallels. The route drawn from Vancouver to Stockholm is also shown on the Mercator projection (Figure 2.8); intermediate points on a route plotted on the gnomonic projection can be used to approximate a great circle route on the Mercator projection.

MERCATOR PROJECTION

The **Mercator projection** (Figure 2.8) is a rectangular grid with meridians shown as straight vertical lines, and parallels as straight horizontal lines. Meridians are evenly spaced, but the spacing between parallels increases with latitude and at 60° is double that at the equator. Closer to the poles, the spacing increases even more, and the map must be cut off at some arbitrary parallel; about 80° N and 65° S are typically used. This change of scale enlarges features near the poles. For example, Greenland appears to be nearly the size of Africa. In fact, the surface area of Greenland is about 2,175,000 km^2, compared with about 30,000,000 km^2 for Africa.

The Mercator projection has several special properties that make it indispensable for navigation. One is that a straight line drawn anywhere on the map is a line of constant bearing or constant compass direction. A navigator can therefore determine the direction to travel by drawing a line between the starting point and destination and measuring the angle with respect to north. Since all meridians on a Mercator projection are true north-south lines, the angle between any meridian and the plotted course will give the compass bearing to follow. Once heading in that compass direction, a ship or airplane would eventually reach its destination.

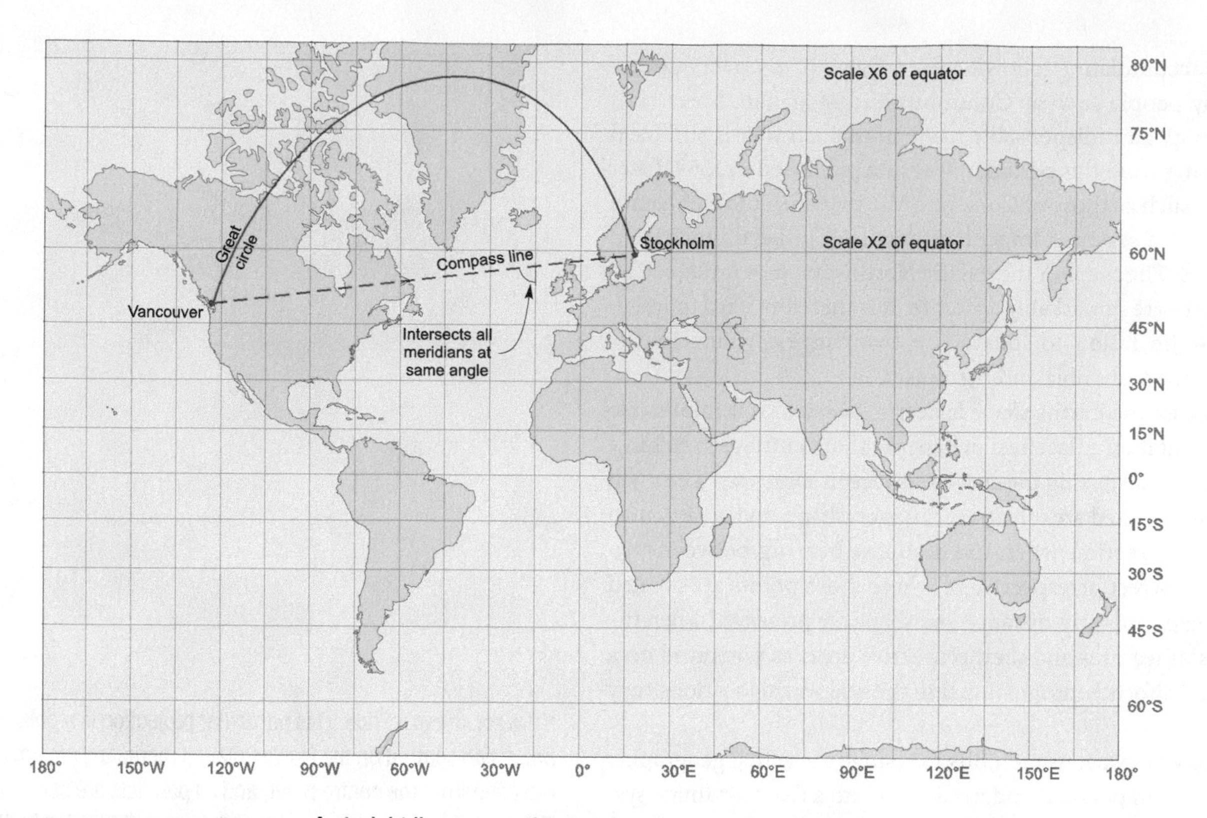

2.8 THE MERCATOR PROJECTION A straight line connecting two locations, such as Vancouver and Stockholm, shows the compass bearing of the course that directly connects them. However, the shortest distance between these points lies on a great circle, which is represented by a curving line on this map projection. Map scale increases rapidly at higher latitudes. At lat. 60°, the scale is double the equatorial scale. At lat. 80°, the scale is six times greater than at the equator. For this reason, a Mercator map is typically truncated at about 80° N and 65° S. Intermediate points along a route plotted on a gnomonic projection (Figure 2.7) can be used to approximate a great circle route on the Mercator projection.

However, this route will not necessarily follow the shortest actual distance between two points, since the shortest path between two points on the globe is always a portion of a great circle. On a Mercator projection, great circle lines are curves (except on the equator) and can falsely seem to represent a much longer distance than a compass line. Using the gnomonic projection in conjunction with the Mercator projection allows a navigator to approximate the great circle route. This is done by transferring intermediate points along the route from the gnomonic projection to the Mercator projection.

Because the Mercator projection shows the true compass direction of any line on the map, it is used to show many features, such as the flow lines of winds and ocean currents, and lines depicting global temperature and pressure patterns.

GOODE PROJECTION

The **Goode projection** uses two sets of mathematical curves to form its meridians. It uses sine curves between the 40th parallels, and beyond the 40th parallels, toward the poles, it uses ellipses. Since the ellipses converge to meet at the pole, the entire globe can be shown. The straight, horizontal parallels make it easy to scan across the map at any given level.

The Goode projection represents areas in correct proportion to the Earth's surface. Because of this, the Goode map is used to show regional distributions of geographical features, such as the soils and vegetation that occur in various parts of the world. However, it distorts shape, particularly in high latitudes. This effect is minimized by separating the map into sectors, each centred on a different vertical meridian (Figure 2.9). This type of split map is called an *interrupted projection.* Although the interrupted projection greatly reduces shape distortion, it does have the drawback of separating parts of the Earth's surface that are actually close together.

Maps are in wide use today for many applications as a simple and efficient way to compile and store spatial information—information associated with a specific location or area of the Earth's surface. However, in the past two decades, maps have been supplemented by more powerful computer-based methods for

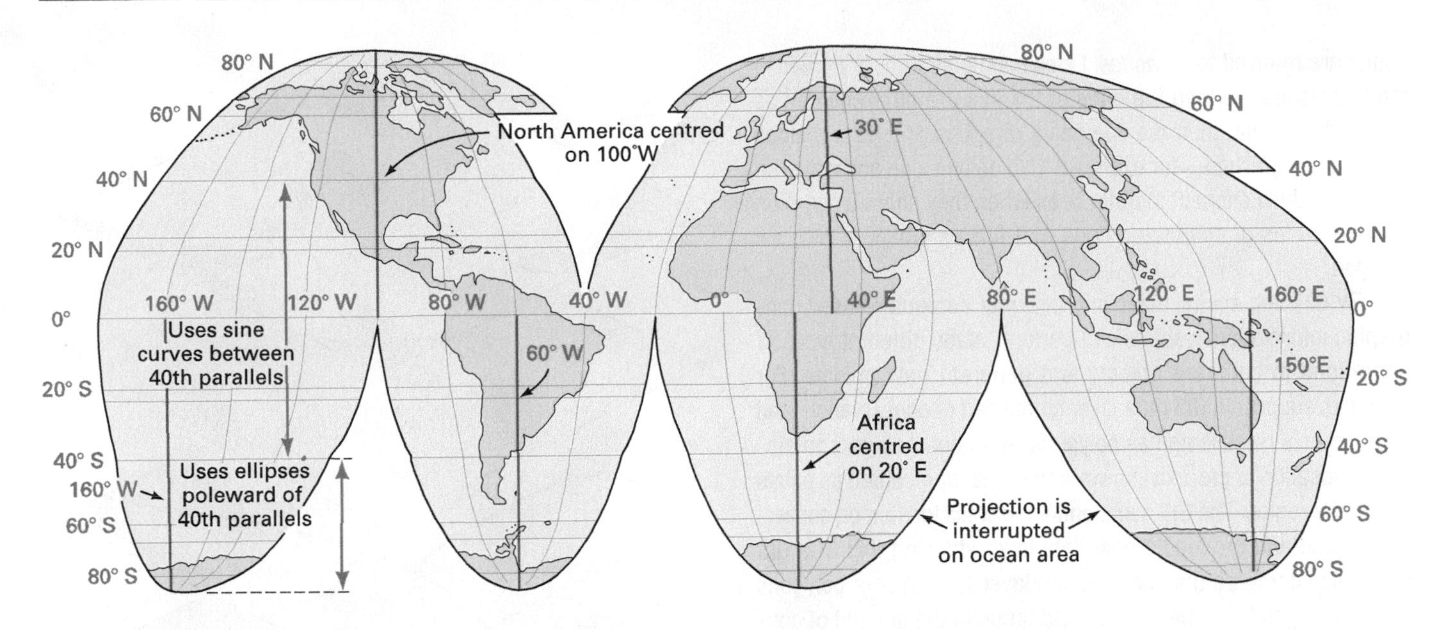

2.9 **THE INTERRUPTED GOODE HOMOLOSINE PROJECTION** The meridians in this projection follow sine curves between lat. 40° N and lat. 40° S, and ellipses between lat. 40° and the poles. Although the shapes of continents are distorted, their relative surface areas are preserved.

GEOGRAPHER'S TOOLS
GEOGRAPHIC INFORMATION SYSTEM

What Is a GIS?

A **geographic information system** contains software to acquire, process, store, query, create, analyze, and display spatial data—pieces of information that are in some way associated with a specific location or area of the Earth's surface. A simple example of a GIS is a map overlay system. Imagine that a planner is deciding how to divide a tract of land into building lots. Appropriate inputs might include a topographic map showing land elevations, a vegetation map showing the type of existing plant cover, a map of existing roads and trails, a map of streams and watercourses, a map of wetland areas, a map of power transmission lines and gas pipelines crossing the area, and so forth. From these, the planner could identify steep slopes, conservation areas, and other attributes before laying out roads, designing bridges, and planning for other essential amenities.

To do this efficiently, the planner needs to be able to overlay the various maps. That's where computer-based geographic information systems come in; they allow the planner to manipulate spatial data for a variety of applications. GIS technology can alleviate problems that might arise, for example, when the input maps are drawn at different scales or when map data need further analysis, such as to identify areas of steep slopes.

Spatial Objects in Geographic Information Systems

Geographic information systems are designed to manipulate spatial objects. A *spatial object* is a geographic area to which some information is attached. This information may be as simple as a place name, or as complicated as a large data table. A spatial object will normally have a boundary, described as a *polygon*, that outlines the object by means of a series of points and lines.

A *point* can be considered a special type of spatial object with no area. A *line* is also a spatial object with no area, but it has two points associated with it, one at each end of the line. These special

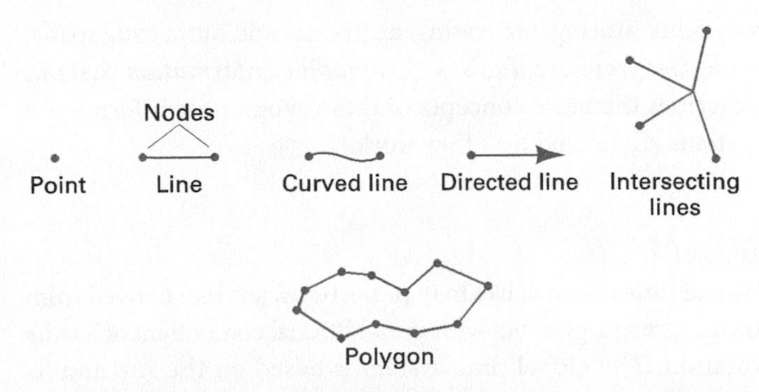

Spatial objects Spatial objects in a GIS can include points, lines of various types, intersecting lines, and polygons.

points are referred to as *nodes.* Lines can be straight or curved. If the two nodes marking the ends of the line are differentiated as the start and end, then the line has a direction and its two sides can be distinguished—for example, land on one side and water on the other. Lines connect to other lines when they share a common node. A series of connected lines that form a closed chain is a polygon.

By defining spatial objects in this way, computer-based geographic information systems can perform many different types of operations to compare objects and generate new objects. For example, suppose a GIS data layer composed of conservation land in a region is represented as polygons, and another layer containing the location of pre-existing water wells is represented as points within the region. The GIS can identify the wells located on conservation land and produce a new data layer showing this. The GIS could also compare the conservation layer to a layer of polygons showing vegetation type, and it could tabulate the amount of conservation land in forest, grassland, brush, and so on. Similarly, it could calculate distance zones around a spatial object to create a map of buffer zones located within, say, 100 m of conservation land.

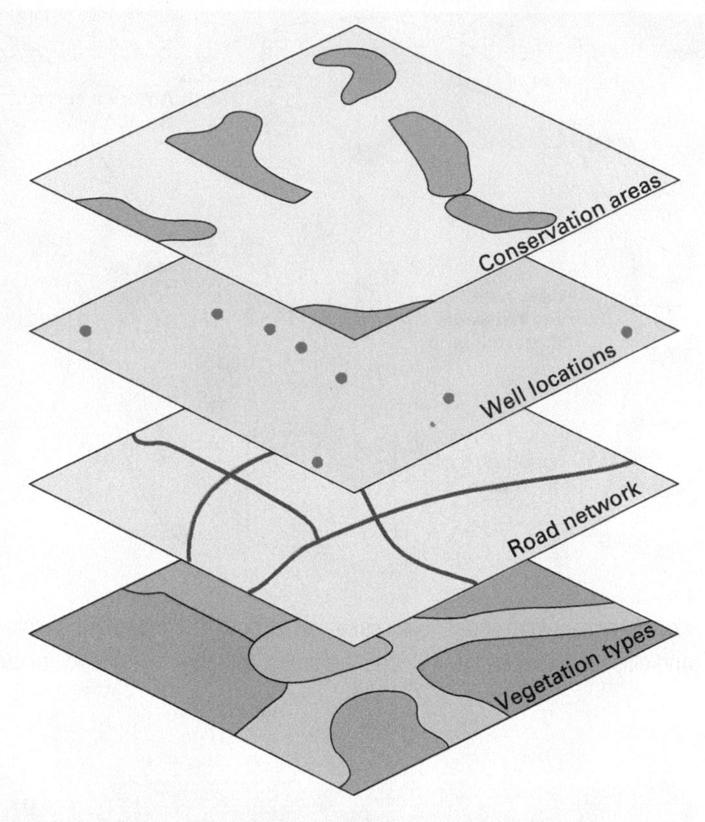

Data layers in a GIS A GIS facilitates overlay of spatial data layers for such queries as "Identify all wells on conservation land."

Key Elements of a GIS

A geographic information system consists of five elements: data acquisition, preprocessing, data management, data manipulation and analysis, and product generation. Each component or process is needed to ensure the functioning of the system as a whole. In the *data acquisition* process, appropriate data are gathered together for the particular application. These may include maps, air photos, or information tables. In *preprocessing,* the GIS converts the assembled spatial data into forms that are compatible with the system to produce data layers of spatial objects and their associated information. The *data management* component creates, stores, retrieves, and modifies data layers and spatial objects, and is essential to the proper operation of all parts of the GIS. The key feature of the GIS is the *manipulation and analysis* component, with which the user asks and answers questions about spatial data and creates new data layers of derived information. The last component of the GIS, *product generation,* produces output in the form of maps, graphics, tabulations, and statistical reports, which are the end products desired by the users.

acquiring, storing, processing, analyzing, and outputting spatial data. *Geographer's Tools • Geographic Information Systems* describes the basic concepts of these geographic information systems (GISs) and how they work.

GLOBAL TIME

Global time systems, like map projections, are also derived from the geographic grid, but with the additional component of Earth's rotation. The global time system is based on the sun and its apparent passage across the sky. In the morning, the sun is low on the eastern horizon. It rises higher as the day progresses, until **solar noon**, when it reaches its highest point in the sky, or *zenith.* After solar noon, the sun's elevation in the sky decreases. By late afternoon, the sun appears low in the sky, and at sunset, it rests on the western horizon. In most locations, the transition from daylight to darkness occurs with each daily rotation of the Earth. The daily cycle is not synchronous throughout the world—it may be a sunny morning in St. John's, Newfoundland, but the residents of Vancouver, British Columbia, are still awaiting dawn.

The starting point for global time systems is the noon meridian—the north-south line along which at any given moment the sun is at its zenith. The noon meridian sweeps around the Earth with every 24-hour rotation. One hour of time is therefore equivalent to 15° of longitude; because of the direction of rotation, places to the east are always ahead.

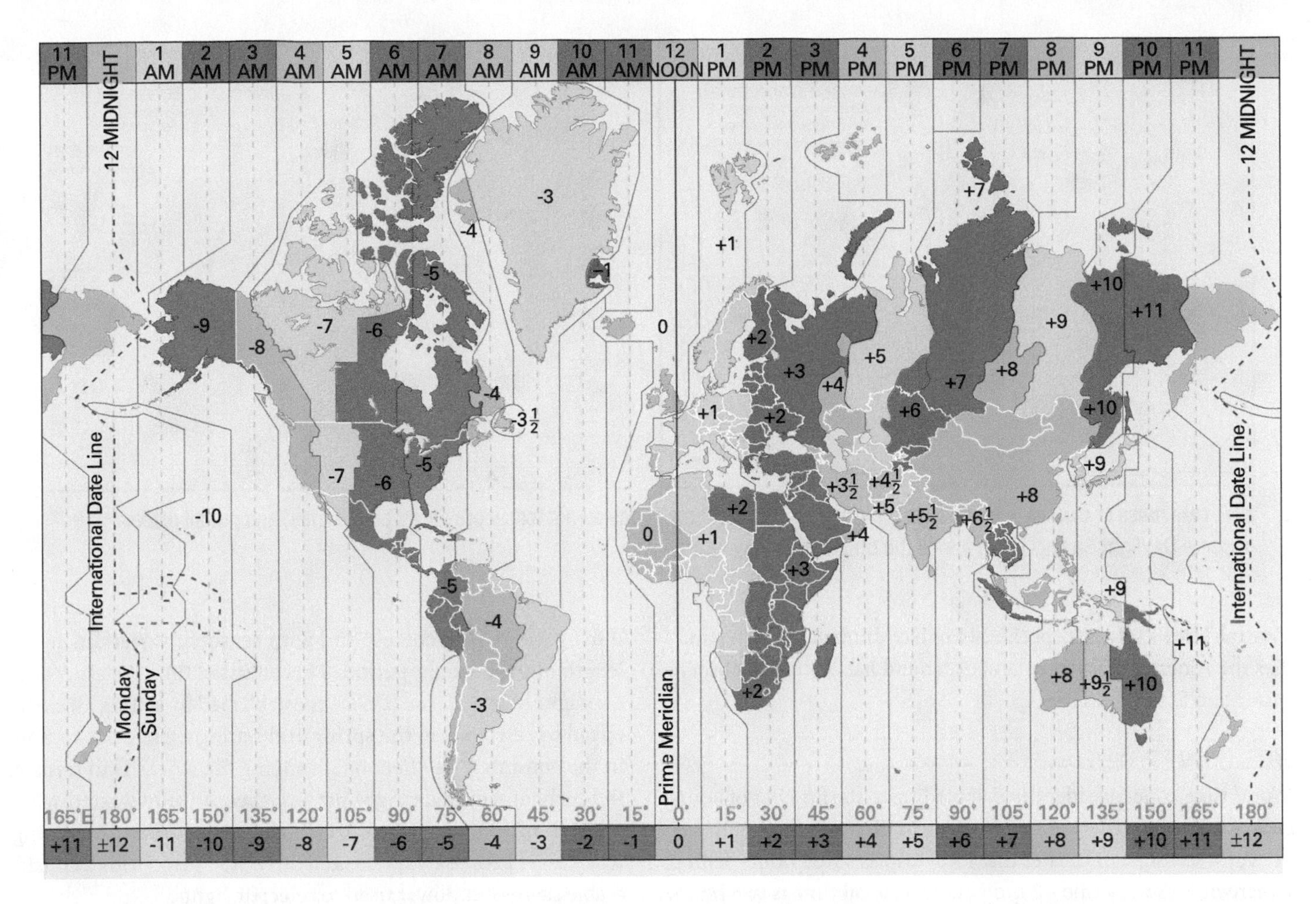

2.10 TIME ZONES OF THE WORLD Dashed lines represent 15° meridians, and bold lines represent 7½° meridians. Alternate zones appear in colour. (After U.S. Navy Oceanographic Office.)

STANDARD TIME

Standard time was first used in Britain in 1840. The world adopted standard time zones in 1884, largely as a result of the efforts of Canadian railway engineer Sir Sandford Fleming. In the **standard time system**, the Earth is divided into 24 **time zones**. All inhabitants within a zone keep time according to the *standard meridian* that passes through their zone. Since the standard meridians are usually 15° apart, the difference in time between adjacent zones is normally one hour.

Theoretically, the standard time system should consist of belts exactly 15° wide, extending to meridians 7½° east and west of each standard meridian. However, this is inconvenient if the boundary meridians divide a province, region, or city into two different time zones. As a result, time zone boundaries are often routed to follow agreed-upon natural or political boundaries. The 24 principal standard time zones of the world are shown in Figure 2.10, together with the time of day in each zone when it is noon at the Greenwich (prime) meridian. The country spanning the greatest number of time zones from east to west is Russia, with eleven zones, but these are grouped into eight standard time zones. China spans five time zones, but runs on a single national time using the standard meridian of Beijing. Canada uses six time zones, and all provinces except Saskatchewan adjust their clocks in spring and autumn in accordance with Daylight Saving Time protocols.

The time zones for Canada in summer and winter are shown in Figure 2.11. Some of the time zone boundaries are conveniently located along provincial boundaries, but there are several exceptions. For example, the Kootenay region of British Columbia sets its clocks to Mountain Standard Time in keeping with the neighbouring province of Alberta. Ontario, Quebec, and Labrador each have two time zones. Most of Nunavut uses Eastern Standard Time; however, Kangiqliniq (Rankin Inlet) uses Central Standard Time and Southampton Island retains Eastern Standard Time year round, as clocks there are not changed in spring and fall. Newfoundland Standard Time is adjusted by 30 minutes to coincide with the half-hour meridian at the provincial capital St. John's. In this way, the time on the clocks is more closely aligned to local solar time. Most of Saskatchewan is in the

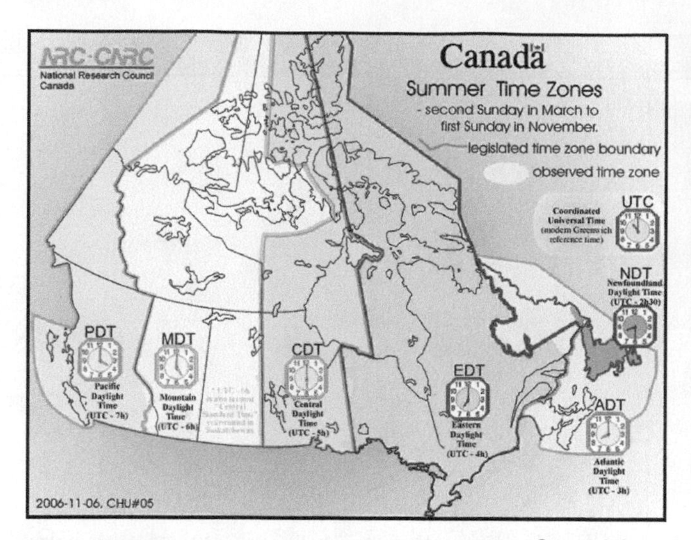

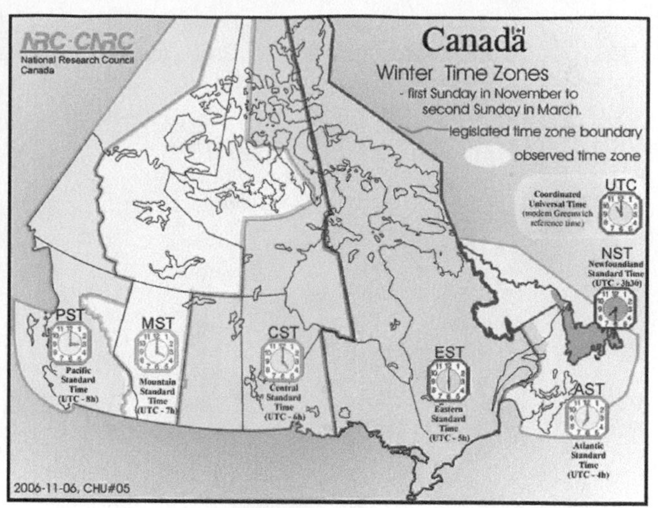

2.11 TIME ZONES OF CANADA DURING SUMMER AND WINTER Several time zones do not follow provincial boundaries, and not all places observe Daylight Saving Time, hence the boundaries vary seasonally.

Central Time Zone, except the city of Lloydminster, which straddles the Alberta-Saskatchewan border and has elected to observe Mountain Standard Time.

WORLD TIME ZONES

World time zones are identified according to the number of hours each time zone differs from the time in Greenwich. A number of −7, for example, indicates that local time is seven hours behind Greenwich Time, while +2 indicates that local time is two hours ahead of Greenwich Time. Because of the historical importance of the Greenwich Observatory, world time was traditionally referenced to Greenwich Mean Time (GMT). Recently, **Coordinated Universal Time (UTC)** has superseded GMT as the global reference. UTC is the legal time standard recognized by all nations and administered by the Bureau International de l'Heure, located near Paris. UTC is a high-precision atomic time standard that is periodically adjusted by *leap seconds* to compensate for discrepancies in the Earth's rotation. It is precisely related to the Earth's angular rotation, rather than to the passage of a 24-hour period.

The relationship between Earth's daily rotation and time has been of fundamental importance in navigation. Prior to the development of modern navigation systems, longitude was established by noting the time on a clock set to GMT when the sun reached its zenith. This established noon local time. By comparing this time to GMT, longitude could be computed knowing that each hour represents a difference of 15° from the prime meridian; each minute of time equals 0.25° and each second 0.004°, or 15 seconds of longitude.

DAYLIGHT SAVING TIME

Many human activities begin well after sunrise and continue long after sunset. In order to correspond more closely with the pace of society, in many parts of the world time adjustments are made during the year to correspond with seasonal variations in the length of the daylight period. This adjusted time system, called **Daylight Saving Time (DST)**, is established by setting all clocks ahead by one hour in the spring and setting them back an hour in the autumn. The effect of advancing the clocks is to transfer the early morning daylight period, theoretically wasted while schools, offices, and factories are closed, to the early evening, when most people are awake and busy. DST also yields a considerable savings in power used for electric lights.

In Canada, DST comes into effect on the second Sunday in March and is discontinued on the first Sunday of November, although it is not observed in some parts of the country. For example, DST is not observed in Saskatchewan, with the exception of Lloydminster in the west of the province and Denare Beach and Creighton in the east. Other places that do not alter their clocks include Fort St. John and Dawson Creek in northeastern British Columbia, Creston in the East Kootenays of British Columbia, and Southampton Island in Nunavut.

DST originated in Germany and Austria on April 30, 1916, and was quickly adopted throughout Europe as a wartime effort to reduce fuel consumption. Manitoba and Nova Scotia introduced DST in 1916, followed by Newfoundland in 1917. Interestingly, several Canadian municipalities implemented Summer Time some years before this date. For example, Moose Jaw, Saskatchewan, introduced Municipal DST in 1912 to run from June 1, until "the end of the summer." DST is now used in many countries, with the exception of those near the equator where the period of daylight changes very little over the course of the year (Figure 2.12). Japan, India, and China are the only major industrialized countries that presently do not observe some form of DST. In the United States, only Arizona and Hawaii do not observe DST. In Australia, Queensland and the Northern Territories do not participate in DST; following a three-year trial

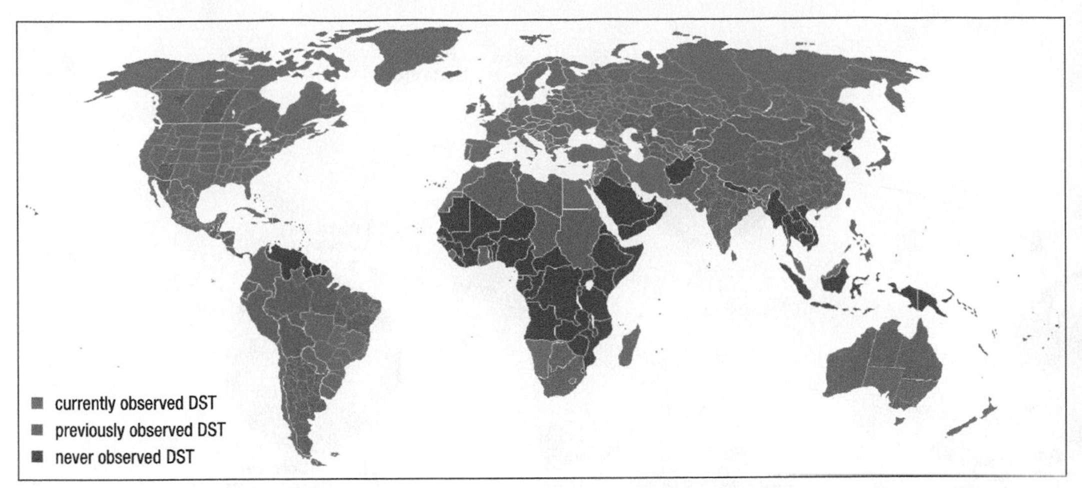

2.12 **DAYLIGHT SAVING TIME** Most industrialized nations in the northern hemisphere have adopted DST; major exceptions are Japan, China, and India. DST is generally not implemented in tropical countries where the period of daylight changes little during the course of the year. (Data from http://www.webexhibits.org/daylightsaving/g.html)

period, DST was also rejected in Western Australia in a referendum held in May 2009.

Variations on DST, including double, single/double, and winter DST, have been proposed on occasion. These are mostly considered when there are concerns about energy consumption, although other benefits, such as reduction in traffic accidents, are also cited. Double DST, in which clocks are advanced two hours, was adopted briefly in Newfoundland in 1988. It was also used year-round in many countries during WWII. Single/double DST results in clocks being moved forward by one hour in winter and two hours in summer; Winter DST is used in conjunction with DST when clocks are advanced by one hour year-round. Both of these options were considered recently in California.

INTERNATIONAL DATE LINE

Counting eastward from the Greenwich meridian, locations on the 180th meridian are 12 hours ahead. Counting westward from the Greenwich meridian, locations on the 180th meridian are 12 hours behind. Since this is the same meridian, an adjustment must be made to resolve this apparent paradox.

Imagine that it is exactly noon on the Greenwich meridian on, say, June 26. At that precise moment, the same 24-hour calendar day covers the entire globe. Time will be progressively later to the east; at the 180th meridian it will be midnight the instant that June 26 is ending. Conversely, to the west of the Greenwich meridian, time becomes progressively earlier; in this case, the time on the 180th meridian is also midnight June 26—but at this instant, June 26 will just be starting. Therefore, it is the same calendar day on both sides of the meridian, but 24 hours apart in time. Only when it is noon at Greenwich is it the same day throughout the world.

Repeating the calculations a few hours later—say at 3 p.m. on June 26 at Greenwich—midnight will occur nine hours to the east of Greenwich at 135° E. Consequently, beyond 135° E it will be early morning on June 27. To the west of the prime meridian, time is behind Greenwich. In Toronto, for example, it would be 10 a.m. on June 26, and in Vancouver it would be 7 a.m. Further west at the 180th meridian it would be 3 a.m. on June 26; immediately across the 180th meridian, it is also 3 a.m., but the date is June 27. Because of this discrepancy, the 180th meridian serves as the **International Date Line**. This means that travelling westward across the date line, calendars must advance by one day. When travelling eastward, calendars turn back by a day. For example, flying westward from Vancouver to Tokyo, an aircraft departing at 2 p.m. on a Tuesday would arrive at 5 p.m. on Wednesday after a flight lasting only 10 hours. On an eastward flight, you may actually arrive the day before you took off.

The International Date Line does not follow the 180th meridian exactly. Like many time zone boundaries, it deviates from the meridian for practical reasons. As shown in Figure 2.10, it has a zigzag offset between Asia and North America, as well as an eastward offset in the South Pacific around New Zealand and several island groups.

THE EARTH'S REVOLUTION AROUND THE SUN

As well as rotating each day on its polar axis, the Earth moves in orbit around the sun. This orbital motion is termed **revolution**. It takes 365.242 days for Earth to complete one orbit of the sun. This is about one quarter day more than a normal calendar year—or *common year*—of 365 days. Every four years, the extra quarter days add up to about one whole day. The 29th day in February in *leap years* corrects the calendar, although further minor corrections are necessary to perfect the system.

Because the Earth traces a slightly elliptical orbit around the sun, the distance between them varies by about 3 percent during each revolution. Earth is nearest to the sun at *perihelion*, which occurs on or about January 3; at this time the distance is 147.5 million km. Earth is farthest from the sun at *aphelion*, on or about July 4, when the distance increases to 152.6 million km. The direction of Earth's revolution around the sun is counter-clockwise, the same direction as Earth's rotation on its axis (Figure 2.13).

The moon also rotates on its axis and revolves around the sun in a counter-clockwise direction. However, the moon's rate

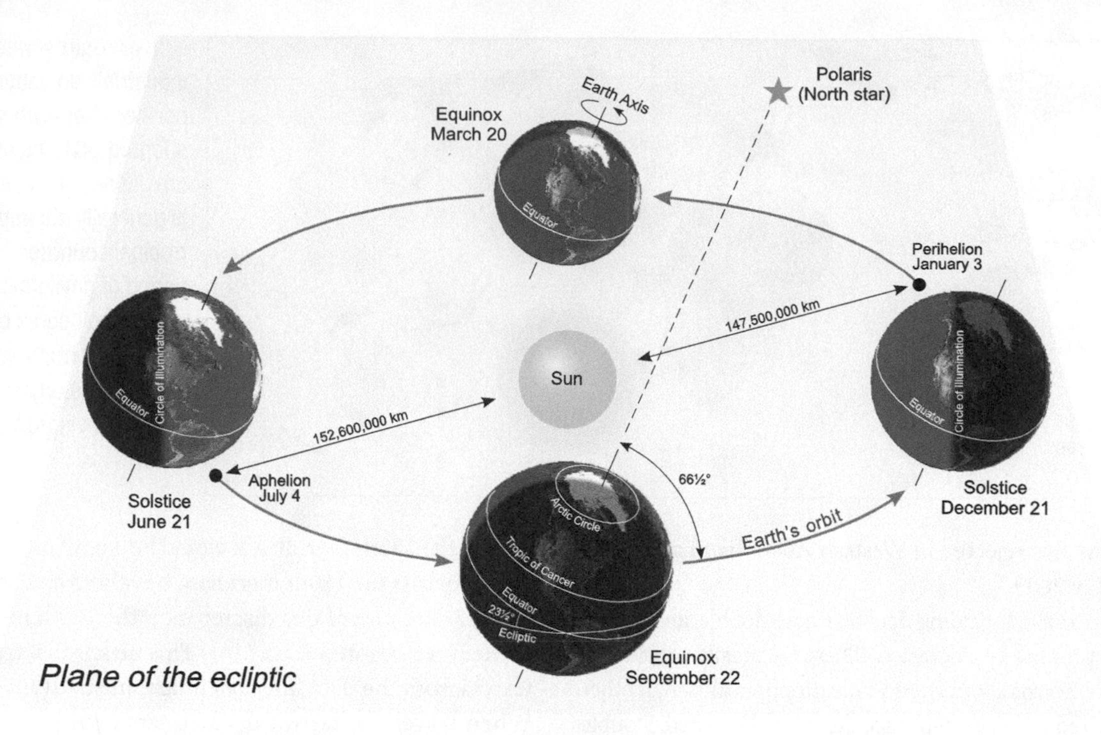

2.13 REVOLUTION OF THE EARTH Viewed from a point over the North Pole, the Earth both rotates and revolves in a counter-clockwise direction. Earth is closest to the sun at perihelion, and furthest from the sun at aphelion (relative distances are greatly exaggerated). (From *Introduction to Weather and Climate* by O.W. Archibold, John Wiley & Sons Canada, Ltd., 2011.)

of rotation is such that one side of the moon is always directed toward the Earth. Thus, astronomers had never seen the far side of the moon until a Soviet spacecraft passing the moon transmitted photos back to Earth in 1959. The phases of the moon are determined by its position in orbit around the Earth during the month (Figure 2.14).

2.14 THE PHASES OF THE MOON The moon orbits Earth every 28 days in a counter-clockwise direction. The new moon occurs when the moon is between the sun and Earth. It waxes through first quarter to full moon when it is on the side away from the sun, then wanes through last quarter to new moon again, and then the cycle repeats.

TILT OF THE EARTH'S AXIS

The seasons arise from the way Earth's axis of rotation is oriented as it revolves around the sun. The Earth's axis is tilted with respect to the **plane of the ecliptic**—the plane circumscribed by the Earth's orbit around the sun. Figure 2.13 shows the plane of the ecliptic as it intersects the Earth. The axis of the Earth's rotation is tilted at an angle of 66½° to the plane of the ecliptic. Note that the direction toward which the axis points is fixed in space—it aims toward Polaris, the North Star. At this stage it can be assumed that neither the angle nor the direction of the axis change as the Earth revolves around the sun, although changes do occur over periods of thousands of years (see Chapter 18).

SOLSTICE AND EQUINOX

The Earth is always divided into two hemispheres with respect to the sun's rays. One hemisphere is illuminated by the sun, and the other lies in darkness (see Figure 2.13). The *circle of illumination*, which separates the illuminated hemisphere from the non-illuminated hemisphere, essentially represents the twilight period at dawn and dusk. Because the direction of Earth's axis of rotation is fixed, the North and South Poles are alternately tilted toward and away from the sun at different points on the orbit. Consequently, the geographic regions encompassed by the illuminated and dark hemispheres change throughout the year.

On or about December 22, the Earth is positioned so that the North Pole is inclined at an angle of 23½° away from the sun, and the South Pole is inclined at the same angle toward the sun. In Canada, this event is usually called the **winter solstice**. But while it is winter in the northern hemisphere, it is summer in the southern hemisphere, so the preferred term is *December solstice* to avoid any confusion. At this time of the year, the southern hemisphere is tilted toward the sun and receives strong solar heating.

Six months later, on or about June 21, the Earth is at the opposite point on its orbit. On this date, known in the northern hemisphere as the **summer solstice** (or *June solstice*), the North Pole is tilted at an angle of 23½° toward the sun and the South Pole is tilted away from the sun. Theoretically, this is the time of maximum solar energy gain for the northern hemisphere, and minimum energy gain for the southern hemisphere.

The **equinoxes** occur midway between the date of the solstices, and at these times the Earth's axial tilt is neither toward nor away from the sun. The *vernal equinox* (spring equinox) occurs on or about March 21, and the *autumnal equinox* (fall equinox) on or about September 22. The orientation of the Earth with the sun is identical during the two equinoxes. Note that the date of the solstices and equinoxes in a particular year may vary by a day or so, since the period of revolution is not exactly 365 days.

EQUINOX CONDITIONS

At the equinoxes, the circle of illumination passes through the North and South Poles (Figure 2.15). Consequently, the sun's rays just touch the Earth's surface at either pole. On these dates, the *subsolar point,* the point on the Earth's surface where the sun at noon is directly overhead, falls on the equator. Here, the angle between the sun's rays and the Earth's surface is 90°, and solar radiation theoretically is at its most intense. At intermediate latitudes, such as 40° N, the rays of the sun at noon strike the surface at a lesser angle. This is the *angle of elevation*, which represents the elevation of the sun above the horizon at noon and is equal to 90° minus the latitude, in this example 90° − 40° (latitude) = 50° (angle of elevation). A *sextant*, a common navigation instrument used by sailors, is based on this relationship. To calculate latitude, a sight is taken at noon to determine the elevation of the sun. The elevation angle is complementary to the *zenith angle*, which is the angular measurement from a point directly above the observer to the sun (zenith angle = 90° − angle of elevation). At the time of the equinoxes, latitude is equal to 90° minus the elevation angle, and so it is the same as the zenith angle. At any other time of the year, a correction has to be applied to compensate for the sun's movement north and south of the equator with the changing seasons (see below).

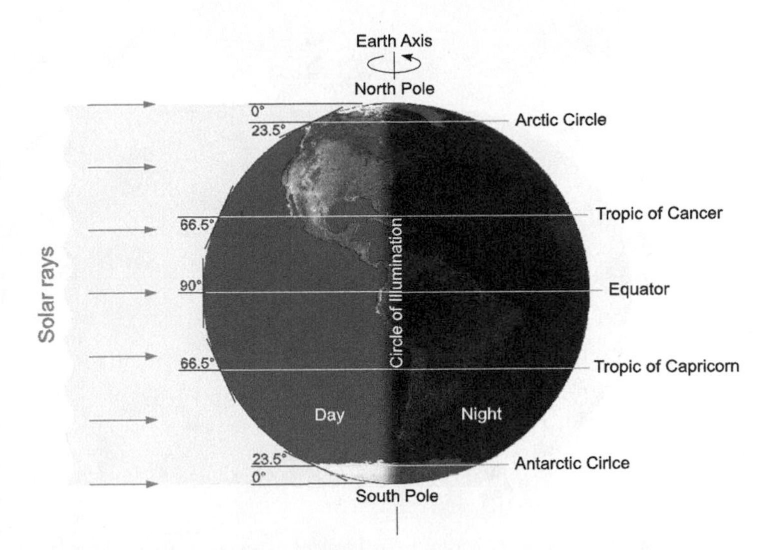

2.15 EQUINOX CONDITIONS At this time, the Earth's axis of rotation is exactly at right angles to the direction of solar illumination. The subsolar point lies on the equator. The sun's rays are at a tangent with Earth's surface at the poles. (From *Introduction to Weather and Climate* by O.W. Archibold, John Wiley & Sons Canada, Ltd., 2011.)

Every point on the Earth's surface receives 12 hours of darkness and 12 hours of daylight at the time of the equinoxes. This is because the circle of illumination passes through the poles, dividing every parallel exactly in half. The word "equinox" is derived from the Latin *aequus* (equal) and *nox* (night). Note that the term daylight is used to describe the period of the day during which the sun is above the horizon. When the sun is not too far below the horizon, scattering of solar rays by atmospheric particles still lights the sky and twilight is observed. At high latitudes during the polar night, the twilight period can last for several hours and provide enough illumination for many human activities.

SOLSTICE CONDITIONS

In both the summer and winter solstices, the circle of illumination passes from the Arctic Circle to the Antarctic Circle (Figure 2.16). Note that during the summer solstice the Earth's orientation puts the entire region north of the **Arctic Circle** in 24 hours of daylight. The situation is reversed in the southern hemisphere, where regions beyond the **Antarctic Circle** are in continuous darkness. During the summer solstice, the subsolar point is at lat. 23½° N, the parallel known as the **Tropic of Cancer**. Because the sun is directly over the Tropic of Cancer at this time, solar energy is theoretically most intense at this latitude.

At the winter solstice, conditions are exactly reversed from those of the summer solstice. Everywhere south of lat. 66½° S lies under the sun's rays for 24 hours, and the subsolar point is now positioned over the **Tropic of Capricorn** at lat. 23½° S.

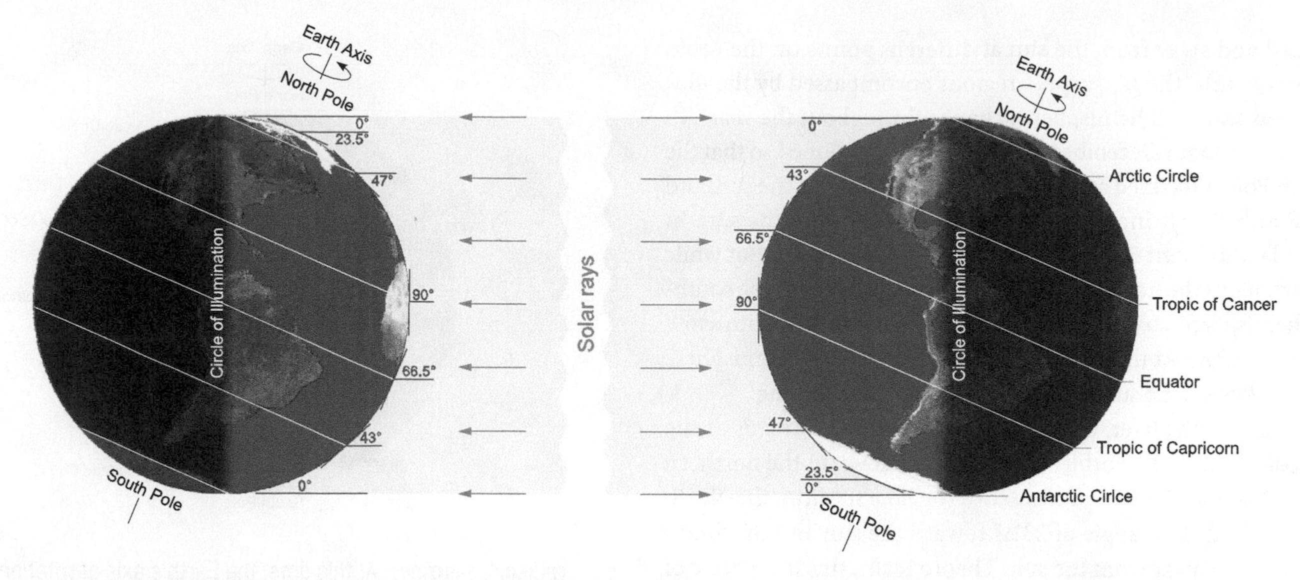

2.16 SOLSTICE CONDITIONS Because of the tilt of the Earth's axis, polar regions experience either 24-hour daylight or 24-hour darkness. The subsolar point lies on the Tropic of Cancer (lat. 23½° N) in June (left) and on the Tropic of Capricorn (lat. 23½° S) in December (right). (From *Introduction to Weather and Climate* by O.W. Archibold, John Wiley & Sons Canada, Ltd., 2011.)

The solstices and equinoxes represent conditions at only four times of the year. During the intervening periods, the subsolar point travels northward and southward in its annual cycle between the Tropics of Cancer and Capricorn. The latitude of the subsolar point is referred to as the sun's **declination** (Figure 2.17). Nautical almanacs contain tables listing these values for every day of the year. Navigators use these tables to adjust sextant readings to establish latitude. By convention, latitudes in the northern hemisphere are positive, and in the southern hemisphere they are negative. Thus, the sun's declination is added or subtracted depending on the season. Using this same convention, the formula for calculating latitude also differs depending on whether the observer is in the northern or southern hemisphere.

2.17 SUN'S DECLINATION The declination of the sun (also referred to as the sun's ephemeris) refers to the latitude where the noon sun is directly overhead (i.e., the zenith angle is 0°). When the subsolar point lies in the northern hemisphere, declination is positive; if it lies in the southern hemisphere, declination is negative.

For observations in the northern hemisphere (i.e., sun is due south at solar noon):

$$\text{Latitude } (°) = 90° \pm \text{sun's declination } (°) - \text{sextant reading } (°)$$

For observations in the southern hemisphere (i.e., sun is due north at solar noon):

$$\text{Latitude } (°) = -90° \pm \text{sun's declination } (°) + \text{sextant reading } (°)$$

A sextant reading near Prince Rupert, British Columbia, of 14° 30′ at noon on January 19 when the sun's declination is –20° 30′, when applied to the northern hemisphere formula, establishes the ship's latitude as:

$$(90° - 20°\ 30' - 14°\ 30') = 54°\ \text{N}$$

Similarly, a sextant reading of 52° 00′ taken at noon on August 9 when the sun's declination is +16° 00′ would give:

$$(90° + 16°\ 00' - 52°\ 00') = 54°\ \text{N}$$

An equivalent reading of 76° 30′ taken on January 19 near Sydney, Australia, and applied to the southern hemisphere formula would give a latitude of:

$$(-90° - 20°\ 30' + 76°\ 30') = -34° \text{ or } 34°\ \text{S}$$

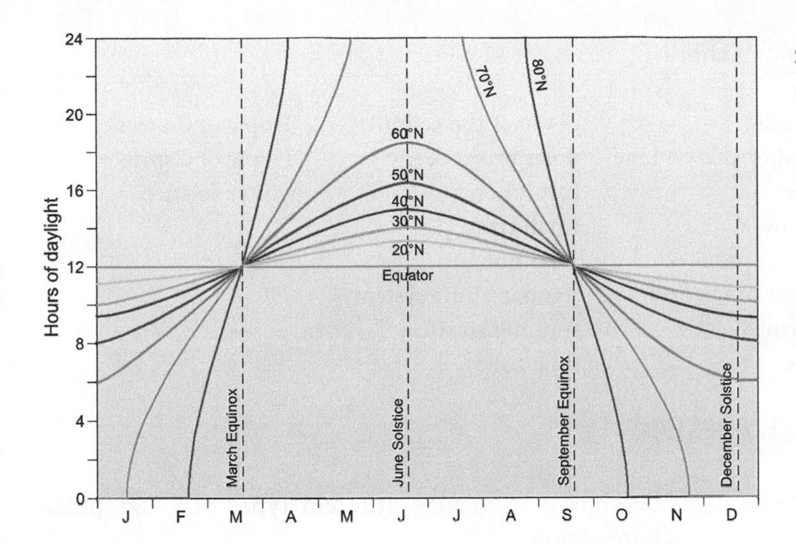

2.18 GLOBAL PATTERNS OF DAYLIGHT Annual variations in hours of daylight length for selected latitudes in the northern hemisphere.

A reading of 40° 00′ on August 9 would give:

$$(-90° + 16° 00' + 40° 00') = -34°, \text{ which again is } 34° \text{ S.}$$

The seasonal cycle is most pronounced in the polar regions. At these high latitudes, the areas that experience 24-hour daylight or 24-hour night shrink as the sun begins to move out of their respective hemispheres, and they grow as it returns. At other latitudes, the length of daylight changes progressively from one day to the next. The difference in duration of daylight and darkness increases with latitude. Only the equator experiences approximately 12 hours of daylight and 12 hours of darkness each day of the year (Figure 2.18).

A Look Ahead

The flow of solar energy to the Earth powers most of the natural processes that occur each day, from changes in the weather to the work of streams in carving the landscape. The next chapter examines in detail solar energy and its interaction with the Earth's atmosphere and surface.

CHAPTER SUMMARY

- The rotation of the Earth on its axis and the revolution of the Earth in its orbit around the sun are fundamental topics in physical geography. The Earth is nearly spherical. It rotates on its *axis* once in 24 hours. The axis of rotation marks the *North* and *South Poles*. The direction of rotation is counter-clockwise.
- The Earth's rotation provides the first great rhythm of our planet—the daily alternation of sunlight and darkness.
- The Earth's axis of rotation provides a reference for the system of location on the Earth's surface—the geographic grid, which consists of meridians and parallels. This system is indexed by the latitude and longitude system.
- Map projections are required to display the Earth's curved surface on a flat map. The polar projection is centred on either pole. The Mercator projection converts the curved geographic grid into a flat, rectangular grid and best displays directional features. The Goode projection distorts the shapes of continents and coastlines, but preserves the areas of land masses in their correct proportion.
- The Earth's rotation is the basis of time. The Earth rotates through 15° each hour. In the standard time system, time is defined by standard meridians that are theoretically 15° apart. Daylight Saving Time is commonly used to advance the clock by one hour. At the International Date Line, the calendar day changes, advancing forward one day for westward travel, and dropping back a day for eastward travel.
- The seasons, the second great rhythm of the Earth, arise from the revolution of the Earth in its orbit around the sun combined with the tilt of the Earth's axis with respect to its orbital plane. The solstices and equinoxes mark the cycle of this revolution. At the June (summer) solstice, the northern hemisphere is tilted toward the sun. At the December (winter) solstice, the southern hemisphere is tilted toward the sun. At the equinoxes, day and night are of equal length.

KEY TERMS

Antarctic Circle
Arctic Circle
Coordinated Universal Time (UTC)
Daylight Saving Time (DST)
declination
Earth rotation
equator
equinox
geographic grid
geographic information system (GIS)
Global Positioning System (GPS)
Goode projection
great circle
International Date Line
latitude
longitude
map projection
Mercator projection
meridian
parallel
plane of the ecliptic
polar projection
revolution
small circle
solar noon
standard time system
summer solstice
time zone
Tropic of Cancer
Tropic of Capricorn
winter solstice

REVIEW QUESTIONS

1. What is the approximate shape of the Earth? How is this known? What is the Earth's true shape?
2. What is meant by Earth rotation? Describe three environmental effects of the Earth's rotation.
3. Describe the geographic grid, including parallels and meridians.
4. How do latitude and longitude determine position on the globe? In what units are they measured?
5. Name three types of map projections and describe each briefly. Give reasons why a different map projection might be chosen to display different types of geographical information.
6. Explain the global time-keeping system. Define and use the terms *standard time, standard meridian,* and *time zone* in your answer.
7. What is meant by the *tilt of the Earth's axis*? How is the tilt responsible for the seasons?
8. What is a geographic information system?
9. Identify and describe three types of spatial objects.
10. What are the key elements of a GIS?

VISUALIZATION EXERCISES

1. Sketch a diagram of the Earth at either the spring or autumnal equinox. Show the North and South Poles, the equator, and the circle of illumination. Indicate the direction of the sun's incoming rays, and shade the night portion of the globe.
2. Sketch a diagram of the Earth at the June (summer) solstice, showing the same features. Also include the Tropics of Cancer and Capricorn, and the Arctic and Antarctic Circles.

ESSAY QUESTION

1. Suppose that the Earth's axis were tilted at 40° instead of 22½°. What would be the global effects of this change? How would the seasons change at your location?

WORKING IT OUT • DISTANCES FROM LATITUDE AND LONGITUDE

Statements of latitude and longitude do not describe distances in kilometres directly. However, for latitude, the conversions from degrees into kilometres can be estimated quite easily. One degree of latitude is approximately equivalent to 111 km of surface distance in the north-south direction. For example, Saskatoon at lat. 52° N is about $52 \times 111 = 5{,}772$ km north of the equator.

East-west distances cannot be converted so easily from degrees of longitude into kilometres because the meridians converge toward the poles. Only at the equator is a degree of longitude equivalent to 111 km (see Table 2.1). At lat. 60° N or S, the distance between meridians is about half that at the equator, or about 56 km. At the poles, the distance is zero.

The formula to determine the length of a degree of longitude is:

$$L_{LONG} = \cos(\text{lat}) \times L_{LAT}$$

where L_{LONG} is the length of a degree of longitude; cos is the trigonometric cosine function; lat is the latitude, in degrees, of the location at which the length is to be calculated; and L_{LAT} is the length of a degree of latitude, that is, 111 km.

For example, a degree of longitude at 20° lat. has a length of:

$$\begin{aligned} L_{LONG} &= \cos(\text{lat}) \times L_{LAT} \\ &= \cos(20°) \times 111 \text{ km} \\ &= 0.866 \times 111 \text{ km} \\ &= 96.1 \text{ km} \end{aligned}$$

This principle can be used to find the distance between two cities located at the same latitude. For example, Vancouver, British Columbia, (49° 16′ N; 123° 06′ W), and Corner Brook, Newfoundland, (48° 57′ N; 57° 57′ W), are separated by 65.15 degrees of longitude.

First, determine the length of a degree of longitude at 49° lat.:

$$\begin{aligned} L_{LONG} &= \cos(\text{lat}) \times L_{LAT} \\ &= \cos(49°) \times 111 \text{ km} \\ &= 0.856 \times 111 \text{ km} \\ &= 72.8 \text{ km} \end{aligned}$$

Then:

$$65.15 \text{ degrees of longitude} \times 72.8 \text{ km} = 4{,}743 \text{ km}$$

QUESTIONS / PROBLEMS

1. Toronto, Ontario (43° 40′ N; 79° 23′ W), and Quito, Ecuador (0° 17′ S; 78° 32′ W), are located on about the same meridian, but they are about 44° lat. apart. What is the approximate distance between these cities?
2. Saskatoon, Saskatchewan, and Oxford, England, are both located at approximately 52° N, but their longitudes are 106° W and 1° W, respectively. What is the approximate distance between the two cities?
3. A map of a region close to the equator shows an area of 1° of latitude by 1° of longitude. About how many square kilometres does the map cover? How does this compare with a 1° by 1° area at lat. 60° N near Churchill, Manitoba?

Find answers at the back of the book.

APPENDIX 2.1 FOCUS ON MAPS

The purpose of this appendix is to provide additional information on the subject of *cartography*, the science of maps and their construction.

More about Map Projections

A *map projection* is an orderly system of parallels and meridians that is used as a base for transferring the features of the spherical Earth to a flat surface. All map projections distort the Earth's shape in some way as a result of cutting, stretching, or other manipulation of the true spatial properties. A simple map projection might consist of a grid of squares or rectangles with horizontal lines representing parallels and vertical lines representing the meridians. Maps that are formed by considering longitude and latitude as a simple rectangular coordinate system are known as unprojected maps. Modern computer-generated world maps often use this type of grid to display data that consist of a single number for each square (see Figure 2A.1). A grid of this kind can represent the approximate spacing of the parallels, but it fails to show how the meridians converge toward the poles. Consequently, scale, distance, area, and shape are all increasingly distorted toward the poles.

Early attempts to find satisfactory map projections used a simple concept. Imagine the continents and a grid of meridians and parallels drawn on a transparent sphere. A light source placed at the centre of the sphere will cast an image of the continents and grid on a surface outside the sphere. Three basic surfaces can be used: flat (or planar), conical, and cylindrical (Figure 2A.2).

The basic principles of a planar projection can be simulated by placing a flat paper disk on the North Pole. The shadow of the grid on this plane surface appears as a combination of concentric circles (parallels) and radial straight lines (meridians). This produces a polar (or zenithal) projection. Now consider a cone resting point up on the sphere. The cone can be unrolled to produce a map that is part of a full circle. This is called a conic projection, in which parallels are arcs of circles, and meridians are radiating straight lines. Similarly, a cylinder touching the sphere at the equator when unrolled produces a cylindrical projection with a rectangular grid. The Mercator projection uses this principle. In practice, each of these types of projections is actually drawn using precise mathematical calculations.

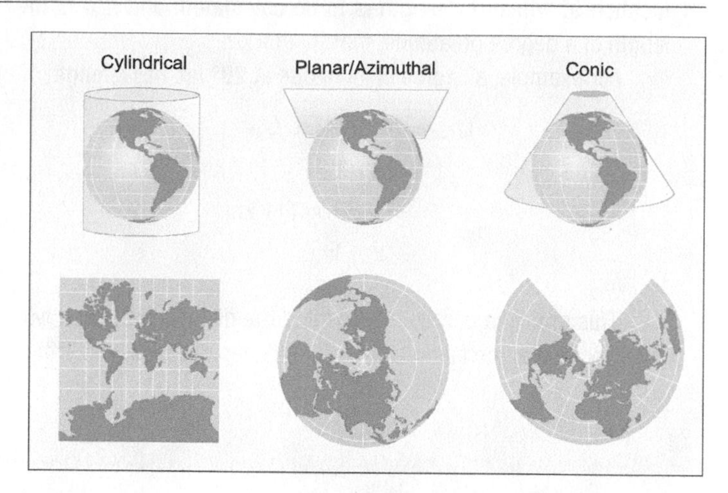

2A.2 Rays from a central light source would cast shadows of a transparent, spherical Earth on different surfaces. The conical and cylindrical surfaces can be unrolled to make flat maps.

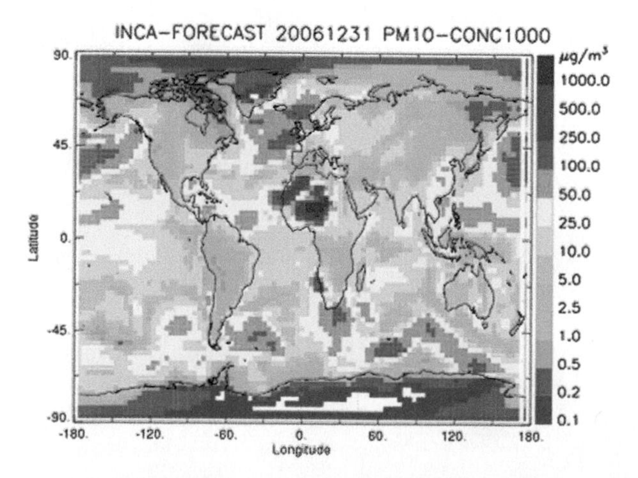

2A.1 A computer-generated map showing the fine particulate matter concentration in the atmosphere on November 28, 2006.

Note that none of these three projection methods can show the entire geographic grid, no matter how large a sheet of paper is used to draw the image. To show the entire Earth, a quite different system must be devised that is entirely based on various mathematical techniques; unlike the planar, conical, and cylindrical projections, the basic principles of these so-called unique projections cannot be portrayed with simple models. The *Goode projection* is an example (see Figure 2.9).

Scales of Globes and Maps

All globes and maps depict the Earth's features at a much smaller size than they really are. Globes are basically scale models of the Earth—the *scale* of a globe represents the ratio between the size of the globe and the size of the Earth. Size in this case refers to some measure of length or distance. For example, a globe 20 cm in diameter would have a scale of 20 cm to 13,000 km (the approximate

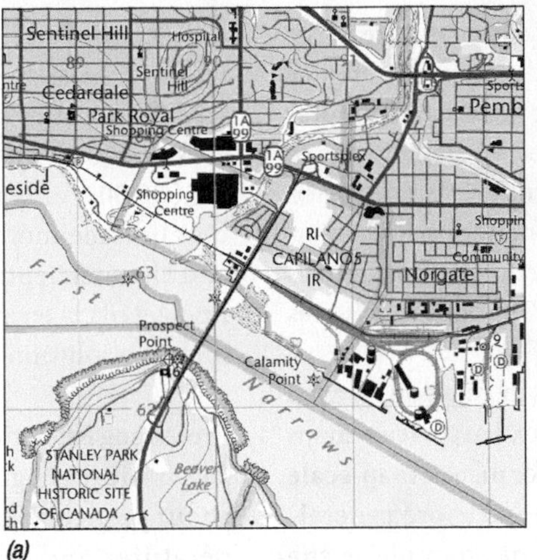

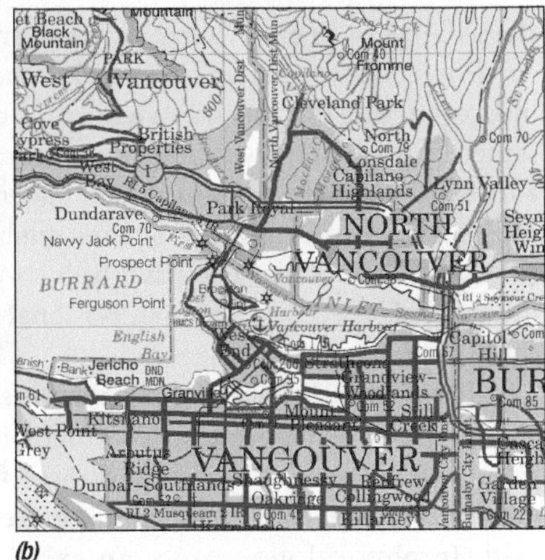

2A.3 In the current National Topographic Series, the Canadian government produces maps at two scales: (a) 1:50,000 and (b) 1:250,000.

diameter of the Earth), which reduces to 1 cm = 650 km. This relationship holds true for distances between any points on the globe.

Scale is more usefully stated as a simple fraction, termed the *representative fraction*, obtained by reducing both the Earth and globe distances to the same unit of measure. The advantage of the representative fraction is that it is entirely free of any specified units of measure, such as metre, kilometre, or mile. Thus, in the previous example, the size of the Earth is converted to centimetres, and the representative fraction is obtained as 20/1,300,000,000, which gives 1/65,000,000 or 1:65,000,000. The fractional notation in which a colon separates the numerator from the denominator is the most common way to express this ratio.

Being a true-scale model of the Earth, a globe has a constant scale everywhere on its surface; however, this is not true of a map projected onto a flat surface. All map projections stretch the Earth's surface in a non-uniform manner so that the map scale changes from place to place. However, on some projections this is negligible if the area being represented is small. It is also possible to select a meridian or parallel—the equator, for example—for which a scale fraction is known so that the amount of distortion can be calculated.

Small-Scale and Large-Scale Maps

When geographers refer to small-scale and large-scale maps, they refer to the value of the representative fraction. For example, a global map at a scale of 1:65,000,000 has a scale value of 0.00000001524, which is obtained by dividing 1 by 65,000,000. A hiker's topographic map at a scale of 1:25,000 has a scale value of 0.00004. Since the global scale value is smaller, it is a small-scale map, while the hiker's map is a large-scale map, even though it represents only a small area (Figure 2A.3). Most large-scale maps carry a *graphic scale,* which is a line marked off into units representing kilometres (Figure 2A.4). Graphic scales make it easy to measure ground distances by marking the positions of two places on the edge of a piece of paper, laying the paper on the graphic scale, and reading the distance directly from it.

Informational Content of Maps

Multipurpose, large-scale maps are capable of portraying a considerable amount of geographic information (see Figure 2A.3a). High-quality maps show this information in a convenient and effective manner using a variety of symbols.

Map symbols associate information with points, lines, and areas. Point symbols consist of various types of dots, such as a

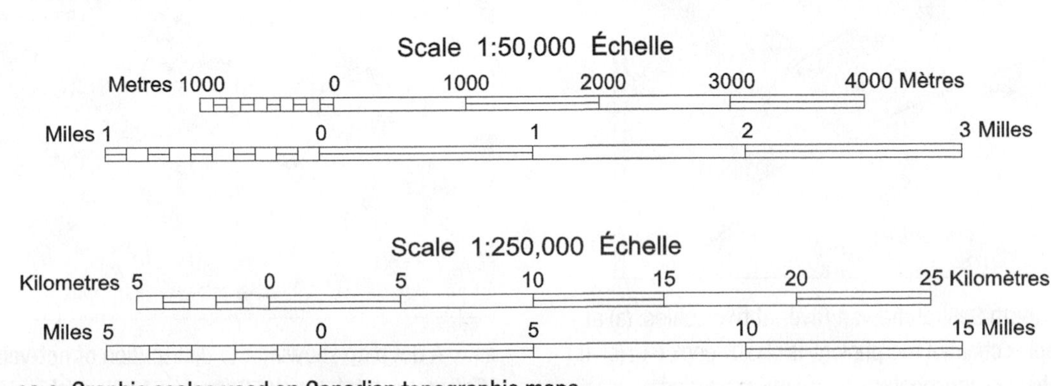

2A.4 Graphic scales used on Canadian topographic maps.

closed circle, open circle, letter, numeral, or a little picture of the object it represents. Linear features like roads are represented by lines of varying widths and styles. Specific areas can be represented by patches, shown simply by a line marking their edge or a distinctive pattern or solid colour. Patterns are highly varied and might consist of tiny dots, stylized symbols, or parallel, intersecting, and wavy lines. The inside back cover of this book shows the symbols used on Canadian National Topographic Series (NTS) maps.

The size and shape of map symbols in relation to map scale is of prime importance in cartography. Large-scale maps can show objects in their true outline form. As map scale decreases, representation becomes more and more generalized. In physical geography, an excellent example is the depiction of a river, such as the South Saskatchewan River. Figure 2A.5 shows the river at two scales. On the larger scale 1:2,000,000 map, the cartographer has shown details of islands and sandbars in the river channel. At a smaller scale of 1:2,000,000, only minimal detail can be shown. As map symbols become more generalized, the details of the river banks and channel bends simplify as well. The level of depiction of fine details is termed *resolution*. Large-scale maps have much higher resolution than small-scale maps.

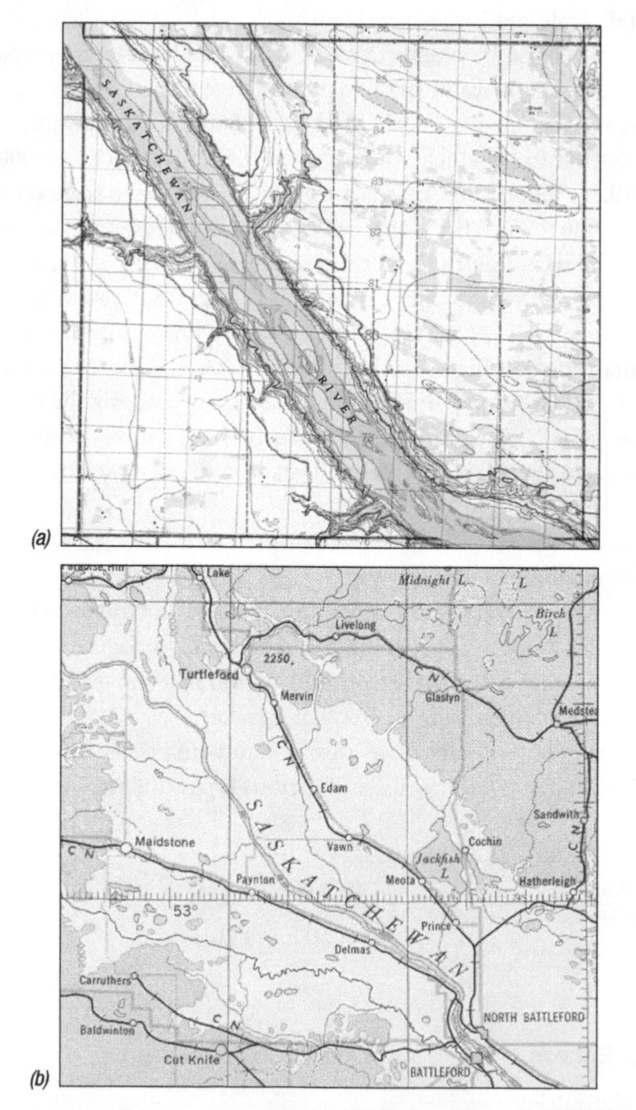

2A.5 Maps of the South Saskatchewan River at two scales: (a) at 1:20,000, the complex channel morphology is clearly depicted; (b) at 1:2,000,000, the channel morphology is greatly simplified.

Presenting Numerical Data on Thematic Maps

In physical geography, much of the information collected about particular areas is in numerical form, such as air temperatures and amounts of rainfall. Another information category consists merely of the presence or absence of an attribute, such as a particular species of plant. In such cases, a dot can mean "present," so that an area of scattered dots develops as the locations are entered (Figure 2A.6). Note that the absence of a dot implies that the attribute is not present. Thus, a dot map—or any map for that matter—should be critically assessed in terms of its accuracy, as this will depend on the thoroughness with which the data were collected. For example, an area may be inaccessible, but this does not guarantee that the attribute is not there.

Some measurements are taken at predetermined locations, such as weather stations, and although the numbers and locations may be accurate, it may be difficult to see the spatial pattern present in the data. For this reason, cartographers often simplify arrays of point values into *isopleth* maps, which use lines of equal values. In drawing an isopleth, the line is routed

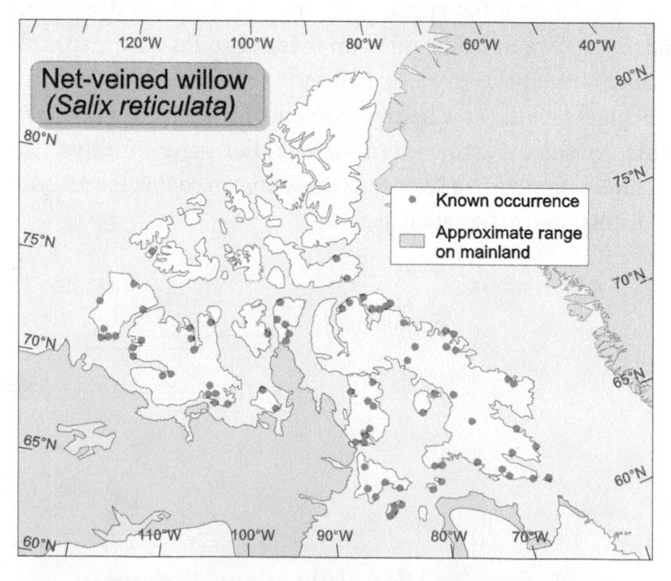

2A.6 A dot map showing the distribution of net-veined willow (*Salix reticulata* L.) in the Canadian Arctic archipelago.

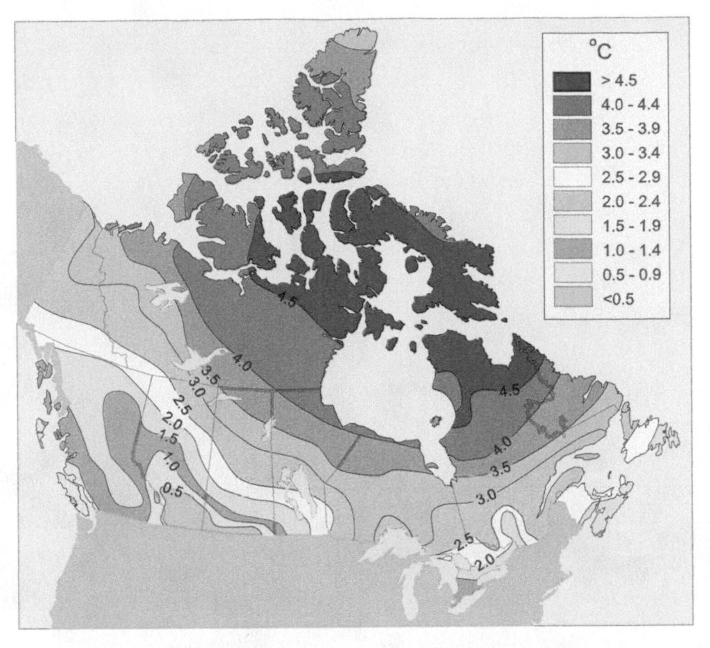

2A.7 This isopleth map shows the temperature departure from normal in Canada during the summer (Jun, Jul, Aug) of 2010. Colours enhance the temperature pattern, with blue signifying cooler than normal areas and yellow indicating warmer than normal areas.

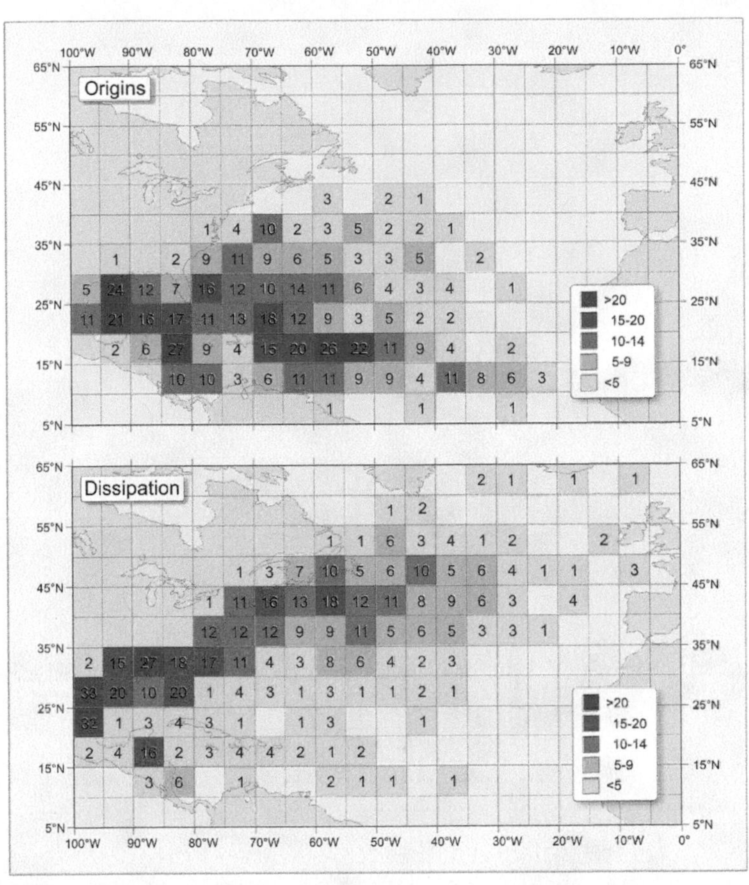

2A.8 The origin and dissipation of hurricanes in the North Atlantic (1886–2008). (Based on Elsner and Kara 1999 *Hurricanes of the North Atlantic*, OUP; from *Introduction to Weather and Climate* by O.W. Archibold, John Wiley & Sons Canada, Ltd., 2011.)

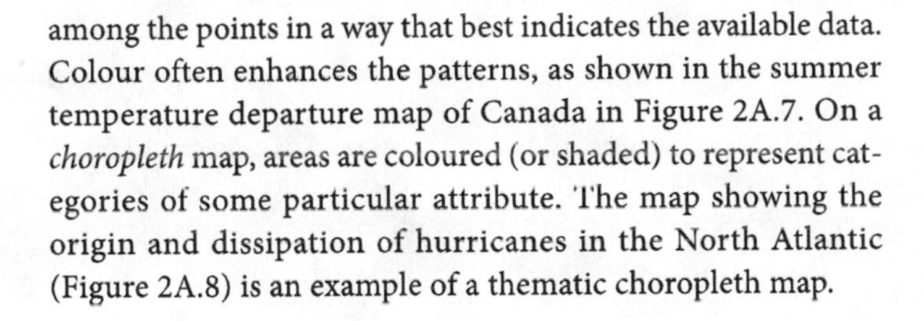
among the points in a way that best indicates the available data. Colour often enhances the patterns, as shown in the summer temperature departure map of Canada in Figure 2A.7. On a *choropleth* map, areas are coloured (or shaded) to represent categories of some particular attribute. The map showing the origin and dissipation of hurricanes in the North Atlantic (Figure 2A.8) is an example of a thematic choropleth map.

QUESTIONS

1. Can the scale of a flat map be uniform everywhere on the map? Do large-scale maps show large areas or small areas?
2. What types of symbols are found on maps, and what types of information do they carry?
3. How are numerical data represented on maps? Identify three types of isopleths. What is a choropleth map?

PART 2

WEATHER AND CLIMATE SYSTEMS

Late-winter sunset over the Churchill River, northern Saskatchewan.

The flow of energy from the sun to the Earth powers a vast and complex system of energy and matter flows within the atmosphere and oceans and at the land surface. Part 2 explores how these flows are linked to weather and climate. ● Five chapters (3 to 7) look at the forces and factors that influence the measurable properties of the atmosphere—those that are

discuss the link between energy flows and global temperature, then review global pressure and circulation patterns. The atmospheric components of the global water cycle are discussed with particular emphasis on evaporation, precipitation, and storm systems. ● The last three chapters in Part 2 are devoted to climate and describe the general weather patterns that different

CHAPTER 3

THE GLOBAL ENERGY SYSTEM

Sunrise. The clouds are illuminated by the rising sun.

The constant flow of energy from the sun drives many of the natural processes that shape the Earth. It is the source of energy for wind, waves, rivers, and ocean currents and is reflected in the rich diversity of climates.

The flows of energy within the atmosphere and between the atmosphere and the Earth's surface depend on the nature and properties of the gases and other atmospheric constituents that envelope the Earth. These have a profound influence on the energy balance of the planet, and so play a fundamental role in the maintenance of local, regional, and global scale weather and climate patterns.

This chapter explains how solar radiation is intercepted by the Earth, flows through the atmosphere, and interacts with the land surfaces and the oceans. Solar energy is not received uniformly over the surface because of the Earth's spherical shape, the tilt of the polar axis, and the constant rotation and revolution of the planet. Regional differences in energy flow rates affect temperature and pressure, which drive the global wind system. Ocean currents are also linked to the redistribution of energy. Although more energy is received in low latitudes than at the poles, the redistribution of energy through oceanic and atmospheric circulation results in a global energy balance; the flow of energy reaching Earth is the same as is returned to space.

EARTH'S ATMOSPHERE

EARTH'S PRIMORDIAL ATMOSPHERE

It is believed that Earth was formed by an aggregation of gas and dust that was part of a *nebula*. The gravitational pull of the growing planet was probably too small to retain any gaseous envelope, and so it is thought that Earth's atmosphere evolved from self-contained secondary sources. Initially, the atmosphere probably consisted of helium (He) and hydrogen (H_2), with traces of ammonia (NH_3) and methane (CH_4). A hydrogen and helium atmosphere is similar to that found on the outer planets and the sun. With time, carbon dioxide (CO_2) began to accumulate, with lesser amounts of nitrogen (N_2), sulphur dioxide (SO_2), and water vapour (H_2O). These gases would have been emitted from ancient volcanoes. This carbon dioxide-dominated atmosphere is similar to present-day conditions on Mars and Venus.

No oxygen (O_2) is, or presumably ever has been, released from volcanoes. However, volcanoes do expel water vapour, and in ancient times this would have resulted in the formation of clouds and rain of high acidity which caused rapid weathering of rocks. Geologic evidence suggests that such conditions had developed by about 3.8 billion years ago. With time, oxygen became more prevalent; evidence of an oxidizing atmosphere is seen in the formation of ancient limestone.

The change from an oxygen-poor to an oxygen-rich atmosphere arose from the photo-dissociation of water molecules by ultraviolet (UV) radiation. Atmospheric oxygen levels produced in this way would have been about 0.001 percent of present concentrations. The subsequent rise in oxygen is attributed to photosynthesis. For primitive life forms to develop, cells would need to have been protected from the lethal UV radiation penetrating the primitive atmosphere; this was possible a few metres below the ocean surface. Once oxygen concentrations reached about 10 percent of current atmospheric levels, sufficient ozone was likely present to shield land surfaces from harmful UV radiation. As a result, organisms rapidly spread onto the land about 400 million years ago.

Terrestrial photosynthesis led to a rapid increase in oxygen. At the same time, atmospheric carbon dioxide levels declined as the gas was progressively incorporated into organic materials and deposited in the ancient coal seams and carbonate rocks. The Earth's atmosphere has remained essentially unchanged since this time, although concerns are now being raised over the impact air pollution is having on the global environment.

TODAY'S ATMOSPHERE

The Earth's atmosphere is a mixture of gases, not a single chemical compound; it is held to the Earth by gravity. Because air is highly compressible, the lower layers of the atmosphere are much denser than those above. Consequently, about 97 percent of the mass of the atmosphere lies within 30 km of the Earth's surface. The chemical composition of air in the lower atmosphere is uniform in terms of the proportions of constituent gases. Pure, dry air consists mainly of nitrogen, about 78 percent by volume, and oxygen, about 21 percent (see Table 3.1). Other gases, such as argon (Ar) and carbon dioxide, as well as water vapour and various pollutants, account for the remaining 1 percent.

Nitrogen (N) is an inert gas that does not readily combine with other substances. It is removed from the atmosphere in precipitation during lightning storms and by specialized nitrogen fixing bacteria in the soil and on the roots of plants. Unlike nitrogen, *oxygen* (O) is chemically active and combines readily with other elements in the oxidation process. Combustion of fuels represents a rapid form of oxidation, while in certain forms of rock weathering oxidation is very slow. Similarly, living tissues require oxygen to convert foodstuffs into energy.

TABLE 3.1 AVERAGE COMPOSITION OF THE ATMOSPHERE UP TO AN ALTITUDE OF 25 K

GAS	CHEMICAL FORMULA	PERCENT VOLUME
Nitrogen	N_2	78.08%
Oxygen	O_2	20.95%
*Water	H_2O	0 to 4%
Argon	Ar	0.93%
*Carbon dioxide	CO_2	0.0360%
Neon	Ne	0.0018%
Helium	He	0.0005%
*Methane	CH_4	0.00017%
Hydrogen	H_2	0.00005%
*Nitrous oxide	N_2O	0.00003%
*Ozone	O_3	0.000004%

*Denotes gases that vary in concentration both spatially and seasonally.

The remaining 1 percent of dry air is mostly argon, a non-reactive gas of little importance in natural processes, although it has many industrial applications. In addition, there is a small amount of *carbon dioxide* (CO_2), which is particularly efficient at absorbing long-wave energy radiated from the Earth's surface. This warms the lower atmosphere, which then reradiates some heat back to the surface through a process called the *greenhouse effect*. Green plants also convert carbon dioxide into carbohydrates in photosynthesis.

Another important component of the atmosphere is *water vapour*, the gaseous form of water. Individual water vapour molecules mix freely throughout the atmosphere, just like the other gases. The concentration of water vapour is highly variable and typically ranges from less than 1 percent to more than 4 percent. Since water vapour, like carbon dioxide, is a good absorber of long-wave radiation, it also plays a major role in warming the lower atmosphere.

The atmosphere contains many other gases, most of which are pollutants, such as sulphur dioxide and halogens. Dusts and fine particles are also present; these originate from volcanic eruptions or other natural processes and contribute to the absorption and scattering of radiant energy. Many of these particulates also play an important role in the formation of precipitation by acting as **condensation nuclei** around which water droplets and ice crystals can form (see Chapter 6).

OZONE IN THE UPPER ATMOSPHERE

A small, but important, constituent of the atmosphere is **ozone** (O_3), a form of oxygen in which three oxygen atoms are bonded together. Ozone absorbs UV radiation, thus shielding plants and animals from its harmful effects. The concentration of ozone in the atmosphere is reported in **Dobson units (DU)**, which measure the density of ozone in a column of atmosphere above a specified area of Earth's surface. A Dobson unit represents the thickness of ozone in the atmosphere that would form if the gas throughout the column was at standard temperature and pressure (i.e., at 0 °C and 101.3 kilopascals, kPa). The thickness of that layer, measured in hundredths of a millimetre, would be the concentration of ozone expressed in Dobson units (Figure 3.1). Ozone is found mostly in the stratosphere, a layer of the atmosphere that lies about 14 to 50 km above the surface (see Chapter 4).

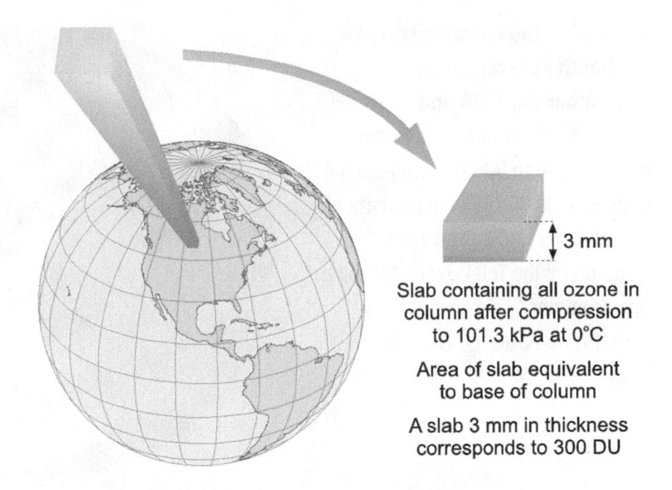

3.1 OZONE IN THE ATMOSPHERE IS MEASURED IN DOBSON UNITS (DU) If the ozone in an air column, compressed to standard temperature and pressure (0 °C at 101.3 kPa), formed a slab 3 mm thick, it would be equivalent to 300 DU.

Gaseous chemical reactions in the stratosphere produce ozone. Oxygen molecules (O_2) absorb UV energy and split into two oxygen atoms (O + O). A free oxygen atom (O) then combines with an O_2 molecule to form ozone (O_3). Once formed, ozone can also be destroyed by UV radiation, which dissociates to form O_2 + O. The net effect is that ozone (O_3), molecular oxygen (O_2), and atomic oxygen (O) are constantly formed, destroyed, and reformed in the ozone layer. When the concentration of ozone is reduced, fewer transformations occur, diminishing the amount of UV absorption.

If UV radiation were to reach Earth's surface at full intensity, exposed bacteria and unprotected plant and animal tissues would be killed or severely damaged. A particular concern of higher levels of UV radiation is the potential increase in skin cancer. Other possible effects include a reduction in crop yields and damage to some forms of aquatic life. The presence of the ozone layer is thus essential to maintaining a viable environment for life on this planet. Certain forms of air pollution can reduce ozone concentrations substantially in some regions of the atmosphere at particular times of the year. *Eye on the Environment • The Ozone Layer—Shield to Life* describes this threat to the ozone layer.

EYE ON THE ENVIRONMENT
THE OZONE LAYER—SHIELD TO LIFE

The release of **chlorofluorocarbons (CFCs)** into the atmosphere poses a serious threat to the ozone layer. CFCs are synthetic industrial chemical compounds containing chlorine (Cl), fluorine (F), and carbon (C) atoms. Although household uses of CFCs in aerosol sprays were banned in most parts of the world in the 1970s, CFCs are still in wide use as cooling fluids in refrigeration systems. When these appliances are disposed of, CFCs can be released into the air.

Molecules of CFCs in the atmosphere are very stable. They move upward by diffusion without chemical change until they eventually reach the ozone layer. There, they absorb UV radiation and decompose into chlorine oxide (ClO) molecules. The chlorine oxide in turn breaks down molecules of ozone, converting them into oxygen molecules. This reduces the concentration of ozone in the ozone layer, which adversely affects its ability to absorb UV radiation. Other gaseous molecules, including nitrogen oxides (NO_x) and bromine oxides (BrO), can also reduce the concentration of ozone in the stratosphere.

However, not all threats to the ozone layer can be attributed to human activity. Aerosols inserted into the stratosphere by volcanic activity can also reduce ozone concentrations. The June 1991 eruption of Mount Pinatubo in the Philippines reduced the average concentration of global ozone in the stratosphere by 4 percent the following year, with reductions over midlatitudes of up to 9 percent.

Studies of the ozone layer based on satellite data during the 1980s showed a substantial decline in total global ozone. By late in the decade, scientists had agreed that ozone reduction was occurring much faster than had been predicted. Compounding the problem of global ozone decrease was the discovery in the mid-1980s of a "hole" in the ozone layer over Antarctica. The ozone hole is defined as the area over Antarctica in which the total ozone concentration is less than 220 Dobson units. This value was selected as ozone concentrations in the region had never previously dropped below this. The ozone hole has steadily increased in area and intensity over the past two decades. Seasonal thinning of

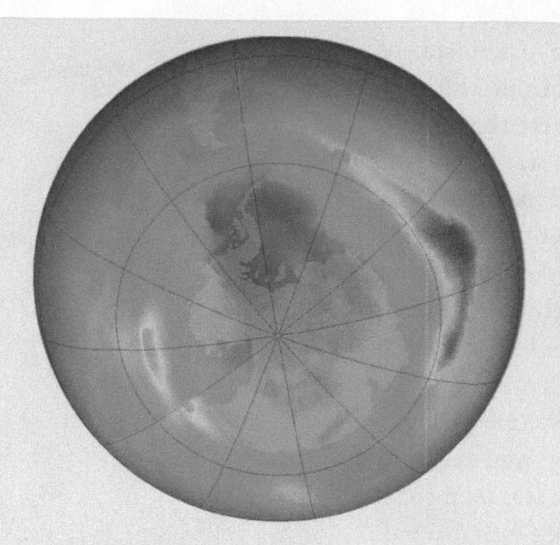

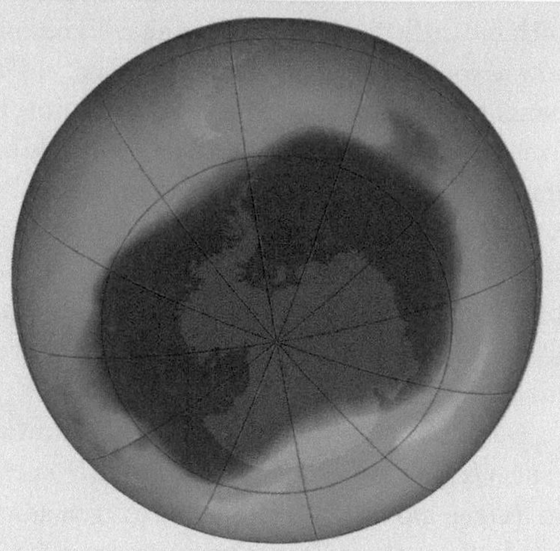

Maps of ozone concentration for the South Pole region on September 24, 1979 and 2006
Prior to 1979, values were in the range of 250 to 300 DU, but plunged to as little as 85 DU in some parts of Antarctica in 2006. Data were acquired by the Total Ozone Mapping Spectrometer (TOMS) on NASA's Earth Probe TOMS-EP Satellite. (Courtesy of NASA.)

the Antarctic ozone layer occurs during spring in the southern hemisphere, with ozone concentration typically reaching a minimum in late September and October of each year.

This thinning occurs after the formation of a polar vortex in the stratosphere in the winter. During the long polar night, this vast whirlpool of trapped air is kept out of the sun, preventing ozone formation. The air in the vortex is very cold, and clouds of crystalline water (ice) or other water-containing compounds form within it. The crystals are important because they provide a surface on which chemical reactions can take place. One of these reactions converts chlorine from a stable form to chlorine oxide, which can rapidly destroy ozone in the presence of sunlight. As spring in the southern hemisphere approaches, the sun slowly illuminates the polar vortex and the chlorine oxide reacts with ozone, reducing ozone concentrations and forming the ozone hole.

On September 24, 2006, the area of the Antarctic ozone hole reached 29.5 million km^2 (see Figure to the left), which equaled the previous record single-day value of September 9, 2000. In addition, the concentration of ozone was only 85 DU in some parts of eastern Antarctica, and was practically absent in the ozone layer between 13 to 21 km above the surface. In this critical layer, a record low of only 1.2 DU was recorded, compared with an average reading of 125 DU earlier in the year.

In the northern hemisphere, conditions for the formation of an ozone hole are not as favourable. An Arctic polar vortex forms, but it is much weaker than the Antarctic vortex and is less stable. As a result, an early-spring ozone hole is not usually observed in the Arctic. However, in 1993, 1996, 1997, and 1999, notable Arctic ozone holes occurred, and atmospheric computer models have predicted more of these events in the future.

Although the most dramatic reductions in stratospheric ozone levels have been associated with Arctic and Antarctic ozone holes, reductions have been noted elsewhere as well—including the northern midlatitudes. A number of studies during the 1990s showed reductions of 6 to 8 percent in stratospheric ozone concentrations occurring over North America.

As the global ozone layer thins, the amount of incoming UV solar radiation reaching the Earth's surface should increase. Recent studies of satellite data have documented this effect. Since 1978, the average annual exposure to UV radiation increased by 6.8 percent per decade along the 55° N parallel. At 55° S, the increase was 9.8 percent per decade.

Responding to the global threat of ozone depletion and its anticipated impact on the biosphere, 20 nations signed the Vienna Convention for the Protection of the Ozone Layer in 1985. In the Montreal Protocol of 1987, they agreed to specific reductions in CFCs. As a result, the concentrations of ozone-depleting substances in the stratosphere are decreasing from a peak reached in 2001 by about 0.1 to 0.2 percent annually. However, many of these substances stay in the atmosphere for more than 40 years. The present estimate is that the ozone hole will not fully recover until about 2065.

AIR POLLUTANTS

An **air pollutant** is an unwanted substance injected into the atmosphere from the Earth's surface by either natural or human activities. Air pollutants include gases, aerosols, and particulates. **Aerosols** are extremely small particles that float freely with normal air movements. **Particulates** are larger, heavier particles that eventually fall back to Earth. Most pollutants generated by human activity arise in two ways. The first is through day-to-day activities, such as driving automobiles. The second is through industrial processes, particularly the combustion of fossil fuels and smelting of mineral ores.

One group of pollutants is known collectively as greenhouse gases. Although relatively scarce in the atmosphere, they are of special importance because of their ability to trap long-wave radiation. (Chapter 4 discusses the link between greenhouse gases and global warming.) The most significant greenhouse gas is carbon dioxide, which although naturally present in the atmosphere, is released in great quantities through combustion of fossil fuels. Methane (CH_4) is also a problem. It is produced by anaerobic processes in bogs and swamps, such as those found throughout the Canadian boreal forest region. However, methane is also produced in rice paddies, by livestock, and by biomass burning when clearing land for agriculture. Nitrous oxide (N_2O) is released into the atmosphere from automobile exhaust and from chemical fertilizers. Chlorofluorocarbons (CFCs), which include aerosol propellants and refrigerants, are also important greenhouse gases.

In addition to N_2O, nitric oxide (NO), nitrogen dioxide (NO_2), and nitrates ($-NO_3$) are also present in the atmosphere; this mixture is usually referred to as NO_X. NO_X reacts in the presence of sunlight with volatile organic compounds (VOCs), such as those released from petroleum refineries. These **photochemical reactions** can cause brown hazes and also produce ozone. Unlike stratospheric ozone, the ozone found in the lowest layers of the atmosphere (tropospheric ozone) is an air pollutant and is damaging to lungs and other living tissues.

Similarly, sulphur oxides (SO_2 and SO_3) are referred to as SO_X. These gases, together with NO_2, readily combine with oxygen and water in the presence of sunlight to form sulphuric and nitric acid aerosols. These aerosols serve as condensation nuclei and acidify the tiny water droplets that coalesce to form acid precipitation. Sulphur gases are mainly produced in smelters and electricity generating stations that burn coal or lower-grade fuel oils. Sulphur dioxide was responsible for the extensive damage that occurred to the vegetation around Trail, British Columbia, and Sudbury, Ontario, as well as in many other parts of the world (Figure 3.2). These types of facilities also supply most of the unwanted particulate matter in the atmosphere. Some of it is coarse soot particles—fly ash—that settle quite close to the source as **fallout**. Particles too small to settle are flushed out of the atmosphere by precipitation in a process called **washout**.

Acidic particles removed from the atmosphere through **dry deposition** form thin dust layers on plants and soils. This dust acidifies any precipitation that subsequently falls, which can have severe environmental consequences. In winter, acid particles mix with snow as it accumulates, and during snow melt, a surge of acid water is released into soils and streams.

The impact of acid deposition on the environment is linked to the neutralizing properties of the soils and surface waters. In dry climates, such as the Canadian Prairies, acidic deposition can be readily neutralized because surface waters are normally somewhat alkaline. Areas where soil water is naturally acidic, such as the forest regions of eastern Canada and parts of Scandinavia, are most sensitive to acid deposition. Acidification of lakes in these regions has been linked to an increase in fish mortality.

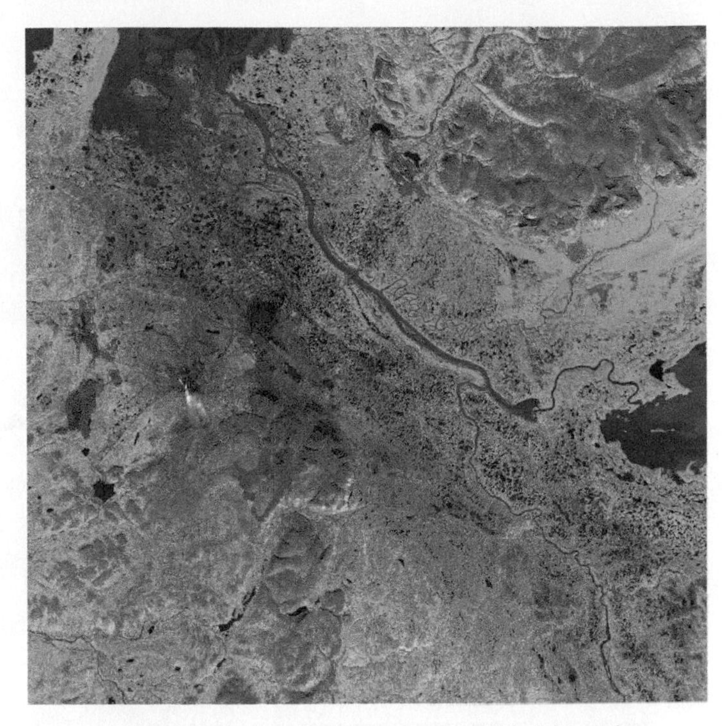

3.2 **AIR POLLUTION FROM INDUSTRIAL SMOKESTACKS** This false-colour satellite image shows the devastating effects of pollutants emitted from mining operations near Noril'sk, Siberia. Shades of pink and purple indicate bare ground where there is little or no vegetation. Greens show predominantly healthy forested areas, and water appears as shades of blue.

THE GLOBAL ENERGY SYSTEM

ELECTROMAGNETIC RADIATION

The study of the global energy system begins with **electromagnetic radiation**—the principal form in which energy is transported from the sun to the Earth. Electromagnetic radiation consists of electric and magnetic fields that simultaneously propagate as waves through space. The waves oscillate together but at right angles to each other, and travel in a direction that is perpendicular to the direction of their oscillation (Figure 3.3).

Electromagnetic radiation is characterized by its wavelength (λ) and frequency (f). Wavelength is the distance separating adjacent wave crests. The general unit of measurement is the *nanometre* (nm), which is 10^{-9} m (0.000000001 m). Micrometres (μm) (10^{-6} or 0.000001 m) are commonly used, while longer wavelengths are conveniently expressed in metres. Frequency refers to the number of waves that pass a point every second, and thus is inversely proportional to wavelength. The SI unit of frequency is the hertz (Hz), which is defined as 1 wave per second. Higher frequency, shorter wavelengths carry more energy than longer wavelengths at lower frequency.

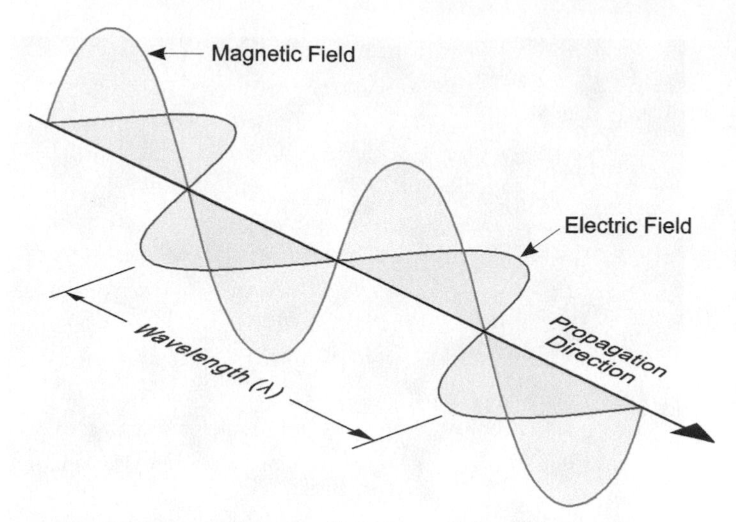

3.3 **ELECTROMAGNETIC RADIATION** Electromagnetic radiation is a collection of energy waves with different wavelengths. Wavelength (λ) is the crest-to-crest distance between successive wave crests. (From *Introduction to Weather and Climate* by O.W. Archibold, John Wiley & Sons Canada, Ltd., 2011.)

Electromagnetic waves of different wavelengths form the **electromagnetic spectrum** (Figure 3.4). At the short wavelength end of the spectrum are *gamma rays* and *X-rays*, which are extremely high-energy forms of radiation. Gamma rays have wavelengths of less than 0.01 nm and are generated in the hottest regions of the universe, such as a *supernova* (exploding star). Gamma radiation has several medical applications in cancer treatment and for sterilization of surgical instruments. X-rays, which range in wavelength from 0.01 to 10 nm, are best known for their penetrating power and application in medical imaging.

Gamma and X-ray radiation grade into **ultraviolet (UV) radiation**, which begins at about 10 nm and extends to 400 nm on the electromagnetic spectrum. UV wavelengths between 100 and 400 nm are usually separated into UVC (100–280 nm), UVB (280–320 nm) and UVA (320–400 nm). It is these wavelengths that are of greatest concern. UVC causes damage to microorganisms, and the small amounts that reach Earth act as a natural purification system that inhibits the growth of pathogens. Overexposure to UVC will cause reddening of the skin and eye damage. UVB is particularly troublesome because it reaches Earth at intensities that can cause cataracts, blindness, and cancer after prolonged exposure. UVA is less dangerous and is most often associated with sun tanning and sunburns.

The **visible light** portion of the electromagnetic spectrum corresponds to the wavelength range 400–700 nm. The visible spectrum covers the wavelengths of greatest intensity produced by the sun. This part of the spectrum produces the various colours, from violet through blue, green, yellow, orange, and red, that can be detected by the human eye. These wavelengths are little affected by Earth's atmosphere, a property which is of great importance in remote sensing as many procedures make use of this so-called *visible window*.

The **infrared** region of the electromagnetic spectrum extends from the visible wavelengths at 700 nm to about 1 mm (1×10^6 nm). Infrared wavelengths include thermal radiation, which though not visible, can be easily sensed as heat given off by a warm object. Four divisions of the infrared region are usually recognized. *Near-infrared radiation (NIR)* ranges from 700 to 1200 nm (1.2 μm). Although the human eye is not sensitive to NIR, image intensifiers, such as night-vision goggles, can be used to rectify this. Healthy plant leaves and green tissues strongly reflect near-infrared light, a property that is used in remote sensing (see Appendix 3.1: Remote Sensing for Physical Geography).

Short-wave infrared radiation (SWIR) lies in the range of 1.2 to 3 μm. In remote sensing, these wavelengths can distinguish

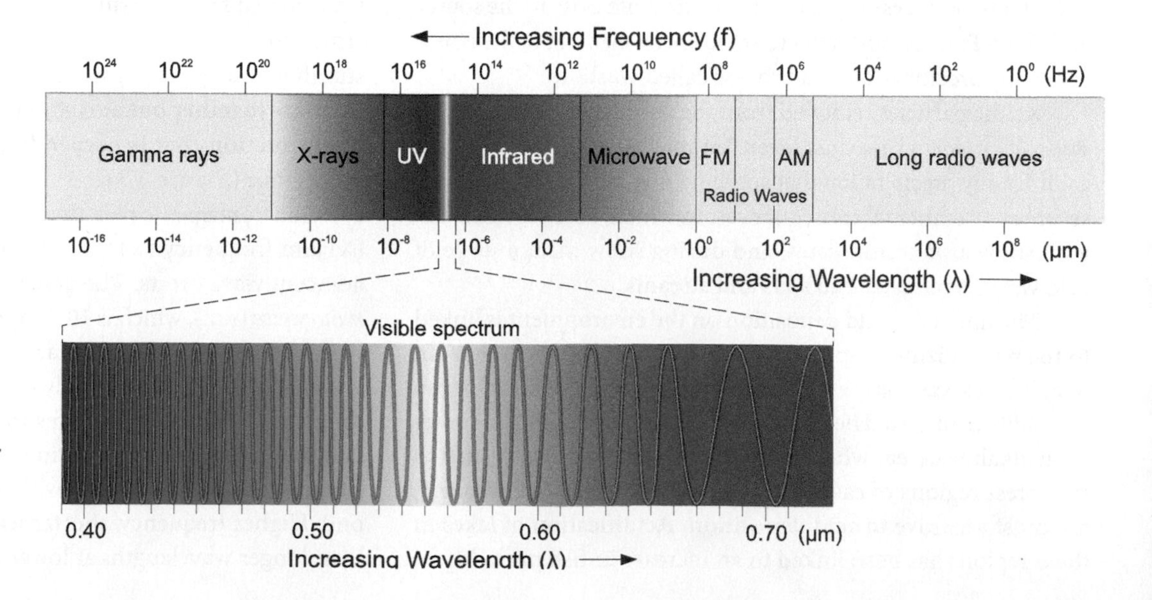

3.4 **THE ELECTROMAGNETIC SPECTRUM** Electromagnetic radiation can exist at any wavelength. By convention, names are assigned to a specific wavelength region. (From *Introduction to Weather and Climate* by O.W. Archibold, John Wiley & Sons Canada, Ltd., 2011.)

3.5 **A THERMAL INFRARED IMAGE** This suburban scene was imaged at night in the thermal infrared spectral region. Red tones indicate the highest temperatures and black tones the coldest. Windows appear red because they are warm and radiate more intensely. The blue tones of the trees show they are relatively cool. The black sky is the coldest.

3.6 **COMPOSITE THERMAL INFRARED IMAGE OF THE CANADIAN ARCTIC** This image was acquired with 10.7 μm sensor on board the polar orbiting NOAA 17 satellite on January 12, 2011. At this time of year, most places north of the Arctic Circle (66½° N) are in darkness.

clouds from snow; they are also useful for differentiating rock types. *Middle-infrared radiation (MIR)* from 3.0 to about 6.0 μm is just within the thermal infrared region of the spectrum. Unlike the NIR and SWIR wavelengths, MIR originates not only from the sun, but is also emitted by hot spots at the Earth's surface, such as lava flows, forest fires, or gas well flares.

Thermal infrared radiation is between 6.0 μm and about 300 μm and includes the heat radiated from objects at temperatures normally found at Earth's surface. Every object at temperatures above absolute zero (0 K or −273.15 °C) radiates energy. The wavelength emitted depends on the temperature of the object. The hotter the object, the shorter the wavelength of emission. Thus thermal infrared radiation is no different in principle from visible light radiation, but because it is beyond the sensitivity of the human eye, it can only be detected with specialized thermal imaging equipment. In this way it is possible to "see" one's environment without visible illumination (Figure 3.5).

Because the amount of radiation emitted by an object also increases with temperature, thermal images can be calibrated for temperature. This has a number of commercial and industrial applications. It is particularly useful at night when only thermal infrared is present. In this way, for example, satellites can continue to monitor temperatures in northern Canada throughout the long, dark Arctic winter (Figure 3.6).

Beyond the infrared wavelengths are microwaves, radar, radio waves, and television transmissions, all of which are essential to life in the modern, high-tech world. Microwaves and radar are also widely used in remote sensing. Radar (**ra**dio **d**etection **a**nd **r**anging) uses wavelengths of 0.8 cm to 1m, which is part of the microwave region of the spectrum. Most sensors are *passive* since they record energy emitted by the sun or thermal sources. Radar, however, is an *active* system in that bursts of energy are generated by the instruments and reflected back to the receiver. The travel time of the outward and reflected signals are used to measure distance. Canada's RADARSAT-2 operates in this way and can acquire detailed images of Earth's surface, day or night, under clear or cloudy conditions (Figure 3.7).

Doppler radar compares the frequency of outgoing and return signals to determine the speed of moving objects. Police radar is a common application of this technology. In a similar way, *Doppler weather radar* uses a succession of pulses that are variously scattered or reflected by raindrops or snowflakes. The strength of the return signal is recorded on a logarithmic scale

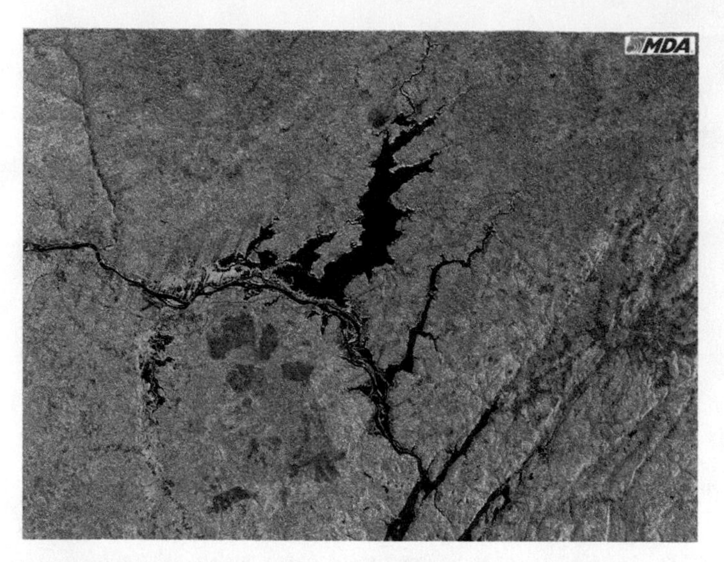

3.7 **RADARSAT-2** This image shows flooding in New Brunswick in May 2008. Completely flooded zones (river, lakes, etc.) are in black. The bright-toned areas indicate zones of flooded vegetation.

that shows decibels of reflectivity (dbZ). The dbZ scale is calibrated to show precipitation intensity.

RADIATION AND TEMPERATURE

Two important physical principles are associated with the emission of electromagnetic radiation. First, there is an inverse relationship between the range of wavelengths that an object emits and the surface temperature of the object. Consequently, the sun, a very hot object, emits radiation predominantly in shorter wavelengths. In contrast, the Earth, a much cooler object, emits radiation at longer wavelengths. Second, hot objects radiate more energy than cooler objects. The flow of radiant energy from the surface of an object is directly related to its absolute temperature raised to the fourth power. Absolute temperature is measured on the Kelvin (K) scale, with absolute zero (0 K) representing the absence of all heat. Because of this relationship, a small increase in temperature results in a large increase in the rate at which radiation is emitted. The *Working It Out • Radiation Laws* feature in the end-of-chapter material describes these two principles in more detail.

SOLAR RADIATION

The sun is a mass of constantly churning gases that are heated by continuous nuclear reactions. It is of average size compared to other stars and has a surface temperature of about 5,800 K (5,500 °C). Like all objects at temperatures above absolute zero, the sun emits energy in the form of electromagnetic radiation. Within the sun's interior, hydrogen is converted to helium at extremely high temperatures and pressures. This process of nuclear fusion produces a vast quantity of energy that finds its way to the sun's surface and travels through space at a speed of about 300,000 km per second. At this rate, it takes about 8.3 minutes for the energy to reach the Earth, at an approximate distance of 150 million km.

No energy is lost as it radiates from the sun, but the rays spread apart; therefore light intensity is inversely proportional to the square of the distance travelled. The intensity of a flow of radiant energy is measured as watts per square metre (W m^{-2}). For Earth, the average intensity of solar energy reaching the top of the atmosphere directly facing the sun is 1,367 W m^{-2}. This represents less than one-billionth of the enormous amount of energy emitted by the sun.

The maximum amount of energy received by Earth is termed **total solar irradiance**; previously this was called the **solar constant** because it was assumed that the quantity of energy emitted by the sun never varied. However, it is now known that solar output varies by about 0.2 percent due to sunspots and associated solar activity. Sunspots develop in regions where convection is reduced by intense magnetic activity. In these regions, the surface temperature of the sun is reduced by as much as 2,000 K. This affects the amount of energy that is emitted by the sun, and also increases the proportion of UV radiation. Total solar irradiance also varies by about 7 percent between perihelion and aphelion due to Earth's elliptical orbit.

Because of Earth's spherical shape, the amount of energy intercepted is considerably less than the total solar irradiance. Assuming there is no atmosphere, the shape of the planet allows only one side to be illuminated at any given time; this reduces total solar irradiance by 50 percent. In addition, total irradiance is measured with respect to a surface that is perpendicular to the path of incoming solar rays. Depending on the season, this only occurs at solar noon between the Tropic of Cancer and the Tropic of Capricorn. At all other latitudes, the intensity of the solar beam is reduced because it is distributed over a greater surface area (Figure 3.8). The net effect is a further reduction in average solar irradiance by an additional 50 percent which, averaged over the entire planet, amounts to approximately 340 W m^{-2}.

EFFECTS OF THE ATMOSPHERE ON SOLAR ENERGY

Energy emitted by the sun is predominantly in the visible UV and infrared wavelengths (Figure 3.9). These wavelengths span the general range 0.2 to 4.0 μm and are collectively referred to as **short-wave radiation**. The uppermost curve for solar radiation shows how the sun would supply solar energy at the top of the atmosphere if it was a perfect radiator or **blackbody**. The actual output, as measured at the top of the atmosphere, is very similar to that expected for a blackbody, except the sun emits less energy in the UV wavelengths than is theoretically possible.

The line showing solar radiation reaching the Earth's surface is quite different from the pattern at the top of the atmosphere.

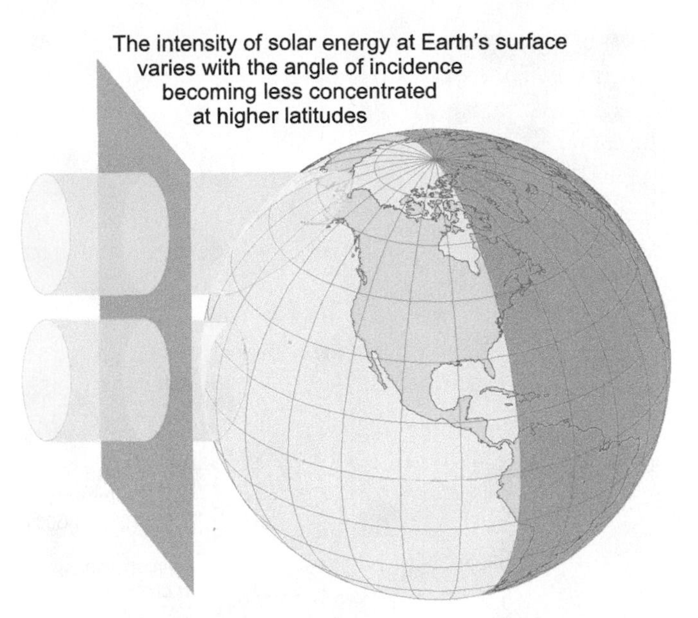

3.8 **SOLAR IRRADIANCE** Because half of the Earth is in darkness and radiant energy from the sun is not spread evenly over the illuminated hemisphere, mean total solar irradiance is effectively reduced from 1,367 W m^{-2} to about 340 W m^{-2}.

As solar radiation passes through the atmosphere, some wavelengths are absorbed, reflected, or scattered as they encounter gas molecules and dust particles. **Absorption** occurs when molecules and particles in the atmosphere intercept radiation at particular wavelengths. Absorption is shown by the various dips in the graph—for example, at about 1.3 and 1.9 µm. These are *absorption bands*, in contrast to the peaks on the graph that are regions of high transmittance and are called *atmospheric transmission bands*, or *atmospheric windows*. The relative effect of specific atmospheric gases on absorption of solar energy is shown in Figure 3.10.

Reflection occurs when a light ray strikes a molecule or particle and bounces off again. When light is reflected, the angle of the incident ray with respect to a line perpendicular to the surface equals the angle of the reflected ray. **Scattering** differs from reflection in that light striking a molecule is redirected in all directions. Reflected and scattered rays may go back toward space or continue toward Earth's surface.

As short-wave solar radiation penetrates the atmosphere, its energy is absorbed or diverted in various ways. Gamma rays, X-rays, and ultraviolet radiation less than 200 nm in wavelength are absorbed by oxygen and nitrogen in the thin outer layers of the atmosphere. Most of the ultraviolet radiation in wavelength range 200 to 300 nm is absorbed by the ozone layer in the upper atmosphere. All of these absorption phenomena are essential for organisms because prolonged exposure to radiation at wavelengths shorter than 300 nm is lethal to living tissue.

Molecules of water vapour, carbon dioxide, and to a lesser extent dust, absorb most of the solar radiation in the red and near infrared part of the spectrum, at wavelengths greater than 700 nm. Absorption of these longer wavelengths is spatially quite variable, because water vapour is not uniformly distributed through the atmosphere. Some direct heating of the lower atmosphere occurs from absorption of the longer infrared wavelengths. However, it is important to note that only about 4 percent of solar irradiance is at wavelengths greater than 2.5 µm, so the amount of warming

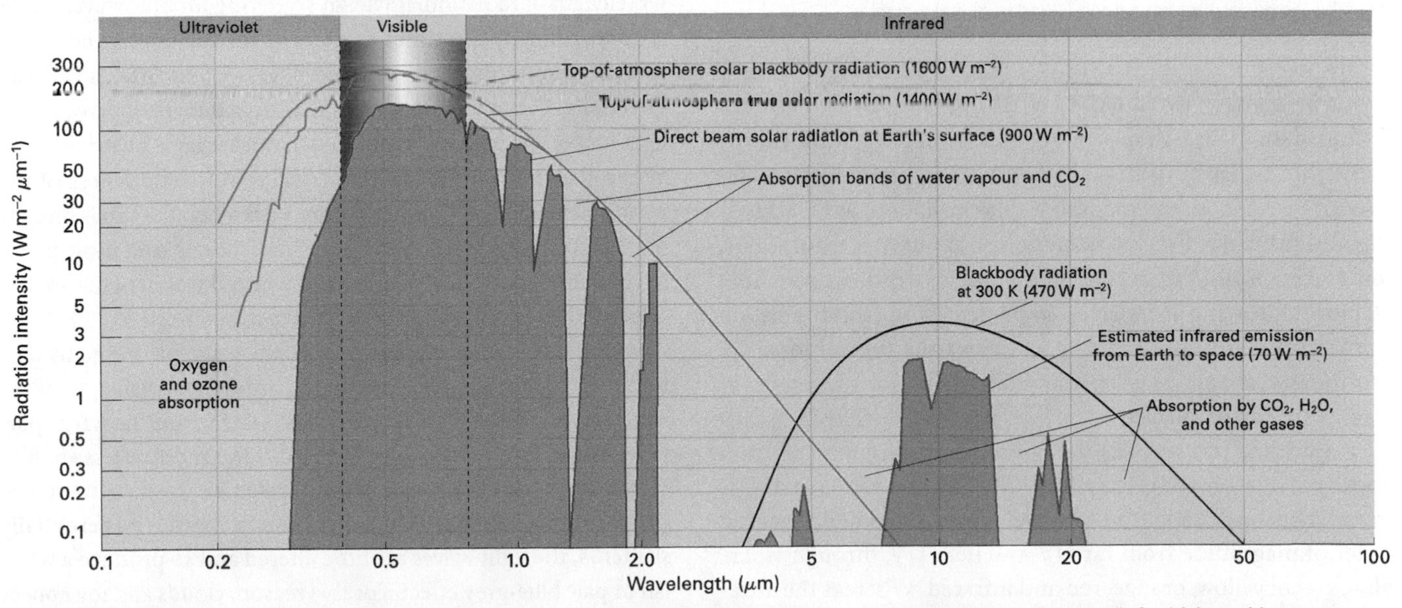

3.9 **SPECTRA OF SOLAR AND EARTH RADIATION** This figure plots both short-wave radiation, which comes from the sun (left side), and long-wave radiation, which is emitted by the Earth's surface and the atmosphere (right side). Note that both scales are logarithmic. (After W. D. Sellers, *Physical Climatology*, University of Chicago Press.)

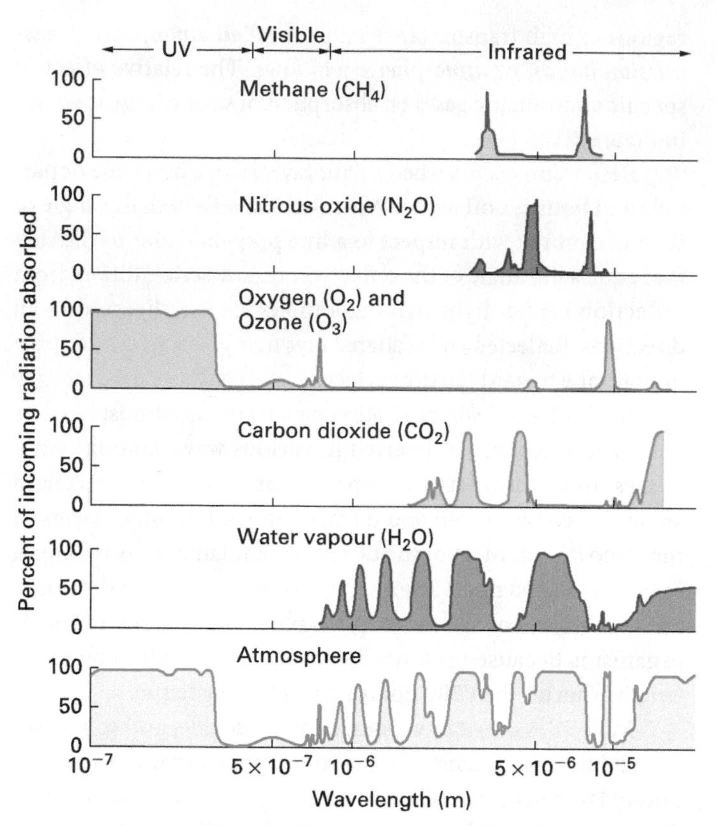

3.10 ABSORPTION OF SOLAR ENERGY BY DIFFERENT ATMOSPHERIC GASES The shorter wavelengths are nearly all absorbed by oxygen and ozone. Water vapour absorbs energy over a wide range of infrared wavelengths, but because of its variable distribution in the atmosphere, the effects are most pronounced in the moist air of tropical regions. Absorption by carbon dioxide, nitrous oxide, and methane is in the longer infrared wavelengths and contributes to atmospheric warming; these are the greenhouse gases that have been linked to global warming.

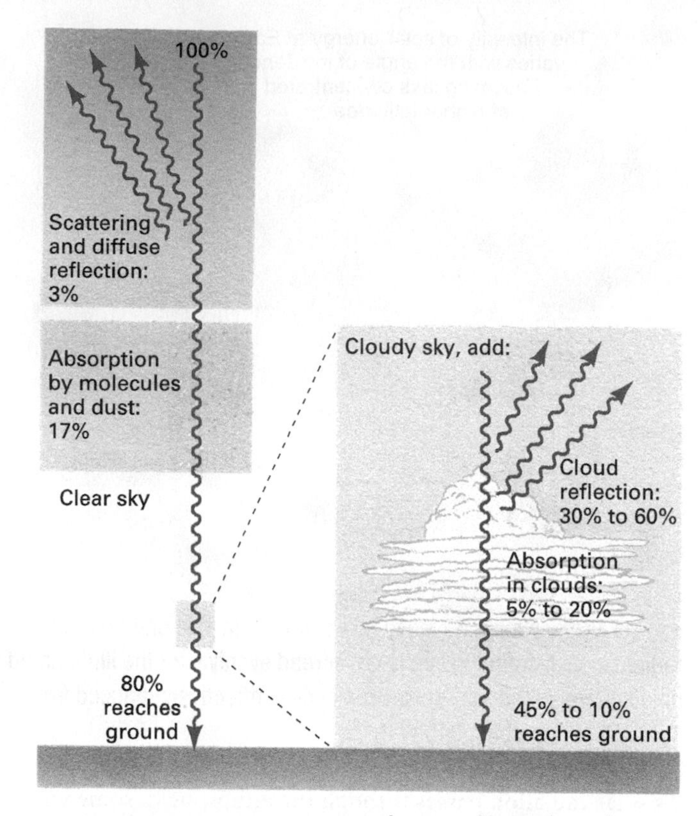

3.11 FATE OF INCOMING SOLAR RADIATION Losses of incoming solar energy are much lower with clear skies (left) than with cloud cover (right). (Copyright © A. N. Strahler.)

is considerably less than what comes from long-wave energy emitted from Earth itself (see below). Absorption accounts for about 17 percent of the incoming solar radiation, while scattering and diffuse reflection account for about 3 percent. Thus, under clear skies, diffuse reflection and absorption total about 20 percent, leaving as much as 80 percent of the solar radiation to reach the ground. Figure 3.11 shows the typical losses of incoming solar energy as it passes through the atmosphere on its way to the Earth's surface.

Scattering increases as radiation moves through the lower and denser atmospheric layers. The degree of scattering is inversely proportional to the wavelength, and so decreases in order of magnitude from far UV and near UV, through violet, blue, green, yellow, orange, red and infrared. Whereas the ultraviolet, violet, and blue bands of the spectrum are scattered in all directions, the longer wavelengths of light pass through more or less unaffected.

Selective scattering, also known as **Rayleigh scattering**, is greatest in the shorter wavelengths. It is caused by atmospheric gases and particles that are smaller in diameter than a particular wavelength of radiation. Rayleigh scattering for blue wavelengths is approximately ten times as great as for the longer red wavelengths of sunlight. The scattered blue light reaches us from all directions, and so the sky appears blue. At sunrise and sunset, sunlight passes obliquely through a much thicker layer of atmosphere. When the sun is on the horizon, the effective thickness of the atmosphere is about 45 times greater than when the sun is directly overhead. This results in maximum atmospheric scattering of violet and blue light, but because the longer red wavelengths are less affected they pass through, and the sky appears red.

Non-selective scattering or **Mie scattering** is caused by particles with diameters greater than ten times the wavelength of the solar radiation; this includes dust and water droplets. Large particles cause all wavelengths of light to scatter more or less equally. However, light is increasingly propagated forward as the size of the particles gets larger. Because no wavelength is preferentially scattered, the light waves are little altered and so produce a whitish or pale blue-grey effect. For this reason, clouds and fog appear white although the constituent water droplets are colourless.

The presence of clouds can greatly increase the amount of incoming solar radiation reflected back to space. The bright,

3.12 **EFFECT OF CLOUD COVER ON THE AMOUNT OF ENERGY RECEIVED AT THE SURFACE** The marked colour contrast between the upper and lower parts of clouds illustrates how much energy is reflected and absorbed.

white surfaces of thick, low clouds are extremely good reflectors of short-wave radiation. Cloud reflection can account for the direct return to space of 30 to 60 percent of incoming radiation (Figure 3.12). Clouds also absorb as much as 5 to 20 percent of the radiation, depending on cloud type and thickness, so under heavy cloud cover as little as 10 percent of the incoming solar radiation may actually reach the ground. The amount and distribution of global cloud cover are therefore important factors affecting the global energy budget.

LONG-WAVE RADIATION FROM THE EARTH

Given the size of Earth and its distance from the sun and assuming there is no atmosphere, under radiative equilibrium the average surface temperature of the planet would be −18 °C. Based on global temperature data for the period 1901 to 2000, the mean temperature for Earth is approximately 14 °C, about 32 °C higher than would be the case without the atmosphere. This increase in temperature is due to greenhouse gases, such as water vapour and carbon dioxide, that reduce the rate at which energy is lost to space. Water vapour is the most important of the greenhouse gases and accounts for about 80 percent of the increase in the mean global temperature above radiative equilibrium.

The temperature of Earth's surface and atmosphere are much lower than the sun's surface. Consequently, Earth radiates less energy than the sun, but in comparatively longer wavelengths. The right side of Figure 3.9 shows the Earth's emission spectrum. The upper line is the radiation curve for a blackbody at a temperature of 300 K (23 °C), which approximates conditions for Earth. At this temperature, radiation ranges from about 3 to 30 μm and peaks at about 10 μm in the thermal infrared region; these emissions are considered **long-wave radiation**.

Beneath the Earth's theoretical blackbody curve is an irregular line that shows upwelling energy emitted by the Earth and its atmosphere, as measured at the top of the atmosphere. Compared with the blackbody curve, much of the Earth's surface radiation is absorbed by the atmosphere, especially at about 6 to 8 μm, 14 to 17 μm, and above 21 μm. Water and carbon dioxide are the primary absorbers of this terrestrial long-wave energy. Such absorption is an important part of the *greenhouse effect*, which helps to warm the Earth's surface. These absorption regions are flanked by regions in which outgoing energy flow is significant at 4 to 6 μm, 8 to 14 μm, and 17 to 21 μm. These are *windows* through which long-wave radiation leaves the Earth and flows to space.

ALBEDO

The proportion of short-wave radiant energy scattered upward by a surface is termed **albedo**. A surface with a high albedo, such as snow or ice (0.50 to 0.85), reflects much or most of the solar radiation that reaches it (Table 3.2), so the amount of energy it absorbs is considerably reduced. A surface with a low albedo, such as black pavement (0.03), absorbs nearly all the incoming solar energy. Since energy absorbed by a surface will heat the air immediately above it by radiation and conduction, temperatures will be warmer over a low albedo surface, such as asphalt, than over a high albedo surface, such as snow.

The albedo of a water surface varies with the calmness of the water and the angle of incidence of the incoming light. Albedo is very low (0.03) for nearly vertical rays on calm water, compared to as high as 0.50 when the sun shines at a low angle. In rough seas, the albedo from near-vertical rays increases because more light hits a sloping surface of a wave. Conversely, the angle of incidence for light coming from just above the horizon can be more or less perpendicular to the wave front, and so lowers the albedo. Reflection from smooth surfaces, such as calm water, is termed *specular reflection*; in this case the light rays are all redirected parallel to each other. For rough surfaces, the light rays are reflected at many different angles. This is termed *diffuse*

TABLE 3.2 ALBEDO (REFLECTIVITY) OF VARIOUS SURFACES

SURFACE	ALBEDO	PERCENT REFLECTION
Fresh snow	0.80–0.85	80–85
Old snow	0.50–0.60	50–60
Sand	0.20–0.30	20–30
Grass	0.20–0.25	20–25
Dry earth	0.15–0.25	15–25
Wet earth	0.10	10
Forest	0.05–0.10	5–10
Water (sun near horizon)	0.50–0.08	50–80
Water (sun near zenith)	0.03–0.05	3–5
Thick cloud	0.70–0.80	70–80
Thin cloud	0.25–0.50	25–50
Earth and atmosphere	0.30	30

reflection and is identical to the way light is scattered in infinite directions by gas molecules in the atmosphere.

Certain orbiting satellites are equipped with instruments to measure the energy levels of short-wave and long-wave radiation at the top of the atmosphere. Data from these satellites have been used to estimate Earth's average albedo. This includes reflection by the Earth's atmosphere as well as all the terrestrial and ocean surfaces, so it includes diffuse reflection by clouds, dust, and atmospheric molecules. The albedo values obtained in this way vary between 0.29 and 0.34. This means that the Earth–atmosphere system directly returns to space about one third of the solar radiation it receives. *Geographer's Tools • CERES—Clouds and the Earth's Radiant Energy System* describes how satellite instruments map and monitor the reflection of solar short-wave radiation and the emission of long-wave radiation by the Earth–atmosphere system.

COUNTER-RADIATION AND THE GREENHOUSE EFFECT

The amount of short-wave energy absorbed by a surface is an important determinant of its temperature. However, a surface is also warmed significantly by long-wave radiation emitted by the atmosphere and absorbed by the ground. Figure 3.13 shows energy flows between the surface, atmosphere, and space. Some of the incoming short-wave radiation from the sun is reflected back to space, but much of it is absorbed and warms the surface.

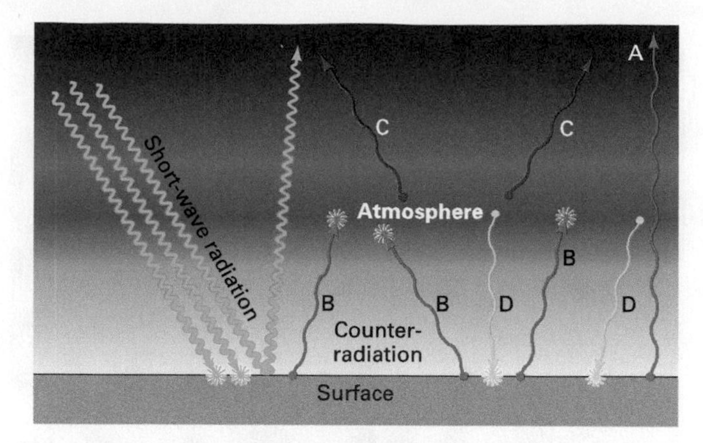

3.13 COUNTER-RADIATION AND THE GREENHOUSE EFFECT Short-wave radiation (left) passes through the atmosphere and is absorbed at the surface, warming the surface. The surface emits long-wave radiation. Some of this flow passes directly to space (A), but most is absorbed by the atmosphere (B). In turn, the atmosphere radiates long-wave energy back to the surface as counter-radiation (D) and also to space (C). The return of outbound long-wave radiation by counter-radiation constitutes the greenhouse effect.

The surface emits long-wave radiation, some of which escapes directly to space (A). The atmosphere absorbs the remainder (B). Although the atmosphere is colder than the surface, it too emits long-wave radiation. It emits this radiation in all directions, some upward to space (C), and some toward the Earth's surface (D). Since this downward flow is the opposite direction to long-wave radiation leaving the surface, it is termed **counter-radiation**. This prevents some of the heat emitted by the Earth's surface from radiating to space.

The amount of counter-radiation depends mainly on the presence of water vapour and carbon dioxide in the atmosphere. Much of the long-wave radiation emitted upward from the Earth's surface is absorbed by these two atmospheric gases. The absorbed energy raises the temperature of the atmosphere, causing it to emit more counter-radiation. Thus, the lower atmosphere, with its long-wave-absorbing gases, acts like a blanket that traps heat. Because liquid water is also a strong absorber of long-wave radiation, thermal insulation by cloud layers is extremely effective. This mechanism, in which the atmosphere traps long-wave radiation and returns some of it to the surface, is termed the **greenhouse effect**.

PLANETARY INSOLATION

Although the flow of solar radiation from the sun remains more or less constant, the amount received on Earth varies both spatially and over time. Average incoming energy flow, measured in watts per square metre over the course of a 24-hour day, is referred to as daily **insolation** (**in**coming **sol**ar radi**ation**). Annual insolation is the sum of these daily energy flows through the year. Assuming that the Earth is a uniformly spherical planet with no atmosphere, the amount of energy intercepted daily at a location depends on two factors: (1) the angle at which the sun's rays strike the Earth, and (2) the length of time the location is directly exposed to the rays. Both of these factors are controlled by the latitude of the location and the time of year.

Insolation is greatest when the sun is directly overhead; the incoming rays are vertical and energy is concentrated on a small part of the Earth. When the sun is lower in the sky, the same amount of solar energy is spread over a greater surface area, which effectively reduces insolation (Figure 3.8). Consequently, insolation at any location increases as the sun gets higher above the horizon. For locations polewards of the tropics, maximum insolation occurs at the summer solstice in each hemisphere. For the northern hemisphere this would be in June, and for the southern hemisphere in December. At all latitudes north and south of the equator, the length of the daylight period increases to a maximum at the summer solstice (see Figure 2.18).

If Earth's axis of rotation were vertical with respect to the path of its orbit around the sun, insolation at any latitude would not change seasonally. However, the tilt of Earth's axis results in the northern and southern hemispheres receiving more direct insolation and for longer periods during their respective high-sun seasons. This leads to profound spatial differences in insolation during the course of the year (Figure 3.14). For example, at 40° N (the approximate latitude of Windsor, Ontario), daily average insolation ranges from about 160 W m^{-2} at the December solstice to about 460 W m^{-2} at the June solstice.

GEOGRAPHER'S TOOLS

CERES—CLOUDS AND THE EARTH'S RADIANT ENERGY SYSTEM

The Earth's global **radiation balance** is the primary determinant of long-term surface temperature. However, this balance is increasingly sensitive to human activities, such as the conversion of forests to pasture or the release of greenhouse gases into the atmosphere. Thus, it is important to monitor the Earth's radiation budget over time as accurately as possible.

For nearly 20 years, Earth's global radiation balance has been the subject of NASA missions, which have launched radiometers—radiation-measuring devices—into orbit around the Earth. These devices scan Earth and measure the amount of short-wave and long-wave radiation at the top of the atmosphere. An ongoing NASA experiment entitled CERES—Clouds and the Earth's Radiant Energy System—is placing a new generation of these instruments in space to monitor the global radiation balance.

The CERES instruments scan Earth from horizon to horizon, measuring outgoing energy flows of reflected solar radiation (0.3 to 5.0 μm), outgoing long-wave radiation (8 to 12 μm), and outgoing total radiation (0.3 to 100 μm). Daily global maps of Earth's upwelling radiation fields are prepared from these observations. The bottom figure shows global reflected solar energy and emitted long-wave energy averaged over the month of March 2000.

The top image shows average short-wave flux, ranging from 0 to 210 W m^{-2}. The largest flows occur over regions of thick cloud cover near the equator, where much of the solar radiation is reflected back to space. In the midlatitudes, persistent cloudiness during this month also shows up as light tones, as do bright tropical deserts, such as the Sahara. Snow and ice surfaces in polar regions are quite reflective, but in March, at about the time of the equinox, the sun's angle is still quite low. Therefore, the amount of radiation the polar regions receive is much less than at the equator or midlatitudes. Consequently, the polar regions don't appear bright in this image. Oceans, especially where skies are clear, absorb much of the incoming solar radiation, and thus show low short-wave fluxes.

The bottom image shows long-wave flux on a scale from 100 to 320 W m^{-2}. Cloudy equatorial regions have low values, showing the blanketing effect of thick clouds that trap long-wave radiation beneath them. Warm tropical oceans in regions of clear sky emit the most long-wave energy. Toward the poles, surface and atmospheric temperatures drop, so long-wave energy emission also drops significantly.

Clouds are important determinants of the global radiation balance. A primary goal of the CERES experiment is to learn more about the Earth's cloud cover, which changes from hour to hour. This knowledge can improve global climate models that predict the impact of human and natural change on the Earth's climates. The most important contribution of CERES, however, is continuous monitoring of the Earth's radiant energy flows. In this way, small, long-term changes induced by human or natural processes can be detected, despite the large spatial and temporal variations in energy flows associated with cloud cover.

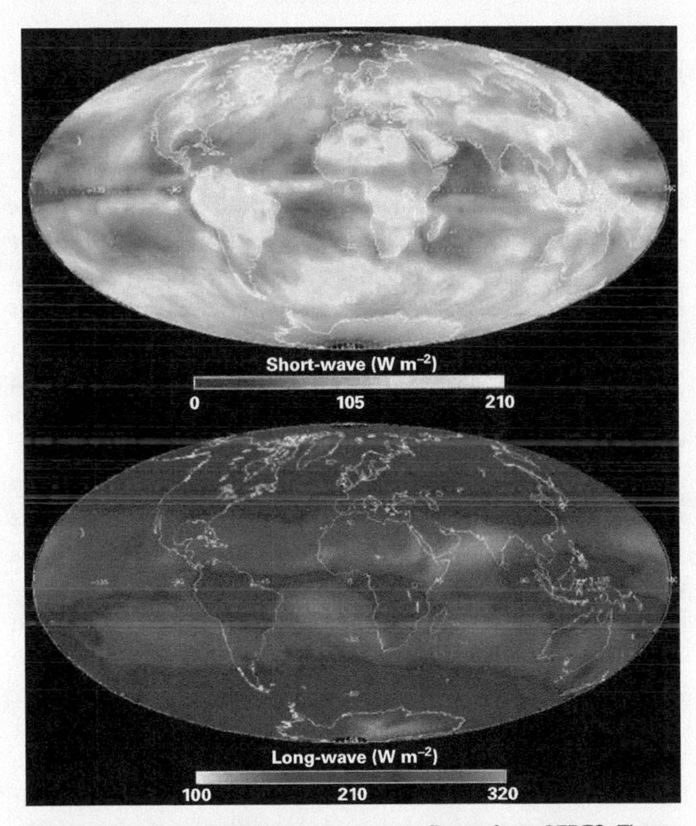

Global short-wave and long-wave energy fluxes from CERES These images show average short-wave and long-wave energy flows from Earth for March 2000, as measured by the CERES instrument on NASA's Terra satellite platform. (Courtesy of NASA.)

Equinox values are about 350 W m^{-2}. Inuvik, Northwest Territories, is situated just north of the Arctic Circle. At this latitude, direct insolation is zero during the December solstice, rising to about 175 W m^{-2} by the March equinox, and reaching 475 W m^{-2} at the time of the June solstice. Note that direct insolation during June is potentially higher at Inuvik than Windsor, and both locations would receive more insolation than the equator, which at this time of the year would expect about 380 W m^{-2}.

At the equator, there are two insolation maxima—one at each equinox—and two minima—one at each solstice. The two insolation

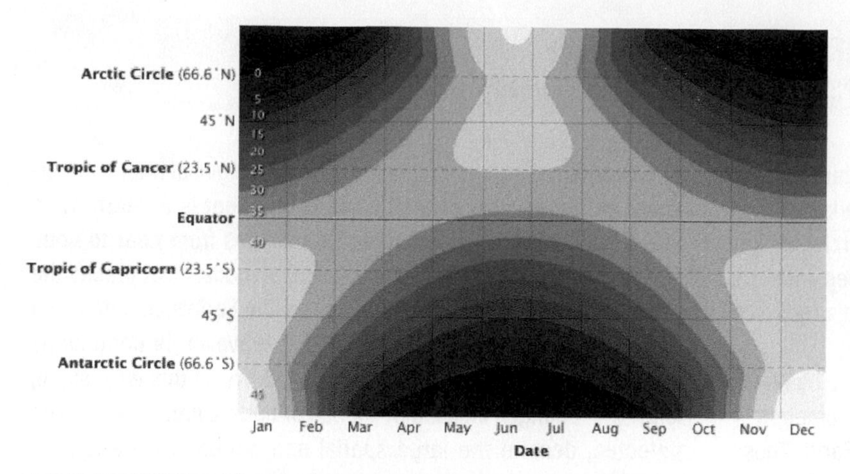

3.14 MEAN DAILY INSOLATION (W m^{-2}) AT THE TOP OF THE ATMOSPHERE THROUGH THE YEAR AS A FUNCTION OF THE DAY OF THE YEAR AND LATITUDE No direct insolation occurs in the white areas, which correspond to the polar night.

peaks are not quite equal and do not occur exactly at the equinoxes. These small variations arise because the Earth's orbit is not exactly circular. Thus, in 2009, insolation at the equator peaked at 440.04 W m^{-2} on March 9 and 433.62 on October 5. This compares to values at the solstices of 412.48 W m^{-2} on December 19 and 386.51 W m^{-2} on June 23. Two daily insolation maxima occur at all latitudes between the Tropic of Cancer (23½° N) and the Tropic of Capricorn (23½° S). However, as either tropic is approached, the two maxima get closer and closer in time, and eventually merge into a single maximum.

Between the Tropic of Cancer and Arctic Circle, and between the Tropic of Capricorn and the Antarctic Circle, insolation follows a wavelike pattern, increasing during the high-sun season and decreasing during the low-sun season. Polewards from the Arctic and Antarctic circles, the sun is below the horizon for at least part of the year, during which daily insolation drops to zero. Daily insolation is greatest at either the North or South Pole during the respective summer solstice in each hemisphere. For example, at the North Pole in 2009, insolation reached 526.51 W m^{-2} on June 20; at the South Pole the maximum value was 561.91 W m^{-2} on December 21.

Latitudinal differences in daily insolation are important because this affects how much solar energy is available to heat the Earth's surface, and so is the most important factor in determining air temperature. Seasonal change in daily insolation at a location is, therefore, a major determinant of climate. Annual insolation is greatest at lower latitudes (Figure 3.15). Despite the long dark winters, the high latitudes still receive a considerable flow of solar energy, with annual insolation at the poles about 40 percent of that at the Equator.

WORLD LATITUDE ZONES

The seasonal pattern of daily insolation can be used as a basis for dividing the globe into broad latitude zones (Figure 3.16). The specified limits are generalized and provide a convenient nomenclature to identify important geographic zones.

The *equatorial zone* encompasses the equator and covers the latitude belt roughly 10° N to 10° S. Here the sun provides intense insolation throughout most of the year, and the daylight and nighttime periods are of roughly equal length. Spanning the Tropics of Cancer and Capricorn are the *tropical zones*, ranging from latitudes 10° to 25° N and S. A marked seasonal cycle exists in these zones, combined with high annual insolation. Moving beyond the tropical zones are transitional *subtropical zones*. These may be conveniently delineated by the latitude belts 25° to 35° N and S. Despite a characteristic seasonal cycle, annual insolation in the subtropical zones is similar to that of the tropical zones.

The *midlatitude zones* lie between 35° and 55° N and S latitude. Here the elevation of the sun changes appreciably during the year. There are pronounced differences in daylight from winter to summer, and seasonal contrasts in insolation are strong. Consequently, these regions can experience a large annual range in temperature.

Bordering the mid-latitude zones on the poleward side are the *subarctic* and *subantarctic zones* at 55° to 60° N and S latitudes. Astride the Arctic and Antarctic Circles from latitudes 60° to 75° N and S lie the *Arctic* and *Antarctic zones*. All of these high-latitude zones have an extremely large yearly variation in day length, resulting in enormous contrasts in insolation from solstice to solstice. The northern and southern *polar zones* occupy the areas between about 75° latitude and the poles, a region characterized by approximately six months of daylight followed by six months of darkness. The polar zones experience the greatest seasonal contrasts of insolation.

SENSIBLE HEAT AND LATENT HEAT TRANSFER

The transfer of short-wave energy from the sun across the vacuum of space is accomplished by radiation, and occurs without the involvement of a physical substance. In the same way, long-wave energy from

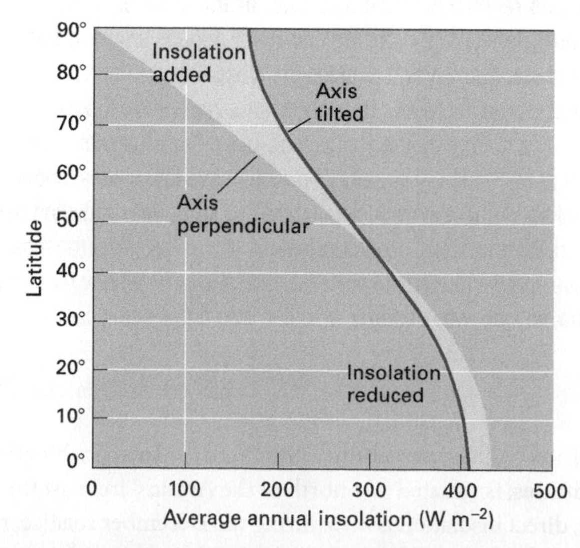

3.15 THE EARTH'S ANNUAL INSOLATION FROM EQUATOR TO POLES The effect of the tilt of the Earth's axis is most pronounced at the poles.

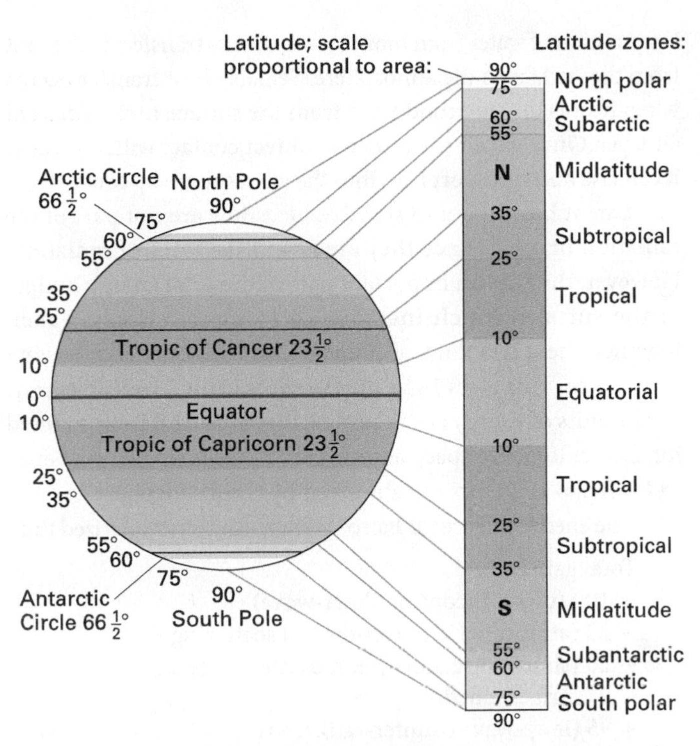

3.16 **WORLD LATITUDE ZONES** A system of latitude zones, based on the seasonal patterns of daily insolation observed over the globe.

the Earth and its atmosphere can escape into space. In addition to radiation, energy transfer in the form of sensible and latent heat is also important in maintaining the global energy balance.

Sensible heat is heat energy held by an object or substance that can be measured by a thermometer. The store of sensible heat increases as the temperature of the object rises. When two objects of unlike temperature contact each other, heat energy moves by **conduction** from the warmer object to the cooler. This type of heat flow is referred to as **sensible heat transfer**. Sensible heat transfer can also occur in fluids by **convection**. In this process, when a fluid is heated, it expands; this lowers its density and so it moves away from the source of heat. Air acts as a fluid in much the same way as water. Convection in the atmosphere creates currents that carry warm air away from the surface, allowing cooler air to descend. Without convection, the radiative equilibrium surface temperature for Earth would be 67 °C. In a similar way, ocean currents also moderate global temperatures by circulating heat away from the surface and bringing cold water up from the depths.

Latent heat is heat that is taken up and stored in the form of molecular motion. It is associated with a change of state of substance from a solid to a liquid, from a liquid to a gas, or from a solid directly to a gas. For example, energy is required to convert liquid water to water vapour. As water evaporates, it absorbs energy from the environment, which accelerates the random motion of the water molecules. The water absorbs the energy in the form of latent heat and uses it to overcome the potential energy stored in the intermolecular bonds in the liquid. This change in state from liquid to vapour occurs without a rise in temperature. However, energy must be supplied to carry out this process, and this is acquired by absorbing heat from the surface or from the atmosphere, which effectively reduces their temperature. In this way, energy is transferred away from the surface in the form of randomly moving water vapour molecules. Conversely, when water vapour reverts to a liquid or solid, the stored latent heat is released to the atmosphere. This is absorbed by the air molecules, which warms the atmosphere.

Latent heat transfer occurs when water evaporates from a moist land surface or an open water surface. This process transfers heat from the surface to the atmosphere. On a global scale, latent heat transfer by movement of moist air provides an important mechanism for transporting large amounts of energy from one region of the Earth to another. The *Working It Out* feature in Chapter 5 provides more information on the amount of energy acquired or released in these processes.

GLOBAL ENERGY BALANCE OF THE EARTH–ATMOSPHERE SYSTEM

Earth constantly absorbs short-wave solar radiation and emits long-wave radiation, and these flows of energy from the sun to the Earth and then back out into space form a complex system involving not only radiant energy, but also energy storage and energy transport in various forms. The pathways of energy flow between the sun, the Earth's surface, and the atmosphere integrate all the dynamic processes that shape the planet. Ultimately, the amount of energy received by the Earth–atmosphere system equals the amount of energy returned to space. Therefore, on a global scale, there is a **radiation balance**.

The global radiation balance is an example of an energy flow system, and can be calculated from the incoming short-wave and outgoing long-wave radiation fluxes. The input flow of solar radiation to Earth is determined from the flow rate and the area that receives the flow. The rate for total solar irradiance is 1,367 W m^{-2} (1.367 kW m^{-2}). The Earth's radius is about 6,400 km. Thus, Earth presents a 6,400 km-wide disk to the sun's rays, with a surface area of 1.29×10^{14} m^2. Therefore, the energy flow from the sun to the Earth is:

$$\text{Flow per Unit Area} \times \text{Area of Earth Presented to Sun} = \text{Sun–Earth Energy Flow}$$

$$1.367\ \text{kW m}^{-2} \times 1.29 \times 10^{14}\ \text{m}^2 = 1.77 \times 10^{14}\ \text{kW}$$

Energy outflows take the form of reflected short-wave radiation and emitted long-wave radiation. Earth reflects about one third of the solar radiation that it receives, or about 1.77×10^{14} kW $\div 3 = 0.59 \times 10^{14}$ kW. The remaining two thirds $(1.77 - 0.59 = 1.18 \times 10^{14}$ kW) is absorbed by the Earth and its atmosphere and is ultimately emitted as long-wave radiation.

An important principle of physics is that energy is neither created nor destroyed, but only transformed. It is therefore possible to follow the initial stream of solar energy and account for its diversion into different system pathways and its conversion into different energy forms. A full accounting of all the energy flows among the sun, the atmosphere, the Earth's surface, and space forms the **global energy balance**.

INCOMING SHORT-WAVE RADIATION

The global energy balance for the Earth-atmosphere system is shown in Figure 3.17; Figure 3.17a shows the fate of the incoming radiant energy flow. For incoming short-wave energy, assume 100 units reach Earth from the sun. Reflection by gas molecules and dust, clouds, and the surface (including the oceans) totals 31 units and represents a combined albedo for the Earth and the atmosphere of 0.31. Short-wave energy absorption in the atmosphere through absorption by molecules, dust, and clouds averages 20 units. Combining this 20 unit loss of incoming solar energy through absorption with the 31 units lost through reflection reduces energy input to Earth's land and water surfaces to 49 units.

ENERGY FLOWS AT EARTH'S SURFACE

The components of outgoing long-wave radiation for the Earth's surface and the atmosphere are shown in Figure 3.17b. Note that total long-wave radiation leaving the Earth's land and ocean surface is equivalent to 114 units, which is 14 percent higher than the 100 units of shortwave radiation originally received from the sun. This increase is due to the supplemental flow of energy to Earth's surface through counter-radiation. After selective depletions, direct solar radiation adds 49 units of short-wave radiation. An additional 95 units of long-wave counter-radiation comes from the atmosphere. The combined input of energy to Earth's surface from these sources is equivalent to 144 units.

Some of this energy flows away from the surface as latent heat (23 units), and a smaller amount is lost as sensible heat (7 units). Evaporation of water from moist soil or oceans transfers latent heat from the surface to the atmosphere. Sensible heat transfer occurs when heat is directly conducted from the surface to the adjacent air layer. Once the air is warmed by direct contact with a surface, it can rise and transport heat into the atmosphere by convection.

Latent heat flow and sensible heat flow are not part of the radiation balance, since they are not in the form of radiation. However, they are an important part of the total energy budget of the surface, which includes all forms of energy. Taken together, these two flows account for 30 units of energy leaving the surface. This gives 144 − 23 (latent heat) − 7 (sensible heat) = 114 units of energy at the surface that must still be accounted for. Direct losses to space account for 12 units of energy; hence 114 − 12 = 102 units are absorbed by the atmosphere.

The energy balance for Earth's surface can be summarized thus:

Total gains are:
+ 100 (direct incoming short-wave)
− 22 (atmospheric reflection and scattering)
− 20 (short-wave absorption by atmosphere)
− 9 (surface albedo)
+ 95 (long-wave counter-radiation) = 144

Total losses are:
− 23 (latent heat)
+ 7 (sensible heat)
+ 12 (long-wave radiation to space)
+ 102 (long-wave radiation to the atmosphere) = 144

ENERGY FLOWS TO AND FROM THE ATMOSPHERE

Earth's atmosphere gains and loses energy by processes that are similar to those operating at the surface. The atmosphere gains energy by absorption of short-wave radiation, amounting to 20 units (17 units by gas molecules and dust and 3 units by clouds).

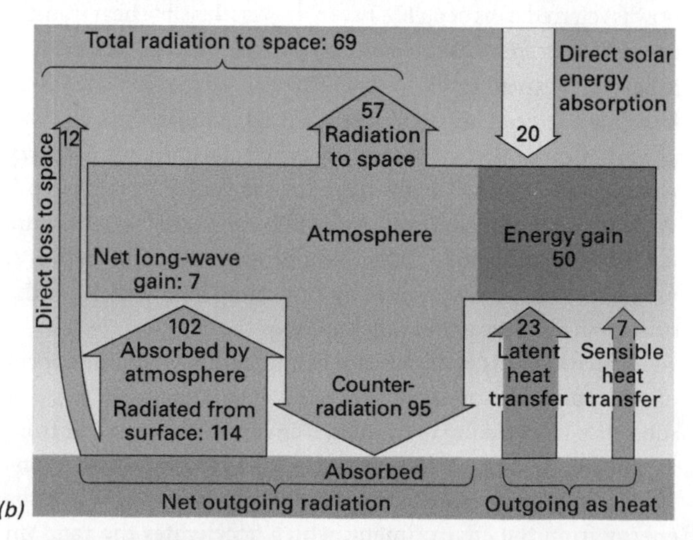

3.17 THE GLOBAL ENERGY BALANCE (a) The fate of incoming solar radiation. (b) Long-wave energy flows occurring between the surface and atmosphere and space. Also shown are the transfers of latent heat, sensible heat, and direct solar absorption, which balance the budget for the Earth–atmosphere system. Values are given as percentages of a total insolation of 100. (Data from Kiehl, J. T., and K. E. Trenberth, 1997: Earth's annual global mean energy budget. *Bulletin of the American Meteorological Society*, 78, 197–208.)

The atmosphere also gains energy from latent heat transfer (23 units) and sensible heat transfer (7 units). Additionally, the atmosphere absorbs 102 units of long-wave radiation emitted by the surface. Together these energy gains total 20 (absorption) + 23 (latent heat) + 7 (sensible heat) + 102 (long-wave) = 152.

A loss of 57 units of long-wave energy is radiated from the atmosphere to space and a loss of 95 units occurs as counter-radiation to the surface. Together these total 57 (long-wave radiation) + 95 (counter-radiation) =152 units, which balances the atmosphere's energy budget.

This analysis illustrates the atmosphere's vital role in trapping heat through the greenhouse effect. Without the 95 units of counter-radiation from the atmosphere, the surface would have only 49 units of absorbed short-wave radiation to emit. The temperature needed to radiate so few units is well below freezing. So without the greenhouse effect, Earth would be much less hospitable.

The global energy budget is determined by combining the surface and atmosphere flows, and can be summarized thus:

Total unit gains to the Earth-atmosphere system are:
+ 49 (short-wave absorption by surface)
+ 20 (short-wave absorption by atmosphere) = 69

Total unit losses to the Earth-atmosphere system are:
– 57 (long-wave radiation to space from atmosphere)
+ 12 (long-wave radiation to space from surface) = 69

CLIMATE AND ENVIRONMENTAL CHANGE

The preceding analysis helps to illustrate how global change might affect Earth's climate. Suppose that clearing forests for agriculture and turning agricultural lands into urban and suburban areas decreases surface albedo. The ground would absorb more energy, raising its temperature. That, in turn, would increase the flow of surface long-wave radiation to the atmosphere. This radiation would be absorbed and then boost counter-radiation, which would probably amplify the warming through the greenhouse effect.

But what if air pollution causes more low, thick clouds to form? Since low clouds increase short-wave reflection back to space, the effect will be to cool the surface and atmosphere. Conversely, an increase in high-altitude clouds would result in greater absorption of long-wave energy compared with the amount of short-wave energy they reflect. This should make the atmosphere warmer and boost counter-radiation, increasing the greenhouse effect. The energy flow linkages between the sun, surface, atmosphere, and space are critical components of the climate system, and understanding them helps to better appreciate the complexities of global climate change.

Many natural feedback processes help to moderate global temperatures. One example is the role played by dimethyl sulphide (DMS). DMS is a volatile sulphur compound found in sea water. It is produced by marine planktonic algae. When this gas gets into the atmosphere, it oxidizes to form substances that promote cloud formation. Emission of DMS by plankton may act as part of a climate control feedback loop because it is related to cloud cover. If an increase in cloud cover due to DMS oxidation increases global albedo, this in turn would cause a reduction in temperature and create an environment that is less favourable to the plankton. Consequently, DMS production would decline, cloud cover would be less, and with lower albedo global temperatures would rise. Such conditions would favour increased planktonic activity, and the cycle would be repeated.

NET RADIATION, LATITUDE, AND THE ENERGY BALANCE

The Earth intercepts solar energy, which when absorbed results in an overall rise in the planetary temperature. This is offset by the reduction in temperature that occurs when energy is radiated to space. Over time, these incoming and outgoing radiation flows must balance for the Earth as a whole. However, incoming and outgoing flows do not have to balance instantaneously at a specific location. At night, for example, there is no incoming radiation, yet the darkened Earth's surface and the atmosphere still emit outgoing radiation.

Solar energy input varies strongly with latitude. In some locations, radiant energy flows in faster than it flows out; elsewhere, radiant energy flows out faster than it flows in. The difference between all incoming radiation and all outgoing radiation is termed **net radiation**. Between about lat. 40° N and lat. 40° S, incoming solar radiation exceeds outgoing long-wave radiation throughout the year; here net radiation is positive, providing an energy surplus (Figure 3.18). From 40° N and 40° S toward the poles, outgoing long-wave radiation exceeds incoming short-wave radiation. At these higher latitudes, net radiation is negative, providing an energy deficit.

The pattern of energy surplus at low latitudes and energy deficit at high latitudes creates a flow of energy from the tropics toward the poles. This energy flow, in the form of sensible and latent heat, occurs through the poleward movement of warm water and warm, moist air in global oceanic and atmospheric circulation. At the same time, cooler water and cooler, drier air moves toward the equator. Without this broad-scale circulation, low latitudes would heat up and high latitudes would cool down until a radiative balance is achieved, resulting in much more extreme temperature contrasts.

Global incoming short-wave radiation exactly balances global outgoing long-wave radiation on an annual basis for the entire Earth–atmosphere system. However, considerable differences occur in net radiation at the regional scale and with the seasons. This ultimately determines how much energy is available in different parts of the world and is an underlying factor in many aspects of global climate.

SOLAR ENERGY

The Earth intercepts solar energy at the rate of 1.77×10^{14} kilowatts per year. This quantity of energy is about 28,000 times as much as all the energy human society currently consumes each year. Thus, an enormous energy source is available for use. In addition to its abundance, solar energy, unlike fossil fuels, does not release carbon dioxide or other pollutants.

Solar energy can be converted directly into electricity using photovoltaic cells. Its use has been limited because of the relatively high cost of manufacturing these solar cells, but the technology is now widely used for roadside emergency telephones and similar applications. Using current technology, arrays of solar cells would need to occupy large areas to produce sufficient energy for general use. Electricity storage is another problem, since the solar cells will not generate power at night.

A second common application of solar energy is heating buildings. Most systems use solar collectors placed on roofs. A typical collector consists of a network of aluminum or copper tubes carrying circulating water. The tubes, painted black, absorb solar energy and use it to heat the water to a temperature of about 65 °C. Several advanced installations have been constructed, which use computer-controlled mirrors to focus the sun's rays onto boilers to produce steam for use in power generators.

Commercial-scale projects have generally employed arrays of flat, moveable mirrors called heliostats to focus the sun's energy onto a central **solar power tower**, such as Solar Tres in Spain. This solar plant was designed to use 2,493 heliostats with a total reflective area of 240,000 m^2. The solar energy is used to heat molten nitrate salt to temperatures in excess of 500 °C. The stored heat is then used during the night to produce steam to generate a continuous supply of electricity.

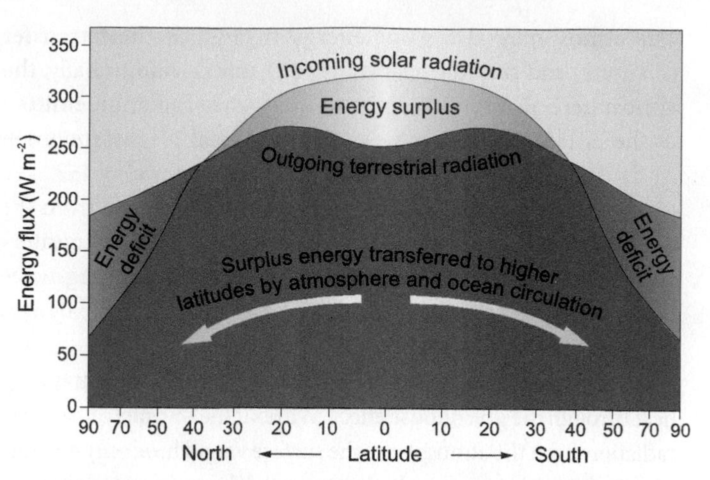

3.18 ANNUAL SURFACE NET RADIATION FROM POLE TO POLE In low latitudes where net radiation is positive, incoming solar radiation exceeds outgoing long-wave radiation; there is an energy surplus, and energy moves toward the poles as latent and sensible heat. Where net radiation is negative, there is an energy deficit, and latent and sensible heat are lost in the form of outgoing long-wave radiation.

A Look Ahead

Earth's energy balance is sensitive to a number of factors that determine how energy is transmitted and absorbed. The net effect is seen in Earth's mean global temperature. However, many other factors are linked to regional temperature patterns. Chapter 4 discusses present-day temperatures at both local and regional scales, and examines how and why they vary from place to place as well as seasonally and over the course of a day. A fundamental aspect of the global energy budget is that energy must be redistributed from regions of net surplus to regions where there is a net deficit. Chapter 5 covers the link between the energy balance and global circulation.

CHAPTER SUMMARY

- The Earth's atmosphere is dominated by *nitrogen* and *oxygen gases*. *Carbon dioxide* (CO_2) and *water vapour*, though lesser constituents by volume, are important because they absorb long-wave radiation and enhance the greenhouse effect.
- Ozone (O_3) is a small but important constituent of the atmosphere and is concentrated in a layer of the upper atmosphere. It is formed from oxygen (O_2) by chemical reactions that absorb ultraviolet radiation, thus sheltering the organisms of the Earth's surface from its damaging effects.
- *Electromagnetic radiation* is a form of energy emitted by all objects. The wavelength of the radiation determines its characteristics. The hotter an object, the shorter the wavelengths of the radiation and the greater the amount of radiation it emits.
- Radiation emitted by the sun includes *ultraviolet, visible, near-infrared, and short-wave infrared radiation. Thermal infrared radiation*, which is emitted by Earth surfaces, is familiar as heat. The atmosphere absorbs and scatters radiation in certain wavelengths. Radiation flows are measured in watts per square metre.
- The Earth continuously absorbs solar short-wave radiation and emits long-wave radiation. In the long run, the gain and loss of radiant energy remain in balance, and the Earth's average temperature remains constant.
- *Insolation*, the rate of solar radiation flow available at a location at a given time, is greater when the sun is higher in the sky. *Daily insolation* is also greater when the period of daylight is longer. Between the tropics and poles, the sun rises higher in the sky and stays longer in the sky at the summer

solstice than at the equinox; daylight hours are shortest during the winter solstice. Near the equator, daily insolation is greater at the equinoxes than at the solstices. Annual insolation is greatest at the equator and least at the poles. However, during the course of a year, the poles still receive 40 percent of the radiation received at the equator.

- The pattern of annual insolation with latitude leads to a natural naming convention for latitude zones: *equatorial, tropical, subtropical, midlatitude, subarctic (subantarctic), Arctic (Antarctic),* and *polar*.
- Latent heat and sensible heat are additional forms of energy. Latent heat is taken up or released when a change of state occurs. Sensible heat is contained within a substance and can be transferred to another substance by *conduction* or *convection*.
- The global energy system includes a number of important pathways by which insolation is transferred and transformed. Part of the solar radiation passing through the atmosphere is absorbed or scattered by molecules, dust, and larger particles. Some of the scattered radiation returns to space as *diffuse reflection*. Land surfaces, ocean surfaces, and clouds also reflect some solar radiation back to space.
- The percentage of radiation that a surface absorbs is termed its albedo. The albedo of the Earth and atmosphere together is about 30 percent.
- The atmosphere absorbs long-wave energy emitted by the Earth's surface, causing the atmosphere to counter-radiate some of that long-wave radiation back to Earth, thereby creating the *greenhouse effect*. Because of this heat trapping, the Earth's surface temperature is considerably warmer than it would be without an atmosphere.
- *Net radiation* describes the balance between incoming and outgoing radiation. At latitudes lower than 40 degrees, annual net radiation is positive. It is negative at higher latitudes. This imbalance creates poleward heat transport of latent and sensible heat in the movement of warm water and warm, moist air, and thus provides the power that drives ocean currents and broad-scale atmospheric circulation patterns.

KEY TERMS

absorption
aerosol
air pollutant
albedo
blackbody
chlorofluorocarbons (CFCs)
condensation nuclei
conduction
convection
counter-radiation
Dobson unit (DU)
dry deposition
electromagnetic radiation
electromagnetic spectrum
fallout
global energy balance
greenhouse effect
infrared
insolation
latent heat
latent heat transfer
long-wave radiation
Mie scattering
net radiation
ozone (O_3)
particulates
photochemical reaction
radiation balance
Rayleigh scattering
reflection
remote sensing
scattering
sensible heat
sensible heat transfer
short-wave radiation
solar constant
solar power tower
Stefan-Boltzmann Law
total solar irradiance
ultraviolet (UV) radiation
visible light
washout
Wien's Law

REVIEW QUESTIONS

1. Identify the three largest components of dry air. Why are carbon dioxide and water vapour important atmospheric constituents?
2. How does the ozone layer protect the life layer?
3. What is electromagnetic radiation? How is it characterized? Identify the major regions of the electromagnetic spectrum.
4. How does an object's temperature influence the nature and amount of electromagnetic radiation it emits?
5. What is the solar constant? What is its value? What units are used to measure it?
6. How does solar radiation received at the top of the atmosphere differ from solar radiation received at the Earth's surface? What are the roles of absorption and scattering?
7. Compare the terms *short-wave radiation* and *long-wave radiation*. What are their sources?
8. How does the atmosphere affect the flow of long-wave energy from the Earth's surface to space?
9. What is the Earth's global energy balance, and how are short-wave and long-wave radiation involved?
10. How does the sun's path in the sky influence daily insolation at a location?
11. What influence does latitude have on the annual cycle of daily insolation? On annual insolation?
12. What is energy? What distinguishes the following forms of energy: kinetic, potential, mechanical, radiant, and chemical?
13. Describe latent heat transfer and sensible heat transfer.
14. Define albedo and give two examples.
15. Describe the counter-radiation process and how it relates to the greenhouse effect.

16. What important physical principle permits the construction of budgets of energy and matter flow?
17. Discuss the energy balance of the Earth's surface. Identify the types and sources of energy flows that the surface receives. Do the same for energy flows that it loses.
18. Discuss the energy balance of the atmosphere. Identify the types and sources of energy flows that the atmosphere receives. Do the same for energy flows that it loses.
19. What is net radiation? How does it vary with latitude?
20. What is the role of poleward heat transport in balancing the net radiation budget by latitude?
21. Why is solar power in principle more desirable than power produced by burning fossil fuels?
22. What are CFCs, and how do they affect the ozone layer?
23. When and where have ozone reductions been reported? Have corresponding reductions in ultraviolet radiation been noted?
24. What is the outlook for the future health of the ozone layer?

VISUALIZATION EXERCISES

1. Sketch the world latitude zones on a circle representing the globe and give their approximate latitude ranges.
2. Sketch a simple diagram of the sun above a layer of atmosphere and the Earth's surface. Draw arrows to indicate flows of energy among sun, atmosphere, and surface. Label each arrow according to the process it represents.

ESSAY QUESTION

1. Discuss how the atmosphere will influence a beam of either (a) short-wave solar radiation entering the Earth's atmosphere heading toward the surface, or (b) a beam of long-wave radiation emitted from the surface heading toward space.

WORKING IT OUT • RADIATION LAWS

The relationship between the flow of energy from an object and that object's surface temperature is described by the **Stefan-Boltzmann Law** as:

$$M = \sigma T^4$$

where M is the energy flow from the surface, in watts per square metre (W m^{-2}); σ is the Stefan-Boltzmann constant (5.67×10^{-8} W m^{-2} K^4); and T is the temperature of the surface, in kelvins. The watt (W) is a measure of energy flow. The symbol K denotes absolute temperature in kelvins.

Wien's Law states that the wavelength of maximum energy emission from an object is related to its surface temperature.

$$\lambda_{MAX} = b/T$$

where λ_{MAX} is the wavelength at which radiation is at a maximum, in micrometres; b is a constant equal to 2,898 μm; and T is the temperature of the surface in kelvins.

Both of these laws apply only to a perfectly radiating surface known as a **blackbody**; that is, one that emits energy at the same rate as it absorbs it. Most natural surfaces are good radiators and do not differ too much from a blackbody.

The following calculation uses these formulae to determine the flow of energy emitted by a surface and then to determine the wavelength at which radiation from the surface is greatest. Assume the temperature of the surface is 450 K (177 °C, a common oven temperature for baking). Using the Stefan-Boltzmann equation, substituting 450 K for T gives:

$$M = \sigma T^4 - (5.67 \times 10^{-8}\ \text{W m}^{-2}\ \text{K}^4) \times (450\ \text{K})^4$$
$$= (5.67 \times 10^{-8}\ \text{W m}^{-2}\ \text{K}^4) \times (4.10 \times 10^{10})$$
$$= 2.325 \times 10^3\ \text{W m}^{-2}$$
$$= 2{,}325\ \text{W m}^{-2}$$

Applying Wien's Law, then

$$\lambda_{MAX} = b/T = 2{,}898\ \mu\text{m} / 450\ \text{K} = 6.44\ \mu\text{m}$$

In comparison, a surface at 293 K (20 °C, approximately room temperature) gives an emission rate of 418 W m^{-2} and a wavelength of maximum emission of 9.89 μm. Thus, the oven surface emits more than five times more energy than the room temperature surface, and the wavelength of maximum emission for the oven is shorter.

QUESTIONS / PROBLEMS

1. Assuming that the sun behaves as a blackbody, what is the flow rate of energy leaving its surface if the surface is at a temperature of 5,950 K (5,677 °C)? At what wavelength will the emitted radiation be greatest?
2. The average surface temperature of the Earth is approximately 15.4°C. Assuming that the surface radiates perfectly, what is the flow rate of energy emitted by the surface? At what wavelength will the radiation be greatest?
3. Using the answers to problems 1 and 2, determine the ratio of the flow rate of energy emitted by the sun's surface to the flow rate of energy emitted by the Earth's surface.
4. Venus has a radius of about 6,050 km. Because it is closer to the sun than the Earth, the solar radiation flow it receives is 1.92 times stronger than that received by the Earth. The atmosphere of Venus consists of dense clouds of carbon dioxide vapour, which reflect about 65 percent of the incoming solar radiation. Draw a diagram of the energy flow system for Venus and calculate the rates of inflow and outflow of short-wave and long-wave radiation.

Find answers at the back of the book.

APPENDIX 3.1 REMOTE SENSING FOR PHYSICAL GEOGRAPHY

Remote sensing refers to gathering information from great distances and over broad areas, usually through instruments mounted on aircraft or orbiting space vehicles. These instruments measure electromagnetic radiation coming from the Earth's surface and atmosphere. The data acquired by remote sensors are typically displayed as images, but are often processed further to provide other types of outputs, such as maps of albedo, condition of the vegetation, or land-cover class. Solar radiation reaching the Earth's surface is concentrated in wavelengths from about 0.3 to 2.1 μm in the visible, near-infrared, and short-wave infrared wave bands, and remote sensors are commonly constructed to measure radiation in all or part of this range. For remote sensing of emitted energy, the object or substance itself is the source of the radiation, which typically is related to its temperature.

Colours and Spectral Signatures

Most objects or substances at the Earth's surface appear coloured to the human eye. The various colours mean that they reflect different wavelengths of light in the visible spectrum. Figure 3A.1 shows typical reflectance spectra for water, vegetation, and soil in the solar wavelength range as they might be viewed through the atmosphere. Note that water vapour in the atmosphere strongly absorbs radiation at wavelengths from about 1.2 to 1.4 μm and 1.75 to 1.9 μm, so it is not possible for a remote sensor to detect surface features at those wavelengths.

Water surfaces are always dark, but are slightly more reflective in the blue and green regions of the visible spectrum. Thus, water appears blue or blue-green to our eyes. Beyond the visible region, water absorbs nearly all short-wave radiation it receives and so looks black in images acquired in the near-infrared and short-wave infrared regions.

Vegetation appears dark green to the human eye, which means that it reflects more energy in the green portion of the visible spectrum while reflecting somewhat less in the blue and red portions. But vegetation also reflects strongly in near-infrared wavelengths, which the human eye cannot see. Because of this property, vegetation will be bright in near-infrared images. This distinctive behaviour of vegetation—appearing dark in visible bands and bright in the near-infrared—is the basis for most vegetation remote sensing.

The soil spectrum shows a steady increase of reflectance across the visible and near-infrared spectral regions. Soil is brighter overall than vegetation and is somewhat more reflective in the orange and red portions. Thus, it often appears brown. (Note that this is just a "typical" spectrum—soil colour can actually range from black to bright yellow or red.)

The pattern of relative brightness within spectral bands is referred to as the *spectral signature* of an object or substance. Spectral signatures can be used to recognize objects or surfaces in remotely sensed images in much the same way that we recognize objects by their colours. In computer processing of remotely sensed images, spectral signatures can be used to make classification maps showing, for example, water bodies, different vegetation types, and soil conditions.

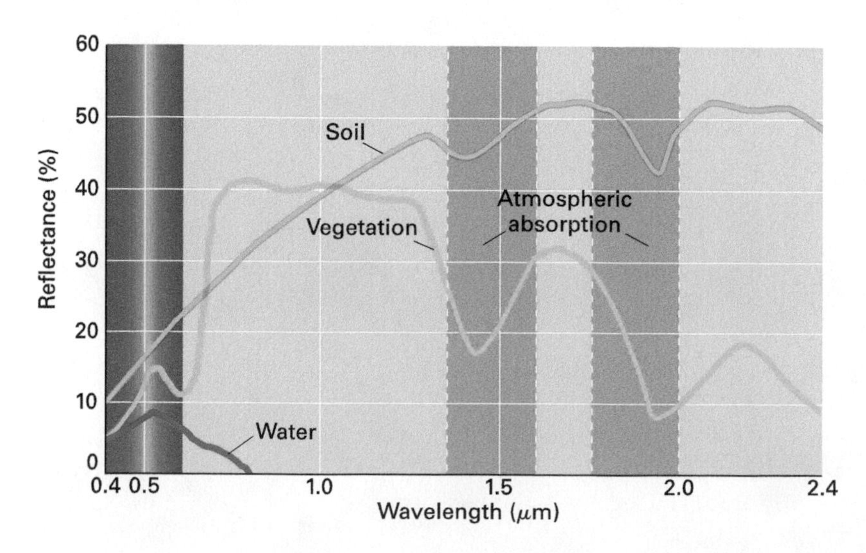

3A.1 Reflectance spectra of water, vegetation and soil.

Aerial Photography

Aerial photography is the oldest form of remote sensing. Air photos have been in general use since the 1930s. Common practice is to have the field of one photograph overlap the next along the plane's flight path so that the photographs can be viewed stereoscopically for a three-dimensional effect. Because of its high resolution (degree of sharpness) and low cost, aerial photography is commonly used in remote sensing.

Aerial photography often uses colour infrared film. This special film is sensitive to near-infrared wavelengths in addition to visible wavelengths. The red tones in an image are produced in response to near-infrared light, green tones by red light, and blue tones by green light. Because healthy, growing vegetation reflects much more strongly in the near-infrared than in the red or green regions of the spectrum, vegetation has a characteristic red appearance (Figure 3A.2).

Thermal Infrared Sensing

Radiation leaving the Earth's surface is concentrated in the thermal infrared spectral region, ranging from about 8 to 12 μm. Besides absolute temperature, the intensity of infrared emissions depends on the *emissivity* of an object or substance. Emissivity is the ratio of the actual energy emitted to that of a blackbody at the same temperature. For most natural Earth surfaces, emissivity is comparatively high—between 0.85 and 0.99.

3A.2 This colour infrared aerial photograph shows part of Bow River Irrigation District near Vauxhall, Alberta. The various red tones represent different types of crops, several of which are irrigated with centre-pivot systems that give the distinctive circular patterns.

Differences in emissivity affect thermal images. For example, two different surfaces might be at the same temperature, but the one with the higher emissivity will look brighter because it emits more energy (Figure 3A.3).

Radar

There are two classes of remote sensor systems: passive and active. *Passive systems* acquire images without providing a source of energy, in much the same way as a camera acquires an image. *Active systems*, such as radar, use a beam of energy as a source, sending the beam toward an object. Part of the energy is reflected back to the source, where it is recorded by a detector.

Radar systems in remote sensing use the *microwave* portion of the electromagnetic spectrum. An advantage of radar systems is that they use wavelengths that are not significantly absorbed by liquid water. This means that radar systems can penetrate clouds to provide images of the Earth's surface in any weather. At some wavelengths, however, microwaves are scattered by water droplets and can produce a return signal sensed by the radar apparatus. This effect is the basis for weather radars, which can detect rain and hail and are used in local weather forecasting.

Surfaces that are more perpendicular to the radar beam will return the strongest echo and therefore appear lightest in tone. In contrast, those surfaces facing away from the beam will appear darkest. This produces an image resembling a three-dimensional model of the landscape illuminated by a light.

Digital Imaging

Modern remote sensing relies heavily on computer processing to extract and enhance information from remotely sensed data. This requires the data to be in the form of a *digital image*. In this format, the picture consists of a large number of grid cells arranged in rows and columns. Each grid cell records a brightness value and is referred to as a *pixel*. Normally, low pixel values represent dark (low reflectance), and high pixel values represent light (high reflectance). The brightness values are usually viewed as a false colour image that is hard to distinguish from a colour photograph.

The great advantage of digital images over images on photographic film is that a computer can process them, for example, to increase contrast, sharpen edges, or assign distinctive colour classes. *Image processing* refers to the manipulation of digital images to extract, enhance, and display the information they contain. In remote sensing, image processing is a broad field that includes many methods and techniques for processing remotely sensed data.

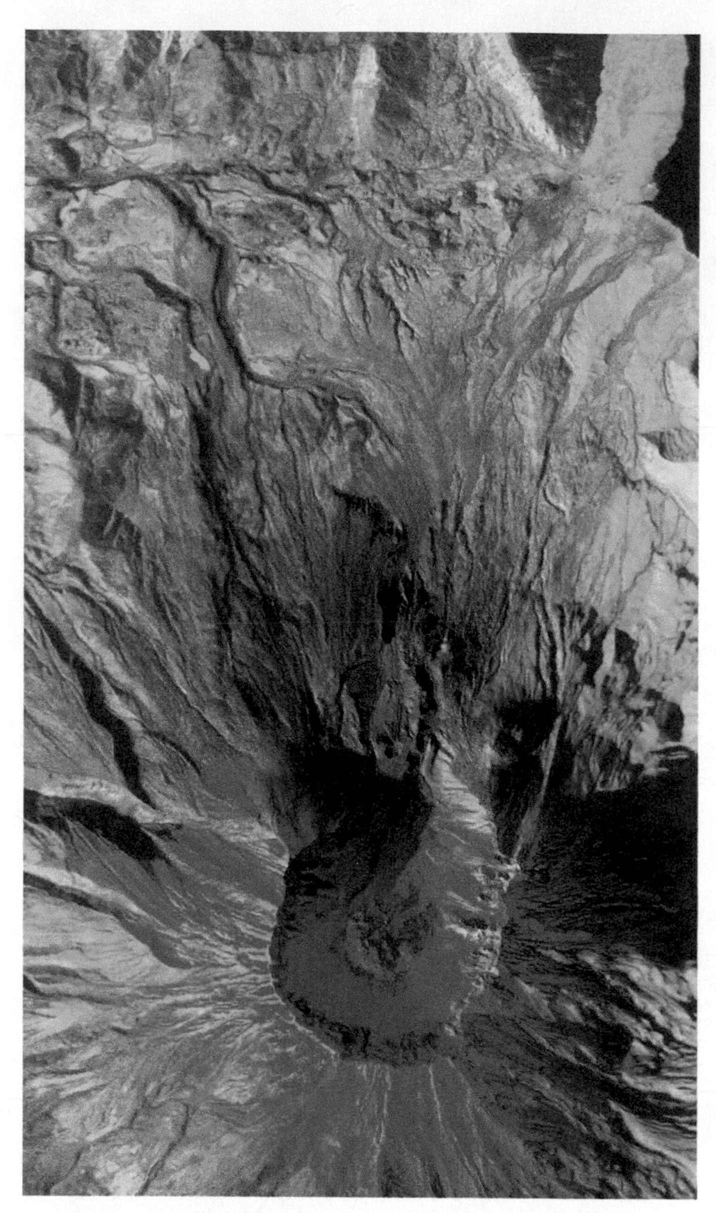

3A.3 **THERMAL INFRARED IMAGE OF MOUNT ST. HELENS** This image was taken by NASA's MODIS/ASTER Airborne Simulator (MASTER) on September 24, 2004. The blue dome at the bottom of the image is the caldera of the volcano. The bulge in the centre of the volcano's caldera is the lava dome, which had been growing for several weeks when this image was taken due to the upward pressure of magma within the volcano.

Orbiting Earth Satellites

With the development of orbiting Earth satellites, remote sensing has expanded into a major branch of geographic research. Because orbiting satellites can image and monitor large geographic areas or even the entire Earth, global and regional studies have become possible. The two general types of Earth-observing satellites follow either sun-synchronous or geostationary orbits.

Sun-synchronous polar orbiting environmental satellites (POES) have an orbit that passes close to the North and South Poles (Figure 3A.4). Because the Earth is not a perfect sphere, the force of gravity is slightly greater at the Equator. When the orbit crosses the Equator, the difference in gravity pushes the satellite slightly eastward. This keeps the satellite in time with the sun, so that its overpasses occur at the same time of day. Typical sun-synchronous satellites take 90 to 100 minutes to circle the Earth and are located at altitudes of about 700 to 800 km above the Earth. As the Earth rotates to the east at a rate of 15° longitude per hour, the satellite's orbit moves to the west on each successive pass. In this way, a polar-orbiting satellite completes 14 to 16 orbits per day, and covers the entire Earth's surface every 16 days.

Satellites in *geostationary orbit*, such as the GOES weather satellites, constantly revolve above the equator. The orbit altitude, about 35,800 km, is set so the satellite makes one revolution in 24 hours in the same direction Earth turns. Thus, the satellite always remains above the same point on the equator. From its high vantage point, the geostationary orbiter provides a view of nearly half of the Earth and its atmosphere.

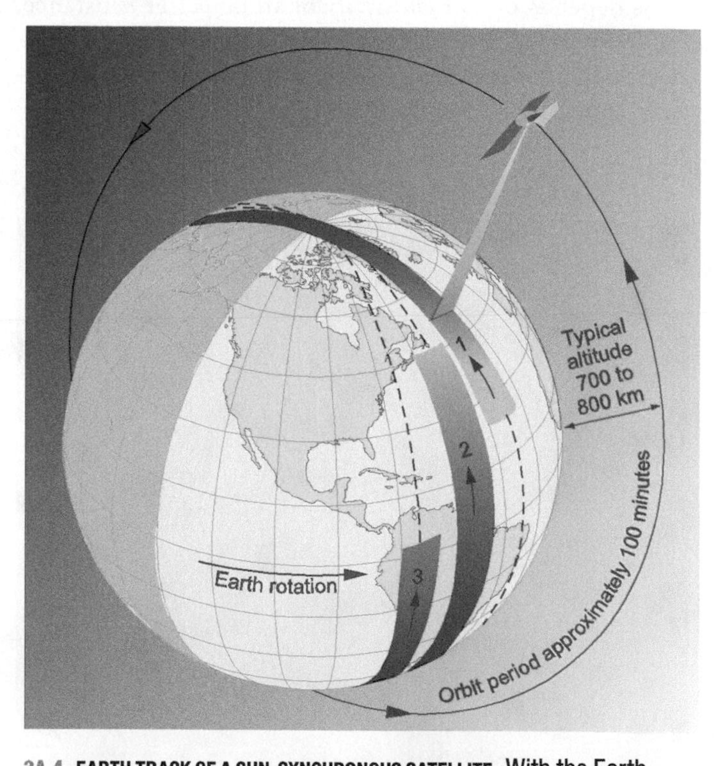

3A.4 **EARTH TRACK OF A SUN-SYNCHRONOUS SATELLITE** With the Earth track inclined at about 80° to the equator, the orbit slowly swings eastward at about 30° longitude per month, maintaining its relative position with respect to the sun. The atmosphere normally contains fewer clouds early in the morning, so Earth-observing satellites, such as Landsat and SPOT, are often set to pass over at about 9:30 a.m. to 10:30 a.m. local time (the height of the orbit above the Earth's surface is greatly exaggerated).

QUESTIONS

1. What is remote sensing? What is a remote sensor?
2. Compare the reflectance spectra of water, vegetation, and a typical soil. How do they differ in visible wavelengths? In infrared wavelengths?
3. What colour is vegetation on colour infrared film? Why?
4. What is emissivity, and how does it affect the amount of energy radiated by an object?
5. Is radar an example of an active or passive remote sensing system? Why?
6. What is a digital image? What advantage does a digital image have over a photographic image?
7. How does a sun-synchronous orbit differ from a geostationary orbit? What are the advantages of each type?

CHAPTER 4 AIR TEMPERATURE

Winter in Jasper National Park, Alberta.

Air temperature affects many aspects of human life, from the clothing we wear to the fuel we use. Air temperature, along with precipitation, is a key determinant of climate. It is, therefore, an important environmental factor that affects many landscape processes, including rock weathering and erosion, soil formation, and the type of plant and animal life that is present in a region.

Is air temperature changing? It is commonly assumed that the Earth is becoming increasingly warmer, and that as a result, sea level is rising, climate boundaries are shifting, and severe weather is becoming more frequent. Global climate change is a controversial issue, but it is generally agreed that human activities, such as forest clearance and air pollution, may have had an impact on global temperatures. This is discussed later in the chapter, in the context of the Earth's natural long-term cycles.

THERMAL ENERGY, HEAT, AND TEMPERATURE

The thermal energy of a substance is the total kinetic energy associated with all its atoms and molecules. In the case of the sun, thermal energy is created from nuclear reactions; in the case of Earth, most thermal energy is acquired from solar radiation, although some is still stored in the molten rocks beneath the planet's surface. A flow of thermal energy to a substance increases its kinetic energy and its temperature rises. Conversely, if a substance loses energy, its temperature falls. Heat is the transfer of thermal energy across a temperature gradient, with energy flowing from warmer to cooler objects. The mechanisms of heat transfer are radiation, conduction, and convection. Earth and its atmosphere gain heat principally by **radiation** from the sun, and lose heat by radiation to space. Heat from deep within Earth's interior is conducted to the surface. **Conduction** also accounts for the vertical transfer of heat in soils causing them to warm during the summer and cool in winter. **Convection** is the principal method of heat transfer in the atmosphere and oceans.

Temperature represents the amount of thermal energy available in a system, and is conveniently defined as a measure of the average kinetic energy associated with the motion of atoms and molecules in a gas, liquid, or solid. The lower limit of temperature is absolute zero (0 K) and represents a theoretical condition in which the kinetic energy is considered to be zero. The temperature of a substance therefore is directly related to the velocity at which its constituent particles are moving. The upper limit of temperature is reached when the atoms and molecules of the substance are moving at their maximum velocity. The SI unit of energy is the joule (J), but in meteorology the common unit of energy is the watt (W), which is equivalent to 1 joule per second.

Air temperature is influenced by several factors. Insolation is of primary importance and varies in response to Earth's daily rotation about its polar axis and the annual revolution of Earth in its solar orbit (see Chapter 2). These two fundamental motions of the Earth determine the daily and annual radiation budgets of the planet, which in turn produce the cycles of air temperature that distinguish day from night and winter from summer. Latitudinal variations in annual and seasonal insolation are reflected in global temperature patterns. Temperatures generally decline with latitude, and seasonal differences are more pronounced at higher latitudes. Comparative data for Windsor, Ontario, (42° 16′ N), and Resolute, Nunavut (74° 43′ N), show these effects (Figure 4.1). At Windsor, mean daily radiation at the top of the atmosphere varies from 145 W m^{-2} in December to 484 W m^{-2} in June; here, mean monthly temperatures range from −4.5 °C in January to 22.6 °C in July, a range of 27.1 °C. At Resolute, mean daily radiation is 0 W m^{-2} in December and January and peaks at 502 W m^{-2} in June when there is continuous daylight. The mean monthly temperature at Resolute varies from −33.2 °C in February to 4.3 °C in July, an annual range of 37.5 °C.

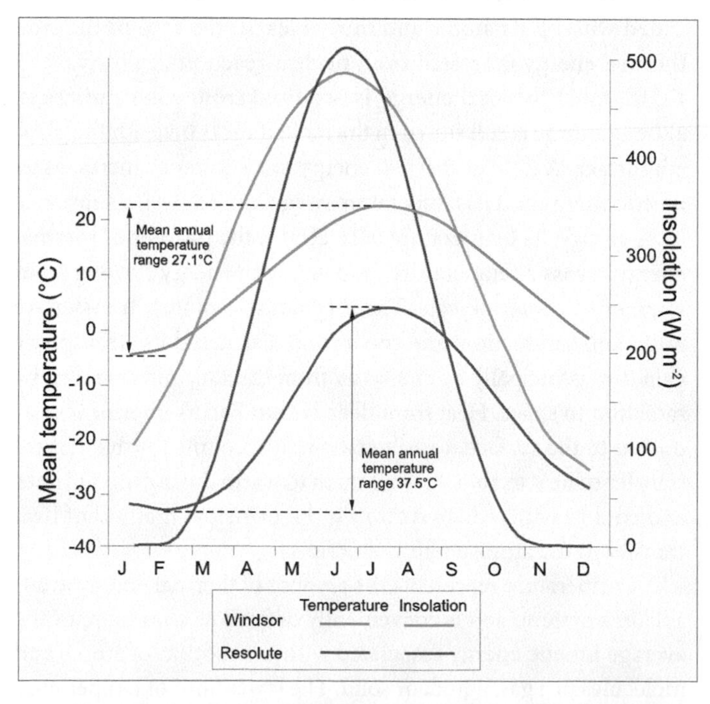

4.1 LATITUDINAL EFFECT OF RADIATION ON TEMPERATURE Solar radiation at the top of the atmosphere and mean temperatures at Windsor, Ontario, and Resolute, Nunavut.

NET RADIATION, ENERGY TRANSFER, AND THERMAL INERTIA

When an object or substance absorbs energy, its temperature increases. The rate of heating is determined not only by the amount of energy that is absorbed over a given period of time, but also by the specific heat capacity of the substance and its density. **Specific heat capacity** (generally referred to as specific heat) is the amount of heat per unit mass required to raise the temperature of a substance by 1 °C. Dry sand, for example, has a specific heat of 0.80 J $g^{-1}K^{-1}$ (joules per gram per degree kelvin). *Density* is defined as the ratio of an object's mass to its volume and is expressed as kg m^{-3}. Note the difference between the terms mass and weight. *Mass* is a measure of the quantity of matter; this is constant all over the universe. *Weight* is proportional to mass, but varies with location because of the force exerted on an object by gravity. Thus, the mass of a rock is constant, but if its weight on Earth is 300 kg, it would weigh only 50 kg on the moon because the force of gravity on the moon is one-sixth that of the Earth.

The ability of surface materials to store or conduct heat can be determined from the volumetric heat capacity and thermal conductivity of the substances. **Volumetric heat capacity** is a measure of how well a substance stores heat and is calculated as specific heat capacity multipled by density. **Thermal conductivity** is the rate at which heat flows within a solid, such as soil. It is determined by the inherent properties of the substance, as well as the temperature gradient that exists between different parts of the substance. The rate at which energy flows into or out of a substance, such as soil, affects its temperature and is intrinsically related to diurnal and seasonal insolation cycles.

The thermal properties of water are very different from those of solid substances. Also, when water evaporates at a surface, the change of state from liquid to vapour removes stored heat, which cools the surface through **latent heat transfer**. This is the latent heat of vaporization. When water condenses at a surface, an equivalent amount of energy is released, warming the surface. This is the latent heat of condensation. Energy transfer also occurs when water freezes and melts, although lesser quantities of heat are involved. The processes are important when comparing temperatures in coastal locations with locations at similar latitudes far inland. Because water heats and cools more slowly than land, air temperatures over water bodies tend to be less extreme than over land surfaces (Figure 4.2). Greater cloud cover in maritime areas compared with less humid interior continental locations can also contribute to the temperature patterns by blocking incoming solar radiation and preventing the escape of long-wave energy to space. The influence of land and sea on temperature is discussed later in the chapter.

Convection is another form of energy transfer important in understanding air temperature patterns. In convection, heat is

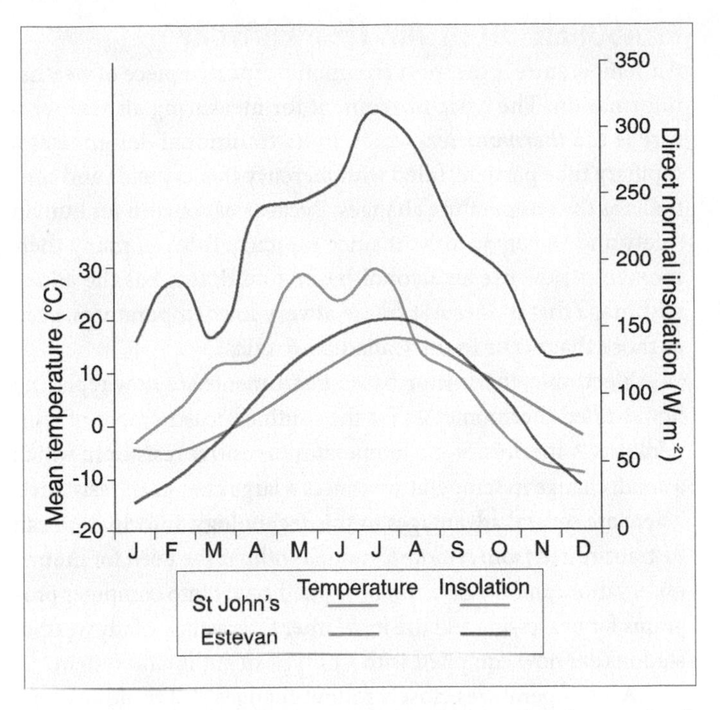

4.2 MARITIME AND CONTINENTAL TEMPERATURE CYCLES The slower rate of heating and cooling of water is an important factor contributing to the smaller annual temperature range at the coastal city of St. John's, Newfoundland, compared to Estevan, Saskatchewan, in the continental interior.

distributed by the mixing of fluids and gases because of differences in density. For example, if air in contact with a heated soil surface is warmed, the gas molecules spread out, making it less dense than the surrounding, unheated air. The warmed air will subsequently rise and transfer heat by convection. Upward and downward air currents can warm or cool the surface and generate turbulence in the atmosphere. Convection is an important energy transfer process in both the atmosphere and the oceans.

THE SURFACE ENERGY BALANCE EQUATION

Since energy is neither created nor destroyed—just transformed—the energy flows occurring at a surface must be in equilibrium. The *surface energy balance equation* for land describes how **net radiation**, latent heat, and sensible heat flows are balanced through conduction and convection. As well as explaining the temperature of the surface, the latent heat transfer term in the equation also describes how water evaporates from the soil.

The surface energy balance equation considers the surface to be a thin boundary layer positioned between the atmosphere and the underlying ground. For daytime conditions, the equation is written as:

$$Q^*\downarrow = Q_G\downarrow + Q_H\uparrow + Q_E\uparrow$$

where:

Q^* is net wave radiation;
Q_G is soil heat flux;
Q_H is sensible heat flux; and
Q_E is latent heat flux.

Arrows indicate the direction of the positive fluxes and would be reversed at night.

The surface is also considered to be an energy transfer layer that changes one type of energy flow to another, but is too thin to hold any heat itself. Thus, the surface energy balance equation can be rewritten as:

$$R_{LONG} - R_{SHORT} + H_{LATENT} + H_{SENSIBLE} + H_{SOIL} = 0$$

where:

R_{LONG} is the net flow of long-wave radiation between the surface and the atmosphere;
R_{SHORT} is the flow of short-wave radiation absorbed by the surface (note that outgoing flows are considered positive and incoming flows are negative);
H_{LATENT} is the latent heat flow between the surface and the atmosphere;
$H_{SENSIBLE}$ is the sensible heat flow between the surface and the atmosphere; and
H_{SOIL} is the sensible heat flow between the surface and the soil.

From the principle of conservation of energy, incoming energy flow must balance outgoing energy flow, so the terms of the equation must add up to zero.

The equation includes two radiation terms. R_{SHORT} is the flow of short-wave radiation to the surface from the sun. This flow includes only radiation that is absorbed by the surface and does not include short-wave radiation that is reflected directly back toward the atmosphere. Because the normal convention is to consider energy flows to the surface as negative, R_{SHORT} is negative during the day. At night, this flow will drop to zero (Figure 4.3). R_{LONG} is the net long-wave radiation. Because it is a net term, it represents the balance between outgoing long-wave radiation emitted by the surface and incoming long-wave radiation from the atmosphere above. Given that the surface is usually warmer than the atmosphere, even at night, the flow will normally be away from the surface, which is taken as positive. The sum of R_{SHORT} and R_{LONG} is the net radiation flow ($Q^*\downarrow$).

H_{LATENT} is the latent heat flow and represents the energy associated with the evaporation of water or sublimation of ice. Latent heat flow ($Q_E\uparrow$) occurs when water vapour diffuses into the atmosphere following a phase change. During the day, latent heat flow will be positive as soil water evaporates. The latent heat flux is regulated by the moisture gradient away from the surface. Without vertical motion, the air immediately above the surface would quickly become saturated and energy transfer would cease. Transfer of latent heat increases during turbulent, windy conditions, when moist air near the surface is continually replaced by drier air. At night, condensation or deposition may occur, yielding dew or frost. If so, latent heat will be released at the surface, providing a heat flow to the surface (negative flow).

$H_{SENSIBLE}$ is the sensible heat flux ($Q_H\uparrow$) that arises when the surface conducts heat to the atmosphere in the boundary layer. Sensible heat is then carried upward by convection. It will be positive when the air is warmed by the surface, which is the normal condition during the day. At night, the surface can become colder than the air in contact with it, so the heat flow may be negative, from the air to the surface.

H_{SOIL} is the flow of heat by conduction from the surface to the soil below ($Q_G\downarrow$). This flow will normally be away from the surface (positive) during the day, as heat is conducted from the warm surface into the soil. At night, however, the surface will normally cool sufficiently for heat to be conducted upward (negative).

MEASUREMENT OF AIR TEMPERATURE

Air temperature is the most commonly reported piece of weather information. The basic instrument for measuring air temperature is the *thermometer*, which in its traditional design uses a capillary tube partially filled with mercury that expands and contracts as the temperature changes. Because of concern for human safety and the environment since mercury is toxic, many thermometers now use an alcohol-based liquid; this has the added advantage that it does not freeze at very low temperatures, such as those that occur in the Canadian Arctic.

Electronic, thermistor-based instruments are now replacing liquid-filled thermometers for the routine measurement of temperature. A thermistor is a temperature sensitive resistor in which a small change in temperature causes a large change in resistance. There are several advantages to this technology. Information can be transmitted from remote locations without the need for human observation, and the data can be input directly into computer programs for processing and use in weather forecasting. Many weather stations are now equipped with this type of automatic system.

Air temperatures closely follow changes and trends experienced at the ground surface and the conventional practice is to locate the instruments in an open area at least 9 m in diameter

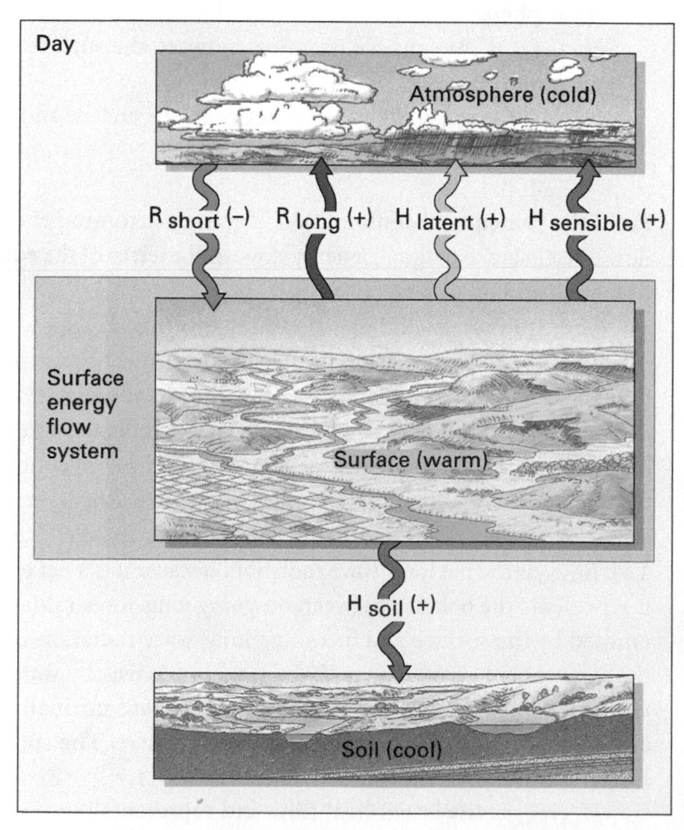

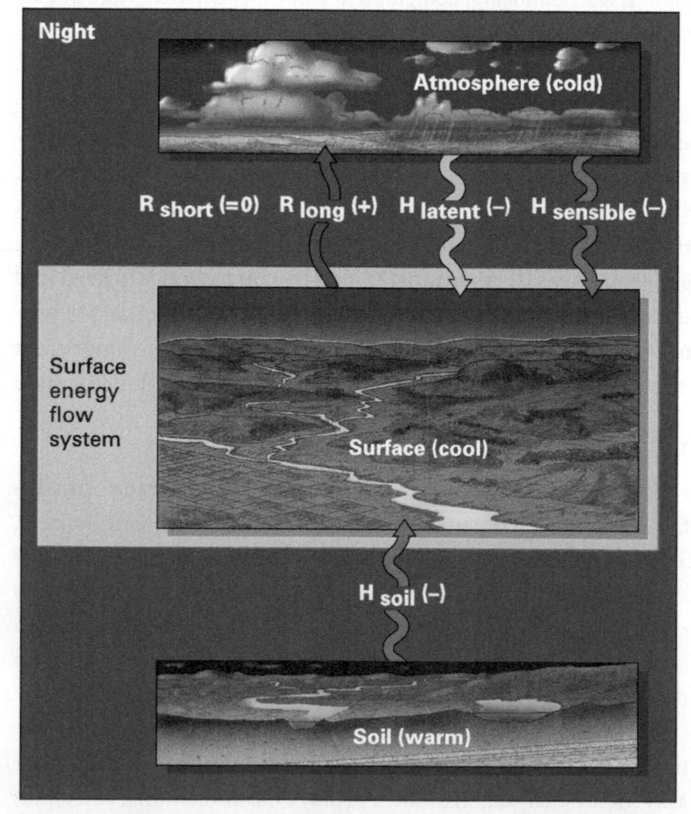

4.3 **SURFACE ENERGY BALANCE EQUATION FOR TYPICAL DAY AND NIGHT CONDITIONS** In this example, the surface is warmer than the atmosphere so long-wave energy remains positive throughout the day and night. However, the rate of long-wave energy flow will vary as the surface warms and cools in response to temporal variations in short-wave energy gain and sensible heat transfer within the soil.

4.4 **TYPICAL ENVIRONMENT CANADA WEATHER STATION** The instrument shelter (shown on the left side of the photo) is designed with louvred sides for ventilation and painted white to reflect solar radiation. It normally houses various thermometers and humidity instruments. The other weather instruments shown in the photo are described in Chapters 5 and 6.

Saskatoon

Current Conditions | More info +

-7 °C	Observed at:	Saskatoon Diefenbaker Int'l Airport		
	Date:	10:00 AM CST Monday 24 January 2011		
	Condition:	Partly Cloudy	Temperature:	-6.7°C
	Pressure:	101.7 kPa	Dewpoint:	-8.5°C
	Tendency:	falling	Humidity:	87 %
	Visibility:	24 km	Wind:	S 9 km/h
	Air Quality Health Index:	2	Wind Chill:	-11

Forecast | More info +

Today	Tue	Wed	Thu	Fri	Sat	Sun
-5°C	-4°C	-2°C	2°C	-4°C	-15°C	-14°C
	-7°C	-7°C	-3°C	-6°C	-16°C	-22°C
30%						

Issued : 5:00 AM CST Monday 24 January 2011

4.5 **TEMPERATURE IS THE MOST COMMONLY REPORTED ATMOSPHERIC PROPERTY** The Environment Canada Meteorological Service of Canada website (http://www.weatheroffice.ec.gc) provides current forecasts as well as archived data for all weather stations in Canada, including historical data from stations that are no longer active.

covered by short grass (Figure 4.4). In regions where grass doesn't grow, such as the desert, a natural bare soil area is used. The temperature values presented in Canadian weather reports are generally measured at a height of 1.25 m above the ground surface. The WMO (World Meteorological Organization) standard allows for instrument heights of between 1.25 and 2 m. Height specifications make no allowance for winter snowpacks.

For standard meteorological measurements, temperature instruments are housed in *instrument shelters*. Traditionally these louvred boxes are painted white and are designed to shade the instruments from the direct rays of the sun. Air circulates freely through the louvres, ensuring that temperatures inside the shelter are the same as the outside air. Smaller radiation shields of similar design are used for modern electronic instruments.

Although some weather stations report temperatures hourly, most stations report only the highest and lowest temperatures recorded during a 24-hour period. Temperature measurements collected by national agencies, such as the Meteorological Service of Canada, are used to generate weather reports and forecasts (Figure 4.5). This agency keeps daily, monthly, and annual temperature statistics for each reporting station in Canada. For many stations, hourly temperature readings are archived, as well as summary data such as daily maximum, minimum, and mean temperature—the average of the maximum and minimum daily values. Monthly summaries of maximum, minimum, and mean temperatures are also available. With sufficient length of record (usually at least 30 years), these statistics, along with others such as precipitation, humidity, and wind, are used to describe the climate at the station and its surrounding area.

THE DAILY CYCLE OF AIR TEMPERATURE

Because the Earth rotates on its axis, incoming solar energy at a location varies throughout each 24-hour period, whereas outgoing long-wave energy remains more constant. During the day, incoming solar radiation normally exceeds outgoing long-wave radiation, so the net radiation balance is positive and the Earth's surface warms. At night, net radiation is negative, and the surface temperature falls as long-wave energy is radiated to space.

Figure 4.6 shows an idealized curve of daily insolation, outgoing long-wave radiation, and air temperature for a typical observation station at approximately lat. 45° N under cloud-free conditions at the equinox. At this time of the year, insolation begins at sunrise (6 a.m.), rises to a peak value at noon, and declines to zero at sunset (6 p.m.). Incoming short-wave radiation drops to zero at night, but long-wave radiation will continue to escape to space. Consequently, during the night, under cloud-free skies, net radiation is negative. This continues into the early morning hours until shortly after sunrise, when net radiation becomes positive then rises sharply to a peak at noon. In the afternoon, net radiation decreases and becomes negative shortly before sunset, and a deficit energy balance persists until morning.

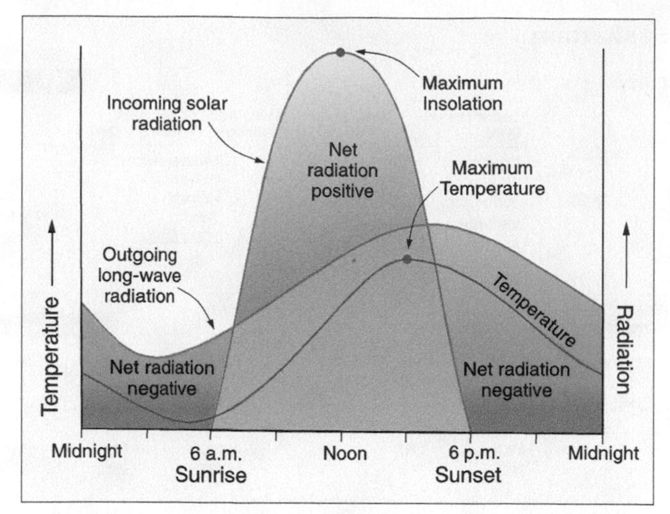

4.6 DAILY VARIATION IN INCOMING SHORT-WAVE SOLAR RADIATION AND OUTGOING LONG-WAVE RADIATION FOR AN IDEALIZED MIDLATITUDE STATION AT THE TIME OF THE EQUINOX The daily cycle of air temperature responds by increasing when net radiation is positive and decreasing when it is negative.

The radiation balance changes over the year depending on the relative duration and intensity of the incoming and outgoing radiation fluxes. At the June solstice, insolation at 45° N typically begins about two hours earlier (4 a.m.) and ends about two hours later (8 p.m.). Total insolation at the June solstice is greater than at the equinox due to higher sun angle and longer period of daylight. Consequently, net radiation is higher and there is a corresponding increase in daily temperatures. At the December solstice, insolation begins about two hours later than at the equinox (8 a.m.) and ends about two hours earlier (4 p.m.). Daily insolation and net radiation are both greatly reduced in the winter and temperatures are correspondingly lower.

The 24-hour cycle of air temperature shows that the minimum daily temperature usually occurs about half an hour after sunrise. Since net radiation is negative throughout the night, heat flows from the ground surface, which in turn cools the near-surface air layer to its lowest temperature. As net radiation becomes positive, the surface warms quickly and transfers heat to the air above. Air temperature rises sharply in the morning hours and continues to rise long after the noon peak of net radiation.

Although air temperature might be expected to rise as long as net radiation is positive, another process begins in the early afternoon on a sunny day. Convection currents develop and mix with the air several hundred metres above the surface, carrying heated air aloft and bringing cooler air down toward the surface. Therefore, the temperature peak usually occurs in mid-afternoon before convection is well established. In Figure 4.6, the peak is shown at about 3 p.m., but it can typically occur between 2 p.m. and 4 p.m., depending on local conditions. Air temperature tends to fall quite rapidly until sunset, then at a slower rate through the night.

Over the course of a year, daily temperatures are highest in summer and lowest in winter. However, temperatures at the September equinox are higher than those in March even though net radiation is more or less the same. Temperatures at these times of the year differ because each reflects conditions in the preceding season. The summer energy surplus persists into the autumn, whereas warming in the spring may be delayed by the reflective properties of the remaining snow cover and the thermal inertia of the frozen soils.

TEMPERATURES CLOSE TO THE GROUND

Although air and ground temperatures show the same general trends, temperatures at the surface are likely to be more extreme. Figure 4.7 presents a series of generalized temperature profiles from about 30 cm below the surface to a height of 1.5 m at five times during an autumn day. At 8 a.m., air and soil temperatures are uniform, as represented by the vertical line on the graph. By noon, the soil surface is slightly warmer than the air, and is considerably warmer by 3 p.m. The soil cools rapidly during the evening and through the night, and by 5 a.m. it is colder than the air.

Thus, daily temperature variation is greatest at the surface, while air temperature at the standard recording height is less variable. Heat from the soil surface moves upward and downward through the soil by conduction, and the deeper soil layers warm and cool over the course of each day. However, within the soil the daily cycle weakens with depth to a point where the temperature does not change during the 24-hour cycle.

Several factors can alter this idealized temperature cycle. Insolation can be considerably reduced due to cloud cover, which will affect both incoming and outgoing energy fluxes. Seasonal

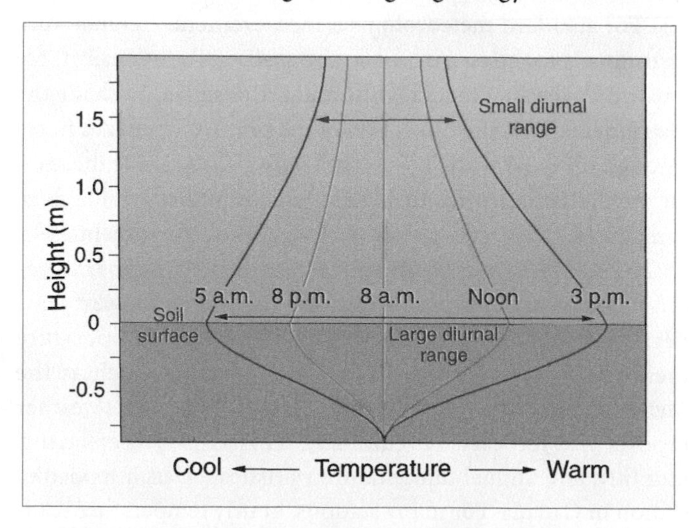

4.7 IDEALIZED DAILY CYCLE OF AIR AND SOIL TEMPERATURES The temperature profile close to the ground surface is shown for five times during a typical sunny autumn day. The daily temperature range decreases with distance from the surface, which is the principal heat source for the air.

and short-term variations in soil moisture will modify the rate at which heat is conducted to and from the soil surface during the day and night. Changes in albedo due to snow cover or from plant growth can also have important effects on the radiation budget and associated temperature cycles.

Whereas latitude has a regular and predictable effect on global insolation, outgoing long-wave energy patterns are more variable because they are dependent on surface conditions. Local and regional topography also influence temperature patterns. For example, the atmosphere is less dense at high elevations, allowing more solar energy to reach the surface during the day. However, lower air density also allows more long-wave energy to escape to space. As a result, the daily temperature range in mountainous areas is generally quite pronounced. Slope aspect—the direction the slope faces—influences the amount of radiation received, with south-facing slopes being normally warmer than north-facing slopes. Radiation intensity and surface warming also increase with slope angle to the point at which incoming solar rays are perpendicular to the surface. However, the slope angle that corresponds to maximum insolation varies seasonally with the sun's declination (see Chapter 2). There is also a diurnal component due to the sun's passage across the sky each day.

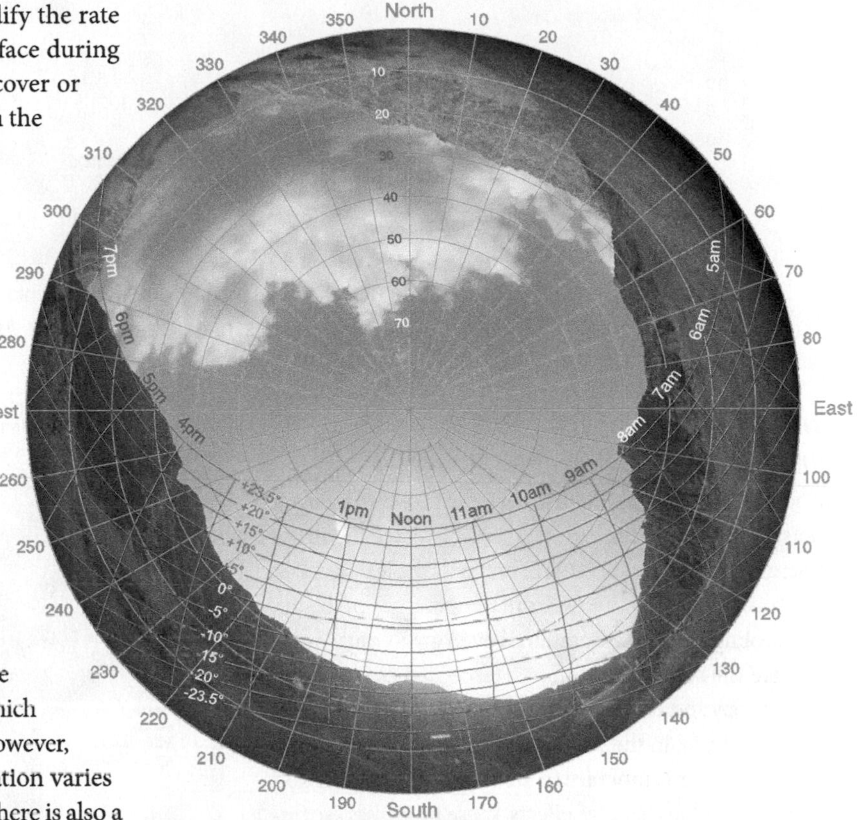

4.8 SOLAR PATH DIAGRAM This solar path diagram for 49° N for a location in the Canadian Rockies shows the effect of local topography on direct-beam insolation at different times of the year. (From *Introduction to Weather and Climate* by O.W. Archibold, John Wiley & Sons Canada, Ltd., 2011.)

Many of the parameters that affect insolation at a given location can be conveniently represented in a **solar path diagram** (Figure 4.8). These show the solar altitude and the solar azimuth at any date of the year and any hour of the day. Theoretically, a solar path diagram is specific for a particular latitude, although in practice these diagrams are generally considered to be sufficiently accurate for use over a range of 5 to 10 degrees. The basic grid is a stereographic projection in which the lines radiating from the centre represent direction and the concentric circles are used to plot the sun angle. Superimposed on this is a series of curved lines running more-or-less horizontally across the diagram that show the declination of the sun at different times of the year. The time of day is indicated by the shorter, roughly vertical lines associated with the declination lines. In open, level terrain the horizon would be represented by the outermost circle (0° elevation), but the prime use of a solar path diagram is to determine how specific site factors will affect incoming radiation. This particular diagram shows a location in the Canadian Rockies. Here the rugged mountain topography has a marked effect on daily sunshine cycles. For example, at the summer solstice, the sun rises above the mountains to the east at about 8:30 a.m. and sets in the west about 4:00 p.m., and the period of direct sunshine lasts for about 7½ hours. In comparison, for level terrain the sun would rise at about 4:30 a.m. and set at 7:30 p.m.; this would extend the period of sunshine to 15 hours. At the winter solstice in this mountain location, the sun never rises above the local horizon, so there is no direct-beam insolation.

Several environmental changes occur with increasing altitude. With fewer air molecules and aerosol particles to scatter and absorb the sun's light, incoming radiation is more intense. Also, there is less CO_2 and water vapour, reducing the greenhouse effect. This generally results in cooler temperatures, especially at night when heat loss is rapid. Elevation has a pronounced effect on the daily temperature cycle; this is apparent in the Andes Mountains of South America, which rise to nearly 7,000 m in places. Daily maximum and minimum temperatures over a 15-day period are shown in Figure 4.9 for stations ranging in elevation from sea level to 4,380 m. The mean daily temperature for this period decreases progressively with elevation, from 16 °C at sea level to −1 °C at 4,380 m. The daily range also increases with elevation, from 6 °C at the coastal town of Mollendo compared to 24 °C at Vincocoya (4,380 m). At Cuzco (3,350 m), the city's heat island effect keeps nighttime temperatures higher than expected and the daily temperature range is also reduced.

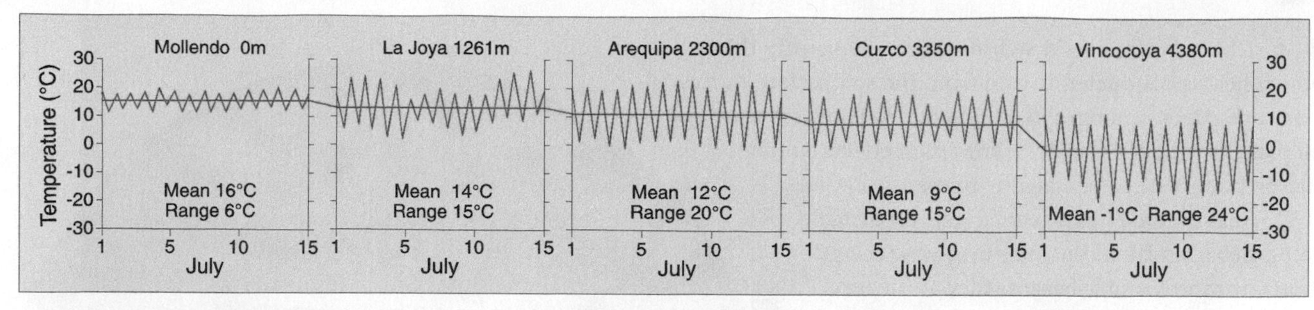

4.9 **THE EFFECT OF ELEVATION ON AIR TEMPERATURE CYCLES** Daily maximum and minimum air temperatures for stations at various elevations in Peru. All data cover the same 15-day observation period in July. As elevation increases, the mean daily temperature decreases and the temperature range increases.

ENVIRONMENTAL CONTRASTS: RURAL AND URBAN TEMPERATURES

In actively growing vegetation, water is taken up by plant roots and moved to the leaves, where it is lost to the atmosphere through **transpiration**. This cools the leaf surfaces, which in turn tends to lower the temperature of the air. The cooling effect of vegetation is even greater in a forest because the solar radiation is intercepted by the deep, leafy canopy and the understory plants. Water also evaporates directly from the soil, again cooling the surface and moderating air temperature. Because evaporation and transpiration have similar effects, these processes are frequently referred to collectively as **evapotranspiration**.

In contrast, water drains from the impervious roofs, sidewalks, and streets of a city. It is channelled into storm sewer systems, where it flows directly into rivers, lakes, or oceans rather than soaking into the ground beneath the city. Little energy is dissipated in evaporation, so these dry surfaces can warm quickly on sunny days. In urban environments, any moderating effect from evapotranspiration is mainly connected with parks, lawns, ornamental trees, and bare soil.

City surfaces are also darker than rural surfaces; asphalt paving and roofing absorb about twice as much solar energy as vegetation. Heat absorption is also enhanced by the many vertical surfaces in a city, which redirects radiation from one to another. Since some radiation is absorbed by each surface it strikes, the network of vertical surfaces tends to trap solar energy and radiant heat more effectively than a simple horizontal surface. Concrete, asphalt, and other materials conduct and store heat better than soil. At night, this heat is conducted back to the surface, keeping nighttime temperatures warmer.

Another important factor in warming the city is fuel consumption and waste heat. In summer, city temperatures increase through the use of air conditioning which pumps heat out of buildings, releasing it to the air. The power used to run the air conditioning systems also generates heat. In winter, heat is directly lost from poorly insulated buildings. Heat contributed from vehicles, as well as greenhouse gases from their exhaust, are additional sources of heat gain.

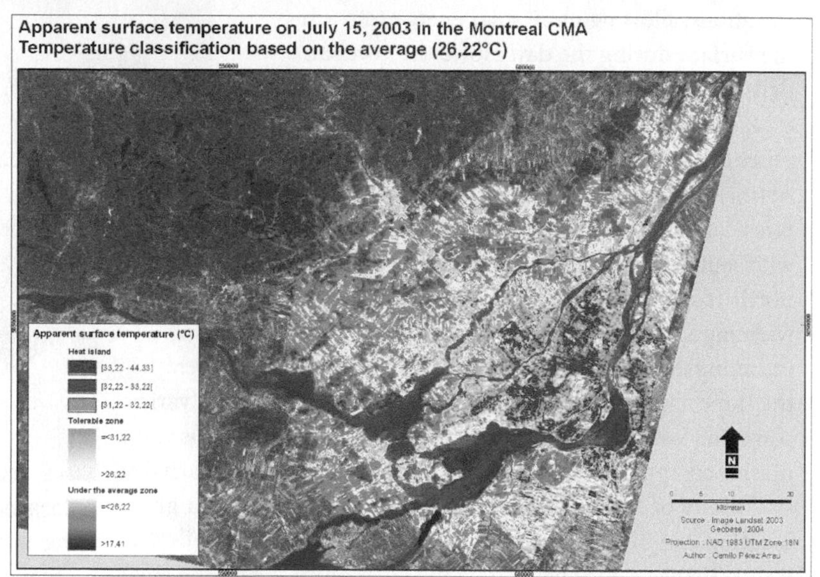

4.10 **THE URBAN HEAT ISLAND OF MONTREAL** Temperatures are highest in the downtown area; parks and rural areas are coolest. The St. Lawrence River also stands out as a cool line in the image.

These effects can increase air temperatures in the central region of a city several degrees above the surrounding suburbs and countryside, an effect known as an urban **heat island** (Figure 4.10). In large cities, the heat island can persist through the night depending on the rate at which heat stored during the daytime is released and dissipated into the atmosphere. Heat islands develop best under clear, calm conditions.

TEMPERATURE STRUCTURE OF THE ATMOSPHERE

The atmosphere is mainly warmed from the surface below, so temperatures can be expected to decrease with altitude. This general condition is termed a **lapse rate**. The **normal temperature lapse rate** measures the drop in temperature in stationary air, averaged for the entire Earth over a long period. On average, temperature drops with altitude at a rate of 6.4 °C per 1,000 m (6.4 °C 1,000 m^{-1}). For example, when the air temperature near the surface is 20 °C, the air at an altitude of 8 km should be about −31 °C. Keep in mind that the normal temperature lapse rate is

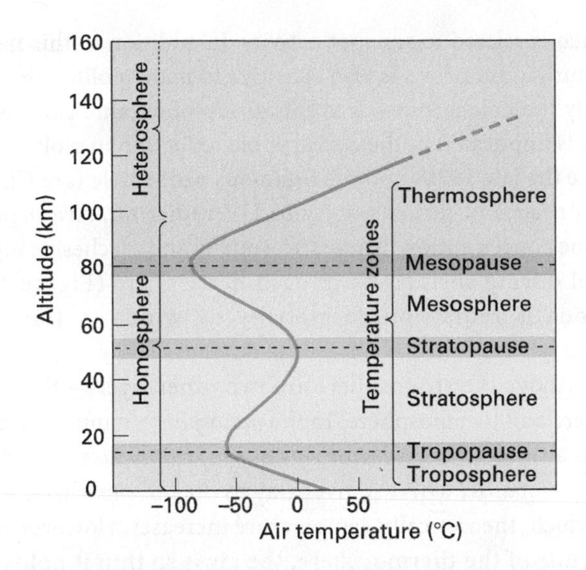

4.11 TEMPERATURE STRUCTURE OF THE ATMOSPHERE Temperature decreases at a mean rate of 6.4 °C 1,000 m^{-1} in the troposphere. This value is termed the normal temperature lapse rate. Temperature increases with altitude in the stratosphere. Above the stratosphere are the mesosphere and thermosphere. The homosphere, which includes all of the atmosphere in which the chemical components are well mixed, ranges from the surface to nearly 100 km. Above this is the heterosphere, where gases tend to be sorted according to their molecular weight.

an average value, and that on any given day the observed lapse rate might be quite different. The actual change in temperature with altitude at a specific time and specific location is termed the **environmental temperature lapse rate**.

For the first 12 km or so above the surface of the Earth, temperature falls with increasing altitude (Figure 4.11). However, at about 12 km (depending on latitude), the temperature trend reverses and the temperature slowly increases with altitude. This property distinguishes two different layers in the lower atmosphere—the troposphere and the stratosphere.

TROPOSPHERE

The **troposphere** is the lowest layer of the atmosphere and contains about 80 percent of its total mass. The troposphere is thickest in the equatorial and tropical regions, where it extends to about 16 km; at the poles it thins to about 6 km. This is partly due to the rotation of the Earth on its axis, but is also linked to thermal expansion in the warmer zones. Most weather phenomena are restricted to this layer. One important feature of the troposphere is that it contains practically all of the water vapour present in the atmosphere. However, because warm air holds more water vapour than cold air and temperature decreases with altitude through the troposphere, water vapour is mainly concentrated below 10 km. Also, the average water content of the atmosphere is highest in the warmer air of tropical and equatorial regions (Figure 4.12).

As well as decreasing in thickness from the low latitudes to the poles, the depth of the troposphere decreases seasonally. Over North America, for example, the height of the tropopause ranges from about 13 km in the southern United States to 7 km in the high Arctic during the winter season, due to thermal contraction of the air column. In summer, the height increases to over 10 km

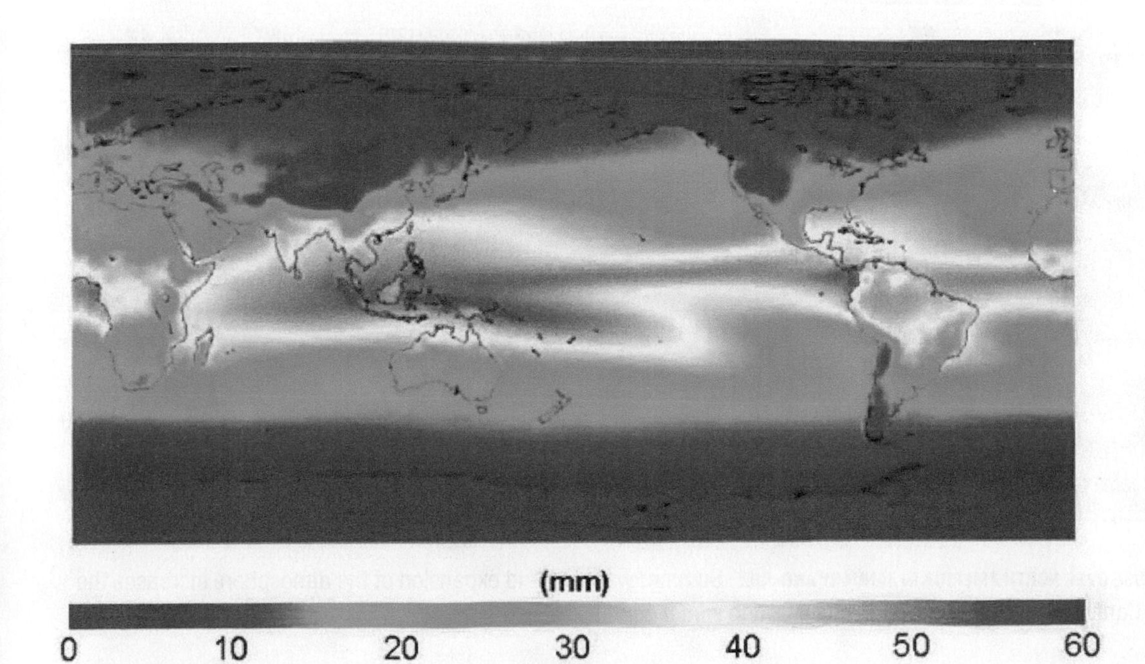

4.12 MEAN ANNUAL WATER VAPOUR CONTENT OF THE ATMOSPHERE Moisture held in the atmosphere is equivalent to about 10 days' supply of precipitation globally. Average water vapour content is highest in lower latitudes where air temperatures are warm throughout the year.

in northern Canada and exceeds 15 km around the Gulf of Mexico (Figure 4.13). The troposphere gives way to the stratosphere at the *tropopause*. Traditionally the tropopause is defined as the lowest level where the lapse rate falls below 2 °C km^{-1} for a vertical thickness of at least 2 km. Beyond the tropopause, the temperature starts to increase with altitude.

STRATOSPHERE AND UPPER LAYERS

Above the tropopause lies the **stratosphere**, in which the air becomes slightly warmer as altitude increases. The stratosphere extends to an altitude of roughly 50 km above the Earth's surface. Because there is little mixing of air between the troposphere and stratosphere, water vapour and dust are practically absent from the stratosphere. However, large thunderclouds will occasionally build into the stratosphere, especially during severe hurricanes. Similarly, powerful volcanic eruptions will inject water and dust directly into the stratosphere. When the water eventually returns to the troposphere, it helps to remove some of the pollutants that also find their way into the stratosphere. Strong, persistent winds—the jet streams—are found in the lower stratosphere; they are linked to the formation and movement of weather systems in the troposphere (see chapters 5 and 7).

Another important feature of the stratosphere is the **ozone layer**, which shields organisms by absorbing the intense UV radiation emitted by the sun. The stratospheric ozone layer contains over 90 percent of the Earth's ozone and is responsible for absorbing 97 to 99 percent of the UV light emitted by the sun. The warming of the stratosphere with altitude is mainly caused by the absorption of solar energy by ozone molecules. Ozone concentrations vary by about 2 percent over an 11-year cycle, which is related to sunspot activity. In addition to this natural perturbation, ozone is also sensitive to many pollutants, especially the halocarbons. It is this group of organic compounds that is implicated in the catastrophic reduction in global ozone since the late 1970s and the infamous ozone hole (see Chapter 3). Because of its sensitivity to UV radiation, stratospheric ozone concentration changes seasonally and reaches its highest level during springtime in each hemisphere (Figure 4.14). Ozone is redistributed globally by winds in the lower stratosphere.

Above the stratosphere are two other layers—the mesosphere and thermosphere. In the *mesosphere*, temperature falls with altitude. This layer begins at the *stratopause* and ends at the *mesopause*, where it gives way to the *thermosphere*, a layer in which, theoretically, temperature increases. However, at the altitude of the thermosphere, the air is so thin it holds very little heat.

Apart from important differences in water vapour and ozone, the composition of the atmosphere is uniform for the first 100 km of altitude, which includes the troposphere, stratosphere, mesosphere, and lower portion of the thermosphere. This region is referred to as the *homosphere*. Above 100 km is the *heterosphere*, a region where gas molecules are progressively sorted into layers by molecular weight and electric charge.

TEMPERATURE INVERSION

During the night, the ground surface radiates long-wave energy to the atmosphere, net radiation becomes negative, and the surface cools; this in turn cools the overlying air. Under clear, calm conditions, cool, dense air will accumulate at the ground surface

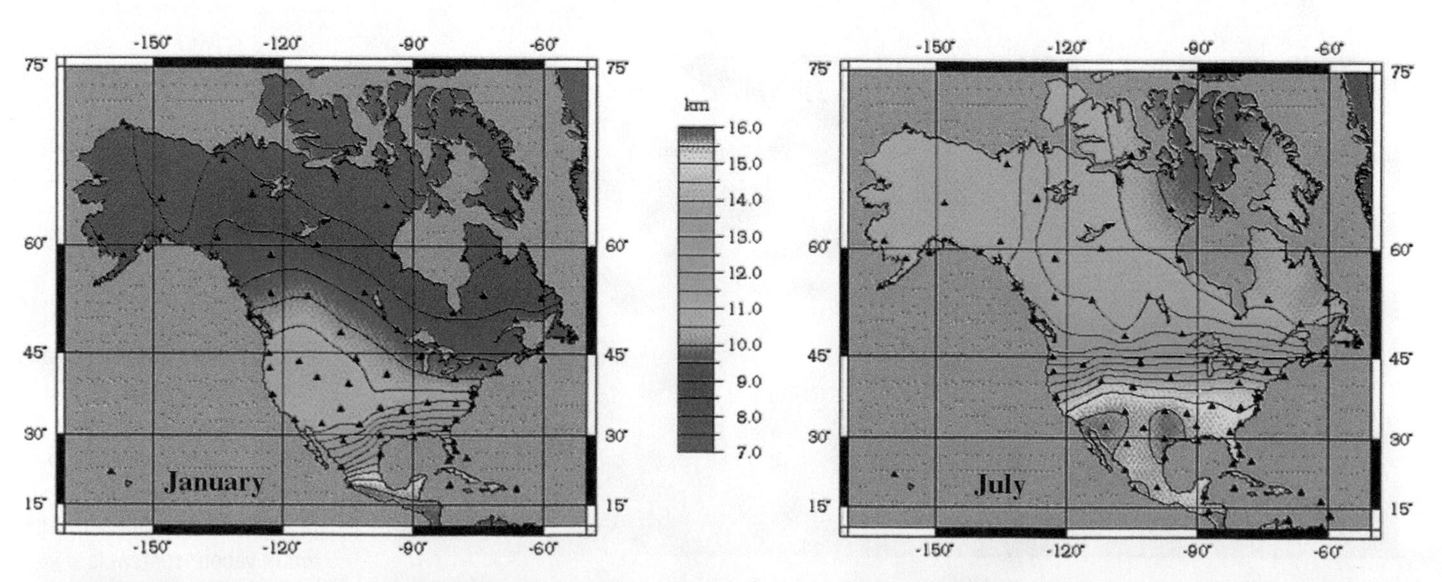

4.13 MEAN HEIGHT OF THE TROPOPAUSE OVER NORTH AMERICA IN JANUARY AND JULY Summer warming and expansion of the atmosphere increases the altitude of the tropopause over Canada by 2,000 to 3,000 m compared to winter conditions.

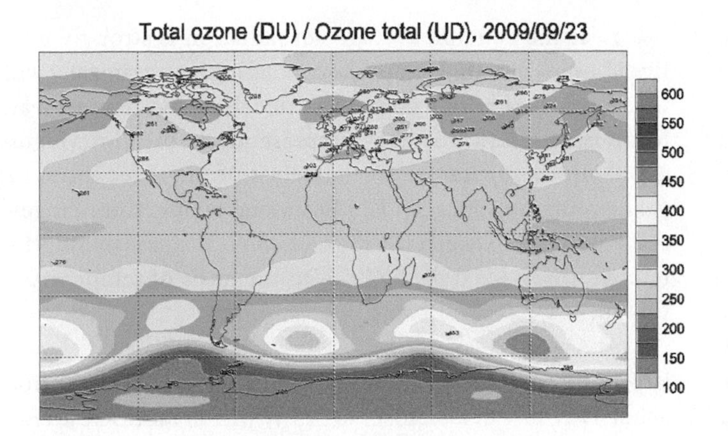

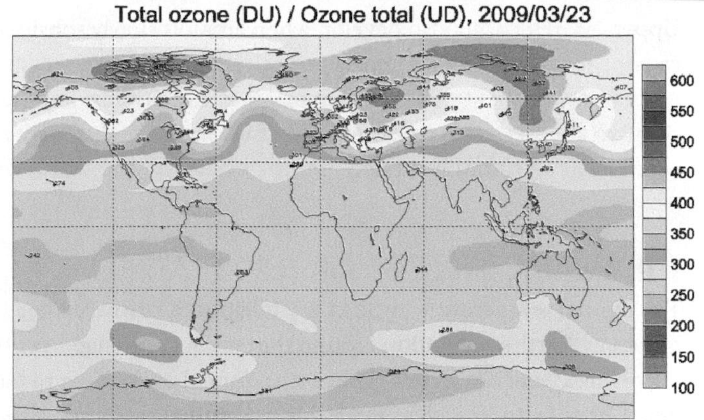

4.14 SEASONAL OZONE CONCENTRATION Stratospheric ozone concentration is now monitored by the solar backscattering ultraviolet (SBUV) instruments on board the NOAA polar orbiting satellites. Because ozone is sensitive to UV radiation, its concentration increases during the dark winter periods in both hemispheres and reaches it maximum during the respective spring seasons.

to produce a **temperature inversion** (Figure 4.15). In this example, a **ground inversion** has formed in which air temperature at the surface is −1 °C. Temperature becomes warmer up to about 300 m; here the lapse rate reverses and the normal decrease in temperature with altitude resumes. As air cools and becomes denser, it can flow gently downward into low lying depressions. Areas that are susceptible to this type of cold air drainage are termed *frost hollows*.

Ground inversions often occur over snow-covered surfaces in winter. Inversions of this type are intense and can extend thousands of metres into the air. They build up over many nights in arctic and polar regions, where the solar heat of the short winter day cannot completely compensate for nocturnal cooling. Inversions can also result when a layer of warm air is carried over colder air by airflow aloft. A characteristic sign of a low-level inversion is smoke spreading across the sky as a thin layer (Figure 4.16). Temperature inversions suppress vertical mixing of the air because normal convective circulation cannot occur. This can lead to an increase in the concentration of pollutants within the thin, stagnant surface air layer.

Temperature inversions are not always ground-based, and neither are they always caused by radiation heat loss during the night. For example, the horizontal transport of air by **advection** can bring a layer of warm air into a region at any height in the atmosphere, and the normal temperature lapse might be interrupted hundreds or even thousands of metres above the surface.

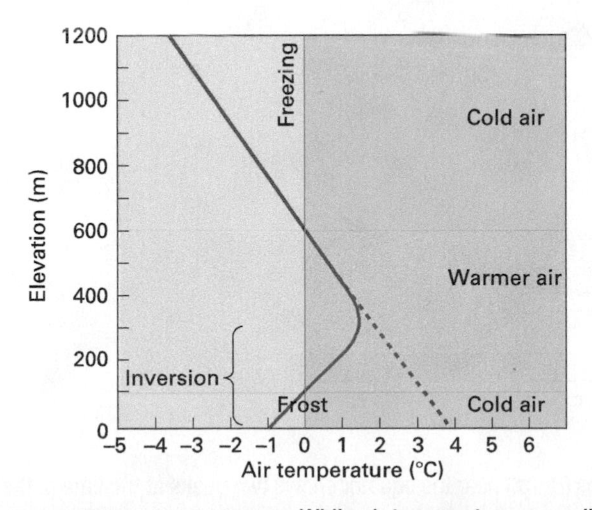

4.15 A GROUND INVERSION WITH FROST While air temperature normally decreases with altitude (dashed line), in a ground inversion temperature increases for some height above the ground before the normal temperature lapse rate is resumed.

4.16 LOW-LEVEL TEMPERATURE INVERSION Smoke from a power station in Saskatoon, Saskatchewan, spreads out across the top of an inversion layer that developed during a cold winter night.

Upper-air inversions also develop when air aloft slowly subsides, causing it to warm by compression (see Chapter 6). **Subsidence inversions** formed in this way can occur at any altitude; if the air sinks to the surface, the same type of weather conditions will occur as in a ground inversion.

THE ANNUAL CYCLE OF AIR TEMPERATURE

The annual temperature cycle at any location is primarily determined by Earth's revolution around the sun in combination with the planet's axial tilt, as this governs the annual cycle of net radiation. Other factors, such as maritime or continental location and cloud cover, also have an important influence.

Near the equator, average net radiation is strongly positive in every month, with peaks coinciding approximately with the equinoxes when the sun is nearly directly overhead. Manaus, Brazil (lat. 3° S) is a representative location (Figure 4.17). Here temperature is relatively uniform throughout the year, averaging about 27 °C. The annual temperature range—the difference between the highest and lowest mean monthly temperatures—is only 1.7 °C. On the basis of temperature, this climate has no seasons.

At Aswan, Egypt (lat. 24° N) the net radiation curve shows a large surplus of energy in every month, with values ranging from about 35 W m^{-2} in December to 125 W m^{-2} in June. The average temperature shows a corresponding cycle, reaching 33 °C in June,

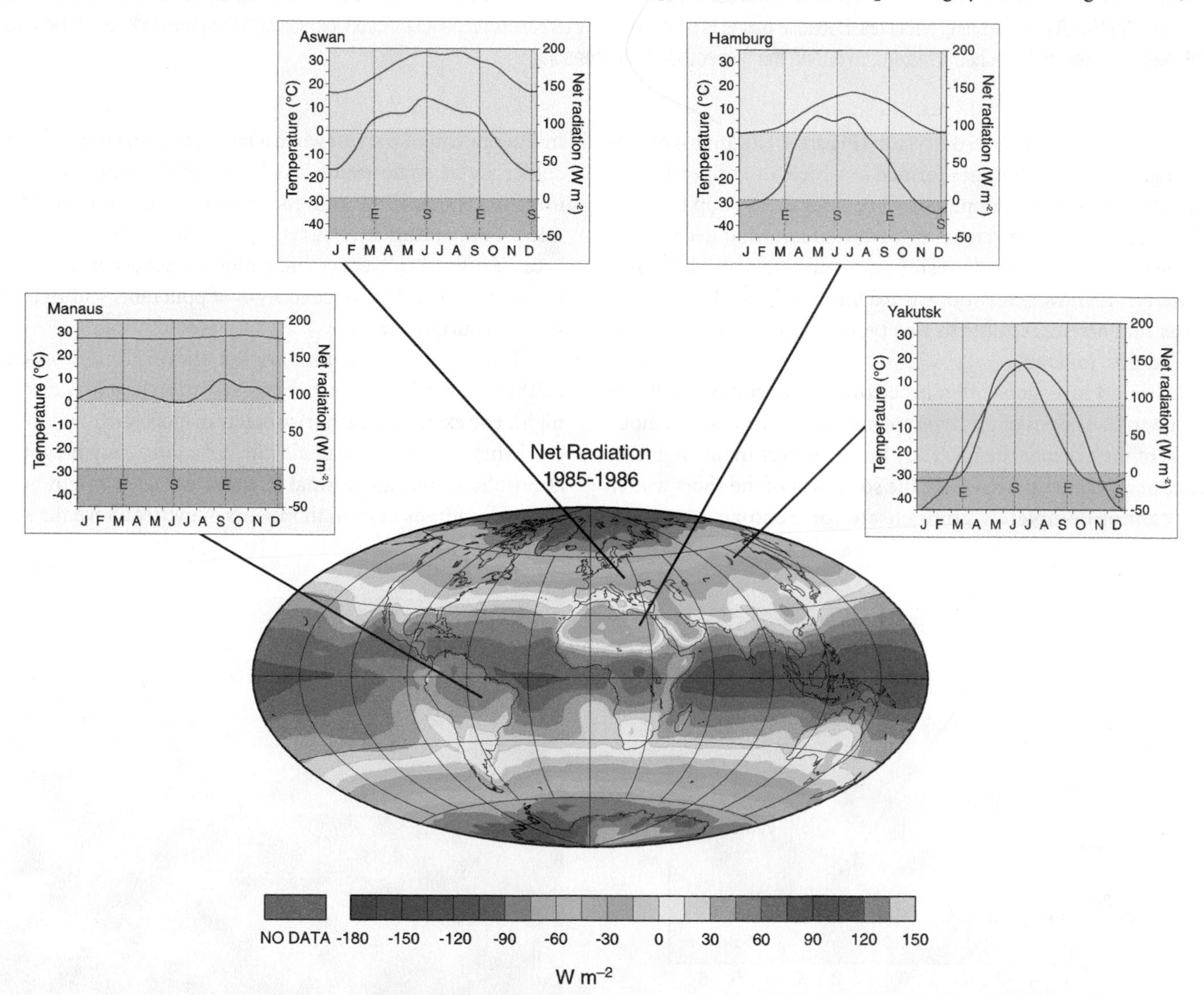

4.17 NET RADIATION AND TEMPERATURE CYCLES Mean monthly net radiation at Manaus (Brazil) near the equator shows two peaks at the time of the equinoxes, although mean monthly temperature remains nearly uniform throughout the year at about 27 °C. Annually, Manaus receives a net radiation surplus of about 70 W m^{-2}. Aswan (Egypt), Hamburg (Germany), and Yakutsk (Russia) all show strong annual temperature and net radiation cycles with summer maxima and winter minima becoming more pronounced with latitude. Aswan at 24° N has an annual net radition surplus of about 35 W m^{-2}. Hamburg has a small net radiation deficit in the winter months. An annual net radiation deficit of about 40 W m^{-2} occurs at Yakutsk.

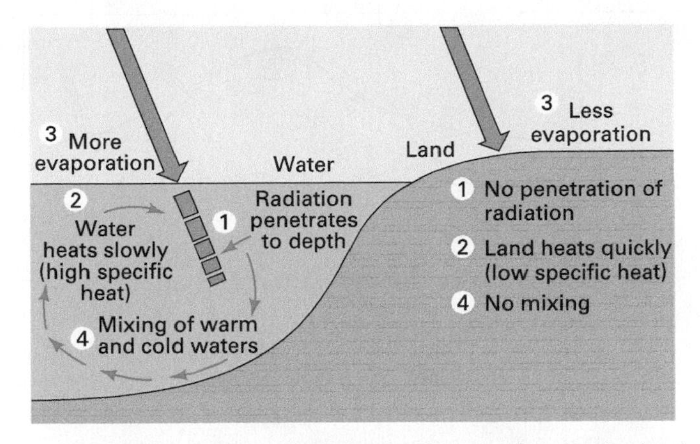

4.18 LAND–WATER CONTRASTS Temperature differences between land and ocean arise because of variations in transparency, specific heat, evaporation, and convectional circulation. As a result, land surfaces heat and cool more rapidly and more intensely than the surface of a deep water body. Temperatures at places located near oceans are consequently more uniform than at places at equivalent latitudes inland.

July, and August, and dropping to 16 °C in December and January. The annual temperature range is 17 °C. Net radiation at Hamburg, Germany (lat. 54° N) is positive from February to October, but a deficit occurs from November to January. The temperature cycle reflects the general reduction in total insolation at this latitude. Temperatures in summer reach a maximum of just over 16 °C and in winter drop to about 0 °C. The annual temperature range is about 17 °C, the same as at Aswan.

Further north at Yakutsk, Russia (lat. 62° N), net radiation is negative during the long, dark winters and a radiation deficit lasts about six months. Mean temperatures in winter are between −35 °C and −45 °C. In mid-summer, when daylight lasts almost 24 hours each day, net radiation rises to a strong peak exceeding values at the other three stations. Air temperature rises quickly after the winter solstice and reaches 17 °C by June. Because of Yakutsk's high latitude and continental interior location, the annual temperature range is very large—exceeding 60 °C.

LAND AND WATER CONTRASTS

The surface of any large body of water heats and cools more slowly than a land surface when both are subjected to the same intensity of insolation, and because of this coastal and continental interior locations experience very different temperature cycles. This is primarily due to four important thermal differences between land and ocean surfaces.

First, incoming solar radiation penetrates to some depth in the ocean, which distributes heat throughout a substantial volume of water (Figure 4.18). In contrast, the ground absorbs incoming short-wave solar radiation only at the surface so the heating effect is more concentrated. Consequently, short-wave radiation will warm the water only slightly, whereas it warms a land surface more intensely.

A second factor is the different specific heats of water and land. Water has a comparatively high specific heat of 4.18 J $g^{-1}K^{-1}$, while a typical value for rock and soil is about 0.80 J $g^{-1}K^{-1}$. Thus, it takes about five times as much heat to raise the temperature of water by 1 °C as it does to raise the temperature of rock. The same will be true for cooling—after losing the same amount of heat, the decrease in temperature of water is less than for rock.

A third difference is the amount of energy that is dissipated through evaporation of water. This is equivalent to 2,259 J $g^{-1}K^{-1}$. Water continually evaporates from the surface of the ocean and so reduces the amount of energy that is available for warming. Land surfaces also can be cooled by evaporation, but only if water is

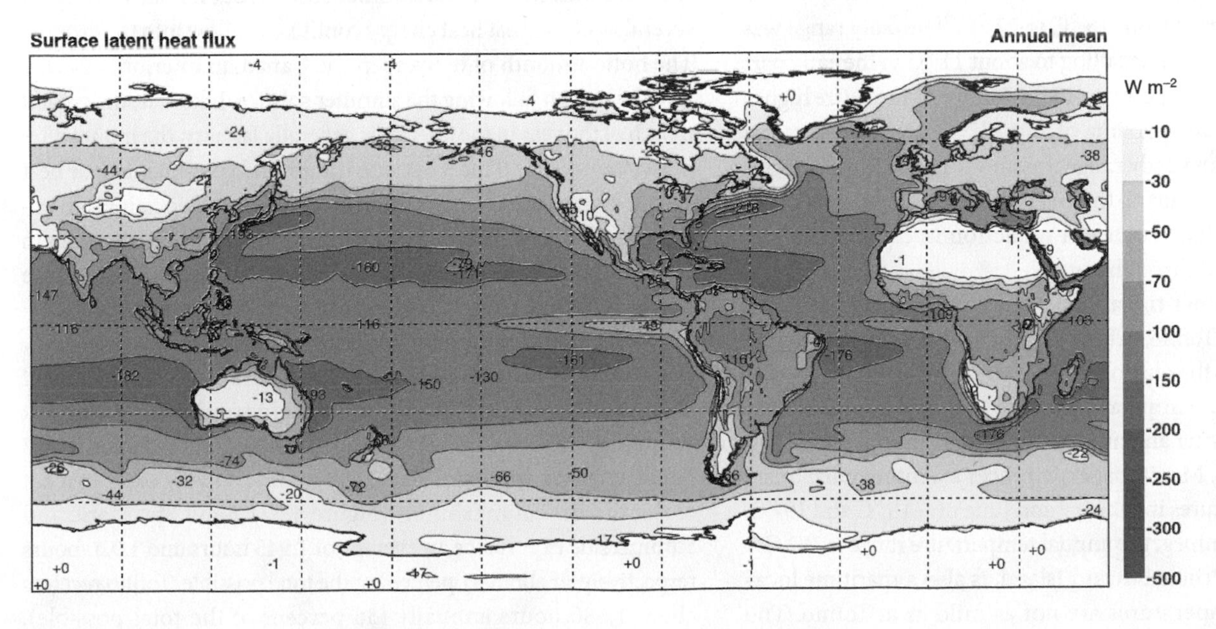

4.19 MEAN ANNUAL SURFACE LATENT HEAT FLUX Energy used in evaporation is greatest in the subtropical oceans and negligible in the polar regions and extreme deserts.

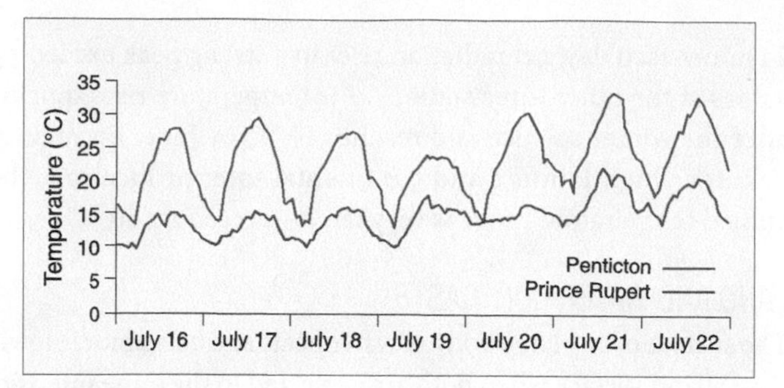

4.20 MARITIME AND CONTINENTAL TEMPERATURES Hourly temperature records for Prince Rupert and Penticton in British Columbia for the period of July 16 to 22, 2006. The daily temperature cycle at Prince Rupert on the Pacific Coast is less pronounced than at Penticton, in the province's interior.

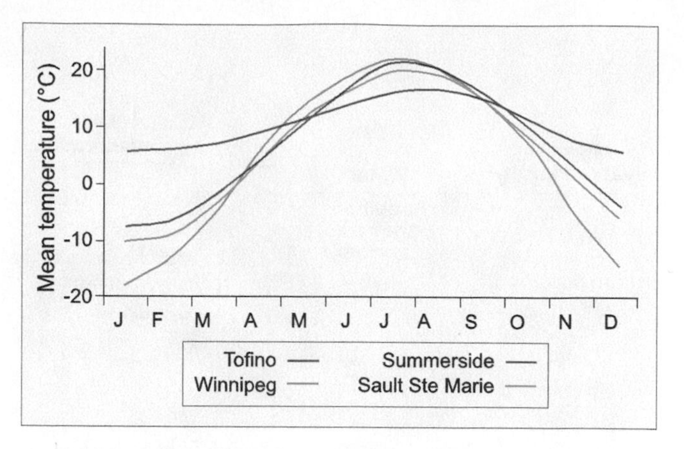

4.21 MEAN MONTHLY TEMPERATURES FOR TOFINO, BRITISH COLUMBIA, WINNIPEG, MANITOBA, SAULT STE. MARIE, ONTARIO, AND SUMMERSIDE, PRINCE EDWARD ISLAND The moderating effect of the nearby ocean is less pronounced at Summerside on the Atlantic coast compared with Tofino, where prevailing winds bring mild Pacific air onshore. Similarly, temperatures at Sault Ste. Marie are influenced by the Great Lakes, so the annual temperature range here is less than at Winnipeg.

present, as is the case for moist soils. When the surface dries, evaporation ceases, and the energy is then converted to heat. Figure 4.19 shows the global surface heat flux, representing the quantity of energy from oceans and continents used for evaporation.

Finally, a warm surface layer in the oceans can mix with cooler water below to produce a more uniform temperature. For open water, the mixing is produced by wind-generated waves and density currents. Clearly, no such mixing occurs on land surfaces. In some localized areas, as much as 150 W m^{-2} is transferred from the surface oceans in this way, but generally the process accounts for less than 50 W m^{-2}. Elsewhere, energy is brought to the surface, notably in the North Pacific and North Atlantic Oceans. Energy transfer by ocean currents is discussed further in Chapter 5.

Comparative data for July 16 to 22, 2006 for Prince Rupert on the west coast of British Columbia and Penticton in the province's interior illustrate how the different properties of land and water affect temperature cycles at coastal and inland stations (Figure 4.20). During that time, the daily maximum temperature at Prince Rupert varied from 15 °C to 22 °C. The daily range was 5 to 6 °C, with temperatures falling to about 11 °C in the early part of the week and later dropping to 15 °C. Temperatures were higher at Penticton, with daily maxima of 25 to 35 °C and daily minima of 15 to 20 °C; here the daily range was about 15 °C. Temperatures at both locations rose during the week, which indicates that other factors were important, such as a reduction in cloud cover as a drier air mass passed through region.

Figure 4.21 shows the annual temperature cycles for four Canadian locations. Tofino, British Columbia, located on Vancouver Island, is exposed to the onshore flow of air from the Pacific Ocean. In January, the mean temperature falls to about 5 °C, and rises to 15 °C in August, giving an annual temperature range of 10 °C. In contrast, Winnipeg, Manitoba, experiences a continental climate with mean temperatures in January and June of −18 °C and 19 °C, respectively; at Winnipeg, the annual temperature range is 37 °C.

Summerside, Prince Edward Island, is also a maritime location. However, temperatures are not as mild as at Tofino. The influence of the Atlantic Ocean at Summerside is reduced because airflow is principally from west to east across the continent. Mean temperature in January is about −7 °C, rising to 19 °C in July. The annual temperature range at Summerside is 26 °C, compared with 10 °C at Tofino. Sault Ste. Marie, Ontario, is located at the eastern end of Lake Superior, with Lake Michigan and Lake Huron close by. The Great Lakes are sufficiently large to moderate the climate of the continental interior. With a January temperature of −11 °C and a July temperature of 17 °C, the annual range at Sault Ste. Marie is 28 °C, compared to 37 °C at Winnipeg.

Another contrast between maritime and continental locations is the timing of maximum and minimum temperatures. Insolation reaches a maximum at the summer solstice, but remains strong for several weeks so that heat energy continues to flow into the ground. The hottest month of the year in the Canadian interior is usually July, the month following the summer solstice. Likewise, the coldest month of the year in the interior is generally January, the month after the winter solstice. This is because the ground continues to lose heat even after insolation begins to increase. At coastal locations, maximum and minimum air temperatures generally occur later than in continental interiors because water bodies heat and cool more slowly; at Tofino, for example, August is the warmest month.

Temperature patterns are affected by many factors, especially cloud cover, which can reduce insolation and also alter the rate of energy loss to space. Total hours of bright sunshine is greatest at Winnipeg, with mean monthly values ranging from 96 hours to 317 hours, totalling approximately 2,400 hours annually or 51 percent of the theoretical maximum (Figure 4.22). Sault Ste. Marie and Summerside experience an average of 1,945 hours and 1,920 hours respectively, or about 41 percent of the total possible. Tofino receives about 1,680 hours annually (36 percent of the total possible),

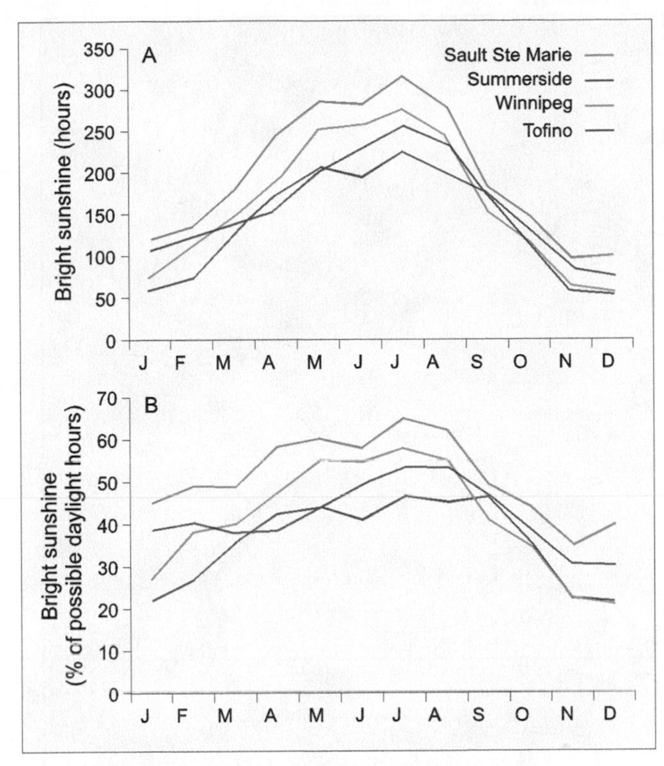

4.22 ANNUAL SUNSHINE AT FOUR CANADIAN LOCATIONS (a) Total hours of bright sunshine and (b) percent of maximum possible sunshine hours at Tofino, British Columbia; Winnipeg, Manitoba; Sault Ste. Marie, Ontario; and Summerside, Prince Edward Island. Less cloud cover in the drier climate of the Prairies accounts for the high number of sunshine hours at Winnipeg compared with other regions in Canada.

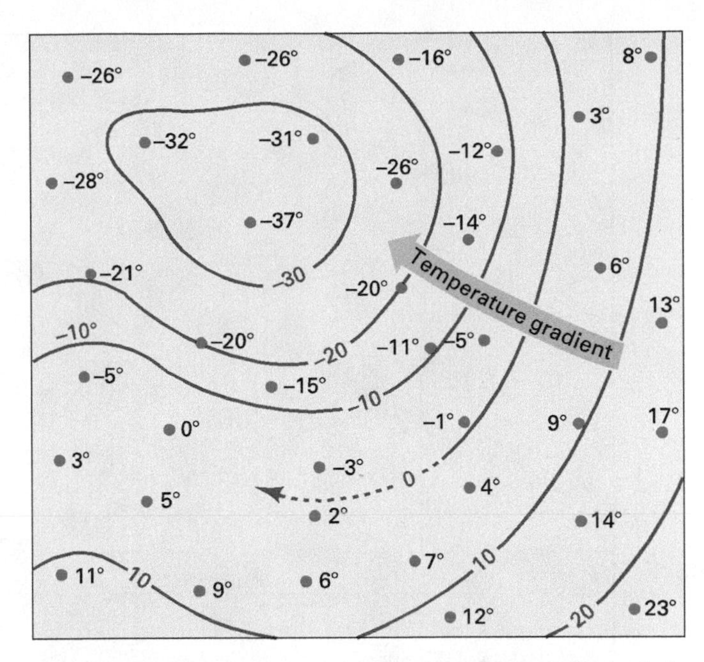

4.23 ISOTHERMS SHOWING SPATIAL VARIATION IN TEMPERATURE Recorded air temperatures are placed at their proper locations on the map. These may be single readings, such as a daily maximum or minimum, or they may be averages of many years for a particular day or month, depending on the purpose of the map. Each line connects points having the same temperature. A directional change in temperature is referred to as a temperature gradient.

ranging from 55 hours in December to 220 hours in July. The shorter period of daylight during the winter months also contributes to the seasonal reduction in sunshine hours.

GLOBAL AIR TEMPERATURE PATTERNS

The distribution of air temperatures on a map is normally shown by **isotherms**—lines drawn to connect locations having the same temperature (Figure 4.23). They are constructed by drawing smooth lines through and between the points of known temperature. Isotherms usually represent 5- or 10-degree differences, but they can be drawn at any convenient temperature interval. Isothermal maps depict broad temperature patterns from which *temperature gradients*—directions along which temperatures change—can be discerned.

Three main factors—latitude, marine-continental location, and altitude—determine the patterns of isotherms at the global scale. The latitudinal decrease in mean annual insolation leads to a corresponding decrease in temperature, with affects the seasonal temperature cycle. Marine-continental influences also produce seasonal temperature contrasts. Marine locations, where prevailing winds come from the oceans, have more uniform temperatures throughout the year as the adjacent oceans tend to lower temperatures in summer and reduce heat loss in winter. Continental stations show a much larger annual temperature range. Ocean currents modify air temperature by bringing water from the cold polar oceans toward the equator or warm tropical water poleward. Similarly, mountain ranges are cooler than the surrounding plains because temperature decreases with elevation.

The combined effect of these factors influences global temperature patterns in several distinctive way (Figure 4.24).

1. **Temperatures decrease from the equator to the poles.** The latitudinal decrease in temperature corresponds closely with the decrease in mean annual insolation. The effect is most pronounced in the southern hemisphere where land areas are less extensive.
2. **Large land masses located in the subarctic and Arctic zones experience extremely low temperatures in winter.** In January, low-temperature centres are strongly developed over North America and Siberia. The high albedo of the snow cover is an important factor keeping winter temperatures low in these regions. Greenland is also a region of low temperature. In addition to the permanent ice cover, the general height of Greenland's ice surface rises to more than 3,000 m.

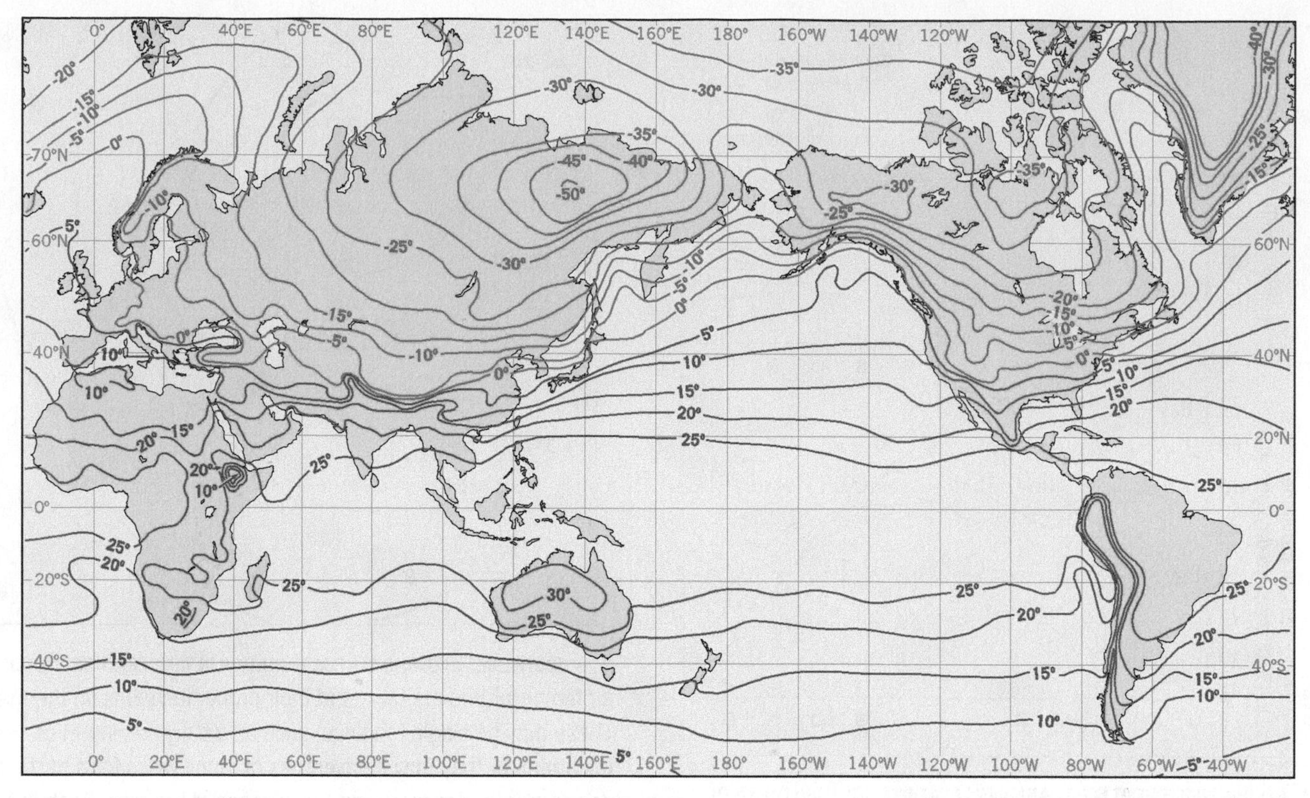

JANUARY

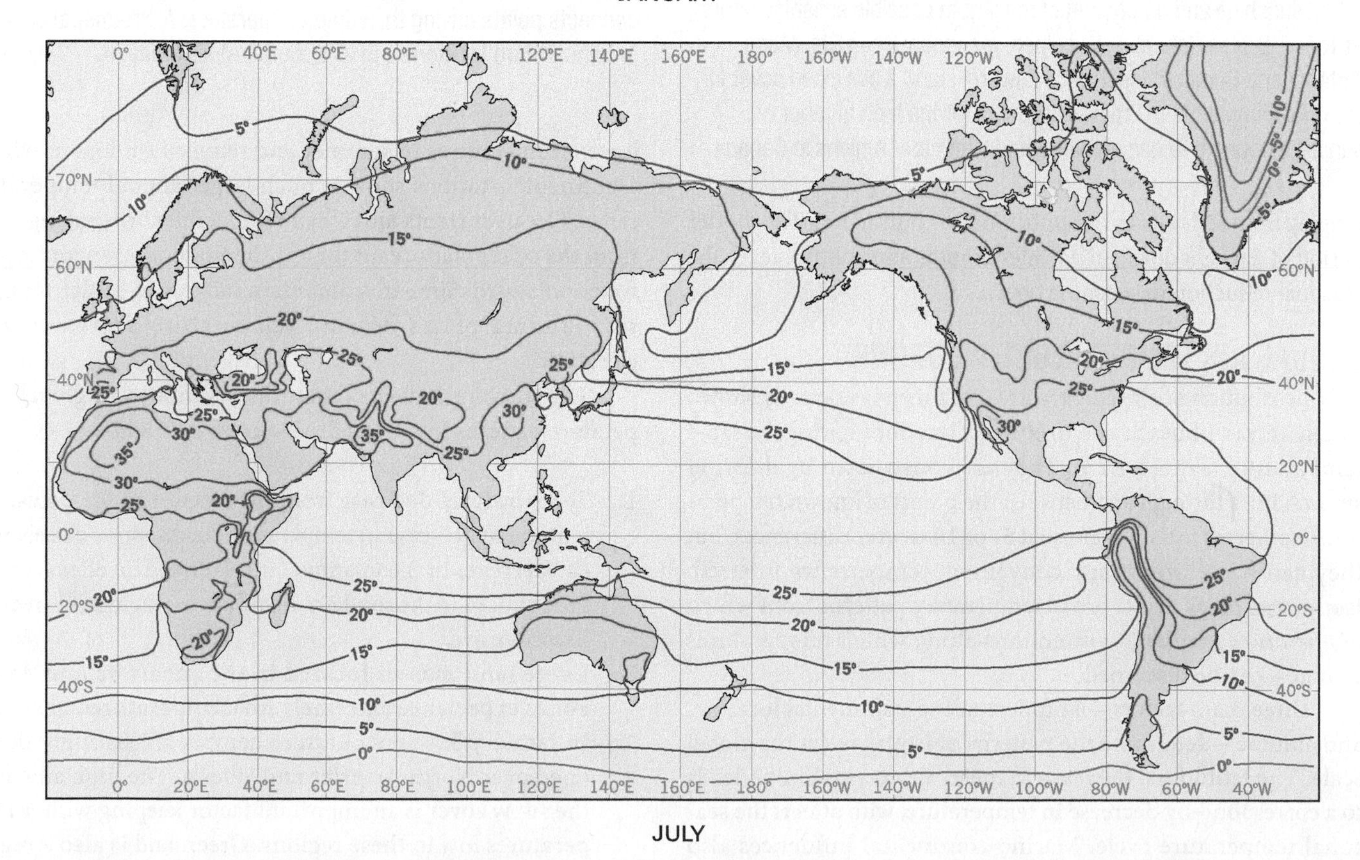

JULY

4.24 MEAN MONTHLY AIR TEMPERATURES (°C) FOR JANUARY AND JULY Temperatures remain relatively constant in tropical regions, but marked seasonal contrasts occur at higher latitudes.

3. **Temperature in equatorial regions change little over the course of the year.** The northern and southern positions of the 25 °C isotherms are separated by a broad zone in which temperature remains above 25 °C in both January and July. Although latitudinal position of the two isotherms moves with the seasons, the equator, with a few minor exceptions, always falls between them. This uniformity of equatorial temperatures is caused primarily by the relatively constant insolation in the region.
4. **Isotherms make a large north–south shift from January to July over continents in the midlatitude and subarctic zones.** In winter, isotherms generally move toward the equator, while in summer, they arch toward the poles. The effect is especially pronounced in North America and Eurasia. For example, the 10 °C isotherm borders the Gulf of Mexico in January, but in July it reaches to the Arctic in northwestern Canada. Isotherms over oceans shift much less. This difference is due to the contrast between oceanic and continental surface properties, which cause continents to heat and cool more rapidly than oceans.
5. **Highlands are always colder than surrounding lowlands.** This effect is illustrated by the Andes Mountains in South America and the highlands of Ethiopia. Similarly, in North America the −5 °C and −10 °C isotherms dip southwards around the Rocky Mountains in winter, indicating that the centre of the mountain range is colder than the adjacent regions. The equatorward bend in the 20 °C and 25 °C isotherms in the western United States in July reflects the decrease in temperature with elevation.
6. **Areas of perpetual ice and snow are always intensely cold.** Greenland and Antarctica contain Earth's two great ice sheets. They are cold for two reasons. First, their surfaces are high, rising to more than 3,000 m in places. Second, the snow surface has a high albedo. Since little solar energy is absorbed, the snow surface remains cold and chills the air above it. The Arctic Ocean, with its cover of floating ice, also maintains its cold temperatures throughout the year (Figure 4.25). However, the cold is much less intense in January in the Arctic than on the Greenland ice sheet, because the ocean water acts as a heat reservoir and moderates the temperature of the ice above.

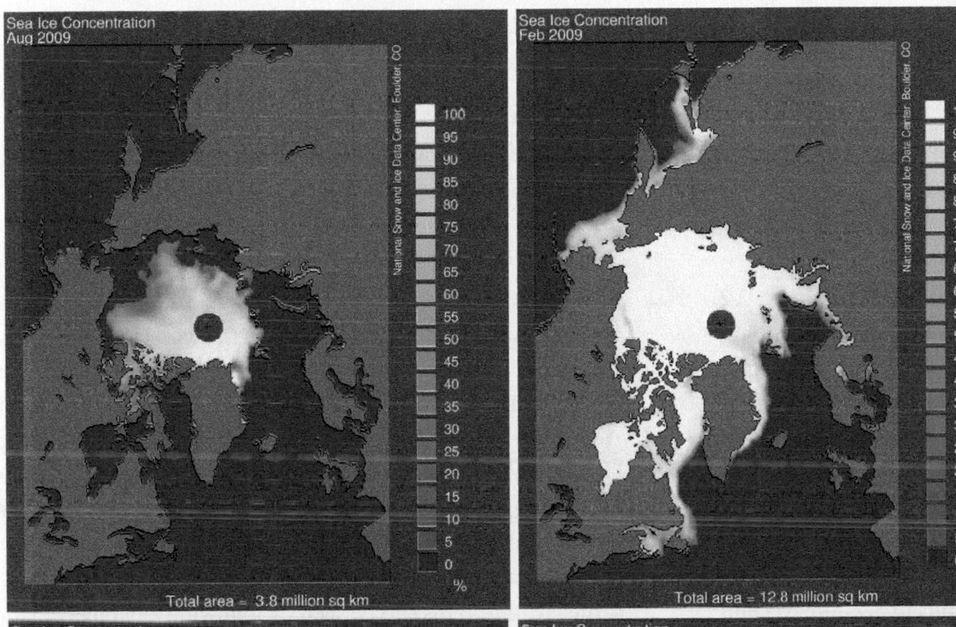

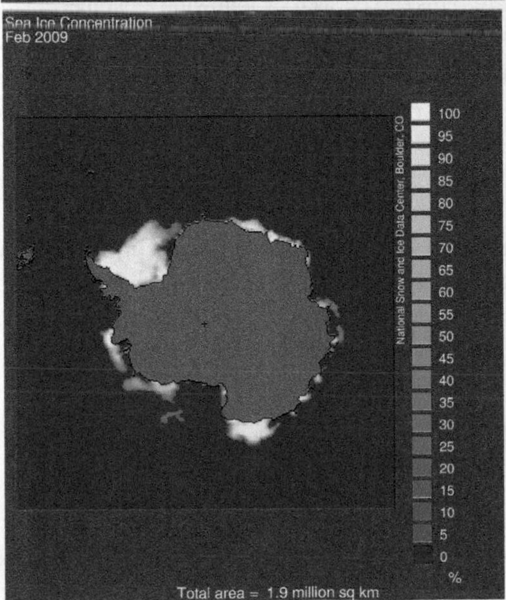

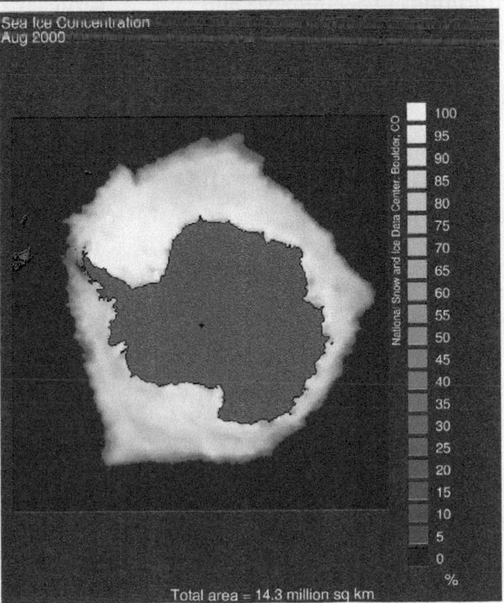

4.25 POLAR SEA ICE The extent of sea ice cover in the Arctic and Antarctic changes with the seasons. In the northern hemisphere, its maximum extent is generally reached in March, shrinking to a minimum in September. The ice seasons are reversed in the southern hemisphere.

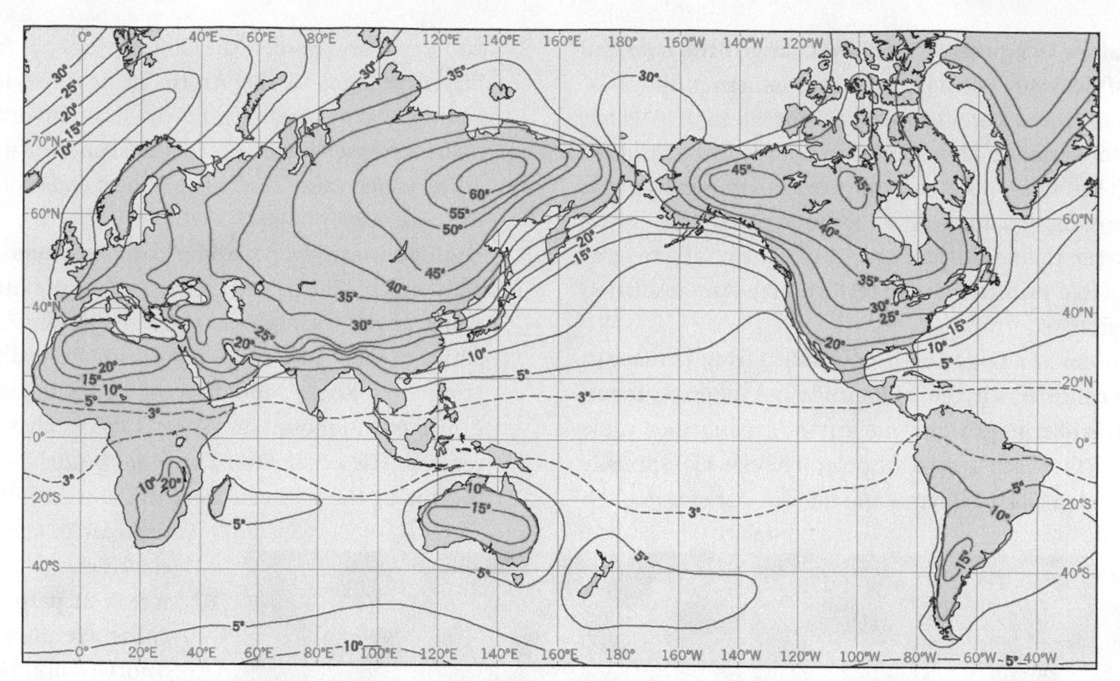

4.26 ANNUAL RANGE OF AIR TEMPERATURE (°C) The isolines show the difference between January and July mean temperatures. Temperature range is largest over land masses at mid- and high-latitudes. The smallest temperature ranges are found in equatorial regions.

7. **Isothermal patterns are associated with ocean currents.** The northward bend in the 20 °C and 25 °C isotherms in January off the west coast of South America is associated with the cold Peru Current which brings cold water north from Antarctica. The Benguela Current has a similar effect off the coast of southwest Africa. The 5 °C January isotherm is positioned farther north over the Atlantic Ocean compared with the Pacific Ocean. This is caused by warm water carried from the general vicinity of the Gulf of Mexico by the North Atlantic Drift, which helps to moderate conditions in Scandinavia and other locations in western Europe.

Figure 4.26 shows the annual range of air temperature; the **isolines** resemble isotherms, but in this map they show the difference between the January and July monthly mean temperatures. Five characteristic patterns emerge.

1. **The annual range increases with latitude, especially over northern hemisphere continents.** This trend is clearly shown for North America and Asia due to the pronounced contrast between summer and winter insolation at higher latitudes.
2. **The largest ranges occur in the subarctic and arctic zones of Asia and North America.** Northeast Siberia and northwest Canada–eastern Alaska are regions of exceptionally large annual temperature range. Here summer insolation is nearly the same as at the equator, while winter insolation is very low.
3. **The annual range is moderately large on land areas near the Tropics of Cancer and Capricorn.** These are principally desert regions, for example the Sahara in North Africa, the Kalahari in southern Africa, and the Gibson Desert in Australia. Dry air and the absence of clouds and moisture allow these continental locations to cool strongly in winter and warm strongly in summer, even though insolation contrasts with the season are not as great as at higher latitudes.
4. **The annual range over oceans is less than over land at the same latitude.** This can be seen by following a parallel of latitude. For example at lat. 60° N, near the west coast of Europe, the range is about 10 °C over the Atlantic Ocean, but increases to 55 °C in central Asia. Similarly, over the Pacific Ocean, the range falls to 20 °C off Alaska and increases to more than 40 °C within North America. Again, these major differences are due to the contrast between land and water surfaces.
5. **The annual range is very small over oceans in the tropical zone.** The temperature range over tropical oceans is less than 3 °C since there is little seasonal variation in insolation near the equator, and water heats and cools more slowly.

GLOBAL CLIMATE CHANGE AND THE GREENHOUSE EFFECT

Global temperatures have risen over the past few decades, but although the cause of this change is not fully understood, there is concern that human activities have had a contributory effect through emissions of greenhouse gases. Because it is an efficient absorber of long-wave radiation, water vapour is considered to be the most important greenhouse gas. However, its presence in the lower troposphere is not significantly affected by emissions from anthropogenic sources. For this reason it is not usually included as a gas implicated in global warming and climate change. CO_2 is also naturally present in the atmosphere, but unlike water vapour its concentration is influenced by many human activities. The concentration of CO_2 has been increasing steadily since the industrial revolution of the mid-eighteenth century, and consequently it is often cited as a major cause of climate warming.

Also of concern are other gases that are normally present in much smaller concentrations—methane (CH_4), nitrous oxide (N_2O), *tropospheric* ozone (O_3), chlorofluorocarbons (CFCs), perfluorocarbons (PFCs), and sulphur hexafluoride (SF_6). Taken together with CO_2, they are known as **greenhouse gases**. Though less abundant, these gases are better absorbers of long-wave radiation than CO_2, and so can readily enhance the greenhouse effect. Apart from *stratospheric* ozone, which has a narrow absorption peak at 9.6 μm, there is little natural absorption of long-wave radiation, so the 8 to 13 μm wavelength band provides the atmospheric window by which most of Earth's outgoing radiation escapes to space. However, CH_4, N_2O, and the CFCs also absorb energy within this window. As concentrations of these gases increase, more long-wave energy is trapped, the greenhouse effect is enhanced, and the potential for anthropogenic global warming increases. The comparative efficiency of gases to enhance the greenhouse effect is termed the **global warming potential (GWP)**. The GWP of a gas is based on its heat-absorbing capacity, as well as the rate at which it is removed from the atmosphere over a given number of years. The GWP of CO_2 is 1; this serves as a reference for comparing other gases (Table 4.1). Some pollutants, such as sulphur hexafluoride (SF_6), are so long-lived that their contribution to the greenhouse effct is essentially permanent.

The increase in global atmospheric CO_2 since the industrial revolution is due mainly to emissions from fossil fuel combustion, gas well flaring, and cement production. Deforestation and other land use changes have also contributed to the rise of CO_2 to its current global mean concentration of 384 ppm (Figure 4.27a). Residence time in the atmosphere for CO_2 is quite variable, but it is considered to remain there for 50 to 100 years. It is absorbed rapidly by growing plants, but then is stored or returned to the atmosphere through various processes operating over a range of time scales. The storage and transfer of CO_2 in the global environment is encompassed in the carbon cycle (see Chapter 20).

Because CO_2 absorbs long-wave radiation, any increase in the atmospheric concentration of this gas will alter the energy balance of the Earth and the global climate system must readjust to a new equilibrium. Anything that can change Earth's energy balance is referred to as a forcing. In this case, the change in CO_2 brings about a change in net radiation, and so this is considered a **radiative forcing**. Radiative forcing is defined by the IPCC (Intergovernmental Panel on Climate Change) *as a measure of the influence a factor has in altering the balance of incoming and outgoing energy in the Earth-atmosphere system*. Each factor is assessed on how it has affected Earth's radiation balance

TABLE 4.1 GLOBAL WARMING POTENTIAL (GWP) OF THE MAIN GREENHOUSE GASES

GREENHOUSE GAS CONCENTRATION IN PPM	ATMOSPHERIC LIFETIME (YEARS)	GWP PER 100 YEAR TIME HORIZON[1]	PRE-1750 TROPOSPHERIC CONCENTRATION	RECENT TROPOSPHERIC CONCENTRATION	INCREASED RADIATIVE FORCING ($W\ m^{-2}$)
Carbon dioxide (CO_2)	Variable	1	278	384.8	1.66
Concentration in ppb					
Methane (CH_4)	12	21	700	1,774	0.48
Nitrous oxide (N_2O)	114	310	270	319	0.16
Tropospheric ozone (O_3)	hours	n.a.	25	34	0.35
Concentration in ppt					
CFC-11 (CCL_3F)	45	4,750	0	244	0.063
CFC-12 (CCl_2F_2)	102	10,900	0	538	0.17
HCFC-22 ($CHClF_2$)	12	1,810	0	206	0.033
HFC-134a (CH_2FCF_3)	14	1,430	0	54	0.0055
Sulphur Hexafluoride (SF_6)	3,200	22,800	0	6.70	0.0029
Carbon tetrachloride	26	1,400	0	89	0.012

[1] The time interval, which is normally 100 years, is an essential part of the rating scheme.

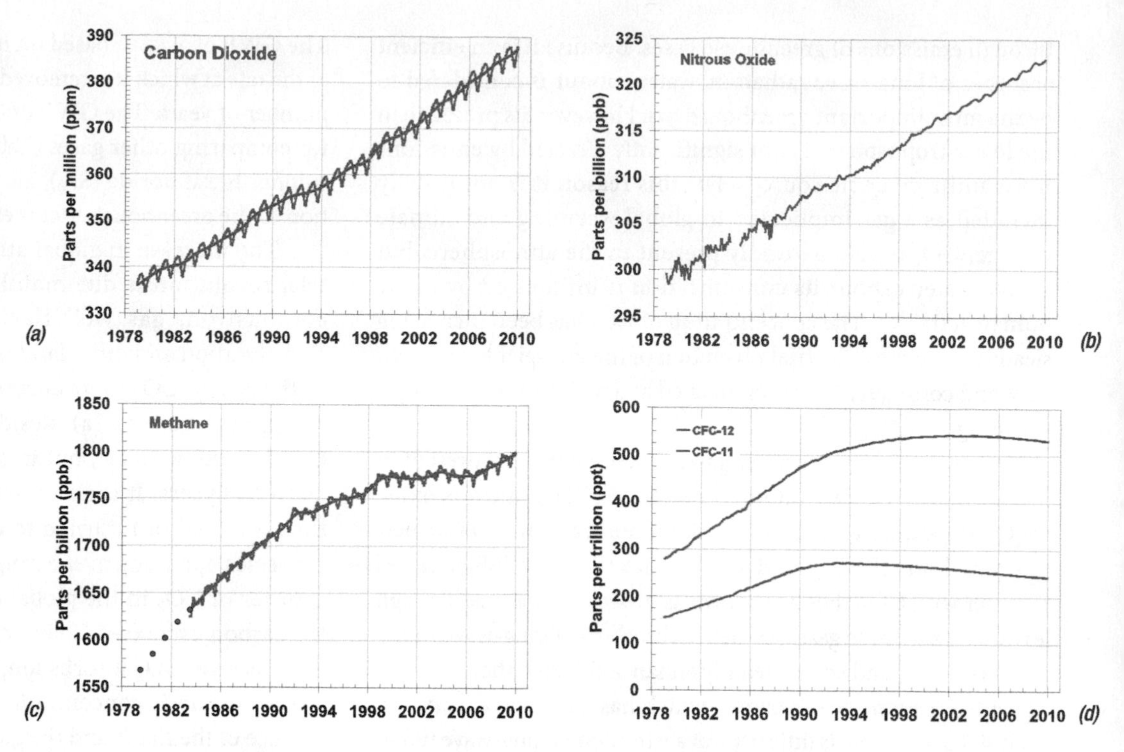

4.27 RECENT TRENDS IN ATMOSPHERIC CONCENTRATION OF GREENHOUSE GASES The concentration of the principal greenhouse gases have all increased since 1978, although the trends vary. Carbon dioxide (CO_2) and nitrous oxide (N_2O) have both increased steadily during this period (a) and (b). Methane (CH_4) concentration has remained relatively steady since the late 1990s (c) while concentrations of the chlorofluorocarbons (CFC-11 and CFC-12) have declined in recent years (d).

compared to pre-industrial conditions—prior to 1750. Radiative forcings are measured in W m^{-2}, the same units as radiation. The radiative forcing for CO_2 is 1.66 W m^{-2}, which is the largest of any forcing agent (Table 4.1).

The present concentration of CH_4 is 1,780 ppb (Figure 4.27c), which is considerably less than CO_2. However, CH_4 is an important greenhouse gas because its GWP is 21 times that of CO_2 and so has a relatively high radiative forcing of 0.48 W m^{-2}. For this reason, the overall contribution of CH_4 to global warming could potentially be as high as 20 percent. CH_4 is produced naturally by anaerobic decomposition in wetlands. Paddy rice production and the digestive processes of cattle have contributed significantly to the rise in CH_4 emissions. Biomass burning, coal mining, and natural gas production also contribute CH_4 to the atmosphere, as do some landfill sites. It is removed from circulation by certain soil bacteria which use the carbon through a process of oxidation.

Atmospheric concentration of N_2O is currently about 320 ppb (Figure 4.27b). The emission of N_2O is part of the global nitrogen cycle (see Chapter 20). Most N_2O enters the atmosphere from soils and the oceans through the process of denitrification under anaerobic conditions. N_2O from agricultural land has increased as more fertilizers are used to increase crop yields. This is of particular concern in tropical regions, since plant growth there is generally limited by a shortage of phosphorus; consequently, additional nitrogen is available for denitrification. Some N_2O originates in the oceans from microbial denitrification of organic matter as it sinks through the water column and from sediments in estuaries and other shallow marine environments. These processes are enhanced by sewage discharge and agricultural runoff into rivers, which then transport nitrogen-rich water to the ocean.

Concentrations of CFC-11 and CFC-12 are currently 244 and 538 parts per trillion, respectively (Figure 4.27d). Although this is considerably lower than CO_2, CH_4, and N_2O, these halocarbons have comparatively long residence times and very high GWPs—4,750 for CFC-11 and 10,900 for CFC-12—so their radiative forcings are high for compounds that are present at such low concentrations. The concentrations of CFC-11 and CFC-12 have decreased slightly in recent years, but have been replaced by other halocarbons which, although having lower radiative forcings, will still affect Earth's radiation balance.

Air samples are collected weekly from about 100 global clean air sites to assess the average total radiative forcing of all the long-lived greenhouse gases. They are also used in the **Annual Greenhouse Gas Index (AGGI)** introduced in 2004 to help clarify the magnitude of the impact of greenhouse gases on the global radiation balance. The five major greenhouse gases—CO_2, CH_4, N_2O, CFC-11, and CFC-12—account for about 96 percent of the direct radiative forcing that is assumed to have occurred since 1750.

As well as affecting the global energy budget by reducing outgoing long-wave radiation, CFCs and other pollutants react with UV light and release chlorine (Cl) or bromine (Br) atoms, which can then cause the breakdown of ozone. This group of compounds is referred to as ozone-depleting substances (ODS).

The chlorine (or bromine) atom that is released from the halocarbon can combine with ozone and break it down to form oxygen and chlorine monoxide (ClO):

$$Cl + O_3 \longrightarrow ClO + O_2$$

ClO can combine with ozone and atomic oxygen to form molecular oxygen:

$$ClO + O_3 \longrightarrow Cl + 2\,O_2$$

$$ClO + O \longrightarrow Cl + O_2$$

However, oxygen atoms that combine with ClO are now no longer available to combine with molecular oxygen to form ozone:

$$O_2 + O \longrightarrow O_3$$

Although chlorine is not abundant in the stratosphere, it is approximately 1,500 times more reactive with oxygen atoms than is the case for reactions between atomic and molecular oxygen. Also, chlorine is continuously regenerated and so acts as a catalyst in these reactions (Figure 4.28). In this way the protective stratospheric ozone layer is progressively destroyed by chlorine atoms derived from the breakdown of CFCs by UV radiation. Current observations indicate a decline in ODSs in the air over Antarctica since the mid-1990s and some reduction in the extent of the ozone hole (see Chapter 3).

The ozone-depleting potential (ODP) assesses an ODS according to the amount of ozone it can destroy. ODP is determined as the ratio of the impact of a substance to that caused by a similar mass of CFC-11. The ODP of CFC-11 is given a value of 1; the ODP of other common ODSs is listed in Table 4.2.

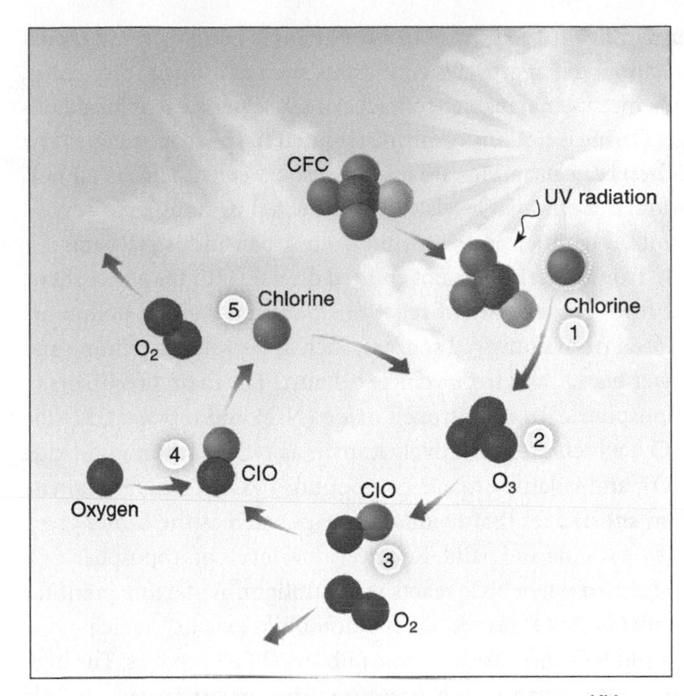

4.28 THE ROLE OF CHLORINE AS AN OZONE DEPLETING SUBSTANCE UV radiation breaks off a chlorine atom (1), which then breaks up an ozone molecule (2), releasing an oxygen molecule and a chlorine monoxide molecule (3). The chlorine monoxide molecule bonds with a free oxygen atom (4), releasing the chlorine atom (5), which is now free to break up another ozone atom. (From *Visualizing the Environment* by Berg, Hager, Goodman, and Baydack, John Wiley & Sons Canada, Ltd., 2010.)

TROPOSPHERIC OZONE

Stratospheric ozone is concentrated in the ozone layer and is considered "good ozone" because it shields the surface from intense UV radiation. However, tropospheric ozone, which represents about 10 percent of total atmospheric ozone, is often referred to as "bad ozone" because of its harmful impacts on vegetation and

TABLE 4.2 OZONE DEPLETION POTENTIAL (ODP) OF COMMON CFCS AND HALOGENS (AFTER WMO *SCIENTIFIC ASSESSMENT OF OZONE DEPLETION, 2006*)

		LIFETIME (YEARS)	ODP
CFC-11 (CCl_3F)	Trichlorofluoromethane	45	1
CFC-12 (CCl_2F_2)	Dichlorodifluoromethane	100	1
CFC-113 ($C_2F_3Cl_3$)	1,1,2-Trichlorotrifluoroethane	85	0.8
CFC-114 ($C_2F_4Cl_2$)	Dichlorotetrafluoroethane	300	1
CFC-115 (C_2F_5Cl)	Monochloropentafluoroethane	1,700	0.44
Halon 1211 (CF_2ClBr)	Bromochlorodifluoromethane	16	7.1
Halon 1301 (CF_3Br)	Bromotrifluoromethane	65	16
Halon 2402 ($C_2F_4Br_2$)	Dibromotetrafluoroethane	20	11.5
CCl_4	Carbon tetrachloride	26	0.73
$C_2H_3Cl_3$	Trichloroethane	5	0.12
CH_3Br	Methyl bromide	0.7	0.51

human health. Exposure to ozone causes coughing and throat irritation and aggrevates conditions such as asthma, bronchitis, and emphysema; regular contact can cause permanent lung damage. Ozone is not uniformly distributed in the troposphere. The highest concentrations are usually associated with urban air pollution, but elevated levels can be detected in rural areas several hundred kilometres downwind from urban-industrial centres.

Tropospheric O_3 is not emitted directly into the atmosphere, but forms by a series of reactions between various pollutants emitted from industrial sources, such as petroleum refining and power plants, and from vehicle exhaust. The main precursors of tropospheric O_3 are nitrogen oxide (NO) and nitrogen dioxide (NO_2) which are collectively known as NO_x, carbon monoxide (CO), and volatile organic compounds (VOC), which originate from substances that readily vaporize, such as the fumes given off by gasoline and paint. Relatively low levels of tropospheric O_3 are formed when NO_x reacts with sunlight. A starting point for tropospheric O_3 formation is automobile exhaust, which emits CO and NO and possibly some unburned fuel vapours. The high temperatures and high pressures that occur in automobile engines are sufficient to combine nitrogen and oxygen:

$$N_2 + O_2 \longrightarrow 2NO$$

NO then reacts with oxygen in the atmosphere to form NO_2:

$$2NO + O_2 \longrightarrow 2NO_2$$

NO_2 is dissociated by strong sunlight to form NO and atomic O, which in turn reacts with molecular O_2 to form ozone:

$$NO_2 \xrightarrow{\text{sunlight}} NO + O$$
$$O + O_2 \longrightarrow O_3$$

O_3 concentration is limited because once formed it can be destroyed by reacting with NO; this generates NO_2 so the cycle is repeated:

$$O_3 + NO \longrightarrow O_2 + NO_2$$

Ground-level O_3 concentrations are now monitored in many cities as part of the **Air Quality Health Index (AQHI)**. The Canadian AQHI incorporates ground-level O_3, NO_2, and two classes of fine particulate matter ($PM_{2.5}$) and (PM_{10}), which are microscopic solid or liquid particles $< 2.5\ \mu m$ and $<10\ \mu m$ in diameter. The AQHI is used to issue "smog alerts" or other advisories when air quality conditions pose potential risks to human health (Table 4.3).

TABLE 4.3 AIR QUALITY HEALTH INDEX FOR BRITISH COLUMBIA. THE AQHI IS CALCULATED BASED ON THE RELATIVE RISKS OF A COMBINATION OF GROUND-LEVEL OZONE, NITROGEN DIOXIDE, AND FINE PARTICULATE MATTER, WHICH ARE KNOWN TO HARM HUMAN HEALTH.

AIR QUALITY HEALTH INDEX LEVELS OF RISK	VALUE	ACCOMPANYING HEALTH MESSAGES FOR AT-RISK POPULATIONS AND THE GENERAL POPULATION	
		AT-RISK POPULATION	GENERAL POPULATION
Low Health Risk	1 – 3	Enjoy your usual outdoor activities. Follow your doctor's advice for exercise.	Ideal conditions for outdoor activities.
Moderate Health Risk	4 – 6	If you have heart or breathing problems and experience symptoms, consider reducing physical exertion outdoors or rescheduling activities to times when the index is lower. Follow your doctor's usual advice about managing your condition.	No need to modify your usual outdoor activities, unless you experience symptoms.
High Health Risk	7 – 10	Children, the elderly, and people with heart or breathing problems should reduce physical exertion outdoors or reschedule activities to times when the index is lower, especially if they experience symptoms. If you have heart or breathing problems, follow your doctor's usual advice about managing your condition.	Anyone experiencing discomfort such as coughing or throat irritation should consider reducing physical exertion outdoors or rescheduling strenuous activities to periods when the index is lower.
Very High Health Risk	Above 10	Children, the elderly, and people with heart or breathing problems should avoid physical exertion outdoors. If you have heart or breathing problems, follow your doctor's usual advice about managing your condition.	Everyone should consider reducing physical exertion outdoors or rescheduling strenuous activities to times when the index is lower, especially if they experience symptoms.

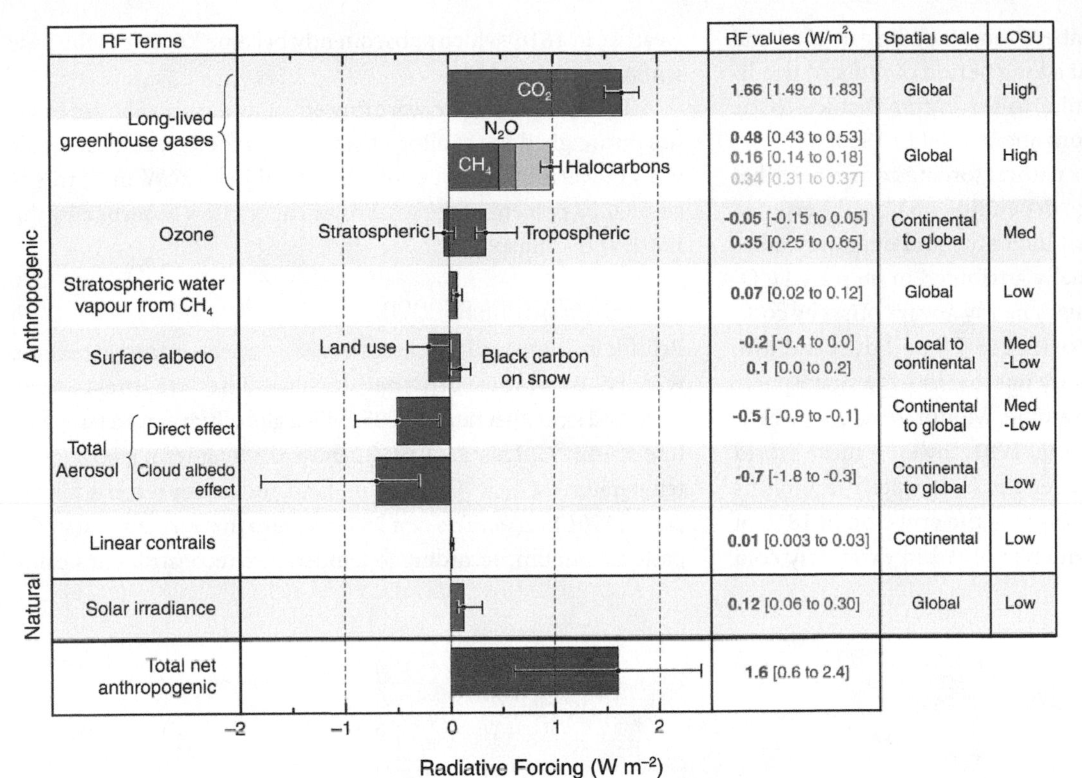

4.29 FACTORS AFFECTING GLOBAL WARMING AND COOLING Greenhouse gases act primarily to enhance global warming, while aerosols, cloud changes, and land-cover alterations caused by human activity act to retard global warming. Natural factors may be either positive or negative.

The effects of anthropogenic factors that can potentially influence global temperatures are shown in Figure 4.29. The three upper bars include the principal greenhouse gases. Although CO_2 is the largest, CH_4, N_2O, the halocarbons, and tropospheric O_3 together potentially contribute a similar amount of warming. Taken as a whole, the total enhanced energy flow to the surface produced by greenhouse gases is about 2.5 W m^{-2}, which is equivalent to 1.25 percent of the solar energy absorbed by the Earth and atmosphere.

Increases in tropospheric aerosols and changes in cloud and surface albedos are primarily a result of human activity. Tropospheric aerosols are produced mainly by fossil fuel burning and are considered a potent form of air pollution. They include sulphate particles, fine soot, and organic compounds. Aerosols act to scatter solar radiation back to space, thus reducing the flow of solar energy available to warm the surface. In addition, they enhance the formation of low, bright clouds that also reflect solar radiation. These, along with other changes in cloud cover caused directly or indirectly by human activity, lead to a significant cooling effect. Land-cover alteration, which can also induce cooling, includes conversion of forested lands to cropland and pastures, which are brighter and reflect more solar energy.

Solar irradiance varies naturally by about 0.12 W m^{-2} and can cause very slight warming and cooling trends. Changes in solar irradiance are principally linked to sunspot activity. In 2008, solar output was at the lowest level in the period since satellite measurements began in the late 1970s (Figure 4.30). The time since the previous solar minimum in 1996 was two years longer than had occurred in other recent cycles. This, in conjunction

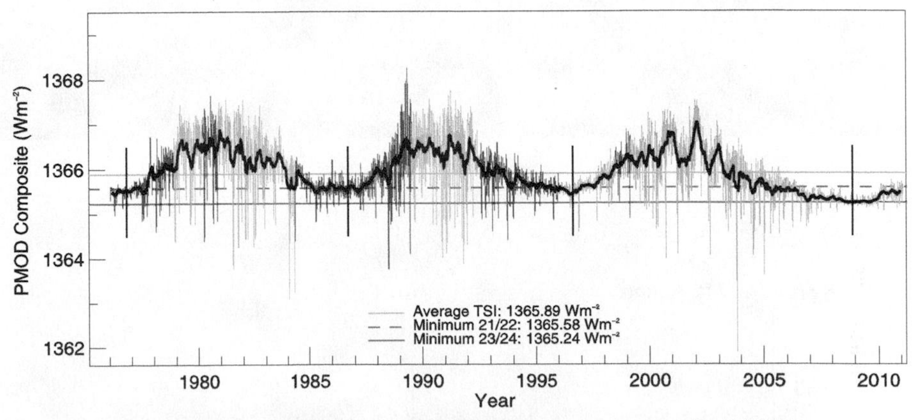

4.30 VARIATIONS IN SOLAR IRRADIANCE FOR THE PERIOD 1976 TO MARCH 2009 (C. Fröhlich, "Evidence of a long-term trend in total solar irradiance," *Astronomy and Astrophysics 501*, L27-L30, 2009, Fig. 1. Reproduced with permission © ESO, and courtesy of Claus Fröhlich, Davos.)

with the fact that sunspot numbers remained low in 2008 and 2009, has led to speculation that a long period of reduced irradiance could be developing, similar to the events that led to the "Little Ice Age" which lasted from about 1250 to 1850.

Volcanic activity is another natural forcing component that can warm or cool Earth's surface depending on how the erupted gases and aerosols interfere with incoming and outgoing radiation. Any warming effect is usually attributed to increased CO_2 in the atmosphere. However, this is nearly always offset by cooling effects caused by sulphur oxides (SO_x) and dust and ash, particularly when these aerosols are injected into the stratosphere during major eruptions. For example, Mount Pinatubo, which erupted in the Philippines in June, 1991, lowered mean world temperatures by about 0.6 °C over the subsequent 15 months (Figure 4.31). Historically important is the eruption in 1815 of Mount Tambora, Indonesia, which resulted in extremely cold weather in 1816, which subsequently became known as the *year without a summer.*

Taken together, the warming effect of the greenhouse gases has outweighed the cooling effect of other factors, and the result is a net warming effect equivalent to about 1.6 W m^{-2}; this is about 0.23 percent of the total solar energy flow absorbed by the Earth and atmosphere.

THE TEMPERATURE RECORD

Reliable instrumented temperature measurements started to become widely available in the 1880s. The warmest year on record since that time is 2005, with a globally averaged temperature of 14.5 °C. This was 0.61 °C above the long-term global mean temperature of 13.9 °C, based on land and ocean records for the period 1901 to 2000. The year 2005 was also the warmest year of the past millennium, according to temperature reconstructions using

(a)

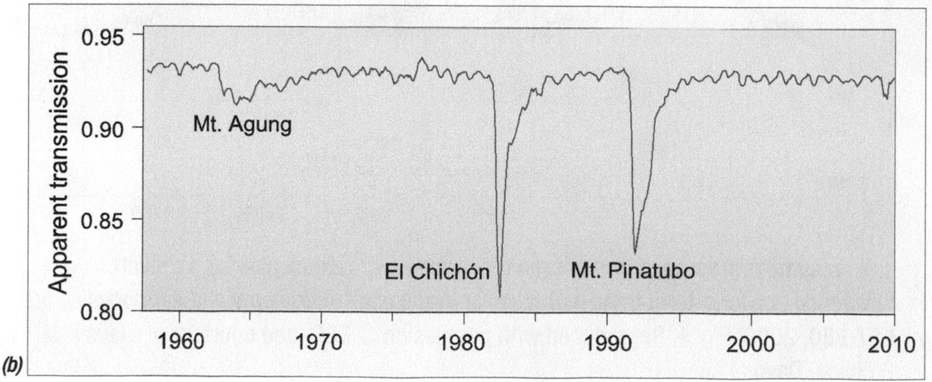

(b)

4.31 ERUPTION OF MOUNT PINATUBO, PHILIPPINES, JUNE 1991 (a) Volcanic eruptions of this magnitude can inject particles and gases into the stratosphere, influencing climate for several years afterward. ***Fly By: 15° 08′ N; 120° 21′ E*** (b) Net solar radiation at Mauna Loa Observatory, relative to 1958, showing the effects of wind transported aerosols associated with major volcanic eruptions.

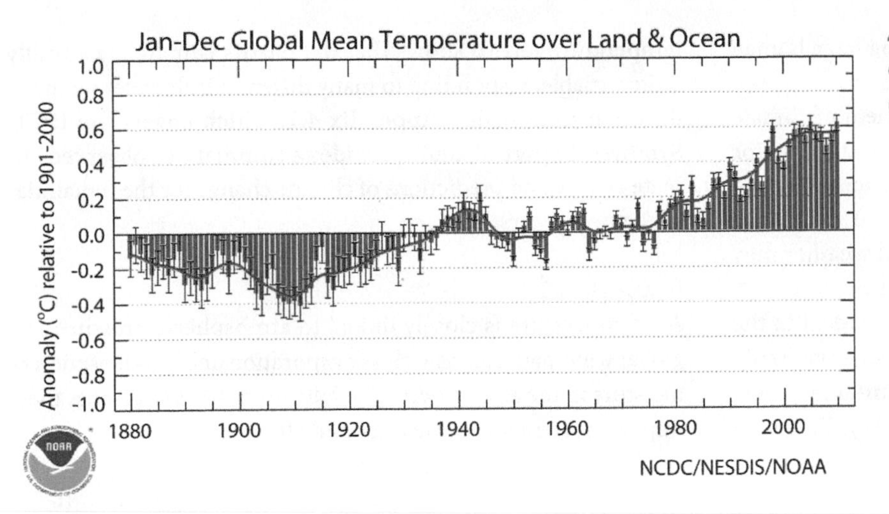

4.32 Annual global mean temperature anomalies over land and ocean for the period 1880 to 2009 compared with the mean value for 1901 to 2000.

tree rings and glacial ice cores. The year 2009 is tied with 1998 as the second warmest year on record, with 2010 ranked third. Thus, 2001 through 2010 all rank among the ten warmest years since 1880, and extend an unbroken period of warmer-than-average global temperatures that extends back to 1977 (Figure 4.32).

Although direct air temperature measurement only began in the middle of the nineteenth century, the record can be extended further back by using **dendrochronology** (tree-ring analysis), a method based on the annual growth rings that develop in trees. In climates where the seasonal conditions are quite different, annual rings are usually clearly defined (Figure 4.33). If growing conditions are good, the annual rings are wide; if poor, they are narrow.

4.33 Tree rings in a cross-section taken from a 50-year-old Douglas fir (*Pseudotsuga menziesii*).

For trees along the timberline in North America, the width is related to temperature—the trees grow better when temperatures are warmer. However, in some locations annual growth is primarily regulated by moisture, and so growth will decrease in drought years. Therefore, it is important to understand the factors that are controlling annual growth patterns at different locations. Since only one ring is usually formed each year, the date of each ring is easy to determine by counting backward from the present. Tree species with long lifespans can be used to develop temperature records that extend back many centuries. In the southwest United States, for example, living and dead bristlecone pines (*Pinus longaeva*) have been used to develop a chronology that extends back 8,500 years.

Careful measurement of the width of tree rings can be used to assess long-term temperature patterns. In 1995, the IPCC concluded that human activity has caused climatic warming. This judgement was based largely on computer simulations of global climate that accounted for the release of CO_2 and SO_2 from fossil fuel burning since the turn of the century. The simulations agreed with the patterns of warming observed over that period, and suggested that, despite natural variability, human influence has been felt in the climate record of the twentieth century. Reports issued in 2001 and 2007 cited increased confidence in this assesment.

Using various projections of the continuing release of greenhouse gases, coupled with computer climate models, the latest IPCC "best estimate" projection is that global temperatures will warm between 1.8 °C and 4.0 °C by 2090–2099 compared to 1980–1999 conditions. This warming trend will be accompanied by other environmental changes. Of particular concern is a rise in sea level, as glaciers and sea ice melt in response to the warming. The predicted rise in sea level is between 18 and 59 cm by the decade 2090–2099. This would place as many as 92 million people at risk from annual flooding. Climate change could also promote the spread of insect-borne diseases, such as malaria. Similarly, a shift in climatic boundaries making some regions wetter and others drier could change agricultural patterns, displacing large human populations as well as affecting natural ecosystems.

A recently discovered effect is the enhancement of climate variability. Events such as very high 24-hour precipitation—for example extreme snowstorms, rainstorms, and ice storms—appear to be occurring more frequently since 1980. More frequent and more intense spells of hot and cold weather may also be indicative of climate change.

Still longer cycles of temperature change are evident in the geologic record, the most dramatic in recent history being the advance and retreat of continental and mountain glaciers during the Late-Cenozoic Ice Age (see Chapter 19). Thus, the temperature record shows that the Earth's climate is naturally quite variable, responding to many different influences on many different time scales. Appendix 4.1, which covers the IPCC Synthesis Report of 2007, provides a summary of observed climate change and predictions of climate change for the remainder of the twenty-first century.

A Look Ahead

Air temperature is closely linked to atmospheric pressure and global wind patterns, as well as evaporation and condensation of moisture in the atmosphere. The following chapters discuss these important elements of weather and climate.

CHAPTER SUMMARY

- Air temperature—measured at 1.25 m above the surface—is influenced by insolation, latitude, cloud cover, surface type, location, elevation, slope, and aspect. The energy balance of the ground surface is determined by net radiation, conduction to the soil, latent heat transfer, and convection to and from the atmosphere.
- Air temperature is measured using a *thermometer* or *thermistor*. Weather stations take daily minimum and maximum temperature measurements. Some measure temperature hourly.
- The two major cycles of air temperature—daily and annual—are controlled by the cycles of insolation produced by the rotation and revolution of the Earth. These cycles induce cycles of net radiation at the surface. When net radiation in the daily cycle is positive, air temperatures increase; when it is negative, air temperatures decrease. This principle also operates through the seasonal cycle and produces annual temperature variations.
- Temperatures of air and soil at or very close to the ground surface are more variable than air temperature measured at the standard height of 1.25 m.
- Surface characteristics affect temperatures. Rural surfaces are generally moist and slow to heat and cool, while urban surfaces are dry and both absorb and release heat readily. This difference creates an urban heat island effect.
- Air temperature normally falls with altitude in the troposphere. The average value of this decrease is the normal temperature lapse rate of 6.4 °C 1,000 m^{-1}. The temperature decrease measured on a specific day may differ from the average rate, and is known as the environmental temperature lapse rate. At the *tropopause*, the altitudinal decrease in temperature ceases and the trend then reverses. In the overlying stratosphere, temperatures increase slightly with altitude, mainly because of absorption of solar energy by ozone.
- Air temperatures observed at mountain locations decrease with elevation, but day–night temperature differences increase with elevation.
- If air temperature in the troposphere increases with altitude, a temperature inversion is present. This can develop on clear nights when the surface loses long-wave radiation to space.
- Annual air temperature cycles are influenced by the annual pattern of net radiation, which depends largely on latitude and cloud cover.
- Maritime or continental location is another important factor influencing temperature. Ocean temperatures vary less than land temperatures because water heats more slowly, absorbs energy throughout a surface layer, and can mix and evaporate freely. Maritime locations that receive oceanic air therefore show smaller ranges of both daily and annual temperature.
- Global temperature patterns for January and July show the effects of latitude, maritime-continental location, and elevation. Equatorial temperatures vary little from season to season. Toward the poles, temperatures decrease with latitude, and continental surfaces in polar regions can become very cold. At higher elevations, temperatures are always colder.
- Isotherms over continents move over a wide latitudinal range with the seasons, while isotherms over oceans move through a much smaller range. The annual range in temperature increases with latitude and is greatest in northern hemisphere continental interiors.
- Earth's global temperature changes from year to year. Within the last few decades, global temperatures have been increasing. CO_2 released by fossil fuel burning is implicated in this warming, as well as the other greenhouse gases, CH_4, CFCs,

O_3, and N_2O. Aerosols scatter sunlight back to space and induce more low clouds, so they tend to lower global temperatures. Solar output and volcanic activity also influence global temperatures.

- Nearly all scientists agree that the human-induced buildup of greenhouse gases has begun to affect global climate. However, natural cycles, such as variations in the sun's output, also provide strong influences. The continued release of large quantities of greenhouse gases at increasing rates is predicted to cause a significant rise in global temperatures that could affect climate zones and raise sea levels.

KEY TERMS

advection
Air Quality Health Index (AQHI)
air temperature
Annual Greenhouse Gas Index (AGGI)
conduction
convection
dendrochronology
environmental temperature lapse rate
evapotranspiration
global warming potential
greenhouse gases
ground inversion
heat island
isolines
isotherm
lapse rate
latent heat transfer
net radiation
normal temperature lapse rate
ozone layer
radiation
radiative forcing
solar path diagram
specific heat capacity
stratosphere
subsidence inversions
temperature inversion
thermal conductivity
transpiration
troposphere
volumetric heat capacity

REVIEW QUESTIONS

1. Identify the important factors that determine air temperature and air temperature cycles.
2. What factors influence the temperature of a surface?
3. How are mean daily air temperature and mean monthly air temperature determined?
4. How does the daily temperature cycle measured within a few centimetres of the surface differ from the air temperature cycle at the standard measurement height?
5. Compare the characteristics of urban and rural surfaces and describe how the differences affect urban and rural air temperatures. Include a discussion of the urban heat island.
6. What are the two lowest layers of the atmosphere, and what is the zone that separates them? How are they distinguished? Name the two upper layers.
7. How and why are the temperature cycles of high mountain stations different from those of lower elevations?
8. Explain how latitude affects the annual cycle of air temperature through net radiation by comparing Manaus, Aswan, Hamburg, and Yakutsk.
9. Why do large water bodies heat and cool more slowly than land masses? What effect does this have on daily and annual temperature cycles for coastal and interior locations?
10. What three factors are most important in explaining the world pattern of isotherms? Explain how and why each factor is important, and what effect it has.
11. Turn to the January and July world temperature maps shown in Figure 4.24. Make seven important observations about the patterns and explain why each occurs.
12. Turn to the world map of annual temperature range in Figure 4.26. What five important observations can you make about the annual temperature range patterns? Explain each.
13. Identify the important greenhouse gases and rank them in terms of their warming effect. What human-influenced factors act to cool global temperature? How?
14. Describe how global air temperatures have changed in the recent past. Identify some factors or processes that influence global air temperatures on this time scale.
15. Consider the surface energy balance during the day. What happens if the surface falls under the shadow of a cloud? How will this affect energy flows to and from the surface? What would you expect to happen to the temperature of the surface?
16. Suppose that on a hot sunny day, a surface layer of soil dries out so that water is no longer available for evaporation. How will this affect energy flows to and from the surface? What will happen to the surface temperature?
17. Air temperature is measured at a standard height (1.25 m) above the surface. Would you expect this air temperature to be warmer or colder than the surface during the day? Why? What about at night? Why?
18. Why has the atmospheric concentration of CO_2 increased in recent years?

VISUALIZATION EXERCISES

1. Sketch graphs showing how insolation, net radiation, and temperature might vary during a 24-hour cycle at a midlatitude station such as Toronto.
2. Sketch a graph of air temperature with height showing a ground inversion. Where and when is such an inversion likely to occur?

ESSAY QUESTIONS

1. Prince Rupert, on British Columbia's north Pacific coast, and Edmonton, Alberta, are at about the same latitude. Sketch the annual temperature cycle you would expect for each location. How do they differ and why? Select one season, summer or winter, and sketch a daily temperature cycle for each location. Again, describe how they differ and why.
2. Many scientists have concluded that human activities are acting to raise global temperatures. What human processes are involved? How do they relate to natural processes? Are global temperatures increasing now? What other effects could be influencing global temperatures? What are the consequences of global warming?

APPENDIX 4.1 THE IPCC REPORT OF 2007

In 1990, the United Nations Intergovernmental Panel on Climate Change (IPCC), a body of scientists nominated by countries throughout the world, issued its First Assessment Report. The report presented information on the scale at which human-induced changes could affect the global climate. It covered the scientific basis for anticipated climate change, the projected impacts of the change, and strategies for mitigating these impacts. This was superseded by reports in 1995 and 2001; the Fourth Assessment Report was released in 2007.

To estimate the degree of climate change throughout the world, IPCC uses several complex global climate models—mathematical models that predict the state of the atmosphere and land and water surfaces at short time intervals over long periods. The models are driven by predicted releases of greenhouse gases under different scenarios of global economic growth and social evolution to the end of the twenty-first century. Because different greenhouse gases are involved, the outcomes are based on CO_2-equivalent emissions, which are calculated by multiplying the emission of a specific gas by its Global Warming Potential (GWP) over the desired forecast period; this is generally up to 2100 (Figure 4A.1).

The accompanying graph of predicted global temperature change as modelled under these scenarios shows there is considerable variation between outcomes (Figure 4A.2). However, it is clear that global temperatures will rise significantly by 2100 based on this analysis.

Some of the panel's more specific findings, taken from *Climate Change 2007: The Scientific Basis, A Report of the Working Group I of the Intergovernmental Panel on Climate Change,* are listed below:

Recent Climate Change

- Global average surface temperature increased by 0.74 °C between 1906 and 2005. The warming trend for the period 1956 to 2005 is nearly twice that for the 100-year period ending in 2005. The temperature increase is widespread over the globe, but greater in the higher northern latitudes. Average temperatures in the Arctic have increased at almost twice the global rate in the past 100 years. Eleven of the 12 years between 1995 and 2006 rank among the 12 warmest years on record (see Figure 4A.3a).
- Satellite measurements show that temperatures in the lower- and mid-troposphere have risen at rates similar to those observed at the surface.
- Land areas have warmed faster than the oceans, even though the oceans have taken up over 80 percent of the heat being added to the climate system. The average temperature of the global oceans has increased to depths of at least 3000 m.

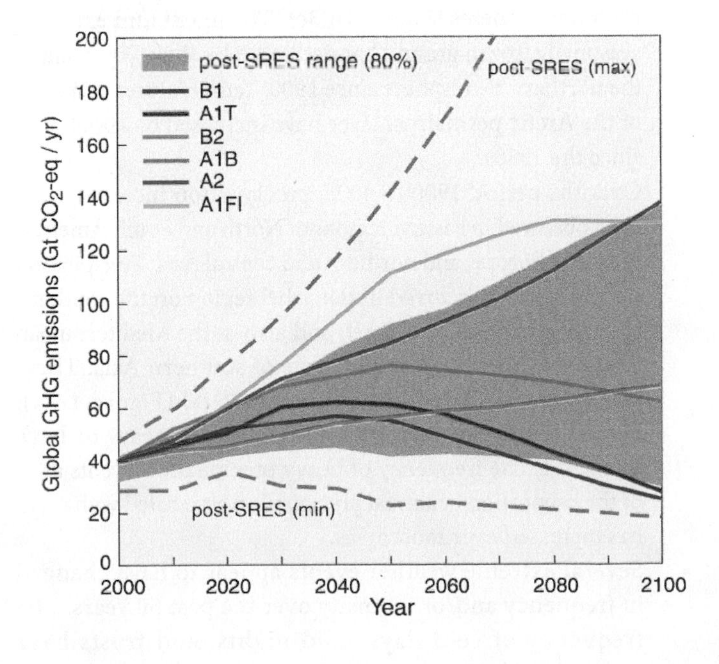

4A.1 GLOBAL GREENHOUSE GAS EMISSIONS (IN Gt CO_2-EQ PER YEAR) IN THE ABSENCE OF ADDITIONAL CLIMATE POLICIES The curves show predicted outcomes for different predictive models under six emissions scenarios. The ***A1*** scenarios describe a world of rapid growth with economic and social convergence among regions, leading to a more uniform world. Scenario ***A1F1*** projects heavy reliance on fossil fuels; ***A1T*** projects a reliance on non-fossil energy sources; and ***A1B*** projects a balance across all sources. The ***A2*** scenario is a more heterogeneous world with less convergence and greater regional isolation. The ***B*** scenarios are similar to ***A1***, but move toward a service and information economy that emphasizes social and environmental sustainability. ***B1*** assumes more global convergence, while ***B2*** assumes more independence among regions. (From IPCC, *Climate Change Report 2007*.)

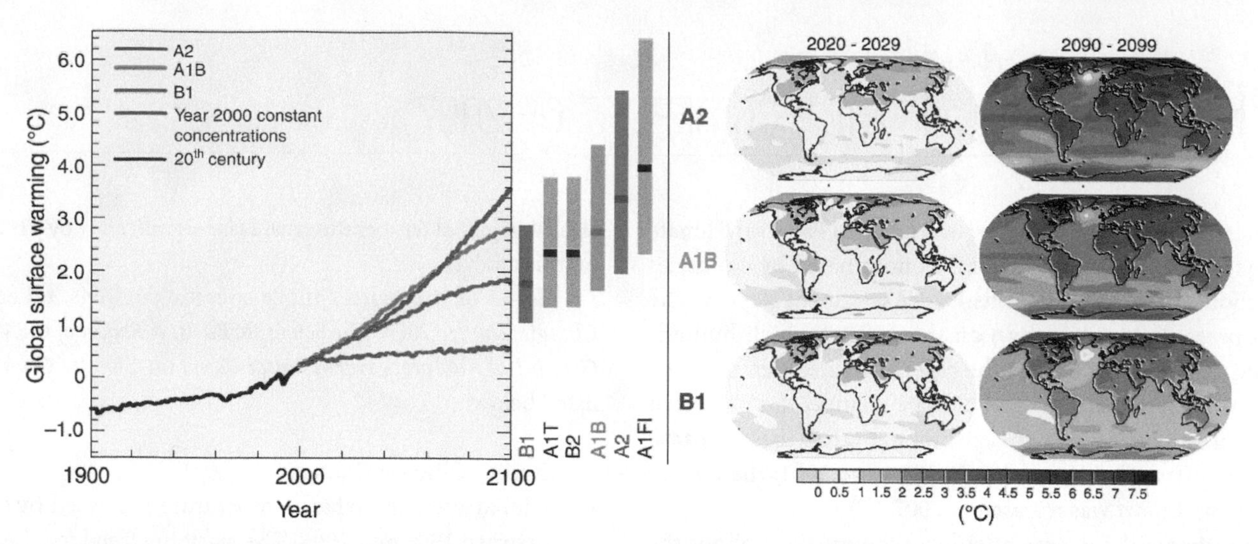

4A.2 ATMOSPHERIC-OCEAN GENERAL CIRCULATION MODEL PROJECTIONS OF SURFACE WARMING Left panel: The coloured lines are multi-model averages of surface warming relative to 1980–1999 for scenarios A2, A1B, and B1, shown as continuation of the corresponding simulations for the twentieth century. The pink line is the condition in which concentrations of greenhouse gases are held at their 2000 values. The vertical bars to the right of the graph indicate the best estimate (solid line within each bar) and likely range in temperature at 2090–2099 relative to 1980–1999 predicted by the six models. Right panel: Projected surface temperature changes for the early and late twenty-first century relative to 1908–1999 for the A2, A1B, and B1 models.

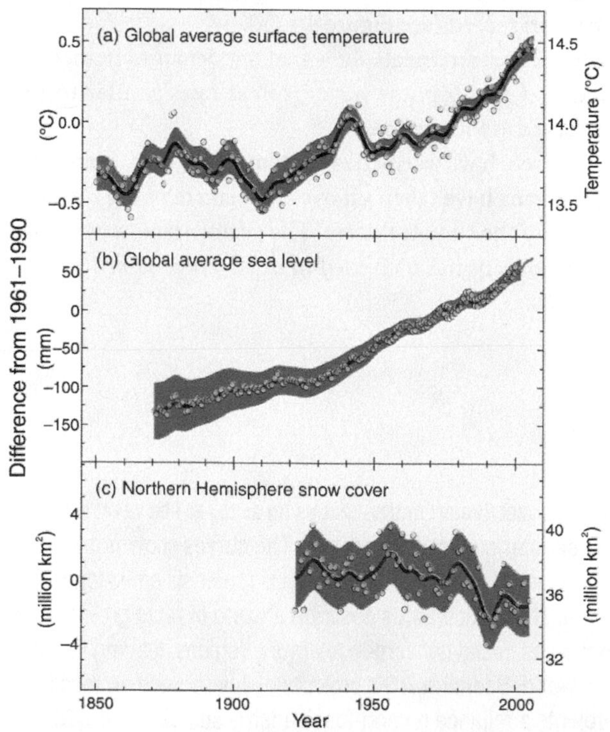

4A.3 OBSERVED ENVIRONMENTAL CHANGES IN THE NORTHERN HEMISPHERE Observed changes in (a) global average surface temperature; (b) global average sea level based on tide gauge and satellite data; and (c) northern hemisphere snow cover for March to April. All differences are compared to averages for the period 1961–1990. Smoothed curves represent decadal averages. The shaded areas are the 10 percent and 90 percent confidence intervals based on known uncertainties in the global datasets.

- Increases in sea level are consistent with warming trends. Global average sea level rose at a mean annual rate of 1.8 mm from 1993 to 2003 (Figure 4A.3b). Since 1993, thermal expansion of the oceans has contributed about 57 percent to this rise, melting of glaciers and ice caps about 28 percent, and losses from the polar ice sheets about 15 percent.
- Satellite data since 1978 indicate that annual average Arctic sea ice extent has shrunk by 2.7 percent per decade, with larger decreases in the summer of 7.4 percent per decade. Mountain glaciers and snow cover on average has declined in both hemispheres (Figure 4A.3c). The maximum extent of seasonally frozen ground has decreased by about 7 percent in the northern hemisphere since 1900. Temperatures at the top of the Arctic permafrost layer have increased by about 3 °C since the 1980s.
- Over the period 1900 to 2005, precipitation increases have been observed in eastern regions of North and South America, northern Europe, and northern and central Asia. Precipitation decreases have occurred in the Sahel region on the southern margin of the Sahara Desert, and also in the Mediterranean lands, southern Africa, and parts of southern Asia. These trends are predicted to continue to 2100 (Figure 4A.4). Globally, the area affected by drought appears to have increased. The frequency of heavy precipitation events and/or the proportion of annual precipitation associated with them has increased over most areas.
- Several extreme weather events appear to have changed in frequency and/or intensity over the past 50 years. The frequency of cold days, cold nights, and frosts have

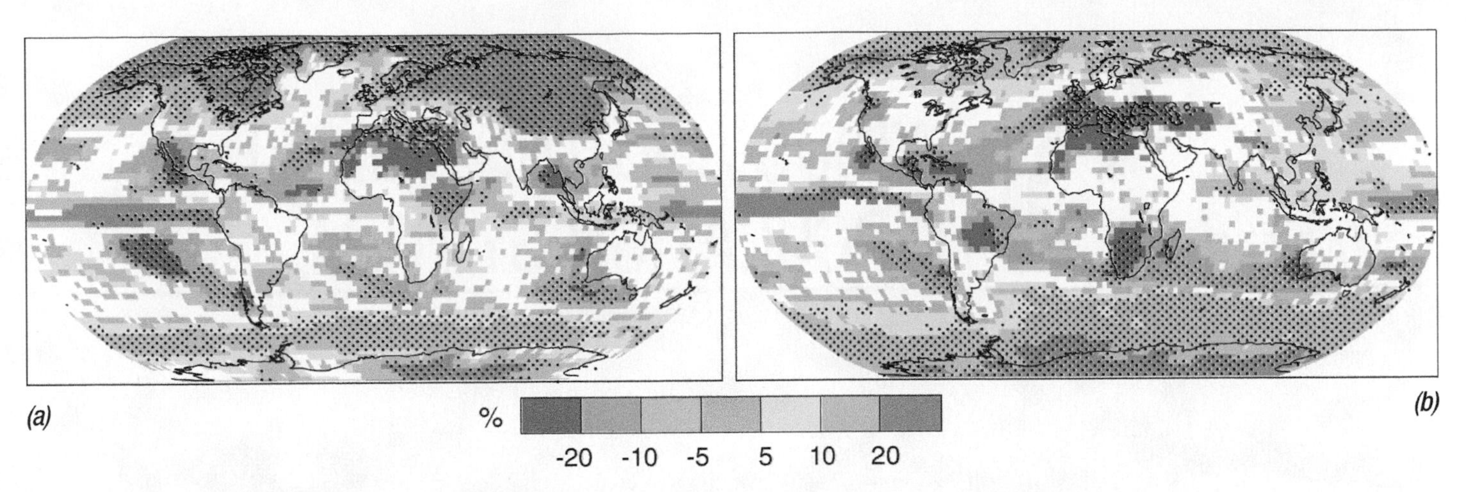

4A.4 PROJECTED CHANGES IN PRECIPITATION FOR THE PERIOD 2090–2099 RELATIVE TO 1980–1999. Values are multi-model averages for (a) December to February and (b) June to August. White areas indicate where agreement between models is less than 66 percent in terms of whether change will be positive or negative. Stippled areas are those where more than 90 percent of the models agree on whether the trend will be positive or negative.

decreased over most land areas, while hot days and hot nights occur more frequently, resulting in more frequent heat waves.

- Intense tropical storm frquency in the North Atlantic appears to have increased since the 1970s.

QUESTIONS

1. What global climate changes have been noted for the last half of the twentieth century in global temperature, snow and ice cover, precipitation, and greenhouse gas concentrations?
2. How is climate predicted to change by the end of the twenty-first century with respect to temperature, precipitation, snow and ice cover, and sea level?

CHAPTER 5

WINDS AND THE GLOBAL CIRCULATION SYSTEM

Modern wind mills at Pincher Creek, Alberta.

Winds are primarily caused by unequal heating of the Earth's atmosphere, which results in differences in pressure and causes air at the surface to move toward the warmer location. For example, sea breezes bring cooler air from the oceans towards the warm land surface during the day. This effect also produces global wind motions, since solar heating of the atmosphere in equatorial and tropical regions is more intense than in mid- and higher latitudes. In response to this difference in heating, global-scale pressure gradients move vast bodies of warm air toward the poles, and cool air is brought toward the equator. The direction of these global wind motions is also influenced by the Earth's rotation.

Another feature of the global circulation system is ocean currents, which are largely driven by surface winds. Ocean currents, like global wind motions, act to move heat energy across parallels of latitude, and are also affected by the Earth's rotation.

ATMOSPHERIC PRESSURE

The Earth is surrounded by a deep layer of air that presses on the solid or liquid surface beneath it. The weight of all the gas molecules above the Earth's surface creates atmospheric pressure. Pressure occurs because air molecules have mass and are constantly being pulled toward the Earth by gravity. The standard unit of pressure is the *pascal* (Pa), which is equal to one newton per square metre (N m^{-2}). One newton is the force required to move an object with a mass of 1 kg so that it accelerates by 1 metre per second (i.e., $1\ N = 1\ kg\ m^{-1}\ s^{-2}$). For meteorological purposes, Environment Canada measures pressure in **kilopascals** (kPa). At sea level, the average pressure of the atmosphere is 101.3 kilopascals. Note that 101.3 kilopascals is equivalent to 1,013 hectopascals (hPa). A hectopascal is therefore the same as a millibar (mb), an older unit of pressure that is still commonly used. Thus, 101.3 kPa equals 1013 hPa or 1013 mb.

MEASURING ATMOSPHERIC PRESSURE

Atmospheric pressure is measured by a **barometer**; because of its high accuracy, the mercury barometer is the standard instrument for weather measurements. It works on the principle that

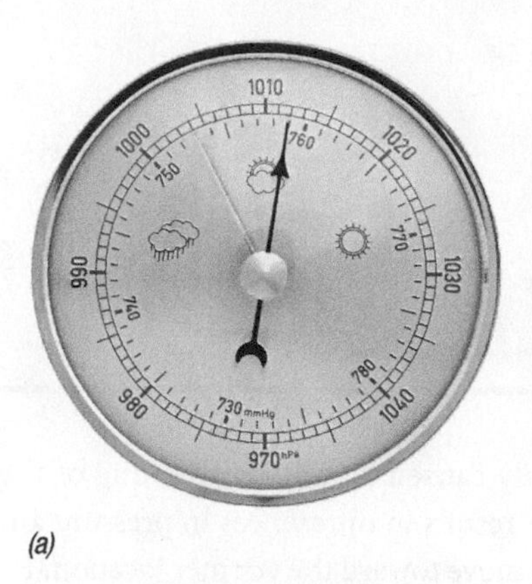

(a)

(b)

(c)

5.1 INSTRUMENTS USED TO MEASURE AIR PRESSURE A common instrument in everyday use is the aneroid barometer (a). The recording barograph (b) provides a graphical trace of pressure fluctuations. Electronic pressure sensors (c) send information directly to a data storage device for processing.

atmospheric pressure forces liquid mercury to rise in a tube from which air has been evacuated.

Another type of barometer is the *aneroid barometer* (Figure 5.1a). It uses a thin-walled, sealed canister from which some air has been removed to establish a partial vacuum. The canister flexes as air pressure changes, and through a mechanical linkage a needle moves across the scale. The aneroid barometer is the most common type of barometer. The same principle is used in a barograph (Figure 5.1b), which records changes in pressure over a period of time by tracing a line on a paper chart. Most weather stations are now equipped with electronic pressure sensors (Figure 5.1c), which comprise a resonator and pressure sensitive diaphragm. Pressure is measured by the change in the resonant frequency of the diaphragm.

Atmospheric pressure at a location varies by only a small proportion from day to day. On a cold, clear winter day, the barometric pressure at sea level might be as high as 103 to 104 kPa, while in the centre of a storm system, pressure might drop to 98 kPa, a difference of about 5 percent.

AIR PRESSURE CHANGES WITH ALTITUDE

Air is compressible, so the mass and density of the atmosphere decrease progressively with altitude above the Earth's surface. As a result, about 95 percent of Earth's atmosphere by mass is found below an altitude of 25 km. About 50 percent of the mass of the atmosphere lies below about 5.5 km altitude, and 99 percent below 30 km. Consequently, atmospheric pressure also decreases with altitude (Figure 5.2). Moving upward from sea level, the decrease in pressure is initially quite rapid, but the rate of change slows progressively with altitude. Aircraft cabins are pressurized for passenger comfort to about 80 kPa, which corresponds to an elevation of about 1,800 m.

Local weather reports usually give station pressure, which depends not only on the condition of the atmosphere, but also on the elevation of the station. Because atmospheric pressure decreases with altitude, for meteorological purposes readings taken at different locations are usually standardized to sea-level so that regional pressure conditions are discernible.

Air pressure is also affected by the decrease in temperature with altitude through the troposphere. The relationship between

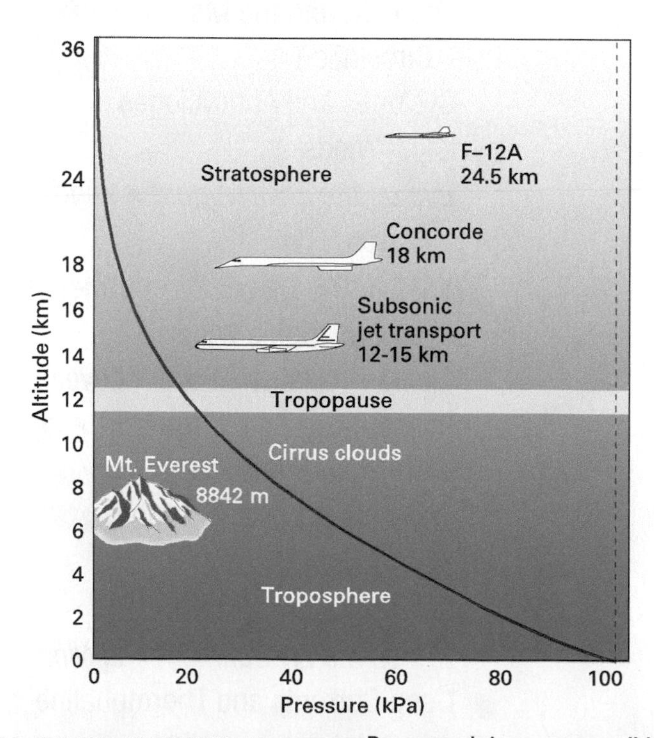

5.2 ATMOSPHERIC PRESSURE AND ALTITUDE Because air is a compressible mixture of gases, atmospheric pressure decreases with altitude above the Earth's surface in response to the reduction in mass of the overlying layers.

air pressure, air temperature, and the density of the air is expressed by the **ideal gas law** as:

$$PV = nRT$$

where P = pressure (in atmospheres; 1 atmosphere = 101.3 kPa)
V = volume (litres)
T = temperature (K)
R = universal gas constant (8.314 J K^{-1} mol^{-1})
n = number of moles (a measure of the amount of substance)

The ideal gas law is derived by combining Boyle's Law, Charles' Law, and Avogadro's Law. Boyle's Law states that the volume of a gas is inversely proportional to pressure when temperature is held constant:

$$P_iV_i = P_fV_f$$

where P_i and P_f = initial and final pressure
V_i and V_f = initial and final volume

Charles' law states that the volume occupied by a gas is proportional to temperature provided pressure is held constant:

$$\frac{V_i}{T_i} = \frac{V_f}{T_f}$$

where V_i and V_f = initial and final volume
T_i and T_f = initial and final temperature

Avogadro's law states the total number of molecules of a gas (n) is proportional to the volume (V) it occupies:

$$V \propto n$$

The relationships expressed in the ideal gas law are important for understanding conditions at higher levels in the troposphere. Meteorological measurements in the troposphere are routinely associated with various **constant pressure surfaces** (Table 5.1). The standard altitudes used by Environment Canada are 850, 700, 500, and 250 hPa. Although these charts resemble surface pressure charts, they differ insofar as the isobars on the surface charts show how pressure changes regionally, whereas the isolines on an upper air chart show regional differences in the altitude of the pressure surface. The unit of measurement for these isolines is decametres (dam)—1 dam equals 10 m—so the isolines are actually contour lines.

The *Working It Out* problems at the end of the chapter explain the physical principles associated with change in atmospheric pressure with height above the surface; similar principles can be applied to the increase in water pressure in the ocean with depth.

TABLE 5.1 APPROXIMATE HEIGHTS AND TEMPERATURES OF STANDARD CONSTANT PRESSURE SURFACES

PRESSURE (hPa)	APPROXIMATE HEIGHT (m)
1,000	100
850	1,500
700	3,000
500	5,000
300	9,000
200	12,000
100	16,000

WIND

Wind is air motion across the Earth's surface and is predominantly horizontal. Air motion that is vertical is known by other terms, such as updrafts or downdrafts. Wind is characterized by direction and velocity. Wind direction can be determined by a *wind vane*. The wind vane faces into the wind, and wind direction is always given as the direction from which the wind is blowing. So, a west wind is one that comes from the west and moves to the east. Wind speed is measured by an *anemometer*. The most common type is the cup anemometer (Figure 5.3). The cups rotate with a speed proportional to that of the wind and are calibrated in metres per second (m s^{-1}). Modern wind vanes and anemometers are normally mounted at a height of 10 m above the surface; they are connected to data storage modules and their measurements fed into computers for analysis.

5.3 ANEMOMETER AND WIND VANE The rotating cup anemometer is a common instrument for measuring wind speed; the wind vane points to the direction from which the wind is blowing. The standard height for wind measurement is 10 m above the surface in an area that is free of obstructions.

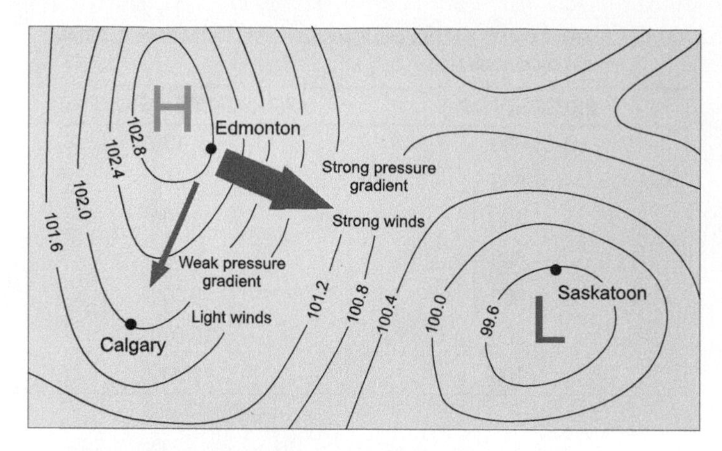

5.4 **ISOBARS AND PRESSURE GRADIENTS** A comparatively strong pressure gradient has developed between the high-pressure region centred over Edmonton and the low-pressure region over Saskatoon. The pressure gradient is weaker between Edmonton and Calgary because of the smaller difference in pressure.

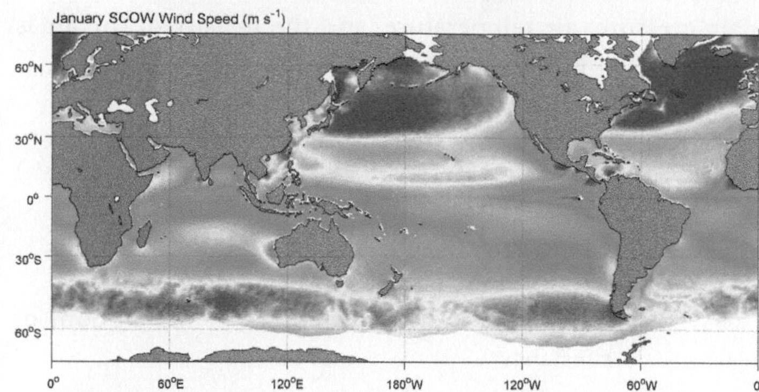

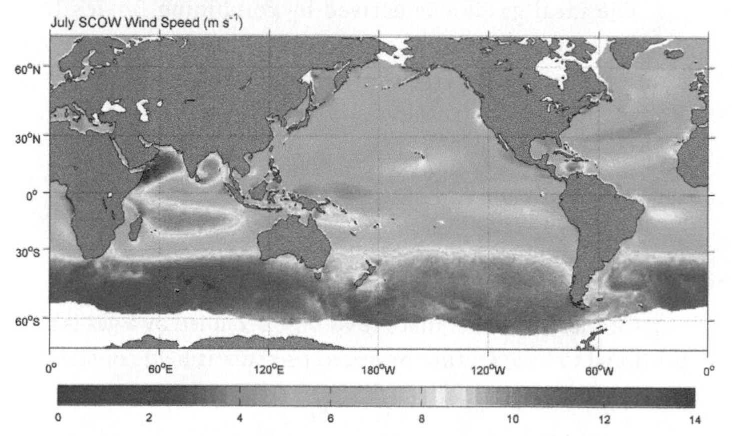

5.5 **GLOBAL WIND SPEEDS** The southern oceans are renowned for their strong winds. The terms roaring forties, furious fifties, and screaming sixties, which are occasionally used to describe the stormy conditions experienced at different latitudes in the southern hemisphere, are a legacy from the days of commercial sailing ships.

WINDS AND PRESSURE GRADIENTS

Wind is caused by differences in atmospheric pressure from place to place with air moving from high to low pressure. This arises from the simple physical principle that any fluid (such as air) subjected to gravity will move until the pressure is uniform. Pressure conditions are shown on a map using isobars—lines connecting places of equal pressure (Figure 5.4). In this example, air pressure at Edmonton is high (H) and the barometer (adjusted to sea-level pressure) reads 102.8 kPa; at Saskatoon, the pressure is low (L), with a barometer reading of 99.6 kPa. The resulting **pressure gradient** produces the *pressure gradient force* that moves air from Edmonton toward Saskatoon. The greater the pressure difference between the two locations, the greater this force will be and the stronger the wind. Thus, the weaker pressure gradient south from Edmonton would be expected to cause less vigorous winds at Calgary where pressure is 102.0 kPa.

Wind speeds are generally higher during the winter when thermal pressure gradients are more intense. In addition, mean wind speeds are higher over the oceans where there is less friction to impede progress (Figure 5.5).

THE CORIOLIS EFFECT AND WINDS

The pressure gradient force moves air from high pressure to low pressure; however, on a global scale, the direction of air motion is also affected by Earth's rotation. This is termed the **Coriolis effect**, and causes air currents to appear to follow gently curving paths as they blow across the Earth's surface. This apparent deflection is to the right in the northern hemisphere and to the left in the southern hemisphere. Deflection is strongest near the poles and decreases to zero at the equator (Figure 5.6). The Coriolis effect does not depend on direction of motion—deflection occurs whether the object is moving toward the north, south, east, or west—and always acts at right angles to the direction of motion. However, the magnitude of the Coriolis effect increases with the speed of the air current. The Coriolis effect also influences the paths followed by ocean currents.

The Coriolis effect is a result of Earth's rotation. Imagine a wind moving north across Alberta (Figure 5.7). It will appear to be deflected to the right because the points of reference on Earth's surface turn counter-clockwise by a distance that increases incrementally with time. In this example, the wind passes over Medicine Hat at time T_1 (position MH_1). By the time it reaches the next point on its path, the Earth has turned to the position indicated by time T_2. This displacement continues as the wind blows northward. The net effect is that the wind appears to follow a curving path. Thus, at Wainright it appears that the wind is blowing from the southwest.

Air in motion near the surface is also subjected to a frictional force due to the roughness of the underlying ground or water. Friction slows the speed of air movement until wind velocity at the surface drops to zero. The frictional force is proportional to

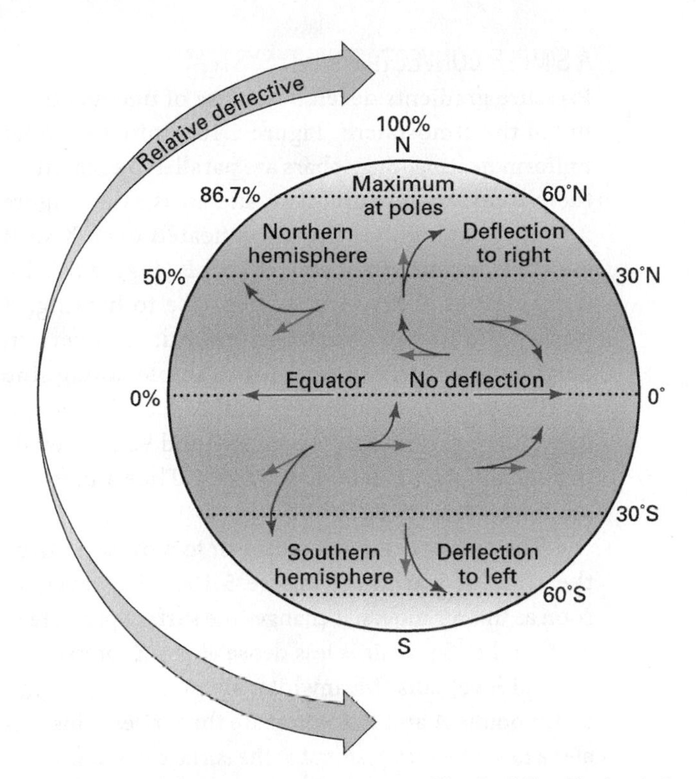

5.6 CORIOLIS EFFECT DIRECTION AND STRENGTH The Coriolis effect acts to deflect the paths of winds or ocean currents to the right in the northern hemisphere and to the left in the southern hemisphere, with respect to their points of origin. Blue arrows show the direction of initial motion, and red arrows show the direction of motion apparent to the observer on the Earth.

the wind speed and always acts opposite to the direction of motion. Because wind speed is reduced by friction, this also reduces the Coriolis effect. The combination of the pressure gradient force, the Coriolis effect, and the force due to friction affects the direction of air motion so that it is toward the low pressure but at an angle to the pressure gradient (Figure 5.8).

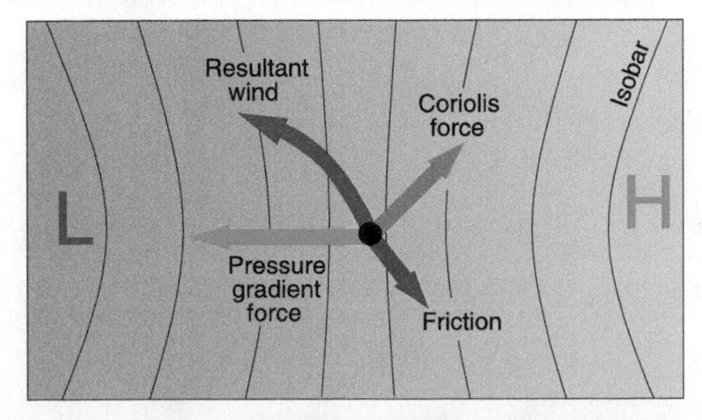

5.8 BALANCE OF FORCES ON A PARCEL OF AIR NEAR THE SURFACE IN THE NORTHERN HEMISPHERE Although the pressure gradient will push a parcel of air toward low pressure, it will be deflected by the Coriolis effect and slowed by friction with the surface. The direction of air motion will be the result of the three forces acting together.

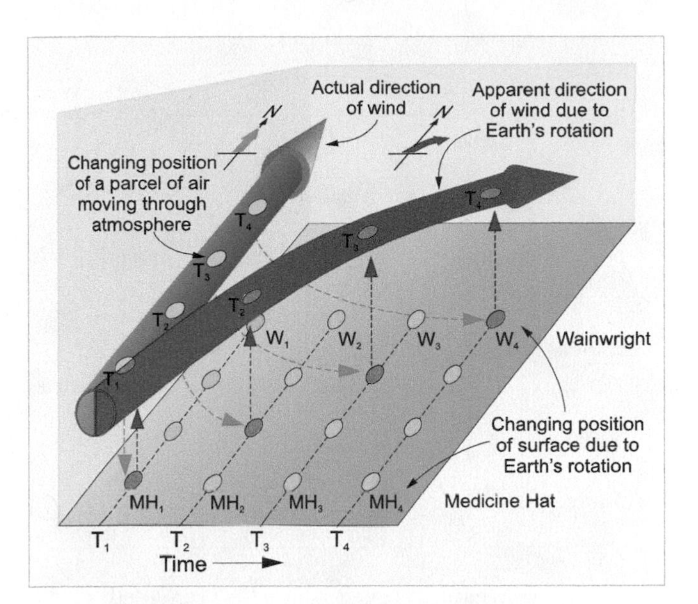

5.7 THE CORIOLIS EFFECT The apparent curving path followed by air moving through the atmosphere arises because of the changing position of reference points on Earth's rotating surface. Here a wind moving north across Alberta appears to be deflected to the right because the points of reference on Earth's surface turn counterclockwise by a distance that increases incrementally with time.

SURFACE WINDS ON AN IDEALIZED EARTH

Figure 5.9 shows the wind and pressure patterns on a hypothetical Earth—one without a complicated arrangement of land and water and without seasonal changes.

Since insolation is strongest when the sun is directly overhead, the surface and atmosphere at the equator will be heated more strongly than other places on this hypothetical planet. A zone of surface low pressure forms what is known as the *equatorial trough*. The heated air rises at the equator, which creates two large-scale convection loops—one in the northern hemisphere and one in the southern hemisphere; these are the **Hadley cells** that form in the northern and southern hemispheres. In each Hadley cell, air rises over the equator and is drawn toward the poles. In addition, the air is turned by the Coriolis effect and so heads westward as well as poleward. The uplifted air descends at about lat. 30°.

Where the Hadley cell circulation descends, surface pressures are high. This produces the **subtropical high-pressure belts**, centred at about lat. 30° in the northern and southern hemispheres. The subtropical high-pressure belts comprise a series of large and stable high-pressure centres. Winds around the subtropical high-pressure centres spiral outward and move toward the equator and also toward the midlatitudes.

The winds moving toward the equator are the strong and dependable *trade winds*. In the northern hemisphere, this is the zone of the *northeast trades*; the equivalent winds in the southern hemisphere are the *southeast trades*. These winds move

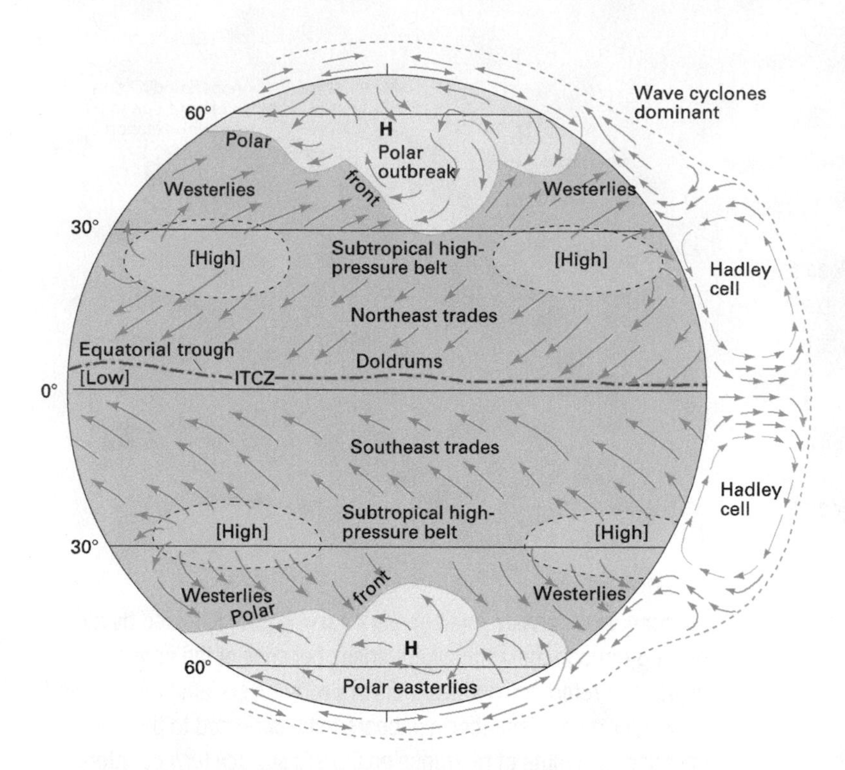

5.9 GLOBAL SURFACE WINDS ON AN IDEALIZED EARTH Global surface winds and pressures are shown on an Earth with no disruption from oceans and continents and no seasonal variation. The cross-section at the right shows general air circulation within the troposphere.

toward the equatorial trough where the air converges and rises into the atmosphere as part of the Hadley cell circulation. This zone of convergence is termed the **intertropical convergence zone (ITCZ)**. Winds in the ITCZ are light and variable because the air is generally rising (the ITCZ is discussed in more detail later in this chapter).

Air spiralling outward from the subtropical high-pressure centres toward the poles produces the zone of vigorous southwesterly winds in the northern hemisphere and northwesterly winds in the southern hemisphere. Between about lat. 30° and 60°, the pressure and wind pattern become more complex. A conflict zone, delineated by the **polar front**, occurs where colder polar air meets the warmer subtropical air. Masses of cool, dry air, termed *polar outbreaks*, move into this zone from the Arctic and Antarctic regions. Many of the weather systems that affect Canada and western Europe originate along the polar front. The resulting low-pressure systems usually bring cloudy, unsettled weather and precipitation as they travel from west to east through these regions (see Chapter 7).

Pressure and winds can be quite variable in the midlatitudes. On average, however, winds are more often from the west, so the region is said to have *prevailing westerlies*. At the poles, the air is intensely cold, which creates centres of high pressure. Winds spiralling outward around these polar high-pressure centres create surface winds generally from an easterly direction, known as the *polar easterlies*.

A SIMPLE CONVECTIVE WIND SYSTEM

Pressure gradients develop because of unequal heating of the atmosphere (Figure 5.10). Initially, under uniform heating, the isobars are parallel to each other, there is no pressure gradient, and air is calm (Figure 5.10a). Imagine that the air is heated over *X* so it becomes warmer than over *A* and *B* (Figure 5.10b). As the air at *X* expands in response to heating, it pushes the isobaric surfaces upward. At a certain height in the atmosphere, for example, along line *A′-B′*, a pressure gradient is induced; at this height over *X*, the pressure is between 97 and 98 kPa, while over *A* and *B*, it is less than 97 kPa. Thus a pressure gradient develops aloft.

The pressure gradient causes air to move away from the zone of warming (Figure 5.10c). However, as soon as this air moves, it changes the surface pressure at *A*, *X*, and *B*. Since air is less dense above *X*, pressure at ground level falls. Meanwhile, air pressure is higher above points *A* and *B*, so pressure there rises. This creates a new pressure gradient at the surface that moves air from *A* and *B* toward *X* (Figure 5.10d). The resulting

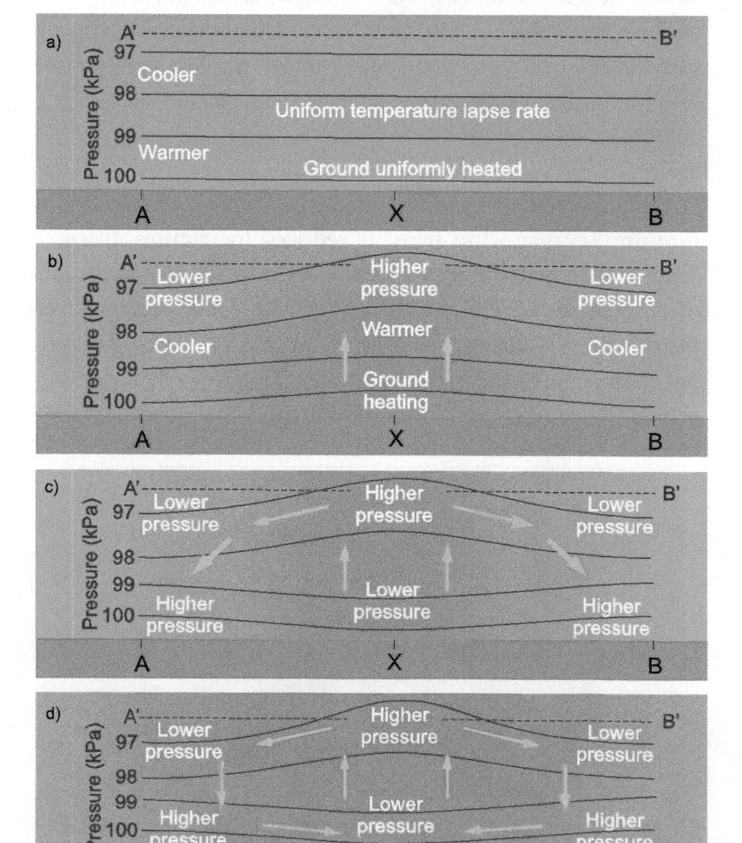

5.10 FORMATION OF A SIMPLE CONVECTION LOOP Heating of the atmosphere over point X creates a pressure gradient that causes a convection loop.

circulation has produced two convection loops (shown by the arrows) in which air at the surface converges at the location of heating (*X*), and diverges at higher altitude above this point.

The principle that heating of the air produces a pressure gradient that causes air to move is important and is associated with the establishment of surface thermal high- and low-pressure zones. In general, low surface pressure is associated with warm air, and high surface pressure with cool air. At the global scale, the equatorial trough is a region of persistent low pressure, whereas polar regions with permanent ice and snow are characterized by high pressure. However, it is important to note that not all surface pressure conditions develop from temperature differences. High- and low-pressure systems can also result from convergence and divergence associated with general global circulation of the atmosphere (see the section *Cyclones and Anticyclones* later in this chapter).

5.11 LAND–OCEAN CONTRASTS BY HEMISPHERE The northern and southern hemispheres are quite different in their distribution of lands and oceans.

GLOBAL WIND AND PRESSURE PATTERNS

In reality, the general circulation of the atmosphere is much more complex because of differences caused by the seasonal passage of the sun across the equator and the corresponding latitudinal shift in the global pressure and wind belts. It is also affected by the different configuration of landmasses and ocean basins in the northern and southern hemispheres (Figure 5.11). The northern hemisphere has two large continental masses separated by oceans and an ocean at the pole. The southern hemisphere is essentially a large ocean with a cold, ice-covered continent at the centre. So whereas the idealized planetary wind system approximates conditions in the southern hemisphere reasonably well, it is less appropriate for the northern hemisphere.

SUBTROPICAL HIGH-PRESSURE BELTS

A prominent feature of the seasonal global pressure patterns are the subtropical high-pressure belts created by the Hadley cell circulation (Figure 5.12). The high-pressure belt in the southern hemisphere conforms well to the hypothetical pattern of Figure 5.9. It has three large high-pressure cells that persist year round over the oceans. A fourth, weaker high-pressure cell forms over Australia in July, as the continent cools during the southern hemisphere winter.

In the northern hemisphere, two large high-pressure cells occur over the subtropical oceans—the Hawaiian High in the Pacific and the Azores High (commonly referred to as the Azores-Bermuda High) in the Atlantic. From January to July, these intensify, move northward, and have a dominant influence on summer weather in North America. Warm, moist air from the Azores High is brought onshore from the Caribbean Sea, producing generally hot, humid weather for the central and eastern United States. On the west coast, dry subsiding air moves out of the continent, bringing fair weather and rainless conditions to places like California. In winter, the two semi-permanent highs weaken and move to the south, allowing colder air from the north and west to penetrate more deeply into the continent.

Whereas the Hadley cells result in transport of air away from the equatorial region, an east-west component to air circulation at low latitudes is provided by the **Walker circulation**. The Walker circulation is best developed over the Pacific Ocean where, under normal conditions, the trade winds bring warm, moist air from South America toward Indonesia. Periodically the trade winds weaken and a series of convective cells develop across the Pacific Ocean, heralding the onset of El Niño events; this is discussed in more detail in *Eye on the Environment* later in this chapter.

THE ITCZ AND THE MONSOON CIRCULATION

Insolation is most intense when the sun is directly overhead. Over the course of the year, the subsolar point migrates north and south of the equator so that it lies at the Tropic of Cancer on the June solstice and at the Tropic of Capricorn on the December solstice. Since Hadley cell circulation is driven by energy input from the sun, the general global pressure and wind belts also shift with the seasons. The seasonal extremes generally occur in January and July because of thermal inertia.

In July, the mean position of the ITCZ lies well north of the equator and reaches its most northerly position over Asia at a latitude of about 35° N, which coincides approximately with the southern edge of the Himalayas (Figure 5.12). The pronounced latitudinal shift over Asia is caused by excessive heating of the landmass in summer. A similar displacement occurs in Africa over the Sahara, but elsewhere the ITCZ is positioned a few degrees north of the equator. In January, the most southerly position of the ITCZ over the continents is about 15° S, but this is quite variable across each landmass. The ITCZ remains north of the equator over the eastern Pacific and across most of the Atlantic Ocean. Maximum seasonal displacement is about 50° of latitude across Asia and the Indian Ocean, compared to less than 5° of latitude in the western Pacific.

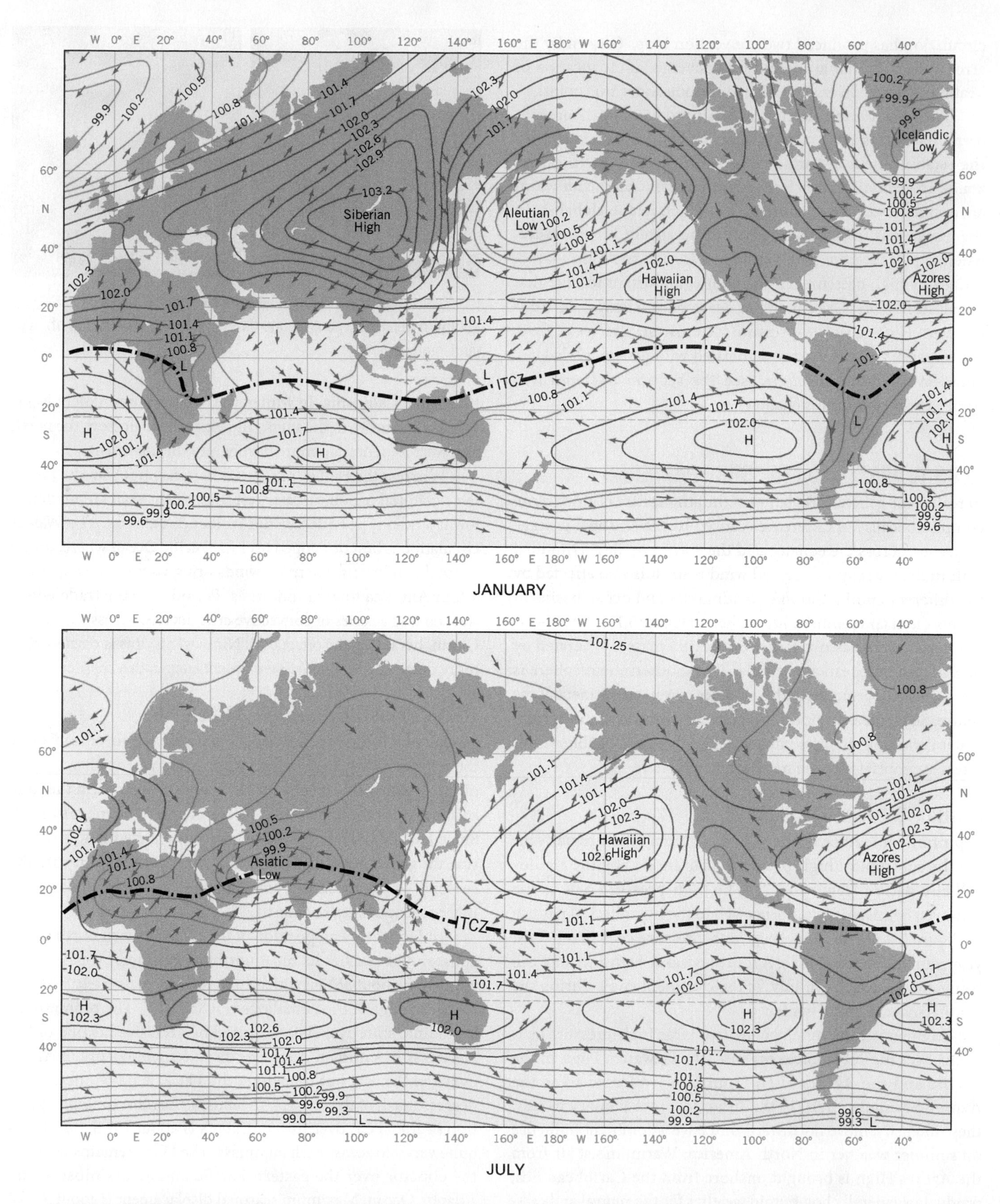

5.12 ATMOSPHERIC PRESSURE MAPS The maps show mean monthly atmospheric pressure and prevailing surface winds for January and July. Pressure (kPa) is reduced to sea level to correct for the elevations of the recording stations. (Data compiled by John E. Oliver.)

In January, an intense high-pressure system, the Siberian High, forms in central Asia due to the extreme cold. In July, the Siberian High is replaced by the Asiatic Low centred over the strongly heated lands of the Middle East. This seasonal change in pressure and the latitudinal movement of the ITCZ create the reversing wind pattern that underlies the **monsoon** climate of Asia (Figure 5.13). The winter monsoon is marked by a strong outflow of dry, continental air from the north across China, Southeast Asia, India, and the Middle East. In the summer, warm, humid air from the Indian Ocean and the southwestern Pacific moves northward and northwestward into Asia, passing over India, Indochina, and China. This airflow is known as the *summer monsoon* and is accompanied by heavy rainfall in southeastern Asia.

An additional element that accentuates the monsoonal airflows in Asia is the diversion of high level air streams north and south of the Himalayas. Intense radiational heating and cooling of the high Tibetan Plateau, at a general elevation of 5,500 to 6,000 m, also strengthens the pressure gradient and the seasonal wind patterns. In addition, the movement of the ITCZ from the southern to northern hemisphere means that the southeast trade winds are subjected to a reversal of the Coriolis effect as they cross the equator. This also contributes to the southwesterly flow of warm moist air from the Indian Ocean, the principal source of moisture for the wet monsoon season (see also Chapter 9).

North America does not have the extreme monsoonal climate experienced in Asia. Even so, in summer, warm, moist air originating in the Gulf of Mexico tends to move northward across the central and eastern part of the United States into Canada. At times, moist air from the Gulf of California also invades the southwestern deserts, causing widespread scattered thundershowers and creating the "Arizona monsoon." In winter, the airflow pattern across North America changes, and dry, continental air moves out from the Arctic, bringing occasional frosts as far south as Florida.

The north–south movement of air across North America is accentuated by the continent's topography. The central lowlands present few barriers to meridional (north–south) airflow. Conversely, the high mountains in the west tend to block the flow of air from the Pacific Ocean. Generally, air from the Pacific is modified by passing over the mountains, where much of its moisture is lost. Similarly, the Appalachian Mountains limit airflow from the Atlantic Ocean, although their effect is much less than the western mountains. However, the Appalachian Mountains do influence the passage of weather systems moving across the continent, deflecting them to the northeast through the Gulf of St. Lawrence (Figure 5.14); many of the heavy rainstorms and snowfalls that affect eastern Canada can be attributed to this.

Over Canada, a high-pressure centre develops in winter favouring the southward flow of cold air. Two centres of low pressure—the Icelandic Low and the Aleutian Low—deepen over the adjacent oceans; these are regions where winter storm systems are formed. In summer, the pressure gradient weakens. The Icelandic Low and the Aleutian Low are displaced by the poleward expansion of the subtropical high-pressure cells; pressure over Canada is lowered as the snow cover melts, the days get longer, insolation increases, and the ground warms.

CYCLONES AND ANTICYCLONES

Low- and high-pressure centres are familiar features on daily weather maps. Unlike the semi-permanent highs and lows displayed on the January and July world maps (Figure 5.12), the daily maps show the pressure systems that move through a region and bring about changing weather conditions. Low-pressure centres, or **cyclones**,[1] are often

[1] Note: *cyclone* is also the name given to intense, low-pressure tropical storms (hurricanes) in Australia, Japan, and the eastern Pacific. The term **depression** is often preferred when describing a low-pressure centre, particularly those which bring cloudy conditions and precipitation in midlatitudes (see Chapter 7).

102.0
102.3
102.6
102.8
103.2
H
102.0
101.7
101.4
101.1
60°
45°
30°
15°
N
0°
JANUARY

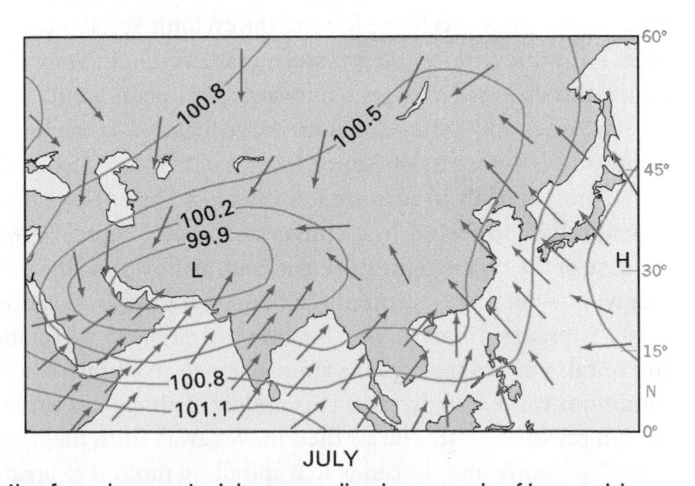

5.13 MONSOON WIND PATTERNS The Asiatic monsoon winds alternate in direction from January to July, responding to reversals of barometric pressure (kPa) over the large continent.

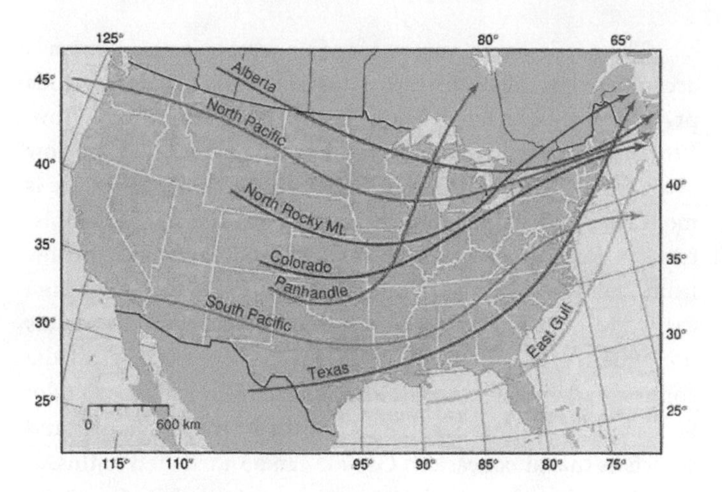

5.14 **DEPRESSION TRACKS IN NORTH AMERICA** Midlatitude low-pressure weather systems move predominantly from west to east and are directed by upper atmosphere circulation patterns, as well as surface topography. They are usually distinguished by region of origin. For example, Alberta Clippers bring unusually cold weather, strong winds, and snow to eastern Canada and the United States. Other examples include the Colorado Lows and Panhandle Hooks, which originate in the foothills of Colorado and in the panhandle regions of northern Texas and western Oklahoma.

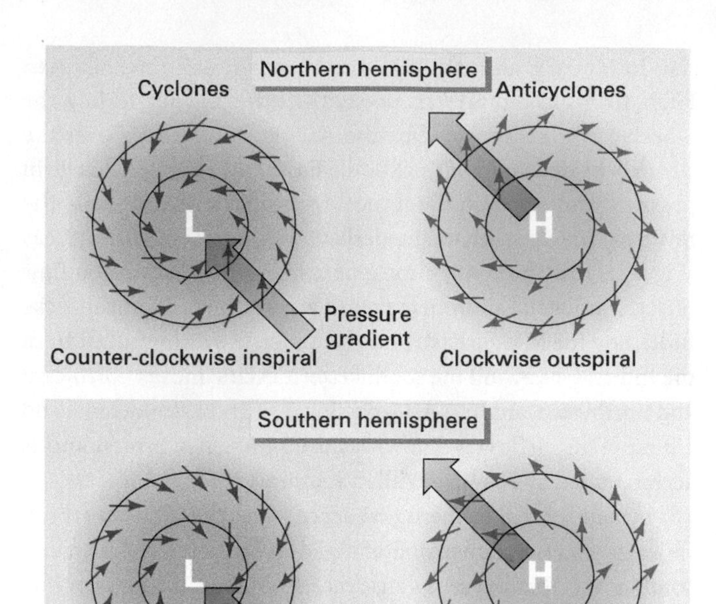

5.15 **AIR MOTION IN CYCLONES AND ANTICYCLONES** Surface winds spiral inward toward the centre of a cyclone but outward and away from the centre of an anticyclone. Because the Coriolis effect deflects moving air to the right in the northern hemisphere and to the left in the southern hemisphere, the imposed curvature reverses from one hemisphere to the other.

associated with cloudy or rainy weather, and high-pressure centres, or **anticyclones**, usually bring fair weather.

These low- and high-pressure centres develop as air converges and diverges in response to regional pressure gradients (Figure 5.15). In a low-pressure system the pressure gradient is inward. The Coriolis effect and friction with the surface cause the surface air to move at an angle across the pressure gradient. This creates a convergent, inward spiralling motion. In the northern hemisphere, the cyclonic spiral is counter-clockwise because the Coriolis effect acts to the right. In the southern hemisphere, the Coriolis effect acts to the left and the cyclonic spiral is clockwise. For anticyclones, the pressure gradient is out from the centre; this creates a divergent, outward spiralling motion.

Cyclones and anticyclones are large features of the lower troposphere—often 1,000 km or more in diameter. The horizontal motion of air in adjacent low- and high-pressure centres is generally connected in a convection loop (Figure 5.16). Because of the regional pressure gradient, airflow near the surface will be out from a centre of high pressure toward a centre of low pressure. Near-surface airflow converges in a cyclone and spirals upward through the atmosphere to an altitude where conditions cause it to diverge. This upflow of air creates a ridge of high pressure aloft. The air then moves away from the ridge of high pressure and descends in a spiralling motion to create the surface anticyclone.

The interconnection between the rising and descending currents by horizontal air flow from the anticyclones to adjacent cyclones near the surface and by diverging air flow from the high pressure ridges aloft is important. Through this dynamic convective loop the development of surface pressure systems is closely

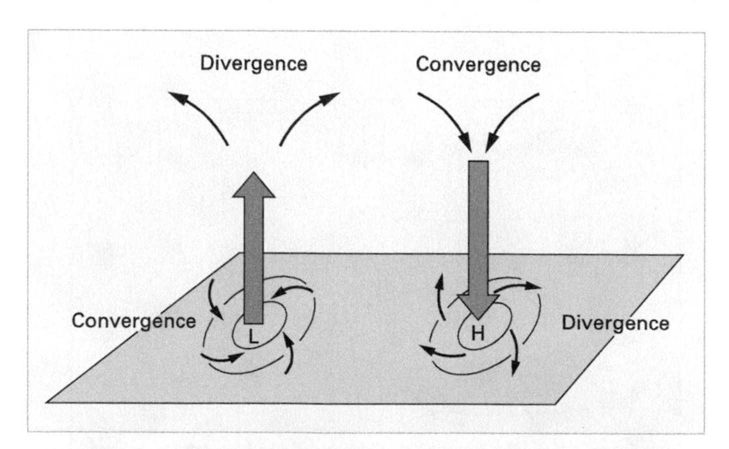

5.16 **DYNAMIC CONVECTION LOOP** A cyclone and anticyclone linked together in a convection loop by horizontal inflow and outflow of air and rising and descending currents. The convective loop is not perfect because air escapes or is drawn into the system as it circulates.

linked to conditions in the upper troposphere, especially the position of the polar front jet stream (see Chapter 7).

LOCAL WINDS

Although air movements at the global scale are important in determining the prevailing winds in a given area, local weather conditions may be influenced by winds that develop in response to small-scale or short-term pressure gradients. Local winds can be divided into two groups. The first includes those that develop only in one area because a local topographic feature or body of water affects air movements. The second includes winds that an area's inhabitants consider to be distinctive, despite the fact that they may originate many hundreds of kilometres away. Sea and land breezes are examples of *local wind* systems that affect relatively small areas and are brought about by localized differences in atmospheric heating and cooling. Mountain and valley winds, drainage winds, and chinooks are also usually classed as local winds because their effects are relatively confined.

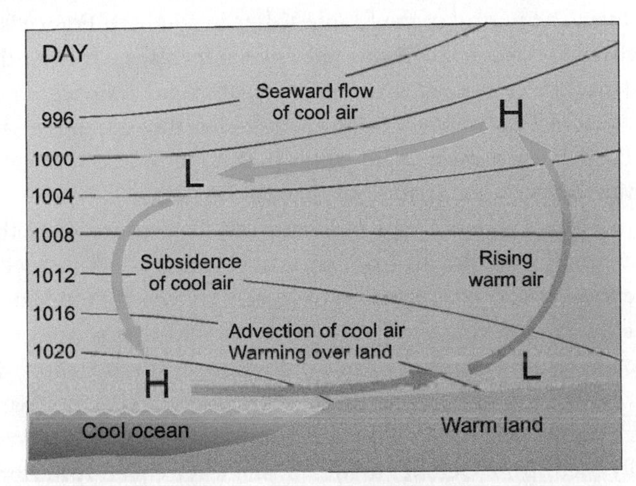

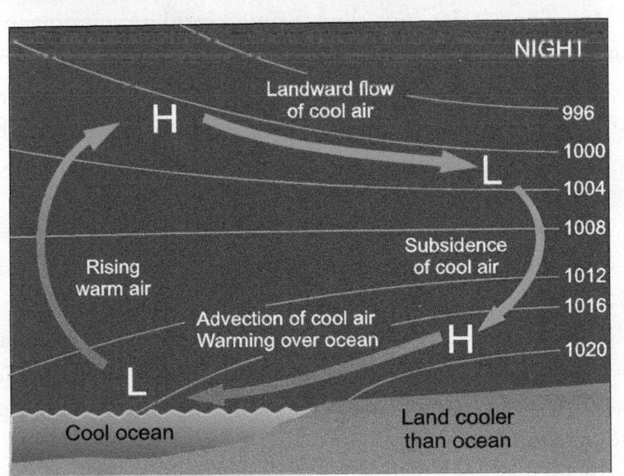

5.17 SEA AND LAND BREEZES The contrast between the land surface, which can heat and cool rapidly each day, and the ocean surface, which has a more uniform temperature, induces pressure gradients to create sea and land breezes.

Sea and *land breezes* are convection loops that develop from localized heating and cooling along shorelines. Because land surfaces heat and cool more rapidly than water, temperature contrasts frequently develop along coastlines. During the day, convective uplift begins when air over the land is warmed by surface heating (Figure 5.17). The *sea breeze*, which usually begins to blow onshore in the late morning, brings cool marine air toward land, and a return flow higher in the atmosphere carries air toward the ocean. The sea breeze can provide a pleasant cooling effect on a hot summer afternoon. At night, in response to radiation cooling over land, the convection loop reverses, and a land breeze develops that blows out to sea. Air flow associated with land and sea breezes is generally about 5 to 10 m s^{-1}. The effect is limited to the lowest 1,000 m or so of the atmosphere and to a relatively narrow zone some 15 to 20 km wide, straddling the shoreline.

This effect is not restricted to ocean beaches, but can be detected around large lakes, such as the Great Lakes. The narrow strip of land that separates the largest of the Great Lakes—Superior, Michigan, and Huron—experiences another effect from the onshore air flow. In summer, the warm air can pick up moisture from the lakes, and a little way inland where the ground is warmer, it rises in gentle convection currents. This can sometimes produce afternoon clouds and light rain (Figure 5.18).

Mountain and *valley winds* change direction in response to the diurnal temperature cycle in much the same way as land and sea breezes. During the morning, as valley slopes begin to heat up, a convection loop forms across the valley (Figure 5.19a). The air rises up the valley sides, spreads across the valley, and sinks back to the surface near the valley's centre. The descending air is warmed by compression, which can help disperse mists and fogs

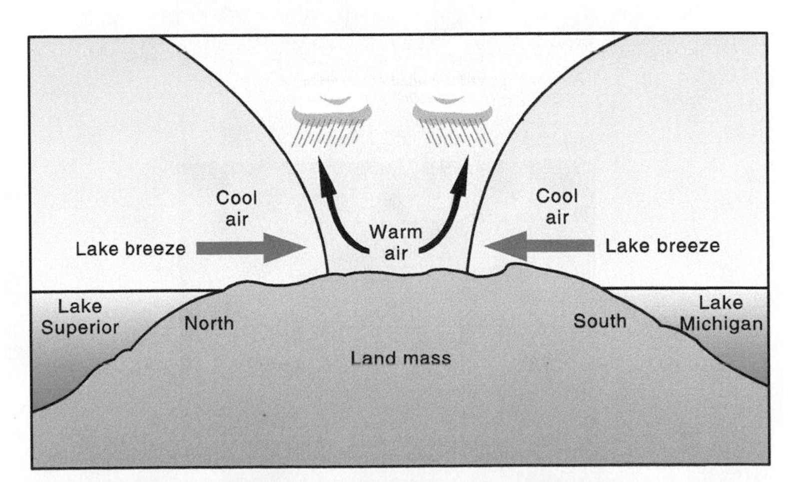

5.18 DEVELOPMENT OF RAIN SHOWERS IN THE GREAT LAKES REGION Convergence of moist air brought onshore by lake breezes rises, cools, and condenses to form rain showers further inland.

that might have formed in low-lying areas during the night (Figure 5.20).

Later in the morning, as the valley floor gets warmer, a longitudinal flow component develops that moves air slowly along the valley toward its head. This is the valley wind. In the evening, the higher slopes begin to cool by radiation, which, in turn, chills the air. This reverses the convection loop (Figure 5.19b) and the cool, dense air begins to move down into the valley as the mountain wind. Air movement is relatively gentle, and small undulations, even bands of forest, can temporarily dam the flow. This may induce a pulsing flow in mountain winds as they move down the slope. During the night, the coldest air will settle at the lowest elevations, and by morning a temperature inversion will often develop. The gentle air currents up and down slopes that are associated with mountain and valley winds help insects move into alpine meadows during the day, where they feed and pollinate the plants before returning to more hospitable sites at night.

Drainage, or *katabatic winds* are similar to mountain winds, except that the downslope flow of cold air is accelerated by gravity. This is because it is much denser than the air it encounters as it flows away from its source region. Drainage winds usually develop over high plateaus and snowfields; this accentuates cooling and creates reservoirs of cold, dense air. The cold air eventually spills over low divides or through mountain passes, down into adjacent lowlands as a strong, cold, dry wind. Drainage winds occur in many mountain regions and go by various local names. The *mistral* of the Rhône Valley in southern France is a well-known example. In extreme environments, such as on the ice sheets of Greenland and Antarctica, powerful drainage winds move across the ice surface and are funnelled through the coastal valleys. These strong, frigid winds pick up loose snow, causing severe blizzard-like conditions that last for days.

5.20 VALLEY MIST Mists and fog that form as cold air settles in a valley floor begin to dissipate within a few hours of sunrise. Clearing usually begins in the centre of the valley where local convection currents descend and the air is warmed adiabatically.

A *Santa Ana* is a type of local wind that occurs when the outward flow of dry air from an anticyclone is combined with the local effects of mountainous terrain. In the United States, the Santa Ana is the hot, dry easterly wind that sometimes blows from the interior desert region of southern California across the coastal mountain ranges to reach the Pacific coast; each event usually lasts for several days. Santa Ana winds typically occur from October to March, and less frequently in April or May. The wind is funnelled through local mountain gaps or canyon floors where it gains great force, with speeds frequently exceeding 80 kilometres per hour (kph). At times the Santa Ana wind carries large amounts of dust. Because this wind is hot, dry, and strong, it can fan wildfires in brush or forest so that they burn out of control (Figure 5.21).

A *Bora* is a cold, dry wind that brings unseasonable conditions to the coastal regions of the Adriatic Sea. The outflow of air is associated with a high-pressure area positioned over the cold, snow-covered mountains of the interior plateau of Croatia. Speeds in excess of 200 kph have been recorded during some of these events. Most boras last for only one day, but some may persist for four or six days.

A *Chinook* is a warm wind that develops under certain conditions when air currents pass over the Rocky Mountains and descend into the eastern foothills and interior plains (Figure 5.22). The descending air is warmed and dried by compression, when subjected to increasing pressure at lower altitudes through the process of adiabatic warming (see

5.19 MOUNTAIN AND VALLEY WINDS (a) During the day, air rises over the heated valley slopes and will flow towards the head of the valley. (b) At night, as the air cools, the circulation reverses and a down-valley flow begins.

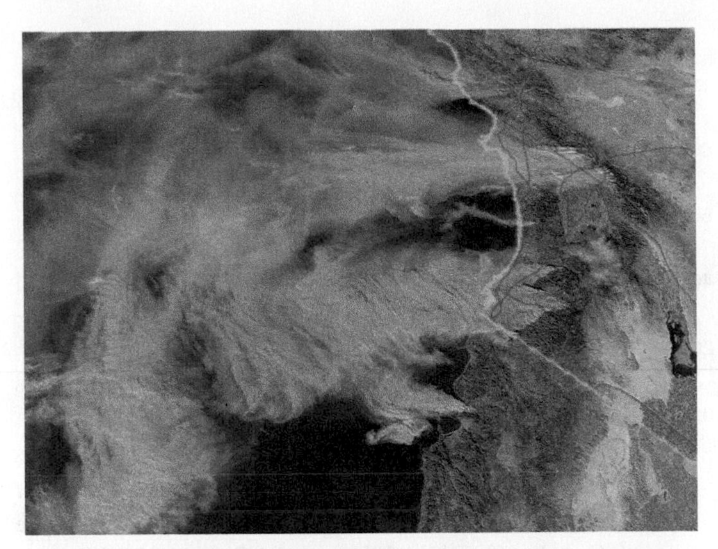

5.21 SANTA ANA This satellite image shows smoke from brush fires carried offshore by the Santa Ana winds blowing out from a high-pressure centre over California.

Chapter 6). In Canada, Chinooks occur most frequently in the Crowsnest Pass region of southwestern Alberta, with places such as Lethbridge experiencing as many as 30 per year. The number of Chinooks decreases northward and eastward; Calgary records about 15 per year and Red Deer about five per year. Occasionally Chinook winds reach southwestern Saskatchewan. In Europe, a similar wind experienced in the Alps is termed the *foehn*, and in Argentina the *zonda* is created by winds descending from the Andes.

The effect of a Chinook is most pronounced in the winter when a dramatic rise in temperature can lead to rapid loss of snow. For example, at Pincher Creek, Alberta, the temperature rose from −23.3°C to 2.2°C between 3 p.m. and 4 p.m. on January 6, 1966. A temperature rise of 10°C to 20°C over the course of a few hours is not uncommon, and is generally accompanied by a fall in humidity and an increase in wind speed (Figure 5.23). The Chinook brings relief from the cold of winter, but the changing weather patterns do have some adverse effects. For example, Chinooks can lead to drought stress in conifers because the warm, dry winds cause rapid loss of water through transpiration that cannot be replaced because the soil is frozen. Several medical complaints, including migraines, asthma, dry skin, and lethargy, have also been linked to Chinook events.

Surface winds are generated when pressure gradient forces cause air to move. Energy of motion—kinetic energy—is stored in the moving air. The stored energy can be extracted by windmills or wind turbines to do other work, such as generate electric power. *Eye on the Environment • Wind Power* describes this source of energy, as well as those of waves and ocean currents.

WINDS IN THE UPPER ATMOSPHERE

From the fundamental physical laws presented earlier in this chapter, it follows that pressure decreases less rapidly with height in warmer air than in colder air. Also important is the fact that insolation is greatest near the equator and least near the poles; this creates a latitudinal temperature gradient. Because of this permanently maintained temperature difference, the isobaric surfaces slope downward from the warmer low latitudes towards the cooler poles, which creates a pressure gradient force. Figure 5.24 shows a generalized cross-section of the atmosphere from 90° latitude (the pole) to about 30° latitude. At height H_1, the pressure at 90° latitude is about 94.0 kPa, while at 30° latitude it is about 95.5 kPa, creating a pressure gradient of 1.5 kPa. However, the pressure difference becomes greater with altitude. At H_2, the pressure at 90° is about 82.5 kPa, while at 30° it is about 89 kPa; at this level the pressure gradient is 6.5 kPa. At H_3, the gradient is about 10 kPa. This increase in the pressure gradient force with altitude is accompanied by progressive strengthening of the winds.

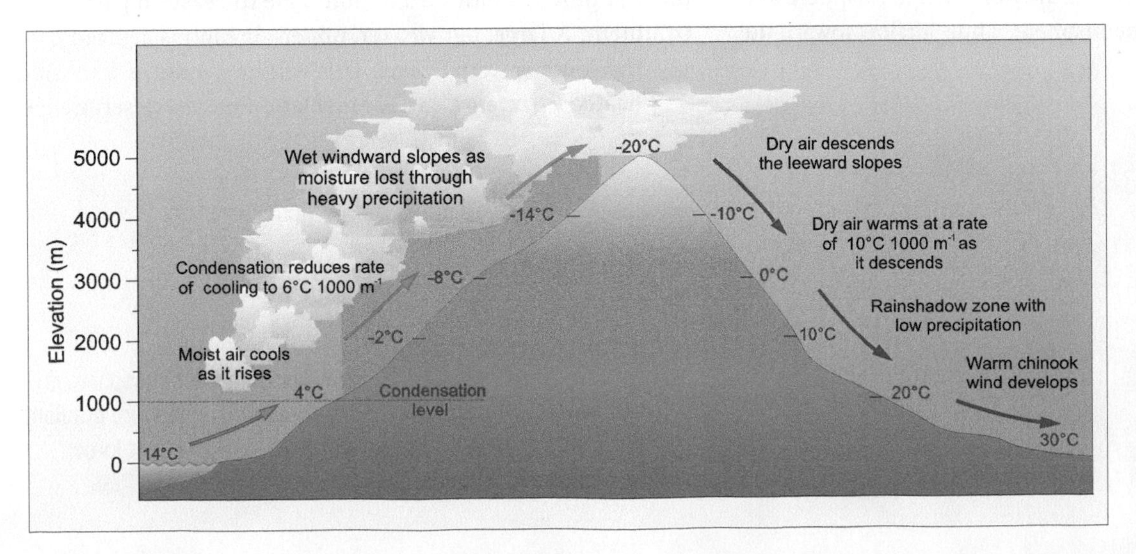

5.22 CHINOOK WINDS Chinook winds can form when moist air is forced over mountain ranges. If the air loses moisture through condensation, it will warm at a faster rate as it descends on the lee side, creating the warm, dry Chinook winds that are a common weather feature in southern Alberta.

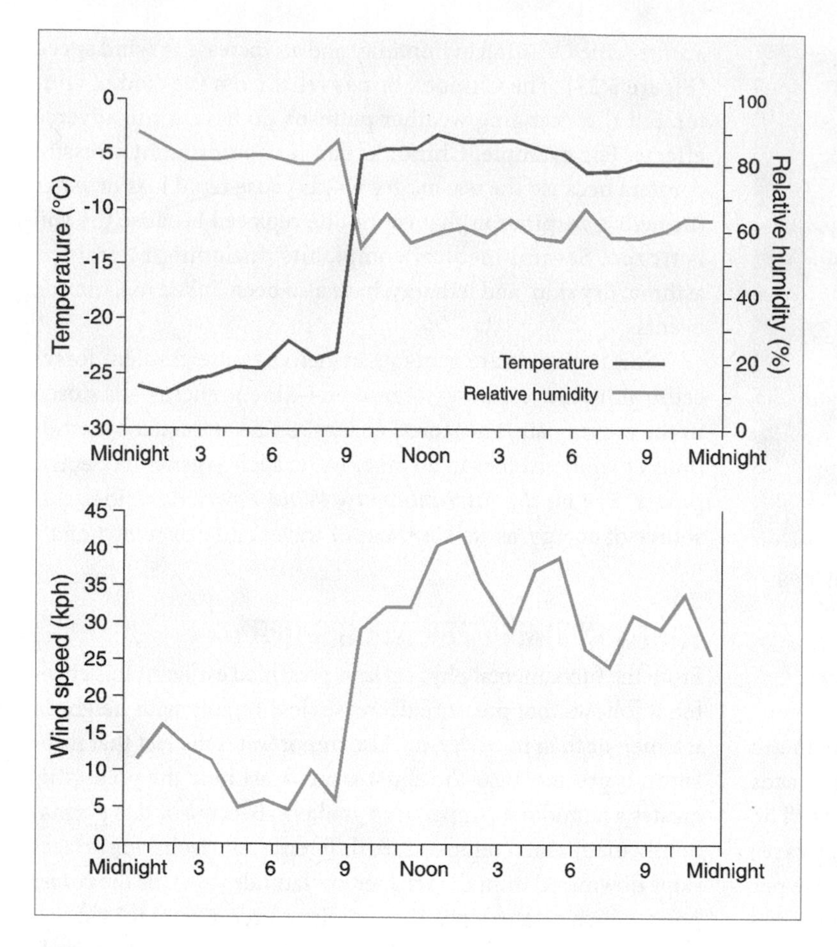

5.23 HOURLY TEMPERATURE, RELATIVE HUMIDITY, AND AVERAGE WIND SPEED AT PINCHER CREEK, ALBERTA, DURING A CHINOOK EVENT ON JANUARY 23, 1964 A rapid rise in temperature accompanied by a fall in relative humidity and gusty winds are the characterisitic weather conditions associated with a Chinook.

THE GEOSTROPHIC WIND

The pressure gradient force generated by the temperature differences between high and low latitudes is termed the **thermal wind**. Note this term applies only to the pressure gradient that is created rather than the actual movement of the air, although the pressure gradient does cause the air to move toward higher latitudes. As the air moves, it is influenced by the Coriolis effect, which will turn it to the right in the northern hemisphere and to the left in the southern hemisphere. Thus, airflow toward the poles in both hemispheres will be subject to a progressively stronger force that is directed toward the east.

Unlike air moving near the Earth's surface, the winds in the upper troposphere are not affected by friction. At these higher altitudes, the horizontal air pressure gradient causes air to accelerate across the isobars from areas of high pressure toward areas of low pressure. The Coriolis effect then deflects the airflow, which in the case of the northern hemisphere is to the right. As the wind gains speed, the Coriolis effect increases in magnitude until it balances the pressure gradient force. At that point, the sum of forces on the moving air is zero, so its speed and direction remain constant. Consequently, the winds blow parallel to isobars (Figure 5.25). This configuration is termed the *geostrophic wind*.

Figure 5.26 shows an upper-air map of North America in late June. The contours show how the altitude of the 50-kPa pressure surface varies in relation to the pressure conditions aloft. The pressure surface dips toward the Earth's surface where the upper-air pressure is lower, and rises where the upper-air pressure is higher. Thus, lower altitude indicates lower pressure, and higher altitude indicates higher pressure.

The map shows a large, upper-air low-pressure centre over Baffin Island. Low pressure in the upper atmosphere indicates cold air, as would be expected at this latitude. The upper-air low centred over the Great Lakes is a mass of cold, arctic air that has moved southward to dominate the eastern part of the continent. A large, but weaker, upper-air high is centred over the southwestern desert. This will be a mass of warm air, heated by intense surface insolation on the desert below.

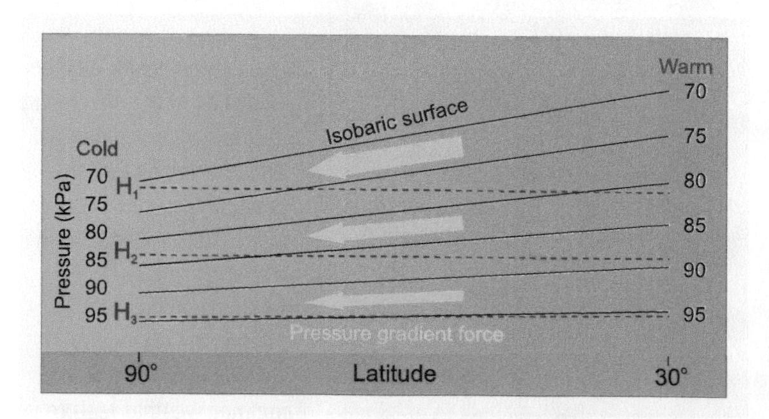

5.24 UPPER-AIR PRESSURE GRADIENT Because the atmosphere is warmer near the equator than at the poles, a pressure gradient force acts to push air toward the poles. The gradient force increases with altitude.

EYE ON THE ENVIRONMENT

WIND POWER

Wind power is an indirect form of solar energy that has been used for centuries. In the Netherlands, the windmill played a major role in pumping water from the polders as they were reclaimed from tidal marshes. In these low, flat areas, streams could not be adapted to waterpower, and windmills were also used to grind grain.

The design of new forms of windmills has intrigued inventors for many decades. The total supply of wind energy is enormous. The World Meteorological Organization has estimated that the combined electrical-generating wind power of favourable sites throughout the world totals about 20 million MW.

Wind turbines, each with a generating capacity ranging from a few hundred KW up to 4.2 MW for large-scale production facilities, have been assembled in groups at favourable locations to form "wind farms." Arranged in rows along ridge crests, the turbines intercept local winds of exceptional frequency and strength. Currently in Canada, wind generators have the capacity to produce about 1,200 MW of electricity. One of the first installations was at Pincher Creek, Alberta. It began operating in 1993 with the construction of a 19-MW wind farm that captured the persistent airflow descending from the Rocky Mountains. There are now more than 100 wind turbines in the area. Canada's largest wind farms are in Ontario and include the Wolfe Island facility near Kingston with an installed capacity of 197.8 MW, the Prince Wind Farm near Sault Ste. Marie (189.0 MW), and the Ontario Wind Power Farm near Kincardine

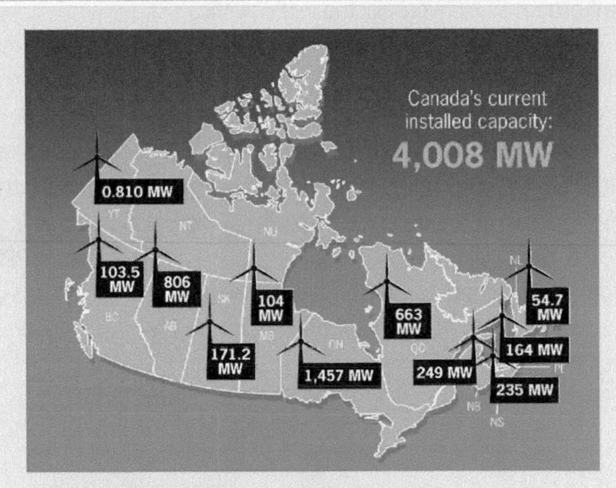

Canada's wind resource is well distributed in rural areas throughout the country.

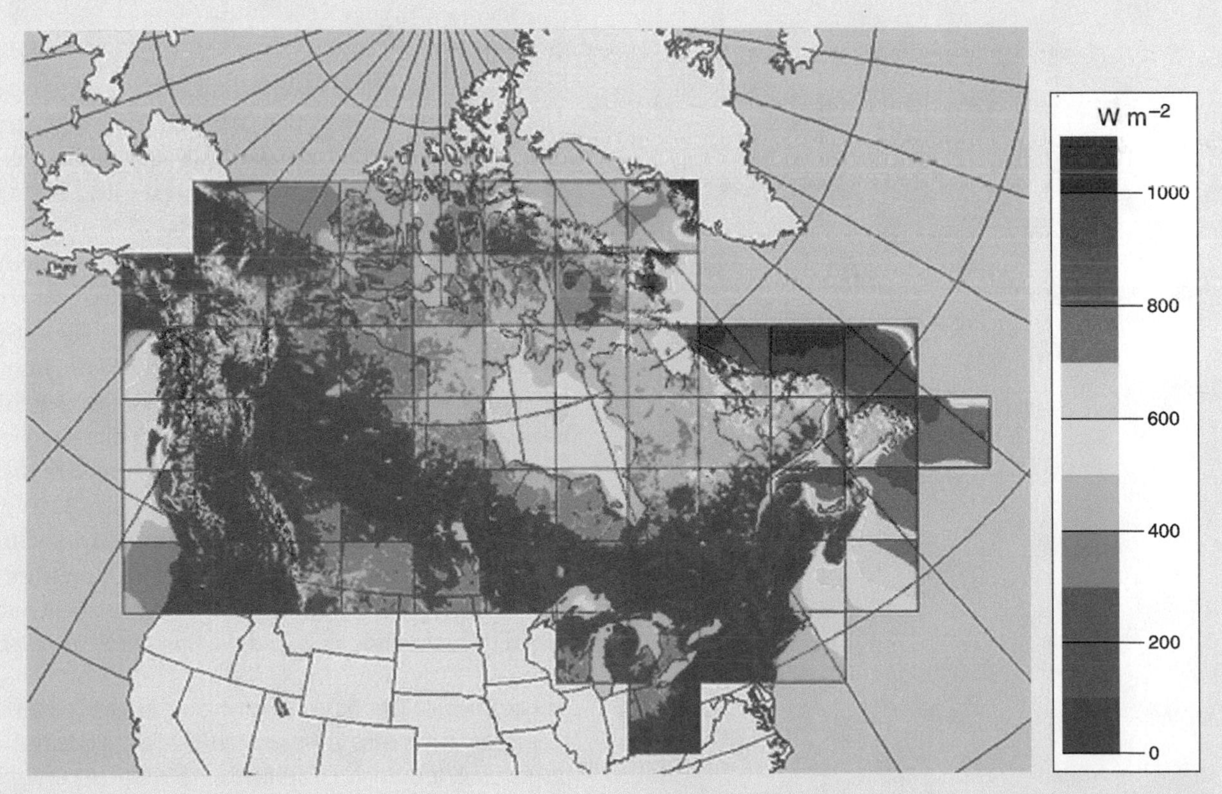

Numerical simulation of potential annual wind energy in Canada, based on wind speeds at 50 m above the surface.

(181.5 MW). Wind energy in Canada currrently generates about 3,250 MW of electricity, which is equivalent to about 1 percent of the country's total electricity demand. This compares to about 60 percent produced by hydroelectric installations.

The figure on the previous page shows the potential for wind energy generation in Canada. The majority of wind turbines currently in use are located at sites where the average wind speed is at least 6 m s^{-1} (22 kph). Wind energy has become competitive with conventional sources of energy, and government support for wind power continues to increase. In Canada, wind energy is promoted through the Wind Power Production Incentive (WPPI), a program designed to reduce greenhouse gas emissions. Its goal is to increase wind power generation to 4,000 MW.

Wave energy is another indirect form of solar energy. Nearly all ocean waves are produced by winds blowing over the sea surface. Water moves in vertical orbits in waves, and energy is extracted from this motion with the use of a floating mechanism tethered to the sea floor. As the device rises and falls, an attachment drives a generator.

Harnessing the vast power of ocean currents, such as the Gulf Stream or the Kuroshio (a similar current that flows along the east coast of Japan), is a prospect that has not gone unnoticed. One plan proposes anchoring a large number of current-driven turbines to the ocean floor. Each turbine would operate below the ocean surface. They would be 170 m in diameter and capable of generating 83 MW of power.

The intensity of the high- and low-pressure centres is inferred from the values and spacing of the surface contours.

The winds generally follow the contours of the 50-kPa surface, showing geostrophic flow. Wind speeds are faster where the contours are closest and the pressure gradient force is strongest. The winds around upper-air lows tend to spiral inward and converge on the low before descending to the surface. Conversely, the winds around upper-air highs spiral outward and diverge from the centre. This outflow is fed by the inward and upward spiral of winds from a surface low-pressure system. Convergence in the upper-air low creates the subsiding air current that contributes to surface anticyclones (see Figure 5.16).

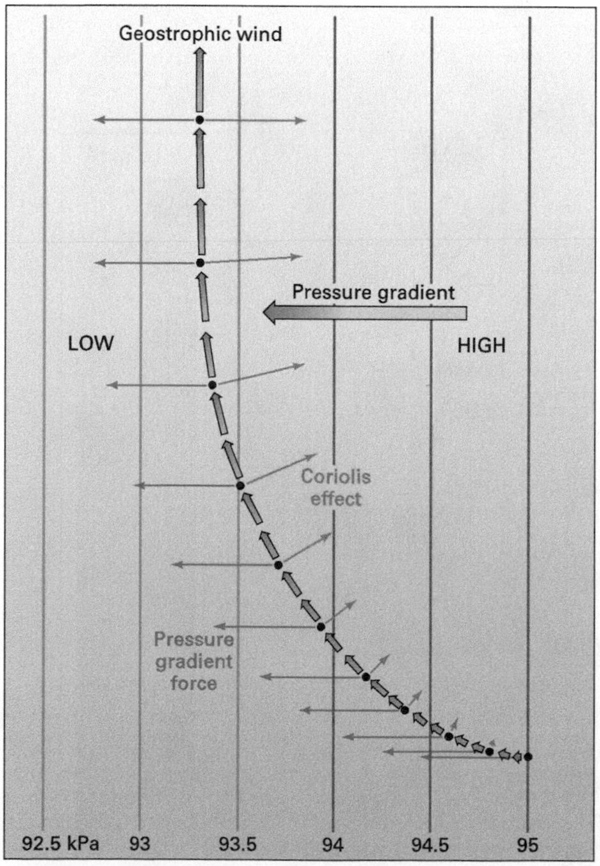

5.25 **GEOSTROPHIC WIND** As air at higher levels in the troposphere accelerates in response to the pressure gradient, it is turned progressively to the right (in the northern hemisphere) until the gradient force and Coriolis effect balance. This produces the geostrophic wind.

GLOBAL CIRCULATION AT UPPER LEVELS

The general pattern of airflow at higher levels in the troposphere comprises four major features—weak equatorial easterlies, tropical high-pressure belts, upper-air westerlies, and a polar low (Figure 5.27).

The pressure gradient force that generates westerly winds in the upper atmosphere is caused by the permanent temperature differences between the tropics and the poles. The *upper-air westerlies* blow in a complete circuit around the Earth, from about lat. 25° almost to the poles. At high latitudes, the westerlies form a huge circumpolar spiral, encircling a great polar low-pressure centre. Toward lower latitudes, atmospheric pressure rises steadily, forming a *tropical high-pressure belt* at lat. 15° to 20° N and S. This is the zone of high-altitude divergence associated with the rising limbs of the Hadley cells. Between these high-pressure ridges is a zone of lower pressure in which the winds are light but generally easterly. These winds are called the *equatorial easterlies*.

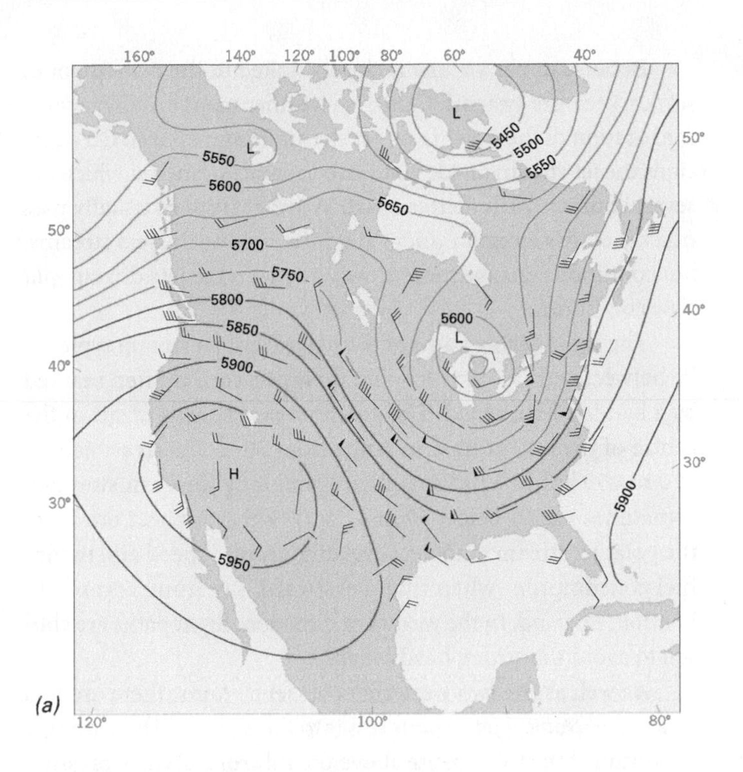

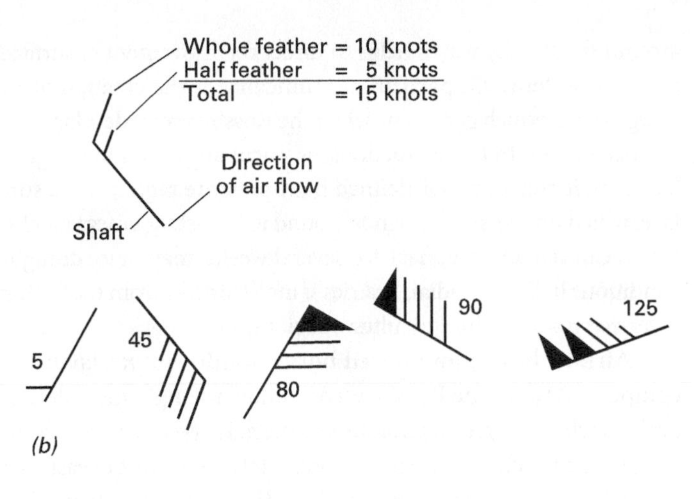

5.26 UPPER-AIR WIND MAP (a) An upper-air map for a late June day. Lines are height contours (m) for the 50-kPa surface. (b) Explanation of wind arrows used on upper-air charts. 1 knot (nautical mile per hour) = 0.514 m s^{-1} = 1.852 kph.

5.27 GLOBAL UPPER-LEVEL WINDS Global circulation in the upper troposphere at mid- and high-latitudes is dominated by strong westerly winds. These sweep to the north or south around centres of high and low pressure in the upper atmosphere. In the equatorial region, a weak easterly wind pattern prevails. (Copyright © A. N. Strahler.)

ROSSBY WAVES, JET STREAMS, AND THE POLAR FRONT

The path of the upper-air westerlies periodically experiences undulations, called **Rossby waves** (Figure 5.28). The number of Rossby waves ranges from three to seven. They form in the contact zone between cold polar air and warm tropical air known as the *polar front*. Although the polar front is a common feature on surface weather maps (see Chapter 6), this boundary zone extends from the Earth's surface to the upper troposphere.

For several days or weeks, the flow configuration may be fairly smooth. Then an undulation develops, and warm air pushes toward the poles while a tongue of cold air moves toward the equator. Eventually, the tongue is pinched off, leaving a pool of cold air far removed from its point of origin. This cold pool may persist for some days or weeks. Because of its cold centre, the air column will be reduced in thickness, resulting in lower pressure and convergence higher in the atmosphere. Air will descend in its core and diverge at the surface, creating a surface high pressure. Conversely, a high-pressure warm air pool will diverge in the upper atmosphere because it is fed by rising air that has converged in a low-pressure system at the surface.

Because the Rossby wave circulation brings warm air toward the poles and cold air toward the equator, it is a primary mechanism of heat transport. It is also the reason weather in the midlatitudes is often so variable, as pools of warm, moist air and cold, dry air alternately pass over midlatitude land masses. In addition, upper air flow around the Rossby waves helps to direct the movement of surface pressure systems. Of particular significance is the creation of an *omega block*, which can occur when the Rossby waves develop pronounced lobes. In these situations, as warm air pushes into higher latitudes, it creates a well-defined high-pressure region at the surface, which diverts surface winds around it. Persistent omega blocks, which can remain stationary for several weeks, may cause drought conditions in the Canadian Prairies if moist airflow from the Pacific is carried north of the agricultural regions.

Airflow has a pronounced north–south—or *meridional*—component when the Rossby waves follow strongly meandering paths, such as in an omega block situation. However, at the beginning of each cycle, airflow is predominately from west to east—or *zonal*. Such variation in flow paths influences the time it takes for surface weather systems to pass through a region and will affect how quickly weather changes at midlatitudes.

Jet streams are important features of upper-air circulation. These narrow currents of fast-moving air extend for thousands of kilometres but are only a few hundred kilometres in width and of limited vertical depth. Jet streams occur at the approximate altitude of the tropopause (10 to 12 km), where atmospheric pressure gradients are strong. This accounts for their development along the polar front where pressure changes rapidly in response to the strong temperature gradient between polar and subtropical air masses. The polar jet streams essentially mark the boundary of the Rossby waves in both the northern and southern hemispheres. Consequently, the position of the polar front jet streams closely follows the changing configuration of the Rossby Waves.

Because upper airflow is closely linked to the movement of surface pressure systems, Canadian weather maps usually include the location of the polar front jet stream (Figure 5.29). On occasions the jet stream divides into two separate branches which are several hundred kilometres apart. Weather systems usually pass quickly across a region under the influence of the jets streams, but conditions change more slowly between the jets during *split flow* conditions.

The meandering paths of the polar front jet stream typically lie between lats. 35° and 65° and their presence is often marked by a band of high cloud. The greatest wind speeds occur in the centre of the jet stream and range from 75 to 125 m s^{-1} (about 270 to 450 kph). Wind speeds are generally lower in summer, sometimes falling below 30 m s^{-1} (100 kph). Aircraft often use the polar jet stream to increase relative ground speed and reduce fuel consumption when flying eastward, say from Toronto to London, England. In the westward direction, flight paths are chosen to avoid the strong headwinds.

As well as the two westerly polar jet streams, there are two smaller *subtropical jet streams* closer to the equator. These occupy a position at the tropopause above the subtropical high-pressure cells at about lats. 20° to 40° in the northern and southern hemispheres; this is the poleward limit of the Hadley cells. The westerly subtropical jet streams are generally not as strong as those at the polar front due to the weaker temperature and pressure gradients that develop between subtropical and tropical air masses. Wind speeds in the subtropical jets are typically slower than for the polar jet with maximum values of 100 to

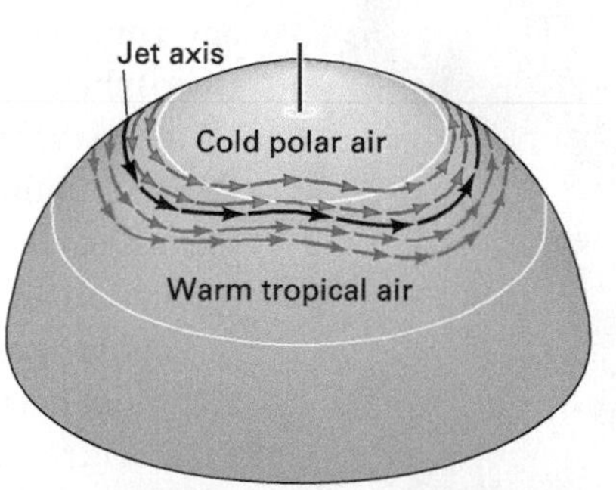

The jet stream begins to undulate.

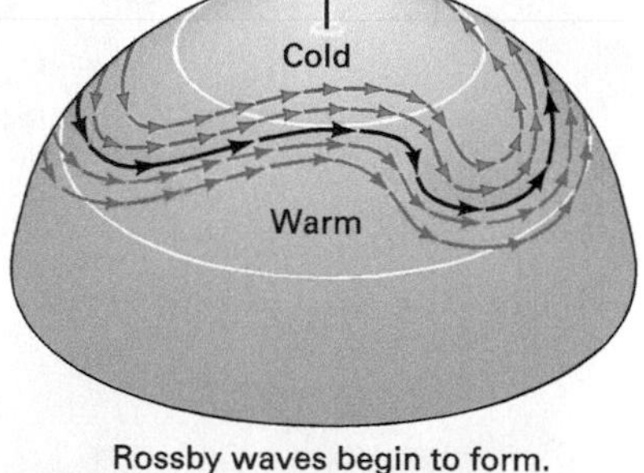

Rossby waves begin to form.

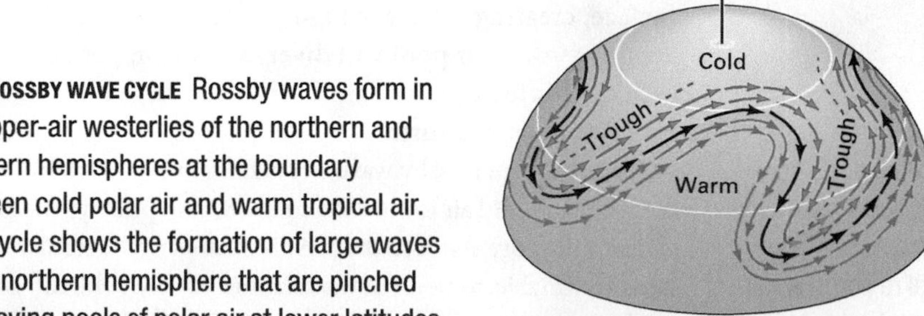

Waves are strongly developed. The cold air occupies troughs of low pressure.

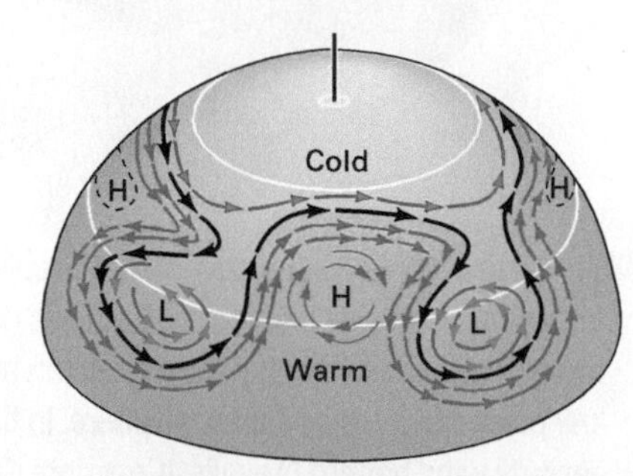

When the waves are pinched off, they form cyclones of cold air.

5.28 ROSSBY WAVE CYCLE Rossby waves form in the upper-air westerlies of the northern and southern hemispheres at the boundary between cold polar air and warm tropical air. This cycle shows the formation of large waves in the northern hemisphere that are pinched off, leaving pools of polar air at lower latitudes. (Copyright © A. N. Strahler.)

110 m s^{-1} (about 360 to 400 kph), but speeds in excess of 135 m s^{-1} (485 kph) have been reported over India.

The subtropical jet stream develops because the poleward flow of air in the Hadley circulation is subject to *angular momentum*. Angular momentum arises when objects (in this case air molecules) follow a rotating path. Angular momentum (L) is equal in magnitude to the *angular velocity* (ω) of the object multiplied by its *moment of inertia* (I), that is $L = I\omega$. Angular velocity measures how fast an object is turning. For air molecules angular velocity is the speed of rotation of the Earth, which varies from 1,700 kph at the equator to 0 kph at the poles. The moment of inertia measures an object's resistance to a change in its rate of rotation. The moment of inertia depends on how the mass of an object is distributed with respect to the axis of rotation. In this case, the distance involved is measured from the axis of rotation to the top of the troposphere. This distance decreases from about 6,400 km at the equator to 10 km near the poles. Consequently, as air moves poleward it gets closer to the axis of rotation.

As air molecules move away from the equator, their speed of rotation decreases with latitude. However, rotational kinetic energy, like any other form of energy, must comply with the principle of conservation of energy. In this case, there must be a balance between the speed of rotation of the object and the torque or force that is causing it to rotate. Therefore, to conserve angular momentum, the velocity of air moving away from the equator must increase. As velocity increases, so too does the Coriolis effect; this results in the air stream being progressively turned until it flows at right angles to the pressure gradient. In the northern hemisphere, for air moving toward the pole, the Coriolis force is directed toward the east; this creates the fast-moving subtropical jet stream that feeds into the descending currents associated with the subtropical highs at the surface.

A third type of jet stream is found at even lower latitudes. Known as the *tropical easterly jet stream*, it runs from east to west—opposite in direction to that of the polar front and subtropical jet streams. The tropical easterly jet stream occurs only in the summer and is limited to the northern hemisphere over Southeast Asia, India, and Africa.

TEMPERATURE LAYERS OF THE OCEAN

Sea surface temperatures vary with latitude and season. In tropical regions, mean annual sea surface temperatures typically range from 25–28°C, decreasing to 15–22°C through the midlatitudes and falling to as low as −3°C in the polar regions (Figure 5.30). Between 30° N and 30° S, the western sides of the ocean basins tend to be warmer due to westward drift imposed by the trade winds. This pattern reverses from about lat. 40° toward the poles; in these regions, the eastern sides of the ocean basins are generally warmer due to the effect of currents such as the North Atlantic Drift. Sea surface temperatures remain relatively constant at high and low latitudes with a seasonal range of only one or two degrees. In midlatitude oceans, a range of 8 to 10°C is not

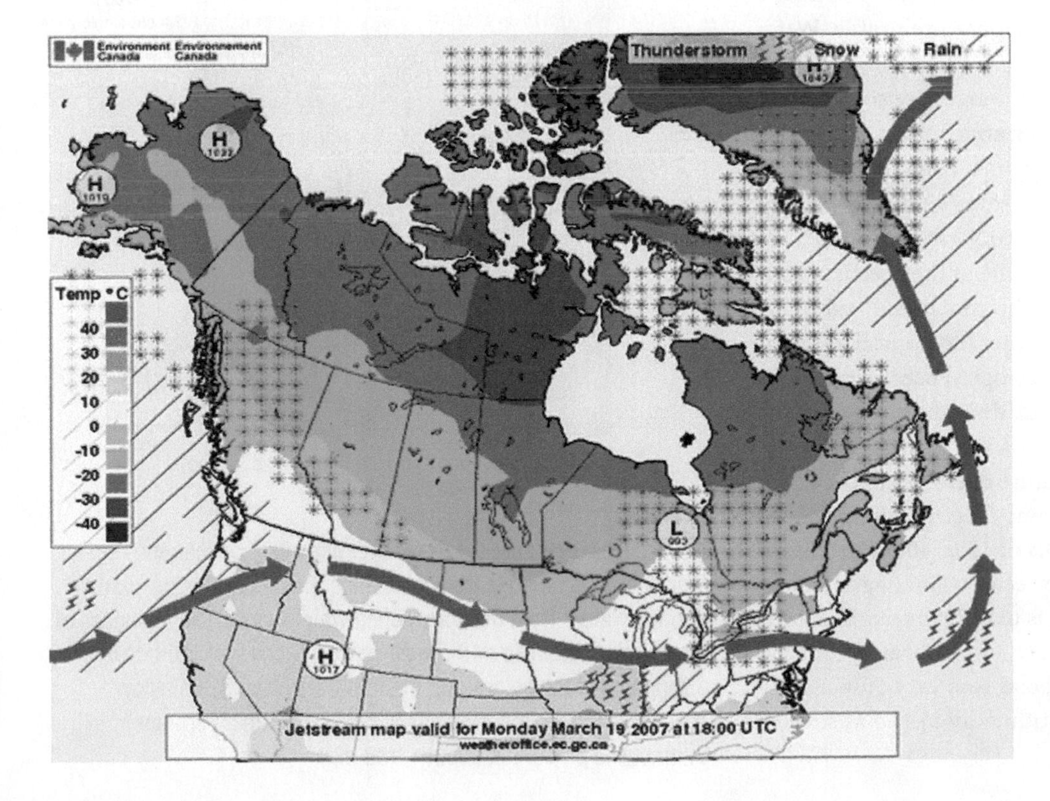

5.29 ENVIRONMENT CANADA'S "WEATHER AT A GLANCE" MAP SERIES The position of the jet stream is marked by red arrows on the daily weather maps produced by Environment Canada. They can be accessed at http://weatheroffice.ec.gc.ca/jet_stream/index_e.html.

5.30 AVERAGE 2010 SEA SURFACE TEMPERATURE Average sea surface temperatures range from about 25 to 30 °C in tropical and equatorial regions, except where ocean currents bring cold water to the region from the polar seas. This is the case along the western coast of South America and southwestern Africa. In the temperate midlatitudes the surface waters are about 15 to 22°C with temperatures decreasing to near 0°C in the polar seas.

uncommon, especially on the western side of the oceans where cold winds blow off of the continents in winter.

The oceans, like the atmosphere, have a layered structure defined according to their vertical temperature profile. Water temperature is generally highest at the surface and decreases with depth because the principal heat sources are solar insolation and heat supplied by the overlying atmosphere. Below this warm layer, temperatures drop rapidly in a zone known as the *thermocline*. Very cold water extends from the thermocline to the deep ocean floor, where temperatures range from 0°C to 5°C. About 90 percent of the total volume of the oceans is contained in the deep water below the thermocline. Figure 5.31 shows representative temperature profiles for tropical, midlatitude (temperate), and polar oceans. In tropical oceans, the warm surface layer is separated from the cold, dense bottom waters by a permanent thermocline. This creates a warm surface layer up to 500 m thick below which temperature decreases to the ocean floor. In midlatitude oceans, a temporary thermocline develops during the summer, usually at depths of 15 to 40 m, but disappears as the water cools, allowing convectional mixing in winter. However, a permanent thermocline is usually present at depths of 500 to 1500 m, which separates the surface water from the deeper ocean. Temperatures in the polar seas vary only slightly with depth because the water is continually mixed as surface layers cool and sink.

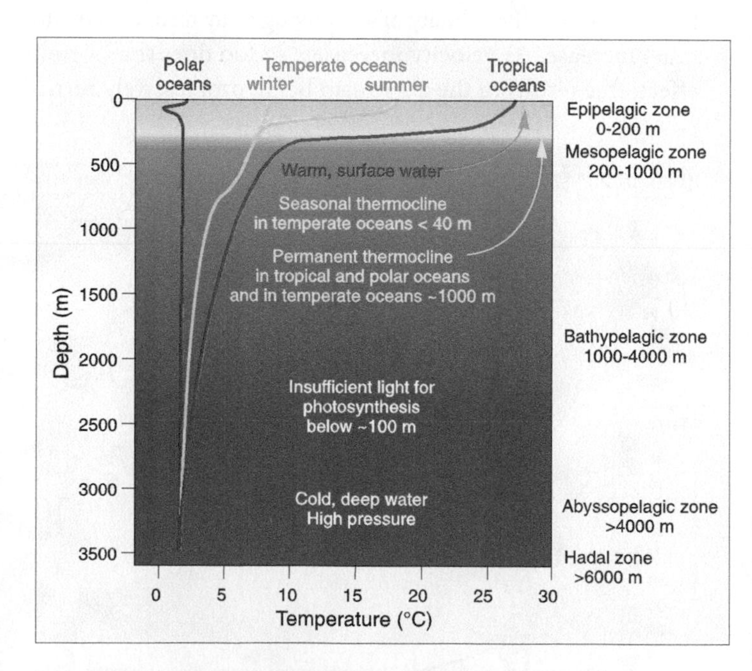

5.31 TEMPERATURE PROFILES OF THE DEEP OCEANS A permanent thermocline is found in tropical oceans that separates the warm surface waters from the cold, dense water below. The thickness of the warm water layer at the surface decreases with latitude and disappears in the polar seas. In temperate oceans, a shallow seasonal thermocline develops in summer, with a permanent thermocline present at about 1,000 m.

OCEAN CURRENTS

Just as the atmosphere has a circulation pattern, so too do the oceans. An *ocean current* is any persistent, dominantly horizontal flow of ocean water, which can occur at the surface or at any depth. Surface currents are driven by prevailing winds. Deep currents are powered by changes in temperature and density occurring in surface waters, causing them to sink. Current systems act to exchange heat between low and high latitudes and are essential in sustaining the global energy balance.

SURFACE CURRENTS

The pattern of surface ocean currents is strongly related to prevailing surface winds (Figure 5.32). Energy is transferred from wind to water by the friction of the air blowing over the water surface. Because of the Coriolis effect, the actual direction of water drift is deflected several degrees from the direction of the driving wind. Ocean currents move warm waters toward the poles and cold waters toward the equator, and consequently, they are important regulators of air temperatures.

The ocean current system includes large circular movements, called **gyres**, which are centred approximately at lat. 30°. The gyres track the movements of air around the subtropical high-pressure cells. The *equatorial currents* with westward flow mark the belt of the trade winds. As the equatorial currents approach land, they are turned poleward along the western margins of the oceans, forming warm currents that flow parallel to the coast. Examples are the Kuroshio Current off Japan and the North Atlantic Drift off the eastern coast of North America. The North Atlantic Drift transports warm water from the Gulf of Mexico to Europe and helps to moderate winter temperatures in Britain and other parts of western Europe. The Russian port of Murmansk, on the Arctic Circle, remains ice-free year round because of this influx of warm water. The northern and southern equatorial currents are separated by an equatorial counter-current.

A slow, eastward movement of water over the zone of the westerlies is named the *west-wind drift*. It covers a broad belt between lat. 35° and 45° in the northern hemisphere and between lat. 30° and 60° in the southern hemisphere. As these currents approach the western margins of the continents, the flow of cool water is deflected along the coasts toward the equator. These currents are often accompanied by *upwelling* along continental margins. In this process, colder water from the ocean depths rises to the surface. Examples of upwelling are the Peru (or Humboldt) Current off the coast of Chile and Peru, the Benguela Current off the coast of southern Africa, and the California Current. Upwelling is important for returning nutrients to the illuminated waters at the ocean surface, where phytoplankton can use them to sustain the marine food cycle.

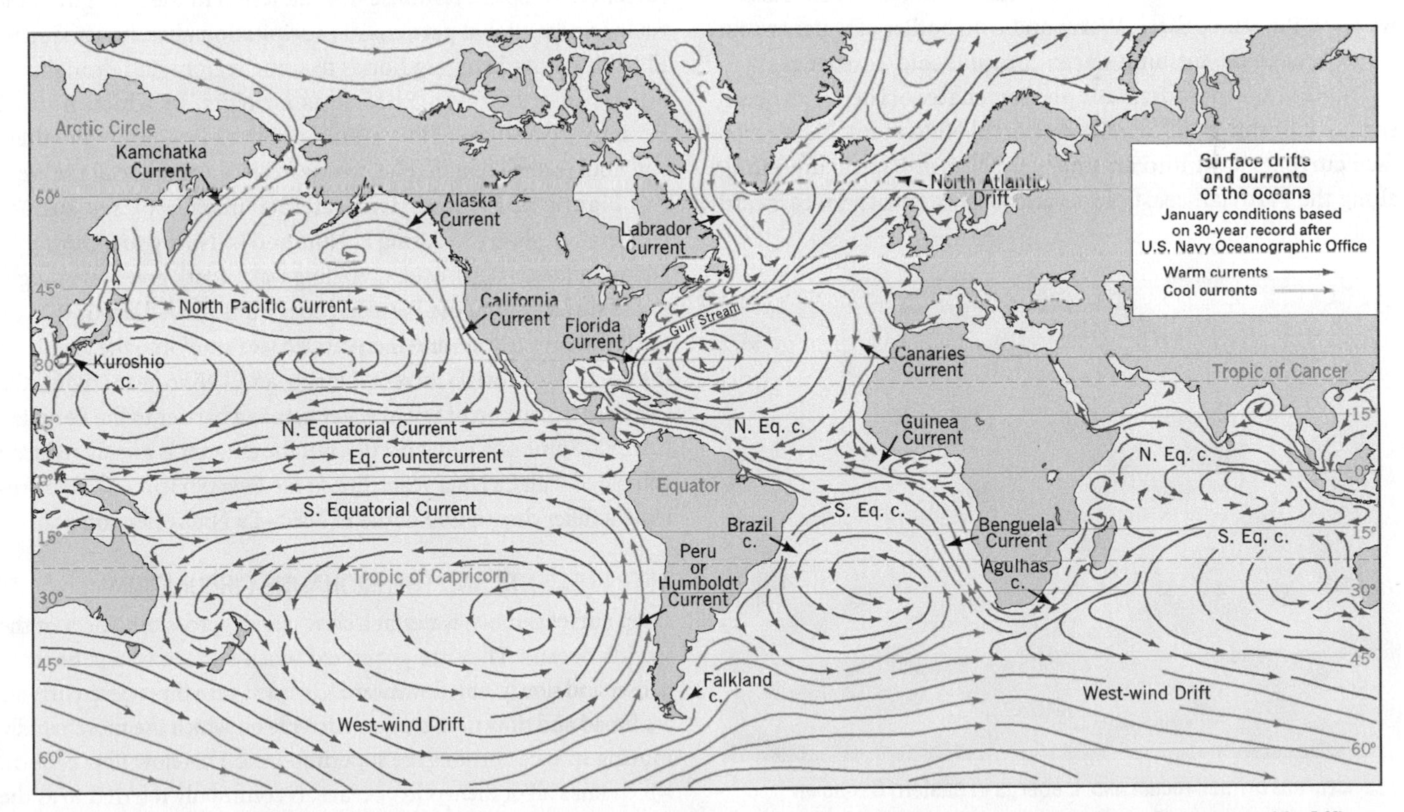

5.32 **JANUARY OCEAN CURRENTS** Surface drifts and currents of the oceans in January. (Based on data from U.S. Navy Oceanographic Office. Redrawn and revised by A. N. Strahler.)

In addition, fogs that form over these cold currents are carried onshore by the prevailing winds and provide much needed moisture to adjacent desert regions.

The west-wind drifts also move water toward the poles to join Arctic and Antarctic circulations. Such is the case in the northeastern Atlantic Ocean, where the North Atlantic Drift spreads around the British Isles, into the North Sea, and along the Norwegian coast. In the northern hemisphere, where the polar sea is largely landlocked, cold currents flow toward the equator along the east coasts of the continents. For example, the Labrador Current flows south along the east coast of Canada. The mixing of cool air associated with the Labrador Current and air warmed by the North Atlantic Drift produces frequent fogs over the Grand Banks off Newfoundland (see Chapter 6). In addition, the Labrador Current carries icebergs from Greenland into Canadian coastal waters until early summer (Figure 5.33).

Figure 5.34 shows a satellite image of ocean temperatures along the east coast of North America for a week in April. The Gulf Stream stands out as a tongue of warm water (red and yellow tones), moving along the coast and heading in a northeasterly direction. Cooler water from the Labrador Current (green and blue tones) hugs the northern Atlantic coast. This current heads south and then turns to follow the Gulf Stream to the northeast. Instead of mixing, the two flows remain quite distinct. The boundary between them shows a wavelike flow, much like Rossby waves in the atmosphere. Warm and cold bodies of water are cut off to float freely, forming warm-core and cold-core rings.

One of the most dramatic phenomena associated with ocean surface currents is *El Niño.* During an El Niño event, Pacific surface currents shift into an unusual pattern. Pacific upwelling along the Peruvian coast ceases and a weak equatorial eastward current develops in response to a reduction in the strength of the trade winds. Global patterns of precipitation also change during El Niño events, bringing floods to some regions and droughts to others. In contrast to El Niño is *La Niña,* in which normal Peruvian coastal upwelling is enhanced, trade winds strengthen, and cool water is carried far westward in an equatorial plume.

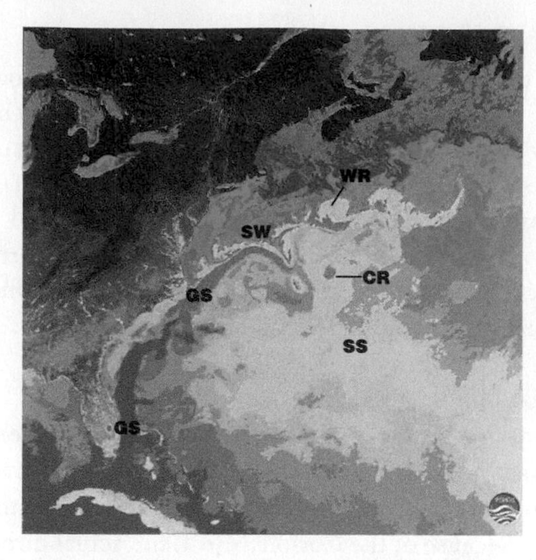

5.34 SEA-SURFACE TEMPERATURES IN THE WESTERN ATLANTIC A satellite image showing sea-surface temperature for a week in April from data acquired by the NOAA-7 orbiting satellite. Cold water appears in green and blue tones; warm water in red and yellow tones. The image shows the Gulf Stream (GS) and its interactions with the cold water of the continental slope (SW) brought down by the Labrador Current, and the warm water of the Sargasso Sea (SS). Other features include a warm-core ring (WR) and a cold-core ring (CR).

5.33 ICEBERGS OFF NEWFOUNDLAND Icebergs in eastern Canadian waters mostly originate from the glaciers of western Greenland, where more than 30,000 are produced each year. They usually are seen from March until July, but are most abundant in late April.

Figure 5.35 shows two satellite images of sea-surface temperature observed during El Niño and La Niña years. During La Niña conditions (left), cold, upwelling water (dark green) is brought to the surface along the Peruvian coast. It is moved northward by the Peruvian Current, and then carried westward into the Pacific by the South Equatorial Current. During an El Niño year (right), the eastward expansion of warm water acts as a barrier to the Peruvian Current. Some upwelling occurs, but the amount is greatly reduced compared with La Niña years. *Eye on the Environment • El Niño* provides a fuller description of the El Niño–La Niña phenomenon.

DEEP CURRENTS AND THERMOHALINE CIRCULATION

Deep currents move water in a slow circuit across the floors of the world's oceans. They are generated when surface waters become denser and slowly sink downward. Coupled with these deep currents are broad and slow near-surface currents on which the more rapidly moving surface currents are superimposed. This slow flow pattern, which links all of the world's oceans, is commonly referred to as the *great ocean conveyor* (Figure 5.36). It is a *thermohaline circulation* process that depends on the change in density of sea water caused by temperature and salinity differences in North Atlantic Ocean waters.

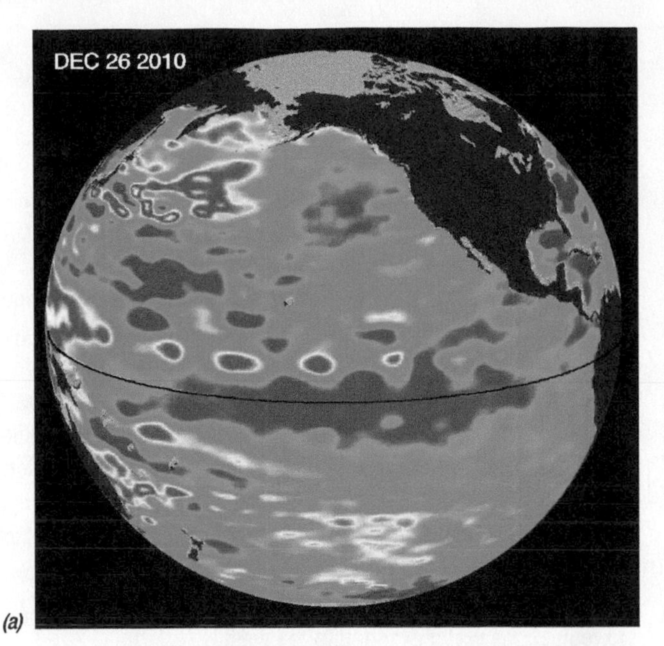

(a)

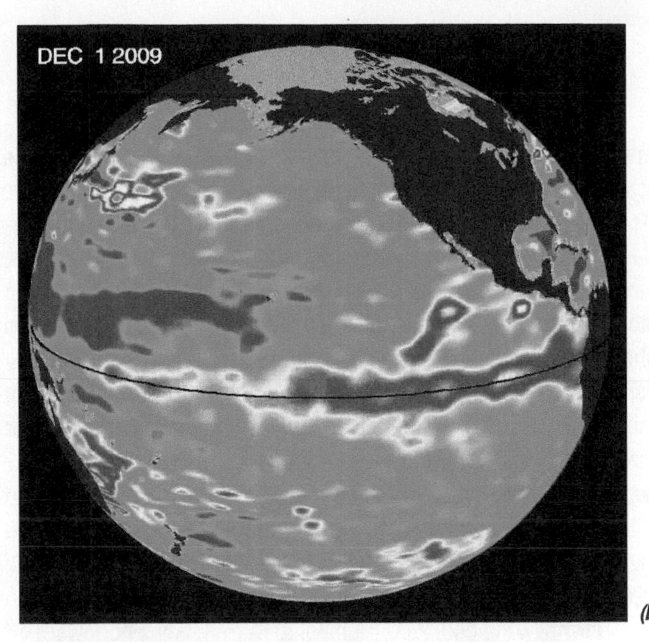

(b)

5.35 **LA NIÑA AND EL NIÑO SEA-SURFACE TEMPERATURE** These images show sea-surface temperatures in the eastern tropical Pacific during (a) La Niña and (b) El Niño years. Blue tones indicate cooler temperatures, while red tones indicate warmer temperatures.

Warm Atlantic surface water moves slowly northward through the equatorial and tropical zones. As this surface layer is warmed, evaporation occurs and the layer becomes saltier, slightly increasing its density. When the water reaches higher latitudes, it loses heat to the atmosphere, and so it becomes colder and even denser. Eventually, along the northern boundary of the North Atlantic, the surface layer becomes dense enough to sink.

Carried along the bottom of the North and South Atlantic Oceans, the cold, dense water eventually reaches the Southern

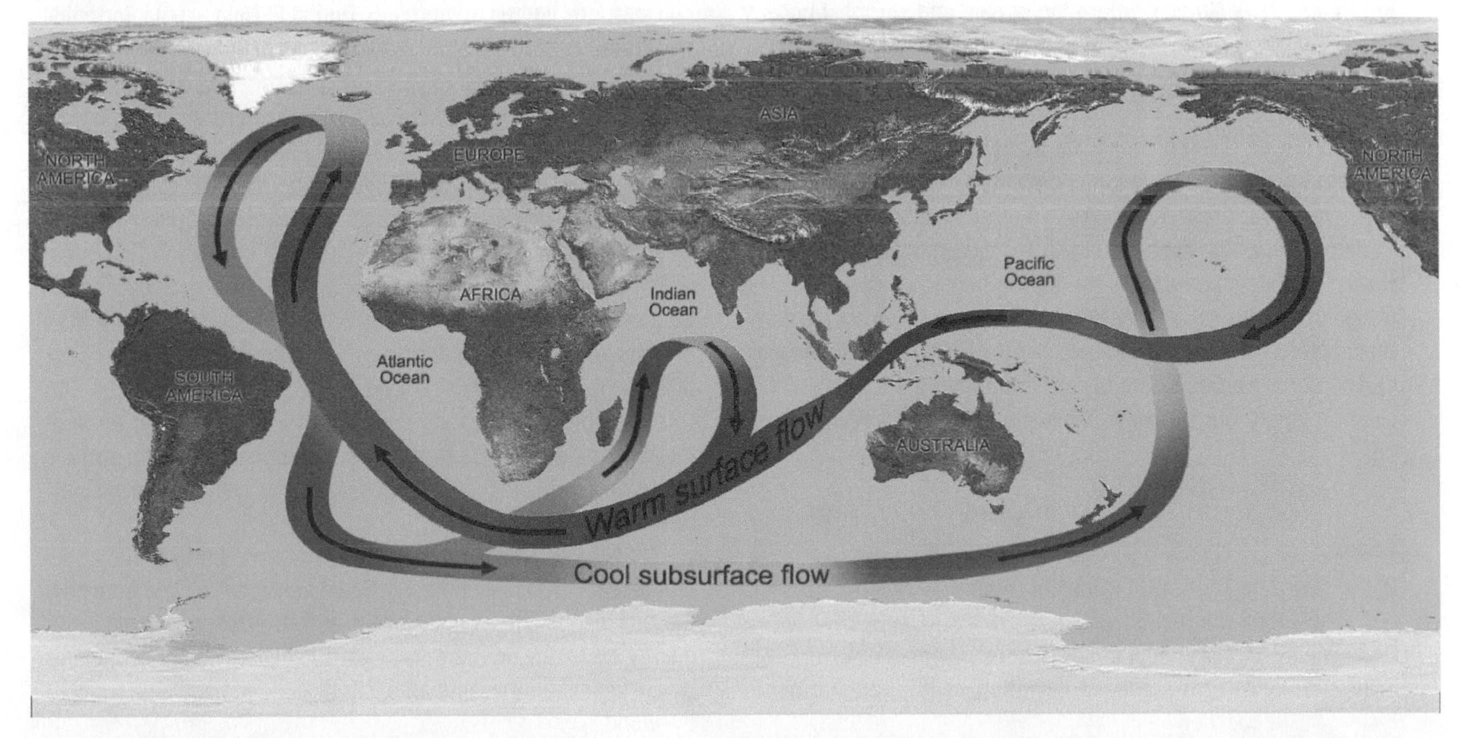

5.36 **OCEAN CIRCULATION** Deep ocean currents, generated by the sinking of cold, dense water of high salinity in the northern Atlantic, circulate sea water in slowly moving coupled loops involving the Atlantic, Pacific, Indian, and Southern oceans.

EYE ON THE ENVIRONMENT

EL NIÑO

At intervals of about three to eight years, a remarkable disturbance of ocean and atmosphere occurs. It begins in the eastern Pacific Ocean and spreads its effects widely over the globe. This disturbance lasts more than a year, bringing droughts, heavy rainfalls, severe spells of heat and cold, or a high incidence of cyclonic storms to various parts of the Pacific and its eastern coasts. This phenomenon is called *El Niño.* The expression comes from Peruvian fishermen, who refer to the *Corriente del Niño*, or the "Current of the Christ Child," in describing an invasion of warm surface water that occurs once every few years around Christmas time, greatly depleting their catch of fish. El Niño occurs at irregular intervals and with varying degrees of intensity. Notable El Niño events occurred in 1891, 1925, 1940–41, 1965, 1972–73, 1982–83, 1989–90, 1991–92, 1994–95, 1997–98, and 2002–03.

Normally, the cool Peru (Humboldt) Current flows northward off the South American coast, then near the equator it turns westward across the Pacific as the South Equatorial Current (see Figure 5.32). The Peru Current is fed by upwelling of cold, deep water, bringing with it nutrients that serve as food for marine life. With the onset of El Niño, upwelling ceases, the cool water is replaced by warm, sterile water from the west, and the abundant marine life disappears.

In an El Niño year, a major change in barometric pressure occurs across the entire stretch of the equatorial zone as far west as southeastern Asia. Normally, high pressure prevails in the eastern Pacific, with low pressure over northern Australia, the East Indies, and New Guinea, where the largest and warmest body of ocean water can be found (see part a in figure to the right). Abundant rainfall occurs in this area during December, which is the high-sun season in the southern hemisphere.

During an El Niño event, the low-pressure system over the western Pacific is less intense and drought conditions replace the heavy rains. Air pressure drops in the equatorial zone of the mid-Pacific, and the high-pressure region in the eastern Pacific weakens. Heavy rainfall accompanies the strengthening of the equatorial trough in the central and eastern Pacific (part b). The shift in barometric pressure patterns is known as the *Southern Oscillation* (part c). The Southern Oscillation is calculated from the monthly differences in air pressure between Tahiti and Darwin, as follows:

$$\mathrm{SOI} = 10\,\frac{[P_{diff} - P_{diffav}]}{SD(P_{diff})}$$

where:

P_{diff} = (average Tahiti MSLP for the month) – (average Darwin MSLP for the month)

P_{diffav} = long-term average of pdiff for the month in question

$SD(P_{diff})$ = long-term standard deviation of P^{diff} for the month in question

The multiplication by 10 is a convention. Calculated in this way, the Southern Oscillation ranges from about −25 to about +25, and the value can be quoted as a whole number. The Southern Oscillation is usually computed on a monthly basis.

Surface winds and currents are affected by this change in pressure. During normal conditions, the strong prevailing trade winds blow westward, causing warm ocean water to move to the western Pacific and "pile up" near the western equatorial low. The characteristic airflow driven westward by the pressure gradient, known as the Walker circulation, results in the accumulation of warm ocean water in the western Pacific (see figure to the right). The ocean surface is some 60 cm higher in the western Pacific as a result of this motion. The westward flow causes normal upwelling along the South American coast, as bottom water is carried up to replace the water dragged to the west.

During an El Niño event, the easterly trade winds slacken with the change in atmospheric pressure. A weak westerly wind flow sometimes occurs, completely reversing the normal wind direction. Without the force of the trade winds to hold them back, warm waters expand eastward causing sea-surface temperatures and sea level to rise off the tropical western coasts of the Americas.

The major change in sea-surface temperatures that accompanies an El Niño can also shift weather patterns across large regions of the globe. Recurring, large-scale weather anomalies caused by changes in pressure and circulation patterns are termed teleconnections (see figures to the right). During El Niño events, torrential rains bring relief to the arid coastal regions of South America, while drier conditions prevail in Australia and the East Indies. Above-normal rainfall is reported in east Africa. In North America, winter storms are generally more intense along the Pacific coast, but winters are milder and snowfalls lighter across the Prairies.

A somewhat rarer phenomenon also capable of altering global weather patterns is *La Niña* (the girl child), a condition that is the antithesis of El Niño. During a La Niña period, sea-surface temperatures in the central and western Pacific Ocean fall to lower than average levels. This happens because the South Pacific subtropical high becomes strongly developed during the high-sun season. The result is abnormally strong southeast trade winds. The force of these winds drags a more-than-normal amount of warm surface water westward, which enhances upwelling along western continental coasts.

El Niño and La Niña show how dynamic Earth really is. As grand-scale, global phenomena, El Niño and La Niña illustrate how the circulation patterns of the ocean and atmosphere, coupled with energy exchange, interact to provide teleconnections capable of producing extreme events affecting millions of people throughout the world.

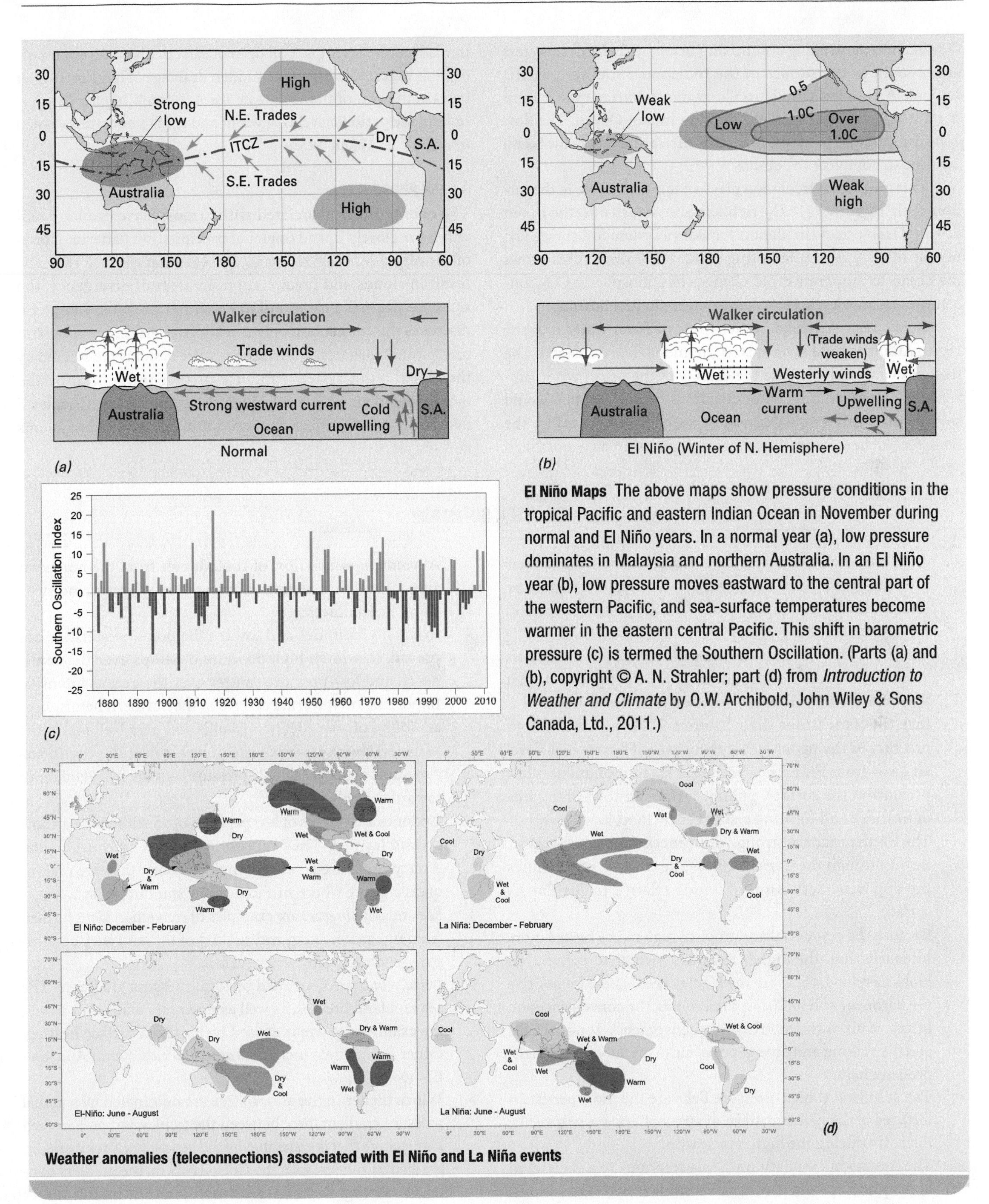

El Niño Maps The above maps show pressure conditions in the tropical Pacific and eastern Indian Ocean in November during normal and El Niño years. In a normal year (a), low pressure dominates in Malaysia and northern Australia. In an El Niño year (b), low pressure moves eastward to the central part of the western Pacific, and sea-surface temperatures become warmer in the eastern central Pacific. This shift in barometric pressure (c) is termed the Southern Oscillation. (Parts (a) and (b), copyright © A. N. Strahler; part (d) from *Introduction to Weather and Climate* by O.W. Archibold, John Wiley & Sons Canada, Ltd., 2011.)

Weather anomalies (teleconnections) associated with El Niño and La Niña events

Ocean. Here, upwelling and mixing occur, and the deep waters are brought to the surface of the Indian and southern Pacific Oceans. A coupled circulation loop moves surface water from the Pacific past Australia and into the Indian Ocean. The flow continues around the southern tip of Africa and enters the South Atlantic to complete the circuit.

Thermohaline circulation plays an important role in the carbon cycle by moving CO_2-rich surface waters into the ocean depths. Deep ocean circulation provides a system for storage and release of CO_2 in a cycle lasting about 1,500 years. This allows the ocean to moderate rapid changes in atmospheric CO_2 concentration, such as those produced by fossil fuel burning.

It has been suggested that inputs of fresh water into the North Atlantic could slow or stop thermohaline circulation. The fresh water would decrease the density of the ocean water, preventing it from sinking; without sinking, the circulation would stop. In turn, this would interrupt a major flow pathway for the transfer of heat from equatorial regions to the northern midlatitudes. The impact on deep ocean circulation from inputs of fresh water released by the sudden drainage of large meltwater lakes at the end of the Cenozoic Ice Age (see Chapter 18) may explain the periodic cycles of warm and cold temperatures noted over the past 12,000 years.

A Look Ahead

The energy flows associated with atmospheric pressure and winds are closely linked to global precipitation patterns. Zones of convergence create rising air currents that characteristically result in clouds and precipitation. In areas of divergence, the skies are normally clear and dry weather prevails. Chapter 6 discusses the connection between atmospheric circulation and precipitation processes. Surface pressure systems are steered by the global wind systems and are further influenced by the movement of air upward through the troposphere. Chapter 7 discusses how this affects the development of weather systems and storms.

CHAPTER SUMMARY

- The term *atmospheric pressure* describes the weight of air pressing on a surface of known area. Atmospheric pressure is measured by a barometer. Atmospheric pressure decreases rapidly as altitude increases.
- Wind occurs when air moves with respect to the Earth's surface. Air motion is produced by pressure gradients that are formed when air in one location is heated to a temperature that is warmer than another. Heating creates high pressure in the upper atmosphere, which moves high-level air away from the area of heating. This motion induces low pressure at the surface, pulling surface air toward the area of heating, and forming a convection loop.
- The Earth's rotation strongly influences atmospheric circulation through the *Coriolis effect*. The Coriolis effect causes the apparent deflection of winds relative to the Earth's surface.
- Because the equatorial and tropical regions are heated more intensely than the higher latitudes, two great convection loops develop—these are the Hadley cells. These loops drive the *northeast* and *southeast trade winds*, the convergence and lifting of air at the intertropical convergence zone (ITCZ), and the sinking and divergence of air in the subtropical high-pressure belts.
- The subtropical high-pressure belts are the most persistent features of the global pattern of atmospheric pressure They intensify during the high-sun season.
- The monsoon circulation of Asia responds to a reversal in atmospheric pressure over the continent with the seasons. A *winter monsoon* flow of cool, dry air from the northeast alternates with a *summer monsoon* flow of warm, moist air from the southwest.
- In the midlatitudes and toward the poles, westerly winds prevail. In winter, high pressure develops over the continents, and low-pressure centres over the oceans intensify. Important low-pressure centres in the northern hemisphere are found off the Aleutian Islands and near Iceland. In the summer, low-pressure centres develop over the continents as oceanic subtropical high-pressure cells intensify and move toward the poles.
- Cyclones are centres of low pressure in which surface airflow spirals inward. They represent regions of *convergence*. Anticyclones are centres of high pressure; they are regions of divergence where surface airflow spirals outward.
- *Sea* and *land breezes* are examples of *convection loops* formed from unequal heating and cooling of the land surface compared with a nearby water surface.
- *Local winds* are generated by local pressure gradients. The sea and land breezes, as well as *mountain* and *valley winds*, are examples of winds caused by localized surface heating. Other local winds include *drainage winds*, *Santa Ana*, and *Chinook winds*.
- Winds higher in the atmosphere are dominated by a global pressure gradient force between the tropics and pole in each hemisphere that is generated by the hemispheric temperature gradient. Coupled with the Coriolis effect, the gradient generates strong westerly *geostrophic winds* in the upper

troposphere. In the equatorial region, weak easterlies dominate the upper-level wind pattern.

- Rossby waves develop in the *upper-air westerlies*, bringing cold, polar air toward the equator and warmer air toward the poles. The *polar-front* and *subtropical* jet streams are concentrated westerly airflows with high wind speeds. The *tropical easterly jet stream* is weaker and limited to Southeast Asia, India, and Africa.
- Tropical and temperate oceans show a warm surface layer, separated from the deep cold water by the *thermocline*. Near the poles, the warm layer and thermocline are absent.
- *Ocean surface currents* are dominated by huge gyres driven by the global surface wind pattern. *Equatorial currents* move warm water westward and then poleward along the eastern coasts of continents. Return flows bring cold water toward the equator along the western coasts of continents.
- *El Niño* events occur when an unusual flow of warm water in the equatorial Pacific moves eastward to the coasts of Central and South America, suppressing the normal northward flow of the Peru Current. This greatly reduces upwelling along the Peruvian coast. El Niño events normally occur on a three- to eight-year cycle and affect global temperature patterns and precipitation in many regions.
- Slow, deep ocean currents are driven by the sinking of cold, salty water in the northern Atlantic. This *thermohaline circulation* pattern involves nearly all the Earth's ocean basins, and also acts to moderate the buildup of atmospheric CO_2 by moving CO_2-rich surface waters to ocean depths.

KEY TERMS

anticyclone
barometer
constant pressure surfaces
Coriolis effect
cyclone
depression
gyre
Hadley cell
ideal gas law
intertropical convergence zone (ITCZ)
jet stream
kilopascal
monsoon
polar front
pressure gradient
Rossby waves
subtropical high-pressure belts
thermal wind
Walker circulation
wind

REVIEW QUESTIONS

1. Explain atmospheric pressure. Why does it occur? How is atmospheric pressure measured and in what units? What is the normal value of atmospheric pressure at sea level? How does atmospheric pressure change with altitude?
2. Describe a simple convective wind system, explaining how air motion arises from a pressure gradient force induced by heating.
3. Describe land and sea breezes. How do they illustrate the concepts of pressure gradient and convection loop?
4. What is the Coriolis effect, and why is it important? What produces it? How does it influence the motion of wind and ocean currents in the northern hemisphere? In the southern hemisphere?
5. Define cyclone and anticyclone. How does air move within each? What is the direction of circulation of each in the northern and southern hemispheres? What type of weather is associated with each and why?
6. What is the Asian monsoon? Describe the features of this circulation in summer and winter. How is the ITCZ involved? How is the monsoon circulation related to the high- and low-pressure centres that develop seasonally in Asia?
7. Compare the winter and summer patterns of high and low pressure that develop in the northern hemisphere with those that develop in the southern hemisphere.
8. What are drainage winds? Give examples of local names that are applied to them.
9. How does global scale heating of the atmosphere create a pressure gradient force that increases with altitude?
10. What is the geostrophic wind, and what is its direction with respect to the pressure gradient force?
11. Describe the basic pattern of global atmospheric circulation at upper levels.
12. What are Rossby waves? Why are they important?
13. Identify five jet streams. Where do they occur? In which direction do they flow?
14. What is the general pattern of ocean surface current circulation? How is it related to global wind patterns?
15. How does thermohaline circulation induce deep ocean currents?
16. Discuss wind power as a source of energy. Include indirect use in tapping waves and ocean currents for energy.
17. Compare the normal pattern of wind, pressure, and ocean currents in the equatorial Pacific with the pattern during an El Niño event.
18. What are some of the weather changes reported for El Niño events?
19. What is La Niña, and how does it compare with the normal pattern?

VISUALIZATION EXERCISES

1. Sketch an idealized Earth (without seasons or ocean–continent features) and its global wind system. Label the following on your sketch: equatorial trough, Hadley cell, ITCZ, northeast trades, polar easterlies, polar front, polar outbreak, southeast trades, subtropical high-pressure belts, and westerlies.
2. Draw four spiral patterns showing outward and inward flow in clockwise and counter-clockwise directions. Use appropriate labels to identify cyclonic and anticyclonic circulation in the northern and southern hemispheres.

ESSAY QUESTIONS

1. An airline pilot is planning a non-stop flight from Vancouver, British Columbia, to Sydney, Australia. What general wind conditions can the pilot expect to find in the upper atmosphere? What jet streams will the aircraft encounter? Will they slow or speed the aircraft on its way?
2. You are planning to take a round-the-world cruise, leaving Montreal, Quebec, in October. Your vessel's route will take you through the Mediterranean Sea to Cairo, Egypt, in early December. Then you will pass through the Suez Canal and Red Sea to the Indian Ocean, calling at Mumbai, India, in January. From Mumbai, you will sail to Djakarta, Indonesia, and then go directly to Perth, Australia, arriving in March. Rounding the southern coast of Australia, your next port of call is Auckland, New Zealand, which you will reach in April. From Auckland, you head directly to Vancouver, British Columbia, your final destination, arriving in June. Describe the general wind and sea conditions you will experience on each leg of your journey.

WORKING IT OUT • PRESSURE AND DENSITY IN THE OCEANS AND ATMOSPHERE

Although it is sometimes useful to think of the Earth's surface as the bottom of a vast "ocean of air," the atmosphere is quite different from the oceans; the oceans are composed of a nearly incompressible liquid, whereas the atmosphere is composed of a mixture of readily compressible gases. Recall that both liquids and gases are classified as fluids.

The pressure at any level in a fluid is created by the weight of the fluid above that level. In the oceans, it's relatively simple to determine that weight. Pressing down on 1 m^2 of ocean at a given depth (D) is the weight of $D \times 1$ m^3 of water. Each cubic metre of water weighs about 1,000 kg. So as a quick rule of thumb, the pressure at any depth in the ocean is:

$$P_{OCEAN} = 1{,}000\text{D} = \text{D} \times 10^3 \text{ kg m}^{-2}$$

where P is the pressure in kilograms per square m (kg m^{-2}) and D is the depth in metres.

For example, the pressure at 500 m depth is about 500,000 kg m^2, which equals 500 $\times$ 10^3 kg m^{-2}. At 5,000 m below the ocean surface, the pressure is about 5,000,000 kg m^{-2}, or 5,000 $\times$ 10^3 kg m^{-2}.

These values are based on a constant density of 1,000 kg m^{-3}. To be precise, there is a slight variation in the density of sea water with pressure, temperature, and salinity. That is, colder, saltier, and deeper water will be denser. At the surface, the density of ocean water is 1,028 kg m^{-3}, while at a depth of 5,000 m the density is 1,051 kg m^{-3}, an increase of about 2.2 percent. In the figure on the facing page, the graph on the left, which plots pressure with depth in the ocean, uses a constant density value of 1,035 kg m^{-3}.

Note that at any depth in the ocean, there will be not only the weight of the water above, but also the weight of the atmosphere above the water. Sea-level pressure is about 100 kPa, which converts to 1,000 kg m^{-2}. Thus, the atmosphere adds the pressure of only about one more metre of depth.

Though small, the density differences in water caused by variations in temperature and salinity can affect the pressure at a given depth enough to create pressure gradients and therefore induce movement of water from higher to lower pressure.

In contrast to the oceans, the atmosphere is readily compressible. What is the weight of a cubic metre of air? There is no rule-of-thumb answer to that question because density varies with pressure. When a given volume of air is compressed, more molecules of the gases that compose air are present, and so the volume will have a greater mass and weigh more. At the Earth's surface, the density of air is about 1.225 kg m^{-3}, while at 11 km, the typical cruising altitude of an airplane, it is about 0.364 kg m^{-3}, or a little less than one third of the surface value.

The difference is nearly all due to pressure, but temperature is also important. At a given pressure, warmer air will be less dense than colder air. Temperature differences create pressure

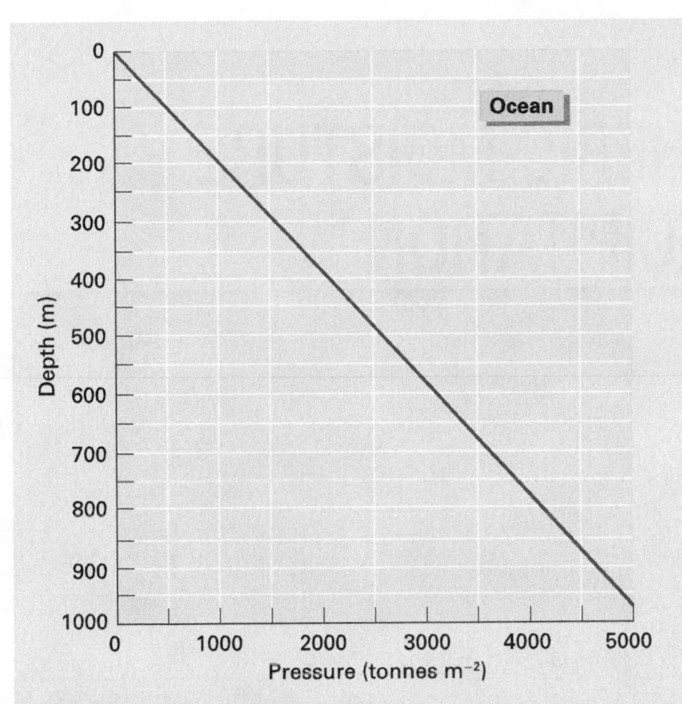

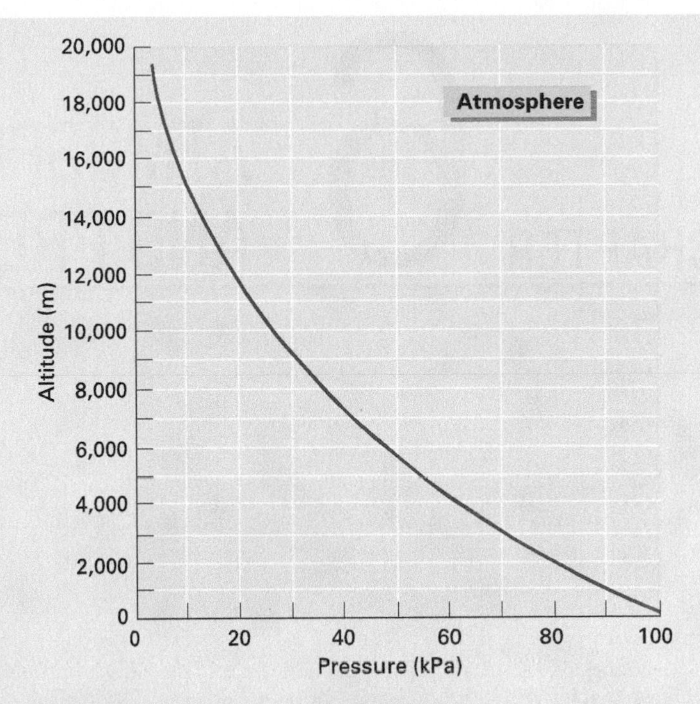

Pressure graphs (left) Graph of pressure with depth in the oceans. (right) Graph of pressure with altitude in the atmosphere.

gradients that induce air movements. The perpetual contrast between the warm atmosphere near the equator and cold atmosphere near the poles sustains the winds that constantly move and interact in the troposphere.

Atmospheric pressure is plotted against altitude in the graph on the right. Note that the graph curves, whereas the graph of pressure against ocean depth is straight. The atmospheric pressure graph curves because air is compressible and its density decreases with altitude.

Atmospheric pressure at any given altitude can be approximated by the following formula:

$$P_Z = 101.3\,[1 - 0.0226_Z]^{5.26}$$

where P_Z is the pressure in kilopascals at height z in kilometres. This formula takes into account the fact that both temperature and pressure decrease with altitude. Note that the value of 101.3 is the surface atmospheric pressure in kilopascals for standard conditions.

For example, to find the pressure at 12 km, which is approximately the altitude of a passenger jet on a transcontinental flight. Substituting for z,

$$\begin{aligned} P_Z &= 101.3 \times [1-(0.0226 \times 12)]^{5.26} \\ &= 101.3 \times (1-0.271)^{5.26} \\ &= 101.3 \times (0.729)^{5.26} \\ &= 101.3 \times 0.189 \\ &= 19.2 \text{ kPa} \end{aligned}$$

QUESTIONS / PROBLEMS

1. A diving pool is 5 m deep. What is the pressure on a person swimming at the bottom of the pool, including atmospheric pressure? What fraction of the pressure is due to the water, and what fraction is due to the atmosphere? Suppose the swimmer is now a deep-sea diver at a depth of 100 m. What proportion of the pressure is due to water and what proportion is due to the atmosphere?
2. The Yellowhead Pass in the Canadian Rockies is at an elevation of 1,110 m, while the summit of nearby Mount Robson is 3,954 m. Assuming a sea-level pressure of 101.4 kPa, what would you expect a barometer to read at these two locations? What percentage of sea-level pressure is each value?
3. In the course of the passage of a hurricane, a weather observer notes a change in barometric pressure from 101.6 kPa to 97.9 kPa, a difference of 3.7 kPa. Using the formula for atmospheric pressure with altitude, at what altitude would you expect to find a pressure of 97.9 kPa under normal conditions?

Find answers at the back of the book.

CHAPTER 6 ATMOSPHERIC MOISTURE AND PRECIPITATION

Rain over Johnson Strait, British Columbia.

This chapter focuses on water in the atmosphere, in its vapour, liquid, and solid forms. Most water enters the atmosphere by evaporation from the oceans. When water evaporates it takes up latent heat, which is later released during condensation. Thus, water plays a key role in the Earth's energy flows. Water returns to the Earth's surface in liquid or solid forms of precipitation. When it falls over land, precipitation provides the water or ice that carves the distinctive landforms that shape the continents.

Whenever air rises through the atmosphere, the decrease in pressure with altitude causes it to expand and cool. If the air is moist, this can lead to widespread condensation and the formation of clouds and precipitation. Air rises as a result of four major situations:

- when winds move air over a mountain barrier;
- when unequal heating at the ground surface creates a parcel of air that is warmer and less dense than the air that surrounds it;
- when convergence takes place, especially along the intertropical convergence zone (ITCZ); and
- when warm air rises over a mass of cooler, denser air to form centres of low pressure that lead to the development of mid-latitude cyclones.

This chapter focuses on the first three processes of uplift. The fourth is associated with the distinctive weather systems that develop along the polar front, and is discussed in Chapter 7.

THREE STATES OF WATER

Water can exist in three states—solid (ice), liquid (water), and gas (water vapour). A change of state from solid to liquid, liquid to gas, or solid to gas requires the input of heat energy (Figure 6.1). This energy, termed *latent heat*, is drawn in from the surroundings and stored within the water molecules. The reverse changes—from liquid to solid, gas to liquid, or gas to solid—require the release of latent heat.

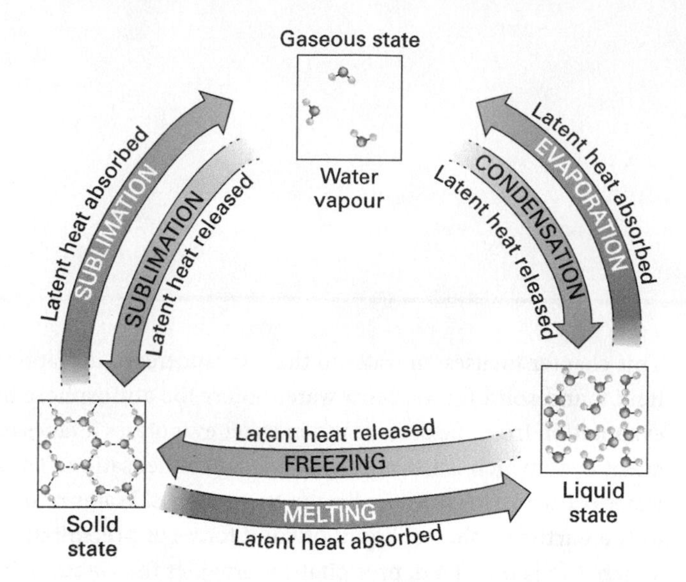

6.1 THREE STATES OF WATER Arrows show the ways that any one state of water can change into either of the other two states. Heat energy is absorbed or released, depending on the direction of change.

Each type of transition is known by a specific name. Melting, freezing, evaporation, and condensation are all familiar terms. **Sublimation** is the direct transition from solid to vapour. A common example would be the way old ice cubes appear to shrink when stored for a long time because of the constant circulation of cold, dry air in the freezer. The Meteorological Service of Canada also calls the reverse process, when water vapour crystallizes as ice, sublimation, although the term **deposition** is also commonly used. It is this process that can cause frost to form on a car windshield on a winter's night.

THE HYDROSPHERE AND THE HYDROLOGIC CYCLE

The **hydrosphere** includes water on the Earth in all its forms. Just over 97 percent of the hydrosphere consists of ocean water. The remaining 2.8 percent is fresh water. The largest reservoir of fresh water is the ice stored in the world's ice sheets and mountain glaciers. This water accounts for 2.15 percent of total global water.

Fresh liquid water is found both above and beneath the Earth's land surfaces. Most of the *subsurface water* is held in deep storage as *groundwater*, at a level where plant roots cannot access it. Groundwater makes up 0.63 percent of the hydrosphere. The remaining portion of the Earth's water (0.02 percent) is distributed in lakes, rivers, the soil, and the atmosphere. Although small in comparison to the total amount of water in the hydrosphere, it is important because it includes the water available for plants, animals, and human use. *Soil water*, which is found in the soil layer within reach of plant roots, comprises 0.005 percent of the global total. *Surface water* is found in streams, lakes, marshes, and swamps. Although most of this surface water is in lakes, about 50 percent is technically classed as saline. An extremely small proportion—about 0.0001 percent—is found in the streams and rivers that flow toward the sea or inland lakes.

The quantity of water held as vapour, cloud water droplets, and ice crystals in the atmosphere is about 0.001 percent of the hydrosphere. This is approximately equivalent to the amount of precipitation that normally falls on Earth over a period of 10 days. Though small, this reservoir of water is of enormous importance. It provides the supply of precipitation that replenishes all freshwater stocks on land. In addition, the transport of water vapour from warm tropical oceans to cooler regions provides a global flow of energy, in the form of latent heat, from low to high latitudes.

The movements of water among the great global reservoirs constitute the **hydrologic cycle** (see Figure 1.11). In the hydrologic cycle, water moves from land and ocean to the atmosphere as water vapour and returns as precipitation. Some water is absorbed into the soil or passes into deeper subsurface storage as groundwater. However, precipitation over land exceeds evaporation and storage, so surplus water also runs off the land to the oceans.

The cycle begins with **evaporation**. In this process, water from ocean or land surfaces changes state from liquid to vapour and enters the atmosphere. Total evaporation is about six times greater over oceans than land. This reflects the immense area of the oceans, which cover more than 70 percent of Earth's surface. The highest evaporation rates occur over tropical oceans. Many land surfaces are not always wet enough to yield much evaporated water, but evaporation is high from wet tropical landmasses, such as the Amazon and Congo basins, where it is supplemented by transpiration from the forest cover. Evaporation is low over tropical deserts, where the rate is similar to that in cold polar regions.

Once in the atmosphere, water vapour can condense into droplets, or directly form ice crystals through sublimation. The water eventually falls to Earth as **precipitation** in the form of rain, snow, or hail. Total precipitation over the oceans is nearly four times greater than over land. Precipitation is generally heaviest along the ITCZ, where the moist trade winds converge to form the rising limb of the Hadley cells (Figure 6.3). Precipitation is much lower in the subtropics, where air descends in the zone of semi-permanent high pressure. In the midlatitudes, precipitation is mainly associated with centres of low pressure that develop along the polar front, although convection is also important. Cold

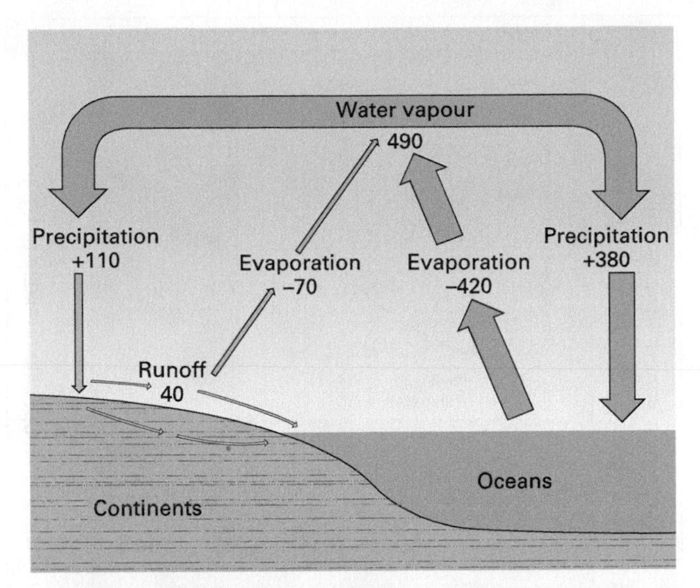

6.2 THE GLOBAL WATER BALANCE The figures give average annual water flows in and out of land areas and oceans. Values are in thousands of cubic kilometres. Global precipitation equals global evaporation. (Based on data of John. R. Mather.)

polar air holds little moisture, and precipitation is generally low in the high latitudes. The effects of mountain barriers, which may increase or decrease precipitation, are superimposed on these general regional patterns. Chapter 8 discusses global precipitation in more detail.

Precipitation that falls directly into the ocean can return immediately to the atmosphere through evaporation. However, on land surfaces, precipitation can follow three pathways. First, it can evaporate and return to the atmosphere as water vapour. Second, it can sink into the soil and then into the underlying rock layers, where it may be temporarily stored as groundwater reserves. Third, precipitation can flow from the land, concentrating in streams and rivers that carry it to a lake or to the ocean. This flow of water is known as surface *runoff*. Chapter 16 discusses the terrestrial components of the hydrologic cycle.

THE GLOBAL WATER BALANCE AS A MATTER FLOW SYSTEM

The Earth and its atmosphere contain a finite amount of water. Consequently, the exchange of water among the atmosphere, oceans, and continents must be in balance at the global scale. The assumption is that the volume of water in the oceans and the overall volume of fresh water on the surface and in subsurface storage remain constant from year to year.

Water moves through the hydrologic cycle primarily as liquid and solid forms in precipitation and as vapour in evaporation. On the continents, water is present in both its liquid and solid forms. The ocean surface is predominantly liquid, but the Arctic Ocean and some parts of the ocean near Antarctica have a cover of sea ice that may be capped with snow.

Flows of water link the atmosphere and the two surface components. For budgeting purposes, the convention is to describe flows to the surface (land or oceans) as positive and flows leaving the surface as negative. Since the cycle is in balance, flows into and out of the atmosphere add up to zero (see Figure 6.2 for flow values). That is,

$$P_{(land + lake)} + E_{(land + lake)} + P_{ocean} + E_{ocean} = 0$$
$$(+110{,}000) + (-70{,}000) + (+380{,}000) + (-420{,}000) = 0$$

where $P_{(land + lake)}$ and P_{ocean} are precipitation flows from the atmosphere to the land and oceans. $E_{(land + lake)}$ and E_{ocean} are evaporation flows from the land and oceans to the atmosphere. Precipitation includes flows of water as rain and snow, as well as the direct sublimation of water vapour as frost. Evaporation includes sublimation of snow and ice to water vapour.

Although the global balance for the atmosphere is zero, there is a positive balance for land (110,000 – 70,000 = + 40,000 km^3), indicating more precipitation than evaporation, and a negative balance for oceans (380,000 – 420,000 = –40,000 km^3), indicating more evaporation than precipitation. Thus, land and lakes gain 40,000 km^3 of water each year, while oceans lose 40,000 km^3 of water. Without a link between land and the oceans, the oceans would eventually empty and all the water would accumulate over

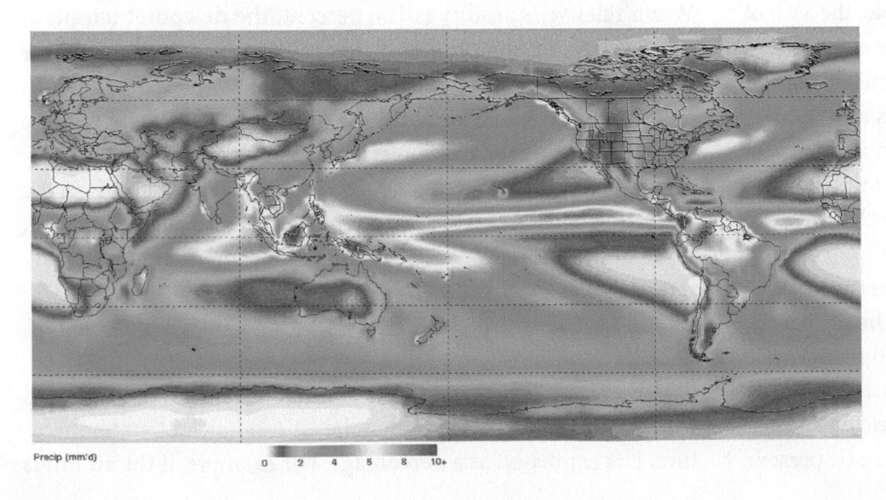

6.3 AVERAGE DAILY PRECIPITATION Daily precipitation is highest in equatorial regions, but considerably less rain falls in areas dominated by the subtropical high-pressure cells, despite the comparatively high water vapour content of the overlying atmosphere. Data from The Global Precipitation Climatology Project (GPCP) (1979–2000).

land. This doesn't happen because a flow pathway connects land and oceans—runoff of water from rivers, glaciers, and ice sheets into the ocean. For the oceans to stay at the same level, and for the same amount of water to be stored on the land year after year, the flow in runoff must equal 40,000 km^3. Current sea level reflects present global temperatures and is sensitive to global warming or cooling insofar as this affects the amount of water in storage as ice.

All matter flow systems need a power source. For the global water flow system, the power source is the sun. Solar energy evaporates water and moves it into the atmosphere, where later it falls to the Earth as precipitation. Gravity plays a further role in moving streams of water and ice off the land and into the oceans as runoff.

HUMIDITY

The amount of water vapour present in the air varies widely from place to place and over time. It is negligible in the cold, dry air of Arctic regions in winter, but can be as much as 4 or 5 percent by volume of air in the warm wet regions near the equator. The amount of water vapour present in the air is commonly reported as **relative humidity**, but this is just one of several terms used in meteorology.

Dalton's law of partial pressure states that the pressure of a mixture of gases is equal to the sum of the pressures of all the constituent gases. For example, if atmospheric pressure is 1,000 hPa and the air is composed of 78 percent nitrogen, 21 percent oxygen, and 1 percent water vapour, then the partial pressures of the gases would be 780 hPa, 210 hPa, and 10 hPa, respectively. The contribution that water vapour makes to the total pressure exerted by the atmosphere is referred to as **vapour pressure**. Assuming the supply of water vapour is not restricted, as is generally the case over the oceans, then the quantity of water vapour in the atmosphere is related to air temperature and increases as temperature increases. However, for any given temperature there is a maximum amount of water vapour that can be held by the atmosphere. When this upper limit is reached, the air is said to be saturated and the pressure exerted by the water molecules under this condition is referred to as **saturation vapour pressure (SVP)**. For example, the SVP of air at 0 °C is approximately 6 hPa, and rises to 17 hPa at 15 °C and 42 hPa at 30 °C (Figure 6.4). Note that a water molecule, comprising two hydrogen atoms and one oxygen atom, has a molecular mass of 18.02. Dry air, composed mainly of nitrogen and oxygen, has an equivalent molecular mass of 28.57. Thus, the total pressure exerted by moist air is actually slightly lower than the pressure exerted by dry air at the same temperature.

Specific humidity (q) compares the actual mass of water vapour held by a parcel of air to the total mass of the air parcel; it is expressed as grams of water vapour per kilogram of air ($g\ kg^{-1}$). Specific humidity therefore is a measure of how much water vapour is available for precipitation. *Saturation specific humidity* (q_s) is the maximum mass of water that can be present

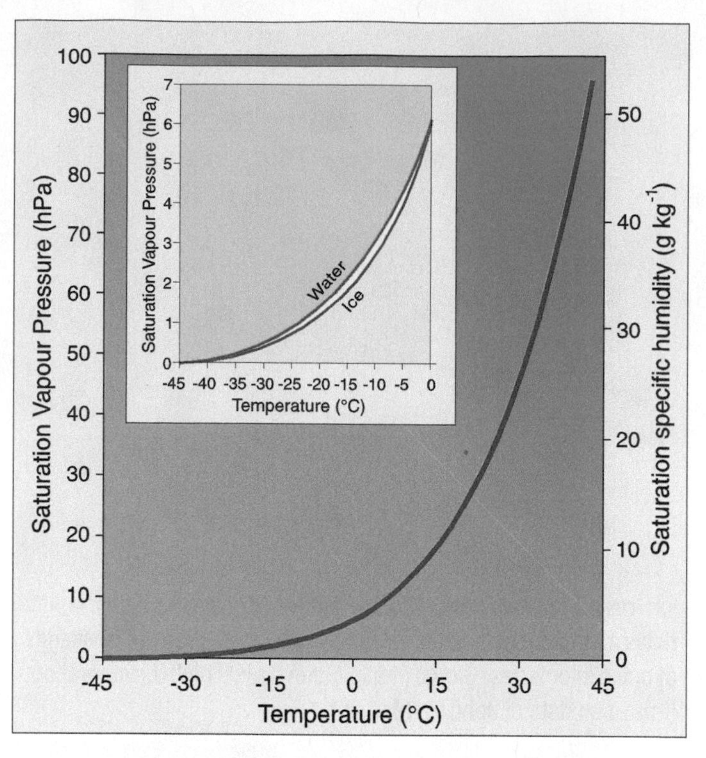

6.4 SATURATION VAPOUR PRESSURE (SVP) AND SATURATION SPECIFIC HUMIDITY (q_s) INCREASE WITH TEMPERATURE At temperatures below freezing, the SVP is greater over super-cooled water than over ice, because more energy is needed to break the molecular bonds in an ice crystal. (From *Introduction to Weather and Climate* by O.W. Archibold, John Wiley & Sons Canada, Ltd., 2011.)

at a given temperature. For example, saturation specific humidity at 10 °C is approximately 7 grams of water vapour, per kilogram of air and rises to 26 $g\ kg^{-1}$at 30 °C (see Figure 6.4).

Another way to describe the water vapour content of air is by the **dewpoint temperature**. This is the temperature to which air must be cooled for saturation to occur. At dewpoint temperature, the air holds the maximum amount of water vapour, and if cooling continues, the water vapour will begin to condense. When relative humidity is 100 percent, the dewpoint temperature and actual air temperature are the same, but at relative humidities below 100 percent the dewpoint temperature is always lower than the actual air temperature. Clouds will begin to form when the temperature of moist air falls below the dewpoint temperature. At ground level, cooling to below dewpoint leads to the formation of dew, or frost when temperatures are below freezing. If the dewpoint temperature is below freezing, then it is usually referred to as the **frost point** (Figure 6.5).

RELATIVE HUMIDITY

Relative humidity compares the amount of water vapour present with the maximum amount that the air can hold at that temperature. It is expressed as a percentage. For example, if the air holds

6.5 HOARFROST Hoarfrost or rime frost forms when the dewpoint temperature is below freezing.

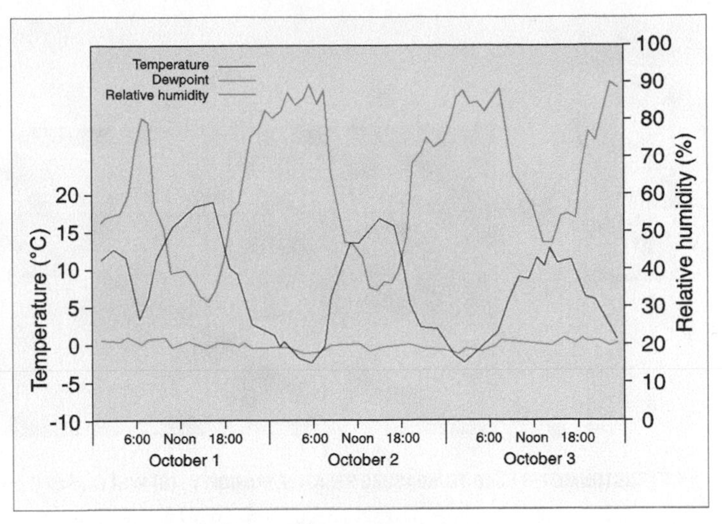

6.6 RELATIVE HUMIDITY, AIR TEMPERATURE, AND DEWPOINT TEMPERATURE FOR KELOWNA, BRITISH COLUMBIA, FOR OCTOBER 1–3, 2006 Relative humidity changes with temperature because the capacity of warm air to hold water vapour is greater than that of cold air, even though the amount of water vapour in the air remains more or less constant.

half the moisture possible for a given temperature, then relative humidity is 50 percent. The general formula is:

$$\text{Relative humidity} = \frac{\text{water vapour content}}{\text{water vapour capacity}} \times 100$$

A change in atmospheric relative humidity can occur in one of two ways. The first is through direct gain or loss of water vapour. For example, if an exposed water surface or wet soil is present, additional water vapour can evaporate into the air, raising the humidity. This process is slow because the water vapour must diffuse upward from the surface into the air above.

The second way is by lowering the temperature of the air. Even though no water vapour is added, relative humidity rises because the air's capacity to hold water vapour is lowered. The existing amount of water vapour then represents a higher percentage of the maximum potential capacity. For this reason, relative humidity is generally highest at night, when air temperatures are lower. For example, at 2 p.m. on October 2, 2006, air temperature at Kelowna was 16.3 °C, relative humidity was 35 percent, and the dewpoint temperature 0.6 °C (Figure 6.6). By 7 a.m. the following morning, air temperature had fallen to 3.9 °C, relative humidity reached 93 percent, and the dewpoint temperature was 2.7°C. From the graph of temperature versus saturation vapour pressure (Figure 6.4), the SVP at 3.9°C is approximately 8 hPa, and at 16.3 °C it is approximately 18 hPa. The SVPs for the recorded dewpoint temperatures of 0.6 °C and 2.7 °C are approximately 6 hPa and 7.5 hPa, respectively. The SVP information can be used to calculate relative humidity as:

$$\text{Relative humidity} = \frac{\text{SVP of dewpoint temperature}}{\text{SVP of air temperature}} \times 100$$

Thus, relative humidity at 2 p.m. on October 2 is:

$$6/18 \times 100 = 33\%$$

and at 7 a.m. on October 3 is:

$$7.5/8 \times 100 = 94\%$$

There was little change in the amount of water vapour present in the air during this three-day period, as indicated by the dewpoint temperature. However, the capacity of the air to hold water vapour increased during the warmer daytime hours, causing relative humidity to fall. With lower air temperature at night, relative humidity increased, and the diurnal cycle repeated.

Humidity can be measured in various ways. The traditional method uses two matched thermometers mounted together side by side in an instrument called a wet-and-dry-bulb thermometer. The wet-bulb thermometer is covered with a cotton sleeve that is kept moist by distilled water drawn up from a reservoir. The dry-bulb thermometer is a standard thermometer (Figure 6.7a). When the thermometers are ventilated, the wet-bulb temperature is lowered by evaporative cooling, provided the air is not saturated. If the air is saturated, then evaporation cannot occur. The drier the air, the greater the rate of cooling, and consequently there is a greater difference in temperature between the two thermometers. A chart is used to determine relative humidity from the wet-bulb and dry-bulb temperatures (Figure 6.8).

The *sling psychrometer* (Figure 6.7b) is a variant of the wet-and-dry-bulb thermometer, but in this instrument the thermometers are ventilated by whirling them around. A record of humidity over a period of time can be obtained using

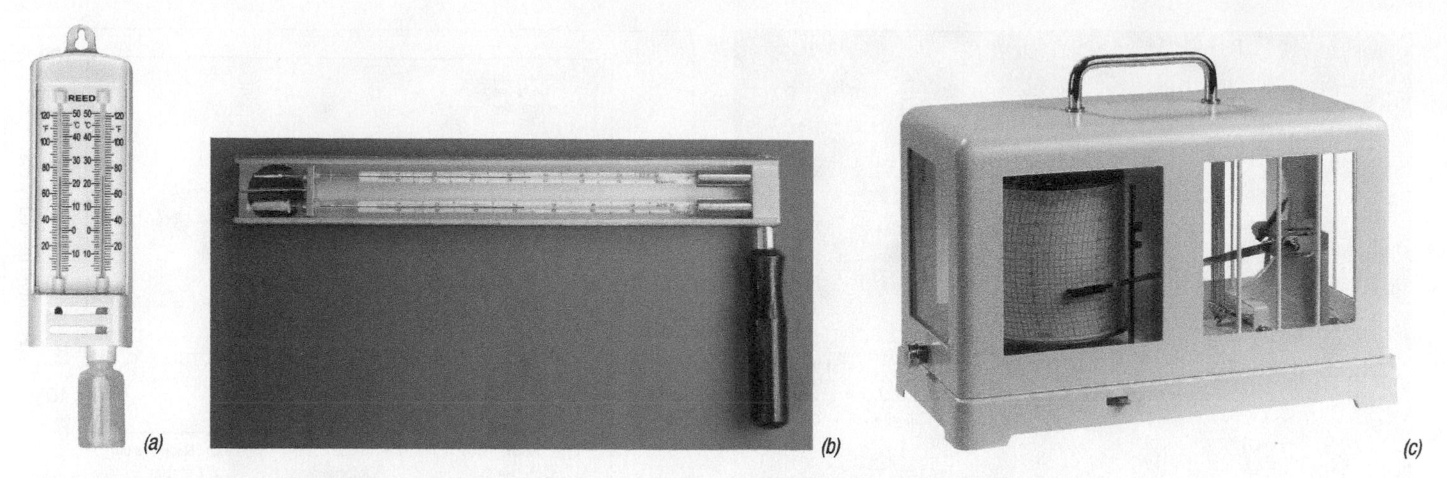

6.7 **INSTRUMENTS USED TO MEASURE RELATIVE HUMIDITY** (a) wet and dry bulb thermometer; (b) sling psychrometer; (c) hair hygrograph.

a *hair hygrograph* (Figure 6.7c). The hair hygrograph works on the principle that human hair changes in length as it absorbs or loses moisture. Modern humidity sensors use electronic circuitry to measure the change in capacitance that occurs when a hygroscopic substance absorbs or releases water in response to changes in relative humidity. The change in moisture affects the ability of the hygroscopic substance to hold an electric charge. Like any electronic sensor, this technology provides data in a form that can be processed directly by computers, and so has become standard equipment for automatic, unmanned weather stations.

THE ADIABATIC PROCESS

If there is sufficient water vapour present in a mass of air, and it can be cooled to the dewpoint temperature, some of it will begin to form water droplets; ice crystals will form if the air is cooled to the frost point temperature. For example, radiative heat loss at night cools the ground surface and the air in contact with it. If air temperature falls below the dewpoint, condensation or sublimation will give rise either to dew or frost. In this way, small patches of mist often develop over lakes and ponds (Figure 6.9). However, radiative heat loss cannot produce the copious condensation necessary for widespread precipitation. Precipitation is formed only when a substantial mass of air experiences a steady drop in temperature to below the dewpoint. For abundant condensation to occur, a parcel of air must be lifted to a higher level in the troposphere. As the air parcel rises, it is cooled *adiabatically*.

DRY ADIABATIC LAPSE RATE

An important principle of physics is that when a gas is allowed to expand, its temperature drops. Conversely, when a gas is compressed, its temperature rises. Heating or cooling that results

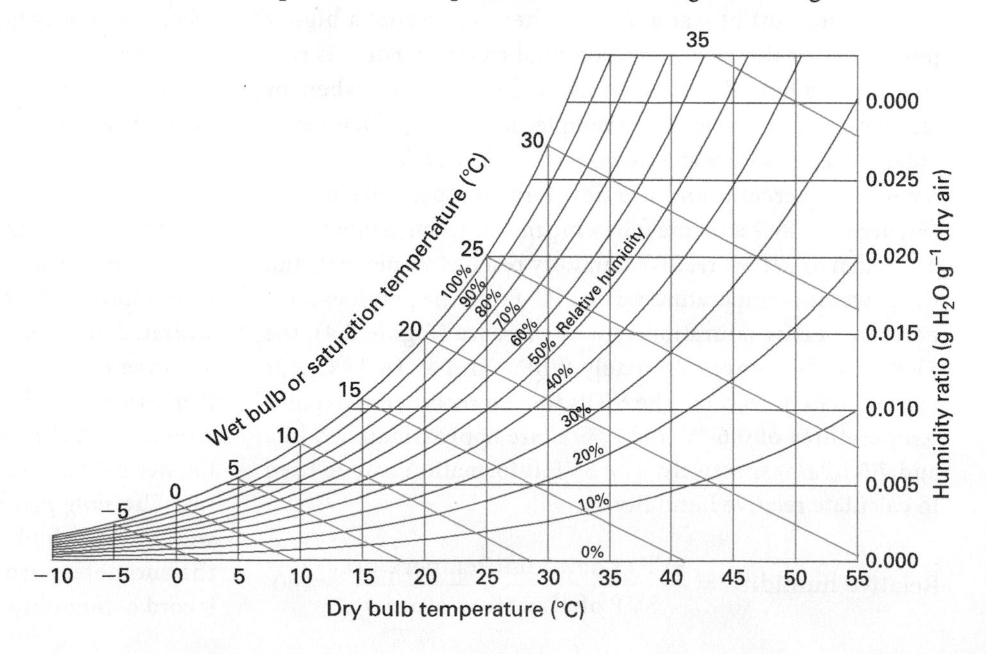

6.8 **PYSCHROMETRIC CHART** A Psychrometric chart is used to determine relative humidity from the wet-bulb and dry-bulb temperatures.

6.9 **TEMPERATURES BELOW DEWPOINT** Mists over lakes and rivers often form through radiative cooling during the night, but nocturnal cooling of this type does not result in precipitation.

solely from a change in pressure is an **adiabatic process**. Because atmospheric pressure decreases with altitude, the pressure exerted on a parcel of air that rises through the troposphere also decreases causing the air parcel to expand. The air molecules must do work as they expand, and this affects the temperature of the air parcel because temperature is a measure of the kinetic energy of the molecules. During expansion, the total amount of energy in the parcel remains the same—none is added or lost. The air parcel can either use this energy to do the work of expansion or to maintain its temperature, but it cannot use it for both. Consequently, when a parcel of air expands, its temperature must drop. Conversely, when a parcel of air sinks, it is compressed by the higher pressure exerted on it closer to the Earth's surface, and so it warms.

If the parcel of air is unsaturated, then its temperature, when displaced vertically, changes at the **dry adiabatic lapse rate (DALR)**. Under these conditions, as the air parcel rises through the atmosphere it will *cool* at a rate of 10 °C per 1,000 m (10 °C 1,000 m^{-1}). If it sinks, it will *warm* at the same rate.

The dry adiabatic rate applies to a mass of air that is moving up or down through the atmosphere. The rate of cooling or heating is always constant and is determined by physical laws. Note that the dry adiabatic lapse rate is quite different from the normal and environmental temperature lapse rates discussed in Chapter 4. These lapse rates refer to the change in temperature of air with altitude in the absence of any vertical motion of the air. The environmental temperature lapse rate will vary from time to time and from place to place, depending on the state of the atmosphere, with the average condition given by the normal temperature lapse rate.

SATURATED (WET) ADIABATIC LAPSE RATE

Assume that an air parcel at sea level has a temperature of 20 °C and is forced to rise upward through the troposphere (Figure 6.10).

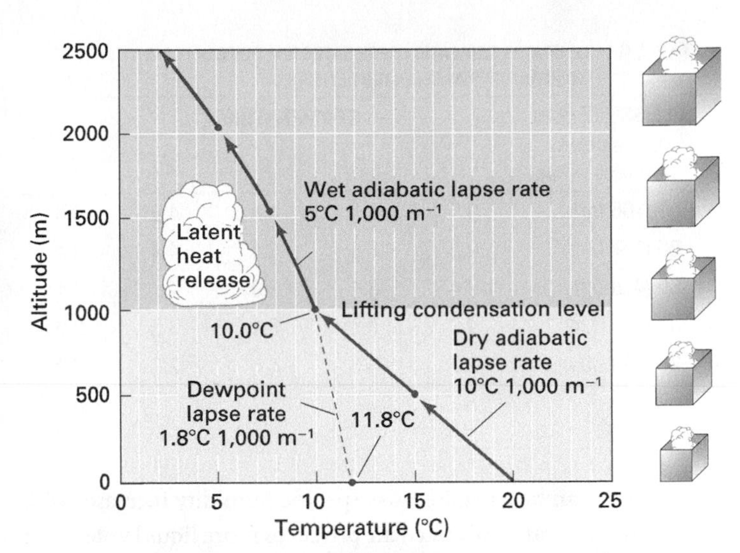

6.10 **ADIABATIC COOLING** An adiabatic decrease in temperature in a rising parcel of air leads to condensation of water vapour and the formation of clouds. (Copyright © A. N. Strahler.)

As it rises and expands, its temperature drops at the dry adiabatic rate of 10 °C 1,000 m^{-1}, so when it reaches 1,000 m, its temperature will have fallen to 10 °C. The dewpoint temperature also changes slightly with altitude at a rate known as the *dewpoint lapse rate*, which is 1.8 °C 1,000 m^{-1}. Given that the initial dewpoint temperature of the air mass was 11.8 °C, at 1,000 m the dewpoint temperature will be 10 °C. This is also the temperature of the lifted air parcel, so condensation will begin at that level. The altitude at which the air reaches this condition is called the **lifting condensation level**. The lifting condensation level can therefore be determined from the initial temperature of the air and its dewpoint. The *Working It Out* feature in the end-of-chapter material shows how to calculate the lifting condensation level given these two lapse rates.

If the air continues to rise above the lifting condensation level, condensation will occur and water droplets will form, producing a cloud. However, the parcel air is now saturated, so if it does continue to rise, latent heat is released at the rate of 2,260 kJ kg^{-1} (see *Working It Out*). This release of latent heat warms the rising air. Consequently, adiabatic cooling is offset by the addition of heat through condensation.

More energy is released through adiabatic cooling than is gained by condensation, so the air will continue to cool as it rises. However, the release of latent heat slows the rate of cooling; this slower rate of cooling for saturated air is called the **saturated (wet) adiabatic lapse rate (SALR)**. The SALR ranges from 4 to 9 °C 1,000 m^{-1}. Unlike the dry adiabatic lapse rate, which remains constant, the wet adiabatic lapse rate is variable because it depends not only on the temperature and pressure of the air, but also its moisture content. For most situations, however, a value of 6 °C

TABLE 6.1 SATURATED ADIABATIC LAPSE RATES FOR DIFFERENT AIR TEMPERATURES AT SELECTED ALTITUDES

PRESSURE (and approximate altitude) (kPa)	TEMPERATURE (°C)				
	−40	−20	0	20	40
100 (100 m)	9.5	8.7	6.8	4.8	3.6
80 (1,950 m)	9.5	8.5	6.4	4.4	3.4
60 (4,200 m)	9.4	8.1	5.9	4.1	3.1
40 (7,200 m)	9.2	7.5	5.1	3.5	2.7
20 (12,000 m)	8.7	6.3	4.0	2.8	2.2

1,000 m^{-1} can be used. Because specific humidity increases with temperature, warm saturated air produces more liquid water than cold saturated air, and so releases more latent heat. Consequently, the wet adiabatic lapse rate is much less than the dry adiabatic lapse rate if the rising air is warm, but is similar if the rising air is cold (see Table 6.1). For example, the SALR for moist air at a temperature of 40 °C originating at 100 m is 3.6 °C 1,000 m^{-1}, compared to 6.8 °C 1,000 m^{-1} if the air temperature is 0 °C, and 9.5 °C 1,000 m^{-1} if its temperature is −40 °C. Note that in Figure 6.10, the wet adiabatic rate is shown as a slightly curving line to indicate that its value changes with altitude.

CLOUDS

Views from space show that clouds cover about half of the Earth at any given time. Clouds are an important variable in Earth's climate system because of the way they control incoming and outgoing radiation. Low clouds are quite thick and so reflect a considerable amount of solar energy; they also absorb energy that is then radiated toward and away from Earth's surface. The net effect of low clouds is to cool the Earth-atmosphere system. Less energy is reflected by the thinner high clouds, but they do absorb some outgoing long-wave radiation. On balance, the high clouds contribute to a net warming of the Earth-atmosphere system (see Chapter 3).

A **cloud** is made up of water droplets or ice crystals suspended in air; these range in diameter from 20 to 50 μm (about 0.002 to 0.005 mm). Most cloud particles form around a tiny centre of solid matter, called a *cloud condensation nucleus (CCN)*. Cloud condensation nuclei have diameters in the range 0.1 to 1 μm. An important source of condensation nuclei is the surface of the sea. When winds create waves, droplets of spray from the crests of the waves are carried upward into the turbulent air. Evaporation of sea water droplets leaves a tiny residue of crystalline salt suspended in the air, which is added to the aerosol load of the atmosphere. Salt aerosols are hygroscopic and strongly attract water molecules. Polluted air over cities is another source of condensation nuclei and can contribute to hazes when in large concentrations. However, even clean air contains sufficient condensation nuclei from natural sources to promote cloud formation. In addition to salt crystals, other natural sources include fine clay particles, soot from grassland and forest fires, and sulphate aerosols produced by marine phytoplankton.

Condensation of water vapour on the nuclei initiates the formation of liquid cloud droplets, and their growth continues as long as water vapour condenses onto them. Liquid droplet growth depends on the humidity of the air, as this controls the rates at which either condensation or evaporation will occur. If the saturation vapour pressure (SVP) in the environment is greater than the SVP in the droplet, the droplet grows due to condensation. If the SVP in the environment is less than the SVP in the droplet, the droplet shrinks through evaporation. After their initial growth, the SVP of a liquid cloud droplet is determined by its *size* and *purity*; these properties affect subsequent growth through the **curvature effect** and the **solute effect**.

The curvature effect is related to the diameter of the droplet, and arises because the surface of a small spherical droplet has greater curvature than the surface of a large sphere. Saturation vapour pressure increases as the droplet size becomes smaller. Consequently, it is easier for water to evaporate from, and harder for water vapour to condense to, a smaller droplet than a larger one. Because smaller droplets lose water at a faster rate than larger droplets, a degree of suspersaturation is required for them to form and grow.

If a cloud is composed of pure water droplets of varying sizes, the smaller ones will tend to evaporate because of the curvature effect. However, the solute effect compensates for this, and allows droplets to develop when air is below saturation. The solute effect occurs because cloud droplets form around condensation nuclei, such as salt crystals, and are rarely composed of pure water. Because equilibrium SVP for solutions is lower than for pure water, condensation begins on these hygroscopic particles before the air is saturated. For small droplets, the SVP is much less than that for pure water. For example, sodium chloride (sea salt) crystals can initiate condensation at relative humidities as low as 80 percent, and this is the principal cause of the haze that often develops in marine areas. Because of the solute effect, droplet growth is dependent on its chemistry: those with a higher salt concentration will grow faster. Through these processes, a range of droplet sizes develops within a cloud, which is important for the subsequent development of precipitation.

Typically, water at the Earth's surface turns to ice when the temperature falls below 0 °C. However, when water is dispersed as tiny droplets in clouds, it can remain in the liquid, *super-cooled* state at temperatures far below freezing. However, freezing will occur if **ice nuclei** are present in the atmosphere. A common form of ice nuclei is fine particles of kaolinite clay; these become active at −10 °C. Hence clouds may consist mostly of water droplets at temperatures down to about −10 °C; however, below that temperature, a mix of water droplets and ice crystals occurs. The coldest clouds, with temperatures below −40 °C, occur at higher altitudes and are formed entirely of ice particles.

CLOUD FORMS

Meteorologists classify the variety of cloud forms into four families distinguished by their height and development as the high, middle, and low clouds, and clouds with strong vertical development (Figure 6.11). The high, middle, and low clouds are further classified on the basis of their general shape.

Two major classes are recognized on the basis of form—stratiform clouds and cumuliform clouds. *Stratiform* clouds are extensive, layered clouds which generally cover large areas. A common type is low-altitude *stratus*, which forms when moist air layers spread horizontally with little vertical uplift. This can happen when a layer of warm moist air passes over a cooler air layer. As the overriding layer is cooled, condensation occurs and a blanket-like cloud cover forms. If the overriding layer is quite moist and some vertical lifting occurs, denser nimbostratus clouds may develop, which may give rise to prolonged rain or snow showers.

Cumuliform clouds are associated with rising parcels of air and are distinguished from stratiform clouds by their stronger vertical development and globular, puffy appearance (Figure 6.12). They may appear alone or in groups, and they rise to different altitudes depending on the prevailing atmospheric conditions. Cumuliform clouds develop when air is forced to rise. In most cases this is because a parcel of air becomes warmer than the surrounding air. As it is buoyed upward, the air parcel is cooled adiabatically, and once it reaches condensation level, a cloud will begin to form. The most common cumuliform clouds are the *cumulus* clouds, which are characteristically seen on sunny days when the ground is heated sufficiently for convection cells to develop in the overlying air. Under these conditions thermals begin to develop, which rise, expand, and cool.

As the air rises, it mixes with the surrounding air and gradually loses its identity. The process continues until eventually a visible cumulus cloud forms. Although air moves upward within a cumulus cloud, air motion is downward on its edges. These subsiding currents are caused by evaporation around the outer edges of the cloud; this cools the air and makes it denser. The downward airflow completes the convection cell started by the thermal.

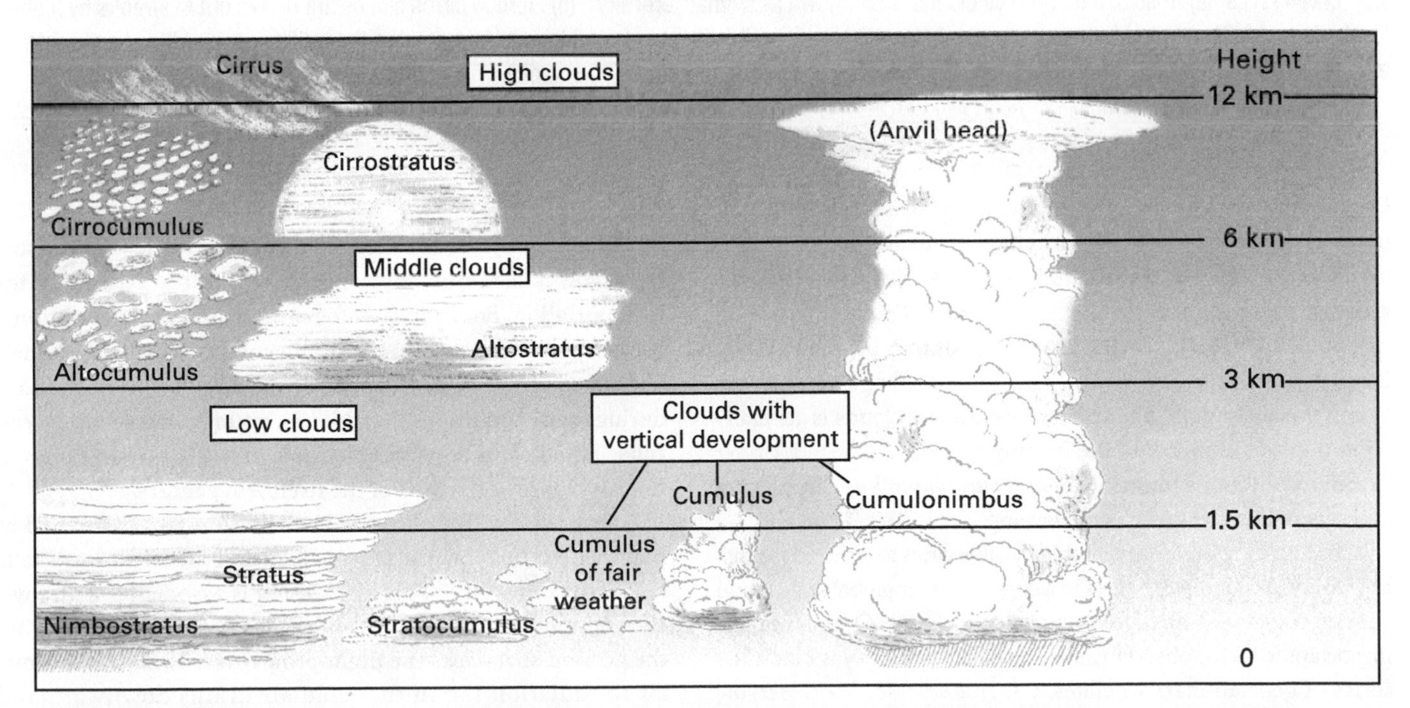

6.11 CLOUD FAMILIES AND TYPES Clouds are grouped into families on the basis of height. Individual cloud types are named according to their form.

6.12 **CLOUD TYPES** (a) Rolls of altocumulus clouds, also known as a "mackerel sky." (b) Fibrous cirrus clouds are drawn out in streaks by high-level winds. (c) Stratocumulus clouds form a broken layer of small puffy clouds and tend to form toward the trailing edge of the warm sector of a midlatitude wave cyclone. (d) Lenticular clouds, a type of altostratus formed when moist air is uplifted as it crosses a range of hills.

Because subsiding air currents counteract the uplift caused by heating, this usually results in the formation of small clouds with blue sky between them. Also, as cumulus clouds grow they shade the ground, which reduces surface heating and further restricts thermal uplift and cloud formation. New clouds may begin to form as the previous clouds drift downwind and thermal uplift is renewed. When the air is moist, the size of the cumulus clouds is generally related to the intensity of the heating that has occurred. Under extreme conditions towering, dense, *cumulonimbus* clouds form and severe thunderstorms will develop.

Individual cloud forms provide visible clues to air motions that are otherwise not discernible. On a continental or global scale, cloud patterns also reveal major features of air circulation. One of the most important tools for observing cloud patterns remotely is the GOES series of geostationary satellites. *Geographer's Tools • Observing Clouds from GOES* shows some examples of GOES images.

FOG

Fog is simply a layer of stratus cloud at or very close to the surface. Fog is a major environmental hazard, especially for transportation. For centuries, fog at sea has been a navigational hazard, increasing the danger of ship collisions and groundings. Highways, railways, and airports are also subject to restrictions during foggy conditions. In addition, polluted fog—**smog**—can cause lethal respiratory problems, as was the case in London, England, in the early part of the twentieth century.

One type of fog, known as *radiation fog*, is formed at night when the temperature of the near-ground air layer falls below the dewpoint. This kind of fog is associated with low-level temperature inversions and typically forms upward from the ground surface as the night progresses. *Valley fog* is similar to radiation fog in that cold air drains downward and collects in low-lying areas, chilling the overlying moist air

GEOGRAPHER'S TOOLS

OBSERVING CLOUDS FROM GOES

Some of the most familiar images of Earth from space are those acquired by the Geostationary Operational Environmental Satellite (GOES) system. Images from the GOES series of satellites and its predecessors have been in constant use by meteorologists and weather forecasters since 1974. The primary mission of the GOES series is to view cloud patterns and track weather systems by providing frequent images of the Earth from a consistent viewpoint in space. This capability assists with forecasting storms and severe weather.

A key feature of the GOES series is its geostationary orbit, which keeps the satellites above the same point on the equator. From this vantage point, the satellites can acquire a constant stream of images of the Earth below. Geostationary satellites are therefore ideally suited for viewing clouds and tracking cloud movements.

Six GOES satellites are currently in use, and together they provide relatively undistorted coverage of Earth's surface in a latitudinal band from about 55° N to 55° S. NOAA's *GOES-West (GOES-11)* at 135° W and *GOES-East (GOES-13)* at 75° W provide coverage for North and South America. Europe's *Meteosat-9* at 0° and *Meteosat-7* at 57.5° E covers Europe and Africa, India's *INSAT-3A* at 93.5° E provides coverage for central Asia, and Japan's *MTSAT-1R* at 140° E covers eastern Asia, Australia, and the western Pacific. *GOES-12* was in on-orbit storage at 105° W since it was launched in May 2006, but in May 2010 it started operations over South America. *GOES-14* was launched on June 24, 2009, and became operational in December of that year, although it is currently classified as a backup satellite. *GOES-15* is expected to be operational in 2015.

From its position above long. 135° W, the GOES-West satellite can track Pacific storm systems as they approach North America and move across western Canada and the United States. The GOES-East satellite at long. 75° W observes weather systems in the eastern part of the continent. GOES-East also observes the tropical Atlantic, identifying tropical storms and hurricanes as they form and move toward the Caribbean Sea and along the east coast.

The present generation of GOES platforms carries two primary instruments—the GOES Imager and the GOES Sounder. The Imager acquires data in five spectral bands, ranging from the visible red to the thermal infrared. It provides images of clouds, water vapour, surface temperature, winds, albedo, fires, and smoke. The Sounder uses 19 spectral channels to observe atmospheric profiles of temperature, moisture, ozone, and cloud height and cover.

Figure 1 was acquired from the GOES-East (GOES-12) satellite on June 8, 2009, when the entire side of the Earth nearest to the satellite was illuminated. The brown and green tones in the image are derived from visible band data. Vegetated areas appear green, while semi-arid and desert landscapes appear yellow-brown. The dark blue primary colour is derived from a thermal infrared band, but is scaled inversely so that cold areas are light and warm areas are dark. Since clouds are bright white in visible wavelengths and are colder than surface features, they appear white. The oceans appear blue because they are dark in the visible bands, but are still comparatively warm.

Figure 1 Earth from GOES-12 The GOES-12 geostationary satellite acquired this image on June 8, 2009.

For weather forecasting, an important tool of the latest generation of GOES imagers is the water vapour image; the images in Figure 2 are from GOES-West and GOES-East on December 19, 2009. The brightest areas show regions of active precipitation—note the numerous convective storms in the equatorial zone and the swirling forms of the midlatitude cyclones in the North and South Pacific. Water vapour images are processed so that areas with the highest atmospheric water vapour appear brightest; dark areas indicate low water vapour content.

Figure 2 Water vapour images acquired by (a) GOES-West (GOES-11) and (b) GOES-East (GOES-12), 2009 In these images, areas of highest atmospheric water vapour content appear the brightest.

(Figure 6.13). *Precipitation fog* forms when rain falls into cold air and evaporates; it often occurs along warm fronts (see Chapter 7). *Steam fog*, a common feature of Arctic coastlines, forms when water evaporates from a warm body of water and condenses in the cold air above.

Another type of fog, *advection fog*, forms when a warm, moist air layer moves over a cold surface. As the warm air layer loses heat to the surface, its temperature drops below the dewpoint and condensation sets in. Advection fog commonly occurs over oceans where warm and cold currents flow toward each other. Condensation occurs as the warm, moist air above the warm current mixes with the colder air over the cold current. The fogs associated with the Grand Banks off Newfoundland are formed in this way; here the cold Labrador Current comes in contact with the warmer waters of the Gulf Stream. Advection fog can also develop on land when a cold air mass mixes with warm, moist air. A similar process gives rise to the *sea fog* that is a characteristic feature along the west coasts of South America and Africa. These sea fogs are caused when warm, moist tropical air is cooled as it passes over the cold Peru and Benguela currents that carry water north from the southern oceans surrounding Antarctica. Similar fogs develop off the coast of California where the air is chilled by the cold California Current (see Figure 6.14).

6.13 **VALLEY FOG** Valley fog forms when moist air in the valley is chilled to below dewpoint by cold air drainage from the upper slopes.

PRECIPITATION

The growth of the droplets in clouds slows down as they get bigger and competition increases for the available moisture. As the surface area of a droplet gets progressively larger, more water must condense to effectively increase its size. The need for more water also slows the condensation rate, which is limited by the speed at which latent heat can be released. Once droplets grow large enough, they will begin to fall through the cloud. As they fall, they may collide with other droplets and coalesce to form larger droplets. The larger droplets also drag smaller droplets down behind them. This is the **collision-wake capture** process, which arises because of the range of droplet sizes naturally present in clouds.

The rate at which the droplet descends depends on its size. For example, in still air a droplet with a diameter of 20 μm has a terminal velocity of 0.01 m s^{-1}, compared with 6.5 m s^{-1} for a droplet 4,000 μm in diameter. However, larger droplets can also break apart as they collide, which increases the number of droplets. The longer a droplet remains in a cloud, the greater the

6.14 **SEA FOG** A layer of sea fog along the coast of California.

opportunity it has to interact with other droplets. Droplet size can therefore increase if strong vertical updrafts develop within the cloud.

In mid- and high latitudes, much of the precipitation originates from ice crystals through the **Bergeron (Bergeron-Findeisen) process**. This *cold-cloud* process depends on the presence of super-cooled water droplets. In the absence of freezing nuclei, small ice needles will form directly from super-cooled water droplets if the temperature in the cloud falls to about −35 °C. Freezing occurs at much higher temperatures if freezing nuclei are present. Once ice crystals have started to form, super-cooled water droplets will automatically freeze to them. An ice crystal will also grow as water vapour is converted directly to ice by deposition. Ice crystals will aggregate into larger snowflakes as they fall through the cloud in a process similar to the collision-wake capture process. If temperatures in the lower atmosphere are near or below freezing, the snowflakes will remain in solid form; otherwise, they melt to produce rain or sleet (Figure 6.15). Precipitation can develop quickly through the Bergeron process; it is the dominant process in large cumulus and cumulonimbus clouds, which usually extend well above the freezing level, even in the tropics.

Cloud droplets grow by condensation and sublimation to diameters of 50 to 100 μm. Droplets formed by coalescence, either in liquid or solid form, can become much larger. A droplet diameter of about 500 μm is characteristic of *drizzle*. For *rain*, the average droplet size is about 1,000 to 2,000 μm, but droplets can reach a maximum diameter of about 7,000 μm (about 7 mm). If they grow bigger than this, they become unstable and disintegrate into smaller droplets while falling. The largest droplets are usually associated with thundershower activity and with the warm, moist clouds in equatorial and tropical zones.

Most precipitation begins as ice crystals and snowflakes that melt as they fall, reaching the Earth's surface as rain. Conversely, raindrops that fall through a cold air layer can produce pellets or grains of ice, or a mixture of snow and rain commonly known as *sleet*. *Ice storms* occur when the ground is frozen and the temperature in the lowest air layer is also below freezing. Under these conditions, rain falling through the layer is chilled and freezes onto ground surfaces as a clear glaze. Ice storms cause great damage, especially to telephone and power lines and to tree limbs,

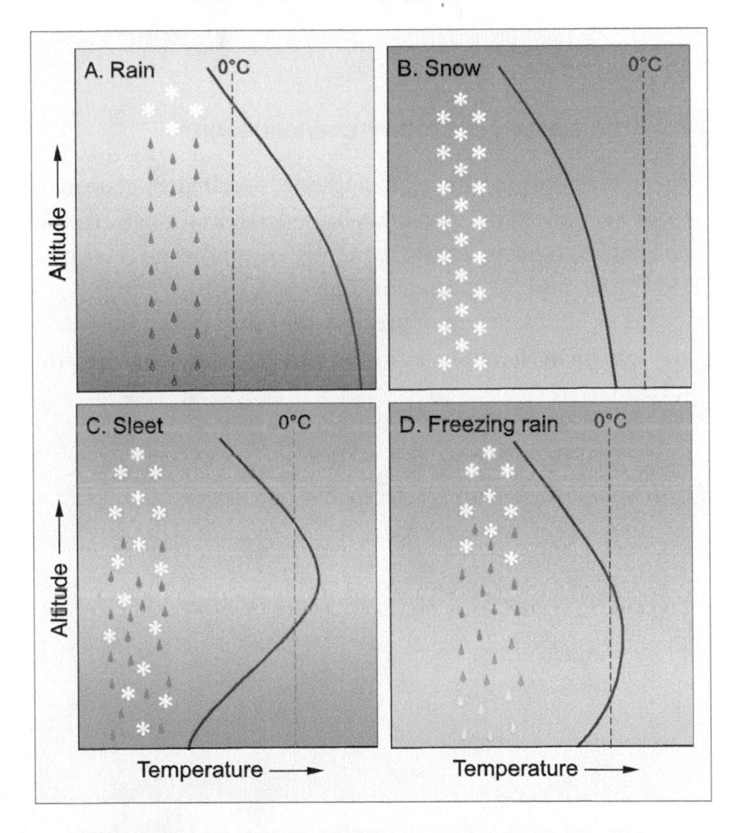

6.15 **RELATIONSHIP BETWEEN AIR TEMPERATURE PROFILE AND PRECIPITATION TYPE** (a) Snow occurs when the falling precipitation passes through a deep cold air layer in the lower troposphere. (b) Sleet falls if a warm air layer allows partial melting to occur. (c) Freezing rain typically occurs when liquid precipitation comes in contact with a thin cold air layer. (d) With air temperatures above freezing, precipitation falls as rain. (From *Introduction to Weather and Climate* by O.W. Archibold, John Wiley & Sons Canada, Ltd., 2011.)

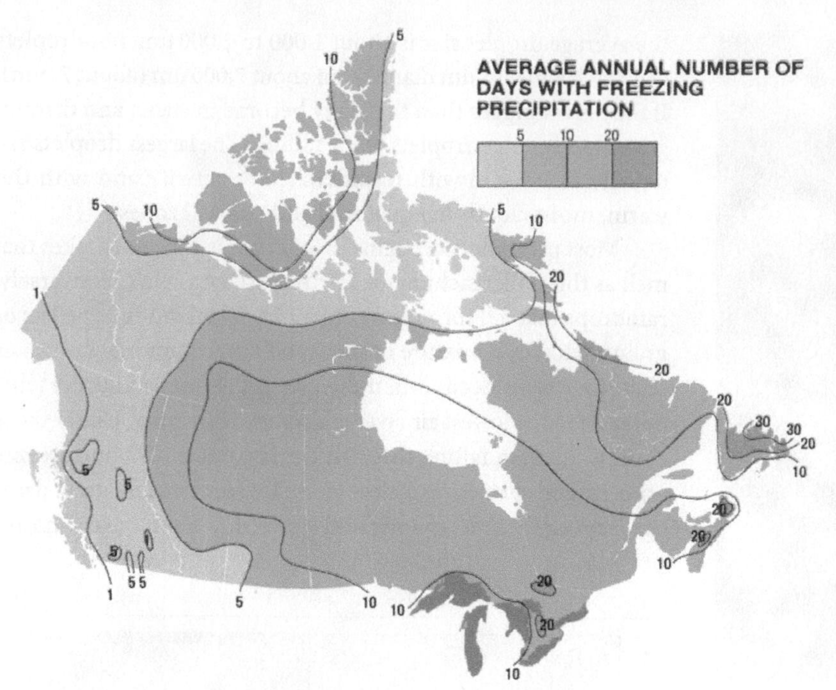

6.16 **MEAN NUMBER OF DAYS WITH FREEZING RAIN IN CANADA**

due to the weight of the ice. In addition, the slippery glaze makes roads and sidewalks extremely hazardous. In Canada, freezing rain tends to occur most frequently in southern Ontario, the Maritimes, and Newfoundland (Figure 6.16).

Hail, another form of precipitation, consists of lumps of ice averaging 5 to 50 mm in diameter, and can be much larger when associated with severe thunderstorms. The largest hailstone recorded in Canada fell in 1973 near Cedoux, Saskatchewan; it weighed 290 grams and was 114 mm in diameter. In 2003, a hailstone with a diameter of 178 mm was reported in Nebraska, U.S. Hailstones are formed from super-cooled water droplets that add layers of ice to ice pellets suspended in the strong updrafts of a thunderstorm. The maximum growth rate occurs at about −13 °C; growth is uncommon below −30 °C, as super-cooled water droplets are not abundant at that temperature. Thus, hail is especially common in midlatitudes during early summer when surface temperatures are warm enough to promote thunderstorm activity but the atmosphere is still relatively cool. Hail is less common in low latitudes, despite frequent thunderstorms, because the tropical atmosphere tends to be warmer. In Canada, hail occurs most frequently in Alberta, British Columbia, and Saskatchewan (Figure 6.17).

On some occasions, wisps or streaks of rain or snow fall from a cloud, but evaporate before reaching the ground. This phenomenon, called *virga*, is commonly observed during summer in the Canadian Prairies (Figure 6.18).

Rainfall is measured with a *rain gauge*; the standard design consists of a narrow, calibrated cylinder with a funnel at the top (Figure 6.19a). These are usually read once each day to provide the total precipitation amount for that day. An alternative

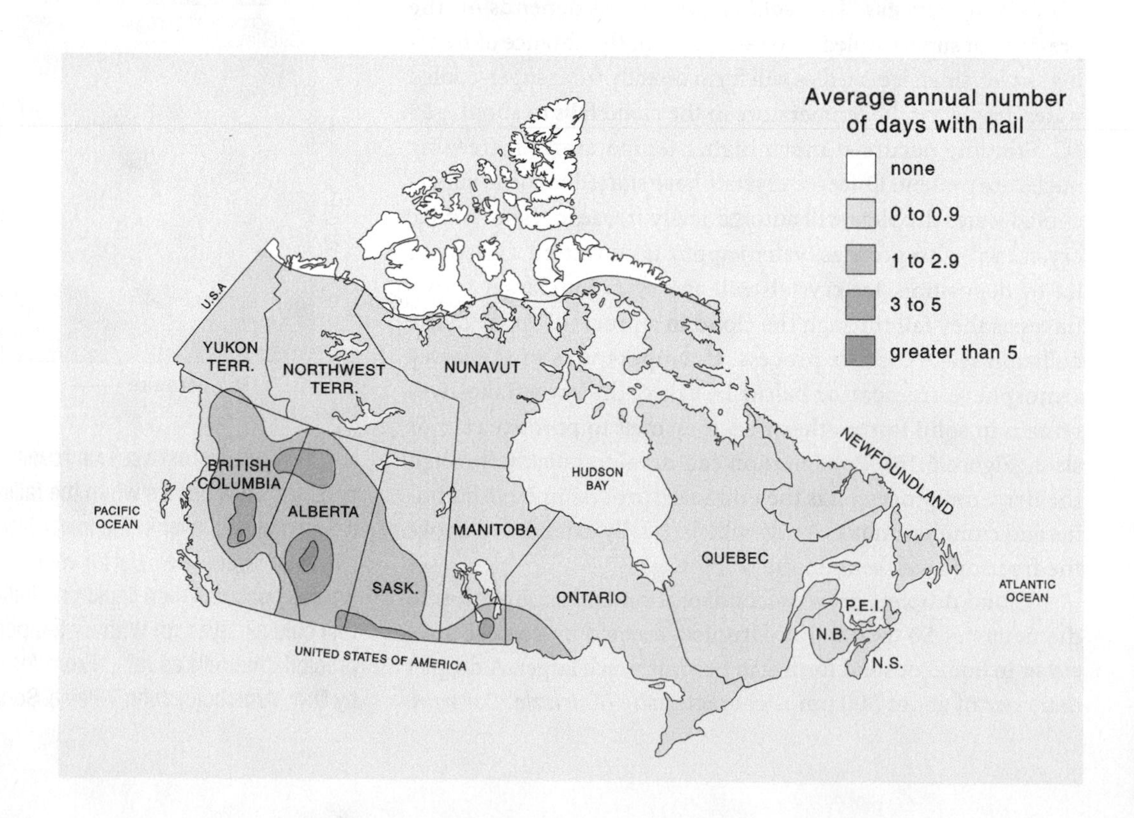

6.17 **MEAN NUMBER OF DAYS WITH HAIL IN CANADA**

6.18 VIRGA Virga are rain streaks that evaporate before reaching the ground.

design is a tipping bucket rain gauge, which consists of two containers mounted on a pivot, beneath a funnel (Figure 6.19b). When one container is filled, it tips and empties, and the second one begins to fill. Each container (or bucket) is calibrated to tip when it contains 0.1 mm of rain. With each tip, a signal is sent to a recording device that provides a record of the time and intensity of the rain.

Snowfall can be measured in a similar way by melting the snow as it falls into a heated rain gauge, but in Canada the standard snowfall instrument is the Nipher snow gauge (Figure 6.19c). It consists of a 10-cm diameter collecting tube surrounded by a trumpet-shaped wind baffle. The snow is melted and measured in a graduated cylinder in the same way as rainfall. This converts the snowfall to an equivalent rainfall amount, allowing rainfall and snowfall to be combined in a single record of precipitation. Ordinarily, a 10-cm layer of snow is assumed to be equivalent to 10 mm of rain. Because snow can persist on the ground for much of the winter, the accumulated depth of snow is often reported for each month. Traditionally, snow stakes—poles calibrated like rulers—have been used to read snow depths, but electronic instruments are now available (Figure 6.19d). Snow depth measurements can be problematic; wind can blow the snow away in some areas, and elsewhere drifting can heap it up to unrepresentative depths. In addition, the snowpack ages over time through settling and melting or sublimation.

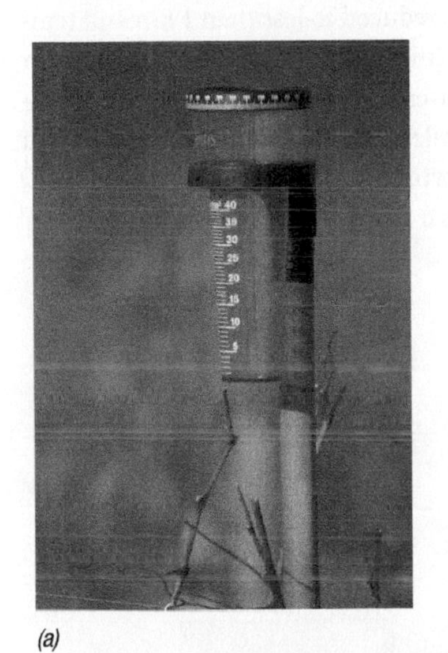

(a)

(c)

(b)

(d)

6.19 PRECIPITATION INSTRUMENTS (a) Standard rain gauge; (b) tipping bucket rain gauge; (c) Nipher snow gauge; (d) ultrasonic sensor used to measure accumulated snow depth.

In Canada, the snow season begins in August in the Arctic islands, and by early December, most of the country is covered. In some areas, such as coastal British Columbia and southern Ontario, the snow cover may last for only a few weeks, but it persists until June or July at high latitudes (Figure 6.20).

The Meteorological Service of Canada defines *blowing snow* as a condition in which horizontal visibility is restricted to 10 km or less because the wind has raised snow particles to a height greater than 2 m. This occurs at wind speeds of 35 to 39 kph. Blowing snow occurs on about 20 days each year in the Prairies and other parts of southern Canada, but in northern communities 60 to 90 days of blowing snow are not uncommon (Figure 6.21). Wind speeds below 35 kph are associated with *drifting snow*, which moves at a height of less than 2 m. *Blizzard* conditions develop at wind speeds above 40 kph, when visibility is reduced to less than 1 km and temperatures are below −10 °C. Blizzards can last for several days. For example, in January 2008, a blizzard in Rankin Inlet continued for 179 hours (about 7½ days) with winds gusting above 100 kph and temperatures with wind chill dropping to −57 °C.

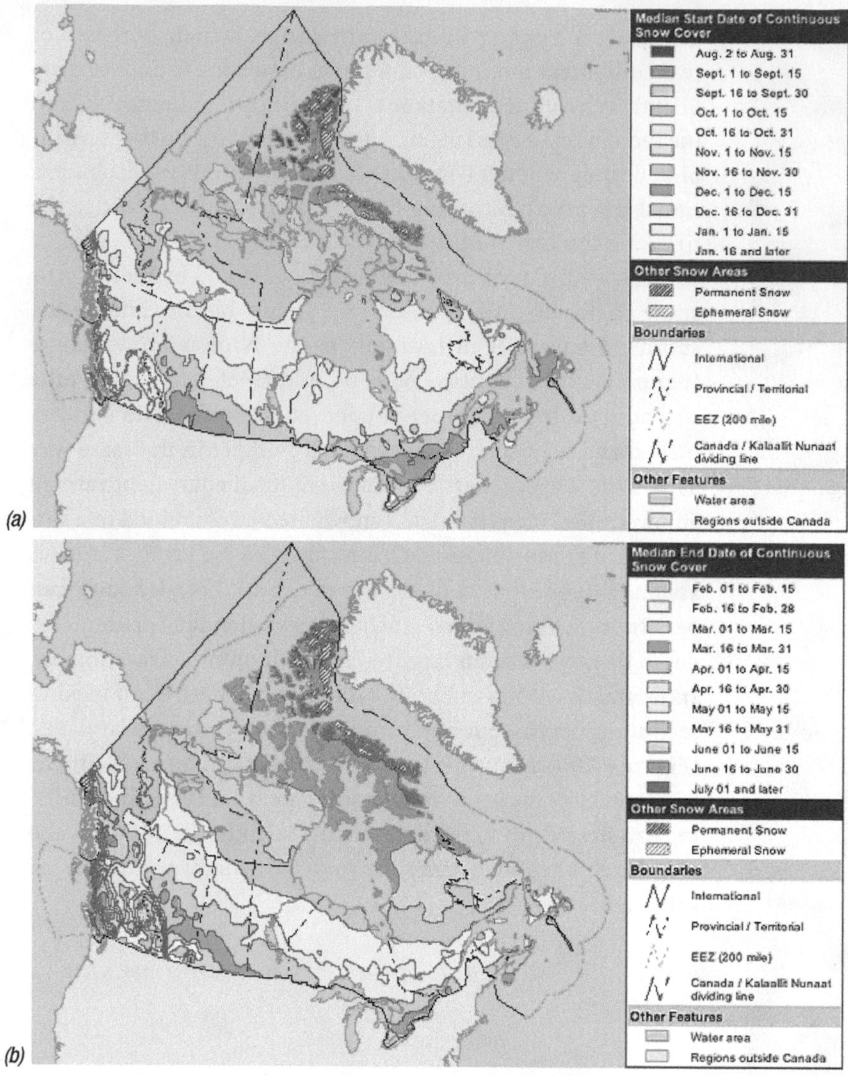

6.20 MEDIAN DATES FOR THE START (A) AND END (B) OF SNOW COVER IN CANADA. The data are based on the first and last dates with 14 consecutive days of snow cover greater than 2 cm in depth.

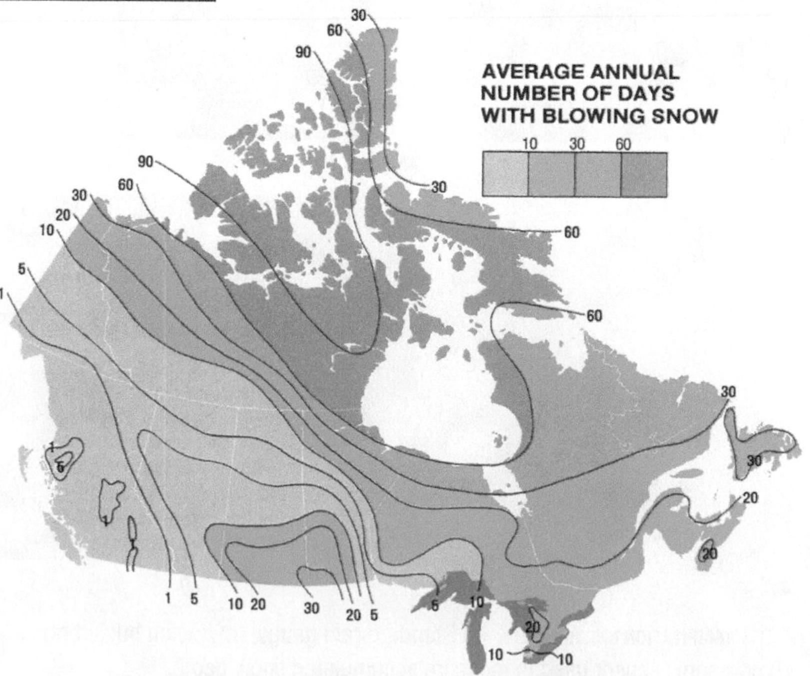

6.21 AVERAGE NUMBER OF DAYS WITH BLOWING SNOW IN CANADA

PRECIPITATION PROCESSES

Condensation and precipitation will occur when a parcel of moist air continues to rise and cool by the adiabatic process past the point where it becomes saturated. A parcel of air can be forced upward through the atmosphere in four ways:

- Air can be forced to rise over a range of hills or mountains; if sufficient moisture condenses, this will result in **orographic precipitation.**
- Air is forced to rise when there is a net inflow due to horizontal **convergence**, as occurs with the northeast and southeast trade winds along the ITCZ.
- During **convection**, a parcel of air rises when it is heated by the underlying surface, making it less dense than the air around it.
- In midlatitudes, air masses of contrasting types meet along frontal zones where colder air tends to be forced under warmer air, forcing it to rise. The process leads to the formation of low-pressure systems, and characteristically produces cloudy skies and precipitation. This is known as *cyclonic (or frontal) precipitation*, and is discussed further in Chapter 7.

OROGRAPHIC PRECIPITATION

Orographic precipitation develops when winds move moist air up and over high terrain. In the example shown in Figure 6.22, moist air arrives at the coast (1) and rises on the windward side of the mountain range. As the air rises, it is cooled at the dry adiabatic rate until it reaches the lifting condensation level. Here condensation sets in, and clouds begin to form (2). In this example, the condensation level is at 600 m, but its altitude depends on the condition of the atmosphere. Cooling now proceeds at the wet adiabatic rate. Eventually, precipitation begins and continues to fall as the air rises further up the slope.

After passing over the mountain summit, the air begins to descend the leeward slopes of the range (3). As it descends it is compressed and warmed, causing cloud droplets and ice crystals to evaporate or sublimate. Initially this takes up latent heat so the descending air warms at the wet adiabatic rate, but eventually the sky clears, and warming continues at the dry adiabatic rate. At the base of the mountain (4), the air is now warmer and drier, since much of its moisture was removed in precipitation. This effect creates a *rain shadow*, and under some conditions in winter, Chinook winds can occur.

Precipitation patterns in western Canada are clearly influenced by the orographic process (Figure 6.23). For example, mean annual precipitation at Salmon Arm, British Columbia, is 551 mm; this rises to 946 mm at Revelstoke as the moist Pacific air begins its ascent over the Rocky Mountains. The highest annual precipitation on this transect is 1,547 mm at Rogers Pass at an elevation of 1323 m. Over 60 percent of the precipitation

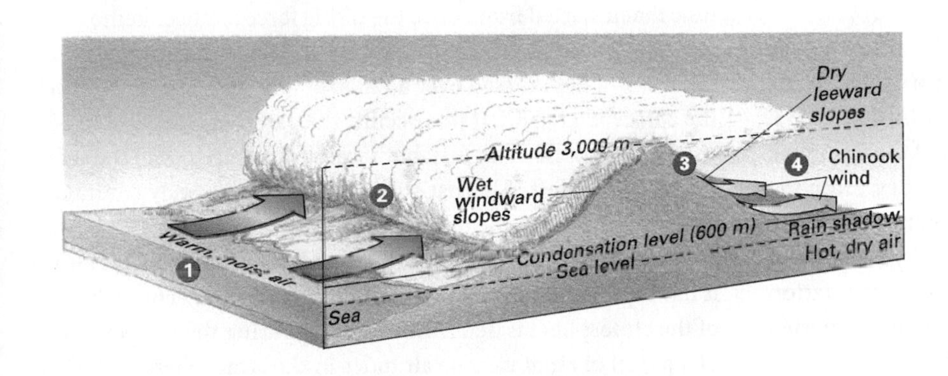

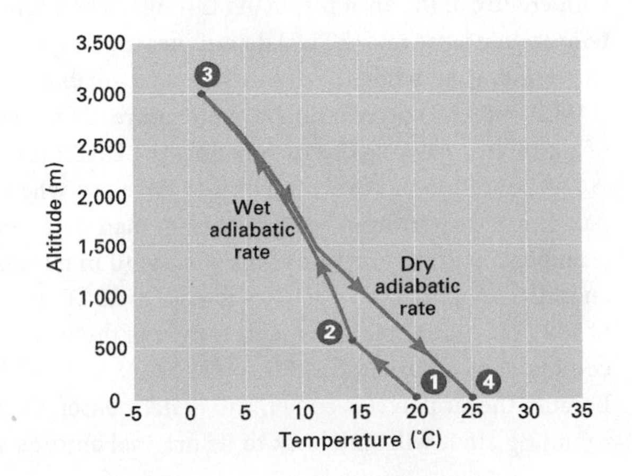

6.22 **OROGRAPHIC PRECIPITATION** The forced ascent of warm, moist oceanic air over a mountain barrier produces orographic precipitation and a rain shadow effect. As the air moves up the mountain barrier, it loses moisture through precipitation. As it descends, the far slope is warmed and may lead to Chinook conditions (see Chapter 5).

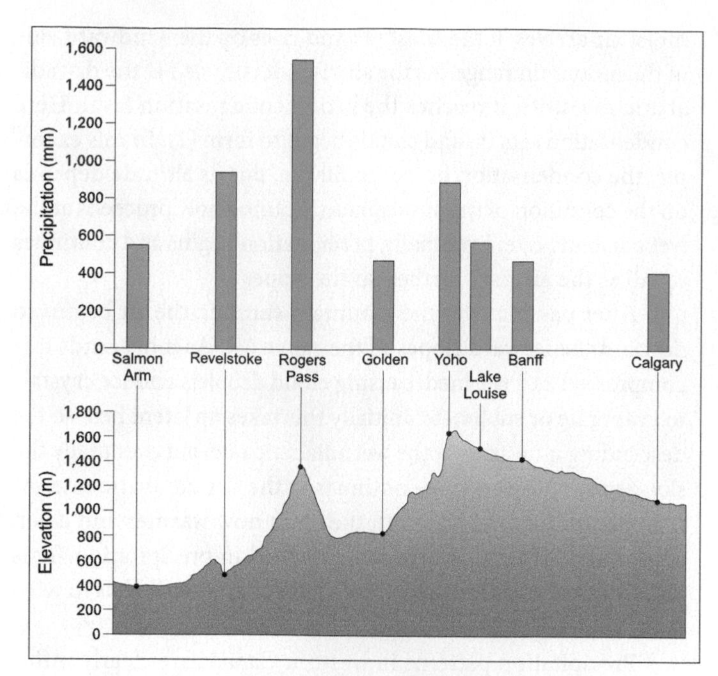

6.23 **OROGRAPHIC PRECIPITATION IN WESTERN CANADA** The effect of mountain ranges on precipitation is strong in western Canada because of the prevailing flow of moist oceanic air. High precipitation occurs on the windward western slopes of the mountain ranges, with drier conditions on the leeward eastern slopes. (From *Introduction to Weather and Climate* by O.W. Archibold, John Wiley & Sons Canada, Ltd., 2011.)

recorded at Rogers Pass comes as snow. Here the extreme daily snowfall is 74 cm (February 12, 1982); this compares to the extreme daily rainfall of 82 mm recorded on July 11, 1983. Golden, British Columbia, situated in the valley of the Columbia River at an elevation of 785 m, is comparatively dry, receiving only 475 mm of precipitation. Much of the remaining moisture in the air is lost as it traverses the crest of the Rocky Mountains, with 884 mm reported at Yoho National Park. Precipitation decreases as the air descends into Alberta, with 413 mm reported at Calgary. Elevation drops from 1,524 m at Lake Louise to 1,084 m at Calgary, with a corresponding change in mean annual temperature from −0.3 °C to 4.1 °C. Some of the decrease in precipitation can therefore be attributed to adiabatic warming as the orographic lifting force is removed.

CONVERGENT PRECIPITATION

Convergence arises from opposing airflows, and at the global scale is mainly associated with the trade winds that carry air toward the ITCZ. Vertical motions associated with convergence are typically rather weak and will usually give rise to cloud layers with little vertical development. However, along the ITCZ, warm, moist tropical air contributes to the rising limbs of the Hadley cells. Once uplift is initiated, energy released through condensation will augment the process, and usually considerable cloud cover and rain develop. Hurricanes are also convergent systems that generate abundant precipitation. In a much less dramatic way, cyclonic precipitation arises from the inflow of air in a low-pressure system (see Chapter 7).

CONVECTIONAL PRECIPITATION

The convection process starts when a surface is heated unequally, resulting in localized pockets of warm air that rise because they are less dense than the surrounding cooler air. As the air pocket rises it is cooled adiabatically, but it will continue to rise as long as it remains warmer than the air around it. When it reaches the dewpoint, condensation occurs, and the rising air becomes visible as a cumulus cloud. The flat base of cumulus clouds marks the lifting condensation level and shows the altitude at which condensation begins. As the cloud grows it can shade the ground, which may reduce surface heating and limit further development. However, activity will begin again if the cloud encounters another warm area as it drifts downwind. Also, if the air is moist, sufficient latent heat can be released to ensure that the uplift will continue; this is an important factor in the formation of thunderstorms.

ATMOSPHERIC STABILITY

A parcel of air that rises may settle back down to the same altitude that it started from, once the lifting force has been removed. For example, a parcel of air might settle back to its original altitude after it has crossed over a range of hills and continues over the adjacent lowland. If this occurs, the parcel of air is said to be **stable**. The parcel of air is stable because it is denser than the surrounding air aloft. Alternatively, the parcel of air may continue to rise even though the initial lifting force is removed; this leads to a condition of instability. The parcel of air is **unstable** because it has become less dense than the surrounding air. The stability of the atmosphere is determined by comparing the temperature of a parcel of air at various altitudes in the atmosphere with the temperature of the air it is moving through. The environmental temperature lapse rate (ETLR) determines the temperature of the surrounding air, while the dry or saturated adiabatic lapse rates (DALR or SALR) determine the temperature of the rising parcel of air. In the following discussion, only the DALR is used.

A condition of **absolute stability** develops when the temperature of the surrounding air is warmer than the air parcel. For example, a parcel of dry air at the surface (0 m elevation) with an initial temperature of 30 °C will cool to 25 °C at an altitude of 500 m (Figure 6.24a). The temperature of the surrounding air cooling at an assumed ETLR of 8 °C 1,000 m^{-1} will be 26 °C. Because the air parcel is cooler, and hence denser, than the surrounding air, it will sink back to its original altitude when the

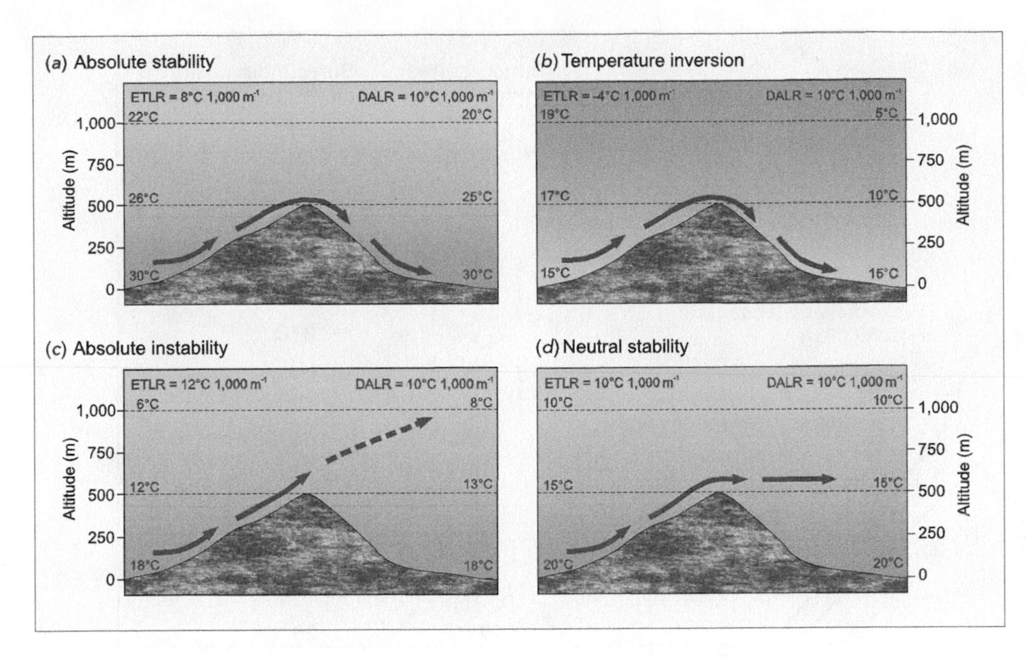

6.24 ATMOSPHERIC STABILITY The stability of an air parcel depends on its density relative to that of the surrounding air, and is primarily determined by its temperature.
(a) Absolute stability occurs when the temperature of a rising air parcel is cooler than its surrounding;
(b) a condition of absolute stability always exists in a temperature inversion; (c) absolute instability occurs when the temperature of a rising air parcel becomes warmer than the surrounding air; (d) under neutrally stable conditions, the temperature of the rising air parcel changes at the same rate as the surrounding air.

lifting force is removed. When stable air is forced to rise, say by passing over a mountain range, it tends to spread out horizontally, and if clouds form, they tend to be relatively thin layers, such as stratus or altostratus clouds. Inversion layers are absolutely stable because at any altitude the surrounding air is always warmer than air that is forced upward. For example, in Figure 6.24b, the temperature of the air forced to rise to 500 m has dropped to 10 °C, compared to 17 °C at the corresponding altitude in the inversion layer (Figure 6.24*b*). Common ways in which a stable atmosphere can develop include raising the temperature of the air in the upper atmosphere through warm advection, or by cooling the surface air by nighttime radiation or cold advection.

Absolute instability develops when the rising air remains warmer and less dense than the surrounding air. For example, the temperature of a parcel of dry air forced to rise to 500 m will cool from 18 °C to 13 °C, whereas the temperature of the surrounding air with an assumed ETLR of 12 °C 1,000 m^{-1} would cool from 18 °C to 12 °C (Figure 6.24c). The air will continue to rise after the initial lifting force is removed because it is now less dense than the surrounding air. Some general ways in which the atmosphere can become unstable include cooling of the air in the atmosphere by cold advection or by radiation cooling, warming of the air at the surface due to daytime heating, or the influx of warm air at the surface. **Neutral stability** (Figure 6.24d) occurs when the ETLR is the same as the DALR; a parcel of air that is forced to rise will always be the same temperature as its surroundings and so will remain at the altitude to which it is forced to rise. In this example the temperature of both the air parcel and the surrounding air is 15 °C at 500 m.

The terms "stability" and "instability" apply to air that is displaced from its original altitude. Thus it is important to note that air which is forced down through the atmosphere would also be considered stable if later it rises and returns to its original altitude. Conversely, if the air continues to sink, it would be considered unstable.

Figure 6.25 provides an example of how the stability of a moist air parcel is determined from the ETLR and the dry and wet adiabatic lapse rates. The temperature of the surrounding air at ground level is 26 °C and it has an ETLR of 12 °C 1,000 m^{-1}. Assume that energy conducted from the underlying warm ground heats a parcel of air by 1 °C to 2 7 °C, and so it begins to rise. At first, this air parcel cools at the DALR. At 500 m, the temperature of the parcel is 22 °C, while the surrounding air is 20 °C. Since it is still warmer than the surrounding air, it continues to rise. At 1,000 m, it reaches the lifting condensation level. The temperature of the parcel is 17 °C.

As the parcel rises above the condensation level, it cools at the SALR, which here is assumed to be 5 °C 1,000 m^{-1}. Now the parcel cools more slowly as it rises. At 1,500 m, the temperature of the parcel is 14.5 °C, while the surrounding air is 8 °C. Since the parcel is still warmer than the surrounding air, it continues to rise. Note that the difference in temperature between the rising parcel and the surrounding air actually increases with altitude. This means that the parcel will be buoyed upward ever more strongly, forcing even more condensation and precipitation.

Whereas absolute stability, absolute instability, and neutral stability could arise with respect to either the DALR or SALR, air

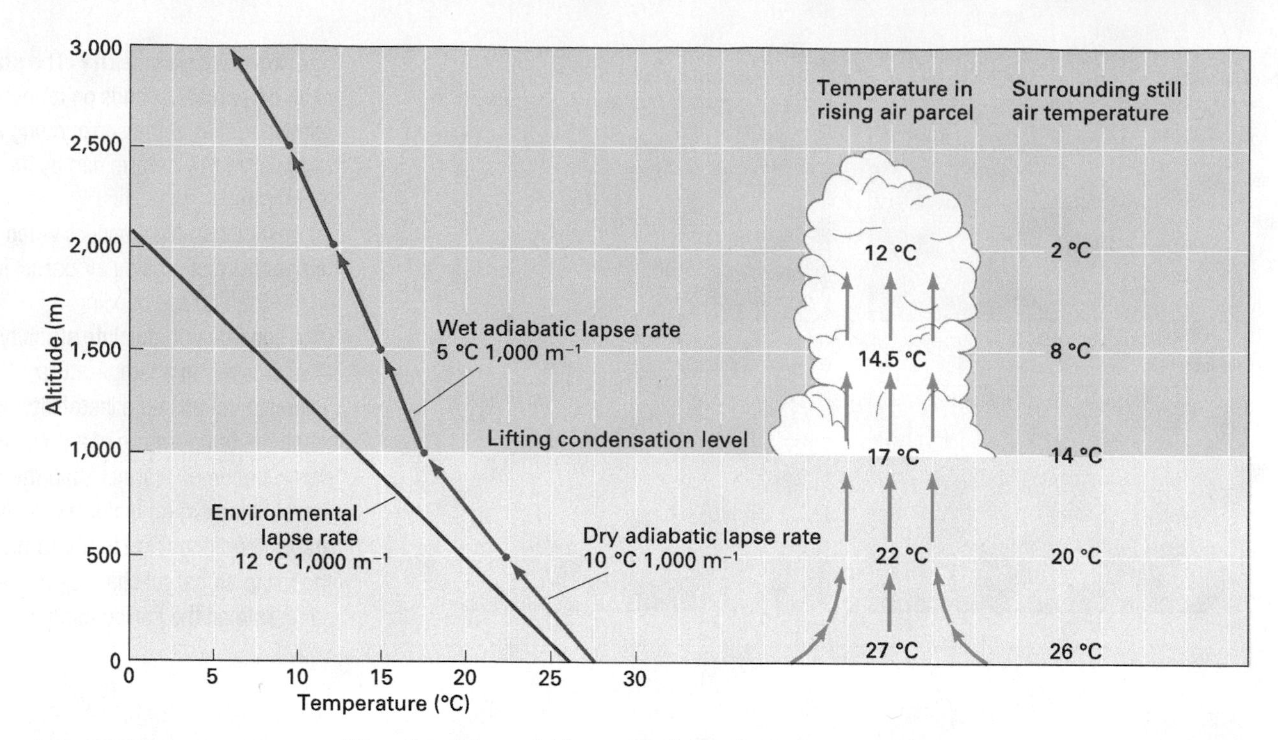

6.25 CONVECTION IN UNSTABLE AIR When the air is unstable, a parcel of air that is heated sufficiently to rise will continue to rise to great heights.

may become conditionally unstable when the ETLR lies between the DALR and SALR. **Conditional instability** is associated with moist air. If the rising air parcel remains unsaturated, it is stable unless it rises above the condensation level. At the condensation level, the air becomes saturated and unstable. Thunderstorms can be triggered when conditionally unstable air rises over mountains.

THUNDERSTORMS

Thunderstorms are intense convective storms associated with tall, dense cumulonimbus clouds in which there are very strong updrafts. The key to the convectional precipitation process is latent heat. When water vapour condenses into cloud droplets or forms ice crystals, it releases latent heat to the rising air parcel. By keeping the parcel warmer than the surrounding air, this latent heat fuels the convection process, driving the parcel ever higher. Once the parcel reaches an altitude where most of its water has condensed, adiabatic cooling is progressively reduced, and less latent heat is released. As a result, the uplift will weaken and eventually cease, and the parcel will dissipate into the surrounding air.

Thunderstorms typically consist of several individual convection cells, each with a distinct life cycle that lasts about 30 minutes (Figure 6.26). Air rises within each cell as a succession of bubble-like air parcels, and intense adiabatic cooling produces precipitation. Precipitation can be in the form of liquid droplets at lower levels, with solid forms at high levels where cloud temperatures are coldest. The uplift rate slows toward the top of the cloud, which in extreme cases may exceed 12 km. A distinctive

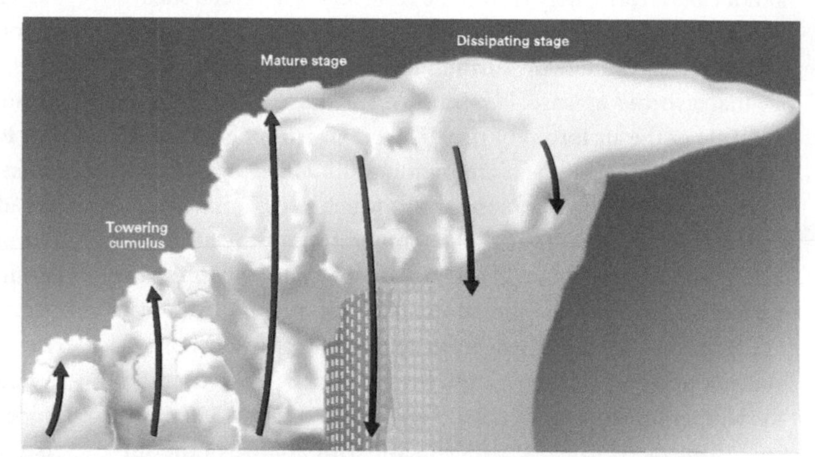

6.26 ANATOMY OF A THUNDERSTORM A thunderstorm begins as a small convective cloud in which air movement is predominantly upward. In the mature stage, a thunderstorm usually consists of several cells of moist condensing air rising upward. This upward movement is countered by downdrafts, which expel rain, hail, and cool air from the storm as it moves forward. In the later dissipating stage, the predominant air movement is downward, moisture begins to evaporate, and the energy to sustain the storm is lost.

anvil shape develops where strong winds drag out the top of the thunderstorm cloud.

Ice particles falling from the top of the cloud act as nuclei for freezing and sublimation at lower levels. Ice crystals can be kept in the upper atmosphere by the strong updrafts, but eventually they fall and coalesce into large raindrops as they melt. The falling raindrops drag the air downward; this feeds a downdraft within the convection cell. As moisture in the cloud is used up, the supply of latent heat is diminished, updrafts become weaker, downdrafts begin to dominate, and the cell dissipates. However, on reaching the surface, the cold downdraft may force warm, moist surface air upward. This rising air then condenses and builds into a new cell. In this way a thunderstorm can continue for several hours. In some cases, the downdrafts emerge from the cloud base with great force to form a gust front or *plough wind*.

The most dangerous type of thunderstorm is the **supercell**, which, as well as causing heavy rain, severe hail, and powerful winds, also has a high probability of tornadoes. Supercells are single cell thunderstorms usually 20 to 50 km in diameter that last for 3 to 4 hours. A distinctive feature of a supercell is that the updraft is subjected to strong wind shear that causes the air to rotate as it rises. This helps to draw the surrounding air into the storm, which contributes to the supercell rather than developing into competing thunderstorms. Outflow is facilitated by strong winds aloft. Transition to a supercell generally occurs when the storm builds into the tropopause and encounters a high level jet with speeds exceeding 50 m s^{-1} (180 kph). The rotational updraft is a unique feature of a supercell and forms the core of the system. Once it is large enough to be detected on Doppler radar, it is termed a **mesocyclone**; at this stage the sky appears as an ominous dark, swirling cloud mass. The downdraft, which originates from the inflow of dry air in the mid-troposphere at altitudes of 3,000 to 6,000 m, descends around the updraft. The updraft and downdraft regions do not interfere with each other as in less-organized multicell cluster storms; this helps to prolong the life of these severe storms.

All thunderstorms characteristically produce heavy, localized rainfall. In Canada, the highest rainfall intensity on record was reported at Buffalo Gap, Saskatchewan, in 1961, where 250 mm of rain fell in less than an hour. In addition to heavy rainfall, many thunderstorms produce hail. Hailstones have a minimum diameter of 5 mm. If they are smaller than that, they are defined as ice pellets. Hailstones are formed by the accumulation of clear and opaque ice layers on ice pellets suspended in the thunderstorm's strong updrafts. The onion-like layering depends on the size and abundance of super-cooled water droplets present in the cloud. Opaque ice forms when small, super-cooled liquid water droplets freeze rapidly on impact and trap air bubbles within the ice. When freezing is slower, air bubbles are able to escape and the ice layer is clear.

Economic losses from hailstorms can be considerable in both urban and rural areas. For example, the hailstorm that hit Calgary on September 7, 1991, caused more than $400 million in damage. Losses to crops are equally devastating. Hail insurance is a necessary expense for many Prairie farmers. On average, hail destroys roughly 3 percent of Prairie crops each year, mostly during a few severe storms. Damage swaths are typically 3 to 20 km wide and 50 to 150 km long. To reduce damage from hail, several cloud seeding experiments have been initiated. Typically, aircraft inject silver iodide crystals into the clouds in an attempt to produce smaller and softer hailstones that might melt as they fall to the ground. The success of these attempts at weather modification is still unclear.

Lightning is another phenomenon that results from convective activity. It occurs when updrafts and downdrafts cause positive and negative static charges to accumulate within different regions of the cloud. Lightning is an electric discharge passing between differently charged parts of the cloud mass or between the cloud and the ground. During a lightning discharge, an electric current as high as 100,000 amperes may develop. This heats the air intensely and makes it expand very rapidly, generating sound waves that are heard as thunder. Most lightning discharges occur within the cloud, but a significant proportion strike land. Most do little or no damage, although electrical supplies are frequently interrupted. However, lightning is responsible for about 50 percent of the forest fires in Canada and economic losses are considerable.

THE THUNDERSTORM AS A FLOW SYSTEM

A thunderstorm is a dramatic and spectacular example of a system in which matter, in the form of air and water, moves from one location to another under the influence of a coupled flow of energy. Figure 6.27 shows (a) the structure of a storm, (b) how water flows and changes state within a storm, and (c) the flow of energy within a storm.

A mature-stage thunderstorm consists of powerful updrafts and downdrafts that distribute energy throughout the cloud mass. The updrafts that are initiated by convectional heating are strengthened by heat released through condensation and by powerful winds aloft (Figure 6.27a). Precipitation in the form of water droplets, ice crystals, and hail are kept aloft by the strong updrafts, but as the precipitation particles grow in size, some begin to fall through the cloud. The drag from the falling precipitation creates downdrafts, which produce strong turbulent outflow at the ground surface. The balance between updrafts and downdrafts in the cloud shifts during the life of the storm. As the energy for the storm is dissipated, normal weather patterns for the region resume.

Water flows in a thunderstorm system (Figure 6.27b) start as rising cells of warm, moist air. They serve to bring in water vapour at low altitude, which condenses to liquid and solid forms as it moves upward. At the top of the cloud, high altitude, through-flowing winds carry away some ice particles. The

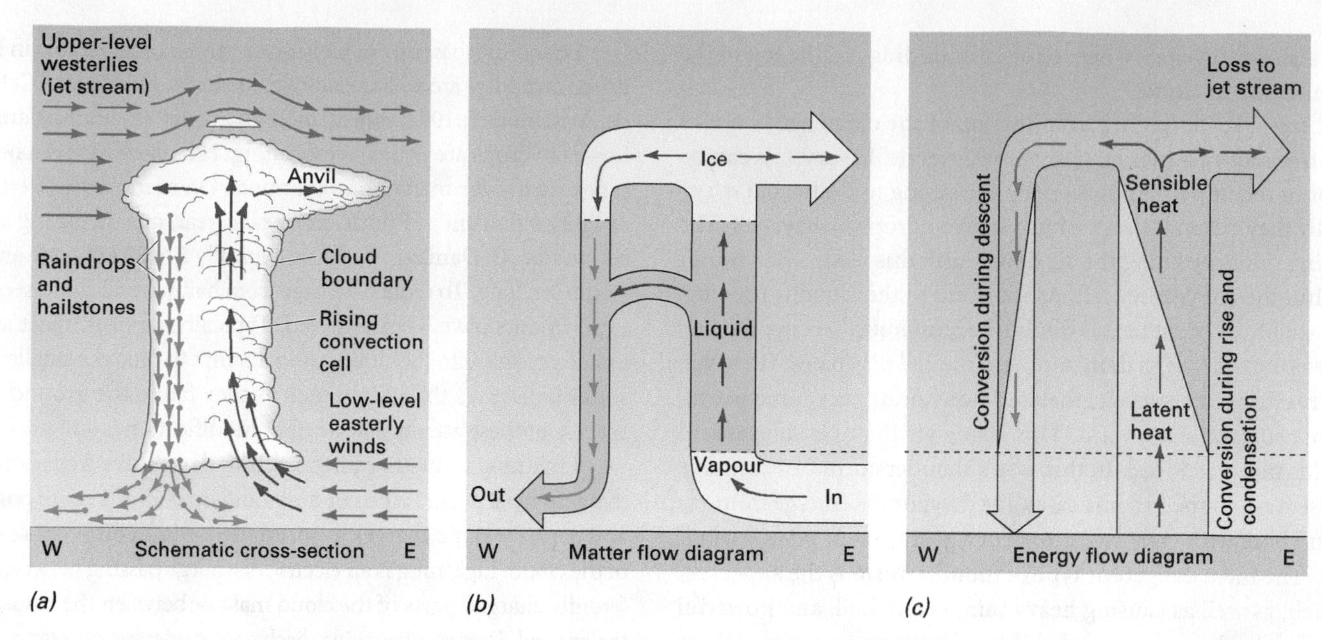

6.27 THE THUNDERSTORM AS A FLOW SYSTEM OF MATTER AND ENERGY (a) Structure of a thunderstorm. (b) Water flows in a thunderstorm. (c) Energy flows in a thunderstorm.

remaining ice crystals and water droplets descend in another part of the cell and strike the ground as rain, hail, or even snow. Thus, water flows upward as vapour, then downward as precipitation.

As the convection cells within the storm cloud move moist air into the atmosphere, condensation converts latent heat to sensible heat, enhancing the convective uplift (Figure 6.27c). At the top of the storm, high-level winds carry a portion of the sensible heat forward and away from the storm. As solid and liquid water particles are carried *downward*, some evaporation and sublimation occur, which convert a portion of the sensible heat released back into latent heat. At the surface, the net effect is that a larger quantity of latent heat is lost than is gained, resulting in generally cooler air. At higher altitudes, the air is warmer than before the storm, owing to the gain in sensible heat.

The power source for the storm is the latent heat in the water vapour that moves upward. Changing from vapour into the liquid or solid state releases the energy that powers the storm. Some of that energy is recovered when solid or liquid water is converted to vapour in the downdraft cell, but a larger portion is dissipated at higher altitudes. Thus, the storm moves heat from the surface to higher altitudes. The ultimate power source for the storm is the sun, which heats the surface and brings about the evaporation that creates the moist air.

MICROBURSTS

The downdraft that accompanies a thunderstorm can sometimes be very intense, and on occasion has been linked to aircraft accidents. Such intense downdrafts are called **microbursts** (Figure 6.28). The downward-moving air flows outward in all directions and is often, but not always, accompanied by rain. An aircraft flying through a microburst first encounters strong headwinds, which may cause a bumpy ride but do not interfere with the aircraft's ability to fly. However, as it passes through the far

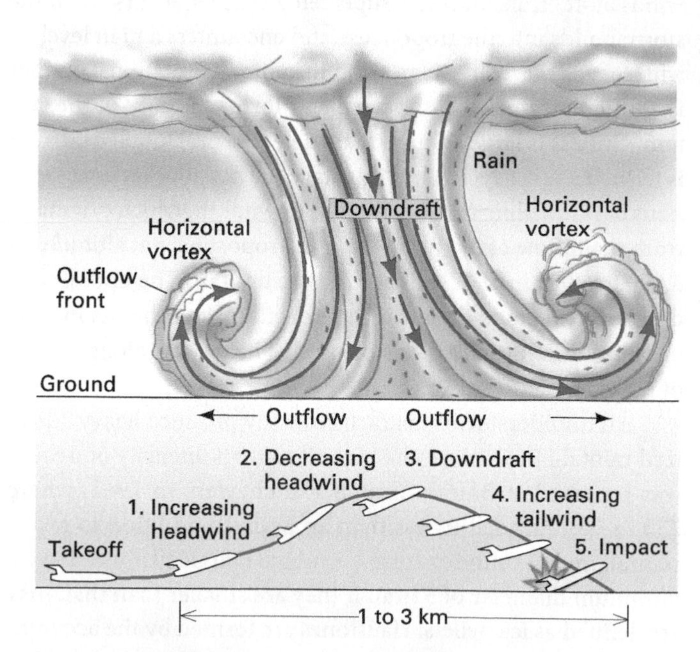

6.28 ANATOMY OF A MICROBURST (top) Schematic cross-section showing a downdraft reaching the ground and producing a horizontal outflow. (bottom) An aircraft taking off through a microburst suffers a loss of lift and could crash. (Adapted from diagrams by Research Applications Program, National Center for Atmospheric Research, Boulder, Colorado.)

side of the microburst, the aircraft encounters a strong tailwind. The lift of the aircraft's wings depends on the speed of the air flowing across them, and the tailwind greatly reduces this speed. If the tailwind is strong enough, the aircraft may be unable to maintain its altitude and could crash.

A Look Ahead

This chapter has shown that no matter what process causes precipitation, latent heat is always released. Because a significant amount of evaporation from the ocean results in precipitation over land, there is a significant flow of latent heat from oceans to land. In addition, global air circulation patterns move this latent heat toward the poles. This creates the conditions necessary for another precipitation process—that associated with the low-pressure centres that move across the midlatitudes. The following chapter discusses these dynamic systems, which are such an integral part of the weather conditions in Canada.

CHAPTER SUMMARY

- Precipitation is liquid or solid water that falls from the atmosphere to the Earth's land or ocean surface. Evaporating, condensing, melting, freezing, and sublimation describe water's changes of state.
- Water moves freely between ocean, atmosphere, and land in the hydrologic cycle. The global water balance describes these flows. The fresh water in the atmosphere and on land in lakes, streams, rivers, and *groundwater* is only a small portion of the total water in the hydrosphere.
- Humidity describes the amount of water vapour present in air. The ability of air to hold water vapour depends on temperature. Warm air can hold much more water vapour than cold air.
- Vapour pressure refers to the contribution that water vapour makes to the pressure exerted by the atmosphere. *Saturation vapour pressure (SVP)* is the maximum amount of water vapour that can be present at a given temperature.
- *Specific humidity* measures the mass of water vapour in a mass of air, in grams of water vapour per kilogram of air. *Relative humidity* measures water vapour in the air as the percentage of the maximum amount of water vapour that can be held at a given air temperature. The dewpoint temperature, at which condensation occurs, can also provide a measure of the moisture content of air.
- The *adiabatic principle* states that when a gas is compressed, it warms, and when a gas expands, it cools. When an air parcel moves upward in the atmosphere, it encounters a lower pressure and so expands and cools.
- The *dry adiabatic lapse rate* describes the rate of cooling with altitude. If the air is cooled below the dewpoint, condensation or sublimation occurs and latent heat is released. This released heat reduces the air parcel's cooling rate with altitude. When condensation or sublimation is occurring, the cooling rate is described as the *saturated (or wet) adiabatic lapse rate*.
- Clouds are composed of water droplets or ice crystals that form on *condensation nuclei*. Clouds typically occur in layers, as *stratiform* clouds, or in globular masses, as *cumuliform* clouds. *Fog* occurs when clouds form at ground level.
- Precipitation from clouds occurs as *rain*, *hail*, *snow*, and *sleet*. When *supercooled* rain falls on a surface at temperatures below freezing, it produces an *ice storm* or *glaze*.
- There are four types of precipitation processes: *orographic*, *convergent*, *convectional*, and *cyclonic*. In orographic precipitation, air moves up and over a topographic barrier. As it moves up, it is cooled adiabatically and rain forms. As it descends on the far side, it is warmed, producing a *rain shadow* effect.
- Convergent precipitation occurs when moist air flows come together and then rise, as is the case along the ITCZ. Convergent precipitation is also associated with hurricanes, and in a much less dramatic way, with cyclonic precipitation.
- In convectional precipitation, unequal heating of the surface causes a moist air parcel to become warmer and less dense than the surrounding air. Because it is less dense, it rises. As it moves upward, it cools, leading to condensation and precipitation.
- *Stability* and *instability* refer to the tendency for air, if displaced vertically, to return to its former altitude. Stable air returns to its original altitude. Air that, when displaced, continues to move away from its original altitude is considered unstable. Comparing the air parcel's temperature with that of the surrounding air determines its stability.
- Instability can lead to thunderstorm formation, producing hail and lightning.

KEY TERMS

absolute instability
absolute stability
adiabatic process
Bergeron (Bergeron-Findeisen) process
cloud
collision-wake capture
conditional instability
convection
convergence
curvature effect
Dalton's law of partial pressure
deposition
dewpoint temperature
dry adiabatic lapse rate (DALR)
evaporation
fog
frost point
hydrologic cycle
hydrosphere
ice nuclei
lifting condensation level
mesocyclone
microburst
neutral stability
orographic precipitation
precipitation
relative humidity
saturated (wet) adiabatic lapse rate (SALR)
saturation vapour pressure (SVP)
smog
solute effect
specific humidity (q)
stable air
sublimation
supercell
thunderstorm
unstable air
vapour pressure

REVIEW QUESTIONS

1. Identify the three states of water and the terms used to describe the various changes of state.
2. What is the hydrosphere? Where is water found on Earth? In what amounts? How does water move in the hydrologic cycle?
3. Define saturation vapour pressure. How is the moisture content of air influenced by air temperature?
4. Define relative humidity. How is relative humidity measured? Sketch a graph showing relative humidity and temperature through a 24-hour cycle.
5. Use the terms *saturation*, *dew point*, and *condensation* to describe what happens when a parcel of moist air is chilled.
6. What is the adiabatic process? Why is it important?
7. Distinguish between dry and wet adiabatic lapse rates. In a parcel of air moving upward in the atmosphere, when do these rates apply? Why is the wet adiabatic lapse rate less than the dry adiabatic rate? Why is the wet adiabatic rate variable?
8. How are clouds classified? Name four cloud families, two broad types of cloud forms, and three specific cloud types.
9. What is fog? Explain how radiation fog and advection fog form.
10. How is precipitation formed? Describe the process for warm and cold clouds.
11. Describe the orographic precipitation process. What is a rain shadow? Provide an example of the rain shadow effect.
12. What is unstable air? What are its characteristics?
13. Describe the convectional precipitation process. What is the energy source that powers this source of precipitation? Explain.
14. Suppose that global climate warming increases evaporation over land and oceans. Assuming the cycle remains in balance (that is, $P_{(\text{land + lake})} + E_{(\text{land + lake})} + P_{\text{ocean}} + E_{\text{ocean}} = 0$), what will be the effect on precipitation?
15. At present, oceans cover about 71 percent of the Earth's surface. Suppose this value were 50 percent. How would the global water balance be affected?
16. Compare and contrast the energy and matter flows in updraft cells and downdraft cells within a convective storm.
17. How would the convective precipitation process be affected by stable surrounding air?

VISUALIZATION EXERCISES

1. Draw a diagram showing the main features of the hydrologic cycle. Include water flows connecting land, ocean, and atmosphere. Label the flow paths.
2. Sketch a diagram showing five pathways of matter flow for water in the global water balance flow system. Indicate their magnitudes on the diagram.
3. Draw a graph of the temperature of an air parcel as it moves up and over a mountain barrier, producing precipitation.
4. Sketch the anatomy of a thunderstorm cell. Show rising bubbles of air, updraft, downdraft, precipitation, and other features.

ESSAY QUESTIONS

1. Water in the atmosphere is important for understanding weather and climate. Write an essay or prepare an oral presentation on this topic, focusing on the following questions: What part of the global water supply is atmospheric? Why is it important? What is its global role? How does the capacity of air to hold water vapour vary? How is the moisture content of air measured? Clouds and fog visibly demonstrate the presence of atmospheric water. How do they form?
2. Compare and contrast orographic and convectional precipitation. Begin with a discussion of the adiabatic process and the generation of precipitation within clouds. Then compare the two processes, paying special attention to the conditions that create uplift. Can convectional precipitation occur in an orographic situation? Under what conditions?

WORKING IT OUT • THE LIFTING CONDENSATION LEVEL

When a parcel of air moves upward, it cools at the dry adiabatic rate of 10 °C 1,000 m^{-1}. This cooling occurs because the parcel is subject to lower atmospheric pressure as it rises, and under the adiabatic principle, it expands and cools. The change in pressure also affects the dewpoint temperature, which falls at a rate of 1.8 °C 1,000 m^{-1}. When the temperature of the cooling air parcel reaches the dewpoint temperature, condensation will begin. The elevation at which condensation begins is referred to as the *lifting condensation level*.

Suppose an air parcel at a temperature of 20 °C is raised 500 m at the dry adiabatic lapse rate. Its temperature will drop by $500/1{,}000 \times 10 = 5$ °C, resulting in a temperature of $20 - 5 = 15$ °C. In equation form, this is:

$$T = T_0 - (H \times R_{dry})$$

where T is the temperature of the parcel; T_0 is the initial temperature of the air; H is the change in altitude in metres; and R_{dry} is the dry adiabatic rate, 10 °C 1,000 m^{-1}.

Assume that the dewpoint of that same parcel of air is 11 °C. At 500 m, the dewpoint temperature will fall by $500/1000 \times 1.8 = 0.9$ °C, and the resulting dewpoint temperature will be $11 - 0.9 = 10.1$ °C. Similarly, we can write:

$$T_d = T_{dew} - (H \times R_{dew})$$

where T_d is the dewpoint temperature of the parcel; T_{dew} is the starting dewpoint temperature; H is the change in altitude; and R_{dew} is the dewpoint lapse rate, 1.8 °C 1,000 m^{-1}.

Condensation will occur when the parcel's temperature reaches the dewpoint; that is, at the altitude at which $T = T_d$. This level can be found from

$$T_0 - (H \times R_{dry}) = T_{dew} - (H \times R_{dew}) \quad (1)$$

After some algebraic rearrangement, the formula becomes:

$$H = \frac{T_0 - T_{dew}}{R_{dry} - R_{dew}}$$

By substituting the values for the dry adiabatic rate and dewpoint lapse rate, the formula becomes:

$$H = 1{,}000 \times (T_0 - T_{dew})/8.2 \quad (2)$$

This gives the lifting condensation level directly in metres. The lifting condensation level for the air parcel described above is:

$$H = 1{,}000 \times (20 - 11)/8.2 = 1{,}000 \times 9/8.2 = 1{,}098 \text{ m}$$

To find the temperature of the air parcel at the lifting condensation level, the equation becomes:

$$T = T_0 - (H \times R_{dry}) = 20 - (1{,}098 \times 10/1{,}000) = 9.0 \text{ °C}$$

QUESTIONS / PROBLEMS

1. What is the lifting condensation level for a parcel of air at an initial temperature of 25 °C and a dewpoint temperature of 18 °C? What is the temperature of the air parcel at that level? Assume that the dry adiabatic lapse rate is 10 °C 1,000m^{-1} and the dewpoint lapse rate is 1.8 °C 1,000m^{-1}.
2. Suppose the same parcel of air is warmed to 30 °C before it begins its ascent. What will be the lifting condensation level for the warmed parcel? What will be its temperature at that level?

Find answers at the back of the book.

WORKING IT OUT • ENERGY AND LATENT HEAT

In the metric system, energy is measured in joules. The *joule* (J) is defined as one newton-metre, which is the energy expended by a force of one newton acting through a distance of one metre. A newton is the force produced by an acceleration of one metre per second squared (m s^{-2}) applied to a mass of one kilogram. The joule thus has units of force multiplied by distance, which is a measure of energy. The joule is related to the watt (W), in that one watt is defined as the flow of one joule of energy per second. From this definition, the energy consumed by a 100-watt light bulb in one second is 100 joules.

The amount of latent heat taken up or released by water when a change of state occurs depends on the type of change and the temperature of the water. For melting and freezing, the energy amounts to about 335 kilojoules per kilogram (kJ/kg) when water changes from a liquid at 0 °C to a solid at the same temperature. For evaporation and condensation, the energy required or released is about 2,260 kilojoules per kilogram, at 100 °C. Energy is also required to heat the water from 0 °C to 100 °C—4.19 kilojoules per kilogram for each Celsius degree, or 419 kilojoules per kilogram for 100°. These values depend partly on atmospheric pressure and apply to sea-level conditions.

To determine the amount of energy required for sublimation from solid to gas, add the amount of heat required first for melting, then for warming to the boiling point, and finally, for evaporating. That is, $335 + 419 + 2{,}260 = 3{,}014$ kilojoules per kilogram. This will also be the amount of energy released when the reverse sublimation process (deposition) occurs, depositing frost.

Evaporation can, of course, occur at temperatures well below 100 °C. For example, the energy required to evaporate a kilogram of liquid water at 25 °C will be the energy required to bring the water to the evaporation point added to the latent heat required to convert it to vapour. For the first quantity, each degree below 100 °C requires 4.19 kilojoules per kilogram. Thus, $100 - 25 = 75$ °C, hence $4.19 \times 75 = 314$ kilojoules per kilogram are needed. To this add the 2,260 kilojoules per kilogram necessary to bring about a change of state from liquid to vapour. Therefore, the total energy required to evaporate a kilogram of liquid water at 25 °C is $314 + 2{,}260 = 2{,}574$ kilojoules per kilogram.

QUESTION / PROBLEM

1. Figure 6.2 shows that about 490 cubic kilometres of water evaporate from land and water surfaces each year. If all this evaporation occurs from a water surface at 15 °C, how much energy is required? (1 cubic metre of water has a mass of about 103 kilograms.)

Find the answer at the back of the book.

CHAPTER 7 WEATHER SYSTEMS

Clearing skies as low stratus clouds begin to dissipate in the warm sector of a wave cyclone.

The Earth's atmosphere is in constant motion, driven by pressure gradients and steered by the Coriolis effect. Air that has acquired characteristic temperature and humidity properties in one location is moved horizontally by the wind to new locations. Vertical motion in the atmosphere favours cloud formation and precipitation when air is lifted and cooled adiabatically; when air subsides, moisture returns to the vapour state through compression and adiabatic warming. In this way, the atmospheric motion influences the day-to-day weather patterns.

Some patterns of wind circulation occur regularly and give rise to recurring weather conditions. For example, travelling low-pressure centres often bring clouds and precipitation; outbreaks of polar air typically give cold, clear weather in winter. These types of recurring circulation patterns and the conditions associated with them are called **weather systems**.

Weather systems range in size from 1,000 km or more, in the case of a large travelling anticyclone, down to less than 1 km for some tornadoes. Weather systems may persist for hours or weeks, depending on their size and strength. Some forms of weather systems—for example thunderstorms, tornadoes, and hurricanes—involve strong winds and heavy rainfall and can be devastating to life and property.

AIR MASSES

An **air mass** is a large body of air with fairly uniform temperature and moisture characteristics. Air masses can be several thousand kilometres across and extend to the top of the troposphere. Each air mass is characterized by a distinctive combination of conditions at the surface, particularly temperature, environmental temperature lapse rate, and humidity. These properties are acquired in *source regions* over which the air moves slowly or stagnates. For example, an air mass over a tropical ocean is characteristically warm and has a high water vapour content. Slowly subsiding air associated with the semi-permanent high-pressure regions over tropical deserts gives rise to hot air masses with low humidity. Very cold air masses with low water vapour content form over snow-covered land surfaces in the Arctic and Antarctica.

TABLE 7.1 **PROPERTIES OF TYPICAL AIR MASSES**

AIR MASS	SYMBOL	SOURCE REGION	PROPERTIES
Maritime equatorial	mE	Warm oceans in the equatorial zone	Warm, very moist; relatively unstable
Maritime tropical	mT	Warm oceans in the tropical zone	Warm and moist; relatively unstable
Continental tropical	cT	Subtropical deserts	Hot and dry; unstable
Maritime polar	mP	Midlatitude oceans	Cool and moist; relatively unstable, especially in winter
Continental polar	cP	Northern continental interiors	Cold and dry in winter; mild and dry in summer; stable
Continental arctic (and continental antarctic)	cA (cAA)	Regions near North and South Poles	Very cold and extremely dry in winter; cool and dry in summer; very stable

Air masses move from one region to another under the influence of pressure gradients and upper-level wind patterns, and are sometimes pushed or blocked by high-level jet stream winds. When an air mass moves to a new area, its temperature and moisture properties will begin to change because they are influenced by the new surface environment.

Air masses are classified on the basis of the latitude where they originate and the nature of the underlying surface of their source regions. Latitude primarily determines surface temperature and the environmental temperature lapse rate of the air mass, while the nature of the underlying surface—continent or ocean—influences its moisture content. Five types of air masses are distinguished with respect to latitude. These are referred to as arctic (A), antarctic (AA), polar (P), tropical (T), and equatorial (E). For the type of underlying surface, two subdivisions are used: maritime (m) and continental (c). Six important air mass types can be identified from these descriptive labels; their general source regions are shown in Figure 7.1, with characteristic properties listed in Table 7.1. Note that the polar air masses (mP, cP) originate in the subarctic latitude zone, not in the polar latitude zone.

Maritime tropical (mT) air masses and maritime equatorial (mE) air masses originate over warm oceans at low latitudes and have quite similar properties. Temperatures characteristically average around 24°C in mT air and about 27°C in mE air. Because specific humidity is about 17 g kg^{-1} in mT air and about 19 g kg^{-1} in mE air, both are capable of producing heavy precipitation. The source regions for continental tropical (cT) air masses

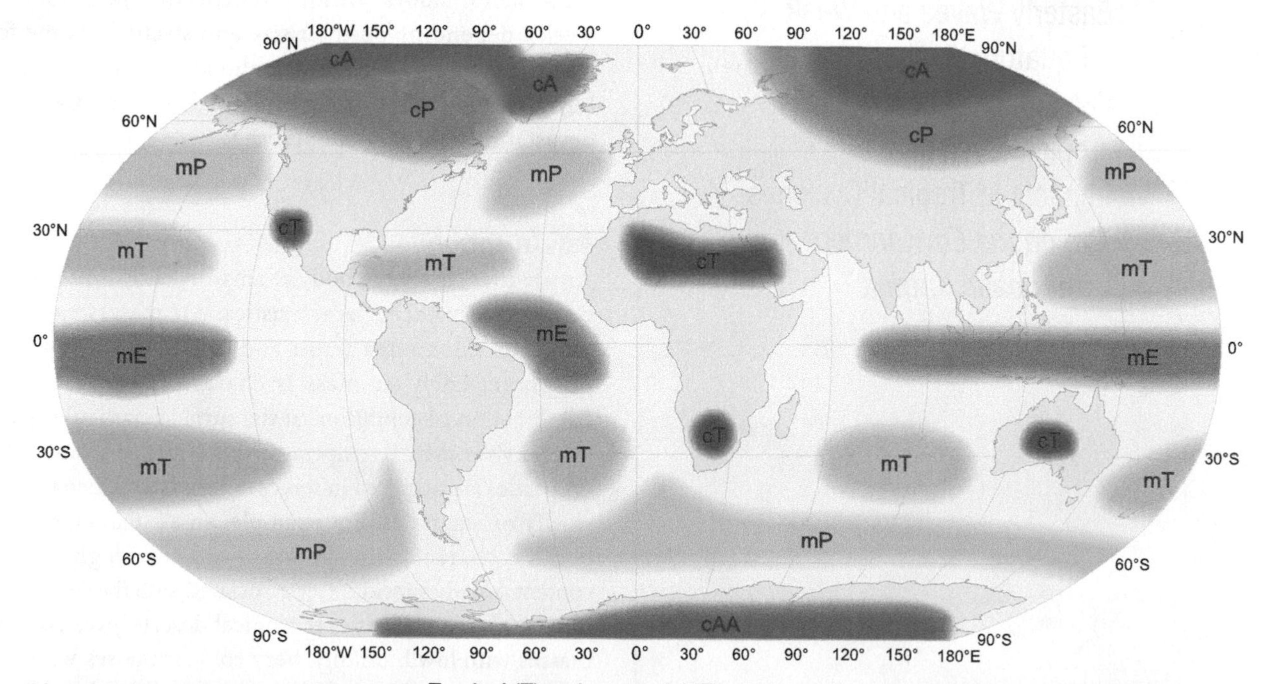

7.1 **GLOBAL AIR MASSES AND SOURCE REGIONS** Tropical (T) and equatorial (E) source regions provide warm or hot air masses, while polar (P), arctic (A), and antarctic (AA) source regions provide colder air masses of low specific humidity. Maritime (m) air masses originating over oceans are moister than those associated with continental (c) land surfaces. (From *Introduction to Weather and Climate* by O.W. Archibold, John Wiley & Sons Canada, Ltd., 2011.)

are subtropical deserts and include places such as North Africa, central Australia, and Mexico. The average temperature in cT air masses is around 24 °C, and because they are warm they hold comparatively large amounts of water vapour; the specific humidity for cT air masses is characteristically about 11 g kg^{-1}. Their general association with the semi-permanent subtropical high-pressure cells causes cT air masses to be stable. The air is heated strongly during the daytime so relative humidity is low, despite its comparatively high water vapour content.

Maritime polar (mP) air masses originate over midlatitude oceans. The average temperature of mP air is around 4°C. Because it is cooler than mT air, mP air holds less water vapour—about 4 g kg^{-1}—and consequently yields less precipitation. Much of the precipitation associated with mP air masses is orographic and occurs over mountain ranges on the western coasts of continents. The amount of precipitation can be substantial in regions where prevailing winds bring a steady supply of mP air. For example, average annual precipitation at Prince Rupert, on the northwest coast of British Columbia, exceeds 3,000 mm.

Continental polar (cP) air masses originate in the subarctic zone over North America and Eurasia. The average temperature for cP air is −11°C, but it is much colder in winter. Moisture content varies seasonally, but specific humidity is comparatively low and averages about 1.4 g kg^{-1}. Continental arctic (and continental antarctic) air mass types (cA, cAA) are extremely cold, with average temperatures of about −40°C. These frigid air masses hold almost no water vapour, so specific humidity is about 0.1 g kg^{-1}.

NORTH AMERICAN AIR MASSES

Figure 7.2 shows the source regions of air masses that have a strong influence on the weather in North America. Continental polar (cP) and continental arctic (cA) air masses are the dominant types affecting North America. The cP air masses originate over north-central Canada; they form tongues of cold, dry air that periodically extend south and east from the source region forming anticyclonic conditions with characteristically cool temperatures and clear skies. Arctic air masses that develop in winter over the frozen Arctic Ocean and adjacent land areas are extremely cold and stable. When an arctic air mass moves southward, it produces a severe cold wave that may be felt as far south as the Gulf of Mexico. Continental polar and continental arctic air masses sometimes cross the Rocky Mountains and bring unseasonably cool conditions to the Pacific coast.

Maritime polar air masses originate over the North Pacific and Bering Strait, in the region of the persistent Aleutian low-pressure centre. These air masses are cool and moist; in winter they tend to become unstable, causing heavy precipitation over the coastal ranges of Alaska, British Columbia, and Washington. Other mP air masses originate over the North Atlantic Ocean. They, too, are cool and moist, and affect Newfoundland and Labrador, the Maritimes, and New England, especially in winter. Occasionally mP air from the Atlantic moves deeper into the continent by way of the Gulf of St Lawrence.

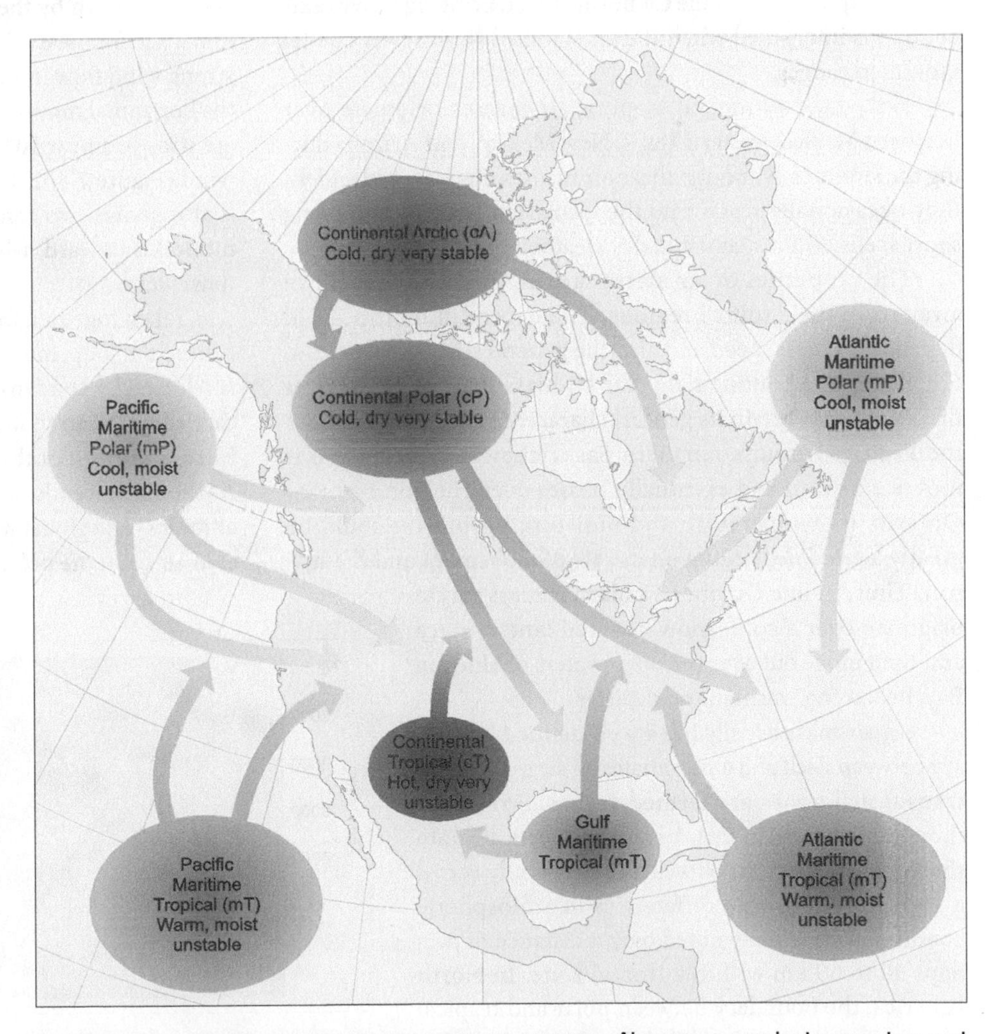

7.2 **NORTH AMERICAN AIR MASS SOURCE REGIONS AND TRAJECTORIES** Air masses acquire temperature and moisture characteristics in their source regions, and then move across the continent. (From *Introduction to Weather and Climate* by O.W. Archibold, John Wiley & Sons Canada, Ltd., 2011.)

Maritime tropical air masses from the Gulf of Mexico most commonly affect the central and eastern United States; mT air is the main opponent of cA and cP over North America. As mT air moves northward, it brings warm, moist, unstable conditions to the eastern part of the continent. Because of these air masses, the weather in summer is often hot and uncomfortably humid; thunderstorms are also expected. Maritime tropical air from the Atlantic Ocean is brought onshore by air flowing out of the subtropical (Azores-Bermuda) high-pressure cell and mostly affects the eastern seaboard of the United States. Many of the hurricanes that move into the continent are associated with this source region.

In the Pacific Ocean, the source region for mT air masses lies within the semi-permanent Hawaiian high-pressure cell. In summer, moist, unstable air masses occasionally penetrate the southwestern desert region, bringing severe thunderstorms to southern California and southern Arizona. In winter, tongues of mT air frequently reach the California coast, bringing heavy rainfall that is intensified when the air is forced to rise over coastal mountain ranges.

Hot, dry continental tropical air masses originate over northern Mexico, western Texas, New Mexico, and Arizona during the summer. Although these air masses do not travel widely, they occasionally reach into the Canadian Prairies and bring short spells of unusually hot, dry weather.

The properties of air masses are modified as they move away from their source regions. The degree of modification depends on their speed and also the general routes they follow. For example, a continental polar air mass that originates over the Yukon may retain its general characteristics if it moves into the Prairies through northern Saskatchewan. However, if it moves eastward and eventually passes over Hudson Bay into Ontario, its temperature and moisture properties could be greatly altered depending on its rate of movement and the season. Thus, in late October a polar air mass might originate over a cold, snow-covered land surface and then move out over the open waters of Hudson Bay, becoming warmer and moister.

An air mass usually has a well-defined boundary between itself and a neighbouring air mass; these transitional zones are termed **fronts**. Fronts are three-dimensional in that they extend vertically into the upper troposphere as well as horizontally across a region. Pronounced differences in atmospheric conditions are usually noted over a distance of perhaps 30 to 60 km within a frontal zone. In North America, the boundary between polar and tropical air masses produces the well-defined *polar front* that is located below the axis of the polar front jet stream (Figure 7.3).

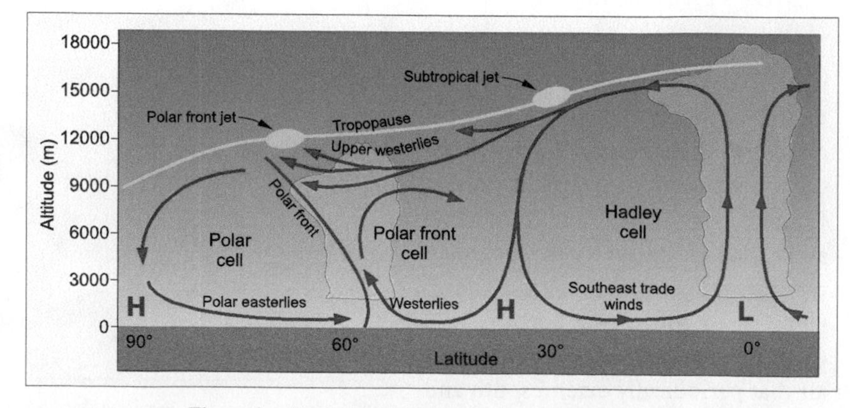

7.3 POLAR FRONT The polar front marks the boundary between polar and tropical air masses. It extends from the surface to the tropopause where it is associated with the jet stream.

POLEWARD TRANSPORT OF HEAT AND MOISTURE

The atmosphere is a fluid layer that transmits much of the sun's energy to the surface of the Earth. Because solar energy does not warm the Earth's surface uniformly, the atmosphere and oceans exhibit a complex circulation pattern of air and water flows, which acts to redistribute absorbed solar radiation more evenly by way of various convection loops. The process is referred to as *poleward heat transport*. In the atmospheric global convection system, warm, moist air rises along the ITCZ and gives rise to the **thermally direct** Hadley cells (see Figure 5.9). The air then subsides and diverges in the subtropical high-pressure belts.

Wind speeds associated with the Hadley cell circulation for a typical day in December are shown in Figure 7.4. The meridional (north-south) component of the wind (Figure 7.4a) shows a series of nested ovals below an altitude of 4 km, with a centre of strong horizontal motion near the surface. This flow is southward, as shown by the direction of the arrow. At about 10 to 15 km, a second nest of isotachs (lines of equal velocity) indicate strong wind flow in a northerly direction. These flows represent the horizontal movement of air in the convection loop. The flows are strongest near 10° N latitude. The vertical component of the wind (Figure 7.4b) shows upward and downward wind speeds; it also shows two centres of wind strength. Near the equator, motion is upward, while at about 15° N latitude, the motion is downward.

Taken together, these four directional flows provide a single convection loop, indicated by the broad arrows. In this loop, heat is released by condensation in rising air near the equator and is carried by the air in motion to the upper part of the troposphere. Here, a portion of the heat is lost by direct radiation to space. As the air descends, some heat escapes toward the poles at lower altitudes. The global wind system is completed by the outflow of cold air from the polar regions. Because this airflow depends on the increase in density that results from loss of energy, the polar cell is also considered thermally direct.

In the mid- and high latitudes, poleward heat transport occurs through the Rossby wave mechanism (see Figure 5.28). Lobes of cold, dry polar or arctic air (cP, cA, and cAA air masses) plunge toward the equator, while tongues of warmer, moister air (mT and mP air masses) flow toward the poles. At their margins, wave cyclones develop, releasing latent heat in precipitation, and providing a heat flow that warms the mid- and higher latitudes well beyond the capabilities of direct solar insolation. The net effect is that heat gathered from tropical and equatorial zones is released in the zone of subsidence associated with the semi-permanent subtropical high-pressure cells, where it can be conveyed into the midlatitudes by the motion of mT and cT air masses.

TRAVELLING CYCLONES AND ANTICYCLONES

Air masses are set in motion by the pressure gradients created by uneven heating or cooling of the atmosphere. They are driven along by the general global wind system and by regional pressure gradients associated with centres of high and low pressure. The interconnected flow between regions of high pressure and low pressure is therefore a fundamental process in the movement of air masses. Because pressure varies regionally over great distances, air masses are moved far their source regions. This type of movement is a characteristic feature of the midlatitude atmosphere, where *travelling cyclones* and *travelling anticyclones* bring changing weather conditions as they pass across a region.

In an anticyclone, divergence and settling cause air to descend and be warmed adiabatically so that condensation does not occur. Skies are clear, except for occasional puffy cumulus clouds that sometimes develop in a moist surface air layer due to weak local convection. Because of these characteristics, anticyclones are often termed *fair-weather systems*. Toward the centre of an anticyclone, the pressure gradient is weak, and winds are light and variable. Travelling anticyclones are found in the midlatitudes. These high-pressure systems are typically associated with clear, dry air that moves eastward and toward the equator.

In a travelling cyclonic system, convergence and upward motion cause air to rise and be cooled adiabatically. If the air is moist, condensation or sublimation can occur; this may lead to **cyclonic precipitation**. Many of these travelling low-pressure systems are weak and pass overhead with little more than a period of cloud cover or light precipitation. However, when pressure gradients are steep and the air is laden with water vapour, strong winds and heavy rain or snow can accompany them. In this case, the disturbance is called a **cyclonic storm**.

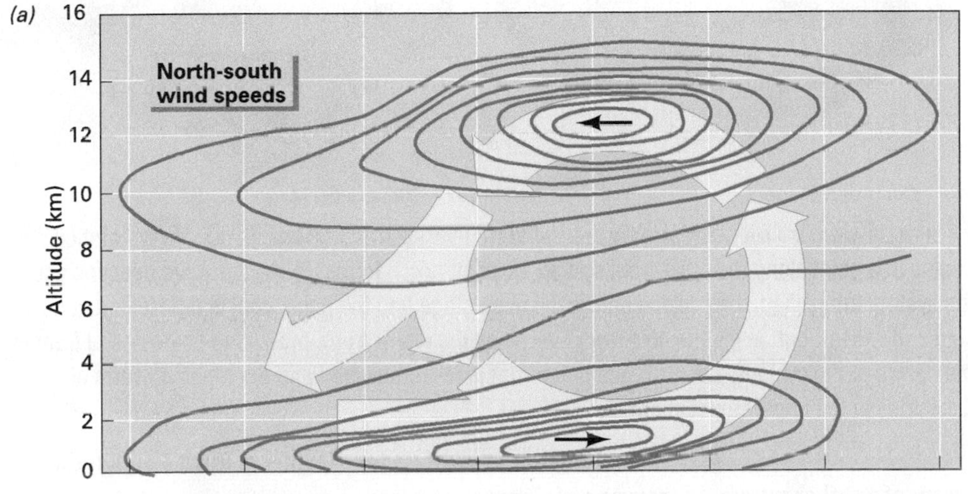

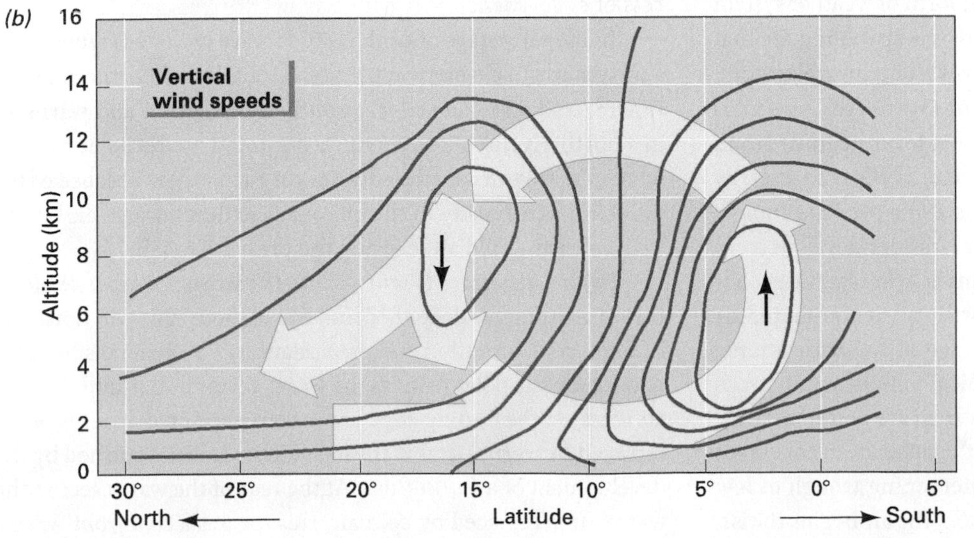

7.4 A CROSS-SECTION OF THE NORTHERN HEMISPHERE HADLEY CELL SHOWING ISOTACHS FOR A TYPICAL DAY IN DECEMBER
(a) Horizontal component of velocity, ranging to about 4 m s^{-1}. (b) Vertical component, ranging to about 0.01 m s^{-1}. Note that the height of the cell is greatly exaggerated. (Copyright © A.N. Strahler.)

Travelling cyclones fall into three types distinguished primarily by their intensity and geographic location. First is the wave cyclone of the midlatitudes, which also affect the arctic and antarctic zones. This type of cyclone ranges in strength from a weak disturbance to a vigorous storm. Several terms are used interchangeably to describe these features, which are variously known as midlatitude cyclones, *wave cyclones*, or *depressions*. Second is the tropical cyclone associated with tropical and subtropical zones. Tropical cyclones range in intensity from mild disturbances to severe *hurricanes* or *typhoons*. In Australia and other parts of the Pacific these intense storms are simply referred to as *cyclones*. The less severe disturbances are termed *easterly waves* because they are moved by the trade winds. The passage of an easterly wave is usually accompanied by light rain showers. A third type is the *tornado*, a small, intense cyclone of enormously powerful winds. The tornado is much smaller in size than other cyclones, and is related to exceptionally strong convectional activity.

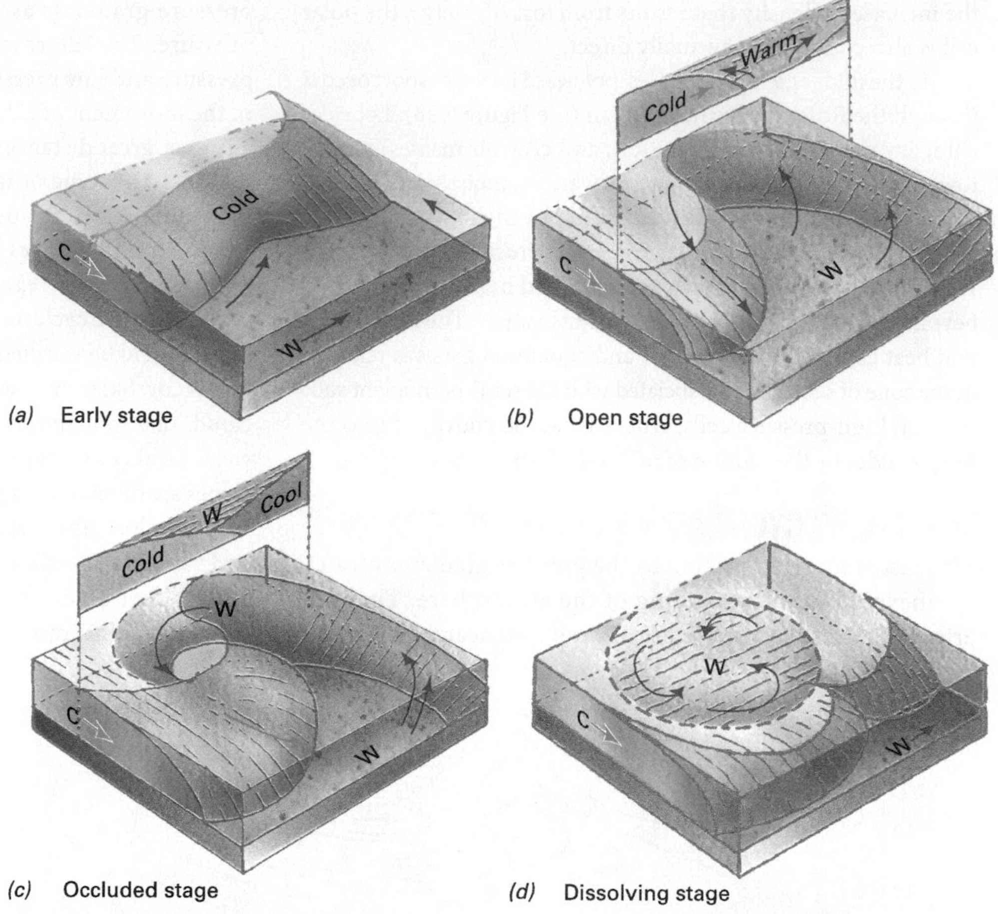

7.5 DEVELOPMENT OF A MIDLATITUDE WAVE CYCLONE IN THE NORTHERN HEMISPHERE In (a), a wave motion begins at a point along the polar front and an incipient notch forms (b). As the wave deepens and the system intensifies, the cyclone enters the open stage and the polar front is differentiated as a cold and warm front; in (c), the cold front overtakes the warm front, producing an occluded front in the centre of the cyclone. Later (d), the polar front is re-established and the mass of warm air dissipates in the upper troposphere.

WAVE CYCLONES

In mid- and high latitudes, the dominant form of weather system is the **wave cyclone**, a large mass of inward-spiralling air that repeatedly forms, intensifies, and dissolves along the polar front. A situation favourable to the formation of a wave cyclone occurs when two large anticyclones lie on either side of the polar front. The boundary between the two high-pressure cells is necessarily a region of lower pressure and so forms a *low-pressure trough*. The anticyclone that originates at higher latitudes develops from cold, dry cP air; the anticyclone that forms at lower latitudes contains warm, moist mT air. In the northern hemisphere, air circulates clockwise out from the centre of an anticyclone. Consequently, airflow from the subtropical high is mainly from the southwest, and winds from the polar high are predominantly from the northeast. Thus, airflow along the polar front converges from opposite directions toward the intervening trough of low pressure. This creates an unstable situation, the air begins to rise, and a midlatitude wave cyclone begins to form through the process of *cyclogenesis*.

In the early stage of a midlatitude wave cyclone (Figure 7.5a), a wave starts to develop on the polarfront. In the northern hemisphere, cold air is turned in a southerly direction and warm air in a northerly direction, producing a notch on the polar front in which the two air masses advance on each other. Because of the differences in density of the opposing airflows, warm air begins to lift over the cold air, causing the pressure to fall.

By the *open stage* (Figure 7.5b), the wave disturbance along the polar front has deepened and intensified. This will normally lead to cloud formation and precipitation. Two general situations develop on the polar front as the warm sector builds into the cold air mass. At the leading edge of the warm sector, cold air is being replaced by warm air; this frontal activity is distinguished by the development of a *warm front*. At the rear of the warm sector, the warm air is replaced by cold air; this creates a *cold front*. Warm

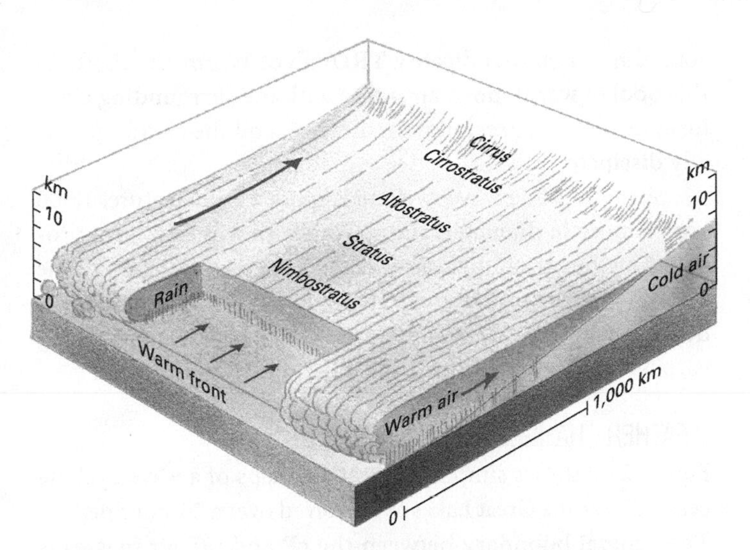

7.6 **WARM FRONT** In a warm front, warm air advances toward cold air and rides up and over it. A notch of cloud is cut away to show where rain is falling from the dense stratus cloud layer. (Drawn by A. N. Strahler.)

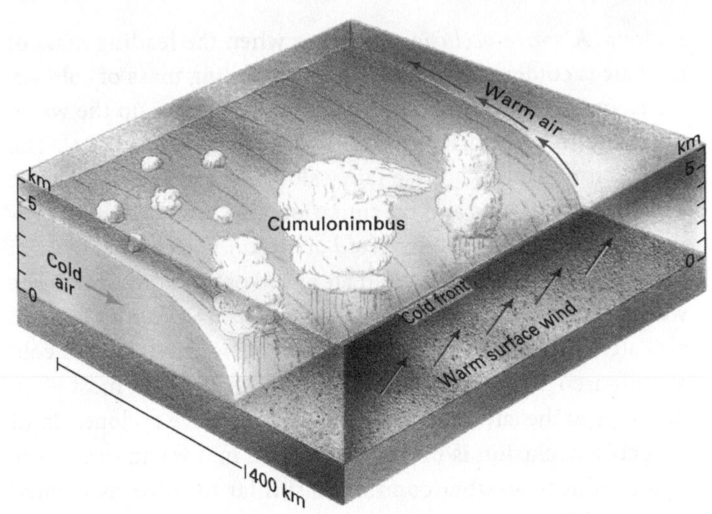

7.7 **COLD FRONT** At a cold front, a cold air mass lifts a warm air mass into the atmosphere. The upward motion can set off a line of thunderstorms. Note that the frontal boundary is actually much less steep than is shown in this schematic drawing. (Drawn by A. N. Strahler.)

air actively rises over the cold air along a warm front, whereas cold air pushes under the warm air along a cold front.

Figure 7.6 shows a **warm front** where warm air is moving over a region of colder air. The cold air mass remains in contact with the ground because it is denser, so consequently the moving mass of warm air is forced to rise over it. The gradient on a warm front is generally about 1:300; this gentle slope results from the cold, denser air in the leading sector being feathered out by friction at the surface as it moves across the region. The rising motion is therefore quite gradual, and a characteristic sequence of clouds will develop if the warm air contains a lot of moisture. Cloud formation will begin with high cirrus clouds and become progressively thicker and lower until the sky is obscured with low stratus clouds. Precipitation often follows, which, although not intense, will generally last for several hours or even days.

A **cold front** develops when a cold air mass advances into the zone occupied by the warm air mass (Figure 7.7). Because the colder air mass is denser, it remains in contact with the ground. The gradient of a cold front is about 1:50, so it is much steeper than a warm front. In this case, the front represents the leading edge of the approaching cold air. Friction at the Earth's surface also affects the advancing cold air so that the base of the cold front curves forward over the warm air. This helps to steepen the gradient of the cold front. Consequently, as the cold air mass moves forward, it forces the warmer air to rise quite abruptly. If the warm air mass is unstable, this rapid uplift will often produce heavy precipitation for a comparatively short time, and possibly thunderstorms, until the cold front has passed. In some cases, severe thunderstorms may develop as a *squall line*—a long line of massive clouds stretching for tens of kilometres along the cold front.

Cold fronts normally move across a region at a faster rate than warm fronts, and where they are in close proximity, a cold front can eventually overtake a warm front. This produces the *occluded stage* of the wave cyclone (Figure 7.5c). The colder air of the faster-moving cold front remains next to the ground, forcing the warm air ahead of it to rise along the **occluded front** (Figure 7.8).

Two types of occlusions are recognized, depending on the relative temperatures of the opposing air masses and the degree of modification they have undergone since they left their source

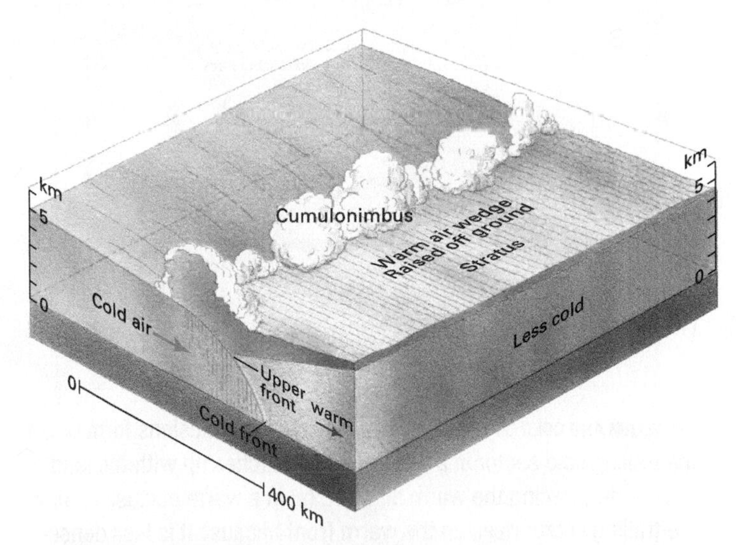

7.8 **OCCLUDED FRONT** In an occluded front, a cold front overtakes a warm front. The warm air is pushed upward, and it no longer contacts the ground. Abrupt lifting by the denser cold air produces precipitation. (Drawn by A. N. Strahler.)

regions. A *warm occlusion* develops when the leading mass of cold air is colder and denser than the trailing mass of cold air. Consequently, the trailing cold air mass will rise up the warm front and lift the warm air off of the surface (Figure 7.9a). The warm air continues to rise fairly slowly over the cold air, and the resulting clouds and precipitation are similar to those produced by a warm front. In a *cold occlusion*, the trailing mass of cold air is denser than the cold air ahead of the warm front. The warm sector air is lifted off of the surface as the trailing cold air catches up with and pushes under the leading mass of cold air (Figure 7.9b). The cold occlusion is the type that most often develops in the later stages of a midlatitude wave cyclone. Uplift in a cold occlusion is more abrupt than in a warm occlusion, and produces weather conditions similar to those associated with cold fronts.

Eventually, the polar front is re-established in the *dissolving stage* (Figure 7.5d), but a pool of warm, moist air is present in the higher troposphere. Once the warm air mass is lifted completely free of the ground, the term TROWAL is sometimes used to indicate a **TRO**ugh of **W**arm air **AL**oft. As this pool of warm moist air mixes with the surrounding air, it loses its identity, precipitation dies out, and the clouds gradually dissipate.

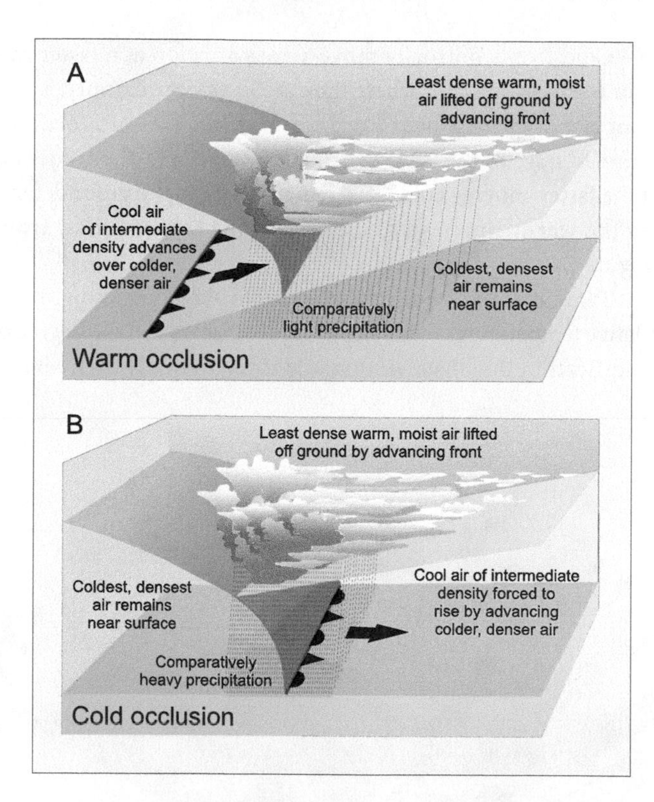

7.9 WARM AND COLD OCCLUSIONS Warm and cold occlusions form when the trailing cold sector in a wave cyclone catches up with the leading cold sector, forcing the warm air aloft. (a) In a warm occlusion, air in the trailing sector rises up the warm front because it is less dense than the colder air in the leading sector. (b) In a cold occlusion, air in the trailing sector is the coldest and densest, which forces the cold air of the leading sector to rise up over the cold front. In both cases, the warm air is the least dense and is forced aloft.

A midlatitude wave cyclone is quite a large feature, 1,000 km or more in diameter, and normally moves eastward, propelled by the prevailing westerlies and the jet stream aloft. The different stages of a wave cyclone, shown in Figure 7.5a–d, develop progressively over the course of several days and bring variable weather conditions as a system tracks eastward.

WEATHER CHANGES WITHIN A WAVE CYCLONE

Figure 7.10 shows simplified weather maps of a wave cyclone centred near the Great Lakes as it evolved over a 24-hour period. The general boundary between the cP and mT air masses is shown by the dashed white line with the structure of the storm indicated by the isobars. Standard line symbols are used to represent the warm and cold fronts and the occluded front. Areas of precipitation are shown in grey.

The map on the left shows the wave cyclone in an open stage, similar to Figure 7.5b. The isobars show that the wave cyclone is a low-pressure centre with inward-spiralling winds. The cold front is pushing south and east, supported by a flow of cold, dry cP air from the northwest filling in behind it. Note that the wind direction changes abruptly as the cold front passes. A sharp drop in temperature behind the cold front is to be expected as the cP air extends into the region. The warm front is moving in a general northeasterly direction, with warm, moist mT air following. The precipitation pattern includes a broad zone of rain near the warm front and the central area of the wave cyclone. A thin band of precipitation extends down the length of the cold front. Cloudy conditions prevail over much of the region occupied by the wave cyclone.

A day later, the wave cyclone has moved rapidly northeastward; its track is shown by the dashed red line. The centre has moved about 1,600 km in 24 hours—a speed of just over 65 kph. The cold front has overtaken the warm front, forming an occluded front in the central part of the disturbance. A high-pressure area, formed by a tongue of cold polar air, has moved into the area west and south of the wave cyclone, and the cold front is now further south and east of its original position. Within the cold air tongue, the skies are clear, but where the warm air mass is lifted well off the ground, there is heavy precipitation.

An idealized relationship between fronts and clouds is shown in Figure 7.11. Along the warm front is a broad layer of stratus clouds. This extends ahead of the front and takes the form of a wedge with a thin leading edge of cirrus. As the wedge thickens, the sequence of clouds seen from the ground changes to altostratus, then stratus, and finally nimbostratus with steady rain, as shown in the cross-section along the line A–A′. Within the sector of warm air, the sky may be partially

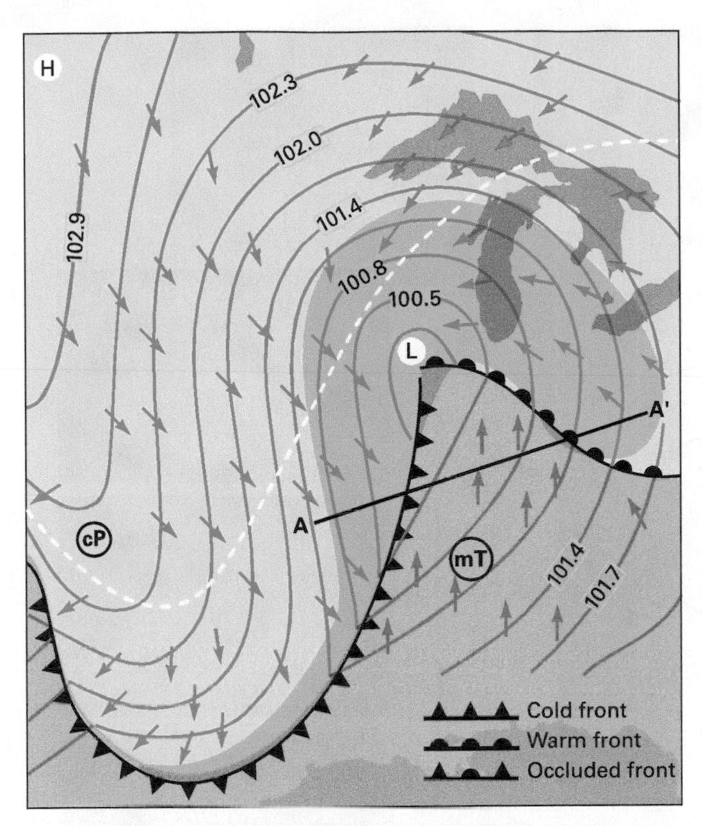

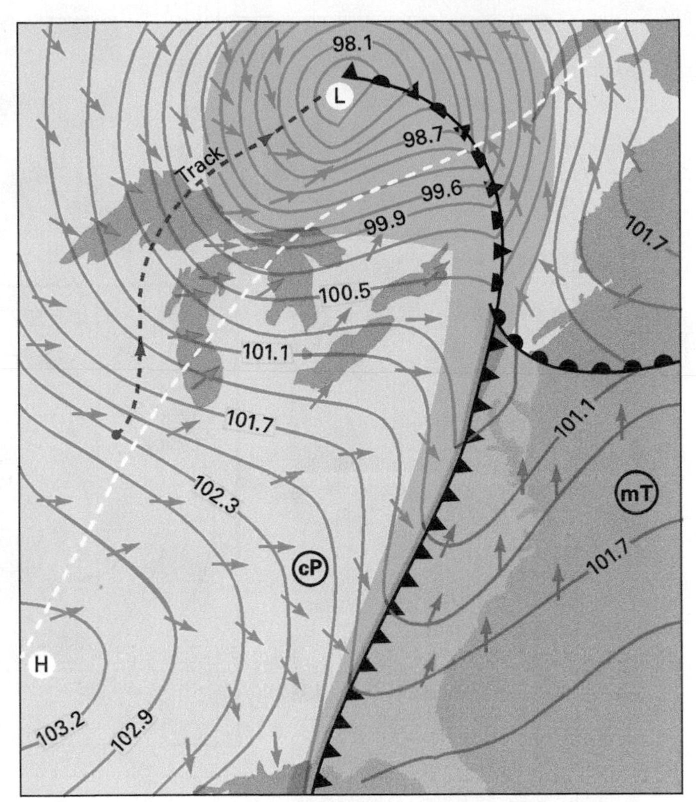

7.10 SIMPLIFIED SURFACE WEATHER MAPS SHOWING CHANGES IN A MIDLATITUDE WAVE CYCLONE OVER A 24-HOUR PERIOD In the open stage (left), cold and warm fronts pivot around the centre of the wave cyclone. In the occluded stage (right), the cold front has overtaken the warm front, and a large pool of warm, moist air has been forced aloft. The dashed white line shows the general boundary between the cP air and mT air and approximates the position of the polar front aloft. The dashed red line shows the route followed by the wave cyclone.

clear with scattered cumulus. Along the cold front, cumulonimbus clouds and associated thunderstorms have developed. These yield heavy rain but only in a comparatively narrow belt, and the skies clear quickly after the cold front passes. Note that the actual conditions that occur during the passage of a wave cyclone will depend on the position of the observer relative to the centre of the weather system, the stage of development of the disturbance, and the specific properties of the air masses involved.

CYCLONE TRACKS AND CYCLONE FAMILIES

Wave cyclones tend to form in certain areas and follow similar routes until they dissolve. In the northern hemisphere, wave cyclones are heavily concentrated in the neighbourhood of the Aleutian and Icelandic Lows. Three or four commonly form in succession, then travel as a chain across the North Pacific and North Atlantic oceans (Figure 7.12). Several groups may develop one after the other, until the opposing air masses have lost their individuality. The secondary disturbances usually form along the trailing cold front associated with the preceding wave cyclone. Thus, in the northern hemisphere, as the polar air pushes out from its source region, the eastward course followed by the successive low-pressure centres is progressively displaced to the south. The sequence terminates in the formation of a broad zonal ridge of high pressure.

The western coast of North America commonly receives wave cyclones that originate in the North Pacific Ocean, but they can also develop over the continent, especially in Alaska, northern Canada, or the eastern Rocky Mountains, as is the case for Alberta clippers and Colorado Lows. Most wave cyclone tracks converge toward the northeast, often passing into the Gulf of St. Lawrence because the general surface airflow over eastern North America is directed by the Appalachian Mountains (see Figure 5.14). Once in the North Atlantic, they may strengthen in the region of the Icelandic Low before moving into Europe.

In the southern hemisphere, storm tracks more often form along a single lane, following the parallels of latitude. The storm tracks are less variable because the extensive ocean surface circling the globe at these latitudes is broken only by the southern tip of South America.

The general eastward movement of wave cyclones is largely controlled by the paths of the Rossby waves and their relationship to the polar front jet streams. The curving path of the Rossby

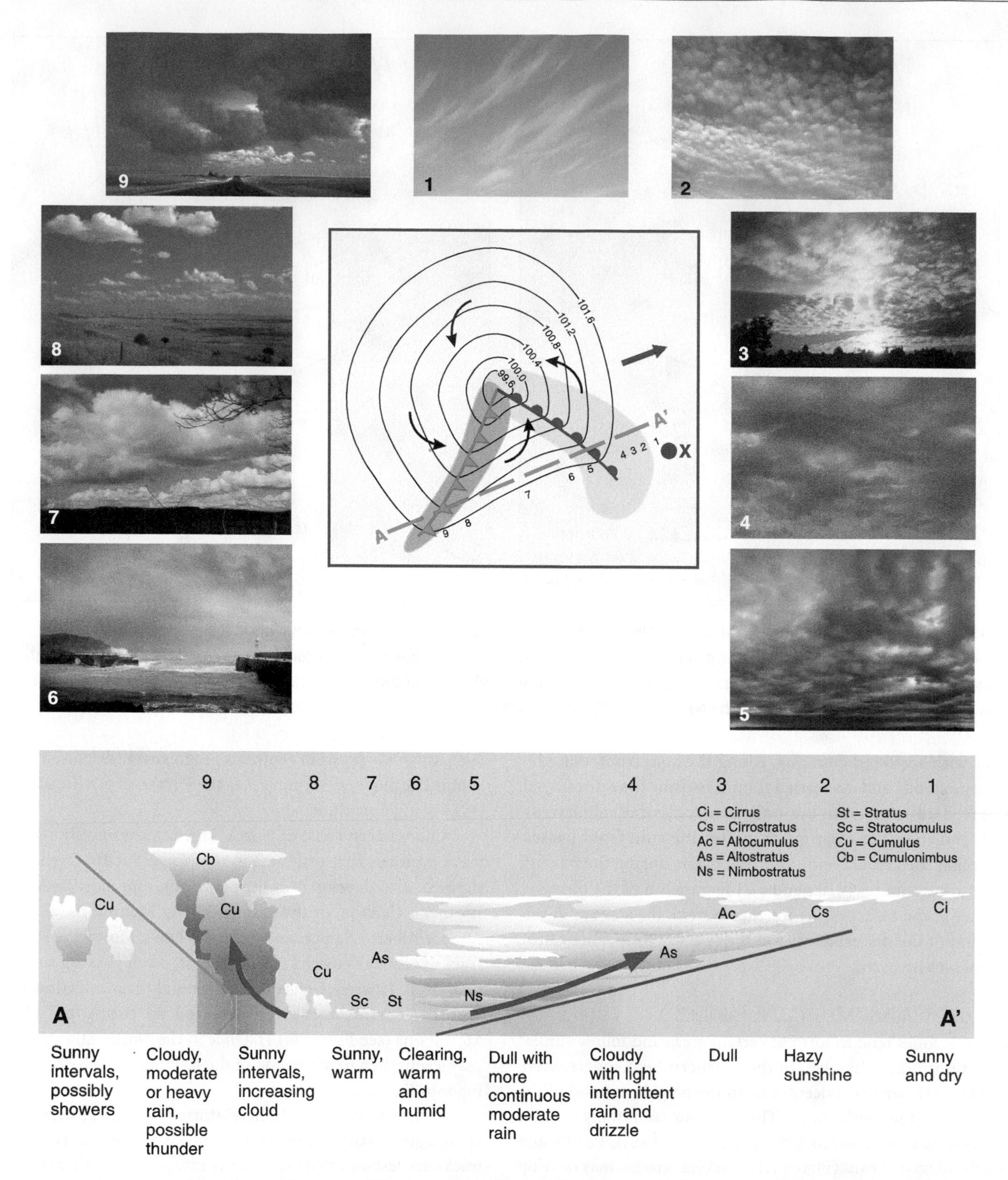

7.11 GENERAL CONDITIONS IN THE OPEN-STAGE OF A MIDLATITUDE WAVE CYCLONE The sequence of clouds and weather conditions associated with a midlatitude wave cyclone depend on the stage of development and the position of the observer relative to the centre of the disturbance. Assuming the wave cyclone moves in the direction shown by the arrow, then the changing conditions at point X would correspond to the sequence shown on line A–A′. The cloud types are (1) cirrus, (2) cirrocumulus, (3) altocumulus, (4) altostratus, (5) nimbostratus, (6) stratus, (7) stratocumulus, (8) cumulus, (9) cumulonimbus.

7.12 DAILY WORLD WEATHER MAP A daily weather map of the world for a given day during July or August might look like this map, which is a composite of typical weather conditions. Note the trains of wave cyclones (lows) in various stages of development moving through the midlatitudes. The closely-spaced circular arrangement of isobars over Florida depicts a hurricane. (After M. A. Garbell.)

waves is a prime factor in wave cyclone development. Air in the atmosphere alternately converges or diverges in the frontal zones, and this is linked to surface pressure conditions by strengthening updrafts and downdrafts in the troposphere (Figure 7.13).

Mountain barriers and ocean surface temperatures also influence the routes followed by wave cyclones and their associated weather patterns. For example, the passage of a wave cyclone across the Rocky Mountains often fails to produce precipitation because orographic lifting removes the moisture on the windward side of the mountains. Similarly, unusually cold water in the north-central Pacific Ocean in winter tends to push storm tracks farther north across North America, whereas warmer surface water in this region tends to displace storm tracks well to the south of their normal position.

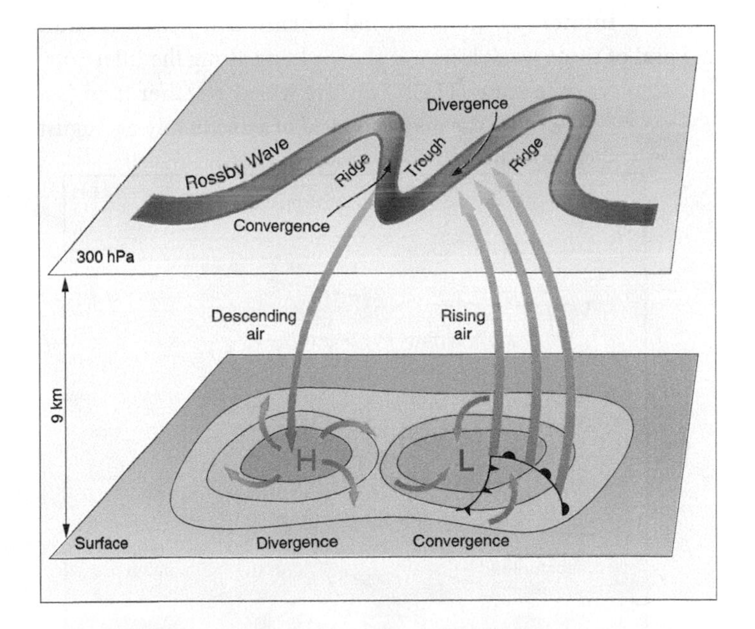

7.13 THE MEANDERING PATTERN OF THE ROSSBY WAVES IN THE ATMOSPHERE IS LINKED TO SURFACE PRESSURE CONDITIONS The fast-moving air alternately converges and diverges as it bends toward and away from the equator, and this complements inflow and outflow of air in surface pressure systems. (From *Introduction to Weather and Climate* by O.W. Archibold, John Wiley & Sons Canada, Ltd., 2011.)

TROPICAL AND EQUATORIAL WEATHER SYSTEMS

Tropical and equatorial zone weather systems show some basic differences from those of the midlatitudes. Upper-air winds are often weak, so air mass movement is slow and gradual. Tropical and equatorial air masses tend to be warm and moist, and consequently clearly defined fronts and large, intense wave cyclones do not develop. On the other hand, strong convectional activity occurs because of the high moisture content of low-latitude maritime air masses. With these conditions, even slight convergence and uplift can be enough to trigger abundant precipitation.

EASTERLY WAVES AND WEAK EQUATORIAL LOWS

One of the simplest forms of tropical weather systems is an **easterly wave**, a slowly moving, convectively active trough of low pressure that is driven westward by the trade winds. The easterly waves that cross the Atlantic Ocean are initiated by instability over northwestern Africa and are associated with the low latitude easterly jet. Once formed, they are directed westward in the troposphere by the winds of the Azores high-pressure cell; they move at a rate of about 20 to 35 kph. Easterly waves first develop in April or May and form about every three or four days until October or November. Approximately two easterly waves per week travel from Africa to North America during hurricane season in the general latitude range 5° to 30° N. They are prevented from moving too close to the equator by the strong convective uplift that generates the rising limb of the Hadley cells. Figure 7.14 shows the upper-troposphere isobars and airflow lines associated with an easterly wave. At the surface, a zone of weak low pressure underlies the axis of the wave. Surface airflow converges on the eastern, or trailing, side of the wave axis. This convergence causes the moist air to be lifted, producing scattered showers and thunderstorms. The rainy period may last a day or two as the wave passes.

Another related weather system is the *weak equatorial low*, a disturbance that forms near the centre of the equatorial trough. Moist equatorial air masses converge on the centre of the low, causing numerous convectional storms and heavy rainfall. Several of these weak lows are shown lying along the intertropical convergence zone (ITCZ) on the world weather map (see Figure 7.12). Because the map is typical of a day in July or August, the ITCZ is shifted well north of the equator. During this season, the rainy monsoon in Southeast Asia is marked by weak equatorial lows, such as the one over Bangladesh.

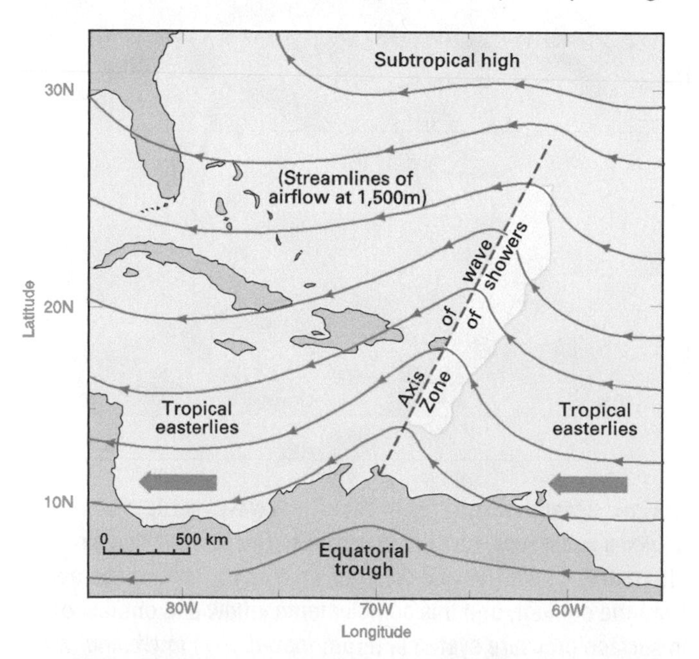

7.14 EASTERLY WAVE An upper-air easterly wave passing over the West Indies at an altitude of 1,500 m (Data from H. Riehl, Tropical Meteorology, New York: McGraw-Hill.)

POLAR OUTBREAKS

Another distinctive feature of low-latitude weather is the occasional penetration of powerful tongues of cold polar air from the midlatitudes into very low latitudes. The leading edge of these *polar outbreaks* is marked by squalls along the cold front, and is followed by unusually cool, clear weather with strong, steady winds. Polar outbreaks are best developed in the Americas. Outbreaks that move southward from the United States into the Caribbean Sea and Central America are called "northers" or "nortes," while those that move north from Patagonia into tropical South America are called "pamperos." One such outbreak is shown as a high-pressure cell over South America on the world weather map in Figure 7.12. A severe polar outbreak may bring subfreezing temperatures to the highlands of South America, severely damaging coffee and other important crops.

TROPICAL CYCLONES

Tropical weather disturbances range from weak tropical lows to powerful and destructive **tropical cyclones** (Table 7.2). A tropical cyclone is known variously as a *hurricane* in the western hemisphere, a *typhoon* in the western Pacific, and a *cyclone* in the Indian Ocean. The World Meteorological Organization classifies a tropical cyclone as "a storm of tropical origin of small diameter with a minimum surface pressure below 95 kPa, with violent winds greater than 125 kph, and accompanied by torrential rain." These storms revolve around a central "**eye**," which is about 30 to 50 km in diameter and characterized by light winds and clear skies. The distinctive pattern of inward-spiralling bands of clouds and a clear central eye make it easy to track tropical cyclones by satellite (Figure 7.15).

TABLE 7.2 U.S. NATIONAL HURRICANE CENTRE TROPICAL CYCLONE CLASSIFICATION

Tropical low	A surface low-pressure system in tropical latitudes
Tropical disturbance	A tropical low generally 200 to 600 km in diameter with weak surface winds and associated clusters of thunderstorms
Tropical depression	A tropical low with maximum sustained wind speed < 63 kph (< 18 m s^{-1}) Also applicable to the decaying stages of a hurricane when wind speed drops below 63 kph
Tropical storm	A warm-core tropical cyclone with maximum sustained surface winds between 63–119 kph (18–33 m s^{-1})
Hurricane	A warm-core tropical cyclone with maximum sustained surface winds > 119 kph (33 m s^{-1})
Major hurricane	A hurricane with maximum sustained surface winds > 180 kph (50 m s^{-1})

7.15 **HURRICANE KATRINA IN THE GULF OF MEXICO, AUGUST 29, 2005** Katrina was designated a Category 5 Hurricane (see Table 7.3), with sustained winds greater than 250 kph. This image comes from the GOES-12 weather satellite.

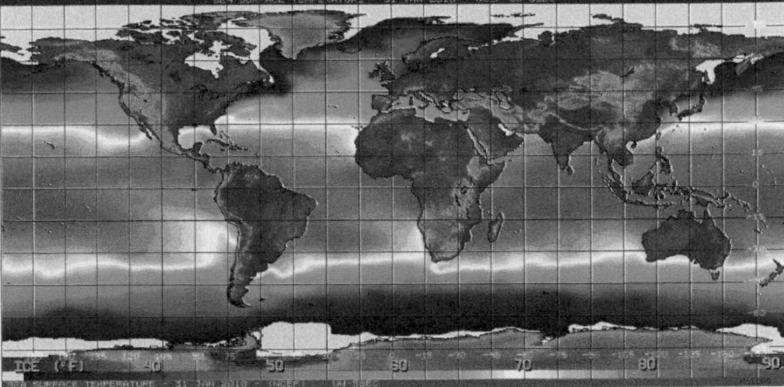

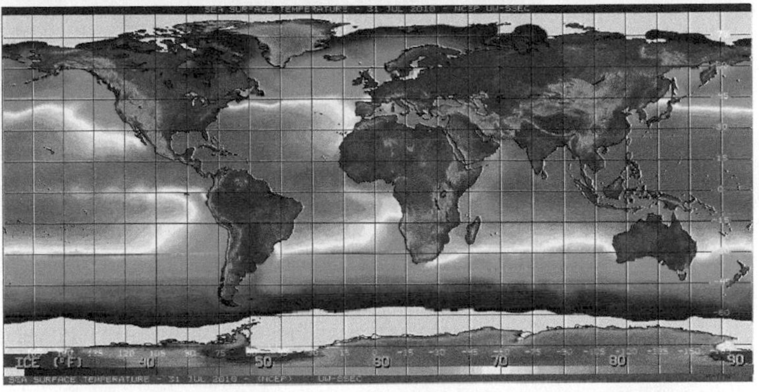

7.16 **SEA-SURFACE TEMPERATURE FOR JANUARY 2010 (TOP) AND JULY 2010 (BOTTOM)** Tropical cyclones form where sea-surface temperatures are above 27 °C. The warmest temperatures are generally found in the western tropical ocean basins where the thermocline is at its deepest. A considerable volume of warm water is therefore available to supply energy for storm development.

Tropical cyclones develop over oceans at 7° to 15° N and S latitudes (see Figure 2A.8 in the appendix to Chapter 2). The poleward limit is determined by the need for sea-surface temperatures that are high enough to ensure sufficient evaporation and release of latent heat to maintain the storm. This approximately correlates with sea-surface temperatures of over 27°C. Typically this condition is met in the western regions of tropical oceans in late summer (Figure 7.16). At latitudes higher than 15°, the cooler sea-surface temperatures do not contribute sufficient energy to the convective process. Although sea-surface temperatures exceed 27°C near the equator, at latitudes below 7° the Coriolis effect is too weak to initiate the storm's circular motion.

Once formed, the storms move westward through the trade-wind belt, often intensifying as they travel (Figure 7.17). In the North Atlantic Ocean, hurricanes travel westward and northwestward, and then turn toward the north and northeast at about 30° to 35° N as they come under the influence of the westerly winds. Storm intensity lessens as they move over cooler ocean surfaces because the source of energy is reduced; storms are further weakened by friction if they move over land. In the trade-wind belt, hurricanes travel at about 10 to 20 kph, although their speeds are more variable in the zone of the westerlies. Tropical cyclones can penetrate into the midlatitudes, bringing severe weather to the southern and eastern coasts of the United States. Heavy rain and strong winds often persist as far north as the Atlantic provinces; at these latitudes, hurricanes are normally downgraded to tropical storms. However, Hurricane Igor struck Newfoundland in September 2010 as a Category 1 storm with strong winds and heavy precipitation, causing a state of emergency to be declared in some parts of the province. Note that tropical cyclones never cross the equator.

An intense tropical cyclone is an almost circular storm of extremely low pressure. The storm's diameter may be 150 to 500 km. Because of the strong pressure gradient, winds spiral inward at high speed, which often exceed 30 to 50 m s^{-1} (100 to 180 kph). Barometric pressure in the storm centre commonly falls to 95 kPa or lower. Ascending air currents give rise to spirals of cumulonimbus clouds that converge on the eye of the storm. Typically about 10 percent of a hurricane is represented by these ascending cloud bands, which are often obscured by a dense overcast of cirrus cloud (Figure 7.18).

The cloud spirals give rise to very heavy rainfall, and as condensation continues, more energy is added to the storm through the release of latent heat. The cloud spirals are hundreds of kilometres long, but only a few kilometres wide. They develop 50 to 80 km apart at the perimeter of the storm, with the distance between adjacent bands decreasing toward the eye. The general

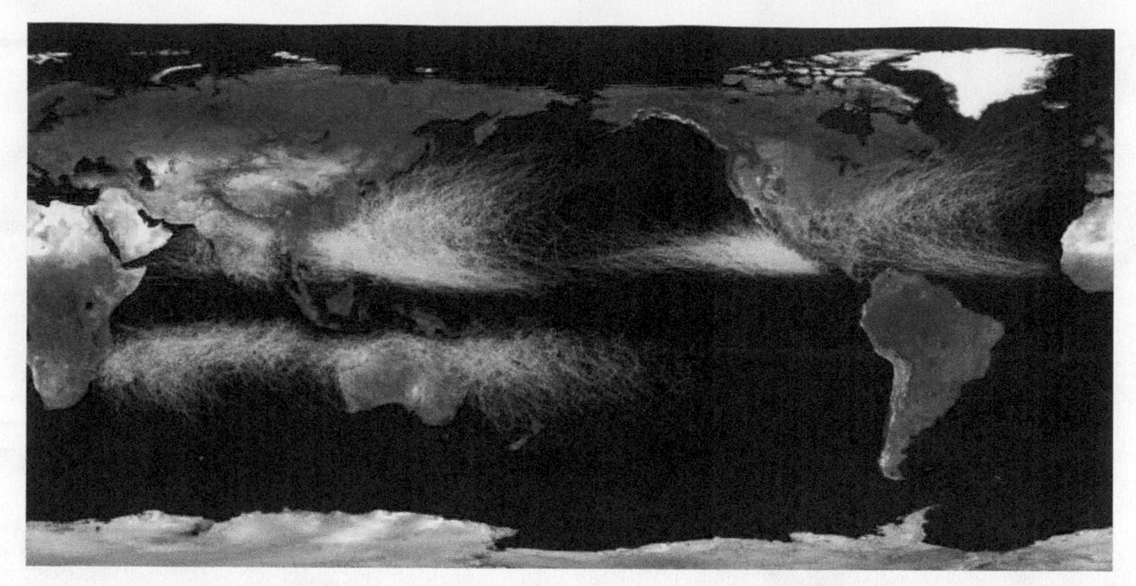

7.17 **TROPICAL CYCLONE TRACKS** Tropical cyclones always form over oceans. In the Atlantic, hurricanes originate off the west coast of Africa and in the Caribbean Sea. Curiously, tropical cyclones do not form in the South Atlantic or southeast Pacific regions, and apart from its northernmost coast, South America is never threatened by them.[1] In the Indian Ocean, cyclones originate both north and south of the equator, and strike India, Pakistan, and Bangladesh, as well as the eastern coasts of Africa and Madagascar. Typhoons of the western Pacific also form both north and south of the equator, moving into northern Australia, Southeast Asia, China, and Japan.

structure of hurricanes is clearly revealed by precipitation radar data collected as part of the Tropical Rainfall Monitoring Mission (TRMM), which began in 1997 (Figure 7.19). As well as the general pattern of the cloud and rain bands, the TRMM precipitation radar shows areas of intense vertical development known as *hot towers*. Hot towers are exceptionally tall cumulonimbus clouds that can grow to heights of 16 to 18 km. However, they are relatively small, ranging in diameter from 1 to 20 km, and short-lived, lasting from 30 minutes to 2 hours.

A characteristic feature of the tropical cyclone is its central eye, in which clear skies and calm winds prevail. The eye is a cloud-free vortex produced by the intense spiralling of the storm, and usually develops when sustained wind speeds exceed 120 kph. The eye is formed through the combined effects of *conservation of angular momentum* and *centrifugal force*. Conservation of angular momentum causes objects to spin faster as they move toward the centre of the hurricane, so wind speeds increase. However, as speed increases, the air molecules are subject to an outward-directed centrifugal force. The centrifugal force increases with the degree of curvature and with faster rotation speeds. At about 120 kph, the strong rotation of air balances the force of the inflow to the centre of the hurricane, which causes the air to ascend and form the eye wall.

The eye wall consists of a ring of tall thunderstorms that produce heavy rains and usually the strongest winds. However, the strong rotation also creates a vacuum and causes some of the air flowing out the top of the eye wall to turn inward and sink to replace air lost from the centre. Hence, in the eye, air descends from high altitudes and is adiabatically warmed. As the eye passes over a site, calm conditions prevail and the sky clears. The eye of the storm passes over fairly quickly, after which the storm strikes with renewed ferocity, but with the wind blowing the opposite direction.

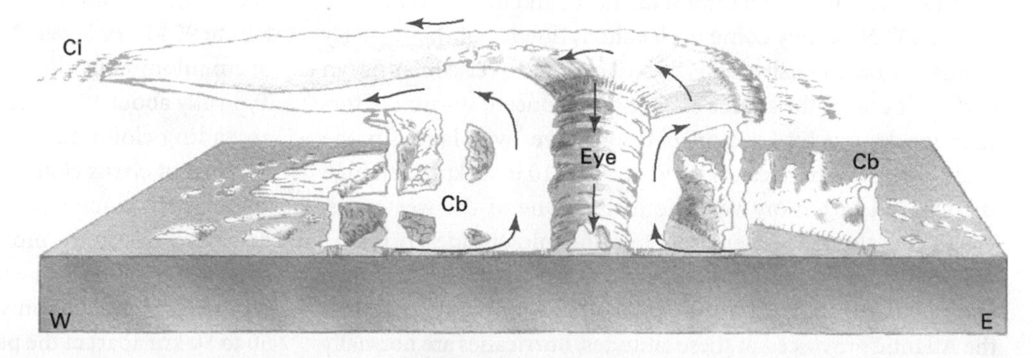

7.18 **ANATOMY OF A HURRICANE** In this schematic diagram, cumulonimbus (Cb) clouds in concentric rings rise through dense stratiform clouds. Cirrus clouds (Ci) fringe out ahead of the storm. The width of this diagram represents about 1,000 km. (Redrawn from NOAA, National Weather Service.)

[1] Only one hurricane has been reported in the South Atlantic Basin, and this occurred when tropical storm Catarina reached hurricane strength in March 2004.

(a) (b)

7.19 TROPICAL RAINFALL MONITORING MISSION (TRMM) DATA FOR HURRICANE KATRINA AS IT REACHED CATEGORY 4 STRENGTH OVER THE GULF OF MEXICO, AUGUST 28, 2005 (a) Horizontal distribution of rain intensity derived from TRMM precipitation radar; (b) vertical structure of the hurricane showing the central eye surrounded by a region of light to moderate rainfall intensity (blue and green) in the eyewall and the heavy rainbands (red). Also shown are two *hot towers* that developed from the energy released through condensation. Some of the hot towers in Katrina extended to altitudes of 16 km.

The Saffir-Simpson scale rates the intensity of tropical cyclones (Table 7.3). The rating is based on the central pressure of the storm, mean wind speed, and height of accompanying **storm surge**, which can raise sea level several metres above normal. Category 1 storms are weak, while Category 5 storms are devastating. The top three hurricanes in the United States ranked in terms of intensity are the Great Labor Day Storm, which occurred in September 1935 with a recorded minimum pressure of 89.2 kPa; Hurricane Katrina in August 2005, 90.2 kPa; and Hurricane Camille in August 1969, 90.9 kPa.

TABLE 7.3 SAFFIR-SIMPSON SCALE OF TROPICAL CYCLONE INTENSITY

CATEGORY	CENTRAL PRESSURE (kPa)	STORM SURGE (m)	MEAN WIND SPEED (kph)
1 Weak	> 98.0	1.2–1.7	119–153
2 Moderate	96.5–97.9	1.8–2.6	154–177
3 Strong	94.5–96.4	2.7–3.8	178–209
4 Very Strong	92.0–94.4	3.9–5.6	210–249
5 Devastating	< 92.0	> 5.6	> 250

Tropical cyclones occur only during certain times of the year. The season for North Atlantic hurricanes runs from May through November, with maximum frequency in late summer or early autumn. In the southern hemisphere, the season is roughly the opposite (Figure 7.20). These situations follow the seasonal migration of the ITCZ and correspond to periods when ocean temperatures are warmest. In a typical year, there are about 60 hurricanes worldwide. About 16 are reported in the Atlantic, 18 in the eastern Pacific, and 25 in the western Pacific. The remainder occur in the Indian Ocean and the tropical oceans of the

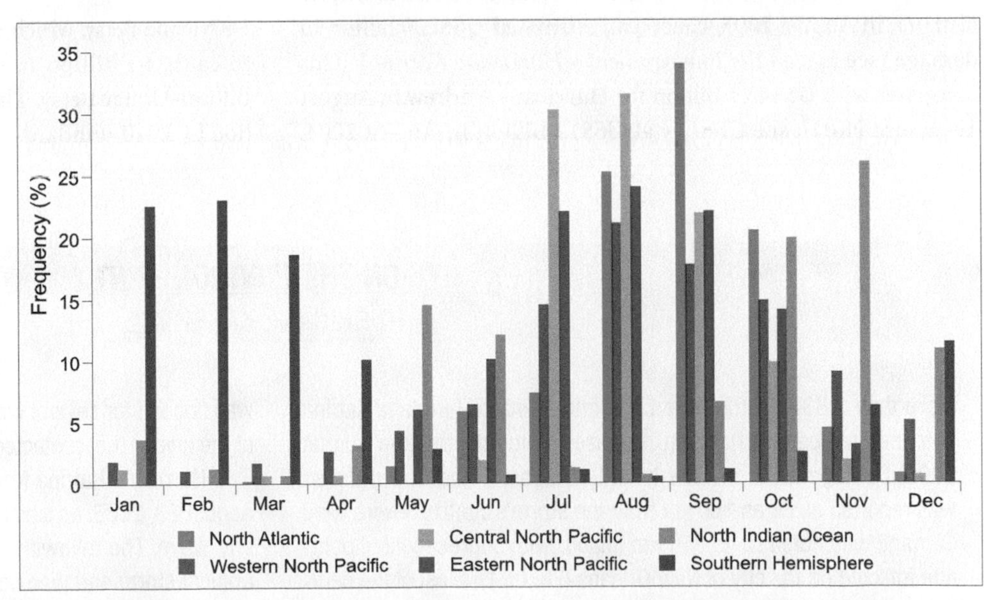

7.20 SEASONAL DISTRIBUTION OF HURRICANE ACTIVITY IN THE VARIOUS TROPICAL OCEAN REGIONS Hurricane activity is predominantly concentrated in the late summer and autumn in the respective hemispheres. (From *Introduction to Weather and Climate* by O.W. Archibold, John Wiley & Sons Canada, Ltd., 2011.)

southern hemisphere, with the exception of the South Atlantic, where only one hurricane has ever been reported.

For convenience, weather forecasters tracking tropical cyclones give them names. They alternate male and female names in an alphabetical sequence, with the exception of ones beginning with Q, U, X, Y, and Z. Each region has its own set of names; in North America one set is used for hurricanes in the Atlantic, and another for typhoons of the Pacific. The names are reused every six years with the exception that the names of storms that have caused significant damage are retired. The 2005 hurricane season was one of the worst on record with four Category 5 storms reported, including the infamous Katrina that devastated parts of New Orleans. A total of 28 tropical storms and hurricanes were reported in 2005. Following on from Tammy, Vince, and Wilma, the Weather Service resorted to the Greek alphabet (Alpha, Beta, Gamma, Delta, Epsilon, and Zeta) for the final six storms of the season, which lasted from June 8 until January 7, 2006. The names Dennis, Katrina, Rita, Stan, and Wilma were retired following the 2005 season.

IMPACTS OF TROPICAL CYCLONES

Tropical cyclones can be tremendously destructive. Islands and coasts feel the full force of the high winds and flooding as tropical cyclones move onshore (Figure 7.21). Densely populated low-lying areas are particularly vulnerable to tropical cyclones. The deadliest tropical cyclone on record struck Bangladesh on November 13, 1970, with at least 500,000 deaths reported. In North America, the worst hurricane in terms of loss of life occurred on September 8, 1900, at Galveston, Texas, with an estimated 6,000 to 12,000 casualties. In terms of cost, Hurricane Katrina, in August 2005, caused an estimated US$200 billion in damage (see *Eye on the Environment • Hurricane Katrina*). This compares with US$43.7 billion for Hurricane Andrew in August 1992, and Hurricane Charley at US$15 billion in August 2004.

7.21 **BAY ST. LOUIS, MISSISSIPPI** The lower photo, taken on August 31, 2005, reveals the devastation caused by Hurricane Katrina, which made landfall two days earlier. The resulting storm surge was estimated at more than 6 m. The arrows provide comparative reference marks.

Hurricane Betsy, which occurred in September 1965, was the first to cause $1 billion in damage, and so is often referred to as Billion-Dollar Betsy. This would be equivalent to about $12 billion by 2010 standards.

EYE ON THE ENVIRONMENT

HURRICANE KATRINA

More than 1,830 deaths have been attributed to Hurricane Katrina and the subsequent flooding it caused when it struck in August 2005. Damage along the low-lying Gulf Coast of the United States was reported as far as 160 km from the storm's centre. Severe wind damage was noted up to 150 km inland, with unprecedented damage inflicted on the city of New Orleans (Figure 1). Most of the damage was caused by the storm surge, which raised the sea level by an estimated 6 m in some locations. About 80 percent of New Orleans was flooded for several weeks because the surge breached several of the levees that protected the city from Lake Pontchartrain.

Hurricane Katrina formed in the Atlantic near the Bahamas on August 23, 2005, as a tropical depression associated with an easterly wave. The following morning the system was upgraded to a tropical storm and was named Katrina. Katrina continued to move westward and attained Category 2 hurricane status as it reached the coast of Florida on the morning of August 25 (Figure 2).

Figure 1 Damage caused by Hurricane Katrina

Katrina weakened to a Category 1 hurricane as it crossed Florida, but quickly regained strength as it entered the Gulf of Mexico. On August 27, Katrina was upgraded to Category 3 and continued to increase in size and intensity. By the following day, it had been upgraded to a Category 5 hurricane. Katrina reached peak strength in the early afternoon, with maximum sustained winds of 280 kph and a minimum pressure of 90.2 kPa. Winds subsided during the night to 190 kph, but increased to 205 kph by about 6 a.m. the following day, when Katrina reached the Gulf Coast as a Category 3 hurricane.

Katrina maintained hurricane strength for about 240 km inland across the state of Mississippi. It was downgraded to a tropical depression in Tennessee, when it was about 1,000 km north of the Gulf of Mexico. Remnants of Katrina produced storms in parts of southern Ontario and Quebec on August 30. In Canada, the storm produced heavy rain and strong wind gusts. Port Colborne, Ontario, received the heaviest rain in the region, with 102 mm falling that day—the first measurable rain in seven days. The storm passed quickly, with only 2.8 mm of rain falling on August 31, followed by another seven-day dry spell. Some localized flooding and fallen trees were reported. The weather system finally dissipated over Labrador on August 31.

Hurricanes follow many paths as they approach North America, and they appear to move farther to the west during El Niño years. Although 2005 was not an El Niño year, it was unusually warm in many parts of the world, and the global annual temperature for combined land and ocean surfaces established a record that exceeded the previous record set in 1998 under the influence of an extremely strong El Niño.

Perhaps for this reason, three Category 5 hurricanes formed in the Atlantic Basin for the first time in a single season. Two of these, Katrina and Rita, passed over the oil-producing region in the Gulf of Mexico. The combined effect of Hurricane Rita's more westerly path and the large diameter of Hurricane Katrina affected most of the 2,900 oil platforms in the region. As a result approximately 150 oil rigs were severely damaged, 36 rigs sank, and several floated free, having broken their moorings.

In preparation for Hurricane Katrina, daily production of oil was reduced by 1.4 million barrels a day to about 5 percent of daily production; natural gas production was reduced by about 35 percent. The impact of Katrina on the oil industry lasted several weeks. Twenty-one oil refineries, which produced almost 50 percent of the petroleum distillates in the United States, were still not functioning in mid-September. Four refineries suffered serious damage, and none of these was at full capacity at the end of 2005.

In addition to loss of production, there was a serious threat of oil pollution. At least 44 oil spills were reported, the largest of which involved about 4 million gallons. Of this total, 960,000 gallons were recovered, 2 million were contained, and 982,000 evaporated. More than 7 million gallons of oil escaped from industrial plants and storage depots, most of which was contained within earthen berms. A major concern was that the oil would have an impact on the Mississippi River. However, only a few minor oil slicks were reported, and these evaporated fairly quickly. There were no reports of major offshore spills into the Gulf of Mexico.

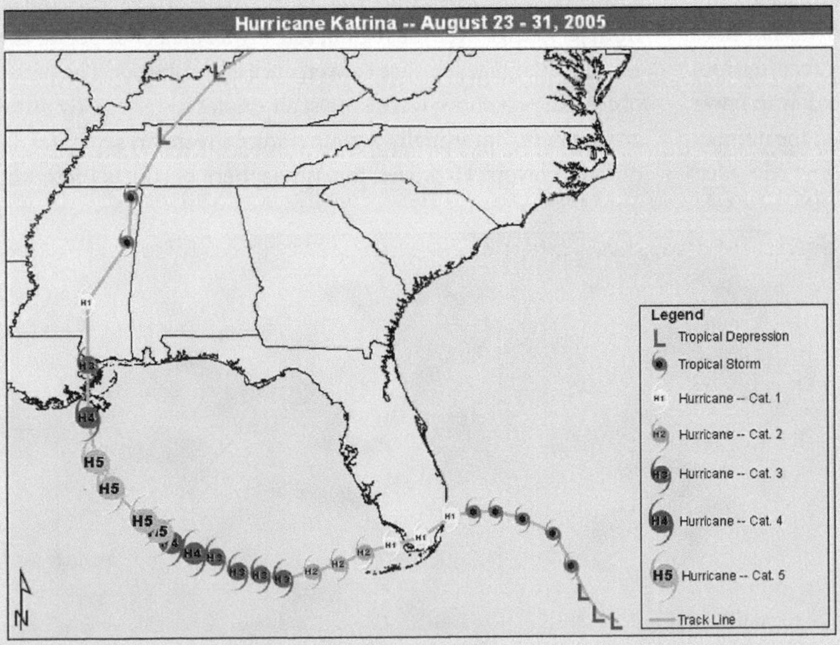

Figure 2 Tracking Hurricane Katrina

THE TORNADO

A **tornado** is a small but intense cyclonic vortex in which air spirals at tremendous speed. Tornadoes generally occur in the midlatitudes between about 30° and 50° (Figure 7.22). Nowhere are they as frequent, or as violent, as in the United States, where 1,000 or more strike annually. In Canada, about 80 tornadoes are reported each year. The tornado season in Canada typically extends from May to September, with most occurring in June or July. Tornadoes are frequently spotted in the Prairie provinces, southern Ontario, and southern Quebec. For example, eight tornadoes were reported in Quebec in 2009. The best-known tornadoes in Canada are the Regina tornado, which occurred on June 30, 1912; the Edmonton tornado (July 31, 1987); and more recently, the Pine Lake tornado, near Red Deer, Alberta, on July 14, 2000. Significant tornado damage has also been reported in Windsor, Sudbury, and Barrie, Ontario. Tornadoes are reported occasionally in British Columbia and the Maritimes. In Europe, the total number of tornadoes is estimated at about 250 annually, with as many as 50 occurring in southern England and 30 in other countries, such as Germany and the Netherlands. Japan reports about 10 tornadoes each year, with one or two noted in India and Bangladesh. In the southern hemisphere, tornadoes are most often reported in Argentina, Australia, and New Zealand.

Tornadoes occur most frequently in the Mississippi valley, which has earned the name "tornado alley" (Figure 7.23). They are most common in the spring and summer, but can occur in any month. In a typical year, tornadoes are reported on about 180 days somewhere in North America. Although records of tornadoes for the United States extend back to the early 1900s, it was not until 1953 that official records began to be collected. Early records from 1916 indicate fewer than 100 tornadoes were reported, with fewer than 200 generally reported annually until 1953, when the number rose above 400. Since then, the number of reported tornadoes has steadily increased (Figure 7.24). The initial increase in tornado frequency is attributed primarily by improved reporting methods.

The tornado season in North America generally includes spring and early summer. At this time of the year, cold, dry air from the High Plains overrides warm maritime air entering the Central Lowlands from the Gulf of Mexico (Figure 7.25). Under these conditions, a wedge of cold dry air at the 70 to 50 kPa level lies directly above the warm surface air. Higher still, at the 30 kPa level, the polar front jet swings over the region, causing an area of upper air divergence that initiates surface convergence and rising air. The presence of cold, dry air above warm, moist air creates an inherently unstable atmosphere, but initially a temperature inversion separates these different layers. However, the atmosphere becomes increasingly

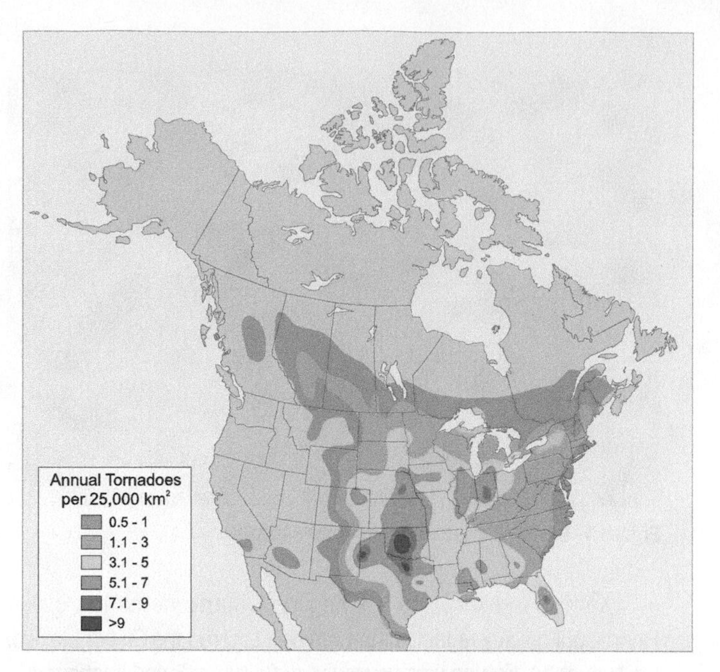

7.23 **TORNADO FREQUENCY IN NORTH AMERICA** Tornadoes occur most frequently in the Great Plains region of the United States, especially in the Mississippi Valley which is commonly referred to as Tornado Alley. (From *Introduction to Weather and Climate* by O.W. Archibold, John Wiley & Sons Canada, Ltd., 2011.)

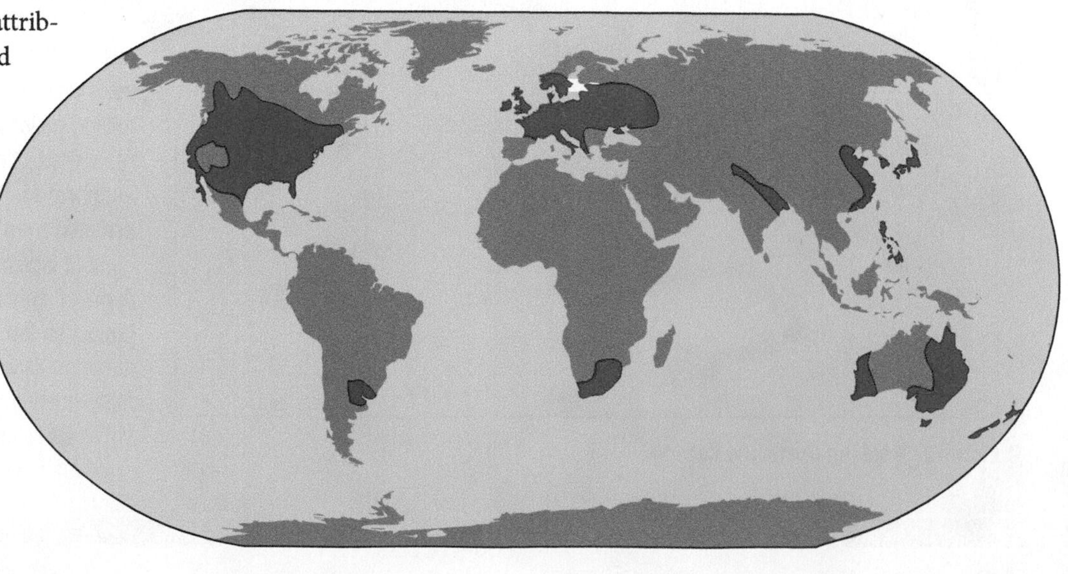

7.22 **TORNADO REGIONS** Apart from North America, tornadoes occur quite frequently in western Europe and Australia, and less commonly in South America, South Africa, and the Far East.

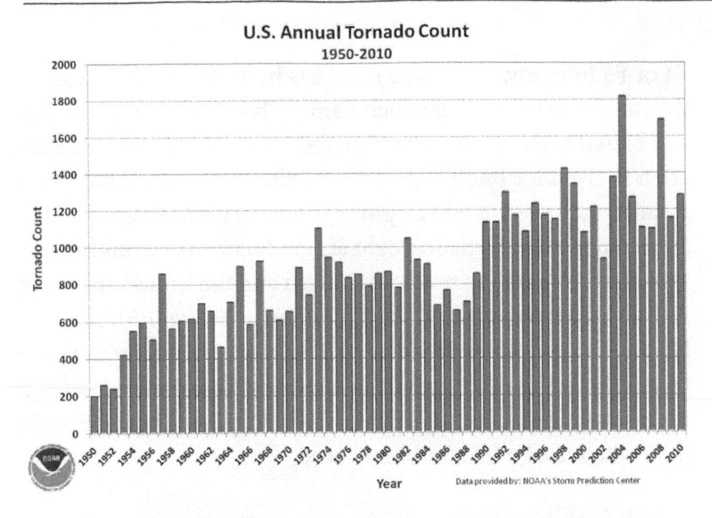

7.24 **NUMBER OF TORNADOES REPORTED IN THE UNITED STATES FROM 1953 TO 2008** The initial increase in tornado frequency is generally explained by improved reporting methods.

unstable as daytime warming continues, with some lifting occurring at frontal zones. Eventually the inversion breaks and convective instability causes the moist surface air to rise rapidly, leading to thunderstorms. The strong wind shear that develops in severe thunderstorms provides the rotation that can spawn a tornado.

Tornadoes form most often in association with intense supercell thunderstorms, with the first stage of tornadogenesis indicated by the development of a mesocyclone due to rotation in the updraft core induced by strong wind shear. A major tornado typically evolves through a series of stages, each of which can be of variable duration. In the initial *dust whirl stage*, a tornado usually takes the form of a swirling dust cloud that extends from the ground to a short, pendant funnel cloud hanging below the base of the mesocyclone cloud mass. Funnel clouds form by condensation of water vapour when the air is cooled adiabatically through expansion as it is taken aloft in the low-pressure vortex. The dust whirl stage is followed by the *organizing stage*, which is often marked by the rapid lowering of the base of the thunderstorm as a rotating wall cloud develops beneath the mesocyclone.

Many tornadoes never develop beyond the organizing stage, but simply lose energy and disappear. Those that continue to develop pass into the *mature stage* in which the vortex becomes nearly vertical and is darkened with dust and debris. A tornado attains its maximum size and strength in the mature stage (Figure 7.26). At this stage, the tornado is generally in the order of 100 to 500 m in diameter with wind speeds ranging from 250 to 350 kph. The tornado writhes and twists as it moves across the countryside, and where it touches the ground it can cause the complete destruction of almost anything in its path. As it loses energy, a tornado will pass into the *shrinking stage* in which the width of the vortex decreases and becomes more tilted. Finally, in the *decay stage* a tornado develops a thin, snaking, rope-like form, but even in its final stages it can still be a serious threat.

Wind speed is never constant within a tornado. Not only do wind speeds vary over time as a tornado strengthens and dissipates, but at any given instant different parts of a tornado can generate winds of different strength. This arises because the tornado itself is advancing across the ground and because the air in the tornado is itself rotating. For example, if a tornado is tracking northeast at 60 kph, and the counterclockwise rotational wind speed is 200 kph, wind speed at its leading and trailing edges would also be 200 kph. However, in the northwestern quadrant, the forward speed of the tornado would oppose the rotational speed, resulting in an effective wind speed of 140 kph. Conversely, in the southeastern quadrant, the effect would be additive, increasing the effective wind speed to 260 kph.

In 1971, the *Fujita (F) scale* was devised to assess the intensity of North American tornadoes on the basis of the damage they caused (Table 7.4). Subsequent analysis of the 1912 Regina, Saskatchewan tornado suggests it reached F4 intensity, with winds possibly

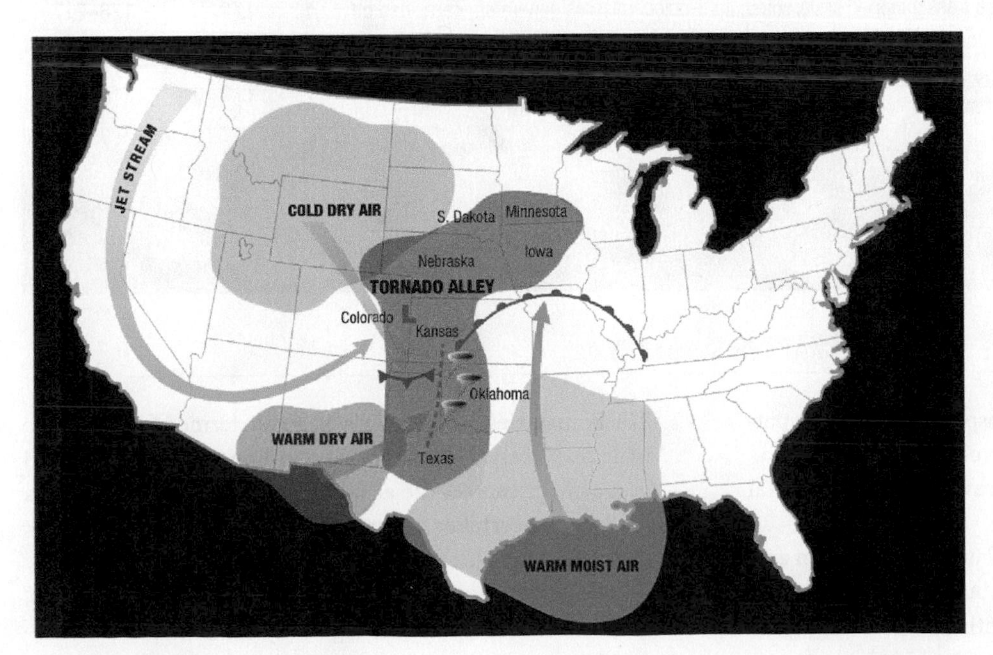

7.25 **TORNADO DEVELOPMENT IN THE UNITED STATES** Tornado development in the United States in the spring and early summer is favoured by the interaction of distinctive air masses, in conjunction with upper air divergence associated with the polar front jet stream aloft.

7.26 **TORNADO** Surrounding the tornado vortex is a cloud of dust and debris carried into the air by violent winds.

exceeding 400 kph. The 1987 Edmonton tornado was similarly rated F4. The first documented F5 tornado in Canada passed through Elie, Manitoba, on June 22, 2007; it remained on the ground for 35 minutes, cutting a 300 m swath with a path length of about 5.5 km. Fewer than 1 percent of Canadian tornadoes reach F4 or F5 intensity. The Fujita scale has been criticized because of the subjective manner in which damage is assessed. In particular, it is biased toward the worst damage, even though no consideration is given for building type or method of construction. As a result, wind speeds could be grossly over- or underestimated. The appearance of the tornado might also influence a person's impression of its strength. As result, in Canada and the United States, the F-scale was replaced in 2007 by the *Enhanced Fujita (EF) scale*. The EF scale incorporates 28 damage indicators, including different types of buildings and trees for which varying degrees of damage are assessed. Because the EF scale is less intuitive than the F-scale, the Fujita scale is still commonly used.

A Look Ahead

Annual cycles of temperature and precipitation in most regions are quite predictable given the changes in wind patterns, air mass flows, and weather systems that occur with the seasons. These recurring annual cycles vary in response to global differences in energy. This creates spatially variable weather patterns and provides the basis for climate and climate classification. Chapters 8 to 10 discuss these topics.

TABLE 7.4 THE FUJITA AND ENHANCED FUJITA SCALES OF TORNADO INTENSITY

FUJITA SCALE			ENHANCED FUJITA SCALE	
SCALE	ESTIMATED WIND SPEED (kph)	TYPICAL DAMAGE	3 SECOND GUST (kph)	
F0	< 120	**Light damage**—some damage to chimneys; branches broken off trees; shallow-rooted trees pushed over; sign boards damaged	EF0	105–137
F1	120–185	**Moderate damage**—surface peeled off roofs; mobile homes pushed off foundations or overturned; moving automobiles blown off roads	EF1	138–177
F2	186–260	**Considerable damage**—roofs torn off frame houses; mobile homes demolished; boxcars overturned; large trees snapped or uprooted; light-object missiles generated; cars lifted off ground	EF2	178–217
F3	261–340	**Severe damage**—roofs and some walls torn off well-constructed houses; trains overturned; most trees in forest uprooted; heavy cars lifted off the ground and thrown	EF3	218–265
F4	341–430	**Devastating damage**—well-constructed houses levelled; structures with weak foundations blown away some distance; cars thrown; large missiles generated	EF4	266–322
F5	431–530	**Incredible damage**—strong frame houses levelled off foundations and swept away; automobile-sized missiles generated and fly through air in excess of 100 m; trees debarked; incredible phenomena occur	EF5	Over 322

CHAPTER SUMMARY

- A weather system is a recurring atmospheric circulation pattern. Weather systems include wave cyclones, travelling anticyclones, tornadoes, easterly waves, weak equatorial lows, and tropical cyclones.
- Air masses are distinguished by the latitudinal location and type of surface of their *source regions*. Air masses influencing North America include those of continental and maritime source regions, and of arctic, polar, and tropical latitudes.
- The boundaries between air masses are termed fronts. These include cold and warm fronts, according to whether the cold or warm air masses are advancing. In an occluded front, a cold front overtakes a warm front, lifting a pool of warm, moist air above the surface.
- *Travelling cyclones* include wave cyclones, tropical cyclones, and the tornado. The *travelling anticyclone* is typically a fair-weather system.

- Wave cyclones form in the midlatitudes at the boundary between cool, dry air masses and warm, moist air masses. In the wave cyclone, a vast inward-spiralling motion produces cold and warm fronts, and eventually an occluded front. Precipitation normally occurs along each type of front.
- Tropical weather systems include *easterly waves* and *weak equatorial lows*. Easterly waves occur when a weak low-pressure trough develops in the easterly wind circulation of the tropical zones, producing convergence, uplift, and shower activity. Weak equatorial lows occur near the ITCZ. In these areas of low pressure, convergence triggers abundant convectional precipitation.
- Tropical cyclones can be very powerful storms. They develop over warm tropical oceans and can intensify to become extensive inward-spiralling systems of high winds with low central pressures. As they move onto land, they bring heavy surf and storm surges.
- Tornadoes are very small, intense cyclones that usually occur as a part of severe thunderstorm activity. The high winds associated with tornadoes can be very destructive.
- Global air and ocean circulation provides the mechanism for *poleward heat transport* by which excess heat moves from the equatorial and tropical regions toward the poles. In the atmosphere, the heat is carried primarily in the movement of warm, moist air poleward, which releases its latent heat when precipitation occurs. In the oceans, global circulation moves warm surface water northward through the Atlantic Ocean. Heated in the equatorial and tropical regions, the surface water loses its heat to the air over the North Atlantic and sinks to the bottom. These heat flows help make high-latitude climates warmer than expected based on solar heating alone.

KEY TERMS

air mass
cold front
cyclonic precipitation
cyclonic storm
easterly wave
eye
front
occluded front
storm surge
thermally direct
tornado
tropical cyclone
warm front
wave cyclone
weather system

REVIEW QUESTIONS

1. Define air mass. What two features are used to classify air masses?
2. Compare the characteristics and source regions for mP and cT air mass types.
3. Identify three weather systems that bring rain in equatorial and tropical regions. Briefly describe each system.
4. Describe the structure of a tropical cyclone. What conditions are necessary for the development of a tropical cyclone? Give a typical path for the movement of a tropical cyclone in the northern hemisphere.
5. Why are tropical cyclones so dangerous?
6. Where did Hurricane Katrina first form, and how did it develop and move?
7. What damage was sustained by south Florida and coastal Louisiana from Hurricane Katrina?
8. What were the maximum wind speeds of Katrina? What other phenomena were associated with this storm?
9. What effect does El Niño have on hurricanes?
10. Describe a tornado. Where and under what conditions do tornadoes typically occur?

VISUALIZATION EXERCISES

1. Identify three types of fronts and draw a cross-section through each. Show the air masses involved, the contacts between them, and the direction of air mass motion.
2. Sketch two weather maps, showing a wave cyclone in open and occluded stages. Include isobars and identify the centre of the cyclone as a low. Lightly shade areas where precipitation is likely to occur.

ESSAY QUESTIONS

1. Compare and contrast midlatitude and tropical weather systems. Be sure to include the following terms or concepts in your discussion: air mass, convectional precipitation, cyclonic precipitation, easterly wave, polar front, stable air, travelling anticyclone, tropical cyclone, unstable air, wave cyclone, and weak equatorial low.
2. Prepare a description of the annual weather patterns that your location experiences through the year. Refer to the general air mass patterns, as well as the types of weather systems that occur in each season.

CHAPTER 8 THE GLOBAL SCOPE OF CLIMATE

Afternoon sun reflecting off the Pacific Ocean at Los Padres National Forest, California, U.S.

The astronomical cycles related to the daily rotation of Earth on its axis, and the annual revolution of Earth around the sun are intrinsically linked to climate. Rotation results in a daily temperature cycle that is produced by hourly changes in insolation and the surface energy balance. The Earth's revolution produces an annual cycle of insolation, which is determined by latitude and solar declination. Both effects are strongly related to the total daily energy flow received at the surface; this in turn influences temperature, evaporation and moisture content of the atmosphere, and global circulation. The dynamic nature of the atmosphere in terms of its spatial and temporal properties is reflected in the diversity of global climates.

In its most general sense, **climate** is the average weather of a region, but except where conditions change very little during the course of the year, such a broad definition can be misleading. Thus, it is preferable to define climate in terms of both average conditions and the variability of those conditions based on long records of daily and seasonal weather patterns. Climate descriptions could include any of the measures that express the state of the atmosphere, such as daily net radiation, barometric pressure, wind speed and direction, cloud cover, and precipitation type and intensity. Although annual climate data are available covering one or two centuries for many parts of the world, most weather stations do not regularly make detailed observations of more than a few basic variables. For this reason, comparisons of world climates generally rely on temperature and precipitation data, and are based on average values for each month, as well as seasonal variation during the year.

Temperature and precipitation strongly influence a region's natural vegetation—for example, forests generally occur in moist regions, and grasslands in drier regions. The natural vegetation cover is often a distinctive feature of a climatic region and typically influences the human use of the area. Temperature and precipitation are also important factors in the cultivation of crop plants. Similarly, the development of soils, as well as the types of processes that shape landforms, are partly dependent on temperature and precipitation. For these reasons, the study of global climates is an important part of physical geography.

KEYS TO CLIMATE

Several of the principles discussed in earlier chapters are helpful in understanding the global scope of climate. First, the annual cycle of air temperature experienced at a weather station is primarily influenced by latitude and location on a continent:

- *Latitude.* Near the equator, temperatures are warm and the annual temperature range is low. Toward the poles, temperatures are colder and the annual range is greater. These effects are produced by the latitudinal variation in the annual cycle of insolation.
- *Coastal-continental location.* Coastal stations show a smaller annual range in temperature, while the range is larger at stations in continental interiors. This effect occurs because of different rates of heating and cooling on land surfaces and oceans.

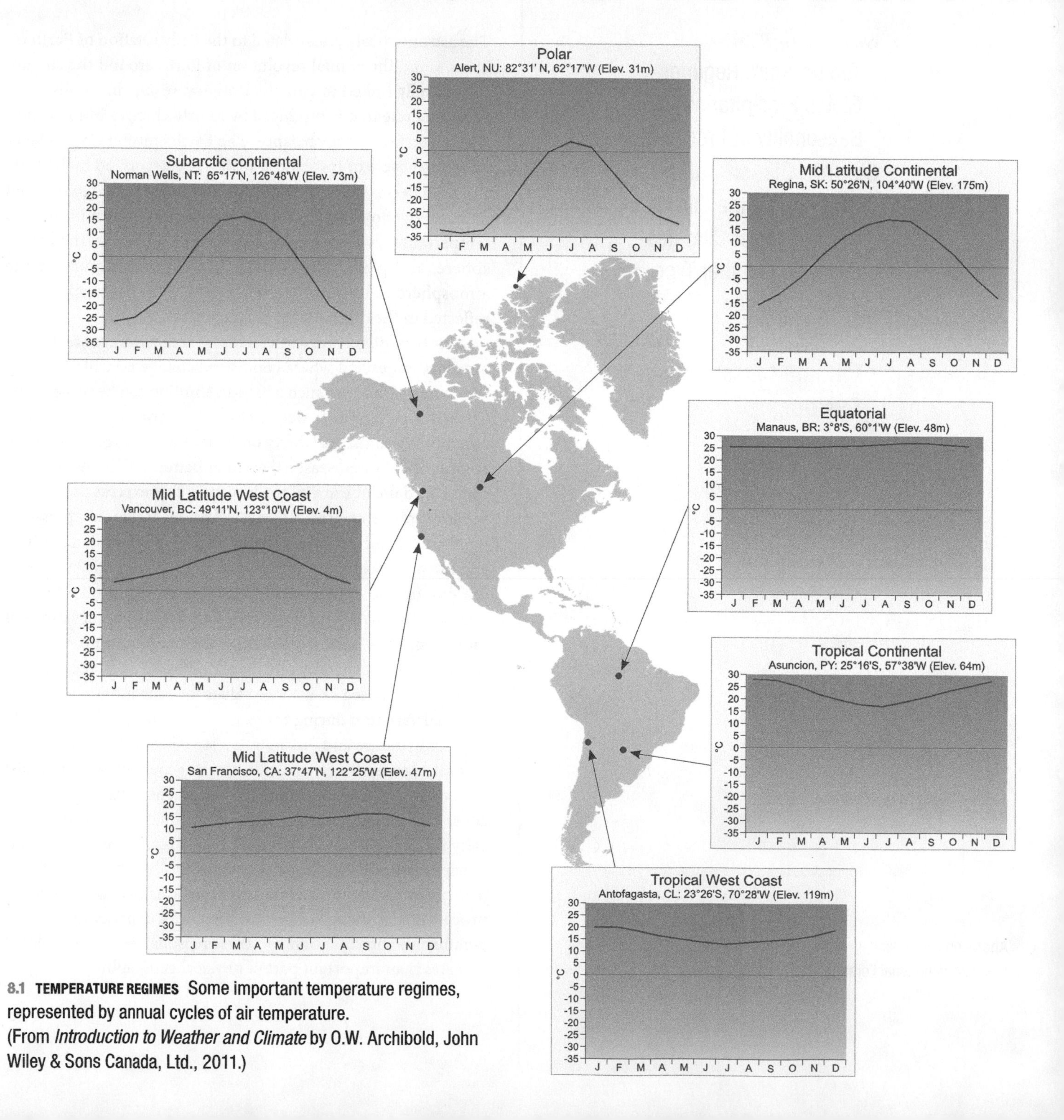

8.1 TEMPERATURE REGIMES Some important temperature regimes, represented by annual cycles of air temperature.
(From *Introduction to Weather and Climate* by O.W. Archibold, John Wiley & Sons Canada, Ltd., 2011.)

Air temperature also has an important effect on precipitation:

- *Warm air can hold more moisture than cold air.* This means that colder regions generally have lower precipitation than warmer regions. Also, precipitation tends to be greater during the warmer months of the temperature cycle.

TEMPERATURE REGIMES

Typical patterns of mean monthly temperatures observed at stations around the world are presented in Figure 8.1. These patterns are referred to as *temperature regimes*—distinctive types of annual temperature cycles related to latitude and location. Each regime is labelled according to its latitude zone: equatorial, tropical, midlatitude, and subarctic. Some labels also describe the location of the weather station in terms of its position on a landmass—"continental" or "west coast."

The equatorial regime (Manaus, Brazil, 3° S) is uniformly very warm. Temperatures are close to 27 °C year-round. There are no pronounced temperature seasons because insolation is similar in all months. In contrast, the tropical continental regime (Ascunsion, Paraguay, 25° S) shows a stronger temperature cycle. Temperatures change from very hot during the high-sun season, to warm during the low-sun season. The situation is quite different at Antofagasta, Chile (23° S), which has a similar insolation cycle as Ascunsion. The tropical west-coast regime at Antofagasta has a weaker annual cycle and no extreme heat. The difference is due to the moderating effects of Antofagasta's maritime location, an effect that is enhanced by the cold Peru Current that flows north along the coast. The moderating effect of a coastal location is also shown in the graphs for the midlatitude west-coast regime—San Francisco, California (36° N), and Vancouver, British Columbia (49° N).

In continental interiors, however, the annual temperature cycle remains strong. The midlatitude continental regime of Regina, Saskatchewan (50° N), and the subarctic continental regime of Norman Wells, Alberta (65° N), show annual variations in mean monthly temperature of about 35 °C and 40 °C, respectively. The polar regime of Alert, Nunavut, (82° N), is characterized by low temperatures year round.

Other regimes can be identified, and because temperature varies continuously, the list could be expanded indefinitely. However, what is important is that (1) annual variation in insolation, which is determined by latitude, provides the basic control of temperature patterns; and (2) the effect of location—maritime or continental—moderates that variation.

The mean temperature for each month is calculated from the average of the daily maximum and minimum temperatures recorded over a long period of time. The Canadian *climate normals,* or average values for climate parameters measured at weather stations, are presently based on the period 1971 to 2000. At the end of each decade, Environment Canada revises the climate normals; the record will next be updated in 2011 to incorporate the period 1981 to 2010. Mean monthly temperature is just one of several temperatures that might be reported from the record of daily maximum and minimum temperatures. A more complete record would include the long-term average maximum and minimum temperatures, as well as the extreme maximum and minimum temperatures (Figure 8.2).

The end-of-chapter feature *Working It Out • Averaging in Time Cycles* demonstrates the power of averaging to reveal distinctive time cycles in data, such as the temperature regimes presented above.

GLOBAL PRECIPITATION

Global precipitation patterns are largely determined by air masses and their movements, which in turn are produced by global air circulation patterns. The general precipitation patterns expected for a hypothetical landmass are shown in Figure 8.3. Five classes of annual precipitation—wet, humid, subhumid, semi-arid, and arid—are recognized on this landmass.

The equatorial zone shows a wet band stretching across the landmass. This represents convectional precipitation in the weak equatorial lows that develop near the ITCZ (see Chapter 7). Note that the wet band widens and extends away from the equator into the tropical zone along the eastern coast. Precipitation in this coastal region is influenced by the trade winds that carry warm, moist mT air masses and tropical cyclones onshore. Humid conditions continue along the east coasts into the midlatitude zones. In these midlatitude regions, general circulation around subtropical high-pressure cells brings mT air masses onshore in the summer, while in winter, wave cyclones bring cyclonic precipitation from the west.

Another important feature of the hypothetical landmass is the pattern of arid and semi-arid regions that stretches from tropical west coasts to subtropical and midlatitude continental interiors.

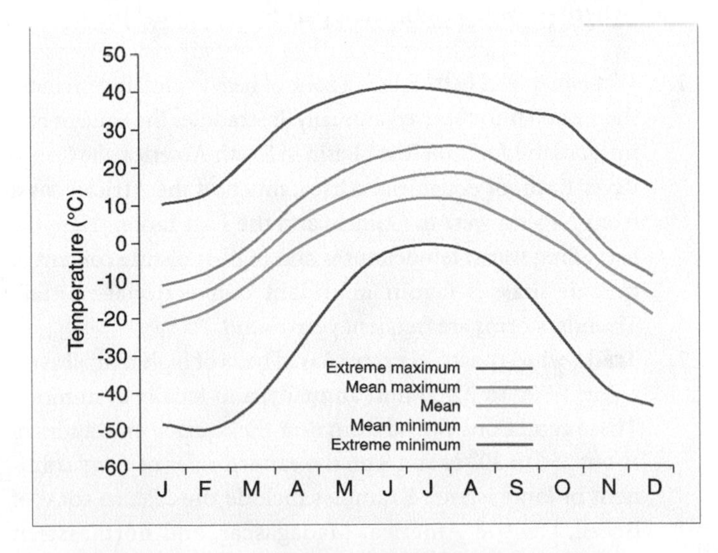

8.2 TEMPERATURE DATA FOR SASKATOON, SASKATCHEWAN, BASED ON ENVIRONMENT CANADA'S 1971–2000 CLIMATE NORMALS

In the tropical and subtropical latitudes, the arid pattern is produced by dry, subsiding air in persistent subtropical high-pressure cells. The arid zone extends eastward and poleward into the semi-arid continental interior where precipitation is limited by the distance moist air masses must travel from their source regions. Rain shadows, created by coastal mountain barriers that block the flow of moist westerly winds from the ocean, may also contribute to dry conditions inland.

A further distinctive feature of the landmass is the pair of wet bands along the west coasts that extend from the midlatitudes into the subarctic zones. The heavy precipitation comes from the moist mP air masses that are carried by the prevailing westerly winds in a succession of wave cyclones; precipitation on the western coast may be augmented by orographic uplift along coastal mountain ranges.

The high-latitude zone is depicted as arctic desert, where precipitation is limited because cold air can hold only a small amount of moisture. In addition, most of the Arctic Ocean is frozen for part of the year; this means that in winter, associated mP air masses are not readily distinguishable from their cP counterparts.

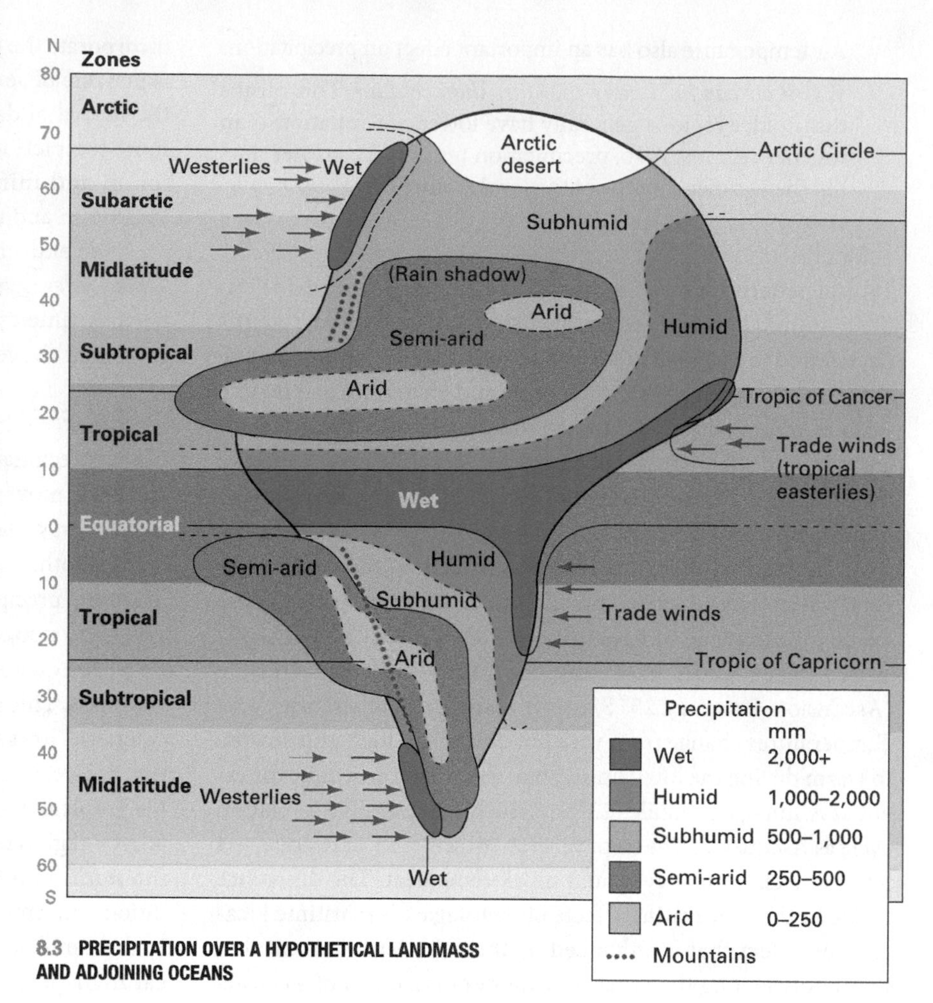

8.3 PRECIPITATION OVER A HYPOTHETICAL LANDMASS AND ADJOINING OCEANS

These general patterns can be seen in Figure 8.4, which shows mean annual precipitation for the world. The map uses **isohyets**, which are lines joining places with the same precipitation. Seven global precipitation regions can be discerned (Table 8.1) and are explained below. Note that the word "rainfall" is used for regions where all or most of the precipitation is rain; "precipitation" is used where snow is a significant part of the annual total.

1. **Wet equatorial belt.** This is a zone of heavy rainfall, with more than 2,000 mm received annually. It straddles the equator and includes the Amazon River Basin in South America, the Congo River Basin of equatorial Africa, much of the African coast from Nigeria west to Guinea, and the East Indies. Here the prevailing warm temperatures and high-moisture content of mE air masses favour abundant convectional rainfall. Thunderstorms are frequent year-round.
2. **Trade-wind coasts.** Narrow coastal belts of high rainfall averaging 1,500 to 2,000 mm annually, and locally even more. These conditions extend from near the equator to latitudes of about 25° to 30° N and S on the eastern sides of every continent or large island. Examples include the eastern coast of Brazil, Central America, Madagascar, and northeastern Australia. The rainfall on these coasts is supplied by moist mT air masses from warm oceans, brought onshore by the trade winds. As they encounter coastal hills and mountains, these air masses produce heavy orographic rainfall.
3. **Tropical deserts.** These arid regions are centred approximately on the Tropic of Cancer and Tropic of Capricorn. The tropical deserts are hot, barren places, with less than 250 mm of rainfall annually and in many places less than 50 mm. They are located under the large, semi-permanent subtropical high-pressure cells in which subsiding cT air masses are adiabatically warmed and dried. These desert climates extend from the west coast of the landmasses out over the oceans. Rain in the tropical deserts is mainly convectional and extremely unreliable.
4. **Midlatitude deserts and steppes.** Farther north, in the interiors of Asia and North America between lat. 30° N and 50° N, are extensive deserts and vast tracts of semi-arid grasslands known as **steppes**. Annual precipitation ranges from less than 100 mm in the driest areas to 500 mm in the moister steppes. The dryness results from the remote location far from oceanic sources of moisture; this may be intensified by the rain shadow effects of mountain ranges. For example, the Cordilleran ranges of Alaska, British Columbia, Oregon, and Washington shield the interior of North America from moist mP air

TABLE 8.1 WORLD PRECIPITATION REGIONS

NAME	LATITUDE RANGE	CONTINENTAL LOCATION	PREVAILING AIR MASS	ANNUAL PRECIPITATION (mm)
1. Wet equatorial belt	10° N to 10° S	Interiors, coasts	mE	More than 2,000
2. Trade-wind coasts (windward tropical coasts)	5–30° N and S	Narrow coastal zones	mT	More than 1,500
3. Tropical deserts	10–35° N and S	Interiors, west coasts	cT	Less than 250
4. Midlatitude deserts and steppes	30–50° N and S	Interiors	cT, cP	100–500
5. Moist subtropical regions	25–45° N and S	Interiors, coasts	mT (summer)	1,000–1,500
6. Midlatitude west coasts	35–65° N and S	West coasts	mP	More than 1,000
7. Arctic and polar deserts	60–90° N and S	Interiors, coasts	cP, cA	Less than 300

masses originating in the Pacific. When these air masses descend into the intermontane basins and interior plains, they are adiabatically warmed and dried and most of the moisture remains in the vapour state.

The mountains of Europe and the Scandinavian Peninsula are less obstructive, but because of the huge area of the Asian continent, much of the moisture in mP air masses from the North Atlantic is lost as they move inland. In addition, any moisture present in the air in summer usually remains as vapour because of high temperatures in the continental interior. In winter, the intense high-pressure region that develops over Siberia further restricts airflow into the continent. Dry steppes give way to more arid conditions in Kazakhstan and eventually merge with the Gobi Desert in Mongolia, where dryness is accentuated by the blocking effect of the Himalayas on moist mT and mE air masses from the Indian Ocean.

The southern hemisphere has too little land in the midlatitudes to produce a true continental desert, but the dry steppes of Patagonia, lying on the lee side of the Andean chain, are similar to the dry interior regions of Oregon and northern Nevada.

5. **Moist subtropical regions**. The moist subtropical regions occur on the southeastern sides of the continents in lat. 25° to 45° N and S where annual rainfall ranges from 1,000 to 1,500 mm. Representative areas in North America include Alabama, Georgia, and South Carolina. In Asia, moist subtropical conditions prevail in the eastern plains of the Yangtze and Yellow Rivers. Smaller areas are found in the southern hemisphere in Uruguay, Argentina, and southeastern Australia. These regions are positioned on the western margins of the oceanic subtropical high-pressure cells. As a result, the lands receive moist mT air masses from the tropical ocean. These areas commonly receive heavy rains from tropical cyclones.
6. **Midlatitude west coasts**. Another distinctively wet location is found on the midlatitude west coasts of all continents and large islands lying between about lat. 35° and 65° in the zone of prevailing westerly winds. In these regions, abundant orographic precipitation occurs as a result of forced uplift of mP air masses. Where the coasts are mountainous, as in British Columbia and Alaska, southern Chile, and Norway, the annual precipitation is more than 2,000 mm. Although not a continental location, the South Island of New Zealand experiences a similar climate because of its situation in the zone of the westerlies. During the ice age, precipitation in these regions fed alpine glaciers that descended to the coast, carving the picturesque deep bays (fjords) that are so typically a part of this climate type.
7. **Arctic and polar deserts**. Northward of the 60th parallel, annual precipitation is generally less than 300 mm, except for the western coastal belts. Cold cP and cA air masses cannot hold much moisture, and consequently, they do not yield large amounts of precipitation. At the same time, however, the relative humidity is high and evaporation rates are low.

SEASONALITY OF PRECIPITATION

Total annual precipitation is useful in establishing the general character of a climate, but variation from month to month through the annual cycle is also very important. A pattern of alternating dry and wet seasons, instead of a uniform distribution of precipitation throughout the year, will affect the natural vegetation, soils, crops, and human use of the land. It also makes a difference whether the wet season coincides with a season of higher or lower temperatures. If the warm season is also wet, growth of both native plants and crops will be enhanced. If the warm season is dry, the stress on plants increases and crops may require irrigation.

Monthly precipitation patterns can be described by three general types: (1) uniformly distributed precipitation; (2) a precipitation maximum during the high-sun season (summer) when insolation is at its peak; and (3) a precipitation maximum during the low-sun or cooler season (winter), when insolation is least. Precipitation patterns can include a wide range of conditions, from little or no precipitation in any month to abundant precipitation in all months.

The monthly precipitation diagrams shown in Figure 8.5 represent the main patterns found in various parts of the world. Uapes (0° N) is a characteristic wet equatorial station

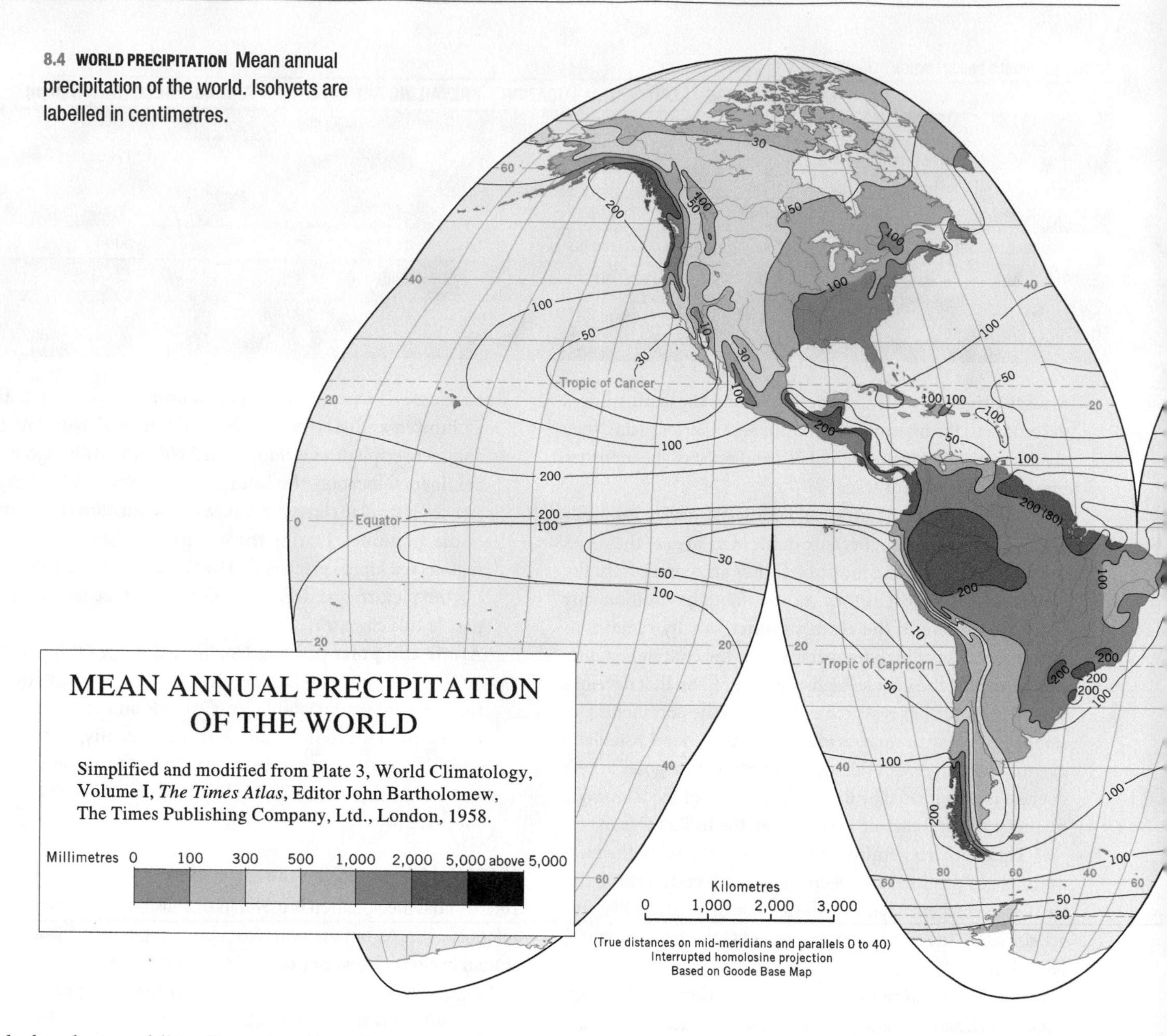

8.4 WORLD PRECIPITATION Mean annual precipitation of the world. Isohyets are labelled in centimetres.

with abundant rainfall in all months, although some months are wetter than others. In contrast, Guaymas, Mexico (28° N), is a tropical desert station where in some months so little rain falls that it scarcely shows on the graph.

Mumbai, India (19° N), and Belo Horizonte, Brazil (20° S), both experience a wet summer season and a dry winter season. Mumbai is an Asian monsoon station, with a large amount of rainfall falling during the high-sun season. Belo Horizonte, with about half the total annual precipitation, is a wet-dry tropical location. At both of these stations, wet-season rainfall is associated with convective activity along the ITCZ.

A summer precipitation maximum also occurs at higher latitudes on the eastern sides of continents. This regime extends from the subtropical zone into the midlatitudes. Charleston, South Carolina (33° N), shows, for example, a relatively dry winter with a marked period of summer rain. In contrast, San Francisco, California (38° N), receives maximum precipitation in winter. This is an example of a Mediterranean climate, which is characterized by dry summers and moist winters. As the name suggests, this climate prevails in the lands surrounding the Mediterranean Sea in southern Spain, southern France, Greece, Turkey, parts of Israel, and Morocco in northern Africa. Southern and central California experience a similar precipitation regime. In the southern hemisphere, this regime occurs in Chile, parts of South Africa, and southwestern Australia.

The summer drought is produced by subtropical high-pressure cells, which intensify and move poleward during the high-sun season. These regions are also influenced by cT air masses from nearby arid regions, such as the southwestern United States. The incursion of cT air is accompanied by very hot, dry conditions. In the low-sun season, the subtropical high-pressure cells move toward the equator and weaken, and

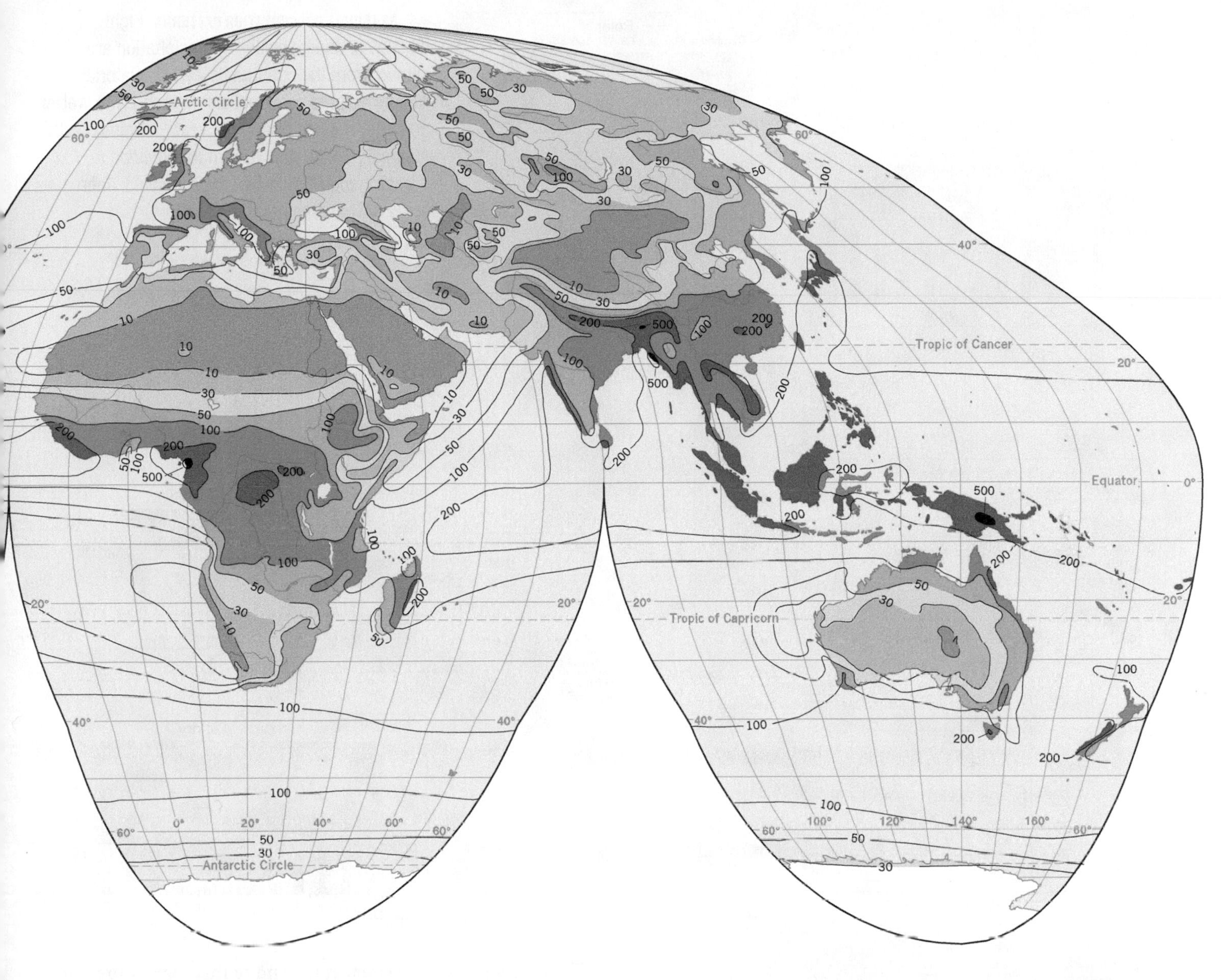

cyclonic storms bring precipitation as they pass through these regions.

The dry-summer, moist-winter cycle extends into coastal areas at higher midlatitudes. However, the difference between summer and winter precipitation becomes less and less marked, and eventually gives way to the marine west-coast type. The marine west-coast regime is characteristic of the Pacific Northwest, from Oregon to Alaska, but occurs as well in Britain and elsewhere in western Europe; Prince Rupert, British Columbia (54° N), is a representative location. Typically, less precipitation is received in summer than in winter. This is partly due to the blocking effects of subtropical high-pressure cells that extend poleward into these regions. Consequently, the polar front shifts further north in summer which reduces the frequency of cyclonic storms. Also, cyclonic storms are less intense in summer because the temperature and moisture contrasts between polar and arctic air masses, and tropical air masses are weaker at this time of the year.

CLIMATE CLASSIFICATION

Mean monthly values of air temperature and precipitation are the two variables that are most commonly used to accurately describe the climate at a weather station and its nearby region. However, both of these variables vary continuously and so can produce an infinite set of conditions which, for the purpose of global comparisons, must be presented in a more generalized way. Various methods have been devised to group climate information into distinctive climate types. In this way, characteristic regional conditions can be described and maps compiled to show their distribution.

In 1918, Vladimir Köppen developed one of the earliest and best-known approaches to organizing the vast amount of world climate information. It is an empirical system based on long-term

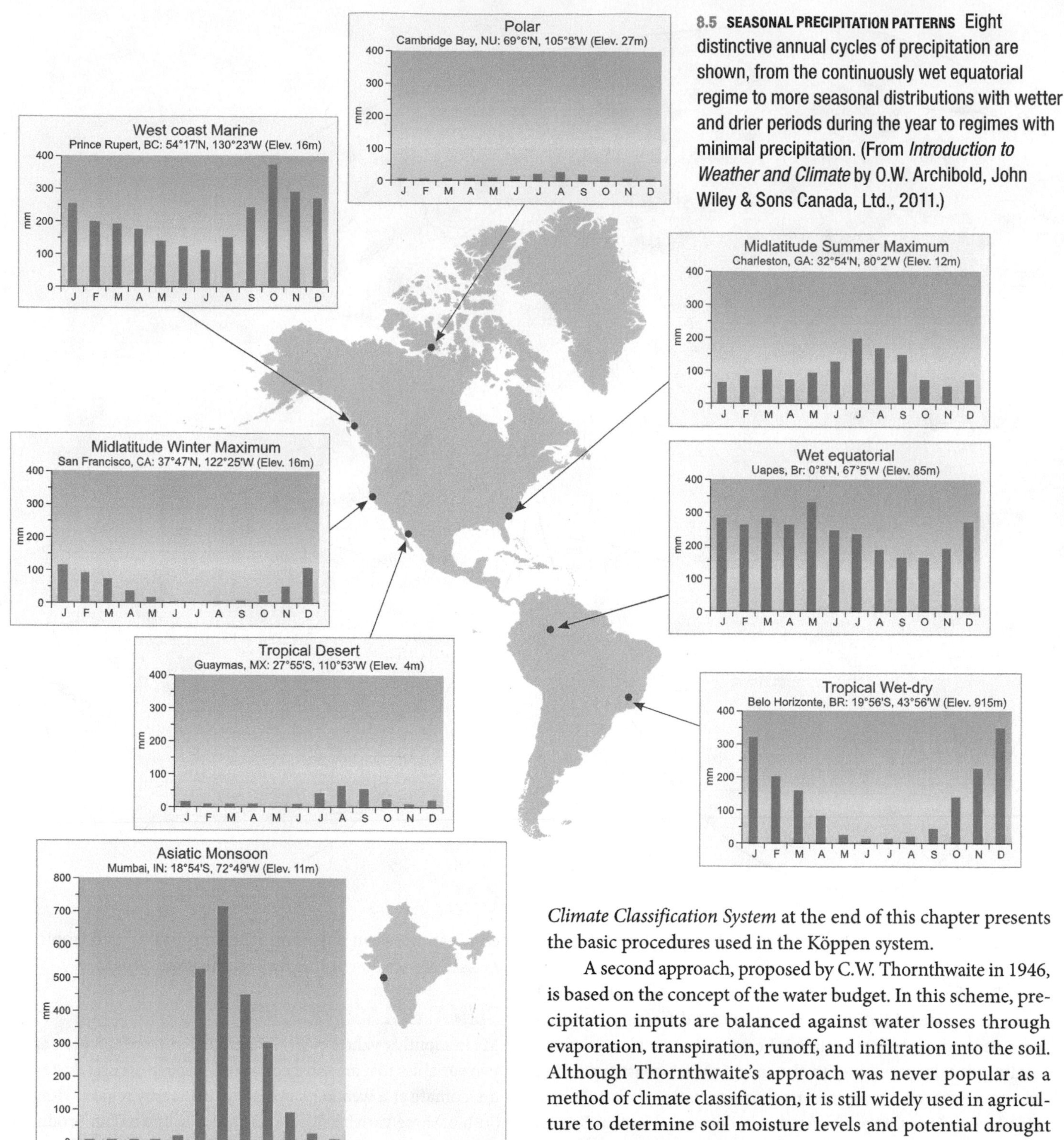

8.5 SEASONAL PRECIPITATION PATTERNS Eight distinctive annual cycles of precipitation are shown, from the continuously wet equatorial regime to more seasonal distributions with wetter and drier periods during the year to regimes with minimal precipitation. (From *Introduction to Weather and Climate* by O.W. Archibold, John Wiley & Sons Canada, Ltd., 2011.)

temperature and precipitation records. The Köppen climate classification system uses a code of letters to group climates into classes according to predefined values of annual and seasonal temperature and precipitation. *Appendix 8.1 • The Köppen Climate Classification System* at the end of this chapter presents the basic procedures used in the Köppen system.

A second approach, proposed by C.W. Thornthwaite in 1946, is based on the concept of the water budget. In this scheme, precipitation inputs are balanced against water losses through evaporation, transpiration, runoff, and infiltration into the soil. Although Thornthwaite's approach was never popular as a method of climate classification, it is still widely used in agriculture to determine soil moisture levels and potential drought problems. Chapter 19 provides additional information on the concept of the water budget.

A more practical approach to climate classification—applied classification—defines climates according to some specified purpose. For example, the construction industry in Canada might divide the country according to high, medium, and low heating costs on the basis of mean winter temperatures; regional differences in roof design might be related to total snowfall loads; or

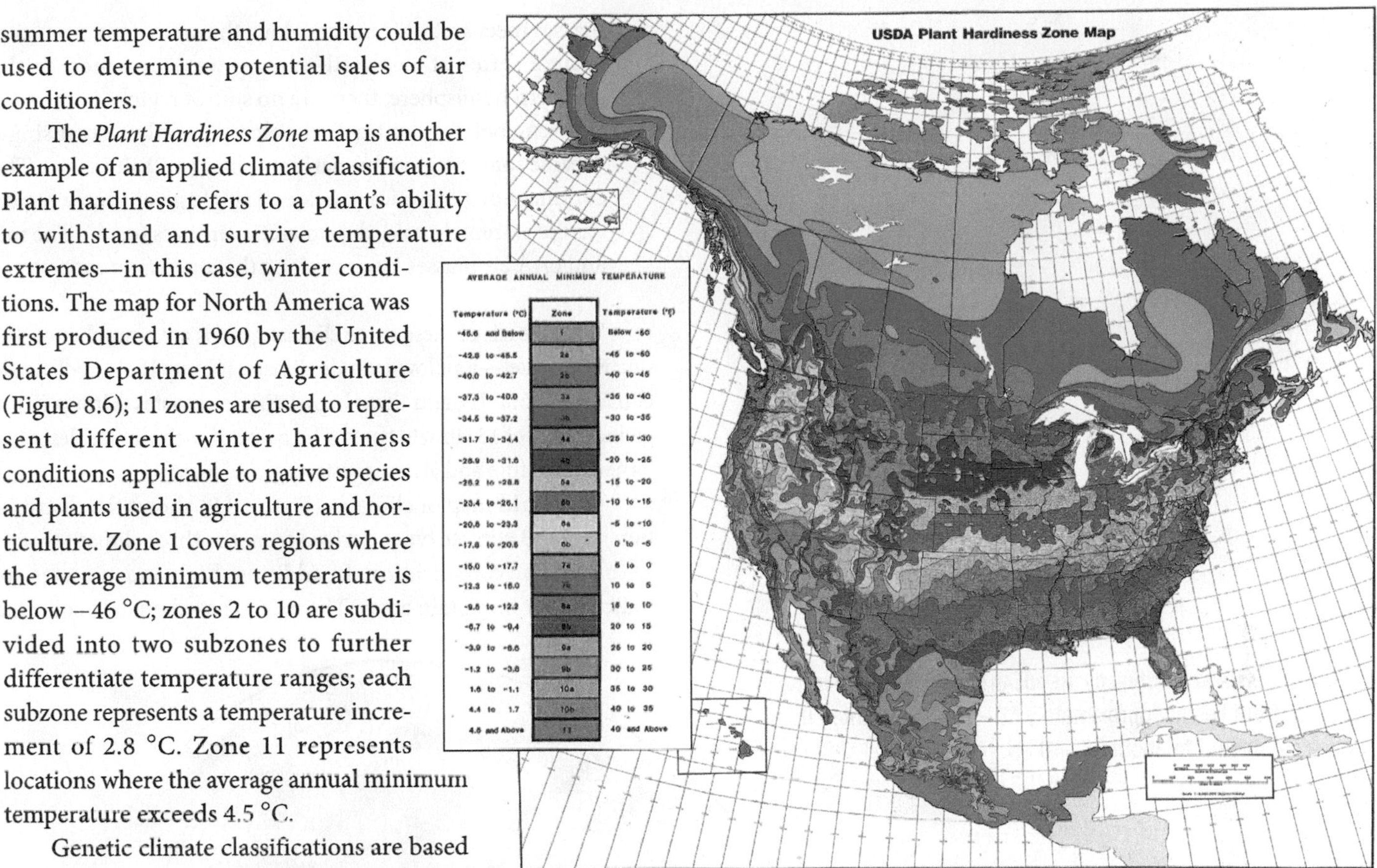

8.6 PLANT HARDINESS ZONES OF NORTH AMERICA. This classification is based on a plant's ability to withstand extreme winter temperatures.

summer temperature and humidity could be used to determine potential sales of air conditioners.

The *Plant Hardiness Zone* map is another example of an applied climate classification. Plant hardiness refers to a plant's ability to withstand and survive temperature extremes—in this case, winter conditions. The map for North America was first produced in 1960 by the United States Department of Agriculture (Figure 8.6); 11 zones are used to represent different winter hardiness conditions applicable to native species and plants used in agriculture and horticulture. Zone 1 covers regions where the average minimum temperature is below −46 °C; zones 2 to 10 are subdivided into two subzones to further differentiate temperature ranges; each subzone represents a temperature increment of 2.8 °C. Zone 11 represents locations where the average annual minimum temperature exceeds 4.5 °C.

Genetic climate classifications are based on the factors that cause the climate of a region. A comprehensive **genetic classification system** must incorporate all the factors that affect the common properties of the atmosphere, particularly temperature and precipitation. In this method, climates are distinguished according to the air masses and frontal zones that affect them. Air masses are classified on the basis of the general latitude of their source regions and the surface—land or ocean—over which they develop (see Chapter 7). Latitude is an important determinant of air mass temperature and can impose pronounced seasonal variation as solar radiation varies over the year. Temperature will also influence the moisture content of an air mass; however, at any given latitude, moisture content is primarily controlled by the distribution of land and ocean. Air mass characteristics therefore reflect the two most important climate variables—temperature and precipitation—and so provide a rational basis for classifying climates.

Air masses acquire distinctive temperature and moisture properties based on the characteristics of their source regions. Adjacent air masses are separated by frontal zones across which the properties of the atmosphere change. However, the position of frontal zones varies with the seasons. For example, the polar-front zone generally lies across the midlatitudes of the United States in winter, but it moves northward to Canada during the summer (Figure 8.7). The seasonal movements of frontal zones greatly influence annual cycles of temperature and precipitation.

Figure 8.8 shows a general zonal subdivision according to air mass source regions. Three broad climate groups are recognized: low-latitude (Group I), midlatitude (Group II), and high-latitude (Group III).

- *Group I: Low Latitude Climates.* The zone of low-latitude climates is dominated by the source regions of continental tropical (cT), maritime tropical (mT), and maritime equatorial (mE) air masses. These source regions are related to the two subtropical high-pressure belts and the equatorial trough at the ITCZ. Air originating in the polar regions very rarely invades regions of low-latitude climates. Easterly waves and tropical cyclones are important weather systems in this climate group.
- *Group II: Midlatitude Climates.* The region of midlatitude climates lies in the polar-front zone—a zone of intense interaction between different air masses. Here tropical air masses moving toward the poles and polar air masses moving toward the equator are in conflict. Wave cyclones are normal features of this climate group.
- *Group III: High-Latitude Climates.* The region of high-latitude climates is dominated by polar and arctic (or antarctic) air masses. In the arctic belt of the 60th to 70th parallels, cP air

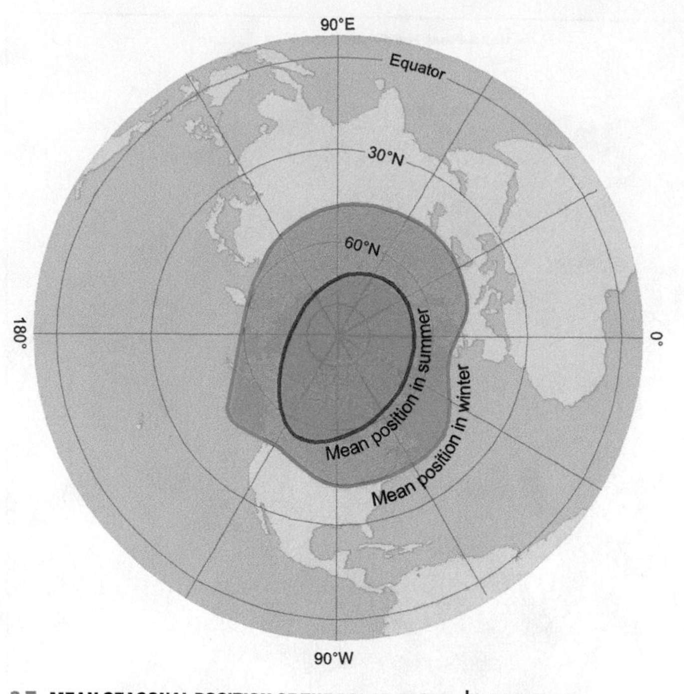

8.7 MEAN SEASONAL POSITION OF THE POLAR FRONT In summer, the polar front lies at approximately 60–65° N; in winter, it moves southward to about 35–40° N.

masses meet arctic air masses along the *arctic-front zone*, creating a series of eastward-moving wave cyclones. In the southern hemisphere, there are no source regions in the sub-antarctic belt for continental polar air—only a great single oceanic source region for maritime polar (mP) air masses. The continent of Antarctica provides a single great source region for the extremely cold, dry antarctic air mass (cAA). The mP and cAA air masses interact along the *antarctic-front zone*.

Within each of these three climate groups are a number of distinctive climates—four low-latitude climates (Group I), six midlatitude climates (Group II), and three high-latitude climates (Group III)—giving a total of 13 climate types. The name of each climate describes its general nature and also suggests its global location.

The world map of climates (Figure 8.9) shows the distribution of the 13 climate types derived from the three climate groups of the genetic classification. In addition, highland climates are indicated for mountain areas. The map is based on data collected

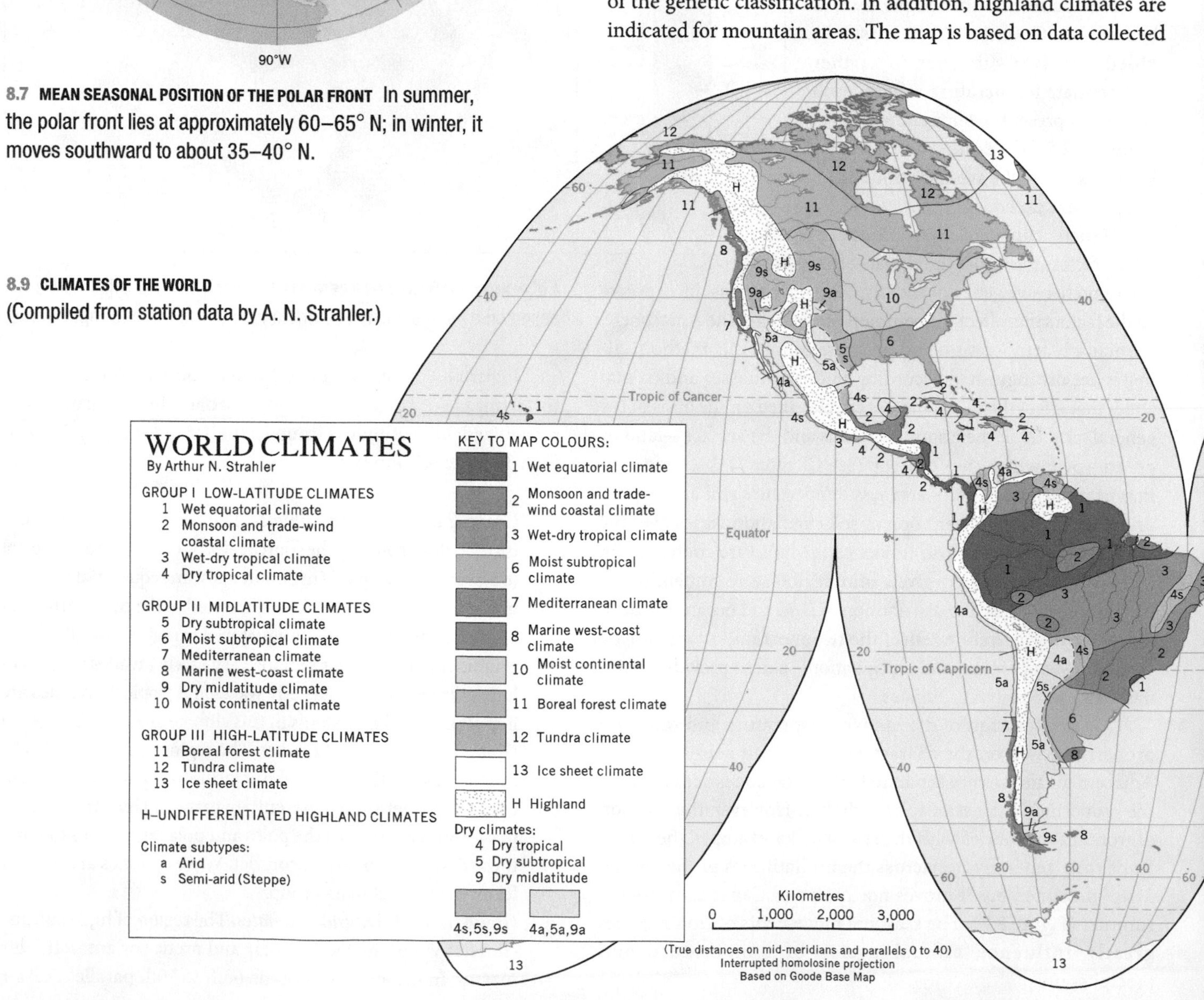

8.9 CLIMATES OF THE WORLD
(Compiled from station data by A. N. Strahler.)

8.8 CLIMATE GROUPS AND AIR MASS SOURCE REGIONS Using the concept of air mass source regions, five global bands associated with three major climate groups can be identified. Within each group is a set of distinctive climates with unique characteristics that are explained by the movements of air masses and frontal zones.

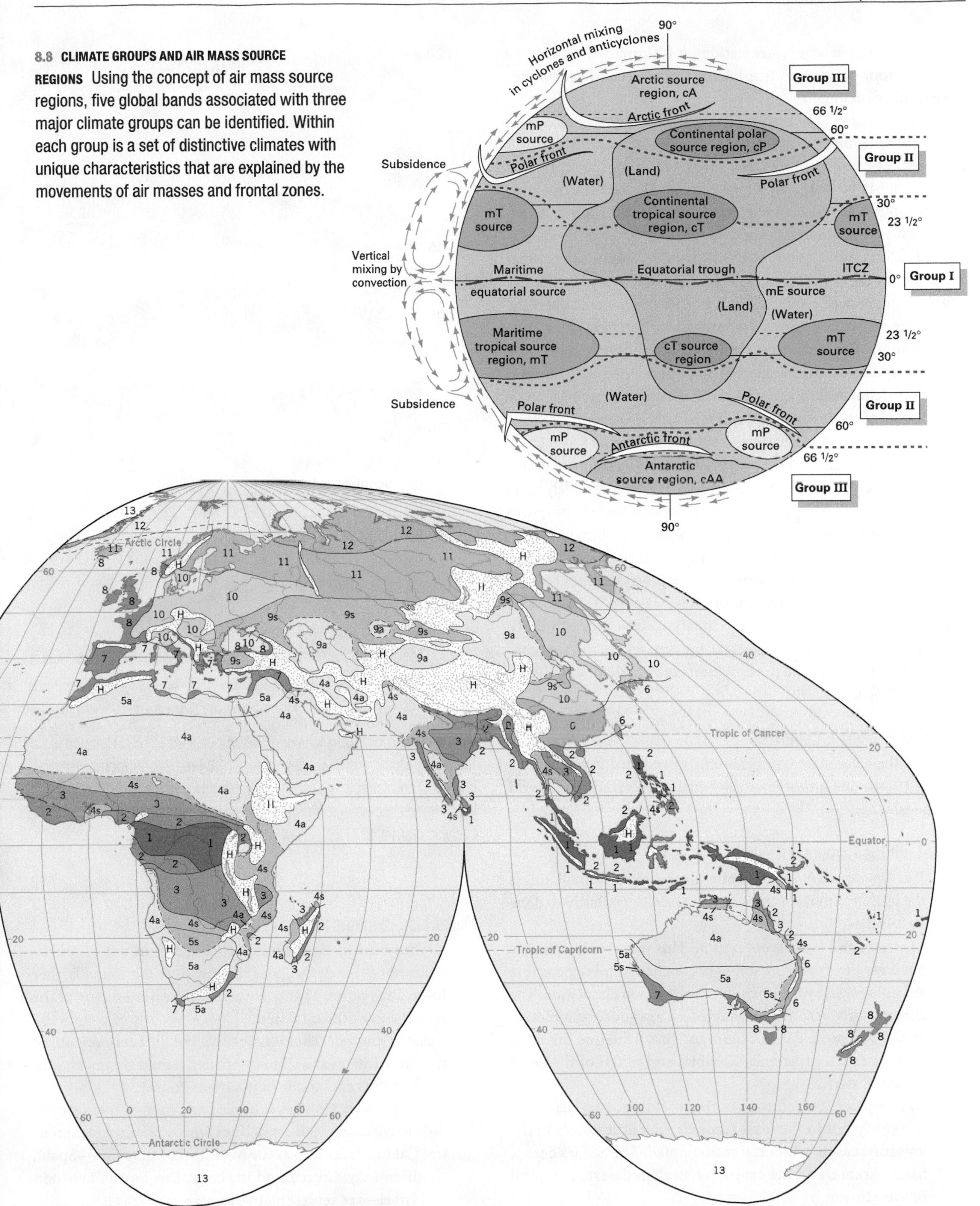

at a large number of weather stations. It is simplified because the climate boundaries are uncertain in many areas where stations are thinly distributed.

A **climograph** can readily portray climate information (Figure 8.10). This example shows the annual cycles of monthly mean air temperature and monthly mean precipitation for Banff, Alberta, located at 51° N. At the top of the climograph, the mean monthly temperature is plotted as a line graph; mean monthly precipitation is shown as a bar graph at the bottom. The annual range in temperature and total annual precipitation are provided. Some climographs also display the declimation of the sun and dominant weather features using picture symbols. For Banff, a dominant feature is the influence of cyclonic storms that bring precipitation in all months.

A more comprehensive method of combining temperature and precipitation into a single diagram was devised by Walter and Leith. The climate data and station information, including altitude and length of the temperature and precipitation records, are plotted in a standardized format for each location (Figure 8.11). These diagrams have undergone only slight modification since they were originally devised in the late 1950s and are still widely used, especially in studies of biogeography. For northern hemisphere stations, the diagram runs from January to December; for southern hemisphere stations, the convention is to start with July and end with June so that the warm season is always in the centre of the diagram. A general indication of water surplus is given by shading the area in which the precipitation curve is drawn above the temperature curve. Conversely, the area where the temperature curve lies above the precipitation curve is shaded to signify a water deficit.

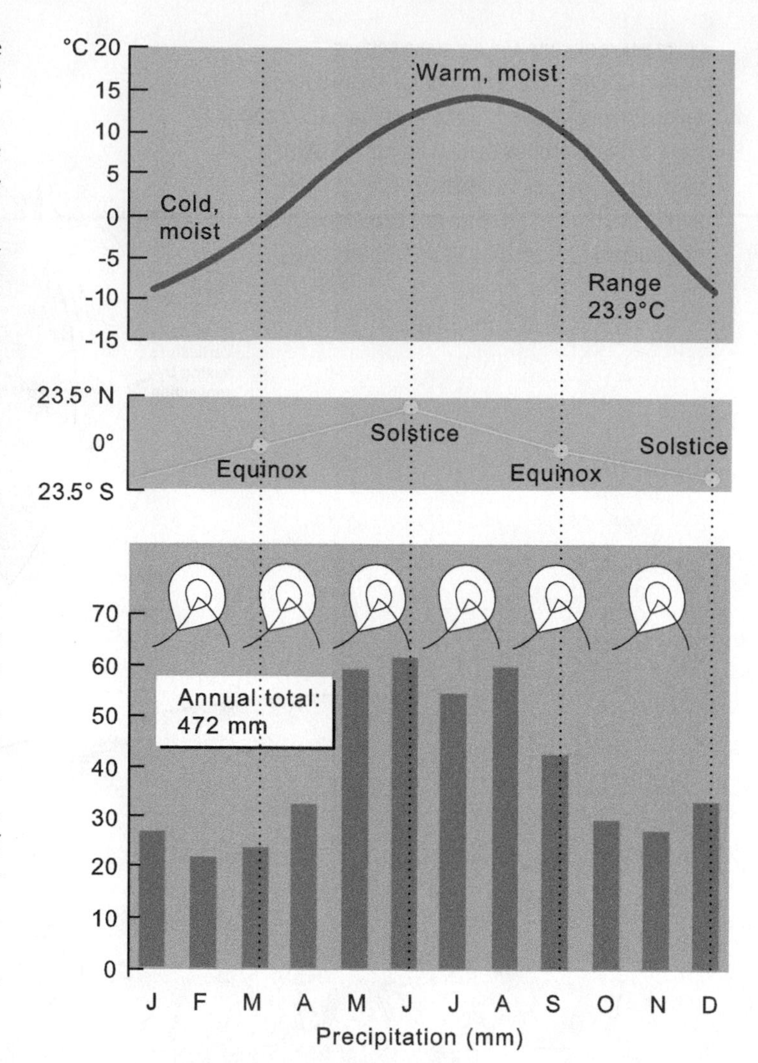

8.10 REPRESENTATIVE CLIMOGRAPH Banff, Alberta, at 51° N, is located in the Rocky Mountains at an elevation of 1,384 m. The temperature range is 23.9°C and reflects the relatively high latitude of this location. Precipitation occurs in all months, and comes predominantly from cyclonic storms.

OVERVIEW OF THE CLIMATES

Each of the 13 climate types is briefly introduced here; chapters 9 and 10 provide a more comprehensive description of world climates.

Low-Latitude Climates (Group I)

- *Wet equatorial.* Warm to hot with abundant rainfall, this is the humid climate of the Amazon and Congo Basins and the East Indies.
- *Monsoon and trade-wind coastal.* This warm to hot climate has a very wet rainy season. It occurs in coastal regions that are influenced by trade winds or a monsoon circulation. The climates of Vietnam and Bangladesh are good examples.
- *Wet-dry tropical.* Parts of India and Brazil fall into this type, as does much of Angola, Zambia, and much of the Sahel region of Africa.
- *Dry tropical.* The climate of the world's hottest deserts—extremely hot in the high-sun season, a little cooler in the low-sun season, with little or no rainfall. The Sahara desert, Saudi Arabia, and the central Australian desert are typical of this climate.

Midlatitude Climates (Group II)

- *Dry subtropical.* Another desert climate, but not as consistently hot as the dry tropical climate, since it is found farther toward the poles. This type includes the hottest part of the American southwest desert.
- *Moist subtropical.* The climate of the southeastern regions of the United States and China—hot and humid summers, with mild winters and ample rainfall year-round.
- *Mediterranean.* Hot, dry summers and rainy winters are the distinctive characteristics of this climate. Southern and central California, as well as the Mediterranean region—Spain, southern Italy, Greece, and the coastal regions of Lebanon and Israel—are representative locations.

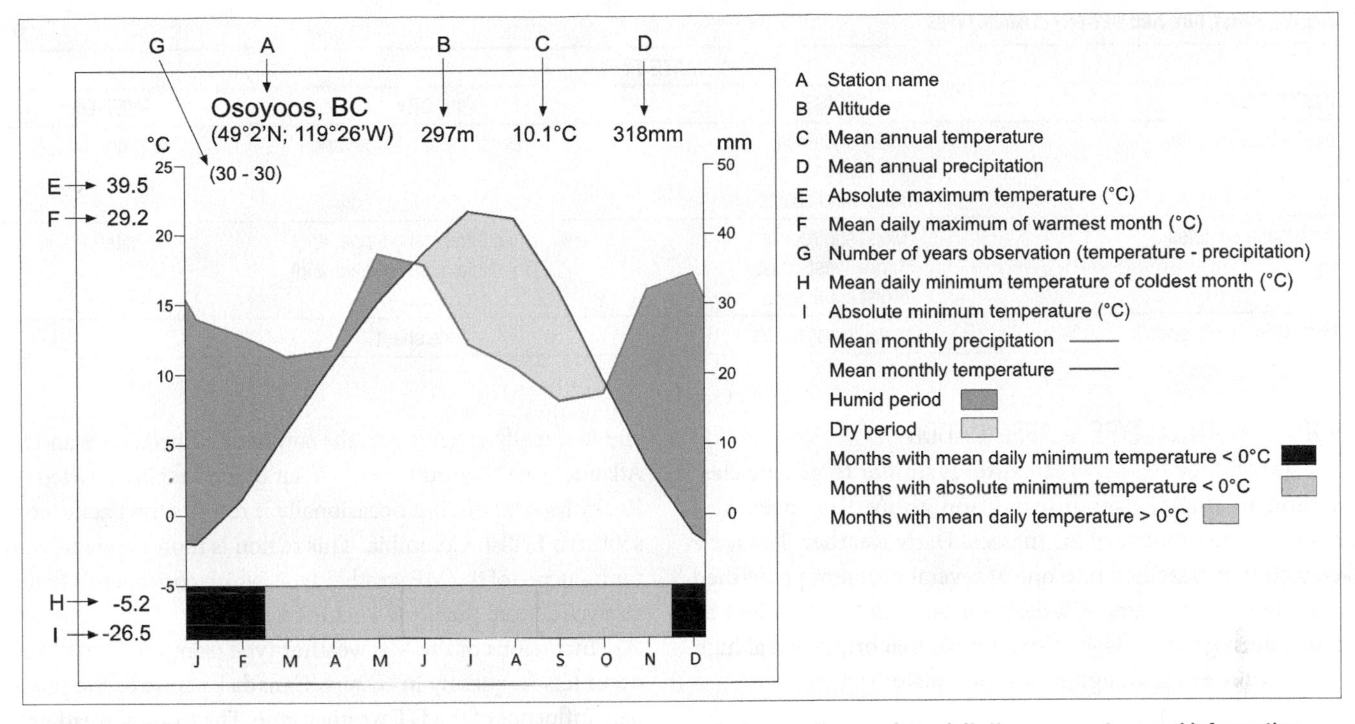

8.11 THE STANDARD WALTER AND LEITH CLIMATE DIAGRAM The diagram presents temperature and precipitation parameters and information about the station. (From *Introduction to Weather and Climate* by O.W. Archibold, John Wiley & Sons Canada, Ltd., 2011.)

- *Marine west-coast.* The climate of the Pacific Northwest—coastal Oregon, Washington, and British Columbia; western Europe; and the South Island of New Zealand. Warm summers and cool winters, with more rainfall in winter, are characteristic of this climate.
- *Dry midlatitude.* This dry climate is found in midlatitude continental interiors. The steppes of central Asia and the Great Plains of North America experience this climate; here it is warm to hot in summer, cold in winter, and has low annual precipitation.
- *Moist continental.* This is the climate of eastern Canada and most of the eastern United States—cold in winter, warm in summer, with ample precipitation through the year.

High-Latitude Climates (Group III)

- *Boreal forest.* Short, cool summers and long, bitterly cold winters characterize this snowy climate. Northern Canada, Siberia, and central Alaska are regions of boreal forest climate.
- *Tundra.* Although this climate has a long, severe winter, temperatures on the tundra are somewhat moderated by proximity to the Arctic Ocean. This is the climate of the coastal arctic regions of Canada, Alaska, Siberia, and northern Scandinavia.
- *Ice sheet.* The bitterly cold temperatures of this climate, found only in Greenland and Antarctica, can drop below −50 °C during the sunless winter months. Even during the 24-hour days of summer, temperatures remain well below freezing.

DRY AND MOIST CLIMATES

Apart from the wet-dry tropical climate and the Mediterranean climate, all of the climate types are classified as either dry climates or moist climates. *Dry climates* are those in which total annual evaporation of moisture from the soil and transpiration from plants exceeds the annual precipitation. Dry climates generally do not support permanently flowing streams. The soil is dry much of the year, and the land surface is clothed either with sparse plant cover—scattered grasses, shrubs, or cacti—or is simply devoid of any plants. This distinction is recognized by two climatic subtypes: the *semi-arid* or steppe climate and the *arid* climate. *Moist climates* are those with sufficient precipitation to maintain the soil in a moist condition through much of the year and to sustain the year-round flow of rivers. Moist climates support forests and prairies.

The wet-dry tropical and Mediterranean climate types alternate between a very wet season and a very dry season. The striking seasonal contrast gives both of them a special character, hence they are designated *wet-dry climates*. Table 8.2 summarizes the moist, dry, and wet-dry climates that occur within the three main climate groups.

TABLE 8.2 MOIST, DRY, AND WET-DRY CLIMATE TYPES

	CLIMATE TYPE		
CLIMATE GROUP	MOIST	DRY	WET-DRY
I: Low-latitude climates	Wet equatorial Monsoon Trade-wind coastal	Dry tropical (steppe; arid)	Wet-dry tropical
II: Midlatitude climates	Moist subtropical Marine west-coast Moist continental	Dry subtropical (steppe; arid) Dry midlatitude (steppe; arid)	Mediterranean
III: High-latitude climates	Boreal forest tundra	Ice sheet	

SYNOPTIC WEATHER TYPE CLASSIFICATION

Synoptic weather type classification is similar to genetic classification in that it uses information about the speed and direction of movement of air masses. Daily weather data for a given station is classified into one of several different predefined *weather types*. For example, a designation of *dry polar* might be assigned during the passage of an air mass that originated at high latitude. Some of the designations used include:

- *DP (dry polar)*. This weather type is analogous to a cP air mass; it is usually associated with the lowest temperatures recorded at any time of year and clear, dry conditions.
- *MP (moist polar)*. Similar to an mP air mass, this weather type usually brings cool, cloudy weather conditions. Note, however, that synoptic weather typing emphasizes the weather conditions rather than air masses. Thus, MP could also apply to a cP air mass that has been greatly modified as it passed over Hudson Bay or the Great Lakes.
- *DT (dry tropical)*. Similar to a cT air mass, this weather type is associated with the hottest and driest conditions that occur at any location. In North America, the DT weather type may arise through advection from its source in the arid southwest or by subsidence of air from aloft.
- *MT (moist tropical)*. This weather type is analogous to an mT air mass and brings warm and very humid weather to a region.

The seasonal frequencies of weather types within North America are indicative of air mass movements throughout the continent and so provide a guide to the climatic conditions experienced. The DP weather type is predominant in northern Canada in January, and frequently extends into the southern United States and to the Atlantic coast. The outflow of DP air to the west is restricted by the Rocky Mountains, but occasionally it reaches the Pacific coast in southern British Columbia. This region is more frequently under the influence of the MP weather type, which corresponds to the mP air masses from the north Pacific Ocean.

Incursions of the MP weather type from the North Atlantic occur less frequently in eastern Canada because of the predominant influence of the DT weather type. The tropical weather types are absent from Canada during January, but incursions of the MT weather type over southeastern and south-central regions of North America occur in July in response to anticyclonic flow associated with the Azores-Bermuda high-pressure system. In Canada, the DT weather type is mainly associated with the Okanagan Valley and similar areas in the interior of southern British Columbia. Less frequently it occurs across the southern Prairie provinces, but only rarely does it extend into eastern Canada.

A Look Ahead

This introduction to global climate has stressed the relationship between climate and the factors that influence annual cycles of temperature and precipitation. Temperature cycles may be uniform, seasonal (continental), or moderated by oceanic influences (marine). Precipitation may be uniform (whether it be scarce in all months or abundant in all months), may have a maximum at the time of high sun, or may have a maximum at the time of low sun. These temperature and precipitation cycles, in turn, are produced by the annual cycle of insolation and by the global patterns of atmospheric circulation and air mass movements. The next two chapters examine the climates of the world in more detail.

CHAPTER SUMMARY

- Climate is the average seasonal weather pattern of a region. Because temperature and precipitation are measured at many stations worldwide, the combined annual patterns of monthly averages of temperature and precipitation can be used to assign climate types.
- *Temperature regimes* are typical patterns of annual variation in temperature. They depend on latitude, which determines the annual insolation cycle, and on location—continental or maritime—which enhances or moderates conditions imposed by annual insolation.

- Global precipitation patterns are largely determined by air masses and their movements, which in turn are produced by global air circulation patterns. The main features of the global pattern of rainfall are the following:
    1. a wet equatorial belt produced by convectional precipitation along the ITCZ;
    2. trade-wind coasts that receive moist flows of mT air from trade winds as well as tropical cyclones;
    3. tropical deserts located under subtropical high-pressure cells;
    4. midlatitude deserts and steppes, which are dry because they are far from oceanic moisture sources;
    5. moist subtropical regions that receive westward flows of moist mT air in the summer and eastward-moving wave cyclones in winter;
    6. midlatitude west coasts, which are subjected to eastward flows of mP air and wave cyclones by prevailing westerly winds; and
    7. polar and arctic deserts, where little precipitation falls because the air is too cold to hold much moisture.
- Annual patterns of precipitation can be described as uniform (ranging from abundant to scarce); high-sun (summer) maximum; and low-sun (winter) maximum.
- Climate classifications are devised to simplify and categorize the diversity of weather conditions that occur worldwide. The most commonly used classification is the empirical system of Köppen, which uses predefined categories based on temperature and precipitation.
- Genetic classifications are based on the factors that cause the observed temperature and precipitation patterns.
- In the genetic classification system, three groups of climate types are distinguished and arranged by latitude. *Low-latitude climates (Group I)* are dominated by mE, mT, and cT air masses, and are largely related to the global circulation patterns that produce the ITCZ, trade winds, and subtropical high-pressure cells. *Midlatitude climates (Group II)* lie in the polar-front zone and are strongly influenced by eastward-moving wave cyclones in which mT, mP, and cP air masses interact. *High-latitude climates (Group III)* are dominated by polar and arctic (antarctic) air masses. Wave cyclones mixing mP and cP air masses along the *arctic-front zone* provide precipitation in this region.
- *Dry climates* are those in which precipitation is largely evaporated from soil surfaces and transpired by vegetation, so that permanent streams cannot be supported.
- Synoptic weather type classification uses the frequencies of air mass types to assess the general conditions that can be expected at a location.

KEY TERMS

climate
climograph
genetic classification system
isohyet
steppe
synoptic weather type classification

REVIEW QUESTIONS

1. Discuss the use of monthly records of average temperature and precipitation to characterize the climate of a region. Why are these measures useful?
2. Why are latitude and location (maritime or continental) important factors in determining the annual temperature cycle of a location?
3. How does air temperature, as a climatic variable, influence precipitation?
4. Describe three temperature regimes and explain how they are related to latitude and location.
5. Identify seven important features on the global precipitation map and describe the factors that produce each.
6. The seasonality of precipitation at a station can generally be described as following one of three patterns. Identify them and provide examples, explaining how each pattern can arise.
7. What are the important global circulation patterns and air masses that influence low-latitude (Group I) climates? What are their effects?
8. What air masses and circulation patterns influence the midlatitude (Group II) climates, and how?
9. Identify the air masses and frontal zones that are important in determining high-latitude climates (Group III) and explain their effects.

VISUALIZATION EXERCISES

1. Victoria, British Columbia, enjoys a midlatitude west-coast location. Sketch a climograph showing monthly temperature and precipitation.
2. Sketch a climograph for Sudbury, Ontario, noting that it has a midlatitude continental location.

ESSAY QUESTION

1. Sketch a hypothetical continent with a shape and features of your own choosing, stretching from about 70° N to 40° S. Add some north-south mountain ranges, positioned where you like. Then select four different locations and describe and explain the annual cycles of temperature and precipitation you would expect at each.

WORKING IT OUT • AVERAGING IN TIME CYCLES

The weather at most stations can vary widely from day to day, or even from hour to hour. For the study of climate, however, analysis focuses on weather in a statistical, or average, sense. Typically, the averages for variables such as daily or monthly temperatures and precipitation are calculated over decades.

Figure 1 shows monthly precipitation for a period of 20 years, from 1986 to 2005, as observed at St. John's, Newfoundland. At first inspection, the graph shows a rhythmic pattern to the precipitation, with sequences of wet months interspersed with drier months. But beyond that general observation, it is difficult to determine the typical annual cycle of precipitation. In fact, it is not easy to determine which month is, on average, the wettest or the driest.

The typical monthly cycle of precipitation can be determined by calculating the average, or mean, of a number of observations—the sum of the observations divided by the number of observations.

The monthly averages for St. John's as listed in Environment Canada's Canadian Climate Normals, 1971–2000, are shown in Figure 2a. Note that the monthly values appear to be much less variable than the record shown for the individual years. Maximum precipitation occurs in October, decreasing to a low in July. The

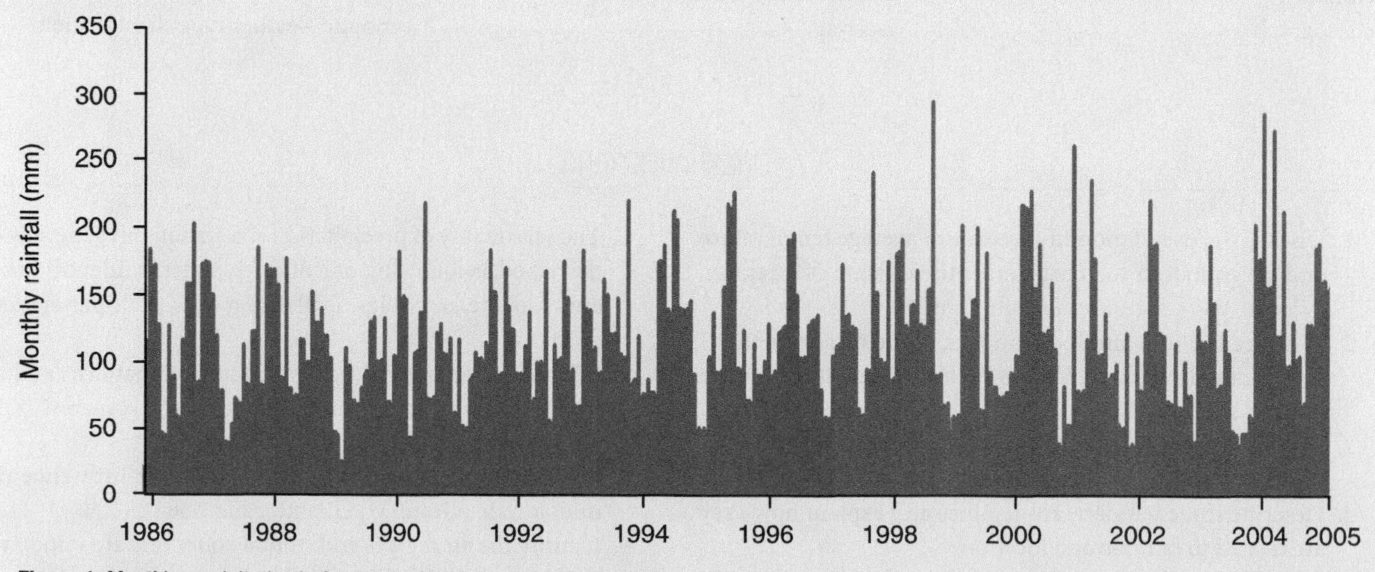

Figure 1 Monthly precipitation in St. John's, Newfoundland, 1986–2005

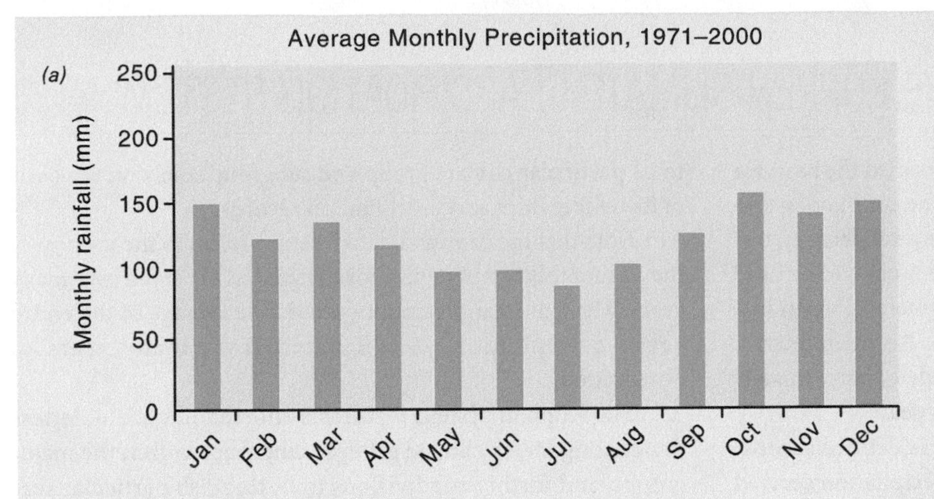

average eliminates the year-to-year variability in the record. Figure 2b shows the average for two five-year periods, 1986 to 1990 and 2001 to 2005. The period 1986 to 1990 was generally drier than the 2001 to 2005 period. This is the case for all months except February, April, June, and August, which received more precipitation during the 1986 to 1990 period.

In general, the longer the period of averaging, the closer the average will come to the long-term average, and the smoother the cycle will be. Thus, averaging is a useful tool for revealing patterns and time cycles in data sequences.

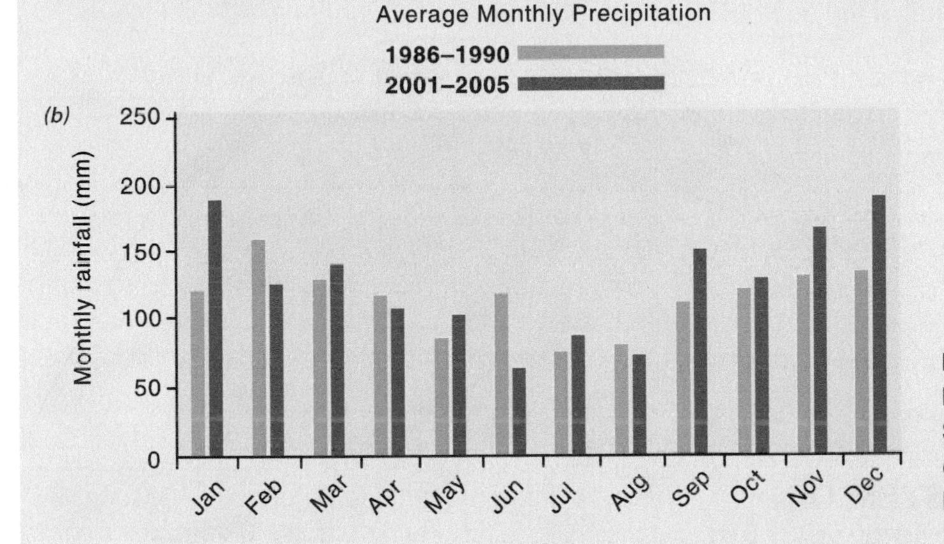

Figure 2 St. John's precipitation Precipitation records and statistics for St. John's, Newfoundland, 48° N, moist continental climate (Data from Environment Canada.)

QUESTIONS / PROBLEMS

1. The table below shows monthly precipitation (mm) at Dryden, Ontario, for 2001 to 2005 and the 1971 to 2000 normals. Find the average for each month of this five-year period.
2. Plot the monthly averages for 2001 to 2005, and then plot the 30-year monthly means. How do the two graphs compare?

	Jan	Feb	Mar	Apr	May	Jun	Jul	Aug	Sep	Oct	Nov	Dec
2001	15.5	16.4	11.7	80.3	124	126.9	106.6	23.1	39.9	85.3	25.7	24.4
2002	20.4	6.2	21.8	68.8	58.7	177.3	45.8	87	92.2	37.5	19.3	16.4
2003	18	15	13.6	24	68.7	80.2	79.2	116.6	204.8	45	16.9	32.7
2004	55.7	16.8	60.2	15	172.3	60.6	58.4	191.4	137.2	84.8	22.5	54.5
2005	39.5	13.4	24.4	45.6	86.1	154.1	95.3	93.8	68.8	43.2	62.4	29.3
1971–2000 Normal	**28.4**	**22.7**	**32.7**	**37**	**67.2**	**105.1**	**98.8**	**85**	**90.4**	**60.7**	**44.8**	**28.7**

Find answers at the back of the book.

APPENDIX 8.1 THE KÖPPEN CLIMATE CLASSIFICATION SYSTEM

Air temperature and precipitation data have formed the basis for several climate classification systems. One of the commonest systems is that of Köppen, devised in 1918. For several decades, this system, with various later revisions, was the most widely used climate classification method among geographers. Köppen was both a climatologist and plant geographer, so his main interest lay in finding climate boundaries that coincided approximately with boundaries between major vegetation types.

Under the Köppen system, each climate is defined according to assigned values of temperature and precipitation derived from annual or monthly climate data. Any station can be assigned to its particular climate group and subgroup solely on the basis of its temperature and precipitation records.

Note that mean annual temperature refers to the average of the 12 monthly temperatures for the year, as observed over many years. Mean annual precipitation uses the average of the entire year's precipitation as calculated from many years of observation.

The Köppen system features a shorthand code of letters designating major climate groups, subgroups within the major groups, and further subdivisions to distinguish particular seasonal characteristics of temperature and precipitation. Five

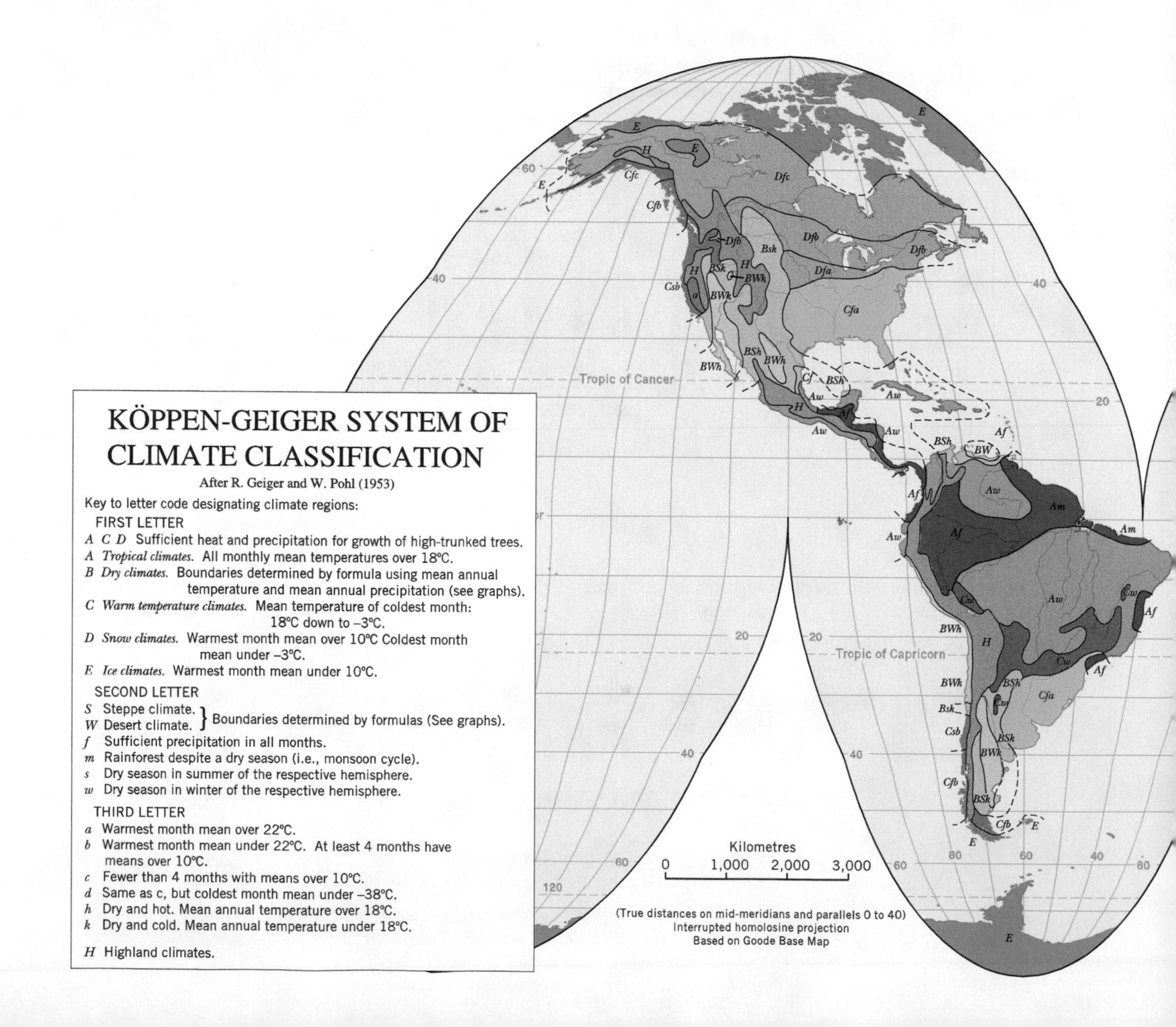

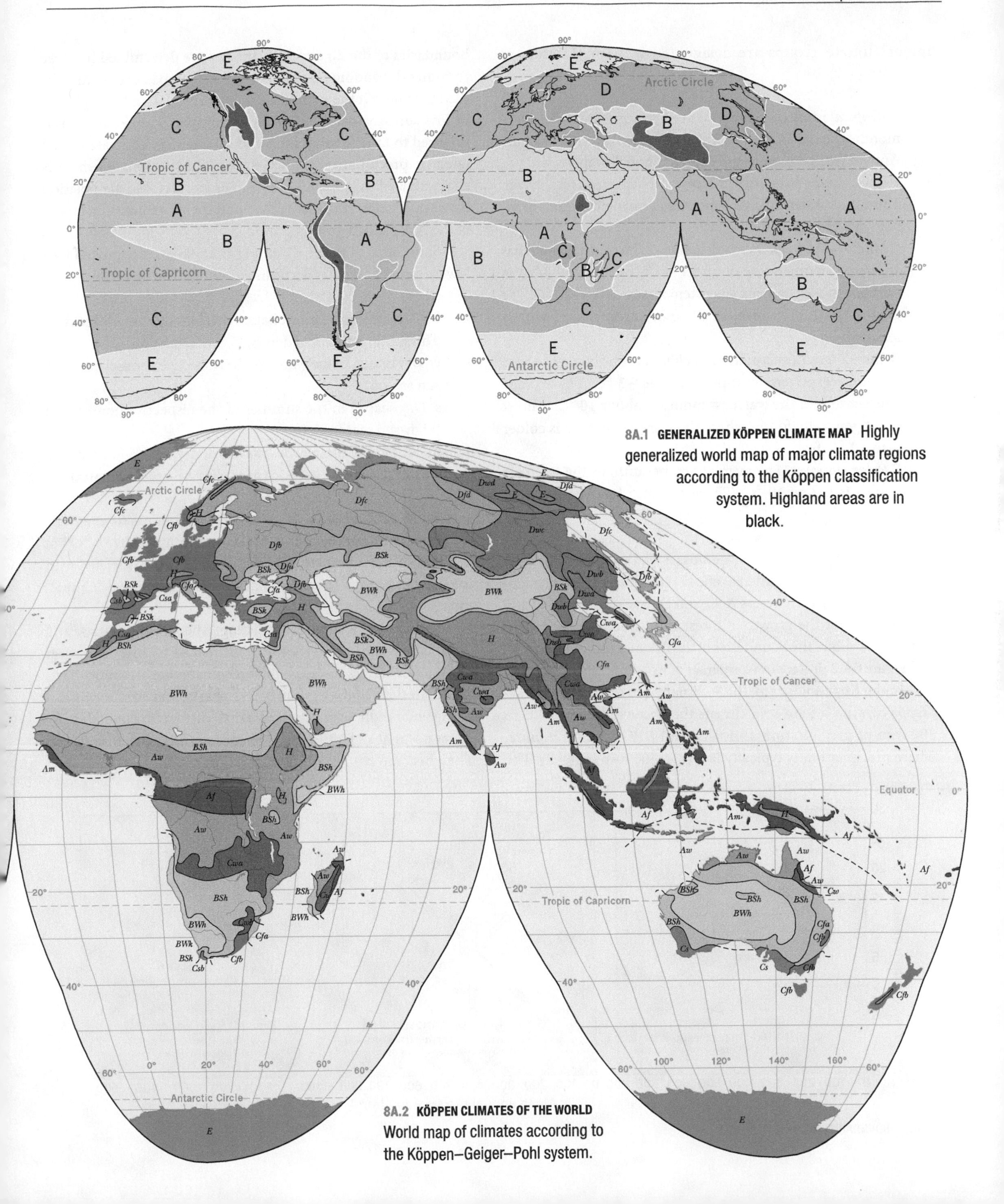

8A.1 GENERALIZED KÖPPEN CLIMATE MAP Highly generalized world map of major climate regions according to the Köppen classification system. Highland areas are in black.

8A.2 KÖPPEN CLIMATES OF THE WORLD World map of climates according to the Köppen–Geiger–Pohl system.

major climate groups are designated by capital letters as follows:

- *A Tropical rainy climates.* The average temperature of every month is above 18 °C. These climates have no winter season. Annual rainfall is large and exceeds annual evaporation.
- *B Dry climates.* Evaporation exceeds precipitation for the year. There is no water surplus; hence, no permanent streams originate in B climate zones.
- *C Mild, humid (mesothermal) climates.* The coldest month has an average temperature of under 18 °C but above −3 °C; at least one month has an average temperature above 10 °C. The C climates therefore have distinct summer and winter seasons.
- *D Snowy-forest (microthermal) climates.* The coldest month has an average temperature of under −3 °C. The average temperature of the warmest month is above 10 °C. (Forest is not generally found where the warmest month is colder than 10 °C.)
- *E Polar climates.* The average temperature of the warmest month is below 10 °C. These climates have no summer season.

Note that four of these five groups (A, C, D, and E) are defined by average temperatures, whereas one (B) is defined by precipitation-to-evaporation ratios. Groups A, C, and D have sufficient heat and precipitation for the growth of forest and woodland vegetation. Figure 8A.1 shows the boundaries of the five major climate groups, and Figure 8A.2 presents a world map of Köppen climates.

In the Köppen system, the B climates are divided into ***BS—Steppe climate***, a semi-arid climate that receives between 400 to 800 mm of precipitation annually, and ***BW—Desert climate***, where precipitation is typically less than 400 mm annually. The boundaries of the BS and BW climates are determined by three formulas depending on whether precipitation is evenly distributed throughout the year or if it is concentrated in the high-sun or low-sun seasons. If rain falls in summer, a larger amount is required to offset evaporation and maintain an equivalent total effective precipitation (Figure 8A.3). The BS climate, which is characterized by grasslands, occupies an intermediate position between the desert climate (BW) and the more humid climates of the A, C, and D groups.

Subgroups of the A, C, and D climates are designated by a lowercase second letter according to the following code:

- f Moist, adequate precipitation in all months, no dry season. This modifier is applied to A, C, and D groups.
- w Dry season in the winter of the respective hemisphere (low-sun season).
- s Dry season in the summer of the respective hemisphere (high-sun season).
- m Rainforest climate, despite short, dry season in monsoon type of precipitation cycle (m applies only to A climates).

Two E climates are recognized. In the ***Tundra climate (ET)***, the mean temperature of the warmest month is above 0 °C but below 10 °C. The ***Perpetual frost climate (EF)*** is associated with ice sheets where mean temperature remains below 0 °C in all months.

Highland (H) climates are distinguished from the arctic and polar climates because they are influenced by high elevation rather than high latitude. Highland climates are also more variable than their high-latitude counterparts, and conditions can change markedly over relatively small distances. Highland climates occur at all latitudes and in coastal and interior locations. They become colder at higher elevations, but have the same

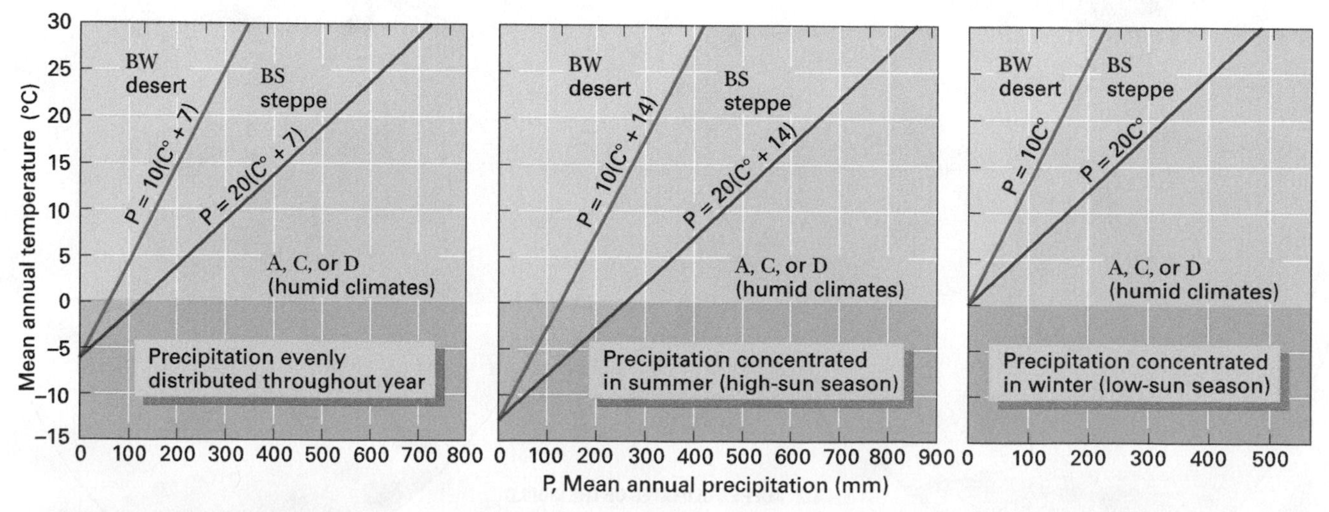

8A.3 BOUNDARIES OF THE B CLIMATES

seasonal precipitation patterns as the regions that surround them. Because of their inherent variability, no subdivisions are used for the Highland climates.

Excluding the Highland climates, twelve distinct climates can be differentiated using the basic Köppen classification.

QUESTIONS

1. Describe the characteristics of tropical Af, Am, and Aw climates as used in the Köppen system, and discuss the factors that give rise to the different precipitation patterns that occur in these regions.
2. Select four climate types that occur in Canada. Describe the characteristics that distinguish them in the Köppen system, and account for their different temperature regimes.

CHAPTER 9 LOW-LATITUDE CLIMATES

Overview of the Low-Latitude Climates
The Wet Equatorial Climate (Köppen *Af*)
The Monsoon and Trade-Wind Coastal Climate (Köppen *Am, Af*)
The Wet-Dry Tropical Climate (Köppen *Aw, Cwa*)
Eye on the Environment • Drought and Land Degradation in the African Sahel
The Dry Tropical Climate (Köppen *BWh, BSh*)
Highland Climates of Low Latitudes

Todd River, Australia, near Alice Springs. The dry channel of the Todd River can be seen at ***Fly By: 23° 44′ 50′ S; 133° 52′ 33′ E.***

OVERVIEW OF THE LOW-LATITUDE CLIMATES

The *low-latitude climates* lie for the most part between the Tropics of Cancer and Capricorn and represent about 50 percent of Earth's surface. They occupy the entire equatorial zone (10° N to 10° S), most of the tropical zone (10° to 15° N and S), and part of the subtropical zone (25–35° N and S). The low-latitude climate region includes the equatorial trough of the intertropical convergence zone (ITCZ), the belt of tropical easterlies (trade winds), and large portions of the oceanic subtropical high-pressure belt. The northeast and southeast trade winds that blow toward the equatorial trough are the steadiest and most reliable of any of the planetary wind belts, and represent the dominant airflow in the low latitudes. However, the trade winds tend to slacken during the respective summer seasons, and are generally weaker in the northern hemisphere.

The low-latitude climates are characteristically warm throughout the year, and are sometimes described as lacking any winter season. Constant warmth is the distinguishing feature of these climates, with the 18 °C winter isotherm used to establish their poleward limits. Whereas temperatures are relatively uniform throughout the tropics, rainfall is much more variable. Near the equator, rainfall is generally abundant and occurs year round, but total rainfall decreases and becomes more seasonal with latitude. In tropical and subtropical latitudes, a sequence of wet and dry seasons characteristically alternate through the year.

The principal low-latitude climate types are distinguished mainly by rainfall amount and distribution. They include the wet equatorial climate, the monsoon and trade-wind climate, the wet-dry tropical climate, and the dry tropical climate (Figure 9.1). Their general characteristics are listed in Table 9.1.

The seasonal distribution of precipitation at these latitudes is largely controlled by the annual movement of the sun north and south of the equator, and the corresponding shift in circulation that accompanies it (Figure 9.2). A zone of convective precipitation moves with the sun as it travels toward its June solstice position at the Tropic of Cancer, culminating in a short wet season of two to three months at this latitude. The wet season gets progressively shorter as the influence of the semi-permanent high-pressure zone becomes more pronounced. Storms

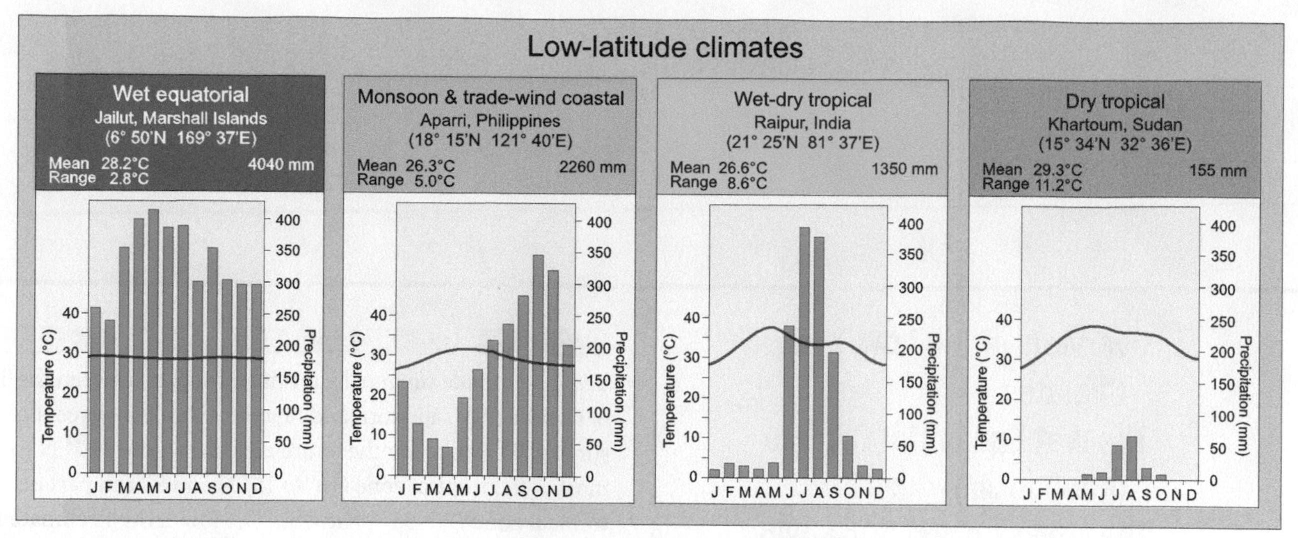

9.1 CLIMOGRAPHS FOR LOW-LATITUDE CLIMATES Temperatures are warm in all months with annual means typically above 26 °C; annual variation is about 3 °C near the equator and increases with latitude. The low-latitude climates are distinguished mainly on the basis of precipitation, with some areas receiving ample rainfall throughout the year and with prolonged dry spells occurring elsewhere.

accompany the sun as it moves toward its winter solstice position at the Tropic of Capricorn. Thus, near the equator, uplift along the ITCZ provides precipitation year round. At latitudes 5° to 15° N and S, the changing declination of the sun results in two short wet seasons alternating with two short dry seasons, and at higher latitudes, the wet-dry tropical climate merges with the dry tropical climate, where precipitation is very limited.

THE WET EQUATORIAL CLIMATE (KÖPPEN *Af*)

The **wet equatorial climate** occurs in areas where the ITCZ is nearby for most of the year. This climate is dominated by warm, moist maritime equatorial (mE) and maritime tropical (mT) air masses that yield heavy convectional rainfall. The wet equatorial climate is found in the latitude range 10° N to 10° S (Figure 9.3); the principal locations include the Amazon lowland of South

TABLE 9.1 LOW-LATITUDE CLIMATES

CLIMATE	TEMPERATURE	PRECIPITATION	EXPLANATION
Wet equatorial	Uniform temperatures, mean near 27 °C.	Abundant rainfall in all months, from mT and mE air masses. Annual total may exceed 2,500 mm.	The ITCZ dominates this climate, with abundant convectional precipitation generated by convergence in weak equatorial lows. Rainfall is heaviest when the ITCZ is nearby.
Monsoon and trade-wind coastal	Temperatures show an annual cycle, with warmest temperatures in the high-sun season.	Abundant rainfall, but with a strong seasonal pattern.	Trade-wind coastal: Rainfall from mE and mT air masses is heavy when the ITCZ is nearby, lighter when the ITCZ moves to the opposite hemisphere. Asian monsoon coasts: dry air flowing southwest in low-sun season alternates with moist oceanic air flowing northeast, producing a seasonal rainfall pattern on west coasts.
Wet-dry tropical	Marked temperature cycle, with hottest temperatures before the rainy season.	Wet high-sun season alternates with dry low-sun season.	Subtropical high pressure moves into this climate in the low-sun season, bringing very dry conditions. In the high-sun season, the ITCZ approaches and rainfall occurs. Asian monsoon climate: alternation of dry continental air in low-sun season, with moist oceanic air in high-sun season, bringing a strong pattern of dry and wet seasons.
Dry tropical	Strong temperature cycle, with intense high temperatures during high-sun season.	Low precipitation. Sometimes rainfall occurs when the ITCZ is near.	This climate is dominated by subtropical high pressure, which provides clear, stable air for much or all of the year. Insolation is intense during high-sun period.

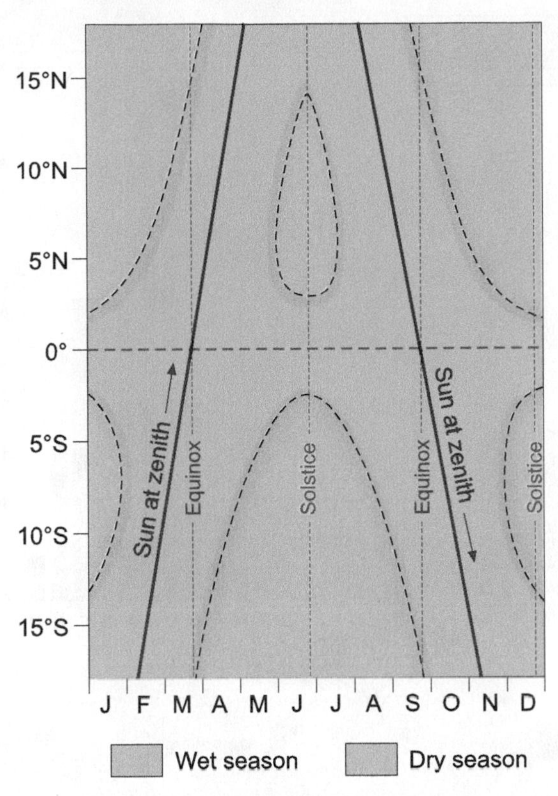

9.2 **ANNUAL DISTRIBUTION OF PRECIPITATION IN LOW-LATITUDE CLIMATES**
Near the equator, rain occurs throughout the year due to convergence along the ITCZ. It is increasingly associated with convectional processes during the high-sun season in the tropics and subtropics. (From *Introduction to Weather and Climate* by O.W. Archibold, John Wiley & Sons Canada, Ltd., 2011.)

America, the Congo Basin of equatorial Africa, and the East Indies, from Malaysia to New Guinea.

Precipitation is plentiful in all months, and the annual total often exceeds 2,500 mm. However, there is usually a seasonal pattern to the rainfall, with a somewhat wetter period occurring during the time of year when the ITCZ migrates into the region. Remarkably uniform temperatures prevail throughout the year. Both mean monthly and mean annual temperatures are typically close to 27 °C. In fact, the diurnal temperature range is normally greater than the annual range.

Iquitos, Peru (3° 45′ S; 75° 15′ W; elevation 126 m), is a representative wet equatorial station; it is situated close to the equator in the broad, low basin of the upper Amazon River. The climograph for Iquitos shows the small annual range in temperature and the high annual rainfall total that is characteristic of the wet equatorial climate. Monthly air temperatures are very uniform; typically, the mean monthly air temperature will range between 26° and 29 °C at low elevation stations in the equatorial zone.

THE MONSOON AND TRADE-WIND COASTAL CLIMATE (KÖPPEN *Am*, *Af*)

The monsoon and trade-wind coastal climate occurs in latitudes from 5° to 25° N and S (see Figure 9.3). Like the wet equatorial climate, the **monsoon and trade-wind coastal climate** has abundant rainfall. However, the rainfall in monsoon and trade-wind coastal regions shows a stronger seasonal pattern. Seasonality is especially pronounced in monsoon climates, as is shown in the climograph for Cochin, India (10° 9′ N; 76° 24′ W; elevation 3 m). In the trade-wind coastal climates, the seasonal rainfall pattern is related to the latitudinal migration of the ITCZ. In places such as Belize City (17° 32′ N; 88° 18′ W; elevation 5 m) and Tamatave, Madagascar (18° 12′ S; 49° 04′ E; elevation 6 m), rainfall increases in the high-sun season, when the ITCZ is nearby. Conversely, in the low-sun season, when the ITCZ has migrated to the other hemisphere, subtropical high pressure dominates and monthly rainfall is less. In general, rainfall is more evenly distributed throughout the year in trade-wind climates than is the case for monsoon climates.

As the name suggests, the monsoon and trade-wind coastal climate results from two somewhat different situations. On trade-wind coasts, moisture-laden maritime tropical (mT) and maritime equatorial (mE) air masses produce rainfall. The trade winds move these air masses onshore across narrow coastal zones backed by highlands. As the warm, moist air passes over coastal hills and mountains, the orographic effect produces shower activity. Rainfall is also intensified by easterly waves, which are more frequent when the ITCZ is nearby. This trade-wind effect is limited to the east coasts of land masses because the trade winds blow from east to west. Trade-wind coasts are found along the eastern margins of Central and South America, the Caribbean Islands, Madagascar, Southeast Asia, the Philippines, and northeast Australia.

The climograph for Belize City (see Figure 9.3) shows that rainfall is abundant from June through November, due to the proximity of the ITCZ. Easterly waves are common in this season, and occasionally, a tropical cyclone will bring torrential rainfall. Following the December solstice, rainfall is greatly reduced, with minimum values in March and April. At this time, the ITCZ lies at its farthest distance from Belize City, and the climate is dominated by subtropical high pressure. Air temperature shows an annual range of 5 °C, with the maximum occurring in the high-sun months.

The coastal precipitation effect also applies to the summer monsoon of Asia, when the monsoon circulation brings mT air onshore. However, the onshore monsoon winds blow from southwest to northeast, so the western coasts of land masses are exposed to this moist airflow. Western India and Myanmar are examples. Moist air also penetrates well inland

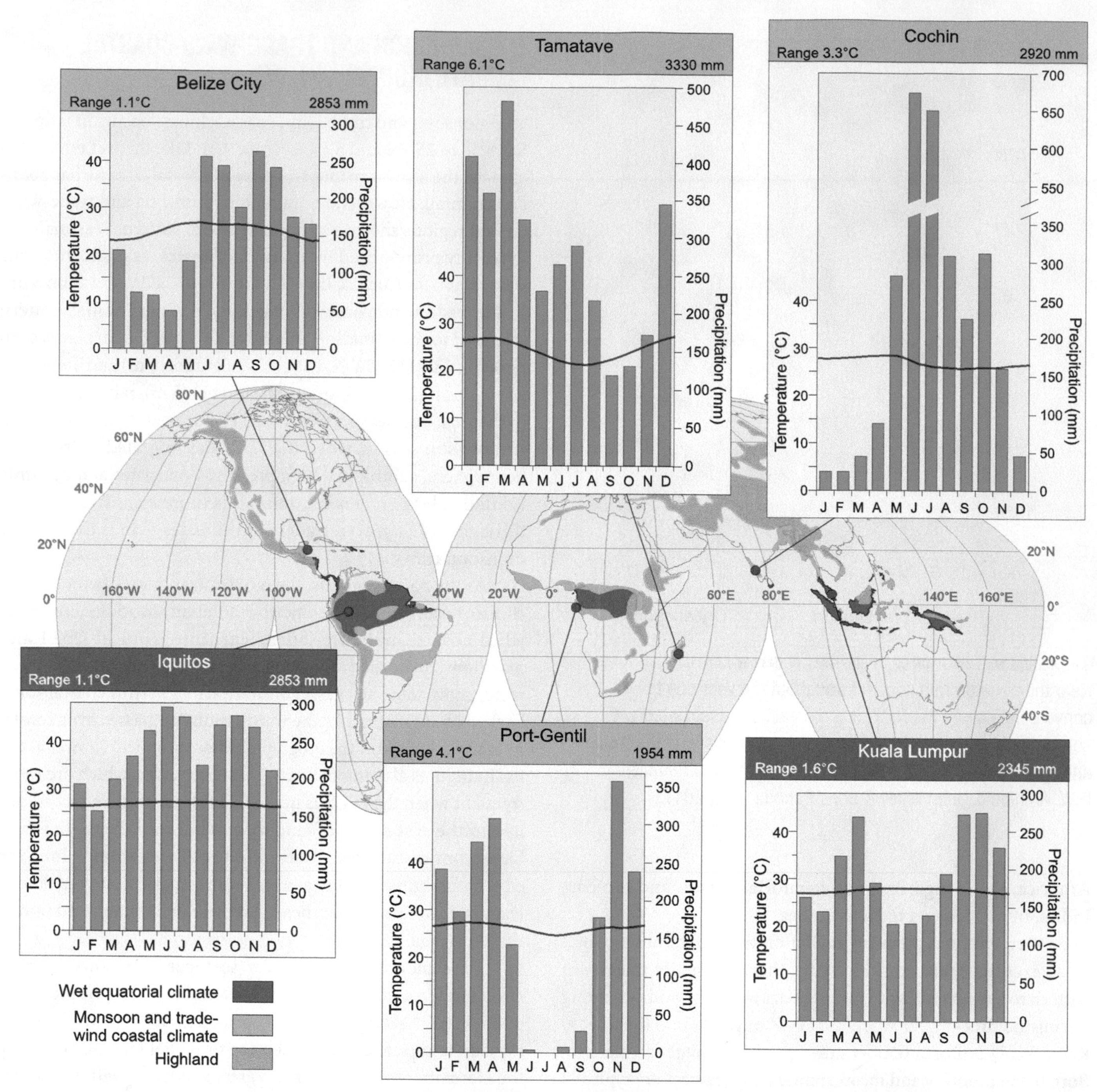

9.3 WORLD DISTRIBUTION OF WET EQUATORIAL AND MONSOON AND TRADE-WIND COASTAL CLIMATES The wet equatorial climate is a seasonless climate with consistently warm temperatures and heavy precipitation in all months. It lies astride the equator in South America, Africa, and Southeast Asia with a greater area in the southern hemisphere. The monsoon climate represents a poleward extension of the wet equatorial climate where temperature and precipitation patterns are more seasonal. The trade-wind coastal climates occur where moist onshore airflows are forced to rise over topographic barriers.

in Bangladesh, bringing the very heavy monsoon rains for which the region is well known.

The southwest monsoon season accounts for more than 75 percent of the annual rainfall over most of India. The typical progression of the summer monsoon over India is shown in Figure 9.4. The dates for the onset and end of the wet season are determined by the respective increase and decrease in rainfall. The actual dates used are the middle dates of consecutive five-day periods during which there is a noticeable rise or fall in precipitation.

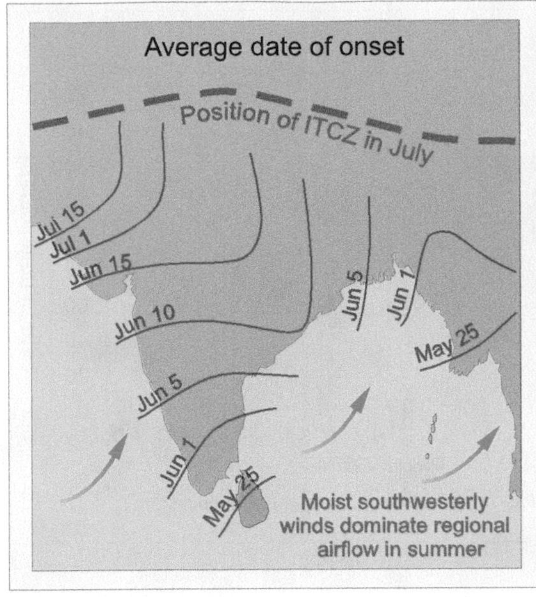

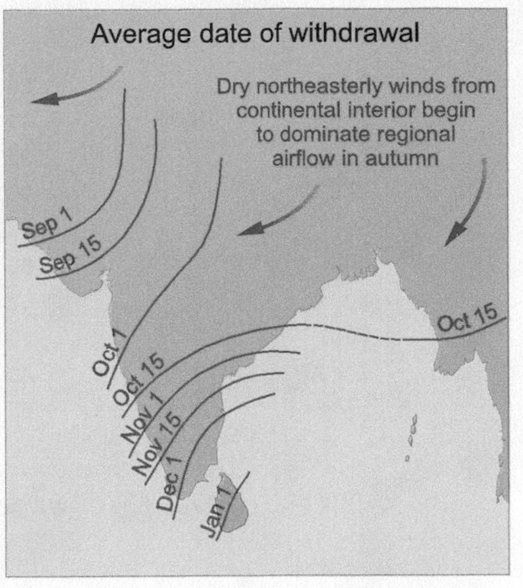

9.4 TYPICAL DATES FOR THE ONSET AND WITHDRAWAL OF THE WET MONSOON OVER INDIA The wet monsoon begins in Sri Lanka in late May and reaches the northwestern part of the Indian subcontinent by mid-July before retreating southwards in early September. The rainy season lasts for about six weeks in the northwest, compared to about seven months in Sri Lanka, bringing much needed moisture to the region.

One of the wettest places on Earth is Cherrapunji, which lies at 1,370 m in the foothills of the Himalayas in northeastern India. The average annual rainfall at Cherrapunji is 11,430 mm; however, in one 12-month period (August 1, 1860 to July 31, 1861), it received 26,461 mm, which still stands as a world record. It also holds the record for maximum rainfall in a single month—9,300 mm in July 1861. In that year, the total rainfall was 22,987 mm with 22,454 mm falling between April and September. The heavy rain period at Cherrapunji is generally from May to September, when the rain-bearing monsoon airflow is forced to rise over the high terrain. Local agriculture is vulnerable to excessive soil erosion from the heavy rainfall. Surprisingly, because of the seasonality of the monsoon rainfall, Cherrapunji can face acute water shortages in some months.

Based on recent records, Mawsynram, located about 10 km from Cherrapunji at an elevation of 1,400 m, is reputedly the wettest place on Earth, with an average annual rainfall of 11,872 mm. Other contenders for highest rainfall include Mount Waialeale (1,598 m) in Kauai, Hawaii; here the average rainfall is 11,685 mm, with rain falling on nearly every day of the year. The maximum annual rainfall recorded for Mount Waialeale is 16,916 mm, received in 1982. At Lloro, Colombia (elevation 159 m), the estimated annual rainfall is 13,299 mm, but this may be erroneous since Quibdo, at an elevation of 37 m and only 22 km away, receives an average rainfall of 8,992 mm, and is officially regarded as the wettest place in South America.

The Asiatic monsoon shows an extreme peak in rainfall during the high-sun period, and a well-developed dry season of two or three months with little rainfall. This characteristic pattern is shown in the climograph for Cochin, India (see Figure 9.3). Cochin lies in the path of the warm, moist southwest winds of the summer monsoon, and monthly rainfall is extreme in both June and July. Conversely, a strongly pronounced low rainfall season occurs at the time of low sun—December through March. Air temperatures show only a weak annual cycle, cooling a little during the rains, so the annual range is small, which is characteristic of all low-latitude climates.

The monsoon climate occurs anywhere where there is a seasonal shift in wind direction, and although characteristically associated with the heavy rainfall in Southeast Asia, similar conditions occur in places such as West Africa and northeastern Brazil. In West Africa, the seasonal shift of the ITCZ is about 15° of latitude (see Figure 5.12). Here, heavy rainfall occurs in the high-sun season, when the ITCZ is nearby; drier conditions prevail in the low-sun season, when the ITCZ is more distant.

In the monsoon and trade-wind coastal climate, the warmest temperatures occur in the high-sun season, just before the arrival of the ITCZ brings clouds and rain. The seasonal reduction in solar radiation caused by the heavy cloud cover affects temperature through the summer. Daytime maximum temperatures are characteristically reduced by one or two degrees, although minimum temperatures at night are generally not affected. Seasonal minimum temperatures occur at the time of low sun.

Generally there is no shortage of moisture in the wet equatorial and monsoon and trade wind coastal climates, and even during the extended dry periods that occur in the monsoon regions, there is sufficient moisture to maintain a forest cover (Figure 9.5). The combination of continually warm temperatures and high soil moisture create the *low-latitude rain forest environment.* Under these conditions, rock decomposition occurs to great depths, which has created a thick soil layer. The soil is typically rich in iron oxides, which impart a deep red colour (see Chapter 19). Most rain forest soils have lost the ability to hold

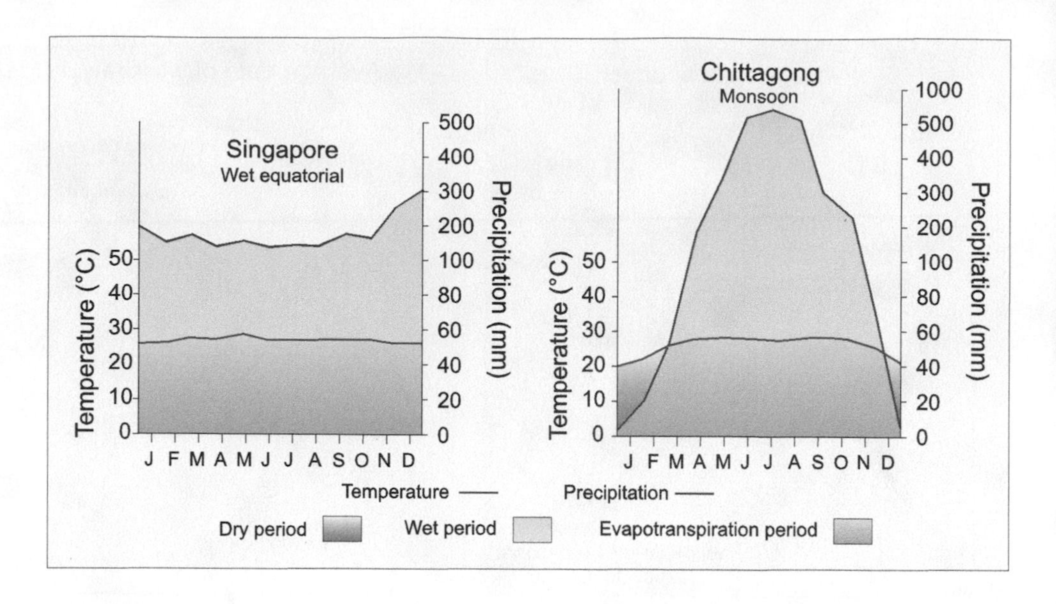

9.5 SIMPLIFIED CLIMATE DIAGRAMS SHOWING SEASONAL MOISTURE AVAILABILTY FOR LOW-LATITUDE RAIN FOREST AND MONSOON FOREST STATIONS Despite the dry period at Chittagong, soil moisture reserves are adequate to offset the lack of rainfall.

the minerals that plants need. However, the native plants have adapted to these infertile soils by rapidly taking up the essential plant nutrients that are released by the decay of fallen leaves and branches before they can be leached from the impoverished soil. The most conspicuous plants are the trees of the evergreen rainforests. These form a multi-layered canopy above a diverse flora that may contain more than 3,000 different species in an area of only a few square kilometres. Chapter 22 discusses in detail the characteristic plants and animals found in the low-latitude rain forest environment.

THE WET-DRY TROPICAL CLIMATE (KÖPPEN *Aw, Cwa*)

In the monsoon and trade-wind coastal climate, the ITCZ moves into and away from the region to produce the seasonal cycles of rainfall and temperature. These cycles become stronger farther from the equator, and the monsoon and trade-wind coastal climate grades into the **wet-dry tropical climate**.

The wet-dry tropical climate is distinguished by very dry conditions during the low-sun season; this alternates with very wet conditions during the high-sun season. During the low-sun season, the ITCZ moves well beyond the wet-dry tropical climate region, and the dominant air masses are of the dry continental tropical (cT) type. In the high-sun season, when the ITCZ is nearby, moist mT and mE air masses dominate. Cooler temperatures accompany the dry season, but give way to a very hot period before the rains begin. Figure 9.6 shows the global distribution of the wet-dry tropical climate, which is found at latitudes of 5° to 20° N and S in Africa and South America, and at 10° to 30° N in Asia. In Africa and South America, the climate occupies broad bands on the poleward side of the wet equatorial and monsoon and trade-wind coastal climates. Because these regions are farther from the ITCZ, less rainfall occurs during the rainy season, and subtropical high pressure can dominate more strongly during the low-sun season. In central India and Indochina, mountain barriers tend to protect the wet-dry tropical climate regions from the warm, moist mE and mT airflows associated with the trade winds and monsoons. These barriers create a rain shadow effect, which reduces rainfall during the rainy season and accentuates the dry season.

Bamako, Mali (12° 32′ N; 7° 57′ W) in West Africa is a representative wet-dry tropical climate station. Here the rainy season begins just after the March equinox and reaches a peak in August, about two months following the June solstice. At this time, the ITCZ has migrated to its most northerly position, and moist mE air masses flow into the region from the ocean lying to the south. Monthly rainfall then decreases as the low-sun season arrives and the ITCZ moves southward. Three months—December through February—are practically rainless. During this season, subtropical high pressure dominates the climate, and stable, subsiding cT air pervades the region. This air originates in the Sahara and brings with it the dry, dusty conditions of the *harmattan*—a wind that blows from the east and northeast, carrying dust across western Africa and into the eastern Atlantic.

The temperature cycle at Bamako is closely linked to both the solar cycle and the precipitation pattern. Air temperature rises in February through May, as insolation increases; this is the warmest period of the year with mean monthly temperatures reaching about 32 °C in April. As soon as the rains set in, the effect of cloud cover and evaporation of rain causes the temperature to fall. By July, temperatures have dropped to about 26 °C and remain more or less at this level until the following March.

A characteristic of the tropical wet-dry climate is its large year-to-year variability in precipitation. In this climate, rainfall

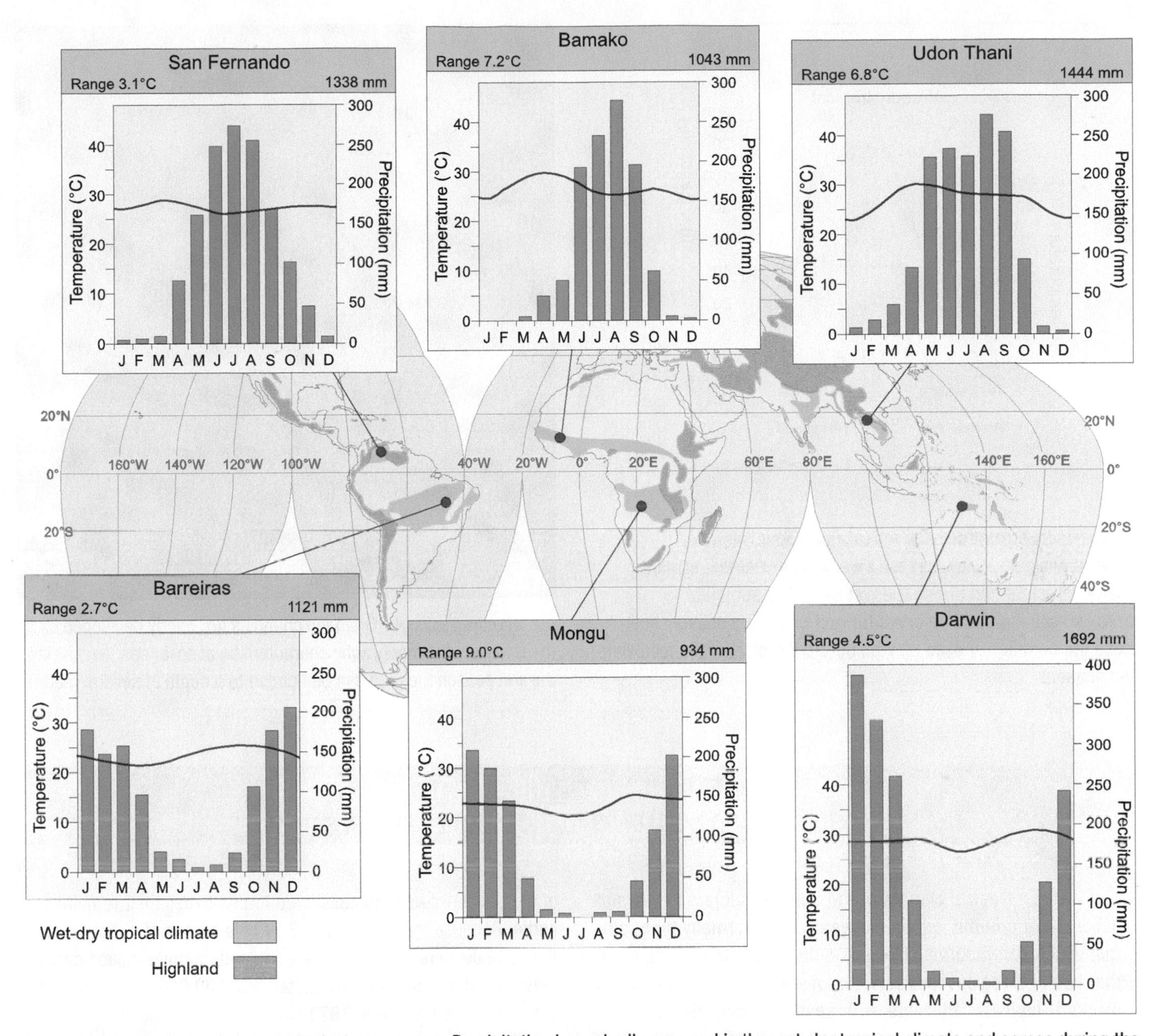

9.6 WORLD DISTRIBUTION OF THE WET-DRY TROPICAL CLIMATE Precipitation is markedly seasonal in the wet-dry tropical climate and comes during the high sun season. The heavy cloud cover lowers temperatures at this time of the year when, theoretically, insulation is at a maximum.

occurs when the ITCZ migrates into the region. In years when the ITCZ does not move as far from the equator, rainfall is greatly reduced. The *Working It Out • Cycles of Rainfall in the Low Latitudes* feature at the end of the chapter shows how year-to-year variation is measured and looks at multi-year cycles at various tropical climate stations.

The wet-dry tropical climate is associated with the savanna environment. Here the native vegetation must survive alternating seasons of very dry and very wet weather (Figure 9.7). River channels that are not fed by nearby moist highland regions are nearly or completely dry in the low-sun dry season. In the rainy season, the rivers become filled with swiftly flowing, turbid water. Often, extensive tracts of land are flooded (Figure 9.8). However, the rains are not reliable, and agriculture without irrigation is hazardous at best. When there is no rain, a devastating famine can ensue. The Sahel region of Africa, discussed further in *Eye on the Environment • Drought and Land Degradation in the African Sahel*, is a region well known for such droughts and famines.

Soils of the savanna environment are similar in their physical characteristics and fertility to those of the rain forest environment—that is, they are largely reddish soils of low fertility.

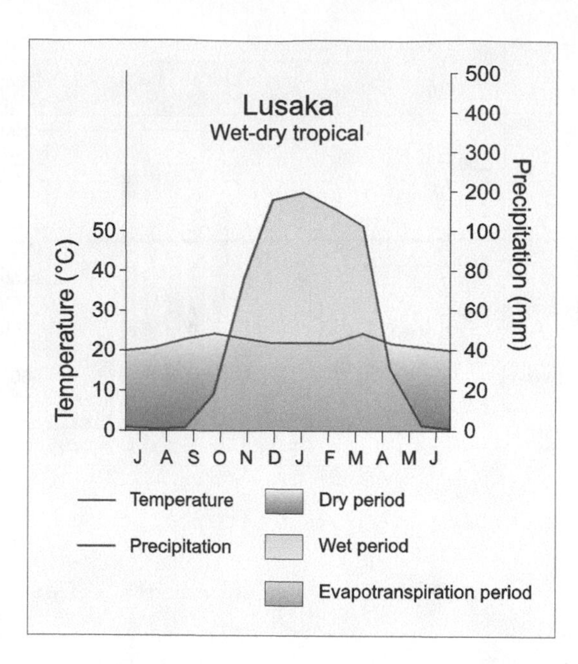

9.7 SIMPLIFIED CLIMATE DIAGRAM FOR LUSAKA, ZAMBIA, SHOWING SEASONAL MOISTURE AVAILABILTY FOR A WET AND DRY TROPICAL CLIMATE STATION Precipitation during the wet season is sufficient to replenish soil moisture reserves for part of the dry season, but these are completely used up well before the start of the following wet season.

9.8 A FLOODED SAVANNA Drainage channels are poorly developed in the relatively flat topography characteristic of savannas, and during the wet season the land can be flooded to a depth of several metres.

EYE ON THE ENVIRONMENT

DROUGHT AND LAND DEGRADATION IN THE AFRICAN SAHEL

The wet-dry tropical climate of the Sahel is subject to years of devastating drought, as well as years of abnormally high rainfall, which can result in severe flooding. Climate records show that two or three successive years of abnormally low rainfall (a drought) typically alternate with several successive years of average or higher than average rainfall. Such variability is a characteristic feature of the wet-dry tropical climate, and the native plants and animals inhabiting this region are well adapted to the rainfall regime.

Africa's wet-dry tropical climate, including the adjacent semi-arid southern belt of the dry tropical climate to the north, provides a lesson on the human impact on a delicate ecological system. Figure 1 shows countries in this perilous belt, called the *Sahel* or *Sahelian zone*. From 1968 through 1974, all these countries were struck by a severe drought, and again from 1983 to 1985. Both nomadic cattle herders and grain farmers share this zone. During the drought, grain crops failed and foraging cattle and goats could find no food to eat (Figure 2). In the worst stages of the Sahel drought, nomads were forced to sell their remaining cattle. Because the cattle were their sole means of subsistence, the nomads were soon desperate for food. Some 5 million cattle perished, and it has been estimated that 100,000 people died of starvation and disease in 1973 alone.

The Sahelian drought of 1968 to 1974 was an example of a phenomenon that, at that time, was called *desertification*—the permanent transformation of the land surface to resemble a desert, largely through activities that led to the destruction of grasses, shrubs, and trees, such as overgrazing and fuel wood harvesting. The term desertification has now been abandoned in favour of *land degradation*. Land degradation accelerates the effects of soil erosion, such as gullying of slopes and accumulation of sediment in stream channels. It also intensifies the removal of soil by wind.

Periodic droughts throughout past decades are well documented in the Sahel, as they are in other regions of the wet-dry tropical climate. Figure 3 shows the percentage departure from

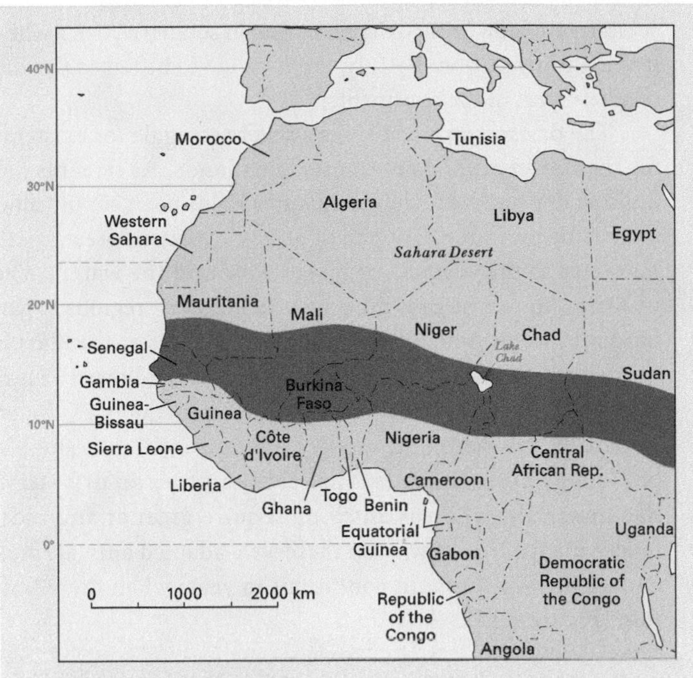

Figure 1 The African Sahel The Sahel, or Sahelian zone, shown in brown, lies south of the great Sahara Desert of North Africa.

Figure 2 Sahelian drought At the height of the Sahelian drought, vast numbers of cattle had perished and even goats were hard pressed to survive. Trampling of the dry ground prepared the region for devastating soil erosion by the rains that eventually ended the drought.

the long-term mean of each year's rainfall in the western Sahel from 1901 through 2009. Note the wide year-to-year variation. Since about 1950, the duration of periods of continuous departures both above and below the mean seem to have increased substantially. The period of sustained high-rainfall years in the 1950s contrasts sharply with a series of severe drought episodes starting in 1971. To obtain an earlier record, scientists have examined fluctuations in the level of Lake Chad. In times of drought, the lake's shoreline retreats, while in times of abundant rainfall, it expands. These changes document periods of rainfall deficiency and excess over some 200 years: the period 1820 to 1840 was below normal; 1870 to 1895 was above normal; and 1895 to 1920 was below normal.

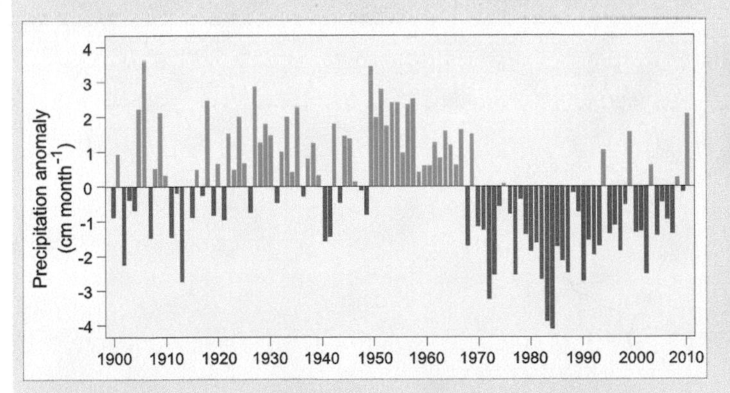

Figure 3 Rainfall in the Sahel Rainfall fluctuations for stations in the western Sahel, 1900–2010, expressed as a monthly average departure from the long-term mean. Data from Joint Institute for the Study of the Atmosphere and Ocean (JISAO), http://www.jisao.washington.edu/data/sahel/sahelprecip19002010.big.gif.

The long-term precipitation record shows that droughts and wet periods are a normal phenomenon in the Sahel. Historically, the landscape was able to recover from the droughts during periods of abundant rainfall. Today, the pressures from the increased human and cattle populations keep the land degraded. As long as these populations remain high, land degradation will become a severe and permanent feature.

Climate models developed in the 1990s suggested that the Sahelian droughts could be explained by natural large-scale climate changes. However, in the early 2000s, these ideas were revised to incorporate the effect of air pollution on cloud cover. It has been suggested that an increase in abundance of sulphate aerosols from pollution results in smaller cloud droplets and increases the albedo of the cloud surfaces. This reduces convective rain activity in the tropics and limits the northward migration of the rains as the sun moves to its July position at the Tropic of Cancer.

The effect of accelerated climate change in the coming decades on wet-dry tropical climate regions is difficult to predict. Some regions may experience greater swings between drought and surplus precipitation. At the same time, these regions may expand toward the poles as a result of intensification of the Hadley cell circulation. Thus, the Sahel region may move northward into the desert zone of the Sahara. However, such changes are highly speculative. At this time, there is no scientific consensus on how climate change will affect the Sahel or other regions of the wet-dry tropical climate.

However, substantial areas of the savanna environment have fertile soils developed and sustained by the slow accumulation of windblown dust from adjacent deserts (Figure 9.9). Equally important are highly productive soils that occur along major through-flowing rivers. Annual flooding of these rivers leaves deposits of fertile silt carried down from distant mountain ranges.

Most savanna plants enter a dormant phase during the dry period, then come into leaf and bloom with the rains. For this reason, the native plant cover is described as rain-green vegetation (Figure 9.10). In the dry season, the grasses turn to straw, and many of the tree species shed their leaves to cope with the drought. As the dry season progresses, there is increasing probability of fires (Figure 9.11).

Fires are especially common in Africa, where they are deliberately set to maintain farmland and grazing areas. The burning area shifts from north to south over the course of the year, as the seasons change (Figure 9.12). Fires burn extensively in the grasslands south of the Sahara Desert in January during the dry season, with little fire activity occurring in southern Africa where it is the rainy season. By July, the fires have shifted to the dry, southern part of the continent.

The dry season also brings a severe struggle for existence for the native animals of the savanna lands. As streams and hollows dry up, the few muddy water holes must supply all of the drinking water. Danger of attack by carnivores greatly increases as the animals congregate around the water holes. In Africa, the herbivores must move across vast regions to find food and water. The annual migration begins at the start of the dry season with the zebra and wildebeest, followed in turn by gazelles and other antelope species. The sequence is determined by the type and quality of food available and the physiology of the animals. The zebra can feed on dry, strawlike materials, whereas antelope require greener and more tender plants. However, even these well-adapted animals must endure unusual drought conditions in years when there is no rain (Figure 9.13).

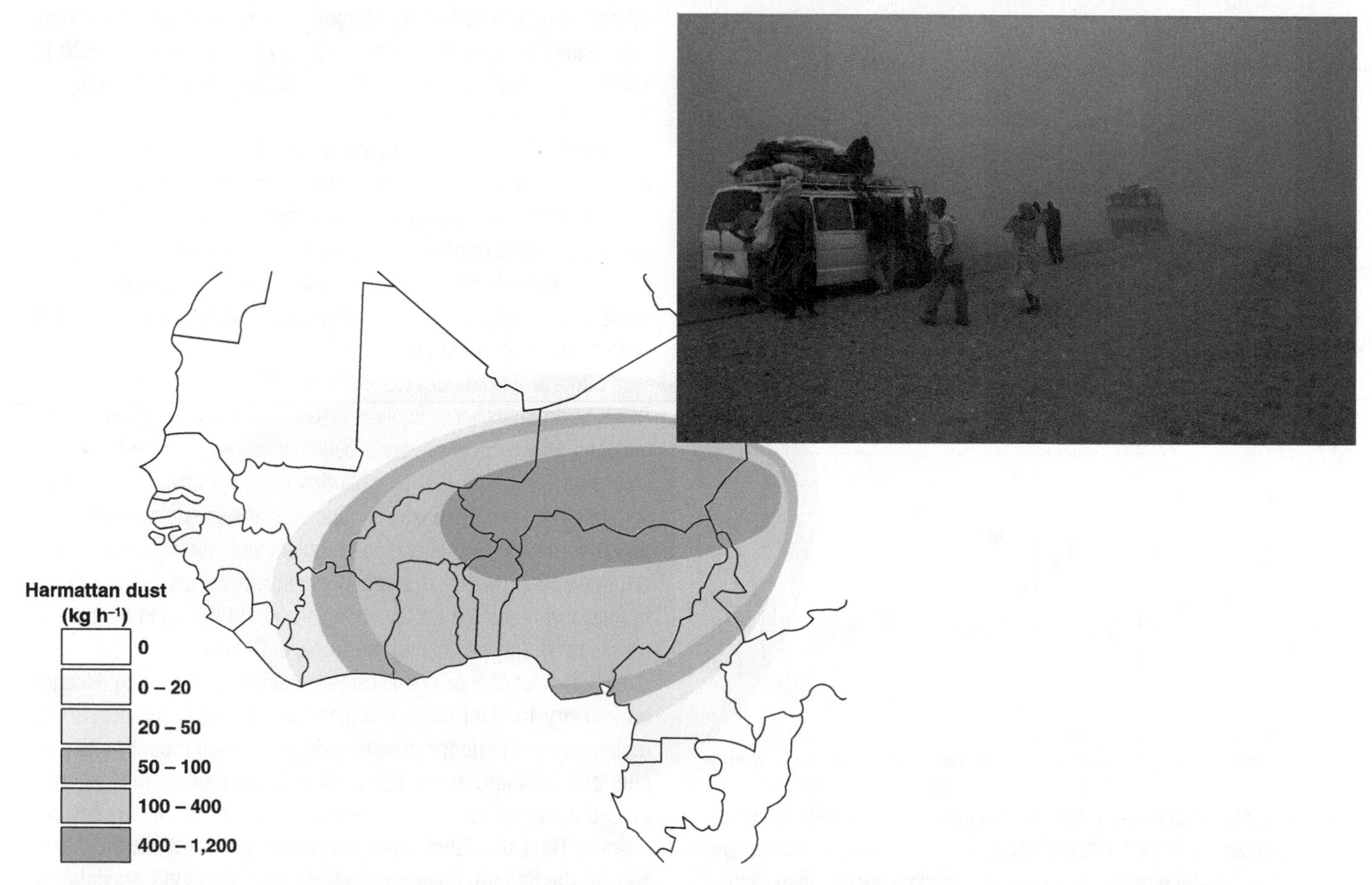

9.9 HARMATTAN WINDS (a) Harmattan dust deposited in West Africa contributes, on average, 3.8 g kg^{-1} of nitrogen, 0.8 g kg^{-1} of phosphorus, and 18.7 g kg^{-1} of potassium to the soil. (b) The large amount of dust in the air can obscure the sky for several days and create what is known as the Harmattan haze.

9.10 **A COMPARISON OF THE WET AND DRY SEASON** Wet and dry seasonal aspects of Mopane (*Colophospermum mopane*) woodland in Zimbabwe.

9.11 **FIRE IN A WOODED SAVANNA IN AUSTRALIA** Fires are a common occurrence in the latter part of the dry season in the savanna region.

(a)

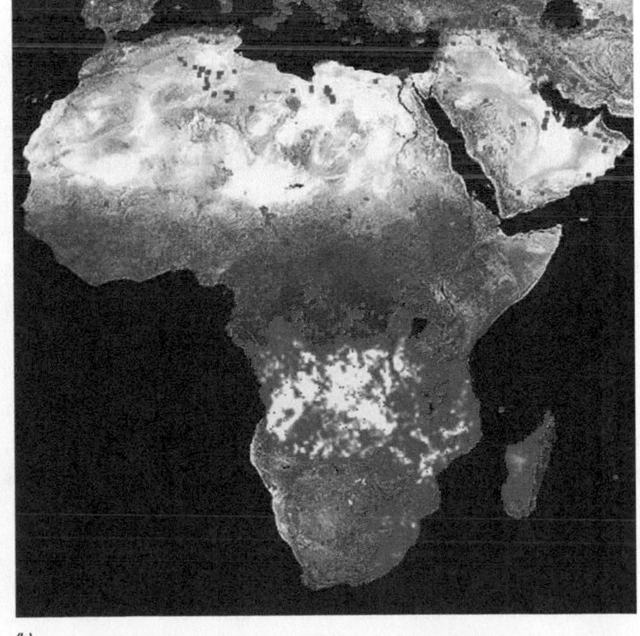

(b)

9.12 **SEASONAL FIRE PATTERNS IN AFRICA DURING JANUARY (a) AND JULY (b) 2005** The images are compiled from fires detected by NASA's Moderate Resolution Imaging Spectroradiometer (MODIS) during consecutive 10-day periods. Red marks occasional fires in a location; yellow signifies frequent fires.

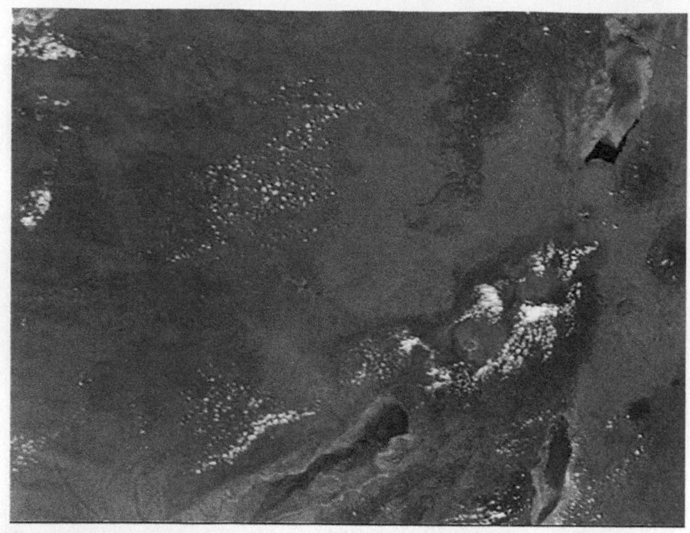

(a) January 12 2006

(b) January 9 2006

9.13 **INTERANNUAL RAINFALL VARIATION** These images from the Moderate Resolution Imaging Spectroradiometer (MODIS) on NASA's Terra satellite compare the landscape of the Serengetti in January 2005 (left) with January 2006 (right), when rainfall totals for the year were only 20 percent of normal in some areas.

THE DRY TROPICAL CLIMATE (KÖPPEN *BWh, BSh*)

The **dry tropical climate** is associated with the subtropical high-pressure cells. Here, strong subsidence and adiabatic warming inhibit condensation. Rainfall is rare and occurs only when unusual weather conditions move moist air into the region. Since skies are clear most of the time, the sun heats the surface intensely, keeping air temperatures high. During the high-sun period, heat is extreme. The record high air temperature for the Earth—57.7 °C—was recorded in this climate type at Al Aziziyah, Libya. During the low-sun period, temperatures are cooler. Given the dry air and lack of cloud cover, the daily temperature range is large.

The driest areas of the dry tropical climate are centred near the Tropics of Cancer and Capricorn, with rainfall increasing toward the equator. Immediately adjacent to the dry tropical regions, precipitation comes during a short rainy season at the time of the year when the ITCZ is nearby. Although much of the world's dry climates consist of extremely *arid* deserts, there are, in addition, broad zones at the margins of the deserts that are best described as *semi-arid.*

Nearly all of the areas of dry tropical climate lie in the latitude range 15° to 25° N and S (Figure 9.14). The largest region extends across the Sahara Desert, through Saudi Arabia and Iran, to the Thar desert belt of northwestern India and Pakistan. This vast desert expanse includes some of the driest regions on Earth. Another large region of dry tropical climate is the desert of central Australia. The west coast of South America, including portions of Ecuador, Peru, and Chile, is also very dry; however, temperatures there are moderated by a cool marine air layer associated with the Peruvian Current that blankets the coast. Temperatures in the coastal deserts of Namibia in southwest Africa are similarly regulated by the Benguela Current.

Bilma, Niger (18° 41′ N; 13° 20′ E), is a representative dry tropical station in the heart of the North African desert. The temperature record shows a strong annual cycle with a very hot period at the time of high sun, when the mean temperature of four consecutive months exceeds 33 °C. Daytime maximum air temperatures are frequently between 43° and 48 °C in the warmer months. There is a comparatively cool season at the time of low sun, when the coolest month averages 17 °C. The climograph shows rainfall is limited to June, July, and August, although traces of rain can occur in all months. In this environment, dew is a significant source of moisture for plants and animals.

Dewfalls are especially important in the coastal deserts that are found along the western edge of South America and southwest Africa. These regions are strongly influenced by cold ocean currents and the upwelling of deep, cold water, which occurs just offshore. The cool water moderates coastal zone temperatures and reduces the seasonality of the temperature cycle. This can be seen in the climographs for Lima, Peru (12° 6′ S; 76° 55′ W), and Walvis Bay, South Africa (22° 58′ S; 14° 38′ E) (Figure 9.14). At Lima, the mean temperature of the warmest month is only 22.7 °C, while at Walvis Bay it is 19.5 °C, despite their tropical locations. Coastal fog is a characteristic feature of these coastal deserts; it is formed when warm moist air is chilled as it moves over the cold currents. The fog provides essential moisture to the region as it drifts onshore. Arica in the Atacama Desert of northern Chile holds the record as the Earth's driest location, with an

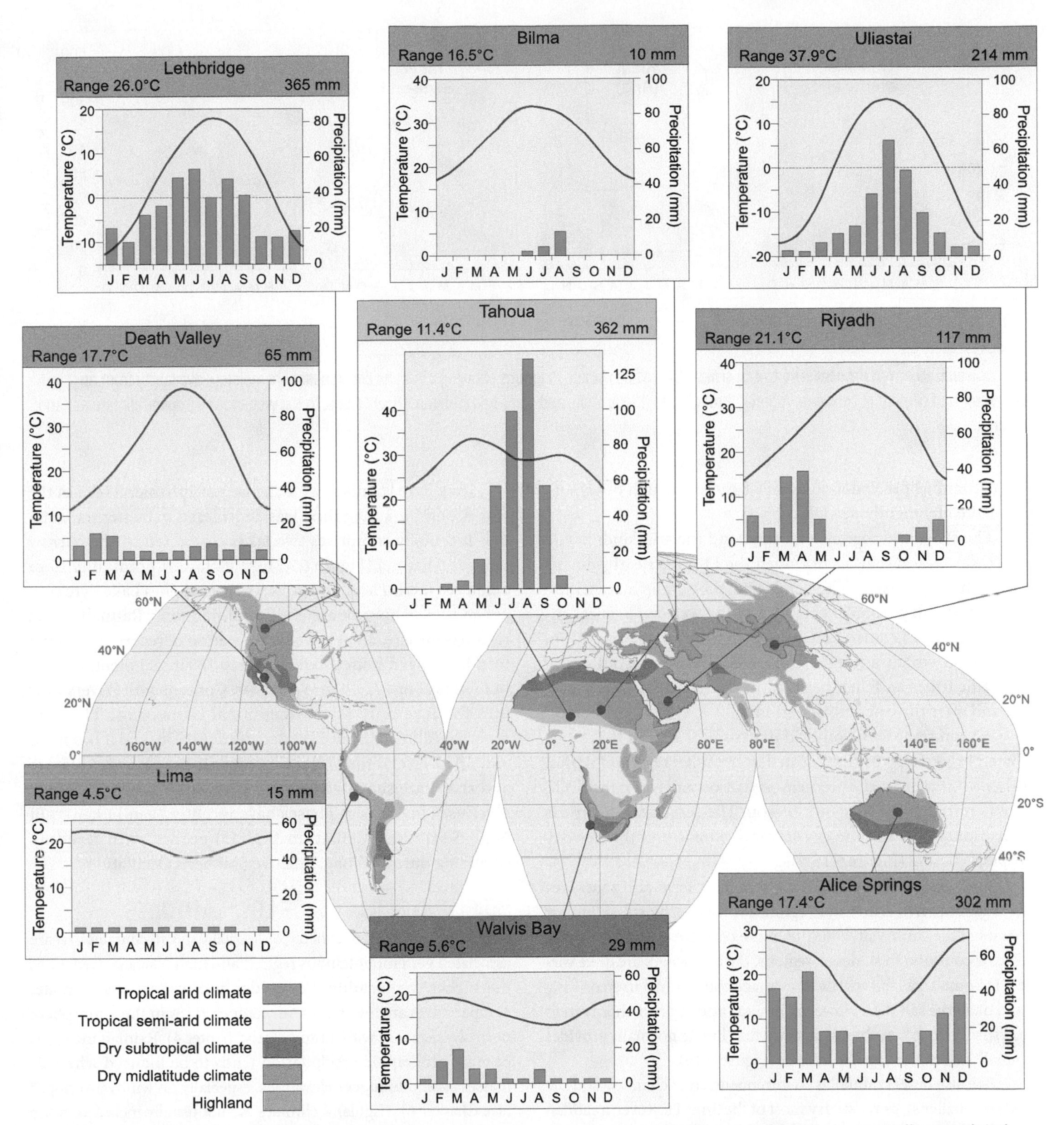

9.14 WORLD DISTRIBUTION OF THE DRY TROPICAL, DRY SUBTROPICAL, AND DRY MIDLATITUDE CLIMATES The arid climates are generally associated with regions of high pressure created by the Hadley circulation cells. The dry subtropical and dry midlatitude climates are poleward and eastward extensions of the dry tropical climate, although at higher latitudes aridity is increasingly related to other factors, especially topographic barriers. Locations with temperate arid climates, such as Lethbridge, are discussed in Chapter 10.

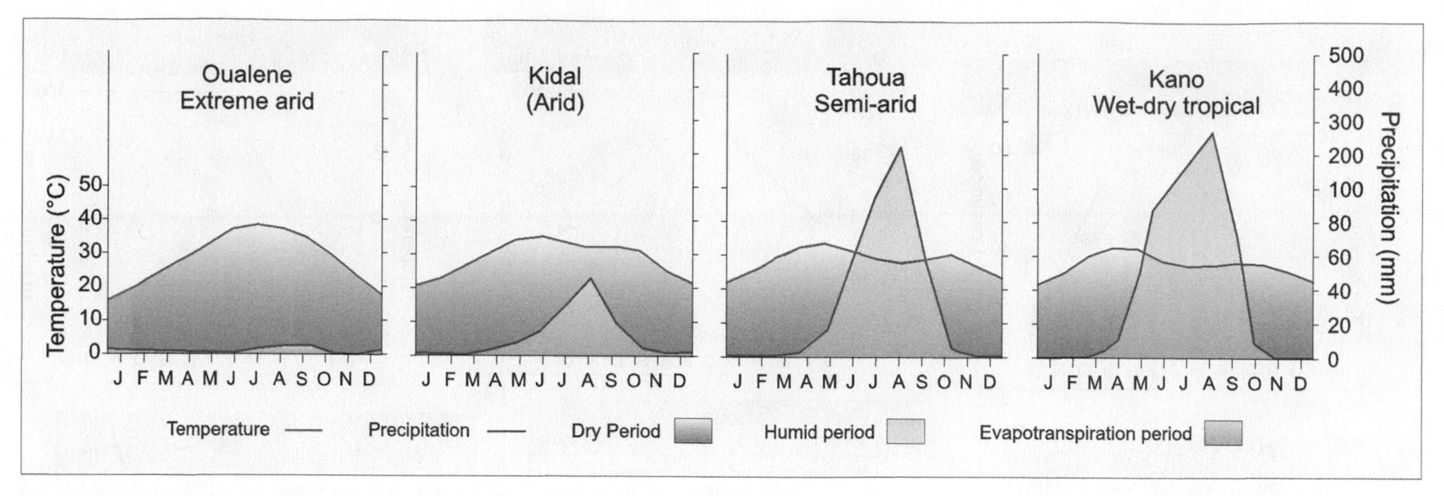

9.15 SEASONAL MOISTURE REGIMES IN NORTHERN AFRICA Simplified climate diagrams showing the transition in seasonal moisture availabilty from an extremely arid climate at Ouallene, Algeria, through arid (Kidal, Mali), and semi-arid (Tahoua, Niger) types, to the wet and dry tropical climate at Kano, Nigeria.

average annual precipitation of only 8 mm, although this is supplemented by nightly dewfalls.

The semi-arid regions that surround the arid lands have a short wet season that occurs when the ITCZ is nearby during the high-sun season. The semi-arid subtype is a transition between the dry tropical climate and the wet-dry tropical climate. Rainfall is sufficient to support the sparse growth of grasses on which animals (both wild and domestic) graze. Nomadic tribes and their herds of animals visit these areas during and after the brief moist period.

A representative semi-arid dry tropical station is Tahoua, Niger (14° 9′ N; 5° 25′ E). Located in the Sahel region of Africa, Tahoua has a distinct rainy season that occurs when the ITCZ moves north in the high-sun season. The semi-arid subtype is transitional between the dry tropical climate and the wet-dry tropical climate (Figure 9.15).

The tropical deserts and their bordering semi-arid zones create a global environmental region dominated by subsiding air masses of the continental high-pressure cells. Despite aridity being a factor common to all desert regions, the landscapes are quite varied. In some areas the arid desert surface consists of barren drifting sand or sterile salt flats. However, in semi-arid regions, thorny trees and shrubs are often abundant, since the climate normally provides a small amount of regular rainfall (Figure 9.16).

Because desert rainfall is so infrequent, river channels and the beds of smaller streams are dry most of the time. However, a sudden and intense downpour can cause local flooding of brief duration that transports large amounts of silt, sand, gravel, and boulders. These events are termed *flash floods* (see Chapter 16). The stream channels often end in flat-floored basins having no outlet. Clay and silt are deposited and accumulate here, along with layers of soluble salts. Shallow salt lakes occupy some of these basins.

Lake Eyre in Australia is the largest ephemeral lake in the world, and for a long time was considered to be permanently dry. It actually comprises two lakes connected by the narrow Goyder Channel (Figure 9.17). At its lowest point, it is 15 m below sea level. The drainage basin that supplies Lake Eyre covers an area of approximately 1.2 million km^2. Rainfall in the region is meagre, although over the period of record the annual total has ranged from about 50 mm to 760 mm. Potential evaporation is about 3,500 mm per year. Consequently, rivers flow into the lake only after exceptional rainstorms. The first reported filling of Lake Eyre occurred in 1938, and it has filled and dried several times since. Because it has no outlet, its size and shape are sensitive to the amount of inflowing water it receives. The wet-dry phases have recently been linked to El Niño-Southern Oscillation (ENSO) events, with flooding occurring during strong, positive Southern Oscillations.

HIGHLAND CLIMATES OF LOW LATITUDES

Highland climates are cool to cold, usually moist, climates that are associated with mountainous regions and high plateaus. Generally, the higher the location, the colder and wetter is its climate. Temperatures are lower since air temperatures in the atmosphere normally decrease with altitude (see Chapter 4). Rainfall increases because orographic precipitation tends to be induced when air masses ascend to higher elevations, especially on windward slopes (see Chapter 6). Highland climates are not usually included in the broad schemes of climate classification, since many small highland areas are simply not shown at the scale of a world map.

The climate of a highland area is usually closely related to that of the climate of the surrounding lowland, particularly in terms of the annual temperature cycle and the occurrence of wet and dry seasons. The climographs for two Indian stations in close

(a)

(b)

(c)

9.16 **DESERT LANDSCAPES** (a) **Sahara desert** Palms and hardy desert trees grow in this sandy region in Morocco. (b) **Great Australian desert** Red colours dominate this desert scene from the Rainbow Valley, south of Alice Springs, Australia. (c) **Baja California desert** Granite boulders, rounded by spheroidal weathering, and cactus are seen here in the Catavina Desert of the northern Baja California Peninsula, Mexico.

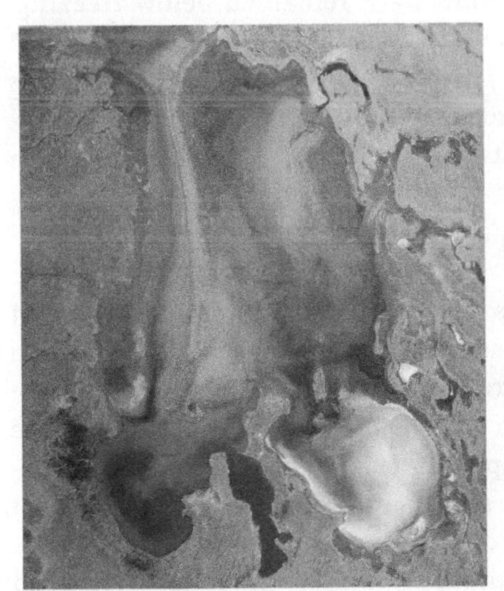

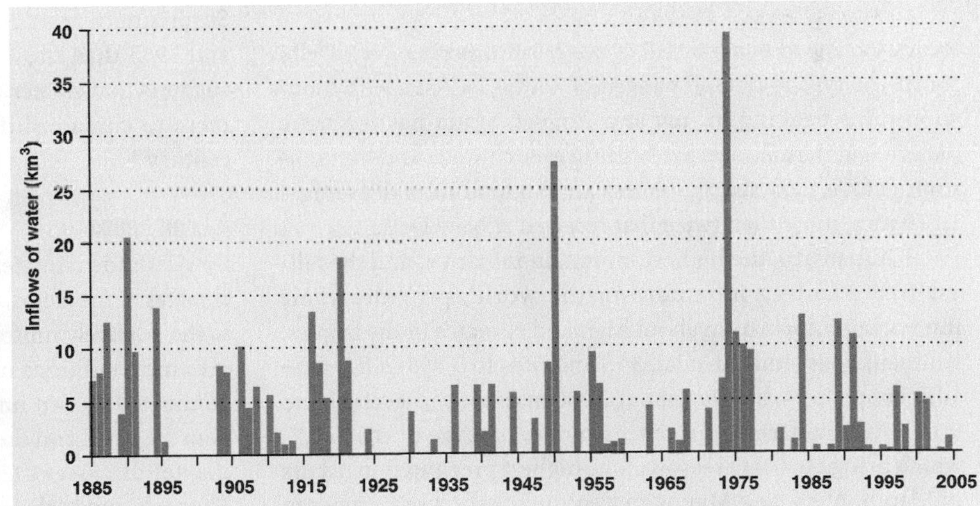

9.17 **LAKE EYRE, AUSTRALIA** Inflows of water occur periodically and have been linked to ENSO events (image taken in June 2009).

geographical proximity provide an example of this effect in the tropical zone (Figure 9.18). New Delhi (28° 43′ N; 77° 18′ E) lies in the Ganges lowland at an elevation of 210 m, and Simla (31° 9′ N; 77° 15′ E) is located at 2,200 m in the foothills of the Himalayas. When the average temperature in the hot season exceeds 32 °C in New Delhi, Simla is enjoying a pleasant 18 °C. Notice, however, that the two temperature cycles are quite similar in appearance, with January being the coolest month at both

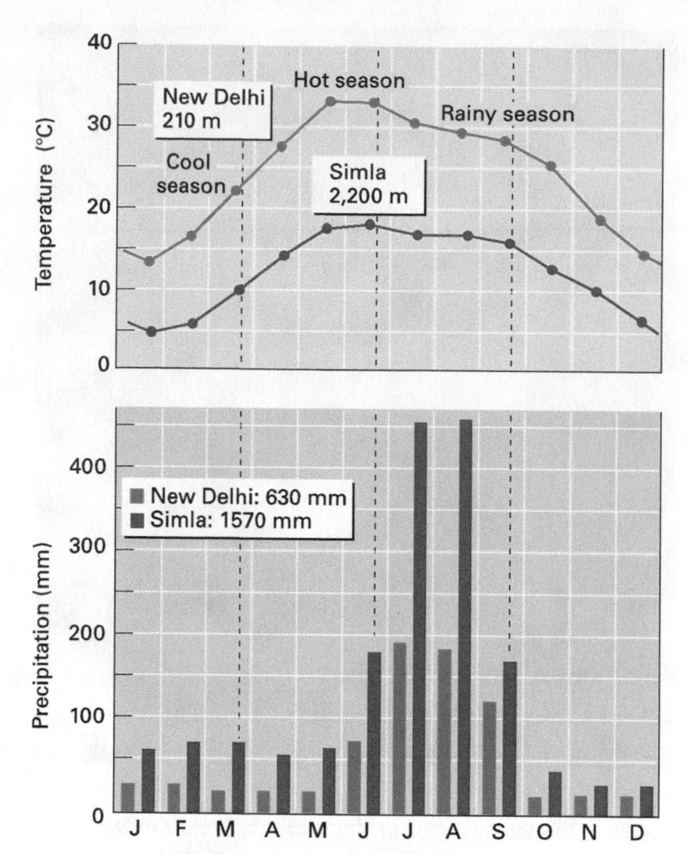

9.18 CLIMOGRAPHS FOR NEW DELHI AND SIMLA Both of these stations are in northern India, but at different elevations. Simla is a welcome refuge from the intense heat of May and June in the Ganges Plain, where New Delhi is located. However, in July and August, Simla is much wetter.

9.19 MOUNT KILIMANJARO, TANZANIA The effect of altitude on tropical climates is dramatically illustrated by Mount Kilimanjaro in Tanzania. *Fly By: 03° 04′ S; 137° 21′ E*

locations. The annual rainfall cycles are also similar. New Delhi shows the typical rainfall pattern of Southeast Asia, with monsoon rains peaking in July and August. Simla has the same pattern, but the amounts are larger in every month, and the monsoon peak is very strong. Simla's annual total rainfall averages 1,570 mm, more than twice that received at New Delhi.

Kilimanjaro, the highest mountain in Africa, and the tallest free-standing mountain in the world, provides some interesting information about highland climates in the tropics. Kilimanjaro is situated at lat. 3° S and rises to 5,895 m from the adjacent plains, which are about 2,000 m above sea level (Figure 9.19). Kilimanjaro is influenced by the passage of the ITCZ, which brings two wet seasons. The highest precipitation occurs in March, April, and May, with a second peak from October to November. The southeastern slopes receive the heaviest precipitation, with totals of about 1,800 mm at 2,400 m elevation. At similar elevations on the north and west slopes, precipitation is about 1,000 mm. Desert-like conditions prevail at the summit, where annual precipitation, in the form of snow, is usually less than 100 mm (rainfall equivalent). Temperatures at the base of the mountain average between 25° and 30 °C, but can drop to −10 °C on the summit. Nighttime frosts are usual at 3,000 m year round.

Kilimanjaro is capped with glaciers, but recent measurements have shown that they have been shrinking very rapidly. Most people attribute this to global climate change, but recent variations in atmospheric moisture patterns have been implicated. In 2003, researchers noted that the ice cover was about 20 percent of what it was in 1912, despite the fact that air temperatures at the summit have remained below freezing. Surprisingly, recession rates were much higher between 1912 and 1953 than those recorded between 1989 and 2003. This suggests that glacier retreat on Kilimanjaro may be an adjustment to climate shifts that occurred in the late nineteenth century.

A Look Ahead

Low-latitude climates include some of the wettest and driest climates of the world, as well as various temperature cycles such as the relatively uniform pattern in equatorial regions to more extreme conditions in the tropical deserts. These climates are dominated by two important features of atmospheric circulation: the ITCZ and the subtropical high-pressure cells. The next chapter discusses the climates of the mid- and high latitudes. The mid- and high-latitude climates are extremely diverse and include regions that experience very wet, very dry, very hot, and very cold conditions at least at certain times of the year. The polar-front boundary and the contrast of warm, moist air masses of the subtropics with colder, drier air masses of the polar regions are the most important factor influencing the temperature and precipitation cycles of the mid- and high-latitude climates.

CHAPTER SUMMARY

- *Low-latitude climates* are located mainly between 30° N and S latitudes, and are controlled by the characteristics and annual movements of the ITCZ and the subtropical highs. These climates range from very moist to very dry, with precipitation distributed throughout the year or exhibiting a pronounced seasonal distribution.
- In the wet equatorial climate, the ITCZ is always nearby, and so rainfall is abundant throughout the year. The annual temperature cycle is very weak; the daily range greatly exceeds the annual range.
- The monsoon and trade-wind coastal climate shows a dry period when the ITCZ has migrated toward the opposite tropic. With the return of the ITCZ, monsoon circulation and enhanced easterly waves provide increased rainfall. Temperatures are highest in the dry weather before the onset of the wet season.
- The wet equatorial and monsoon and trade-wind coastal climates are associated with the *low-latitude rain forest environment.*
- The wet-dry tropical climate has a very dry period at the time of low sun and a wet season at the time of high sun, when the ITCZ is nearby. Temperatures peak strongly just before the onset of the wet season. The *savanna environment* is associated with this climate type.
- Extreme year-to-year variability in rainfall in the tropical wet-dry climate provides a constant threat of extreme drought, especially in the Sahel region of western and north-central Africa. The drought not only causes famine and disease, but also triggers land degradation through overgrazing and fuel wood harvesting.
- The dry tropical climate is dominated by subtropical high pressure and is often nearly rainless. Temperatures are very high during the high-sun season. The *semi-arid* subtype of this climate has a short wet season and supports sparse grasslands; the *arid* subtype represents the tropical desert environment. In the desert, streams flow after heavy rainstorms, which often produce *flash floods.*
- Highland climates of low latitudes normally show a similar seasonality to lowland climates of nearby regions, but are cooler and wetter.

KEY TERMS

dry tropical climate
highland climates
monsoon and trade-wind coastal climate
wet-dry tropical climate
wet equatorial climate

REVIEW QUESTIONS

1. Why is the annual temperature cycle of the wet equatorial climate so uniform?
2. The wet-dry tropical climate has two distinct seasons. What factors produce the dry season? The wet season?
3. Describe how seasonal climate patterns affect the savanna environment.
4. How do the arid (a) and semi-arid (s) subtypes of the dry climates differ?
5. How do low-latitude highland climates differ from their counterparts at low elevation?
6. What is meant by the term *land degradation?* Provide an example of changes that have occurred in a region that has experienced land degradation.
7. Examine the Sahel rainfall graph. Compare the pattern of rainfall fluctuations for the periods 1901 to 1950 and 1951 to 2009. How do they differ?

VISUALIZATION EXERCISE

1. Sketch the temperature and rainfall cycles for a typical station in the monsoon and trade-wind coastal climate. What factors contribute to the seasonality of the two cycles?

ESSAY QUESTIONS

1. The ITCZ moves north and south with the seasons. Describe how this movement affects the four low-latitude climates.
2. Compare and contrast the low-latitude environments of Africa.

WORKING IT OUT • CYCLES OF RAINFALL IN THE LOW LATITUDES

The cycle of monthly precipitation shown in the climographs for low-latitude stations is basically controlled by the sun's declination cycle. This causes a rhythmic variation in the amount of insolation a location receives, which is related to latitude and to the time of the year. The insolation cycle produces an annual pattern of precipitation that is revealed by taking monthly averages over a long period (see *Working It Out • Averaging in Time Cycles* in Chapter 8).

Although this annual cycle is strong at most locations, there are also other cycles that influence rainfall over a sequence of years, rather than from month to month. This may give rise to several "wet" years followed by several "dry" years in a repeating sequence.

Rainfall cycles that span several decades are shown in the three graphs to the right. The graphs present the total rainfall for the wettest month at each location for a sequence of 46 years. The low-latitude stations that have been selected are (a) Padang, Sumatra, in the wet equatorial climate region where the wettest month is November; (b) Mumbai, India, which experiences a monsoon tropical climate and where the wettest month is July; and (c) Abbassia, near Cairo, Egypt, a dry tropical climate location where most rain falls in January. Notice the difference in vertical scale of the Abbassia record.

The data are presented in two ways, using bars and a continuous line that smoothes the rainfall totals. Many of the peak rainfalls appear to be followed by a period of low rainfall. This suggests that there is a rainfall cycle with a length, or period, of between two and three years. This cycle is particularly pronounced at Padang and Mumbai, especially in the early part of the sequence.

Another way to describe a cycle is by its *amplitude*. For a smooth wavelike curve, amplitude is the difference in height between a crest and the adjacent trough. That difference is expressed for these data in millimetres of rainfall. The varying amplitude of the curve is a measure of the *variability* of the precipitation. The difference in amplitude of the cycles at Padang and Mumbai suggests that rainfall at Mumbai has higher variability.

Variability can be measured by the *mean deviation* of all the values in the record. This first requires the calculation of the deviation of each value; this is the difference between the value and the mean of all the values, taken without respect to sign.

Deviation is defined as:

$$D_i = \left| P_i - \bar{P} \right|$$

where D_i is the deviation of P_i, the ith observation in the series; $\bar{P}$ is the mean of the series (see *Working It Out • Averaging in Time Cycles* in Chapter 8). and the two vertical bars enclosing the right side of the equation indicate absolute value. The absolute value is the value taken without respect to sign and is always considered to be positive.

The mean deviation ($\bar{D}$) is:

$$\bar{D} = \frac{1}{n}\sum_{i=1}^{n} D_i = \frac{1}{n}\sum_{i=1}^{n} \left| P_i - \bar{P} \right|$$

where n is the number of observations.

Calculation of the mean and mean deviation for each station for their respective wettest months yields the following results:

Station	Mean (mm)	Mean deviation (mm)
Padang	514.0	129.0
Mumbai	610.0	207.0
Abbassia	6.7	7.2

Of the three stations, Mumbai has the largest mean value for the wettest month at 610 mm for the period of record. Padang has a corresponding mean of 514 mm. Compared with these two stations, Abbassia is very dry, with a mean rainfall for its wettest month of 7 mm. The mean deviations of these stations show that Mumbai has the largest year-to-year fluctuation at 207 mm; Padang is next, followed by Abbassia with a value of 7.2.

Note that if only mean deviation is considered, it appears that rainfall in its wettest month is more or less constant at Abbassia. However, in many years, Abbassia actually receives no rainfall during the wettest month, while in other years more than 30 mm falls.

Comparing the mean deviation with the mean gives a measure of *relative variability*.

Station	Relative Variability
Padang	$129.0 \div 514.0 = 0.25$
Mumbai	$207.0 \div 610.0 = 0.34$
Abbassia	$7.2 \div 6.7 = 1.07$

In other words, the mean deviation during the wettest months is 25 percent of the mean at Padang, 34 percent at Mumbai, and 107 percent at Abbassia. By this measure, Abbassia's rainfall is the most variable, while Padang's is the least variable.

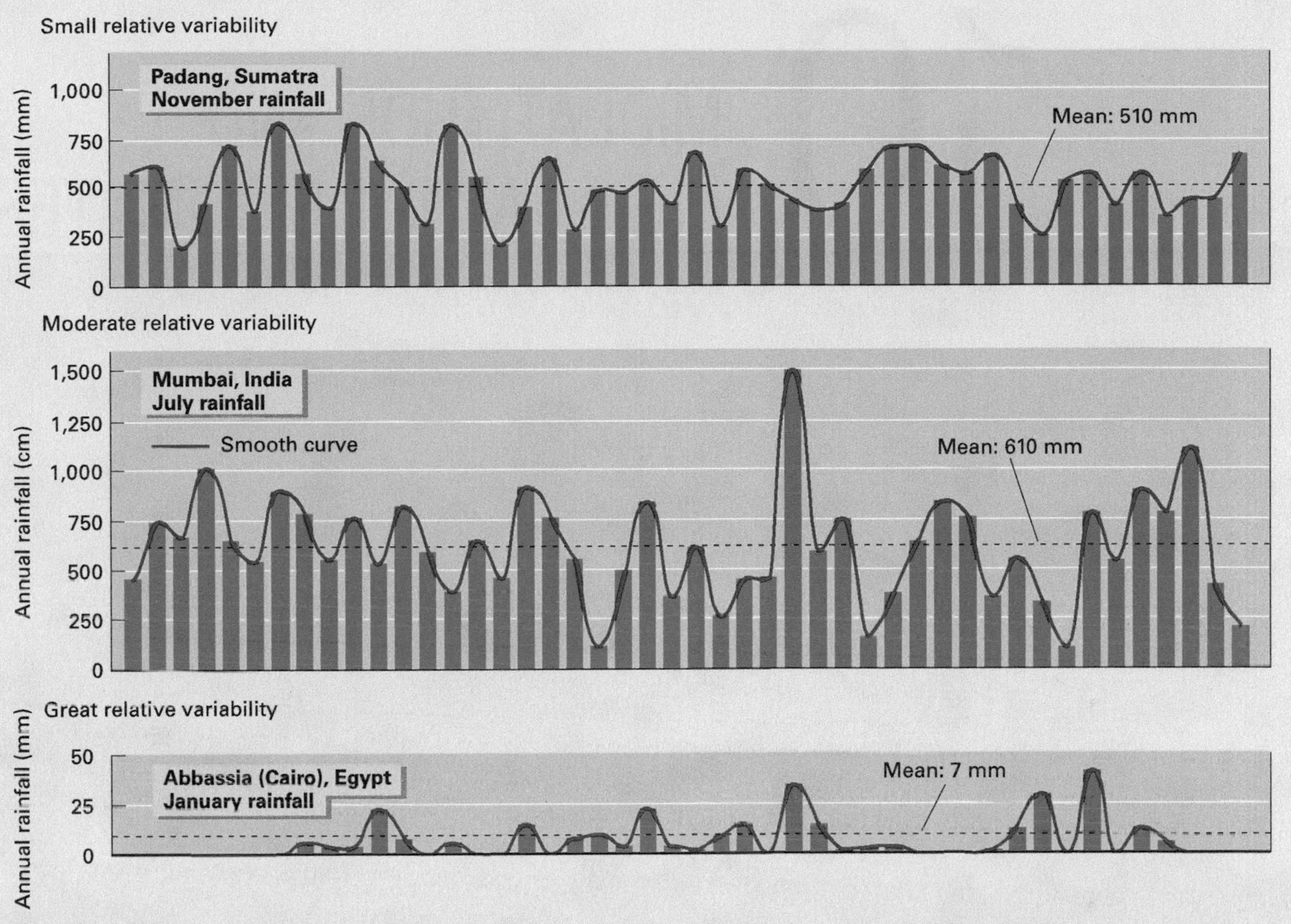

Rainfall in the wettest month at three tropical wet-dry climate stations These data are for Padang, Sumatra; Mumbai, India; and Abbassia, Egypt. They show a tendency for rainfall in the wettest month to rise and fall in a two- to three-year cycle. Large variability is a characteristic of monthly rainfall in many low-latitude climates.

QUESTIONS / PROBLEMS

1. The data below are measurements of total precipitation for November for the years 1985 to 1994 at San Juan, Puerto Rico. (November is the wettest month during this 10-year period.) Find the mean deviation of these data.

Year	Precipitation (mm)
1985	115
1986	149
1987	190
1988	144
1989	126
1990	135
1991	156
1992	304
1993	110
1994	211

2. Calculate the relative variability by taking the ratio of the mean deviation to the mean. How does the result compare with those of the wettest months of Padang (Sumatra), Mumbai (India), and Abbassia (Egypt)?

Find answers at the back of the book.

CHAPTER 10

MIDLATITUDE AND HIGH-LATITUDE CLIMATES

Alpine meadow in Mount Rainier National Park, Washington, U.S.

This chapter continues examining global climates, focusing on midlatitude and high-latitude regions. It explores how latitude, continental location, and the movement of air masses and fronts can explain the temperature and precipitation cycles of the individual climates.

OVERVIEW OF THE MIDLATITUDE CLIMATES

The *midlatitude climates* extend from the subtropical zone to approximately 55° N and S latitudes. Along the western fringe of Europe, midlatitude climates also continue into the subarctic zone, reaching to the 60th parallel. Unlike the low-latitude climates, which are almost equally distributed between northern and southern hemispheres, nearly all of the midlatitude climate area is located in the northern hemisphere. In the southern hemisphere, the land area at latitudes higher than the 40th parallel is so small that the climates are dominated by the great Southern Ocean and do not develop the continental characteristics of their counterparts in the northern hemisphere.

The midlatitude climates of the northern hemisphere lie in a broad zone of intense interaction between two groups of unlike air masses (see Figure 8.8). Maritime tropical (mT) air from the subtropical zone enters the midlatitudes, where it meets with maritime polar (mP) and continental polar (cP) air along the discontinuous and shifting polar front.

In terms of prevailing pressure and wind systems, the midlatitude climates include the poleward halves of the great subtropical high-pressure systems and much of the belt of prevailing westerly winds (see Figure 5.12). As a result, weather systems, such as travelling low-pressure systems and their associated fronts, characteristically move from west to east. This predominantly zonal airflow influences the distribution of climates across the North American and Eurasian continents.

The interaction between warm, moist air masses and cooler, drier air masses produces climates that are quite variable from day to day. The *Working It Out • Standard Deviation and*

Coefficient of Variation feature at the end of this chapter uses monthly precipitation from stations in two different midlatitude climates to show how variability is measured.

The midlatitude climates range from those with strong wet and dry seasons to those with precipitation that is more or less uniformly distributed through the year (Figure 10.1). Temperature cycles also are quite varied. The windward west coasts experience low annual ranges. In contrast, annual ranges are large in the continental interiors. Table 10.1 summarizes the important features of these climates.

THE DRY SUBTROPICAL CLIMATE (KÖPPEN *BWh, BWk, BSh, BSk*)

The **dry subtropical climate** is a poleward extension of the dry tropical climate, but because of the midlatitude location, the annual temperature range is greater. There is a distinct cool season at the time of low sun, which is accentuated by invasions of cold continental polar (cP) air masses that also bring frontal storms to the region. As is the case for the dry tropical climate, the dry subtropical climate type has both arid and semi-arid subtypes.

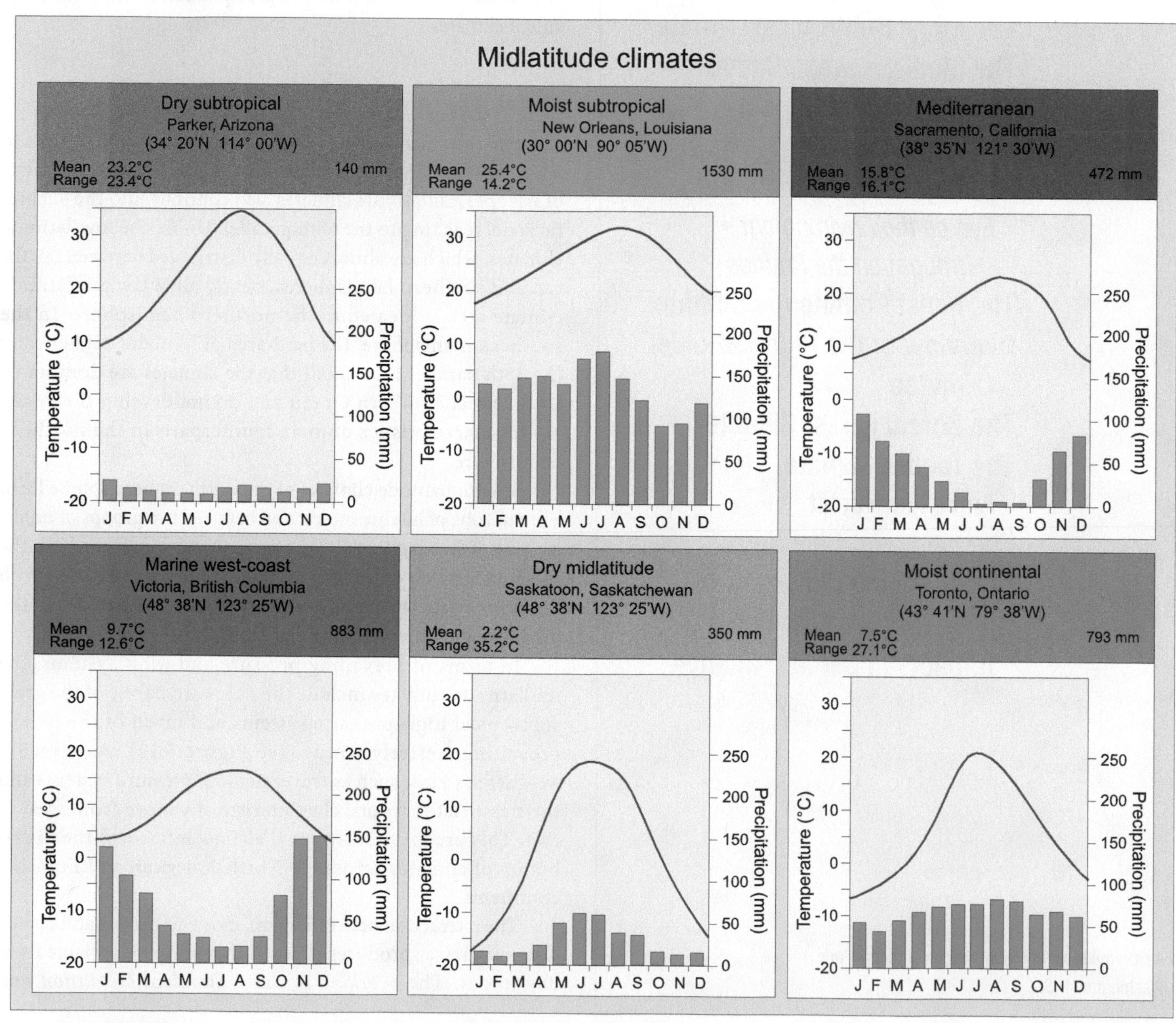

10.1 CLIMOGRAPHS FOR THE SIX MIDLATITUDE CLIMATES The midlatitudes exhibit several temperature and precipitation regimes, which contribute to the distinctive climates of this region.

TABLE 10.1 MIDLATITUDE CLIMATES

CLIMATE	TEMPERATURE	PRECIPITATION	EXPLANATION
Dry subtropical	Temperatures show a distinct cool or cold season at low-sun period.	Precipitation is low in nearly all months.	This climate lies poleward of the subtropical high-pressure cells and is dominated by dry cT air most of the year. Rainfall occurs when moist mT air reaches the region, either in summer monsoon flows or in winter frontal movements.
Moist subtropical	Temperatures show a strong annual cycle, but with no winter month below freezing.	There is abundant rainfall, cyclonic in winter and convectional in summer. Humidity is generally high.	The flow of mT air from the west sides of subtropical high-pressure cells provides moist air most of the year. cP air may reach this region during the winter.
Mediterranean	Temperature range is moderate, with warm to hot summers and mild winters.	Precipitation shows an unusual pattern of wet winter and dry summer. Overall, the area is drier when nearer to subtropical high pressure.	The poleward migration of subtropical high-pressure cells moves clear, stable cT air into this region in the summer. In winter, cyclonic storms and polar frontal precipitation reach the area.
Marine west-coast	Temperature cycle is moderated by marine influence.	There is abundant precipitation, but with a winter maximum.	Moist mP air, moving inland from the ocean to the west, dominates this climate most of the year. In the summer, subtropical high pressure reaches these regions, reducing precipitation.
Dry midlatitude	This area has a strong temperature cycle with a large annual range. Summers are warm to hot; winters are cold to very cold.	Precipitation is low in all months, but usually shows a summer maximum.	This climate is dry because of its interior location, far from mP source regions. In winter, cP dominates. In summer, a local dry continental air mass develops.
Moist continental	Summers are warm and winters are cold, with three months below freezing. The area has a very large annual temperature range.	There is ample precipitation, with a summer maximum.	This climate lies in the polar-front zone. In winter, cP air dominates, while mT invades frequently in summer. Precipitation is abundant, cyclonic in winter and convectional in summer.

The global distribution of the dry subtropical climate was shown previously in Figure 9.14. In North America, this climate occurs in the Mojave and Sonoran deserts of the American southwest and northwest Mexico. In South America, the dry subtropical climate extends south into Argentina. Elsewhere in the southern hemisphere, it is found in southern Australia and southern Africa. In North Africa, the dry subtropical climate stretches as a discontinuous band north of the Sahara desert from Morocco, through Algeria, Libya, and Egypt, into Syria and northern Iraq.

The climograph for Yuma, Arizona (32° 39′ N; 114° 36′ W), illustrates the arid subtype of the dry subtropical climate (Figure 10.2). Monthly temperatures show a strong seasonal cycle with hot summers and cool winters. Here the mean temperature in July, the warmest month, is 33.7 °C, compared to 13.2 °C in December, the coldest month, giving an annual temperature range of 20 °C. Freezing temperatures can be expected at night in December and January. Annual precipitation, which totals about 85 mm, is low in all months, but has peaks in late winter and late summer. Midlatitude depressions produce higher rainfalls from December through March. May and June are nearly rainless, followed by the August maximum caused by the invasion of maritime tropical (mT) air masses, which bring thunderstorms to the region.

The shift from a dry tropical to a dry subtropical climate type is gradual; however, the effect of the cooler, moister winters in midlatitudes becomes increasingly apparent in the landscapes with the establishment of denser and more varied plant covers.

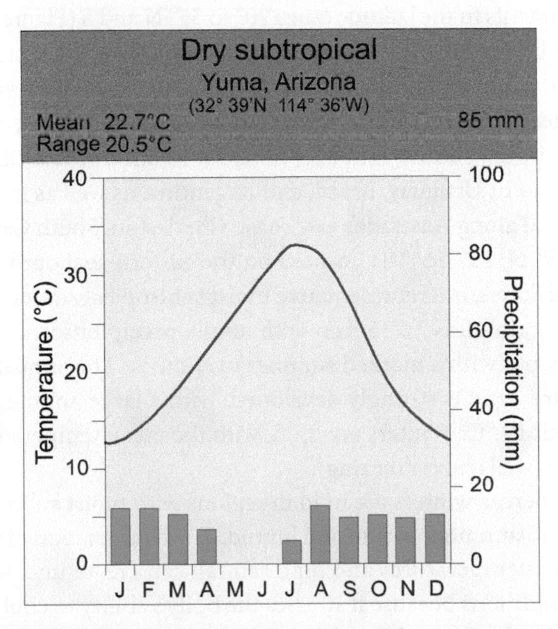

10.2 DRY SUBTROPICAL CLIMATE Yuma, Arizona, has a strong seasonal temperature cycle. Compare with Lima, Peru (Figure 9.14).

Although many soils in subtropical deserts are naturally saline, they may support an open cover of trees and shrubs. Deep-rooting species help to mitigate the problem of salinity by keeping the water table many metres below the surface.

Replacing native species with shallow-rooted crops can increase soil salinity through the process of *dryland salinization*. The rise in the water table that results when shallow-rooted crops are planted brings dissolved salts upward into the surface soils where it accumulates through evaporation. *Irrigation salinization* occurs when irrigation water soaks through the soil and causes the water-table to rise; deterioration in the quality of the water supply exacerbates the problem.

THE MOIST SUBTROPICAL CLIMATE (KÖPPEN *Cfa*)

Circulation around subtropical high-pressure cells provides a flow of warm, moist maritime air onto the eastern side of continents (see Figure 5.12). This flow of mT air dominates the **moist subtropical climate**. In this climate, summer is usually a time of abundant rainfall, much of it convectional. Occasional tropical cyclones augment summer precipitation. There is also a strong monsoon effect in Southeast Asia, with much more rainfall in summer than in winter. Summer temperatures are warm, with persistent high humidity.

Winter precipitation in the moist subtropical climate is similarly plentiful, and is associated with the passage of midlatitude cyclones. Invasions of cP air masses are frequent in winter, bringing spells of below-freezing temperatures. However, no winter month has a mean temperature below 0 °C.

The moist subtropical climate is found on the eastern sides of continents in the latitude range 20° to 35° N and S (Figure 10.3). In the United States, this climate covers most of the southeast, from the Carolinas to eastern Texas. Another extensive area includes southern China, Taiwan, and the southernmost part of Japan. In the southern hemisphere, moist subtropical climates are found in Uruguay, Brazil, and Argentina, as well as in a narrow band along Australia's east coast. Charleston, South Carolina (32° 47′ N; 79° 56′ W), located on the eastern seaboard of the United States, is a representative moist subtropical station. Total annual rainfall is 1,235 mm with ample precipitation in every month, but with a marked summer maximum. The annual temperature cycle is strongly developed, with a large annual range of about 19 °C. Winters are mild, with the mean temperature in January well above freezing.

Whereas winters are mild in regions with moist subtropical climates, summers are hot and humid. The combination of warm summer temperatures and high rainfall can create uncomfortable conditions because it stresses the body's ability to cool itself. Several indices have been developed to describe seasonal conditions based on human perceptions.

The comfort index or humidex (Figure 10.4) is derived from temperature and relative humidity readings, and is subjectively based on people's apparent comfort or discomfort under given conditions. This index provides an "apparent temperature"; that is, what the air temperature feels like when the effect of humidity is taken into account. The comfort index makes use of the fact that humans maintain body temperature within proper physiological limits by perspiration, which has a cooling effect as the water evaporates from the surface of the skin. High humidity levels will reduce the rate of evaporation and so limit this cooling process. Unfavourable humidex conditions can lead to heat exhaustion or heat stroke. Oppressive conditions can occur throughout much of North America, but are especially frequent in the moist subtropical regions of the United States.

Other derived temperature indices include cooling degree days and heating degree days, both of which are widely used in engineering and the building trades. Cooling degree days represent the cumulative number of degrees in a month or year in which the mean temperature is above 18.3 °C. The reference temperature of 18.3 °C is generally considered to be a comfortable inside temperature for a building. Cooling degree days are therefore used to assess the potential need for air conditioning. Heating degree days are similarly calculated from a base temperature of 18.3 °C, but in this case the number of days that the mean air temperature is lower than 18.3 °C is used. In the southeastern United States, annual cooling degree days typically exceed 2,400, indicating a high potential energy demand for air conditioning. Conversely, the mild winter reduces potential demand for heating with the annual heating degree day index in the general range of 1,500 to 2,000 (Table 10.2); selected Canadian locations are listed for comparison.

The abundant rainfall of the moist subtropical environment supplies rivers and streams throughout the year; there is no period of water shortage (Figure 10.5). Flooding is not uncommon when torrential rains accompany the passage of tropical cyclones through this region. Also, in areas where the natural forest cover has been removed, excessive runoff has led to severe soil degradation (Figure 10.6), and the heavy sediment loads carried by the rivers seriously impair water quality. Reforestation and other forms of erosion control are used extensively in an effort to combat the problem.

THE MEDITERRANEAN CLIMATE (KÖPPEN *Csa, Csb*)

The **Mediterranean climate** is located along the west coasts of continents in the latitude range 30° to 45° N and S (Figure 10.7). It is distinguished by its annual precipitation cycle, which has a wet winter and very dry summer. This is caused by the poleward movement of the subtropical high-pressure

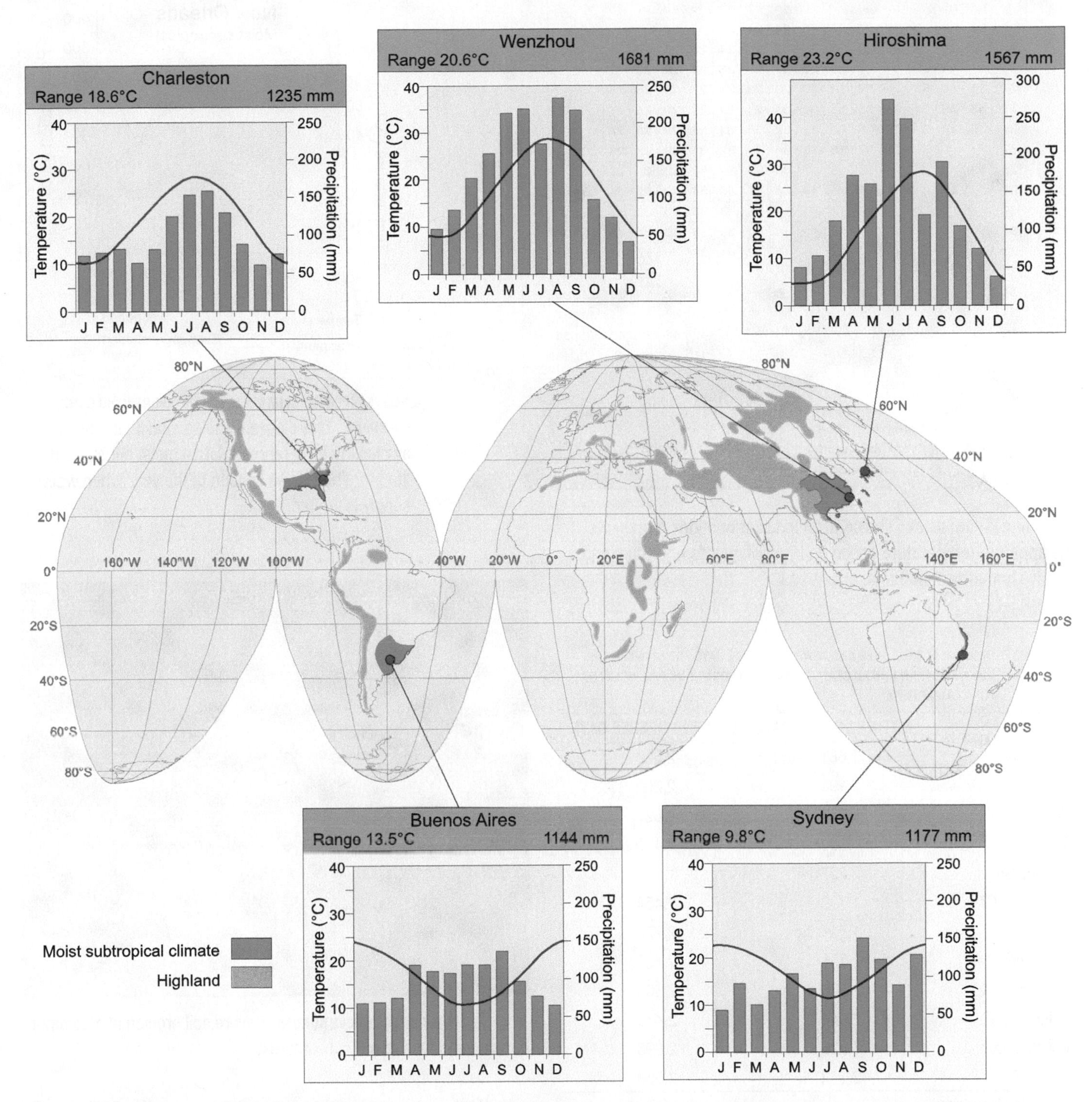

10.3 WORLD DISTRIBUTION OF THE MOIST SUBTROPICAL CLIMATE The moist subtropical climate is found on the eastern sides of continents; in general, the latitude range is 20° to 35° N and S. The combination of warm temperatures and high humidity due to plentiful precipitation can sometimes lead to uncomfortable conditions in the summer months.

Relative humidity (%)

Temperature (°C)	100	95	90	85	80	75	70	65	60	55	50	45	40	35	30	25	20
21	29	29	28	27	27	26	26	24	24	23	23	22					
22	31	29	29	28	28	27	26	26	24	24	23	23					
23	33	32	32	31	30	29	28	27	27	26	25	24	23				
24	35	34	33	33	32	31	30	29	28	28	27	26	26	25			
25	37	36	35	34	33	33	32	31	31	31	29	28	28	27			
26	39	38	37	36	35	34	33	32	31	31	29	28	28	27			
27	41	40	39	38	37	36	35	34	33	32	31	30	29	28	28		
28	43	42	41	41	39	38	37	36	35	34	33	32	31	29	28		
29	46	45	44	43	42	41	39	38	37	36	34	33	32	31	30		
30	48	47	46	44	43	42	41	40	38	37	36	35	34	33	31	31	
31	50	49	48	46	45	44	43	41	40	39	38	36	35	34	33	31	
32	52	51	50	49	47	46	45	43	42	41	39	38	37	36	34	33	
33	55	54	52	51	50	48	47	46	44	43	42	40	38	37	36	34	
34	58	57	55	53	52	51	49	48	47	45	43	42	41	39	37	36	
35		58	57	56	54	52	51	49	48	47	45	43	42	41	38	37	
36			58	57	56	54	53	51	50	48	47	45	43	42	40	38	
37					58	57	55	53	51	50	49	47	45	43	42	40	
38							57	56	54	52	51	49	47	45	43	42	40
39									56	54	53	51	49	47	45	43	41
40										57	54	52	51	49	47	43	41
41											56	54	52	50	48	46	44
42												56	54	52	50	48	46
43													56	54	51	49	47

20-29	No discomfort
30-39	Some discomfort
40-45	Great discomfort
>45	Dangerous; possible heat stroke

10.4 THE COMFORT INDEX The comfort index or humidex represents the apparent temperature based on prevailing air temperature and relative humidity conditions.

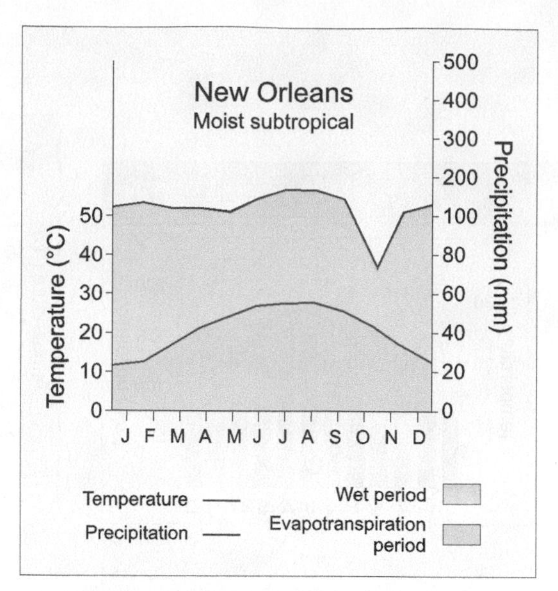

10.5 SIMPLIFIED CLIMATE DIAGRAM FOR NEW ORLEANS, LOUISIANA This diagram shows seasonal moisture availability for a representative moist subtropical station. There is no season of the year when water supply is limited.

TABLE 10.2 COOLING DEGREE DAYS AND HEATING DEGREE DAYS FOR CITIES IN THE SOUTHEASTERN UNITED STATES. CANADIAN LOCATIONS ARE INCLUDED FOR COMPARISON.

	HEATING DEGREE DAYS	COOLING DEGREE DAYS
Mobile, AL	1,681	2,539
Montgomery, AL	2,194	2,252
Jacksonville, FL	1,354	2,627
Tampa, FL	591	3,482
Atlanta, GA	2,827	1,810
Savannah, GA	1,799	2,454
New Orleans, LA	1,417	2,773
Shreveport. LA	2,251	2,405
Charleston, SC	2,005	2,306
Columbia, SC	2,044	2,475
Houston, TX	1,525	2,893
Dallas, TX	2,370	2,568
Canadian locations		
Vancouver, BC	2,926	44
Winnipeg, MB	5,777	186
Toronto, ON	4,066	252
Halifax, NS	4,367	104
Yellowknife, NWT	8,256	41
Resolute, NU	12,526	0

10.6 THE EFFECT OF EXCESSIVE RUNOFF Severe soil erosion is a common hazard in moist subtropical climates.

cells during the summer season, which brings dry cT air into the region. In winter, moist mP air masses bring low pressure systems and ample precipitation.

In terms of total annual precipitation, the Mediterranean climate type varies from arid to humid, depending on location. Generally, the closer an area is to the tropics, the stronger the influence of the subtropical high-pressure cells, and thus the drier the climate. Annual precipitation ranges from about 275 to

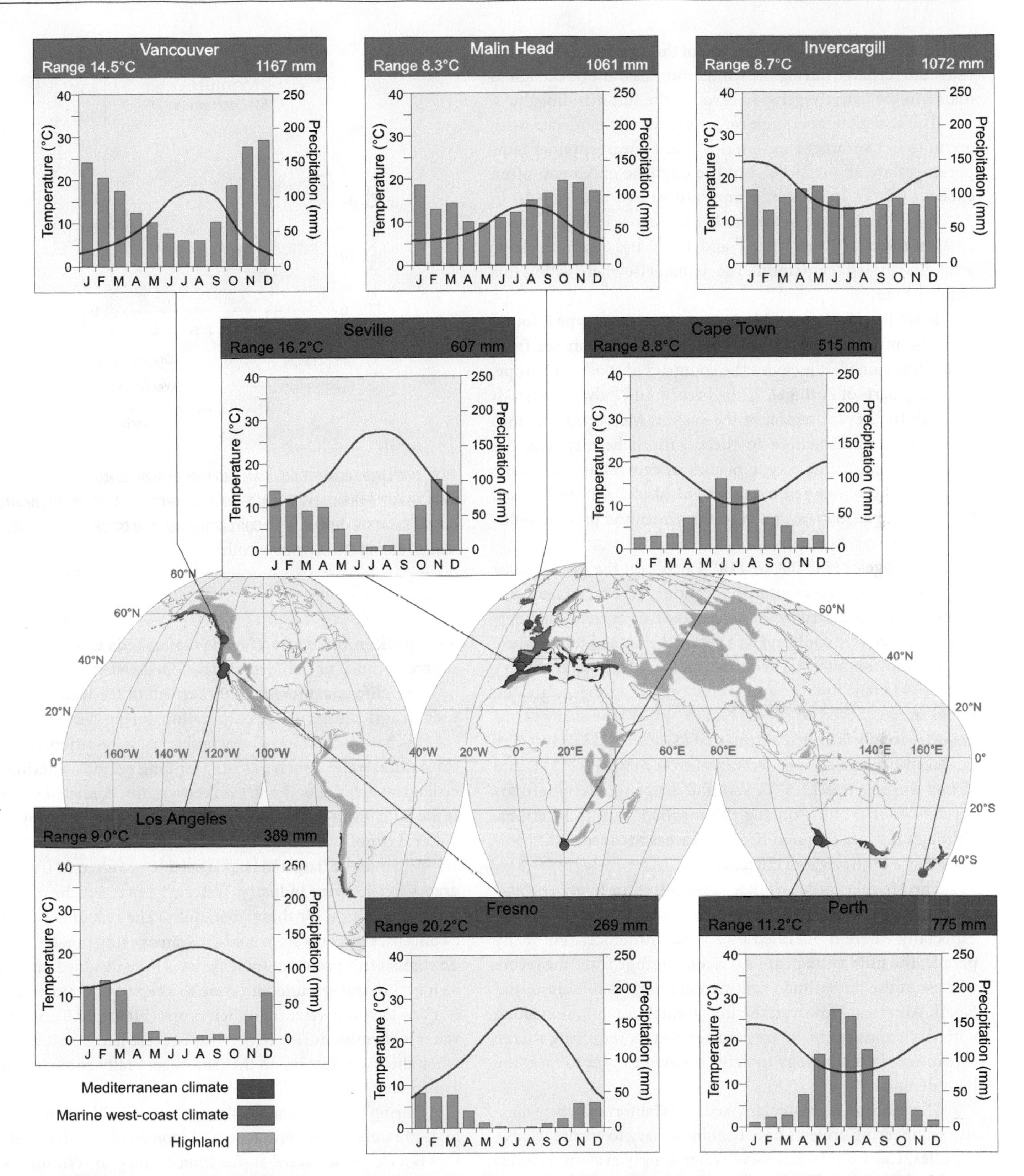

10.7 WORLD DISTRIBUTION OF THE MEDITERRANEAN AND MARINE WEST-COAST CLIMATES Both of these climate types are found on the west coast of the continents. The Mediterranean climate with its distinctive dry summer period is replaced poleward by the wet marine west-coast climate, although here too, precipitation usually exhibits a winter maximum.

900 mm, with more than 65 percent of the total falling in winter. Moisture recharge during the winter months must be enough to maintain the native vegetation through the summer drought.

The annual temperature range is generally moderate, with warm to hot summers and mild winters. Mean summer temperatures are about 25 °C, but the daytime maximum often exceeds 35 °C. In winter, temperatures average from 10 to 12 °C, but there is an occasional risk of frost in higher areas. Coastal zones between lat. 30° and 35° N and S typically have a smaller annual temperature range than elsewhere within this climate type.

More than half of the Mediterranean climate type is found in a discontinuous belt around the Mediterranean Sea from which it is named. It includes the countries of southern Europe, including parts of Portugal, Spain, France, and Italy, and extends through the Levant region of the eastern Mediterranean into North Africa. Elsewhere in the northern hemisphere, the Mediterranean climate type occurs in central and southern California. Locations in the southern hemisphere include coastal Chile, the Cape Town region of South Africa, and parts of southern and western Australia.

Los Angeles (34° 00′ N; 118° 15′ W) on the Pacific coast of California is a representative location. The annual temperature cycle is moderated by the cold California current. Summers are comparatively cool, and as a result the annual temperature range is only about 9 °C. In California's Central Valley, the dry summers persist, but the annual temperature range is greater. For example, at Fresno (36° 43′ N; 119° 47′ W) the average temperature in July is 27.3 °C compared to 20.6 °C at Los Angeles; in December, the average temperature at Fresno is 7.3 °C and at Los Angeles it is 13.8 °C. Rainfall drops to nearly zero for three or four months during the summer at both locations, although fogs are frequent in coastal areas because of the cooling effect of the offshore current.

The Mediterranean climate is attractive for human habitation, mainly because of year-round pleasant temperatures, especially where moderated by coastal influences. For many people, the mild winters are a welcome refuge from the severe winters of the midlatitude continental interiors of Eurasia and North America. However, the low annual precipitation along with dry summers make fresh water scarce, requiring a large investment in technology to deliver enough water to meet the high demand (Figure 10.8).

The problem is particularly acute in California, where massive engineering projects have been necessary to ensure reliable supplies. California's extensive water supply system includes reservoirs, groundwater basins, and interregional conveyance facilities in the form of irrigation canals and pipelines, all of which are needed to reduce short-term water shortages.

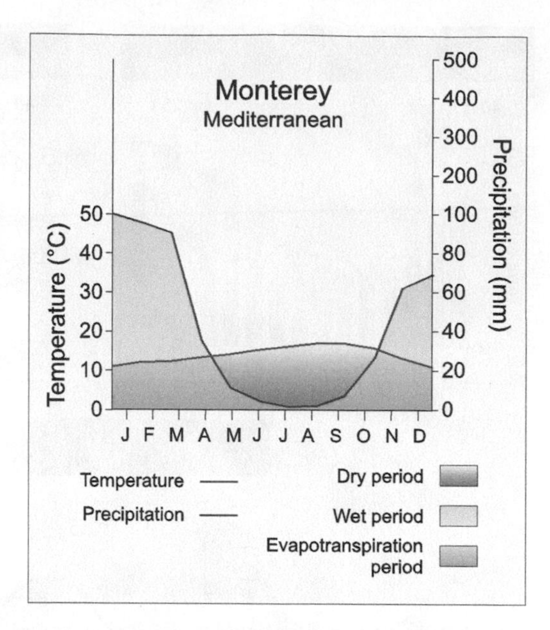

10.8 SIMPLIFIED CLIMATE DIAGRAM SHOWING SEASONAL MOISTURE AVAILABILITY FOR MONTEREY, CALIFORNIA, A REPRESENTATIVE MEDITERRANEAN CLIMATE STATION Irrigation is commonly used to offset the lack of rainfall during the summer months.

Snowpack in the Sierra Nevada watersheds is an important source of much of California's developed water supply.

In California, more than 90 percent of the water supply is used in agriculture, where evaporation during the hot summer months has resulted in salinity problems. In addition, many of the natural water courses are dry for long periods, altering the ecological balance of the stream ecosystems. A related concern is the high levels of chemicals carried into the streams by agricultural runoff.

Much of the irrigated land is used to grow citrus fruits and grapes for the wine industry. Water supply is not the only factor that is critical for these operations. The risk of frost is also a concern, especially from cold air drainage into low-lying sites. To combat this problem, some farmers have installed machines to mix the near-ground air layers to keep temperatures above 0 °C, or to spray water onto their crops (Figure 10.9). Spraying water is effective during light frosts because latent heat released when the water freezes on the fruit is generally enough to prevent injury.

During summer, the persistent flow of stable air out of the high-pressure centres brings several months of hot, dry weather. Fire is a frequent hazard at this time of the year. Wildfire is an integral part of all Mediterranean environments, with an average frequency in most regions of 20 to 30 fires per year. The relatively dense canopy of fine stems and the volatile oils in the leaves of

10.9 THE RISK OF FROST (a) Wind machines disperse frost pockets that may develop in areas susceptible to cold air drainage. (b) Latent heat released during freezing can prevent frost injury in sensitive crops.

many of the native species make the vegetation very flammable. The fires burn fiercely and tend to spread rapidly, especially where they are fanned by hot, dry winds such as the Santa Ana (see Chapter 5). Fires as large as 10,000 hectares are frequently reported. However, the native vegetation is well adapted to fire, and most plants renew their growth from buried rootstocks or germinate from fire resistant seeds (Figure 10.10).

Another weather-related hazard in Mediterranean climate regions is excessive soil erosion that can occur when the rains return in winter. This can be a serious problem when the vegetation cover has been burned off during the previous summer. Also, overgrazing, particularly by goats, has increased the probability of soil loss, especially in southern Europe. During the rainy winter months, large quantities of coarse mineral debris are swept down the slopes and carried long distances by the flooding streams.

Rainfall in the Mediterranean climate can be quite variable from year to year. Sometimes the weather patterns that provide winter precipitation fail to appear, leading to drought, while in other years, precipitation may be much greater than normal. In California, attempts have been made to link annual rainfall to large-scale climatic patterns associated with the Southern Oscillation Index (SOI) and El Niño events, and with the Pacific Decadal Oscillation (PDO). The PDO index is calculated monthly from sea surface temperatures in the North Pacific. It has a time cycle of 20 to 30 years in which ocean temperatures shift between warm and cool phases. This, in turn, affects the track and properties of air masses moving into California from the Pacific Ocean. However, neither the SOI nor the PDO provides a reliable signal of rainfall patterns in this region, although there is a tendency for precipitation to increase during El Niño years and during warm phases of the PDO.

10.10 FIRE IS A FREQUENT THREAT IN MEDITERRANEAN ECOSYSTEMS (a) Cork oak is well protected by its thick bark, which in this specimen has been recently stripped for commercial use. (b) New branches sprouting in fire-damaged Eucalyptus in Australia. (c) Regrowth from the undamaged root tissue of burned shrubs in California.

THE MARINE WEST-COAST CLIMATE (KÖPPEN *Cfb, Cfc*)

The **marine west-coast climate** occurs in midlatitude west coasts where prevailing westerlies bring moist air onshore from the adjacent oceans. Precipitation comes mainly from low-pressure systems. Where the coast is mountainous, annual precipitation is increased substantially by the orographic effect. Precipitation is plentiful in all months, but there is often a distinct winter maximum. In summer, the poleward extension of the subtropical high-pressure system leads to a reduction in rainfall. The annual temperature range is relatively small for midlatitudes. The marine influence keeps winter temperatures mild, compared with inland locations at equivalent latitudes.

The general latitude range of this climate is 35° to 60° N and S (see Figure 10.7). Locations include the Pacific northwestern coast of North America, the British Isles, and parts of western Europe. In the southern hemisphere it is associated with New Zealand, the southernmost parts of Australia, particularly Tasmania, and the Chilean coast south of 35° S.

Vancouver, British Columbia, is a typical marine west-coast station. Annual precipitation at Vancouver averages 1,167 mm, with most of it falling during the winter months. Depending on latitude, much of the winter precipitation can fall as snow, but this will usually melt fairly quickly because of the mild temperatures. An average snow depth of 1 cm is reported at Vancouver in December, January, and February. Mean temperatures in the winter months remain above freezing; the January mean is 3.3 °C. August is the warmest month with a mean temperature of 17.6 °C, giving an annual temperature range of 14.3 °C. However, frost is not uncommon in Vancouver; below freezing temperatures have been recorded from October to April, with extreme minimum temperatures falling to −17.8 °C. Similarly, unusually warm summers occur; the extreme maximum temperature at Vancouver is 33.3 °C.

Water is an abundant natural resource in all marine west-coast environments (Figure 10.11). The mountainous terrain is also suited to hydroelectric power, and many of the rivers have been dammed to use the enormous water surpluses that run to the sea. One of the most ambitious projects in Canada was the Kenny Dam. It was built in the 1950s to impound and reverse the flow of the Nechako River to supply electricity to the Alcan smelter in Kitimat, British Columbia. The water now passes through a 16-km tunnel under Mount Dubose to the power-generating facility.

Despite control structures on many of the rivers, unusually heavy or prolonged precipitation events can result in flooding in low-lying areas and cause disruption of roads and railways (Figure 10.12). In addition, high turbidity can lead to problems with water quality. This was the case in November 2006, when about a million residents in British Columbia's Lower Mainland region were instructed to boil their drinking water.

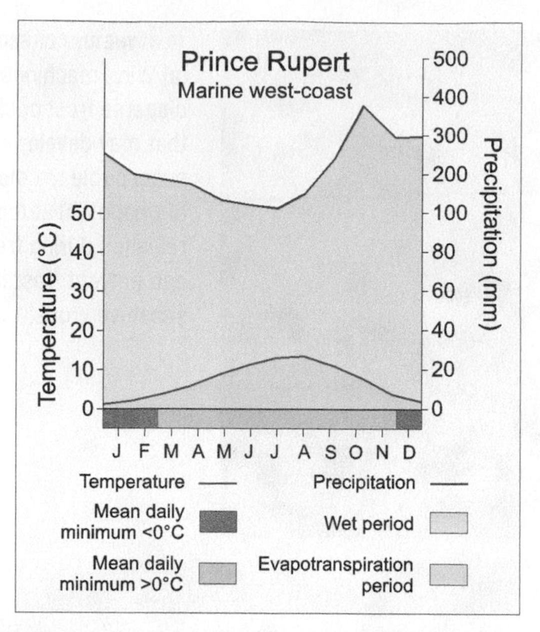

10.11 SIMPLIFIED CLIMATE DIAGRAM SHOWING SEASONAL MOISTURE AVAILABILITY AT PRINCE RUPERT, A REPRESENTATIVE MARINE WEST-COAST CLIMATE STATION Water is an abundant natural resource; in these regions there is a year-round water surplus.

10.12 THE DESTRUCTION CAUSED BY FLOODING Heavy rains commonly cause flooding and infrastructure damage in the wet west-coast marine environment. Flood plain mapping has helped to minimize flood losses in susceptible areas.

In western North America, southern Chile, and the South Island of New Zealand, much of the land is mountainous and has been heavily scoured by glaciers. Hence, the soils are poorly developed. Dense needle-leaf forests of fir, cedar, hemlock, and spruce flourish in the wet mountainous coastal areas of British Columbia (Figure 10.13). They are important lumber trees and are considered one of the greatest timber resources in the world. Much of the world's old-growth forests have been destroyed, and the rain forest of the Pacific Northwest is one of the few regions

10.13 NEEDLE-LEAF FOREST Douglas fir, hemlock, and cedar grow to great sizes in protected environments. Ferns and mosses provide a lush ground cover.

that still contain relatively undisturbed areas. Clear-cutting is the harvest method most widely used in British Columbia. Trees are removed from patches of forest some 50 to 60 hectares in size. Regulations are in place to minimize the environmental impact of clear-cutting—for example, fish-bearing rivers and streams should be protected by riparian leave strips. However, such forest protection regulations are sometimes difficult to enforce. Logging on steep slopes in this wet environment causes soil erosion problems, and slumping and landslides are common. Helicopter logging is carried out in more inaccessible regions to minimize environmental problems from road construction.

THE DRY MIDLATITUDE CLIMATE (KÖPPEN *BWk, BSk*)

The **dry midlatitude climate**, at a latitude range of about 35° to 50° N, is limited almost exclusively to the interior regions of North America and Eurasia. Typically the region lies in the rain shadow of mountain ranges. Maritime air masses are effectively blocked out much of the time. In winter, the climate is dominated by cP air masses; in summer, a dry continental air mass of local origin is dominant. Summer rainfall is mostly convectional and is caused by occasional invasions of maritime air. The temperature cycle is strongly developed, with a large annual range. Summers are warm to hot, followed by cold to very cold winters.

The largest expanse of the dry midlatitude climate is in Eurasia; it stretches from the southern republics of the former Soviet Union to the Gobi desert and into northern China (see Figure 9.14). In the centre of this region lie true deserts of the *arid* climate subtype where precipitation is very low. The dry western interior regions of North America, including the Great Basin, Columbia Plateau, and the Great Plains, are of the *semi-arid* subtype. In the southern hemisphere, a small area of dry midlatitude climate is found in southern Patagonia, near the tip of South America.

The arid subtype of the dry midlatitude climate is restricted to the driest of midlatitude continental interiors. In North America, this cold desert environment supports a cover of sagebrush and associated low woody shrubs. Some of the drier valleys in the interior of British Columbia are of this subtype (Figure 10.14a). The semi-arid or steppe subtype of the dry midlatitude climate occurs

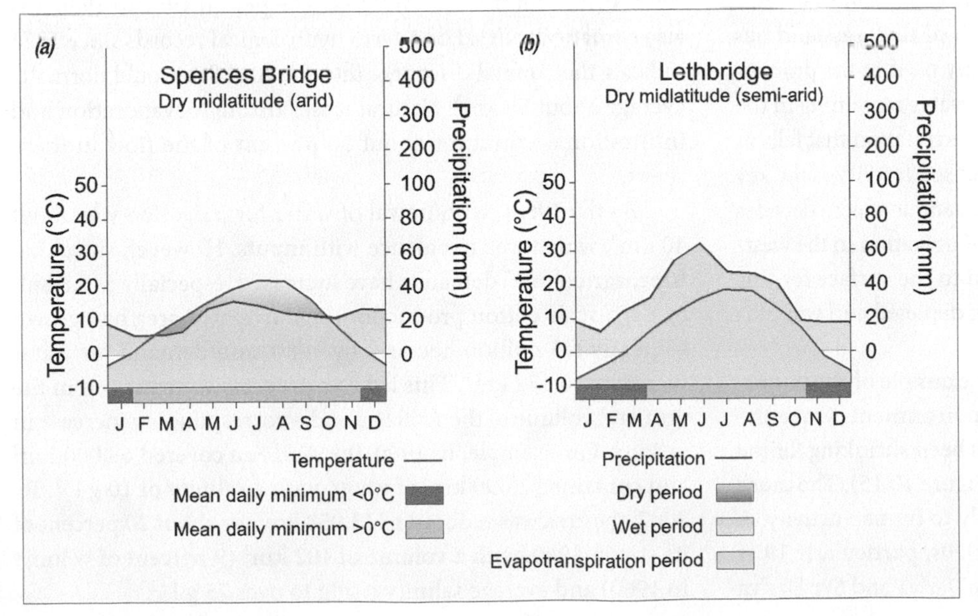

10.14 SIMPLIFIED CLIMATE DIAGRAMS SHOWING SEASONAL MOISTURE AVAILABILTY FOR DRY MIDLATITUDE STATIONS (a) The prolonged water deficit at Spences Bridge, British Columbia, is characteristic of the arid subtype; (b) Lethbridge, Alberta, is characteristic of the semi-arid steppe subtype in which a water deficit develops in the latter part of the summer. Note also the periods of frost that distinguish the dry midlatidude climates from their subtropical and tropical counterparts.

more extensively and is found in large regions of interior North America and central Asia. The dominant natural vegetation comprises hardy perennial grasses that can tolerate the effects of low annual precipitation, combined with a strongly continental thermal regime, such as occurs at Lethbridge, Alberta (Figure 10.14b).

Lethbridge, Alberta (49° 42′ N; 112° 50′ W) experiences a semi-arid climate because it lies in the rain shadow of the Rocky Mountains. Total annual precipitation here is 386 mm. Most of this comes in the form of convectional summer rainfall. Winter snowfall is light, averaging 130 cm, and is reported on about 50 days each year. Average snow depth at month-end varies from 1 to 6 cm between October and March. The temperature cycle has a large annual range, with warm summers and cold winters. January, the coldest month, has a mean temperature of −8 °C, although the record low is −43 °C, which occurred in 1950. The warmest months are July and August, when the average daily maximum temperature is about 25.5 °C, with record highs above 39 °C.

Annual precipitation in the dry midlatitude climate is quite variable and there may be large differences in precipitation from year to year. Drought is an ever-present threat in the dry midlatitude climates, and its occurrence has serious implications for the region. When dry conditions prevail for several years, the drought may be intense. This was the case in the mid-1930s when a series of drought years caused severe hardships in the central and western Great Plains of North America. During the drought, many centimetres of soil were removed from the barren, unprotected fields. Exceptionally intense dust storms transported the finer material out of the region, while the coarser silt and sand particles accumulated in drifts along fence lines and around buildings. The affected area became known as the Dust Bowl. *Eye on the Environment • Drought on the Prairies* examines climate conditions during recent drought episodes in the Canadian Prairies.

In North America and Eurasia, much of the grassland has been converted to cropland, which is mainly used in the production of spring wheat. The crops depend on water remaining in the soil from winter rains and snows and the precipitation that falls in late spring and early summer. Soil water, not soil fertility, is the key to wheat production over these vast regions, and in recent decades there has been a great increase in the use of irrigation in the western Great Plains. Groundwater is pumped to the surface for this purpose, but the reserves are rapidly being depleted and will ultimately be gone.

The fate of the Aral Sea is a dramatic example of the importance of water in the dry midlatitude environment. Once the world's fourth largest lake, the Aral Sea has been shrinking for the last 40 years because of reduced inflows (Figure 10.15). The cause is partly climatic, but is also linked directly to human activity. A series of dry years that occurred in the 1970s, particularly 1974 to 1975, lowered discharge from the Amu Dar'ya and Syr Dar'ya rivers that drain into the Aral Sea. The period 1982 to 1986 was also climatically dry. Long-term hydrological records since 1926 indicate that annual discharge into the Aral Sea would normally average about 55 km^3. Natural losses through evaporation and infiltration account for about 50 percent of the flow in these rivers.

10.15 THE ARAL SEA Once the world's fourth largest lake, the Aral Sea has shrunk considerably in the past few decades and much of its former basin is now a dry, salty plain. The satellite image shows the area of the Aral Sea in 1960 (black border) and the lake in 2009, which now consists of four disconnected small lakes (the green areas plus a very shallow grey area in the centre). The image also shows dust blowing across the centre of the old lake. The effect on native terrestrial and aquatic ecosystems has been dramatic; commercial fishing has been abandoned. ***Fly By: 45° 05′ N; 59° 24′ E***

By the 1960s, withdrawal of water for irrigation was about 40 km^3, which was in balance with inputs. However, since that time, agricultural demands have increased, especially as a result of large-scale cotton production. The irrigated area had grown to nearly 6.5 million hectares by 1980, and demand for water increased to 132 km^3. This led to a progressive reduction in the area and volume of the Aral Sea and a corresponding increase in salinity. For example, in 1960, the Aral Sea covered 68,000 km^2 and contained 1,090 km^3 of water, with a salinity of 10 g l^{-1}. By 2007, the area was reduced to 13,958 km^2, or about 20 percent of its size in 1960, with a volume of 102 km^3 (9 percent of volume in 1960) and average salinity rising to over 75 g l^{-1}.

EYE ON THE ENVIRONMENT

DROUGHT ON THE PRAIRIES

Drought on the Canadian Prairies is the most important factor that limits crop yield, and is a recurring threat to agriculture. The long-term precipitation record for Indian Head, Saskatchewan, (see Figure 1), shows that precipitation was significantly lower than normal in the 1930s, as was the case throughout much of the American Great Plains region. However, droughts of equal or greater magnitude have been recorded in the early 1960s, 1997, 2001, and 2003.

The winter of 2001–02 was warmer and drier than normal on the Prairies, resulting in limited snow accumulation; as a result, low soil moisture conditions continued into the following growing season. Soil moisture for crops is conveniently expressed in terms of the agricultural year, which for the Prairies runs from September to the following August. The agricultural year includes the precipitation for soil moisture recharge after harvest, the winter precipitation available for soil moisture recharge, and growing

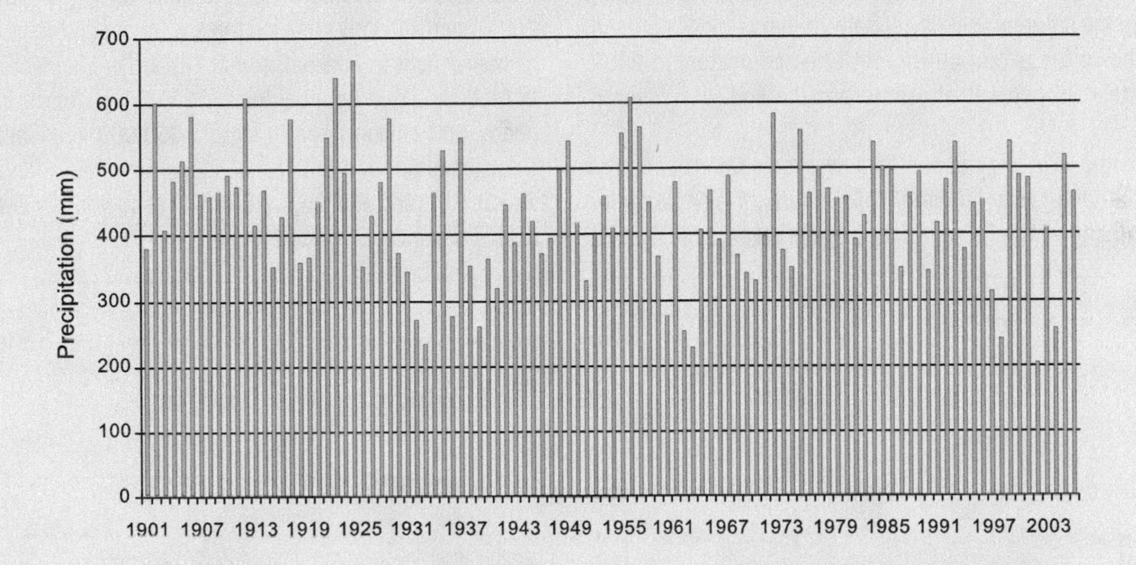

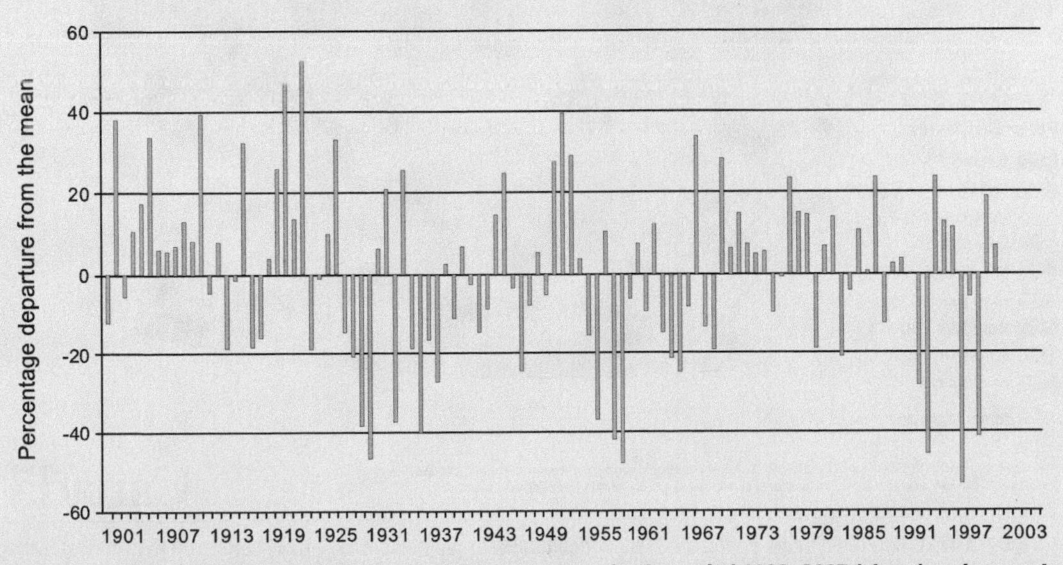

Figure 1 Annual precipitation for Indian Head, Saskatchewan, for the period 1900–2005 (above) and percentage departure from the mean (below).

season precipitation. The maximum extent of the drought was assessed on August 6, as rainfall after this date comes too late to improve yields. More than 65 percent of the prairie cropland was affected by moderate, severe, or record drought conditions, although some areas in southern Alberta and southwestern Saskatchewan actually received higher than average precipitation.

Annual crops such as wheat and barley are susceptible to drought at any time during their growth and development. However, forage grasses are more sensitive to moisture deficiencies in the fall and spring. Thus, drought conditions in one year have a pronounced carry-over effect into the following year. Poor grass growth due to forage drought leads to reduced stocking rates during the summer and hay supply problems for winter feed. Figure 2 shows the extent of drought-affected pasture in 2002. Approximately 60 percent of prairie pastureland was severely affected.

Hydrologic drought caused by low precipitation and high water demand affects livestock operations. Water supplies declined in central Alberta, central Saskatchewan, and in some areas of western Manitoba during the summer of 2002, and at the peak of the drought, more than 70 percent of the farms were affected.

Although the cyclic occurrences of drought on the Prairies cannot be prevented, improved farming practices, such as zero tillage, help to reduce its impact. Zero tillage has two general benefits—the stubble left from harvested grain crops forms an effective snow trap, and by seeding directly into the stubble, there is less bare soil exposed to the wind.

Dust storms still occur on the Prairies, although their frequency and magnitude are now reduced because of improved farming practices. In addition, the Prairie Shelterbelt Centre at Indian Head distributes trees to farmers to create wind breaks. In 2005, approximately 4 million trees were used to establish 220 km of shelter belts in Saskatchewan, 121 km in Manitoba, and 78 km in Alberta. Since its inception in 1901, the centre has shipped nearly 600 million trees to almost 650,000 applicants, and has established nearly 4,500 km of shelter belts. An additional benefit is that the trees planted in 2005 will sequester an estimated 1.5 million tonnes of CO_2 by 2054.

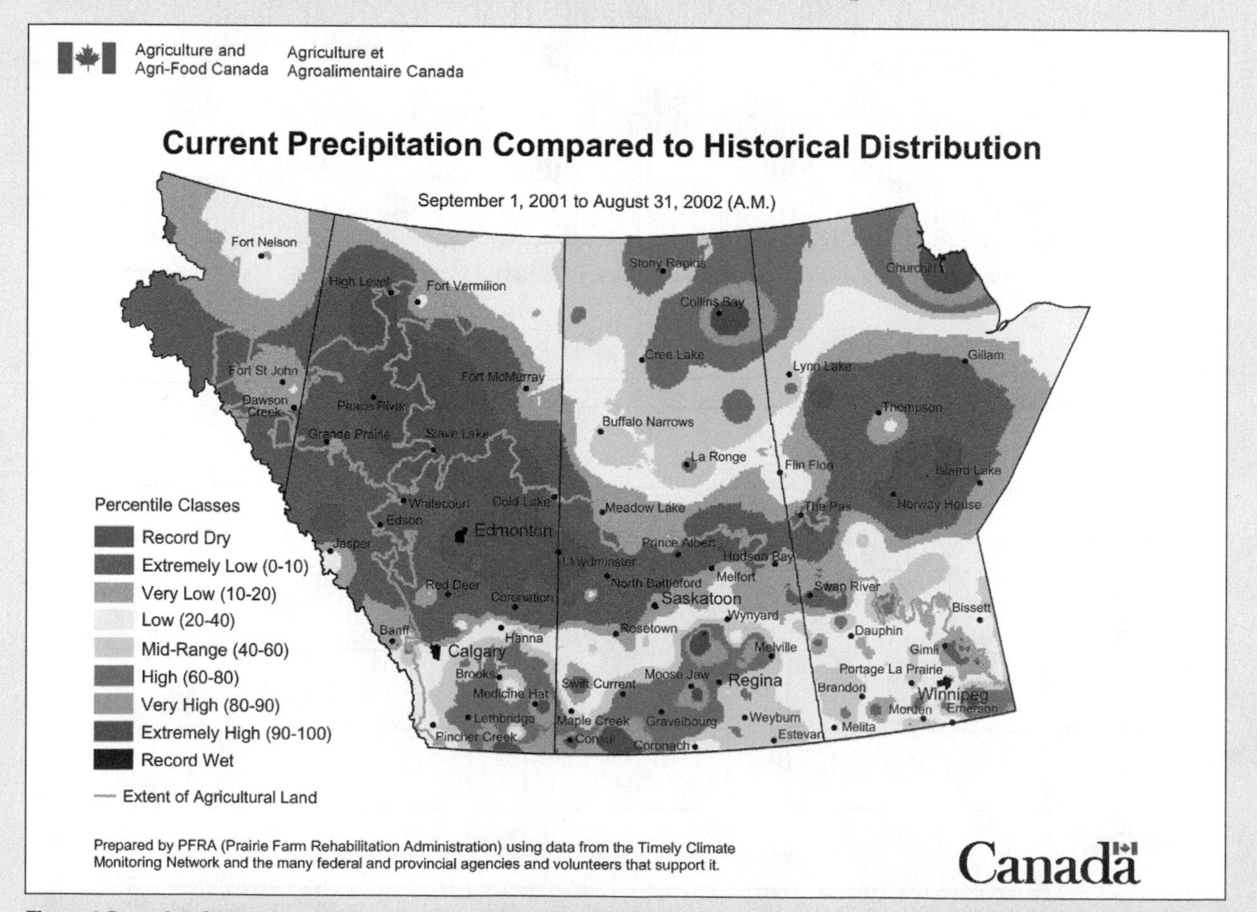

Figure 2 Drought of 2002 Precipitation during the 2002 agricultural year (September to August) compared to historical average conditions. The values are expressed as percentages of the long-term mean.

The environmental impact has been considerable. An estimated 43 million metric tonnes of salt are carried from the dried-out seabed each year and deposited over an area of 150,000 to 200,000 km^2. Many native plant communities have disappeared. The forests, which once occupied the river deltas and shallow margins of the Aral Sea, were especially vulnerable to the 3- to 8-m drop in groundwater that has occurred. Loss of habitat has resulted in the elimination of all but 38 of the original 173 animal species that once occupied the area. The aquatic ecosystems have been similarly devastated, with 20 of the 24 native fish species disappearing, together with the loss of the commercial fishery.

THE MOIST CONTINENTAL CLIMATE (KÖPPEN *Dfa, Dfb, Dwa, Dwb*)

The **moist continental climate** is restricted to the northern hemisphere, occurring at lat. 30° to 55° N in central and eastern parts of North America and Eurasia, and at lat. 45° to 60° N in Europe (Figure 10.16). In North America, the moist continental climate region stretches across southern Canada from eastern Manitoba to the Maritime provinces, and extends through the central and eastern United States as far south as Tennessee. The moist continental climate is also characteristic of central and eastern Europe as far east as the Ural Mountains. In Asia, it is found in northern China, Korea, and Japan. These regions lie in the polar-front zone; seasonal

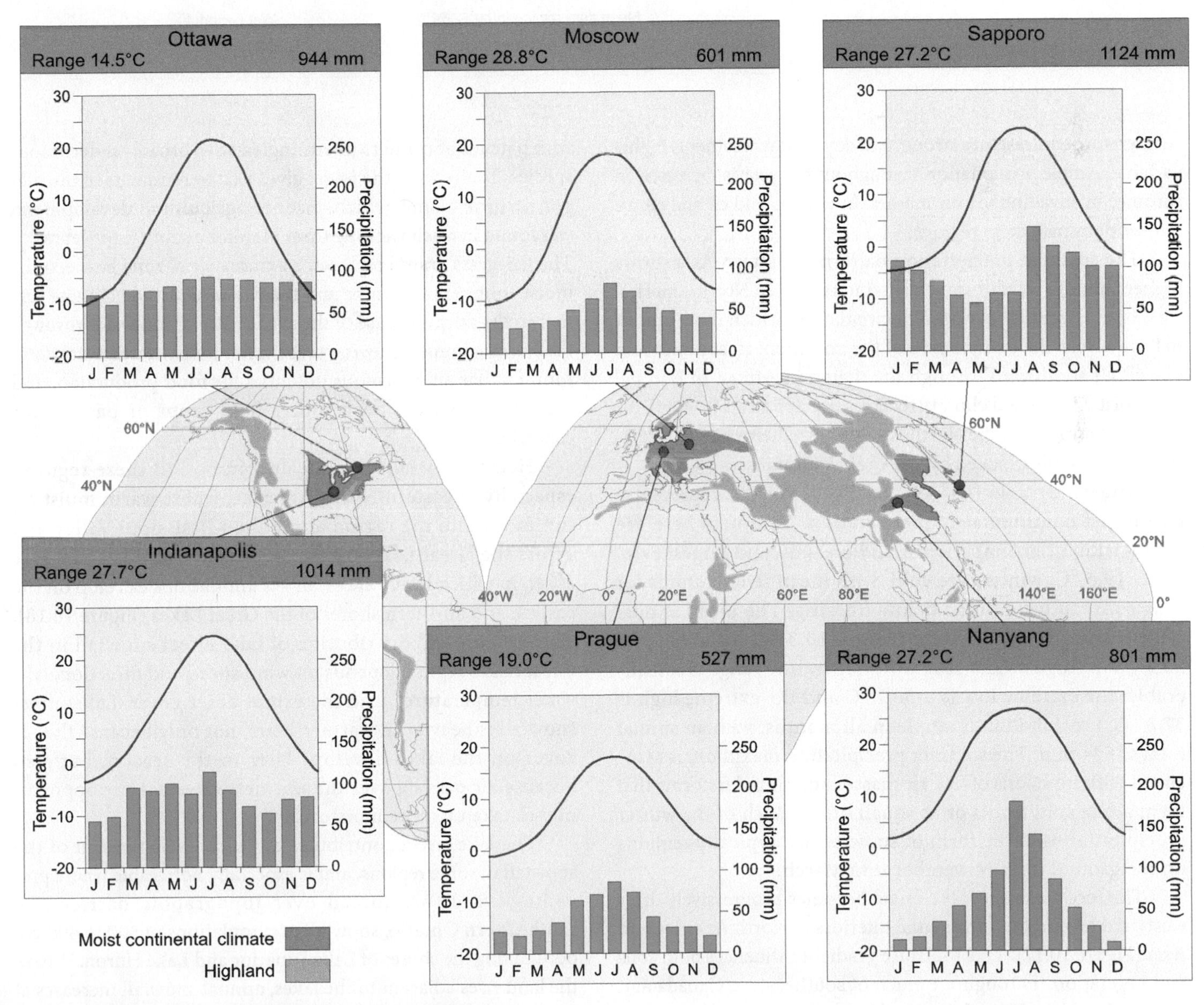

10.16 WORLD DISTRIBUTION OF THE MOIST CONTINENTAL CLIMATE

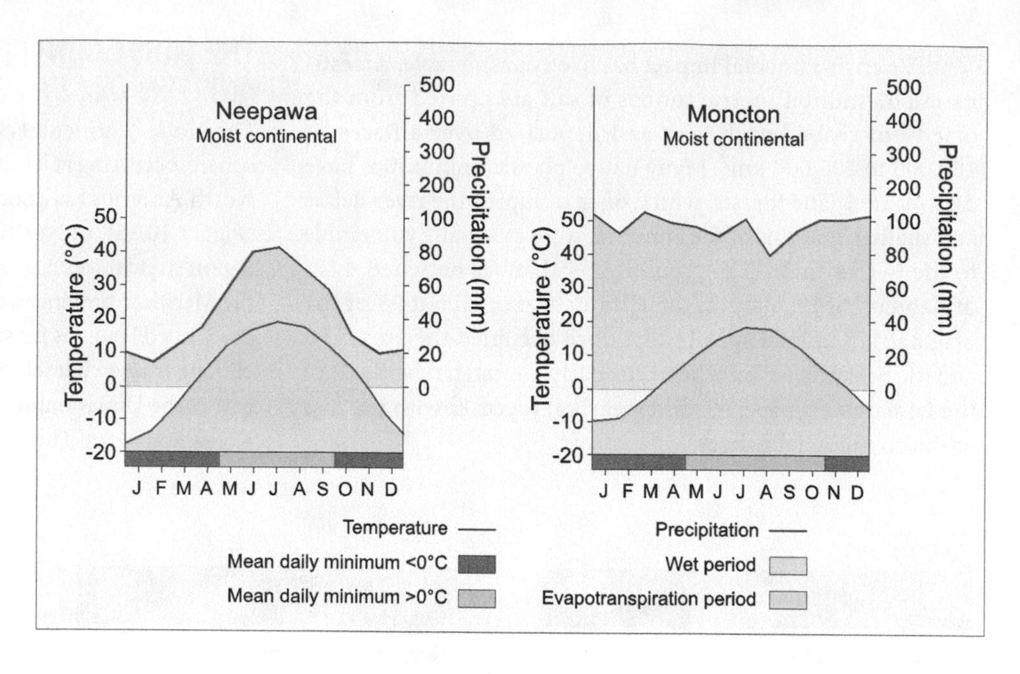

10.17 SIMPLIFIED CLIMATE DIAGRAMS SHOWING SEASONAL MOISTURE AVAILABILTY FOR MOIST CONTINENTAL CLIMATE STATIONS In Canada, precipitation decreases from east to west across the region and a water deficit can occur as soil moisture reserves are used over the summer. Moisture reserves at Neepawa, Manitoba, are generally adequate to meet crop demands, but are considerably lower than the moisture surplus recorded at Moncton, New Brunswick.

temperature contrasts are strong, and day-to-day weather is highly variable. Ample precipitation throughout the year is increased in summer by invading mT air masses. In winter, cold cP and cA air masses dominate these regions.

The seasonal precipitation pattern in eastern Asia shows higher summer rainfall and drier winters than in North America. This is an effect of the monsoon circulation, which moves moist mT air across the eastern side of the continent in summer and dry cP air southward through the region in winter. In Europe, the moist continental climate receives precipitation from mP air masses coming from the North Atlantic, but their eastward progress is eventually blocked by the Ural Mountains.

Ottawa, Ontario (45° 19′ N; 75° 40′ W), is a representative moist continental climate station. Summers here are warm, with mean temperatures in June, July, and August averaging 19.6 °C; winters are cold, with mean temperatures for four consecutive months below freezing. The mean annual temperature range is large, from −10.8 °C in January to 20.9 °C in August. The absolute temperature range is considerable; the extreme low is −36.1 °C and the extreme high is 37.8 °C. Precipitation is ample in all months, with an annual total of 944 mm. The summer precipitation maximum is associated with invasions of mT air masses and thunderstorms that form along cold fronts or as squall lines. Much of the winter precipitation is in the form of snow, which typically remains on the ground from November into March.

The moist continental climate becomes progressively drier westward toward the continental interiors of North America and Asia (Figure 10.17). This moisture gradient influences both soils and vegetation. Throughout much of southeastern Canada and the northeastern United States, the vegetation is mixed forest, with patches of conifers intermingled with broadleaf deciduous species. To the west, the forest gives way to remnants of the *tall-grass prairie* biome, which, prior to agricultural development, was found in a belt running from Manitoba south into Nebraska. The tall-grass prairie represents a transitional zone between the moist continental climate and the dry midlatitude climate further to the west. Because of the availability of soil water through the warm summer growing season, the moist continental environment has an enormous potential for food production, and most regions have been under field crops or pasture for centuries.

Heavy snowfalls are not uncommon in these regions, especially in eastern North America, where warm moist air can move into the region along the Mississippi Valley and across the Great Lakes. One consequence of this is the *lake effect*, in which heavy snowfall accumulations develop on the eastern and northern shores of the Great Lakes (Figure 10.18). The amount and distribution of lake-effect snowfall in the Great Lakes region depends on wind speed and direction, lake water temperatures, and the extent of ice cover. Lake-effect snowfall is heavier in warmer years, not only because the ice cover on the lakes develops later in the season, but also because air passing over the warmer water has the opportunity to take up more moisture.

The lake effect contributes as much as 50 percent of the snowfall in some regions, and is especially noticeable where prevailing winds are forced over topographic barriers. In southeastern Ontario, snowfall accumulations of 300 to 400 cm occur along the shores of Lake Superior and Lake Huron. Where the land rises adjacent to the lakes, annual snowfall increases at a rate of about 60 cm per 100 m of elevation. Major storms can

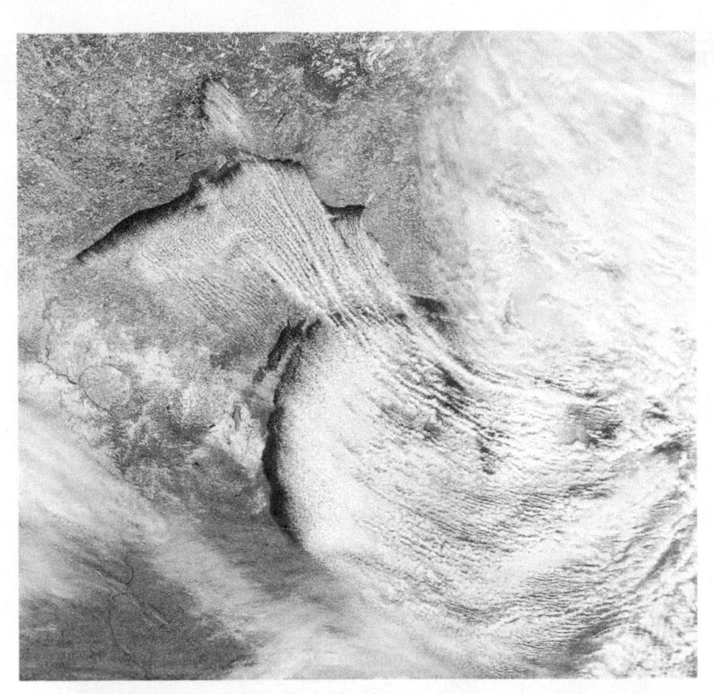

10.18 **THE LAKE EFFECT CAUSED BY WINDS BRINGING MOIST AIR FROM THE GREAT LAKES CREATES PRONOUNCED SNOW BELTS IN SOUTHEASTERN ONTARIO** The clear skies on the northern and western margins of Lake Superior and Lake Michigan show a striking contrast with cloudy conditions on the southern and eastern shores in this image acquired December 5, 2000. Drier air from the northwest picks up moisture from the ice-free lakes as it is carried over them in the prevailing northwesterly winds, with subsequent cooling to dewpoint, and resulting in condensation and precipitation.

drop up to 100 cm of snow in these areas, and the accompanying winds can cause heavy drifting.

Ice storms—or freezing rain—are another hazard in eastern Canada and New England. They typically develop when a warm air mass invades the region and overruns a shallow layer of cold air at the surface. Ice storms can cause considerable damage and inconvenience. They are especially disruptive for all methods of transport, and interrupt power supplies when transmission lines break under the weight of the ice or trees collapse on them. Eastern Canada can expect about 15 ice storms each year, which typically last for only a few hours. Maintenance crews equipped with de-icing materials can usually deal with them quite quickly, and the power transmission infrastructure can normally cope with storms that deposit 50 to 60 mm of ice.

In January 1998, a storm of unprecedented magnitude caused considerable damage and was directly linked to the deaths of at least 25 people. The storm lasted from January 4 to 10, and in that time deposited more than 100 mm of ice in some areas. It affected a relatively narrow band from eastern Ontario to Nova Scotia and bordering areas in the northeastern United States. Millions of people were without power for days or weeks until

10.19 **ICE STORMS** Ice storms are an expected feature of moist continental climates and can cause considerable damage.

the power grid was repaired (Figure 10.19). Damage was estimated at $4 billion to $6 billion.

In addition to the material damage caused by the large numbers of trees that collapsed under the weight of the ice, the ice storm impacted the agriculture sector since many of those trees were fruit trees or sugar maples. The Quebec maple sugar industry was devastated.

OVERVIEW OF THE HIGH-LATITUDE CLIMATES

The *high-latitude climates* are mainly restricted to the northern hemisphere. They occupy the northern subarctic and arctic latitude zones. Only the ice sheet climate of the polar zones is present in both hemispheres. Table 10.3 provides an overview of the three high-latitude climates.

The dynamic wind pattern at high latitudes brings mP air masses from the northern oceans into conflict with cP and cA air masses on the continents. The Rossby wave system transports lobes of warmer, moister air toward the poles in exchange for colder, drier air that moves toward the equator. The result of these processes is frequent wave cyclones produced along a discontinuous and constantly fluctuating arctic-frontal zone. In summer, tongues of mT air occasionally reach subarctic latitudes to interact with polar air masses and yield significant amounts of precipitation.

THE BOREAL FOREST CLIMATE (KÖPPEN *Dfc, Dfd, Dwc, Dwd*)

The **boreal forest climate** is a continental climate with long, bitterly cold winters and short, cool summers. The latitude range for this climate type is from 50° to 70° N. In North America, the boreal forest climate stretches from central Alaska,

TABLE 10.3 **HIGH-LATITUDE CLIMATES**

CLIMATE	TEMPERATURE	PRECIPITATION	EXPLANATION
Boreal forest	This region has short, cool summers and long, bitterly cold winters, with the greatest annual range of all climates.	Annual precipitation is small, falling mostly in the summer months.	This climate falls in the source region for cold, dry stable cP air masses. Travelling cyclones, more frequent in summer, bring precipitation from mP air.
Tundra	There is no true summer, but there is a short mild season. Otherwise, temperatures are cold.	Annual precipitation is small, falling mostly during the mild season.	The coastal arctic fringes occupied by this climate are dominated by cP, mP, and cA air masses. The maritime influence keeps winter temperatures from falling to the extreme lows of interiors.
Ice sheet	All months are below freezing, with the lowest global temperatures on Earth experienced during the Antarctic winter.	Very low precipitation, but snow accumulates since temperatures are always below freezing.	Ice sheets are the source regions of cA and cAA air masses. Temperatures are intensely cold.

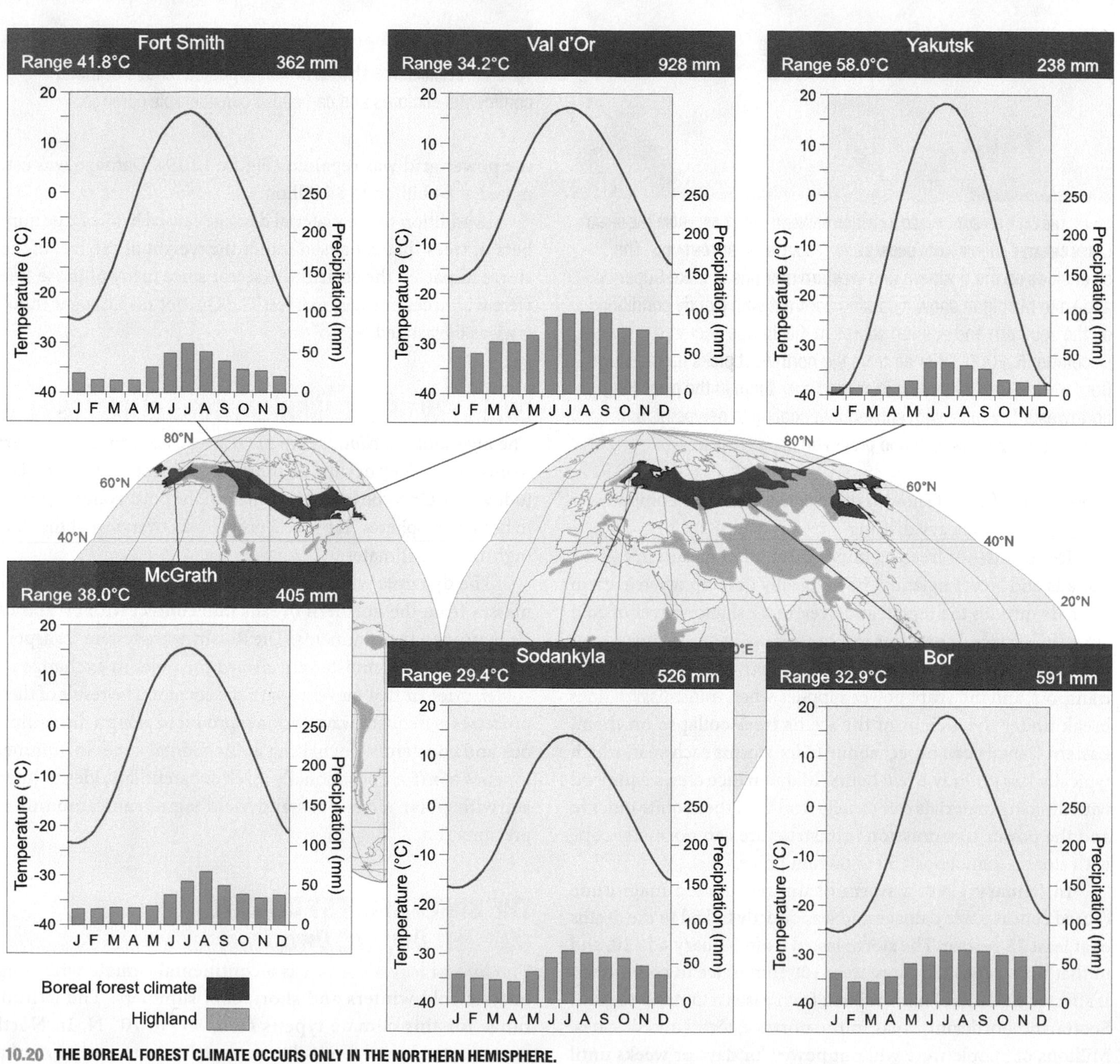

10.20 **THE BOREAL FOREST CLIMATE OCCURS ONLY IN THE NORTHERN HEMISPHERE.**

across the Yukon and Northwest Territories, to Labrador on the Atlantic coast (Figure 10.20). In Europe and Asia, it extends from the Scandinavian Peninsula eastward across Siberia to the Pacific. The boreal forest climate represents the source region for cool cP air masses, but invasions of cold, dry, stable cA air masses are frequent at these latitudes.

The annual temperature range in the boreal forest climate is the largest of any climate type and is especially pronounced in Siberia. Yakutsk, Russia (62° N) is generally considered to be the coldest city in the world. The mean annual temperature ranges between −39.5 °C in January and 18.6 °C in July. In February 1984, the record low temperature dropped to −64.4 °C, with a record high of 38.3 °C in July 2001—an extreme range of 102.7 °C. Extending the record for Yakutsk back to 1885 gives an all-time record low for the northern hemisphere of 67.8 °C. Precipitation increases substantially in summer, when maritime air masses penetrate the continent bringing cyclonic systems to the region. However, over large areas in western Canada and Siberia, annual precipitation is low, much of it coming as snow.

Fort Smith, Northwest Territories (60° 01′ N; 111° 58′ W), is a representative boreal forest climate station. Monthly mean air temperatures are below freezing for seven consecutive months. The summers are short and cool. Total precipitation for the year averages 362 mm and comes with a marked summer maximum; the four months June through September account for more than half of the annual total. Winter snowfall remains over frozen ground for about five months, and accumulates to depths of 40 to 50 cm by February. The snow melts quickly in the spring, the meltwater, together with the increase in rainfall over the summer, being sufficient to maintain a positive soil moisture budget (Figure 10.21). Mean monthly temperatures range from −24.1 °C in January to 16.5 °C in July, giving an annual range of 40.6 °C. The extreme range at Fort Smith is 89.2 °C, based on the record low of −53.9 °C in February 1947 and 35.3 °C in August 1981.

The Pleistocene ice sheets shaped much of the boreal forest climate region, and lakes and bogs are common features in the landscape. Most of the area is covered with coniferous forests that are exploited for pulp, paper, and lumber. The forests of Europe and Asia consist of closely-related species, each of which is well adapted to the harsh environment and to periodic disturbance from fire. Each year several thousand fires burn in the coniferous forests of Canada, affecting about 1 percent of the forest area. The average interval between successive fires at a given location is about 60 years, but this varies according to site conditions. For example, jack pine (*Pinus banksiana*) stands are especially prone to fire because they are generally associated with well-drained sandy soils.

Many fires are started by people, but about 50 percent are caused by lightning strikes. Natural fires are typically preceded by rainless periods lasting at least one to two weeks, accompanied by high temperatures and low humidity. The probability of these conditions determines the fire climate. The fire climate is strongly seasonal at higher latitudes, and most fires occur in summer when long days and strong winds promote rapid drying. In Canada, the danger of forest fires is based on the Fire Weather Index (FWI) that represents the intensity of a fire in terms of

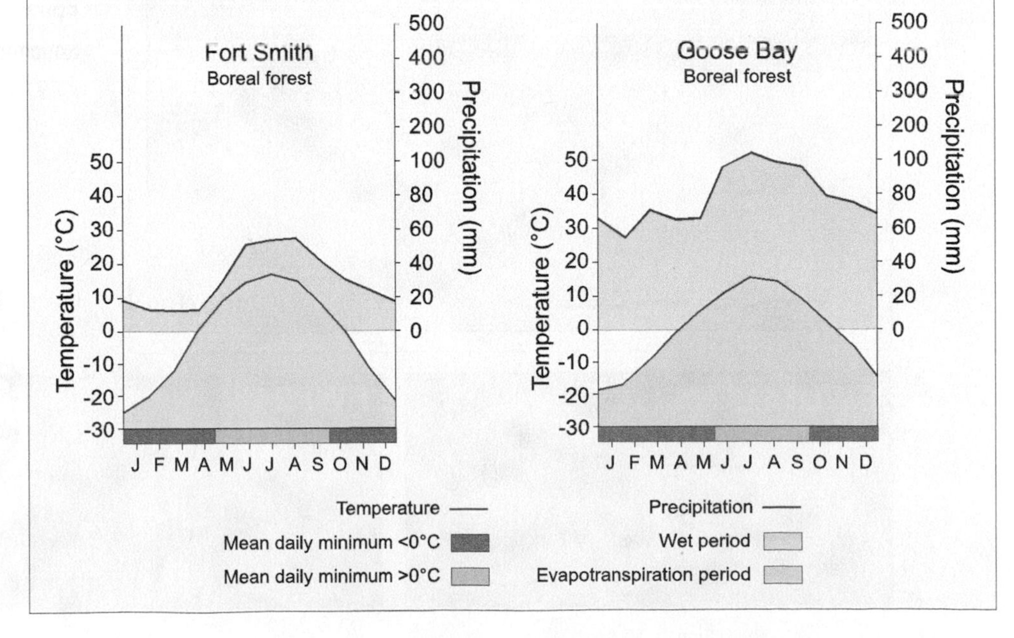

10.21 SIMPLIFIED CLIMATE DIAGRAMS SHOWING SEASONAL MOISTURE AVAILABILTY FOR BOREAL FOREST CLIMATE STATIONS Typically there is no moisture deficit in the boreal forest climate type in Canada. However, precipitation increases from west to east across the region, so the moisture surplus is less at Fort Smith, Northwest Territories (left), than at Goose Bay, Labrador (right).

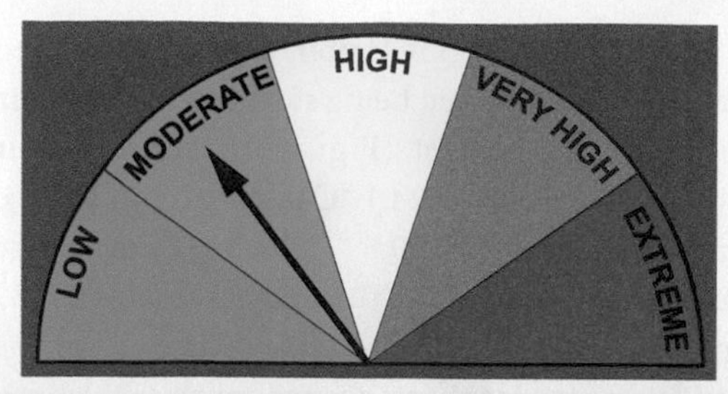

LOW	New fires are not likely to start. If started, they spread very slowly or may go out on their own (existing fires may continue to smoulder if burning in deep, dry fuel layers).
MODERATE	Creeping or gentle surface fires can start but are easily controlled by ground crews.
HIGH	Vigorous surface fires can start and will require heavy equipment for containment.
VERY HIGH	Intense surface fires and developing tree crown fires can readily occur and will challenge fire suppression forces, which will need to attack the fires with air tankers and water bombers.
EXTREME	Burning conditions are considered explosive, and attempts to control the fast-spreading, high-intensity fires are not possible until the severity of the situation diminishes.
NIL	No calculations are performed due to snow cover, cold weather, or lack of combustible material.

10.22 FIRE DANGER CLASSIFICATIONS The risk of fire as determined by the Fire Weather Index is displayed at the entrance to most national parks.

energy output. Several parameters determine the index, including temperature, rainfall, relative humidity, and wind speed. The calculations also account for the amount of litter on the forest floor and its moisture content (Figure 10.22).

Typical average values for the Fire Weather Index range between 2 and 10, but this can rise dramatically depending on weather conditions during a particular fire event (Figure 10.23). The FWI is generally highest in northern Alberta and Saskatchewan, and decreases eastward in response to the generally wetter climate conditions. The wet climate of coastal British Columbia also leads to relatively low average FWI ratings, but in the dry interior exceptionally high ratings are common. Once a fire starts, its intensity will depend not only on the amount and condition of the fuel, but also on weather conditions and topography. For example, intense crown fires often develop during the afternoon when wind speeds and temperatures increase. Low fire intensities are characteristic of surface fires that consume litter on the forest floor as well as shrubs and small trees. These may become more intense during gusty periods, but fall back to the surface when wind speeds drop. High intensity crown fires will develop under more extreme conditions.

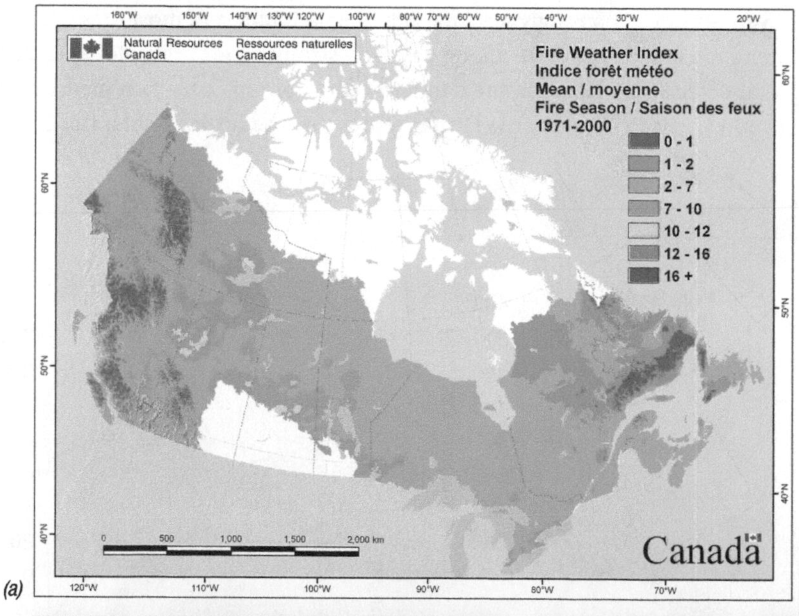

10.23 FIRES ARE A COMMON HAZARD IN THE BOREAL FORESTS CLIMATE REGION
(a) Mean Fire Weather Index (FWI) during fire season in Canada for 1971–2000. (b) Fires become more intense as FWI increases, as shown in this sequence ranging from FWI 14 to FWI 34. These are similar to fires anticipated under moderate, high, and very high fire danger ratings.

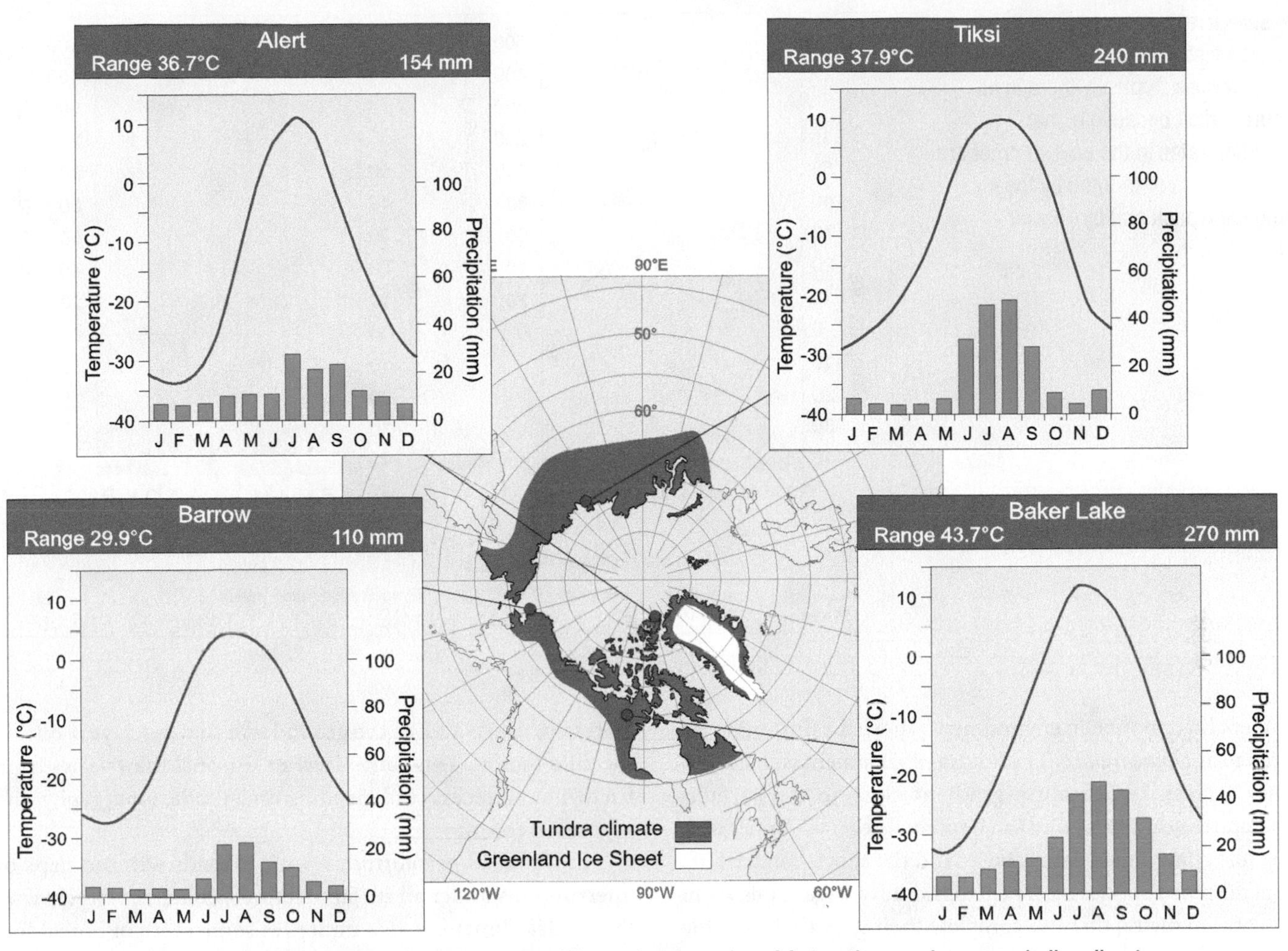

10.24 **THE TUNDRA CLIMATE REGION OF THE NORTHERN HEMISPHERE** The peninsula region of Antarctic experiences a similar climate.

THE TUNDRA CLIMATE (KÖPPEN *ET*)

The **tundra climate** is dominated by polar (cP, mP) and arctic (cA) air masses. Winters are long and severe. The nearby oceans moderate winter temperatures, preventing from falling to the extreme lows found in the continental interior. There is a short mild season. Precipitation generally increases in the summer, when more moisture is available from the ice-free polar seas.

The tundra climate surrounds the Arctic Ocean and extends across the island region of northern Canada (Figure 10.24). It includes the Alaskan north slope, the Hudson Bay region, and the Greenland coast. In Eurasia, this climate type occupies the northernmost fringe of the Scandinavian Peninsula and the Siberian coast. The Antarctic Peninsula also experiences a tundra climate. The latitude range for this climate is 60° to 75° N and S, except for the northern coast of Greenland, where tundra occurs at latitudes beyond 80° N.

Baker Lake, Nunavut (64° 18′ N; 96° 05′ W), is a representative tundra climate station. In this part of Canada, summer is short and mild with mean monthly temperatures remaining above freezing for only four months. Average temperature in July, the warmest month, is 11.4 °C, although a temperature of 33.6 °C was recorded in July 1989. Winters are long and very cold, with average temperatures in January dropping to −32.3 °C. The extreme daily minimum at Baker Lake is −50.6 °C, recorded in January 1975. Generally, July and August are the only months when the average minimum temperature does not drop below 0 °C. However, below-freezing temperatures have been reported in all months. Annual precipitation is 270 mm, with a pronounced late summer maximum; about half of the yearly total falls in July, August, and September. This is associated with the increased moisture content of maritime air masses due to warmer air temperatures and increased evaporation from the ocean surface following the melting of the sea ice. Consequently, there is a moisture surplus year round (Figure 10.25).

Arctic tundra lying beyond the arctic treeline describes both an environmental region and a type of vegetation dominated by herbaceous plants, such as grasses and sedges, together with mosses and lichens, and in some places small woody shrubs.

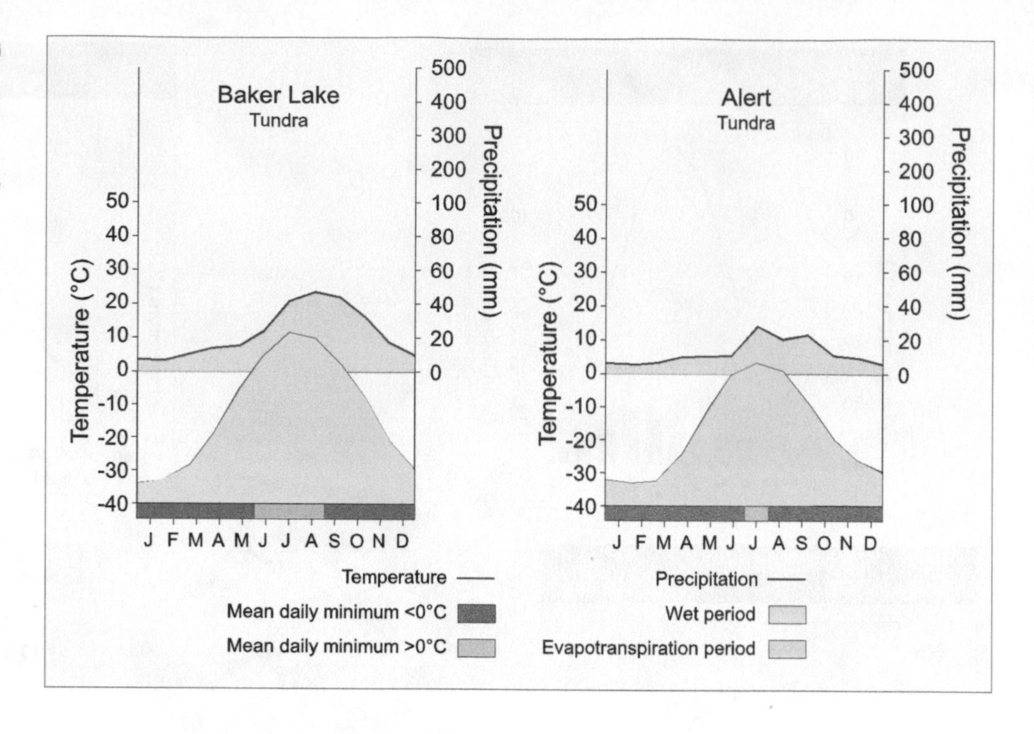

10.25 SIMPLIFIED CLIMATE DIAGRAMS SHOWING SEASONAL MOISTURE AVAILABILTY FOR TUNDRA CLIMATE STATIONS Typically there is no moisture deficit because higher evaporation rates in the cool summer are offset by water contributed by the late melting snowpack and by heavier precipitation.

Equivalent high mountain environments above the timberline are referred to as *alpine tundra*. The tundra environment is characterized by a steep temperature gradient close to the surface. Temperatures near the ground can be several degrees warmer than the air just a few centimetres above, and the short, compact life form of the native species is adapted to take advantage of this warm microzone, allowing them to complete their growth during the brief summer.

The tundra is a treeless zone that borders the open lichen woodland at the extreme range of the boreal forest. The transition zone is marked by stunted, deformed trees that have lost their branches due to abrasion from ice crystals carried in the bitter winds. In some places the forest-tundra boundary is fairly distinct. In North America it coincides approximately with the 10 °C isotherm of the warmest month and with the median position of the arctic front that separates cool arctic air masses from milder air originating in the Pacific.

ARCTIC PERMAFROST

Soils of the arctic tundra are poorly developed and consist of freshly broken mineral particles and varying amounts of partially decomposed plant matter. Peat bogs are numerous. Because the soil remains permanently frozen not far below the surface, the summer thaw saturates the soil with water, especially in areas where drainage is limited by lack of slope.

Perennially frozen ground, or **permafrost**, occurs throughout the tundra region and much of the adjacent boreal forest. Normally, a top layer of the ground will thaw each year during the mild season. This *active layer* of seasonal thaw varies from 0.6 to 4 m in thickness, depending on latitude, topography, and vegetation cover.

Continuous permafrost, which extends without gaps or interruptions under all surface features, coincides largely with the tundra climate. It also underlies some areas of boreal forest in northern Canada and Siberia, especially where the tree cover is very open. *Discontinuous permafrost*, which occurs in patches separated by frost-free zones under lakes and rivers, is common throughout the boreal forests of North America and Eurasia, and occurs sporadically even near the southern limits of the forest.

The general distribution in Canada is related to latitude, with continuous, discontinuous, and isolated permafrost encountered in a southward progression. However, there are some deviations from this simple pattern. For example, the zone of continuous permafrost swings southward along the coast of Hudson Bay, compressing the zones in northern Ontario. This is due to persistent sea ice offshore for much of the summer, which lowers summer air temperatures near the coast.

Permafrost is sensitive to disturbance, so regulations have been developed to minimize the impact of mining and exploration activities in these regions. The impact of climate change is especially serious for coastal environments in this climate zone. Not only are they susceptible to rising sea level and loss of ice cover, but much of the coastline is sensitive to slumping and erosion. Chapter 18 discusses permafrost in more detail.

THE ICE SHEET CLIMATE (KÖPPEN *EF*)

The **ice sheet climate** coincides with the source regions of arctic (A) and antarctic (AA) air masses, situated on the vast ice sheets of Greenland and Antarctica and over the polar sea ice of the Arctic Ocean. The mean annual temperature is much lower than in any other climate, and no monthly mean exceeds 0 °C. Strong temperature inversions, caused by radiation loss from the surface, develop over the ice sheets. In Antarctica and Greenland, the high elevation of the ice sheet surface intensifies the cold (Figure 10.26). Temperatures in the interior of Antarctica are far lower than those at any other place on Earth. The record low temperature of −88.3 °C was recorded in 1958 at Vostok, located about 1,300 km from the South Pole at an altitude of about 3,500 m. Conditions are less extreme at McMurdo Sound because this station is located near sea level on the Ross Sea coast. Strong cyclones with blizzard winds are frequent. Precipitation, almost all occurring as snow, is very low, but accumulates because of the continuous cold.

Because of low monthly mean temperatures throughout the year over the ice sheets, this environment is practically devoid of vegetation. Some algae and bacteria survive in the surface layers of the ice and in locally heated volcanic openings. The few species of animals found on the ice margins, such as polar bears in the northern hemisphere and penguins in the south, are associated with the marine habitat. Concern has been raised for both of these species as a result of significant changes in the ice conditions in recent years.

OUR CHANGING CLIMATE

In Chapter 8, climate was described as a region's long-term average seasonal conditions, based on mean monthly temperature and precipitation observed at weather stations. However, climate change has affected present weather conditions and will likely become more pronounced in the coming years. As noted in *Appendix 4.1 • The IPCC Report of 2007*, recent human activity appears to have contributed to warmer temperatures, which in turn have reduced snow and ice cover and raised sea levels. Precipitation, enhanced by greater evaporation, has increased in mid- and high latitudes, but has decreased in subtropical regions. The variability of weather has also increased, with extreme precipitation events reportedly occurring more frequently than in the past.

For most of North America, temperatures in the twenty-first century are predicted to rise significantly, bringing warmer winters and hotter summers. Although precipitation will increase in many regions, higher temperatures will cause more evapotranspiration,

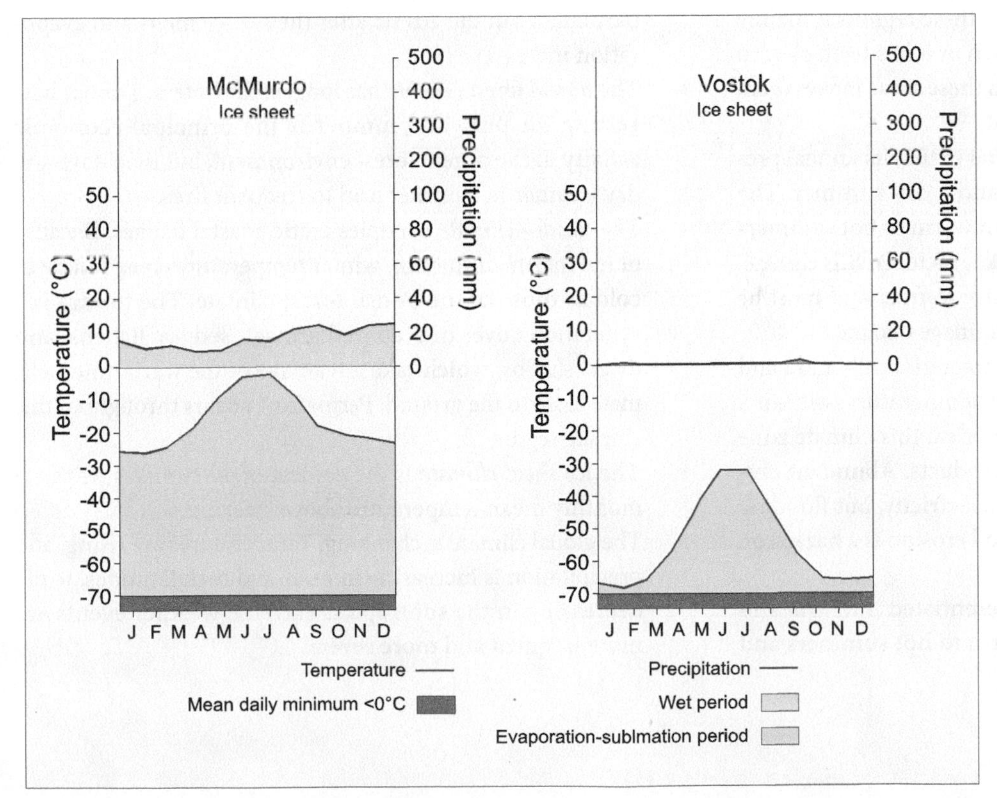

10.26 SIMPLIFIED CLIMATE DIAGRAMS FOR REPRESENTATIVE ICE SHEET CLIMATE STATIONS IN ANTARCTICA Precipitation is minimal in this cold climate where below-freezing temperatures occur year round. However, the permanent snow and ice cover ensures there is no moisture deficit. Temperatures are less extreme at McMurdo Sound (77° 53′ S; 166° 48′ W) coastal station at an elevation of 2 m, compared to Vostok (78° 27′ S; 106° 52′ E) further inland and at an elevation of 3,420 m.

with the probability that summer droughts will be more pronounced. Meanwhile, more frequent extremes of precipitation will enhance flooding and storm damage. *Appendix 10.1 • Regional Impacts of Climate Change* summarizes climate change predictions and impacts for nine regions in North America.

A Look Ahead

The Earth's climates are remarkably diverse, ranging from the hot, humid, wet equatorial climate to the bitterly cold and dry ice sheet climate at high latitudes. Many other climates are recognized between these extremes, each with distinctive characteristics, such as dry summers or dry winters, or uniform or widely varying temperatures. Climate exerts strong controls on landform processes, such as those related to flowing water, glaciers, and wind and wave action. Soil formation and soil properties also show close relationships with regional climate, as do world biomes. These features of the physical landscape are discussed in parts 4 and 5.

CHAPTER SUMMARY

- *Midlatitude climates* are varied, since they lie in a broad zone of intense interaction between tropical and polar air masses.
- The *dry subtropical climate* is dominated by subtropical high pressure and resembles the dry tropical climate; however, it has a larger annual temperature range and a distinct cool season. Agriculture is possible with adequate water supply, but this has led to serious soil degradation through the processes of dryland salinization and irrigation salinization.
- Abundant rainfall with a summer maximum is a characteristic of the *moist subtropical climate*. The temperature cycle of this climate type includes cool winters with spells of below-freezing weather and warm, humid summers. This climate is often uncomfortably hot and humid in the summer. The natural vegetation cover in these regions is mainly broadleaf deciduous forest, but much of it has been cleared for agriculture. The heavy rainfall in these areas makes them prone to severe soil erosion.
- The *Mediterranean climate* is unusual in that its annual precipitation cycle has a wet winter and a dry summer. The temperature range is moderate, with warm to hot summers and mild winters. Water supply is a key factor in this climate. Some crops such as fruit are sensitive to frost and must be protected in areas where cold air drainage occurs.
- The *marine west-coast climate* is characteristically mild and moist. The marine influence keeps temperatures within a narrow annual range. In North America, this climate zone is a source of many diverse forest products. Abundant precipitation is used to generate hydroelectricity, but flooding is common in low-lying areas and soil erosion is a hazard on steep slopes.
- The *dry midlatitude climate* is differentiated into *arid* and *semi-arid* subtypes. Both have warm to hot summers and cold to very cold winters. The semi-arid subtype typically has relatively fertile soils with a sparse cover of grasses. Wheat is the dominant crop, and cattle are also important. Because rainfall is highly variable, drought is a recurring hazard and has led to dust-bowl conditions in the past.
- The *moist continental climate* has ample precipitation with a summer maximum. Summers are warm and winters are cold. Heavy snowfalls and ice storms commonly occur in this climate and can cause significant damage and disruption of day-to-day activities.
- *High-latitude climates* are comparatively dry since temperatures are low and the air holds little moisture. The precipitation generally occurs during the short warm period, particularly in the Arctic after the sea ice melts and evaporation increases.
- The *boreal forest climate* has long, cold winters. Timber harvesting for pulp and lumber is the principal economic activity in the boreal forest environment, but long days and dry weather in summer lead to frequent fires.
- The *tundra climate* occupies arctic coastal fringes. Because of the marine influence, winter temperatures may not be as cold as those of the boreal forest climate. The tundra is a vegetation cover of scattered grasses, sedges, lichens, and dwarf shrubs, which take advantage of the warm microclimate close to the ground. Permafrost occurs throughout this climate region.
- The *ice sheet climate* is the coldest of all climates, with no monthly mean temperature above freezing.
- The global climate is changing. Temperatures are rising, and precipitation is increasing in mid- and high latitudes, while decreasing in the subtropics. Extreme weather events are more frequent and more severe.

KEY TERMS

boreal forest climate
dry midlatitude climate
dry subtropical climate
ice sheet climate
marine west-coast climate
Mediterranean climate
moist continental climate
moist subtropical climate
permafrost
tundra climate

REVIEW QUESTIONS

1. What climate type is associated with the Mojave Desert? Describe how this climate type arises and how it affects the desert environment.
2. Both the moist subtropical and moist continental climates are found on eastern sides of continents in the midlatitudes. What are the major factors that determine their temperature and precipitation cycles? How do these two climates differ?
3. What natural vegetation types are associated with the moist subtropical forest environment? What weather characteristics in these areas sometimes make them uncomfortable in the summer?
4. Both the Mediterranean and marine west coast climates are found on the west coasts of continents. Why do they experience more precipitation in winter than in summer? How do the two climates differ?
5. Describe human use of the Mediterranean climate environment and discuss the advantages and limitations the climate places on these uses.
6. Contrast the natural vegetation types and human use of the land in the marine west-coast environments of North America and Europe.
7. The dry midlatitude environment is one of great agricultural importance. Why is this? What natural hazards can occur and how do farmers attempt to minimize their impact?
8. How does drought impact agriculture on the Canadian Prairies?
9. The moist continental climate region is often subject to severe winter weather. Explain the nature, causes, and impacts of such events.
10. The subtropical high-pressure cells influence several climate types in the midlatitudes. Identify the climates and describe the effects of the subtropical high-pressure cells on them.
11. Both the boreal forest and tundra climate are climates of the northern regions; however, areas of tundra typically extend to the Arctic Ocean, and the boreal forest is located further inland. Compare these two climates, taking into account coastal-continental effects.
12. Describe two types of permafrost and discuss how these different characteristics develop.
13. What is the coldest climate on Earth? How is the annual temperature cycle of this climate related to the cycle of insolation?

VISUALIZATION EXERCISES

1. Sketch climographs for the Mediterranean climate and the dry midlatitude climate. What are the essential differences between them? Explain why they occur.
2. Imagine that South America is inverted so that the southern tip is now at about lat. 10° N and the northern end (Venezuela) is at about 55° S. The Andean mountain chain would still be on the west side. Sketch this continent and draw possible climate boundaries, using your knowledge of global air circulation patterns, frontal zones, and air mass movements.

ESSAY QUESTIONS

1. Discuss how the polar front and the air masses that come in conflict in this zone affect the temperature and precipitation cycles of the midlatitude and high-latitude climates.
2. Compare and contrast the differing climate regions of Canada using Vancouver, Saskatoon, Toronto, and Yellowknife as examples.

WORKING IT OUT • STANDARD DEVIATION AND COEFFICIENT OF VARIATION

Although the mean, or average, monthly precipitation and temperature are important determinants of climate, the variability from one year to the next is also informative. In *Working It Out • Cycles of Rainfall in the Low Latitudes* in Chapter 9, the mean deviation was calculated as a measure of variation around the mean value of the data; the ratio of the mean deviation to the mean was used as a measure of relative variability.

A more common measure of variation is the *sample standard deviation*, which is defined for a sample of n precipitation values, where $i = 1, \ldots, n$, as:

$$S_P = \sqrt{\frac{1}{n-1}\sum_{i=1}^{n} D_i^2}$$

where S_P is the sample standard deviation of the precipitation and D_i is the deviation of the ith observation from the sample mean. To find the sample standard deviation, take the deviation for each precipitation value and square it. Note that any deviations that are negative will become positive after squaring. Thus, there is no need to take the absolute value of the deviation, as was done in *Working It Out • Cycles of Rainfall in the Low Latitudes*. Next, add the squared deviations, and divide that amount by $n - 1$—that is, one less than the number of samples. Finally, take the square root of the result.

The set of graphs show an example of the sample standard deviation for Penticton, British Columbia, and Charlottetown, Prince Edward Island. Graphs (a) and (b) present monthly precipitation for the two stations for a 10-year period from 1996 to 2005. Penticton lies in the dry interior of the province and experiences a mean annual precipitation of 330 mm. Charlottetown, with a moist continental climate, has mean annual precipitation of 1,297 mm.

By comparing the two graphs, it appears that Charlottetown has greater variation, largely because many months are much wetter than those in Penticton. Graph (c) compares the monthly standard deviations for each sample. In all months, the value is larger for Charlottetown, which strengthens the impression gained from the comparison of graphs (a) and (b). The standard deviation for the sample of annual totals, shown in the last pair of bars, is larger than the monthly values for both locations since the annual total itself is always larger than those of individual months.

However, the standard deviation does not measure the relative variation—that is, the variation with respect to the mean. Graph (*d*) shows the *coefficient of variation* by month. The coefficient of variation for a sample is defined as the ratio of the standard deviation to the mean. That is:

$$CV_P = \frac{S_P}{\overline{P}}$$

where CV_P is the coefficient of variation for precipitation. Relative variation of monthly precipitation at Penticton is greater than at Charlottetown in all months. Note that the coefficients of variation for the annual totals (the last pair of bars in graph (d) are smaller than those of the monthly averages, showing that the relative variation of the total—that is, for the entire year—is less.

The mean, standard deviation, and coefficient of variation of a sample are values that give basic information about the sample and its variability. They are referred to as *sample statistics* and are widely used in many branches of science.

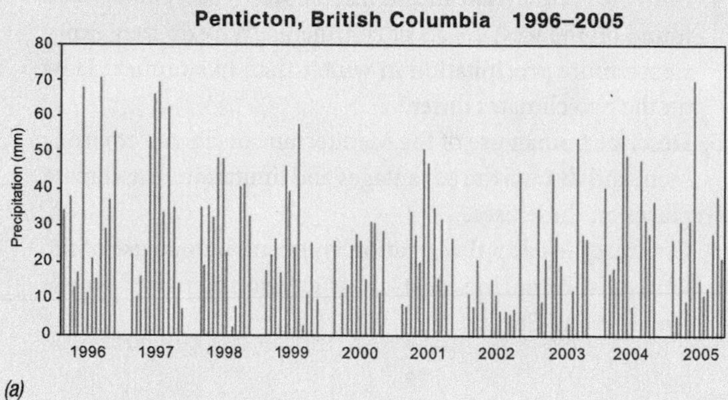

(a)

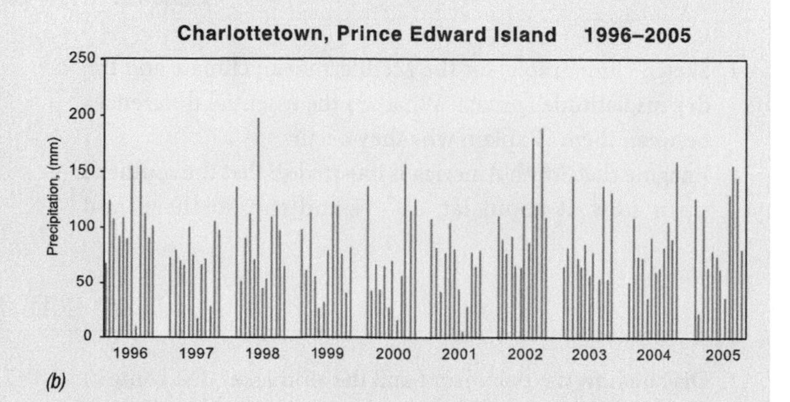

(b)

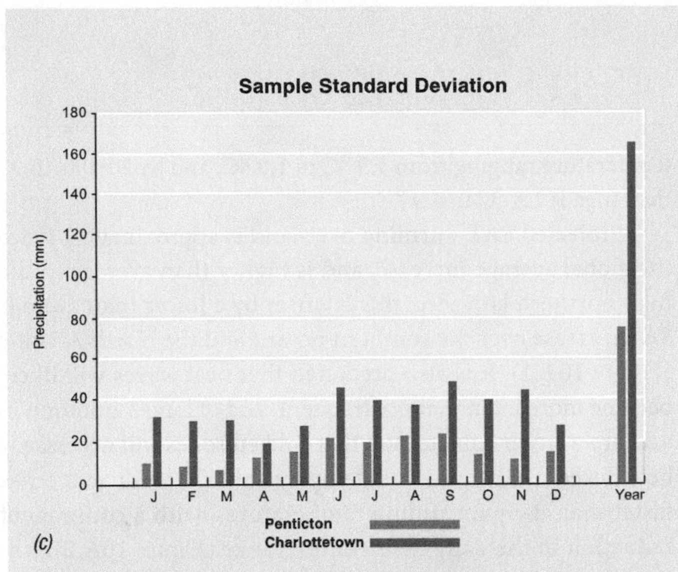

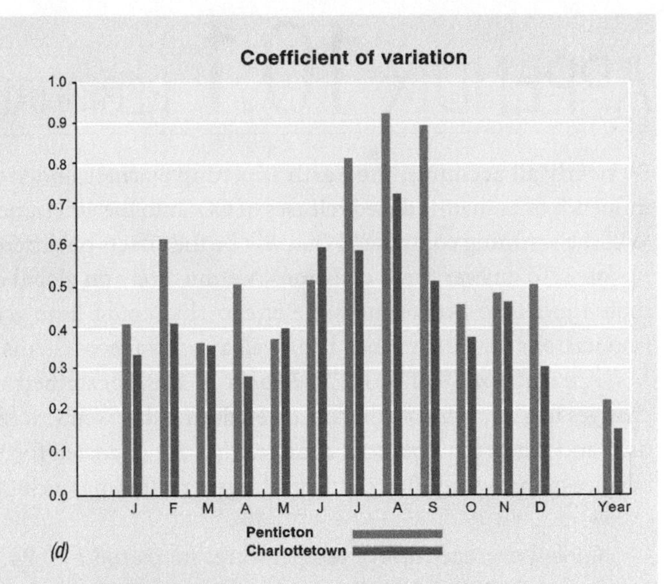

Precipitation data and statistics for Penticton, British Columbia, and Charlottetown, Prince Edward Island.

QUESTION / PROBLEM

1. The following data are monthly precipitation (mm) for January and July at Gander, Newfoundland. For each month, find the sample standard deviation and the coefficient of variation. Which month has the largest standard deviation? Which has the highest coefficient of variation?

Year	January	July
1996	78	141
1997	118	52
1998	65	108
1999	109	79
2000	190	64
2001	53	97
2002	116	72
2003	75	95
2004	135	62
2005	150	118

Find answers at the back of the book.

APPENDIX 10.1 REGIONAL IMPACTS OF CLIMATE CHANGE

By nearly all accounts, the Earth is getting warmer under the influence of human-induced releases of CO_2 into the atmosphere. Will the warming continue? What will be the effects on different regions? To answer these questions, we must rely on global climate models to predict possible effects that could have great implications for human society as well as natural ecosystems.

Appendix 4.1 • The IPCC Report of 2007 described the changes in global climate that occurred during the twentieth century and listed the potential changes that are expected for the balance of the twenty-first century. The general findings included:

- Global average surface temperatures increased by 0.74 °C between 1906 and 2005.
- The warming trend for the period 1956 to 2005 is nearly twice that for the 100-year period ending in 2005.
- Average Arctic temperatures have increased at almost twice the global rate in the past 100 years.
- Land areas have warmed faster than the oceans.
- Global average sea level rose at an average annual rate of 1.8 mm from 1993 to 2003.
- Since 1978, the extent of the annual average Arctic sea ice has shrunk by 2.7 percent per decade.
- Mountain glaciers and snow cover on average have declined in both hemispheres.
- Over the period 1900 to 2005, precipitation increases have been observed in eastern regions of North and South America, northern Europe, and northern and central Asia.
- Globally, the area affected by drought appears to have increased.
- Several extreme weather events appear to have changed in frequency and/or intensity over the past 50 years.

The following discussion is based on the Intergovernmental Panel on Climate Change (IPCC) report published in 2007—*Climate Change 2007: The Physical Science Basis; Contribution of Working Group I to the Fourth Assessment Report of the Intergovernmental Panel on Climate Change.* It uses the A1B scenario. Using this non-mitigation scenario, there is general agreement that global mean surface air temperature (SAT) will continue to increase over the twenty-first century as a result of increased greenhouse gas emissions. Results from complex Atmosphere-Ocean General Circulation Models (AOGCMs) suggest an average warming of 0.64 °C to 0.69 °C from 2011 to 2030, compared to conditions in 1980 to 1999. However, the projected rise in temperature becomes more variable as lead time is extended. By 2046 to 2065, the different models predict a rise in temperature ranging from 1.3 °C to 1.8 °C, and by 2090 to 2099 the range is 1.8 °C to 4.0 °C.

Projected SAT warming over land is approximately twice the global average increase, and is higher than average in the high northern latitudes; this is offset by a lower than average SAT increase over the southern oceans and the North Atlantic (Figure 10A.1). It is also predicted that heat waves will likely become more intense, more frequent, and of longer duration in a future warmer climate, and that cold episodes will necessarily be reduced. Daily minimum temperatures are predicted to rise faster than daily maximum temperatures, with a consequent reduction in the daily temperature range (Figure 10A.2). It is

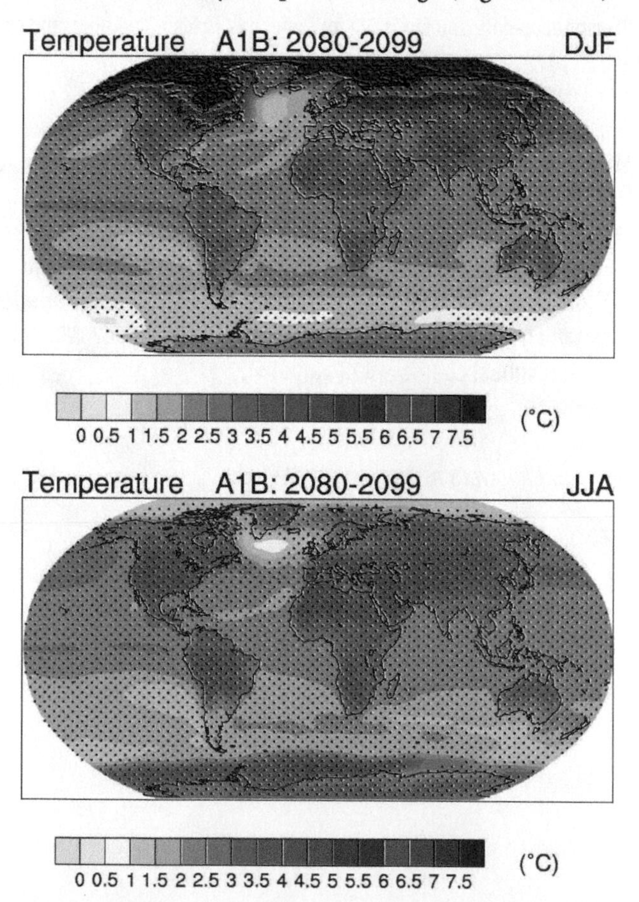

10A.1 MULTI-MODEL MEAN CHANGE IN SURFACE AIR TEMPERATURES FOR THE NORTHERN HEMISPHERE WINTER (TOP) AND NORTHERN HEMISPHERE SUMMER (BOTTOM) BASED ON THE A1B SCENARIO This scenario assumes rapid economic growth, a global population that peaks in mid-century, and rapid introduction of new technologies, leading to a balance of fossil fuel intensive and non-fossil energy sources. Changes are given as mean values for the period 2080–2099 relative to 1980–1999. Stippling denotes areas where there is poor agreement between models.

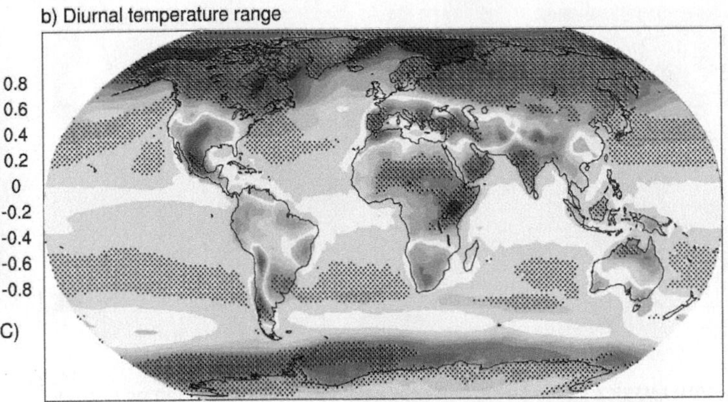

10A.2 MULTI-MODEL MEAN CHANGE IN DIURNAL TEMPERATURE RANGE BASED ON THE A1B SCENARIO Changes are given as annual means for the period 2080–2099 relative to 1980–1999. Stippling denotes areas where there is considerable disagreement between models.

also predicted that heat waves will occur more frequently in all parts of the world (Figure 10A.3a). The general warming trend will likely decrease the number of frost days throughout the mid- and high latitudes, which consequently will lengthen the growing season in these regions (Figures 10A.3b and 10A.3c).

All of North America is expected to warm during the twenty-first century, with annual mean warming likely to exceed the global mean (Figure 10A.4). Warming in northern areas is likely to be greatest in winter, with temperatures possibly 7 to 10 °C higher than present. In the southern part of the continent, the seasonal temperature change will likely be greatest in summer and be in the order of 3.5 to 4 °C. Apart from the Arctic, the coastal regions will experience the lowest overall temperature increase.

The general global increase in temperature will be accompanied by higher rates of evaporation and a higher concentration of water vapour in the atmosphere. This is expected to alter the amount and distribution of cloud cover (Figure 10A.5). A general increase in high-level cloud cover is predicted for all latitudes. Conversely, mid-level cloud cover tends to decrease at all latitudes. The change in low-level cloud cover is more variable, with increases predicted in equatorial regions and at higher latitudes, but a decrease expected in subtropical areas, especially in the northern hemisphere. The overall effect is that most of the lower and middle latitudes will likely experience a decrease in total cloud cover, whereas more cloud is expected over high latitudes. While clouds do not always produce precipitation, there appears to be some spatial correlation between the predicted changes in cloud cover and precipitation.

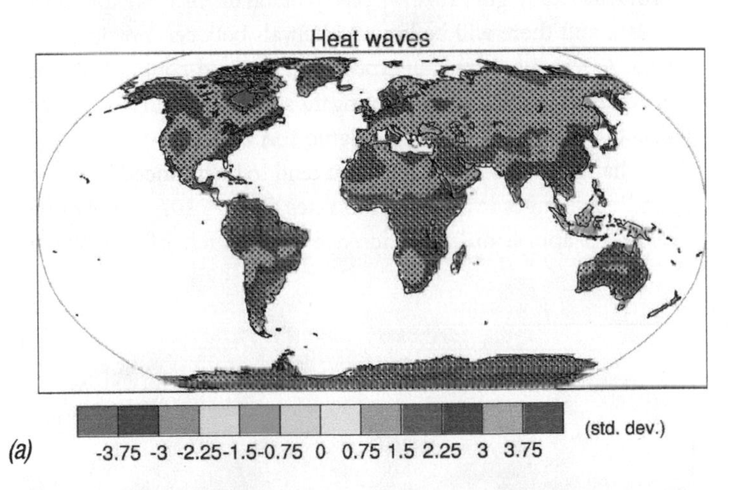

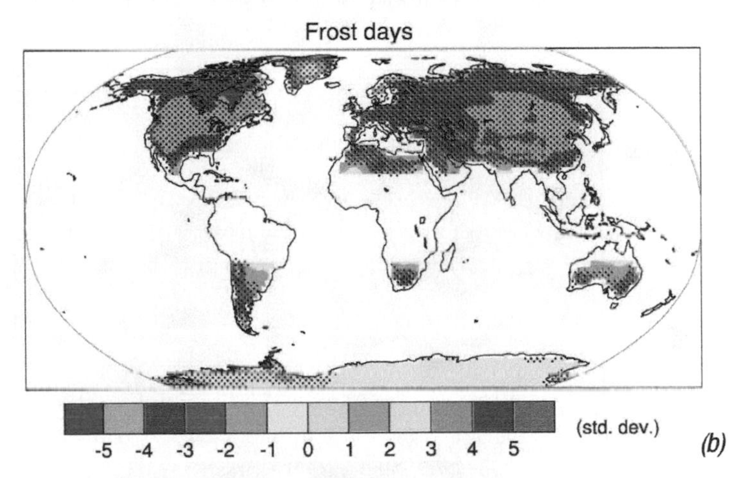

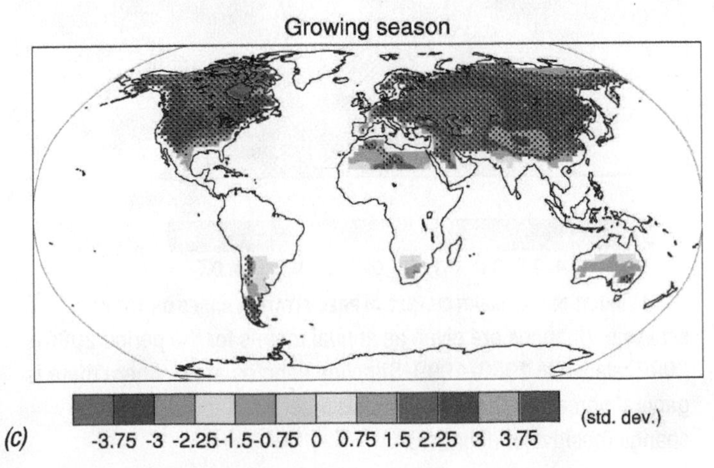

10A.3 PROJECTED CHANGE IN OCCURRENCE OF TEMPERATURE RELATED EVENTS (a) Globally averaged change in heat waves (defined as the longest period of the year of at least five consecutive days with maximum temperatures at least 5 °C higher than the mean temperature for the corresponding dates); (b) globally averaged change in frost days (defined as days in which the absolute minimum temperature is below 0 °C); (c) length of growing season (defined as the period between the first spell of five consecutive days with mean temperatures above 5 °C and the last such spell). The maps are based on the A1B scenario, with changes given in terms of the standard deviation above and below the mean for the period 2080–2099 relative to 1980–1999. Stippling denotes areas where there is general agreement between models.

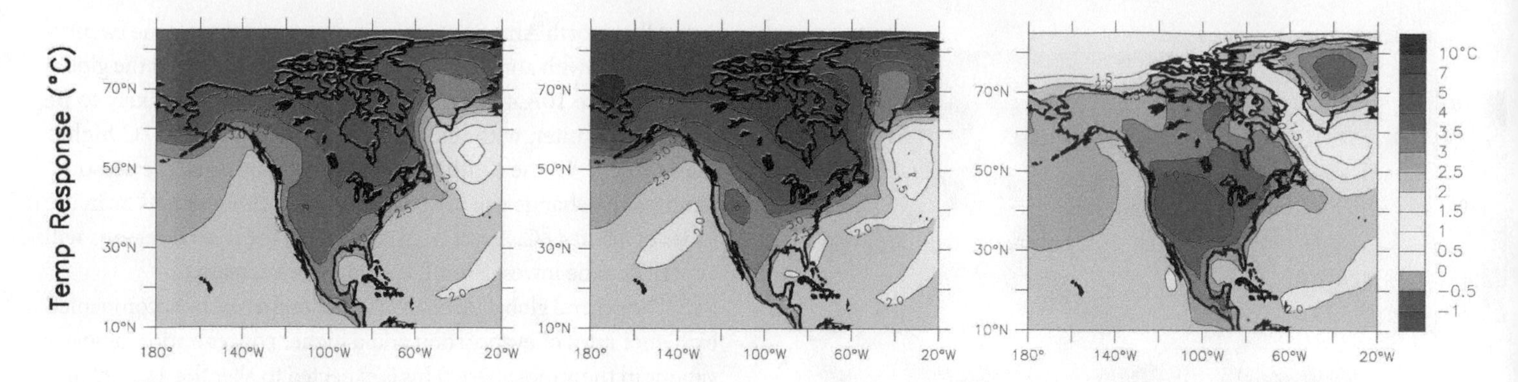

10A.4 AVERAGE PREDICTED ANNUAL AND SEASONAL TEMPERATURE CHANGE OVER NORTH AMERICA BASED ON THE A1B SCENARIO Changes are given as annual means for the period 2080–2099 relative to 1980–1999.

Under a warmer climate scenario, the models predict that precipitation will increase in wet tropical areas and at high-latitudes; a decrease in precipitation is predicted in the subtropics and lower mid-latitudes (Figure 10A.6). The general pattern suggests that precipitation will increase by about 20 percent in most high latitude regions, as well as eastern Africa, central Asia, and the equatorial Pacific Ocean. The change in precipitation over the ocean in the latitudinal band between 10° N and 10° S accounts for about half of the predicted global increase. Decreases in precipitation by as much as 20 percent are predicted in the Mediterranean Basin, the Caribbean, and along subtropical western coasts. In general, precipitation over land is predicted to increase by about 5 percent and over the oceans by 4 percent. The net change over land therefore accounts for 24 percent of the global mean increase in precipitation. A warmer climate is expected to result in more intense precipitation events because warm air has the ability to hold more moisture. This effect will likely be most pronounced in the middle and high latitudes, especially in Europe and North America (Figure 10A.7a); precipitation intensity is expected to increase and there will be longer intervals between precipitation events. As a result, there is an expectation that mid-continental areas will experience general drying during the summer, leading to a greater risk of drought in those regions (Figure 10A.7b).

Changes in global precipitation tend to be balanced by corresponding changes in evaporation rates (Figure 10A.8a). Annual average evaporation will increase over much of the ocean,

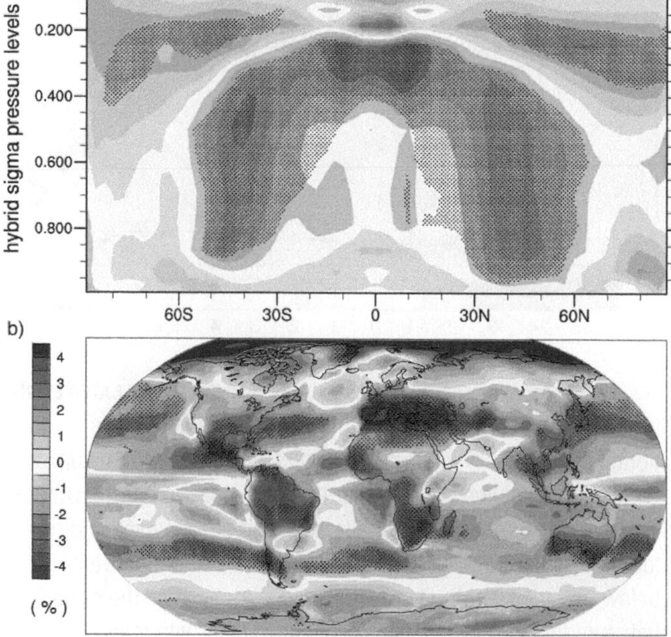

10A.5 MULTI-MODEL MEAN CHANGE IN (a) ZONAL MEAN CLOUD COVER (%) SHOWN AS A CROSS-SECTION THROUGH THE ATMOSPHERE, AND (b) TOTAL GLOBAL CLOUD COVER (%) BASED ON THE A1B SCENARIO Changes are given as annual means for the period 2080–2099 relative to 1980–1999. Stippling denotes areas where there is considerable disagreement between models.

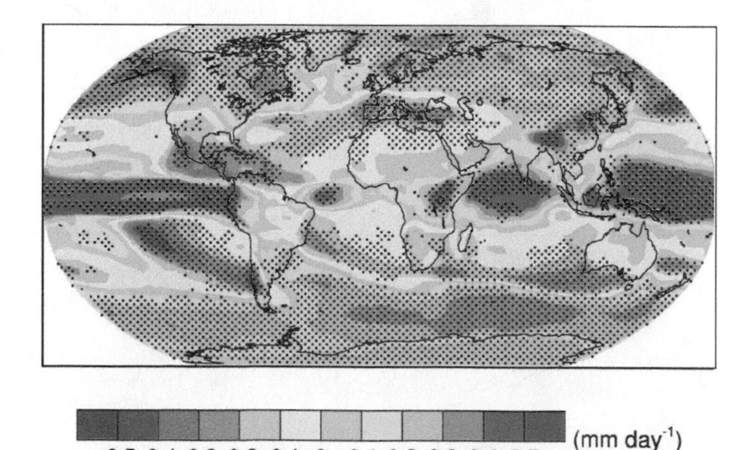

10A.6 MULTI-MODEL MEAN CHANGE IN PRECIPITATION BASED ON THE A1B SCENARIO Changes are given as annual means for the period 2080–2099 relative to 1980–1999. Stippling denotes areas where there is general agreement between models in terms of the sign of the change (positive or negative).

with spatial patterns generally related to the predicted change in sea surface temperature. Over land, changes in precipitation will lead to changes in soil moisture and surface runoff. Runoff is predicted to be considerably reduced in southern Europe and central America. An increase in runoff is expected in places such as southeast Asia, east Africa, and western Canada (Figure 10A.8b). Associated changes in soil moisture suggest decreases across most regions, especially in the subtropics, the Mediterranean Basin, and the southern United States and Mexico (Figure 10A.8c). Increases in soil moisture will be most pronounced in Africa and central Asia.

In North America, precipitation is expected to increase over most of the continent (Figure 10A.9). The annual increase could be as much as 20 percent in the north, and seasonally as high as 30 percent in the winter. The increase during winter is attributed to the northward displacement of the westerly winds and intensification of the Aleutian Low. This is expected to increase orographic precipitation on the windward slopes of the western mountains. Precipitation is predicted to decrease annually and in all seasons in the southwestern part of the continent due to enhanced subsidence attributed to an intensification of regional anticyclonic conditions. Summer precipitation is predicted to decrease by 5 to 15 percent across much of central and southern parts of North America, and perhaps increase by about 10 percent in northern Canada. Warmer temperatures in summer will increase evaporation, and soil moisture is likely to drop despite the higher precipitation in winter over most of the continent.

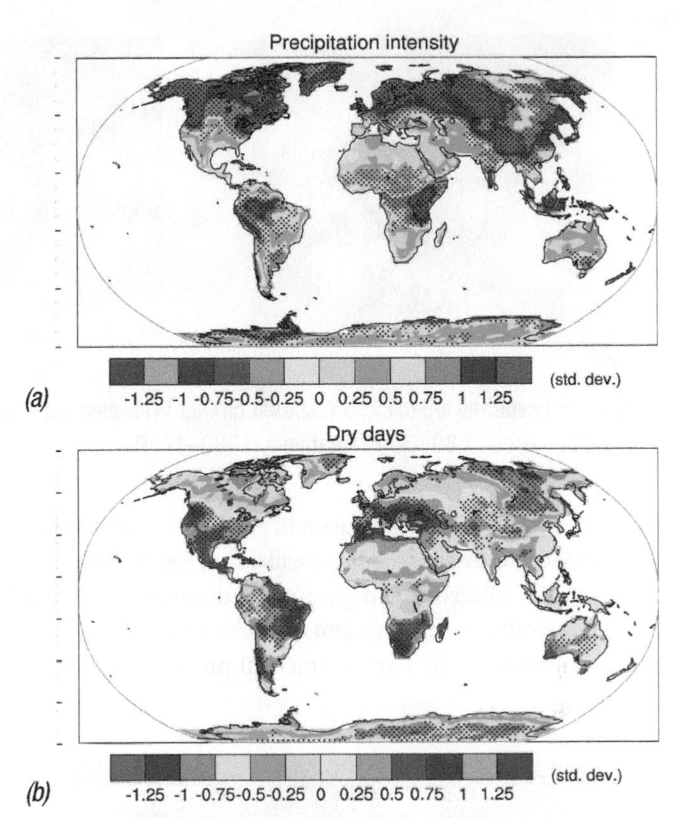

10A.7 PROJECTED CHANGE IN PRECIPITATION EVENTS (a) Globally averaged change in precipitation intensity (defined as the annual total precipitation divided by the number of wet days); (b) globally averaged change in dry days (defined as the annual maximum number of consecutive days without precipitation). The maps are based on the A1B scenario with changes given in terms of the standard deviation above and below the mean for the period 2080–2099 relative to 1980–1999. Stippling denotes areas where there is general agreement between models.

(a) Evaporation

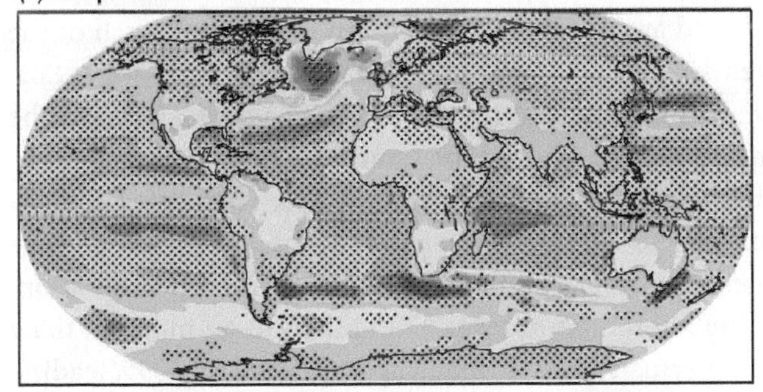

(b) Soil moisture

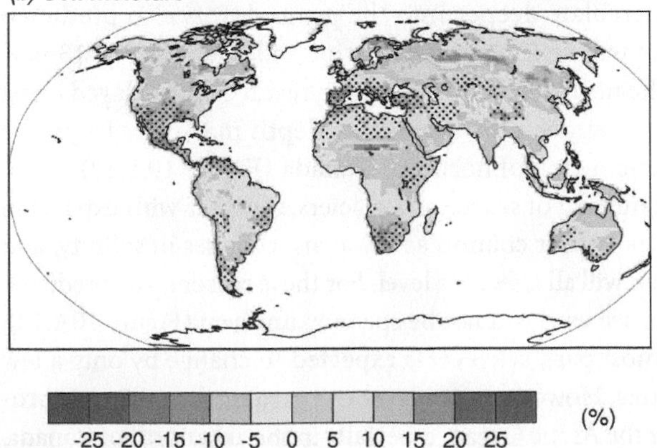

(c) Runoff

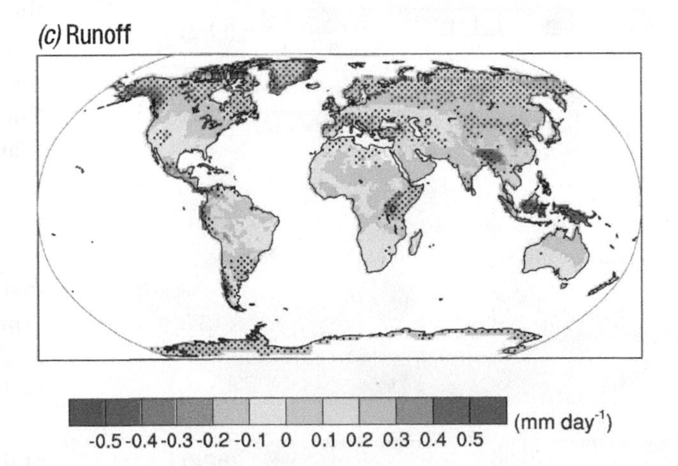

10A.8 MULTI-MODEL MEAN CHANGE IN (a) EVAPORATION, (b) SOIL MOISTURE, AND (c) SURFACE RUNOFF, BASED ON THE A1B SCENARIO Changes are given as annual means for the period 2080–2099 relative to 1980–1999. Stippling denotes areas where there is general agreement between models in terms of the sign of the change (positive or negative).

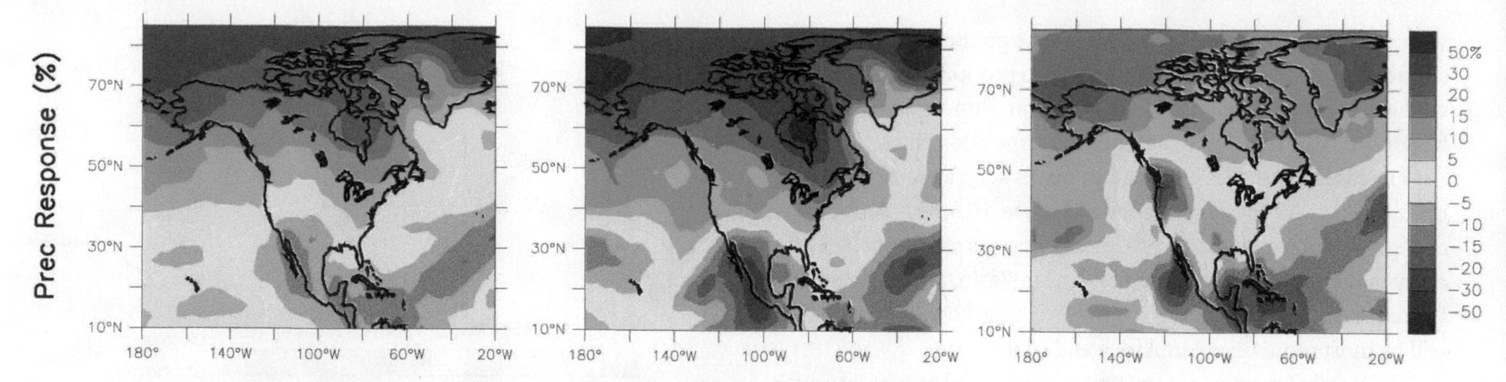

10A.9 AVERAGE PREDICTED ANNUAL AND SEASONAL CHANGE IN PRECIPITATION OVER NORTH AMERICA BASED ON THE A1B SCENARIO Changes are given as annual means for the period 2080–2099 relative to 1980–1999.

Projections of sea level pressure and circulation patterns suggest that spatial and seasonal changes will occur. Sea level pressure simulations show a decrease at high latitudes during both summer and winter in both the northern and southern hemispheres. This is mainly compensated by semi-permanent pressure increases at midlatitudes, particularly across the southern Pacific and Indian Oceans, and by a seasonal increase in pressure in the Mediterranean Sea (Figure 10A.10). Such a change is expected to cause an expansion of the Hadley circulation and a poleward shift in the midlatitude storm tracks. This would also contribute to the increased precipitation predicted at high latitudes and the decrease in the subtropics; it would also strengthen the midlatitude westerlies and help to bring moist air to the continental interiors.

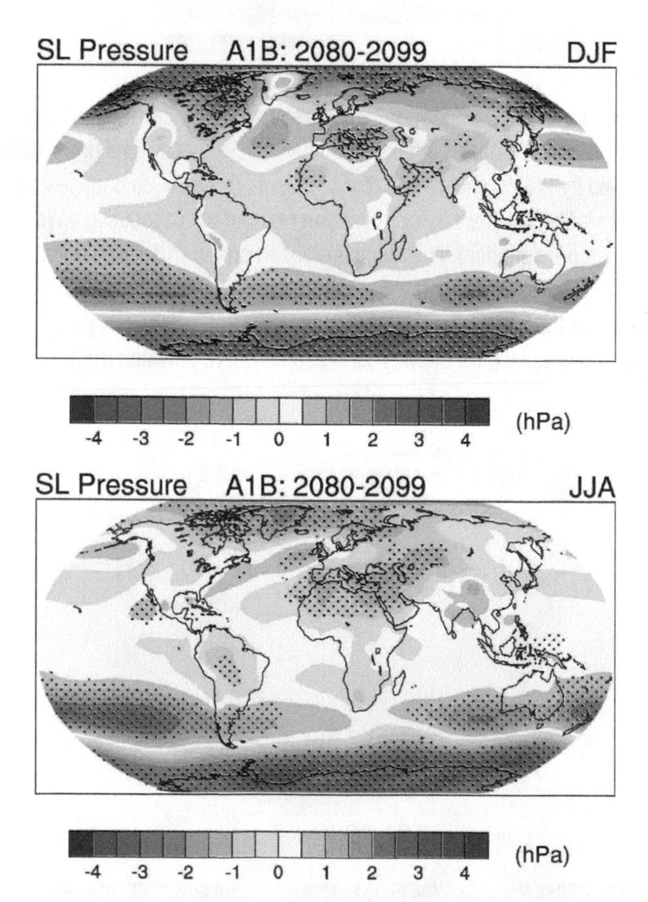

10A.10 MULTI-MODEL MEAN CHANGE IN SEA LEVEL PRESSURE FOR THE NORTHERN HEMISPHERE WINTER (TOP) AND NORTHERN HEMISPHERE SUMMER (BOTTOM) BASED ON THE A1B SCENARIO Changes are given as annual means for the period 2080–2099 relative to 1980–1999. Stippling denotes areas where there is poor agreement between models.

The projected warming trend for the twenty-first century is greatest at high latitudes, and is expected to greatly reduce the summer ice area. Antarctic sea ice cover is expected to decrease at a slower rate than the Arctic ice pack (Figure 10A.11). This is largely attributed to the rate at which heat is dissipated into the deep ocean. In the Southern Ocean, heat is transported away from the surface; in the Arctic, much of the heat remains in the upper 1 km of the water column. In addition, more heat will be transported into the Arctic Basin by the thermohaline conveyor. At the same time there will be accelerated ablation of the Greenland ice cap. This is expected because the summer melt rate will be greater than any increase in ice thickness that might accrue from higher winter precipitation. Warming at high latitudes is also linked to an increase thaw depth in permafrost by as much as 40 percent by 2099, leading to a reduction in soil moisture as snow melt and rain will be able to percolate deeper into the ground. It is also predicted that there will be a general reduction in snow cover of 13 percent in the northern hemisphere as a result of the delayed onset of snow in winter. However, snow depth may increase in the high Arctic region of northern Canada (Figure 10A.12).

The melting of sea ice and glaciers, together with expansion of the ocean water column as it warms, changes in salinity, and circulation will all affect sea level. For these reasons, the predicted change in sea level will not be spatially uniform (Figure 10A.13). For the most part, sea level is expected to change by only a few centimetres. However, a sea level rise greater than 0.2 m is projected for the Arctic Ocean, especially in the area north of Canada. A second zone of pronounced sea level rise extends across the

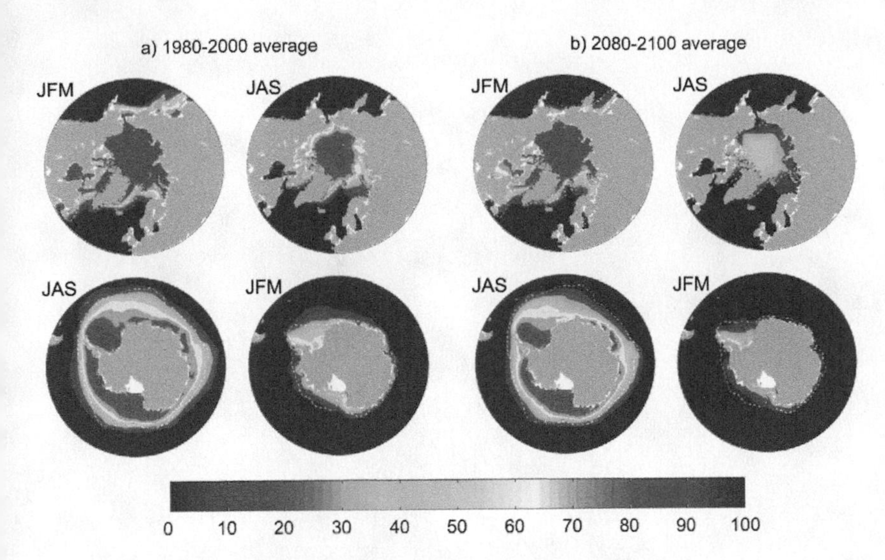

10A.11 CHANGE IN MEAN SEASONAL SEA ICE CONCENTRATION IN THE ARCTIC OCEAN AND ANTARCTICA IN 2080–2100 COMPARED TO CONDITIONS IN 1980–2000 The projection is based on the A1B scenario. The dashed white line indicates the present limit of 15 percent sea ice concentration.

Indian Ocean. A fall in sea level by more than 0.2 m is suggested for the Southern Ocean.

These observations show that climate change will have major impacts on North America. To reduce the costs to society and the environment, a two-pronged strategy needs to be implemented. The first strategy is abatement: emissions of CO_2 and other greenhouse gases need to be lowered by reducing the use of fossil fuels. This will require both active energy conservation and a shift to alternative energy sources, such as solar, wind, and wave power. Water conservation will also be important in many regions as water supplies dwindle and demand increases. The second strategy is mitigation. Substantial investments in infrastructure will be needed, such as the construction of new coastal structures where sea levels are rising. In agriculture, new crop varieties and cropping practices will have to be developed. In nearly all economic endeavours, the costs of dealing with a greater risk of extreme weather events will need to be factored in.

Whatever the course of global warming, human society will be best served by anticipating and planning for a changing future, rather than by reacting to it as it occurs.

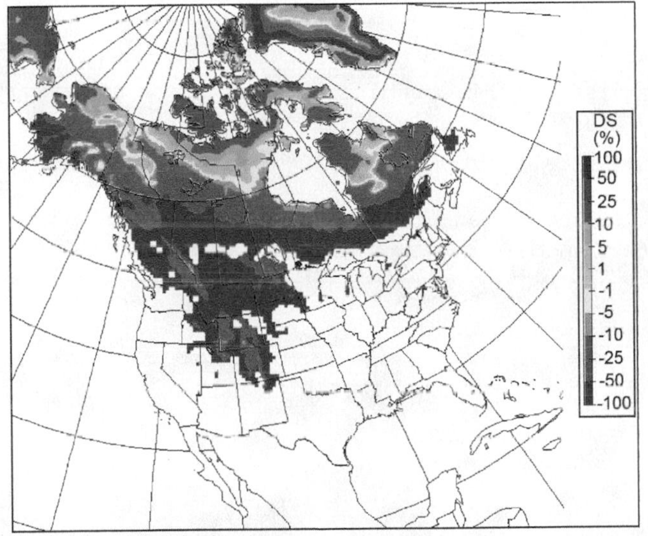

10A.12 PERCENT SNOW DEPTH CHANGE IN MARCH (CALCULATED ONLY WHERE SNOW AMOUNTS EXCEED 5 MM WATER EQUIVALENT) AS PROJECTED BY THE CANADIAN REGIONAL CLIMATE MODEL UNDER A2 SCENARIO.

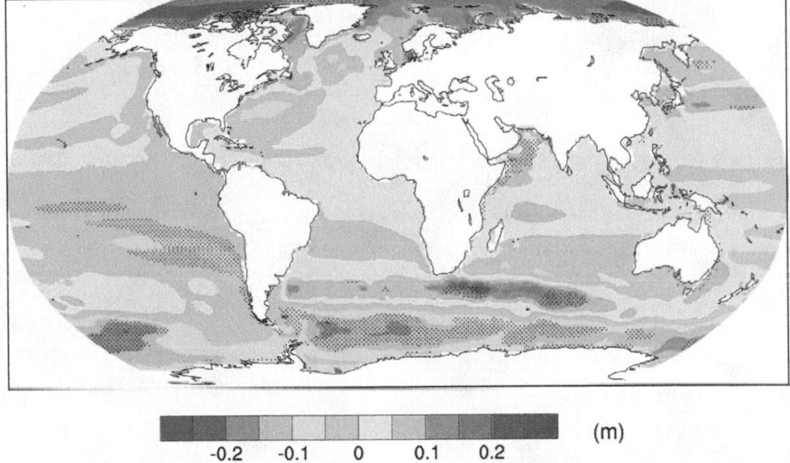

10A.13 LOCAL SEA LEVEL CHANGE DUE TO OCEAN DENSITY AND CIRCULATION PATTERNS RELATIVE TO THE GLOBAL AVERAGE The projection is based on the A1B scenario. Changes are given as annual means for the period 2080–2099 relative to 1980–1999. Stippling denotes areas where there is poor agreement between models.

QUESTIONS

1. Identify the major changes in the North American climate that are anticipated for the remainder of the twenty-first century and describe their general impact.
2. Identify the major changes in the climate of the northern hemisphere that are anticipated for the remainder of the twenty-first century and describe their general impact.
3. Describe how the response of ocean regions to anticipated climate change is expected to differ from continental areas in a) the northern and b) the southern hemisphere.

PART 3

SYSTEMS AND CYCLES OF THE SOLID EARTH

Wave Rock, Western Australia. This granite outcrop has been undercut by weathering and erosion from flowing water. Its face is about 15 m high and is stained by streaks of black algae.

Part 3 examines the systems and cycles of the solid Earth. Unlike surface energy and matter flows, such as heat and precipitation driven by solar radiation, the solid Earth systems are driven by the planet's internal energy. These two great systems of energy and matter flow also operate on very different time scales. The surface energy flow systems are influenced by the rhythms that arise from the daily rotation of the Earth on its axis and its annual revolution around the sun. However, the cycles associated with the solid Earth systems are typically measured in thousands to millions of years. ● Chapter 11 begins Part 3 by presenting the basic materials of the solid Earth—rocks and minerals—and some principles of their formation. The continuous production and conversion of one type of rock into another is described by the cycle of rock transformation, in which Earth materials are constantly recycled over geologic time. Forces deep within the planet drive part of the rock cycle. This energy source causes the vast tectonic plates that form the Earth's crust to converge, collide, split, and separate. Because of this, the configuration of continents and ocean basins slowly changes. Chapter 12 discusses this slow, inexorable movement, called plate tectonics. The present pattern of plates and their motions explains the location of many geologic phenomena, such as earthquakes and volcanoes. These types of geologic activity are the subject of Chapter 13.

CHAPTER 11

EARTH MATERIALS AND THE CYCLE OF ROCK TRANSFORMATION

The Spectrum Range, Mount Edziza Provincial Park, British Columbia.

The study of the solid Earth begins by examining the minerals and rocks found at or near the surface. The solid Earth materials are constantly formed and reformed over geologic time in the rock cycle that ultimately depends on energy within the planet and from the sun. A few radioactive elements provide a nearly infinite source of energy that migrates slowly outward from the Earth's interior; the most notable are uranium (U), thorium (Th), and potassium (K). Chapter 12 discusses radiogenic heating and radioactive decay. In addition to radioactive energy, a second source of internal energy is the residual heat generated by compression of cosmic debris when the Earth was created some 4.6 billion years ago. Both of these sources are important for powering the cycles of the solid Earth.

THE CRUST AND ITS COMPOSITION

The outermost layer of the planet is the *Earth's crust*. It varies in thickness from 8 to 80 km and comprises the continents and ocean basins. The thickness of the crust is generally less than 10 km under the oceans and averages about 40 km in most continents (Figure 11.1). Approximately 10 percent of the crust

11.1 THICKNESS OF EARTH'S CRUST The map uses a 10-km contour interval (plus the 45-km contour line) and was created directly from the 5-degree-by-5-degree gridded crustal model CRUST 5.1.

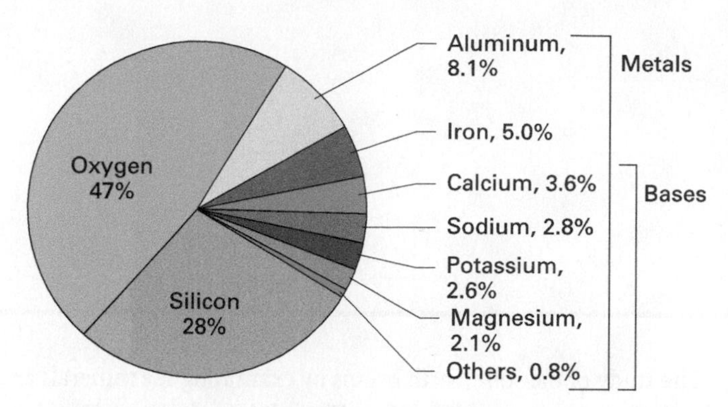

11.2 CRUSTAL COMPOSITION The eight most abundant elements in the Earth's crust, measured by percentage of weight.

exceeds 50 km, with the greatest thickness reported in the Himalayas, where it is about 80 km.

Figure 11.2 shows the eight most abundant elements of the Earth's crust by weight. The predominant elements are oxygen (O) and silicon (Si), which together account for approximately 75 percent of the crust. Aluminum (Al) and iron (Fe) account for about 8 percent and 5 percent respectively. Four other metallic elements occur at abundances greater than 2 percent. Three of these, calcium (Ca), potassium (K), and magnesium (Mg), are essential nutrients for plant and animal life. All other elements together account for less than 1 percent of the total weight of the crust.

ROCKS AND MINERALS

The elements of the Earth's crust are usually bonded together to form minerals. A **mineral** is a naturally occurring, inorganic substance that has a definite chemical composition and a characteristic atomic structure. This gives each mineral its distinctive properties, such as colour, lustre, and hardness. Most minerals have a crystalline structure that reflects the internal arrangement of its composing atoms; for example, crystals of lead sulphide (PbS) are commonly found as cubes whereas quartz (SiO_2) may form translucent columns (Figure 11.3). Large crystals will only form under ideal conditions, and in most minerals, such as quartz, orthoclase feldspar ($KAlSi_3O_8$), and biotite [$K(Mg,Fe)_3(AlSi_3O_{10})(OH,F)_2$], the crystals are so small that they combine into more or less uniform masses (Figure 11.4).

11.3 MINERAL CRYSTALS (a) Lead sulphide is commonly found in its crystalline form as cubes. (b) Quartz can form large six-sided translucent columns.

Minerals are combined into **rock**. Although many types of rocks are recognized on the basis of their different compositions, physical characteristics, and ages, they are all assemblages of minerals in the solid state (Figure 11.5). Most rock in the Earth's crust was formed many millions of years ago, but rock is also being currently formed, such as when active volcanoes emit lava that solidifies on contact with

11.4 COMMON ROCK FORMING MINERALS (a) Quartz is a common mineral, and because it is very resistant to weathering, it is the main component of beach and river sand (note the contrast in form with the larger crystals shown in Figure 11.3); (b) Feldspar is the most abundant mineral in the Earth's crust and is found in most types of rock; (c) Biotite is usually found in thin flexible sheets that can be easily peeled apart.

11.5 GRANITE The constituent minerals in this igneous rock can be readily identified by their different colours. In this rock sample, the white grains are quartz, feldspar is pink, and biotite is black.

the atmosphere or ocean. Rocks in the Earth's crust fall into three major classes based on the nature in which they are formed: igneous, sedimentary, and metamorphic. Figure 11.6 shows the general distribution of these major rock classes in North America.

Igneous rocks are solidified from mineral matter in a high-temperature molten state. The mineral grains in igneous rocks are tightly interlocked, and the rock is normally very strong. The molten rock, or **magma**, cools and solidifies between about 600 and 1,200 °C, either below the surface as *intrusive rocks*, such as granite, or on the surface as *extrusive rocks*, such as volcanic lava.

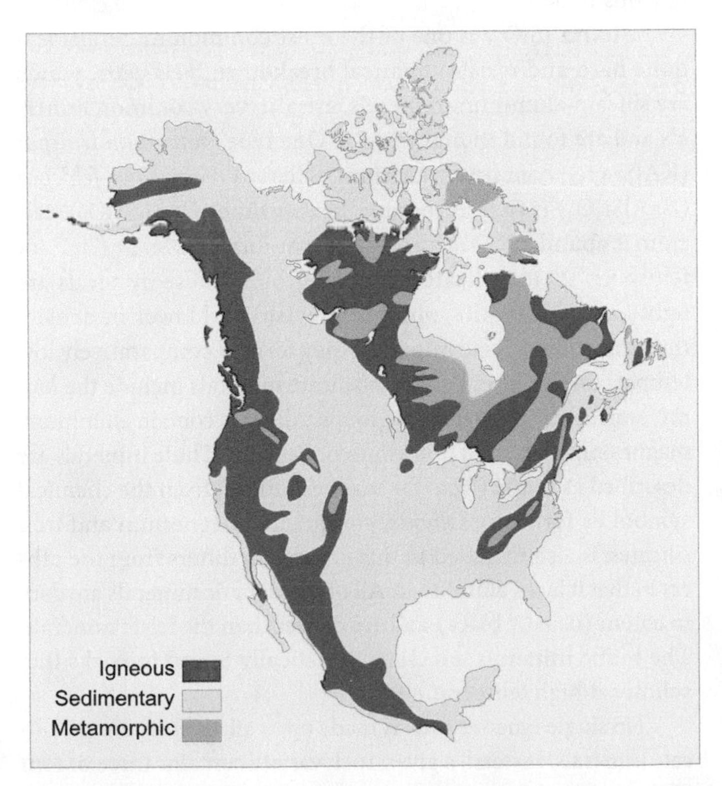

11.6 GENERAL DISTRIBUTION OF THE IGNEOUS, SEDIMENTARY, AND METAMORPHIC ROCKS IN NORTH AMERICA

11.7 GRANITE ROCK OUTCROP IN STRAWAMUS CHIEF PROVINCIAL PARK, NEAR SQUAMISH, BRITISH COLUMBIA This granite outcrop originally formed many kilometres below the surface and has been exposed by removal of the overlying rock.

Although intrusive rocks are formed at some depth beneath the Earth's crust, they can be exposed at the surface as overlying rock layers are eroded over millions of years (Figure 11.7).

Sedimentary rocks are formed from layered accumulations of mineral particles, derived mostly by weathering and erosion of pre-existing rock. The various chemical, physical, and biological changes that occur as layers of sediment transform into rocks are collectively referred to as *diagenesis*. The resulting rock layers are called **strata** (Figure 11.8). Sedimentary rocks are the most common exposed rock types and cover about 75 percent of the Earth's surface. However, they account for only about 5 percent of the total mass of the Earth's rocks because they are only found within a comparatively thin layer over igneous and metamorphic rocks. Unlike igneous and metamorphic rocks, sedimentary rocks often contain *fossils*, making them invaluable for interpreting geologic history.

11.8 SEDIMENTARY ROCK STRATA EXPOSED IN THE ERODED GULLIES AND COULEES OF THE BADLANDS NEAR DRUMHELLER, ALBERTA The rocks are sandstones and mudstones, interlaced with layers of shale and coal.

11.9 METAMORPHIC ROCKS IN THE HARDANGERVIDDA MOUNTAIN PLATEAU IN WESTERN NORWAY This glaciated region of southern Norway is composed of ancient metamorphic rocks, including gneiss, quartzite, and phyllite.

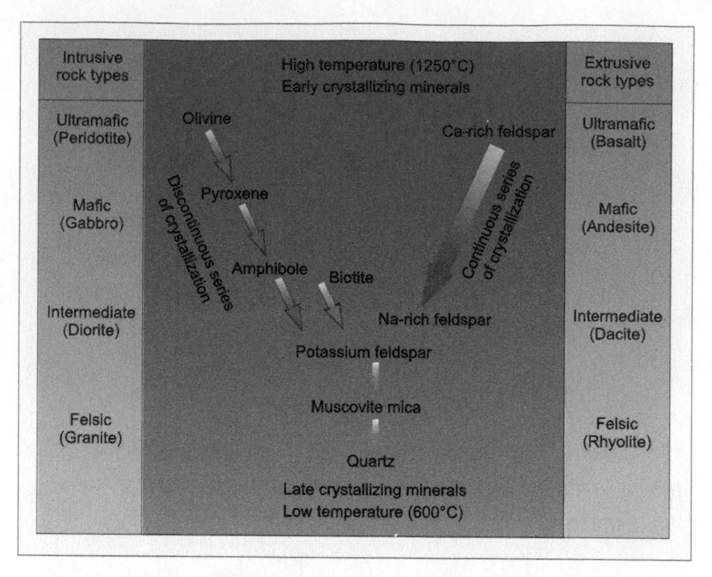

11.10 BOWEN'S REACTION SERIES Bowen's reaction series represents the sequence of mineral crystallization from basaltic magma, as suggested by experimental evidence.

Metamorphic rocks are formed from igneous or sedimentary rocks that have been physically or chemically changed, usually by the subsequent application of heat and pressure during episodes of significant crustal movement (Figure 11.9). The process of **metamorphism** creates new minerals and rock structures. Certain minerals such as garnet, for example, are formed in this way, and can provide information about the temperatures and pressures under which specific metamorphic rocks are created. The formation process can partially or completely reset the radiometric decay sequence so it can be difficult to determine the age of metamorphic rocks. In North America, exposed metamorphic rocks occur most extensively in the Canadian Shield.

IGNEOUS ROCKS

Igneous rock crystallizes from hot, molten magma that originates deep below the surface of the Earth. The temperature required to melt rock varies widely depending on composition. For the common *silicate minerals*, melting occurs at temperatures ranging from 600 °C for quartz, mica, and potassium-rich feldspar, to about 1,250 °C for calcium-rich feldspar. The molten rock migrates upward from the magma chambers through fractures in older solid rock and eventually solidifies within or on top of the Earth's crust. Magma that is injected into existing rock masses and remains surrounded by older, pre-existing rock is called **intrusive igneous rock**. The process of injection into existing rock is called *intrusion*. Intrusive rocks solidify below the Earth's surface. Where magma reaches the surface, it emerges as **lava**, which solidifies to form **extrusive igneous rock**. The sequence of crystallization at temperatures below about 1,200 °C is described by *Bowen's reaction series* (Figure 11.10). The chemical composition of the magma and the rate of cooling are the main determinants of rock composition and crystal size.

The *silicate minerals* that form igneous rocks are chemical compounds that contain silicon and oxygen atoms, combined with various proportions of other elements, particularly aluminum, iron, calcium, sodium, potassium, and magnesium. Figure 11.11 shows seven of the most common silicate minerals and the various combinations found in a selection of igneous rocks.

Quartz (SiO_2) is one of the most common minerals. It is quite hard and resists chemical breakdown. *Feldspars*, which are silicate-aluminum minerals, are also very common minerals and are found in many rocks. One type, *potassium feldspar* ($KAlSi_3O_8$), contains potassium, whereas *plagioclase feldspar* ($NaAlSi_3O_8$) is rich in sodium. In anorthite ($CaAl_2Si_2O_8$), calcium is abundant. Quartz and feldspar form the **felsic** ("fel" for feldspar; "si" for silicate) mineral group. These minerals are light in colour (white, pink, or greyish) and lower in density than the other silicate minerals; they form at comparatively low temperatures. Other common silicate minerals include the *biotite*, *amphibole*, and *pyroxene* groups. All three contain aluminum, magnesium, iron, and potassium or calcium. These minerals are described as **mafic** ("ma" for magnesium; "f" from the chemical symbol Fe for iron). *Olivine*, comprising magnesium and iron silicates, is also included in this group, but differs from the others in that it lacks aluminum. All of these mafic minerals are dark in colour (usually black) and are denser than the felsic minerals. The mafic minerals are characteristically found in rocks that solidify at high temperatures.

No single igneous rock is made up of all seven common silicate minerals. Instead, a given rock variety contains three or four of those minerals as its major constituents (see Figure 11.11).

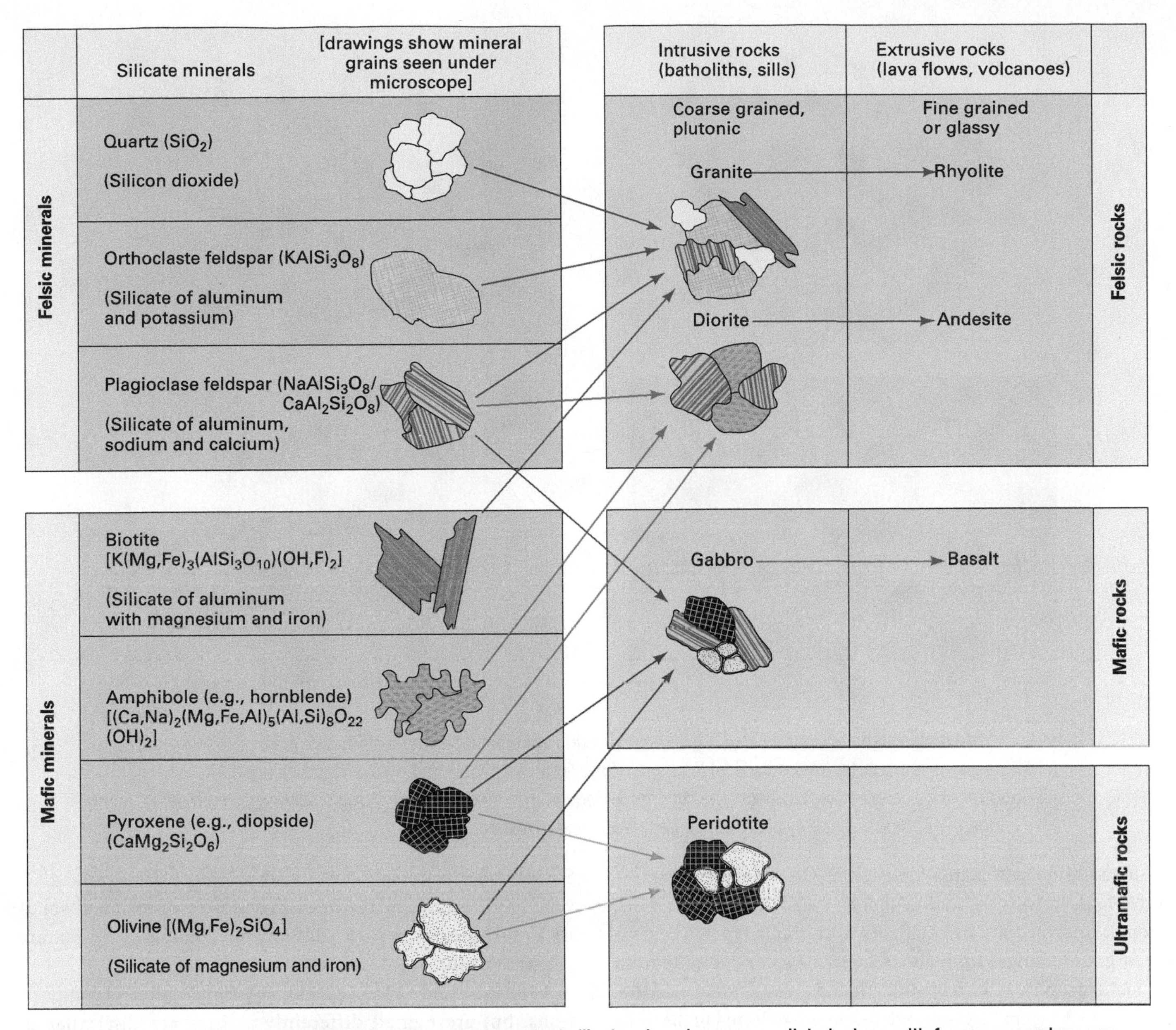

11.11 SILICATE MINERALS AND IGNEOUS ROCKS Only the most important silicate mineral groups are listed, along with four common igneous rock types. The patterns shown for mineral grains indicate their general appearance through a geological microscope, and do not necessarily indicate the relative amounts of the minerals within the rocks.

Granite is one of the most common intrusive igneous rocks. Typical **granite** consists mainly of potassium feldspar, with lesser amounts of quartz and Na-plagioclase, and some mica and amphibole. The actual composition of granite can be quite varied, with quartz, for example, ranging between 20 and 60 percent, and feldspar ranging between 10 and 65 percent. Because most of the volume of granite is of felsic minerals, granite is classified as a *felsic igneous rock*. Granite is generally composed of a mixture of white, pinkish, and black grains, giving it an overall light grey or pink colour (Figure 11.12a). *Diorite*, an intermediate rock, lacks quartz and is mostly composed of Na-plagioclase (60 percent) and secondary amounts of amphibole and pyroxene; it is also a light-coloured felsic rock (Figure 11.12b).

Gabbro, in which the major mineral is pyroxene (60 percent), also contains a substantial amount of Na-plagioclase (20 to 40 percent), and there may also be some olivine (0 to 20 percent). As the mafic minerals, pyroxene, and olivine are dominant, gabbro is classed as a *mafic igneous rock*. Gabbro is darker in colour than the felsic rocks (Figure 11.12c). *Peridotite* is composed mainly of olivine (60 percent) and pyroxene (40 percent). Peridotite is classed as an *ultramafic igneous rock*, and is darker than the mafic types (Figure 11.12d).

11.12 **INTRUSIVE IGNEOUS ROCK SAMPLES** (a) Granite – dark grains are amphibole and biotite; pink grains are feldspars and light grains are quartz; (b) Diorite – feldspar grains are light; amphibole and pyroxene grains are dark; (c) Gabbro – dark colour of rock reflects the high percentage of pyroxene with paler Na-plagioclase also present; (d) Peridotite – composed mainly of dark-coloured olivine and pyroxene.

In addition to affecting their colour, the change in composition of igneous rock also causes an increase in density, from felsic, through intermediate and mafic, to ultra-mafic types. The density of granite ranges from about 2.6 to 2.7 g cm^{-3}, diorite from 2.8 to 3.0 g cm^{-3}, gabbro from 2.7 to 3.3 g cm^{-3}, and peridotite from 3.1 to 3.4 g cm^{-3}. These differences are reflected in the principal rock layers that make up the solid Earth, with the least dense layer (mostly felsic and intermediate rocks) near the surface and the most dense layer (ultramafic rocks) deeper in the Earth's mantle (see Chapter 12).

INTRUSIVE AND EXTRUSIVE IGNEOUS ROCKS

Although both intrusive and extrusive rock can solidify from the same original body of magma, their outward appearances are quite different. Intrusive igneous rocks cool very slowly—over hundreds or thousands of years—and, as a result, develop large, readily seen crystals. Granite and diorite are good examples of *coarse-textured* intrusive igneous rocks.

Extrusive rock, which cools rapidly, is *fine-textured*, and the individual crystals can only be seen through a microscope. Most lava solidifies as a dense, uniform rock with a dark, dull surface. Sometimes lava cools to form shiny *obsidian* or volcanic glass (Figure 11.13a). If the lava contains dissolved gases, it may solidify to form *scoria*, a rock with a frothy, bubble-filled texture (Figure 11.13b).

Intrusive and extrusive rocks have similar mineral compositions, but are named differently as they are dissimilar in appearance. The distinctive chemical compositions of the intrusive rocks granite, diorite, and gabbro are also found in extrusive rocks (see Figure 11.11). Thus, **rhyolite** is the name for lava of the same composition as granite; **andesite** is lava with the composition of diorite; and **basalt** is lava with the composition of gabbro. Note that peridotite has no extrusive counterpart. Rhyolite and andesite are pale greyish or pink in colour, whereas basalt is black. Andesite and basalt are the two most common types of lava. Basalt is the most common rock of the oceanic crust.

A body of intrusive igneous rock is called a **pluton**. Granite typically accumulates in enormous plutons, called *batholiths* (Figure 11.14). As the hot fluid magma rises, it may melt and incorporate some of the older rock lying above it. A single batholith may extend down several kilometres and be many thousands of square kilometres in area.

(a)

(b)

11.13 EXTRUSIVE IGNEOUS ROCK SAMPLES (a) The smooth, glassy appearance of this obsidian or volcanic glass is acquired when gas-free lava cools very rapidly. (b) A specimen of scoria, a form of lava containing many small holes and cavities produced by gas bubbles.

Several other types of igneous intrusions are recognized. A *sill*, a flat, layer-like body, is formed when magma forces its way between pre-existing rock layers (Figure 11.15). A second kind of intrusion is the *dike*, a vertical, wall-like mass formed when a rock fracture is forced open by magma. Dikes tend to be of intermediate texture because the magma cools relatively close to the Earth's surface. These vertical fractures often conduct magma to the surface; magma entering small, irregular, branching fractures in the surrounding rock solidifies in a branching network of thin *veins*.

CHEMICAL ALTERATION OF IGNEOUS ROCKS

The minerals in igneous rocks are formed in magma chambers beneath the Earth's surface and begin to crystallize as the molten rock cools and solidifies. Extrusive igneous rocks are formed at the surface, but even deep-seated intrusive rocks can be uplifted millions of years after they were formed and exposed through erosion of overlying crustal rocks. The temperature and pressure at the surface is very different from the environment in which the rocks were formed. The minerals become unstable under these new conditions and slow chemical changes begin to weaken the rock structure. Once exposed, igneous rock minerals may also be altered by water and dissolved gases.

The processes that cause **mineral alteration** are collectively referred to as *chemical weathering*. Weathering also includes

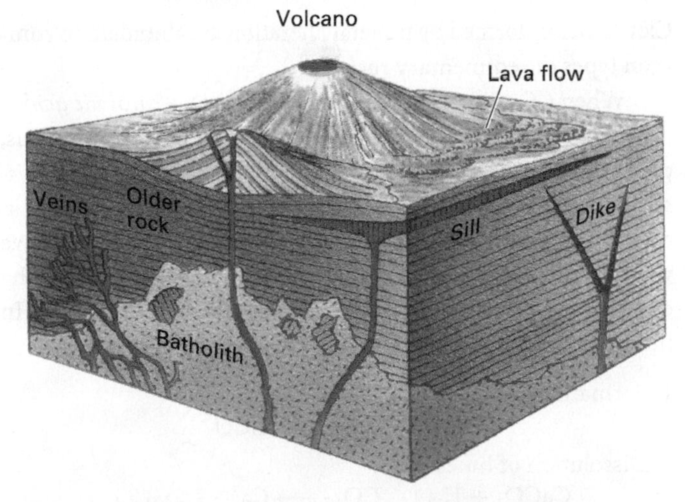

11.14 VOLCANIC ROCK FORMATIONS This block diagram illustrates various forms of intrusive igneous rock plutons, as well as an extrusive lava flow.

11.15 A DOLERITE SILL IN NORTHEASTERN SPITSBERGEN, NORWAY The sill appears as a dark layer of rock where it has been injected into older layers of lighter-coloured sedimentary rock.

11.16 PHYSICAL WEATHERING Fragmentation of a rock outcrop exposes a greater surface area to chemical weathering processes.

physical forces, like frost action, that break up rock into small fragments and separate the component minerals, grain by grain. Rock fragmentation caused in this way is termed *mechanical weathering*. Mechanical weathering accelerates mineral alteration because it greatly increases the surface area that is exposed to chemically active solutions (Figure 11.16). Chemical and mechanical weathering processes are closely linked to prevailing climatic conditions, as the principal factors involved in both processes are temperature and moisture. In the case of chemical weathering, decomposition is accelerated by warm, wet conditions. Mechanical weathering is most active in moist, temperate environments, especially where temperatures frequently fluctuate above and below freezing (Figure 11.17). The distinctive landforms produced by weathering are discussed in Chapter 14.

There are three main processes involved in chemical weathering: oxidation, hydrolysis, and dissolution. Many silicate minerals exposed at the surface undergo *oxidation* when metallic elements, such as iron, magnesium, and aluminum react with oxygen dissolved in water. With oxidation, the metallic elements bond completely with oxygen to form *oxides*. Iron is very susceptible to oxidation and is readily converted to its oxidized ferric form (Figure 11.18). For example, iron silicate, such as in pyroxene, would oxidize to iron oxide in the form of haematite:

$$\underset{\text{(pyroxene)}}{4FeSiO_3} + O_2 \longrightarrow \underset{\text{(hematite)}}{2Fe_2O_3} + \underset{\text{(silica)}}{4SiO_2}$$

Other readily oxidized mineral elements include magnesium and aluminum. Once formed, the oxides tend be very stable under normal environmental conditions. Quartz is a good example. This long-lasting mineral oxide is found abundantly in many types of rocks and sediments.

Although silicate minerals do not generally dissolve in water, some undergo chemical alteration through *hydrolysis*. This process is not merely a soaking or wetting of the mineral, but rather it is a true chemical change that produces a different mineral

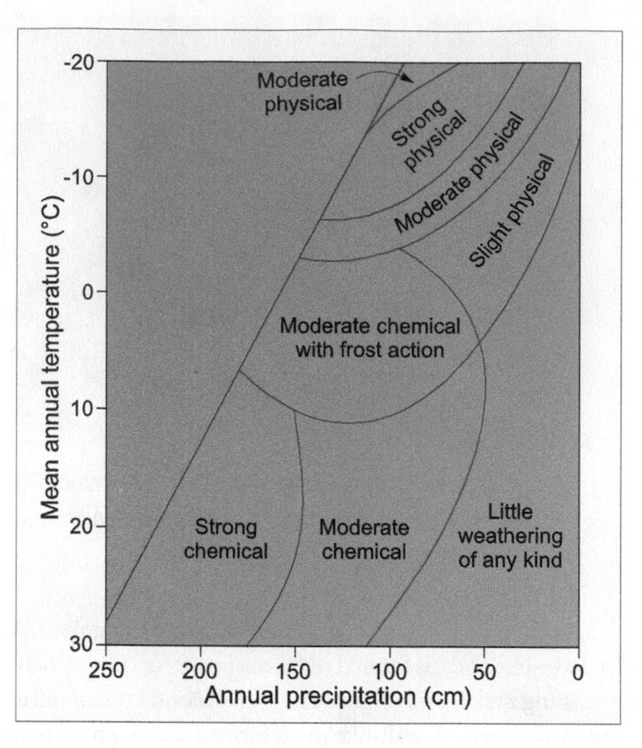

11.17 CLIMATE AND WEATHERING The relative importance of chemical and mechanical weathering processes is closely linked to general climate conditions. (After L.C. Peltier, 1950.)

compound in which water is added to the mineral structure itself. The *secondary minerals* produced by hydrolysis are stable and long-lasting, like the products of oxidation. **Clay minerals**, such as kaolinite, are typical products of hydrolysis:

$$\underset{\text{orthoclase}}{4KAlSi_3O_8} + \underset{\text{hydrogen ions}}{4H^+} + \underset{\text{water}}{H_2O} \longrightarrow \underset{K^+_{(aq)}}{4K^+} + \underset{\text{Kaolinite (clay)}}{2Al_2Si_2O_5(OH)_4} + \underset{\text{silica}}{8SiO_2}$$

Clay minerals formed by mineral alteration are abundant in common types of sedimentary rocks.

When CO_2 dissolves in water, a weak acid—*carbonic acid*—is formed. Carbonic acid can dissolve certain igneous minerals, particularly potassium feldspar and biotite, by the process of *dissolution*. Generally dissolution occurs as a preliminary stage in the hydrolysis of igneous rocks. Dissolution is most effective when rocks are rich in calcium carbonate ($CaCO_3$), and is the characteristic weathering process in sedimentary limestone. In this case, the chemical weathering process can be written as:

1. Formation of carbonic acid

$$H_2O + CO_2 \longrightarrow H_2CO_3$$

2. Dissolution of limestone

$$CaCO_3 + H_2O + CO_2 \longrightarrow Ca^{2+} + 2HCO_3$$

As a result of chemical weathering, the iron-bearing silicates (olivine, pyroxene, amphibole, and biotite) break down to form clay

11.18 **CHEMICAL WEATHERING** This rock's surface has been chemical weathered by the process of oxidation.

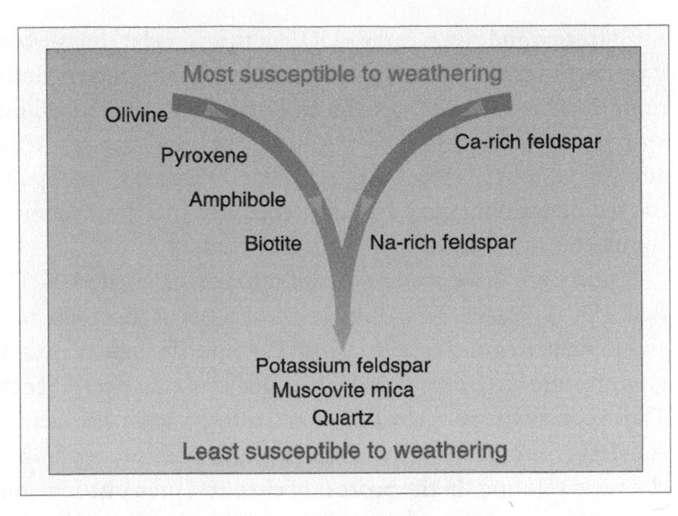

11.19 **GOLDICH STABILITY SERIES** The sequence of chemical weathering of rock minerals is inversely related to the temperature at which they form. Minerals that form at high temperature are the least stable under conditions prevailing at Earth's surface.

minerals and iron oxide. The feldspars weather into clay minerals and ionic forms of potassium (K^+), sodium (Na^+), and calcium (Ca^{++}). Mica reacts to form clay minerals and potassium ions (K^+). Because quartz is very stable, it remains as SiO_2. Interestingly, the rates of chemical weathering are the inverse of Bowen's Reaction Series. This is because minerals that form at higher temperatures are the most unstable under conditions at the Earth's surface. This inverse relationship is referred to as the Goldich Stability Series (Figure 11.19).

SEDIMENTS AND SEDIMENTARY ROCKS

Sedimentary rocks are formed from mineral particles derived from pre-existing rock of any of the three rock classes—igneous, metamorphic, or sedimentary—and from compaction of organic materials, as well as through chemical precipitation. For example, granite can weather to yield grains of quartz and particles of clay minerals derived from feldspars, thus contributing sand and clay to a sedimentary rock. Sedimentary rocks include rock types with a wide range of physical and chemical properties (see Table 11.1).

In the process of mineral alteration, solid rock is weakened, softened, and fragmented, yielding particles of many sizes and mineral compositions. When transported by a fluid medium, such as water or air, these particles are known collectively as **sediment**. Dissolved mineral matter in solution is also an important source of material for rocks that form by evaporation, such as the potash deposits mined in Saskatchewan.

TABLE 11.1 SOME COMMON SEDIMENTARY ROCK TYPES

SUBCLASS	ROCK TYPE	COMPOSITION
Clastic (Composed of rock and/or mineral fragments)	Sandstone	Cemented sand grains
	Siltstone	Cemented silt particles
	Conglomerate	Sandstone containing pebbles of hard rock
	Mudstone	Silt and clay, with some sand
	Claystone	Clay
	Shale	Clay, breaking easily into flat flakes and plates
Chemically precipitated (Formed by chemical precipitation from sea water or salty inland lakes)	Limestone	Calcium carbonate, formed by precipitation on sea or lake floors
	Dolomite	Magnesium and calcium carbonates, similar to limestone
	Chert	A microcrystalline form of quartz formed when underground fluids replace sedimentary materials with silica (see organic chert below)
	Evaporites	Minerals formed by evaporation of salty solutions in shallow inland lakes or coastal lagoons
Organic (Formed from organic material)	Coal	Rock formed from peat or other organic deposits; may be burned as a mineral fuel
	Chalk	Shells of microscopic marine organisms composed of calcium carbonate
	Coquina	Clastic shell fragments
	Chert	Shells of microscopic marine organisms composed of Silica
	Coral	Calcareous exoskeletons of colonial marine algae

Streams and rivers carry sediment to lower elevations where it can accumulate. The most favourable sites for sediment accumulation are shallow sea floors bordering continents, but sediments are also deposited in valleys, lakes, and marshes. Wind and glacial ice also transport sediment. Over long time spans, the sediments undergo physical or chemical changes, becoming compacted and hardened to form sedimentary rock.

There are three major types of sedimentary materials. The first type is **clastic sediment**, which consists of inorganic rock and mineral fragments called *clasts*. Examples include the quartz grains found in the sand on a river bed or on an ocean beach. The second type is **chemically precipitated sediment**, which consists of inorganic mineral compounds precipitated from a chemical solution. In the process of chemical precipitation, ions in solution combine to form solid mineral matter; the deposits of salts around some prairie sloughs are an example. Coral reefs are usually considered to be this type of sediment, as coral organisms precipitate calcareous exoskeletons that contribute directly to the growth of the reef. The third class is **organic sediment**, which consists of the tissues of plants and animals, accumulated and preserved after death. This might include a layer of peat in a bog, or the hard parts of organisms, such as sea shells.

Transported sediment is eventually deposited and gradually accumulates in approximately horizontal strata, or *beds*. Individual strata are separated from those below and above by surfaces called *bedding planes*, which represent interruptions in time, sediment source, or rate of sedimentation within the sequence of deposition. Adjacent strata are most conspicuous when they exhibit different properties, such as colour or grain size. Strata of very different compositions can occur alternately, one above the other (see Figure 11.8).

CLASTIC SEDIMENTARY ROCKS

Clastic sediments are derived from fragments of any of the rock groups, and thus may include a wide range of minerals. Silicate minerals are the most important, both in original form and as altered by oxidation and hydrolysis. Because quartz is abundant and resistant to alteration, it is the most important component of clastic sediments. Second in abundance are fragments of unaltered fine-grained parent rocks. Feldspar and mica are also commonly present, as well as the clay minerals that are derived from them.

The range of particle sizes in a clastic sediment determines how easily and how far the particles are transported. Finer particles are more easily held in suspension in the fluid, whereas coarser particles tend to settle out. In this way, a separation of sizes, called *sorting*, occurs. Sorting determines the texture of the deposited sediment and of the sedimentary rock derived from it. The finest clay particles do not settle out unless they clump together. This process, called *flocculation*, normally occurs when river water carrying clay mixes with the salt water of the ocean.

When sediments accumulate in thick sequences, the lower strata are subjected to the pressure produced by the weight of the overlying layers. This compacts the sediments, squeezing out excess water. Cementation occurs as dissolved minerals recrystallize in the spaces between mineral particles. Silicon dioxide (SiO_2) is slightly soluble in water, and so the cement is often a form of quartz, called *silica*, which lacks a true crystalline form. Calcium carbonate ($CaCO_3$) is another common cementing agent. Compaction and cementation harden the sediments into sedimentary rock.

Important varieties of clastic sedimentary rock are distinguished by the size of their particles. They include sandstone, conglomerate, mudstone, claystone, and shale. **Sandstone** is formed from fine to coarse sand, and mostly consists of quartz (Figure 11.20). Very coarse-grained sedimentary rock containing numerous rounded pebbles is called *conglomerate* (Figure 11.21).

Muddy sediments with a high percentage of fine silt and clay can harden to form *mudstone*. Compacted and hardened clay layers become *claystone*. Sedimentary rocks derived from mud are commonly layered in such a way that they easily break apart into small flakes and plates. The rock is then described as

11.20 **SANDSTONE STRATA** This photo of sandstone rock formations from Arizona shows individual layers, or strata, that make up the rock. The layers were originally deposited within a tract of moving sand dunes.

11.21 CONGLOMERATE This type of sedimentary rock is composed of pebbles that have been rounded through transport in flowing water, and subsequently cemented together by finer sediment and recrystallization of dissolved minerals.

11.22 BURGESS SHALE TRILOBITES Marine trilobites are one of many well-preserved groups of marine invertebrates found in the Burgess Shale in British Columbia. Other fossils include early crabs, sponges, and marine worms.

being *fissile* and is given the name **shale**. Shale, the most abundant of all sedimentary rocks, is composed largely of clay minerals. The compaction of muddy sediments to form mudstone and shale results in a considerable loss of volume as water is driven out of the clay.

In 1909, a remarkable fossil bed was discovered in the Burgess Shale in Yoho National Park in the Canadian Rockies. The Burgess Shale fossils are exceptionally well preserved and represent a unique diversity of Cambrian animal fossils dating from 545 to 525 million years ago (Figure 11.22). The sediment was deposited in a deep-water basin adjacent to an enormous algal reef that was hundreds of metres high. It is thought that many of the fossilized organisms were living on mud banks close to the reef, but frequent slumping of the unstable sediment carried them into deeper waters, where they were subsequently preserved. Because of its importance, UNESCO declared the Burgess Shale a World Heritage Site in 1981.

CHEMICALLY PRECIPITATED SEDIMENTARY ROCKS

Under favourable conditions, mineral compounds are deposited by evaporation from saline waters of the oceans or from lakes. For example, surface deposits of sodium sulphate, in the form of Glauber's salt ($Na_2SO_4 \cdot 10H_2O$), occurs in natural brines and crystal deposits in alkaline lakes in southern Saskatchewan (Figure 11.23). Sedimentary minerals and rocks deposited from concentrated solutions are called **evaporites**. The potash deposits mined in Saskatchewan are contained in the Prairie Evaporite Formation. Potash refers to potassium compounds and potassium-bearing materials, the most common being sylvite (potassium chloride). The Saskatchewan

11.23 GREAT SALT LAKE, UTAH Minerals, including potassium sulphate (K_2SO_4), magnesium chloride ($MgCl_2$) and sodium chloride (NaCl) are deposited through evaporation of the saline lake waters of the Great Salt Lake. Nearby are the Bonneville Salt Flats which developed as former Lake Bonneville dried up in post-glacial times. At its maximum extent, Lake Bonneville was about the same area as Lake Michigan and 300 m deep. ***Fly By:** 41° 13′ N; 112° 44′ W*

deposits originated in the Devonian Period about 400 million years ago, and now lie at depths of more than 1,000 m. Gypsum ($CaSO_4 \cdot 2H_2O$) is another example of an evaporite; the gypsum deposits in Nova Scotia were similarly formed when a large water body evaporated as a result of climatic warming and drying 330 million years ago.

One of the most common sedimentary rocks formed by chemical precipitation is **limestone**, composed largely of the mineral *calcite* ($CaCO_3$). Marine limestone accumulated in thick layers in many ancient seas in past geologic eras. Limestone is often rich in fossils, but the fossil content is a minor part of most deposits. Most of the calcium begins as microscopic needle-like crystals of aragonite (a variety of calcite) produced by cyanobacteria (blue-green algae), which accumulate on the seabed when the organisms die and decompose. Over time, aragonite re-crystallizes to calcite, and so is usually present only in younger limestone. A closely related rock also formed by chemical precipitation is *dolomite*, composed of calcium magnesium carbonate [$CaMg(CO_3)_2$]. Unlike other sedimentary rocks, such as sandstone and shale, dolomite does not form on the surface of the Earth at the present time. It is thought to be formed from calcite rich limestone that undergoes chemical change following deposition, probably as a result of intrusion by magnesium rich groundwater in warm, tropical ocean environments. Limestone and dolomite are grouped together as the *carbonate rocks*; they are relatively dense rocks, white, pale grey, or even black in colour. The Rocky Mountains are formed mainly of limestone originally deposited in inland seas; Dolomite Peak (2,782 m) near Banff is one of the few places where dolomite is found in the Canadian Rockies (Figure 11.24).

Sea water also yields sedimentary layers of silica in a hard, non-crystalline form called *chert*. Chert is a variety of sedimentary rock, but it also commonly occurs combined with limestone (cherty limestone). *Chalk*, a soft, white, porous form of limestone, is a marine deposit and consists almost entirely of microscopic, single-celled algae called coccolithophores. These marine organisms are covered with ornate calcium carbonate plates, and their remains are referred to as coccoliths. Massive chalk deposits hundreds of metres thick, which accumulated in the Cretaceous Period (144 to 66 million years ago), are almost entirely composed of these calcareous coccoliths (Figure 11.25). Chalk deposits reflect a time of extensive warm oceans and are indicative of climates much warmer than today.

HYDROCARBON COMPOUNDS IN SEDIMENTARY ROCKS

Hydrocarbon compounds are a very important component of some sedimentary deposits. Although hydrocarbons occur both as solids, for example peat and coal, and as liquids and gases, such as petroleum and natural gas, only coal qualifies physically as a rock. At various times in the geologic past, plant remains accumulated on a large scale, accompanied by sinking of the area and burial of the compacted organic matter under thick layers of inorganic clastic sediments. *Coal* is the end result of this process, with individual coal seams interbedded with shale, sandstone, and limestone strata.

Oil deposits and *natural gas* are also of organic origin, but they are classed as mineral fuels rather than minerals. Oil and gas deposits commonly occupy open, interconnected pores in a thick sedimentary rock layer, such as a porous sandstone. The simplest arrangement of strata favourable to trapping oil and gas

11.24 LIMESTONE The Canadian Rockies are mostly formed of limestone that was originally deposited in inland seas. Shown here is Dolomite Peak, near Banff, Alberta.

11.25 CHALK CLIFFS The Etretât cliffs of Normandy, France, are composed of soft limestone, known as chalk.

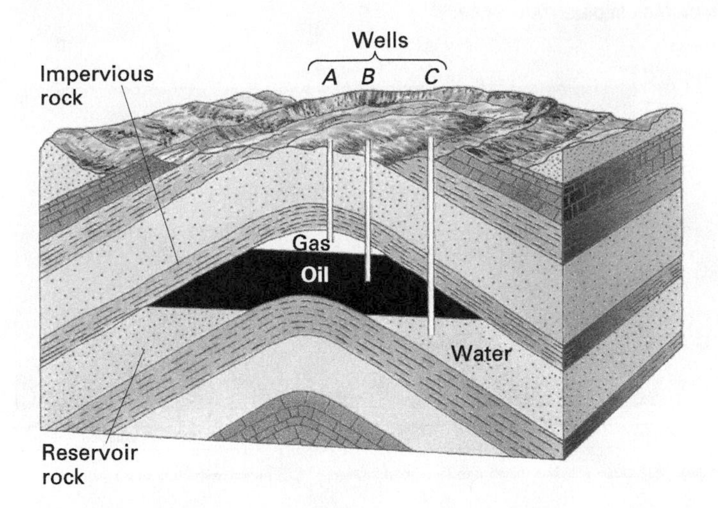

11.26 **TRAPPING OF OIL AND GAS** Idealized cross-section of an oil pool on a dome structure in sedimentary strata. Well *A* will draw gas, well *B* will draw oil, and well *C* will draw water. The cap rock is shale; the reservoir rock is sandstone. (Copyright © A.N. Strahler.)

is upward arching sedimentary rocks in a formation known as an anticline (Figure 11.26). Shale forms an impervious cap rock. A porous sandstone beneath the cap rock serves as a reservoir. Due to their different densities, natural gas occupies the highest position, with the oil below that and water at the bottom. In other forms of hydrocarbon deposits, the oil-bearing materials are in the form of solid bituminous kerogen in *oil shales*, or as viscous bitumen in *oil sands*. Because neither of these forms is free-flowing, the oil has to be extracted by expensive secondary processing.

Oil shales are fine-grained sedimentary rocks containing relatively large amounts of organic matter from which significant amounts of hydrocarbons can be extracted by destructive distillation. Oil shale is derived from sediments deposited in a wide variety of environments, including freshwater lakes, coastal swamps, and shallow marine areas. Most of the organic matter comes from algae, although remains of land plants are sometimes present. Oil shale generally contains enough hydrocarbon materials that the rock will burn without further processing. Oil shale extraction has been carried on for centuries using both underground- and surface-mining methods. However, unlike conventional oil deposits that are pumped directly from the ground, oil shale must undergo industrial retorting before the oil can be recovered. In this process, the oil shale is heated to about 350 °C in the absence of oxygen, causing the kerogen to liquefy and separate from the rock. In the past this was done after extraction, but in situ retorting is now being carried out experimentally using electric heaters placed in deep holes drilled into the shale. The released fluids are then pumped to the surface for further refining. The increase in the price of conventional oil has led to renewed interest in oil shale extraction. Oil shale is found in many parts of the world, but the largest reserves are found in the United States in the Green River Formation, which covers parts of Wyoming, Utah, and Colorado. In Canada, the oil shales near Moncton, New Brunswick, have the greatest commercial potential.

Tar sands, or oil sands, are formed where bitumen occupies pore spaces in layers of sand or porous sandstone. The bitumen remains immobile in the enclosing sand and will flow only when heated. Tar sands are found in many countries, but by far the biggest deposits occur in Canada and Venezuela. Perhaps the best known deposits are the Athabasca Tar Sands at Fort McMurray, Alberta. The Athabasca Tar Sands cover an area of 140,000 km^2; the oil-bearing strata are typically 40 to 60 m thick, and occur 1 to 3 m below the surface, beneath a thin overburden of peat. The tar sands are extracted by surface mining; the first mine in the area became operational in 1967.

The typical composition of the Athabasca Tar Sands is about 10 percent bitumen, 83 percent sand, 4 percent water, and 3 percent clay. Consequently, it must undergo pre-processing before it can be used in conventional oil refineries. This involves the removal of the sand and minerals through the addition of hot water and agitation in a separator to release the bitumen, which then floats to the surface. Bitumen is much thicker than conventional crude oil, so it must be mixed with lighter petroleum, or be chemically modified, before it can be sent by pipeline for upgrading into synthetic crude oil. About two tons of tar sands are required to produce the equivalent of one barrel (about one eighth of a ton) of conventional oil. Over 90 percent of the bitumen in the tar sands can now be recovered. However, all these processes take large amounts of energy and water, and emit CO_2. After oil extraction, the spent sand and other waste materials are returned to the mine, allowing the site eventually to be reclaimed. Currently there is some concern about the environmental quality of this mining process, especially in terms of toxic effluents and wildlife habitat.

The conventional oil deposits and the tar sands in Alberta both formed under similar environmental conditions, basically through the gentle heating under pressure of sediments rich in marine plankton over a long period of time. It is generally assumed that bacteria removed most of the oxygen and nitrogen from the organic debris under anaerobic conditions in stagnant depressions in the ancient seas that covered the region in the Middle Devonian Period, 390 million years ago. The heat and pressure from the subsequent accumulation of sedimentary rock layers resulted in the reorganization of the residual carbon and hydrogen to form the hydrocarbons, which were subsequently absorbed into the porous rocks. The different types of oil are produced by varying degrees of biodegradation and are related to the temperature in the reservoir rocks. Biodegradation ceases at temperatures of 80 °C. This accounts for the progression from comparatively non-biodegraded light oils at greater depths under the thicker sedimentary rocks in western Alberta, to more viscous bituminous forms in the east.

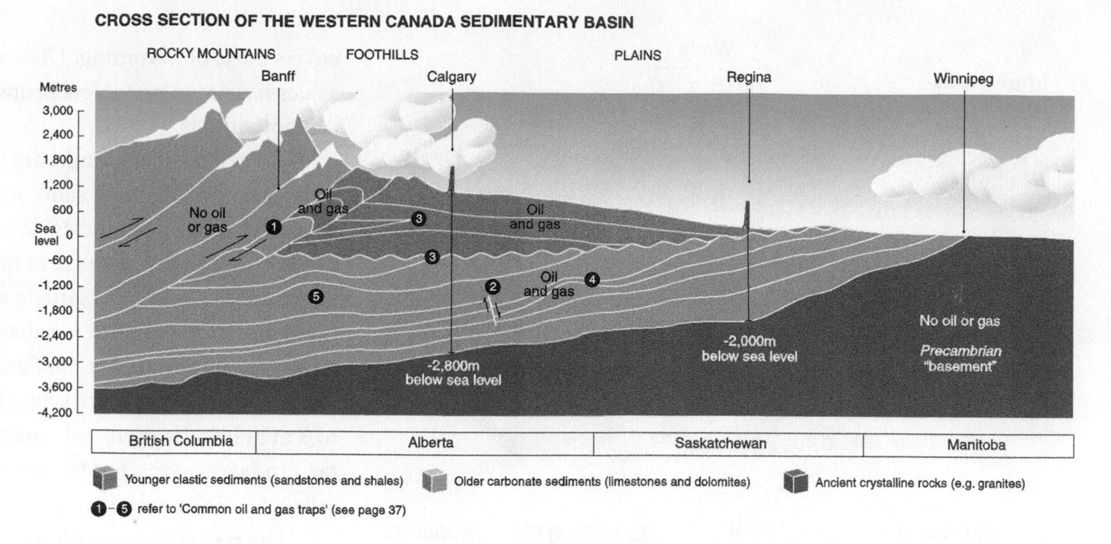

11.27 GENERAL STRUCTURAL GEOLOGY OF OIL-BEARING STRATA IN WESTERN CANADA Lighter oils are generally found in western Alberta and become increasingly viscous toward the east and north, where biodegradation was more intense.

Subsequent formation of the Rocky Mountains, which occurred between 70 and 40 million years ago, forced some of the oil northward, where it was absorbed into sands deposited by ancient rivers to form the oil sands (Figure 11.27).

The **fossil fuels** have taken millions of years to accumulate. However, industrial societies are consuming them at a very rapid rate. These fuels are non-renewable resources. Once they are gone there will be no more, because the quantity produced by present geologic processes is scarcely measurable in comparison to the quantity stored through geologic time. The continuing rise in the price of oil has increased development of non-conventional oil reserves. However, non-conventional oil production has considerable environmental costs due to increased energy requirements needed for extraction and processing, and the associated increase in greenhouse gas production.

METAMORPHIC ROCKS

Any type of igneous or sedimentary rock may be altered by the tremendous pressures and high temperatures that accompany crustal deformation processes, such as the creation of mountain ranges. The result is rock so changed in texture and structure that it is reclassified as metamorphic rock. Mineral components of the parent rock are, in many cases, reconstituted into different mineral varieties. Recrystallization of the original minerals can also occur. Five common metamorphic rocks are slate, schist, quartzite, marble, and gneiss (see Table 11.2).

Slate is formed from shale that is heated and compressed by mountain-building forces. This fine-textured rock splits readily into thin plates, which are commonly used in roofing. With increased heat and pressure, slate changes into **schist**. Schist exhibits foliation, which consists of thin, rough, irregularly curved planes in the rock (Figure 11.28). These are evidence of *shearing*—a stress that pushes the layers sideways, like a deck of cards. Schist is distinguished from slate by the coarse texture of the mineral grains, the abundance of mica,

TABLE 11.2 SOME COMMON METAMORPHIC ROCK TYPES

Slate	Shale exposed to heat and pressure results in a hard rock that readily splits into flat sheets
Schist	Shale exposed to more intense heat and pressure causes some recrystallization and shows evidence of shearing
Quartzite	Sandstone that is "welded" by a silica cement, resulting in a very hard rock composed almost entirely of quartz
Marble	Limestone exposed to heat and pressure, resulting in larger, more uniform crystals
Gneiss	Rock produced when clastic sedimentary rocks or intrusive igneous rocks are subjected to intense heat and pressure deep within the Earth

and occasionally the presence of scattered large crystals of newly formed minerals, such as garnet.

The metamorphic equivalent of conglomerate and sandstone is **quartzite**, which is formed by the fusion of quartz-rich sediments through the heat and pressure associated with tectonic

11.28 SCHIST SAMPLE A specimen of foliated schist.

activity. Limestone, after undergoing metamorphism, becomes *marble*. During formation, calcite in the limestone is reconstituted into larger, more uniform crystals. Bedding planes are obscured, and masses of mineral impurities may be drawn out into swirling bands.

Gneiss forms either from igneous or sedimentary rocks that have been in close contact with intrusive magmas. A single description will not fit all gneisses because they vary considerably in appearance, mineral composition, and structure. One variety of gneiss is strongly banded into light and dark layers or lenses; others can have very pronounced wavy folds (Figure 11.29). The segregation of light and dark minerals is indicative of intense metamorphism, where temperatures as high as 600 to 700 °C have resulted in mineral migration. These bands differ in mineral compositions and are clearly seen in *migmatite* that is formed under extreme temperatures and pressures through distortion and deformation of rock in a plastic state.

11.29 **BANDED GNEISS** This rock surface at South Uist, Scotland, exposes banded gneiss.

GEOGRAPHER'S TOOLS

IDENTIFYING ROCKS USING REMOTE SENSING TECHNIQUES

In 1999, the Japanese-built ASTER instrument (Advanced Spaceborne Thermal Emission and Reflection Radiometer) was launched aboard NASA's Terra satellite platform. ASTER acquires multi-spectral images in three different waveband regions: visible and near-infrared, short-wave infrared, and thermal infrared. Minerals and rocks have distinctive spectral signatures in near- and short-wave infrared domains, and images at these wavelengths are useful for geologic mapping.

A spectral region of particular interest is the thermal infrared, especially the wavelengths from 8 to 12 μm. Here the energy sensed is the radiant energy emitted by substances as a function of their temperature. However, some types of rocks and minerals emit more energy of a specific wavelength at a given temperature than others. In this way, they too can be assigned spectral signatures in the thermal infrared region.

Figure 1 shows reflectance plotted against wavelength for two common minerals: quartz and calcite. Also shown are the spectral locations of the ASTER wavebands. In the left graph, both quartz and calcite are quite bright in ASTER's two visible bands (band 1 – green and band 2 – red). They appear white to the eye, with calcite,

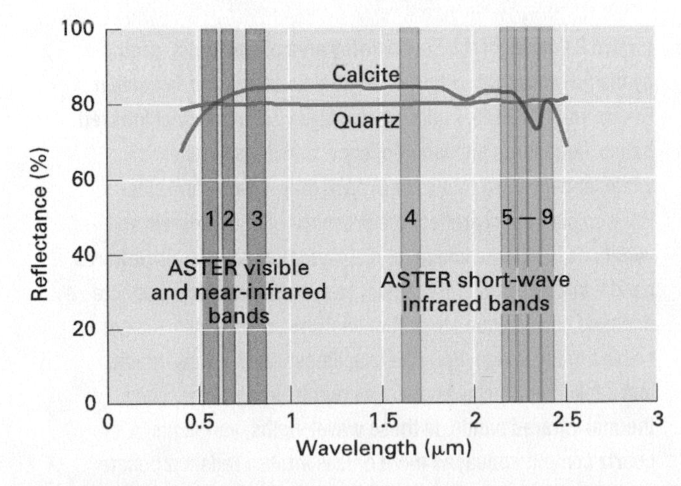

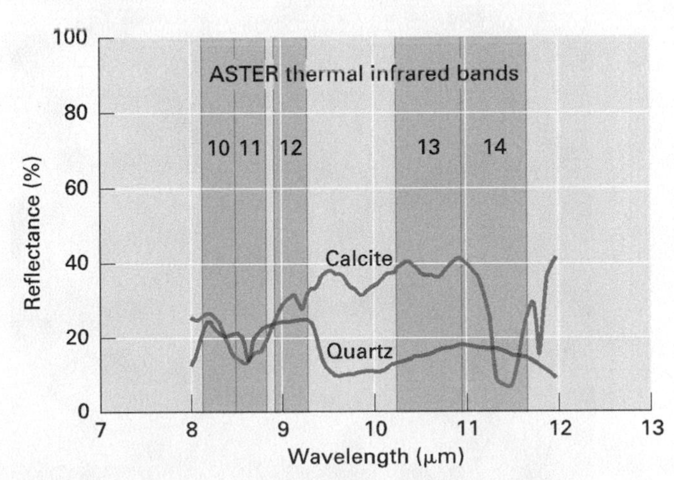

Figure 1 Reflectance spectra for quartz and calcite Wavebands in which ASTER acquires images are identified as numbered bands. Left, the visible and near-infrared spectral region; right, the thermal infrared spectral region.

which is slightly brighter in the red band, appearing to be a "warmer" white. In the short-wave infrared, however, calcite is significantly darker in the ASTER band 8 than in surrounding bands 7 and 9. The right graph shows spectral reflectance in the thermal infrared spectral region. Calcite is brightest in the ASTER band 13 and darkest in the adjacent band 14. Quartz has a distinctive signature as well; it is about 40 percent brighter in bands 10 to 12 than in 13 and 14. Thus, both minerals also have unique spectral signatures.

Colour images in the three ASTER waveband regions are shown in Figure 2. In the infrared image (a), the different rock types appear in tones ranging from brown, grey, and yellow, to blue. Image (b) displays the ASTER short-wave infrared bands 8, 6, and 4 as blue, green, and red. These bands distinguish carbonate, sulphate, and clay minerals. Limestone is yellow-green, while the clay mineral kaolinite appears in purplish tones. In (c), the ASTER bands 10, 12, and 13 are shown in blue, green, and red, respectively. Here, rocks rich in quartz appear in red tones, while carbonate rocks are green. Mafic volcanic rocks are visible in purple tones.

By comparing the three images, it is clear how the spectral information acquired by ASTER has the ability to make geologic mapping faster and easier. By using ASTER data, geologists can not only understand a region's geologic history, but also identify the geologic settings in which valuable mineral ores may be found.

Figure 2 Images of the Saline Valley area of California, acquired by the Advanced Spaceborne Thermal Emission and Reflection Radiometer (ASTER). Image (a) displays visible and near infrared bands. Vegetation appears red, snow and dry salt lakes are white, and exposed rocks are brown, grey, yellow, and blue. Rock colours mainly reflect the presence of iron minerals and variations in albedo. Image (b) displays short wavelength infrared bands. In this wavelength region, clay, carbonate, and sulphate minerals have unique absorption features, resulting in distinct colours in the image. For example, limestone is yellow-green, and kaolinite-rich clay appears as purple. Image (c) displays thermal infrared bands. In these wavelengths, variations in quartz content appear as more or less intense reds, carbonate rocks are green, and volcanic rocks with high proportions of magnesium and iron are purple.

THE CYCLE OF ROCK TRANSFORMATION

Igneous, sedimentary, and metamorphic rocks are transformed from one to another in a continuous cycle over the course of many millions of years. Igneous rock forms when pre-existing rock of any class melts and this magma subsequently cools and solidifies. Through weathering and erosion, pre-existing rock breaks down and the fragments accumulate in layers that eventually become compacted, forming sedimentary rock. Heat and pressure, without complete melting, convert igneous and sedimentary rock to metamorphic rock. These combined processes constitute a single system that creates and recycles Earth materials from one form to another over geologic time. This system is referred to as the **cycle of rock transformation**, or the rock cycle (Figure 11.30).

Two general environments are considered: a low-temperature, low-pressure surface environment, and a high-temperature, high-pressure environment deep within the Earth. The surface environment is the site of rock weathering and erosion, sediment transport, and sediment deposition. In this environment, igneous, sedimentary, and metamorphic rocks are exposed to air and water. Their minerals are altered chemically and broken free from the parent rock, yielding sediment. The sediment accumulates in basins, where deeply buried layers are compressed and cemented into sedimentary rock.

Sedimentary rock forced down into the deep environment is heated by energy released through the slow decay of radioactive elements and also subjected to high confining pressure. Here, it is transformed into metamorphic rock. Pockets of magma also form in the deep environment and move upward, melting and incorporating surrounding rock as they rise. Upon reaching a higher level, magma cools and solidifies, becoming intrusive igneous rock. It may also emerge at the surface to form extrusive igneous rock. The creation of either form of igneous rock perpetuates the rock cycle.

The rock cycle has been active since the Earth became solid and internally stable, continuously forming and reforming rocks of all three major classes. Not even the oldest igneous and metamorphic rocks found thus far are the "original" rocks of the Earth's crust, for they were recycled eons ago. The loops in the rock cycle are powered by a number of sources, ranging from solar energy to internal, radiogenic heat.

POWERING THE CYCLE OF ROCK TRANSFORMATION

For the underground part of the rock cycle, the main power source is *radiogenic heat* that is slowly released by the radioactive decay of unstable isotopes. (See the *Working It Out* feature on radioactive decay in Chapter 12 for more details.) Isotopes of uranium (^{238}U, ^{235}U), thorium (^{232}Th), and potassium (^{40}K), along with the daughter products generated by their decay, are the source of nearly all of this heating.

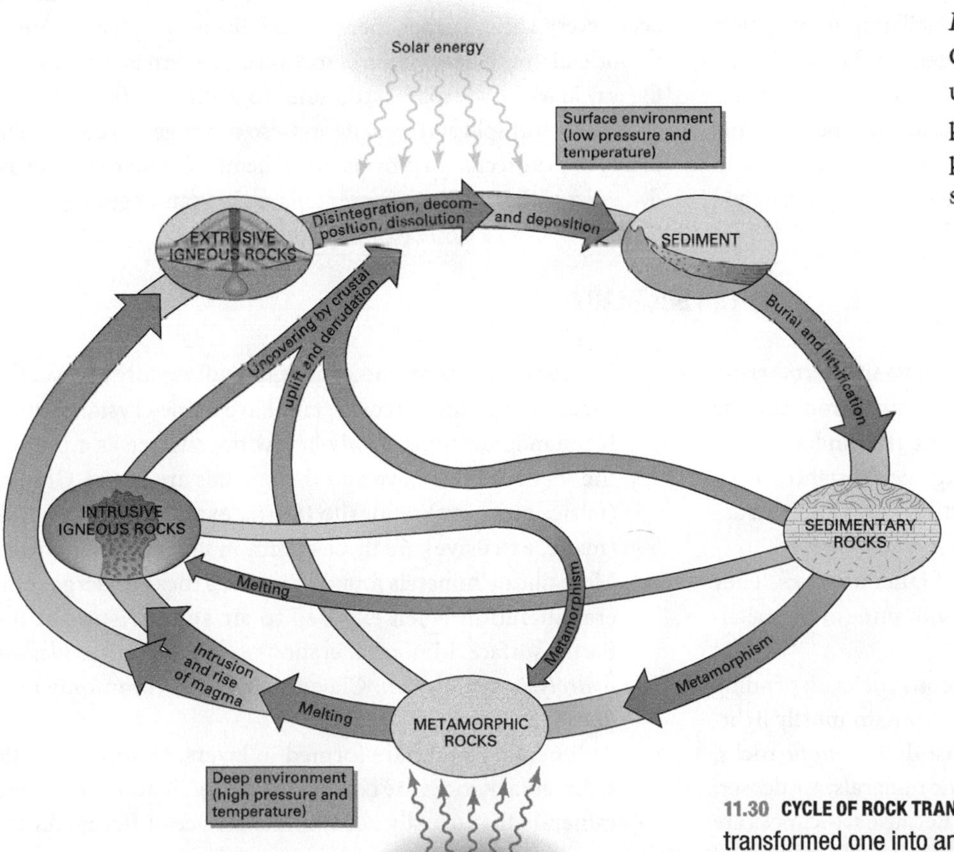

11.30 CYCLE OF ROCK TRANSFORMATION The three classes of rock are transformed one into another by weathering and erosion, melting, and exposure to heat and pressure, through a continuous process of rock formation and destruction.

Figure 11.31 plots temperature against depth below Earth's surface, and shows that temperature increases rapidly at first then more slowly with depth. Much of the radiogenic heat is liberated within the uppermost 100 km or so. This observation fits with chemical analyses, which show that isotopes of uranium, thorium, and potassium are most abundant in the upper layers of continents. There is enough radiogenic heat to keep the Earth layers below the crust close to the melting point, and thus provide the power source for the heating required to form metamorphic and igneous rocks from pre-existing rocks.

A secondary power source for the rock cycle is gravity. As sedimentary layers are buried more and more deeply, the pressure on the underlying strata rises steadily at a rate of about 1,000 atmospheres every 30 km (Figure 11.31). High pressure at depth brings mineral grains into very close contact. Water, often containing dissolved silica or calcium carbonate, is forced through the sediments, leaving deposits of these minerals to cement the grains in place; in this way, the density of the rock increases.

The above-ground portion of the cycle of rock transformation is dependent on the movement of air, water, and glacial ice to transport rock particles. The energy for these processes comes mainly from the sun. Solar energy, and the unequal heating of the Earth's surface that it provides, is the primary energy source for the motions of the atmosphere and oceans. A lesser, but still important, power source for atmospheric and oceanic movements is the Earth's rotation, which transfers momentum to the fluid atmosphere and ocean. The tidal forces of the sun and moon also act to keep the ocean and atmosphere in motion. The force of gravity on the particles eventually brings them to sedimentary basins, where they accumulate and the underground portion of the rock cycle begins.

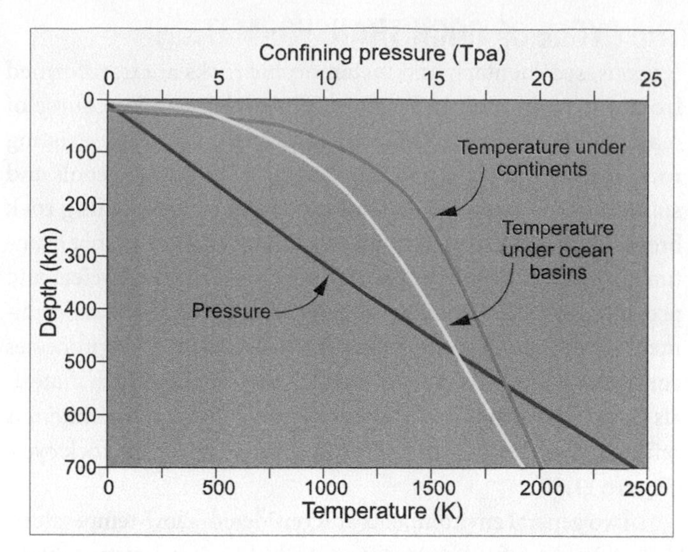

11.31 CHANGE IN TEMPERATURE AND PRESSURE WITH DEPTH FROM THE EARTH'S SURFACE Temperature increases rapidly beneath the continents and ocean basins to about 100 km, at which depth the rate of increase slows. Pressure decreases at a steady rate with depth.

A Look Ahead

This chapter has described the minerals and rocks of the Earth's surface and the processes of their formation. These processes do not occur everywhere. Rather, there is a definite geographic pattern to the formation and destruction of rocks. This pattern is controlled by the way in which the solid Earth's brittle outer layer is fractured into great plates that split and separate and also converge and collide. The theory of plate tectonics provides the scheme for understanding the dynamics of the Earth's crust over millions of years of geologic time. Chapter 12 takes a closer look at plate tectonics.

CHAPTER SUMMARY

- The elements oxygen and silicon dominate the *Earth's crust.* Metallic elements, which include aluminum, iron, and the base elements, account for nearly all the remainder.
- Minerals are naturally occurring, inorganic substances. Each has an individual chemical composition and atomic structure. Minerals are combined into rock.
- *Silicate minerals* make up the bulk of igneous rocks. They contain silicon and oxygen, together with some of the metallic elements.
- There are three broad classes of igneous rocks, depending on their mineral content. *Felsic* rocks contain mostly light-coloured felsic minerals and are least dense; *mafic* rocks, containing mostly dark-coloured mafic minerals, are denser; and *ultramafic* rocks are most dense. Because felsic rocks are least dense, they are generally found in the upper layers of the Earth's crust. Mafic and ultramafic rocks are more abundant in the layers below.
- If magma erupts on the surface to cool rapidly as lava, the rocks formed are extrusive and have a fine crystal texture. If the magma cools slowly below the surface as a pluton, the rocks are intrusive and the crystals are larger. Granite (felsic, intrusive), andesite (felsic, extrusive), and basalt (mafic, extrusive) are three common igneous rock types.
- Most silicate minerals found in igneous rocks undergo mineral alteration when exposed to air and moisture at the Earth's surface. Mineral alteration occurs through *oxidation, hydrolysis,* or *solution.* Clay minerals are commonly produced by mineral alteration.
- Sedimentary rocks are formed in layers, or strata. Clastic sedimentary rocks are composed of fragments of rocks and minerals that usually accumulate on ocean floors. As the layers are buried more and more deeply, water is pressed out and particles are cemented together. Sandstone and shale are common examples.

- Chemical precipitation also produces sedimentary rocks, such as limestone. *Coal, petroleum,* and *natural gas* are hydrocarbon compounds occurring in sedimentary rocks as mineral fuels.
- Metamorphic rocks are formed when igneous or sedimentary rocks are exposed to heat and pressure. Shale is altered to *slate* or *schist*, sandstones become *quartzite*, and intrusive igneous rocks or clastic sediments are metamorphosed into *gneiss*.
- In the cycle of rock transformation, rocks are exposed at the Earth's surface, and their minerals are broken free and altered to form sediment. The sediment accumulates in basins, where the layers are compressed and cemented into sedimentary rock. Deep within the Earth, the heat of radioactive decay melts pre-existing rock into magma, which can move upward into the crust to form igneous rocks that cool at or below the surface. Rocks deep in the crust are exposed to heat and pressure, forming metamorphic rock. Mountain-building forces move igneous, sedimentary, and metamorphic rocks to the surface, providing new material for weathering and breakup, thereby completing the cycle.

KEY TERMS

andesite
basalt
chemically precipitated sediment
clastic sediment
clay mineral
cycle of rock transformation or rock cycle
evaporites
extrusive igneous rock
felsic
fossil fuels
gneiss
granite
igneous rocks
intrusive igneous rock
lava
limestone
mafic
magma
metamorphic rocks
metamorphism
mineral
mineral alteration
organic sediment
pluton
quartz
quartzite
rhyolite
rock
sandstone
schist
sediment
sedimentary rocks
shale
strata

REVIEW QUESTIONS

1. What is Earth's crust? What elements are most abundant in the crust?
2. Define the terms *mineral* and *rock*. Name the three major classes of rocks.
3. What are silicate minerals? Describe two groups of silicate minerals.
4. Name four types of igneous rocks and arrange them in order of increasing density.
5. How do igneous rocks differ when magma cools (a) at depth and (b) at the surface?
6. How are igneous rocks chemically weathered? Identify and describe three processes of chemical weathering.
7. What is sediment? Define and describe three types of sediments.
8. Describe two processes that produce sedimentary rocks, and identify at least three important varieties of clastic sedimentary rocks.
9. How are sedimentary rocks formed by chemical precipitation?
10. What types of sedimentary deposits consist of hydrocarbon compounds? How are they formed?
11. What are metamorphic rocks? Describe at least three types of metamorphic rocks and the conditions under which each formed.
12. Identify and describe the sources of energy that power the cycle of rock change.

VISUALIZATION EXERCISES

1. Sketch a cross-section of the Earth showing the following features: batholith, sill, dike, veins, lava, and volcano.
2. Sketch the cycle of rock change and describe the processes that act within it to form igneous, sedimentary, and metamorphic rocks.

ESSAY QUESTION

1. Granite is exposed at the Earth's surface. Describe how mineral grains from this granite might be released, altered, and eventually become incorporated in a sedimentary rock. Summarize the processes that would incorporate the same grains in a metamorphic rock.

CHAPTER 12 THE LITHOSPHERE AND THE TECTONIC SYSTEM

The Earth's Structure
- The Earth's Interior
- The Lithosphere and Asthenosphere

The Geologic Time Scale

Major Relief Features of the Earth's Surface
- Relief Features of the Continents
- Relief Features of the Ocean Basins

Mid-Oceanic Ridge and Ocean Basin Features
- Continental Margins

The Global System of Lithospheric Plates

Plate Tectonics
- Plate Motions and Interactions
- Orogenic Processes
- Continental Rupture and New Ocean Basins

Continents of the Past

The outpouring of basaltic lava in Hawaii.

The outlines of the continents on a globe or map are so unique that they can be readily identified. These outlines are the result of forces within the Earth that have acted over millions of years. Various types of evidence suggest that the shapes and positions of the landmasses have changed significantly during the long history of the Earth. The theory describing the changing configuration of the continents through time is called *plate tectonics*.

Plate tectonic theory maintains that the Earth's outermost solid layer consists of huge, rigid plates that float on a layer of plastic rock. These plates are in slow, constant motion, powered by energy sources deep within the planet. When two plates come together, one may be forced under the other in a process called **subduction**. In this case, part of the subducted plate can melt, creating pockets of magma that move upward to the surface to create volcanic mountain chains and islands. Plates can also collide without subduction, forming mountain zones where rocks are fractured, folded, and compressed into complex structures. When crustal plates are split apart, rift valleys and ultimately ocean basins are created. Lava that moves upward to fill these widening gaps form submerged mountain chains such as the mid-Atlantic ridge. Thus, ocean basins expand with rifting and contract with subduction, while continents grow by collision and shrink by fracturing.

Because plate motions are powered by forces within the Earth, they are independent of surface conditions. However, processes that affect climate, vegetation, and soils are dependent on the major relief features and Earth materials that tectonic activity creates. Thus, knowledge of plate tectonics is important for understanding the global patterns of the Earth's landscapes and, ultimately, human activity.

THE EARTH'S STRUCTURE

Direct observation of the Earth's interior is limited to deep-drilling operations. The first of these was *Project Mohole*; it was carried out from a deep-sea drilling ship in the Gulf of Mexico. Drilling was abandoned in 1966, after penetrating 183 m into the seabed at a depth of 3,500 m below sea level. Later, from 1968 to 1983, the research vessel *Glomar Challenger* retrieved rock samples from more than 19,000 cores from different oceans in what was known as the *Deep-Sea Drilling Project*. The deepest borehole reached 1,741 m. More recently, in 1993 the research drill ship *JOIDES Resolution* drilled to 2,111 m in the Pacific Ocean near Costa Rica; the rocks at this depth were principally of plagioclase, olivine and pyroxenes associated with igneous dikes. The temperature at the bottom of the borehole was 180 °C. Several deep-drilling projects have also occurred on land. The Kola Superdeep Borehole, carried out by the former Soviet Union, is the deepest at 12,261 m, and penetrated about a third of the way through the **continental crust**.

Indirect evidence of Earth's internal structure comes from studying seismic activity associated with earthquakes. Seismic waves travel away from the focus in all directions, and can be detected by *seismographs* around the world. The timing and patterns of seismic waves following an earthquake vary according to how they are reflected from deep layers within the Earth, and show that the planet's internal structure is far from uniform. Earthquakes are discussed further in Chapter 13. Deep within the Earth is the central core that is surrounded by several layers or shells, each with its own distinctive properties. The densest matter is found within the core, with density decreasing progressively in each of the overlying layers.

THE EARTH'S INTERIOR

The Earth is an almost spherical body, approximately 6,400 km in radius. Its structure dates back to the planet's earliest history, when it was formed by accretion from a mass of gas and dust orbiting the sun. The centre of the Earth is occupied by the **core**, which is about 3,500 km in radius (Figure 12.1). The sudden change in behaviour of earthquake waves when they reach the core suggests that it consists of two zones. The outer core, which accounts for about 31 percent of Earth's mass, has the properties of a liquid; the inner core is solid and represents about 2 percent of Earth's mass. Even though temperature increases toward the Earth's centre, the inner core remains solid because it is under very high pressure. Earth's magnetic field is believed to originate from a dynamo effect set up by convection currents within the molten outer core in conjunction with Earth's rotation. Based on various geophysical data, it has long been inferred that the core consists mostly of iron, with some nickel. The core temperature is estimated at 3,000 °C to 5,000 °C.

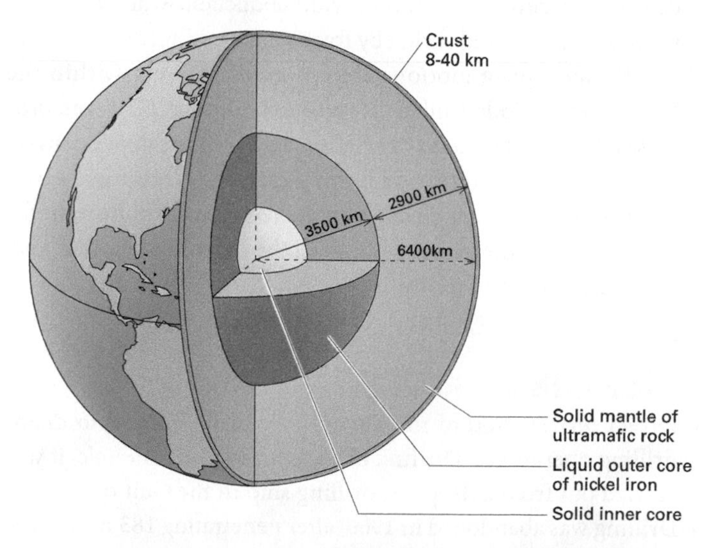

12.1 THE EARTH'S INTERIOR This cutaway diagram of the Earth shows the inner core of iron, which is solid, surrounded by a liquid outer core. The mantle that surrounds the core is a thick inner layer of ultramafic rock. The crust is too thin to be shown at the correct scale.

Enclosing the metallic core is the **mantle**, a rock shell about 2,900 km thick, which accounts for about 50 percent of the Earth's mass. Seismic data indicate that mantle rock is composed of magnesium and iron silicates similar to the ultramafic igneous rock peridotite. Temperatures in the mantle range from about 2,800 °C near the core to about 1,800 °C near the crust, and are maintained by the release of energy through radioactive decay of uranium, thorium, and potassium isotopes. The temperature gradient causes slow convection currents to develop within the mantle.

Several zones are recognized within the mantle. The lowest zone is referred to as the D′ layer, or Gutenburg discontinuity, and represents a transition layer 200 to 300 km thick between the core and the lower mantle. The lower mantle extends from 2,900 km to about 650 km below the surface. From 650 to 400 km there is another transition region; this is sometimes referred to as the middle mantle or fertile layer because it is a source of basaltic magma. The fertile layer is overlain by the upper mantle, which has a maximum thickness of about 400 km. While the lower mantle is relatively homogeneous because of continuous mixing, the upper mantle is more structured, and numerous subdivisions are recognized from seismic velocity measurements and other geophysical properties. The upper mantle is mainly made up of highly viscous olivine and pyroxenes.

The outermost and thinnest of the Earth's layers is the **crust**. It is formed mainly of igneous rock, but also contains substantial proportions of metamorphic rock and a comparatively thin upper layer of sedimentary rock. Two crustal types are distinguished. **Oceanic crust**, which accounts for about 0.1 percent of Earth's mass, can be up to 10 km thick under the oceans, but is absent beneath the continents. **Continental crust** accounts for about 0.4 percent of Earth's mass and is present only in continental areas where it is generally 30 to 40 km in thickness (see Figure 11.1). The base of the crust is sharply defined where it contacts the mantle. This contact is detected by the way in which earthquake waves abruptly change

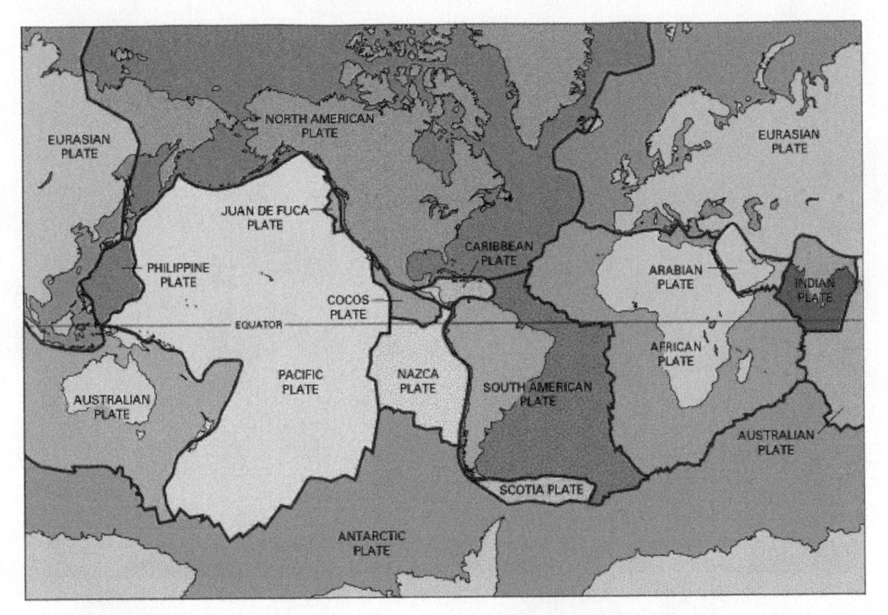

12.2 **WORLD MAP OF LITHOSPHERIC PLATES** Continental plates, such as the North American Plate and Eurasian Plate consist primarily of granitic-type rock. Oceanic plates, such as the Pacific Plate consist mainly of basaltic-type rock. Earthquakes and volcanic activity are common along plate boundaries.

velocity. The boundary between the crust and mantle is called the Mohorovičić discontinuity, or Moho, after the scientist who discovered it in 1909.

THE LITHOSPHERE AND ASTHENOSPHERE

The **lithosphere**, a zone of rigid, brittle rock, includes not only the crust, but also the cooler, upper part of the mantle. It ranges in thickness from 60 to 150 km, and is thickest under the continents. Some tens of kilometres deep within the Earth, the brittle condition of the lithospheric rock gives way gradually to a plastic layer named the **asthenosphere**, where temperatures reach 1,400 °C. The density of the asthenosphere increases from 3.7 to 5.5 g cm^{-3} with depth. However, at still greater depth in the mantle, the strength of the rock material increases again. Thus, the asthenosphere forms a soft layer between the "hard" lithosphere above and a "strong" mantle rock layer below.

Because the asthenosphere is soft and plastic, the rigid lithosphere, consisting of **lithospheric plates**, can easily move over it. A single lithospheric plate can be as large as a continent and can move independently of the plates that surround it (Figure 12.2). Like great slabs of floating ice, lithospheric plates can separate from one another at one location, while elsewhere they may collide and push up mountain ranges. Along the collision zones, one plate moves under the edge of its neighbour as it is drawn down into the mantle. The global structure of lithospheric plates is discussed later in this chapter.

THE GEOLOGIC TIME SCALE

It is generally believed that the solid Earth originated some 4,600 million years ago. This long period of time can be considered in two different ways from a geological perspective. The *chronostratic* approach to geological time is based on the relative age of rock types and events, and uses fossil assemblages and an idealized set of sedimentary rock strata as its reference. The *chronometric* approach establishes the absolute age of earth materials from radiometric and other dating methods.

The foundation of chronostratic dating is the *geologic column* that is made up of all the layers of sedimentary rock that have formed at some time in the geologic past. The relative age of the rocks is established according to two basic assumptions. The first assumption, known as the *Law of Uniformity*, states that "the present is the key to the past." This implies that processes observed today likely operated in the same way in past geologic times. The second assumption is known as the *Law of Superposition*, and states that "in any undisturbed sequence of sedimentary rocks, the lowest stratum was deposited first and is the oldest, and the uppermost stratum was deposited last and is the youngest." A complete set of the different rock strata does not exist anywhere on Earth. However, the entire **geologic column** can be established by cross-referencing many different locations (Figure 12.3).

The geologic column represents a vertical cross-section through the Earth's crust, with the youngest sedimentary rocks at the surface and the oldest sedimentary rock deposited on the lower crystalline basement rocks. Differences in sedimentary rock types and fossil sequences are used to subdivide the geologic column into the familar units of the *geologic time scale* (Table 12.1). The major divisions in the geologic time scale are termed **eons**. All Earth materials and events older than 542 million years (Ma) are assigned to the *Precambrian*. The Precambrian was originally considered to be devoid of fossils, but several primitive life forms are now recognized; the most common are *stromatolites*, which are cyanobacteria preserved in laminated, metamorphosed sediments. The Precambrian is divided into the Proterozoic Eon (542 to 2,500 Ma), the Archean Eon (2,500 to 3,800 Ma), and the Hadean Eon (3,800 to 4,600 Ma). From 542 Ma to the present is covered in the *Phanerozoic*

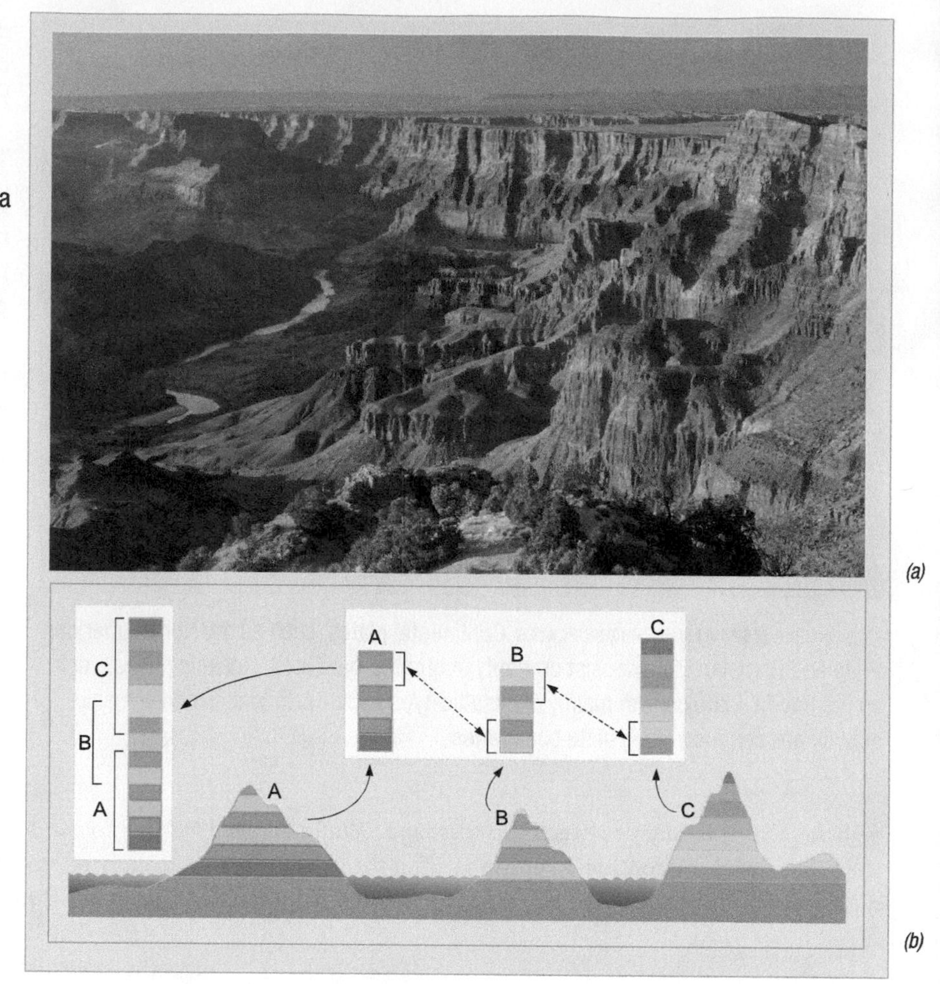

12.3 THE COMPLETE GEOLOGIC COLUMN IS ESTABLISHED BY CROSS-REFERENCING BETWEEN DIFFERENT LOCATIONS
(a) The Grand Canyon. Even with an average depth of 1.6 km, the exposed rock strata in the Grand Canyon represent only a fraction of the geologic column. The oldest sedimentary rock layers at 525 Ma are the Tapeats Sandstones, and the youngest are in the Kaibab Formation at 270 Ma. (b) The geologic column is pieced together by compiling rock sequences gathered from numerous locations.

Eon. The Phanerozoic is subdivided into three *eras*—the *Paleozoic*, *Mesozoic*, and *Cenozoic*; these eras are subdivided into *periods* and epochs.

The Cenozoic Era is particularly important in terms of the continental surfaces. Nearly all of the landscape features seen today have developed in the past 65 million years, even though the rock strata may be considerably older and many previous landforms have long since disappeared. The Cenozoic Era is comparatively short in duration. It is divided into two periods: the Tertiary (sometimes subdivided into the Paleogene and Neogene periods) and the Quaternary (includes the Pleistocene Epoch—popularly known as the Ice Age—and Holocene, which covers only the past 11,500 years).

Various techniques are used to establish the chronometric ages of rocks within the geologic time scale. One of the most important techniques is *radiometric dating*, which establishes the age of certain rock-forming minerals using the using principles of radioactive decay. The *Working It Out • Radioactive Decay and Radiometric Dating* feature at the end of the chapter describes this process in more detail.

MAJOR RELIEF FEATURES OF THE EARTH'S SURFACE

About 29 percent of the Earth's surface is land, with the remaining 71 percent comprising the oceans. However, if the seas were to drain away, broad sloping areas lying close to the continental shores would be exposed. These continental shelves are covered by shallow water, less than 150 m deep (Figure 12.4). Beyond the continental shelves, the ocean floor drops rapidly to depths of thousands of metres. This suggests that the oceans extend over the margins of the continents. If these submerged areas are considered part of the continents, land area would increase to 35 percent, with a reduction in the area of the ocean basins to 65 percent. These revised figures more correctly represent the relative proportions of continents and oceans.

RELIEF FEATURES OF THE CONTINENTS

Broadly viewed, continents can be subdivided into two basic regions: active belts where mountain-building processes are still occurring and shield areas composed of old, stable rock. The mountain ranges in the active belts grow either through volcanism or by tectonic activity. *Volcanism* can result

TABLE 12.1 GEOLOGIC TIME SCALE

EON	ERA	PERIOD		EPOCH	DURATION (Ma)	START (Ma)	OROGENIES	EVOLUTION OF LIFE FORMS
Phanerozoic	Cenozoic	Quaternary		Holocene	0.01	0.01		
				Pleistocene	2	2		Human genus
		Tertiary	Neogene	Pliocene	3	5		
				Miocene	18	23		Hominoids
			Paleogene	Oligocene	11	34	Alpine ends	Whales
				Eocene	23	57	Cordilleran ends	Bats
				Paleocene	9	66	Alpine begins	Mammals
	Mesozoic	Cretaceous			80	146		Flowering plants
		Jurassic			54	200		Dinosaurs (extinct)
		Triassic			51	251		Turtles
	Paleozoic	Permian			48	299		Frogs
		Carboniferous			60	359	Alleghany	Conifers; higher fishes
		Devonian			57	416	Caledonian/ Alleghany	Vascular plants; primitive fishes
		Silurian			28	444		
		Ordovician			44	488		
		Cambrian			54	542		Invertebrates
Precambrian	Proterozoic				1,958	2,500		
	Archean				1,300	3,800		
	Hadean				800	4,600		

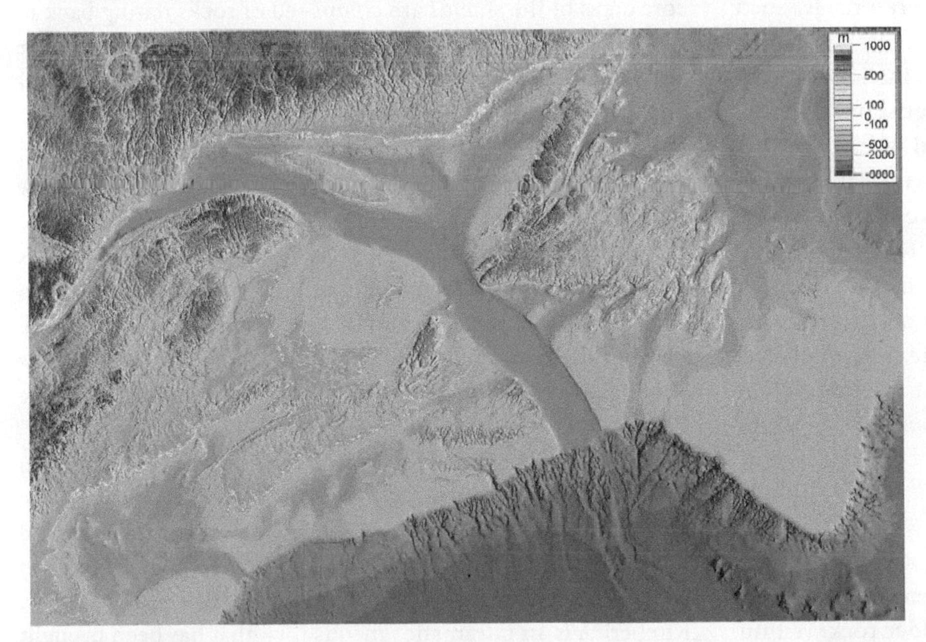

12.4 THE CONTINENTAL SHELF IN THE VICINITY OF CANADA'S ATLANTIC PROVINCES The broad shelf southeast of Newfoundland is the Grand Banks, a regionally important fishing zone.

in massive accumulations of volcanic rock by extrusion of magma, as seen in the Cascade Mountains of western North America. **Tectonic activity** involves the breaking, bending, and upthrusting of the Earth's crust due to internal forces associated with the collision of the lithospheric plates.

Zones of Active Mountain Building

Active mountain-building belts are narrow zones usually found along the margins of lithospheric plates (Figure 12.5). These belts, sometimes referred to as *alpine chains*, are characterized by high, rugged mountains, such as the Rocky Mountains of North America, formed 40 to 70 million years ago, and the Alps of Central Europe. Active alpine chains are characterized by their broadly curving arrangement; each curved section is called a **mountain arc**.

The mountain arcs form two principal mountain belts, the *circum-Pacific belt* that rings the Pacific Ocean Basin, and the

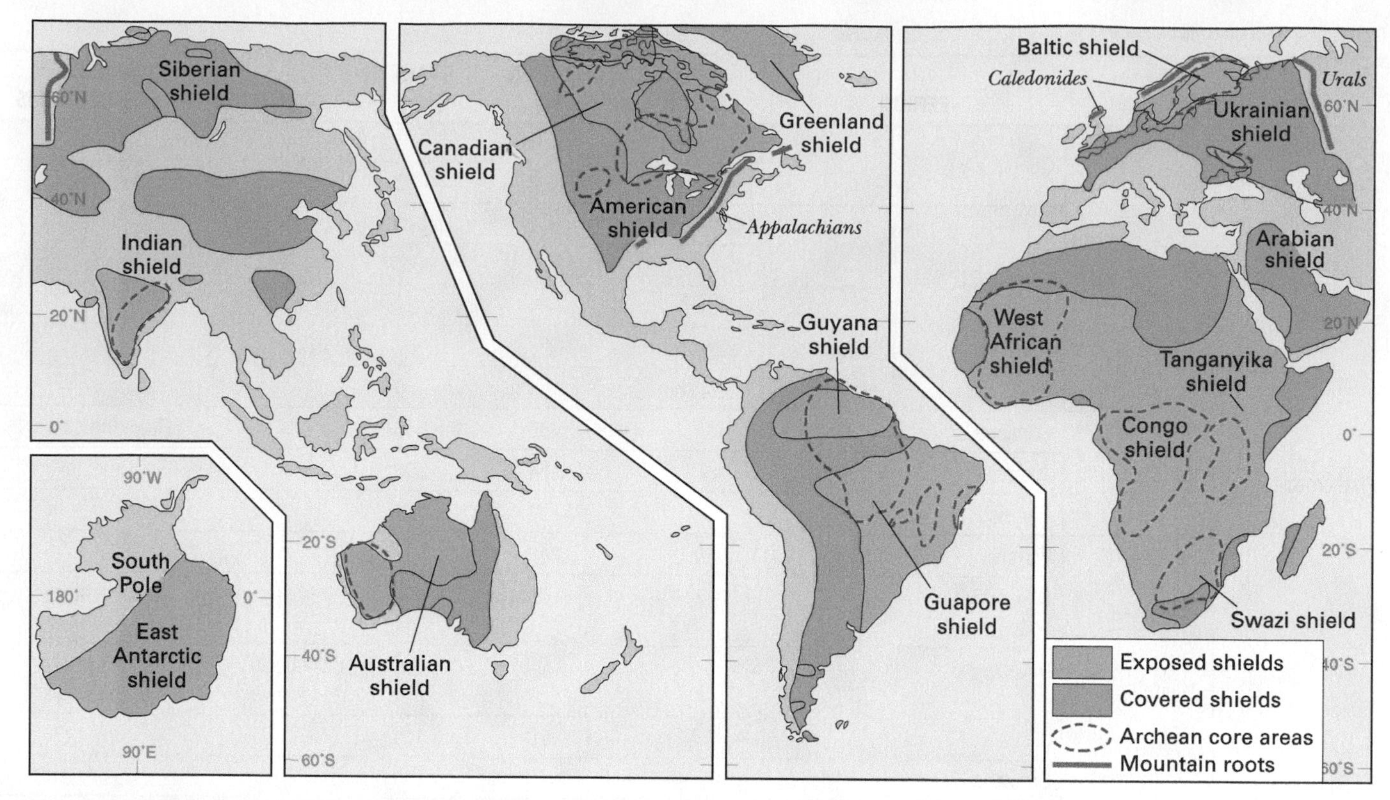

12.5 CONTINENTAL SHIELDS In this generalized world map of exposed and covered continental shields, continental centres of the early Precambrian Time (Archean) are shown by a broken red line. The heavy brown lines show mountain roots of the Caledonian and Hercynian orogenies. In North America, this is usually referred to as the Appalachian orogeny.

Eurasian-Indonesian belt that includes the Himalaya Mountains. The circum-Pacific belt includes the Cordilleran ranges such as the Coast Mountains and Cascade Mountains in North America and the Andes in South America. In the western part of the Pacific Basin, the mountain arcs well offshore from the continents. Here they form partly submerged **island arcs** that include the Aleutian Islands, the Kuril Islands, Japan, and the Philippines. Island arcs are the result of volcanic activity. Isolated volcanoes may rise above the sea to form smaller islands amongst the larger islands. The second chain of major mountain arcs that forms the Eurasian-Indonesian belt runs from the Atlas Mountains of North Africa, through the European Alps and the ranges of the Near East and Iran, to join the Himalayas. It then continues through Southeast Asia into Indonesia, where it meets the circum-Pacific belt.

Zones of Inactive Stable Crust

Belts of recent and active mountain building account for only a small portion of the continental crust. The remainder consists of comparatively inactive regions of much older rock. Within these stable regions are two types of crustal structure: continental shields and mountain roots. **Continental shields** are low-lying continental surfaces, beneath which are complex arrangements of ancient igneous and metamorphic rocks. Some core areas of the shields are composed of rocks dating back to the Archean Eon, 2,500 to 3,800 million years ago. These ancient cores are exposed in some areas, but are covered in others (Figure 12.5).

Exposed shields include very old rocks, mostly from Precambrian time, and have a complex geologic history. These regions occur extensively in Scandinavia, South America, Africa, Asia, India, Australia, and Canada. The exposed shields are typically regions of low hills and low plateaus, although in some areas large crustal blocks have been recently uplifted. Many thousands of metres of rock have eroded away from these shields over the past 500 million years throughout the Phanerozoic.

The Canadian Shield is the greatest area of exposed Archaean rock in the world. It is mostly composed of metamorphic base rocks from the Precambrian, many of which contain substantial mineral deposits, including nickel, copper, and gold. Diamonds are also mined from kimberlite in the Northwest Territories. Kimberlite is an ultramafic igneous rock that has been brought to the surface from great depths in volcanic pipes and dikes. Despite being repeatedly uplifted and eroded, the Canadian Shield has remained almost entirely above sea level throughout

12.6 CANADIAN SHIELD This outcrop of hard metamorphic rock has been shaped and smoothed by the action of ice and water.

its long history, but was heavily glaciated during the Late-Cenozoic Ice Age. The result is a low-relief, poorly drained, lake-studded area with extensive bedrock exposures (Figure 12.6). Together with the covered shield areas to the west and south, the Canadian Shield comprises the North American *craton*, or ancient base of the continent.

Covered shields are areas of the continental shields that are covered by younger sedimentary layers deposited during the Phanerozoic Era. These strata accumulated at times when the shields were inundated by shallow seas. Marine sediments covered the ancient shield rocks in thicknesses ranging from hundreds to thousands of metres. The shield areas were then broadly arched upward to become land surfaces again. Erosion has since removed large sections of their sedimentary cover, but it still remains intact over vast areas. The shallow waters of Hudson Bay are presently receiving sediments carried by numerous rivers in what appears to be a renewed cycle of activity.

Ancient Mountain Roots

Remains of older mountain belts lie within the continental shields in many places. These *mountain roots* are mostly formed of Paleozoic and early Mesozoic sedimentary rocks that have been intensely bent and folded, and in some locations changed into metamorphic rocks—for example, slate, schist, and quartzite. Thousands of metres of overlying rocks have been removed from these old tectonic belts, so that only the lowest structures remain. Roots appear as chains of long, narrow ridges, rarely rising more than a thousand metres above sea level.

One important system of mountain roots was formed in the Paleozoic Era, about 400 million years ago. Called the Caledonides, it forms a highland belt across the northern British Isles and Scandinavia, and is also present in the Canadian Atlantic provinces and in New England (see Figure 12.5). Originally a high alpine mountain chain created by the collision of two lithospheric plates, the Caledonides have since been worn down to a series of subdued mountain ranges and hills.

Three separate periods of mountain building occurred in North America as a result of collisions between ancient landmasses. The first took place between 420 and 280 million years ago and gave rise to the ancient Appalachian Mountains. The last phase, part of the Hercynian Orogeny, occurred when the combined continents of Europe and Africa, known as Gondwana, collided with the North American landmass, known at the time as Laurentia. The tectonic processes that formed the Appalachian Mountains were also responsible for the formation of the Ural Mountains between Europe and Asia. These processes ended about 250 million years ago, when most of the smaller continents had come together to form the single supercontinent of Pangaea. The ancient Appalachian mountain range was eroded to an almost level plain, but a period of renewed uplift during the Cenozoic Era resulted in the general configuration seen today.

RELIEF FEATURES OF THE OCEAN BASINS

The relief features of ocean basins are quite different from those of the continents. The crustal rock of the ocean floors consists almost entirely of basalt, which is generally covered by a comparatively thin accumulation of sediments. Age determinations of the basalt and its overlying sediments show that the oceanic crust is, geologically, quite young. Much of the oceanic crust is less than 60 million years old, although in some cases it has been dated to 135 million years. This is much younger than most of the continental crust, some of which is several billion years old.

MID-OCEANIC RIDGE AND OCEAN BASIN FEATURES

The major relief feature of ocean basins is a central ridge, which in the case of the Atlantic Ocean divides the basin approximately in half (Figure 12.7). This *mid-oceanic ridge* consists of submarine hills that rise gradually to a rugged central zone. In the centre of the ridge is a narrow, elongated depression known as the *axial rift*. The location and form of the axial rift suggest that this is a region where the crust is being pulled apart. This is confirmed by the fact that the age of the rocks on both sides of the ridge increases symmetrically with distance from the rift.

The mid-oceanic ridge and its principal branches can be traced through the world's ocean basins for a total distance of about 60,000 km (Figure 12.8). Beginning in the Arctic Ocean, the Mid-Atlantic Ridge divides the Atlantic Ocean Basin from Iceland to the South Atlantic. Turning east, it runs past Africa into the Indian Ocean as the Southeast Indian Ridge. Near the centre of the Indian Ocean it divides to form the Mid-Indian Ocean Ridge that runs north into the Red Sea, with the Southeast Indian Ridge continuing eastward between Australia and Antarctica. The mid-oceanic ridge crosses the southern Pacific Ocean as the

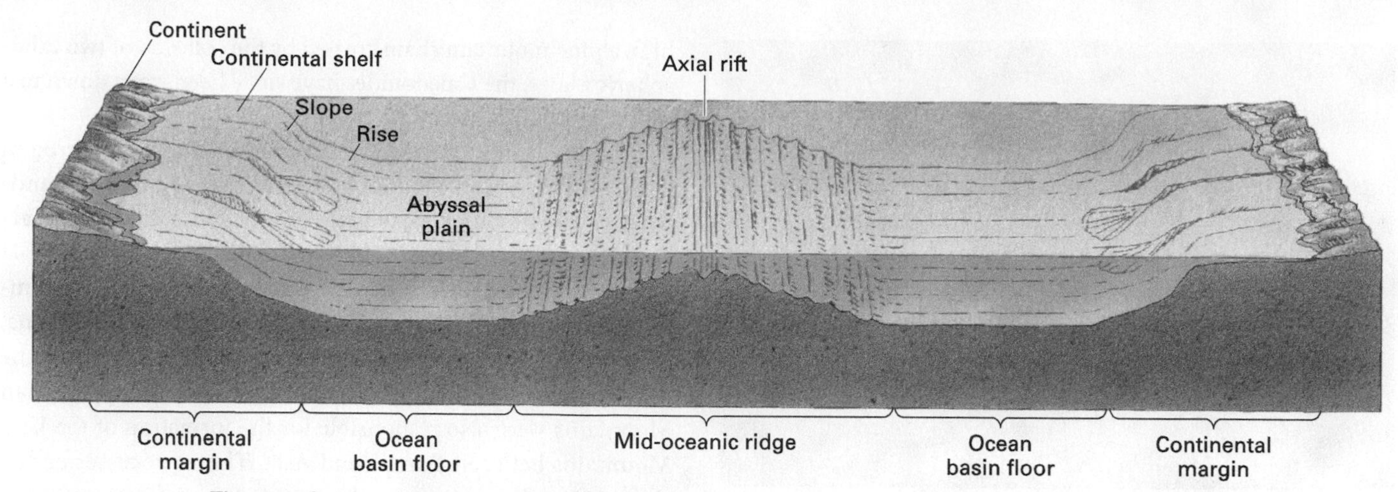

12.7 OCEAN BASINS The main feature of the North and South Atlantic Ocean basins is the mid-oceanic ridge, a region where the crust is separating and moving apart.

Pacific Antarctic Ridge, then runs past South America as the East Pacific Rise, and terminates near North America at the head of the Gulf of California. A disconnected section of the system forms the Juan de Fuca Ridge along the coast of British Columbia.

The mid-oceanic ridge is found at a depth of about 2,000 m below the surface of the oceans, but in some locations it is above sea level and creates islands, such as Iceland, the Azores, and Tristan da Cunha (35° 30′ S; 12° 15′ W) in the South Atlantic. On either side of the mid-oceanic ridge are broad, flat *abyssal plains* that form the deep *ocean basin floor* at an average depth below sea level of about 5,000 m. The level topography of the abyssal plains consists of fine sediment that has settled slowly and evenly from ocean water above. The sediment is derived from debris eroded from the continents, transported to the coasts

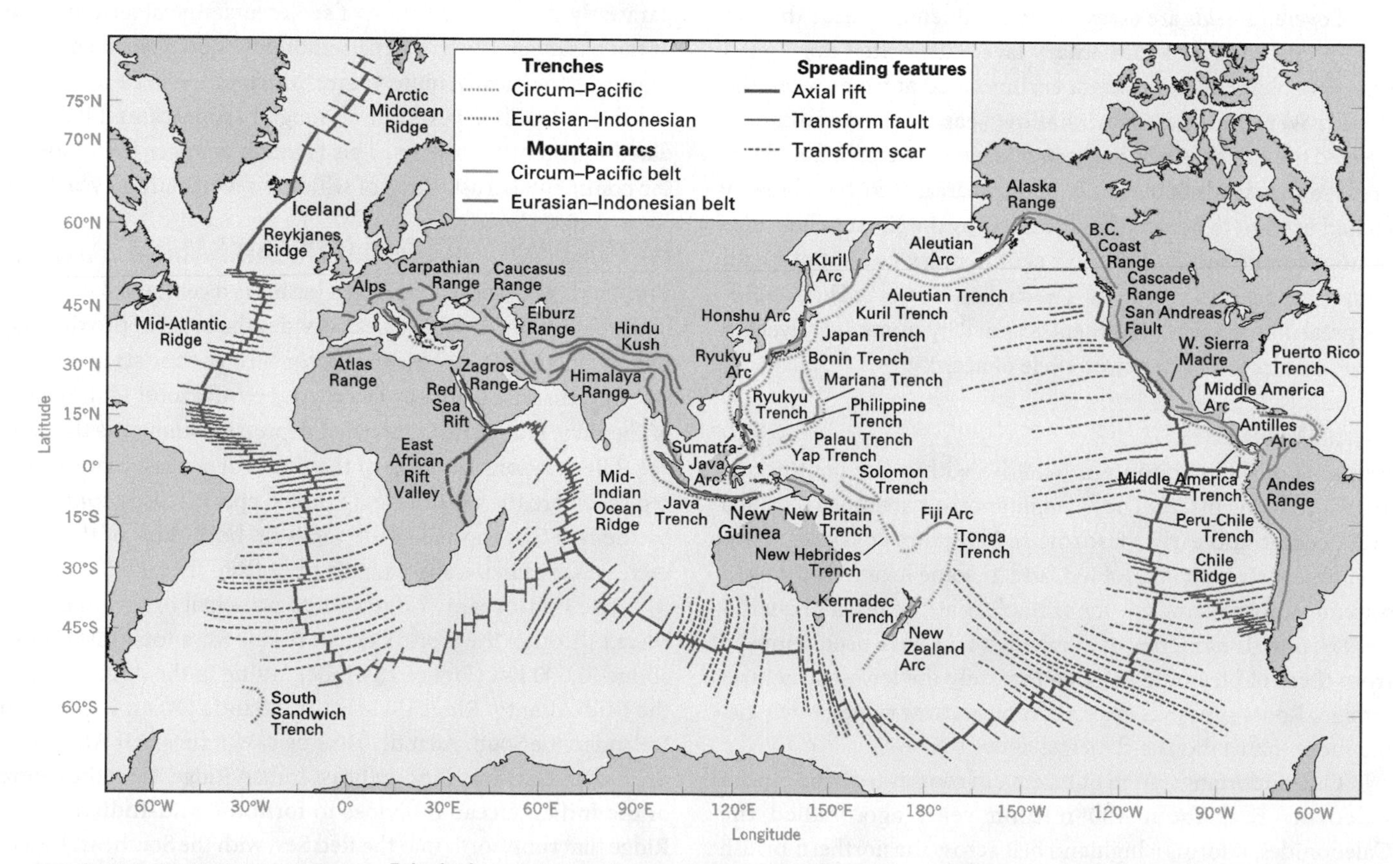

12.8 TECTONIC FEATURES OF THE WORLD Principal mountain arcs, island arcs, ocean trenches, and the midoceanic ridge. (Midoceanic ridge map copyright © A. N. Strahler.)

by rivers, and then dispersed into the oceans by *turbidity currents*. These dense tongue-like flows of suspended sediment move rapidly down the *continental slope* at the edge of the *continental shelf* like underwater avalanches. The average depth of the sediment is about 5 km and is sufficiently thick to cover irregularities in the basaltic rock of the ocean floor. In places where the sediments are thinner, gently sloping abyssal hills of volcanic origin may rise 1,000 m or so above the abyssal plains.

The deepest parts of ocean basins are the *ocean trenches* that mark the positions of subduction arcs where oceanic crust is being forced down into the mantle. Of the 22 ocean trenches that have been located, 18 are in the Pacific Ocean, three are in the Atlantic Ocean, and one is in the Indian Ocean. The deepest is the Mariana Trench in the Pacific Ocean, about midway between Indonesia and Japan; it is 2,540 km in length, descends to 11,033 m, and is 69 km in width. The Puerto Rico Trench, which descends to 8,648 m, is the deepest point in the Atlantic Ocean; the Java Trench, the deepest point in the Indian Ocean, descends to 7,725 m.

CONTINENTAL MARGINS

The *continental margin* is the narrow zone where oceanic lithosphere is in contact with continental lithosphere (see Figure 12.7). As the continental margin is approached from the deep ocean, the ocean floor begins to slope gradually upward, forming the *continental rise*, then becomes much steeper on the *continental slope*. The top of this slope marks the edge of the *continental shelf*, a gently sloping platform with vertical relief of less than 20 m. The average width of the continental shelves is about 80 km, but this is quite variable. It is narrowest where a continent abuts the forward edge of an advancing oceanic plate, such as off the coast of Chile. Shallow seas overlie areas of continental shelf, such the North Sea; the Grand Banks off Newfoundland is also part of the continental shelf (see Figure 12.4). The continental rise, slope, and shelf together form the continental margin.

The symmetrical arrangement of a mid-oceanic ridge with ocean basin floors on either side nicely fits the North Atlantic and South Atlantic Ocean basins. It also applies rather well to the Indian Ocean and Arctic Ocean basins. The margins of these symmetrical basins are described as *passive continental margins* because they have not experienced strong tectonic and volcanic activity during the last 50 million years. This is because the continental and oceanic lithospheres that join at a passive continental margin are part of the same lithospheric plate, and they move away from the axial rift as a single unit.

Although a mid-oceanic ridge is present across much of the southern Pacific Ocean with ocean basin floors on either side, unlike the Atlantic Ocean the margins of the Pacific Basin are characterized by mountain arcs or island arcs with deep *oceanic trenches* offshore. These trenched ocean-basin edges are referred to as *active continental margins*. Here, oceanic crust is bent downward and forced under continental crust or, as in the case of island arcs such as Japan and Indonesia, under other oceanic crust. As well as creating trenches, this type of crustal interaction induces volcanic activity.

THE GLOBAL SYSTEM OF LITHOSPHERIC PLATES

The Earth's crust consists of distinctive lithospheric plates of various sizes; seven of these are considered to be major or primary plates (see Figure 12.2). The Pacific Plate is the largest primary plate and covers about 103.3 million km^2; the smallest is the South American Plate, covering 43.6 million km^2. Several secondary plates are also recognized, including the Caribbean Plate (3.3 million km^2), Cocos Plate (2.9 million km^2), and the Juan de Fuca Plate (250, 000 km^2), which lie adjacent to the North American Plate (75.9 million km^2). More than 50 minor or tertiary plates have also been identified, such as the Galapagos microplate with an area of only 12,000 km^2.

The Pacific Plate occupies much of the Pacific Ocean Basin, and consists almost entirely of oceanic lithosphere. Along most of the western and northern margins there is a subduction boundary in which crustal plates are being forced down into the mantle. The eastern and southern edge is mostly a spreading boundary where material from the mantle rises toward the surface and forms an axial rift. A small area of continental lithosphere that is also part of the Pacific Plate makes up the coastal portion of California and all of Baja California. The boundary of the Pacific Plate in California is marked by the San Andreas Fault.

The North American Plate includes most of the continental lithosphere of North America, Greenland, and parts of Siberia. It also extends into the Atlantic Ocean, where its eastern boundary is formed by the Mid-Atlantic Ridge. Consequently, western Iceland is also included as part of the North American Plate, whereas eastern Iceland is part of the Eurasian Plate. It is believed that Iceland arose from the sea floor some 20 million years ago, and because of its position astride the Mid-Atlantic Ridge, this volcanically active region continues to spread apart at a rate of about 5 cm a year. The southern boundary between the Caribbean and Cocos Plates is marked by faults and trenches. The western boundary with the Pacific Plate is defined by the San Andreas Fault in California and the Queen Charlotte Fault system, which runs parallel to the coasts of British Columbia and Alaska. All of these areas are prone to earthquakes, such as the one that occurred in Haiti in 2010 (see Chapter 13).

The general configuration of the South American Plate is similar to the North American Plate. The western part includes the continent of South America, and to the east consists entirely of oceanic lithosphere as far as the Mid-Atlantic Ridge. For the most part, the western edge of the South American Plate is a subduction boundary, where oceanic lithosphere of the relatively young Nacza Plate (15.6 million km^2) is being forced beneath the continental lithosphere. The southern edge is marked by a fault zone along the secondary Scotia Plate (1.6 million km^2) and the major Antarctic Plate (60.9 million km^2) further east. The eastern edge, formed by the spreading boundary along the Mid-Atlantic Ridge, separates the South American Plate from the African Plate (61.3 million km^2).

The Eurasian Plate (67.8 million km^2) is mostly continental lithosphere, but a belt of oceanic lithosphere fringes it on the west and north. The easternmost part of Asia, including the Kamchatka Peninsula, is part of the North American Plate, but the nature of the boundary in this region is uncertain. The African Plate has a central core of continental lithosphere nearly surrounded by oceanic lithosphere. The Red Sea, which separates the African Plate from the lesser Arabian Plate (5 million km^2), is an active rift zone. A similar tectonic boundary formed by the East African Rift zone separates the African Plate from the secondary Somali Plate (16.7 million km^2).

The Austral-Indian Plate is mostly oceanic lithosphere but contains two cores of continental lithosphere: the Australian Plate (47 million km^2) and the Indian Plate (11.9 million km^2). Recent evidence shows that these two continental masses are moving independently and may properly be considered parts of separate plates. The Antarctic plate forms a central core of continental lithosphere completely surrounded by oceanic lithosphere. It is almost completely enclosed by a spreading plate boundary where the adjacent plates are moving away from the pole.

PLATE TECTONICS

The place where two plates meet is a plate boundary. Different types of boundaries are recognized depending on how the adjacent plates are moving in relation to each other. The motion of lithospheric plates and how they interact at their boundaries is collectively called **plate tectonics**. Many of Earth's large-scale structural and topographic features, or primary landforms, are recognized by the deformation that occurs when moving lithospheric plates collide.

PLATE MOTIONS AND INTERACTIONS

Lithospheric plates move about the Earth's surface at a rate of about 5 to 10 cm a year. It is generally postulated that plate movement is caused primarily by immense convection currents generated by heat energy, derived from radioactive decay of unstable isotopes in the crust and mantle. Plate movement is also linked to density differences within the mantle due to rock chemistry and mineral composition. Gravitational sliding away from the higher topography along the centre of a ridge may also contribute to the spreading motion through the process termed *ridge push*. As the molten rock rises beneath the oceanic lithosphere, the sea floor lifts, and the plates are moved laterally away from the spreading axis (Figure 12.9). Plate movement is generally fastest in zones of subduction. Here too, gravity is considered to contribute to plate movement by exerting a downward frictional pull through a process known as *slab suction*.

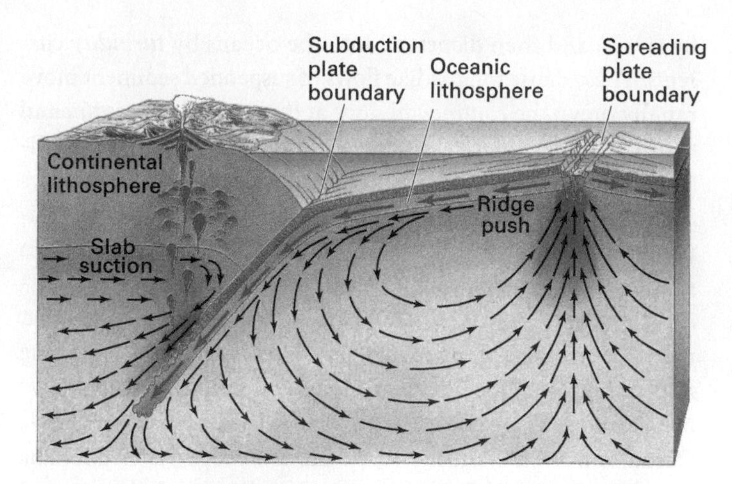

12.9 **MANTLE FLOW SYSTEM AND HEAT EXCHANGE** Mantle flow (black arrows) transports heat to the Earth's surface, where it is lost through volcanic activity associated with zones of subduction and mid-oceanic ridges. The crustal plates move away from the zone of upwelling magma (red arrows) and are eventually carried down to the mantle in the subduction zones. (Copyright © A. N. Strahler.)

The motion of the oceanic plate exerts a drag on the surrounding asthenosphere, which creates a horizontal flow current. At the same time, the plate undergoes cooling, becomes denser, and sinks steadily deeper. The process in which one plate is carried beneath another is called **subduction**; the descending slab is heated and softened, and is eventually incorporated into the mantle.

This convective system brings hot material from the asthenosphere toward the surface, where heat flows to the oceanic and atmospheric layers. It is a one-way energy system that depends on a radiogenic heat source, which is slowly being depleted. As time passes, the plate tectonic system will lose power and the rate of circulation will steadily slow. In the distant future, the Earth's tectonic system will cease, and there will be no more earthquakes or volcanic eruptions.

Both oceanic and continental lithosphere can be thought of as "floating" on the soft asthenosphere. The zone of separation, or *rift*, between the lithospheric plates along the axis of a mid-oceanic ridge is continually filled with magma rising from the mantle. The magma appears as basaltic lava in the floor of the rift and quickly hardens. At greater depth under the rift, magma solidifies into *gabbro*, an intrusive rock of the same composition as basalt. Together, the basalt and gabbro continually form new oceanic crust. This type of boundary between plates is called a divergent or *spreading boundary*.

Converging plate boundaries, where subduction is taking place, results in the formation of deep ocean trenches; these are zones of intense tectonic and volcanic activity. Because the oceanic plate is comparatively thin and dense—in contrast to the thick,

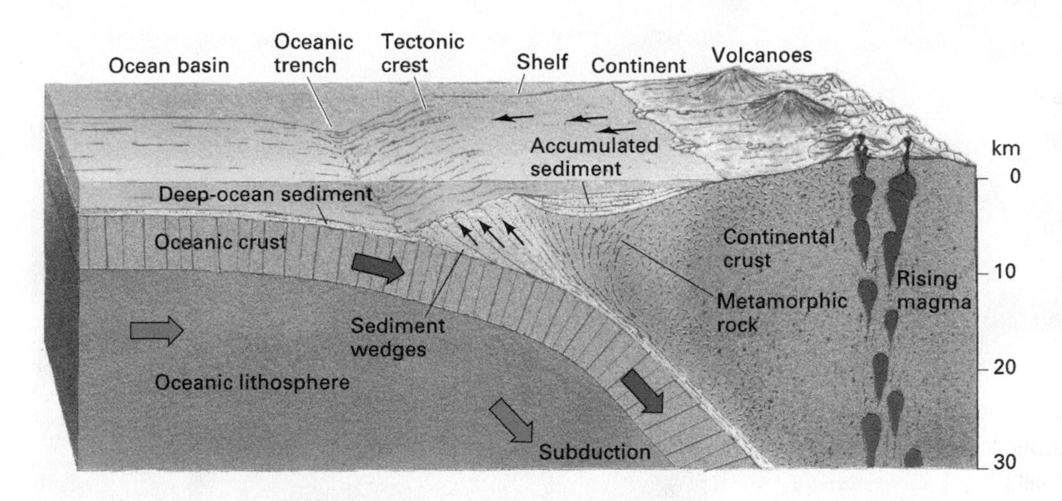

12.10 TYPICAL FEATURES OF AN ACTIVE SUBDUCTION ZONE Sediments scraped off the moving plate form tilted wedges that accumulate in a rising tectonic mass. Near the mainland is a shallow trough in which sediment brought from the land accumulates. Metamorphic rock forms above the descending plate. Magma rising from the top of the descending plate reaches the surface to build a chain of volcanoes.

buoyant continental plate—the oceanic lithosphere bends down and plunges into the soft asthenosphere. The leading edge of the descending plate is cooler and therefore denser than the surrounding hot asthenosphere. Consequently, once subduction has begun, the plate sinks under its own weight. The plate is gradually heated by the surrounding hot rock and eventually softens. The under portion, which is the same composition as mantle rock, is simply incorporated into the mantle as it softens.

An oceanic trench is also a region where sediment accumulates. The sediment comes from two sources. Deep-ocean sediment comprising fine clay and ooze is carried along on the moving oceanic plate. From the continent comes terrestrial sediment in the form of sand and mud brought by streams to the shore, and then carried by currents seaward into the deep water (Figure 12.10). In the bottom of the trench, both types of sediment are intensely deformed and dragged down with the moving plate. The deformed sediment is scraped off the plate and shaped into wedges that ride up, one over the other, on steep fault planes. The wedges accumulate at the plate boundary, forming an *accretionary prism* in which metamorphism takes place. This forms new continental crust of metamorphic rock and builds the continental plate outward.

The accretionary prism is of relatively low density and tends to rise, forming a *tectonic crest*. The tectonic crest is submerged in Figure 12.10, but in some cases it forms an island chain, or a *tectonic arc*, paralleling the coast. Between the tectonic crest and the mainland is the shallow *fore-arc trough*. Terrestrial sediment accumulates in the basin-like structure of the fore-arc trough, the base of which continually subsides under the added load. Where the sea floor in the trough becomes flat and shallow, a type of continental shelf is formed. Sediment carried across the shelf moves down the steep outer slope of the accretionary prism as *turbidity currents*. The sediments eventually come to rest with the heaviest, coarsest grains settling first, grading to the finest, lightest grains above. When compacted and lithified, such materials are referred to as *turbidites*.

The descending lithospheric plate enters the asthenosphere where intense heating of the upper surface melts the oceanic crust. Basaltic magma is formed, which rises into the thick continental lithosphere. As it rises, the magma melts material from the base of the crust; this is subsequently incorporated into the magma, changing its chemical composition to produce andesite. Some magma cools and forms deep-seated plutons, while some may eventually reach the surface and form chains of volcanoes parallel to the deep oceanic trenches where subduction is occurring. Such is the case in the Andes of South America, from which andesite gets its name. The chain of volcanoes that runs from Alaska into northern California was also formed in this way (Figure 12.11).

A third type of lithospheric plate boundary occurs where two lithospheric plates are in contact, but one plate merely slides past the other (Figure 12.12). This is a *transform boundary* where movement of the plates causes neither separation nor convergence. The fault plane along which motion occurs is a nearly vertical fracture extending down into the lithosphere. The San Andreas Fault is one of the best-known examples of a transform fault boundary. It can be followed for about 1,000 km from the Gulf of California to Cape Mendocino on the Pacific coast, where it heads out to sea. The San Andreas Fault

12.11 CASCADE MOUNTAIN VOLCANOES In the foreground is Mount St. Helens, which erupted in 1980. Mount Rainier can be seen in the background.

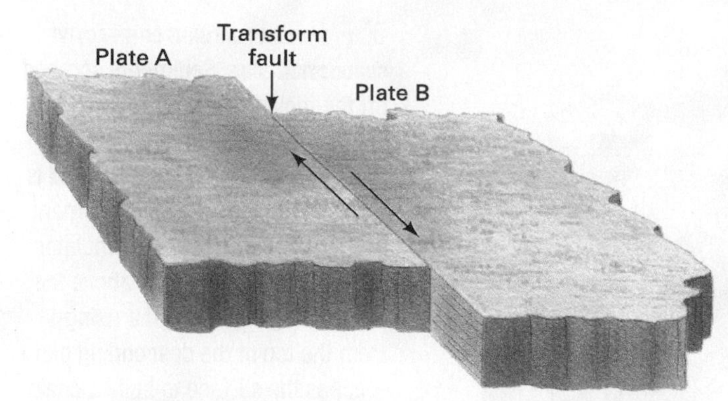

12.12 **TRANSFORM BOUNDARY** A transform boundary involves the horizontal motion of two adjacent lithospheric plates, one sliding past the other. (Copyright © A. N. Strahler.)

marks the boundary between the Pacific Plate, which is moving toward the northwest, and the North American Plate (Figure 12.13). Throughout the many kilometres of its length, the San Andreas Fault appears as a straight, narrow scar. In some places, this widens to a steep-sided valley, and elsewhere it becomes a low scarp.

OROGENIC PROCESSES

Prominent mountain masses and mountain chains (other than volcanic mountains) are elevated by one of two basic tectonic processes: *extension* and *compression*. Extensional tectonic activity occurs where oceanic plates pull apart or where a continental plate undergoes breakup. As the crust thins, it fractures producing rift valleys and block mountains. Compressional tectonic activity occurs at converging plate boundaries. The result is often an alpine mountain chain consisting of intensely deformed rock strata. The strata are tightly compressed into wavelike structures called *folds*. The entire deformed rock mass produced by compression is called an *orogen*, and the event that produced it is an *orogeny*.

Fold Belts

Folding usually shows a developmental sequence known as an *alpine fold system*. The wavelike shapes imposed on the strata consist of alternating arch-like upfolds, called **anticlines**, and trough-like downfolds, called **synclines** (Figure 12.14). Originally each mountain crest would have been associated with the axis of an anticline, with intervening valleys lying over the axis of a syncline. Some of the anticlinal arches may be partially removed by erosion processes, creating step-like skylines such as those that are seen throughout the Canadian Rockies (Figure 12.15).

Fault Landforms

Faulting may accompany folding if the brittle rocks of the Earth's crust yield to unequal stresses by fracturing. Faulting is accompanied by some displacement of the rocks along the line of fracture, or *fault plane*. Major faults often extend a great horizontal distance, and most also extend down into the crust for at least several kilometres. Faulting occurs in sudden slipping movements ranging perhaps from a few millimetres to several metres. The interval between successive movements may be several years or decades, but over long time spans, the accumulated displacements can amount to tens or hundreds of kilometres. In some places, clearly recognizable sedimentary rock layers are offset on opposite sides of a fault, allowing accurate measurement of the amount of displacement (Figure 12.16).

12.13 **THE SAN ANDREAS FAULT IN SOUTHERN CALIFORNIA** The fault is marked by a narrow trough. ***Fly By:*** *35° 11′ N; 119° 44′ W*

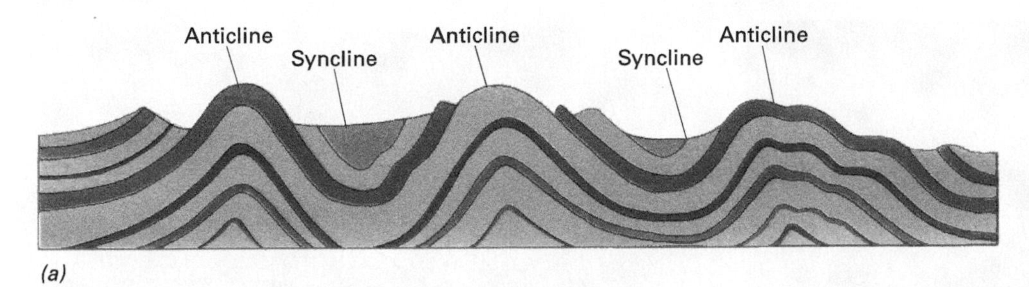

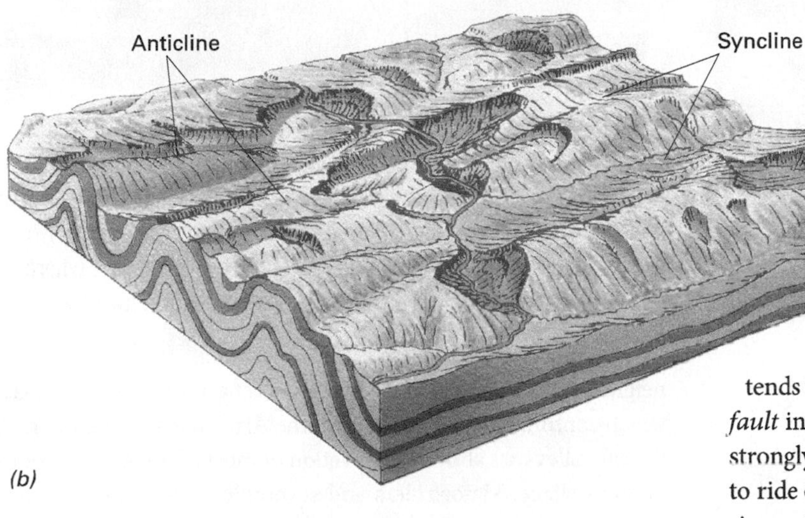

12.14 ANTICLINES AND SYNCLINES Structural diagrams showing (a) a typical cross-section and (b) the landscape developed on folded strata that is characteristic of the Canadian Rockies and other fold mountain areas, such as the Alps and Himalayas. (After E. Raisz.)

One common type of fault associated with crustal rifting is the **normal fault** in which the plane of slippage, or fault plane, is steeply inclined, and the rocks on one side are displaced relative to the other. This may arise either through the land on one side being lowered, or *downthrown*, or because movement has occurred on both sides of the fault. If the land is raised, or *upthrown*, the process is referred to as a **reverse fault**. Both normal and reverse faults can produce a steep, straight, cliff-like feature or *fault scarp*. Fault scarps range in height from a few metres to a few hundred metres and are often many kilometres in length.

Reverse faults produce fault scarps similar to those of normal faults, but the possibility of landslides is greater because the upthrust side tends to overhang the downthrust side. The *low-angle overthrust fault* involves predominantly horizontal movement, usually in strongly folded rocks, which eventually yields, causing one mass to ride over the adjacent rocks (Figure 12.17). In this way, extensive rock slabs called *thrust sheets* or *nappes*, up to 50 km wide, can be carried tens of kilometres across the underlying rock.

Normal faults are not usually isolated features. They commonly occur in multiple arrangements, often as intersecting sets of parallel faults, where land may drop down to form a *graben* or rise as a *horst* (Figure 12.18). The resulting topography is a series of trenches and plateaus. On a regional scale this can create *block mountains*, as in the Basin and Range district of the western

12.15 FOLDED STRATA (a) Complex folding in the Sullivan River area of the Rocky Mountains in British Columbia. (b) At 3,954 m, Mount Robson in British Columbia is the highest peak in the Canadian Rockies. The compressed rock in the axis of a syncline is usually more resistant to denudation, and many of the higher peaks in fold mountains are formed because of this.

12.16 **DISPLACED STRATA** Two normal faults are visible in this road cut in Canberra, Australia.

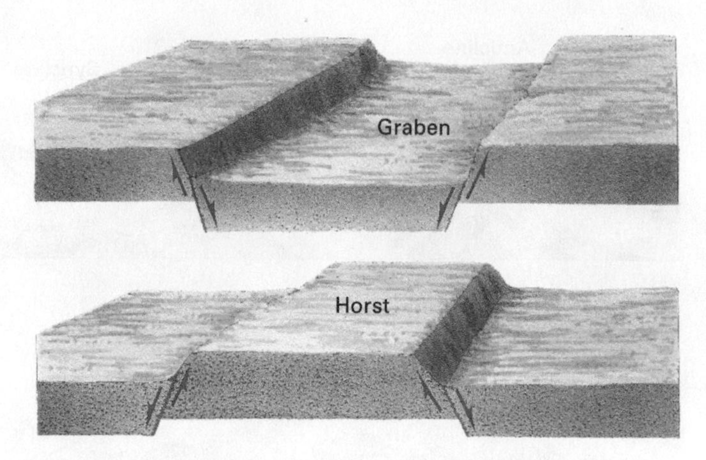

12.18 **INITIAL LANDFORMS OF NORMAL FAULTING** A graben is a downthrown block, often forming a long, narrow valley. A horst is an upthrown block, forming a plateau, mesa, or mountain. The width of the blocks typically increases from top to bottom. This type of rift structure is an extensional feature associated with regions where continental plates are pulling apart. (Copyright © A.N. Strahler.)

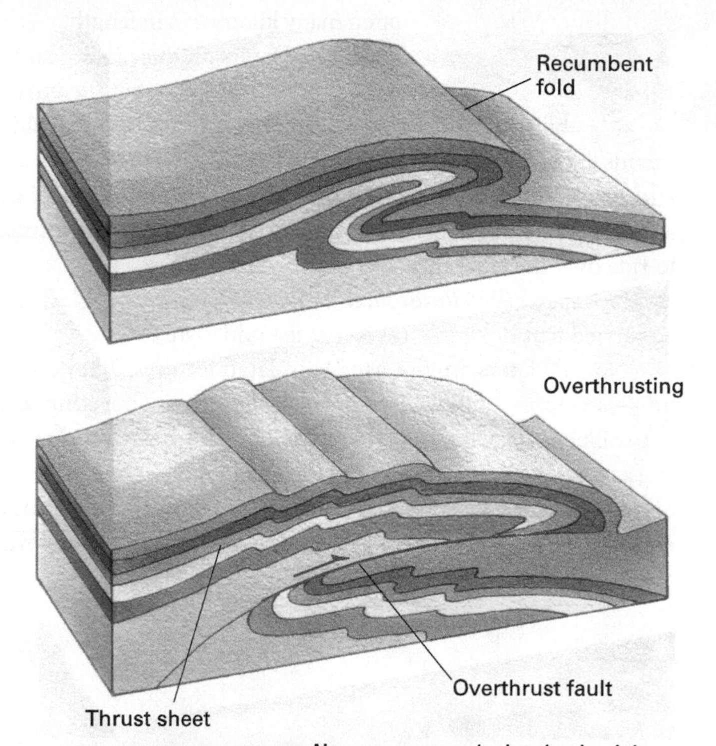

12.17 **THE FORMATION OF A NAPPE** Nappes commonly develop in alpine fold systems where intense folding has occurred along an overthrust fault. (Based on diagrams by A. Heim, Geologie der Schweiz, vol. II-1, Tauschnitz, Leipzig, 1922.)

United States. The East African Rift Valley system, which extends some 3,000 km from the Red Sea southward through the Afar region of Ethiopia to the Zambezi River, illustrates the process on a continental scale (Figure 12.19).

The East African Rift Valley system consists of a number of graben-like troughs. Each is a separate rift valley ranging in width from about 30 to 60 km, in which blocks have slipped down between neighbouring blocks, caused by movement of the Somalian Plate which continues to pull away from the African Plate. The floors of the rift valleys are above the elevation of most of the African continental surface. Major rivers and several long, deep lakes—Lake Malawi and Lake Turkana, for example—occupy some of the valley floors. The sides of the rift valleys typically consist of multiple fault steps (Figure 12.20), and sediments derived from the high plateaus fill the valley floors. Two stratovolcanoes, Mount Kilimanjaro (5,985 m) and Mount Kenya (5,199 m) have built up close to the Rift Valley.

Where two continental lithospheric plates converge along a subduction boundary, they must collide because the impacting masses are too thick and too buoyant to allow either plate to slip under the other. The result is an orogeny that causes complex folding and faulting of the crustal rocks. A mass of metamorphic rock is formed between the joined continental plates, welding them together and terminating further tectonic activity along that collision zone. The collision zone is named a **continental suture** (Figure 12.21).

Continent–continent collisions occurred in the Cenozoic Era along a great tectonic line that marks the southern boundary of the Eurasian Plate. The line begins with the Atlas Mountains of North Africa, and runs across the Aegean Sea region into western Turkey. It appears again in the Zagros Mountains of Iran, where it can be followed discontinuously into the Himalayas. Each segment of this collision zone represents the contact of different northward-moving plates against the single and relatively immobile Eurasian Plate. The European segment containing the Alps was formed when the African Plate collided with the Eurasian Plate. The Persian segment resulted from the collision of the Arabian Plate, and the Himalayan segment represents the collision of the Indian portion of the Austral-Indian Plate.

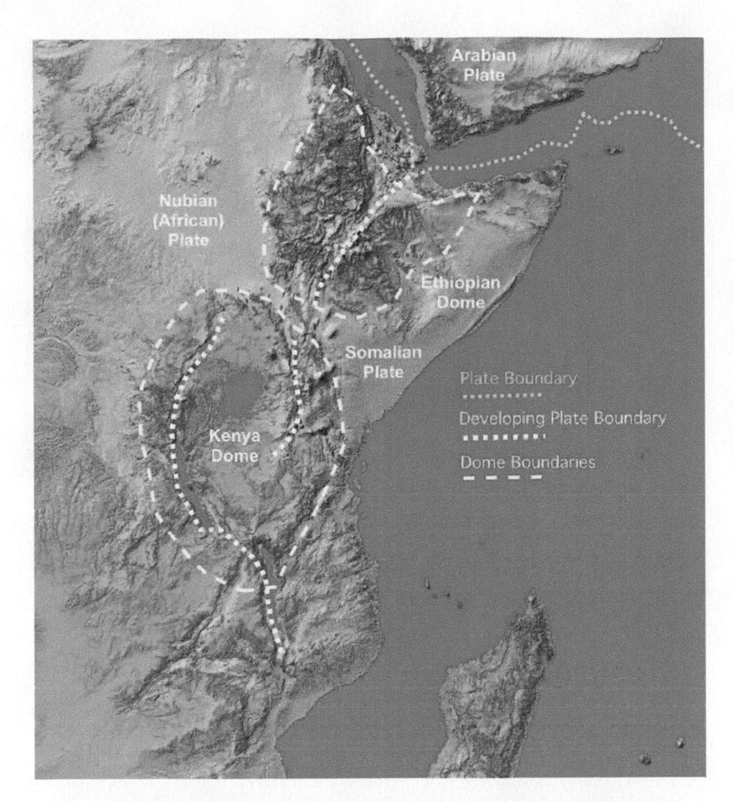

12.19 **EAST AFRICAN RIFT VALLEY** This complex geological region of East Africa marks the boundary between the African and Somalian plates. To the north, where they meet the Arabian Plate, water from the Indian Ocean now forms the Gulf of Aden and the Red Sea.

12.20 **RIFT VALLEY WALL IN NORTHERN KENYA** The multiple fault scarps of several horst blocks within the graben give the landscape a stepped appearance.

Collisions between continental plates have occurred many times since the late Precambrian time. Several ancient sutures have been identified in the continental shields. For example, the Ural Mountains, which divide Europe from Asia, formed in this way toward the end of the Paleozoic Era.

CONTINENTAL RUPTURE AND NEW OCEAN BASINS

Passive continental margins are formed when a single plate of continental lithosphere breaks apart. This process is called

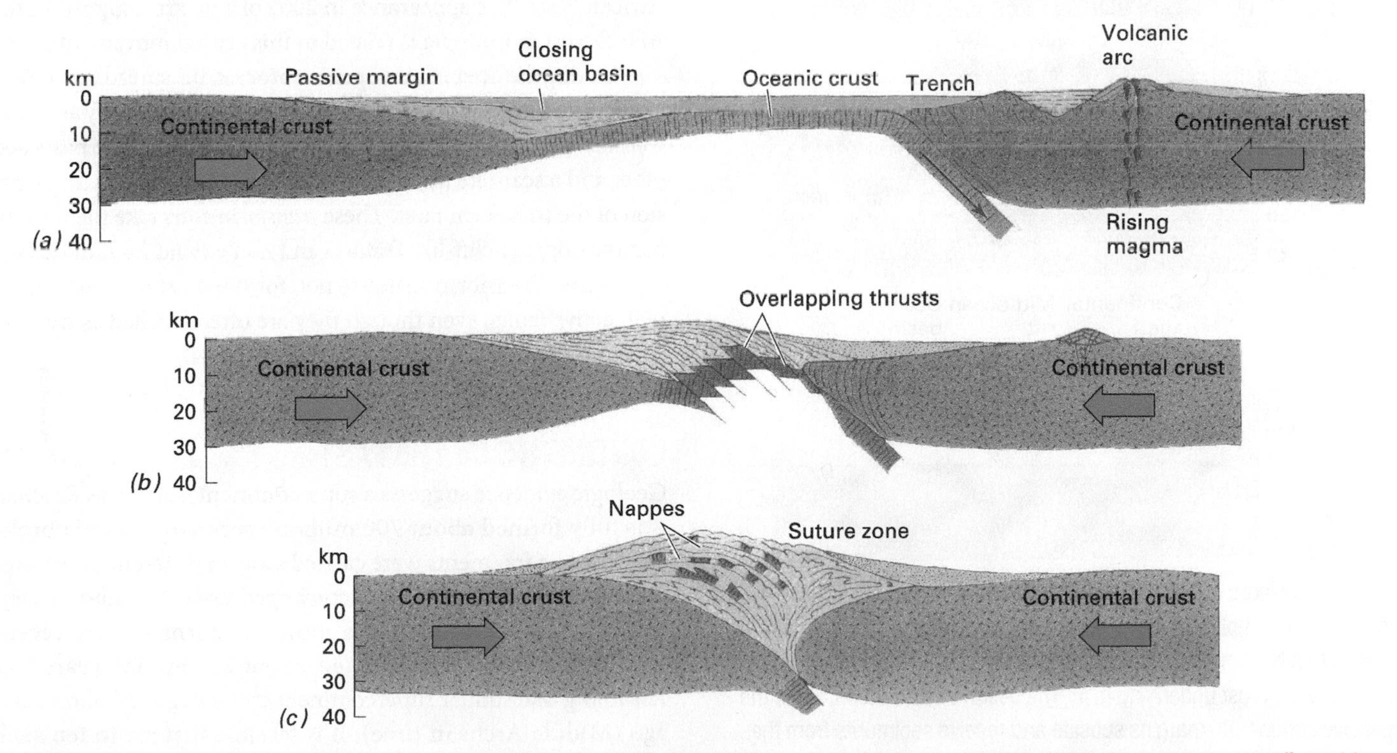

12.21 **CONTINENT–CONTINENT COLLISION AND THE FORMATION OF A SUTURE ZONE WITH NAPPES** Metamorphic activity in the suture zone eventually welds the continental crusts together into a single unit. (Copyright © A. N. Strahler.)

continental rupture (Figure 12.22) and starts when the crust is lifted and fractured, creating a region of upthrown block mountains and intervening basins. Eventually a long narrow *rift valley* appears. The widening rift valley is continually filled with magma that solidifies to form new crust in the floor. Crustal blocks on either side slip down along a succession of steep faults, creating a mountainous landscape. The Rift Valley system of East Africa is a notable example of this stage of continental rupture. As separation continues, the rift valley widens and opens to the ocean to form a narrow sea with a spreading plate boundary running down its centre. Rising magma in the central rift produces new oceanic crust and lithosphere. Continued widening of the basin results in a large ocean with the continental lithosphere moved far apart.

The youngest region of continental rupture is in the area of the Red Sea and the Gulf of Aden (Figure 12.23). Formation of the narrow Red Sea began about 30 million years ago and continues today at a rate of about 1 cm per year. As shown in Figure 12.19, this region is a triple junction of three spreading boundaries created by the motion of the Arabian Plate pulling away from the African Plate. The appearance in 2005 of a 60 km-long rift in the Afar Desert in Ethiopia is related to this crustal movement.

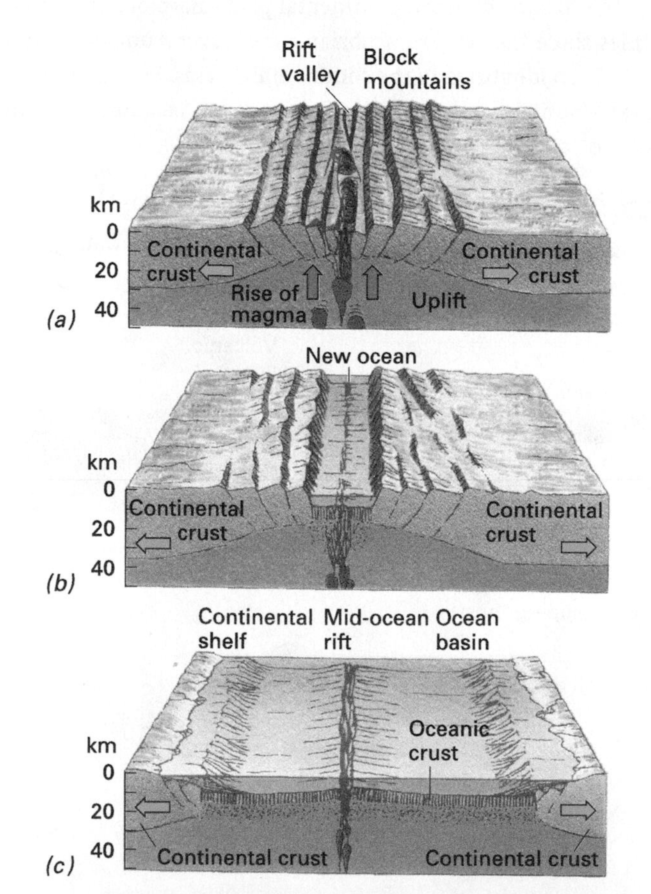

12.22 CONTINENTAL RUPTURE AND THE OPENING UP OF A NEW OCEAN BASIN (a) The crust is uplifted and stretched apart, causing it to break into blocks that become tilted on faults. (b) A narrow ocean is formed, with a new oceanic crust underlying it. (c) The ocean basin widens, while the passive continental margins subside and receive sediments from the continents. The vertical scale is greatly exaggerated to emphasize surface features. (Copyright © A. N. Strahler.)

12.23 THE RED SEA AND GULF OF ADEN IN NORTHEASTERN AFRICA The Afar Triangle is the tan-coloured region near the centre of the image bordered on the north by the Red Sea and to the east by the Gulf of Aden. In this region of northern Ethiopia, three lithospheric plates are pulling apart. This has resulted in a complex mixture of continental and volcanic rock, where oceanic-type flood basalts have emerged through the thinning crust.

During the ocean basin opening process, the spreading boundary develops a series of offset breaks that are connected by an active transform fault. As spreading continues, the offsets slide past each other and a scar-like feature forms on the ocean floor as an extension of the transform fault. These *transform scars* take the form of narrow ridges or cliff-like features and may extend for hundreds of kilometres. Transform scars are not, for the most part, associated with active faults, even though they are often labelled as fracture zones on maps of the ocean floor (Figure 12.24).

CONTINENTS OF THE PAST

Geologic evidence suggests a supercontinent, known as *Rodinia*, was fully formed about 700 million years ago. Rodinia broke apart and its fragments were carried away in different directions. Later these ancient continents converged toward a common centre, where they collided and joined to form a more recent supercontinent, called ***Pangaea***, about 200 million years ago. Assuming that similar supercontinent cycles began 3 billion years ago (Middle Archean time), it is feasible that six to ten such events could have occurred. Such a repeating cycle is called the Wilson Cycle, and can be considered in six stages (Figure 12.25).

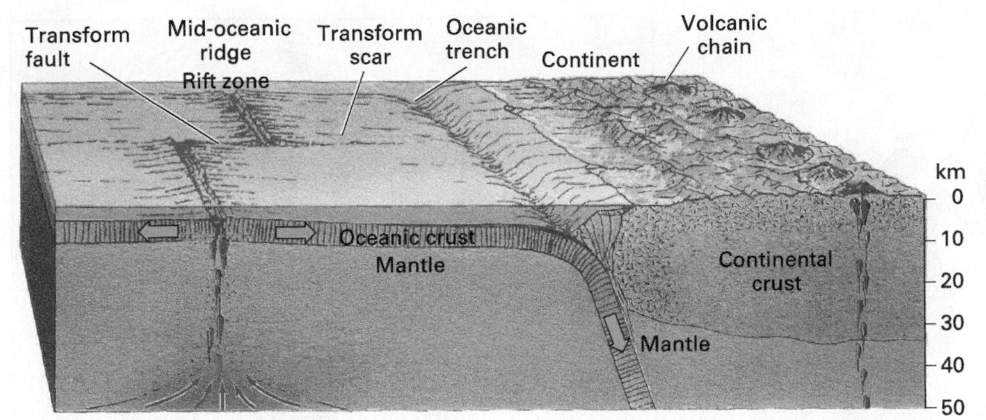

12.24 TRANSFORM FAULTING ALONG MID-OCEANIC RIDGES Transform faults characteristically result in mid-oceanic ridges developing as a series of offset segments rather than a single continuous feature. The extent of this process is shown in Figure 12.8.

- *Stage 1.* Embryonic ocean basin. The Red Sea, which separates the Arabian Peninsula from Africa, is an active example.
- *Stage 2.* Young ocean basin. The Labrador Basin, a branch of the North Atlantic lying between Labrador and Greenland, is an example of this stage.
- *Stage 3.* Old ocean basin. This includes the entire vast expanse of the North and South Atlantic Oceans and the Antarctic Ocean. Passive margin sedimentary wedges have become wide and thick.
- *Stage 4a.* The ocean basin begins to close as continental plates collide with it. New subduction boundaries begin to form, for example the Tonga Trench between the Pacific and Austral-Indian Plates, north of New Zealand.
- *Stage 4b.* Island arcs rise and grow into great volcanic island chains. These presently surround the Pacific Plate; the Aleutian arc is an example.
- *Stage 5.* Closing continues. Formation of new subduction margins close to the continents is followed by arc–continent collisions. The Japanese Islands represent this stage.
- *Stage 6.* The ocean basin has finally closed with a collision orogen, forming a continental suture (see Figure 12.21c). The Himalayan orogen is a recent example, with activity continuing today.

The hypothesis of a repeating cycle of supercontinents is now a fundamental theme in geology. Although modern plate tectonic theory is only a few decades old, the concept of a breakup of an early supercontinent into fragments that drifted apart dates back to the nineteenth century. Almost as soon as good navigational charts showing the continents became available, geographers became intrigued with the close correspondence in outline between the eastern coast of South America and the western coast of Africa. Credit for the first full-scale scientific hypothesis of the breakup of a single large continent belongs to Alfred Wegener, who offered geologic evidence as early as 1915 that the continents had once been united and had drifted apart. He postulated that the supercontinent, Pangaea, existed about 300 million years ago in the Carboniferous Period.

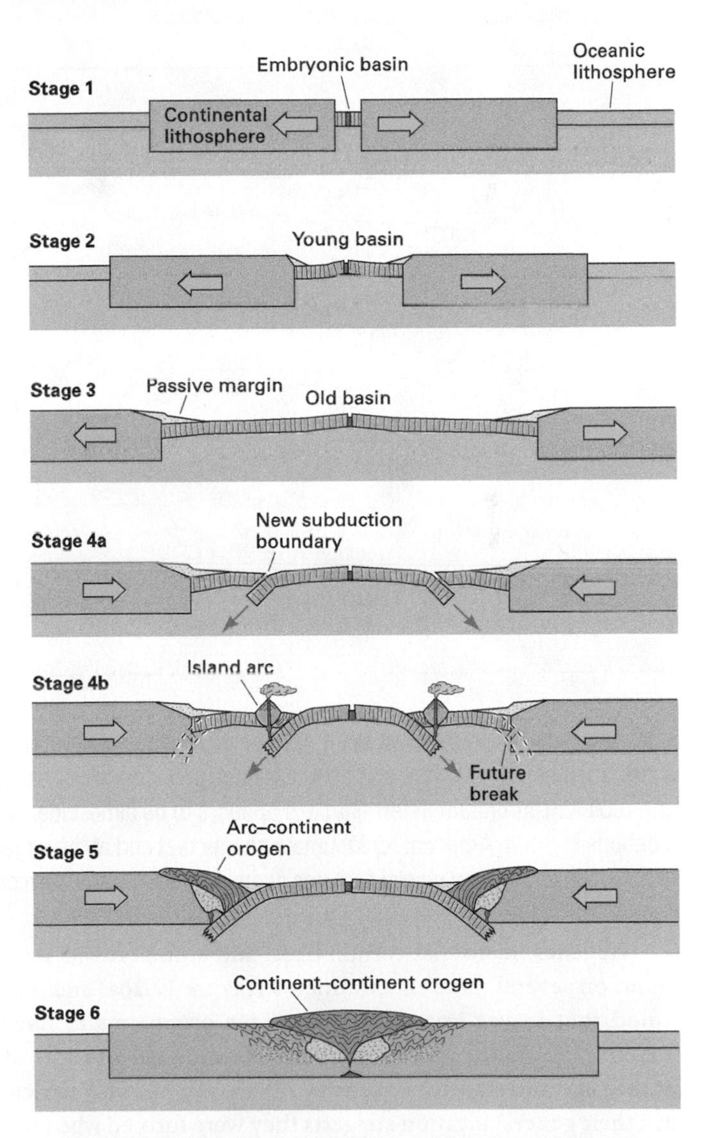

12.25 THE WILSON CYCLE Schematic diagram depicting the six stages of the Wilson Cycle. The diagrams are not to true scale. (From Plate Tectonics, copyright © Arthur N. Strahler, 1998. Used with permission.)

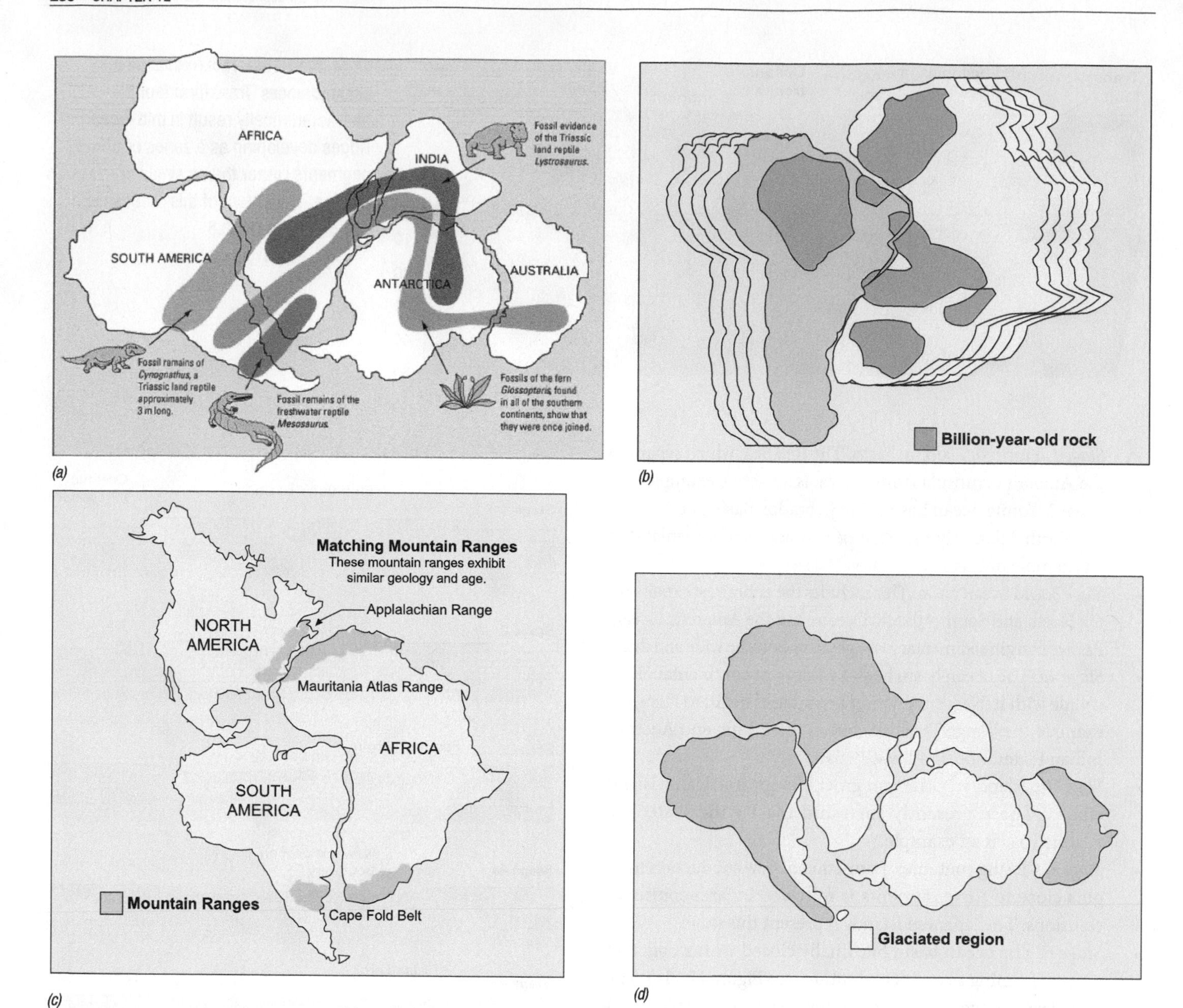

12.26 EVIDENCE FOR CONTINENTAL DRIFT (a) The distribution of several fossil organisms suggests the continents were at one time joined. The distribution of present plant and animal groups provides additional evidence. (b) The distribution of ages obtained by potassium-argon and rubidium-strontium determinations appears to be almost identical for rocks in West Africa and those in potentially contiguous locations in South America. (c) Mountain chains that end at the edge of one continent can be traced to another continent. (d) The distribution of Permian glacial features suggests that the southern continents shared contemporaneous episodes of glaciation.

Wegener noted that certain plant and animal fossils were found on several different continents (Figure 12.26a) and reasoned that it was impossible for these organisms to have travelled across the oceans. Similarly, broad belts of rocks of similar age and structure occur in Africa and South America, and their general location suggests they were formed when the continents were joined (Figure 12.26b). Other commonalities are evident in mountain chains that exhibit similar features in South America and South Africa; in North America, Africa, and Europe; and in Greenland and Europe (Figure 12.26c). Finally, there is evidence that a continental ice sheet covered parts of South America, southern Africa, India, and southern Australia in the Permian Period, about 300 million years ago (Figure 12.26d). Such a glacial episode is most likely if the oceans were much less extensive than they are today.

However, Wegener's explanation of the physical process that separated the continents was weak, and the theory of continental drift was strongly criticized on valid physical grounds. Wegener

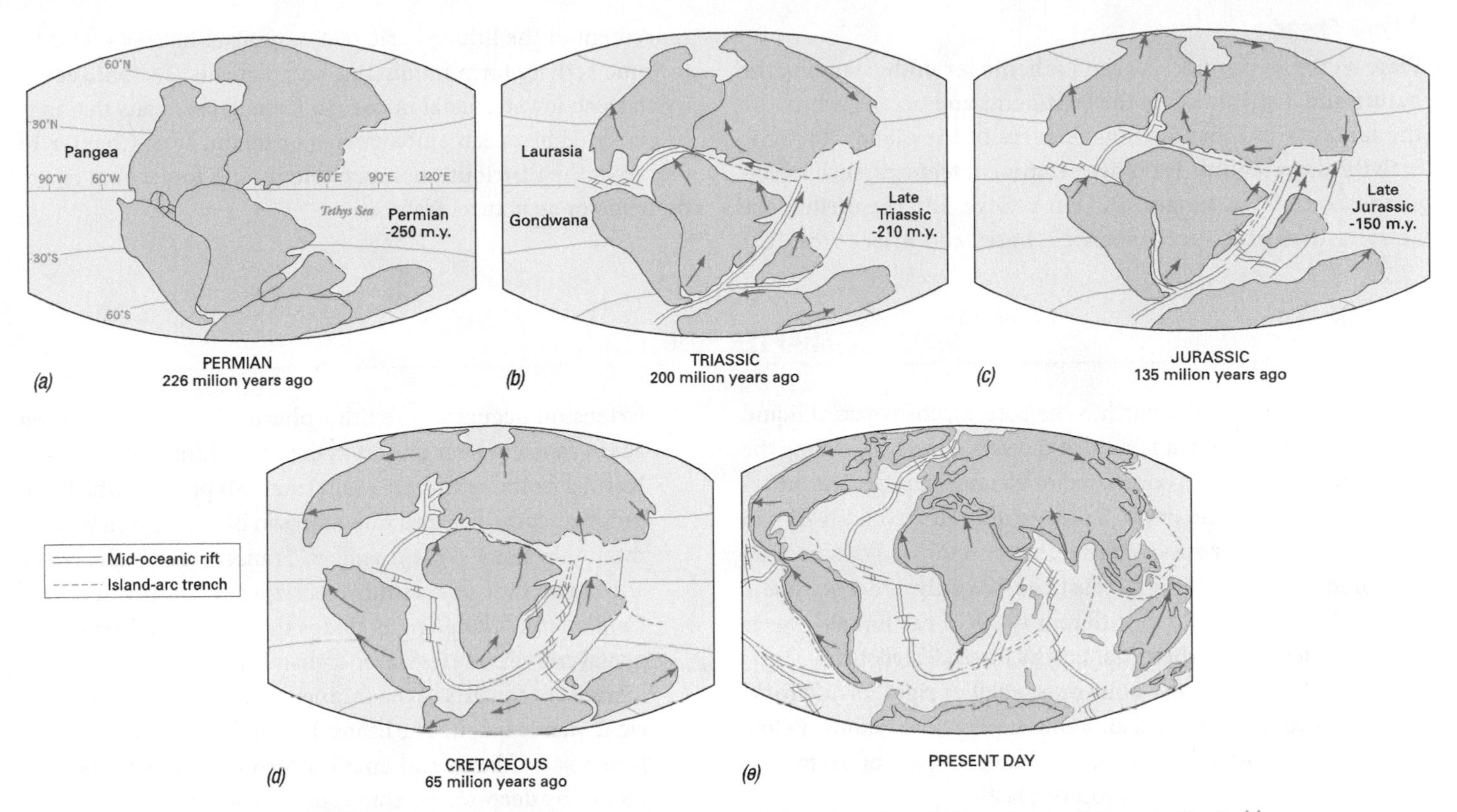

12.27 PANGAEA'S BREAKUP The breakup of Pangaea is shown in five stages. Inferred motion of lithospheric plates is indicated by arrows. (Redrawn and simplified from maps by R. S. Dietz and J. C. Holden, Jour. Geophysical Research, vol. 75, pp. 4943–4951, Figures 2 to 6. Copyright © American Geophysical Union. Used with permission.)

had proposed that a continental layer of less dense rock had moved like a great floating "raft" through a "sea" of denser oceanic crustal rock. Contemporary geologists were able to demonstrate through established principles of physics that this proposed mechanism was impossible.

In the 1960s, however, seismologists showed beyond doubt that the thick lithospheric plates are in motion, both along the mid-oceanic ridges and beneath the deep offshore trenches. Geophysicists also used paleomagnetic data in crustal rock to conclude that the continents had moved great distances apart. Approximately 50 years later, Wegener's scenario was validated, but only by applying a mechanism to the process that was unknown at his time.

The continents are moving today. Data from orbiting satellites have shown that rates of separation, or convergence, between two plates are on the order of 5 to 10 cm per year (50 to 100 km per million years). At that rate, global geography must have been very different in past geologic eras than it is today. Many continental riftings and plate collisions have taken place over the past 2 billion years. Single continents have fragmented into smaller ones, while at other times small continents have merged to form large ones.

Modern reconstructions of the global arrangements of past continents have been available since the mid-1960s. In the Permian Period of the Mesozoic Era, about 250 million years ago (Figure 12.27a), Pangaea lay astride the equator and extended nearly to the poles. Regions that are now North America and western Eurasia were located in the northern hemisphere, forming *Laurasia*. Regions that are now South America, Africa, Antarctica, Australia, New Zealand, Madagascar, and India occupied a region south of the equator; together they formed *Gondwana*. Subsequent maps (Figure 12.27b–d) show the breaking apart and dispersal of Laurasia and Gondwana to yield their modern components and locations (Figure 12.27e).

Note that North America moved into higher latitudes, helping to close off the Arctic Ocean as a largely landlocked sea. This change may have been a major cause of the Late-Cenozoic ice advances about 2 million years ago (see Chapter 18). Similarly, the Indian Peninsula started out from a near-subarctic southerly location in Permian times, and moved northeast to collide with Asia in the northern subtropics. In the future, the Atlantic Ocean Basin will become much wider, and the westward motion of the Americas will cause a reduction in the width of the Pacific Basin.

A Look Ahead

Plate tectonics provides a grand scheme for understanding the nature and distribution of the continents and oceans, which are the largest and most obvious features of the planet. Tectonic activity also accounts for other prominent features, such as volcanoes, mountain ranges, and rift valleys, and the earthquake activity that often accompanies them, and arises from the movement of the lithospheric plates. Crustal masses raised by tectonic activity form mountains and plateaus. Tectonic activity can also lower crustal masses to form depressions that may be occupied by ocean embayments or inland seas. Chapter 13 discusses the distribution, morphology, and forces that create these major structural features.

CHAPTER SUMMARY

- At the centre of the Earth is the core, a dense mass of liquid iron and nickel that is solid at the very centre. Enclosing the core is the mantle, composed of ultramafic rock. The outermost layer is the crust. Continental crust consists of two zones: a lighter zone of felsic rocks on top of a denser zone of mafic rocks. Oceanic crust consists only of denser, mafic rocks. The activity of different types of earthquake waves provides details about the Earth's internal structure.
- The lithosphere, the outermost shell of rigid, brittle rock, includes the crust and an upper layer of the mantle. Below the lithosphere is the asthenosphere, a region of the mantle in which mantle rock is soft or plastic.
- Geologists trace the history of the Earth through the geologic time scale. *Precambrian time* includes the Earth's earliest history. It is followed by three major divisions: the *Paleozoic, Mesozoic,* and *Cenozoic* eras.
- Continental masses consist of active belts of mountain-building and inactive regions of old, stable rock. Mountain-building occurs by *volcanism* and *tectonic activity. Alpine chains* include mountain arcs and island arcs. They occur in two principal mountain belts—the circum-Pacific and Eurasian-Indonesian belts.
- Continental shields are regions of low-lying igneous and metamorphic rocks. They may be *exposed* or *covered* by layers of sedimentary rocks. Ancient *mountain roots* lie within some shield regions.
- The ocean basins are marked by a *mid-oceanic ridge* with its central *axial rift*. This ridge occurs at the site of crustal spreading. Most of the *ocean basin floor* is *abyssal plain*, covered by fine sediment. As *passive continental margins* are approached, the *continental rise, slope,* and *shelf* are encountered. At *active continental margins*, deep *oceanic trenches* lie offshore.
- The two basic tectonic processes are compression and extension. Both processes can lead to the formation of mountains.
- Compression occurs where lithospheric plates are colliding. At first, the compression produces folding—*anticlines* (upfolds) and *synclines* (downfolds). If compression continues, folds may be overturned, and eventually *overthrust* faulting can occur.
- Extension occurs where lithospheric plates are spreading apart, causing them to split as the crust thins and fractures. Normal faults commonly result from this process, which can produce upthrown and downthrown blocks that may form mountain ranges or *rift valleys*. Transcurrent faults occur where two rock masses move horizontally past each other.
- *Continental lithosphere* includes the thicker, lighter continental crust and a rigid layer of mantle rock beneath. *Oceanic lithosphere* consists of the thinner, denser oceanic crust and rigid mantle below. The lithosphere is fractured and broken into a set of large and small lithospheric plates that are moved by deep-seated convection currents.
- Where plates move apart, a *spreading boundary* occurs. At *converging boundaries*, plates collide. At *transform boundaries*, plates move past one another on a *transform fault*. There are seven major lithospheric plates: Pacific, North American, South American, Eurasian, African, Austral-Indian, and Antarctic.
- When oceanic lithosphere and continental lithosphere collide, the denser oceanic lithosphere plunges beneath the continental lithospheric plate in a process called subduction. A trench marks the site where the plate plunges down. Some subducted oceanic crust melts and rises to the surface, producing volcanoes. Under the severe compression that occurs with continent–continent collision, the two continental plates weld together in a zone of metamorphic rock named a continental suture.
- In *continental rupture*, extensional tectonic forces move a continental plate in opposite directions, creating a *rift valley*. Eventually, the rift valley widens and opens to the ocean, and new oceanic crust forms as the spreading continues.
- Plate movements are thought to be powered by *convection currents* in the plastic mantle rock of the asthenosphere.
- During the Permian Period, the continents were joined in a single, large supercontinent—Pangaea—that broke apart, leading eventually to the present arrangement of continents and ocean basins.

KEY TERMS

anticlines
asthenosphere
continental crust
continental shields
continental suture
core
crust
eons
geologic column
island arcs
lithosphere
lithospheric plates
mantle
mountain arc
normal fault
oceanic crust
Pangaea
plate tectonics
reverse fault
subduction
synclines
tectonic activity

REVIEW QUESTIONS

1. Describe the Earth's inner structure from the centre outward. What types of crust are present? How are they different?
2. How do geologists use the term *lithosphere?* What layer underlies the lithosphere, and what are its properties? Define the term *lithospheric plate.*
3. More recent geologic time is divided into three eras. Name them in order from oldest to youngest. How do geologists use the terms *period* and *epoch?* What age is applied to time before the earliest era?
4. What proportion of the Earth's surface is ocean? Land? Do these proportions reflect the true proportions of continents and oceans? If not, why not?
5. What are the two basic subdivisions of continental masses?
6. What term describes belts of active mountain building? What are the two basic processes by which mountain belts are constructed? Provide examples of mountain arcs and island arcs.
7. What is a continental shield? How old are continental shields? What two types of shields are recognized?
8. Describe how compressional mountain building produces folds, faults, overthrust faults, and thrust sheets (nappes).
9. Briefly describe the Rift Valley system of East Africa as an example of normal faulting.
10. What role does gravity play in the motion of lithospheric plates?
11. Name the seven great lithospheric plates. Identify an example of a spreading boundary by general geographic location and the plates involved. Do the same for a converging boundary.
12. Describe the process of subduction as it occurs at a converging boundary of continental and oceanic lithospheric plates. How is the continental margin extended? How is subduction related to volcanic activity?
13. How does continental rupture produce passive continental margins? Describe the process of rupturing and its various stages.
14. What are transform faults? Where do they occur?
15. What is the Wilson Cycle of plate tectonics?
16. Provide a brief history of the idea of "drifting continents."
17. What was Wegener's theory on "continental drift?" Why was it opposed at the time?
18. Briefly summarize the reconstructed spatial history of the continents, beginning with the Permian Period.

VISUALIZATION EXERCISES

1. Sketch a cross-section of an ocean basin with passive continental margins. Label the following features: mid-oceanic ridge, axial rift, abyssal plain, continental rise, continental slope, and continental shelf.
2. Identify and describe two types of lithospheric plates. Sketch a cross-section showing a collision between the two types. Label the following features: oceanic crust, continental crust, mantle, oceanic trench, and rising magma. Indicate where subduction is occurring.
3. Sketch a continent–continent collision and describe the formation process of a continental suture. Provide a present-day example where a continental suture is forming, and give an example of an ancient continental suture.

ESSAY QUESTION

1. Suppose astronomers discover a new planet that, like the Earth, has continents and oceans. They dispatch a reconnaissance satellite to photograph the new planet. What features would you look for, and why, to detect past and present plate tectonic activity on the new planet?

WORKING IT OUT • RADIOACTIVE DECAY AND RADIOMETRIC DATING

The Earth's interior is heated in various ways, including the spontaneous decay of naturally occurring radioactive isotopes of certain elements. The number of positively charged particles, or *protons*, contained in an element's nucleus determines its properties. For example, the element uranium (U) has 92 protons. Nuclei of atoms also contain neutrons. The number of protons defines the atomic number of the element, and this, combined with the number of neutrons, determines the *atomic mass number*. Some elements are found in forms with different mass numbers. These forms are known as *isotopes*. For example, the most common isotope of uranium is ^{238}U, a form with an atomic mass of 238; another form is ^{235}U.

A key to understanding radioactivity is that certain isotopes are *unstable*, meaning that the composition of the isotope's nucleus can experience an irreversible change. Such a change is known as *radioactive decay*. The result is that an atom of one element may be transformed into an atom of another. For example, the uranium isotope ^{238}U decays to form ^{234}Th, an isotope of the element thorium. This new isotope is known as a *daughter product*. Often the new isotope created will also be unstable and will decay into yet another isotope of a different element. As this process continues, the result is a decay chain of daughter products that eventually ends in the formation of a stable isotope. For example, ^{238}U ultimately forms the stable lead isotope ^{206}Pb.

The significance of radioactive decay is that it provides an internal source of heat for the Earth—a source that accounts for the melting of solid rock to form magma, and thus the creation of igneous rocks. The decay of ^{238}U is only one of several radioactive decay chains that are important in heating the Earth from within. Other chains begin with ^{235}U, thorium-232 (^{232}Th), and potassium-40 (^{40}K).

The rate of an unstable isotope's decay is measured by its *half-life*—the time period in which the number of atoms of the isotope will be reduced by half. For example, the half-life of ^{238}U is 4.47 billion years, meaning that one gram of ^{238}U will be reduced to half a gram after that length of time. In another 4.47 billion years, this half gram will diminish to a quarter gram, and so forth.

The graph shows an example of the decay of ^{40}K. This unstable potassium isotope has a half-life of 1.28 billion years. The *y*-axis shows the proportion of the original quantity remaining after the elapsed time, shown on the *x*-axis. This curve has a *negative exponential* form. When ^{40}K decays, one of two products is produced: calcium-40 (^{40}Ca) or, less frequently, argon-40 (^{40}Ar). Both ^{40}Ca and ^{40}Ar are stable isotopes and undergo no further decay.

Suppose that a quantity of ^{238}U is present in an igneous rock at the time it was formed early in the Earth's history. These atoms of ^{238}U will slowly transform themselves to ^{206}Pb at a rate that depends on the half-life of ^{238}U. For example, the ratios of particular elemental isotopes can be used in *radiometric dating* to determine the time of formation of a rock, or in some cases, the time when metamorphism occurred using the following equation:

$$t = \frac{1}{k}\ln\left[\left(\frac{D}{M}\right) + 1\right]$$

where t = time of decay and k = a constant related to the half-life by the expression:

$$k = 0.693/\mathrm{H}$$

where H = the half-life, and 0.693 is the natural logarithm of 2. For the case of ^{238}U, $k = 0.693/4.77 = 0.145$.

M = the number of mother (^{238}U) atoms
D = the number of daughter (^{206}Pb) atoms

Assume that a chemical analysis of a mineral shows that for every two atoms of ^{238}U, one atom of ^{206}Pb is present. The ratio of daughter to mother atoms is $D/M = 1/2 = 0.5$, and since $k = 0.693/4.47 = 0.145$ using units of billions of years (b.y.) then:

$$t = \frac{1}{0.145}\ln(0.5 + 1)$$
$$= \left(\frac{0.405}{0.145}\right) = 2.61\ b.y.$$

The practice of radioactive dating makes it possible to age rocks, based on careful analysis of the concentration of the radioactive isotopes in certain minerals. The oldest rocks so far discovered are ancient gneisses in northwestern Canada. They contain zircon crystals that are 3.96 billion years old. Ancient shield rocks from Antarctica have been dated at 3.93 billion years, and samples from Greenland at 3.80 billion years.

During Earth's earliest history, continental crust was being vigorously formed and reformed, so rocks and minerals from these times are no longer available for analysis. However, some types of meteorites have radiometric ages of about 4.6 billion years, and these also have the same mixture of naturally occurring lead isotopes as the Earth's rocks. This latter fact leads geochemists to suspect strongly that these meteorites were formed from the same primordial matter as the Earth, thus dating the formation of Earth at about 4.6 billion years ago.

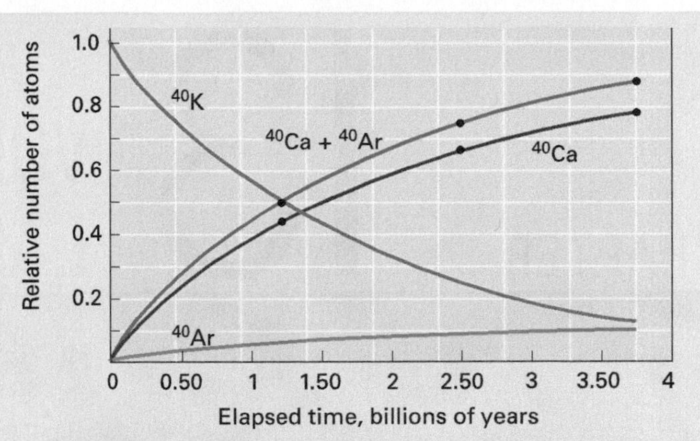

Exponential decay and growth curves for ^{40}K, ^{40}Ca, and ^{40}Ar
(Copyright © A. N. Strahler.)

QUESTIONS / PROBLEMS

1. The half-life of potassium-40 (^{40}K) is 1.28 billion years. What percent of a quantity of ^{40}K will remain after 1 billion years? 3 billion years?
2. The half-life of thorium-232 (^{232}Th) is 14.1 billion years. What percent of a quantity of pure ^{232}Th will remain after 5 billion years? 10 billion years? 15 billion years?
3. The unstable isotope carbon-14 (^{14}C) has a half-life of 5,730 years. How many years will it take for a pure sample of ^{14}C to reduce to 0.1 (10 percent)?
4. A geochemist analyzes a sample of zircon obtained from an igneous rock and observes that the ratio of ^{206}Pb to ^{238}U atoms is 0.448. Using the formula, what is the age of the sample?
5. The geochemist also determines that the atomic ratio of ^{207}Pb to ^{235}U is 9.56 for the same sample. For this decay process, the half-life is $H = 7.04 \times 10^8$ years. Calculate k and then determine the age from the D/M ratio of 9.56. Is it consistent with the age obtained from the $^{206}Pb/^{238}U$ ratio?

Find answers at the back of the book.

CHAPTER 13

VOLCANIC AND TECTONIC LANDFORMS

Mount Bromo, Indonesia, in the early morning

The two previous chapters on the Earth's materials and plate tectonics described how forces within the Earth create major structural features. These same forces create landforms with distinctive characteristics on a regional scale, as well as readily identifiable individual features in local landscapes. Previously it was shown how stresses from compression generated by plate motions can throw rock strata into wavelike forms, such as those seen in the Canadian Rockies. Similarly, faults mark zones in which rock layers break and move past one other, often producing uplifted mountain blocks next to deep valleys. This chapter continues the overview of the landscape features created by forces within the Earth. Here the features produced when upwelling magma spews forth explosively from vents and fissures, creating steep-sided volcanoes, are described. As well, many tectonic processes are accompanied by strong earthquakes; these too are discussed.

LANDFORMS

Landforms are the natural physical features found on the Earth's surface, such as mountains, canyons, beaches, and sand dunes. Their creation, subsequent transformation, and ultimate removal from the landscape attest to the dynamic nature of the planet. The scientific study of the processes that shape landforms is covered in **geomorphology**. The shapes of continental surfaces reflect the balance between two sets of forces. Forces internal to the Earth fold, fracture, and warp crustal materials through volcanic and tectonic processes. At the Earth's surface, other processes and forces, such as flowing water, wave action, glacial ice, and wind, modify the landscape by removing, transporting, and depositing mineral matter. Each geomorphological process creates unique and distinctive landforms; these are discussed in chapters 14 to 18.

Landforms can be generally divided into two basic groups: initial landforms and sequential landforms. *Initial landforms* are directly produced by volcanic and tectonic activity. The energy to produce the initial landforms is generated by the decay of radioactive minerals in the crustal and mantle rocks. The resulting landforms include volcanoes and lava flows, as well as rift valleys and elevated mountain blocks in zones of recent crustal deformation. Landforms, such as river valleys and sand dunes,

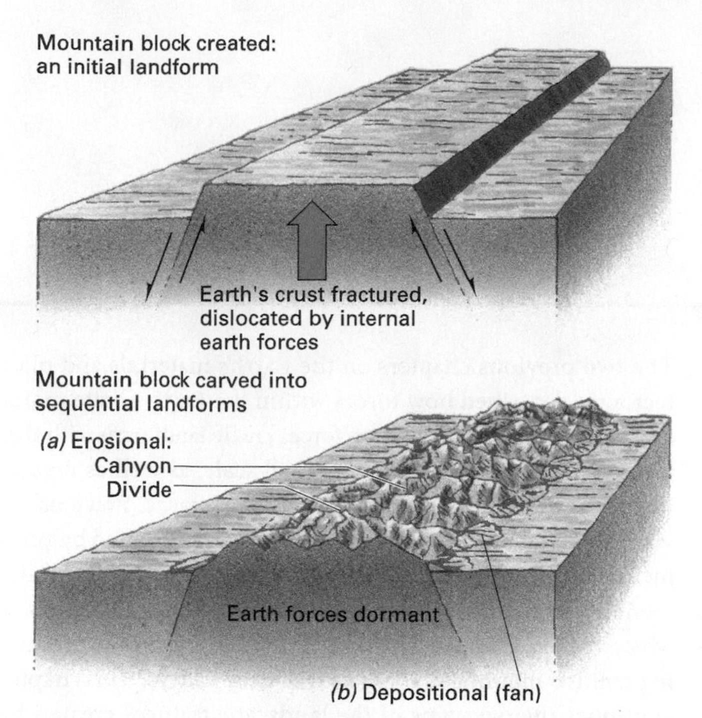

13.1 **INITIAL AND SEQUENTIAL LANDFORMS** An initial landform is created, here by tectonic activity, then carved into sequential landforms. (Drawn by A. N. Strahler.)

13.2 **MOUNT ST. HELENS** This stratovolcano in the Cascade Mountain Range in southwestern Washington erupted violently on the morning of May 18, 1980, emitting a great cloud of condensed steam, heated gases, and ash from the summit crater. Within a few minutes, the plume had risen to a height of 20 km. *Fly By: 46° 12′ N; 122° 11′ W*

which are shaped by processes and agents of denudation, belong to the group of *sequential landforms*, which develop after the initial landforms have been created (Figure 13.1).

VOLCANIC ACTIVITY

Magma extruded by volcanic activity (*volcanism*) can form imposing mountain ranges comprised of volcanic peaks and accumulated lava flows. A **volcano** is a conical or dome-shaped initial landform built of lava and ash emitted from a constricted vent in the Earth's surface. The magma rises in a narrow, pipe-like conduit from a magma reservoir. Upon reaching the surface, magma may pour out in tongue-like lava flows, or it may be violently ejected in the form of solid fragments driven skyward under the pressure of confined gases. Ejected fragments, ranging in size from huge boulders to fine dust, are collectively called *tephra* (Figure 13.2). The type of lava and the amount of tephra ejected during eruptions determines the size and shape of a volcano.

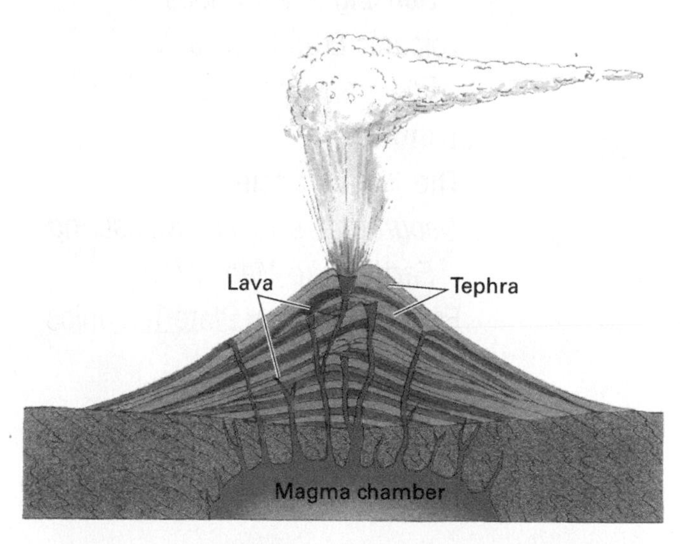

13.3 **ANATOMY OF A STRATOVOLCANO** Idealized cross-section of a stratovolcano with feeders from the magma chamber beneath. The steep-sided cone is built up from layers of lava and tephra. (Copyright © A. N. Strahler.)

STRATOVOLCANOES

The nature of volcanic eruptions, whether explosive or subdued, depends on the viscosity of the magma. Felsic lava (rhyolite and andesite) is highly viscous; it is thick and sticky, and resists flow. Consequently, volcanoes of felsic composition typically have steep slopes as the lava does not usually flow far from the vent. When the volcano erupts, tephra falls on the area surrounding the crater and contributes to the structure of the cone. Included in the tephra are volcanic bombs, which are solidified masses of lava that can be the size of large boulders.

The inter-layering of sluggish streams of felsic lava and eruptions of tephra produces **stratovolcanoes**, which are sometimes referred to as *composite volcanoes*, or composite cones, since they are formed from layers of ash and lava (Figure 13.3). Fine examples of stratovolcanoes include Mount Baker (Figure 13.4),

13.4 **MOUNT BAKER** Located in the Cascade Mountain Range of Washington, U.S., Mount Baker is composed of layers of ash and lava. It is the second most active volcano in this range after Mount St. Helens and rises to a height of nearly 3,300 m.

13.5 **NUÉE ARDENT** A cloud of hot, dense volcanic ash, emitted by the Soufrière Hills volcano, courses down this hillside on the island of Montserrat in the Lesser Antilles. ***Fly By: 16° 43′ N; 62° 11′ W***

Mount Rainier, and Mount Shasta in the Cascade Mountain Range; Mount Fuji in Japan; and Mount Mayon in the Philippines. Their tall cones usually become steeper toward the summit, where a bowl-shaped depression—the *crater*—is located. The crater is the volcano's principal vent. Felsic lava holds large amounts of gas under high pressure, and usually produces explosive eruptions. Fine volcanic dust from these eruptions can rise high into the troposphere and stratosphere, travelling hundreds or thousands of kilometres before settling to the Earth's surface.

Another important form of emission from explosive stratovolcanoes is a cloud of white-hot gases and fine ash. These intensely hot clouds, called *nuée ardentes* or ash flows, travel rapidly down the flank of a volcanic cone, searing everything in its path. Mount Pelée, on the Caribbean island of Martinique, erupted in this way in 1902, destroying the city St. Pierre and reportedly killing all but two of its 30,000 inhabitants. A similar eruption of the Soufrière Hills volcano on Montserrat in 1997 left the southern two-thirds of the island uninhabitable (Figure 13.5).

Calderas

One of the most catastrophic of natural phenomena is a volcanic explosion so violent that it destroys the entire central portion of the volcano. Vast quantities of ash and dust are emitted and fill the atmosphere for many hundreds of square kilometres around the volcano. A great central depression, named a **caldera**, remains after the explosion. Although some of the upper part of the volcano is blown outward in fragments, the depth of the caldera is also affected by subsidence as material settles back into the cavity following the explosion.

Krakatoa, a volcanic island in Indonesia, exploded in 1883, leaving a huge caldera. Great seismic sea waves (*tsunamis*) generated by the explosion killed many thousands of people living in low coastal areas of Sumatra and Java. About 25 km^3 of rock was blown out of the crater during the explosion. Vast quantities of gas and fine particles of dust were carried into the stratosphere and contributed to the rosy glow of sunrises and sunsets seen around the world for several years afterward.

A classic example of a caldera produced in prehistoric times is Crater Lake, Oregon (Figure 13.6). The former volcano, named Mount Mazama, is estimated to have risen 1,200 m higher than the present caldera rim. The great explosion and collapse occurred about 7,000 years ago. The lake that presently occupies the caldera is over

13.6 **CRATER LAKE** Crater Lake, Oregon, is a water-filled caldera marking the remains of the summit of Mount Mazama, which exploded about 7,000 years ago. Wizard Island (right) built up on the floor of the caldera after the major explosive activity had ceased. It is an almost perfectly shaped cone of cinders capping small lava flows. ***Fly By: 42° 56′ N; 122° 06′ W***

8 km in diameter and almost 600 m deep, making it the seventh deepest lake in the world. Ash layers associated with this eruption are reported in British Columbia, Alberta, and Saskatchewan, as well as the western United States, where they provide important stratigraphic information in archaeological studies.

Stratovolcanoes and Subduction Arcs

Most of the world's active stratovolcanoes lie within the circum-Pacific mountain belt. Andesitic magmas rise beneath volcanic arcs of active continental margins and island arcs. For example, the volcanic arc of Sumatra and Java lies over the subduction zone between the Australian Plate and the Eurasian Plate. Active subduction of the Pacific Plate beneath the North American Plate is associated with the Aleutian volcanic arc and the chain of volcanoes found in the Cascade Mountains of northern California, Oregon, and Washington. This merges with the Garibaldi volcanic belt of southern British Columbia, where the small Juan de Fuca Plate is similarly being forced under the North American Plate. Volcanic activity in Central America is linked with the subduction of the Cocos Plate. In South America, the subduction zone between the Nazca and South American Plates accounts for the stratovolcanoes that occur in various segments of the Andes Mountains.

SHIELD VOLCANOES

In contrast to thick, gassy felsic lava, mafic lava (basalt) is usually fluid, has a low viscosity, and holds little gas. As a result, eruptions of basaltic lava are much less violent, with the lava often travelling long distances, spreading out in thin layers (Figure 13.7). Large basaltic volcanoes are typically broad rounded domes with gentle slopes; they are called **shield volcanoes**. Hawaiian volcanoes are of this type. Shield volcanoes are also found in association with mid-oceanic ridges, such as those of Iceland, the Azores, Ascension, and Tristan da Cunha in the Atlantic. Iceland is constructed entirely of basalt. Here, active volcanoes, such as Mount Hekla, are composed of basaltic flows, dikes, and sills which are incorporated within or on older basaltic rocks when magma emerges from deep within the spreading rift. Smaller shield volcanoes occur in northern California and Oregon, where they are typically 5 to 7 km in diameter, with heights of 450 to 600 m.

The shield volcanoes of the Hawaiian Islands are characterized by smooth, gently rising slopes that flatten near the top, producing a broad-topped mountain (Figure 13.8). Domes on the island of Hawaii rise to summit elevations of about 4,000 m, but because they have grown from the seabed, their accumulated height is more than twice that. They range in diameter from 16 to 80 km at sea level, and up to 160 km across at the submerged base. Most of the lava flows from *fissures*, long, gaping cracks on the flanks of the volcano. Hawaiian lava domes have a wide, steep-sided central depression that may be 3 km or more in diameter

13.7 **LAVA FLOWS FROM THE KILAUEA VOLCANO** Hot, fluid basaltic lava typically flows over existing surfaces, which progressively adds to the size and height of shield volcanoes. ***Fly By: 19° 25′ N; 155° 17′ W***

and several hundred metres deep. These large depressions are a type of collapsed caldera. Molten basalt is sometimes seen in the floors of deep pit craters that occur on the floor of the central depression or elsewhere over the surface of the lava dome.

Hotspots, Sea-Floor Spreading, and Shield Volcanoes

The chain of Hawaiian volcanoes was created by the motion of the Pacific Plate over a *hotspot*—a plume of upwelling basaltic magma from very deep within the Earth's mantle. The Galapagos Islands are another example of this process. As the hot mantle rock rises, magma

13.8 **BASALTIC SHIELD VOLCANOES IN HAWAII** Mauna Loa, as seen from the summit of Mauna Kea, shows the characteristic form of a shield volcano. Both volcanoes rise to over 4,000 m and provide high elevation sites for observatories used to gather climate data.

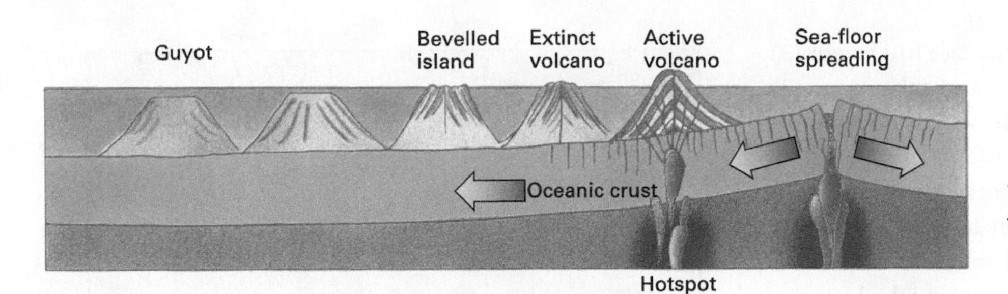

13.9 VOLCANIC CHAIN A chain of volcanoes is formed by an oceanic plate moving over a hotspot. (Copyright © A. N. Strahler.)

forms in bodies that melt their way through the lithosphere and reach the sea floor. Each major pulse of the plume sets off a cycle of volcano formation. However, the motion of the oceanic lithosphere eventually carries the volcano away from the location of the deep plume, and so it becomes extinct. Erosion processes wear the volcano away, and ultimately it becomes a low island. Continued attack by waves and slow settling of the island reduces it to a coral-covered platform. Eventually only a sunken seamount, or *guyot*, exists (Figure 13.9).

The life cycle of a basaltic shield volcano that grows upward from the abyssal ocean may take it from more than 5,000 m below the sea surface to a height greater than 4,000 m above sea level. In this way a huge mass of igneous rock can accumulate to form a volcano at a location where volcanic activity has not previously occurred. A unique feature of this type of volcano is that it rests on oceanic lithosphere that is moving steadily over the asthenosphere. The chain of Hawaiian volcanic domes is a good example. However, this chain lies far away from the spreading boundary of the Pacific Plate, which would be a natural source of magma. Geophysical evidence has shown that there is a *hotspot* beneath the islands—a persistent source of heat that provided the pulses of rising magma that built the Hawaiian chain, island by island, as the Pacific Plate moved over it.

In the first stage of the life cycle (Figure 13.10), magma rising from the hotspot on the ocean floor forms a low

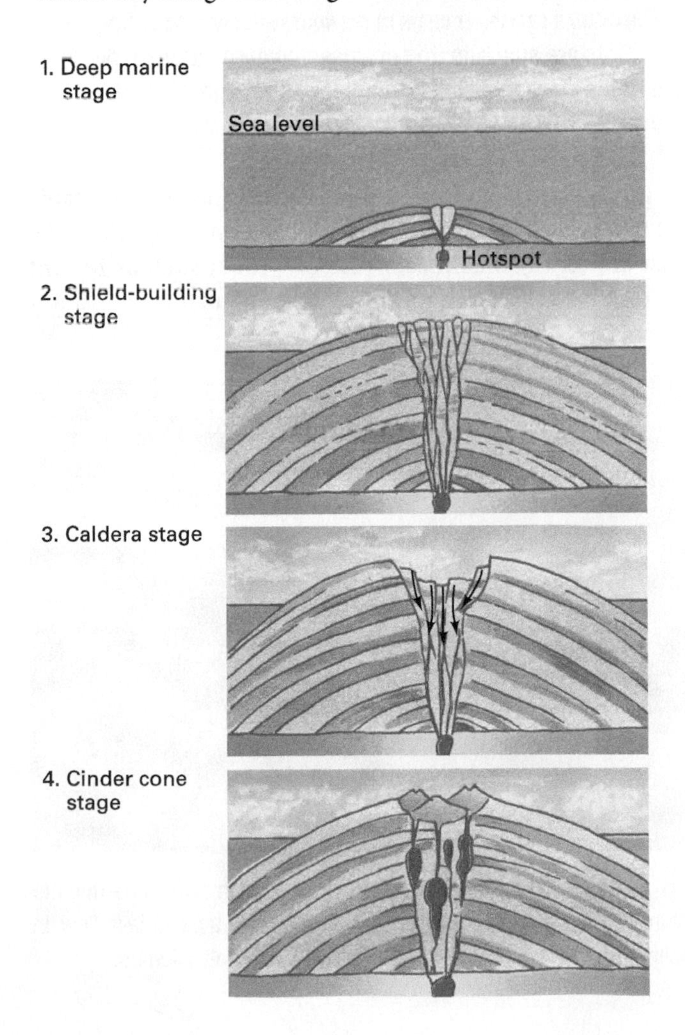

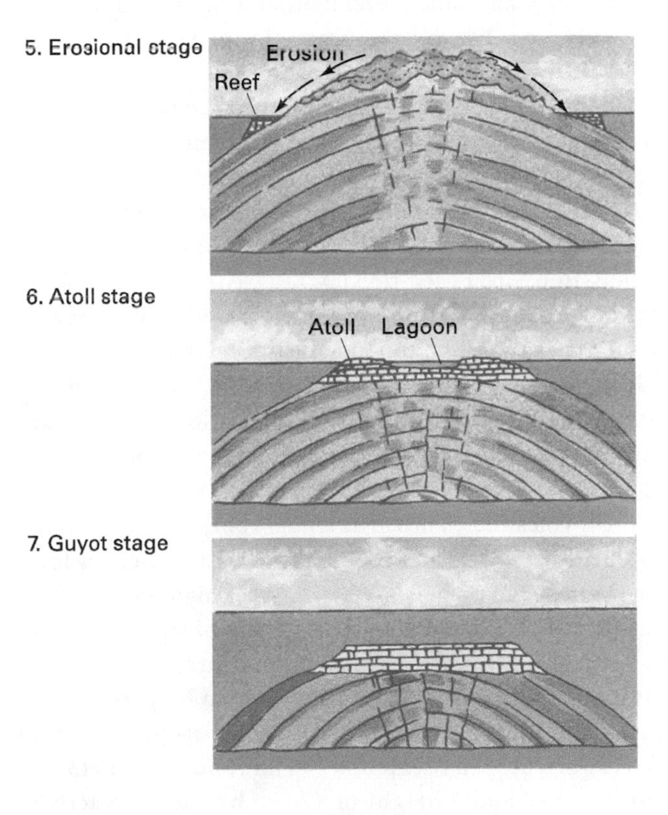

13.10 LIFE CYCLE OF A TYPICAL HAWAIIAN VOLCANO (Adapted from Figure 6.1 from *Volcanoes in the Sea* by Macdonald, Abbott, & Peterson, 1983, University of Hawaii Press, p. 147.)

basaltic lava dome (1). After the dome reaches full height (2), a central caldera forms through collapse and subsidence (3). A post-caldera stage (4) then follows in which large cinder cones fill the caldera and renew some of the original mass. Eventually, dormancy sets in and erosional processes lower the mountain's height, while waves cut back the coast (5). Fringing coral reefs that form on the dome become broader. The volcano is now fully extinct. No longer pushed upward by upwelling magma, the oceanic crust holding the volcano steadily subsides, further lowering its height. Erosional bevelling finishes at the atoll stage (6), when a thick layer of reef corals and related lagoon sediments forms. In the final stage (7), continued crustal subsidence drowns the reef corals, leaving the residual guyot.

Each major pulse of the plume sets off a volcanic cycle, with each volcano eventually moving away from the hotspot by the motion of the lithospheric plate. This produces a long trail of sunken islands and guyots (Figure 13.11). The Hawaiian trail, trending northwestward, is 2,400 km long, and includes a sharp bend to the north caused by a sudden change of direction of the Pacific Plate. This distant leg consists of the Emperor Seamounts. Several other long trails of volcanic seamounts cross the Pacific Ocean Basin. They, too, follow parallel paths that reveal the plate motion.

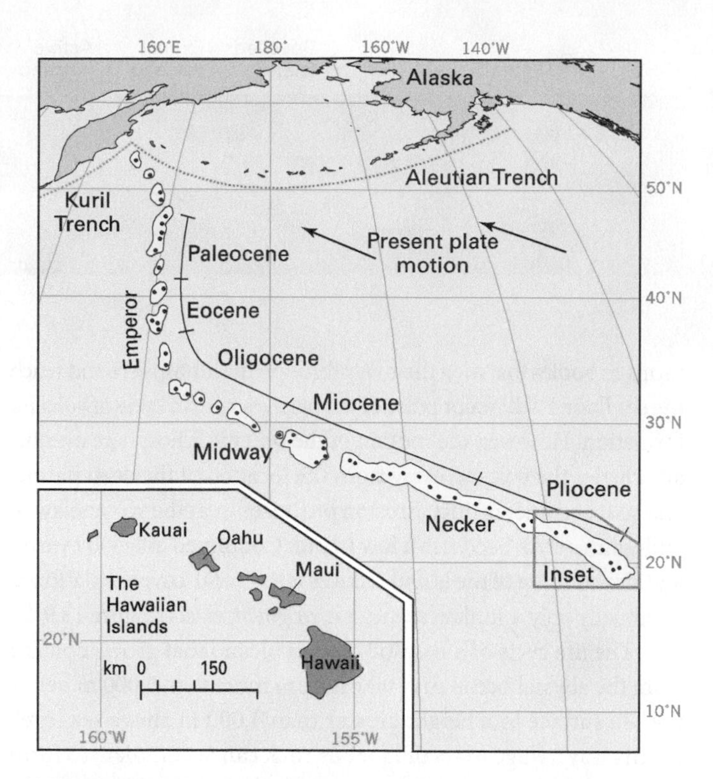

13.11 HAWAIIAN SEAMOUNT CHAIN IN THE NORTHWEST PACIFIC OCEAN BASIN Dots are summits; the enclosed coloured area marks the base of the volcano at the ocean floor. (Copyright © A. N. Strahler.)

Where a mantle plume lies beneath a continental lithospheric plate, the hotspot may generate an enormous volume of basaltic lava that emerges from numerous vents and fissures and accumulates layer upon layer over an extensive area. The basalt may ultimately become thousands of metres thick; such accumulations are called *flood basalts.* An important North American example is found in the Columbia Plateau region of southeastern Washington, northeastern Oregon, and western Idaho. Here, basalts from the Cenozoic Era cover an area of about 130,000 km^2. Individual basalt flows are exposed along the walls of river gorges as cliffs with conspicuous vertical joint columns (Figure 13.12).

Small volcanoes, known as *cinder cones,* are often found with shield volcanoes and basaltic flows. Cinder cones form when frothy basalt magma is ejected under high pressure from a narrow vent, producing tephra. The rain of tephra accumulates around the vent to form a roughly circular hill with a central crater. Cinder cones rarely grow to heights of more than a few hundred metres. One of the best-known cinder cones is Parícutin in Mexico. It began to form in 1943 and eventually reached a height of 424 m by the time activity ceased in 1952.

Although there is evidence of ancient volcanic activity throughout Canada, comparatively young volcanoes (less than 5 million years old) are found only in British Columbia and the Yukon. They are extensions of the Cascades Volcanic Belt that runs through Washington, Oregon, and California. Most are stratovolcanoes, some of which, such as Mount Nazko, have erupted in the past 7,000 to 8,000 years. Mount

13.12 FLOOD BASALTS Basaltic lava flows exposed in cliffs bordering the Columbia River in Washington. Each set of cliffs is a major lava flow. In cooling, vertical cracks form in the lava, creating tall columns.

Meager, in the Garibaldi Volcanic Belt, erupted about 2,350 years ago and is considered the most recent explosive eruption in Canada, although several cinder cones have erupted in the past 500 years. The most recent eruption reported in Canada was the Iskut-Unuk River cinder cone event that occurred in British Columbia in 1904. This area is close to Mount Edziza (2,787 m), a shield volcano similar to those in Hawaii. Mount Edziza began to form 4 million years ago, with the last basalt flow occurring 10,000 years ago; about 30 cinder cones have since built up on its flanks (Figure 13.13).

Hot Springs and Geysers

Where hot rock material is near the Earth's surface, it can heat nearby groundwater to high temperatures. When it reaches the surface, the heated groundwater provides *hot springs* (Figure 13.14). At some places, jet-like emissions of steam and hot water, at temperatures not far below the boiling point, occur at intervals from small vents producing *geysers* (Figure 13.15).

Hot springs and geysers are found wherever there are active or recently dormant volcanoes, for example, the North Island of New Zealand, Chile, Italy, Iceland, Japan, and California. In Canada, the best-known locations are in Alberta, British Columbia, and the Yukon. In the 1880s, commercial development of hot springs began in Banff, which, because of its scenic mountain location, has become a world tourist attraction. British Columbia also has a significant number of commercial hot springs. Water emerges at a temperature of about 50 °C in Canadian hot springs, so spas must mix it with cooler water before use.

The heat from masses of igneous rock close to the surface in areas of hot springs and geysers provides a source of energy for electric power generation, and numerous geothermal power stations are in operation. Geothermal facilities in 24 countries currently supply enough electricity to meet the needs of some

13.14 **MAMMOTH HOT SPRINGS** Small terraces ringed by mineral deposits hold steaming pools of hot water as the spring cascades down the slope. This example of geothermal activity is from Yellowstone National Park, Wyoming. ***Fly By: 44° 28′ N; 110° 50′ W***

13.13 **CINDER CONE** A cinder cone built up on the flank of Mount Edziza, a large shield volcano in northern British Columbia.

13.15 **CASTLE GEYSER** An eruption of Castle Geyser in Yellowstone National Park, Wyoming.

60 million people (Table 13.1). Some of the oldest installations are in Italy, where geothermal plants have been operating for more than 100 years.

In Canada, drilling is underway at Mount Meager, located about 170 km north of Vancouver, to confirm its potential to become Canada's first commercial geothermal power station. Temperatures as high as 225 °C at depths of 800 m have been recorded in test borings at this site, with a projected capacity of 6.4 MW. Abandoned mines can also be used as geothermal reservoirs, as has been demonstrated in Springhill, Nova Scotia. In Yukon, geothermal energy is used to prevent water pipes from freezing. *Eye on the Environment • Geothermal Energy Sources* provides more information about this resource.

As new oceanic crust is formed along the mid-oceanic ridges, it is strongly fractured through deformation, thermal contraction, or earthquakes. Sea water seeps into these fractures, becomes heated, and begins to leach minerals from the rock. As water temperature rises, dark-coloured, mineral-enriched water and steam are expelled as *black smokers* through chimney-like vents built up of sulphide minerals (Figure 3.16). The principal elements found in black smokers are manganese, iron, cobalt, nickel, copper, and lead, which can precipitate onto the ocean floor as nodules up to 50 cm in diameter. Although commonly referred to as manganese nodules, their general composition is a mixture of manganese (15 to 30 percent), iron (7 to 15 percent), nickel (1 to 2 percent), copper (1.5 percent), and cobalt (0.3 percent). The nodules build up in layers around a small nucleus derived from diatoms, fragments of fish bones, and particles of marine clay or rock chips. Radiometric dating indicates that the nodules grow by just a few millimetres in a million years. Seabed nodules represent a vast store of mineral wealth, but as yet no commercial mining has begun.

TABLE 13.1 COUNTRIES WITH GEOTHERMAL POWER GENERATION

COUNTRY	INSTALLED CAPACITY (MW)
Australia	0.2
Austria	1.2
China	28
Costa Rica	163
El Salvador	151
Ethiopia	7.3
France (Guadeloupe)	15
Germany	0.2
Guatemala	33
Iceland	202
Indonesia	797
Italy	791
Japan	535
Kenya	129
Mexico	953
New Zealand	435
Nicaragua	77
Papua New Guinea (Lihir Island)	6
Philippines	1,930
Portugal (Sao Miguel Island)	16
Russia	79
Thailand	0.3
Turkey	20
United States	2,564
Total	**8,933 MW**

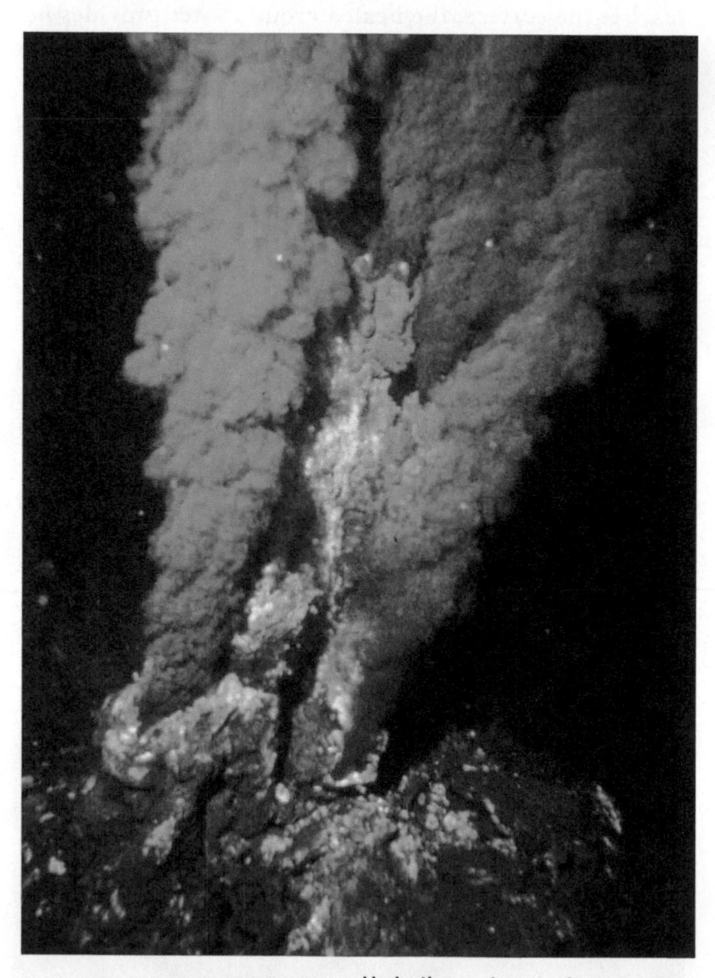

13.16 DEEP-SEA HYDROTHERMAL VENT Hydrothermal vents, known as black smokers, are associated with mid-oceanic ridges. These chimney-like structures are composed of sulphides that are deposited on volcanic rocks where plumes of hot, mineral-rich water are vented onto the ocean floor. Despite the almost total lack of illumination at these depths, hydrothermal vents support unique ecosystems that are dependent on the mineral deposits for their source of energy.

EYE ON THE ENVIRONMENT

GEOTHERMAL ENERGY SOURCES

Geothermal energy is energy in the form of sensible heat that originates within the Earth's crust and makes its way to the surface by conduction. Heat may be conducted upward through solid rock, or carried up by circulating groundwater that is warmed at depth before returning to the surface. Concentrated geothermal heat sources are usually associated with igneous activity, but deep zones of heated rock and groundwater that are not directly related to igneous activity also exist.

Observations made in deep mines and bore holes show that the rock temperature increases steadily with depth and attains very high values in the upper mantle. Heat within the Earth's crust and mantle is produced largely by radioactive decay, and the basic energy resource it provides can be regarded as limitless on a human scale. It might seem simple enough to obtain all our energy needs by drilling deep holes at any desired location into the crust and letting the hot rock turn injected fresh water into steam, which could then be used to generate electricity. Unfortunately, at the depths usually required to furnish the needed heat intensity, crustal rock tends to close any cavity by rupture and slow flowage. Thus, geothermal locations occur where special conditions have caused hot rock and hot groundwater to lie within the range of conventional drilling methods.

Figure 1 Geothermal power plant This electricity-generating power plant at Wairakei, New Zealand, runs on steam produced by superheated groundwater. Steam pipes in the foreground lead to the plant.

Natural hot water and steam were the first type of geothermal energy source to be developed; at present they account for nearly all production of geothermal electrical power. Wells are drilled to tap the hot water. When it reaches the surface, the water turns to steam under the reduced pressure of the atmosphere. The steam is fed into generating turbines to produce electricity, then condensed in cooling towers (see photo). The resulting hot water is usually released into surface stream flow, where it may create a thermal pollution problem. The larger steam fields have sufficient energy to generate at least 15 megawatts of electric power, and a few can generate 200 megawatts or more.

In certain areas, the intrusion of magma has been recent enough that the solidified igneous rock of a batholith is still very hot at depths of perhaps 2 to 5 km. Rock in this zone may be as hot as 300 °C and could supply an enormous quantity of heat energy. The planned development of this resource includes drilling into the hot zone and then shattering the surrounding rock by hydrofracture—a method using water under pressure that is widely practised in petroleum development. Surface water would then be pumped down one well into the fracture zone and heated water pumped up another well. Although some experiments have been conducted, this heat source has not yet been exploited in any practical way.

On a smaller scale, groundwater at 10 to 20 °C can be used directly or with heat pumps to heat buildings. At present, more than 30,000 buildings in Canada are heated in this way. In Nova Scotia, water from the flooded Springhill coal mine is used for this purpose. Annual energy savings compared with conventional sources amount to about 600,000 kWh—and the process is non-polluting.

GLOBAL PATTERN OF VOLCANIC ACTIVITY

Figure 13.17 shows the locations of volcanoes that have been active within the last 12,000 years. Many volcanoes are located along subduction boundaries where lithospheric plates are converging (see Figure 12.2). The "ring of fire" around the Pacific Rim is associated with subduction zones running in an almost unbroken sequence from the southern tip of South America to Central America, along the Pacific northwest coast of North America through the Aleutian Islands to Japan, the Philippines, and the East Indies, and finally terminating south of New

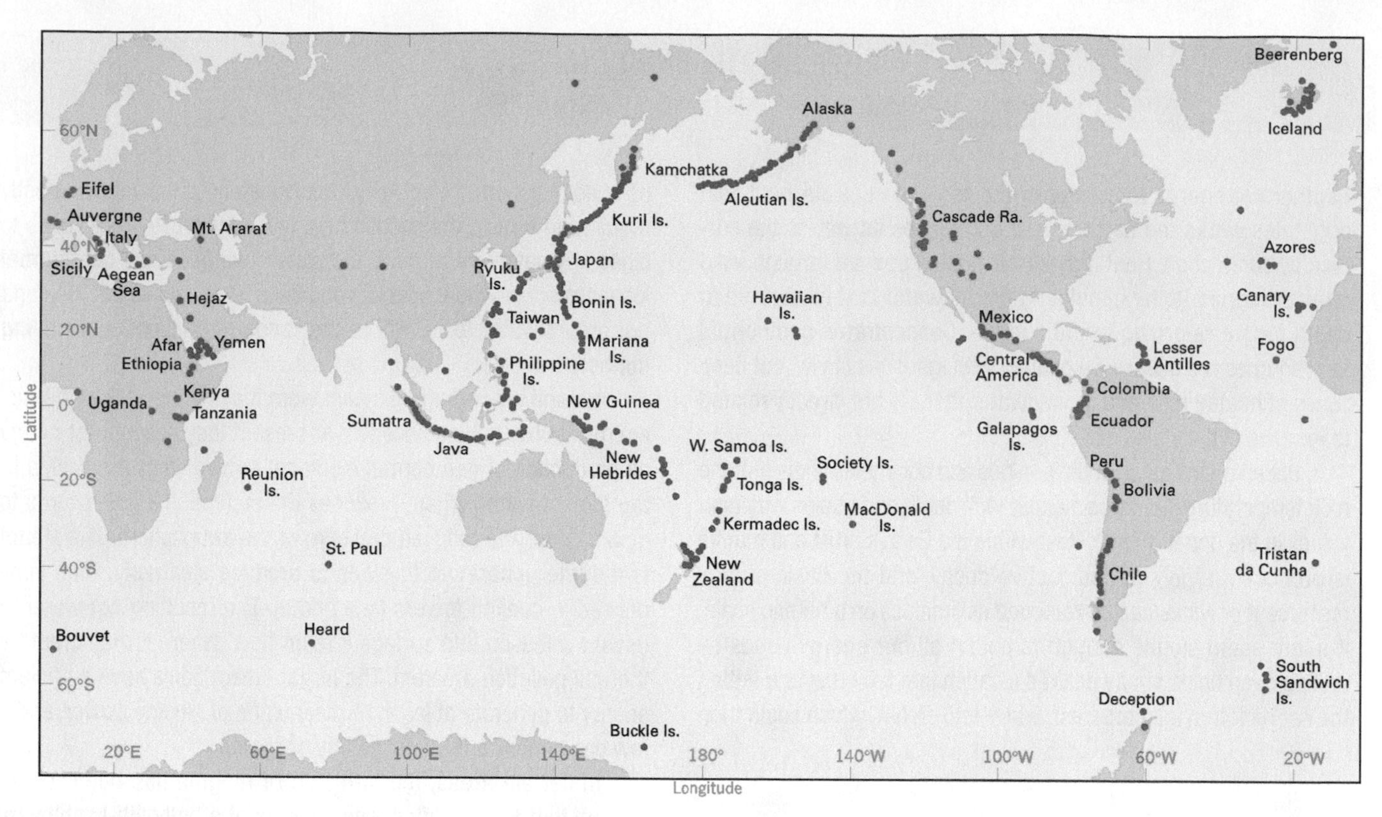

13.17 VOLCANIC ACTIVITY OF THE EARTH Dots show the locations of volcanoes known or believed to have erupted within the past 12,000 years. Each dot represents a single volcano or cluster of volcanoes. (From data of NOAA. Copyright © A. N. Strahler.)

Zealand (Figure 13.18). Volcanoes such as Vesuvius, Etna, and Stromboli in the Mediterranean region are similarly found near converging plate boundaries. Other volcanoes, including those in Iceland, are located on or near mid-oceanic rifts. Rifting has also produced volcanic regions in East Africa, including Mount Kilimanjaro and Mount Kenya. Hotspot activity is represented in Hawaii and in several other islands in the Pacific, such as the Galapagos Islands, Easter Island, Tahiti, and the Society Island group.

Hotspots associated with mantle plumes also occur on some of the islands around Antarctica, including the Kerguelen Islands, Crozet, and Bouvet. Volcanic activity is found in two areas on the Antarctic continent. Large stratovolcanoes, such as Mount Erebus (3,794 m) and Mount Melbourne (2,732 m), occur in the southwesterly part of the continent around the Ross Sea. The second group, located more to the west, consists of shield volcanoes including Mount Waesche (3,293 m), Mount Takahe (3,460 m), and Toney Mountain (3,595 m). Stratovolcanoes are also found on the Antarctic Peninsula, which is part of a continental margin arc associated with the boundary of the Antarctic and Scotia Plates; this continues as an oceanic island arc through South Orkney, South Georgia, and the South Sandwich Islands.

Images of volcanoes taken using different remote sensing techniques are presented in *Geographer's Tools • Remote Sensing of Volcanoes*.

13.18 VOLCANIC ERUPTION ON THE SMALL UNINHABITED ISLAND OF HUNGA HA'APAI IN THE TONGA ISLAND GROUP IN MARCH 2009 Continuing eruptions will lead to the further emergence of this underwater volcano associated with the subduction zone between the Pacific and Austral-Indian Plates.

GEOGRAPHER'S TOOLS
REMOTE SENSING OF VOLCANOES

Remote Sensing of Volcanoes

Mount Vesuvius erupted in A.D. 79, burying the nearby Roman city of Pompeii. In recent history, major eruptions of Mount Vesuvius were recorded in 1631, 1794, 1872, 1906, and 1944. The image of Mount Vesuvius shown in Figure 1 was acquired by the Advanced Spaceborne Thermal Emission and Reflection Radiometer (ASTER) on September 26, 2000. Spatial resolution is 15 × 15 m. Red, green, and blue colours in the image have been assigned to near-infrared, red, and green spectral bands, respectively. Vegetation appears bright red, and urban areas are in blue and green tones. The magnitude of development around the volcano shows that the impact of a major eruption would be catastrophic.

Mount Fuji lies about 100 km southwest of Tokyo. Although it has the symmetry of a simple cone, it is actually a complex structure with two former volcanic cones buried within its outer form. Mount Fuji is considered an active volcano; it last erupted in 1707. The image in Figure 2 was acquired by the Shuttle Radar Topography Mission's interferometric synthetic aperture radar (InSAR) on February 21, 2000. This type of radar sends simultaneous pulses of radio waves toward the ground from two antennas spaced 60 m apart. Very slight differences in the return signals can be related to the ground height. The vertical scale is doubled for visualization, so Mount Fuji and surrounding peaks appear twice as steep as they actually are.

Figure 1 Mount Vesuvius imaged by ASTER ***Fly By: 40° 49′ 20′ N; 14° 25′ 30′ E*** (Image courtesy NASA/GSFC/MITI/ERSDAC/JAROS and U.S./Japan ASTER Science Team.)

Popocatepetl (5,470 m), located only 65 km from Mexico City, last erupted on December 18, 2000. The image shown in Figure 3 was acquired by Landsat-7 on January 4, 1999, at a spatial resolution of 30 × 30 m. Snow and ice flank the summit crater. Canyons carved into the volcano lead away from the summit. The lower slopes are thickly covered with vegetation, which appears green in this image.

Figure 2 Mount Fuji imaged by the Shuttle Radar Topography Mission (Image courtesy NASA/JPL/NIMA.)

Figure 3 Popocatepetl imaged by Landsat-7 ***Fly By: 19° 01′ N; 98° 37′ W*** (Image courtesy Ron Beck, EROS Data Center.)

VOLCANIC ERUPTIONS AS ENVIRONMENTAL HAZARDS

Volcanic eruptions and lava flows are severe environmental hazards, often taking a heavy toll on plant and animal life and devastating human habitations. Loss of life and destruction of towns and cities are frequent in the history of peoples who live near active volcanoes. Devastation might occur in several ways: from the relentless advance of lava flows engulfing whole cities; from showers of ash, cinders, and volcanic bombs; from clouds of incandescent gases that descend the volcano slopes; and from violent earthquakes associated with volcanic activity. For people living along low-lying coasts, there is the additional peril of great seismic sea waves, generated by eruptions of undersea or island volcanoes that may have occurred hundreds of kilometres away.

In 1985, an explosive eruption of Nevado del Ruiz, a volcano in the Colombian Andes, caused the rapid melting of ice and snow at the summit. Mixing with volcanic ash, the water formed a variety of mudflow known as a *lahar*. Rushing down slopes at speeds of up to 145 kph, the lahar was channelled into a valley on the lower slopes, where it engulfed the town of Armero and killed more than 20,000 people.

Scientific monitoring techniques are reducing the toll of death and destruction from volcanoes. By analyzing the gases emitted from the vent of an active volcano, as well as the minor earthquakes and local land tilting that precede a major eruption, scientists have successfully predicted periods of volcanic activity. Extensive monitoring of Mount Mayon and the Mexican volcano Popocatepetl to predict recent eruptions has led to evacuations that saved thousands of lives. However, not every volcano is well monitored or predictable.

Despite their potential for destructive activity, volcanoes are a valuable natural resource in terms of recreation and tourism. British Columbia's Garibaldi Provincial Park preserves volcanic mountain vistas in a vast wilderness area of snow-covered peaks and swift rivers. Hawaii's Volcanoes National Park recognizes the natural beauty of Mauna Loa and Kilauea—their breathtaking displays of molten lava dramatically illustrate many igneous processes. Similarly, Mount Rainier, Mount Lassen, and Crater Lake in the Cascade Mountain Range are long-established national parks.

EARTHQUAKES

An **earthquake** is a motion of the ground surface, ranging from a faint tremor to a wild movement capable of shaking buildings apart (Figure 13.19). This type of movement can be produced by volcanic activity or when magma rises or recedes within a volcanic chamber, but most result from movements along the boundaries of lithospheric plates. Earthquakes can arise in these regions through various processes. In some cases they are triggered by the compressional forces that occur where plates collide. Other earthquakes are generated in subduction zones where descending plates force layers of sediment on the ocean floor against an overlying plate. Many earthquakes are associated with zones of rifting, where brittle continental crust is fractured as it is pulled apart and where plate boundaries slide past each other along transcurrent faults (see Chapter 11). Earthquakes can also be produced by volcanic activity or when molten rock rises or recedes within a magma chamber.

13.19 EARTHQUAKE DEVASTATION Damage in Port-au-Prince, Haiti, following a 7.0-magnitude earthquake on January 12, 2010. The earthquake was followed by 30 aftershocks, some of which were almost as strong. The focus of the earthquake was at a depth of 13 km, with the epicentre 25 km WSW of Port-au-Prince.

Typically, tectonic forces slowly bend the rock at the plate boundaries over many years. When a critical point is reached, the rocks on opposite sides move in different directions to relieve the strain. A large quantity of energy is instantaneously released in the form of seismic waves, which shake the ground. The waves move outward in widening circles from a point of sudden energy release, called the *focus*, and gradually lose energy as they travel outward in all directions.

Earthquakes produce four basic types of waves, two of which—P waves and S waves—travel deep within the Earth's interior, and two of which—Rayleigh waves and Love waves—travel near the surface (Figure 13.20a). These different wave types travel out from the centre of the earthquake in all directions. The centre of an earthquake is determined by the time of arrival of the different wave types at three locations on the surface. The point on the Earth's surface directly above the focus of an earthquake is referred to as the **epicentre**.

The faster of the two deep-seated wave types is called the primary or **P wave**. The P waves propagate by alternately pushing and pulling the rock, and can travel through both solid and molten rock material, as well as the water of the oceans. The slower waves are called secondary or **S waves**. The S waves create a transverse motion perpendicular to the direction of propagation. This can be a vertical or horizontal movement, either of which shears the rock at right angles to the direction of propagation. S waves cannot propagate through liquids and so do not develop in Earth's outer core; hence the inference that this region is essentially molten iron. Similarly, S waves do not travel through magma chambers in the mantle.

The speed of P and S seismic waves depends on the density and properties of the rocks through which they pass. In most earthquakes, the P waves travel through the crust at 5 to 7 km s^{-1} and at about 8 km s^{-1} in the mantle and core. These are followed by the shaking and twisting motion of the S waves, which propagate at about 3 to 4 km s^{-1} through the crustal rocks, 4.5 km s^{-1} in the mantle, and 2.5 to 3.0 km s^{-1} in the solid inner core. S waves are especially damaging to buildings and infrastructure.

Love waves spread out and contract horizontally through the surface rock layers at right angles to the direction of propagation. The amplitude of the Love wave motion decreases with depth; their velocity ranges from 2.2 to 4.4 km s^{-1}, depending on frequency. Generally the low frequency waves propagate at higher velocities and also penetrate to greater depths. Rayleigh waves are also **surface waves**; they are similar to the ripples that travel across the surface of a lake. Their motion is both in the direction of propagation and perpendicular to that direction, so they tend to create an elliptical pattern around the centre of the earthquake. Rayleigh waves cause earth materials to follow a circular path in exactly the same way a wave rolls through water (see Chapter 17); also like ocean waves, their amplitude decreases with depth. The speed of travel of Rayleigh waves is about 2.0 to 4.2 km s^{-1} and depends on frequency; as with the Love waves, depth of propagation is inversely related to frequency.

Because seismographs do not detect S waves on the opposite side of the Earth from an earthquake's point of origin, it has been concluded that the core is liquid. The core's size has been determined by the distance at which S waves are detected. As P waves move through the Earth's interior, they refract or bend, with abrupt changes in direction occurring at the boundary between different layers. P waves entering the outer core are bent toward the Earth's centre so they reach a region opposite the earthquake's point of origin. Thus a **shadow zone** separates the P waves that only pass through the mantle from the P waves that pass through the mantle and the core. The shadow zone forms a ring that extends from about 105° to 140° away from the point of origin (Figure 13.20b). The fact that very weak P waves are felt in the shadow zone suggests that the inner core is solid.

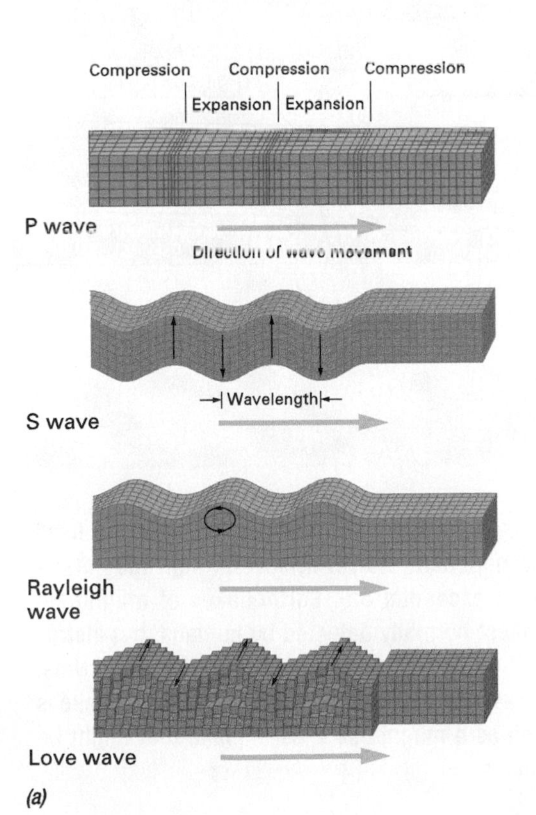

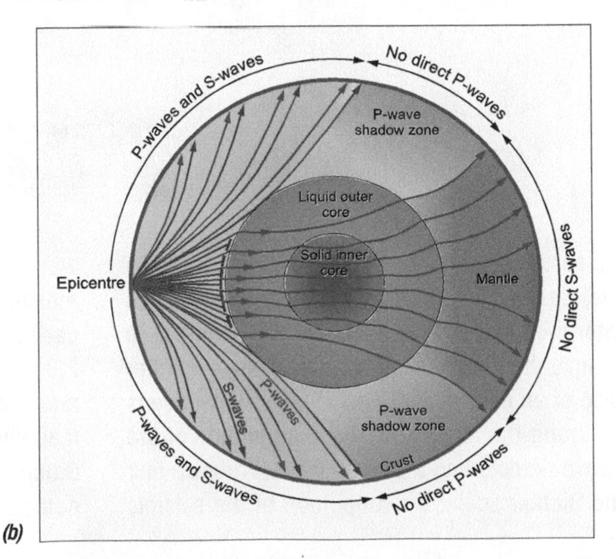

13.20 MAJOR TYPES OF EARTHQUAKE WAVES (a) P waves, which have the highest velocity, vibrate parallel to the direction of motion in a pulse-like manner; in S waves, movement is perpendicular to the direction of propagation and causes a twisting, heaving motion; Rayleigh waves travel in the same manner as waves in water, causing earth materials to follow in a circular path. Love waves cause transverse motion near the surface perpendicular to the direction of propagation. (b) The existence of a shadow zone where neither P nor S waves arrive provides evidence that a core region exists within the Earth. The inability of S waves to pass through the core suggests that at least the outer core is liquid.

THE RICHTER SCALE

The *Richter scale* was designed in 1935 to assess the local magnitude (M_L) of earthquakes. The numbers on the Richter scale range from 0 to 9, but there is really no upper limit. For each whole unit increase (say, from 5.0 to 6.0), the amplitude of the earthquake wave increases by a factor of 10 and the quantity of energy released increases by a factor of 32. A value of 9.5 recorded in the Chilean earthquake of 1960 remains the highest to date. The Richter scale is based on the amount of ground shaking, as measured on a seismograph. It is one of three common measures of earthquake magnitude.

The oldest measure, the *Mercalli scale* devised in 1902, is based on the amount of damage caused by an earthquake and human response to it. The Mercalli scale is descriptive and originally included 10 categories; it was subsequently modified to a 12-point scale that ranges from I (most people do not notice, animals may be uneasy) to XII (all structures are destroyed).

The *Moment magnitude scale (MMS)*, introduced in 1979, is based on the movement that occurs on a fault, not on how much the ground shakes during an earthquake. The MMS estimates the force (M_W) of the earthquake in terms of the torque or movement it created through its twisting and bending motion. Like the Richter scale, the MMS is logarithmic. However, the two scales cannot be directly compared because they are based on different characteristics of an earthquake (Table 13.2). Although the MMS is now used to estimate earthquake magnitude, the Richter scale is still a popularly used reference for earthquakes of less than 3.5 magnitude.

Several million earthquakes occur each year, many of which are undetected because they occur in very remote areas or are too small to be measured on seismographs outside of the immediate area. On average about 75 major earthquakes of magnitude 6.0 or greater occur worldwide each year (Table 13.3). This number remains relatively constant, but the number of lesser earthquakes has increased, mainly because they can now be detected by the growing network of sensitive seismographs. *Geographer's Tools • Measuring Earthquake Magnitude* provides more information on the Richter scale and other methods of assessing earthquake magnitude.

TABLE 13.2 COMPARATIVE EARTHQUAKE MAGNITUDES ON THE RICHTER MAGNITUDE (M_L) AND MOMENT MAGNITUDE (MMS) SCALES

EARTHQUAKE	RICHTER SCALE	MOMENT MAGNITUDE SCALE
New Madrid, MO, 1812	8.7	8.1
San Francisco, CA, 1906	8.3	9.2
Chile, 1960	8.5	9.6
Alaska, 1964	8.4	9.2
Northridge, CA, 1994	6.4	6.7

TABLE 13.3 FREQUENCY OF EARTHQUAKES OF DIFFERENT MOMENT MAGNITUDE (M_W) WORLDWIDE

MAGNITUDE (M_W)	NUMBER OF EARTHQUAKES PER YEAR	DESCRIPTION
> 8.5	0.3	Great
8.0–8.4	1	
7.5–7.9	3	Major
7.0–7.4	15	
6.6–6.9	56	
6.0–6.5	210	Destructive
5.0–5.9	800	Damaging
4.0–4.9	6,200	Minor
3.0–3.9	49,000	
2.0–2.9	300,000	
0–1.9	700,000	

GEOGRAPHER'S TOOLS
MEASURING EARTHQUAKE MAGNITUDE

The magnitude of earthquakes is most commonly assessed by the *Richter scale*, devised in 1935 by seismologist Charles F. Richter, and later modified in 1956. The **Richter magnitude scale** (M_L)—or simply the Richter scale—uses a single number to quantify the size of an earthquake. Earthquake magnitude is obtained by calculating the logarithm of the amplitude of the largest seismic wave detected on a seismograph. For each unit of increase in the Richter scale, the amplitude of the seismic wave increases by a factor of 10. The scale has neither a fixed maximum nor a minimum; however, several high-magnitude earthquakes have exceeded 8.9. Earthquakes of magnitude 2.0 are the smallest normally detected by humans, but instruments can detect quakes as small as −3.0 on the scale. Note, magnitude −3.0 on a logarithmic scale means an earthquake is 0.0001 as strong as a magnitude 2 earthquake that might be detected by humans.

The **moment magnitude scale** (M_W), introduced in 1979, is based on the concept of the **seismic moment** (M_0) and provides a measure of the amount of energy released by an earthquake. The seismic moment combines the amount of movement on the fault with the area of the fault that ruptured and the type of rock involved. The rigidity of the rock incorporates two components: stress (or force acting on the rock) and strain (deformation that occurs within the rock). Thus:

$$\text{Moment} = \text{Rock rigidity} \times \text{Fault Area} \times \text{Slip Distance}$$
$$M_0 = \mu Ad$$

The moment magnitude is calculated as:

$$M_W = \frac{2}{3}\left(\log_{10}\frac{M_0}{\mathrm{N\cdot m}} - 9.1\right)$$

where M_0 is the seismic moment.

The constants in the equation are chosen so that estimates of moment magnitude roughly correspond with estimates based on the Richter magnitude scale. Thus, the relationship between the Richter scale and energy release is given by the following equation:

$$\log_{10} E = 4.8 + 1.5M_L$$

where E is the energy in joules and M_L is the Richter magnitude scale. By taking the exponent of both sides, the expression can be written as:

$$E = 10^{(4.8+1.5M_L)} = 10^{4.8} \times 10^{1.5M_L}$$

The expression $10^{1.5M_L}$ means that each unit in the Richter scale increases by a factor of $10^{1.5\times 1}$ or 31.6. Thus, if the Richter scale number increases by 1 unit, say from 4 to 5, then the energy release will increase by a factor of $10^{1.5\times 1}$ or 31.6, so that about 32 times more energy is released. For a two-unit increase, say from 4.0 to 6.0, the energy released will be $10^{1.5\times 2} = 10^3 = 1{,}000$ times as large (see Figure 1). One advantage of the moment magnitude scale is that there is no value beyond which all large earthquakes are given the same magnitude. For this reason, moment magnitude is now the preferred scale for estimating the magnitude of large earthquakes. However, the Richter scale is generally used for earthquakes with a magnitude of less than 3.5.

Whereas earthquakes in North America are rated according to their magnitude, elsewhere in the world the preference is still to use a scale of intensity. The *Rossi-Forel scale*, devised in the latter part of the nineteenth century, is generally regarded as the first of these descriptive scales. The 1873 version used 10 categories. This scale was used until 1902 when the 10-point Mercalli scale was introduced. The Mercalli scale was changed to a 12-point scale and translated into English in 1931. It is now commonly referred to as the Modified Mercalli Scale.

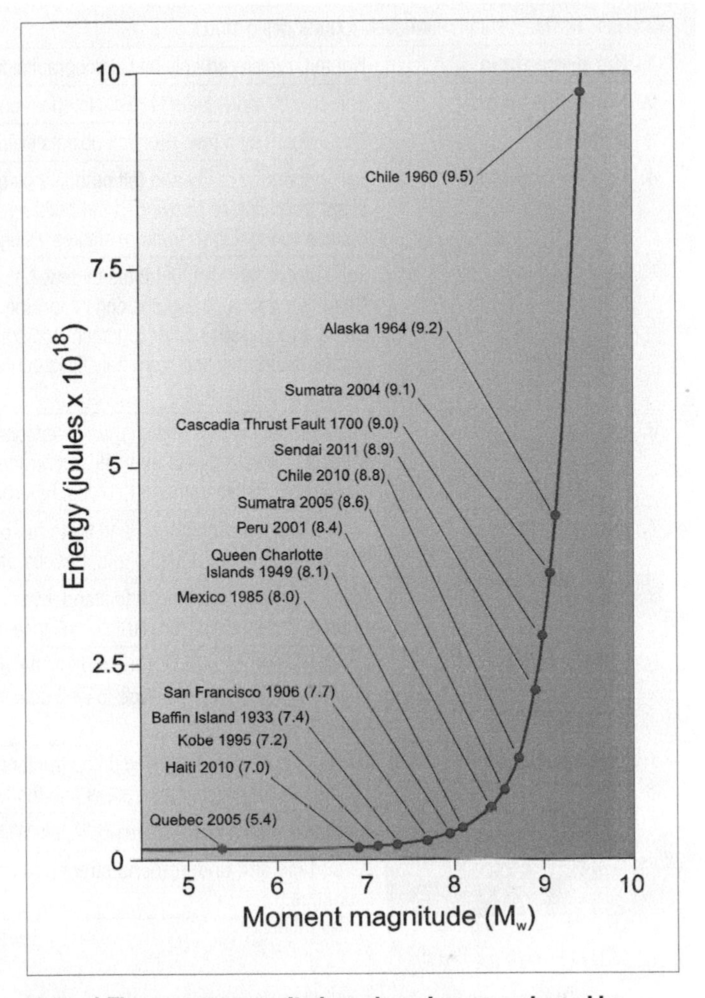

Figure 1 The moment magnitude scale and energy released by earthquakes

A newer scale, the *Medvedev-Sponheuer-Karnik (MSK) scale*, came into widespread use in the 1970s and is still the standard reference in Russia, throughout the former Soviet Union, India, and Israel. The MSK scale incorporates natural landscape conditions, including landslides and tsunamis, into several of its categories, as shown in Table 1. The *European Macroseismic Scale (EMS 98)*, the latest version of which was developed in 1998, is based largely on the principles used in the MSK scale.

The Japan Meteorological Agency seismic intensity (shindo) scale uses 10 categories, ranging from 0 to 7 (two levels are included in categories 5 and 6), as shown in Table 2. It describes the intensity of an earthquake mainly in terms of the building damage incurred. Real-time earthquake reports in Japan are reported from a network of 180 seismographs and 600 seismic intensity meters, and include information on how hard the ground shakes during the event.

TABLE 1 **THE MEDVEDEV-SPONHEUER-KARNIK (MSK) SCALE**

1. Not perceptible	Not felt, registered only by seismographs. No effect on objects. No damage to buildings.
2. Hardly perceptible	Felt only by individuals at rest. No effect on objects. No damage to buildings.
3. Weak	Felt indoors by a few. Hanging objects swing slightly. No damage to buildings.
4. Largely observed	Felt indoors by many and felt outdoors by only very few. A few people are awakened. Moderate vibration. Observers feel a slight trembling or swaying of the building, room, bed, chair, etc. China, glasses, windows, and doors rattle. Hanging objects swing. Light furniture shakes visibly in a few cases. No damage to buildings.
5. Fairly strong	Felt indoors by most, outdoors by few. A few people are frightened and run outdoors. Many sleeping people awake. Observers feel a strong shaking or rocking of the whole building, room, or furniture. Hanging objects swing considerably. China and glasses clatter together. Doors and windows swing open or shut. In a few cases, window panes break. Liquids oscillate and may spill from fully filled containers. Animals indoors may become uneasy. Slight damage to a few poorly constructed buildings.
6. Strong	Felt by most indoors and by many outdoors. A few people lose their balance. Many people are frightened and run outdoors. Small objects may fall and furniture may be shifted. Dishes and glassware may break. Farm animals may be frightened. Visible damage to masonry structures, cracks in plaster. Isolated cracks on the ground.
7. Very strong	Most people are frightened and try to run outdoors. Furniture is shifted and may be overturned. Objects fall from shelves. Water splashes from containers. Serious damage to older buildings, masonry chimneys collapse. Small landslides.
8. Damaging	Many people find it difficult to stand, even outdoors. Furniture may be overturned. Waves may be seen on very soft ground. Older structures partially collapse or sustain considerable damage. Large cracks and fissures open up, rockfalls.
9. Destructive	General panic. People may be forcibly thrown to the ground. Waves are seen on soft ground. Substandard structures collapse. Substantial damage to well-constructed structures. Underground pipelines rupture. Ground fracturing, widespread landslides.
10. Devastating	Masonry buildings destroyed, infrastructure crippled. Massive landslides. Water bodies may be overtopped, causing flooding of the surrounding areas and the formation of new water bodies.
11. Catastrophic	Most buildings and structures collapse. Widespread ground disturbances, tsunamis.
12. Very catastrophic	All surface and underground structures completely destroyed. Landscape generally changed, rivers change paths, tsunamis.

TABLE 2 **JAPAN METEOROLOGICAL AGENCY SEISMIC INTENSITY (SHINDO) SCALE**

MAGNITUDE	CLASSIFICATION	EFFECTS	PEAK GROUND ACCELERATION
7	Ruinous earthquake	In most buildings, wall tiles and windowpanes are damaged and fall. In some cases, reinforced concrete-block walls collapse.	Greater than 4 s^{-2}
6+	Violent earthquake	In many buildings, wall tiles and windowpanes are damaged and fall. Most un-reinforced concrete-block walls collapse.	3.15–4.00 s^{-2}
6–	Violent earthquake	In some buildings, wall tiles and windowpanes are damaged and fall.	2.50–3.15 s^{-2}
5+	Severe earthquake	In many cases, un-reinforced concrete-block walls collapse and tombstones overturn. Many automobiles stop due to difficulty in driving. Occasionally, poorly installed vending machines fall.	1.40–2.50 s^{-2}
5–	Severe earthquake	Most people try to escape from danger; some find it difficult to move.	0.80–1.40 s^{-2}
4	Strong earthquake	Many people are frightened. Some people try to escape from danger. Most sleeping people awake.	0.25–0.80 s^{-2}
3	Weak earthquake	Felt by most people in the building. Some people are frightened.	0.08–0.25 s^{-2}
2	Light earthquake	Felt by many people in the building. Some sleeping people awake.	0.025–0.08 s^{-2}
1	Slight earthquake	Felt by only some people in the building.	0.008–0.025 s^{-2}
0	Insensible	Imperceptible to people.	Less than 0.008 s^{-2}

EARTHQUAKES AND PLATE TECTONICS

Most seismic activity occurs primarily near lithospheric plate boundaries (Figure 13.21). The greatest intensity of seismic activity is found along converging plate boundaries where oceanic plates are undergoing subduction. Strong pressures build up at the downward-slanting contact of the two plates. They are relieved by sudden fault slippages that generate earthquakes of large magnitude. This mechanism explains the great earthquakes experienced in Japan, Alaska, Central America, Chile, and other zones close to trenches and volcanic arcs.

Similar examples can be identified on the Pacific coast of Mexico and Central America, where the subduction boundary of the Cocos Plate lies close to the shoreline. The earthquake that devastated Mexico City in 1985 was centred in the deep trench offshore. During that event, two great shocks in close succession, the first of magnitude 8.1 and the second of 7.5, damaged cities along the Pacific coast. Although Mexico City lies inland, about 300 km from the earthquake epicentre, it experienced intense ground shaking of underlying saturated clay formations, with a resulting death toll of some 10,000 people.

The magnitude 7.0 earthquake that struck Port-au-Prince, Haiti, in January 2010 was also related to tectonic activity along a plate boundary. Here the North American Plate meets the Caribbean Plate in a series of transcurrent faults, with a subduction zone lying further to the east where it plunges under the Caribbean Plate (Figure 13.22). However, in the vicinity of the Bahamas a portion of the American Plate known as the Bahamas Platform is resisting subduction. This has resulted in the development of major fault systems. The northern fault boundary of the Caribbean Plate, the Septentrional Fault, extends from the eastern coast of the Dominican Republic across the north of Haiti, past Cuba toward the Yucatan Peninsula. A second fault, the Enriquilla-Plantain Garden Fault, runs from the Cocos Plate through Guatemala, then crosses the Caribbean Sea north of Jamaica to Haiti. Port-au-Prince lies almost directly on this fault, which appears to be the main region where strain relief occurs as the two lithospheric plates move past each other.

Transcurrent faults on transform boundaries that cut through the continental lithosphere, such as in Haiti, occur in many other parts of the world. These are regions of frequent seismic activity, with moderate to strong earthquakes. In North America, the best example is the San Andreas Fault (Figure 13.23). Similar activity is frequently reported along the North Anatolian Fault in Turkey, where the Persian Subplate is moving westward at its boundary

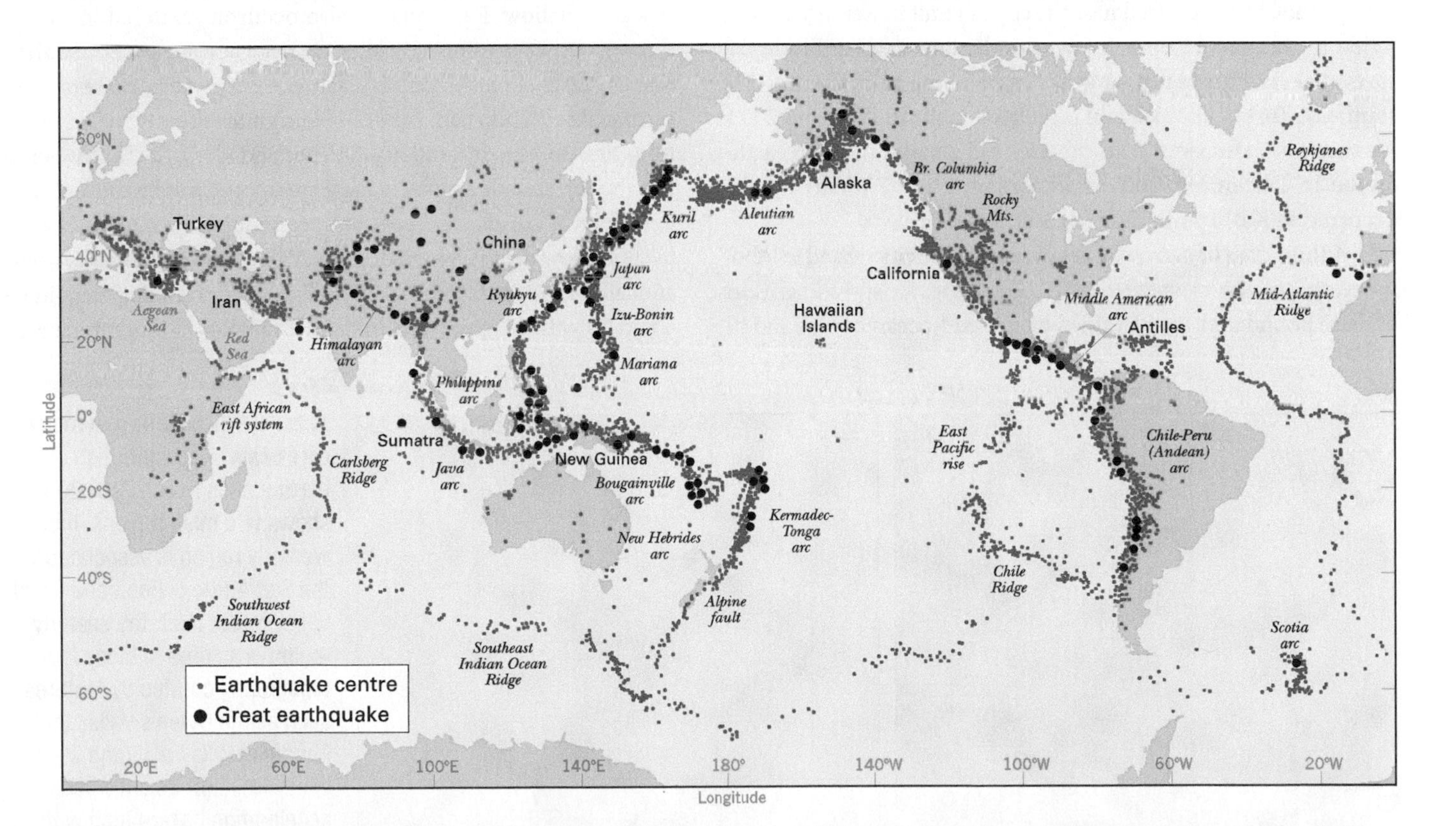

13.21 EARTHQUAKE LOCATIONS Earthquakes originating at depths of 0 to 100 km during a seven-year period are shown by red dots. Each dot represents a single location or a cluster of centres. Black circles identify centres of earthquakes of magnitude 8.0 or greater during an 80-year period. The map clearly shows the pattern of earthquakes occurring at subduction boundaries. (Compiled by A. N. Strahler from U.S. government data. Copyright © A. N. Strahler.)

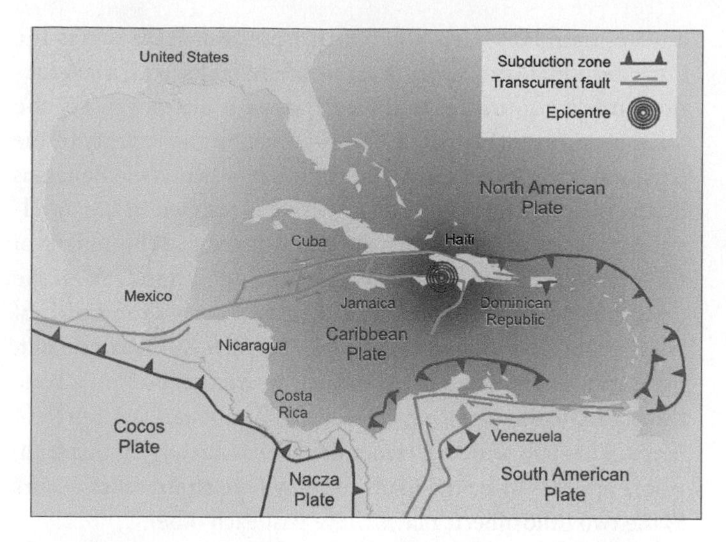

13.22 GEOLOGIC SETTING OF HAITI The capital city Port-au Prince was devastated in January 2010 by a shallow earthquake, caused by the movement in a major fault zone separating the North Atlantic and Caribbean Plates. This region experiences frequent earthquakes.

13.24 EARTHQUAKES IN CANADA Based on long-term records, the strongest and most frequent earthquakes occur in western Canada, but they are also common in the St. Lawrence region and across the Arctic. Earthquakes are rarely reported in the Prairie provinces.

with the European Plate. In August 1999, a major earthquake with a magnitude of 7.4, centred near the city of Izmit in western Turkey, killed more than 15,000. A few months later, a magnitude 7.2 earthquake occurred not far away on the same fault. Central and southeast Turkey shook again on related faults in 2002 and 2003. Further east, southwest Iran experienced a magnitude 6.6 earthquake in December 2003, and a magnitude 7.6 earthquake occurred in Kashmir in October 2005.

A third class of narrow zones of seismic activity related to lithospheric plates occurs where the plate boundaries are spreading. Most of these boundaries are identified by the mid-oceanic ridge and its branches. Earthquakes of this type are generally moderate in intensity and shallow. Earthquakes also occur at scattered locations throughout the continental plates, far from active plate boundaries. In many cases, no active fault is visible, and the geologic cause of these earthquakes is uncertain. For example, significant earthquakes occur from time to time in southern Quebec and along the St. Lawrence River valley. Seismic activity in these areas appears to be concentrated in regions of crustal weakness, with earthquakes occurring at relatively shallow depths down to 30 km. About 450 earthquakes occur annually in eastern Canada, with perhaps four exceeding magnitude 4 and 30 with magnitude 3 to 4. Earthquakes of magnitude 5 or

(a)

(b)

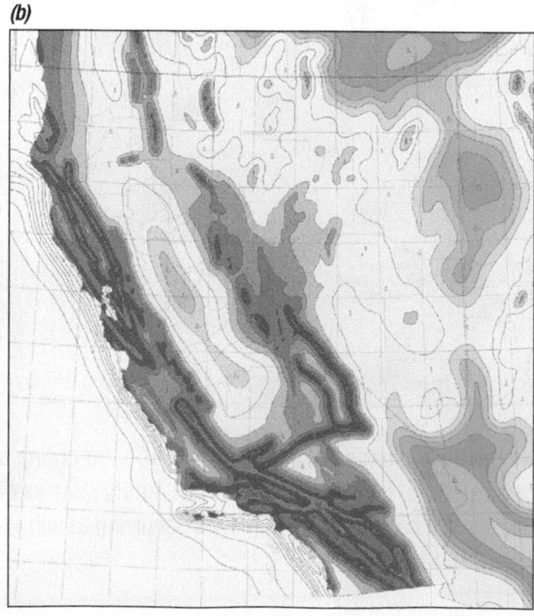

13.23 EARTHQUAKE INFORMATION FOR CALIFORNIA (a) Distribution of earthquakes from 1990–2006 shows two main regions: the westerly region is associated with the San Andreas Fault and Great Valley thrust fault; the easterly region is an area of block faulted topography created by features such as the Owen's Valley fault, Panamint Hills Fault, and Death Valley fault. (b) Seismic peak acceleration hazard map with a 2 percent probability that it will be exceeded during a 50-year period. Peak acceleration is the maximum change in velocity of a particle on the ground.

higher are predicted to occur here about three times per decade; the most recent magnitude 5 earthquake was in June 2010.

In Canada, the strongest earthquakes occur most frequently in British Columbia, particularly off Vancouver Island and the Queen Charlotte Islands, Yukon, and the Northwest Territories (Figure 13.24). Of the 1,000 or so earthquakes that occur in this region each year, several are magnitude 5 or greater. The frequency of earthquakes along the west coast is related to the complex structure associated with the small Juan de Fuca Plate. Along its western margin, the Juan de Fuca Plate is separated from the Pacific Plate by a spreading rift zone. Its eastern margin is marked by the Queen Charlotte Fault, a transcurrent fault where it slides northwards against the North American Plate, and by the Cascade Subduction Zone, where it is moving under the North American Plate at a rate of 2 to 5 cm per year.

In recent years, the largest earthquakes on record in Canada are ones of magnitude 8.1 and 7.4 off the Queen Charlottes in 1949 and 1970, respectively. In 2004, a magnitude 6.7 earthquake occurred on Vancouver Island, which previously had recorded a magnitude 7.3 earthquake in 1946 and one of magnitude 7.0 in 1918. The strongest earthquake reported in eastern Canada is a 7.3 event on the Grand Banks in 1929. An earthquake of magnitude 7.4 occurred in 1933 on Baffin Island. Historically, the largest earthquake in Canada reportedly occurred in 1700, and is estimated to have had a magnitude of 9.0. It was associated with the undersea Cascadia thrust fault that runs in the Pacific Ocean from the vicinity of Vancouver Island to the waters off California.

The probability of an earthquake occurring can be calculated from past records, and is used to determine the seismic hazard for any region in Canada (Figure 13.25). The map is based on the probability of tremors that are of sufficient magnitude to damage one- and two-storey single-family dwellings. In regions of high hazard, earthquakes have a 30 percent chance of causing significant damage every 50 years, compared to a 5 to 15 percent chance in regions of moderate hazard, and less than 1 percent chance in regions of lowest hazard.

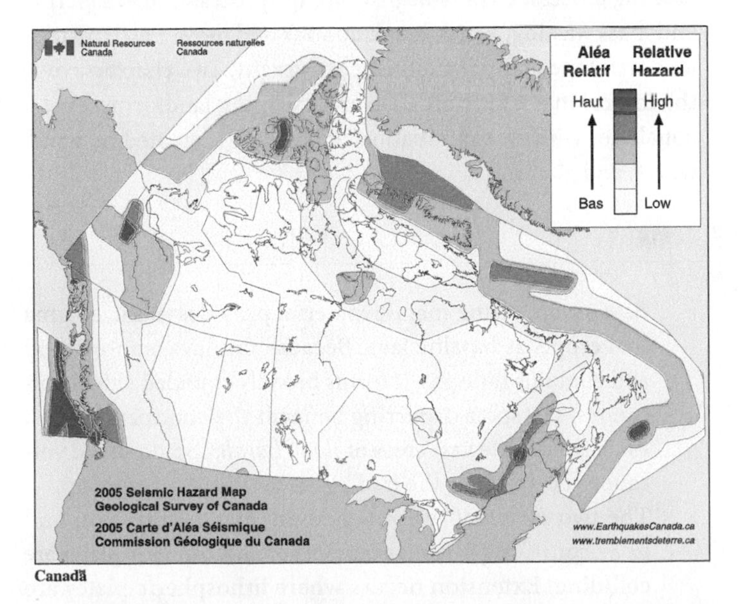

13.25 SEISMIC HAZARD MAP FOR CANADA This map shows the probability of the occurrence of earthquakes of sufficient magnitude to damage one- and two-storey single-family dwellings.

SEISMIC SEA WAVES

An important environmental hazard often associated with a major earthquake centred on a subduction plate boundary is the *seismic sea wave*, or **tsunami**. A succession of these waves is often generated in the ocean by a sudden movement of the sea floor at a point near the earthquake source. The waves travel over the ocean in ever-widening circles, but they are not perceptible at sea in deep water (Figure 13.26).

When a tsunami arrives at a distant coastline, it causes a rise in water level. Normal wind-driven waves, superimposed on the heightened water level, strike places inland that are normally above their reach. Few have been as devastating as the Boxing Day (Asian) Tsunami of 2004. It was triggered by the undersea Sumatra-Andaman earthquake, the epicentre of which was located off the west coast of Sumatra. The magnitude of the earthquake is officially rated at between 9.1 and 9.3, making it the second-largest earthquake recorded on a seismograph. Seismic activity lasted for about 10 minutes, and the vibrations were sufficiently powerful to cause the Earth to vibrate on its axis by more than a centimetre. Tsunamis with waves of up to 30 m spread throughout the Indian Ocean, inundating coastal communities in places such as Indonesia, Sri Lanka, India, and Thailand. The estimated death toll exceeded 275,000 people.

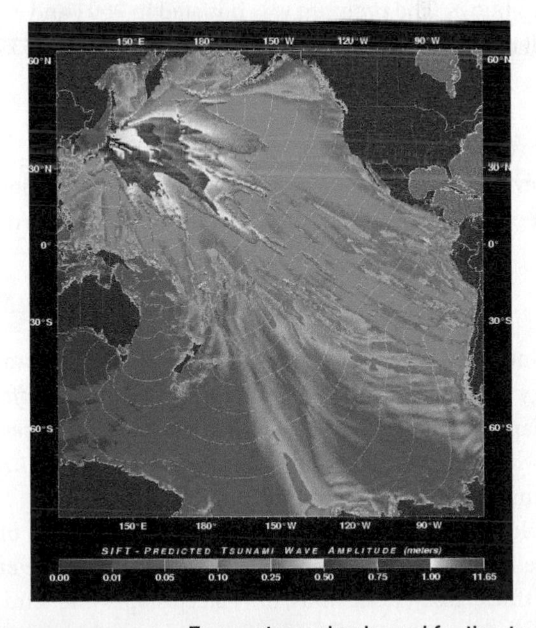

13.26 TSUNAMI TRAVEL TIME Forecast warning issued for the tsunami generated by the 8.9 magnitude earthquake that occurred about 130 km east of Sendai, Japan, on March 11, 2011. Severe damage was caused in coastal areas inundated as a wave exceeding 10 m came ashore.

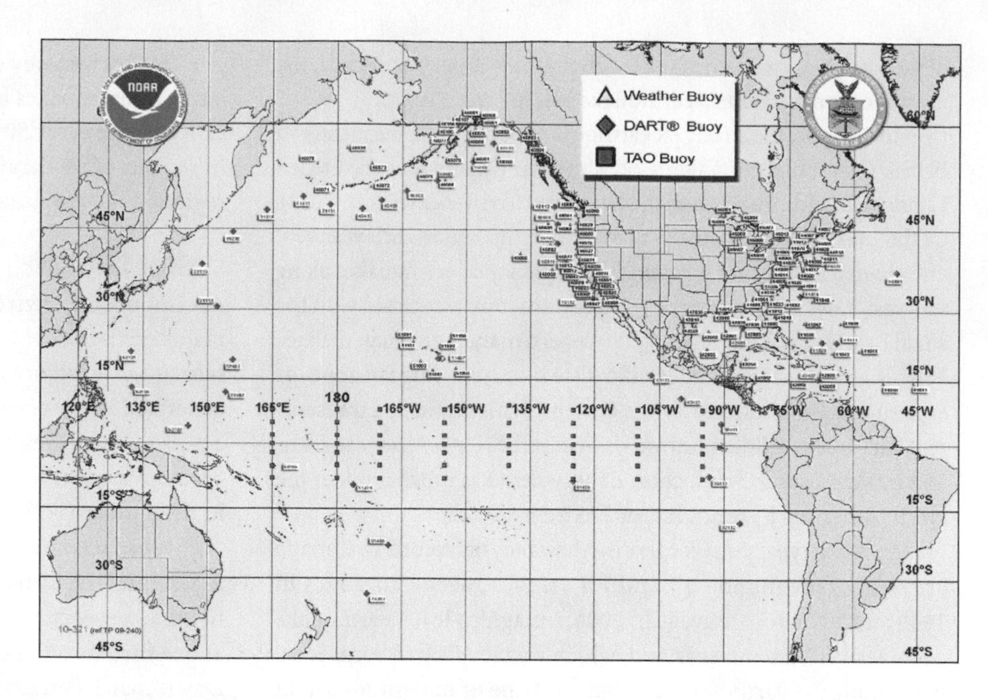

13.27 LOCATION OF DART BUOYS The DART buoys provide tsunami monitoring capabilities throughout the entire Pacific and Caribbean basins, with additional buoys deployed in the Atlantic and Indian Oceans. Each buoy consists of a pressure recording device capable of detecting tsunamis as small as 1 cm. Data are relayed via a GOES satellite link to ground stations for immediate dissemination to National Oceanic and Administration's (NOAA) Tsunami Warning Centers.

The need to provide early warnings of tsunamis led to the establishment of the Pacific Tsunami Warning Center (PTWC) in Hawaii in 1949. A second centre at Palmer, Alaska, was established in 1967 to serve the west coast of North America. Currently, the PTWC, representing 26 countries, is responsible for monitoring seismological and tidal stations throughout the Pacific Basin. The PTWC operational centre issues tsunami warnings and information about potentially tsunamogenic earthquakes to all nations in the Pacific. The present global monitoring system is based on the deployment of Deep-ocean Assessment and Reporting of Tsunami (DART) buoys. The program was initiated in 2001 and expanded to a full network of 39 stations in March 2008 (Figure 13.27).

A Look Ahead

The previous three chapters have surveyed the composition, structure, geologic activity, and initial landforms of the Earth's crust. The rocks and minerals that make up the Earth's crust and core are continuously recycled and transformed in a process that has occurred over some 4 billion years of geologic time. Powered by radiogenic heat stored in the Earth's interior, the cycle of rock transformation is part of the mechanism of plate tectonics that intricately links the geographical distribution of mountain ranges, ocean basins, and continents. Initial landforms, which result directly from volcanic and tectonic activity, occur primarily at the boundaries of spreading or colliding lithospheric plates.

Following this survey of the Earth's crust and the geologic processes that shape it, attention now turns to other landform-creating processes. Part 4 begins with the processes of weathering and mass wasting, which break up rock and move Earth materials downslope under the influence of gravity. Two chapters cover the importance of running water in shaping landforms. Part 4 concludes with an examination of landforms created by wind, waves, and glacial ice.

CHAPTER SUMMARY

- Landforms are the surface features of the land; geomorphology is the scientific study of landforms. *Initial landforms* are shaped by volcanic and tectonic activity, while *sequential landforms* are sculpted by agents of denudation, including running water, waves, wind, and glacial ice.
- Volcanoes are landforms created by the eruption of lava at the Earth's surface. Stratovolcanoes, formed by the emission of thick, gassy, felsic lavas, have steep slopes and tend to have explosive eruptions that can form *calderas*. Most active stratovolcanoes lie along the Pacific Rim, where subduction of oceanic lithospheric plates is occurring.
- At *hotspots*, rising mantle material provides mafic magma that erupts as basaltic lava. Because this lava is more fluid and contains little gas, it forms broadly rounded shield volcanoes. Hotspots occurring beneath the continental crust can also provide vast areas of *flood basalts*. Some shield volcanoes occur along the mid-oceanic ridge.
- The two forms of tectonic activity are compression and extension. Compression occurs where lithospheric plates are colliding. Extension occurs where lithospheric plates are spreading apart.

- Earthquakes occur when rock *layers,* bent by tectonic activity, suddenly fracture and move. The sudden motion at the fault produces earthquake waves that shake and move the ground surface in the adjacent region.
- Earthquakes produce four basic types of waves, two of which—P waves and S waves—travel deep within the Earth's interior, and two of which—Rayleigh waves and Love waves—travel near the surface.
- Large earthquakes occurring near developed areas can cause great damage. Most severe earthquakes occur near plate collision boundaries.
- Seismic activity creates tsunamis, which can inundate coastal areas with waves of up to 30 m. They are particularly devastating in the island nations of the Pacific and Indian Oceans.

KEY TERMS

caldera
earthquake
epicentre
geomorphology
landform
moment magnitude scale
P wave
Richter magnitude scale
S wave
seismic moment
shadow zone
shield volcano
stratovolcano
surface wave
tsunami
volcano

REVIEW QUESTIONS

1. Distinguish between initial and sequential landforms. How do they represent the balance of power between internal Earth forces and the external forces of denudation agents?
2. What is a stratovolcano? What is its characteristic shape, and why does that shape occur? Where do stratovolcanoes generally occur and why?
3. What is a shield volcano? How is it distinguished from a stratovolcano? Where are shield volcanoes found, and why? Give an example of a shield volcano. How are flood basalts related to shield volcanoes?
4. How can volcanic eruptions become natural disasters? Be specific about the types of volcanic events that can devastate habitations and extinguish nearby populations.
5. Describe the stages in the life cycle of a Hawaiian-type basaltic shield volcano.
6. What is a hotspot? What produces it? How is it related to the life cycle of a volcano?
7. What is the ultimate source of geothermal power? Where would you go to find a geothermal power source, and why would it occur there?
8. How is geothermal energy extracted? What environmental concerns arise in this process?
9. What is an earthquake, and how does it arise? How are the locations of earthquakes related to plate tectonics?
10. Name the different types of earthquake waves that travel at depth within the Earth. How do their paths differ and what can be deduced from this?
11. Name the different types of near-surface earthquake waves. What effect do they have on surface structures, and why?
12. Describe a tsunami, including its origin and effects.

VISUALIZATION EXERCISES

1. Sketch the sequence of stages in the life of a volcano associated with a hotspot.
2. Draw sketches to show the form and structure of composite and shield volcanoes. Indicate on a map similar to Figure 13.17 the general areas where each type is found.

ESSAY QUESTIONS

1. Write a hypothetical news account of a volcanic eruption. Select a type of volcano—stratovolcano or shield—and a plausible location. Describe the eruption and its effects as it was witnessed by observers. Include any details you need, but be sure they are scientifically correct.
2. Describe the geographic distribution of earthquakes in Canada and discuss the factors that underlie this distribution.
3. Discuss the different ways of assessing an earthquake's magnitude and intensity, and describe the advantages and disadvantages of each approach.

4 PART SYSTEMS OF LANDFORM EVOLUTION

Chapters In Part 4

Part 4 focuses on matter and energy-flow systems that shape the surface of the land. These systems operate largely on time scales, ranging from hundreds of years to a few million years. Systems of landform evolution are mostly powered by gravity. Gravity releases potential energy that has been stored in positioning solids and fluids above some base level, which for most land areas is sea level. Potential energy is ultimately derived from solar energy, which places water on the continents through precipitation, and the Earth's internal heat, which generates forces that uplift masses of rock and soil above the base level. ● Chapter 14 discusses the influence of rock structure on landforms created by geomorphic processes acting at the surface. The first is weathering, which leads to the disintegration of rock. Mass movement, in which weathered material moves downslope under the influence of gravity (as in landslides and earth flows), is also discussed. ● Chapters 15 and 16 focus on water at or near the land surface, in lakes, rivers, and as groundwater, and the role of flowing water in creating landforms through erosion and deposition; a discussion of lake hydrology is also included. ● Chapter 17 covers the two landform-creating agents powered not by gravity, but by the action of wind and waves. It focuses on the geomorphic features of desert regions and coastlines. ● Chapter 18 concludes Part 4 by examining landforms created by glacial ice, both as high mountain glaciers and as vast continental ice sheets. It also discusses the distinctive landscapes created in regions of permafrost where ground ice persists year round.

CHAPTER 14

ROCK STRUCTURE, WEATHERING, AND MASS WASTING

Talus cones in the Canadian Rockies Frost action has caused these steeply inclined limestone strata to shed angular rock fragments that accumulate as talus cones.

Denudation wears down all rock masses exposed at the land surface, with the rate of removal depending on the type of rock involved. Generally, weaker rocks will underlie valleys, and the most resistant rocks will form hills, ridges, and uplands. In this way the pattern of landforms on a landscape can reveal the type of rock, as well as the general geological structure that is beneath. Repeated episodes of uplift followed by periods of denudation can bring to the surface rocks and structures that formed deep within the crust. In this way, even ancient mountain roots may eventually appear at the surface and be subjected to denudation. The subsequent removal of overlying rock layers can also expose batholiths and other igneous rock structures produced by magmas that cooled at depth.

This chapter discusses the sequential landforms that arise on various rock structures, and describes the weathering and mass wasting processes that modify them. Once rock is exposed at the surface, it is subject to mineral alteration (see Chapter 11). This is an essential part of the cycle of rock transformation and results in the disintegration of the rock and the production of **sediment**. This chapter also discusses the features produced by weathering and how the resulting particles move downhill under the force of gravity, a process called *mass wasting*.

ROCK STRUCTURE AS A LANDFORM CONTROL

As denudation progresses, landscape features develop according to patterns of **bedrock** composition and structure. Under similar climatic and topographic conditions, rock denudation rates are generally related to rock type. The rate of removal is generally fastest for sedimentary rock, and is progressively slower for metamorphic and igneous types. Denudation rates are related to rock hardness, but other lithological factors, such as fracturing, jointing, and permeability are also important. Measurements of denudation rates are based on the concentration of cosmogenic radionuclides that form in rocks and soils through continual exposure to cosmic radiation; beryllium (^{10}Be) is most commonly used for this purpose. For sedimentary rocks, ^{10}Be analysis suggests a rate of denudation in the order of 18.5 m per million years, compared to 12.6 m per million years for metamorphic rocks and 8.3 m per million years for igneous rocks.

Igneous rocks, because of their comparatively low denudation rates, are commonly represented in upland regions. Metamorphic rocks are also generally resistant to denudation. Shale is a weak rock that erodes easily and is often associated with lower terrain. Sedimentary rocks, particularly sandstones with high quartz content, are typically resistant to denudation and form ridges or uplands. Based on ^{10}Be analysis, the rate of denudation for quartz is about 2.3 m per million years. However, limestone, subjected to solution by carbonic acid in rainwater and surface water, erodes more rapidly, and so would be more likely to form valleys in humid climates. In arid climates, however, limestone is a resistant rock and often forms ridges and cliffs.

Although different landforms can be expected to develop on specific rock types, rock structure, as determined by tectonic activity, is also important. Geologic structures include horizontal strata, gently sloping strata, and strata that are folded or warped into domes, as well as rocks that have been fractured into fault blocks or affected by igneous intrusions or volcanic activity (Figure 14.1). These various structures greatly influence how different rock types are exposed to denudational agents. *Geographer's Tools • Remote Sensing of Rock Structures* provides satellite images of landforms associated with distinctive rock structures.

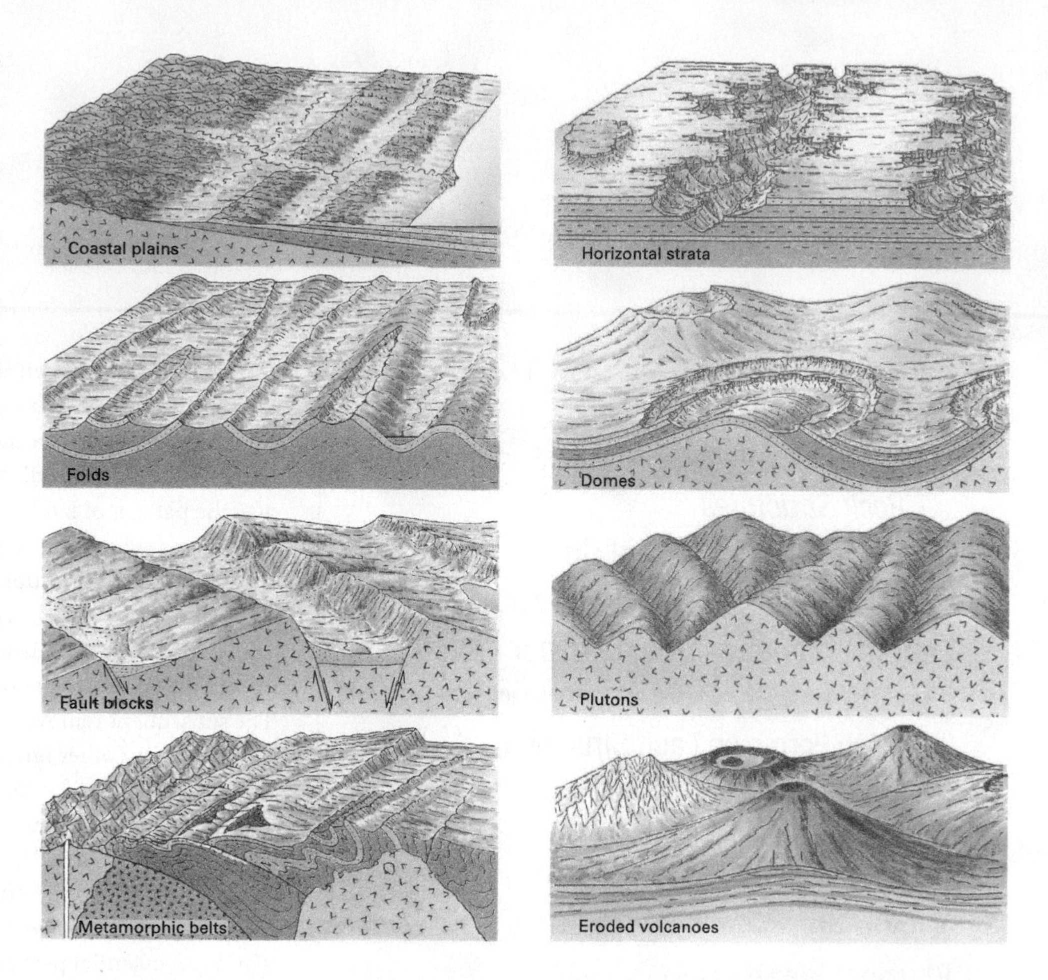

14.1 LANDFORMS ASSOCIATED WITH VARIOUS ROCK STRUCTURES The rugged terrain associated with sedimentary rock strata that has been bent by tectonic forces is very different from the subdued landscapes formed on horizontally bedded rock masses. Igneous intrusions and metamorphic belts also produce distinctive landforms. (Drawn by A. N. Strahler.)

STRIKE AND DIP

Most rock layers are not flat, but rather are tilted at an angle as is readily seen in the bedding layers of many sedimentary rock strata (Figure 14.2). The tilt and orientation of a rock layer is measured against a horizontal plane. For example, Figure 14.3 shows a water surface resting against tilted sedimentary strata. The angle formed between the rock and the horizontal plane is termed the **dip**. The amount of dip is stated in degrees, ranging from 0° for a horizontal rock plane to 90° when the rock plane is vertical. In this example, the dip is 65°. The line of intersection between the inclined rock plane and a horizontal plane gives the **strike** of the rock, which by convention is always

14.2 FOLDED ROCK STRATA A light snow cover enhances the tilted rock strata of this anticlinal structure in the Front Range of the Canadian Rockies near Banff, Alberta.

GEOGRAPHER'S TOOLS
REMOTE SENSING OF ROCK STRUCTURES

With its 30 × 30 m spatial resolution and 175 × 175 km image size, the Landsat satellite provides a perspective from space that captures a large region in a single view. This scale is ideal for revealing the relationship between landforms and rock structure.

Figure 1 shows the ridge-and-valley country of south-central Pennsylvania in a colour-infrared image. The land surface has zigzag ridges formed by bands of hard quartzite. The strata were crumpled into folds during a continental collision that took place more than 200 million years ago to produce the Appalachian Mountains. In this image, the red colour depicts natural vegetation, while the blue colours identify mainly agricultural fields. Condensation trails from passing aircraft formed the cloud bands.

Kauai (Figure 2) is the oldest and most heavily dissected of the shield volcanoes that make up the island chain of Hawaii. The radial drainage pattern of streams and ridge crests leading away from the central summit is a primary feature of this image. The intense red colours of the northern and western slopes identify lush vegetation. The west side of the island is much drier; in this colour infrared image, the bright green corresponds to exposed rocks.

The Brandberg Massif (Figure 3), a dome-shaped plateau in Namibia, covers an area of approximately 650 km^2. This granitic pluton is about 120 million years old and rises 1,900 m above the surrounding arid plains, with its summit at 2,573 m above sea level. The steep slopes are devoid of soil and much of the surface consists of relatively unbroken granite sheets that radiate from the centre.

The Manicouagan Impact Structure (Figure 4) in Quebec originated some 214 million years ago from the collision of an asteroid about 5 km in diameter. The original diameter of the crater was about 100 km, but scouring by subsequent glacial activity has greatly altered its form. The central area is composed of igneous and metamorphic rocks. Evidence of impact comes from shatter-cones—conical fracture surfaces ranging in size from a few centimetres to over a metre—and the presence of glass-like maskelynite produced from altered feldspars. The unusual annular lake was formed following construction of a hydroelectric dam.

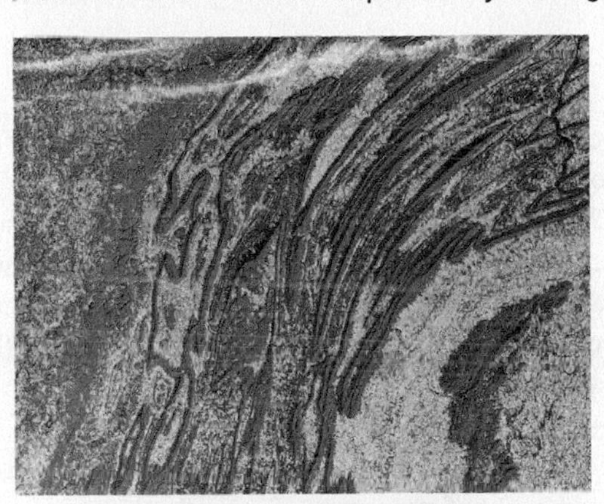

Figure 1 Ridge-and-valley landscape

Figure 2 Kauai, Hawaii

Figure 3 The Brandberg Massif in Namibia

Figure 4 The Manicouagan Impact Structure region in Quebec This image was acquired on June 1, 2001, by the Multi-Imaging SpectroRadiometer (MISR) nadir (vertical-viewing) sensor aboard NASA's Terra platform.

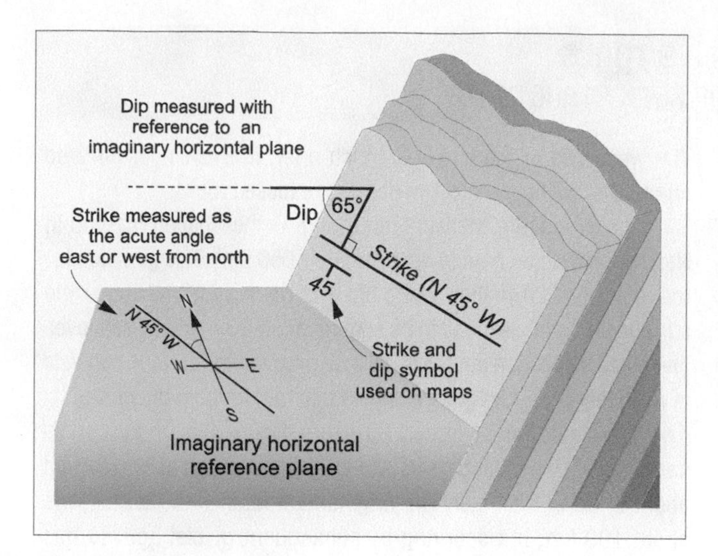

14.3 STRIKE AND DIP In reality, there is rarely a convenient water surface available. Therefore, strike is defined as the compass direction, relative to north, of the line formed by the intersection of a rock layer with an imaginary horizontal plane.

14.4 MONUMENT VALLEY Mesas and buttes have developed in the shale, siltstone, and sandstone in Monument Valley on the borders of Utah and Arizona. The extensive Colorado plateau that formerly occupied this region has been dissected into a series of mesas and buttes. Rockfalls accumulate at the base of the cliffs. Much of the eroded sediment has been carried away by water and wind, leaving a thin sandy soil that supports a sparse cover of mainly coarse grasses and sagebrush. ***Fly By: 37° 03′ N; 110° 06′ W***

measured as an acute angle to the east or west of north. The compass quadrant bearing method is often used. In this example, using this method, the strike trending from northeast to southwest is given as N 45° W.

LANDFORMS OF HORIZONTAL STRATA

Extensive areas of the ancient continental shields are covered by thick sequences of sedimentary rock that were deposited in shallow inland seas at various times over the past 600 million years, since the Precambrian Era. Uplifted with little disturbance other than minor crustal warping or faulting, these areas became continental surfaces underlain with horizontal strata. Subsequent denudation of horizontal strata under different climates gave rise to a series of distinctive landforms.

In arid climates, the normal sequence of landform development on horizontal strata is for a sheer rock wall, or *cliff*, to form at the edge of a resistant rock layer. At the base of the cliff is an inclined slope, which flattens out into a plain beyond. Erosion strips away successive rock layers, leaving a broad platform capped by hard rock layers. Such a platform is usually called a **plateau**. Cliffs retreat as near-vertical surfaces because storm runoff and channel erosion rapidly wash away the weak clay or shale formations exposed at the cliff base. When undermined, the rock in the upper cliff face repeatedly breaks away along vertical joints and fractures.

Cliff retreat produces a **mesa**, essentially a small plateau bordered on all sides by steep rock faces. Mesas represent the remnants of a formerly extensive layer of resistant rock. Continued retreat of the surrounding cliffs reduces the area of a mesa, but it retains its flat top. Eventually, it becomes a small steep-sided feature known as a **butte** (Figure 14.4). Resistant rock layers, outcropping at different levels, characteristically produce a series of steps on the flanks. Further erosion may produce a single, tall column before the landform is totally consumed.

LANDFORMS OF FOLDED ROCK STRATA

Areas underlain by gently upwarped or downwarped rock layers form domes and basins. More intense folding produces the wave-like strata of anticlines and synclines. A **sedimentary dome** is a circular or oval structure in which strata have been forced upward at its centre (Figure 14.5). Sedimentary domes can occur where igneous intrusions at great depth have caused the overlying sediments to bow upward. In other cases, upthrusting on deep faults may have been the cause. When a sedimentary dome wears away,

14.5 DOME EROSION Erosion of sedimentary strata from the summit of the Richat Structure in the Sahara Desert of Mauritania has produced a series of concentric, outward dipping cuestas. The partially eroded sandstones and limestones form a sawtooth pattern of massive triangular slabs generally referred to as flatirons.

strata are usually initially removed from the summit region, where greater bending has weakened the rock; this exposes the older underlying strata. The eroded edges of the dipping strata form sharp-crested sawtooth ridges called *hogbacks*.

Deep erosion of simple folds produces a **ridge-and-valley landscape**. Weaker rocks, such as shale and limestone, erode away. This leaves hard strata, such as sandstone or quartzite, to stand in bold relief as long, narrow ridges. Initially, the uparched anticlines will form ridges and the downwarped synclines will form valleys. However, the geometry of the folds will contribute to the differential resistance of the rock mass; strata in a syncline tend to be compressed, while in an anticline, extensional forces may weaken and fracture the rock. Where rock was weakened as it was bent upward, it can erode away to form an *anticlinal valley* (Figure 14.6a). Conversely, a *synclinal mountain* can form where the rock has been strengthened by compression. These ridge-like forms develop when a resistant rock type is exposed at the centre of a syncline and the surrounding weaker rock is stripped away (Figure 14.6b).

Simple folds are continuous and even-crested, hence they produce ridges and valleys that are approximately parallel and continue for great distances. In some fold regions, however, the axis of the folds is not parallel to the surface, so they rise or descend to form *plunging anticlines* and *plunging synclines*. Plunging folds give rise to a zigzag line of ridges (Figure 14.7).

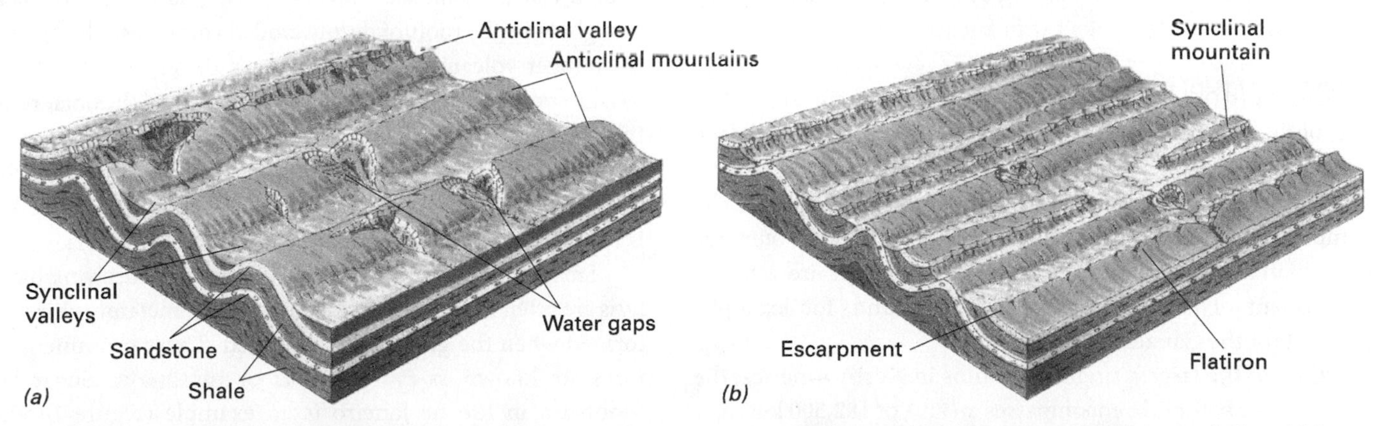

14.6 STAGES IN THE EROSIONAL DEVELOPMENT OF FOLDED STRATA (a) Erosion exposes a highly resistant layer of sandstone or quartzite, which controls much of the ridge-and-valley landscape. (b) Continued erosion partly removes the resistant formation, but reveals another below it, adding to the complexity of the the system of parallel ridges, valleys, and escarpments. (Drawn by A. N. Strahler.)

14.7 PLUNGING FOLDS Folded strata with crests that plunge downward give rise to zigzag ridges following erosion, as illustrated by Sheep Mountain near Lovell, Wyoming. The ridge in the centre of the photograph is a plunging anticline.
Fly By: 44° 39′ N; 108° 12′ W

14.8 EROSION OF AN UPLIFTED FAULT BLOCK The near-vertical face of Mount Yamnuska in Alberta marks a fault scarp along which the rock has been uplifted.

LANDFORMS DEVELOPED ON OTHER GEOLOGICAL STRUCTURES

EROSION FORMS ON FAULT STRUCTURES

Active normal faulting produces a sharp surface break—a *fault scarp*—that can create a rock cliff hundreds of metres high (Figure 14.8). Erosion quickly modifies a fault scarp, but because the fault plane extends down into the bedrock, its effect on erosional landforms persists over a long geologic time span. A fault scarp, produced by movement along a fracture in the Earth's crust, is distinct from a *fault-line scarp*, which is formed by differential erosion of rocks weakened by faulting, or where faulting brings softer rocks in contact with harder rocks.

EXPOSED BATHOLITHS

Batholiths, huge intrusions of igneous rock that are formed deep below the Earth's surface, are sometimes uncovered by erosion of the overlying rock materials. Because batholiths are typically composed of resistant rock, they erode into hilly or mountainous uplands. Batholiths of granitic composition are a major component of ancient shields and are found, for example, throughout the Canadian Shield.

One of the largest single batholiths in North America, the Coast Range batholith, encompasses an area of 182,500 km^2 that extends from northwestern Washington State to the Yukon. The bedrock is mainly granite that was intruded in the Middle Jurassic and early Tertiary Periods, 170 to 45 million years ago. Between the intrusions are zones of metamorphosed, folded, and faulted volcanic and sedimentary rock. The granitic rocks are thought to be the roots of deeply eroded volcanoes. The presence of younger volcanoes, such as Mount Baker in northwestern Washington and Mount Garibaldi in British Columbia, reflect the volcanic origin of the landscape. Batholiths are typically associated with rich mineral deposits. For example, the Guichon Creek batholith near Kamloops, although only about 1,000 km^2, is the principal copper ore reserve in British Columbia.

Small bodies of granite projecting from underlying batholiths are often found surrounded by ancient metamorphic rocks formed when the granite was intruded. These prominent features are known as **monadnocks** or **inselbergs**. Sugar Loaf Mountain in Rio de Janeiro is an example (Figure 14.9). A monadnock develops because it is more resistant to erosion than the bedrock of the surrounding area. The distinctive rounded form develops as the outermost layers of rock break apart and slide off to expose the new layer within (see Weathering below).

14.9 EXFOLIATION DOME Exfoliation domes, such as Sugar Loaf Mountain, Brazil, form when sheets of granite are fractured by expansion as confining pressure is relieved through removal of overlying rock. The loosened slabs fall away to form an apron around the base of the dome.

14.10 VOLCANIC NECK Shiprock in New Mexico is a volcanic neck enclosed by a weak shale formation. The peak rises about 520 m above the surrounding plain. In the background and to the right, wall-like dikes extend far out from the central peak.

DEEPLY ERODED VOLCANOES

The distinctive initial landform produced by a stratovolcano is the volcanic cone. When the volcanic episode ends and the volcano becomes dormant or extinct, a lake usually occupies the caldera and the cone is dissected by streams. With continued erosion, an extinct stratovolcano may gradually be reduced to a remnant *volcanic neck* formed of lava that solidified in the pipe of the volcano. Radiating from it are wall-like dikes created from magma that filled radial fractures around the ancient volcano. Shiprock, New Mexico, is a classic example of a volcanic neck with radial dikes (Figure 14.10). Shiprock rises 520 m above the desert plain. It is mainly composed of volcanic breccia that formed about 1,000 m below the surface; six dikes radiate out from the neck. They are composed of alkali feldspar, biotite, and pyroxenes (see Chapter 11). Radiometric dating indicates that the dikes formed about 27 million years ago.

A similar landform of volcanic origin is Devils Tower, Wyoming, which rises 382 m above the sedimentary rocks that surround it (Figure 14.11). This steep-sided igneous feature may also be an erosional remnant of a volcanic neck, although its exact origin is subject to debate. Like Shiprock, it is made of magma that solidified at a shallow depth below the surface; it is dated at 40 million years. The resistant, fine-grained extrusive rock is rich in alkali-feldspar. The distinctive, predominantly hexagonal columns formed as the rock cooled and contracted.

14.11 DEVILS TOWER, WYOMING This igneous remnant is formed of fine-grained rock that cooled and contracted to give a distinctive columnar structure. It rises 382 m above the surrounding plain.
Fly By: 44° 5′24′ N; 104° 42′55′ W

Shield volcanoes, such as those in Hawaii, show erosion features that are quite different from those of stratovolcanoes. In the humid climate of Hawaii, streams have cut deep, narrow valleys that open out into steep-walled amphitheatres. Over time, the original surface of the shield volcano will be completely dissected, leaving a rugged mountain mass composed of sharp-crested divides and canyons (see *Geographer's Tools • Remote Sensing of Rock Structures*).

IMPACT STRUCTURES

An interesting group of landscape features are those that have originated from the impact of meteors or other extraterrestrial bodies. One of the best examples is the Barringer Meteorite Crater, also known as Meteor Crater, near Flagstaff, Arizona (Figure 14.12). The crater is 1.2 km in diameter and 200 m deep. It was formed by the impact of a nickel-iron meteorite approximately 49,000 years ago. The typical impact speed of a meteorite, about 50,000 kph, produces a crater that is 10 to 20 times larger than the meteorite's diameter. Most craters are bowl-shaped depressions around which the ejected material is deposited. Rocks beneath a large crater are significantly altered, and usually consist of a shallow layer of breccia made up of coarse, angular fragments of the original rocks. The impact fractures the bedrock to great depths. Metamorphism through impact melting also occurs.

About 200 impact structures have been identified on Earth, 29 of which are in Canada (Figure 14.13). The largest on record is at Sudbury, Ontario, with an approximate diameter of 250 km. This impact is dated at 1,850 million years. The Manicouagan impact site (see *Geographer's Tools • Remote Sensing of Rock Structures*) is about 100 km in diameter and formed about 214 million years ago. Another large crater formed about 65 million years ago near Chicxulub, Mexico, from the impact of a meteorite that is estimated to have been about 10 km in diameter. At the time of its formation, the crater measured nearly 100 km in diameter and was 14 km deep. However, subsequent collapse of the perimeter wall increased the crater's diameter to more than 150 km. It is possible that this impact event caused a significant change in Earth's climate and is perhaps linked to the extinction of the dinosaurs.

14.12 BARRINGER METEORITE CRATER NEAR FLAGSTAFF, ARIZONA The nickel-iron meteorite that created this crater is estimated to have been 50 m in diameter and was travelling at about 46,800 kph. Its impact weight was about 150,000 tonnes. Some fragments of the meteorite have been recovered, but most of it appears to have vaporized on impact.

14.13 EARTH IMPACT SITES The distribution of known impact sites comes from a program that began in 1955 at the Dominion Observatory in Ottawa. The Planetary and Space Science Centre at the University of New Brunswick currently manages the program.

WEATHERING

Weathering is the general term applied to the combined action of all processes that cause rock to disintegrate physically and decompose chemically due to conditions at or near the Earth's surface. Two general types of weathering are recognized. In *physical weathering*, rocks are fractured and broken apart without chemical alteration due to processes such as frost action expansion and contraction caused by changes in temperature, and pressure from roots. In *chemical weathering*, rock minerals are transformed from those that were stable when the rocks were initially formed into those that are stable under temperature, pressure, and moisture conditions found at the Earth's surface. Chemical weathering processes include oxidation, hydrolysis, and solution (see Chapter 11). Weathering leads to the production of **regolith**—a surface layer of rock particles that lies above solid, unaltered rock. Weathering, in conjunction with gravity, also creates a number of distinctive landforms.

PHYSICAL WEATHERING

Physical weathering, also known as *mechanical weathering*, produces regolith by the action of forces strong enough to fracture the rock. Fracturing occurs when stress is exerted along zones of weakness in the rock. The size and shape of the residual fragments is ultimately determined by the macro and microstructure of the rock itself. Stresses sufficient to cause rock to fracture arise in various ways, including frost action, salt-crystal growth, unloading, and wedging by plant roots.

FROST ACTION

One of the most important physical weathering processes in cold climates is *frost action* caused by the repeated growth and melting of ice crystals in pore spaces and cracks in a rock. In contrast to most other substances, water expands when it freezes, resulting in an increase in volume of about 10 percent. Over the course of many freeze–thaw cycles, such expansion can fragment even extremely hard rocks. Frost action and ice-crystal growth produce a number of conspicuous effects and several distinctive landforms in all cold winter climates. Features caused by frost action and the buildup of ice below the surface are particularly visible in the tundra climate and above the timberline in high mountains (see Chapter 18).

Almost all bedrock is permeated by systems of fractures called *joints*, which often develop when rock that has been subjected to heat and pressure subsequently cools and contracts. Joints generally occur in a series of parallel and intersecting planes, creating surfaces of weakness along which weathering processes can act to separate the rock into blocks. Since there is no relative movement of rock along the joints, they cannot be considered faults. Joints are important to the physical weathering of rocks because they allow water to penetrate the rock, which can then exert force through freeze–thaw activity and salt-crystal growth. The characteristic layering or stratification in sedimentary rocks develops through the accumulation of successive layers of sediment in contact along bedding planes. Each bedding plane constitutes another natural route for the potential entry of water. Bedding planes and strata are often cut at right angles by sets of joints, and the process of *block separation* can loosen the resulting blocks when comparatively weak stresses are applied (Figure 14.14).

Much stronger forces are needed to create fractures in solid igneous or metamorphic rock, so water may only be able to penetrate the contact surfaces between exposed mineral grains. Here, the water can freeze and exert forces strong enough to separate the grains. This form of breakup is called *granular disintegration* (see Figure 14.14). The end product is a fine gravel or coarse sand. The effects of frost action can be seen in all climates that have a winter season, but it is most effective in areas where temperatures fluctuate above and below freezing many times each year.

14.14 BEDROCK DISINTEGRATION Joint-block separation and granular disintegration are two common forms of bedrock disintegration. (Drawn by A. N. Strahler.)

In high mountains, frost action on barren cliffs detaches rock fragments that fall to the base of the cliff. These loose fragments, known as *talus*, accumulate to form *talus slopes*. Where the cliffs are notched by narrow ravines that funnel the rock fragments into separate tracks, each chute will supply a single *talus cone* so a series of cones develop side-by-side along the cliff base (Figure 14.15a). Talus slopes are generally unstable; simply walking across the slope or a large rock fragment dropping from the cliff above can easily set off a sliding or rolling motion within the surface layer of rock fragments. Where bedrock is exposed on knolls and mountain summits away from steep slopes, large angular rock fragments created by frost action can accumulate in a layer that completely blankets the bedrock beneath (Figure 14.15b). These expanses of broken rock are termed blockfields or *felsenmeer* ("rock sea").

In fine-textured soils and sediments composed largely of silt and clay, soil water freezes in layers or lens-shaped bodies. As the ice layers thicken, they heave the overlying soil upward. Prolonged frost heaving can produce minor irregularities and small mounds on the soil surface. A rock fragment at the surface can sometimes conduct heat from the soil to the cold night air, causing perpendicular ice needles to grow beneath it and raise it above the surface (Figure 14.16). A similar process acting on a rock fragment below the soil surface can eventually push the fragment to the surface. Frost action is a dominant process in arctic and high mountain tundra environments, where it is a factor in the formation of a wide variety of unique landforms (see Chapter 18).

(a)

(b)

14.15 FROST ACTION Ice-crystal growth within the joint planes and other zones of weakness in rock can cause it to split apart. (a) Talus cones are common features in mountain environments and are formed where frost-loosened rock fragments accumulate at the base of a slope. (b) Blockfields or felsenmeer develop when rock split by frost action is not transported from the site by gravity or other agents.

SALT-CRYSTAL GROWTH

The weathering of rock by salt crystals is similar to the process of ice-crystal growth. *Salt-crystal growth* operates extensively in dry climates and is responsible for many of the niches, shallow caves, rock arches, and pits seen in sandstone formations. During long drought periods, *capillary action* moves groundwater to the surface of the rock. The water is drawn upward through fine openings and passages in the rock by surface tension. As water evaporates from the porous outer zone of the sandstone, tiny crystals of salts, such as halite (NaCl), calcite ($CaCO_3$), or gypsum ($CaSO_4{\cdot}2H_20$), are left behind. Over time, the force created by the growth of these crystals results in grain-by-grain breakup of the sandstone, which crumbles into sand and is swept away by wind and rain.

14.16 ICE NEEDLES The formation of ice needles in soils and fine sediments can dislodge small particles and push stones to the surface. The process is particularly common in subarctic regions and gives rise to a variety of patterned ground features.

Rock lying close to bedding planes or the base of a cliff is especially susceptible to breakup by salt-crystal growth because here the groundwater seeps downward and outward to reach the rock surface (Figure 14.17). Salt-crystal growth occurs naturally in arid and semi-arid regions. In humid climates, abundant rainfall dissolves the salts and carries them away, rendering them ineffectual as a weathering agent.

UNLOADING

Unloading is a process that relieves the confining pressure on a rock mass. It occurs when the removal of overlying material brings a rock mass nearer to the surface. Rock formed at great depth beneath the Earth's surface is in a compressed state because of the pressure created by the mass of material above it. As this is slowly worn away, the pressure decreases, allowing the rock beneath to expand slightly. This causes the rock to crack in layers that are more or less parallel to the surface,

14.17 **NICHE FORMATION AT CANYON DE CHELLY, NEW MEXICO** In dry climates, water seeps slowly from bedding planes and the bases of cliffs. Salt-crystal growth separates the grains of permeable sandstone, breaking them loose and creating niches.

creating a type of jointing called *sheeting structure*. In massive rocks like granite or marble, thick curved layers or shells of rock successively break free from the parent mass below through the process of **exfoliation**.

If a sheeting structure forms over the top of a single large mass of rock, it produces a rounded *exfoliation dome* from which the rock peels away in layers several metres thick (Figure 14.18). The rounded shape is best maintained where fractures are not present within the exposed rock mass, allowing weathering processes to act uniformly on its surface. Rock breakup is mainly due to the reduction in pressure that occurs as overlying rock is removed. This causes the granite to expand and fracture into large sheets that eventually fall away. Exfoliation domes are among the largest of the landforms shaped primarily by weathering. They form the spectacular scenery of Yosemite Valley, California, and Sugar Loaf Mountain, Brazil (see Figure 14.9).

14.18 An exfoliation dome is formed by the shedding of slabs of rock from the surface of a non-stratified rock outcrop through the release of confining pressure as overlying material is removed. This example is Enchanted Rock in Texas.

Well-known examples in Canada can be seen in Stawamus Chief Provincial Park near Squamish, British Columbia (see Figure 11.7). Smaller intrusions may be reduced to a collection of granite boulders known as a **kopje**. They are especially common in arid regions, such as southern Africa and Australia. Similar rock outcrops formed by weathering of hill summits in humid climates, such as in Britain, are called *tors* (see the discussion of hydrolysis).

OTHER PHYSICAL WEATHERING PROCESSES

Most rock-forming minerals expand when heated and contract when cooled under normal diurnal temperature cycles. This can create powerful disruptive forces on rocks in deserts and other areas where the surface is subjected to intense daytime heating alternating with cool nights. Daily temperature changes can cause the breakup of a surface layer of rock already weakened by other agents of weathering.

Another mechanism of rock breakup is the growth of plant roots, which can wedge joint blocks apart (Figure 14.19). Even fine rootlets can cause the loosening of small rock fragments and mineral grains. However, the primary role of plants in weathering is biochemical rather than biophysical, and is associated mainly with acidic exudates from mosses and lichens that colonize bare rock surfaces. As well, these small plants enable a film of water to be retained on the surface, which might then be subject to freeze–thaw cycles, or more importantly, initiate chemical alteration. Microorganisms have also been linked to weathering through

14.19 Roots exert a strong wedging action as they permeate cracks and crevices in rock, which can eventually prise it apart.

their ability to oxidize iron and sulphur under anaerobic conditions, converting it to secondary sulphates that expand into rock fractures.

CHEMICAL WEATHERING AND ITS LANDFORMS

Chemical weathering is a process of mineral alteration (see Chapter 11). The dominant processes of chemical change affecting silicate minerals are oxidation, hydrolysis, and carbonation. Oxidation and hydrolysis change the chemical structure of primary minerals, turning them into secondary minerals that are typically softer and bulkier and therefore more susceptible to erosion and mass movement. Carbonation results from the dissolving action of carbonic acid, causing minerals to be removed in solution.

Chemical reactions proceed more rapidly at higher temperatures; water also plays an important role in chemical weathering. Consequently, chemical weathering is most effective in warm, moist climates. Air quality is also a factor, and in many parts of the world industrial emissions have caused pronounced changes in regional precipitation chemistry. Precipitation chemistry has been monitored in Canada since 1972, with as many as 28 sites operating as part of the CAPMoN (Canadian Air and Precipitation Monitoring Network) program. CAPMoN monitors non-urban air quality. The sites are chosen to be regionally representative of areas that are beyond the effect of local sources of air pollution. The majority are in eastern Canada, where transboundary pollution is a concern. In urban areas, air is commonly polluted by sulphur and nitrogen oxides. When these gases dissolve in rainwater, the result is *acid precipitation*. Acid rain and snow can dissolve limestone and chemically weather other types of building stone more rapidly than natural precipitation. The result can be damaging to buildings, sculptures, and tombstones.

HYDROLYSIS AND OXIDATION

Decomposition by hydrolysis and oxidation changes rock-forming minerals into clay minerals and oxides. In the warm, humid climates of the equatorial, tropical, and subtropical zones, prolonged hydrolysis and oxidation has resulted in the decay of igneous and metamorphic rocks to depths of up to 100 m. The resulting clay-rich material is soft and erodes easily.

In dry climates, hydrolysis weathers exposed granite to produce many interesting boulder and pinnacle forms. Although rainfall is infrequent, water penetrates the granite along the interfaces between crystals of quartz and feldspar. Chemical weathering of these surfaces then breaks individual crystal grains away from the main mass of rock, leaving rounded forms (Figure 14.20a). The grain-by-grain breakup forms a fine desert gravel consisting largely of quartz and partially decomposed feldspar crystals.

Tors characteristically develop in humid, temperate climates where hydrolysis acts upon jointed granite masses. In these environments, it is generally thought that weathering is initiated below ground. Water permeating the joints causes the breakdown of feldspars into clays. Organic acids leached from decomposing plant matter may contribute to the process leading to the progressive widening of the joints. Frost action can also help to shatter the rock, which is slowly exhumed as overlying material is stripped to expose the fractured rock masses (Figure 14.20b).

CARBONATION AND SOLUTION

Chemical weathering through *acid action* is mainly associated with *carbonic acid* (H_2CO_3) formed when CO_2 dissolves in water. Rainwater, soil water, and stream water all normally contain dissolved CO_2. Whereas rainwater acquires CO_2 from the atmosphere, percolating water is primarily acidified as it moves through the soil and reacts with CO_2 generated by decomposition of organic material. Carbonic acid slowly interacts with feldspars and other types of minerals to form carbonates; this process of **carbonation** is an important intermediate stage in the weathering of igneous rocks. Carbonate sedimentary rocks,

(a)

(b)

14.20 (a) **Chemical weathering in igneous rocks by hydrolysis** The Devils Marbles, a collection of rounded granitic boulders in northern Australia, was produced by the spheroidal weathering of silicate minerals through replacement of potassium and other minerals by the hydrogen (H^+) and hydroxyl (OH^-) ions of water as water seeped through the zones of weakness in the rock mass. (b) **Weathering of an exhumed, granite intrusion: Haytor, Dartmoor, United Kingdom** Weathering continues to enlarge the joints in the rock mass. In this moist, temperate climate, the process of joint enlargement begins while the rock mass is below the surface.

such as limestone and marble, are particularly susceptible to acid action. In this case, the mineral calcium carbonate ($CaCO_3$) dissolves and is carried away in **solution** in stream water.

Carbonic acid reaction with limestone produces many interesting surface forms, mostly of small dimensions. Outcrops of limestone often show cupping, grooving, and fluting in intricate designs. In some places deep grooves and high wall-like rock fins are produced (Figure 14.21). The solution of limestone is particularly rapid in moist tropical climates and can produce an intricate landscape of spires that

(a)

(b)

14.21 CHEMICAL WEATHERING ON A LIMESTONE OUTCROP NEAR PORT ALBERNI, BRITISH COLUMBIA (a) Cupping and grooving produces intricate patterns on exposed limestone surfaces. (b) Vertical grooves produced by the dissolving action of water on a limestone cliff face in British Columbia.

14.22 **DIFFERENTIAL RATES OF WEATHERING OF LIMESTONE** (a) Grooved surface produced by joint enlargement on a limestone pavement in the moist, temperate climate of North Yorkshire, UK. (b) A "forest" of limestone pinnacles, many of which are over 100 m tall, in the humid tropical climate of Tsingy de Bemaraha, a UNESCO World Heritage Site in Madagascar. Mean annual rainfall here is about 1,500 mm with an average temperature of 28 °C, which favours the accelerated limestone solution.

are often more than 100 m in height (Figure 14.22). Carbonic acid in groundwater can also dissolve limestone to produce underground caverns.

LIMESTONE SOLUTION BY GROUNDWATER

In moist climates, the slow flow of groundwater in the saturated zone can dissolve limestone below the surface, producing large underground caverns. Some caverns collapse, causing the ground above to sink and a unique type of **karst** landscape to develop.

Limestone caverns are interconnected subterranean cavities in bedrock formed by the corrosive action of circulating groundwater on limestone (Figure 14.23). In stage 1, the action of carbonic acid is concentrated in the saturated zone just below the water table. The solution process forms many kinds of underground landforms, such as tunnels, chambers, and vertical chimneys. Subterranean streams carrying the products of solution flow in the lowest tunnels, eventually emerge and join with surface streams and rivers.

In stage 2, the stream valley deepens and the water table is lowered to a new position. The cavern system formed previously is now in the unsaturated zone, and a new series of caverns develops deeper below the surface. Where percolating water can evaporate, the dissolved carbonates are deposited as *travertine* on exposed rock surfaces in the caverns. Encrustations of travertine take many forms: stalactites (hanging rods), stalagmites (upward pointing masses), columns, and drip curtains (Figure 14.24).

Where limestone solution is very active, a karst landscape with many unusual landforms develops. Karst topography refers to any limestone area where sinkholes occur and small surface streams are non-existent. A *sinkhole* is a surface depression in a region of cavernous limestone (Figure 14.25). Some sinkholes

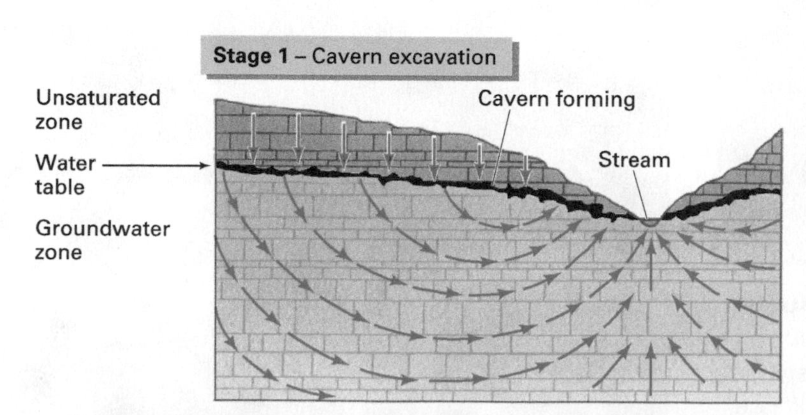

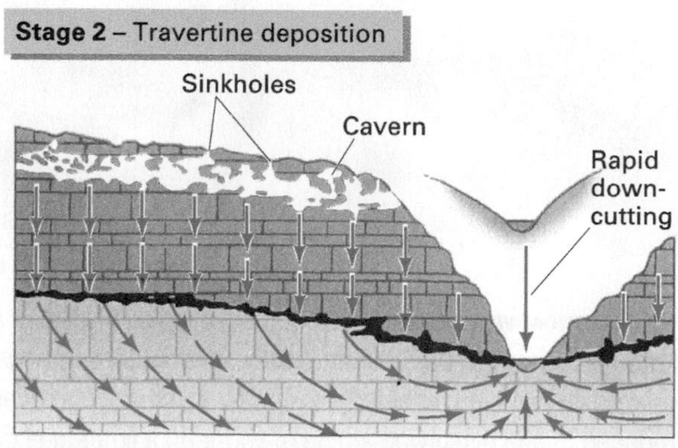

14.23 **CAVERN DEVELOPMENT** Limestone dissolves at the top of the groundwater zone. When rapid erosion of streams lowers the water table, caverns result in the unsaturated zone. (Copyright © A. N. Strahler.)

14.24 TRAVERTINE DEPOSITS IN A CAVERN Travertine deposits in the Papoose Room of Carlsbad Caverns in New Mexico. Deposits include stalactites (slender rods hanging from the ceiling), stalagmites (upward pointing rods), and sturdy columns, all formed when dripping groundwater evaporates, leaving travertine deposits of calcium carbonate behind.

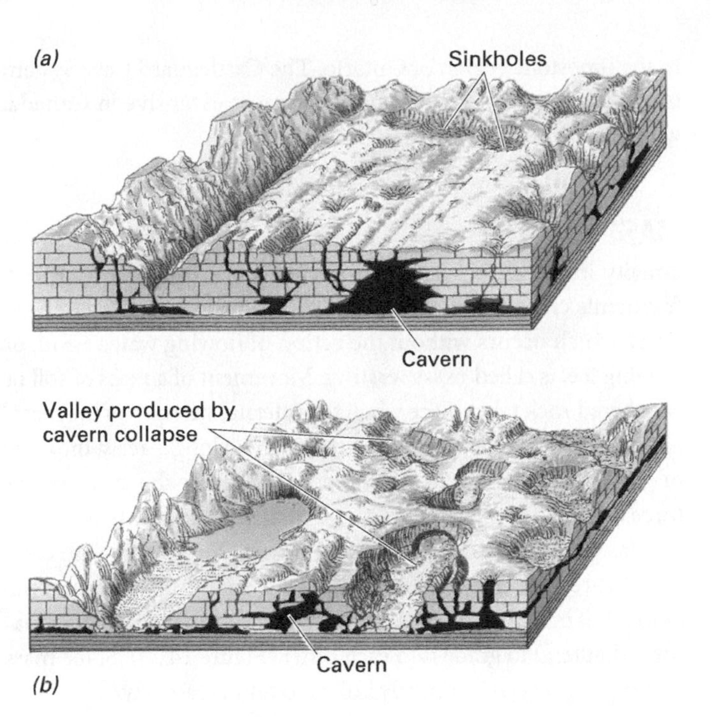

14.26 EVOLUTION OF A KARST LANDSCAPE (a) Rainfall enters the cavern system through sinkholes in the limestone. (b) Extensive collapse of caverns reveals surface streams flowing on shale beds beneath the limestone. Some parts of the flat-floored valleys can be cultivated. (Drawn by Erwin Raisz. Copyright © A. N. Strahler.)

are filled with soil washed from nearby hillsides, while others are steep-sided, deep holes. They develop where the limestone is more susceptible to weathering, or where an underground cavern near the surface has collapsed.

The early stage of karst landscape development is characterized by numerous funnel-like sinkholes (Figure 14.26). Later, the caverns collapse, leaving open, flat-floored valleys. Important regions of karst or karst-like topography are the Dalmatian coast of Croatia, the Mammoth Cave region of Kentucky, the Yucatan Peninsula, and parts of Cuba and Puerto Rico. In southern China and west Malaysia, the karst landscape is dominated by steep-sided, conical limestone hills or towers 100 to 500 m high (Figure 14.27). The towers are sometimes capped by beds of more resistant rock or impure limestone that dissolve more slowly. They are often riddled with caverns and passageways. In Canada, the western mountains have numerous caves, while smaller examples exist

14.25 SINKHOLES IN SOUTH ISLAND, NEW ZEALAND

14.27 TOWER KARST Limestone towers near Guilin (Kweilin), Guanxi Province, southern China.

in the limestone region of Ontario. The Castleguard Cave System in Alberta's Rocky Mountains is the most extensive in Canada, with a length of more than 20 km.

MASS WASTING

Gravity induces the spontaneous downslope movement of rock fragments created by weathering. This movement to lower elevations, which occurs without the action of flowing water, wind, or moving ice, is called **mass wasting**. Movement of a mass of soil or weathered rock takes place when the internal strength of the material declines to a critical point where it can no longer resist the force of gravity. The failure of weathered rock under the ever-present force of gravity occurs over a range of time scales from very slow, imperceptible movement to sudden slope failures, often with catastrophic results. The combination of different time scales and the amount of material involved results in a variety of mass wasting features that tend to grade into each other (Figure 14.28). Some mass wasting events can be directly linked to human activity.

Kinds of earth materials:	Rock (dry)	Regolith, soil, alluvium, clays + water	Water + sediment
Physical properties:	Hard, brittle, solid	Plastic substance	Fluid
Kinds of motion:	Falling, rolling, sliding	Flowage within the mass	Fluid flow

Forms of mass movement: Very slow ← - - - → Very fast

ROCK CREEP
TALUS CREEP
SOIL CREEP
SOLIFLUCTION
LANDSLIDES:
BEDROCK SLUMP
ROCKSLIDE
EARTHFLOW (slump or flowage)
MUDFLOW
STREAM FLOW
ROCKFALL
ALPINE DEBRIS AVALANCHE
DEBRIS FLOOD

14.28 PROCESSES AND FORMS OF MASS WASTING

SLOPES

The inclination of a surface can be expressed in three ways. The change in elevation over a given distance provides a ratio of rise over run, such as 1 in 20. Alternatively, slope can be given as the angle measured in degrees with respect to the horizontal; for example, a 1 in 20 gradient corresponds to a slope of 3 degrees. Grade expresses rise over run as a percentage; in this case, 1/20 is equivalent to a 5 percent grade. On a typical hill, slope soil and regolith blanket the bedrock. The thicknesses of these surficial materials are quite variable and depend on the type and calibre of material, as well the gradient of the slope. Although soil rarely extends deeper than 1 or 2 m, residual regolith overlying decayed and fragmented rock can be more than 100 m thick. Soil and regolith may be absent in some places, exposing outcrops of bedrock (Figure 14.29).

Residual regolith derived directly from the rock beneath moves slowly down the slope. Accumulations of regolith at the foot of a slope are called *colluvium*. Beneath the valley floor are layers of regolith, called **alluvium**, which is sediment transported and deposited by streams. The source of alluvium is regolith from hill slopes many kilometres upstream. Colluvium and alluvium are referred to as transported regolith to distinguish them from residual regolith which forms *in situ*.

SOIL CREEP

On most slopes, soil and regolith move extremely slowly downhill through the process of **soil creep**. Evidence of soil creep is shown in a variety of ways (Figure 14.30). In some layered rocks, such as shale or slate, the edges of vertically oriented strata seem

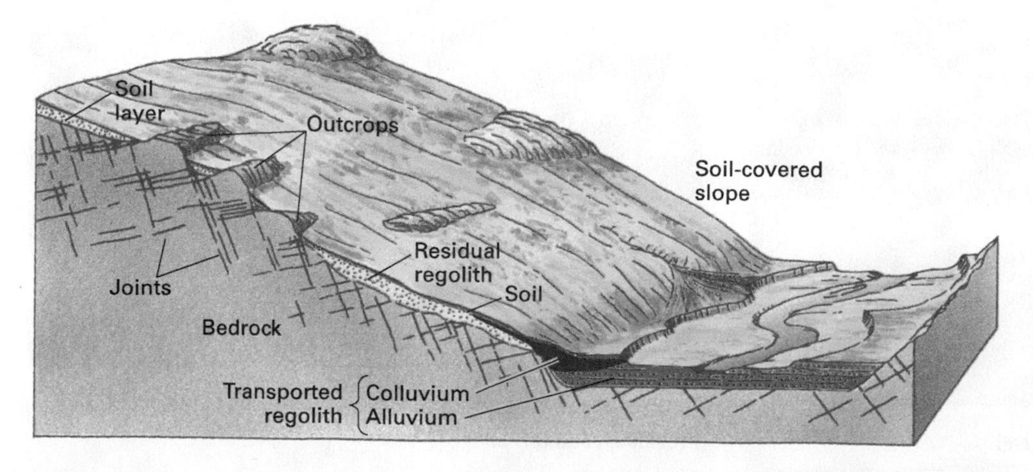

14.29 SOIL, REGOLITH, AND OUTCROPS ON A HILL SLOPE Colluvium accumulates at the foot of the slope, while alluvium lying on a valley floor is transported and deposited by flowing water. (Drawn by A. N. Strahler.)

(a)

(b)

14.30 **INDICATORS OF SOIL CREEP** The slow, downhill creep of soil and regolith appears in many ways on a hillside. (a) Apparent curving occurs in sedimentary strata through displacement of small blocks of rock. (b) Posts become tilted and misaligned.

to "bend" in a downhill direction. This is not true plastic bending, but it is the result of downhill creep of many rock pieces that have separated along small joint cracks. Joint blocks can sometimes be found downslope from the outcrop. Soil creep causes fence posts and utility poles to lean downslope or even shift measurably out of line, and retaining walls can buckle and break.

14.31 **TERRACETTES** The step-like surface on this grassy slope covering a chalk escarpment at Westbury in England is formed by the imperceptible downslope movement of soil particles.

Soil creep occurs through the effect of some process of soil disturbance acting under the influence of gravity. Alternate wetting and drying of the soil, growth of ice needles and ice lenses, heating and cooling of the soil, trampling and burrowing by animals, and shaking by earthquakes all produce some disturbance of the soil and regolith. Because gravity exerts a downhill pull each time the particles are disturbed, they gradually work their way downslope. Small, step-like ridges called **terracettes** are commonly formed through soil creep on grassy slopes (Figure 14.31). Terracettes are generally less than one metre wide, with fairly level treads arranged more or less parallel to each other in curving lines across the slopes.

EARTH FLOWS AND ROTATIONAL SLUMPS

In humid climates, a mass of moist soil and fine regolith may move down a steep slope in the form of an **earth flow** (Figure 14.32a) that may affect a few square metres or cover an area of a few hectares. The rate at which an earth flow develops can be slow or rapid, depending on the texture of the material, moisture content, and angle of slope. Shallow earth flows are common on grass-covered and forested slopes that have been saturated by heavy rains. At the top of a flow, the displaced material leaves a curved, wall-like scarp. Often the soil farther from the scarp moves more rapidly, leaving a series of steps if the soil mass breaks apart as it slips downward. At the bottom, the sodden soil flows in a sluggish mass that piles up in ridges or lobes to form a bulging toe.

A *slump* usually occurs when water percolates deep into a mass of unconsolidated material, resulting in it dropping along a concave slip plane. It involves slow movement along a surface of weakness. Slump units slip over each other as intact blocks

(a)

(b)

14.32 **EARTH FLOWS AND SLUMPS** (a) Earth flows develop in soils and fine regolith on slopes that have been moistened or saturated with water, leaving crescent-shaped scars on the hillside, with small step-like terraces and a hummocky, bulging toe further downslope. (b) Slumping often occurs where an unconsolidated material, usually moistened to some depth by percolating water, is undercut by wave action or flowing water, such as here on the banks of the South Saskatchewan River near Saskatoon.

usually exhibiting some backward rotation. River banks and coastal cliffs are subject to slumping when flowing water or wave action undercuts the base (Figure 14.32b). Slumping also occurs on over-steepened banks in road cuts.

Earth flows often block highways and railroad lines, usually during heavy rains. Generally the flow rate is slow, so the flows are not life-threatening. However, damage to buildings, pavement, and utility lines can be severe where construction has taken place on unstable soil slopes. Slumps are also a hazard to construction, and expensive control structures are needed to try and ameliorate the problem. One special form of earth flow has proved to be a major environmental hazard along the St. Lawrence River and its tributaries in Quebec. The Leda clays of the Ottawa Valley are particularly susceptible. This type of flow involves horizontally layered sediments composed of fine clays and silts that form low, flat-topped terraces adjacent to rivers and lakes. These soft deposits spontaneously turn into a near-liquid state when subjected to a shock or disturbance. As a result, a surface layer 6 to 12 m thick and some 600 to 900 m across begins to move downslope (Figure 14.33).

These types of clay-rich sediments are referred to as *quick clays*. They are derived from glacio-marine deposits formed in shallow saltwater bays of the Champlain Sea toward the end of the Late–Cenozoic Ice Age. The clays initially settled on the bed of the sea as thin layers saturated with saline water. The ionic charges in the salt water help to bind the clay into a cohesive layer. However, with regional uplift following deglaciation, the quick clay beds were raised above sea level, and fresh groundwater replaced the salt water. A mechanical shock, such as an earth tremor, then caused the structure to collapse. Because of the large proportion of water present (from 45 to 80 percent by volume), the saturated clay behaves like a liquid. The moving clay settles downward and breaks into step-like masses and ultimately reaches the river, into which it pours as a disordered flow of mud.

MUDFLOW AND DEBRIS FLOOD

One of the most spectacular forms of mass wasting and a potentially serious environmental hazard is the **mudflow**. These streams of muddy fluid pour swiftly down canyons in mountainous regions. In deserts, rainfall intensity during local thunderstorms can exceed the infiltration capacity of the soil. The sparse vegetation offers little protection, and as the water runs down the slopes, it forms a thin mud that flows to the

14.33 **EARTH FLOW AT ST. JUDE, QUEBEC** This earth flow in St. Jude, Quebec, occurred in May 2010 in an area underlain by Leda clays, a glacio-marine quick clay that is notorious for mass movements of this type. In this case, the liquified clay layer flowed a short distance into the valley of the Salvail River, causing the overlying regolith and soil to collapse.

canyon floors and then follows the stream courses. As it flows, it picks up additional sediment, becoming thicker and thicker until it is too thick to flow further. Large boulders are carried along, buoyed up in the mud, which engulfs and destroys roads, bridges, and houses in the canyon floor.

Mudflows vary in consistency, from a viscous fluid similar to freshly mixed concrete to fast-flowing, water-saturated mixtures of mud and rock. The more fluid type of mudflow is called a *debris flow*. Debris flows can travel many kilometres down valleys, often at high speed, destroying or damaging everything in their paths through impact or burial under layers of sediment that can be several metres thick. Mudflows commonly occur on the slopes of erupting volcanoes. Heavy rain or melting snow turn freshly fallen volcanic ash and dust into a muddy slurry that flows quickly down the flanks of the volcano, forming what is termed a *lahar* (Figure 14.34). Herculaneum, a city at the base of Mount Vesuvius, was destroyed by a mudflow during the eruption of A.D. 79. At the same time, the neighbouring city of Pompeii was buried under volcanic ash.

LANDSLIDES AND ROCKFALLS

A **landslide** is the rapid sliding of large masses of bedrock or regolith. Wherever mountain slopes are steep, there is a possibility of large, disastrous landslides. In mountainous regions, towns and villages built on the floors of steep-sided valleys have been destroyed when millions of cubic metres of rock have suddenly collapsed without warning. A classic example of an enormous, disastrous landslide is the Turtle Mountain slide, which took place near Frank, Alberta, in 1903 (see Figure 1.5). A huge mass of limestone slid from the face of Turtle Mountain, descended to the valley, then continued up the low slope of the opposite valley side until it came to rest as a great sheet of bouldery rock debris. The Hope slide in British Columbia is another well-known example.

Landslides, unlike earth flows and mudflows, are triggered by earth tremors or sudden rock failures rather than by heavy rains or floods; they can also result when excavation or river erosion oversteepens a slope. Landslides can range from *rockslides* of jumbled bedrock fragments to *bedrock slumps* in which most of the bedrock remains more or less intact as it moves. The high speed at which rockslides travel down a mountainside is thought to be due to the presence of a layer of air trapped between the slide and the ground surface. The compressed air reduces frictional resistance and may prevent the rubble from touching the surface beneath.

Large landslides tend to occur only sporadically and usually in thinly populated mountainous regions. However, small slides and rockfalls repeatedly block highways and railway lines. A rockfall involves a comparatively small quantity of rock falling from a cliff face, where the rock is weakened by joints or other discontinuities in the rock mass. Once it has fallen free, the rock mass continues

14.34 LAHAR FROM MOUNT ST. HELENS Melted snow following the March 19, 1982, eruption of Mount St. Helens carried ash and debris down the flank of the volcano to form the dark-coloured lahar.

14.35 ROCKFALL A rockfall created a scar in the face of this cliff formed in the thinly bedded sandstone of the Kaibab Plateau near the Grand Canyon, Arizona. A debris cone has accumulated at the base of the rockfall.

down slope by bouncing, sliding, or rolling, during which the material may be broken into smaller fragments before accumulating in a debris apron at the foot of the cliff (Figure 14.35). It resembles a talus slope in appearance, except that the calibre of the material is generally larger and more variable. Unlike a rockfall, talus slopes represent the accumulated debris of numerous minor events, often involving only a single rock fragment that has been loosened by freeze–thaw action, root-wedging, and other weathering process. Rapid mass wasting events such as landslides and rock falls are similar to avalanches, which occur when masses of snow break free from accumulated snowpacks in mountainous terrain. Avalanches are triggered by several factors that cause zones of weakness to develop in the snow mass; these include temperature gradients within the snowpack, size and shape of the snow crystals, and snow density. Unlike avalanches, sudden rock failures are comparatively rare events, but when they do occur they release a vast amount of gravitational energy; this is discussed in the *Working It Out • The Power of Gravity* feature at the end of this chapter.

A Look Ahead

This chapter examined how rock structure and lithology control the general topography of a region as expressed in the initial landforms created by tectonic and volcanic processes. Subsequent changes in the landscape occur through the subaerial processes of weathering, mass wasting, and erosion. In the weathering process, rock near the surface breaks up into smaller fragments and often alters in chemical composition. In the mass wasting process, weathered rock and soil move downhill in slow or sudden movements. The landforms of mass wasting are produced by gravity acting directly on soil and regolith. Gravity also powers another landform-producing agent flowing water. Flowing water, in the form of rivers and streams, is the principle agent of erosion and is discussed in the next two chapters. The first deals with water in the hydrologic cycle, in soil, and in streams. The second deals specifically with how flowing water erodes rock and regolith and deposits transported sediment to create a variety of distinctive fluvial landforms.

CHAPTER SUMMARY

- Differences in rock composition and crustal structure are related to landform development.
- Where rock layers are arched upward into a dome, erosion produces a circular arrangement of rock layers outward from the dome's centre. Resistant strata form *hogbacks*, and weaker rocks form lowlands.
- In fold belts, the sequence of *synclines* and *anticlines* brings a linear pattern of rock layers to the surface. Resistant strata form ridges, and weaker strata form valleys.
- Faulting provides an initial surface along which rock layers move—the *fault scarp*. This feature can persist as a fault-line scarp long after the initial scarp is gone.
- Exposed batholiths are often composed of uniform, resistant igneous rock. They erode to form a dendritic drainage pattern. Monadnocks of intrusive igneous rock stand up above a plain of weaker rocks.
- Stratovolcanoes produce lava flows that initially follow valleys but are highly resistant to erosion. At the last stages of erosion, all that remains of stratovolcanoes are necks and dikes.
- Meteor craters that originated from the impact of extraterrestrial meteorites and asteroids are found throughout the world. Dust clouds resulting from such impacts might be associated with climate perturbations and changes in the Earth's plant and animal life.
- Weathering is the action of processes that cause rock near the surface to disintegrate and decompose into regolith. Mass wasting is the spontaneous downhill motion of soil, regolith, or rock under gravity.
- Physical weathering produces regolith from solid rock by breaking bedrock into pieces. *Frost action* breaks rock apart by the repeated growth and melting of ice crystals in rock fractures and *joints*, as well as between individual mineral grains. In mountainous regions of vigorous frost, fields of angular blocks accumulate as *felsenmeers*. Slopes of rock fragments form *talus cones*. In soils and sediments, needle ice and ice lenses push rock and soil fragments upward. *Salt-crystal growth* in dry climates breaks individual grains of rock free. *Unloading* of the weight of overlying rock layers can cause some types of rock to expand and break loose into thick shells, producing *exfoliation domes*. Daily temperature cycles in arid environments are thought to cause rock breakup. Wedging by plant roots also forces rock masses apart.
- Chemical weathering results from mineral alteration. Igneous and metamorphic rocks can decay to great depths through *hydrolysis* and *oxidation*, producing a regolith that is often rich in clay minerals. *Carbonic acid* action dissolves limestone. In warm, humid environments, basaltic lavas can also show features of solution weathering produced by acid action.
- Mass wasting occurs on slopes that are mantled with regolith. Soil creep is a process of mass wasting in which regolith moves down slope almost imperceptibly under the influence

of gravity. In an earth flow, water-saturated soil or regolith slowly flows downhill. Quick clays, which are unstable and can liquefy when subjected to a shock, have produced earth flows in previously glaciated regions. A mudflow is much swifter than an earth flow. It follows stream courses, becoming thicker as it descends and picks up sediment. A watery mudflow with debris ranging from fine particles to boulders is called a *debris flow*. A landslide is a rapid sliding of large masses of bedrock, sometimes triggered by an earthquake.

KEY TERMS

alluvium
bedrock
butte
carbonation
chemical weathering
dip
earth flow
exfoliation
inselbergs
karst
kopje
landslide
mass wasting
mesa
monadnocks
mudflow
physical weathering
plateau
regolith
ridge-and-valley landscape
sediment
sedimentary dome
soil creep
solution
strike
terracettes
weathering

REVIEW QUESTIONS

1. Why is there often a direct relationship between landforms and rock structure? How are geologic structures that formed deep within the Earth exposed at the surface?
2. Which of the following types of rocks—shale, limestone, sandstone, conglomerate, and igneous rocks—tends to form lowlands? Uplands?
3. How are the tilt and orientation of a natural rock plane measured?
4. How are domes formed? What type of drainage pattern would a dome have and why?
5. How do ridge-and-valley landscapes arise? Describe their similarities to and differences from landforms associated with sedimentary domes.
6. What type of landform(s) might be associated with a fault? Why?
7. Do shield volcanoes erode differently from stratovolcanoes? How and why?
8. What does the term *weathering* mean? What are the types of weathering?
9. Define the terms *regolith, bedrock, sediment*, and *alluvium*.
10. How does salt-crystal growth break up rock? Give an example of a landform that arises from salt-crystal growth.
11. What is an exfoliation dome, and how does it form? Provide an example.
12. Name three types of chemical weathering. Describe how a chemical weathering process often alters limestone.
13. Define mass wasting and identify the processes it includes.
14. What is soil creep, and how does it arise?
15. What is an earth flow? What features distinguish it as a landform?
16. Compare and contrast earth flows and mudflows. How does a landslide differ from an earth flow?

VISUALIZATION EXERCISES

1. Sketch a ridge-and-valley landscape, identifying the following features: syncline, anticline, synclinal valley, anticlinal mountain, anticlinal valley.
2. Define the terms *regolith, bedrock, sediment*, and *alluvium*. Sketch a cross-section through a part of the landscape showing these features and label them on the sketch.
3. Copy or trace Figure 14.28; then identify and plot on the diagram the mass movement associated with each of the following locations: Turtle Mountain, Mount St. Helens, St. Lawrence Valley, and Herculaneum.

ESSAY QUESTIONS

1. A landscape includes a range of lofty mountains elevated above a dry desert plain. Describe the processes of weathering and mass wasting that might be found on this landscape and identify their location.
2. Imagine yourself as the newly appointed director of public safety and disaster planning for your province. One of your first jobs is to identify locations where potential disasters associated with mass wasting threaten human populations. Where would you look for mass wasting hazards and why?
3. Imagine the following sequence of sedimentary strata starting at the top with sandstone and underlain with shale, limestone, and shale. What landforms would you expect to develop in this structure if the sequence of beds is (a) flat-lying in an arid landscape; (b) slightly tilted as in a coastal plain; (c) folded into a syncline and an anticline in a fold belt; and (d) fractured and displaced by a normal fault? Use sketches in your answer.

WORKING IT OUT • THE POWER OF GRAVITY

Directly or indirectly, gravity causes many of the processes that shape the landscape. By acting to move mineral matter and water downhill in various mass wasting processes, such as earth flows, mudflows, and landslides, gravity works over a span of time and acts as a source of power.

Gravitation is the simultaneous attraction that occurs between two physical bodies. The strength of the attraction depends on the mass of each body and the distance between them. The Earth's attraction on an object near the Earth's surface is called *gravity*. This acceleration acts on the mass of the object to produce a force according to the simple relationship:

$$F = m \times g$$

where F is the force, m is the mass of the object, and g is the acceleration due to gravity (about 9.8 m s^{-2}). Mass is measured in kilograms, acceleration is measured in metres per second squared, and force is in newtons (N).

Suppose that gravity actually does some work; that is, it moves a mass through a distance. The work is measured as force multiplied by distance:

$$W = F \times d$$

where W is the work and d is the distance. The force is measured in newtons, the distance in metres, and the work done is measured in joules (J).

Power describes the rate at which work is accomplished: that is, work per unit of time, or:

$$P = \frac{W}{T}$$

where P is the power, W is the work, and T is the time. Work is measured in joules (J), time is measured in seconds (s), and power is measured in watts (W). Recall from Chapter 3 that work and energy have the same units and that energy is defined as the ability to do work. Thus, power is a measure of both a rate of energy flow and a rate at which work is done.

An example of the power of gravity is the Madison Slide—a catastrophic landslide that took place near Yellowstone National Park in 1959. A rock body with a volume of about 28 million cubic metres (28×10^6 m^3) moved a vertical distance downward of about 500 m into the valley of the Madison River, attaining a velocity of about 150 kph. Propelled by its own inertia, the flow then moved uphill on the far side of the river valley for a vertical distance of about 120 m. In this slide, the potential energy of the position of the mass above the valley bottom was converted to kinetic energy during the slide, and the kinetic energy was dissipated as heat produced by friction during the sliding motion.

In moving downhill for 500 m then uphill another 120 m, the slide converted potential energy to kinetic energy and friction. On the uphill leg, some kinetic energy converted back to potential energy because the mass moved upward against the pull of gravity. So the net release of potential energy produced by a change in vertical distance was equivalent to $500 - 120 = 380$ m.

How much energy was released? Or alternatively, how much work was done? The formulas to the left require information on the slide's mass, acceleration, and distance. Of these, only the mass is unknown. However, the volume of the landslide is 28×10^6 m^3; what is missing is the density—the mass of the rock per unit of volume. Assume that the rock has the density of granite, which is about 2.7×10^3 kg m^{-3}. Then calculate the mass as follows:

$$m = 28 \times 10^6 \text{ m}^3 \times (2.7 \times 10^3 \text{ kg m}^{-3})$$
$$= 7.56 \times 10^{10} \text{ kg}$$

Gravity will act on this mass with a force of:

$$F = m \times g = 7.56 \times 10^{10} \text{ kg} \times (9.8\text{m s}^{-2})$$
$$= 7.41 \times 10^{11} \text{ N}$$

This force is applied through a net distance of 380 m, so the work done is:

$$W = F \times d = 7.41 \times 10^{11} \text{ N} \times 380\text{m}$$
$$= 2.82 \times 10^{14} \text{ J}$$

The friction of the slide converted this work to heat. The amount of work done is rather similar to the amount of electrical energy consumed in a day in the United States, which is about 2.33×10^{14} joules per day.

To calculate the power expended by gravity in the slide, it is necessary to know how long the motion took. With some simple assumptions about the angles of the mountain slopes and the magnitude of friction in the earth flow, a reasonable guess for the time required is about 70 seconds. The power applied by gravity in the downward motion would then be:

$$P = W/T = (2.82 \times 10^{14} \text{ J})/70\text{s} = 4.02 \times 10^{12} \text{ W}$$

Expressed in watts, the average rate of U.S. electric power consumption is about 2.70×10^9 W. The rate of energy release in the Madison Slide is therefore about 1,500 times greater than the rate at which the entire United States consumes electricity. This large number demonstrates gravity's awesome power to shape and reshape the landscape.

QUESTIONS / PROBLEMS

1. The Madison Slide reached a velocity of at least 150 kph. What was the kinetic energy of the slide at that point? Recall from Chapter 3 that the formula for kinetic energy is $E = \frac{1}{2}mv^2$, where m is the mass (kg) and v is the velocity (m s^{-1}). How does this value compare with the total energy released in the slide?
2. Physics predicts the velocity, v (m s^{-1}), of a falling body in the absence of friction to be $v = \sqrt{2gd}$, where g is the acceleration of gravity (m s^{-2}) and d is the distance of the fall from rest (m). What is the velocity of an object falling without friction for a distance of 500 m? How does this compare with the velocity observed for the Madison Slide? Why are the two values different?

Find answers at the back of the book.

CHAPTER 15

THE CYCLING OF WATER ON THE CONTINENTS

The narrow gorge of the Ubaye River in Provence, France.

Fresh water on land and beneath the Earth's surface accounts for about 3 percent of the hydrosphere's total water. Most of this fresh water is locked into ice sheets and mountain glaciers. Groundwater accounts for a little more than 0.5 percent. Although this is a small fraction of the global water supply, it is still many times larger than the amount of fresh water in lakes, streams, and rivers, which account for only 0.03 percent of the entire water resources of the Earth. Groundwater can be found almost everywhere on land that receives rainfall. In contrast, fresh surface water varies widely in abundance. In many dry regions, streams and rivers are non-existent for most or all of the year.

Chapter 6 discussed atmospheric moisture and precipitation, describing the part of the hydrologic cycle in which water evaporates from ocean and land surfaces and then falls to the Earth as rain or snow. A portion of the precipitation returns directly to the atmosphere through evaporation. Another portion travels downward, moving through the soil under the force of gravity to become part of the groundwater reserves. Following underground flow paths, this subsurface water eventually emerges to become surface water, or it may emerge directly into the ocean. A third portion flows over the ground surface as runoff. This water collects in streams, which eventually conduct the flow to the ocean.

This chapter traces the parts of the hydrologic cycle that include both the subsurface and surface pathways of water flow. Most soils are capable of absorbing the water from light or moderate rainfalls by **infiltration**. In this process, water enters the small natural passageways between irregularly shaped soil particles, as well as the larger openings in the soil surface. A mat of decaying leaves and stems breaks the force of the raindrops and helps to keep these openings clear.

The precipitation that infiltrates the soil is temporarily held in the soil layer as soil water, occupying the *soil water belt*. This water can return to the surface and then to the atmosphere through direct evaporation from the soil and through transpiration from vegetation. The combined process is termed *evapotranspiration*.

When rain falls too rapidly to pass into the soil, **runoff** occurs as a thin layer at the soil surface by the process of **overland flow**. In periods of heavy, prolonged rain or rapid snow melt, overland flow feeds directly into streams. Overland flow also occurs when soil that

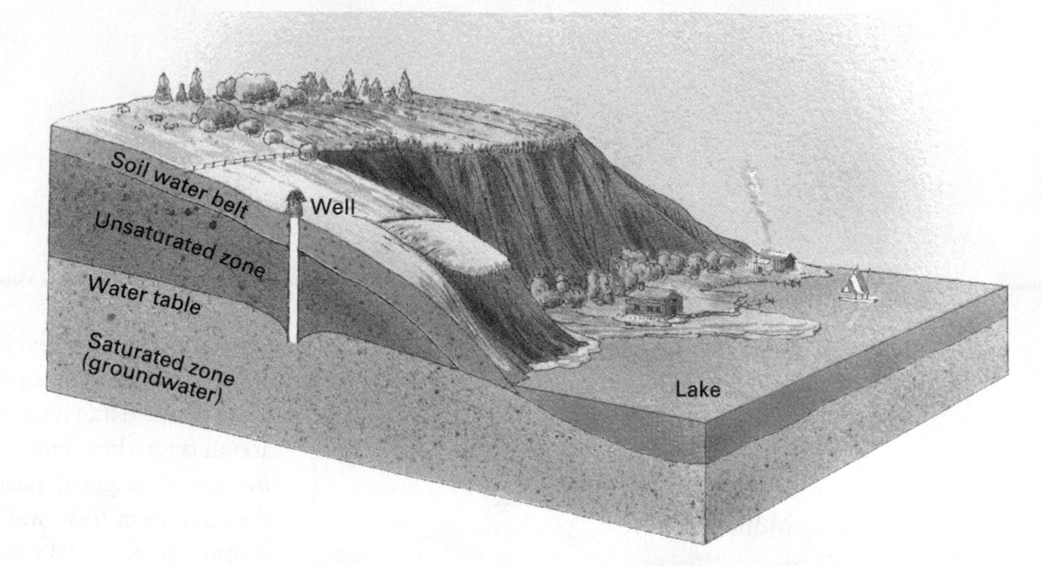

15.1 ZONES OF SUBSURFACE WATER Water in the soil water belt is available to plants. Water in the unsaturated zone percolates downward to the saturated zone of groundwater, where all pores and spaces are filled with water.

is already saturated receives additional rainfall or snow melt. Since soil openings and pores are already full, water cannot infiltrate the soil and percolate to deeper layers. Under these conditions, nearly all of the surplus water will become runoff. This chapter discusses how precipitation contributes to the groundwater reserves and how surplus water becomes organized into drainage systems. Chapter 16 describes the geomorphological role of running water and the landforms produced by stream erosion, sediment transport, and deposition. Processes associated with groundwater and river systems are part of the study of hydrology.

GROUNDWATER

Water from precipitation can continue to flow downward beyond the soil water belt. This slow movement under the influence of gravity is called *percolation*. Eventually, the percolating water reaches the **groundwater**—the zone where the pore spaces in the bedrock or regolith are completely filled with water. The **water table** marks the top of the *saturated zone* (Figure 15.1). In the *unsaturated zone* above the water table, the pore spaces are not fully saturated. Here, water is held in thin films adhering to mineral surfaces (see Chapter 19). The upper part of this zone is the *soil water belt*.

Groundwater moves slowly along deep flow paths, eventually seeping into streams, ponds, lakes, and marshes, where the land surface dips below the water table (Figure 15.2). *Perennial streams*, which flow throughout the year, derive much of their water from groundwater seepage.

THE WATER TABLE

The position of the water table can be mapped in detail in areas where there are many wells. This is done by plotting the height of the water in the wells and noting the change in elevation from

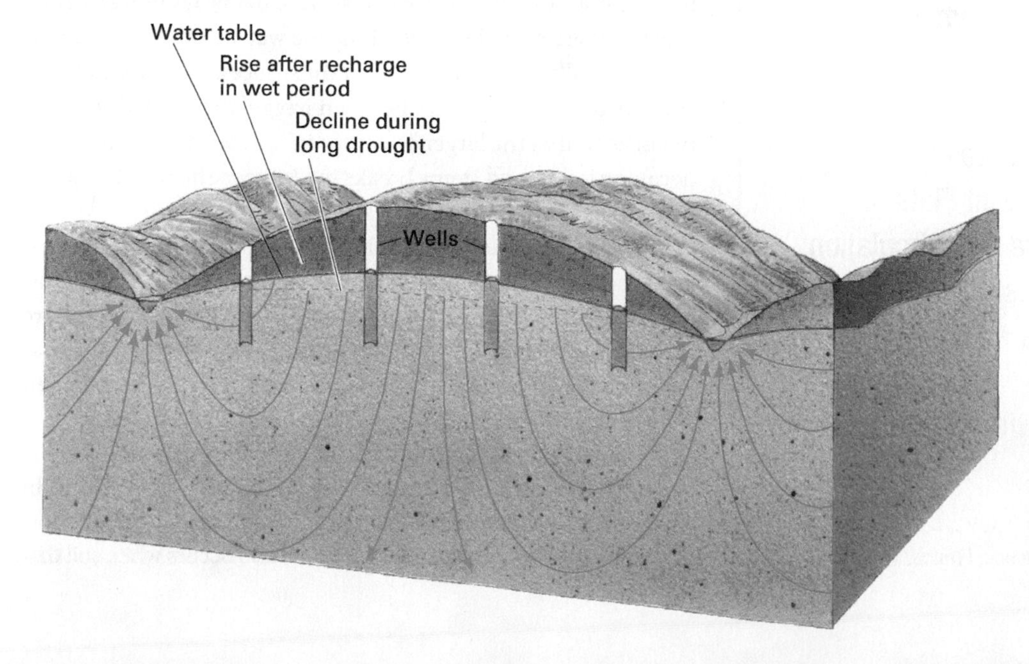

15.2 WATER TABLE The configuration of the water table surface conforms broadly with the land surface above it. It varies in response to prolonged wet and dry periods. Groundwater flow paths circulate water very slowly to deep levels, and eventually it seeps into streams.

one well to the other. The water table is highest under hills and other elevated land surfaces and drops in valleys. Lakes, ponds, and marshes occur where the water table intersects the surface, and on hillsides water may emerge as *springs*.

The undulating configuration of the water table is linked to the very slow flow rate of subsurface water through the pores and openings in the regolith and bedrock. Downward percolation tends to raise the water table, while seepage into lakes and streams draws off groundwater and lowers the water table. During periods of high precipitation, the water table rises under hilltops; in drier periods, the water table falls (see Figure 15.2).

The direction of flow at any point depends on the direction of the pressure exerted on the water. Gravity always exerts a downward pressure, while the difference in the height of the water table between hilltop and streambed produces a lateral pressure. These directional pressures cause groundwater to flow in curving paths. Water that enters the hillside midway between the hilltop and the stream flows almost directly toward the stream. However, water that reaches the water table midway between two streams flows almost straight down to great depths before curving and rising upward. Progress along these deep paths is very slow, whereas flow near the surface is much faster. Flow is most rapid close to a stream where the flowlines converge. Over time, the level of the water table tends to remain stable, and the flow of water released to streams and lakes balances the flow of water percolating into the water table.

AQUIFERS

Sedimentary layers often exert a strong control over the storage and movement of groundwater. For example, clean, well-sorted sand, such as that found in beaches and dunes, can hold an amount of groundwater equal to about one-third of its volume. Consequently, a bed of sand or sandstone is often a good *aquifer*, in that it contains abundant, freely flowing groundwater. In contrast, beds of clay and shale are relatively impermeable and hold little water; they are known as *aquitards* or *aquicludes*. In areas where impervious rock is overlain by porous rock, groundwater will move freely through the aquifer and across the top of the aquitard, possibly emerging as *springs* along the valley (Figure 15.3). A *perched water table* may occur above the main water table if a small lens of impervious rock is present.

15.3 GROUNDWATER IN HORIZONTAL STRATA Water flows freely within the sandstone aquifer, but very slowly in the shale aquiclude. A lens of shale creates a perched water table above the main water table. (© A. N. Strahler.)

When an aquifer is situated between two aquitards, groundwater in the aquifer may be under pressure and so will flow freely from an **artesian well** drilled into the water-bearing strata (Figure 15.4). Precipitation on the hills where the porous sandstone is exposed provides water that saturates the aquifer, filling it to that elevation. Since the elevation of the well that taps the aquifer is below that of the range of hills feeding the aquifer, hydrostatic pressure causes water to rise in the well.

PROBLEMS OF GROUNDWATER MANAGEMENT

Rapid withdrawal of groundwater has seriously affected the environment in many places. Increased urban populations and industrial developments require larger water supplies—needs that cannot always be met by building new surface water reservoirs. To fill these demands, vast numbers of wells using powerful pumps draw huge volumes of groundwater to the surface, greatly altering nature's balance of groundwater discharge and recharge.

In dry climates, agriculture is heavily dependent on irrigation water from pumped wells, especially since major river

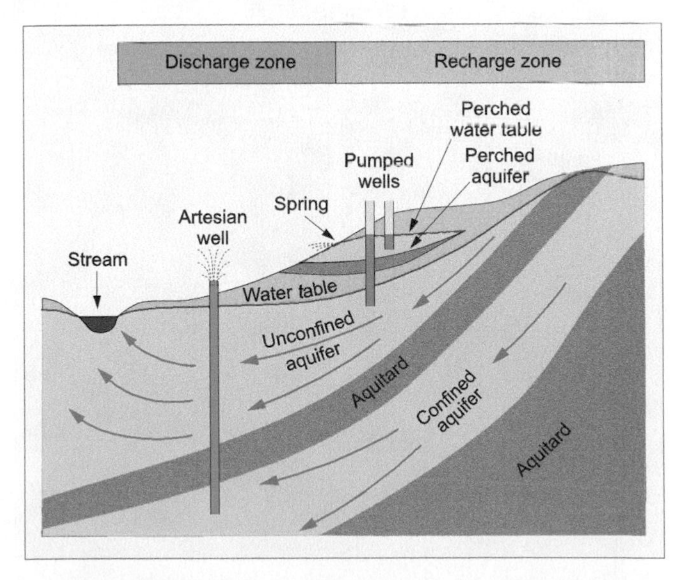

15.4 ARTESIAN WELL A layer of sandstone between impervious layers provides a source of water under pressure that flows naturally to the surface.

systems are likely to be already fully used for irrigation. Wells are also convenient water sources. They can be drilled within the limits of a given agricultural or industrial property, and can provide immediate supplies of water without the need to build expensive canals or aqueducts.

Drilled wells are usually lined with metal casings that exclude impure near-surface water and prevent the walls from caving in and clogging the borehole. Near the lower end of the hole, in the groundwater zone, the casing is perforated to admit the water. The yield of a single drilled well ranges from as low as a few hundred litres per day in a domestic well, to thousands of cubic metres per day in a large industrial or irrigation well.

Water Table Depletion

When water is pumped from a well, the water level will inevitably drop. At the same time, the surrounding water table lowers in the shape of a downward-pointing cone, called the *cone of depression* (Figure 15.5). The difference in height between the tip of the cone and the original water table is the *drawdown*. The cone of depression may extend out more than 15 km from a heavily pumped well. Where many wells are in operation, their intersecting cones produce a general lowering of the water table.

Water table depletion often exceeds the rate at which infiltrating water can move downward to recharge the saturated zone. In dry regions, much of the groundwater for irrigation is drawn from wells driven into thick aquifers of sand and gravel. These deposits are often recharged by the seasonal flow of water originating from snow melt high in adjacent mountains. Fanning out across the dry lowlands, the streams lose water, which sinks into the sands and gravels and eventually percolates to the water table below. Excessive demand on groundwater reserves can essentially exhaust this natural resource because it is renewable only over very long time periods. Water table depletion can also result in subsidence, a sinking of the land in response to the removal of water from underlying sediments. This problem has plagued a number of major cities that rely heavily on groundwater wells for their water supplies.

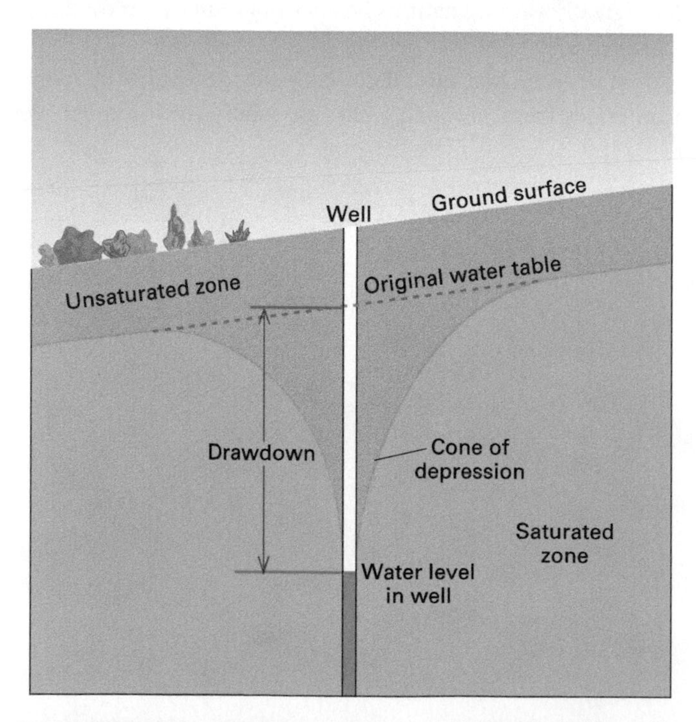

15.5 DRAWDOWN AND CONE OF DEPRESSION IN A PUMPED WELL As the well draws water, the water table is depressed in a cone with the well at its centre. (© A. N. Strahler.)

Contamination of Groundwater

Contamination of wells by pollutants that infiltrate the ground and reach the water table can be a major environmental problem arising from groundwater withdrawal. The contamination of the groundwater supply in the community of Walkerton, Ontario, illustrates this issue. On May 12, 2000, heavy rain washed a virulent strain of *E. coli* bacteria from cattle manure into the soil near the community. It then percolated down to the groundwater, contaminating the shallow well that provided water for the town. Inadequate testing of the water drawn from the well failed to detect the highly toxic bacteria, and the polluted water was distributed in the town's water supply network.

The first cases of illness were reported on May 17, but an advisory warning to residents that they must boil water to kill the bacteria before drinking and cooking with it wasn't issued until four days later. The first death as a result of the contaminated water occurred on May 22. By May 31, six more people had died and more than 1,000 people in the small community fell seriously ill. Eventually the water supply system was cleaned of the contamination. This involved, among other measures, a house-by-house disinfection of the water pipes. This tragedy was a wake-up call. By late July, the Ontario Ministry of the Environment had compiled a list of 131 municipalities in the province with "inadequate" water supply facilities. Before the end of August 2000, the provincial legislature had enacted new laws to ensure that drinking water is adequately treated throughout the province. Thirty percent of Canadians draw their water supply from groundwater reserves. Efforts by federal and provincial agencies continue to address what is an ongoing issue, especially in small, remote communities across the country.

There are other potential sources of groundwater contamination. Disposal of solid wastes poses a major environmental problem in developed countries because their advanced industrial economies provide an endless source of garbage. Traditionally, these waste products were trucked to the town dump, and often burned there in continuously smouldering fires that emitted foul smoke and gases. The partially consumed residual waste was then buried under earth.

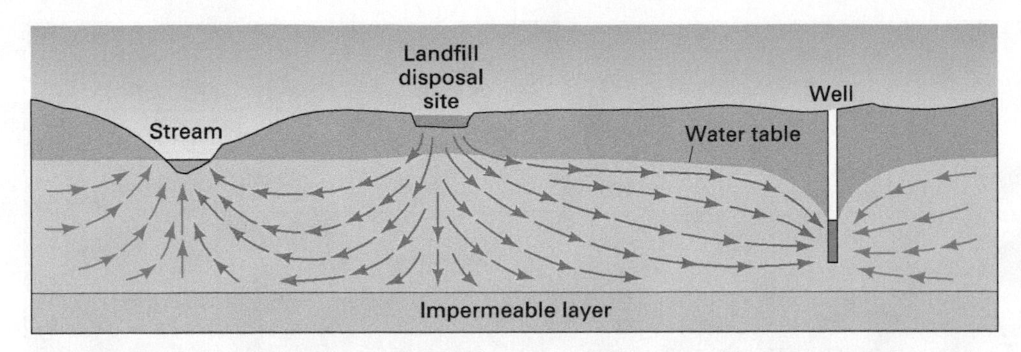

15.6 MOVEMENT OF POLLUTED GROUNDWATER Polluted water, leached from a waste disposal site, moves toward a supply well and a stream. (© A. N. Strahler.)

In recent decades, a major effort has been made to improve solid-waste disposal methods. One method is high-temperature incineration, but this can lead to air pollution. Another is the sanitary landfill method, in which layers of waste are continually buried, usually by sand or clay available on the landfill site. In this way, the waste is situated in the unsaturated zone. However, it can react with rainwater that infiltrates the ground surface. The percolating water can pick up a wide variety of chemical compounds from the waste and carry them down to the water table (Figure 15.6). Here the contaminated water follows the groundwater flow paths and can potentially flow toward a well, rendering the water supply unfit for human consumption. Contaminated water may also move toward nearby valleys, polluting the streams that flow there.

A common source of contamination in coastal wells is *salt-water intrusion*. Since fresh water is less dense than salt water, a layer of salt water from the ocean may lie under a coastal aquifer. As the aquifer is depleted, the level of the salt water rises and eventually reaches the well from below, making the well water unusable.

SURFACE WATER

Surplus water that does not infiltrate into the ground runs off the land surface and ultimately reaches the sea. Surface water flows in rivers and streams. In general usage, rivers describe large watercourses, and streams smaller ones. However, *stream* is also a scientific term describing the channelled flow of any amount of surface water.

OVERLAND FLOW AND STREAM FLOW

Runoff that flows down the slopes of the land in broadly distributed sheets is overland flow. In contrast, *stream flow* is the flow of water along a narrow channel confined by lateral banks. Overland flow can take several forms. Where the soil or rock surface is smooth, the flow may be a continuous thin film, called *sheet flow*. Where the ground is rough or pitted, flow may take the form of a series of tiny rivulets connecting one water-filled hollow with another (see Chapter 16). On a grass-covered slope, overland flow is subdivided into countless tiny threads of water passing around the stems, and is normally not apparent except perhaps during a heavy and prolonged rainstorm. On heavily forested slopes, overland flow can pass entirely concealed beneath a thick mat of decaying leaves.

Overland flow eventually contributes to a stream, which is a deeper, more concentrated form of runoff. A **stream** is a long, narrow body of flowing water that occupies a trench-like depression, or channel, and moves to lower levels due to gravity. The **channel** of a stream is a narrow trough. The force of the flowing water shapes the trough to the most effective form for moving the volume of water and sediment supplied to the stream. A channel may be very narrow, only a few centimetres across, or, in the case of large rivers, several kilometres wide.

STREAM DISCHARGE

The flow of water in a stream is affected by channel geometry. Because of friction between the water and the banks, velocity is nearly zero along each bank and increases to a maximum near the stream's centre (Figure 15.7a). Velocity also increases from the bottom of the channel to the water surface (Figure 15.7b), so

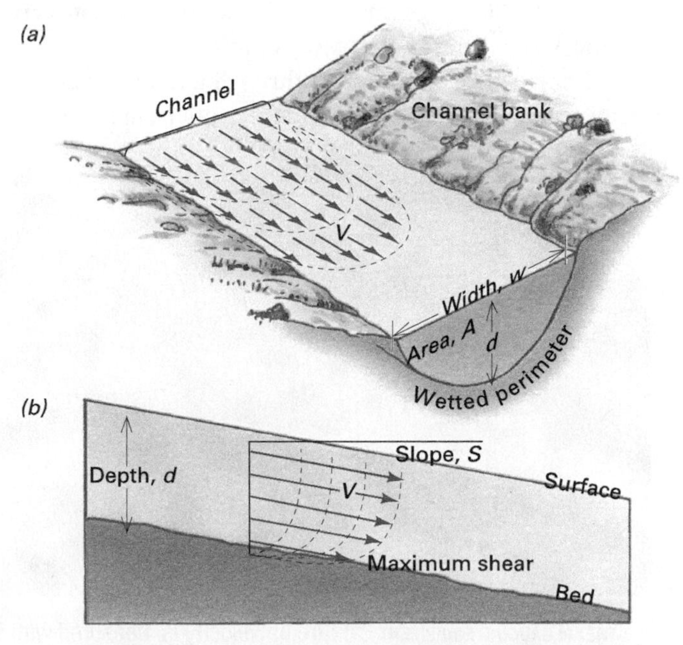

15.7 CHARACTERISTICS OF STREAM FLOW The flow velocity is greatest in the middle (a) and near the top (b) of a stream. (© A. N. Strahler.)

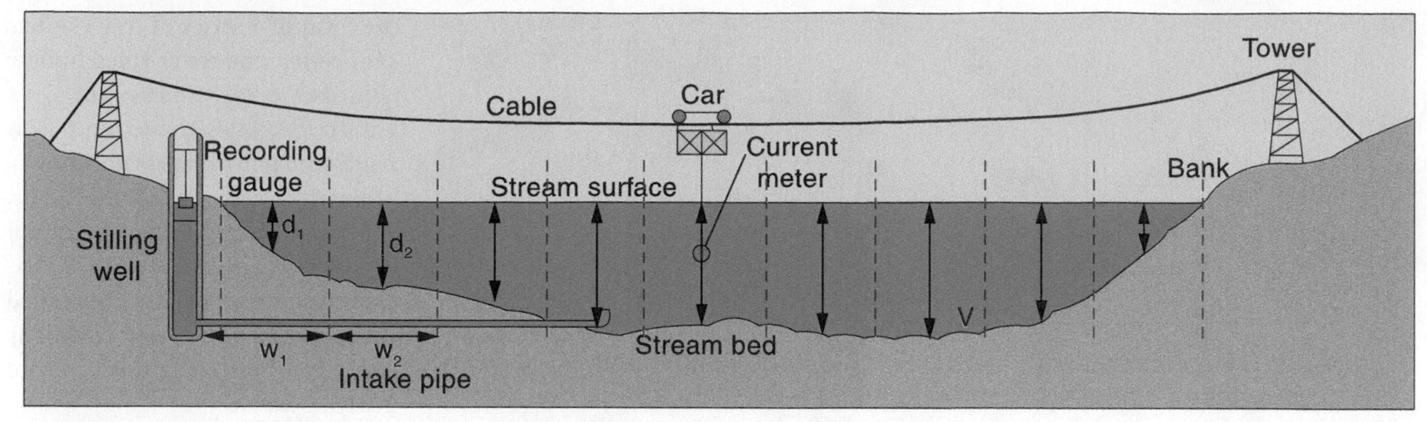

15.8 GAUGING A STREAM The discharge in a river is determined by lowering a current meter at various points across the channel to measure the mean velocity (V) and multiplying that by the area of each sampled increment—depth (d) × width (w). The total discharge is the sum of the discharges in each increment across the entire channel. For a large river, as illustrated here, the operator must work from a cable car or from a bridge or boat; small streams are gauged by wading.

that maximum flow is generally along the centre line of the channel. Stream flow at a given location is measured by its **discharge**, which is usually expressed as the volume of water in cubic metres (m^3) passing a point in the river every second.

The discharge of a stream depends on the mean water velocity and cross-sectional area of the stream. For example, if a stream's mean velocity at a certain point is 2 m s^{-1} and its cross-sectional area is 20 m^2, then in one second, $2 \times 20 = 40$ m^3 of water flow past that point. This relationship is expressed by the equation:

$$Q = A \times V$$

where Q is the discharge (m^3 s^{-1}), A is the cross-sectional area (m^2), and V is the mean velocity (m s^{-1}).

In any given vertical section through a river, the average velocity occurs at about six-tenths of the depth from the surface. By convention, the discharge of a river is measured with a current meter at this depth in a number of equally spaced points across the river (see Figure 15.8). A current meter has a propeller or cups that rotate in the flowing water (Figure 15.9a). Each point is considered to be representative of a small cross-section of the stream of known width and depth. The discharge in each section (width × depth × mean velocity) is added to the discharge in the other sections to obtain the total discharge, Q, of the river at that location on its course.

The discharge of streams and rivers changes from day to day. Records of daily discharges provide important information for planning the development and distribution of surface waters, as well as in designing flood-protection structures. Variations in discharge over time can be determined by carrying out repeated measurements. However, an easier technique is to derive discharge from measurements of the height of the water surface (called

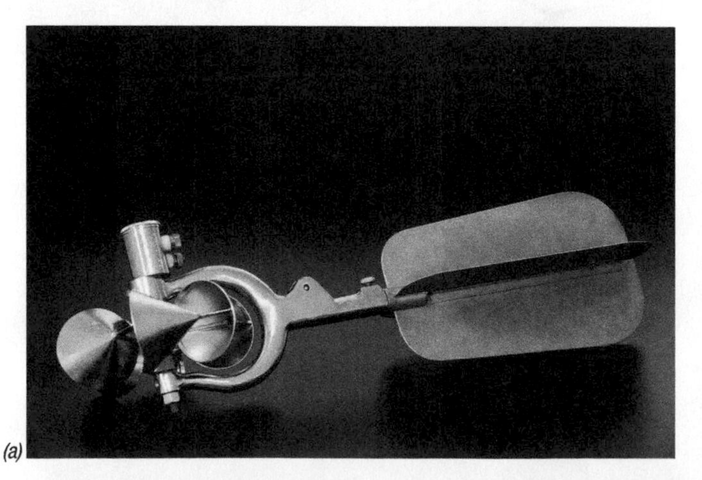

(a)

(b)

15.9 STREAM GAUGING EQUIPMENT (a) Stream velocity is measured with a current meter by counting electonically the number of revolutions that occur in a given period of time. (b) A float activated stage recorder is used to measure fluctuations in the depth of a stream. The white float is suspended in the stilling well from the pulley at the right of the chart drum.

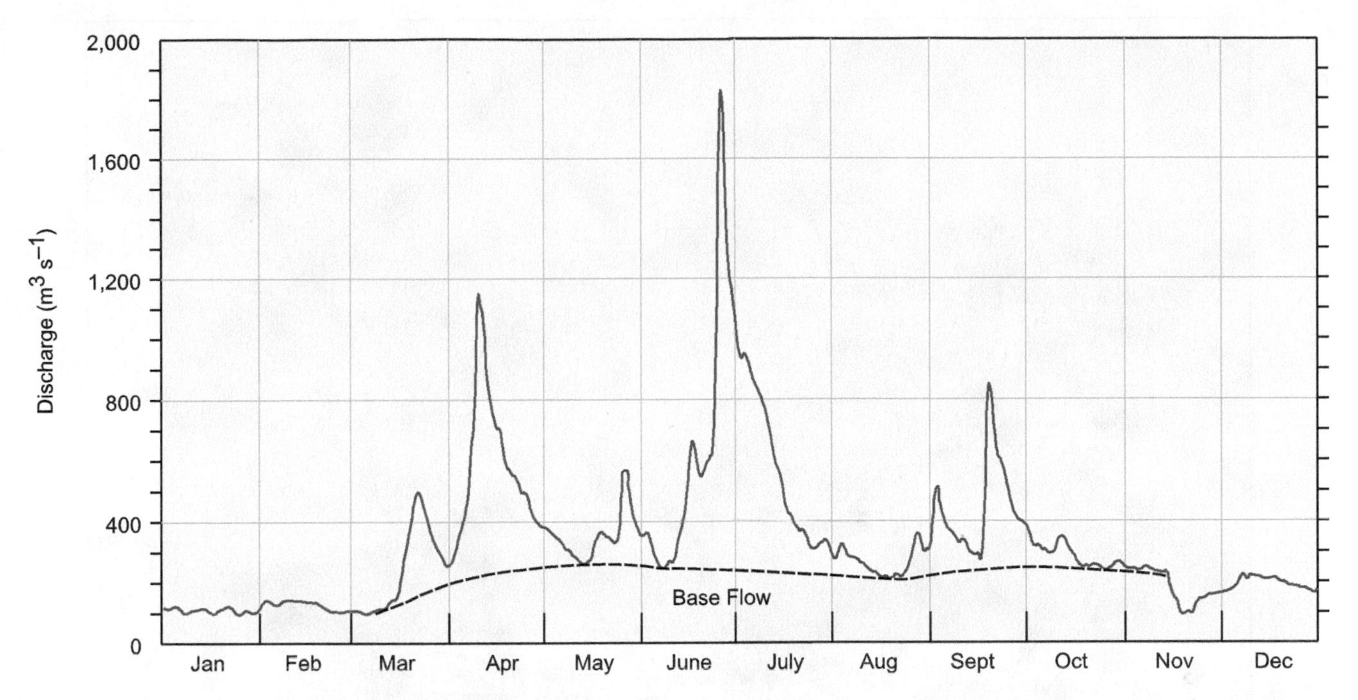

15.10 HYDROGRAPH OF THE NORTH SASKATCHEWAN RIVER AT PRINCE ALBERT, SASKATCHEWAN, IN 2005 The peaks in March and April are caused by snow melt on the Prairies. The peak in June is associated with snow melt in the Rocky Mountains. Other peaks are caused by rainfall in the drainage basin. Base flow is water supplied by groundwater seepage. (Data from the Water Survey of Canada.)

the *stage* of the river), since stage increases with higher discharges. A simple device to measure stage consists of a *staff gauge*, essentially a ruler, permanently anchored in the river. Many staff gauges have been replaced by automated *stage recorders*, which collect a continuous record of changes in stage. The traditional stage recorder is a float-driven device that is mounted in a *stilling well* located in or beside the river, with an intake to ensure the stage in the well is the same as in the river (Figure 15.9b). Changes in the stage of the river are recorded as the float rises and falls with the water level. The relationship between stage and discharge is plotted as a *rating curve* from measurements of both parameters taken under different flow conditions. Once the rating curve has been compiled for the river, the subsequent values for discharge can be determined for any measured stage and a **hydrograph** of the variation of discharge or stage through time can be plotted (Figure 15.10).

Canada's rivers drain to the three oceans that border the country: the Atlantic, the Pacific, and the Arctic (including Hudson Bay). The Arctic drainage basin is by far the largest, at 7,444,000 km^2. The Atlantic (1,520,000 km^2) and the Pacific (1,009,000 km^2) are much smaller. A small portion of Canada's water from southern Alberta and Saskatchewan (21,000 km^2) flows to the Mississippi River and on to the Gulf of Mexico. The largest rivers in Canada are the St. Lawrence, which drains from the Great Lakes, and the Mackenzie River, which flows from Great Slave Lake to the Arctic Ocean (Figure 15.11). Both have an average annual discharge of less than 10,000 $m^3 s^{-1}$. Hudson Bay receives 30,900 $m^3 s^{-1}$ annually from the numerous rivers flowing into it.

The discharge of most rivers increases downstream as a natural consequence of the way streams and rivers combine to deliver runoff and sediment to the oceans (Figure 15.11). The gradient also changes in a downstream direction. The general rule is the larger the cross-sectional area of the stream, the lower the gradient. Great rivers, such as the Mackenzie, Mississippi, and Amazon, have gradients so low that they can be described as "flat." For example, the water surface of the lower Mississippi River falls in elevation about 3 cm for each kilometre downstream.

The slope (S) or the gradient of a channel represents the difference in elevation between two points divided by the distance between them, and is related to the stream's velocity. Pools are located at various points along a stream where the slope is low and the water moves slowly. Between the pools are stretches of rapids or riffles, where the slope is steeper and the water moves more swiftly (Figure 15.12). Since no new tributaries enter this stretch of the stream, discharge (Q) remains constant. If the slope becomes steeper and the velocity (V) increases, the equation $Q = A \times V$ indicates that the cross-sectional area (A) of the stream must decrease, since Q is unchanged. In pools, the slope is low, velocity is slow, and cross-sectional area is large. In the riffles, the situation is reversed—slope is steep, velocity is high, and cross-sectional area is reduced.

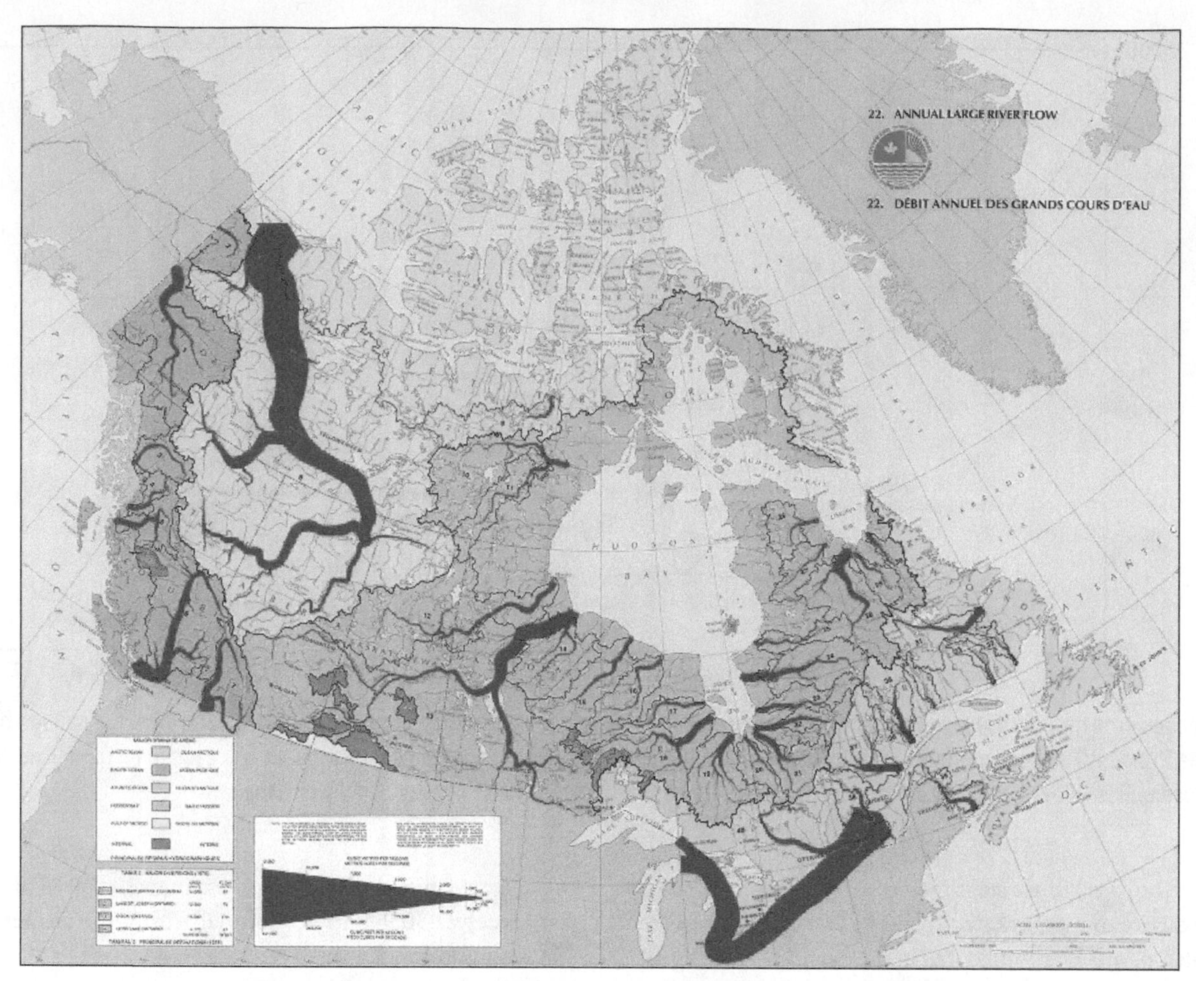

15.11 DISCHARGE OF CANADIAN RIVERS Widths of the rivers as drawn are proportional to mean annual discharge. (From Natural Resources Canada.)

As a stream flows down its course, the potential energy due to the higher elevation of water in the stream's upper reaches converts into kinetic energy. The process is driven by gravity, which supplies energy to overcome the friction between the water and the riverbed. This generates turbulence in the water, and the power to erode and transport the sediment load of the river. Energy is also dissipated as heat (warming the water), and to a lesser extent as sound. The heat is eventually lost by conduction, radiation, or evaporation to the air above, and by conduction into the channel walls.

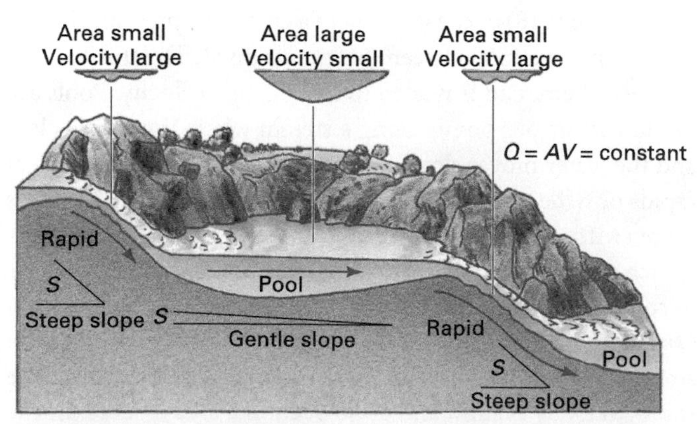

15.12 CHANNEL MORPHOLOGY Mean velocity, cross-sectional area, and slope change in the pools and rapids of a stream with uniform discharge. (© A.N. Strahler.)

DRAINAGE SYSTEMS

As runoff moves to lower levels and eventually to the sea, it becomes organized into a **drainage system**. The system consists of a branched network of stream channels, as well as the adjacent ground surfaces that contribute overland flow to them. Between the channels on the crests of ridges are *drainage divides*, which mark the boundary between slopes that contribute water to different streams or drainage systems. The entire system is bounded by an outer drainage divide that outlines a more-or-less pear-shaped **drainage basin** or **watershed** (Figure 15.13). Each fingertip tributary receives runoff from a small area of land surface surrounding the channel. This water flows downstream and merges with runoff from other small tributaries. The drainage system thus provides a converging mechanism that funnels overland flow into small channels and then to streams and rivers.

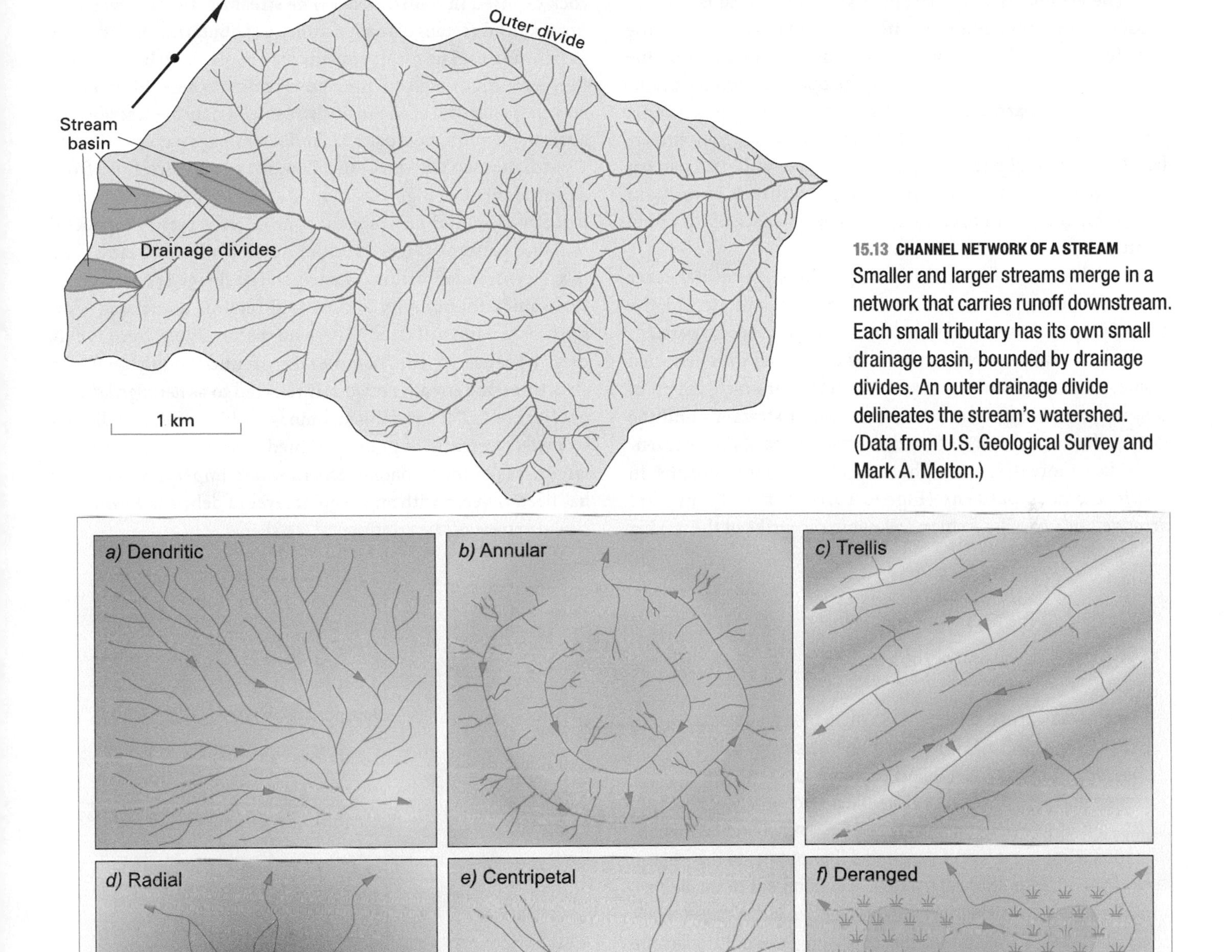

15.13 CHANNEL NETWORK OF A STREAM Smaller and larger streams merge in a network that carries runoff downstream. Each small tributary has its own small drainage basin, bounded by drainage divides. An outer drainage divide delineates the stream's watershed. (Data from U.S. Geological Survey and Mark A. Melton.)

15.14 DRAINAGE PATTERNS Distinctive stream drainage patterns develop in response to local geologic factors. Drainage patterns are classified on the basis of their form and channel density. (a) Dendritic drainage is best developed in areas where nearly horizontal sedimentary strata give rise to relatively gently sloping terrain and uniform lithology, which favours the development of this tree-like, branching pattern. (b) Annular drainage is associated with dome structures. (c) Trellis drainage is characteristic of linear folded strata. (d) Radial drainage flows out from a central peak, such as a volcano. (e) Centripetal drainage flows into a central depression; this is often the case in arid regions where much of water evaporates to form saline lakes or salt flats. (f) Deranged drainage occurs in areas such as the Canadian Shield where thin soils cover gently sloping terrain, underlain with impervious bedrock. Such areas are poorly drained with extensive swamps, bogs, or muskeg.

The arrangement of stream channels in a drainage basin most often resembles a tree with many small streams joining together to form larger tributaries, which eventually combine into a single river that carries all of the outflow from the drainage basin. This arrangement of stream channels is termed a *dendritic drainage* pattern; it is best developed in regions underlain by horizontal strata (Figure 15.14a). As well, this pattern also develops on metamorphic and igneous rock of uniform crystalline composition, and signifies a general lack of structural control.

In addition to dendritic drainage, there are several other types of stream patterns that are variously related to the underlying rock type and structure. *Annular drainage* patterns (Figure 15.14b) are characteristically associated with dissected domes. This produces a more-or-less circular arrangement of major streams cut into the exposed softer strata around the dome. Shorter tributaries flow down the slopes of the intervening ridges more or less at right angles to the bigger streams. In *trellis drainage* patterns (Figure 15.14c), the streams are arranged in a parallel fashion following the strike of the weaker rock exposed in folded strata. The stream network develops a pronounced rectangular appearance as tributaries tend to meet at right angles. The major streams may make sharp bends where they have cut though an intervening ridge. Where this is a general characteristic of the major streams, it may be described as a *rectangular drainage* pattern. *Radial drainage* patterns (Figure 15.14d) develop where streams flow away in all directions from a central point of high elevation. Radial drainage is characteristically observed on volcanic cones. Conversely, *centripetal drainage* patterns (Figure 15.14e) develop where streams flow into a central depression. Centripetal drainage is common in fault block topography and in arid regions associated with saline lakes and salt flats. Where no discernible pattern is evident because of the complexity of underlying geologic structures, the stream network is referred to as *deranged drainage* (Figure 15.14f). Deranged drainage, which is often indicated by extensive tracts of poorly drained land, is characteristic of large areas of the Canadian Shield where impervious bedrock has been covered with an irregular layer of debris following the recent retreat of the continental ice sheets.

GEOGRAPHER'S TOOLS
STREAM NETWORKS AS TREES

In mathematics, a branching system of lines and points is called a "tree," and it is part of the discipline called *topology*. A map of a drainage system, such as presented in Figure 1, shows the tree-like form of a typical network of stream channels. It consists of two kinds of information. First is the set of connected line segments that represent the stream channel system. A second set of connected dashed lines consists of all the drainage divides—those topographic crests that direct the overland flow onto adjacent slopes. When the divides are mapped, they delineate individual drainage basins. Thus, a network of basins can be considered a topological set.

Topology recognizes curving and twisting lines, and also assumes that both lines and points have no breadth or thickness; they are drawn that way only for ease of representation. The basic terminology of topology is shown in Figure 2. What is usually called a "point" is a *node*; a "line" is an *arc* (or *edge*). Nodes are positioned at both ends of an arc. Arcs can form a

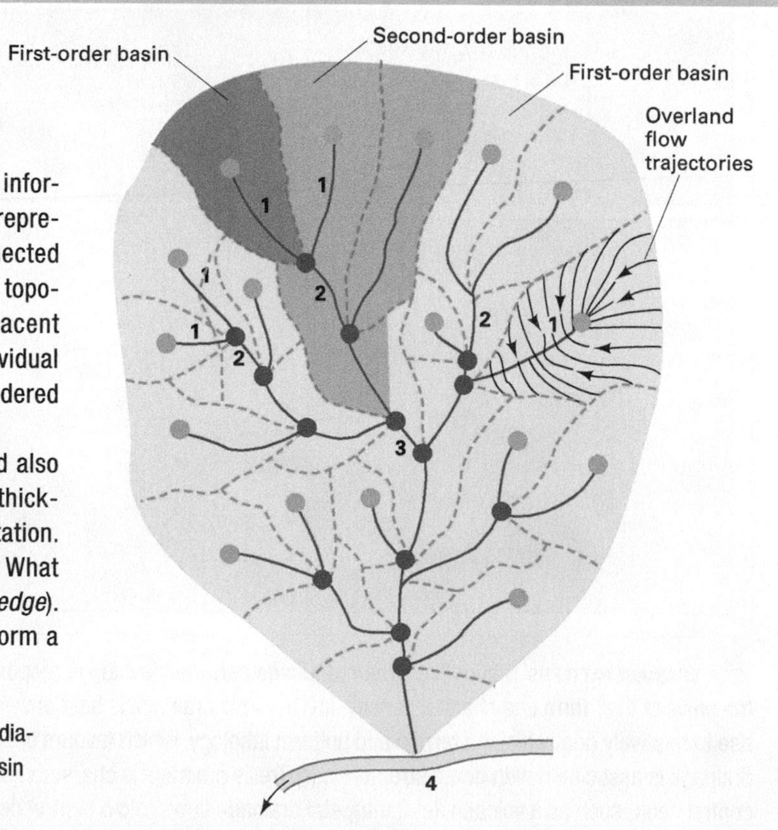

Figure 1 Schematic map of a third-order drainage basin This diagram shows both the stream channel system and the drainage basin network. (© A. N. Strahler.)

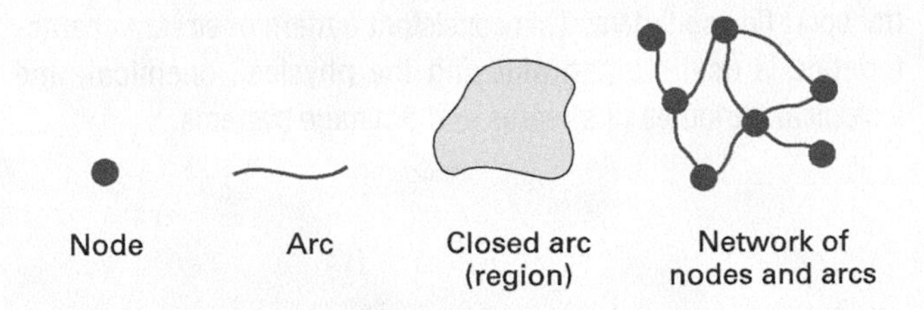

Figure 2 Basic forms used in topology

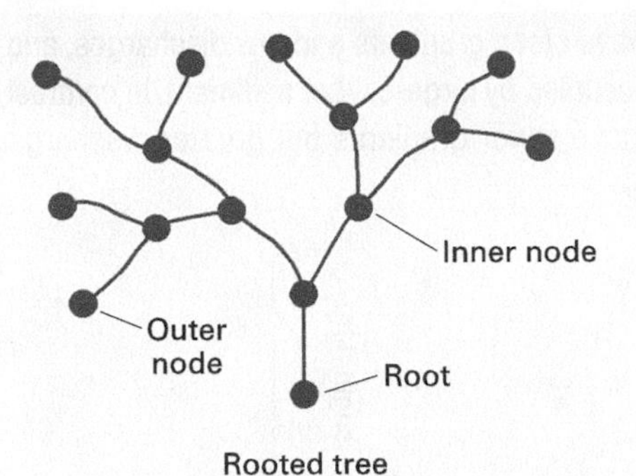

Figure 3 Topological diagram of a rooted tree

closed loop containing a *region*. An assemblage of arcs connected by nodes is called a *network*. All these forms can lie in a space of any dimension, although here only topologies in real spaces of either two dimensions (planes) or three dimensions (volumes) are considered.

Figure 3 is a formal topological diagram of a type of network called a *rooted tree*. It consists solely of nodes and arcs and recognizes two types of nodes: *inner nodes* and *outer nodes*. Each inner node connects with three arcs, and each outer node connects with only one inner node. A third kind of node, called the *root*, serves as a starting point for the construction. This topological system serves well as a model not only of a stream system, but also of several kinds of biological systems, among them the branching patterns in trees and shrubs, venation of leaves, and even the respiratory pathways of the human lung. In these natural systems, a gaseous or fluid agent (water, sap, air) flows into or out of the system.

In Figure 1, the rooted tree represents a map of a natural stream system. It shows not only the network, but also the area of ground surface that encloses each stream arc. This surface slopes down from its perimeter—a drainage divide—to meet each stream channel, and thus provides the channel with its water and sediment. In topology it is called a *region*, consisting of an area bounded by a closed loop of nodes and arcs. In this case, the loop bounding the region is composed of the arcs and nodes of the drainage divide network. All of these features—stream tree, drainage divide network, and regions that contribute flow to the channels—lie above a reference level and constitute a three-dimensional system. They are projected upon a flat base in the figure, so the map is described as *planimetric*.

Geomorphologists use a convenient ordering system for the arcs and nodes within a tree. A single arc with nodes at each end is called a *stream segment*. The outermost (terminal) segments are designated as *first-order segments* (or *first-order streams*). Where two first-order segments join, they generate a *second-order segment*, and so forth, until they reach the single root segment. That terminal node may join an existing stream channel of the same or higher order, but it could also terminate at the shoreline of a lake or ocean. For each stream segment, there is an enclosing drainage basin of the same order. Figure 1 illustrates an example of a *first-order basin*, showing the overland trajectories of surface runoff.

Given time, the streams in the flow network will adjust their form and slope so that all of the precipitation that falls on them passes along to the outlet. In this state, each stream channel segment has the channel form and gradient necessary to denude its drainage area and to transport erosional debris to the next segment.

First-order streams are perennial streams, which carry water all year. When two first-order streams come together they become a second-order stream. Some adjustments must be made to the channel geometry and flow rates, since below this junction the discharge will (on average) have doubled. When two second-order streams come together, they form a third-order stream, and so on, with discharge doubling each time. Thus, when a stream of a given order joins one of the same order, the downstream order is raised by one. However, if a stream joins a higher-order stream, its order remains unchanged. Stream ordering provides a method of measuring the relative size of streams. These range from the smallest first-order streams to eleventh-order streams for large drainage systems, such as the Amazon River. Over 80 percent of the total length of Earth's rivers and streams are first- and second-order.

In the case of streams, the topological ordering of stream segments provides a useful and informative way to study fluvial systems and the way that flowing water erodes landscapes. Morphometric analysis of stream networks has shown that the relationship between stream number, mean length, and mean basin area is approximately linear on semi-log plots (Figure 4). Thus, first-order streams characteristically tend to be numerous and short in length. In contrast, higher order streams are less common, but are longer and have lower gradients. Several other characteristics show similar relationships. For example, first-order streams have steep gradients and low discharges, and their channels are occupied by large-calibre sediment. In contrast, high-order streams have lower gradients but greater discharges, and they transport fine sediment. This consistent pattern of stream characteristics is useful when studying the physical, chemical, and biological attributes of streams and drainage systems.

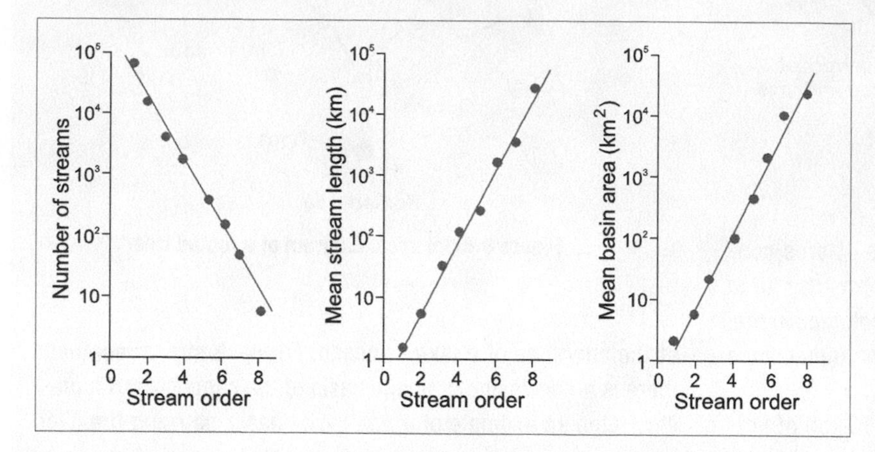

Figure 4 Stream morphometry Semi-logarithmic plots of many attributes of stream networks show a linear relationship with stream order, as shown in this hypothetical drainage system.

STREAM FLOW

A stream's discharge increases in response to a period of heavy rainfall or snow melt. However, this response is delayed, as the movement of water into stream channels takes time. The length of delay depends on a number of factors, the most important of which is the size of the drainage basin feeding the stream. Larger drainage basins show a longer delay.

Figure 15.15 shows a stream hydrograph for Sugar Creek, Ohio, covering a four-day period that included a rainstorm that lasted for 12 hours. The average total rainfall over the watershed of Sugar Creek was about 15 cm. The rainfall and runoff graphs show that prior to the onset of the storm, Sugar Creek was carrying a small discharge. This flow, which is supplied by the seepage of groundwater into the channel, is called *base flow*. After the heavy rainfall began, several hours elapsed before the stream gauge began to show a rise in discharge. This delay indicates that the channel network was acting as a temporary reservoir. At first, the channels were receiving inflow more rapidly than they could pass down the channel system to the stream gauge.

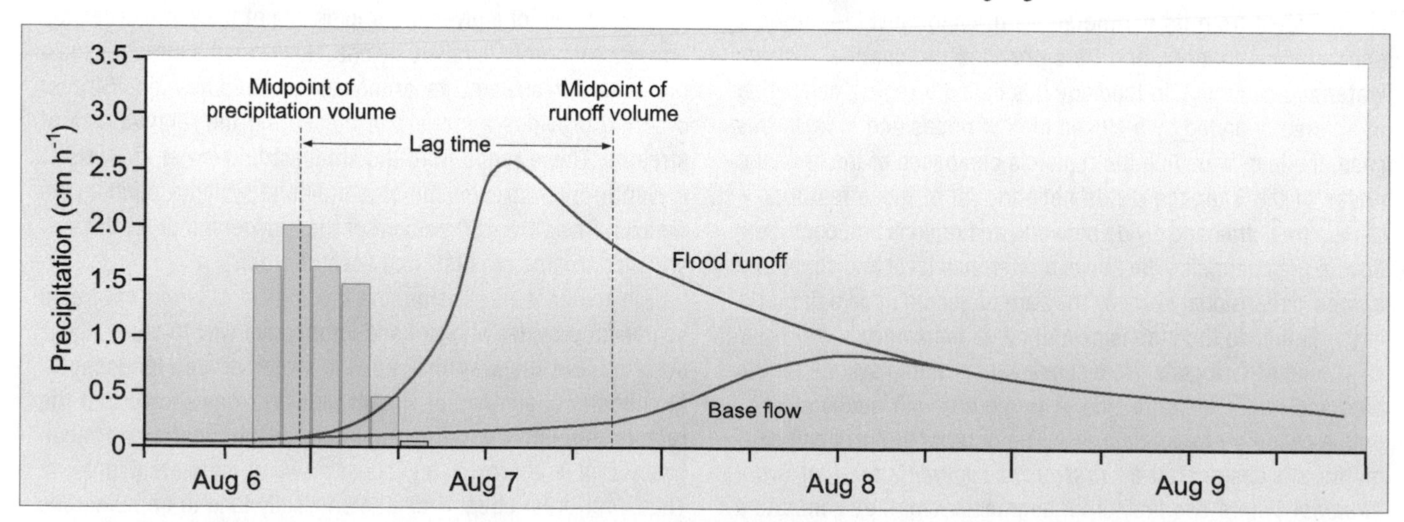

15.15 SUGAR CREEK HYDROGRAPH Precipitation and stream flow were recorded for this 800 km² drainage basin in Ohio following a heavy summer rainstorm. The bar graph shows precipitation rates based on two-hour intervals. (After Hoyt and Langbein, *Floods*, Princeton University Press.)

Lag time is the difference between the time when half the precipitation has occurred and the time when half the runoff has passed downstream. In Sugar Creek the lag time was about 21 hours, with the peak flow reached almost 18 hours after the rainfall began. Note also that the stream's discharge rose much more abruptly than it fell. In general, the larger the watershed, the longer the lag time between peak rainfall and peak discharge, and the more gradual the rate of decline of discharge after the peak has passed. Another typical feature of a flood hydrograph is the slow but distinct rise in the amount of discharge contributed by base flow.

THE ANNUAL FLOW CYCLE OF A LARGE RIVER

In regions of wet climates, where the water table is high and normally intersects the important stream channels, the hydrographs of larger streams clearly show the effects of two sources of water: base flow and direct input, including overland flow. This is illustrated by the hydrograph of the North Saskatchewan River at Prince Albert, Saskatchewan (see Figure 15.10). This river drains a large watershed of 131,000 km^2 on the Alberta and Saskatchewan Prairies. Its headwaters are high in the snowfields and glaciers of the Rocky Mountains to the west. Discharge during winter, from December to March, is low because almost all the precipitation falls as snow and is stored in the snowpack covering the drainage basin. At this time, base flow from groundwater contributes almost all of the discharge to the river—about 100 m^3 s^{-1}.

Snow melt on the Prairies in March and April provides the first discharge peak flows of the year. Contribution from the melt of the large snowpack in the headwaters is delayed until June due to the lag time along the 1,000 km of river channel between the mountains and Prince Albert, and due to the later melt at higher elevations in the mountains. However, both factors contribute to groundwater storage and raise the base flow to about 300 m^3 s^{-1} through the summer and into the autumn. Rainstorms, especially in August and September, cause secondary peaks in discharge and contribute to maintaining the base flow. In November, a cold snap at the onset of winter generally sets ice on the river and reduces the discharge.

Rivers with headwaters in high mountains have flow characteristics that are especially desirable for irrigation and for flood prevention. The higher ranges serve as snow storage areas, slowly releasing winter and spring precipitation in early or midsummer. As summer progresses, melting moves to successively higher levels. In this way, a continuous, sustained river flow is maintained. Water managers and farmers on the Prairies have recently become concerned that the shrinking of glaciers in the Rockies in a warming climate of the twenty-first century will result in a decline in discharge in large rivers, causing water shortages in this vital agricultural region.

RIVER FLOODS

A **flood** is defined as the condition that exists when the discharge of a river cannot be accommodated within its normal channel. As a result, the water spreads over the adjoining ground. Most rivers of humid climates have a *flood plain*, a broad belt of low, flat ground bordering the channel on one or both sides that is flooded periodically. Flooding usually occurs in the season when abundant surface runoff combines with the effects of a high water table to supply more water than the channel can carry.

Inundation that occurs almost every year is also considered a flood, even though its occurrence is expected and does not prevent the cultivation of crops after the water has subsided. Annual flooding does not interfere with the growth of dense forests, which are widely distributed over low, marshy flood plains in all humid regions of the world. The Water Survey of Canada, which provides a flood-warning service, designates a specific river stage at a given place as the *flood stage*. Above this critical level, inundation of the flood plain will occur. Higher discharges of water, the rare and disastrous floods that may occur as seldom as once in 30 or 50 years, inundate ground lying well above the flood plain, as was the case in many parts of Australia in 2011.

Flash floods are characteristic of streams draining small watersheds with steep slopes. These streams have short lag times—perhaps only an hour or two—so when an intense rainfall event occurs, the stream quickly rises to a high level. The flood arrives as a swiftly moving wall of turbulent water, sweeping away buildings and vehicles in its path. In drier regions, the flood water sweeps great quantities of coarse rock debris into the main channel, producing debris floods (see Chapter 14). In forested landscapes, flood water can uproot trees and carry considerable quantities of soil, rocks, and boulders downstream. Because flash floods often occur too quickly to warn affected populations, they can also cause significant loss of life.

FLOOD PREDICTION

The magnitude of a flood is usually measured by the peak discharge or highest stage of a river during the period of flooding. The flood history of any river shows that large floods occur less frequently than smaller ones; that is, the greater the discharge or higher the stage, the less likely the flood. In Canada, the Water Survey of Canada (WSC) monitors water levels in rivers using a network of about 2,500 gauges, many of which are operated in co-operation with provincial governments and other agencies. Historical data from any of the stations, as well as "real time" data

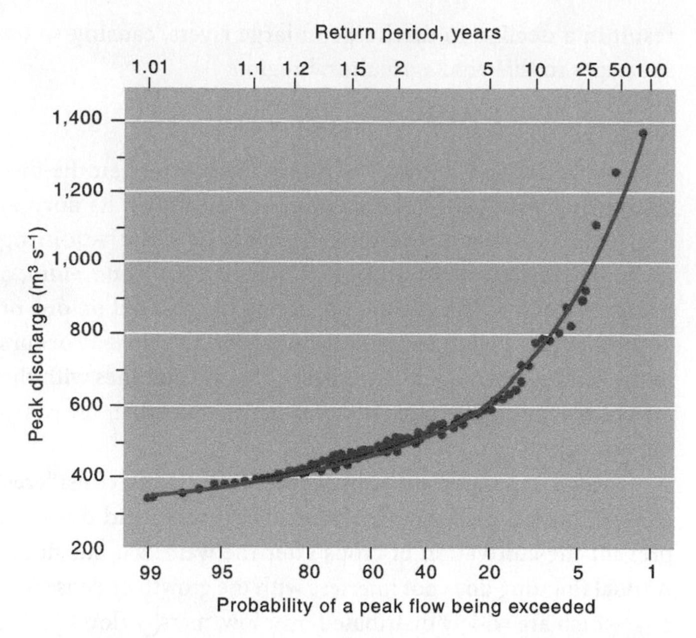

15.16 FLOOD FREQUENCY DATA FOR THE LILLOOET RIVER AT PEMBERTON, BRITISH COLUMBIA Each dot is a measured maximum yearly discharge in an 88-year record. (Data from the Water Survey of Canada.)

from many stations, are available from the WSC website (www.wsc.ec.gc.ca). These data help scientists understand a river's previous behaviour, and are important for monitoring changes in stream flow to predict the onset of a flood in order to alert emergency services in time to deal with its impact.

In nature, extreme events are uncommon, with those of great magnitude occurring only very infrequently. For example, records of the frequency of floods on the Lillooet River in British Columbia show that the most severe floods can be expected about every 50 to 100 years (Figure 15.16). Each dot in this graph plots the maximum average daily discharge of the river recorded in a particular year (that is, the peak flow, or largest flood, each year). The lower horizontal scale indicates the probability that the given discharge will be equalled or exceeded in a given year. For example, a 20 percent discharge is interpreted to mean "a discharge of this magnitude (about 600 $m^3 s^{-1}$) can be expected to be equalled or exceeded in 20 out of 100 years." The numbers on the upper horizontal scale show the *recurrence interval* (or *return period*). This value is the probability divided into 100. For example, the return period for the probability of 20 percent is $100/20 = 5$ years. This means that, over a long period of time, a maximum annual flood of about 600 $m^3 s^{-1}$ can be expected once every five years. Note that the recurrence interval is another way of expressing a probability. Thus, if a river experiences a "five-year flood" in one year, it does not mean that the next flood of the same magnitude will occur five years later. There is nothing to prevent two five-year floods from occurring in successive years.

The recurrence interval is calculated by reordering the peak discharges for each year from greatest to least, assigning a rank of 1 to the largest value, 2 to the next largest, and so on. The recurrence interval is determined for each flow from its rank, using the formula:

$$I = (N + 1)/R$$

where I is the recurrence interval in years, N is the number of years in the record, and R is the rank of the value. Suppose that a flow is the largest in a set of 44 years of observations. Its recurrence interval is given by $I = (N + 1)/R = (44 + 1)/1 = 45/1 =$ 45 years. For a flow that is the fifth largest, the recurrence interval would be $I = (44 + 1)/5 = 45/5 = 9$ years.

Another tool used to present the flood history of a river is the flood expectancy graph (Figure 15.17). The graph for the Fraser River at Mission, British Columbia, illustrates a large river responding to spring floods, with highest water levels in June and low levels in February and March. The red line indicates the flood stage, in this case 7 m. Note that daily stage values can greatly exceed the monthly averages shown on the graph; this could cause frequent flooding for periods shorter than a month. However, because the flood plain in the lower reaches of the Fraser is heavily populated as part of the city of Vancouver and outlying municipalities, large dikes have been built to protect against floods of up to 15 m.

The drainage basin of the Saint John River at Fredericton, New Brunswick, is less than one-fifth the size of the Fraser, and its discharge is about half. Both rivers have similar annual stage patterns except that spring snow melt runoff peaks in April in New Brunswick; in British Columbia, runoff is delayed due to the later melt in the high mountains. A flood stage of 6 m on the Saint John River represents the onset of overbank inundation. Damaging floods do not occur until the river's stage exceeds about 6.5 m, which may happen in both April and May.

Canadian federal and provincial agencies coordinate flood forecasting from 10 regional centres across the country. When there is a threat of a flood, forecasters analyze precipitation patterns and the progress of high waters moving downstream. By examining the flood history of the rivers concerned, they develop specific flood forecasts. These are supplied to communities within the associated district, which usually covers one or more large watersheds. Flood warnings are publicized by every

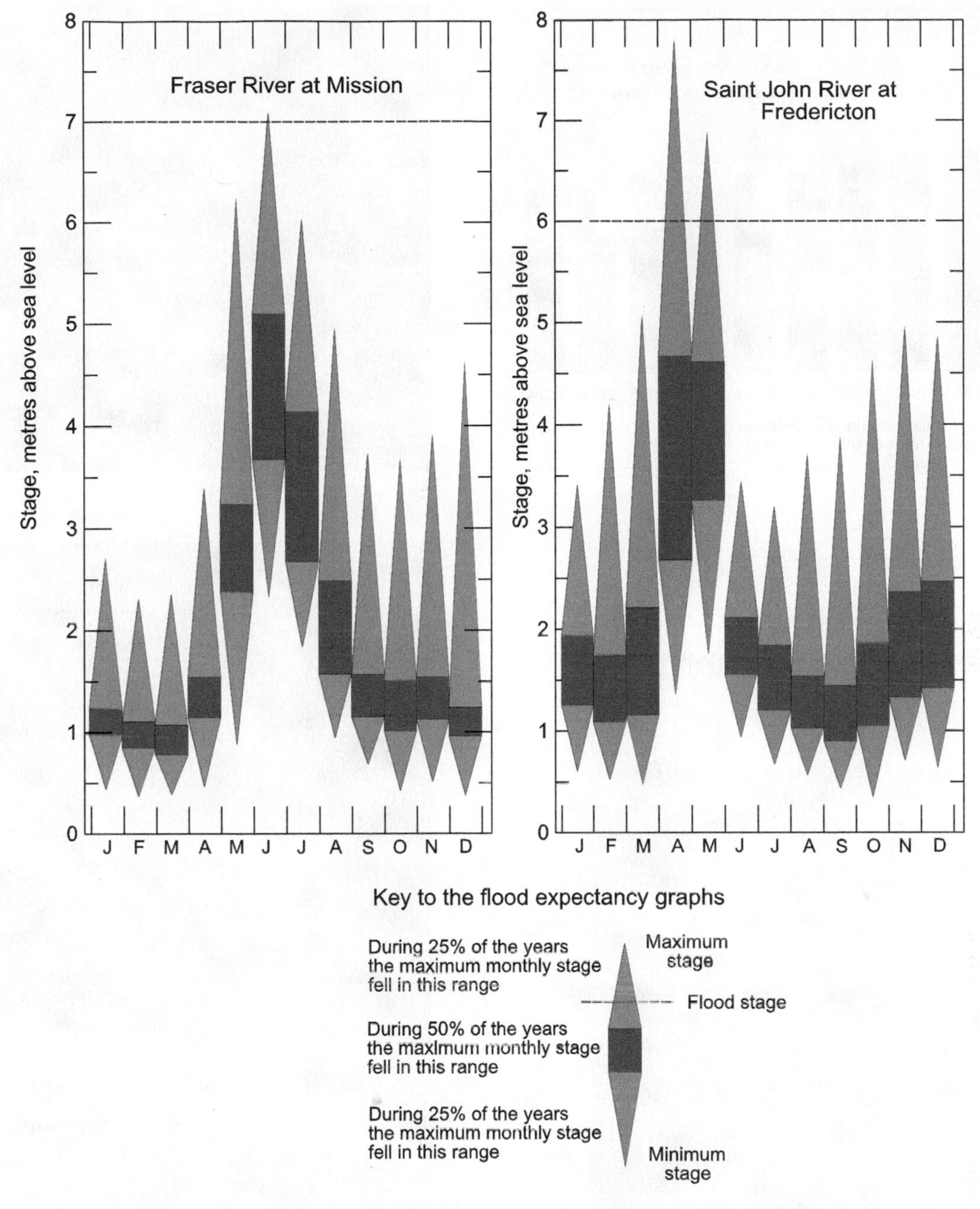

15.17 FLOOD EXPECTANCY GRAPHS Maximum and minimum monthly stages of the Fraser River at Mission, British Columbia, and the Saint John River at Fredericton, New Brunswick.

possible means, and various agencies co-operate closely to plan the evacuation of threatened areas and the removal or protection of property.

THE RED RIVER FLOODS

The Red River region of southern Manitoba is an area that is periodically subject to severe flooding. The Red River flows northward from the United States, principally from Minnesota and North Dakota, through the city of Winnipeg, and then on to Lake Winnipeg in central Manitoba. Its course is over the former bed of glacial Lake Agassiz, the largest ice-marginal lake in North America, which was created by the retreating continental glaciers (see Chapter 18). The broad, flat plain of this ancient lake floor provides an extensive area easily flooded by the rivers that flow across it.

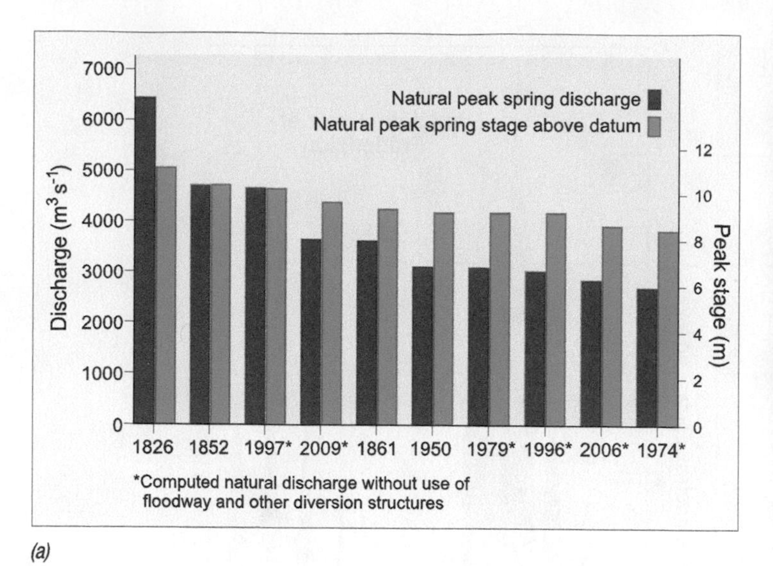

(a)

(b)

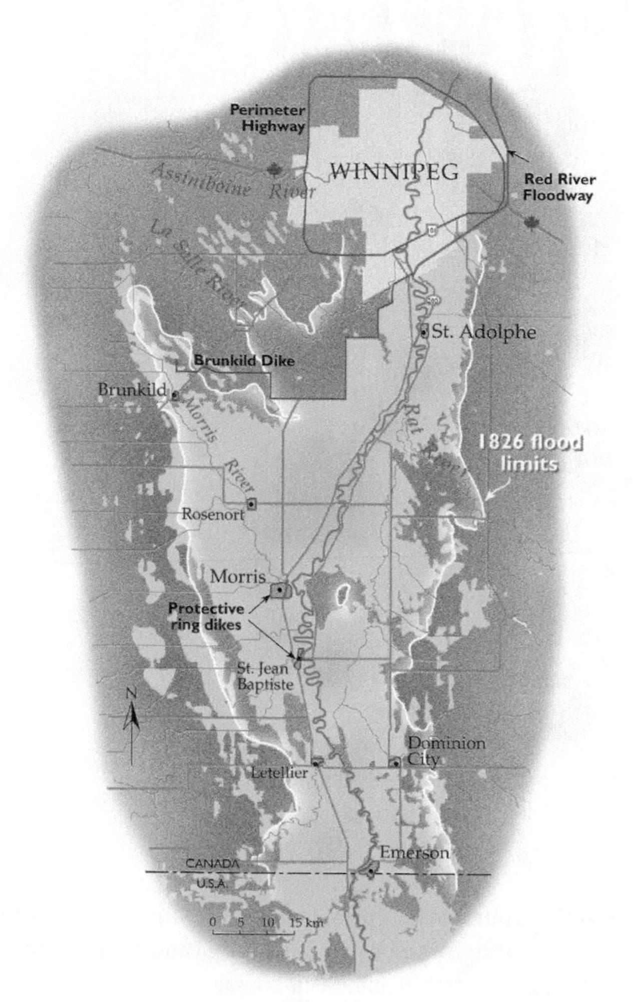

(c)

(d)

15.18 FLOODING IN MANITOBA (a) The graph shows the top 10 Manitoba floods since 1800. (b) The Red River Floodway (on the right) rejoins the Red River. The purpose of the floodway is to divert large floods around the city and so protect the urban centre. (c) The map of southern Manitoba shows the extent of flooding of the Red River in 1826 and 1997. (d) The ice jam in Selkirk, Manitoba, on the Red River as seen from the air, with flood waters approaching on April 9, 2009.

The largest of the floods in the recorded history of Manitoba occurred in 1826 (Figure 15.18a). A more recent flood in 1950, although not reaching the same level and discharge, was important because the region was more heavily populated by then and this event caused the largest evacuation of people in Canadian history. As a result of the 1950 flood, the 47 km-long Red River Floodway (Figure 15.18b) was completed in 1968 as part of a flood protection scheme, which also included diversion of the Assiniboine River to Lake Manitoba. The purpose was to divert flood waters around the city of Winnipeg, and so protect the urban area from flooding.

Several major floods have occurred since the floodway was constructed. The largest flood since 1852 occurred in 1997. With a return period of 110 years, this event, which was named the "flood of the century," severely tested the emergency measures that had been put in place. Four factors caused the flood. First, heavier than normal rains during the autumn of 1996 saturated the soil, leaving little opportunity for the ground to absorb the following spring's snow melt. Second, snowfall in the winter of 1996–97 was greater than average, especially in the upper portions of the drainage basin. Third, a spring blizzard on April 5, shortly before the flood began, dumped another 50 cm of snow on the drainage basin. Fourth, the basin experienced one of the most rapid snow melts on record, which proceeded from south to north. Consequently, peak flows on the tributaries coincided with the progress of the main flood on the Red River as it moved north toward Winnipeg.

Flooding began on April 20, and the flood crested in Winnipeg on May 3. Although most of the previously constructed dikes around smaller communities held, about 1,850 km^2 of land was flooded and 22,000 people were evacuated from rural areas (Figure 15.18c). The Red River Floodway greatly reduced the impact of the flood in Winnipeg itself, although some 6,000 urban residents were also evacuated. Damage from the flood was estimated at $150 million, with another $200 million in agricultural losses.

The flood of 2009 was the second highest on the Red River since 1852, but because of its extended duration, the total discharge during this event makes it the largest flood on record. On April 16, 2009, the city of Winnipeg declared a state of local emergency as flood waters crested at 6.85 m, only 0.6 m below the 1997 crest. At various times the city had more than 30 flood pumping stations in operation, together with as many as 27 temporary pumps and 210 sewer control gates and numerous temporary dikes built from sandbags. Without the Floodway and other control works, the flood crest would likely have reached 9.9 m. The extent of the flooding was 1,000 km^2, about half of the area inundated in 1997.

The 2009 flood resulted in part from record high soil moisture levels at the time of freeze-up in 2008, heavy snowpacks in the part of the drainage basin in the United States, and significant ground frost that increased spring runoff. In addition, unusually heavy and persistent river ice conditions caused severe ice jamming north of Winnipeg (Figure 15.18d), as well as interfering with flood control structures. Ice typically moves freely at the Floodway inlet when flows reach 1,000 to 1,150 m^3 s^{-1}. In 2009, ice remained immobile until flows reached about 2000 cm^3 s^{-1}, resulting in the unprecedented decision to operate the Floodway while it still had a solid ice cover at both the inlet and outlet structures, as well as within the city of Winnipeg. Despite these setbacks, the combination of permanent and temporary flood control methods is estimated to have saved $10 billion in flood damage in the city of Winnipeg, $700 million in damage to 13 smaller communities in the Red River Valley, and $100 million in damage to farms and isolated residences.

LAKES

A **lake** is a body of standing water found within continental margins that is enclosed on all sides by land. The Caspian Sea, which contains 78,700 km^3 of water, is by far the largest lake on Earth. Other large lakes include Lake Baikal in central Asia (23,615 km^3), Lake Tanganyika in the Great Rift Valley of Africa (18,940 km^3), and Lake Superior in central North America (12,251 km^3). Although these are immense water bodies, lakes range in size down to the smallest ponds containing only a few hundred cubic metres of water. Most lakes contain fresh water, although some are saline, with salinities much higher than the ocean.

Lakes are dynamic environments. They respond to seasonal changes in climate and hydrology, and in turn influence the land around them. Lakes modify climate by moderating surrounding air temperatures and adding water vapour to the local atmosphere. *Lacustrine*, or lake-related, sediments, which settle year after year, are important repositories of environmental and paleoenvironmental information. Lakes are also important to humans' needs. The science of *limnology* studies the physical, chemical, and biological processes of lakes.

Lakes are especially important in Canada, which contains more than half of the total surface area of lakes in the world. Canada has nearly half a million lakes with surface areas greater than 1 km^2, covering 8.4 percent of the country's land area. In Canada, lacustrine processes and landforms are extremely important. North America's Great Lakes are a unique resource of immense value to both Canada and the United States. *Eye on the Environment • The Great Lakes* provides more information about the formation of the Great Lakes, their effects on local climate, and their degradation by water pollution.

EYE ON THE ENVIRONMENT
THE GREAT LAKES

The Great Lakes—Superior, Huron, Michigan, Erie, and Ontario—along with their smaller bays and connecting lakes, form a vast network of inland waters in the heart of North America. They contain 23,000 km^3 of water—about 18 percent of all the fresh surface water on the Earth. Of the Great Lakes, Lake Superior is by far the largest. In fact, the volume of the other Great Lakes combined would not fill its basin. Less than 1 percent of the Great Lakes waters is renewed annually through precipitation, runoff, and infiltration. The net basin water supply is estimated to be 500 billion litres per day, which is equal to the discharge into the St. Lawrence River.

The total population of the Great Lakes basin is approximately 42 million, of which 36 million—8 million Canadians and 28 million Americans—live in large metropolitan areas. The lakes are an essential resource for these people, and great demands are put on them for drinking water, as well as power generation, manufacturing, transportation, agriculture, and recreation. In 2004, daily water withdrawal was 164 billion litres, of which 95 percent was returned to the lakes. The bulk of the water is used in power generation (Figure 1).

The Great Lakes play an important role in moderating the climate of their region. Because water bodies heat and cool more slowly than adjacent land, autumn and winter temperatures near large lakes are warmer, while spring and summer temperatures are cooler. For the Great Lakes, the difference may be as much as 5 to 10°C, depending on the prevailing winds. As a result, the frost-free period near the lakes is up to 50 days longer compared to more distant locations. To the east of each of the lakes are snow belts where average snowfall ranges from 200 to 350 cm per year, about twice the amount that would occur otherwise. This lake-effect snowfall is the result of evaporation from the relatively warm, ice-free water surface of the lake, followed by condensation in cold, winter air.

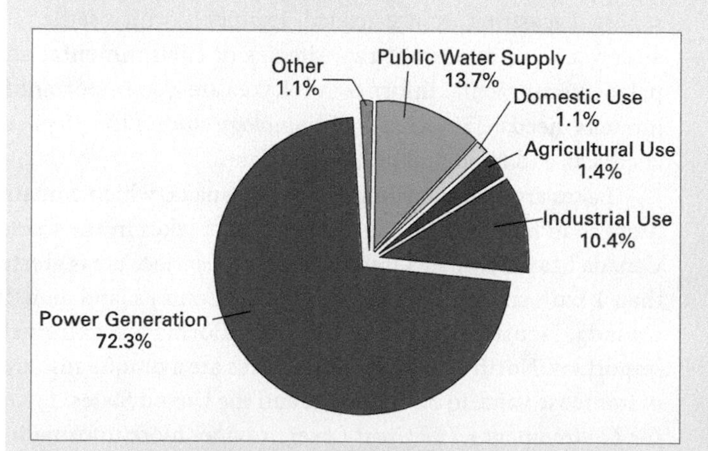

Figure 1 Water withdrawals by use, 2004 Data from *State of the Great Lakes 2009* report.

Ice forms on the Great Lakes each winter. Although Lake Erie is the southernmost of the Great Lakes, it is normally more than 90 percent ice covered in February because its volume is small and it has a limited ability to store heat. Lake Superior (at 75 percent), Lake Huron (70 percent), Lake Michigan (50 percent), and Lake Ontario (25 percent) are less ice covered on average, although the amounts vary greatly from year to year. For example, during the strong El Niño winter of 1983 when temperatures in eastern North America were several degrees warmer than normal, ice cover ranged from 45 percent in Lake Huron to just over 10 percent in Lake Ontario. With global climate change, the ice cover on the lakes is predicted to decrease, and by the last decades of this century, ice may not form at all on some of them.

The sediments of the Great Lakes provide a record of human settlement in their drainage basins. Analysis of sediment core samples shows that logging and clearing for agriculture caused accelerated erosion beginning in the nineteenth century. The input of pollutants associated with industrialization and urbanization increased sharply through the twentieth century.

By the 1960s and 1970s, excess nutrients, especially phosphorus from fertilizer, sewage, and detergents, had entered Lake Erie in amounts sufficient to cause a large increase in the lake's biological productivity. The large amount of biomass produced by this "fertilization" depleted the available oxygen dissolved in the water as it decayed. Suffocation threatened much of the aquatic life in the lake. The Great Lakes Water Quality Agreement between Canada and the United States, passed in 1972 and renewed in 1978, was the first step taken to remediate pollution of the lakes, and was an important recognition of the environmental degradation they have suffered.

Pollution caused by a large number of substances from many sources continues to be a problem in the Great Lakes, even though vigilance has reduced the impact of many pollutants through documentation of their presence and clean-up efforts. Persistent organic pollutants are one group of particular concern. These include substances, mainly of industrial origin, that are long lasting, highly mobile in the aquatic system, and toxic in very small amounts. Many of these compounds accumulate up the food chain as predators consume contaminated prey (Figure 2). For example, the concentration of polychlorinated biphenyls

(PCBs) in the algae of Lake Ontario has been measured at 0.02 ppm. This very small amount is equivalent to about 0.5 mL in volume. However, this low concentration increases as it moves up the food chain through the process of biomagnification—the eggs of herring gulls contain as much as 24 ppm. This amount is still small, but it is 10 million times greater than the concentration in the lake water, and is sufficient to cause deformities and high mortality in the chicks.

Since 1998, the United States Environmental Protection Agency and Environment Canada have coordinated a biennial assessment of the ecological health of the Great Lakes ecosystem using a consistent set of environmental and human health indicators. In 2008, the overall status of the Great Lakes Ecosyem was rated as mixed because some conditions or areas were good but others were poor. Extracts from the *State of the Great Lakes 2009* and the *Nearshore Areas of the Great Lakes 2009* reports are presented in Appendix 15.1, which shows how this rating was determined.

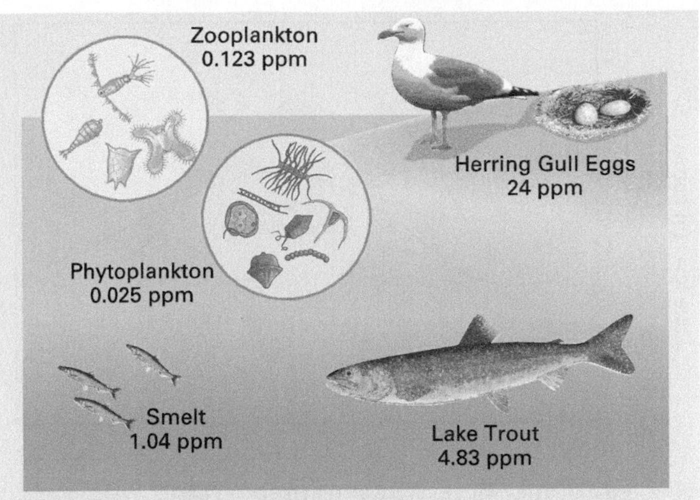

Figure 2 Concentration of PCBs in the Great Lakes food chain The degree of concentration of PCBs in each level of the Great Lakes aquatic food chain increases from aquatic plants to top predators. The eggs of fish-eating birds, such as herring gulls, have the highest levels. (From the U.S. Environmental Protection Agency.)

ORIGINS OF LAKES

The basins that contain lakes originate from many different processes acting on the Earth's surface. Tectonic activity creates rift valleys where some of the deepest lakes on the Earth are found. Meteorite impact and volcanic eruptions form craters that also may contain lakes. Lakes are created on the flood plains of rivers by fluvial erosion and deposition, as well as by the solution of limestone in karst regions. Landslides can block rivers to create lakes in river valleys, and rivers are often dammed artificially to create reservoirs for water supply, hydroelectric power generation, and a variety of other human uses.

In Canada, most lakes originated during the Late-Cenozoic Ice Age when continental ice sheets eroded depressions in areas of weaker bedrock throughout the Canadian Shield. Lakes were also formed in the sedimentary deposits left behind by the melting ice (see Chapter 18).

Although lakes seem permanent from a human perspective, they are normally short-lived features on the geologic time scale. Lakes that have stream outflow will gradually drain as their outlets erode to lower levels. Also, lake basins fill with sediment carried by inflowing streams and with organic matter, either transported or produced *in situ* by plant and animals. As a group, the lakes of Canada are particularly young because the ice sheets that formed them retreated only about 13,000 years ago. The present retreat of alpine and arctic glaciers in response to a warming climate has resulted in the formation of new lakes.

WATER BALANCE OF A LAKE

Lake levels vary, usually rising in spring and early summer and dropping in late summer and fall. Lake levels also rise after a period of heavy rainfall and fall after several weeks of dry weather. Thus, a lake is a repository of water that responds to changing inputs and outputs. It receives water as precipitation and inflow from streams, and loses water from evaporation and outflow. The lake's volume, and therefore the lake level, change in response to variations in these inputs and outputs.

The water balance of a lake can be expressed by the simple equation:

$$\Delta V = P + I - E - O$$

where ΔV is the change in the volume of water in the lake, P is the precipitation that falls on the lake, I is the inflow of streams and groundwater to the lake, E is evaporation from the lake surface, and O is outflow from the lake. The time period over which the water balance is taken can vary. Often the quantities are taken on an annual basis and so are expressed in rates such as cubic kilometres per year. They can also be expressed as flow rates in metres per second, much like stream flow.

Changes in lake levels often follow cycles on different time scales. Annual cycles of precipitation and temperature modulate the lake's inputs and outputs and create regular seasonal variations in lake levels. On a decadal time scale, lake levels rise and fall in response to sequences of cool and wet years followed by

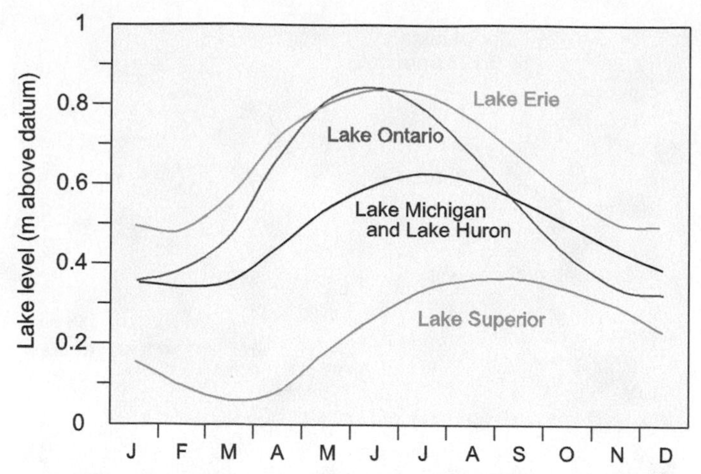

15.19 MEAN MONTHLY WATER LEVELS OF THE FIVE GREAT LAKES Water levels show a strong annual cycle. Data are mean monthly water levels over about 100 years and are shown with respect to the 1987 International Great Lakes Datum for each lake. (Data from the Canadian Hydrographic Survey.)

warm and dry years. Over hundreds to thousands of years, lakes can expand, shrink, or dry up completely in response to climate change. The Great Lakes, for example, exhibit a pronounced annual cycle, with lowest lake levels in winter and highest levels in late spring and summer following the input of snow melt (Figure 15.19). During the warm summer, stream flow decreases and evaporation increases. Starting in July, mean water levels fall steadily through the remainder of the year.

The lake levels not only fluctuate seasonally, but also change significantly from year to year. In Lake Ontario, for example, the 1930s was a period of warmer and dryer weather, and water levels were the lowest on record (Figure 15.20). The cooler and wetter conditions of the 1940s and 1950s resulted in higher than normal levels. In 1958, the Moses-Saunders Dam was completed near Cornwall, Ontario, and within a few years the long-term cyclical pattern in water level on Lake Ontario was no longer evident. Since the 1990s, levels have been declining, in part due to management of water levels at dams on the St. Lawrence Seaway and perhaps in part as an early response to a warming climate.

SALINE LAKES AND SALT FLATS

In moist climate regions with abundant precipitation, a lake fills until water escapes at the lowest point on its rim, which forms an outlet. However, in dry regions, inflows are smaller, and if balanced by evaporation, no outlet is necessary. In this case, the water balance equation for the lake lacks the outflow term, and changes in level are determined by inflow and precipitation (gain) and evaporation (loss). If inflow and precipitation increase, the lake level rises. At the same time, the lake surface increases in area, allowing a greater evaporation rate, thus achieving a new balance. Similarly, if the region becomes drier, reducing input and increasing evaporation, the water level falls.

Lakes without outlets commonly accumulate salts. Dissolved solids are brought into the lake by streams that usually begin in distant highlands where a water surplus exists. Since evaporation removes only pure water, the salts remain behind and the salinity of the water slowly increases. *Salinity* refers to the abundance of certain common ions in the water. Eventually, salinity levels reach a point where salts precipitate as solids. Notable examples of saline lakes include the Caspian Sea, Dead Sea, Aral Sea, and Great Salt Lake in Utah. In Canada, small saline lakes are found on the Prairies and in the interior of British Columbia (see Figure 11.23).

In regions where climatic conditions consistently favour evaporation over input, the lake may disappear. A shallow basin covered with salt deposits, known as a *salt flat* or *dry lake*, remains. On rare occasions, these flats are covered by a shallow

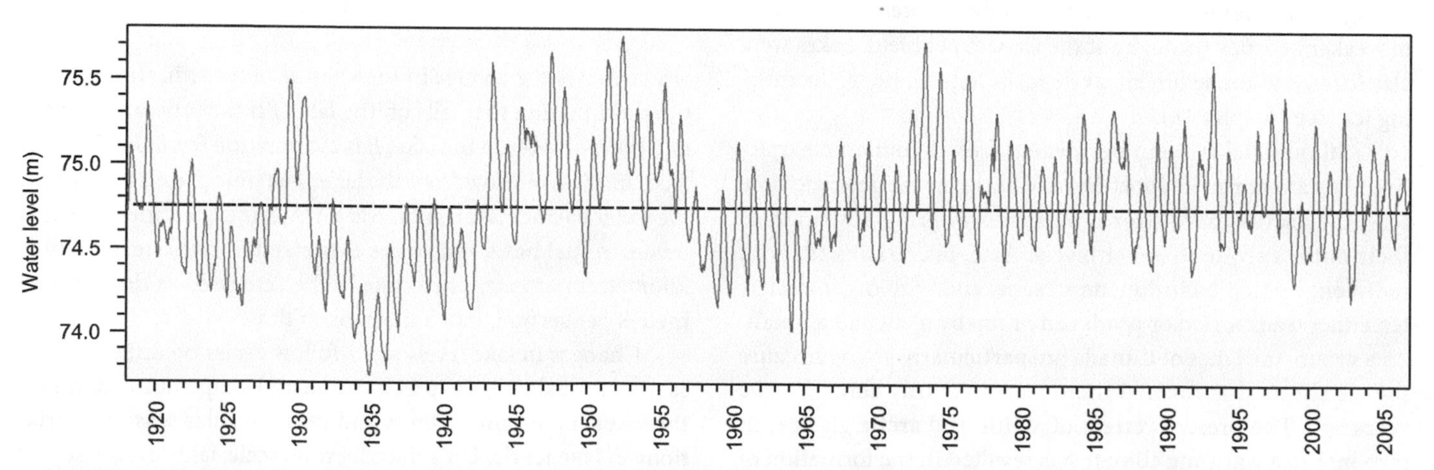

15.20 MONTHLY AVERAGE LEVELS OF LAKE ONTARIO Levels of Lake Ontario vary both seasonally and from year to year, as variations in climate change the parameters of the water balance. (Data from the Canadian Hydrographic Survey.)

layer of water brought in by flooding streams originating in adjacent highlands. The Aral Sea has become saline and has greatly reduced in volume through human activity. The two major rivers that feed this vast central Asian water body have been largely diverted to irrigate agricultural lands. With its inflow greatly reduced, the lake has shrunk and become increasingly salty (see Figure 10.15).

LAKE TEMPERATURE AND CIRCULATION

A lake receives most of its thermal energy from the sun, and so at latitudes where there is a strong seasonal cycle of insolation, there is a corresponding cycle in the behaviour of lakes. Short-wave solar radiation is more readily absorbed by water than by other natural surfaces (see Chapter 2). The albedo of water varies depending on the angle at which the sun's rays strike the lake surface and the amount of sediment in the water, but the average value for most lakes is less than 10 percent. Thus, water absorbs more than 90 percent of the solar radiation it receives. Lakes lose heat energy in several ways. First, the long-wave energy radiates from the water surface at a rate that depends on its temperature. For example, a water surface at a temperature of 18 °C loses about 400 W m^{-2} in radiant energy flux. Cooling through latent heat flux by evaporation is also significant, and is normally greatest in summer and autumn when the lake is warmest. The lake can also gain or lose heat by sensible heat transfer. The net result of these energy gains and losses is that midlatitude lakes experience strong temperature cycles.

The water in lakes is not normally motionless; rather, it is continually moved by winds, which generate waves and currents that mix it. The degree of mixing, however, depends on the temperature structure of the lake. Like the oceans, the absorption of solar energy creates density differences in the water column of a lake (see Chapter 5). This keeps the warmest water with the lowest density at the surface. The warm, upper zone is called the *epilimnion*, and below this is the cold, denser water of the *hypolimnion*. Summer temperature in the epilimnion varies from about 25 to 30 °C in midlatitude lakes, depending on the size of the lake and date. Temperature in the hypolimnion of larger lakes can remain near 4 °C, the temperature of maximum density.

Because there is a large seasonal difference in solar energy in the midlatitudes, the temperature and density profiles in lakes change over the course of the year. In summer, wind-induced waves and currents create a strong circulation in the epilimnion. The hypolimnion experiences a much weaker circulation that is driven by friction with currents at the bottom of the epilimnion. In autumn, the temperature of the surface water cools, sinks, and mixes into the epilimnion. Eventually, the temperature of the epilimnion drops to that of the hypolimnion. Now the barrier caused by the density difference is gone, and the lake is able to circulate through all its depth. This *overturn* of the lake distributes not only water, but also suspended solids and dissolved gases throughout the water. For example, oxygen is carried down from the surface waters, replenishing the supply for fish and other organisms resident in the deeper parts of the lake.

As winter sets in, the upper part of the lake cools below 4 °C. The near-surface water becomes slightly less dense, remains on top, and continues to cool, forming a layer of ice. The layer below the ice has a temperature of between 0 to 4 °C. Because this layer is less dense than the water below it, a weak-layered structure forms, which isolates the upper and lower water masses. The ice cover also shelters the lake from wind-driven waves and currents, so mixing ceases for the winter. In spring the ice melts, and the temperature throughout the lake becomes nearly uniform again. Another period of overturn occurs and water, suspended solids, and dissolved gases circulate freely throughout the lake.

Lakes that experience overturn twice a year, in the spring and autumn, are referred to as *dimictic*. This term applies to most lakes in climates where the lake surface freezes, such as Canada and the northern United States. Lakes that are less than about 10 m deep may not develop a hypolimnion, since substantial solar energy penetrates to that depth. In these lakes, mixing occurs throughout the year, and they are referred to as *polymictic*. In the Arctic, lakes may not become warm enough to establish a thermal structure in summer, and so spring and autumn mixing come together in one short period of mixing in the warm season when the lake ice melts. These lakes are *monomictic*.

SEDIMENT IN LAKES

Lakes receive water and sediment from inflowing streams and rivers, as well as from runoff down slopes adjacent to lake shores. When streams and rivers enter the lake, the velocity of the stream water slows, depositing coarse-grained sediment, typically sand and gravel, to form a delta. However, fine-grained silts and clays remain suspended in the stream water as it moves into the lake. While some suspended sediment can exit through a lake's outlet, most eventually settles out in layers on the lake bottom. In most lakes only a few millimetres of sediment accumulate each year. Biological material produced in lakes is also deposited on the lake floor, where it only partially decomposes. The remains of the smallest organisms, such as *phytoplankton* and *zooplankton*, dominate this organic material.

Lakes can also make their own sediment. This occurs when chemical compounds dissolved in inflowing streams and groundwater precipitate and settle to the lake floor. This precipitation is common in regions underlain by limestone and other carbonate rocks. The limestone that makes up the bedrock is relatively soluble, and the dissolved load of calcium carbonate ($CaCO_3$) in streams and groundwater entering the lake is high. The solubility of $CaCO_3$ in water is inversely related to temperature. Cold water can carry more dissolved $CaCO_3$ than warm water can. When carbonate-rich water in the epilimnion warms in the summer, the dissolved $CaCO_3$ will precipitate, forming tiny solid particles. Several of the Great Lakes, including Lake Michigan and Lake Ontario, experience this *whiting* of their waters about every other year on average. The fine-grained precipitate is deposited around the margins of lakes in beds and banks of *marl*, a soft, white, carbonate-rich mud. Some of the $CaCO_3$ re-dissolves as the lake waters cool, but some accumulates on the lake floor.

The nature and composition of the sediment that accumulates on the lake bottom can reveal much about the environment of the lake and its surroundings. For example, the amount of inorganic sediment the lake receives is greater in wet years as inflowing streams and rivers increase their sediment loads. Temperatures and nutrient concentrations in the epilimnion affect the relative amounts of various species of phytoplankton and zooplankton, and thus the types and amount of organic sediment they produce. Plant pollen falling into the lake is often preserved in lake-bottom sediments, and the mixture of pollen types can indicate the general climate of the region. In some cases, information on water temperature can be determined from the atomic composition of precipitated carbonate or silica based on the ratio of two oxygen isotopes found in the precipitate.

Thus, the sediments deposited on the lake floor provide a record of the environmental history of the lake and its drainage basin. In some cases, even seasonal patterns can be seen in this record. For example, if the rivers flowing into the lake introduce a pulse of clastic sediment every year during spring melt, or if a plankton bloom occurs every summer, then annual layers, called *varves*, may be preserved in the sediments (Figure 15.21). Like annual tree rings, the number of varves can determine the age of the lake. Each varve can provide information on the events that happened during that year, and comparison with the other varves can determine how the lake's environment changed over time. Many lakes have sedimentary records that go back for tens of thousands of years, permitting the environmental history of the region to be reconstructed over long time scales. In the glaciated terrains of Canada and the northern United States, for example, lake sediments reveal where and when the climate warmed and how forests migrated northward following the retreat of the continental ice sheets.

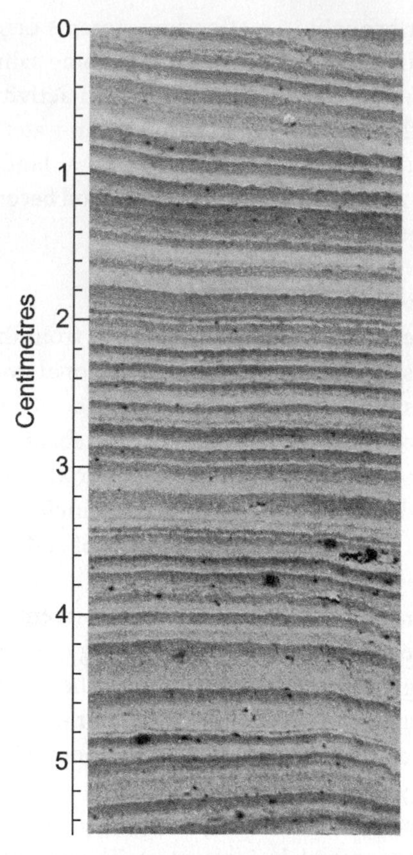

15.21 **LACUSTRINE SEDIMENTARY RECORD** This photograph illustrates a portion of a core of sediment from Ape Lake in the Coast Mountains of British Columbia. Each light–dark couplet, which represents the sediment deposited in one year, is called a varve. The dark sediment is clay deposited in the quiet, ice-covered lake during winter when inflow was small. The light sediment is silt and fine sand deposited in summer when inflowing streams were more active. Notice that the thickness and appearance of each varve is different, illustrating the variability in the climate and hydrology that determines sediment delivery to the lake.

SURFACE WATER AS A NATURAL RESOURCE

Fresh surface water is a basic natural resource essential to human agricultural and industrial activities. Runoff held in reservoirs behind dams provides water supplies for great urban centres. When diverted from large rivers, it provides irrigation water for highly productive lowlands in dry regions. Runoff is also used in hydroelectric power generation where the gradient of a river is steep, or routes of inland navigation where the gradient is gentle. Today's heavily industrialized society requires enormous supplies of fresh water for its sustained operation. Urban dwellers consume water in their homes at rates of 150 to 400 litres per person per day (Figure 15.22). Much of this water is obtained from surface water supplies.

Unlike groundwater, which represents a large water storage body, fresh surface water in the liquid state is generally stored only in small quantities. An exception would be large lakes, such as the Great Lakes system. The global quantity of available groundwater is about 20 times larger than that stored in freshwater lakes, and the water held in streams is only about one one-hundredth of that in lakes. Because of small natural storage capacities, surface water can be drawn only at a rate comparable with its annual renewal through precipitation. Dams provide useful storage capacity for runoff that would otherwise escape to the sea, but once the reservoir is full, water

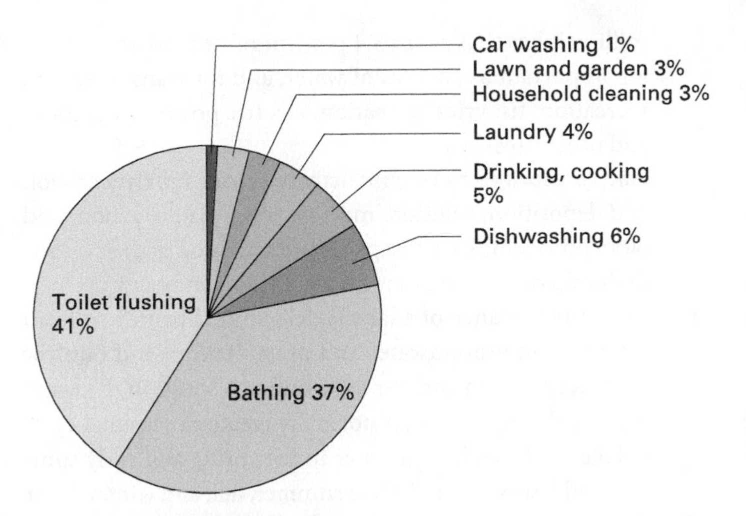

15.22 DOMESTIC WATER USE This chart shows how an average home in Canada uses water. (Environment Canada.)

use must be scaled to match the natural supply rate averaged over the year. Development of surface water supplies causes many environmental changes, both physical and biological, and these must be taken into account when planning for future water developments.

Because most large rivers with a steep gradient do not have falls, dams are necessary to create the vertical drop required to spin the turbines of electric power generators. Hydroelectric power is comparatively inexpensive, non-polluting, and renewable. Most large dams serve a second purpose of providing fresh water for urban use and irrigation. Some 45,000 dams have been constructed worldwide, which collectively hold back water in reservoirs with a combined surface area of 0.66×10^6 km^2. This compares with a surface area of 1.5×10^6 km^2 for the world's freshwater lakes. The Rogun Dam on the Vakhsh River in Tajikistan is the world's highest, at 335 m. Canada's highest dam, the Mica Dam located 135 km north of Revelstoke in British Columbia, is the twelfth highest in the world, at 243 m; it was commissioned in 1973.

The gross reservoir capacity of the Mica Dam is 24,670 million cubic metres, and the lake that has formed behind it extends some 130 km, with a surface area of about 115 km^2. The Caniapiscau Reservoir, with a surface area of 4,318 km^2, is the largest in Canada. Located on the Caniapiscau River in Quebec, its waters flow into the James Bay hydroelectric complex. It is about half the area of the world's largest reservoir, Lake Volta in Ghana, which covers 8,500 km^2 and is 520 km in length.

The Three Gorges reservoir in China is projected to be the world's biggest hydroelectric facility; it will produce about eight times more power than that generated at Niagara Falls. The reservoir began filling in 2003, and will occupy the present position of the scenic Three Gorges area. The initial depth of the reservoir was 135 m, but this will eventually be increased to 175 m; it will contain 39 billion cubic metres of water covering an area of 1,084 km^2, and extend more than 650 km upstream from the dam. The Three Gorges project has had profound ecological impacts as a result of submergence of factories, mines, and waste dumps. It has also caused the displacement of 1.2 million people following the flooding of 13 cities, 140 towns, and about 1,500 villages. Natural hazards, such as landslides, have also increased, and of even greater concern is the possibility that the weight of impounded water will induce earthquakes. The dam also has profound biological implications.

The loss of scenic and recreational resources behind dams, as river valleys, rapids, and waterfalls are drowned, is of major concern. Dam construction also destroys ecosystems adapted to the river environment. In addition, deposition of sediment behind a dam can rapidly reduce the water holding capacity of the reservoir. Depending on the sediment load of the inflowing rivers, many dams are likely to be almost completely filled within a century or so of completion. This reduces the reservoir's ability to provide consistent supplies of water and hydroelectric power. Thus, it is not surprising that new dam projects can meet with stiff opposition from concerned local citizens' groups and national environmental organizations.

A Look Ahead

This chapter focused on water, including the precipitation that runs off the land and flows into the sea, and the drainage system of a stream that conducts this flow of runoff. The smallest streams catch runoff from slopes, carrying the runoff into larger streams. The larger streams, in turn, receive runoff from their side slopes and pass their flow on to still larger streams. However, this cannot happen unless the gradients of slopes and streams allow the water to keep flowing downhill. This means that the landscape has been shaped and organized into landforms that are an essential part of the drainage system. The shaping of landforms within the drainage system occurs as running water erodes the landscape, and that is the subject of the next chapter.

CHAPTER SUMMARY

- The fresh water on land surfaces accounts for only a small fraction of the Earth's water. Since it is produced by precipitation over land, it depends on the continued operation of the hydrologic cycle for its existence. The soil layer plays a key role in determining the fate of precipitation by diverting it in three ways: to the atmosphere through *evapotranspiration*, to groundwater through *percolation*, and to streams and rivers as runoff.

- Groundwater occupies the pore spaces in rock and regolith. The water table marks the upper surface of the *saturated zone* of groundwater, where pores are completely full of water. Groundwater moves in slow paths deep underground, recharging rivers, streams, ponds, and lakes by seeping upward and thus contributing to runoff.
- Wells draw down the water table and, in some regions, lower the water table more quickly than it can be recharged. Groundwater contamination can occur when precipitation percolates through contaminated soils or waste materials. Landfills and dumps are common sources of groundwater contaminants.
- Runoff includes *overland flow*, moving as a sheet across the land surface, and *stream flow*, which is confined to a channel in streams and rivers. Streams and rivers are organized into a drainage system that moves runoff from slopes to channels and from these into larger ones. The discharge of a stream measures the flow rate of water moving past a given location. Discharge increases downstream as tributary streams add more runoff.
- The hydrograph plots the discharge of a stream at a location through time. Since it takes time for water to move down slopes and into progressively larger stream channels, a *lag time* separates peak discharge from peak precipitation. In larger streams, the lag time will be longer and the rate of decline of the discharge after it peaks will be more gradual. Because urbanization typically involves covering ground surfaces with impervious materials, urban streams exhibit shorter lag times and higher peak discharges. Annual hydrographs of streams from wet regions show an annual cycle of *base flow* on which discharge peaks related to individual rainfall episodes are superimposed.
- Floods occur when river discharge increases and its usual channel can no longer contain the flow. Water spreads over the *flood plain*, inundating low fields and forests adjacent to the channel. When discharge is high, flood waters can rise beyond normal levels to inundate nearby areas of development, causing damage and sometimes taking lives. *Flash floods* occur in small, steep watersheds and can be highly destructive.
- Lakes are standing bodies of water, ranging in size from the Caspian Sea to small ponds. Canada has nearly half a million lakes that cover 8.4 percent of its land area. Lakes are important as sources of water, and for transportation, recreation, fisheries, irrigation, electric power generation, and other uses.
- Lakes are formed by tectonic activity, volcanism, river erosion and deposition, solution, mass wasting, glacial action, and other processes. They disappear when erosion of their outlets drains them, or when they fill with sediment or dry up.
- The water balance of a lake is determined by inflow from streams and groundwater and precipitation, and outflow from evaporation and loss at its outlet. Change in the water balance through the year normally creates a seasonal cycle of lake levels, with high water in the spring and early summer and low water in the late summer, fall, and winter. Over decades, sequences of wet and dry years cause lake levels to respond by rising and falling.
- Where lakes occur in inland basins without outlets, they are often saline. When climate changes, such lakes can dry up, creating *salt flats* or *dry lakes*.
- Midlatitude lakes experience a strong annual cycle of thermal heating. Coupled with the density change of water with temperature, warm-season heating creates a warm *epilimnion* atop a cold *hypolimnion*. In *dimictic* lakes, wind-driven mixing occurs twice a year as lake water temperature becomes uniform just before the surface water freezes over and just after it thaws. *Monomictic* lake water mixes once per year during a short warm-season thaw. *Polymictic* lakes are shallow and mix throughout the year.
- Lakes accumulate sediment from several sources. Inflowing rivers provide suspended sediment that settles to the lake bottom. *Phytoplankton* and *zooplankton* contribute organic sediment. Precipitation of calcium carbonate can create a *whiting* of lake water and leave *marl* in lake beds. The sediments on a lake bottom provide a record of the lake's history and prior environments.
- Groundwater and surface water resources are essential for human activities. Human civilization is dependent on abundant supplies of fresh water for many uses. But because the fresh water of the continents is such a small part of the global water pool, use of water resources takes careful planning and management.

KEY TERMS

artesian well	drainage basin	groundwater	lake	stream
channel	drainage system	hydrograph	overland flow	water table
discharge	flood	infiltration	runoff	watershed

REVIEW QUESTIONS

1. How and under what conditions does precipitation reach groundwater?
2. How do wells affect the water table? What happens when pumping exceeds recharge?
3. How is groundwater contaminated? Describe how a well might become contaminated.
4. Define discharge (of a stream) and the two quantities that determine it. How does discharge vary in a downstream direction? How does gradient vary in a downstream direction?
5. How are the cross-sectional area and velocity of a stream related to the flow?
6. What two types of flow occur in a stream, and how do they differ?
7. What is a drainage system? How are slopes and streams arranged in a drainage basin?
8. Define the term *flood*. What is the flood plain? What factors are used in forecasting floods?
9. What factors combined to make the Red River floods of 1997 and 2009 so large? What kinds of engineering structures may be constructed to reduce the size and effect of floods?
10. How are lakes formed? Identify typical processes of lake formation. What processes cause lakes to disappear?
11. Write an equation for the water balance of a lake and define the terms involved. How does the water balance influence the level of a lake?
12. Under what circumstances does a lake become saline? How does the water balance equation for a saline lake differ from that of a freshwater lake?
13. Explain how solar energy penetrates and warms a lake. What portion of the lake is most intensely heated? Explain the formation of a layered structure in a midlatitude lake in the warm season.
14. Describe how the density of water depends on temperature. How does water temperature, and therefore density, affect mixing in a lake?
15. At what times of the year and by what process does the water of a midlatitude lake mix completely? Why do some lakes only mix once per year? Why are some lakes always well mixed?
16. Identify the sources of sediment that accumulates at the bottom of a lake. How can the sediments found on the bottom of a lake be used to reconstruct the history of a lake and its environment?
17. How is surface water used as a natural resource? How do dams affect the stream environment?

VISUALIZATION EXERCISES

1. Sketch a cross-section through the land surface showing the position of the water table. Indicate flow directions of subsurface water motion with arrows. Include the flow paths of groundwater. Be sure to include a stream in your diagram. Label the saturated and unsaturated zones.
2. Why does water rise in an artesian well? Illustrate with a sketched cross-sectional diagram showing the aquifer, aquicludes, and the well.
3. Sketch the annual cycle of water level of a midlatitude lake in a continental climate, such as Lake Ontario, and describe the factors that affect it.

ESSAY QUESTIONS

1. A thundershower causes heavy rain to fall in a small region near the headwaters of a major river system. Describe the flow paths of that water as it returns to the atmosphere and ocean. How do engineered structures affect flows?
2. Imagine yourself as a recently elected mayor of a small city located on the banks of a large river. What issues might you be concerned with that involve the river? In developing your answer, choose and specify some characteristics for this city—such as its population, its industries, its sewage systems, and the present uses of the river for water supply or recreation.

APPENDIX 15.1 STATE OF THE GREAT LAKES 2009

Since 1998, the United States Environmental Protection Agency and Environment Canada have coordinated a biennial assessment of the ecological health of the Great Lakes ecosystem using a consistent set of environmental and human health indicators. The various indicators used are organized into nine categories: Coastal Zones and Aquatic Habitats (shoreline communities), Invasive Species, Contamination, Human Health, Biotic Communities, Resource Utilization, Land Use-Land Cover, and Climate Change. The status of each indicator is assessed according to five categories together with current trends (Table 15A.1).

Shoreline communities

Water levels are regulated in Lake Superior and Lake Ontario and are less variable than in the other Great Lakes. Natural lake level fluctuations significantly impact wetland vegetation, and in Lake Ontario this has resulted in coastal wetlands of low plant species diversity. Similar problems have been noted in several other types of shoreline habitat. For example, more than 90 percent of Great Lakes *alvars*, open habitats occurring on flat limestone bedrock, have been destroyed or substantially degraded, but these areas are now recognized as important habitats for rare plants and animals.

Non-native species invasion

Based on conditions in the 1840s, non-native species in the region now include about 185 aquatic and 160 terrestrial plants and animals. About 10 percent of non-native species are considered to be invasive and negatively impact ecosystem health. The presence of invasive species can be linked to many current ecosystem challenges, including the decline in organisms of the lower food web, fish and waterfowl diseases, and excessive algal growth. Shipping continues to be a major concern for the introduction and spread of invasive species.

Managing the impact of harmful invasive species once they are established is a major challenge. The sea lamprey (*Petromyzon marinus*), a lethal parasite of large fish, initially invaded the Great Lakes Region in the 1920s. They arrived by way of the Welland Canal, which allowed them to bypass Niagara Falls, and by the 1940s they were abundant in all the Great Lakes. Decades of control measures have reduced the number of sea lamprey by over 90 percent compared to the population peak in the early 1960s.

Zebra mussels (*Dreissena polymorpha*) are another problem species that was accidentally introduced to the Great Lakes, probably on the hulls or in the ballast water of ships arriving from Europe. Zebra mussels feed voraciously on phytoplankton and small zooplankton which previously were an important part of the food chain for the native fishes. Zebra mussels were first discovered in 1988 in Lake St. Clair, between Lake Erie and Lake Huron. Their free-swimming larvae attach themselves to hard surfaces, such as rock, metal, fibreglass, and even native shellfish. The mussels grow especially well in fast-flowing water and have become an enormous problem in water intake pipes at power stations, industrial sites, and irrigation and drinking water facilities.

The *State of the Great Lakes 2009* report concludes that the Great Lakes still face a major challenge with respect to the sea lamprey, and that they will continue to be extremely vulnerable to introductions of new invasive species (Figure 15A.1).

Contamination

The Great Lakes continue to be a receptor of contaminants from many different sources, such as municipal and industrial

TABLE 15A.1 ASSESSING STATUS AND TRENDS OF THE GREAT LAKES ECOSYSTEM INDICATOR CATEGORIES

INDICATOR STATUS		
■	Good	The ecosystem component is presently meeting ecosystem objectives or otherwise is in acceptable condition.
■	Fair	The ecosystem component is currently exhibiting minimally acceptable conditions, but it is not meeting established ecosystem objectives, criteria, or other characteristics of fully acceptable conditions.
■	Poor	The ecosystem component is severely negatively impacted and it does not display even minimally acceptable conditions.
■	Mixed	The ecosystem component displays both good and degraded features.
■	Undetermined	Data are not available or are insufficient to assess the status of the ecosystem component.
CURRENT TRENDS		
→	Improving	Information provided shows the ecosystem component to be changing toward more acceptable conditions.
◆	Unchanging	Information provided shows the ecosystem component to be neither getting better nor worse.
←	Deteriorating	Information provided shows the ecosystem component to be departing from acceptable conditions.
?	Undetermined	Data are not available to assess the ecosystem component over time, so no trend can be identified.

Source: *State of the Great Lakes 2009 Highlights report.*

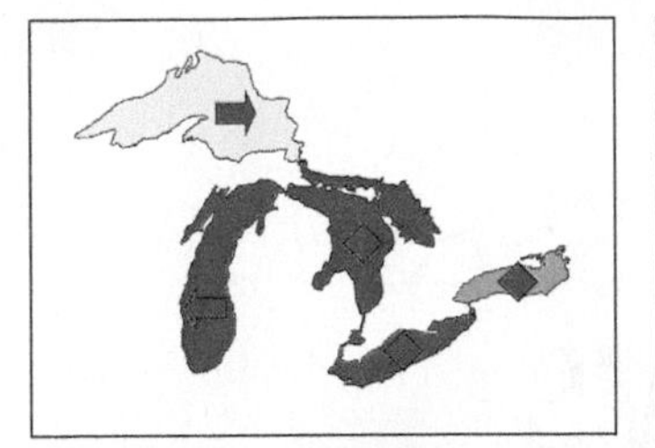

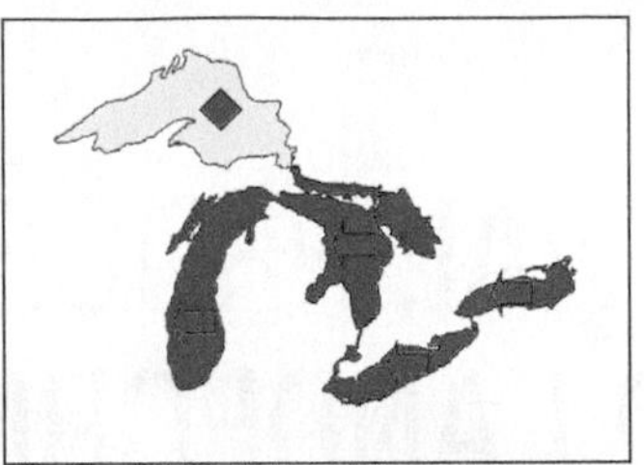

15A.1 STATUS AND TREND OF SELECTED INVASIVE SPECIES INDICATORS IN THE GREAT LAKES IN 2008 The left map shows sea lamprey in the lakes; the right map shows aquatic invasive species. (See Table 15A.1 for explanation of colours and symbols.) Source: *State of the Great Lakes 2009 Highlights Report.*

wastewater, air pollution, contaminated sediments, runoff, and groundwater. However, the release of targeted bioaccumulative toxic chemicals has declined significantly from the peak period in past decades and, for the most part, no longer limit the reproduction of fish, birds, and mammals. Concentrations of contaminants in the open waters are low, and many contaminants are still declining. However, concentrations are higher in some local areas near the shore, such as in bays.

Colonial waterbirds, such as the herring gull, are fish-eaters and usually considered top-of-the-food web predators. They are excellent bioaccumulators of contaminants, and are often among the species with the greatest pollutant levels in an ecosystem. They also breed on all the Great Lakes. Overall, most contaminants in herring gull eggs have declined 90 percent or more since the monitoring began in 1974, but recently the rate of decline has slowed. More physiological abnormalities in herring gulls still occur at Great Lakes sites than at cleaner reference sites away from the Great Lakes basin.

Since the 1970s, concentrations of historically regulated contaminants such as polychlorinated biphenyls (PCBs), dichloro-diphenyl-trichloroethane (DDT), and mercury have generally declined in most monitored fish species. Concentrations of other regulated and unregulated contaminants, such as chlordane and toxaphene, vary in selected fish communities, and these concentrations are often lake-specific. Overall, there has been a significant decline in these contaminant concentrations. However, the rate of decline is slowing, and in some cases concentrations are even increasing in certain fish communities.

Excessive inputs of phosphorus to the lakes from detergents, sewage treatment plants, agricultural runoff, and industrial discharges can result in nuisance algae growth. Efforts that began in the 1970s to reduce phosphorus loadings have been largely successful. However, in some locations, phosphorus loads may be increasing again, and an increasing proportion of the phosphorus is a dissolved form that is biologically available to fuel nearshore algal blooms. The status and trends of phosphorus can be quite different in the nearshore waters compared to the offshore waters of each lake.

The *State of the Great Lakes 2009* report concludes that contamination is generally declining in all the Great Lakes, but water quality is especially compromised by phosophorus. Lake Superior is the least contaminated of the Great Lakes due to its size, upstream position, relatively low population density around its shoreline, and lesser industrial activity compared to its neighbours (Figure 15A.2).

Climate Change

The Great Lakes region is currently experiencing shorter winters and warmer annual average temperatures. If this continues, the projected effects include an overall increase in air and water temperatures and a corresponding decrease in lake ice cover. Heavy rain and snow and extreme heat events are also expected to occur more frequently. Such changes will likely alter lake snowpack density and evaporation rates, and will affect water quality.

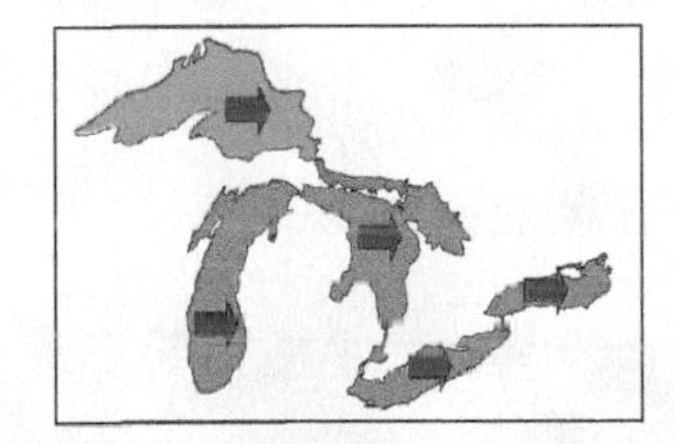

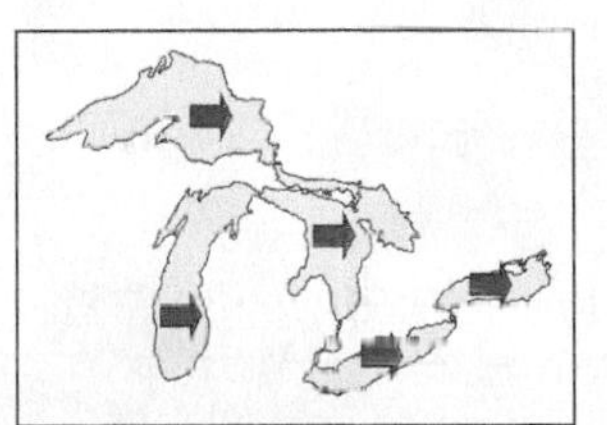

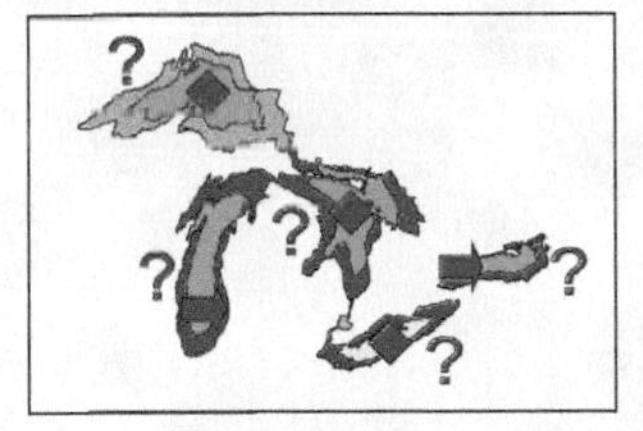

15A.2 STATUS AND TREND OF SELECTED CONTAMINATION INDICATORS IN THE GREAT LAKES IN 2008 The top left map shows contaminants in waterbirds, the top right map shows contaminants in whole fish, and the bottom map shows phosphorus concentrations and loadings. (See Table 15A.1 for explanation of colours and symbols.) Source: *State of the Great Lakes 2009 Highlights Report.*

QUESTIONS

1. What concerns about environmental quality have been raised for the Great Lakes, particularly Lake Erie and Lake Ontario?
2. What are the main sources of pollutants that affect water quality in the Great Lakes? What effects do they have on aquatic species? How are organisms that feed on these aquatic species affected?
3. How does water quality reflect human activity in the watershed? What control measures might be considered to help reduce the impact of these activities?
4. How has each of the lakes in the Great Lakes ecosystem been affected by human activities? What contributes to the differences in how each of the lakes has been affected by human activity?

CHAPTER 16 FLUVIAL PROCESSES AND LANDFORMS

Horton River, Northwest Territories.

Running water has sculpted most of the world's land surface. Moving as a sheet across the land, water picks up particles and carries them down slopes into a stream channel. When rainfall is heavy, streams and rivers swell, lifting large volumes of rock debris and transporting it downstream. In this way, running water erodes mountains and hills, carves valleys, and deposits sediment. This chapter describes the work of running water and the landforms it creates.

Running water is one of four fluid agents that erode, transport, and deposit mineral and organic matter. The other three are waves, glacial ice, and wind. The total action of all processes that wear away exposed rock is termed *denudation*. The resulting sediments are then transported to the sea or to closed inland basins. Denudation causes an overall lowering of the land surface. If left unchecked, it would eventually reduce a continent to a nearly featureless, level surface at sea level. However, tectonic activity continually elevates continental crust well above the oceans. The result is that running water, waves, glacial ice, and wind always have material available to mold into a diversity of landforms.

FLUVIAL PROCESSES AND LANDFORMS

Landforms shaped by running water are described as **fluvial landforms**; they develop from the **fluvial processes** of overland flow and stream flow. Weathering and mass wasting (see Chapter 14) operate in concert with overland flow, providing the rock and mineral fragments that are carried into stream systems.

Fluvial landforms and fluvial processes dominate continental land surfaces across the world. Throughout geologic history, glacial ice has been present only in mid- and high latitudes and in high mountains. Landforms made by wind action are mostly found in arid regions, although even there fluvial action is important. The action of waves and currents is restricted to the narrow contact zone between oceans and continents.

EROSIONAL AND DEPOSITIONAL LANDFORMS

All agents of denudation perform the geological activities of erosion, transportation, and deposition. When an initial landform such as an uplifted crustal block is created, it is immediately subjected to denudation, and particularly to the force of flowing water. Valleys are formed where fluvial agents erode away rock. Between the valleys are ridges, hills, or mountain summits, representing the remaining parts of the crustal block that running water has not yet removed. These are sequential landforms, which because they are shaped by progressive removal of the bedrock mass, are also *erosional landforms*. Fragments of soil, regolith, and bedrock are transported and deposited elsewhere to make an entirely different set of surface features, collectively referred to as *depositional landforms*. Typical landforms created by fluvial deposition include fans built of rock fragments at the mouth of a ravine, and flood plains that form on the floor of river valleys.

SLOPE EROSION

Fluvial action starts with **overland flow**, which consists of a thin film of water or tiny rivulets that move across the ground. In humid climates, overland flow begins when precipitation exceeds the storage capacity of the surface materials (Figure 16.1). Some precipitation, whether rain or snow, will be retained by the vegetation and returned to the atmosphere through evaporation or sublimation. Excess rain and melting snow eventually reach the ground, either as **throughfall** that drips from the vegetation or as **stemflow** that trickles along branches and down stems and

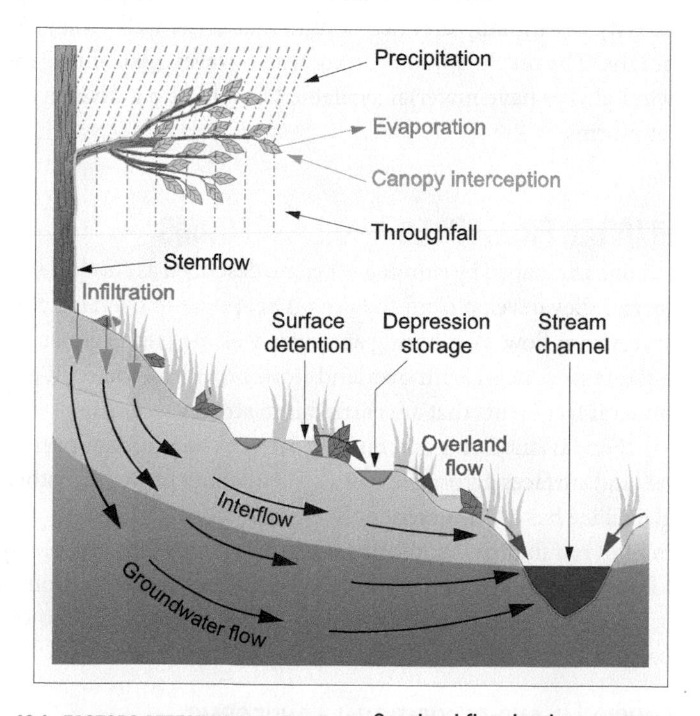

16.1 **FACTORS AFFECTING SURFACE FLOW** Overland flow begins as soon as infiltration, surface detention, and the water retention capacity of plant surfaces are exceeded. Local and regional topography and underlying geology control the ability of the water to become organized into channels and drainage systems.

trunks. If the soil is dry, the water will infiltrate into the soil until the infiltration capacity is reached, at which point the surplus will begin to move downslope as a film of water.

The time required to initiate overland flow depends on the **antecedent conditions**, which determine the amount of water in the soil prior to a precipitation event. If there has been little rain for several days or weeks, then most of the water will infiltrate the dry soil. The downslope flow of water will also be impeded by surface irregularities. Some will also be held in small depressions and behind obstructions, such as little twigs and leaves, through **surface retention** and **depression storage**. Once the storage capacities of the soil and surface depressions are exceeded, the water will begin to move downslope in a process called *infiltration excess overland flow*. If the soil is fully saturated from a previous rainstorm and depression storage is filled, then runoff will occur immediately when another rainstorm begins through the process of *saturation overland flow*.

Saturation overland flow produces a rapid downslope flow response similar to what occurs as *Hortonian overland flow*, or *unsaturated overland flow*. Hortonian overland flow develops when rainfall intensity exceeds the rate at which water can be absorbed by the soil, even though the soil is dry. Hortonian overland flow is widely accepted as the principle stormflow generation process in thinly vegetated watersheds in arid and semi-arid regions under intense rainfall, especially where surface sealing can limit soil infiltration capacity. Hortonian overland flow is also applicable to impervious surfaces, such as exposed bedrock or paved surfaces in urban environments.

As it moves across the surface, overland flow will entrain small particles of mineral matter, beginning the first stage in fluvial erosion (Figure 16.2). At this stage, the water is not concentrated in a well-defined channel, and movement of particles occurs as **sheet erosion**. The effects of sheet erosion are often hard to distinguish, especially in vegetated areas, because soil loss is so gradual. Early signs of sheet erosion include bare areas and puddles of water forming as soon as rain falls. Although

16.2 **OVERLAND FLOW** The transport of soil particles by overland flow across this saturated ground is the initial stage in fluvial erosion.

only a thin layer of soil is being removed, the cumulative effect over time causes significant degradation.

Runoff occurs whenever excess water on a slope cannot be absorbed into the soil or trapped on the surface. Regionally, this is determined by annual precipitation, evaporation rates, and terrain conditions including surface gradients, vegetation cover, and soil type. In Canada, runoff is highest in the coastal areas of British Columbia, and is also high along the Rocky Mountains and in the Atlantic provinces (Figure 16.3). It is lowest in the Prairie provinces and across the Arctic. The seasonal nature of Canada's climate typically results in high runoffs in the spring when the subsoil is still frozen and the overlying layers are quickly saturated by snow melt.

As the water moves across the surface, it exerts a tractive force on the loose soil particles, which become entrained in the flow. The size of the particles that can be transported depends on the gradient of the surface, the volume of water that moves across it, and the degree to which the particles are bound by plant roots or held down by a mat of leaves. Soil particles carried in runoff typically range from about 0.001 to 1.0 mm in diameter. Under normal rainfall intensities, larger particles tend to settle out after moving only short distances, but finer materials suspended in the water can travel much further. In addition to solid materials, the water will carry dissolved mineral matter in the form of ions produced by acid reactions or direct solution.

A steeper slope increases the erosive power of water because it flows more rapidly across the surface. Slope length is also important since a longer slope allows more runoff to accumulate and so increases the degree of scouring. The ability of soils to resist erosion is a measure of soil erodibility, and is based on the physical characteristics of each soil. Generally, well-structured soils with faster infiltration rates and higher levels of organic matter have a greater resistance to erosion. Soils with sand, sandy loam, and loam textures tend to be less erodible than silt, very fine sand, and clay soils.

Properties that determine erodibility, such as soil aggregation and shear strength, show systematic seasonal variation and are strongly affected by climatic factors, including rainfall distribution and frost action. *Shear strength* is the maximum stress that can be sustained by a soil before the particles slide over each other along a failure surface. Shear strength changes considerably over short time periods, especially in response to soil moisture conditions. In fine-textured soils, shear strength is mainly determined by cohesive forces. *Cohesion* is an intermolecular attractive force that acts between particles and binds them together. Cohesion is substantial for clay-size particles, but is virtually non-existent in sands in which the coarser grains are held by friction. In moist soil, water films create a negative pore water pressure; that is, the water films produce suction, which increases cohesion by drawing particles together. However, in saturated soil, positive pore water pressures can develop, which push particles apart. As well as reducing cohesion, positive pore water pressures reduce friction by reducing the number and area of points of contact between grains.

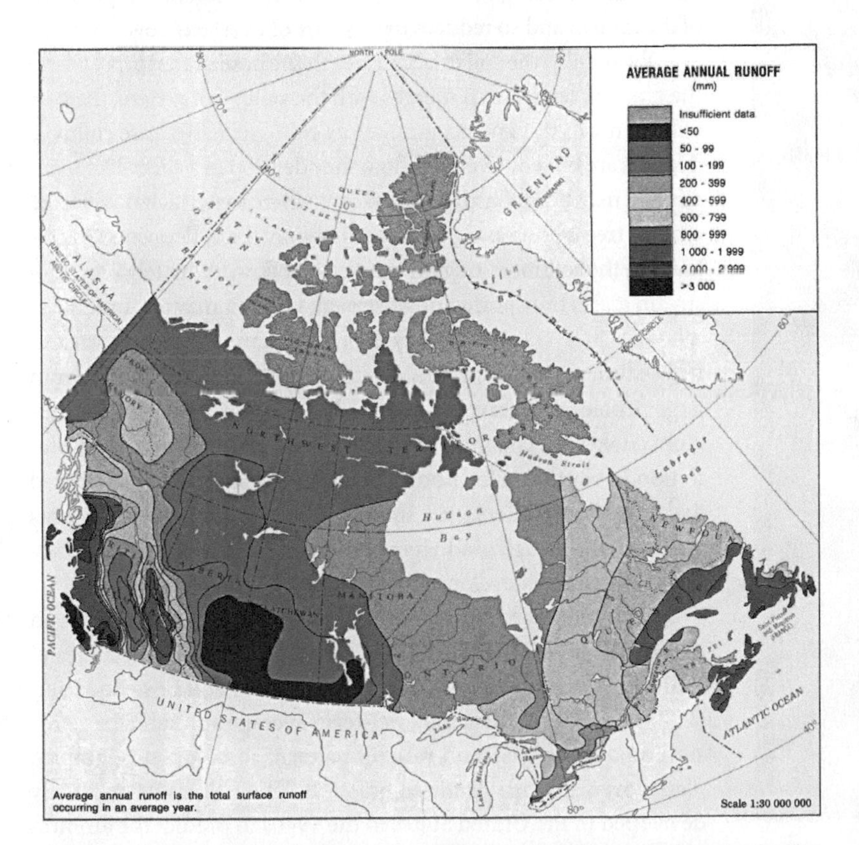

16.3 AVERAGE ANNUAL RUNOFF IN CANADA Runoff is highest in the coastal areas of British Columbia, the Atlantic provinces, and in the western mountains. Runoff is low in the Prairie provinces due to the drier climate and more permeable substrates.

The ongoing removal of soil is part of the natural process of denudation, and occurs everywhere that precipitation falls on land. In general, the erosion rate is slow enough that a layer of soil is maintained on the land surface. Under equilibrium conditions, some soil is removed each year, but it is replaced by material freshly weathered from the underlying regolith. Soil scientists refer to this continual replacement process as the *geologic norm*.

ACCELERATED EROSION

The rate of soil erosion may be greatly increased by human activities or by natural events of unusual magnitude; this results in a state of *accelerated erosion*. The surface soil is removed much faster than it can be formed, progressively exposing the underlying soil layers (Figure 16.4). Accelerated erosion is often initiated by the removal of the plant cover by fire, logging, or other disturbances. No foliage remains to intercept rain, and the protection afforded by a cover of fallen

16.4 ACCELERATED SOIL EROSION On steep slopes or where the vegetation cover has been disturbed, soil can be removed at a faster rate than it forms.

16.5 SOIL EROSION BY RAIN SPLASH When large raindrops land on a wet soil surface, they produce miniature craters. Grains of clay and silt thrown into the air as the soil surface is disturbed are then available to seal up the soil pores.

leaves and stems is removed. Consequently, the raindrops fall directly onto the mineral soil.

The impact of falling raindrops on bare soil lifts and moves soil particles. This process, called *rain splash erosion* (Figure 16.5), can disturb many tonnes of soil during a single rainstorm. Particle movement by rain splash is usually greatest during short-duration, high-intensity rainstorms. Runoff water can easily remove lighter soil particles, such as silt and fine sand, which have been detached in this way. On a sloping ground surface, rain splash slowly shifts the soil downhill. It also can break up soil aggregates. Consequently, detached particles of fine clay and silt can begin to seal up the pores in the soil, making it much less able to absorb water. Reduced infiltration, in turn, increases the depth of overland flow generated by a given amount of rain, and so intensifies the rate of soil erosion.

The removal of vegetation also reduces the resistance of the ground surface to erosion through overland flow. For example, on a grassy slope, overland flow causes little soil erosion, because the energy of the moving water is dissipated in friction with the grass stems. Similarly, on a forested slope, overland flow is countered by the surface layer of leaves, twigs, roots, and fallen tree trunks. Without such a cover, the eroding force is applied directly to the bare soil surface, easily dislodging grains and sweeping them downslope. As well as absorbing the force of erosion, vegetation intercepts some of the rainfall and so reduces the volume of overland flow.

Eventually, the soil materials reach the base of the slope, where the gradient lessens as it merges with the valley floor. Here, the particles come to rest and accumulate as **colluvium**. Because colluvial deposits are built by overland flow, they develop in a sheet-like manner and may be inconspicuous, except where material has collected around tree trunks and other obstructions. If not deposited as colluvium, the sediment carried by overland flow ultimately reaches a stream and is transported downstream, where it may accumulate as **alluvium** in layers on the valley floor. Alluvium refers to any stream-laid sediment deposit; it could include particle sizes ranging from large pebbles and gravel to sands, silts, and fine clays. Coarse alluvium chokes the channels of small streams and can cause the water to flood over the valley floors. The finest particles are distributed widely by flood waters, and, in this way, extensive alluvial deposits accumulate through gradual increment. Typically, alluvial soils are fertile and highly developed for agriculture.

The quantity of sediment that overland flow removes from an area in a given period of time is called *sediment yield*. Estimates of sediment yield are provided by the Universal Soil Loss Equation (USLE), which predicts the long-term average annual rate of erosion on a slope based on rainfall pattern, soil type, topography, plant cover, and management practices. The USLE was originally developed in the United States in the 1940s to predict the amount of soil loss that results from sheet or rill erosion on agricultural land; it has since been modified and is now referred to as the Revised Universal Soil Loss Equation (RUSLE). Five major factors are used to calculate the soil loss for a given site. Each factor is the numerical estimate of a specific condition that affects the severity of soil erosion, all of which can vary considerably due to prevailing weather conditions. Therefore, the values obtained from the RUSLE are representative of long-term average conditions.

The RUSLE is written as:

$$A = R \times K \times LS \times C \times P$$

where A = estimated average soil loss in tonnes per hectare per year (tonnes ha^{-1} a^{-1}); R = *rainfall-runoff erosivity factor* based on the average erosive force of the annual rainfall (R is calculated for rainfall events > 12.7 mm as the product of the total kinetic energy of each storm and its maximum 30-minute intensity); K = *soil erodibility factor* that measures the average soil loss for a specified type of unvegetated soil on a "standard" slope 22.1 m in length with a gradient of 9 percent (soil particle size is the principal factor affecting K, but structure, organic matter, and permeability also contribute); LS = *slope-length factor* that represents a ratio of soil loss under specified conditions to a "standard" slope (length 22.1 m; gradient 9 percent) (the steeper and longer the slope, the higher the risk for erosion); C = *cover management factor*, used to determine the effectiveness of soil and vegetation management systems on erosion rates, that compares the soil loss from land under a specific crop and management system to the corresponding loss from continuously fallow and tilled land; and P = *support practice factor* that reflects the impact of practices designed to reduce the amount and rate of the water runoff and thus reduce the amount of erosion.

Figure 16.6 gives annual average sediment yield and runoff by overland flow from several types of upland surfaces in northern Mississippi. Notice that both surface runoff and sediment yield decrease greatly with the increased effectiveness of protective vegetation cover. Sediment yield from cultivated land undergoing accelerated erosion is more than 10 times greater than that from pasture, and about 1,000 times greater than that from an established pine plantation.

In drier areas, such as the Canadian Prairies, the grass cover is normally sufficient to slow the pace of erosion. However, the natural equilibrium is highly sensitive and easily upset. Depletion of the plant cover by fire or by grazing can easily increase erosion. These sensitive, marginal environments require careful use because they lack the potential to recover rapidly from accelerated erosion once it has begun.

RILLING AND GULLYING

Accelerated soil erosion is a constant problem in cultivated regions with a substantial water surplus. Where the land is steeply sloping, runoff from torrential rains can cut more pronounced channels through the process of *rill erosion*. In this way, a series of closely spaced channels can be scored into the soil over a short period of time (Figure 16.7a). If the rills are not destroyed by soil tillage, they may join together into larger channels. These rapidly deepen and soon become *gullies*—steep-walled trenches whose upper ends grow progressively upslope (Figure 16.7b). A rugged, barren topography results when accelerated soil erosion is allowed to proceed unchecked.

Erosion due to a high rate by overland flow occurs naturally in some semi-arid and arid lands where the process creates *badland topography*. Easily eroded clay formations underlie badlands. Erosion rates are too fast to permit plants to take hold, so no soil can develop. The result is a maze of small stream channels with steep gradients. The landscape of Alberta's Dinosaur Provincial Park is an example of badland topography (Figure 16.8). The exposed horizontal strata at this site date from the Late-Cretaceous Period; the sand and mud deposits were left by rivers that flowed some 75 million years ago. The present landscape dates from the end of the Pleistocene Epoch (10,000 to 15,000 years ago) when meltwater streams issuing from the retreating ice sheets began to carve the distinctive landscape. Final sculpting of the many hills and *hoodoos* is attributed to the effects of flash floods in this semi-arid region.

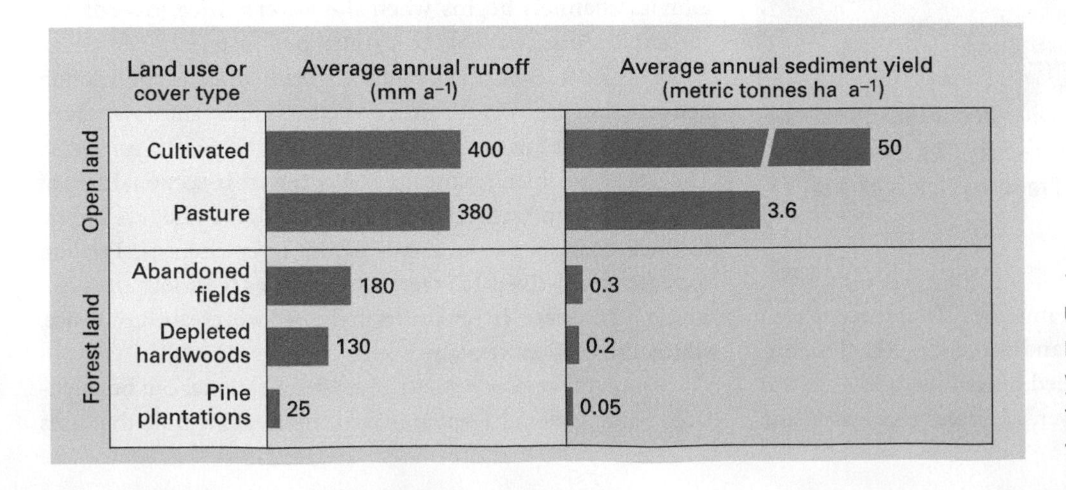

16.6 RUNOFF AND SEDIMENT YIELD IN UPLAND SURFACES IN NORTHERN MISSISSIPPI Both runoff and sediment yield are much greater for open land than for land covered by shrubs and forest. (Data from S. J. Ursic, U.S.D.A.)

(a)

(b)

16.7 **RILLS AND GULLIES** (a) Rills can develop quickly on soil-covered slopes when the vegetation cover is disturbed. They can be a problem in agricultural areas if heavy rains occur before the crop is well developed. (b) If left unchecked, runoff is progressively concentrated in the rills and leads to more pronounced erosion and the formation of deep gullies.

16.8 **BADLANDS** Gullies cut into soft sedimentary rock have created the dissected topography of Dinosaur Provincial Park in Alberta.

THE WORK OF STREAMS

The geomorphological work of streams consists of three related activities: erosion, transportation, and deposition. In this context, "stream" refers to any channelled water flow, and the word is used interchangeably with "river." **Stream erosion** is the progressive removal of mineral material from the floor and sides of the channel, whether bedrock or regolith. **Stream transportation** consists of movement of the eroded particles dragged over the stream bed, suspended in the body of the stream, or held in solution as ions. **Stream deposition** is the accumulation of transported particles on the stream bed and flood plain or on the floor of a lake, where it may be temporarily held until it is carried to the oceans. Erosion cannot occur without some transportation taking place, and the transported particles must eventually come to rest. Thus, erosion, transportation, and deposition are three interrelated phases in landscape development.

STREAM EROSION

Streams erode in various ways, depending on the force of the flowing water, the nature of the channel materials, and the tools carried by the current. The force of flowing water is derived from its mass and acceleration due to gravity (9.81 m s^{-2}). By extension, the work that a stream performs is expressed as force $\times$ distance. Thus, a stream performs a certain amount of work (expressed in joules) to move a pebble a given distance. The ability to do work is a measure of the stream's energy. The water molecules in a stream acquire potential energy due to the height of the land surface above sea level. Potential energy is dissipated as the water flows downhill, but the river system gains kinetic energy as it flows. Kinetic energy dislodges and transports materials in the stream channel or is converted to heat generated by friction between the water molecules.

The force of the flowing water not only sets up a dragging action on the bed and banks, but also causes particles to strike against these surfaces. Dragging and impact can easily erode alluvial materials, such as gravel, sand, and silt. This form of erosion, called *hydraulic action*, can excavate enormous quantities of channel material in a short time. Where banks are undermined, masses of material will slump into the river. There, the particles quickly separate and become part of the river's load. This process of bank caving is an important source of sediment during high river stages and floods. Erosion of alluvial channels begins when the water's force exceeds the threshold shear stress of the channel bed materials.

The force of flowing water is greatly affected by friction between the water and the channel bed and sides, and so is related to channel shape according to the length of the *wetted perimeter*. The wetted perimeter is the portion of the cross-sectional area of a river that is in contact with the channel substrate (see Figure 15.7); it changes as the depth of the river changes. Friction increases near a riverbed where rock particles of various sizes create a rough surface. Irregularities in the bed will create turbulence, which also reduces velocity.

The average flow velocity in a stream channel can be calculated from *Manning's equation*, an empirical formula that uses channel gradient and the roughness of the sand, gravel, and cobble surfaces over which the water flows. It is written as:

$$V = \frac{1.49}{n} R^{2/3} S^{1/2}$$

where V is the velocity (m s^{-1}), R is the hydraulic radius in metres calculated by dividing the cross-sectional area of the channel by the wetted perimeter, S is the gradient of the energy line in metres per metre (m m^{-1}), and n is the coefficient of roughness, specifically known as Manning's n. Manning's roughness coefficient is a function of friction along the channel bed. If the flow is deep relative to the bed material size, Manning's n can be estimated by:

$$n = 0.048\, D_{50}^{\,1/6}$$

where D_{50} (in metres) is the median size of the bed materials (i.e., half of the material is larger than this diameter). Typical values for n in mountain streams with beds of cobbles and large boulders range from 0.050 to 0.070, while those with gravel, cobbles, and few boulders range from 0.04 to 0.05, compared with 0.028 to 0.035 for rivers greater than 30 m wide. Table 16.1 presents general values for n in natural and artificial channels.

TABLE 16.1 VALUES FOR MANNING'S *n* IN VARIOUS NATURAL AND ARTIFICIAL CHANNELS

CHANNEL BED CONDITIONS	*n*
Natural Channel	
Clean and Straight	0.030
Major Rivers	0.035
Sluggish, Deep Pools	0.040
Excavated Earth Channel	
Clean	0.022
Gravelly	0.025
Weedy	0.030
Artificially Lined Channel	
Asphalt	0.016
Brickwork	0.015
Clay Tile	0.014
Concrete, Finished	0.012
Concrete, Sewer	0.015
Concrete, Unfinished	0.014
Wood, Planed	0.012
Wood, Unplaned	0.013
Corrugated Metal	0.024

(After V.T. Chow Open Channel Hydraulics, McGraw-Hill, 1959.)

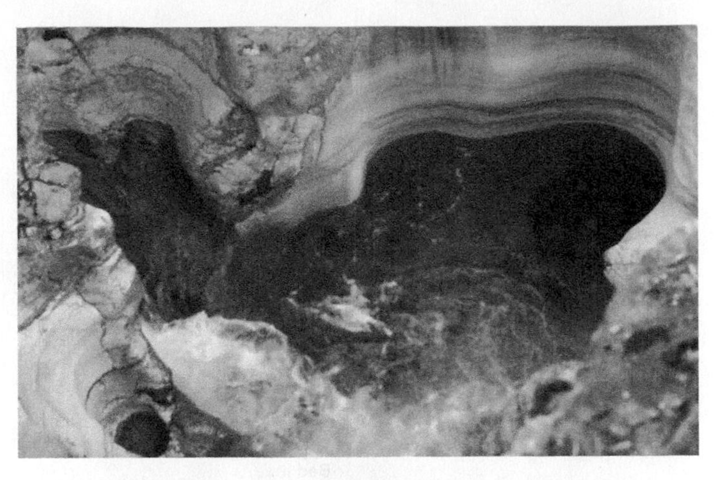

16.9 POTHOLES These potholes have been deeply incised in the limestone bedrock of Maligne Canyon, near Jasper, Alberta.

Mineral particles of any size that are carried by a swift current will strike against the rock floor and walls of a channel. This process of mechanical wear of channel materials is termed *abrasion* and is the principal method of deepening and widening a river channel in bedrock. The transported fragments also become chipped, broken, and rounded, and are reduced in size through *attrition*. The rolling of cobbles and boulders over the stream bed further crushes and grinds the smaller fragments to produce a wide assortment of grain sizes, which are then sorted by the flowing water. As a result, channel materials are generally coarser in the upstream sections of a river and get progressively finer downstream.

Abrasion is effective in bedrock that is too strongly consolidated to be eroded by simple hydraulic action. Potholes are distinctive erosional features associated with abrasion. They begin to form when a shallow declivity or indentation in the bedrock of a stream bed traps stones that are spun around by the flowing water, eventually carving deep cylindrical depressions (Figure 16.9).

In addition to the mechanical processes of scour and abrasion, flowing water can also remove some types of bedrock, such as limestone and dolomite, through chemical solution. This process is called *corrosion* and arises when the rock is slowly dissolved by the river water. River channels in limestone often develop cupped and fluted surfaces due to corrosion.

STREAM TRANSPORTATION

Stream load refers to the materials carried or transported by a stream. Stream transport occurs in three ways, depending on particle size and the energy of the flowing water (Figure 16.10). Finer particles, such as clay and silt, are carried in *suspension*; the turbulent motion of the flowing water holds these materials within the water column. This fraction of the transported matter

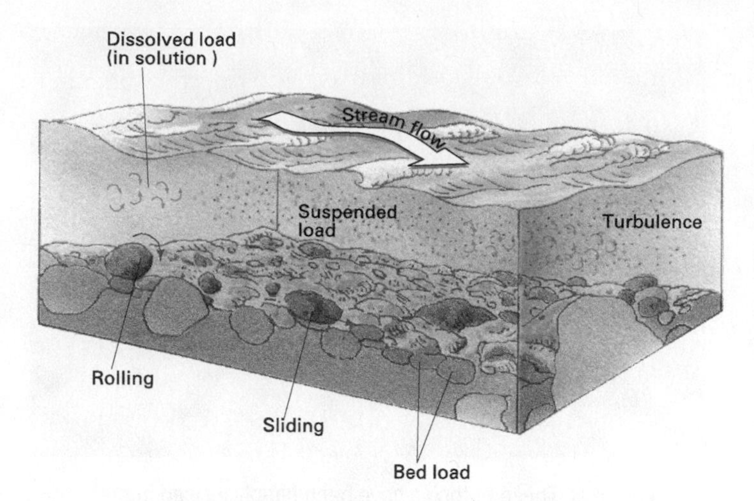

16.10 SEDIMENT TRANSPORT Streams transport materials in three ways: coarse particles move along the channel bed as bed load; finer sediment may be carried in the water column as suspended load; and materials carried in ionic form compose the dissolved load.

is the *suspended load* and normally accounts for the bulk of the material conveyed by a river. Sand, gravel, and larger particles move as *bed load* by rolling, bouncing, or sliding along the channel floor. *Dissolved matter* is transported invisibly in the form of chemical ions. All streams acquire some dissolved ions through mineral alteration of the channel surfaces or of the particles in their loads (Figure 16.11). An additional, albeit rather minor, method of stream transport is *flotation*. Although mainly restricted to buoyant organic matter such as wood, mineral particles can be rafted on ice by flotation.

Particles begin to move once the force of flowing water exceeds the critical shear stress of the bed materials. The force necessary for entrainment is called *fluid drag*, which is determined by the water's density and velocity. Fluid drag exerts a horizontal force that pushes against the particles, as well as a vertical component due to turbulence that lifts the particles. When these forces overcome friction and cohesion between particles, the channel sediment will begin to move. Once the particle is lifted, it is subject to gravity and so will have a tendency to settle back to the bed.

Fine sands about 1 mm in diameter tend to move most readily and can be entrained at flow velocities as low as 20 cm s^{-1} (Figure 16.12). Critical flow velocities increase for finer sediments, such as clays, because cohesion binds the particles together into larger aggregates. Similarly, high flow rates are required to transport coarser sands and gravels because of their greater mass.

A river carries as much as 90 percent of its load in suspension, and the largest river systems discharge considerable amounts of sediment into the oceans. For example, the annual sediment load of the Amazon River is calculated at 1,200 million tonnes, India's Brahmaputra-Ganga system at 1,100 million tonnes, and the Mississippi River at 159 million tonnes. Another river with a high suspended sediment load is the Huang He, or Yellow River, in China, so named because of the vast amount of yellow silt it carries in suspension. The Huang He's drainage basin has an area of about 865,000 km^2. Based on the amount of sediment that the river transports, the annual denudation rate is estimated at about

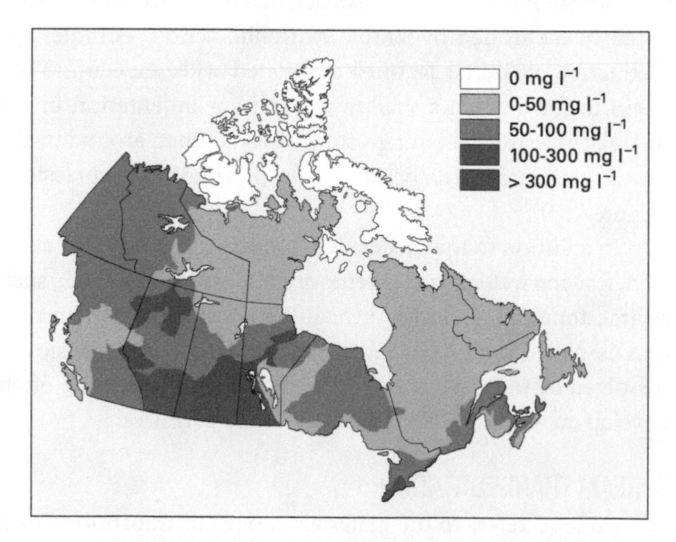

16.11 TOTAL DISSOLVED LOAD FOR CANADIAN DRAINAGE SYSTEMS The calcareous nature of glacial tills in western Canada contributes to the high dissolved load carried by rivers draining from the Prairie provinces. Dissolved loads are less in rivers associated with the Canadian Shield. (Hydrological Atlas of Canada, 1978.)

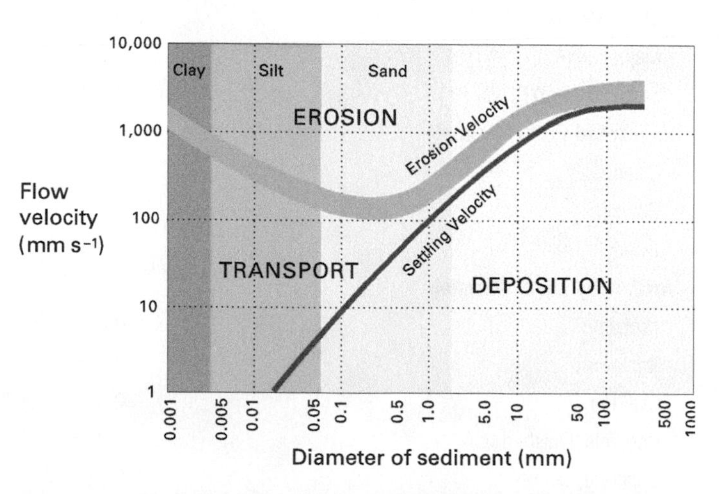

16.12 RELATIONSHIP BETWEEN FLOW VELOCITY AND PARTICLE TRANSPORT IN STREAMS Higher velocities are required to entrain and transport larger calibre materials because of their greater mass. Transport of fine silts and clays also requires a relatively high velocity because they tend to bind together through cohesion. Flowing water most readily moves sand-sized grain approximately 0.3 mm in diameter.

16.13 The Huang He River in China is heavily laden with sediment that is acquired as it flows through Loess Plateau, an area of easily eroded silt.

3,000 tonnes km^{-2}, most of which is removed from the easily eroded Loess Plateau through which it flows (Figure 16.13).

In Canada, the river with the highest average annual suspended sediment load is the Mackenzie, at 100 million tonnes. Much of the sediment carried by the Mackenzie River is derived from unconsolidated materials left by the ice sheets or from sediment supplied from tributaries arising in the wetter climates of the western mountains. In contrast, the rivers of the drier Prairie provinces carry much lower sediment loads, while those in eastern Canada transport relatively little sediment because they flow mainly over bedrock that is more resistant to erosion (Figure 16.14).

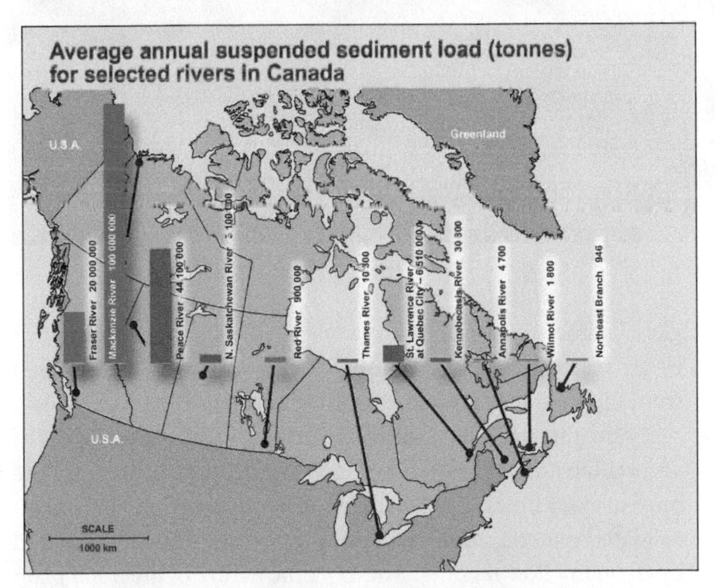

16.14 **AVERAGE ANNUAL SUSPENDED SEDIMENT LOAD (IN TONNES PER YEAR) FOR SELECTED RIVERS IN CANADA** The heavy sediment loads characteristic of western rivers can be attributed to high rainfall and significant areas of unconsolidated materials within the drainage basins. In eastern Canada, most rivers flow over resistant bedrock.

CAPACITY OF A STREAM TO TRANSPORT LOAD

The maximum solid load of debris that a stream can carry at a given discharge is a measure of *stream capacity*. This load is usually recorded in units of metric tonnes per day passing downstream at a given location. Total solid load includes both bed load and suspended load. A stream's capacity to carry suspended load increases as its velocity increases, because the swifter the current, the more intense the turbulence. The capacity to move bed load also increases with velocity and is related to the third or fourth power of the velocity. In other words, when a stream velocity doubles in times of flood, its ability to transport bed load is increased from eight to 16 times. Consequently, most of the conspicuous changes that occur in the channel of a stream are associated with flood events. The *Working It Out • River Discharge and Suspended Sediment* feature at the end of this chapter shows how the suspended sediment load of a stream increases with discharge.

STREAM DEPOSITION

Where a stream flows in a channel of hard bedrock, the channel's form cannot deepen quickly in response to rising waters, and may not change much even during extreme flood events. However, a stream flowing in a channel that is cut into thick layers of silt, sand, and gravel is very sensitive to flow rates and sediment load, and continually adjusts to changes in the supply of water and rock waste from upstream sources. If water flow increases, alluvial channels may widen and deepen. As the flow slackens or more sediment is available, perhaps because of slumping along the banks of the channel upstream, the stream will deposit material on the bed.

When bed load increases and exceeds the transporting capacity of the stream, the sediment accumulates as *river bars* of sand, gravel, and pebbles. These deposits raise the elevation of the stream bed, a process called **aggradation**. The development of distinctive channel forms reflects this type of adjustment to flow regime and sediment load. The three basic channel patterns generally recognized are straight, meandering, and braided. These develop in response to the amount and calibre of the sediment supply, the gradient of the stream, and the stability of the channel as determined by its resistance to erosion. Because these parameters are all highly variable, a wide variety of channel forms are possible.

An example of a straight channel is a mountain stream flowing swiftly over bedrock, with alternating sections of rapids and deeper pools (Figure 16.15a). Even where a stream's banks are relatively straight, the channel along which the water flows typically swerves around gravel bars. The course of a meandering channel consists of a series of broad, sweeping curves (Figure 16.15b). The spacing between adjacent meanders increases with the size of the river. In a braided stream, the water flows in numerous channels

(a)

(b)

(c)

16.15 CHARACTERISTIC CHANNEL FORMS (a) Many mountain streams have straight channels with sections of pools and rapids as the water flows over the coarse bed materials. (b) The sweeping curves of a meandering channel usually develop best on a flood plain. (c) Braided streams consist of numerous channels that divide and reunite in response to varying rates of sediment transport and deposition.

that separate and reunite (Figure 16.15c). The channel geometry of a braided stream changes continually as varying flow rates deposit and scour sediment.

AGGRADATION AND FLOOD PLAIN DEVELOPMENT

Sediment introduced at any point in a stream will be gradually spread along its whole length and slowly build up the land surface through aggradation. In addition to sediment acquired by stream erosion, overland flow carries material to streams in the normal process of slope degradation. Higher amounts may be delivered to a stream following disturbance of the plant cover, for example, after a forest fire or through logging or poor farming techniques. Other causes of aggradation are related to major changes in global climate. For example, many rivers in Canada are supplied by glacier meltwater that is heavily laden with sediment released from the ice under present climatic conditions (Figure 16.16).

Accumulated alluvium has filled many valleys to depths of several tens of metres to form a flood plain across which the present river flows. The river channel is now formed in unconsolidated materials that can be easily and continually reworked by changing flow regimes. The dynamic nature of the flood plain landscape is exemplified by the changing shape and position of **meanders**. Meanders are associated with **alluvial rivers** that transport sediment in channels that are free to migrate across the flood plain. The shapes and positions of meanders change over time as the outer banks erode and deposition builds up the inner

16.16 GLACIAL MELTWATER STREAMS FLOWING FROM THE LLEWELLYN GLACIER Meltwater streams are heavily laden with sediment. Where the streams flow into mountain lakes, the very fine sediment can remain suspended in the water, giving it a milky blue colour.

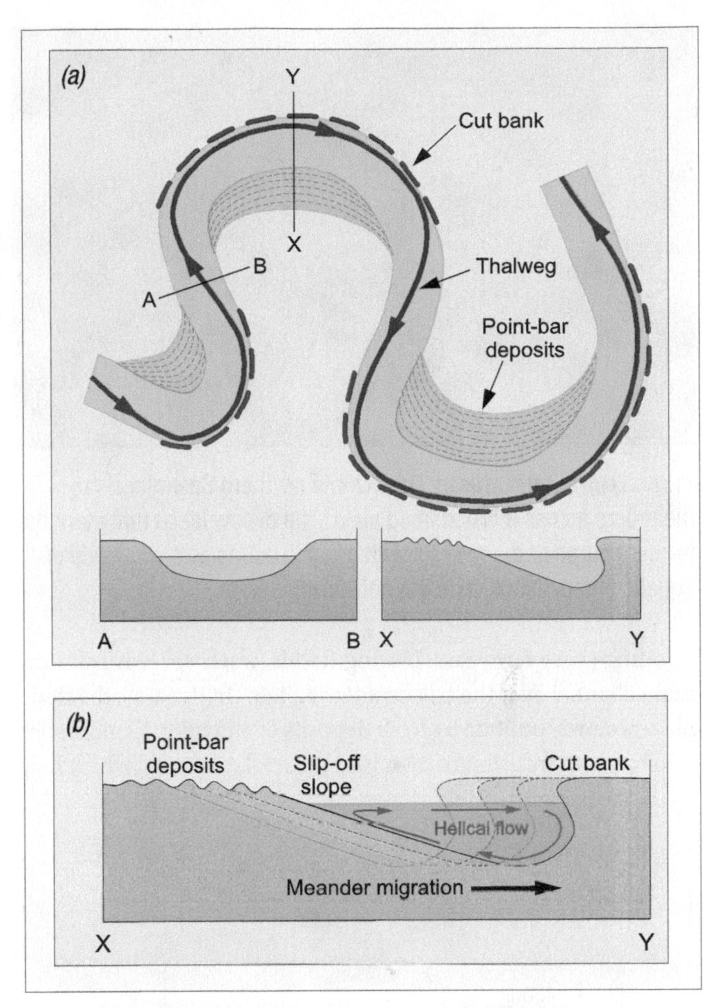

16.17 MEANDER BEND (a) Idealized map of a meandering stream showing general sites of deposition and erosion on the inner and outer banks. (b) Cross-profile of a meander bend of a large alluvial river. Helical flow causes undercutting of the outer bank of the meander and deposition of point bars on the inner bank, and results in the lateral migration of the meander.

banks (Figure 16.17a). The different rates of erosion and deposition are related to the variable flow rates within the curving channel. The deepest and fastest flowing water is represented by the **thalweg**, which swings back and forth across the channel as the course of the river changes.

As well as flowing along the river, water in a meandering channel also is subject to **helical flow** so that it follows a spiralling path, much like a cork crew (Figure 16.17b). Thus, not only is the current slowest on the inside of each bend, but water and sediment is also carried toward the bank by helical flow. This results in the accumulation of long, curving deposits of sediment called *point bars*. On the outside of the bend, the bank is undercut and collapses. In this way, the configuration of the meander changes and a narrow neck develops, which is eventually cut through. This shortens the river course and leaves a meander loop to form a *cutoff* (Figure 16.18). This is quickly followed by deposition of silt and sand across the ends of the abandoned channel, producing an *oxbow lake*. The oxbow lake gradually fills with fine sediment brought in during floods and with organic matter produced by aquatic plants. With time, oxbow lakes convert into swamps, but their identity is normally retained by subtle changes in topography and plant cover (Figure 16.19).

In time, the meandering river widens the **flood plain**, so that broad level areas lie on both sides of the river channel. A river flowing across a flood plain has a low channel gradient and, in unregulated systems, will characteristically experience overbank

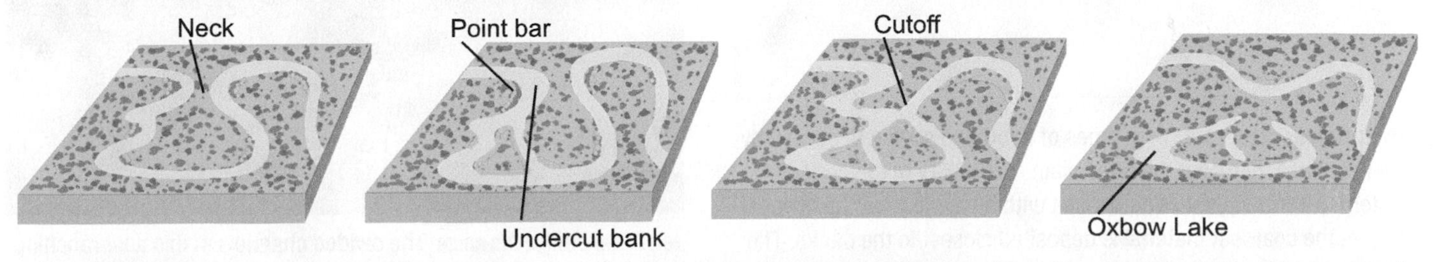

16.18 DEVELOPMENT OF AN OXBOW LAKE Changes in the course of a meandering river due to differences in erosion and deposition on the inner and outer banks will eventually produce a cutoff with the former meander loop left as an oxbow lake.

16.19 **FLOOD PLAIN FEATURES** This river in northern Saskatchewan meanders across a broad flood plain. The oxbow lakes that mark the former course of the river fill with organic debris as a sequence of aquatic plants establish in the non-flowing water.

flooding every few years. During floods, water spreads from the main channel over the adjacent flood plain. In this way, the flood plain is slowly built up by fresh deposits of alluvium. Flood waters also bring an infusion of dissolved mineral nutrients, which help

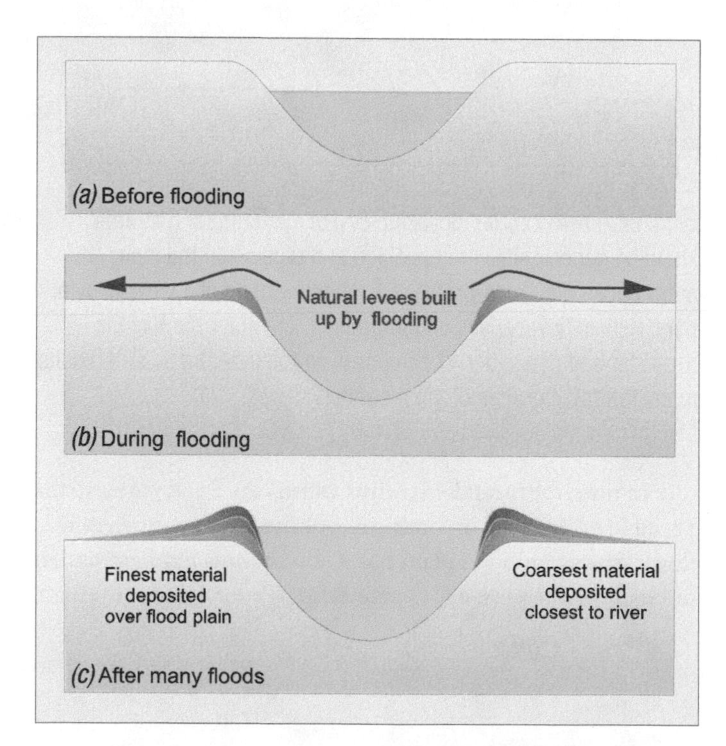

16.20 **NATURAL LEVEES** During times of flood, the sediment carried by rivers is deposited over the flood plain. The ability of the flood waters to carry sediment decreases with distance from the channel so that the coarsest material is deposited closest to the banks. The banks of the channel are slowly built up after many flood cycles, allowing the depth of the river to increase.

16.21 **A BREACHED LEVEE** The Mississippi River rushes through a break in a levee near Elsbury, Missouri, causing major flooding.

retain the high natural fertility of flood plain soils. As the current in the overbank flow slackens, the coarser particles are deposited in a zone adjacent to the channel. The result is an accumulation of higher land on either side of the channel known as a **natural levee** (Figure 16.20). Levees are artificially reinforced in many areas, however they can still be overtopped or even breached during exceptional flood events (Figure 16.21).

The flood plain extends to a point where land elevation precludes flooding, and its boundary is usually marked by a line of *bluffs*. Between the levees and the bluffs is lower ground, called the *backswamp*. Rivers will often divide into several channels and rejoin as they cross a flood plain to produce a pattern that is similar to that of braided streams. These are referred to as *anabranching rivers*, and are distinguished from braided streams by the relative permanence of their channels (Figure 16.22). Each channel in an anabranching

16.22 **ANABRANCHING RIVER** The divided channels in this anabranching river are semi-permanent features that are separated by well-vegetated levees and sandbars.

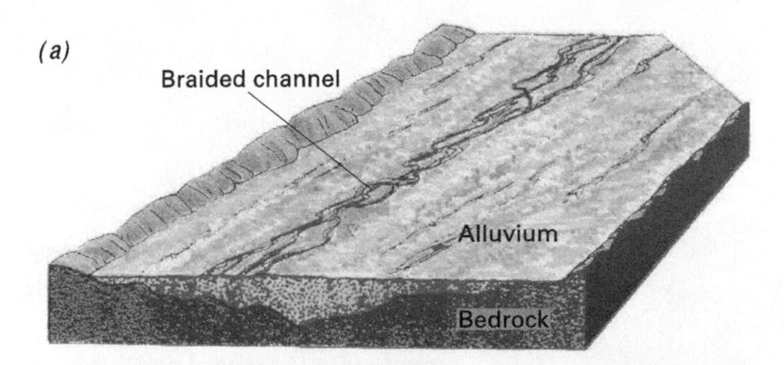

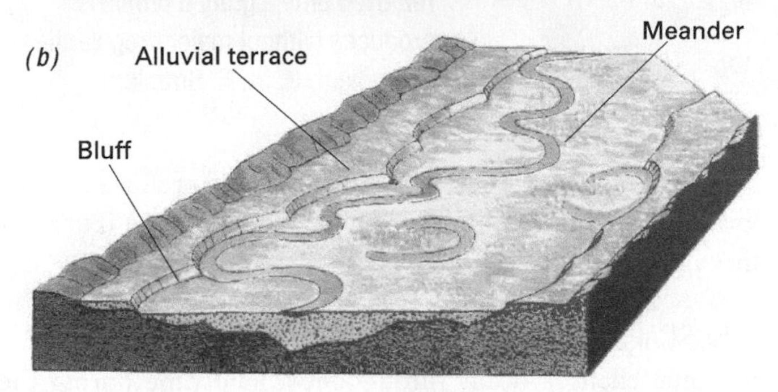

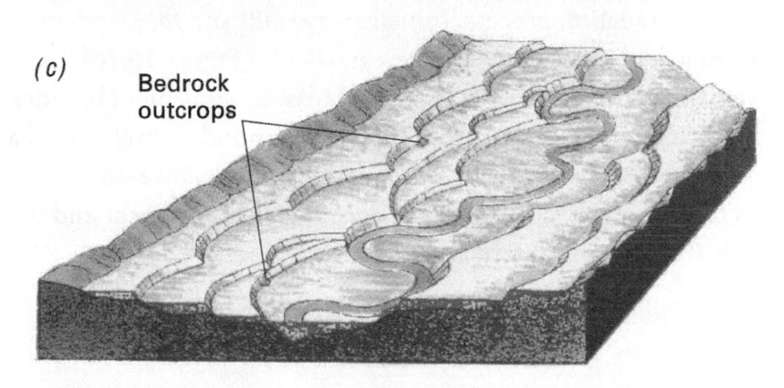

16.23 ALLUVIAL TERRACE FORMATION Alluvial terraces can form if the removal of alluvial fill in a valley is prevented by the greatly reduced rate of lateral erosion when the river encounters bedrock. (Drawn by A. N. Strahler.)

16.24 The coarse sediments underlying these terraces on the River Bran in Scotland are exposed where the slopes are being eroded on the outer bank of the meander.

river is generally bordered by a well-defined natural levee, which characteristically is stabilized by trees and shrubs. The islands that form are quite extensive, so adjacent channels are more widely separated than in braided streams.

If the source of sediment is cut off or greatly diminished and the volume of water is unchanged, less of the stream's energy is needed for transport. This increases the energy of the flowing water and results in channel scour. Where the stream gradient is gentle, this renewed degradation will result in lateral cutting and subsequent growth of meanders. This process gradually excavates the valley alluvium until the channel encounters hard bedrock, which limits the rate of removal of the remaining sediment (Figure 16.23).

A river can create a sequence of terraces when it moves back and forth across the valley, progressively cutting into deeper alluvial fill. As each terrace is cut into the original sediments, the river leaves a relic flood plain at a higher elevation. Each flat terrace surface is separated from the adjacent terrace by a steep slope, with the lowest terrace dropping down to the active channel. If a river cuts evenly into the alluvial sediments, terraces form at the same elevation on both sides of the valley. These are called *paired terraces. Unpaired terraces* occur when the stream encounters material on one side of the valley that does not erode easily. This leaves a terrace on one side with no corresponding terrace on the resistant side. Regardless of their configuration, all **alluvial terraces** in a river valley are created through renewed down-cutting due to *stream rejuvenation* (Figure 16.24).

STREAM GRADATION

Most major stream systems have experienced thousands of years of runoff, erosion, and deposition. Over time, the gradients of the stream segments that make up the drainage network tend to adjust so that they carry the average load of sediment that they receive from slopes and inflowing channels. If more sediment accumulates each year in the stream channel than can be carried away, the surface of the channel will build up and the slope of this segment of the stream will increase. But a steeper gradient increases stream velocity, allowing the stream to carry more sediment. Eventually, the slope will reach a point at which the stream carries away the sediment that it receives. If sediment flow to a stream segment decreases, the stream will gradually erode its channel. This will reduce the gradient and also reduce the stream's ability to transport sediment. In time, a new equilibrium will be established between channel slope, flow velocity, and sediment load. Since every channel segment experiences this process, eventually the whole stream will tend toward a

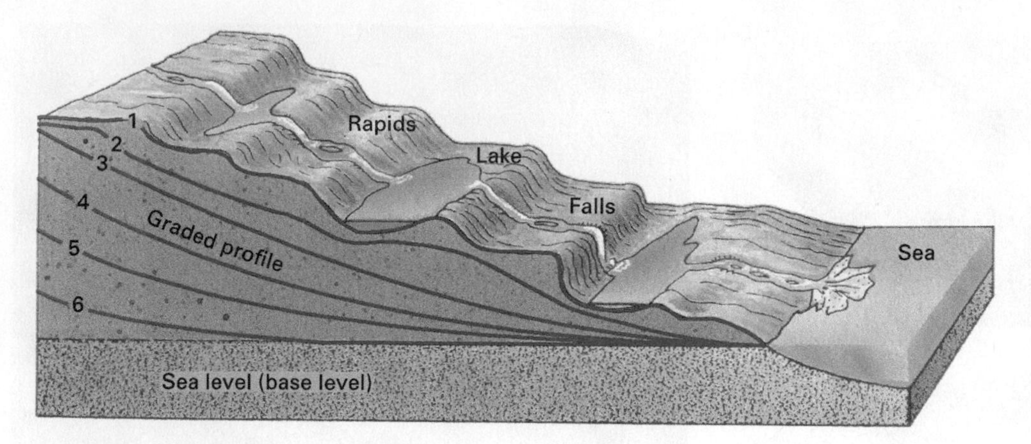

16.25 STREAM GRADATION In this schematic diagram, the initial channel consists of a succession of lakes, falls, and rapids, which are progressively removed until a graded profile is produced without major irregularities in gradient. (© A. N. Strahler.)

coordinated network that carries the sediment load contributed by the drainage basin. A stream in this equilibrium condition is called a **graded stream**.

Figure 16.25 shows how a graded stream might develop on a newly uplifted landscape. A series of longitudinal *stream profiles* plots the elevation of the stream with distance from the sea at different times. At first (profile 1), the stream is not graded, with large variations in its slope profile. Water accumulates in shallow depressions to form lakes, which overflow from higher to lower levels. Numerous rapids and waterfalls occur along its course. As time passes, fluvial action slowly erodes the landscape. As each stream segment seeks its own equilibrium slope, the stream profile becomes more regular (profile 2) and achieves a smooth *graded profile* by stage 3. Over time, the graded profile is steadily lowered through further erosion of the landscape (curves 4 through 6).

LANDSCAPE EVOLUTION OF A GRADED STREAM

Regional changes occur throughout a landscape during the stream gradation process. Initially *waterfalls* and *rapids* occur in stretches of a stream with steep gradients (Figure 16.26). Flow velocity at these points is greatly increased, abrasion of bedrock is intense, and the falls are cut back. The ponded stretches of a stream begin to fill with sediment and are later lowered in level as the outlets are cut down. In time, the lakes disappear and the falls transform into rapids.

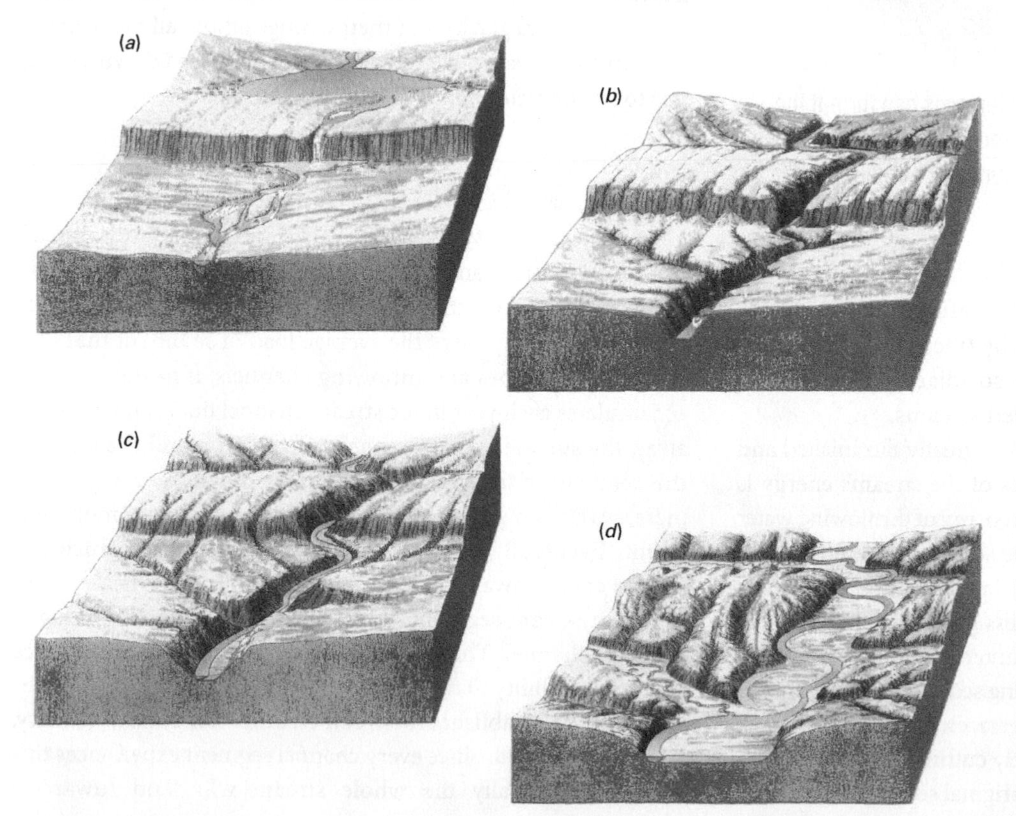

16.26 EVOLUTION OF A GRADED STREAM AND ITS VALLEY (a) A stream is established on a land surface dominated by landforms of recent tectonic activity. (b) Gradation is in progress. The lakes and marshes are drained. The gorge is deepening, and the tributary valleys are extending. (c) The principal river attains a graded profile. Flood plain development is beginning, and the widening of the valley is in progress. (d) The flood plain has widened to accommodate meanders. Flood plains now extend up tributary valleys. (Drawn by E. Raisz; © A. N. Strahler.)

16.27 **DEEPLY INCISED CANYON** This section of the Thompson River in British Columbia has cut a deep canyon into the limestone bedrock. Rapids indicate that the channel is not graded.

In the early stages of gradation, the capacity of the stream exceeds the load supplied to it, so little or no alluvium accumulates in the channels. Abrasion continues to deepen the major channels, resulting in steep-walled *gorges* or *canyons* (Figure 16.27). Weathering and mass wasting of canyon walls contribute an increasing supply of rock material. Also, overland flow supplies debris to the developing branches of the stream network.

As the rapids diminish, erosion reduces the gradient of the channel segment so the slope more closely approximates the average gradient of that section of the stream. At the same time, branches of the main stream are extended into higher parts of the original land mass. These carve out many small drainage basins. Thus, the original tectonic landscape gradually transforms into a fluvial landform system.

The continual change in the landscape results in an increase in the load supplied to the streams in their upper reaches. However, the gradient of the main channel slowly decreases and consequently its capacity to move bed load also decreases. In this way, the load supplied to the fluvial system eventually matches its capacity to transport it, and the major river achieves its graded profile. The first indication that a river has attained a graded condition is the beginning of flood plain development.

16.28 **POINT BAR** The deposition of sediment as a point bar on the inner bank of the meander shows how the river has migrated laterally as it cuts into the steep valley side.

This is initiated when the river begins to migrate laterally, cutting into the side slopes flanking its channel as a sinuous course develops. Point bars form as material is deposited on the inside of each bend, which progressively widens to produce an area of low ground (Figure 16.28).

As lateral cutting continues, the width of the flood plain increases and the channel develops the sweeping meandering course. In this way, the flood plain widens into a continuous belt of flat land in the valley floor. Flood plain development reduces the frequency with which the river attacks and undermines the adjacent valley wall. Weathering, mass wasting, and overland flow can then act to reduce the steepness of the valley side slopes (Figure 16.29). As a result, in a humid climate, the valley's gorge-like aspect gradually disappears and eventually gives way to an open valley with soil-covered slopes protected by vegetation.

The time required for a stream to reach a graded condition and erode a broad valley is usually many millions of years.

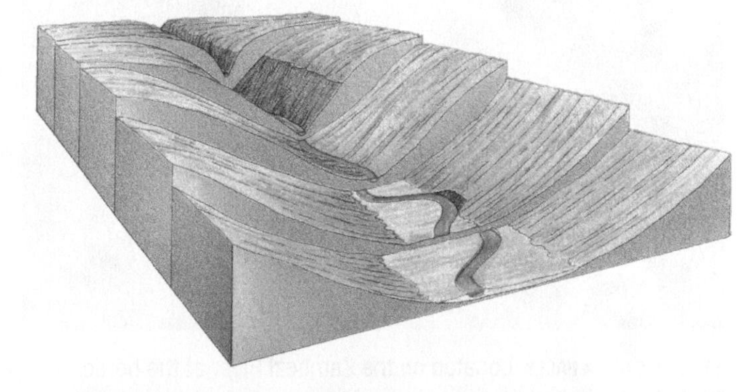

16.29 **EVOLUTION OF SIDE SLOPES** Following stream gradation, the valley walls become gentler in slope, and soil and weathered rock cover the bedrock. (From W. M. Davis; © A. N. Strahler.)

However, similar landforms may develop over a much shorter period when other factors influence river channel development. For example, the South Saskatchewan River near Saskatoon meanders across a broad flood plain and exhibits many of the features associated with the later stages of fluvial landscape development. The general topographic setting here was primarily created by continental ice sheets, which retreated from this part of Canada only about 12,000 years ago, leaving a predominantly glacial landscape upon which fluvial features have been superimposed.

16.30 VICTORIA FALLS Located on the Zambezi River at the border of Zimbabwe and Zambia in southern Africa, Victoria Falls formed where a fault created a zone of weakness in the bedrock. ***Fly By:* *17° 55′ 30′ S; 25° 51′ 25′ E***

WATERFALLS

The stream gradation process smoothes the profile of a stream by draining lakes and removing falls and rapids (see Figure 16.25). Thus, large waterfalls on major rivers are comparatively rare. Faulting and dislocation of large crustal blocks have caused spectacular waterfalls on several rivers, for example, the Victoria Falls on the Zambezi River in East Africa (Figure 16.30). Here, the shattered rocks along a fault have created a zone of weakness that the river has eroded.

Another class of large waterfalls involves new river channels resulting from past glacial activity. Erosion and deposition by vast moving ice sheets greatly disrupted drainage patterns in Canada and other northern regions, creating lakes and shifting river courses to new locations. Niagara Falls, which consists of the larger Horseshoe Falls (about 800 m wide) and the American Falls (about 325 m wide), is a good example (Figure 16.31). The formation of Niagara Falls began about 12,500 years ago when water from the early Lake Erie drained across the Niagara Escarpment at a point near Lewiston, New York, some 12 km from the present falls. The course of the Niagara River makes a right angle bend at the Whirlpool, where it joins an older drainage route that was filled with glacial materials deposited from the former ice sheet.

The Niagara Escarpment is formed from a gently inclined layer of hard Silurian limestone (approximately 400 million years old), beneath which lies more easily eroded shale. The Niagara River has gradually eroded the edge of the limestone layer to produce the steep gorge marked by Niagara Falls at its head. The height of the falls is currently 52 m, although the river has a clear drop of only 21 m before reaching a mass of rock that collapsed in 1954. Niagara Falls is receding at a rate of about 30 cm per year. This is slower than might occur naturally because as much as 75 percent of the water is diverted for hydroelectric power generation.

16.31 NIAGARA FALLS Niagara Falls formed where the river passes over the eroded edge of a massive limestone layer. ***Fly By:* *43° 04′ 41′ N; 79° 04′ 32′ W***

16.32 **UNPAIRED RIVER TERRACES** Multiple terraces have formed along the Thompson River near Ashcroft, British Columbia, due to renewed downcutting following rejuvenation.

ENTRENCHED MEANDERS

When tectonic activity causes regional uplift of a land surface, the gradient of pre-existing stream channels increases; this *rejuvenates* the river system, which continues to erode the bedrock with renewed energy. Initially this may produce a steep-walled gorge on either side of which lies the former flood plain. At first the former flood plain is represented as a flat terrace above the level of the river, but it is progressively dissected and stripped of its alluvial deposits (Figure 16.32).

Slow but considerable uplift can lead to the formation of *incised meanders*. Two types of incised meanders are generally recognized, depending on the cross-profile of the valley that is formed. *Entrenched meanders* are symmetrical in cross-section with little difference in gradient on the opposing valley sides. Entrenched meanders develop where incision has been rapid and can result in a winding gorge-like feature (Figure 16.33). When uplift of the land occurs more slowly, incision is less rapid and the river can cut laterally as well as vertically. This produces asymmetrical *ingrown meanders*.

Although ingrown and entrenched meanders are not free to shift in the same way as meanders do on alluvial flood plains, they can enlarge slowly and produce cutoffs. In this way, areas of high land are isolated from adjacent surfaces of similar elevation by deep abandoned river channels and the newly formed shortened stretches of river. Under unusual circumstances, where strong, massive bedrock is present, a meander cutoff leaves a *natural bridge* formed where the narrow meander neck is finally breached (Figure 16.34).

16.33 **ENTRENCHED MEANDERS** The Goosenecks of the San Juan River in Utah are deeply entrenched river meanders in horizontal sedimentary rock layers. The canyon, carved from sandstone and limestone, is about 370 m deep. ***Fly By: 37° 12′ N; 109° 59′ W***

16.34 **NATURAL BRIDGE** This natural bridge in Utah was formed when the neck of an entrenched meander was pierced by the river that formerly flowed in this dry region.

THE GEOGRAPHIC CYCLE

The Earth's landscapes range from mountain regions of steep slopes and rugged peaks, to regions of gentle hills and valleys, to extensive, nearly flat plains. Given that flowing water is the principal agent of landmass denudation in humid climates, one way to view these landscapes is as stages in a cycle that begins with

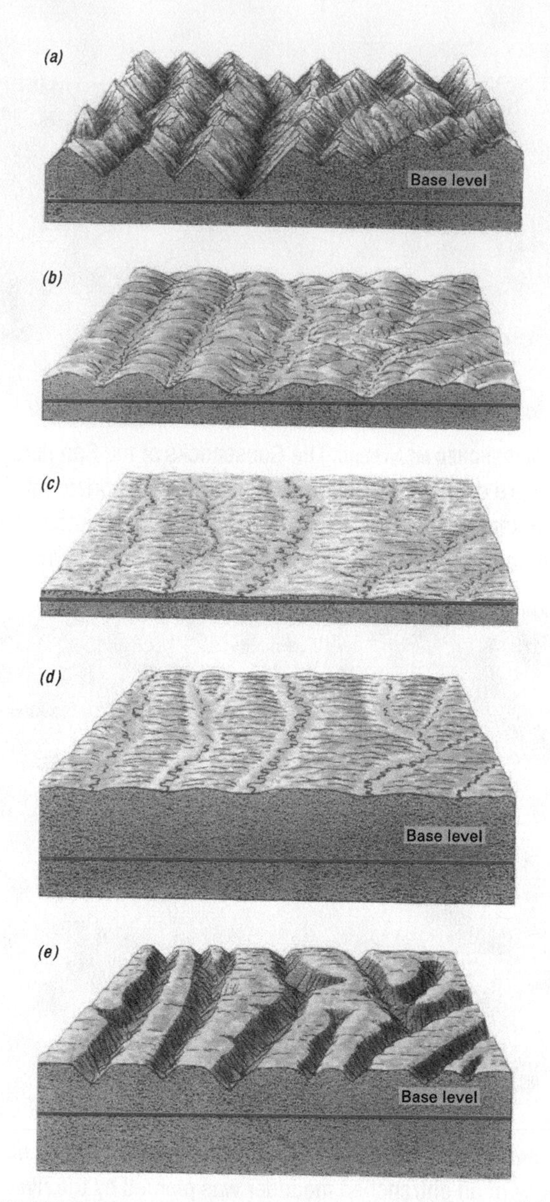

16.35 THE HUMID CYCLE OF LANDMASS DENUDATION (a) In the youthful stage, topographic relief is great, slopes are steep, and the erosion rate is rapid. (b) In the mature stage, relief is greatly reduced, slopes are gentle, and the erosion rate is slow. Soils are thick over the broadly rounded hill summits. (c) In old age, after many millions of years of fluvial denudation, a peneplain forms. Slopes are very gentle, and the landscape is an undulating plain. Flood plains are broad, and the stream gradients are extremely low. The entire land surface lies close to base level. (d) The peneplain is uplifted. (e) Streams erode a new system of deep valleys in the phase of land-mass rejuvenation. (Drawn by A. N. Strahler.)

rapid uplift and follows with erosion by streams. This sequence of development, first described by William Morris Davis in 1899 in what he termed the **geographic cycle**, assumes that landscapes are a function of structure, process, and time. In the *humid cycle of landmass denudation*, the initial landscape, referred to as the *youthful stage*, consists of many drainage basins and their branching stream networks. The region is rugged, with steep slopes and high, narrow crests (Figure 16.35a).

After initial uplift, the main streams draining the region establish a graded condition. They transport rock debris out of each drainage basin at the same average rate as the debris is being contributed from the land surfaces within the basin. Continuous removal of debris from the land surface causes a steady decline in average elevation. Lowering of the land surface is accompanied by a reduction in the average gradients of all streams. As the sharp peaks and gorges of the youthful stage give way to rounded hills and broad valleys, the geographic cycle enters the *mature stage* (b).

With time, the gradients of the streams and valley-side slopes of the drainage basins are gradually reduced. In theory, the ultimate goal of the denudation process is to create a featureless plain at sea level. Sea level is assumed to be at a lower elevation than the landmass, and so represents the lowest level, or *base level*, of fluvial denudation. Because the rate of denudation slows as elevation decreases, the land surface approaches base level at a progressively slower pace. Consequently, the ultimate goal is never reached. Instead, after the passage of some millions of years, the land surface is reduced to a gently rolling surface of low elevation, called a **peneplain** (c). With the evolution of the peneplain, the landscape has reached *old age*.

The development of a peneplain requires a high degree of crustal and sea-level stability for a period of many millions of years. One region that has been cited as a possible example of a contemporary peneplain is the Amazon-Orinoco Basin of South America. This vast region is a stable continental shield of ancient rock with very low relief.

Inevitably, crustal deformation will occur, and this may uplift the peneplain (d). The base level is now far below the land surface. Streams begin to cut into the landmass and carve deep, steep-walled valleys (e). This process is called *rejuvenation*. With the passage of many millions of years, the landscape will again be carved into the rugged stage shown in (a), and the cycle will repeat.

While Davis's idealized geographic cycle is useful for understanding landscape evolution over very long periods, it does little to explain the diversity of the features observed in real landscapes. Most geomorphologists now approach landforms and landscapes from the viewpoint of *equilibrium*. This approach considers a landform to be the product of forces acting upon it, including both forces of uplift and denudation, with the characteristics of the rock material playing an important role. For example, steep slopes and high relief are found where the underlying rock is strong and highly resistant to erosion.

Another problem with Davis's geographic cycle is that it applies only where the land surface is stable over long periods. However, crustal movements are frequent on the geologic time

16.36 FLASH FLOOD A flash flood created by a thunderstorm has filled this desert channel in the Tucson Mountains of Arizona with fast-flowing, turbid waters. A distant thunderstorm produced the runoff.

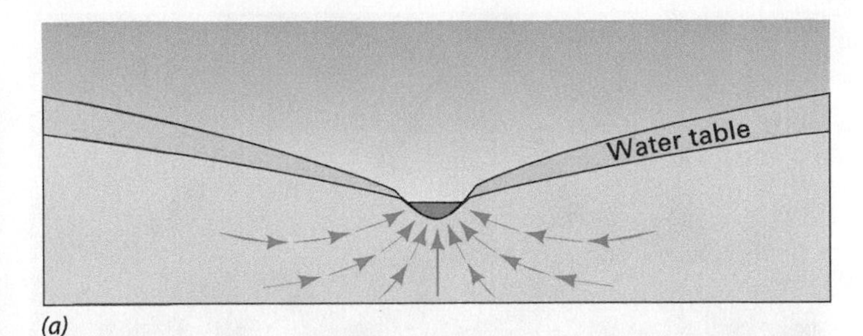

(a)

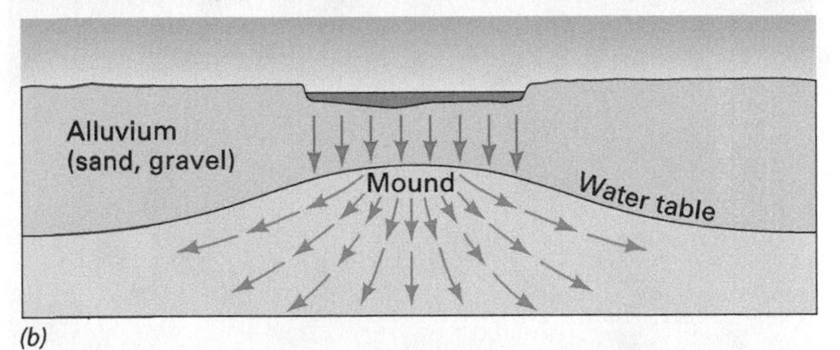

(b)

16.37 GROUNDWATER AND STREAM FLOW In humid regions (a), a stream channel receives groundwater through seepage. In arid regions (b), stream water seeps out of the channel and into the water table below. (© A. N. Strahler.)

scale, and few regions of the Earth's surface remain untouched by tectonic forces for very long. As erosion strips layer upon layer of rock from the continental lithosphere, the landmass becomes lighter and is buoyed upward by the underlying aesthenosphere. The process of crustal rise in response to unloading is known as *isostatic compensation*. Thus, a more appropriate model of landscape development is one in which tectonic processes operate continuously and erosional processes adjust accordingly, rather than as a sudden event followed by denudation.

FLUVIAL PROCESSES IN ARID CLIMATES

Although deserts have low precipitation, running water actually forms many landforms in these arid regions. A specific location in a dry desert may experience heavy rain only once in several years and, at these times, the stream channels carry water and work as agents of erosion, transportation, and deposition. In a few minutes, a dry channel transforms into a raging flood of muddy water heavily charged with rock fragments (Figure 16.36). Fluvial processes are especially effective in shaping desert landforms because the sparse vegetation cover offers little or no protection to the land surface, allowing large quantities of rock debris to be washed into the streams from the barren slopes.

An important contrast between arid and humid regions is the way in which water enters and leaves a stream channel. In a humid region, groundwater moves steadily toward the channel and seeps into the stream bed, producing a permanent stream (Figure 16.37a). In arid regions, the water table normally lies far below the channel floor, and a stream flowing across a gravel or sand plain will lose water by seepage (Figure 16.37b). Loss of water by seepage and evaporation rapidly reduces stream discharge in alluvium-filled valleys of arid regions. As a result, aggradation occurs and braided channels are common. Streams of arid regions are often short and end in alluvial deposits on the floors of shallow, dry lakes.

ALLUVIAL FANS

One common landform built by braided, aggrading streams is the **alluvial fan**, which develops as a low cone of sands and gravels (Figure 16.38). Alluvial fans can be many kilometres wide. The apex, or central point of the fan, lies at the mouth of a canyon or ravine and builds out onto an adjacent plain. Fans are built by streams carrying heavy loads of coarse rock waste from a mountain or upland region. The braided channel shifts constantly, but its position is fixed at the canyon mouth. The lower part of the channel, below the apex, sweeps back and forth, which accounts for the semicircular fan form that radiates downslope away from the apex. Often adjacent fans coalesce to form a *bajada*, or apron of coarse sediment along the foot of a mountain scarp.

In regions of internal drainage, where there is no outflow to the sea, streams respond only to the local base level determined by their elevation relative to the floor of the enclosed basin. Each arid basin becomes a closed system for the transport of sediment. In this situation, only the hydrologic system is open, with water entering as precipitation and leaving as evaporation. Consequently, fine sediment and precipitated salts accumulate in the basin to form a level salt flat or **playa**. In some playas, shallow water forms a salt lake.

16.38 ALLUVIAL FAN EXTENDING ONTO A DESERT PLAIN In places, alluvial fans coalesce to form a *bajada* or apron of sediment along the base of the highlands. The canyons from which they originate have carved deeply into an uplifted fault block.

A Look Ahead

The varied physical landscapes of the world reflect the influence of fluvial processes acting on different rock types, as running water is by far the most important agent of denudation. The three remaining agents of denudation are wind, waves, and glacial ice. Wind is particularly effective in arid climates, where it can readily pick up fine sediment from the barren surface and use the transported grains to abrade exposed bedrock surfaces. Similar processes operate along shorelines where wind can move beach sand inshore and deposit it as coastal dunes. However, in the coastal regions the principal geomorphic processes arise from the action of waves and currents, both of which are linked to wind. Chapter 17 discusses the role of wind in desert and coastal environments. Water in its solid form as glaciers and icesheets has had a dramatic effect on Earth landscapes. Now found only at high altitudes in mountain glaciers and ice fields and at high latitudes as great ice sheets, ice cover was much more extensive in recent geologic times. The work of ice during the Late-Cenozoic Ice Age had a profound effect on landscapes in many parts of the world, including nearly all of Canada. The erosional and depositional features created by ice action are discussed in Chapter 18.

CHAPTER SUMMARY

- This chapter covered the landforms and land-forming processes of running water, one of the four active agents of *denudation*. Like other geomorphological agents, running water erodes, transports, and deposits rock material, forming both *erosional* and *depositional* landforms.
- The work of running water begins on slopes as *overland flow*, which consists of thin films of water or tiny rivulets that move across the ground once precipitation exceeds the storage capacity of the surface materials. Overland flow moves soil particles downslope, producing colluvium that accumulates at the base of the slope. Mineral particles that enter stream channels and are later deposited are referred to as alluvium. In most natural landscapes, *soil erosion* and soil formation rates are more or less equal, a condition known as the *geologic norm*. *Badlands* are an exception in which natural erosion rates are very high.
- The work of streams includes erosion, transportation, and deposition. Where stream channels are carved into soft materials, large amounts of sediment can be obtained by *hydraulic action*. Where stream channels flow on bedrock, channels are deepened only by the abrasion of bed and banks through the impact of large and small mineral particles. Both the *suspended load* and *bed load* of rivers increase greatly as velocity increases. Velocity, in turn, depends on gradient.
- There are three basic channel forms. Straight channels tend to develop on steep terrain in a stream's headwaters; meandering channels develop sweeping curves where the channel gradient is very low; braided streams with multiple channels are common in streams that have heavy sediment loads.
- When provided with a sudden influx of rock material, for example by glacial action, streams build up their beds by aggradation. When that supply of material ceases, streams resume downcutting, leaving behind alluvial terraces.
- Large rivers with low gradients that move immense quantities of sediment are called alluvial rivers. The meandering

of these rivers forms distinctive landforms, such as *cutoff* meanders and *oxbow lakes*. Alluvial rivers are sites of intense human activity. Their fertile flood plains support agricultural crops and provide easy transportation routes.

- Over time, streams become graded—their gradients are adjusted to move the average amount of water and sediment supplied to them from slopes. Lakes and *waterfalls*, created by tectonic, volcanic, or glacial activity, are short-lived events, geologically speaking, that give way to a smooth, *graded* stream *profile*. Grade is maintained as landscapes erode toward *base level*.
- Denudation of the Earth's landscape is primarily by flowing water. The geographic cycle of landmass denudation in humid climates was an early concept that suggested landform development begins with tectonic uplift, followed by a sequence of features that are produced mainly by fluvial erosion. The cycle progresses through youth and maturity to old age, by which time the land surface has been levelled to a peneplain. The cycle is renewed when the land is once again uplifted.
- The geographic cycle has been superceded by the equilibrium approach, in which landforms are viewed as the product of uplift and denudation—continuous processes that act on rocks of varying resistance to erosion.
- Although rainfall is scarce in deserts, running water is still very effective in producing fluvial landforms. Desert streams, subject to flash flooding, build alluvial fans at the mouths of canyons. Water sinks into the fan deposits, creating local groundwater reservoirs. Fine sediments and salts carried by streams accumulate in playas, from which water evaporates, leaving sediment and salt behind.

KEY TERMS

aggradation
alluvial fan
alluvial river
alluvial terrace
alluvium
antecedent conditions
colluvium
depression storage
flood plain
fluvial landforms
fluvial processes
geographic cycle
graded stream
helical flow
meander
natural levee
overland flow
peneplain
playa
sheet erosion
stream deposition
stream erosion
stream load
stream
transportation
stemflow
surface retention
thalweg
throughfall

REVIEW QUESTIONS

1. List the four flowing substances that serve as agents of denudation.
2. Describe the process of slope erosion. What is meant by the geologic norm?
3. Contrast the two terms *colluvium* and *alluvium*. Where on a landscape would you look to find each one?
4. What special conditions are required for badlands to form?
5. When and how does sheet erosion occur? How does it lead to rill erosion and gullying?
6. In what ways do streams erode their bed and banks?
7. What is stream load? Identify its three components. In what form do large rivers carry most of their load?
8. How is velocity related to the ability of a stream to move sediment downstream?
9. How does stream degradation produce alluvial terraces?
10. Describe the evolution of a fluvial landscape according to the geographic cycle. What is meant by rejuvenation?
11. What is the equilibrium approach to landforms? How does it differ from interpretations based on the geographic cycle?
12. Why is fluvial action so effective in arid climates, considering that rainfall is scarce? How do streams in arid climates differ from streams in moist climates?
13. Describe how playas and alluvial fans form an arid desert landscape.

VISUALIZATION EXERCISES

1. Compare erosional and depositional landforms. Sketch an example of each type.
2. What is a graded stream? Sketch the longitudinal profile of a graded stream and compare it with the profile of a stream draining a recently uplifted set of landforms.
3. Sketch the flood plain of a graded, meandering river. Identify key landforms on the sketch. How do they form?

ESSAY QUESTIONS

1. The North Saskatchewan River rises in the Rocky Mountains, crosses the foothills, flows through the agricultural regions of Alberta and Saskatchewan, and finally reaches the sea. Describe the fluvial processes and landforms you might expect to find on a journey along the river from its headwaters to Hudson Bay.
2. What would be the effects of climate change on a fluvial system? Choose either the effects of cooler temperatures and higher precipitation, or warmer temperatures and lower precipitation.
3. What is meant by the humid cycle of landmass denudation? Describe the processes and stages that were originally incorporated into this cycle. How do modern views differ from those proposed earlier?

WORKING IT OUT • RIVER DISCHARGE AND SUSPENDED SEDIMENT

A river carries more sediment during a flood than during periods of normal flow. This happens for a number of reasons.

- First, as discharge increases, more water is available to carry sediment downstream.
- Second, as discharge increases, so does the velocity of the stream flow; and as the velocity of the water flow increases, so does the intensity of the turbulence that keeps suspended sediment in motion. Thus, more sediment can be carried in a volume of swiftly moving water compared to slower moving water.
- Third, as the water rises and increases in velocity, the force of the stream flow on the bed increases. The stream then erodes sediment in its bed, scouring and deepening the channel, a process that also increases the sediment available for transportation.
- Fourth, the heavy precipitation that causes a flood generates overland flow that drags more sediment into upstream river channels. Thus, more sediment is available for transport.

Figure 1 plots discharge against the suspended sediment load for the Oldman River at Lethbridge, Alberta. The discharge is measured in cubic metres per second ($m^3\ s^{-1}$) and the suspended sediment load in metric tonnes per day. The points are daily measurements collected from April to October 1990. Figure 1 shows that as discharge increases, so does suspended sediment. For example, when the discharge increases from 10 to 100 $m^3\ s^{-1}$, the suspended sediment load increases from about 8 to 800 tonnes per day. Thus, when discharge increases by a factor of 10, sediment load increases by a factor of about 100.

This type of increase follows a power function of the form:

$$y = ax^b$$

which gives a straight line on a log-log plot.

In the case of the relationship between suspended sediment load and discharge for the Oldman River, the measurements come close to fitting a line with the equation:

$$S = 0.015\ Q^{2.30}$$

where S is the suspended sediment load (tonnes per day) and Q is the discharge ($m^3\ s^{-1}$).

The rapid increase in suspended load with increasing discharge is characteristic of nearly all rivers. Consequently, much more sediment can be carried downstream during a single small flood than over many months of normal river flow. In addition, a large flood may move vast volumes of sediment that has been accumulating in the channel and flood plain, waiting for a rare and extreme event.

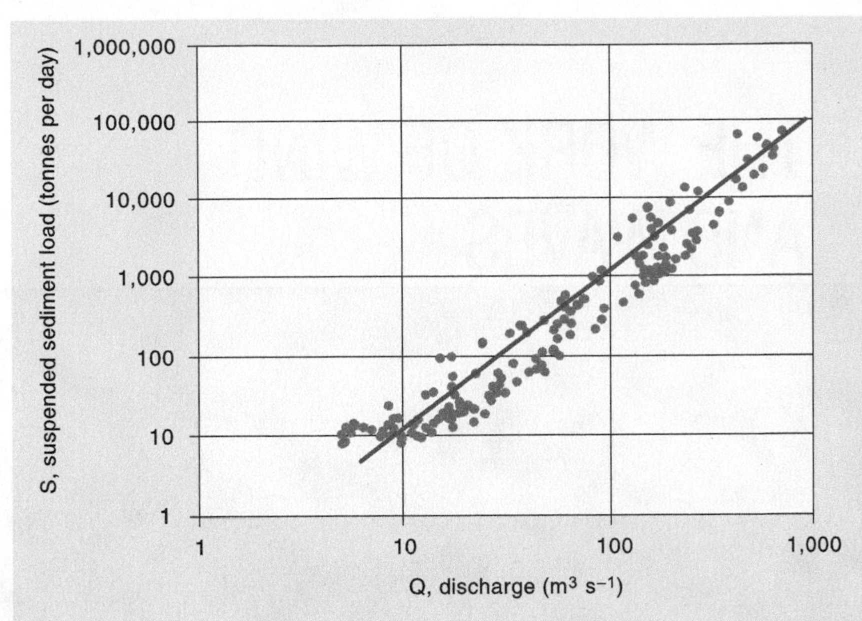

Figure 1 Sediment load and discharge The increase in suspended sediment load with an increase in discharge is plotted on logarithmic scales. The dots are daily observations for April to October 1990, for the Oldman River at Lethbridge, Alberta. (Data from the Water Survey of Canada.)

QUESTION / PROBLEM

1. What suspended load would you project for the Oldman River at Lethbridge when the discharge is 50 $m^3\ s^{-1}$? 500 $m^3\ s^{-1}$? If the discharge of the Oldman River doubles, what is the effect on its suspended sediment load? For this river, $S = 0.015\ Q^{2.30}$ where S is the suspended sediment load (tonnes per day) and Q is the discharge ($m^3\ s^{-1}$).

Find answers at the back of the book.

CHAPTER 17 THE WORK OF WIND AND WAVES

Arid landforms—a flat-topped mesa.

The rotation of the Earth on its axis, combined with unequal solar heating, produces a system of global winds in which air moves as a fluid across the Earth's surface. This air motion produces a frictional drag that can transport surface materials. When winds blow across areas not protected by vegetation, they can pick up fine particles and carry them long distances. Winds can also move coarser particles, such as sand, which may then be molded into dunes. When winds blow over broad expanses of water, they generate waves that expend their energy at coastlines, eroding rock, moving sediment, and creating distinctive landforms. In coastal regions, wave action is complemented by the tides—the rhythmic rise and fall in sea level generated by the gravitational attraction of the sun and moon. Tides are especially important in estuaries.

THE WORK OF WIND IN TERRESTRIAL ENVIRONMENTS

The geomorphic work of wind in terrestrial environments is collectively referred to as **aeolian** processes, and includes erosion, transport, and deposition of Earth materials. The ability of winds to carry out geomorphological work is related to the density of air, the velocity at which it moves, and the size and quantity of mineral particles that are being transported. Wind is generally not powerful enough to remove material from the surface of moist soils, or from soils protected by vegetation. However, it is an effective geomorphic agent in arid and semi-arid regions where vegetation is sparse and there is an abundant supply of unconsolidated surface materials. Wind is also effective in coastal environments, where beaches provide abundant supplies of loose sand.

Wind velocities increase rapidly with height above the surface. Close to the ground, there is a thin zone in which air movement is negligible. This zone of little or no wind is called the *laminar sublayer* and its thickness depends on the roughness of the surface. For soils, the depth of the laminar sublayer is about 0.03 of the average diameter of mineral grains. For example, the zone would be 1 mm thick on a surface covered with pebbles with an average diameter of 30 mm. In the laminar sublayer, air flows parallel to the surface. This *laminar flow* is replaced by *turbulent flow* in the lowest few metres of the atmosphere, with air currents rising at a velocity that is approximately one-fifth of the average wind speed. Hence, the thickness of the stationary air layer at the ground surface, coupled with the average velocity of upward motion in the overlying air, affects the wind's ability to transport sediment.

To move Earth materials, winds must exceed a threshold velocity, which varies according to the size of the mineral particles (Figure 17.1). **Threshold velocity**, the velocity required to entrain a particle of a given diameter, increases as the square root of particle size. Thus larger, heavier particles require a higher wind velocity to move them. However, clay particles tend to become cohesively bonded into larger aggregates. As a result, the threshold velocity required to entrain clay is similar to that of sands. For this reason, the optimum particle size for wind erosion is about 0.08 mm—the size of very fine sand and silt.

The ability of wind to erode is termed **wind erosivity** and is expressed mathematically as:

$$E = V^3\rho$$

where E is erosivity, V is wind velocity, and ρ is air density (1.2 kg m^{-3}).

Wind erosivity is an exponential function of wind velocity. Hence, if the wind velocity doubles, from say 5 to 10 m s^{-1}, erosivity increases 8-fold; an increase in wind speed from 5 to 15 m s^{-1} would result in a 27-fold increase in erosivity. Thus, the action of wind in removing and transporting sediment is facilitated by strong, regular prevailing winds.

Sand grains, up to 2 mm in diameter, are normally the largest particles that can be kept in suspension by strong winds of around 14 m s^{-1} (50 kph). Particles of this size will quickly return to the surface as the wind speed drops. The maximum rate at which a wind-transported particle settles through the air is called the *terminal fall velocity*, which can be calculated from Stoke's Law as:

$$w = \frac{2(\rho_p - \rho_a)gr^2}{\rho\mu}$$

where w is terminal fall velocity, ρ_p is density of the particle, ρ_a is density of air, g is acceleration due to gravity, r is radius of the particle, and μ is viscosity of the air. Viscosity is related to temperature and is important in terms of aerodynamic drag, which is the mechanical force that opposes motion; values range from about 1.45 kg m^{-1} s^{-1} at −40 °C to 1.98 kg m^{-1} s^{-1} at 40 °C. The terminal fall velocity for clay is less than 0.0001 m s^{-1}; for silt, it ranges from 0.001 to about 0.08 m s^{-1}; and for coarse sand, it can be as high as 10 m s^{-1}. Hence, the larger the particle, the greater the wind speed required to keep it moving above the ground surface.

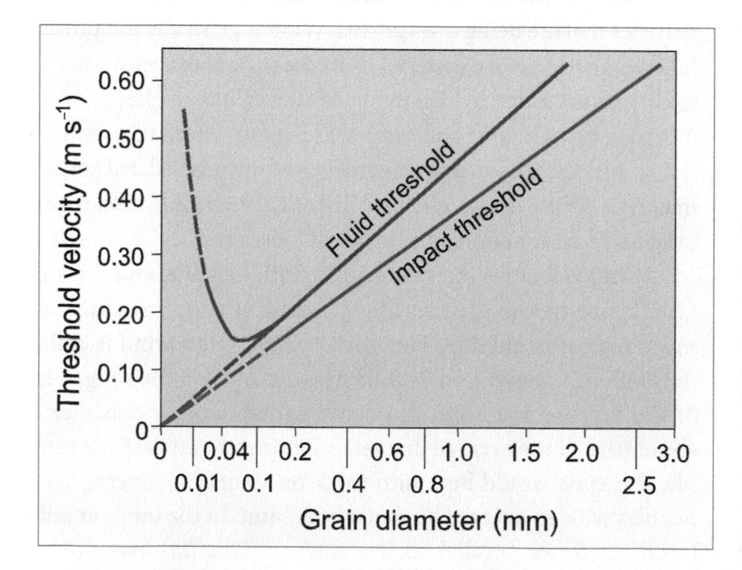

17.1 **GEOMORPHIC WORK OF WIND** Threshold velocities for wind erosion, transportation, and deposition for surface materials of different diameters.

TRANSPORT BY WIND

Wind transports Earth materials in different ways depending on their size and mass. The smallest and lightest particles are held in **suspension** in the atmosphere. Materials carried in suspension are typically less than 0.2 mm in diameter. Once aloft, wind can transport these materials great distances from their source, before precipitation or coagulation removes them from the atmosphere. Severe windstorms can keep larger particles aloft in turbulent eddies. Most of the silty loess deposits found extensively in North America, Europe, and China originated from wind-transported debris left by the ice sheets of the Late-Cenozoic Ice Age (see Chapter 18).

When wind reaches a critical velocity of about 4.5 m s^{-1}, larger particles such as sand become entrained in the airflow and begin to roll through **traction**. Strong winds can move small pebbles in this way. Traction accounts for about 20 to 25 percent of the material moved by wind. Wind alone normally cannot lift sand grains from the surface, but some are thrown into the air through collision. Once the sand particles are airborne, they follow curving paths determined by the horizontal wind velocity and the force of gravity. This process, termed **saltation**, accounts for 75 to 80 percent of sediment transport by wind.

Saltation usually lifts sand grains a centimetre or so above the surface, and moves them downwind a few centimetres at a time in a series of little jumps (Figure 17.2). The rate at which the particles move is about 30 to 50 percent of the wind's speed. When a particle strikes the surface, the force of impact is transferred to another particle, which may in turn be lifted into the air or moved forward through **creep**.

Wind blowing across loose sand creates a rippled surface of crests and troughs that are perpendicular to the wind direction (Figure 17.3). The distance between adjacent crests corresponds to the average distance travelled by airborne grains during saltation. Sorting occurs in these microtopographic features, with the

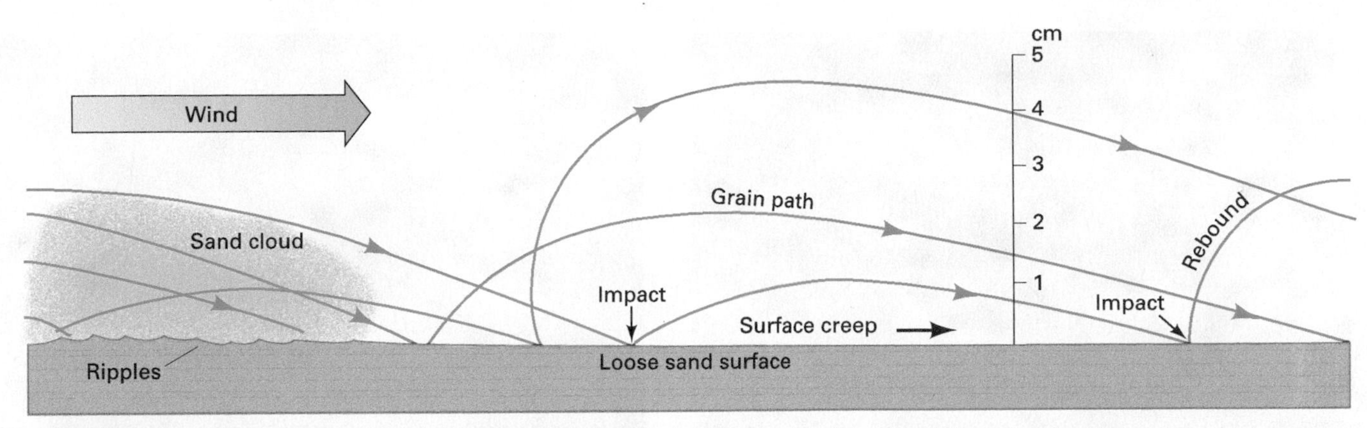

17.2 **SALTATION** Sand particles travel in a series of little jumps. (After R. A. Bagnold, *The Physics of Blown Sand and Desert Dunes*, London: Methuen, 1941.)

17.3 **SAND RIPPLES** The rippled surfaces of sand deposits form at right angles to the prevailing winds; the strength of the wind determines the scale of these features.

coarsest materials collecting at the crests. This is opposite to the sorting that occurs in dune formation, where the coarsest materials generally settle in the troughs.

DUST STORMS

Even in severe sandstorms, sand particles generally move within a metre of the surface, and rarely higher than 2 m. However, strong, turbulent winds blowing over barren surfaces can lift finer materials hundreds of metres, forming a dense **dust storm** (Figure 17.4). In semi-arid grasslands, a dust storm is generated where cultivation or grazing has stripped ground surfaces of protective vegetation cover. The bouncing motion of soil particles under strong winds disturbs the soil surface, and with each impact, fine dust is released and carried aloft by turbulence.

A severe dust storm approaches as a dense cloud extending from the surface to heights of several thousand metres. Light intensity within dust clouds can decrease considerably, and visibility can diminish to a few metres. Winds blowing out of deserts, such as the Sahara, are often heavily laden with fine particles (Figure 17.5). The harmattan, for example, is a hot, dusty wind that affects West Africa as prevailing wind direction changes with the seasonal movement of the ITCZ.

17.4 **DUST STORM** A cloud of fine dust sweeps across this savanna plain in eastern Kenya.

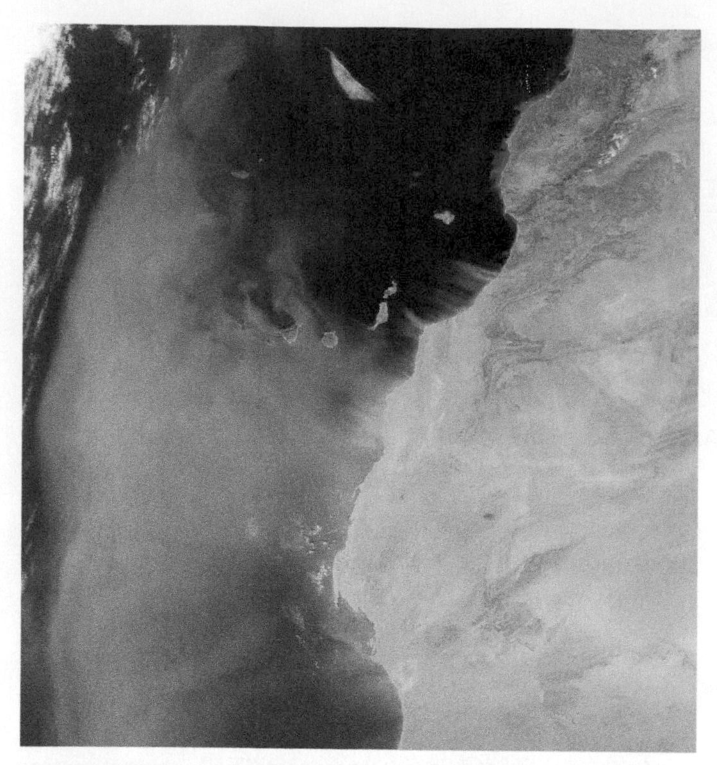

17.5 **ATMOSPHERIC DUST TRANSPORT** Dust from the Sahara region of North Africa is carried westward over the Azores and across the Atlantic Ocean.

17.6 **BLOWOUT IN A SAND DUNE** The dune surface is susceptible to deflation as soon as the protective vegetation cover is removed.

EROSION BY WIND

Dust picked up by the wind is removed from the surface through **deflation**; this is the fundamental process of *aeolian erosion*. As well as lifting particles into the air, deflation also moves loose material along the ground. In dry climates, much of the ground surface is susceptible to deflation because of the sparse vegetation cover. In the same way, materials are removed from dry riverbeds, beaches, and areas of recently formed glacial deposits.

Induced deflation often occurs when short-grass cover in a semi-arid region is cultivated. Plowing disturbs the natural soil surface, and in drought years, when vegetation dies out, the unprotected soil is easily eroded by wind. Much of the Great Plains region of North America has suffered from this type of land degradation, particularly during the severe drought years of the 1930s. Human activities in very dry regions contribute measurably to the creation of high dust clouds. In the Thar Desert of northwest India and Pakistan, the continual trampling of fine-textured soils by grazing animals and humans produces a semi-permanent dust cloud that sometimes extends to a height of 9 km.

A typical landform produced by deflation is a shallow depression called a *deflation hollow*. These subtle topographic features may range from a few metres to a kilometre or more in diameter, and are usually only a few metres deep. In dune environments where the fragile plant cover has been disturbed, a similar process can form a **blowout** (Figure 17.6).

Wind deflation is selective. Finer particles, such as silt, are lifted and raised into the air, but sand grains travel within a metre or two of the ground. Gravel fragments and rounded pebbles can be rolled or pushed over flat ground by strong winds , but they do not travel far and become lodged in hollows or between other large grains. Deflation will remove fine particles, leaving coarser, heavier materials behind as *lag deposits*. In this way, rock fragments ranging in size from pebbles to small boulders are progressively concentrated on the surface, and eventually form what is known as **desert pavement** (Figure 17.7).

Desert pavement acts as an armour that effectively protects the finer particles from deflation, although the stony surface is easily disturbed. Almost half of the Earth's desert surfaces are stony deflation zones. A surface of this kind in the Sahara is called a reg. In Australia, these landforms are known as gibber plains. The Gobi Desert of central Asia gets its name from the rounded pebbles, or gobi, that are found there. The term **hammada** describes a desert surface from which wind has removed most of the regolith, leaving only bedrock surfaces scattered with large rocks.

These stony and rocky landscapes are very different from the extensive sand seas or *ergs* that form in the Sahara and other desert environments. Sand sheets are level to gently undulating areas covered with sand grains that are too large to be moved by saltation under the prevailing wind conditions. They account for approximately 40 percent of aeolian depositional surfaces. One of the largest, the Selima Sand Sheet in southern Egypt and northern Sudan, occupies 60,000 km^2.

A dark, blackish-purple shiny stain, called *desert varnish*, is often found on the surfaces of desert rocks that are no longer etched by sand grains. Chemically, desert varnish consists mainly of manganese and iron oxides and clay minerals. It was previously thought that the shiny coating was formed by colonies of bacteria

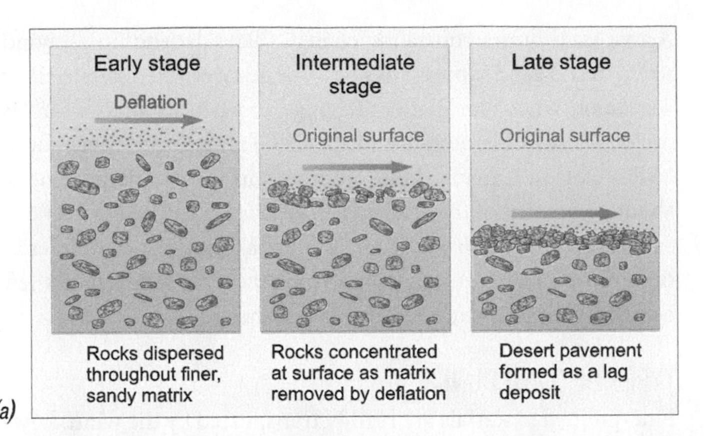

(a)

(b)

17.7 DESERT PAVEMENT (a) Desert pavement forms in arid regions as coarser rock framents become progressively concentrated at the surface through removal of finer materials. (b) Desert pavement in Arizona is formed of closely fitted rock fragments.

17.8 VENTIFACTS Rocks etched by windblown particles in the Dry Valleys of McMurdo Sound, Antarctica.

that preferentially concentrated manganese in their tissues. However, recent studies suggest that silica dissolved from other minerals is responsible for the glaze that forms over the rocks.

17.9 PEDESTAL ROCKS Pedestal, or mushroom, rocks are undercut by the abrasive action of windblown sand. The upper part of the rock is above the zone of abrasion, and its rough appearance contrasts with the smoother rock at the base.

Particles moved by the wind are effective at wearing away exposed rock surfaces, and wind **abrasion** produces many distinctive features in arid environments. At the smallest scale, rock surfaces are pitted and grooved by particles driven against them by the wind. Rocks that have been shaped, and sometimes polished, by the abrasive action of wind are called *ventifacts* (Figure 17.8). They are generally formed from hard, fine-grained materials. Smaller stones commonly have flattened surfaces or facets. The facets are cut in sequence, and as the stone becomes undercut, it topples over and exposes another surface to the wind. Often they develop three curved facets, giving them a somewhat triangular appearance. Stones that acquire this form are called *dreikanters*. At a larger scale are undercut *pedestal rocks* (Figure 17.9). *Yardangs*, tens of metres high and several kilometres long, are formed where wind abrasion has sculpted outcropping bedrock.

DEPOSITIONAL LANDFORMS ASSOCIATED WITH WIND

Wind-transported particles remain suspended until the wind velocity drops. Settling will also occur when particles collide and stick together, which increases their mass, or when precipitation washes them from the atmosphere. In arid areas, deposition is initiated primarily by small obstructions, such as plants and stones, which increase surface roughness and reduce wind velocity. Small mounds of sand will form as particles settle out on the lee side of such obstructions (Figure 17.10). Once formed, the mound will further disrupt airflow and, with a sufficient supply of sand, it may grow into a large dune.

17.10 **DUNE FORMATION** With a sufficient supply of sand, these small mounds formed in the shelter of desert plants in Morocco could grow into a large dune system.

SAND DUNES

A **sand dune** is any mound of loose sand molded by the wind. Dune sand is most commonly composed of quartz, which is extremely hard and largely immune to chemical decay. Large sand dunes require an abundant source of sand; for example, a sandstone formation that weathers easily to release individual grains, or perhaps a beach supplied with abundant sand from a nearby river mouth.

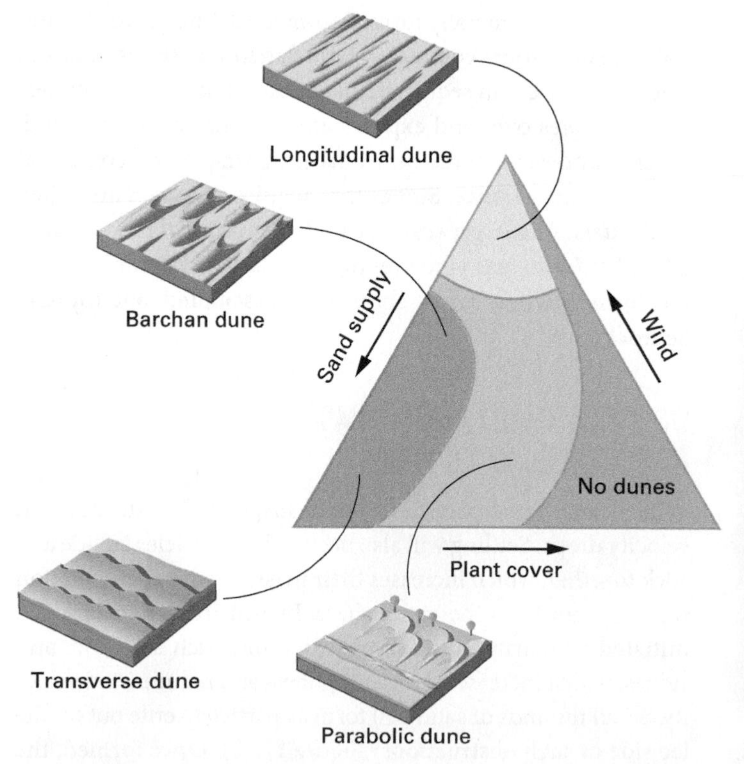

17.11 Dune types formed through the interaction of wind strength, sand supply, and vegetation.

Active sand dunes constantly change shape depending on wind strength and sand supply. They become inactive when stabilized by vegetation, when wind patterns change, or when a source of sand is no longer available. Active and stabilized sand dunes cover about 26,000 km^2 in Canada. Most are found in the Prairie provinces; about 45 percent of the total area occurs in Alberta, 36 percent in Saskatchewan, and 10 percent in Manitoba. The dune complexes in the Prairie provinces originated from the exposed deposits left behind by the retreating continental ice sheets.

TYPES OF SAND DUNES

Fine-grained materials are readily transported by the wind, however it requires much higher wind velocities to become mobile when coarse sand and pebbles are present. The type of sand dunes that form in a region therefore depends principally on the abundance of available sediment, wind strength, and variability of wind direction. Rock type and vegetation cover, insofar as they affect the amount and mobility of sediment, are also important. Variations in each of these factors will result in the formation of different basic forms of sand dunes (Figure 17.11).

The **barchan dune** is a common type of sand dune that develops perpendicular to the wind. These isolated dunes usually rest on a flat, pebble-covered surface and tend to be arranged in chains extending downwind from the sand source. The life of a barchan dune may begin as a sand drift on the lee side of some obstacle, such as a small hill, rock, or shrub. Once a sufficient mass of sand has formed, it begins to move downwind, taking on a distinctive crescent shape with the horns, or points, directed downwind (Figure 17.12). On the upwind side of the dune, the sand slope is gentle and smoothly rounded. On the downwind side, within the crescent, is the steep *slip face*.

17.12 **MIGRATING BARCHAN DUNES IN SKELETON COAST NATIONAL PARK, NAMIBIA** This aerial view shows a large barchan dune that in time will migrate from left to right. ***Fly By: 23° 51′ S; 15° 07′ E***

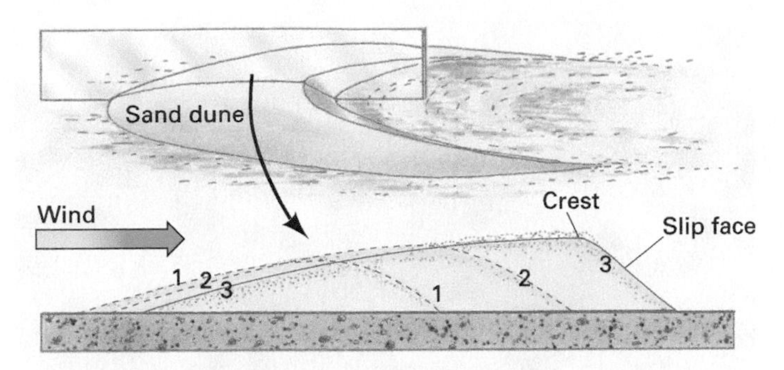

17.13 **GROWTH OF A DUNE AND DEVELOPMENT OF A SLIP FACE** The gradient on the upwind slope is normally 10 to 12°, compared with about 35° on the steep downwind slope. (After R. A. Bagnold.)

Windblown sand moves up the gentle upwind side of the dune by saltation or creeps to form a high, sharp crest (Figure 17.13). From this crest, the grains move on to the steep slip face of the dune, where they come to rest. The slip face maintains a more or less constant angle, which is known as the *angle of repose*. However, the accumulation of wind-carried sand grains steepens the slip face until it becomes unstable and the outermost layer of sand slides down the dune slope, restoring the angle of repose. For loose sand, this angle is about 35°.

The rate at which sand moves over a dune surface increases exponentially with wind velocity. Sand transport is negligible at wind velocities below 20 kph, and even at 50 kph it amounts to less than 500 kg per day. Although the downwind progress of dunes is imperceptible, over the course of a year, a barchan dune can migrate as much as 25 to 30 m.

Where sand is so abundant that it completely covers the underlying consolidated rock, dunes take the form of wavelike ridges separated by trough-like furrows. These dunes are called **transverse dunes** because, like ocean waves, their crests are aligned at right angles to the direction of the dominant wind (Figure 17.14). The sand ridges have sharp crests and are asymmetrical; the gentle slope is on the windward side and the steep slip face on the lee side. Deep depressions lie between the dune ridges. Transverse dunes are most often associated with extensive *sand seas* and develop where enormous quantities of sand are supplied by material weathered from sandstone formations. The surface of a sand sea usually develops very large-scale undulations known as *draas*. These features, which are only a few metres high but several kilometres across, provide the base upon which transverse dunes are superimposed.

Parabolic dunes are characterized by a curving dune crest that is bowed outward in the downwind direction; thus, curvature in these dunes is the opposite of barchan dunes. A common type of parabolic dune is the *coastal blowout dune*, which forms adjacent to beaches. Here, large supplies of sand are available, and prevailing winds blow the sand toward land. Deflation forms a saucer-shaped depression, heaping the sand in a curving ridge. The landward side develops a steep slip face that advances over the lower ground and buries the vegetation (Figure 17.15).

On semi-arid plains where winds are strong, groups of parabolic dunes develop to the lee of shallow deflation hollows. Sand that is caught by low bushes accumulates as a broad, low ridge. These dunes have no steep slip faces and are relatively immobile. However, in some cases, the dune ridge migrates downwind, drawing the dune into a long, narrow form with parallel sides, called a **hairpin dune**.

Longitudinal dunes (or *sief dunes*) develop where sand supply is more limited. These form as long, narrow ridges parallel to

17.14 **TRANSVERSE DUNES IN THE RUB' AL KHALI (EMPTY QUARTER) OF THE ARABIAN PENINSULA** In this largely uninhabited region, prevailing winds from the north have molded the abundant sand supply into transverse dunes. ***Fly By:*** *19° 14′ N; 53° 04′ E*

17.15 **ACTIVE PARABOLIC DUNES NEAR PASCO, WASHINGTON**

17.16 **LONGITUDINAL DUNES** These parallel dune ridges are in the Empty Quarter of the Arabian Peninsula. ***Fly By: 19° 17′ N; 48° 02′ E***

the direction of the prevailing wind. Longitudinal dunes may be many kilometres long; they cover vast areas of tropical and subtropical deserts in Africa, the Arabian Peninsula, and Australia (Figure 17.16).

Some dunes acquire more elaborate forms. For example, the radially symmetrical *star dunes* are pyramidal sand mounds, with slip faces on three or more sides, which radiate from a high central point (Figure 17.17). Star dunes tend to accumulate in areas with multidirectional wind regimes and grow in height rather than move laterally. They have been used for centuries as landmarks in the otherwise featureless landscape of the Sahara and Arabian Deserts.

COASTAL FOREDUNES

A narrow belt of dunes in the form of irregularly shaped hills and depressions typically develops on the landward side of sand beaches. These **coastal foredunes** normally bear a cover of beach grass and a few other species of plants that have adapted to the

17.17 **STAR DUNES IN SAUDI ARABIA'S EMPTY QUARTER** Radially symmetrical star dunes can attain heights of about 500 m and are the tallest form of dunes. ***Fly By: 19° 17′ N; 48° 20′ E***

17.18 **COASTAL FOREDUNES, QUEEN'S COUNTY, PRINCE EDWARD ISLAND** Beach grass thriving on coastal foredunes traps drifting sand to produce a dune ridge.

severe environment (Figure 17.18). The sparse plant cover traps sand moving onshore from the adjacent beach. As a result, the foredune ridge is built up several metres above high-tide level.

Foredunes form a protective barrier for the tidal lands which develop behind them. In a severe storm, waves may cut away the upper part of the beach, and then attack the foredune barrier. Usually the barrier is not breached, except perhaps along coasts subject to high storm surges during hurricanes. Between storms, the beach is rebuilt, and in due time, wind action restores the dune ridge.

Where the dune vegetation is disturbed, a blowout can rapidly develop and extend as a trench through the dune ridge. Storm waves may then funnel through the gap, spreading sand onto the tidal marsh or lagoon behind the dune. If badly eroded, the gap can become a new tidal inlet for ocean water.

LOESS

In several midlatitude areas of the world, extensive regions are covered by deposits of wind-transported silt that have settled out from dust storms over many thousands of years. This material, known as **loess**, generally has a uniform yellowish colour and is deposited as a homogeneous mass with no apparent stratification. Because it is unconsolidated, loess tends to break away along vertical cliffs wherever it is exposed. It is also easily eroded by running water, and is subject to rapid gullying when the protective vegetation cover is removed.

Loess is important to world agriculture; it forms fertile soils particularly suited to the cultivation of grains. The thickest deposits of loess are in northern China, where layers 30 to 100 m thick are common. Extensive loess deposits are also found in central Europe and Argentina. Large areas of the prairie region of Indiana, Illinois, Iowa, Missouri, Nebraska, and Kansas are underlain by loess, ranging in thickness from 1 to 30 m.

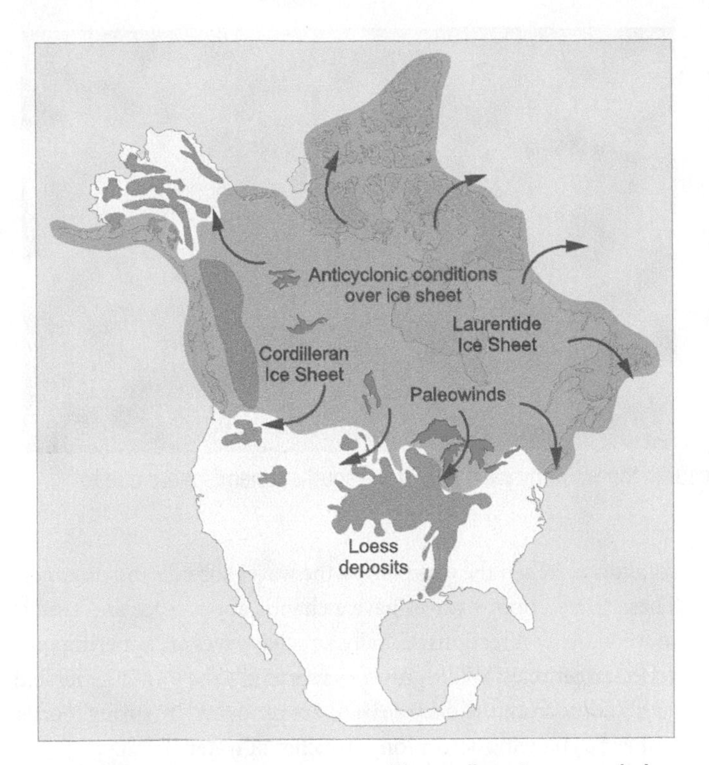

17.19 **LOESS DEPOSITS IN NORTH AMERICA** Fine sediment was carried away from barren areas adjacent to the continental ice sheet by anticyclonic airflow over the cold ice surface.

17.20 **THE LOESS PLATEAU** At elevations ranging from 1,200 to 2,000 m, the plateau covers an area of 380,000 km^2 in north-central China. The yellow silt is easily eroded and contributes to the heavy sediment load of the Huang He River, which flows through this region (see Figure 16.13).

Extensive deposits also occur in Tennessee and Mississippi in areas bordering the lower Mississippi River flood plain, as well as in the Palouse region of northeastern Washington and western Idaho.

The American and European loess deposits are directly related to the continental glaciers of the Late-Cenozoic Era. At the time when ice covered much of North America and Europe, a generally dry winter climate prevailed in the land bordering the ice sheets. Strong winds blew southward over the bare ground, picking up silt from the flood plains of braided streams that carried the meltwater from the ice (Figure 17.19). Loess deposits formed in this way are referred to as "glacial loess" to distinguish them from "desert loess" deposits, such as those of north-central China.

Desert loess deposits originate in arid climates where large quantities of fine silt are derived by abrasion of quartz grains through wind action. The Chinese loess deposits have been linked to tectonic activity that began some 40 million years ago and culminated in the formation of the Himalaya Mountains and the Tibetan Plateau. Intensive development of this highland region about 25 million years ago had a marked effect on the climate of central Asia, which became progressively drier as the flow of moist air from the Indian Ocean was cut off. The dry, winter monsoon winds transported dust from this newly formed desert region, eventually depositing it to form the Loess Plateau (Figure 17.20).

THE WORK OF WIND IN COASTAL ENVIRONMENTS

WIND AND WAVES

The energy in an ocean wave is derived from the wind and is proportional to the square of its height. Thus, a 4-m wave has $4^2 = 16$ times more energy than a 1-m wave. The stored energy in a wave is dissipated when it strikes the shore; 1-m waves dissipate about 10 kWh of energy (36×10^6 J h^{-1}) along a wave front 1 m in length. Wavelength, the distance between adjacent wave peaks, is not related to energy.

When the wind blows across an initially calm surface, local pressure differences generated by small eddies begin to disturb the water surface and generate small ripples, or *cat's paws*. As the ripples grow, the wind begins to exert a shear stress on the windward side, and also creates a pressure differential across the growing waves, which furthers their development. The rougher the surface, the greater the amount of energy that is transferred to the water.

The size and strength of an ocean wave is determined by wind speed and direction, and by the **fetch** or distance of open water over which the wind acts. The amount of energy transferred to the water is proportional to the fourth power of the wind speed. A longer fetch and greater wind speed produces more powerful waves. Waves will eventually reach their maximum development when winds blow consistently from the same direction for a sufficient period of time. This condition is termed a *fully developed sea*. The largest waves are found in the southern hemisphere at latitudes of 40° to 60° S. Here the westerly winds blow across an uninterrupted expanse of ocean, and average wave height is 5 m. In equatorial regions where wind speeds are low, the average height of ocean waves is 1 to 2 m (Figure 17.21).

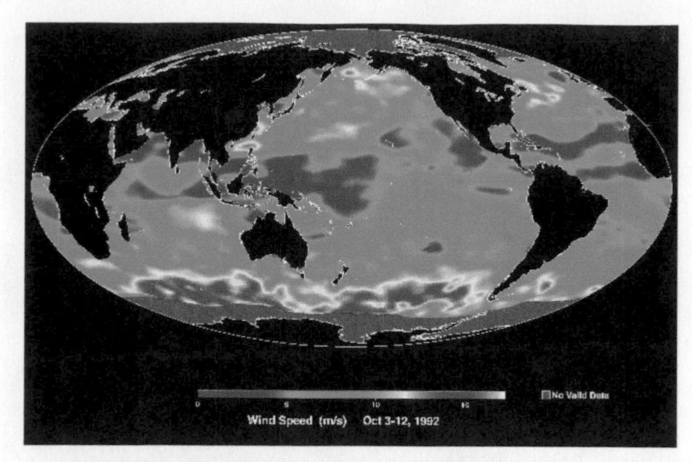

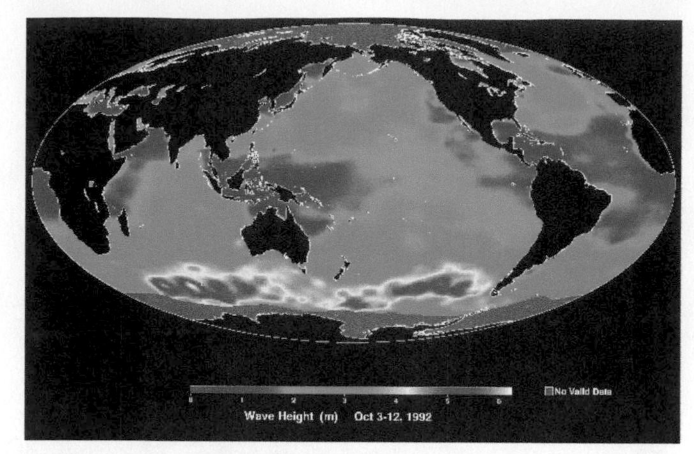

17.21 **AVERAGE WIND SPEED ($m\ s^{-1}$) AND WAVE HEIGHT (m)** The largest waves occur in the westerly wind belt of the southern hemisphere due to persistent, strong winds and extensive fetch.

The condition of the sea is the basis for the descriptive Beaufort scale, originally devised in 1805. As currently used, the Beaufort scale is somewhat modified, with the addition of wind speeds and wave heights (Table 17.1).

Waves are classified into two types: *gravity waves* and *swells.* Gravity waves are short-lived waves created by local wind conditions. When the wind drops, the waves subside and disappear. These short, choppy waves have a chaotic appearance and tend to move in many directions. Usually, smaller waves are superimposed on the larger ones. While gravity waves tend to be locally generated, swells come from the deep sea and are produced by strong, persistent winds blowing across long stretches of water.

TABLE 17.1 **THE BEAUFORT SCALE IS AN EMPIRICAL SCALE BASED ON THE OBSERVABLE EFFECTS THAT WINDS OF DIFFERENT STRENGTH HAVE ON THE OCEAN SURFACE (ON LAND, THE EFFECTS ON VEGETATION AND OTHER FEATURES ARE USED)**

FORCE	WIND ($km\ h^{-1}$)	WORLD METEOROLOGICAL OFFICE (WMO) CLASSIFICATION	APPEARANCE OF WIND EFFECTS	
			ON WATER	ON LAND
0	Less than 2	Calm	Sea surface smooth and mirror-like	Calm, smoke rises vertically
1	2 – 6	Light Air	Scaly ripples, no foam crests	Smoke drift indicates wind direction, wind vanes are still
2	7 – 11	Light Breeze	Small wavelets, crests glassy, no breaking	Wind felt on face, leaves rustle, vanes begin to move
3	12 – 19	Gentle Breeze	Large wavelets, crests begin to break, scattered whitecaps	Leaves and small twigs constantly move, light flags are extended
4	20 – 30	Moderate Breeze	Small waves, 0.3–1.2 m, become longer, numerous whitecaps	Dust, leaves, and loose paper are lifted, small tree branches move
5	31 – 39	Fresh Breeze	Moderate waves, 1.2–2.4 m, take a longer form, many whitecaps, some spray	Small trees in leaf begin to sway
6	40 – 50	Strong Breeze	Larger waves, 2.4–4.0 m, whitecaps common, more spray	Larger tree branches move, whistling in wires
7	51 – 61	Near Gale	Sea heaps up, waves 4.0–6.1 m, white foam streaks off breakers	Whole trees move, resistance felt walking against wind
8	62 – 74	Gale	Moderately high (4.0–6.1 m) but waves of greater length, edges of crests begin to break into spindrift, foam blows in streaks	Trees swaying, some resistance when walking into wind
9	75 – 87	Strong Gale	High waves (6.1 m), sea begins to roll, dense streaks of foam, spray may reduce visibility	Slight structural damage occurs, slates blow off roofs
10	88 –102	Storm	Very high waves (6.1–9.1 m) with overhanging crests, sea white with densely blown foam, heavy rolling, lowered visibility	Rarely experienced on land, trees broken or uprooted, considerable structural damage
11	103 – 117	Violent Storm	Exceptionally high waves (9.1–13.7 m), foam patches cover sea, visibility more reduced	
12	117+	Hurricane	Air filled with foam, waves over 13.7 m, sea completely white with driving spray, visibility greatly reduced	

Swells do not diminish when the wind dies, and can travel thousands of kilometres from their origin. For this reason, sizeable waves can still be seen, even when there is little or no wind. Swell amplitude is low so the water surface does not become steeply inclined, but wavelength is long with adjacent wave crests sometimes several hundred metres apart. Because of their long wavelength, swell waves appear to be slow-moving; the wave period, the time for successive crests to pass, is typically 10 to 15 seconds in swells. Wind-generated gravity waves tend to be relatively steep and typically last for only 4 to 6 seconds. They appear to be fast-moving because their crests are closely spaced, but this is not the case (Table 17.2).

In the open sea, waves travel through water but do not push the water forward as they pass. As a wave arrives, it lifts the water molecules, which then move forward, down, and back so that each particle completes a circle. The orbital motion near the surface sets off smaller and slower circling motions below it, which eventually ceases at a depth of about half the wavelength (Figure 17.22). Waves are divided into deep- and shallow-water forms depending on whether the lower orbital motion is affected by the seabed. As waves approach shallower water, the friction with the bottom causes the wave form to steepen and eventually collapse forward onto the shore.

Table 17.2 **GENERAL CHARACTERISTICS OF WAVES** Wave period is the time between successive peaks; wavelength is the distance between successive peaks; apparent speed is the rate at which each peak passes; and depth is the depth to which the effects of the wave can be felt. These data are for pure wave forms, which is rarely the case at sea.

WAVE PERIOD (seconds)	WAVELENGTH (m)	APPARENT SPEED* (kph)	DEPTH (m)
2	6.1	11.1	3.0
3	14.0	16.7	7.0
4	25.0	22.2	12.5
5	39.0	27.8	19.5
6	56.1	33.3	28.0
7	76.5	38.9	38.1
8	100.0	44.4	50.0
9	126.5	50.0	63.1
10	156.1	55.6	78.0
11	189.0	61.1	94.5
12	224.6	66.7	112.4
13	263.7	72.2	132.0
14	306.0	77.8	153.0
15	351.1	85.2	175.6

*Note, water moves in a circular motion in a wave, but does not move forward.

THE WORK OF WAVES

In coastal geomorphology, the term **shoreline** refers to the shifting line of contact between water and land. The broader term **coastline** includes the shallow water zone in which waves perform their work, as well as beaches and cliffs shaped by waves, and coastal dunes. The coastlines of large lakes exhibit many features and processes similar to those of marine coastlines.

The most important agent shaping coastal landforms is wave action. Waves travel across the deep ocean with little loss of energy until they reach shallow water and feel the drag of the sea floor, which eventually causes them to collapse and surge up the beach. The energy in the breaking wave is expended primarily in the constant churning of mineral particles on the shore.

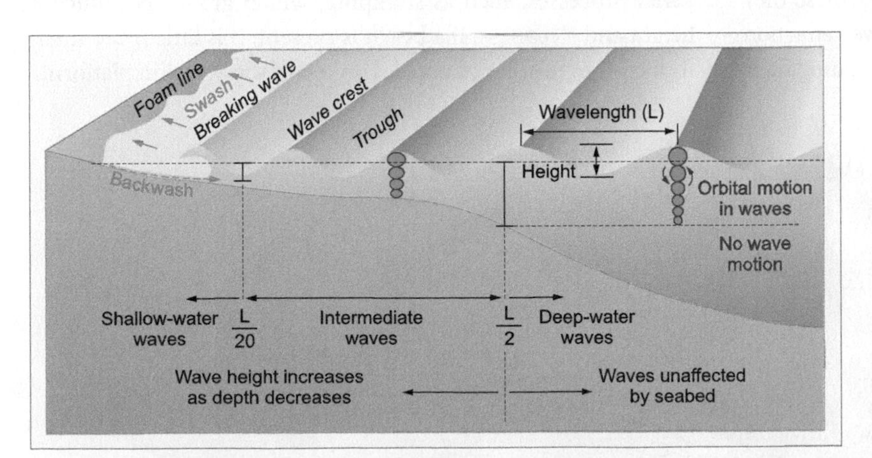

17.22 **DEEP-WATER WAVES AND SHALLOW-WATER WAVES** Deep-water waves appear to carry water forward as they pass. In fact, the water molecules move in a circular motion and maintain their general position. Closer to shore, drag on the seabed causes waves to steepen, fall forward, and rush up the beach.

MARINE EROSION

Marine erosion is caused in four general ways. Earth materials can be dislodged through *hydraulic action* by the force of the waves alone or in combination with air trapped and compressed by the weight of the falling water. *Abrasion* occurs when the waves pick up rock fragments and propel them against the shoreline. The rock fragments are gradually rounded and reduced in size through *attrition* as they grind against each other. Finally, the weakly acidic nature of sea water can dissolve certain rocks, such as limestone and chalk, by *corrosion* (or solution).

Where the coastline is composed of unconsolidated materials, these can be easily dislodged by the force of the forward-moving water. In this way, the shoreline may recede rapidly to form a steep bank, or *marine scarp* (Figure 17.23). If the coastline is made of harder materials, erosion is most pronounced where deep water offshore allows waves to break at the foot of rocky headlands with little loss of energy through friction with the seabed. The impact of the waves causes the headland to be undercut at its base. A **wave-cut notch** marks the level where wave action is most intense (Figure 17.24). As the size of the wave-cut

17.23 **RETREATING SHORELINE, RHODE ISLAND** This marine scarp of Pleistocene sediments is being rapidly eroded, threatening the historic Southeast Lighthouse.

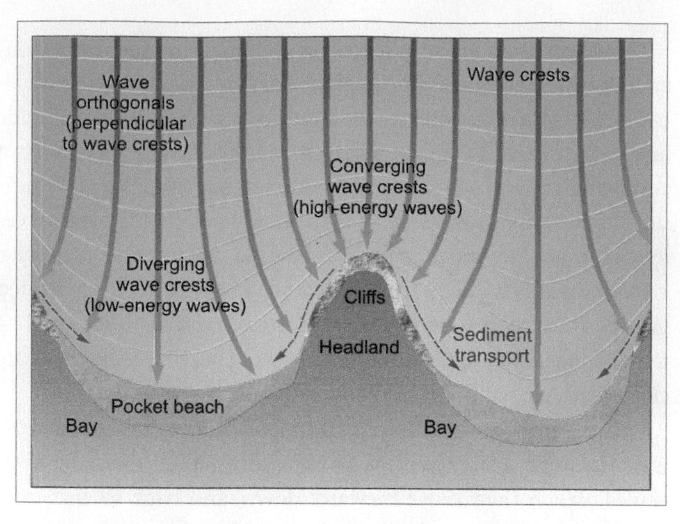

17.25 **WAVE REFRACTION** Waves approaching a shoreline are affected by the depth of water offshore. Deeper water extends further into the bays, causing the waves to bend so that their energy is concentrated on the headland. Lower energy waves continue into the bays, where sand is deposited as pocket beaches.

notch increases, the unsupported rock above collapses to form a *marine cliff*. The collapsed rock provides temporary protection from the waves, but it is eventually worn down and removed, and undercutting continues. In this way, the cliff erodes toward the shore, maintaining its form as it retreats.

The most resistant material will form a headland that extends farther out to sea. Softer rock on either side of the headland erodes to form bays where sediment will accumulate as a *pocket beach*. As the headland becomes more exposed to the waves, the erosion rate increases. However, the approaching waves are also influenced by the configuration of the seabed. The part of the wave in shallower water is slowed by frictional drag, causing the wave crests to bend so they become parallel to the coastline. This process of wave refraction results in waves converging on the headlands, concentrating their energy and accentuating erosion (Figure 17.25). In this way, waves will attack the headland from both sides, creating enlarged *sea caves*. The waves may pierce the headland to form a *marine arch*. Further erosion weakens the arch, causing it to collapse. This process leaves a *stack*, which in turn will be toppled by wave action (Figure 17.26).

As the sea cliff retreats, continued wave action forms an *abrasion platform* (Figure 17.27). Abrasion continues to erode and widen this almost-level rock pavement. However, because the water is shallow on the abrasion platform even at high tide, much of the wave energy is dissipated as friction; in this way the base of the cliff is protected from further erosion, and the rate of retreat slows. The form of the cliff is then increasingly influenced by subaerial processes, such as slumping, which gradually reduce its height and steepness. If a beach is present, it is little more than a thin layer of pebbles or cobbles on top of the abrasion platform.

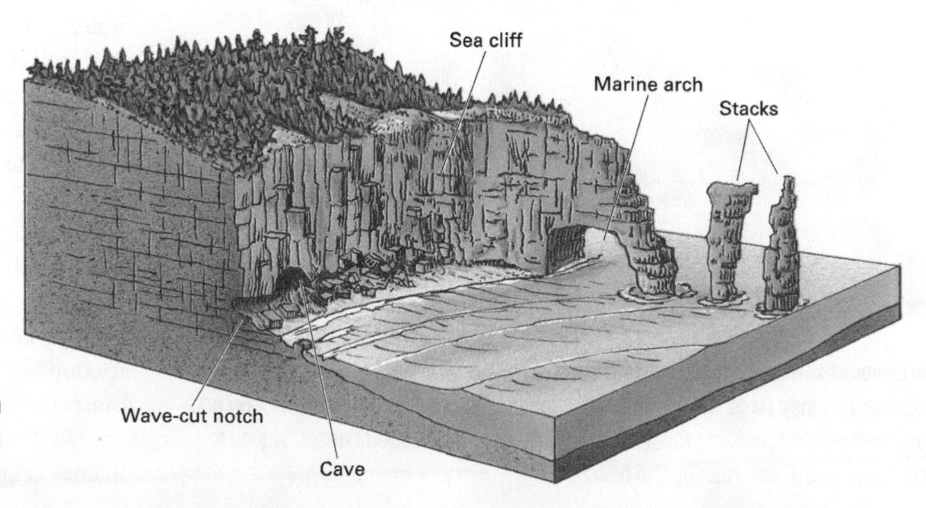

17.24 **LANDFORMS OF SEA CLIFFS** Wave action undercuts the marine cliff, maintaining its form. (Drawn by E. Raisz.)

17.26 **MARINE EROSION** In a typical sequence, a headland is first undercut from both sides, creating an arch, and then a stack forms when the arch collapses. This sea stack, on the west coast of the Orkney Islands northwest of Scotland, has a small abrasion platform at its base and a pronounced wave-cut notch; it has also been pierced by wave action to form a small arched base.

17.27 **A WAVE-CUT ABRASION PLATFORM EXPOSED AT LOW TIDE** As the wave-cut platform gets wider, the energy of the approaching waves is dissipated in the shallow water and eventually the base of the cliff is protected from further erosion. Sub-aerial processes, such as slumping, then reduce the angle of the cliff.

MARINE TRANSPORT AND DEPOSITION

Much of the material that is deposited in coastal environments is actually supplied by rivers that discharge their sediment loads into the sea. For example, the Fraser River carries about 20 million tonnes of suspended sediment annually, the Mackenzie River about 100 million tonnes, and the St. Lawrence about 6.5 million tonnes. In comparison, the average annual sediment load delivered to the Gulf of Mexico by the Mississippi River is more than 150 million tonnes. Added to this is the rock debris derived directly from marine erosion. Once in the sea, waves and currents transport the finer materials, which are temporarily deposited as a **beach** or some form of sandbar. Although many beaches are formed of particles of fine to coarse quartz sand, some are composed of rounded pebbles or cobbles. Still others are formed from fragments of volcanic rock, as in Hawaii, or even from shells and broken coral.

The landward and seaward flow of water generated by breaking waves moves beach materials laterally on the beach face as *beach drift* (Figure 17.28a). When a breaker collapses, a foamy, turbulent sheet of water washes up the beach slope. This *swash* is a powerful surge that moves materials toward land. The water returns to the sea as the *backwash*. The direction in which waves advance toward a shore is mainly determined by their point of origin relative to the coastline. Waves typically approach the shore obliquely. After the wave has spent its energy, the backwash flows perpendicularly down the beach slope, and particles come to rest some distance from their initial position. Over time, this repeated wave action transports a considerable quantity of sediment.

Beaches form where sand is in abundant supply, and although they may appear to be permanent features, the sand is continually moving along the beach face. In addition, more powerful waves will gouge the sand and carry it offshore, where they deposit it as a sandbar. Smaller constructional waves bring it back to the beach. In this way, a beach may retain a fairly stable configuration over many years. As well, this back-and-forth movement transports sediment along the coast in the shallow water offshore.

Waves approaching a shoreline at an angle set up a current parallel to the shore (Figure 17.28b). When wave and wind conditions are favourable, this *longshore current* is capable of carrying sand along the seabed in a process called *longshore drift*. Beach drift and

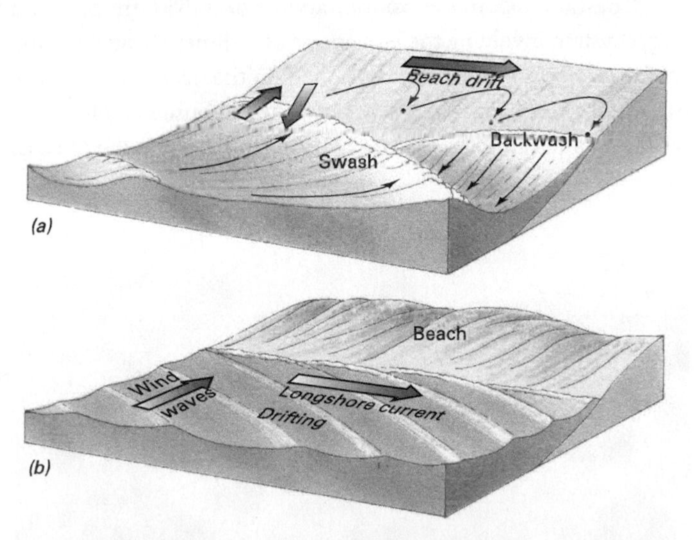

17.28 **COASTAL SEDIMENT TRANSPORT** (a) Swash and backwash move particles along the beach as beach drift. (b) Longshore drift arises from the longshore current produced by the curvature that develops along the wave crests as they advance toward the beach. These processes are collectively known as littoral drift.

17.29 **COASTAL DEPOSITION** Sandspits are a common depositional feature in coastal environments. This example is on the western end of Gialug Island in the Torres Strait, which separates Australia from New Guinea. *Fly By: 10° 36′ S; 142° 09′ E*

longshore drift act together to move particles in the same direction; the combined process is called **littoral drift**.

Littoral drift is a constructive process that shapes shorelines in two ways. Materials are moved along the beach in the direction of the prevailing winds. Where a bay exists at the end of a relatively straight stretch of coastline, the sand is carried out into open water to form a *sandspit* (Figure 17.29). As the sandspit grows, it can form a barrier, or *bar*, across the mouth of the bay. Beaches typically disappear at headlands and promontories where wave action is intense, and littoral drift quickly moves sediment away. Wave refraction around headlands helps to transport sediment into coves where it accumulates as crescent-shaped *pocket beaches* (Figure 17.30).

Coastal erosion, deposition, and littoral drift are part of a larger system involving the movement of sediment brought to the sea by rivers and ultimately deposited on the ocean floor. At the mouth of a river, a beach will broaden in response to an influx of sediment and remain wider for some distance along the coast as the extra sediment is transported in the littoral current. Beaches eventually end where a submarine canyon interrupts the flow of sand in the littoral current. Offshore submarine canyons cross the continental shelf and extend out to great depths. Sediment cascades into these canyons where it forms undersea fans and cones along the continental rise. Many submarine canyons were cut by streams during the Late-Cenozoic Ice Age, when sea level was as much as 200 m lower than present.

17.30 **POCKET BEACHES** On this embayed coastline near Mendocino, California, sediment is carried from eroding headlands into the bays to form pocket beaches. *Fly By: 39° 18′ N; 123° 48′ W*

Draining off sand into submarine canyons is particularly important along the Pacific coast of North America. On the Atlantic coast, the continental shelf is wider and there are fewer submarine canyons. A notable exception is the Hudson Canyon at the southwest end of Long Island, New York. This difference in submarine topography results in beaches along the eastern seaboard being generally wider than those on the west coast, with sandy, barrier islands offshore (Figure 17.31). On the east coast, sand is lost primarily during hurricanes and other storms that gouge the beaches and carry the sand seaward. Sand may also be carried over barrier islands by storm waves where it settles out on the landward side as *washover deposits.*

The flow of sediment from rivers is an important part of this so-called *coastal sediment cell.* Many coastal communities have experienced beach recession because the rivers that feed their coastlines have been dammed, trapping sediment upstream and keeping it out of the flow system. With less sediment flowing through the coastal system as a whole, less is available for storage on the beach, and shorelines are more rapidly cut back.

LITTORAL DRIFT AND SHORE PROTECTION

When sand arrives at a particular section of the beach at a greater rate than it is carried away, the beach widens through the process of **progradation**. A beach will prograde if wave action decreases, depriving littoral drift of its energy. The reverse process, **retrogradation**, occurs when sand is removed more rapidly than it is brought in, narrowing the beach. Along stretches of shoreline affected by retrogradation, the beach may be seriously depleted or even disappear entirely. In some circumstances, structures can be installed to cause progradation. For example, *groynes*—barriers of large rocks, concrete, or wood, built at right angles to the shoreline—installed at close intervals along the beach will trap sediment moving along the shore as littoral drift (Figure 17.32).

THE TIDES

THE PHYSICS OF TIDES

An ocean tide refers to the cyclical rise and fall of the ocean's surface as a result of the gravitational potential of the moon and sun. Tide propagation and amplitude are influenced by the rotation of the Earth and the shapes and depths of the ocean basins. The gravitational force exerted by a celestial body such

17.31 **BARRIER ISLAND RUNNING PARALLEL TO THE EAST COAST OF NORTH CAROLINA** Waves that overtop the dune line carry sands and silts across barrier islands during storms that are deposited on the landward side as washover deposits.

17.32 **GROYNES CONSTRUCTED ACROSS A BEACH** Groynes are strong barriers, usually constructed from wood or stones, that are designed to reduce longshore transport of sand along exposed beaches.

as the moon or sun is proportional to its mass, but inversely proportional to the square of the distance. The equation is:

$$F = G\frac{m_1 m_2}{r^2}$$

where F is the magnitude of the gravitational force, G is the gravitational constant in newtons (6.67×10^{-11} N m^3 kg^{-1} s^{-2}), m_1 is the mass of the first object (i.e., the Earth) in kilograms, m_2 is the mass of second object (i.e., the moon) in kilograms, and r is the distance between objects in metres.

The mass of the sun is 1.98892×10^{30} kg, which is considerably greater than the moon's mass of 7.36×10^{22} kg. The mean distance between the sun and the Earth is 149.5 million km, compared with 385,000 km between the moon and the Earth. Hence, the gravitational force of the sun on the Earth is about 178 times that of the moon. However, tides are not produced by the absolute pull of gravity exerted by the sun and moon, but by the differences in the gravitational fields, or *gravity gradient*, produced by these two bodies across the Earth's surface.

The gravity gradient changes at a rate that is inversely proportional to the cube of the distance. Because the moon is closer to the Earth, its gravitational force field varies much more strongly than that of the sun. Consequently, the gravitational component of the sun's **tide-generating force** is about 46 percent of that of the moon. Given that the Earth's diameter is 12,680 km, a point on the Earth's surface nearest the moon is approximately 378,660 km away, compared with 391,340 km for a point on the opposite side of the Earth. This means that the gravitational force of the moon is about 3.5 percent greater on the side of the Earth nearer the moon.

In addition to gravity, the tide-generating force includes a centrifugal component caused by the revolution of the Earth around the sun. The centrifugal force experienced by all points on the Earth is the same in magnitude and direction, whereas the gravitational force exerted by the moon always points to the centre of the moon (Figure 17.33).

Earth's rotation on its axis does not alter the net effect of Earth's revolution around the sun in terms of the gravitational and centrifugal forces. The only effect that rotation has on the tides is that it moves the entire tide-generating force field around the Earth each day. This relates to another important concept associated with tides—that of the **geoid**. The geoid refers to the shape that the Earth would assume if it were entirely covered

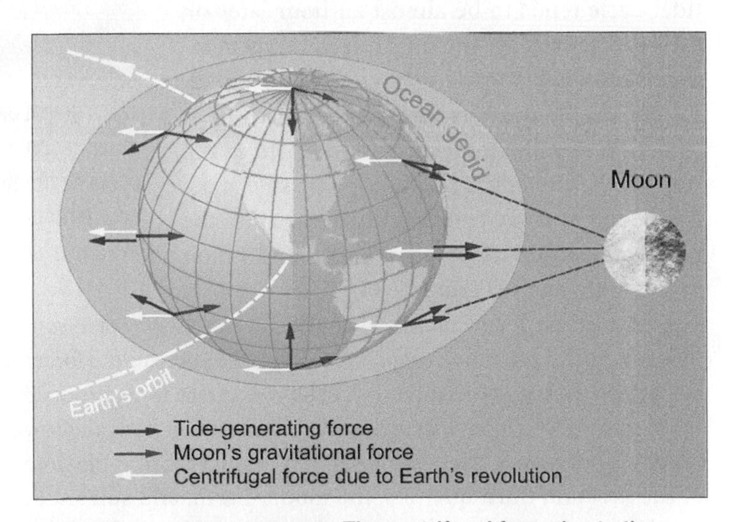

17.33 **THE TIDE-GENERATING FORCE** The centrifugal force due to the Earth's revolution around the sun is the same for all locations, whereas the gravitational force exerted by the moon is directed toward the centre of the moon. The tide-generating force is the result of these two components. Note: the tidal effect on the ocean geoid is greatly exaggerated in this figure.

with water and responding to the forces acting upon it. In this hypothetical situation, sea level would represent the height of the water created by the resultant forces created by revolution, gravitational attraction, and rotation. The geoid is deformed as the forces acting upon it change.

The lunar gravitational force is greatest at the point on the Earth directly beneath the moon (the sub-lunar point), and consequently, the ocean surface is pulled and rises. On the side facing away from the moon (the antipodal point), the moon's gravity is at its weakest, which effectively allows the water to move away from the Earth. Assuming that the Earth is completely covered with water and responded instantly to the changing forces acting upon it, this would create what is called the **equilibrium tide**. Because the equilibrium tide has two bulges, one on each side of the Earth, there would be two high tides and two low tides each lunar day. This is called a **semidiurnal lunar tide** and has a period of 12.42 hours.

TIDE CYCLES AND AMPLITUDES

The moon revolves around the Earth once every 27.3 days in the same direction as the Earth rotates. The combined effect of these relative motions is that it takes 24 hours and 50 minutes for a point on the Earth's surface to return to the same position relative to the moon. Thus, the tidal cycle tends to be almost an hour later on each successive day. The sun's role in tide generation is most apparent over the course of a month. As the moon revolves around the Earth, it is sometimes on the same side of the Earth as the sun, at other times it is on the opposite side of the Earth, and midway between these extremes, the relative positions of the sun, Earth, and moon form a right angle (Figure 17.34).

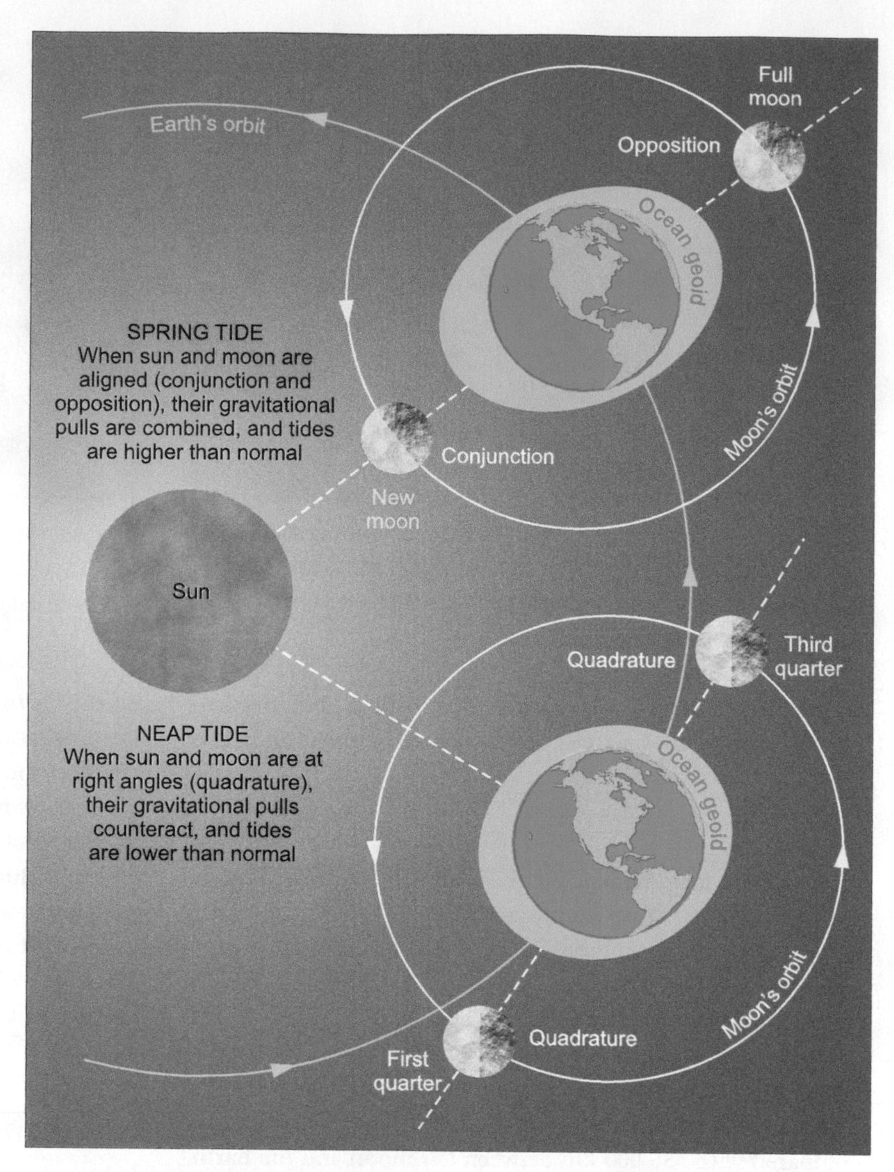

17.34 **EFFECT OF THE SUN AND MOON ON TIDES** At new moon and full moon, the sun's gravitational force is added to that of the moon, creating spring tides. Neap tides occur at the first and third quarters when the gravitational force of the sun counteracts that of the moon. Note: the tidal effect on the ocean geoid is greatly exaggerated in this figure.

At the time of new moon, when the sun and moon are in *conjunction*, and at full moon (*opposition*), their tide-generating forces are acting in the same direction. The net effect is to produce **spring tides**, which have the greatest range between high and low water. About a week after the new moon or full moon, at the time of the first and third quarters, the moon, Earth, and sun are in *quadrature*. At this time, the peaks of the lunar tides tend to coincide with the troughs of the solar tides and so counteract each other. This produces **neap tides**, in which the tidal range between high and low water is at its lowest (Figure 17.35). Additional variation in the tidal ranges at specific locations arises over the course of the year as the sun migrates 23½° north and south of the equator. Because the moon's axis is inclined at 5°, the sub-lunar point can move by an angular distance of 23½° ± 5° north and south of the equator during the year. But note that this cycle of variation from 18½° to 28½° takes 18.6 years to complete.

Ideally the gravitational and centrifugal forces acting on the geoid would generate two high tides and two low tides each day. In reality, this is complicated by several factors, including the shape and distribution of the ocean basin, friction, the Coriolis effect, and inertia. As a result, tidal regimes are classified into three general types: diurnal, semidiurnal, and mixed tides (Figure 17.36).

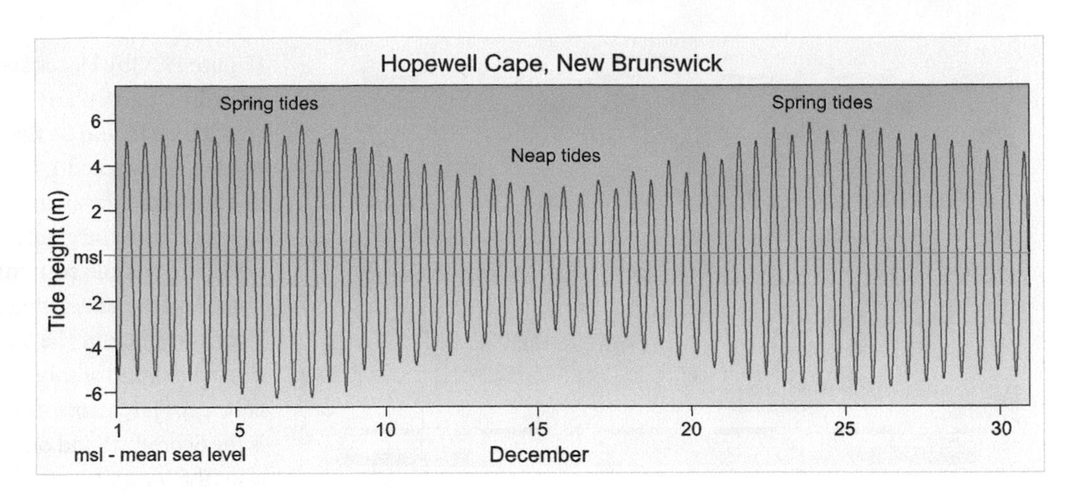

17.35 SPRING TIDES AND NEAP TIDES The effect of the relative positions of the sun and moon during each lunar month is illustrated by the tides at Hopewell Cape on the Bay of Fundy, New Brunswick. The data are for 2010.

Diurnal tides have one high tide and one low tide in each tidal day, for example in the Gulf of Mexico, the southern parts of the Pacific Ocean, and around Southeast Asia (Figure 17.37). *Semidiurnal tides* have two high and two low tides each tidal day, as occur around northern Canada, Europe, southern Africa, and New Zealand. *Mixed tides* have two high tides and two low tides each tidal day, but the tidal ranges during these events are not equal. Mixed tides occur along the west coast of Canada and the United States, the Mediterranean Sea, south of Australia, and in the East Indies.

TIDAL CURRENTS

In bays and estuaries, the changing tide sets in motion flows of water known as *tidal currents*. When the tide begins to fall, an *ebb current* sets in and carries water seaward. This flow ceases about the time when the tide is at its lowest point. A landward current, the *flood current*, starts as the tide begins to rise. Ebb and flood currents perform several important functions along a shoreline. The currents that flow in and out of bays through narrow inlets are very swift and can scour the inlet. This keeps the inlet open, despite the tendency of littoral drift to close the inlet with sand.

Tidal currents also carry large amounts of fine silt and clay in suspension. These sediments come from streams or from bottom mud that has been agitated by storm waves. The fine sediments settle on the floors of some bays and estuaries and gradually accumulate as tidal mud flats. Mud flats are alternately flooded and exposed by the tides, and only specially adapted salt-tolerant plants can grow under these conditions. The mud flats slowly build up to approximately the level of high tide to become a *salt marsh*, and eventually develop a thick layer of peaty soil at the surface. Tidal currents continue to flow through a salt marsh by means of a complex network of winding tidal streams.

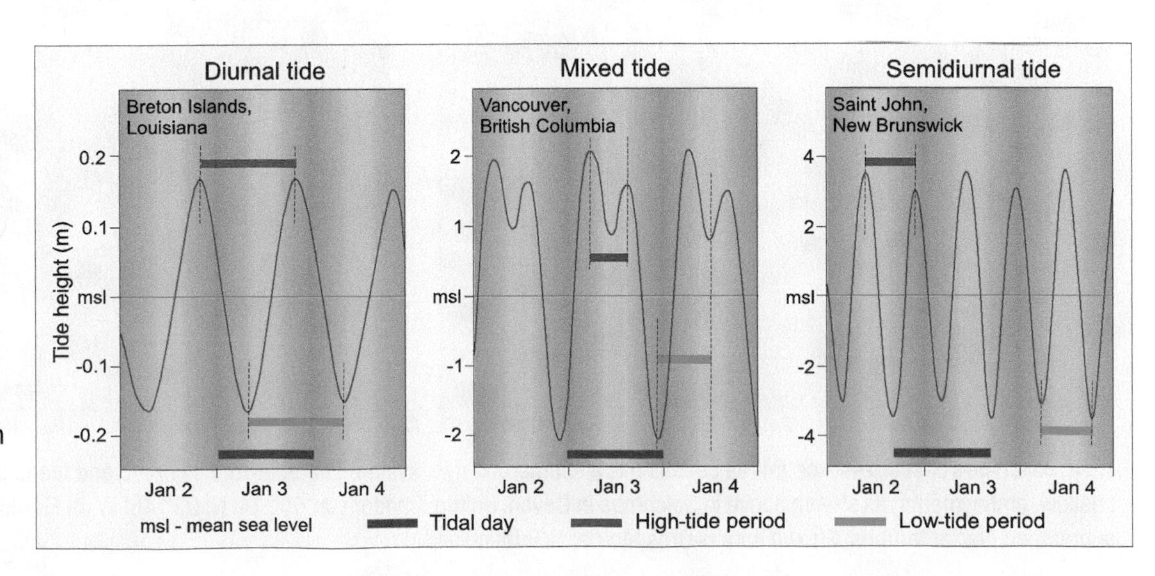

17.36 DIURNAL, MIXED, AND SEMIDIURNAL TIDE CYCLES Diurnal tides have one high tide and one low tide in every 24-hour period, so the tide period measured between consecutive high or low tides is the same as the tidal day. Mixed tides and semidiurnal tides have two high and two low tides each tidal day, so the tide period is approximately 12 hours. The data are for 2010.

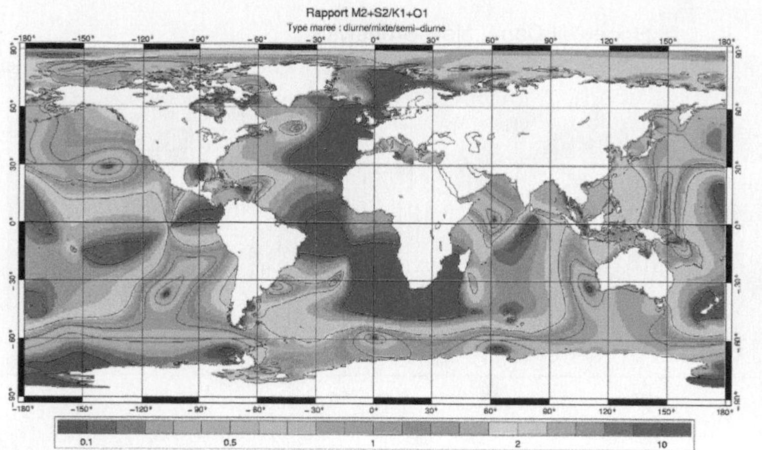

17.37 **GLOBAL TIDE PATTERNS** Depth and configuration of the ocean basins are important determinants of tidal regimes.

TYPES OF COASTLINES

COASTLINES OF SUBMERGENCE

A number of distinctive coastlines have developed as a result of various tectonic and geomorphological processes. One group of coastlines is formed by **submergence**. One of the most common of these is the *ria coast*, a deeply embayed coast created by the submergence of valleys previously carved by rivers (Figure 17.38a). The attack of waves forms cliffs on the exposed seaward sides of the adjacent headlands, and sediment accumulates in the form of beaches at the heads of bays. Because of the gently sloping gradient of the surrounding land further inland, most rias are fringed by tidal mud flats.

A *fjord coast* occurs where steep-walled glacial troughs have been submerged to produce long, narrow inlets of the sea (Figure 17.38b). Fjords are found around much of the Canadian coastline, particularly in British Columbia, throughout the Arctic islands, and on the coast of Newfoundland and Labrador. Many fjords exceed 1,000 m in depth, and characteristically are deepest some distance from the seaward ends. This is thought to result from the greater erosive power of the ice in the landward and middle sections, possibly because the glaciers were floated off the underlying bedrock as they moved out to the sea. This buoyancy of the ice might also account for the shallow *threshold* found at mouths of many fjords that limits water circulation. Fjords can extend many kilometres inland; the longest is the Scoresby Sund on Greenland, stretching for 350 km (it is also the deepest, at 1,500 m). The second largest is the Sognefjord in Norway, at 204 km. Because sediment rapidly sinks into the deep water, beaches are rare in fjords.

Tectonic activity can also produce shorelines of submergence; a *fault coast* is formed if a down-faulted zone is flooded. Faulting can also weaken the rock so it is more susceptible to erosion. Point Reyes, California, is an example of a fault coast. Here the San Andreas fault reaches the Pacific coast, its position being marked by an usually straight length of coastline **(Fly By: 38° 10′ N; 122° 55′ W)**.

EMERGENT COASTLINES

Coastlines of **emergence** form when underwater landforms are exposed by a drop in sea level or when the Earth's crust is elevated. Raised shorelines formed through tectonic uplift develop as a result of faulting or by isostatic rebound following the melting of ice sheets. Along rocky coasts, the former abrasion platforms will form terrace-like features, known as *marine terraces* or *raised beaches*, when elevated above the current sea level.

(a)

(b)

17.38 **COASTLINES OF SUBMERGENCE** (a) Ria coastline is a submerged river system characterized by broad and branching, rather sinuous and shallow, embayments, as shown above in Salcombe in Devon, United Kingdom, at 50° 14′ N; 03° 46′ W (b) Fjords are steep-sided and deep-submerged glacial troughs. Shown here is Gros Morne, Newfoundland.

Because it is no longer eroded at its base, the steep cliff line is gradually reduced by flowing water and other sub-aerial processes, such as slumping (Figure 17.39a).

Where the offshore slope is very gentle, a slowly emerging coastal plain can form a *barrier-island coast* as sand accumulates a short distance offshore (Figure 17.39b). Behind the barrier island, there is normally a *lagoon*, a broad expanse of shallow water that is filled in places with tidal deposits. A characteristic feature of barrier islands is the presence of gaps, known as *tidal inlets*. Strong currents flow alternately landward and seaward through these gaps as the tide rises and falls. In heavy storms, the barrier island may be breached by new inlets.

(a)

(b)

17.39 **EMERGENT COASTLINES** (a) The level terrace is a raised beach formed from a wave-cut platform that ran back to the old cliff line. The angle of the cliffs along the earlier shoreline has decreased by sub-aerial denudation, and a low cliff has developed along the active shoreline. (b) Barrier islands form offshore from sands and gravels deposited by longshore currents. ***Fly By: 34° 39′ N; 76° 31′ W***

NEUTRAL COASTLINES

Neutral coastlines develop when new land is built out into the ocean. For example, **delta** coasts form where sediment-laden rivers enter the sea. A rapid reduction in velocity of the current as it pushes out into the standing water allows the sediment to settle from the water. The coarser sand and silt particles are deposited first, while the fine clays continue out into deeper water. When the fresh water comes in contact with salt water, the finest clay particles coagulate and sink to the seabed.

Deltaic deposits exhibit distinctive layering (Figure 17.40). Sediment initially settles into the shallow sea water to form *bottomset beds*. With time, these beds are covered by steeply inclined *foreset beds*, which form when the river loses its power to transport its load because flow is no longer confined by a channel. The *topset beds* form a superficial cover of horizontal layers as the shoreline continues to prograde. Typically, the river channel divides and subdivides into lesser channels called *distributaries*.

Deltas can have a variety of outlines. The Nile Delta (Figure 17.41a) is triangular in shape. In areas where tides and wave action are less pronounced, a delta may take on a multi-lobed shape referred to as a bird's-foot delta. The Mississippi Delta with its long, branching fingers growing out into the Gulf of Mexico, is of this type (Figure 17.41b). Sediment discharge from the Mississippi amounts to about 1 million metric tonnes per day. Delta growth is often rapid, ranging from 3 m per year for the Nile to 60 m per year for the Mississippi.

A second form of neutral shoreline is the *volcano coast* that builds up through deposition of ash and lava (Figure 17.41c). In most cases, the volcano is partly constructed below water level. Low cliffs develop as wave action erodes the fresh deposits.

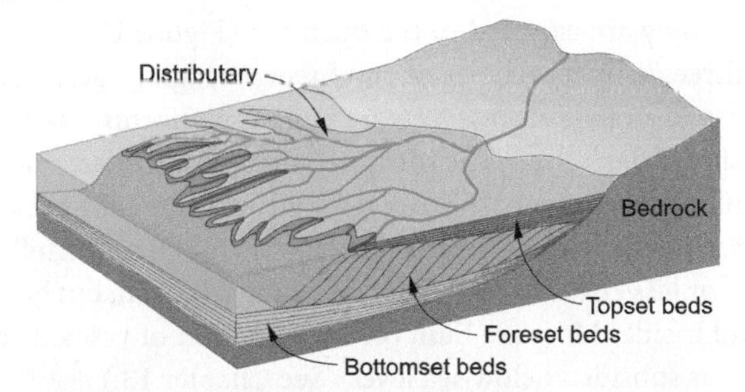

17.40 **THE CREATION OF DELTAS** Deltas form at river mouths in lakes and oceans where sediment is deposited in a characteristically layered sequence in response to the rapid reduction in current velocity. The bottomset beds are composed of the fine material carried in suspension before settling to sea bed. Foreset beds, developed from coarser materials, build the shoreline seaward. Topset beds are made up of mostly of coarser bed load sediments which are spread over the surface due to shifts in position and flow regime of the branching network of distributaries.

17.41 **NEUTRAL COASTLINES** (a) The Nile Delta is a classic triangular shape. (b) The Mississippi Delta is an example of a bird's-foot delta, in which the distributary channels build levees as they extend into open water. (c) A volcanic coast builds into the sea from accumulations of lava or ash from successive eruptions.

Beaches are typically narrow and steep, and are composed of fine particles of dark extrusive rock.

ORGANIC COASTLINES

Coral-reef coasts are unique in that new land is constructed by organisms. Corals secrete rocklike deposits of mineral carbonate. As coral colonies die, new ones are built on top of them, accumulating as limestone **coral reefs**. Coral fragments are torn free by waves, and the pulverized fragments accumulate as sandy beaches.

Coral-reef coasts occur in warm, tropical, and equatorial waters between lat. 30° N and 25° S. For vigorous coral growth, the sea water must remain warm (above 20 °C), well-aerated, and free of suspended sediment. For this reason, corals thrive where they are exposed to the open sea (Figure 17.42). There are three distinctive types of coral reefs: fringing reefs, barrier reefs, and atolls. *Fringing reefs* are built as platforms attached to the shore. *Barrier reefs* lie offshore and are separated from the mainland by a lagoon. Narrow gaps occur at intervals in barrier reefs, through which water returns to the open sea. **Atolls** are more or less circular coral reefs enclosing a lagoon, but have no island inside. Most are built on a foundation of volcanic rock that has subsided below sea level. (See Chapter 13.)

Coral reefs serve as protective barriers to coastlines, reducing the effects of storm waves and surges, as well as affording some protection from tsunamis. However, many coral reefs appear to be at or near their upper temperature limits. The projected rise in ocean temperatures through global climate change could have serious consequences for coral reefs. A warmer climate would also cause a rise in sea level, which would have a dramatic effect on coastlines in many parts of the world. *Appendix 17.1 • Global Climate Change and Coastal Environments* documents the causes and effects of sea-level rise more fully.

A Look Ahead

This chapter has described the processes and landforms associated with wind action, either directly or through wind-driven ocean waves. The power source for wind action is the sun, which also is ultimately responsible for powering the last active geomorphological agent, glacial ice. By evaporating water from the oceans and returning it to the continents as snow, solar energy creates the masses of ice that form mountain glaciers and polar ice sheets. Compared with wind and water, glacial ice

17.42 **FRINGING REEFS** Coral reefs fringe the Island of Moorea, Society Islands, in the South Pacific. The island is a deeply dissected volcano with a history of submergence. Tahiti lies in the background. A good example can be seen at 18° 52′ S; 159° 47′ W.

moves much more slowly. Glaciers transport sediment like a vast conveyor belt, depositing it at the ice margin. By plowing its way over the landscape, glacial ice also shapes the local terrain—loosening and removing rock and plastering sediments underneath its mass. This slow but steady action is very different from that of water, wind, and waves, and produces another set of distinctive landforms. Chapter 18 discusses the impact of glacial ice.

CHAPTER SUMMARY

- This chapter has described the landforms of wind and waves, both of which are indirectly powered by the Earth's rotation and the unequal heating of its surface by the sun.
- Wind is a landform-creating agent that acts by moving sediment. Deflation occurs when wind removes mineral particles—especially silt and fine sand—which can be carried long distances. Deflation creates blowouts in semi-arid regions and lowers desert surfaces. In arid regions, deflation produces dust storms.
- Sand dunes form when a source, such as a sandstone outcrop or a beach, provides abundant sand that can be moved by wind action. Barchan dunes are arranged individually or in chains leading away from the sand source. Transverse dunes form a sand sea of wavelike forms arranged perpendicular to the wind direction. Parabolic dunes are arc-shaped—coastal blowout dunes are an example. Longitudinal dunes lie parallel with the wind direction. Coastal foredunes are stabilized by dune grass and help protect the coast against storm wave action.
- Loess is a surface deposit of fine, wind-transported silt sometimes hundreds of metres in thickness, often with steep, almost vertical exposures. Loess is easily eroded by wind and water. In eastern Asia, the silt forming the loess deposits was transported by winds from extensive interior deserts located to the north and west. In Europe and North America, the silt was derived from fresh glacial deposits during the Pleistocene Epoch.
- Waves act at the shoreline—the boundary between water and land. Waves expend their energy as breakers, which erode hard rock into marine cliffs and produce distinctive landforms such as abrasion platforms, caves, marine arches, and sea stacks. Features are more subdued in softer materials and typically develop coastal plain landscapes.
- Beaches, usually formed of sand, are shaped by the swash and backwash of waves, which continually work and rework beach sediment. Wave action produces littoral drift, which moves sediment parallel to the beach. This sediment accumulates in bars and sandspits, which further extend the beach. Depending on the nature of longshore currents and the availability of sediment, shorelines can experience progradation or retrogradation.
- Tidal forces cause sea level to rise and fall rhythmically, which produces tidal currents in bays and estuaries. Tidal flows redistribute fine sediments, which can accumulate with the help of vegetation to form salt marshes.
- Coastlines of submergence result when the land sinks below sea level or sea level rises rapidly, such as ria and fjord coasts. Coastlines of emergence include barrier-island coasts and delta coasts. Along some coasts, rapid uplift has occurred, creating raised shorelines and marine terraces. Coral reef coasts occur in regions of warm tropical and equatorial waters.
- Global sea level is predicted to rise in the twenty-first century due to both volume expansion of warmer sea water and increased melting of glaciers and snowpacks. Future rises may be very costly to human society as estuaries are displaced, islands are submerged, and coastal zones are subjected to frequent flooding.

KEY TERMS

abrasion
aeolian
atoll
barchan dune
beach
blowout
coastal foredune
coastline
coral reef
creep
deflation
delta
desert pavement
dust storm
emergence
equilibrium tide
fetch
geoid
hairpin dune
hammada
littoral drift
loess
longitudinal dune
neap tide
parabolic dune
progradation
retrogradation
saltation
sand dune
semidiurnal lunar tide
shoreline
spring tide
submergence
suspension
threshold velocity
tide-generating force
traction
transverse dune
wave-cut notch
wind erosivity

REVIEW QUESTIONS

1. What is the energy source for wind and wave action? How do these occur?
2. What is deflation, and what landforms does it produce? What effects do dust storms have at local and regional scales?
3. How do sand dunes form? Describe and compare barchan dunes, transverse dunes, star dunes, coastal blowout dunes, parabolic dunes, and longitudinal dunes.
4. What is the role of coastal dunes in beach preservation? How are coastal dunes influenced by human activity? What problems can result?
5. Define the term loess. What is the source of loess, and how are loess deposits formed?
6. How are waves formed and what determines their size and strength?
7. What landforms can be found in areas where bedrock meets the sea?
8. What is littoral drift, and how is it produced by wave action?
9. Identify progradation and retrogradation. How can human activity influence retrogradation?
10. What causes the tides? Why are there several different types of tidal regimes?
11. What key features identify a coastline of submergence? Identify and compare the two types of coastlines of submergence.
12. Under what conditions do barrier-island coasts form? What are the typical features of this type of coastline? Provide and sketch an example of a barrier-island coast.
13. How are marine terraces formed?
14. What conditions are necessary for the development of coral reefs? Identify three types of coral-reef coastlines.

VISUALIZATION EXERCISES

1. With north at the top of the page, and assuming a north wind, sketch the shapes of the following types of dunes: barchan, transverse, parabolic, and longitudinal.
2. Describe the features of delta coasts and their formation. Sketch and compare the shapes of the Mississippi and Nile deltas. Why are they different?

ESSAY QUESTIONS

1. Wind action moves sand close to the ground in a bouncing motion, whereas it lifts and carries silt and clay longer distances. Compare landforms and deposits that result from wind transportation of sand with those that result from wind transportation of silt and fine clay particles.
2. Consult an atlas to identify a good example of each of the following types of coastlines: ria coast, fjord coast, barrier-island coast, delta coast, coral-reef coast, and fault coast. For each example, provide a brief description of the key features you used to identify the coastline type and describe how each is formed.

APPENDIX 17.1 GLOBAL CLIMATE CHANGE AND COASTAL ENVIRONMENTS

Global climate change over the remainder of the twenty-first century will have major impacts on coastal environments, primarily through a rise of sea level estimated at 9 to 88 cm between now and 2100. Some of this rise will be due to simple thermal expansion of the upper layers of the oceans as they warm. The melting of glaciers and snowpacks as air temperatures rise will also contribute to this change. Sea-level rise will have effects ranging from displacement of shorelines to enhanced coastal erosion. Depending on the amount of rise, some low-lying islands will disappear along with their inhabitants. Other expected effects include a decrease in sea-ice cover, and changes in salinity, wave activity, and ocean circulation.

The global heat content of the ocean has been increasing since at least the late 1950s, resulting in a change in sea-surface temperature of approximately 0.3 °C, about half the increase in the average land-surface temperature. Between 1870 and 1993, sea level rose by about 150 mm, an average rate of 1.7 mm per year (Figure 17A.1). Since 1993, sea level has risen a further 60 mm at an average rate of 3.3 mm per year. In the 1950s, arctic sea ice in summer covered more than 8.0 million km^2; in September 2009, it measured 5.1 million km^2 (Figure 17A.2). In recent years the rate of decrease has increased, and now averages about 9.1 percent per decade. Since 1970, El Niño episodes, which affect ocean circulation as well as the intensity of severe storm tracks, have become more frequent, more intense, and more persistent than during the previous 100 years.

What changes are in store? By 2100, sea surface temperature is predicted to increase between 1.5 and 2.6 °C compared to 1980 to 1999 conditions. The projected rise in sea level over this period ranges from 38 to 43 cm, depending on which greenhouse gas scenario is used. Sea-level rise will not be uniform. Modifying factors of waves, currents, tides, and offshore topography can magnify the rise, depending on the location. For example, some models predict doubled rates of sea-level rise for the North American Pacific coast and the western Arctic shoreline. However, all models agree that snow and ice cover will continue to decrease, and mountain glaciers and icecaps will continue their dramatic retreat. Tropical cyclone peak wind and peak precipitation intensities will increase, leading to higher storm surges (Figure 17A.3). El Niño impacts on flood and drought will be exaggerated.

Coastal Erosion

These predicted changes do not bode well for coastal environments. Most coastal erosion occurs in severe storms, when high winds generate large waves and push water up onto the land in storm surges. Global climate change will likely increase the frequency of high winds and heavy precipitation events, thus amplifying the effects of severe storms. More frequent and longer El Niño events will increase the severity and frequency of Pacific storms on the North American coast. During intervening La Niña events, Atlantic hurricane frequency and

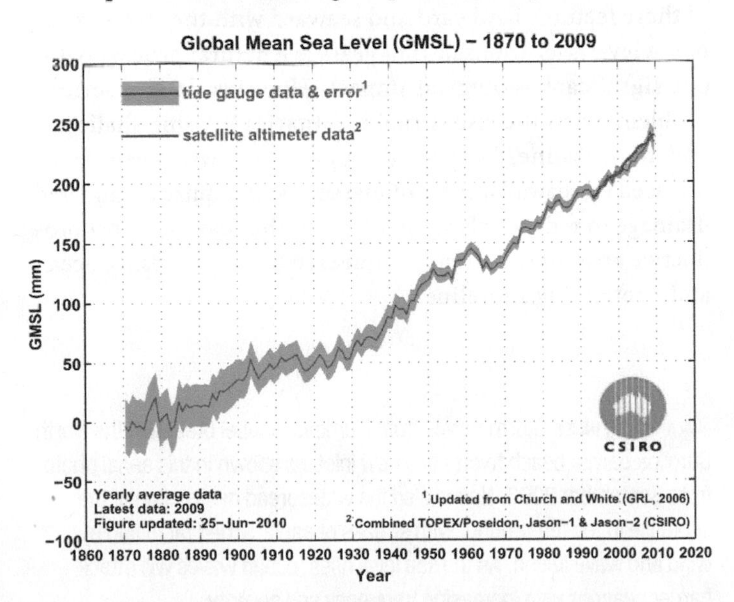

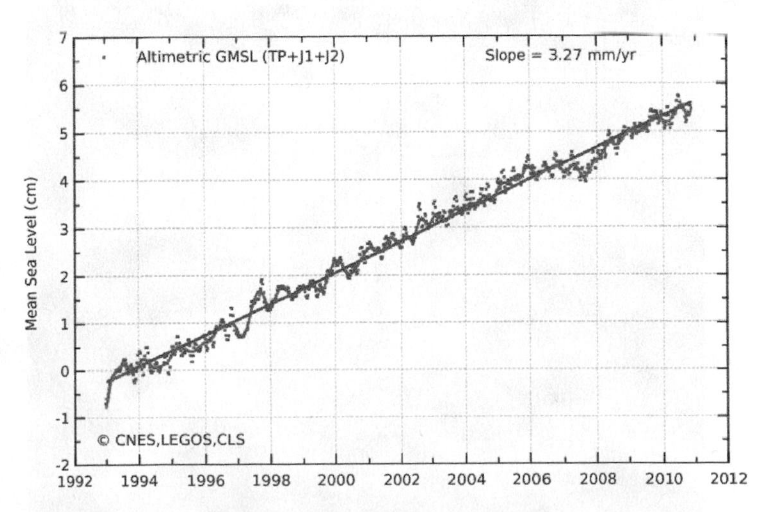

17A.1 The chart on the left shows historical sea-level data derived from coastal tide gauge records. The chart on the right shows the average sea level since 1993 derived from global satellite measurements.

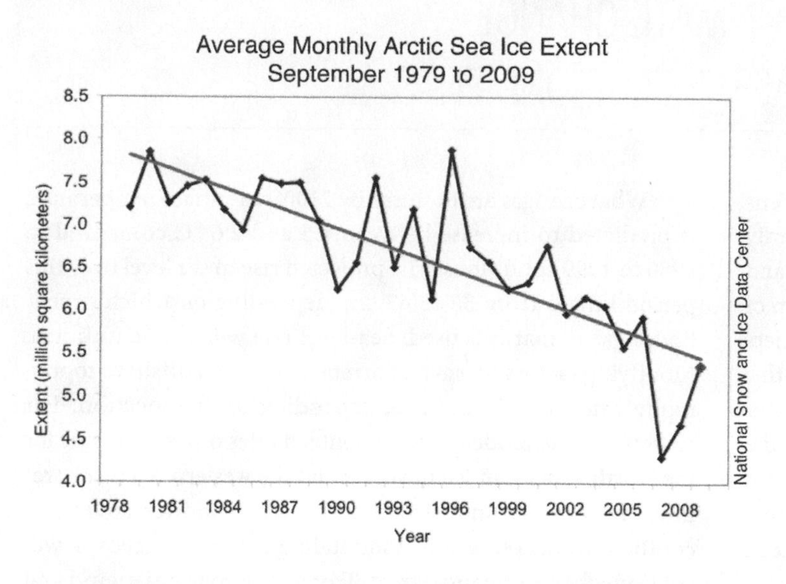

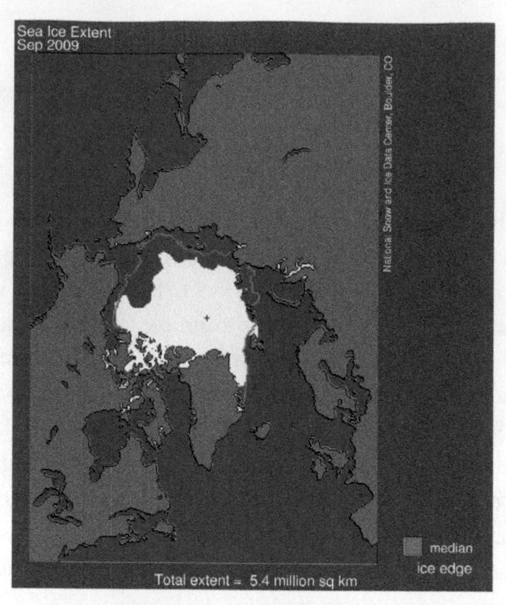

17A.2 Arctic sea ice reaches its minimum extent each September. The graph on the left charts the average September extent from 1979 to 2009, derived from satellite observations. The illustration on the right shows the Arctic sea ice minimum extent for 2009, which was the third-lowest in the satellite record.

intensity will increase, with higher risk of damage to structures and coastal populations (Figure 17A.4). Most of the world's sandy shorelines have retreated over the past century; around the Gulf of Mexico, reported rates of recession since the 1970s have ranged from 2.6 m to 12.0 m per year. One model suggests that shoreline recession will be in the order of 50 to 200 times the rise in sea level.

Sea-level rise enhances coastal erosion in storms, but in the long term, its effect is to push beaches, salt marshes, and estuaries landward. In an unaltered landscape, the migration of these features landward and seaward with the rise and fall of sea level over thousands of years is a natural process without significant ecological impact. However, it is a serious problem when the rise is rapid. Estuaries become shallower and more saline, beaches disappear and will need to be replaced by seawalls, and salt marshes will require engineered drainage to reduce inland flooding. In this way, the most productive areas of the coast are squeezed between a rising ocean and a retreating shoreline.

17A.3 **A NEW INLET** Storm waves from Hurricane Isabel breached the North Carolina barrier beach to create a new inlet, as shown in this aerial photo from September 2003. Notice also the widespread destruction of the shoreline in the foreground, with streaks of sand carried far inland by wind and wave action. As the sea level rises, ocean waves will attack barrier beaches with increasing frequency and severity.

17A.4 **COASTAL EROSION** Wave action has undermined the bluff beneath these two buildings, which collapsed onto the beach below.

Subsidence and Sea-Level Rise

Land subsidence is a contributing factor to the impact of sea-level rise. In an unaltered environment, rivers bring fine sediment to the coastline that settles in estuaries and is also carried by waves, currents, and tides into salt marshes and mangrove swamps. As the fine sediment accumulates, it slowly compacts, forming rich, dense layers of silt and clay mixed with fine organic particles. However, many coastlines are now fed by rivers that have been dammed at multiple points along their courses; this reduces the amount of fine sediment brought to the coast. Without a constant inflow of new sediment, coastal wetlands slowly sink as the older sediments that support them become compacted. This subsidence increases the effects of sea-level rise. Changes in river flows also affect estuaries as a result of changes in the frequency of severe precipitation events and summer droughts.

Delta coasts are especially sensitive to sediment starvation and subsidence. Here, subsidence rates can reach 2 cm per year. The Mississippi River has lost about half of its natural sediment load (Figure 17A.5), and sediment transport by such rivers as the Nile and Indus has been reduced by 95 percent. Extracting ground water from deltas also increases subsidence. The Chao Phraya delta near Bangkok and the old deltas of the Huang and Changjing rivers in China are important examples.

According to recent estimates, sea-level rise and subsidence could cause considerable losses of the world's coastal wetlands by the 2080s. Losses following a 36 cm rise in sea level are predicted at 33 percent of the 1999 area, and a rise of 72 cm would lead to a 44 percent loss. Losses would be most severe in areas of low tidal range, such as the Atlantic and Gulf Coasts of North and Central America, around the Mediterranean and Baltic Seas, and many small islands. Coupled with losses directly related to human activity, coastal wetlands could decrease by 30 percent. This reduction would have major effects on commercially important fish and shellfish populations, as well as other organisms in the marine food chain.

Coral Reefs

Coral reefs, like coastal wetlands, perform important ecological functions, such as harbouring marine fish and nursing their progeny. Coral reefs rank among the most diverse and productive of communities, and some contain more species than rainforests. They also serve to protect low-lying tropical islands from storm waves and surges. However, more than half of the total area of living coral reefs is considered to be threatened by human activities, ranging from water pollution to coral mining.

It appears that simple sea-level rise will not be a factor, because healthy coral reefs are able to grow upward at a rate equal to or greater than the projected sea-level rise. However, the increase in sea-surface temperature that will accompany global warming is of major concern. Many coral reefs appear to be at or near their upper temperature limits. When stressed by a rise in temperature, a lot of corals respond by "bleaching." Coral bleaching has been observed on reefs since the 1980s (Figure 17A.6). In this process, they expel the algae that live symbiotically inside their structures, leaving the coral without colour.

17A.5 **MISSISSIPPI DELTA MARSHLAND** Rising sea level and increasing subsidence have endangered marshlands of many river deltas. Shown here is a marshland of the Mississippi Delta in Louisiana.

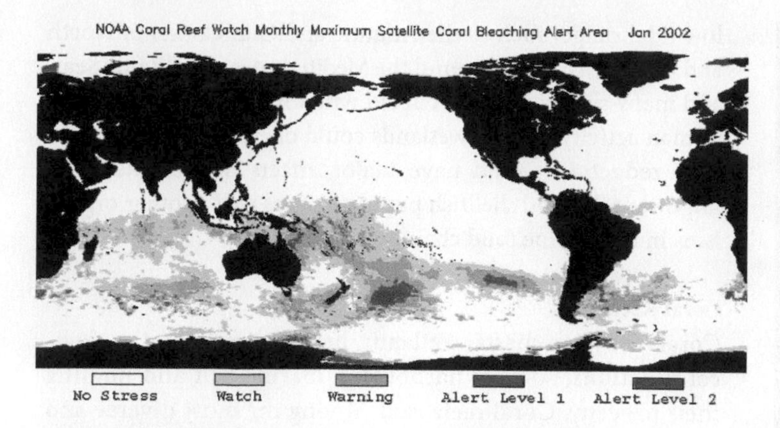

17A.6 **CORAL BLEACHING** Corals are vulnerable to bleaching when the sea surface temperature exceeds the temperature they would normally experience in the hottest month. Here are data for the month of January in 2002, 2005, and 2010. At the No Stress and Watch levels, no bleaching has occured. At the Warning level, bleaching is possible, at Alert Level 1 bleaching is likely, and at Alert Level 2 coral mortality is likely.

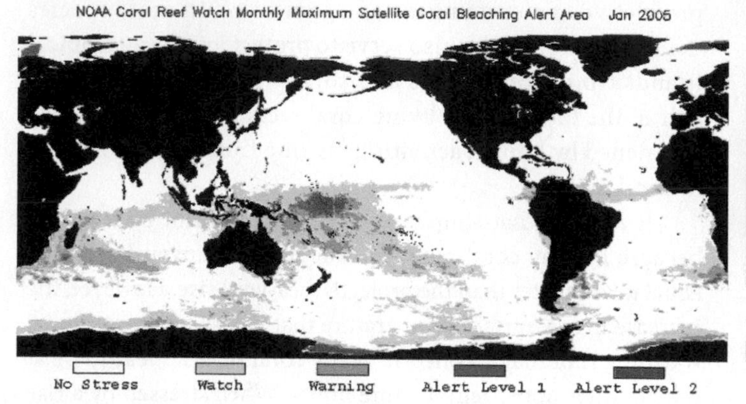

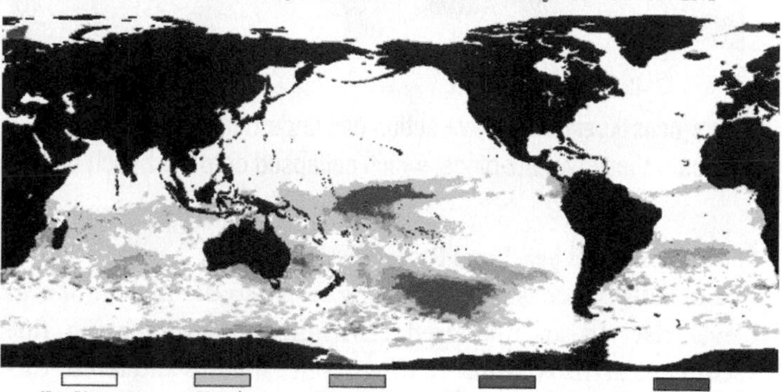

However, the algae are necessary for the coral's continued survival. The bleaching may be temporary if the stress subsides, but if permanent, the corals die. Major episodes of coral bleaching have been associated with strong El Niño events, where water temperatures increased by at least 1 °C.

Another concern is the effect of increased atmospheric CO_2 levels on corals. With higher concentrations of atmospheric CO_2, more of the gas dissolves in sea water, causing the water to become slightly less alkaline. This in turn shifts the solubility of calcium carbonate, making it less available to the corals to use in building their skeletal structure. Whether this will create yet another source of stress for coral reefs is still the subject of study.

High-Latitude Coasts

Many pristine stretches of arctic shoreline are threatened by global warming (Figure 17A.7). In these regions, the shoreline is sealed off for much of the year by sea ice. The shoreline is also buttressed by permanent ground ice that bonds unconsolidated sediment into a rocklike mass. This frozen ground is resistant to wave action and thus helps to protect the shoreline from erosion. In some areas, massive ice beds underlie major portions of the coastline.

With global warming, the shoreline is less protected by these mechanisms. Sea ice melts earlier and returns later, increasing the season of wave action. Greater expanses of open sea allow larger waves to build. Ground ice thaws to a greater depth, releasing more surface sediment and allowing waves to scour the shore more deeply. Near-shore sediments, stored in cliffs or bluffs, also thaw and are released to shoreline processes. Global warming is expected to be especially severe at high latitudes. Rapid coastal recession under wave attack is already reported for many ice-rich coasts along the Beaufort Sea.

It is apparent that global climate change will have major impacts on coastal environments, with broad implications for ecosystems and natural resources. It will take careful management to reduce these impacts on both human and natural systems.

17A.7 ARCTIC SHORELINE A pristine arctic shoreline, such as this one on the Beaufort Sea, will be subjected to rapid change as the global climate warms. Thawing of ground ice will release beach sediments, and the early retreat of sheltering sea ice will expose the shoreline to enhanced summer wave attack (see Chapter 18).

QUESTIONS

1. Review the observed and predicted changes in global climate that will affect coastal environments.
2. Identify the global change factors that will affect coastal erosion and describe their impacts.
3. How have human activities induced land subsidence in wetlands? Why are the wetlands of delta coasts particularly at risk?
4. What impact will global climate change have on coral reefs?
5. Why is global warming expected to cause coastal recession of arctic shorelines?

CHAPTER 18 GLACIER SYSTEMS, THE LATE-CENOZOIC ICE AGE, AND PERMAFROST

Mount Angel Glacier, Columbia Icefields, British Columbia.

Much of northern North America and Eurasia was covered by massive sheets of glacial ice as recently as 1 million years ago. As a result, glacial ice has played a dominant role in shaping landforms of large areas in midlatitude and sub-arctic zones. Glacial ice still exists today in two great accumulations—the Greenland and Antarctic ice sheets—and in many smaller masses throughout the Arctic and in high mountains.

The glacial ice sheets of Greenland and Antarctica strongly influence the radiation and heat balance of the globe. They reflect much of the solar radiation they receive, and their permanently cold surfaces help drive the system of meridional heat transport. In addition, these enormous ice accumulations represent stored water in the solid state and so are important components of the global water balance. When the volume of glacial ice increases, such as occurred during the Late-Cenozoic Ice Age, the sea level must fall. Conversely, when ice sheets melt away, the sea level rises. Today's coastal environments evolved during the rising sea level that followed the melting of the ice sheets of the last Ice Age.

This chapter concludes with a description of the landforms and geomorphic processes found in arctic and subarctic regions. These features are created primarily by the freezing and thawing of water, acting in concert with gravity.

GLACIERS

Where a great thickness of ice exists, the pressure makes the ice at the base lose its rigidity and become plastic. This allows the ice mass to respond to gravity, slowly spreading out over a larger area or moving downhill. On steep mountain slopes, the ice can also move by sliding. Movement is an important characteristic of a *glacier*, which represents a large natural accumulation of land ice affected by present or past motion.

Glacial ice accumulates when the average winter snowfall exceeds the amount of snow and ice lost in summer by **ablation**. Ablation occurs through sublimation, melting, and evaporation. When winter snowfall exceeds summer ablation, a layer of snow is added each year to what has already accumulated. As the snow compacts by melting and refreezing, it turns into granular ice that is gradually compressed into hard crystalline ice by overlying layers. When the ice mass is so thick that the lower layers become plastic, outward or downhill flow starts, and the ice mass is now an active glacier.

Glacial ice forms where temperatures are low and most or all precipitation falls as snow. These conditions can occur both at high latitudes and at high elevations, even in tropical and equatorial zones. In the Andes, the snowline at the equator is at an elevation of 4,600 m, compared with 5,000 m on Mount Kilimanjaro and Mount Kenya in Africa. In the Himalayas, the snowline is generally about 5,200 m. The higher snowline is attributed to warm summer conditions associated with the continental location of this Asian massif. The snowline in the southern Canadian Rockies is at about 2,700 m, similar to the European Alps, and drops to about 1,000 m in the Yukon. Orographic precipitation encourages the growth of glaciers that flow from high-elevation collecting grounds to lower elevations, where temperatures are warmer and the ice disappears through ablation.

Typically, mountain glaciers are long and narrow because they occupy former stream valleys. This distinctive type is called an **alpine glacier** (Figure 18.1a). Smaller glaciers that persist in mountain basins are called *cirque glaciers* (Figure 18.1b). Others, called *hanging glaciers*, end as icefalls when they no longer extend down to the main valley (Figure 18.1c). A regionally extensive ice accumulation in a mountain region that feeds several glaciers is known as an *icefield*. The Columbia Icefield, covering some 325 km^2 of mountainous terrain, straddles the border of Alberta and British Columbia (Figure 18.1d).

18.1 **CANADIAN GLACIER TYPES** (a) Salmon Glacier, British Columbia, is confined between the walls of the valley it has carved. (b) A cirque glacier on Bylot Island, Nunavut, is located in a mountain basin. (c) Angel Glacier, a hanging glacier on Mount Edith Cavell in Jasper National Park, ends in an icefall. (d) The Columbia Icefield covers 325 km^2 of mountainous terrain between Alberta and British Columbia.

In arctic and polar regions, prevailing temperatures are low enough that snow can accumulate over broad areas, eventually forming a vast **ice sheet** several thousand metres thick. The largest ice sheets are in Greenland and Antarctica, but smaller examples occur on Ellesmere Island, Baffin Island, and elsewhere in the Canadian Arctic. The Greenland Ice Sheet has an area of 1.7 million km^2 and occupies close to 90 percent of the entire island of Greenland. Its surface has the form of a broad, smooth dome and only a narrow, mountainous coastal strip of land is exposed. The Antarctic Ice Sheet covers 13 million km^2. It is thicker than the Greenland Ice Sheet, with maximum accumulations of 4,000 m. At some locations, long tongues extend from ice sheets as *outlet glaciers* or **tidewater glaciers** that reach the sea at the heads of fjords. Huge masses of ice break off from the floating edge of the glacier and drift out to open sea with tidal currents to become icebergs. An important glacial feature of Antarctica is the presence of great plates of floating glacial ice, called *ice shelves*. Ice shelves are fed by the ice sheet, but they also accumulate new ice through the compaction of snow. No ice shelves exist in the Arctic Ocean. Here, ice occurs as various forms of floating *sea ice*.

ALPINE GLACIERS

An alpine glacier begins in a bowl-shaped depression, called a **cirque**, and occupies a sloping valley between steep rock walls. Snow collects in the **zone of accumulation** where layers of snow are compacted and recrystallized. Freshly fallen snow has a density of 0.05 to 0.07 g cm^{-3}. The snowflakes quickly lose their hexagonal structure and form snow grains that are converted into *firn*, with densities ranging from 0.4 to 0.8 g cm^{-3}. The transition time depends on conditions. In midlatitudes, conversion can occur over the course of a few days or weeks, depending on snowfall and temperature regimes. At high latitudes, cold, dry conditions may prolong the process for a year or more. The lower part of the glacier lies in the **zone of ablation** where the ice wastes away. The **equilibrium line** marks the boundary between the accumulation zone and the ablation zone. Losses in the ablation zone occur quite rapidly, and the transported rock debris is released as the ice melts away at the *terminus* of the glacier.

Although the uppermost layer of a glacier is brittle, the plastic ice beneath flows slowly downhill in response to gravitational forces generated by the accumulated weight of snow and ice. Movement occurs through internal deformation of ice under compressive stress, which causes slippage within and between ice crystals. Glaciers with steep ice-surface gradients flow faster than those with gentler gradients since the shear stress generated in response to gravity is proportionately greater. Flow rates are generally highest at the equilibrium line, where the ice is thickest. Slower rates occur in the upper parts of the accumulation zone where the ice is colder and more rigid, and near the snout where the ice thins rapidly.

Changes in gradient of the underlying rock surfaces also influence glacier movement. Where the gradient is steeper, a glacier will accelerate and thin; this is called *extending flow*. Areas of extending flow are usually characterized by large tensional cracks, known as **crevasses**. Conversely, *compressive flow* occurs in response to a reduction in gradient. Here, the ice thickens and **pressure ridges** deform the surface. The nature of the underlying surface can also affect the rate of movement. Easily deformable rock, such as shale, favours basal sliding. In this process, the ice moves downhill, lubricated by meltwater and mud at its base. The thin layer of water that separates the ice from the underlying bedrock comes from pressure melting of the basal ice against obstacles. As the ice passes the rock obstruction, it refreezes through the process of *regelation*, and in this way, releases latent heat that contributes to melting of adjacent ice. Faster movement also occurs across impermeable rock surfaces where meltwater is retained at the base of the ice.

Rates of ice flow in glaciers generally range from 3 to 300 m annually. The highest rates, exceeding 12 km per year, are currently recorded on the glaciers of the West Greenland Ice Sheet. Flow rate changes vertically within the ice; usually it is most rapid near the surface in the midline of the glacier where there is less friction with the valley floor and walls. The faster-moving glaciers tend to occur in areas of high snow accumulation. Temperature is also important. Warmer ice not only deforms more easily than cold ice, it also usually generates more meltwater at its base. Hence, temperate glaciers typically flow more rapidly than those in polar regions. The absence of water at the base of glaciers in very cold polar regions limits the effectiveness of basal sliding. In extreme cases, the ice freezes onto the bedrock and basal flow ceases; under these conditions, movement is mainly by internal deformation. Where glaciers encounter weakly consolidated materials, movement can occur through bed deformation. Bed deformation is particularly effective in regions where underlying materials are saturated with water; this further reduces their strength, allowing the glacier to move with little resistance.

Flow rates are generally relatively constant, but some glaciers experience episodes of very rapid movement. A period of cool, snowy weather can produce a *kinematic wave* or bulge in the glacier surface due to an increase in accumulation. Flow accelerates in response to the greater mass of ice. Glacial *surges* likely develop in response to an increase in meltwater beneath the ice, which enhances basal sliding. In one dramatic example, the snout of the Bruarjokull glacier in Iceland surged forward 45 km at a rate of 5 m per hour.

A Glacier as a Flow System of Matter and Energy

An alpine glacier is a system of coupled matter and energy flows. The pathways of the matter flow system are shown in Figure 18.2a. In the zone of accumulation, snowfall provides input to the glacier as it becomes compacted and recrystallized. During warm periods, some snow is lost to melting, evaporation, and sublimation, but the annual balance favours accumulation. When the glacier reaches lower elevations, it loses more water. The balance shifts at the equilibrium line to a net loss of ice over the year, and the glacier enters the zone of ablation. With further descent, the loss rate increases as mean annual temperature rises. The glacier shrinks rapidly until no ice remains at the terminus.

The glacier flow system reaches a state of equilibrium in which the excess water in the form of ice in the zone of accumulation is balanced by the loss of water in the zone of ablation. For the most part, the glacier's shape and appearance remain unchanged, although there may be some small variation from year to year. However, this balance is easily upset by changes in the average annual rates of accumulation or ablation. If temperatures cool and snowfall increases, more ice will form in the zone of accumulation, and the equilibrium line will move to a lower elevation. The glacier will take longer to melt, so the terminus will extend farther down the valley. Eventually, the glacier will reach a new state of equilibrium, larger, thicker, and terminating farther from its source. If conditions were to become warmer or drier, the changes would reverse, producing a new equilibrium state in which a thinner glacier terminates at a higher elevation.

As the glacier flows downhill, it transports rock debris, which grinds away the underlying bedrock. The glacier's transport of debris can be treated as a separate open flow subsystem powered by gravity (Figure 18.2b). Another power source for this subsystem is solar energy. Solar energy provides water at high elevations through precipitation, thus contributing potential energy as an input. Gravity powers the downhill motion of the ice. In this motion, potential energy is converted to kinetic energy, and the kinetic energy is dissipated as heat generated by friction. The surface of a glacier is subjected to short-wave heat flows from solar radiation. This input provides a positive net radiation balance during the day. Most of the energy flow melts the ice, leaving less energy available to warm the overlying air through sensible heat transfer. Also, less energy remains to sublimate the snow and evaporate the meltwater. This reduces the flow of latent heat to the atmosphere, making the local climate colder. In this way, a glacier also acts as a thermal flow subsystem (Figure 18.2c).

Matter flow system

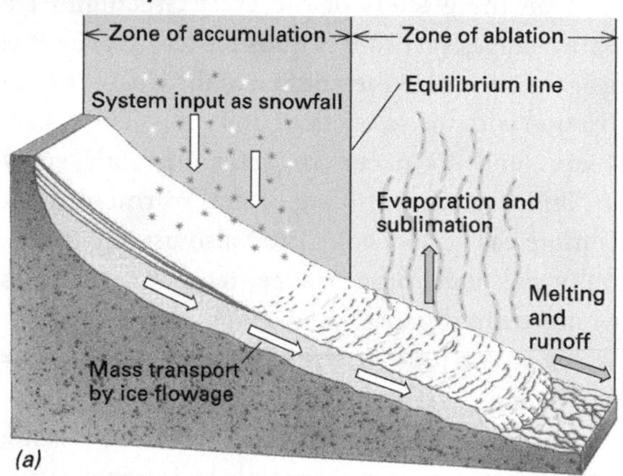

Gravity flow energy system

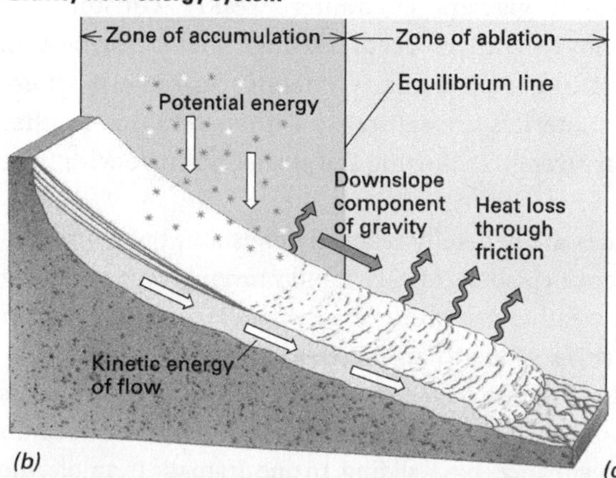

Thermal flow energy system

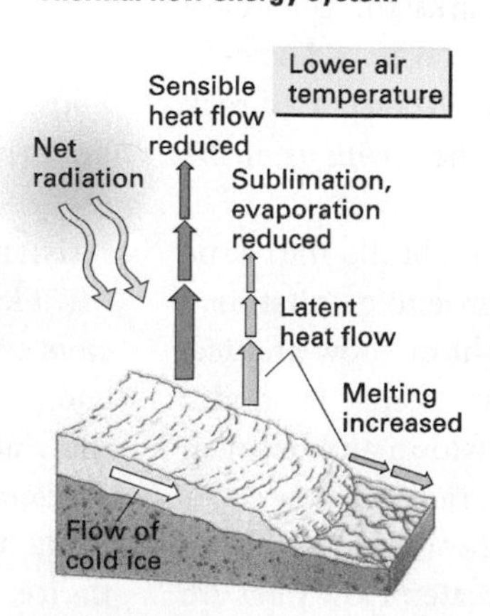

18.2 MATTER AND ENERGY FLOW IN AN ALPINE GLACIER SYSTEM (© A.N. Strahler.)

Landforms Made by Alpine Glaciers

Glacial ice normally contains abundant rock fragments, ranging from huge angular boulders to pulverized rock flour. Some of this material is eroded from the rock floor through *abrasion*, caused when the rock fragments scrape and grind against the bedrock over which the ice moves. In alpine glaciers, rock debris is also derived from material that slides or falls from valley walls onto the ice. Larger fragments of hard rock are the most effective abrasives. Abrasion increases with the speed of ice movement, as this controls the number of rock fragments that pass over the bedrock and the rate at which glaciers carry away eroded materials. The manner in which rocks are moved by glaciers is also a factor. Rocks that are rolled along under the ice are more effective at weakening bedrock than rocks that move by sliding. Erosion also occurs because of the great pressure that develops beneath a moving ice mass. This is sufficient to cause failure in weak rocks such as shale. Once the rock is weakened, moving ice can lift out blocks of bedrock by *plucking*.

The combined actions of abrasion and plucking smooth and chip the bedrock. Abrasion by very fine particles can polish the surface of the bedrock; more pronounced scratches or **striations** are cut by larger particles (Figure 18.3). As the calibre of the transported material increases, grooves and furrows may be formed. Crescentic gouges or **chattermarks** can also be chipped out of the bedrock. A glacier eventually deposits the rock debris it has obtained at its terminus, where the ice melts.

Mountains are eroded and shaped by glaciers, with the resultant landforms exposed as the ice melts away. The process begins with an initial landscape sculpted by weathering, mass wasting, and fluvial processes (Figure 18.4). The dissected highlands are covered with a thick layer of regolith and soil, giving them a comparatively smooth, rounded appearance. At the height of glaciation, much of the original land surface is buried beneath hundreds of metres of ice. The ice grinds and shapes the underlying bedrock as it moves down the slopes to produce an array of new and distinctive glacial landforms, which are exposed as the ice disappears when the climate ameliorates.

A climatic change results in the accumulation of snow in the heads of the higher valleys. Bowl-shaped cirques are carved by the grinding motion of the ice and by intense frost shattering of the bedrock under the masses of compacted snow. Glaciers fill the valleys and are integrated into a system of tributaries that feed the main glacier. The cirques grow steadily larger. Their rough, steep walls replace the smooth slopes of the original highland mass. Where two cirque walls approach from opposite sides, a jagged, knifelike ridge, or *arête*, is created (Figure 18.5a). When three or more cirques grow together, the remnant highland is left as a sharply pointed *pyramid peak*, or *horn* (Figure 18.5b).

18.3 **GLACIAL ABRASION** At the smallest scale, glacial abrasion can polish a bedrock surface until it is quite smooth. Scratches or striations can develop where debris is dragged over the surface, and crescentic chattermarks may occur in hard rock types from the impact of boulders under the ice.

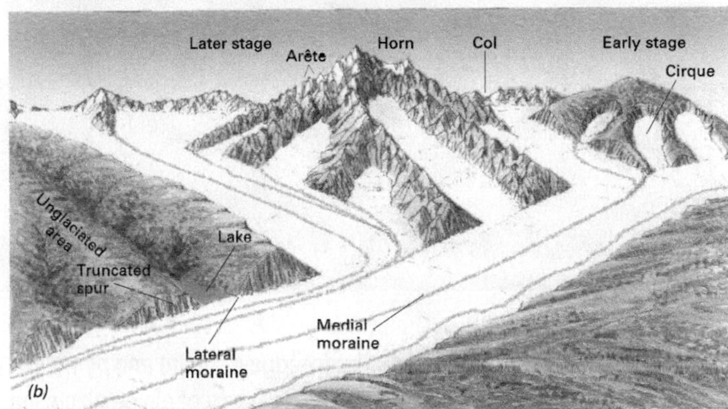

18.4 **LANDFORMS PRODUCED BY ALPINE GLACIERS** (a) Before glaciation, the region has smoothly rounded divides and narrow, V-shaped stream valleys. (b) After glaciation has been in progress for thousands of years, new erosional forms develop. (c) The disappearance of the ice exposes a system of glacial troughs. (Drawn by A. N. Strahler.)

(a)

(b)

18.5 GLACIAL EROSION The arête or knife-edge ridge (a) and pyramid peak or horn (b) are distinctive erosional features of alpine glaciation.

(a)

(b)

18.6 GLACIAL MORAINES (a) Debris carried at the margin of a glacier forms lateral moraines. When two ice streams converge, their lateral moraines form a medial moraine. Two medial moraines appear as dark lines on the surface of this glacier in Kluane National Park, Yukon. ***Fly By: 60° 43′ N; 138° 50′ W*** (b) Terminal moraines deposited as ridges at the snout of a glacier mark the farthest advance of the ice. ***Fly By: 59° 47′ N; 139° 57′ W***

Debris carried in, on, or under the ice is eventually deposited as ridges or piles of rock in various configurations in the glaciated landscape. A **lateral moraine** is a debris ridge formed along the edge of the ice adjacent to the trough wall. Where two ice streams join, marginal debris is dragged as a narrow band on the ice surface. This midstream feature is called a *medial moraine* (Figure 18.6a). At the terminus of a glacier, rock debris accumulates in a **terminal moraine**, which forms an embankment that curves across the valley floor (Figure 18.6b). Even though the terminus of a glacier may be stationary, the ice still delivers rock debris to the snout. As the ice melts away, this debris is released and accumulates as a ridge across the valley. A terminal moraine marks the farthest advance of the glacier. Similar mounds of material may occur upstream from the terminal moraine. These *recessional moraines* mark the positions where the snout of a receding glacier remained stationary long enough to produce a depositional ridge.

A valley is constantly deepened and widened as a glacier passes along it. After the ice has disappeared, a deep, steep-walled **glacial trough** remains. The trough typically has a U-shaped cross-profile (Figure 18.7a). U-shaped troughs carved by tributary glaciers generally have a smaller cross-section and are less deeply eroded. Because the floors of the tributary troughs lie high above the base of the main trough, they are called *hanging troughs*. Streams that later occupy these troughs provide scenic waterfalls that cascade over their lips (Figure 18.7b). In the smaller troughs and cirques, the bedrock is unevenly excavated, so their floors contain basins and steps. Small lakes, called *tarns*, occupy the cirque basins (Figure 18.7c). Major troughs sometimes hold large, elongated *ribbon lakes*, also known as *finger lakes* (Figure 18.7d), but the majority have deep alluvial fills deposited by streams, heavily laden with rock fragments, that flowed from the receding ice fronts (see Figure 16.15c). In coastal areas, deeply excavated glacial troughs are usually flooded with sea water as the ice recedes, creating **fjords**. Fjords are found mainly along mountainous coasts between lat. 50° and 70° N and S, where glaciers were nourished by heavy orographic snowfall (see Chapter 17).

(a) (b)

(c)

(d)

18.7 GLACIAL VALLEYS AND GLACIAL LAKES (a) Glacial troughs have distinctive U-shaped cross-profiles. (b) Hanging troughs develop where less powerful tributary glaciers join a more deeply eroded trough. (c) Tarns are small, rounded lakes that occupy the floors of cirques. (d) Long ribbon or finger lakes form in the floor of glacial troughs. Numerous examples can be found near ***60° 24′ N; 131° 16′ W.***

EYE ON THE ENVIRONMENT
GLACIER RETREAT—A GLOBAL OVERVIEW

Glaciologists use the term *mass balance* to describe the difference between the amount of ice that is added to a glacier and the ice that a glacier loses, averaged over a period of time. Since the end of the Little Ice Age in 1850, glaciers worldwide have generally been in retreat, with a marked reduction recorded in the last 20 to 30 years (Figure 1). This negative trend in glacier mass balance is associated with climate changes linked to variations in both temperature and snowfall. Monitoring in the Cascade Mountains of Washington has concluded that glacier retreat is associated with a 25 percent decline in winter snowpack since 1946 and a 0.7 °C increase in summer temperatures during the same period.

North America

For the past two decades, laser elevation measurements of many glaciers indicate glacial thinning is occurring throughout western North America. In Waterton-Glacier National Parks, which encompasses 4,080 km^2 of mountainous terrain in Alberta, British Columbia, and northwestern Montana, all of the mountain glaciers have receded dramatically since 1850, the date recognized as the end of the Little Ice Age in this region. For example, the Illecillewaet Glacier in British Columbia has retreated 2 km since 1887. Computer analysis suggests that if atmospheric warming continues at the present rate, all glaciers in Glacier National Park will be eliminated by 2030, and even under present conditions, they are predicted to disappear by 2100. Although most glaciers in North America are currently receding, the Taku Glacier in Alaska has advanced about 6.5 km since 1890.

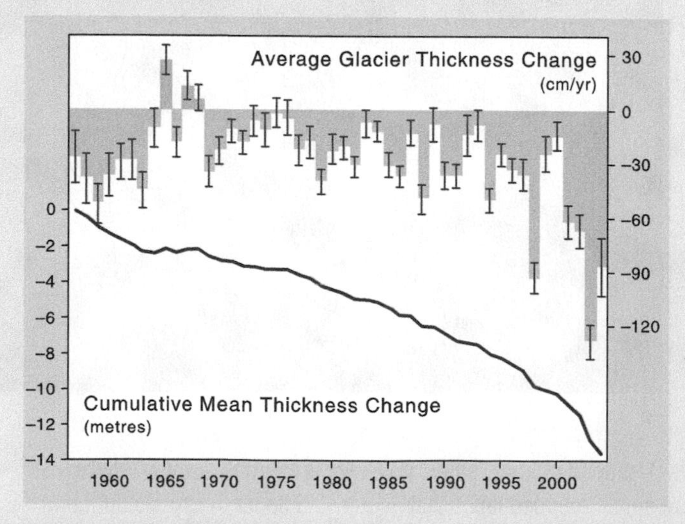

Figure 1 Glacier mass balance The volume of glacier ice has steadily declined worldwide since the late 1960s.

Europe

The glaciers of Europe are in a general state of retreat. In Switzerland, a 20 percent reduction in glacier area has been recorded since 1980. The Aletsch Glacier, the largest in Switzerland, has retreated 2,600 m since 1880, of which 800 m occurred in the past 25 years. Only about 5 percent of the glaciers in the Italian and Swiss Alps are currently advancing. This marks a dramatic reduction compared to the 1980s, when nearly 90 percent of the glaciers in the Italian Alps and more than 60 percent of the glaciers in the Swiss Alps were reportedly advancing. The present situation is similar to conditions in the 1950s.

In Scandinavia, the majority of glaciers are also retreating. Norway has experienced a significant decrease since 2000, attributed to several consecutive years of little winter precipitation and record warm summers in 2002 and 2003. In Iceland, some glaciers have reportedly receded by up to 2 km in the past 30 years.

Africa

In Africa, glaciers are found only on Mount Kilimanjaro (5,895 m), Mount Kenya (5,199 m), and the Ruwenzori Mountains, which rise to 5,109 m on the borders of Uganda and Zaire. The ice cover on the summit of Kilimanjaro has decreased dramatically since 1912. Little ice was reported in 2005, and it is predicted that the summit will be ice-free by 2025. The Ruwenzori Mountains also show evidence of glacier retreat. However, because these mountains are supplied by moisture from the Congo region, retreat will likely be slower than on the isolated peaks of Mount Kilimanjaro and Mount Kenya.

Himalayas

The formation and growth of pro-glacial lakes at the termini of many glaciers in the Himalayas indicate glacier melting and retreat. Studies in China have concluded that 95 percent of the Himalayan glaciers have been retreating since 1990. Regional examples include the Khumbu Glacier, the main route up Mount Everest, which has retreated 5 km since 1953; the Rongbuk Glacier on the north side of Mount Everest, which currently is retreating at a rate of 20 m per year; and the Gangotri Glacier in India, which has retreated 30 m per year since 2000.

Andes

In the northern Andes, most glaciers are less than 1 km^2 in size. Measurements from the Chacaltaya Glacier in Bolivia indicate that it lost 65 percent of its volume and 40 percent of its thickness between 1992 and 1998; it is predicted to disappear by 2015.

Farther south in Peru, glaciers cover an area of 725 km^2. From 1977 to 1983, general glacial recession was reported at 7 percent in this region. Recent studies on the Quelccaya Ice Cap, the largest in Peru, show significant retreat and thinning, with the rate of retreat on one glacier measured at 155 m per year from 1995 to 1998. The glaciers in Patagonia at the southern end of South America are reportedly receding at a faster rate than any other region in the world. Here, ice has retreated by as much as 1 km since the early 1990s.

New Zealand

In New Zealand, mountain glaciers have shown accelerated rates of recession since 1920, and in the past three decades, pro-glacial lakes have appeared behind the terminal moraines of several glaciers.

THE LATE-CENOZOIC ICE AGE

The present ice sheets in Greenland and Antarctica provide an impression of the landscape that would have existed over much of Canada at the peak of the **Late-Cenozoic Ice Age** (or simply the **Ice Age**) that began in late Pliocene time, perhaps 2.5 to 3.0 million years ago. The Ice Age is associated with the last three epochs of the Cenozoic Era: the Pliocene, Pleistocene, and Holocene (see Table 12.1). Previously, geologists associated the Ice Age with the Pleistocene Epoch, which began about 1.6 million years ago. However, new evidence obtained from deep-sea sediments shows that the Ice Age began earlier than this.

A period during which continental ice sheets grow and spread outward over vast areas is known as a **glaciation**. Glaciations are associated with a general cooling of average air temperatures. At the same time, ample snowfall must persist over the growth areas to allow the ice masses to build in volume. When the climate warms or snowfall decreases, ice sheets become thinner and begin to shrink. Eventually, the ice sheets may melt completely. This period is called a *deglaciation*. Following a deglaciation, but before the next glaciation, is a period in which a milder climate prevails; such a period is referred to as an **interglaciation** or **interstadial**. A succession of alternating glaciations and interglaciations, spanning a period of several million years, constitutes an *ice age*.

The last four major ice advances of the Late-Cenozoic Ice Age in North America were traditionally known as the Nebraskan, Kansan, Illinoisan, and Wisconsinan glaciations, but because numerous advances are now known to have occurred during this period, only the Wisconsinan stage is retained in recent chronologies (Table 18.1). Different names are applied to contemporaneous episodes in Europe and other parts of the world. In the *Wisconsinan Glaciation*, 12,000 to 110,000 years ago, ice sheets covered much of North America, northern Europe, and northern Asia (Figure 18.8). In addition, extensive ice covers developed in the Andes Mountains in South America, the South Island of New Zealand, and less extensively in Australia. It is generally thought that Earth is presently experiencing an interglaciation that started about 18,000 years ago.

GLACIATION DURING THE ICE AGE

During the Wisconsinan Glaciation, most of Canada was engulfed by the vast Laurentide Ice Sheet that was centred in the general vicinity of what is now Hudson Bay (see Figure 17.19). Ice spread into the United States, covering most of the land lying north of the Missouri and Ohio rivers, and extending eastward into New England. Alpine glaciers from the western mountains coalesced to form the Cordilleran Ice Sheet that spread to the

TABLE 18.1 CHRONOLOGY OF THE RECENT GLACIAL AND INTERGLACIAL EPISODES

TIME (YRS BP)	EPOCH	EVENT
0-150	Holocene	Present Interstadial
150-750		Little Ice Age (Neoglaciation)
750-1,100		Medieval Warm Period
5,000-7,000		Mid-Holocene Warm Period (Climatic Optimum)
8,200-7,900		8.2 Kiloyear Event
11,500-8,900		Pre-Boreal and Boreal (Early Hypsithermal)
12,900-11,500		Younger Dryas
12,900-110,000	Late Pleistocene	Wisconsinan Glaciation Frequent alternation between Interstadials (warm periods) and Heinrich Events (cold periods)
115,000-130,000	Pleistocene	Sangamon (Interglacial)
750,000-1,650,000	Middle Pleistocene	Many ice advances
1650,000-2,480,000	Early Pleistocene	Many ice advances
>2,480,000	Pliocene	Warm climate

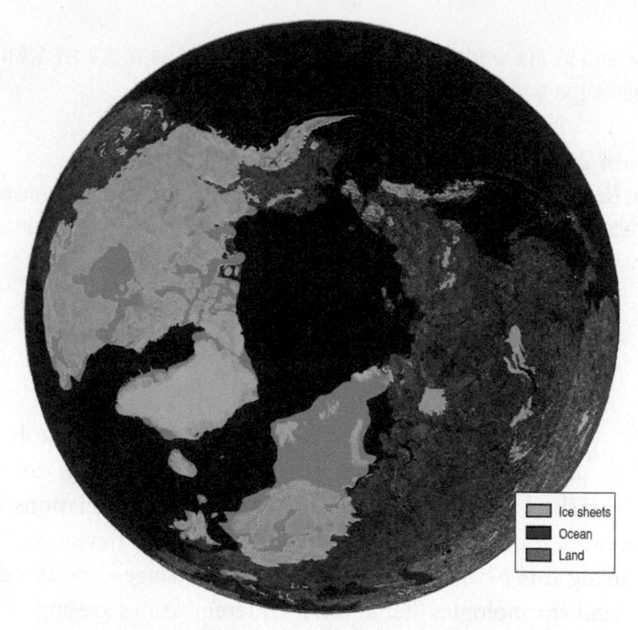

18.8 During the Wisconsinan Glaciation, ice sheets covered much of North America, northern Europe, and northern Asia. In addition to Antarctica, the principal regions in the southern hemisphere affected by the Wisconsin Glaciation were the Andes Mountains and the South Island of New Zealand.

Pacific shores and met the Laurentide sheet lying to the east. A small area south of Lake Superior escaped inundation by the Wisconsinan ice. Known as the Driftless Area, it was apparently bypassed by the glacial lobes that moved on either side.

At the maximum extent of the Wisconsinan ice sheets, sea level was as much as 200 m lower than today, exposing large areas of continental shelf throughout the world. The lower sea level allowed the ice sheets to extend far beyond the present coastlines. These vast ice sheets were several kilometres thick and their great mass depressed the Earth's crust by hundreds of metres at some locations. With deglaciation, the overlying ice load was removed and the crust began to move upward. This crustal movement is known as *isostatic rebound.*

18.9 RAISED BEACHES ON BAFFIN ISLAND Isostatic rebound created a series of raised beaches in this high Arctic site following the melting of the Laurentide Ice Sheet.

Important evidence of isostatic rebound is the position of ancient shorelines, now raised above sea level by the crustal motion. Sea level rose rapidly during the last deglaciation and the oceans flooded over areas where the crust remained depressed. Wave action at the shoreline created beaches that were slowly elevated as the crust rebounded, leaving them hundreds of metres above present sea level. Many elevated shorelines are found today in the coastal areas around Hudson Bay (Figure 18.9). Radiocarbon ages for marine shells and driftwood show that these features developed over the past 14,000 years.

Hudson Bay and the Baltic Sea are two of the largest *epicontinental seas* that presently extend into continental land masses. Both are centred in areas of active isostatic rebound (Figure 18.10). In Hudson Bay, maximum uplift is presently about 100 m. In the Baltic region, the land has risen by as much as 275 m. The crust in these regions is expected to continue to rise, and eventually these bodies of water will be much smaller or disappear entirely. Gravity measurements predict another 100-m rise in the Hudson Bay region. Isostatic rebound is most rapid immediately following the melting of the ice and slows exponentially over time. Around Hudson Bay, the rebound rate has decreased from 2 to 3 m per century about 8,000 years ago to less

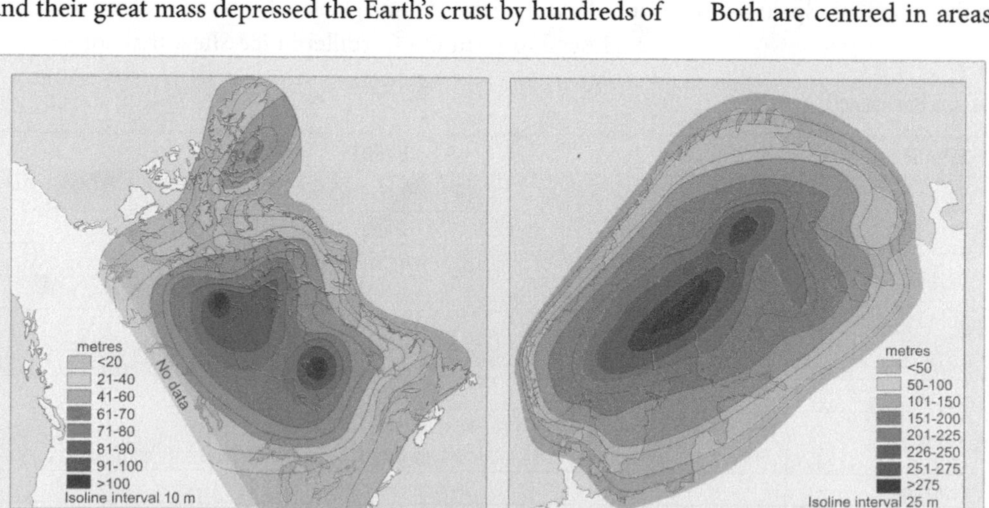

18.10 ISOSTATIC REBOUND (a) Hudson Bay and (b) the Baltic Sea are located in areas that were greatly depressed by the weight of continental ice sheets.

than 10 cm per century at present. Thus, about half of the total uplift had occurred by about 7,000 years ago.

LANDFORMS MADE BY ICE SHEETS

Like alpine glaciers, ice sheets are highly effective agents of erosion, and several landscape features are common to both environments. For example, the slowly moving ice scrapes and grinds away solid bedrock, leaving behind smoothly rounded rock masses. These show countless grooves and striations trending in the general direction of ice movement (Figure 18.11). Sometimes the scratches are so fine that the rock is polished to a smooth, shining surface. More pronounced shaping of solid bedrock by the moving ice can form a *roche moutonnée* (Figure 18.12). The side of the rock that the ice approached is usually smoothly rounded, while glacial plucking on the downstream side leaves the rock irregular and blocky.

Erosion by the ice sheets created many hundreds of lake basins. These range in size from the numerous small lakes scattered across the Canadian Shield (Figure 18.13) to Great Bear Lake, Northwest Territories, which with a surface area of 31,153 km^2, makes it the world's eighth largest lake. Drainage is generally poor on the Canadian Shield as the surface consists of rocky, ice-smoothed hills

18.11 GLACIAL ABRADED GROOVES AND STRIATIONS Rock fragments dragged across bedrock cause abrasions ranging in scale from fine scratches that effectively polish the rock, to deep gouges and grooves.

18.12 A GLACIALLY ABRADED ROCHE MOUTONNÉE Glacial action abrades the rock into a smooth form as it rides over the rock summit, then plucks bedrock blocks from the lee side, producing a steep, rocky slope.

18.13 LAKES ON THE CANADIAN SHIELD Numerous lakes fill glacially scoured bedrock depressions throughout the Canadian Shield.

with an average relief of only 30 m. Between the lake-filled basins, the land is often boggy *muskeg* where the water table is very near the surface. In some areas, the ice sheets excavated linear depressions where the bedrock was weak or where the flow of ice was channelled by the presence of a valley. The Finger Lakes of western New York state were produced in this way (Figure 18.14).

DEPOSITS LEFT BY ICE SHEETS

The term **glacial drift** includes all varieties of rock debris deposited as a result of continental glaciation. About 25 percent of North America is covered with drift, which globally accounts for 8 percent of the Earth's land surface. Two major types of drift are distinguished according to the degree of sorting and layering that occurred during deposition. **Till** or *non-stratified drift* is made up

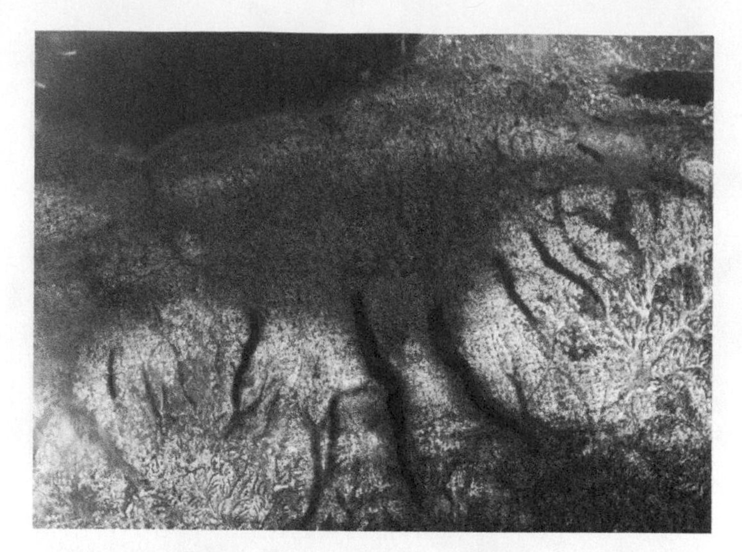

18.14 **NEW YORK'S FINGER LAKES** The steep, roughly parallel valleys and hills of the Finger Lakes region were shaped by ice sheets that were as much as 3.5 km deep. Pre-existing river valleys that were scoured into deep troughs are now filled with lakes. The two largest lakes, Seneca and Cayuga, are so deep that their lakebeds are below sea level. (Image acquired by the International Space Station December 4, 2004.)

of a mixture of rock fragments, ranging in size from clay to boulders, that is deposited directly from the ice without subsequent water transport. **Stratified drift** consists of layers of clays, silts, sands, or gravels that were deposited by meltwater streams or settled out in bodies of water adjacent to the ice.

Although the positions of the ice sheet margins fluctuated many times, there were long periods when they were essentially stable, and this affects the form and composition of the residual drift. When the rate of ice ablation balances the amount of ice brought forward by the ice sheet, the ice margin remains stationary and allows thick deposits of drift to accumulate. Rock fragments are transported by ice sheets in much the same way as in alpine glaciers, but on a vaster scale. Debris is carried forward and deposited in the ablation zone at the ice margin, where it will pile up if not subsequently removed. During the recessional phase when the ice sheet starts to disappear, some of the debris is sorted by meltwater and redeposited as various stratified features (Figure 18.15).

NON-STRATIFIED GLACIAL DEPOSITS

Ablation till is formed when glacial ice melts in a stagnant marginal zone. The rock fragments held in the ice are lowered to the underlying ground surface where they form a layer of debris

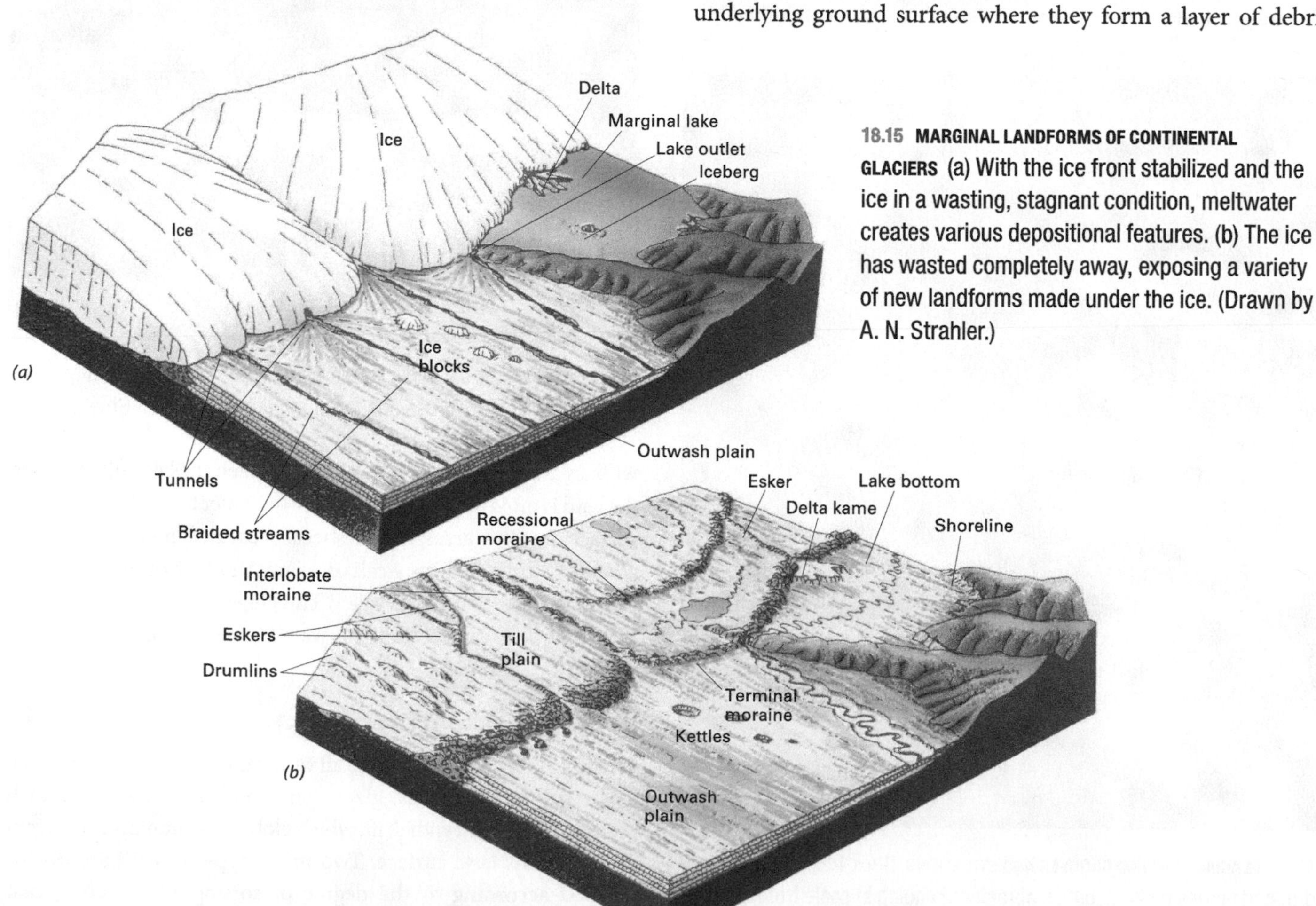

18.15 **MARGINAL LANDFORMS OF CONTINENTAL GLACIERS** (a) With the ice front stabilized and the ice in a wasting, stagnant condition, meltwater creates various depositional features. (b) The ice has wasted completely away, exposing a variety of new landforms made under the ice. (Drawn by A. N. Strahler.)

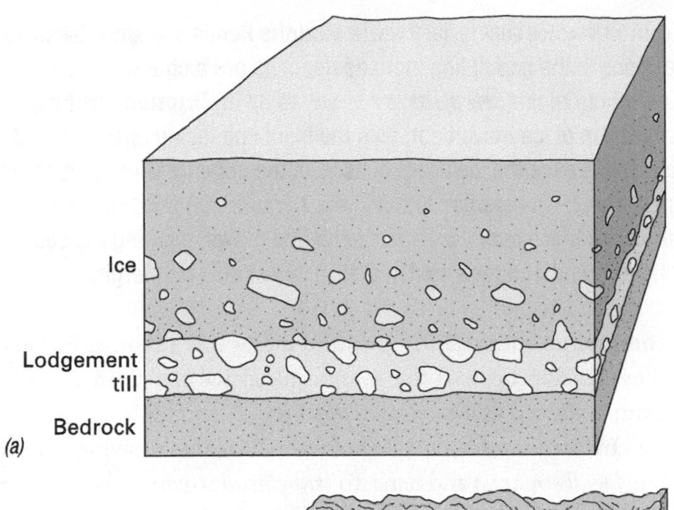

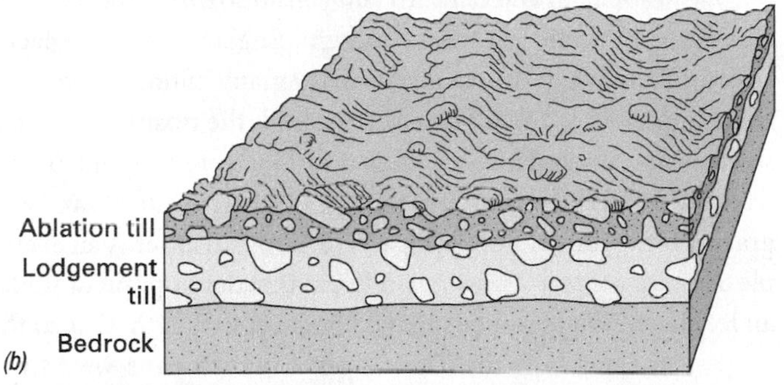

18.16 GLACIAL TILLS (a) As ice passes over the ground, sediment and coarse rock fragments are pressed together to form a layer of lodgement till. (b) When the overlying ice stagnates and melts, the rock debris it contains is left as the residual deposit of ablation till. (Copyright © A. N. Strahler.)

(Figure 18.16). Ablation till shows no sorting, and often consists of a mixture of sand and silt with many angular stones and boulders. Beneath this residual material, there may be a basal layer of dense *lodgement till*, consisting of clay-rich debris previously dragged forward beneath the moving ice. Where ablation till forms a thin, more-or-less even cover, it is called a *ground moraine*.

The thickness of glacial till deposits depends on the amount of debris in the ice, the pre-existing topography, and the length of time that the ice margin stayed at a given location. Glacial till usually does not form prominent landscape features, but tends to obscure, or entirely bury, the landscape that existed before glaciation. Where the deposits are thick and smoothly distributed, they form a level *till plain*; till plains are widespread throughout the central lowlands of the United States and southern Canada. In parts of Alberta and Saskatchewan, drift thickness varies from 300 m in a few pre-glacial valleys to zero on some of the higher rock outcrops, and has contributed to the subdued relief of the Canadian Prairies.

Thicker deposits of glacial till that accumulate in a narrow zone at the ice margin form a *terminal moraine* comprising heaps of unsorted debris. After the ice has disappeared, the moraine appears as a belt of irregular hills. Small lakes may occupy depressions within this undulating landscape of mounds and ridges; such terrain is often referred to as *knob-and-kettle* topography. Terminal moraines reflect the shape of the ice margin, which consisted of a series of great *ice lobes*, each with a curved front. Where two lobes come together, the moraines curve back and fuse together into a single *interlobate moraine*. During the recessional phase, the ice sheet begins to shrink because conditions no longer favour ice accumulation. If the ice front pauses for some time at various positions during its retreat, morainal belts similar to the terminal moraine form. As with alpine glaciers, these belts are called recessional moraines. They run roughly parallel with the terminal moraine, but are often thin and discontinuous. Unsorted glacial till covers areas between the morainal ridges. The numerous *sloughs* found across Alberta, Saskatchewan, and Manitoba are mostly found in *kettle holes* that formed as ice lenses in or under glacial drift eventually melted away, leaving surface depressions that later filled with water (Figure 18.17a).

A **drumlin** is another common feature formed from glacial till; these smoothly rounded, oval hills have the general shape of an inverted spoon (Figure 18.17b). Drumlins invariably lie in a zone behind the terminal moraine. They often occur in groups or swarms and may number in the hundreds—the term "basket of eggs topography" is occasionally used to describe drumlin fields. The long axis of each drumlin parallels the direction of ice movement, with the steeper, broader end facing the oncoming ice. The length of a drumlin is thought to increase in response to ice velocity and the period of time the flowing ice acted upon it; an increase in sediment strength will likely reduce the length of a drumlin as the material is less easily extruded.

Drumlins seem to have formed under moving ice by a plastering action in which layer upon layer of sediment were spread and shaped into their final forms. However, the origin of drumlins is not well understood, and several theories have been put forward to explain them. Some believe that drumlins began as pre-existing features that were reshaped by erosion during a re-advance of the ice sheet. A second theory proposes that till was deposited and molded behind an obstacle or in a cavity under the ice. Alternatively, drumlins might be sculpted into streamlined forms by catastrophic meltwater discharge events beneath the ice. One widely accepted hypothesis suggests that deformable sediment flowed onto and around cores of less-deformable material. Whether the sediment was deposited or deformed would depend on pore water pressure, with deposition occurring if water could drain from the material and deformation occurring if drainage was prevented.

STRATIFIED GLACIAL DEPOSITS

In front of the ice margin, a smooth, gently sloping *outwash plain* or *sandur* accumulates from stratified drift left by braided streams issuing from the ice. Outwash plains are built of layer upon layer of sands and gravels distributed and sorted by the

(a)

(b)

(c)

18.17 LANDFORMS OF CONTINENTAL ICE SHEETS (a) **Prairie sloughs** Ponds and small lakes that formed in depressions in the undulating moraine deposits are a characteristic feature of the prairie pothole region. ***Fly By: 53° 04′ N; 104° 55′ W*** (b) **Drumlin** The form of drumlins shows the direction of ice movement, with the blunt end facing upstream and tapering downstream; in this case the ice direction is from the upper to lower part of the photograph. Drumlins are well-represented around Peterborough, Ontario. ***Fly By: 48° 17′ N; 78° 12′ W*** (c) **Esker** An esker is a curving ridge of sand and gravel, marking the bed of a river of meltwater flowing underneath a continental ice sheet near its margin.

continuous lateral shifting of the meltwater streams. Because they develop beyond the margin of an ice sheet, an outwash plain is classed as a *proglacial deposit*.

In some cases, large meltwater streams flowing within (*englacial streams*) and beneath (*subglacial streams*) the ice sheet emerge from tunnels at the ice margin. Englacial and subglacial streams form when the ice is stationary many kilometres behind the margin. After the ice has disappeared, the positions of such streams are marked by long, sinuous ridges of sediment. These features are called **eskers** and represent the deposits of sand and gravel on the tunnel floors (Figure 18.17c). An esker is an example of an *ice-contact deposit*, a landform that develops on or under an ice sheet. Eskers can be many kilometres in length. One of the longest in Canada is the Thelon esker, which runs west from Dubwant Lake in the Northwest Territories for about 800 km.

When ice sheets advance toward higher ground, outlet valleys may become blocked with ice. Under these conditions, meltwater can collect in marginal glacial lakes that form along the ice front (see Figure 18.15a). Meltwater streams draining from the ice are able to build *glacial deltas* into these marginal lakes. Once filled to capacity, glacial lakes will begin to drain along the lowest available channel, which is progressively deepened, allowing even more water to drain from the lake. Because glacial lakes were continuously supplied with enormous volumes of meltwater, the outlet flow was able to cut broad valleys called *spillways* or *overflow channels* (Figure 18.18). Overflow channels are prominent landscape features that are usually several kilometres across. The streams that now flow in them are no longer supplied with vast amounts of meltwater, and because they are much smaller than the valley they occupy, they are referred to as *underfit* or *misfit streams*.

After the ice has disappeared and the proglacial lake has dried, several distinctive features remain in the landscape. Exposed lake beds remain as *glaciolacustrine plains* underlain by soils that are predominantly clay; these plains often contain extensive areas of marshland. For example, the level terrain near Winnipeg, Manitoba, is a legacy of former glacial Lake Agassiz (Figure 18.19). Lake Agassiz covered much of Manitoba, as well as parts of Ontario, Saskatchewan, Minnesota, and North Dakota. It formed about 12,000 years ago from glacial meltwater that was prevented from flowing northward by remnants of the Laurentide

18.18 GLACIAL SPILLWAY The broad Qu'Appelle Valley in Saskatchewan is one of several overflow channels found in the Canadian Prairies. These valleys were cut by meltwater draining from proglacial lakes through the action of rivers that were much larger than the ones that presently occupy them. *Fly By: 40° 34′ N; 103° 23′ W*

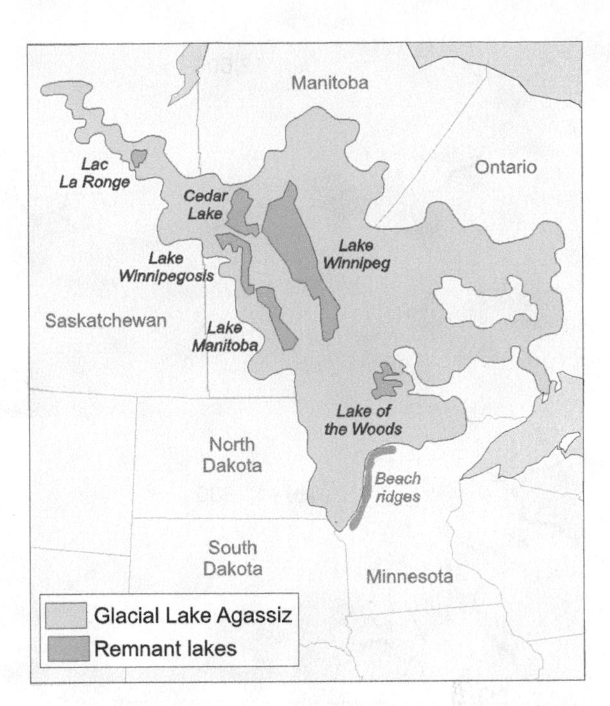

18.19 GLACIAL LAKE AGASSIZ The largest of the ice marginal lakes, glacial Lake Agassiz stretched from Saskatchewan to Ontario into South Dakota and Minnesota. Stratigraphic evidence of Glacial Lake Agassiz is found over an area of approximately 950,000 km^2, although the lake never covered this entire area at any one time.

Ice Sheet. Old shorelines indicate that the size and depth of Lake Agassiz varied as the climate periodically warmed and cooled.

In many locations, the sediments deposited in former glacial lakes contain **varves**. Varved sediments consist of alternating layers of clay and silt, usually only 1 to 2 cm thick (see Figure 15.21). The layers reflect seasonal deposition in the lakes at the ice margin. When temperatures warmed in the high-sun season, meltwater streams began to flow and brought sediment into the ice-free lake. Turbulence in the lake permitted only coarser sediments to settle out of the water. When the streams stopped flowing in the low-sun season, the finer clays were able to settle through the still waters beneath the ice cover on the lake.

The present Great Lakes are considered to have formed from proglacial lakes that developed at the time of the Wisconsinan deglaciation. The Great Lakes occupy a region of old sedimentary rocks. During the Ice Age, the advancing ice sheets scoured and lowered the surface by as much as 500 m. Drainage for this region was southward to the Gulf of Mexico at the time when the last continental glacier began its retreat, about 18,000 years ago. About 12,500 years ago, lakes formed in the ice-free portions of the scoured sedimentary basin, the largest being Lake Algonquin. Dammed by the ice mass to the north, the lakes reached levels many tens of metres higher than the present Great Lakes. At this time, drainage from Glacial Lake Algonquin was through the St. Clair–Detroit River system to the basin of Early Lake Erie. Early Lake Erie drained eastward via Niagara Falls into Glacial Lake Iroquois, and then to the Hudson River through the Mohawk Valley (Figure 18.20a). The retreat of the ice sheet was interrupted by a short re-advance (Figure 18.20b). Water levels subsequently dropped as an outlet to the northeast was uncovered through Early Lake Nipissing and the Ottawa River to the Champlain Sea. By 10,000 years before present (BP), the current form of the Great Lakes was beginning to appear. Former Lake Chippewa expanded to form Lake Michigan, with former Lake Stanley eventually becoming Lake Huron (Figure 18.20c). Soon after, the Ottawa River drainage route was closed due to isostatic rebound. The resulting rise in lake levels allowed outflow to shift to the present route from southern Lake Huron, through Lakes Erie and Ontario, to the St. Lawrence River.

INVESTIGATING THE LATE-CENOZOIC ICE AGE

The study of Ice Age glacial history progressed significantly in the 1960s, when core samples began to be collected from the deep ocean floor (see Chapter 12). Deep-sea cores reveal a long history of alternating glaciations and interglaciations going back at least 2 million years, and possibly 3 million years BP. The cores show that more than 30 glaciations occurred in Late-Cenozoic time, spaced at intervals of about 90,000 years. Many of these events were accompanied by sudden reversals in Earth's magnetic field. Studies of the composition and chemistry of the layers within the core samples has provided additional information about ancient temperature cycles in the atmosphere and ocean.

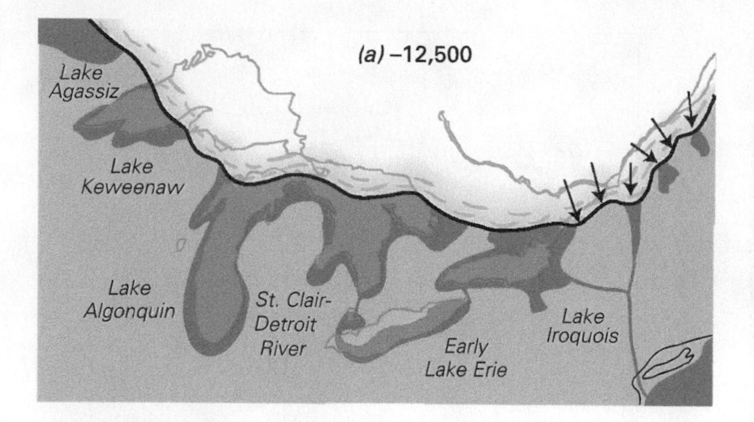

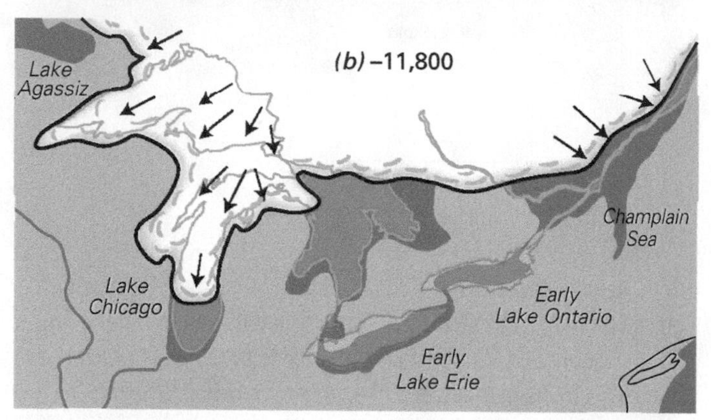

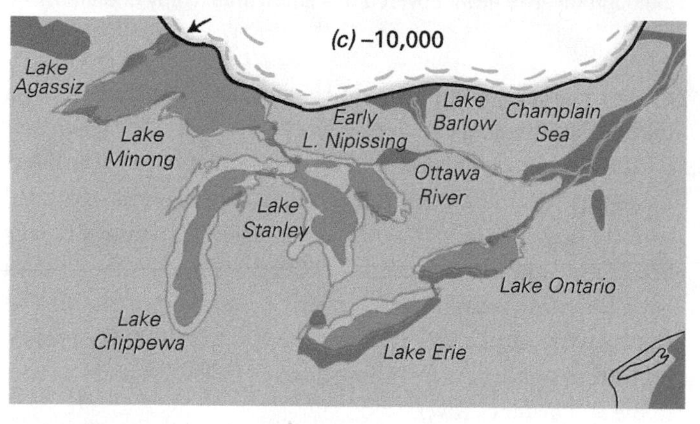

18.20 EMERGENCE OF THE GREAT LAKES Three stages in the evolution of the Great Lakes as the continental ice sheets retreated at the end of the last glacial period.

Earlier periods of extensive glaciation are recorded in Earth's geologic record, although much of the evidence has been lost through subsequent tectonic activity and erosion and deposition. Widespread glaciation in southern Africa, South America, India, Australia, and Antarctica occurred at the end of the Paleozoic Era some 250 million years ago. The distribution of these Permian glaciation features has provided information on continental drift (see Chapter 12). Glacial episodes have also been identified in South America from the Middle Paleozoic (350 million years ago), and in Africa from the Early Paleozoic (500 million years ago). Various regions in the northern hemisphere show glacial features dating from the end of Precambrian, about 600 million years ago. Ancient glaciations dated at 850 million, 1.2 billion, and 2.2 billion years have been noted in India, Australia, Africa, and South America.

CAUSES OF THE LATE-CENOZOIC ICE AGE

Three principal causes are postulated for the Ice Age cycle of glaciations and interglaciations. First is a change in the relative positions of continents on the Earth's surface as a result of plate tectonic activity. Second is an increase in the number and severity of volcanic eruptions. Third is a reduction in the sun's energy output and other aspects of the radiation budget, based on cyclical changes in Earth's orbit and collectively known as the *astronomical hypothesis*.

About 250 million years ago in the Permian period, only the northern tip of Eurasia projected into the polar zone. However, as the Atlantic Basin opened up, North America moved westward and northward to a position opposite Eurasia, with Greenland between them (see Figure 12.28). These plate motions brought enormous landmasses into the high northern latitudes, which began to enclose the polar ocean. It is postulated that this would have greatly reduced the flow of warm ocean currents into the polar ocean and favour the growth of ice sheets. Since the polar ocean was ice covered much of the time, average air temperatures in high latitudes would have decreased enough to allow ice sheets to grow on the surrounding continents. In addition, Antarctica moved southward during the breakup of Pangaea to a position over the South Pole, where it was ideally situated to develop a large ice sheet. The uplift of the Himalayan Plateau might also have modified weather patterns sufficiently to trigger the Ice Age.

The second theory suggests volcanic activity increased in the Late-Cenozoic. Volcanic eruptions produce dust veils that linger in the stratosphere and reduce the intensity of solar radiation reaching the ground (see Chapter 3). Temporary cooling of near-surface air temperatures follows such eruptions. Although the geologic record shows periods of high levels of volcanic activity in the Miocene and Pliocene epochs (from 23 to 2 million years ago), their role in initiating the Ice Age has not been convincingly demonstrated on the basis of current evidence.

CAUSES OF GLACIATION CYCLES

Several timing and triggering theories have been proposed for the glacial cycles of the Ice Age. The most widely accepted of these is the **astronomical hypothesis**, which is based on well-established motions of the Earth in its elliptical orbit around the sun (see Chapter 3). Perihelion, the point in the orbit where the Earth is nearest the sun, presently occurs about January 3; aphelion, when the Earth is farthest from the sun, occurs about July 4. Astronomers have observed that the orbit slowly rotates on a

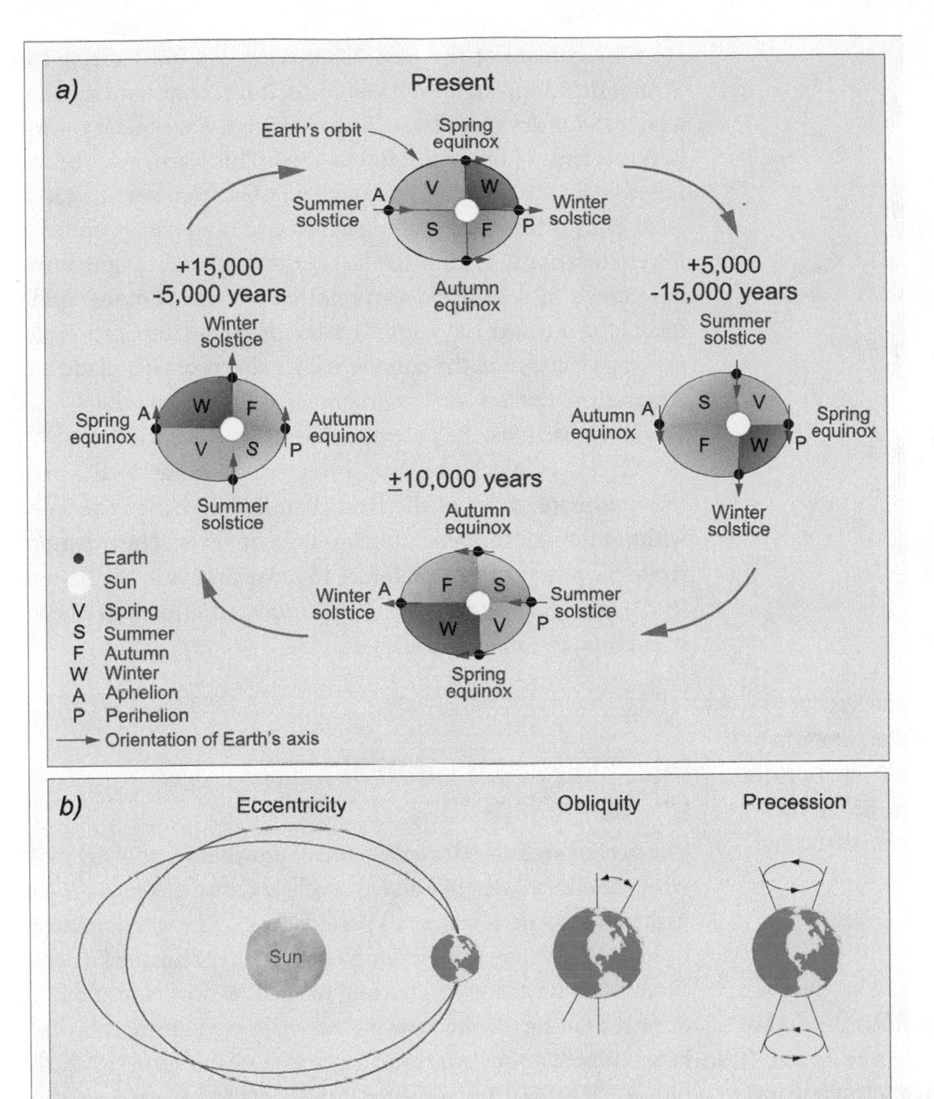

18.21 ASTRONOMICAL FACTORS AFFECTING EARTH'S ORBIT (a) Apsidal precession arises because Earth's orbit slowly rotates around the sun due to the gravitational forces exerted on the planet. This changes Earth's distance from the sun. For example, perihelion presently coincides approximately with the winter solstice, but in 5,000 years the spring equinox will occur at perihelion, followed in turn by the summer solstice (10,000 years) and autumn equinox (15,000 years). (b) Eccentricity, obliquity, and axial precession refer to cyclical changes with respect to the shape of Earth's orbit, axial tilt, and orientation of the polar axis. (From *Introduction to Weather and Climate* by O.W. Archibold, John Wiley & Sons Canada, Ltd., 2011.)

108,000-year cycle in response to changing gravitational forces exerted by the sun and moon, and the planets. This phenomenon, known as *apsidal precession*, shifts the absolute time of perihelion and aphelion by a small amount each year. At present, the northern hemisphere winter occurs during perihelion when the Earth's orbital speed is greatest; this effectively shortens the length of the northern hemisphere winter season by about 4.5 days compared to summer. Over the next several millenia, the northern hemisphere winters will gradually become longer because the orbital position of the winter solstice will no longer occur near the time of perihelion (Figure 18.21a). However, any cooling trend could be offset by several other changes that are known to occur in Earth's orbital properties.

The shape of Earth's orbit varies on cycles ranging from 92,000 to 400,000 years, becoming more and less elliptical over time. The varying shape of Earth's orbit is referred to as **eccentricity** (Figure 18.21b). Eccentricity changes the distance between the Earth and the sun, and will affect the amount of solar energy the Earth receives at each point of the annual cycle. The Earth's axis of rotation also experiences cyclic motions. The tilt of the axis or **obliquity** varies from about 22 to 24 degrees over a 41,000-year cycle. Obliquity affects seasonal insolation by changing the angle at which solar rays strike the Earth's surface. In addition, Earth's axis slowly progresses through a circular path that changes its orientation over cycles of 22,000 to 24,000 years. At present, the axis points to Polaris, the North Star, but in 13,000 years, Vega in the constellation of Lyra will be approximately at the north celestial pole. This change in the direction of the axis in space is called **precession**, or *axial precession* to distinguish it from apsidal precession.

As a result of these cycles in axial orientation and solar revolution, the annual insolation experienced at each latitude changes from year to year. The combined effect, standardized in terms of insolation received at the time, of the summer solstice at lat. 65° N and extended back through time, is called the Milankovitch curve (Figure 18.22). The dominant cycle of the curve has a period of about 40,000 years; however, every second or third peak seems to be higher. Peaks at about 12,000, 130,000, 220,000, 285,000, and 380,000 years ago have been associated with rapid melting of ice sheets and the onset of deglaciations, as revealed by other dating methods involving ancient ice cores and deep-lake sediment cores. It is now generally agreed that cyclic changes in insolation, operating through complex interactions among the atmosphere, ocean, and continental surfaces, were the primary cause of the glaciation cycles during the Ice Age.

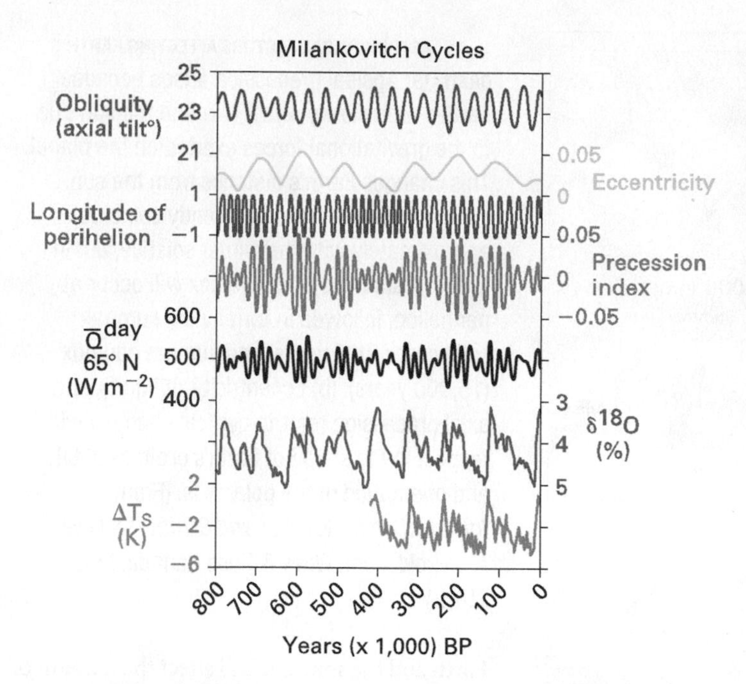

18.22 THE MILANKOVITCH CURVE The Milankovitch curve shows the combined effects of cyclical variations in Earth's orbital parameters on the receipt of solar energy, as indicated by the solar forcing curve showing fluctuations in summer daily insolation at lat. 65° N. The curve is named after Milutin Milankovitch, the astronomer who first calculated it in 1938.

HOLOCENE ENVIRONMENTS

The elapsed time span of about 10,000 years since the end of the Wisconsinan Glaciation is called the *Holocene Epoch.* It began with a rapid warming of ocean surface temperatures. Continental climate zones then quickly shifted poleward, and plants soon re-established in glaciated areas.

Three major climatic periods occurred during the Holocene Epoch leading up to the last 2,000 years. These periods are inferred from studies of changes in vegetation observed in fossil pollen and spores preserved in post-glacial bogs. The Boreal stage that followed the initial ice retreat was characterized by pine, spruce, and fir, species that are now associated with the boreal forests in midlatitude regions. In North America, the cool Boreal stage persisted from about 9,000 to 7,500 years BP. A general warming accompanied by moister conditions gave rise to the Atlantic stage, noted for temperatures somewhat warmer than they are today, which peaked about 8,000 years BP. The vegetation in the Atlantic stage was dominated by deciduous forest species, such as oak and beech. With the onset of drier conditions, the Atlantic stage was replaced by the Sub-boreal stage about 4,500 years BP. The transition to the present Sub-Atlantic stage occurred about 2,000 years BP.

The climate of the past 2,000 years can be described in greater detail through available historical records and detailed analysis of proxy records, such as tree rings. A secondary warm period occurred between 1100 to 1200. This warm episode was followed by the Little Ice Age, which lasted from about 1450 to 1850. During this time, valley glaciers advanced and extended to lower elevations. Within the last century, global temperatures have slowly increased and are projected to increase more rapidly for at least the next 100 years. This has profound implications for existing glaciers and the continental ice sheets of Greenland and Antarctica. *Eye on the Environment • Ice Sheets and Global Warming* discusses the potential impact on these ice sheets.

Cycles of glaciation and interglaciation, as well as the lesser climate cycles of the Holocene Epoch, have proceeded without human influence for millions of years. They demonstrate the power of natural forces to cause dramatic oscillations in climatic conditions, and they confound efforts to understand human impact on climate.

PROCESSES AND LANDFORMS OF THE PERMAFROST REGIONS

The treeless arctic tundra environment provides some insight into what conditions were like across much of Canada toward the end of the Pleistocene. During this period, the climate was dominated by long, cold winters and ground temperatures remained below 0 °C throughout the year. Thawing to shallow depths would have occurred during the short, warm season, leaving the moist surface layer vulnerable to mass wasting and water erosion. With the return of freezing temperatures, ice formation exerted a strong mechanical influence, creating a distinctive set of landforms and landform-creating processes known collectively as the *periglacial system.* The periglacial system is governed by the flow of energy into and out of the ground's surface layer, which seasonally changes water from ice to liquid then back to ice. The expansion and contraction of water as it changes state, coupled with the pressure that ice crystals can exert as they grow, provides the mechanism for the movement of mineral particles that characterizes periglacial processes.

Ground with temperatures perennially below 0 °C is called **permafrost**, while the ice that is commonly present in pore spaces or as lenses is known as **ground ice**. A distinctive feature of permafrost terrain is the shallow surface layer, called the *active layer*, which freezes and thaws each year. This layer ranges in thickness from about 15 cm to 4 m, depending on latitude and the nature of the terrain. Below the active layer is the *permafrost table*, the upper surface of the perennially frozen zone. Water stays within the active layer because it cannot drain into the underlying permafrost. However, water can move laterally,

EYE ON THE ENVIRONMENT

ICE SHEETS AND GLOBAL WARMING

The Antarctic Ice Sheet holds 91 percent of the world's ice, and if it were to melt entirely, mean sea level would rise by about 40 m. The melting of the Greenland Ice Sheet, which holds most of the remaining volume of land ice, would release additional water. Although global warming will accelerate glacial melting and thinning of the ice at the edges of ice sheets, increased precipitation, held by warmer air over the ice sheets, should produce a net growth in ice sheet thickness. Thus, unless the warming is much greater than anticipated, it seems most unlikely that global warming will melt these ice sheets completely. However, other factors may be at work.

In 1993 and 1998, NASA researchers measured the height of the southern half of the Greenland Ice Sheet using an aircraft-mounted laser altimeter. They discovered that marginal thinning had reduced the volume of the ice sheet by about 42 km^3 in that five-year period. The thinning was primarily on the eastern side of the ice sheet. On the western side, melting was more in balance with new snow accumulation. Although the present loss of this ice volume is not large enough to have much effect on global sea level, the substantial flow of fresh meltwater into the ocean could influence the thermohaline circulation of the northern Atlantic (see Chapter 5), affecting temperatures in Europe.

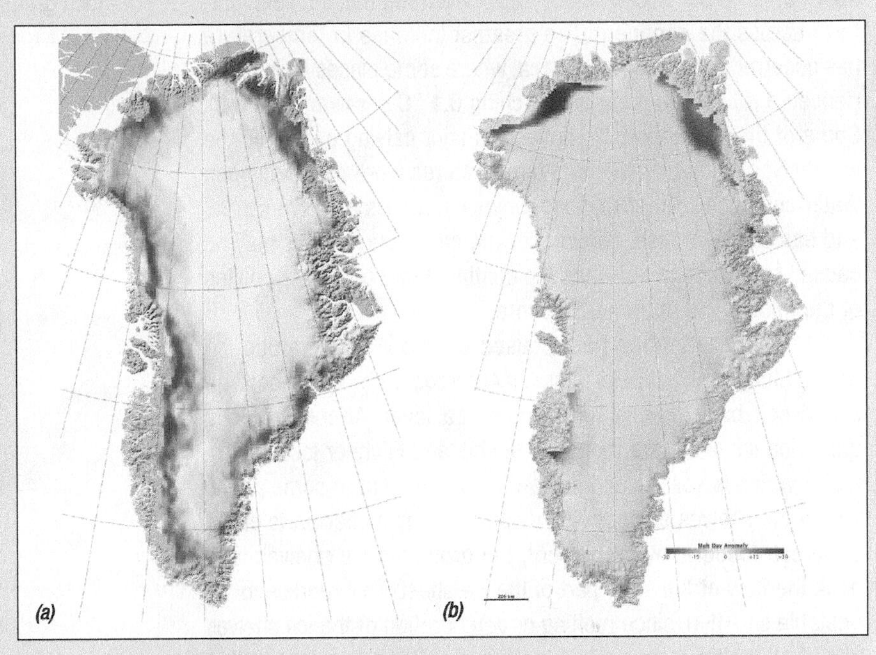

Figure 1 Ice melt in Greenland The number of days when melting occurred on the Greenland Ice Sheet in 2007 (a) and 2008 (b) compared to the average number of melt days for the preceding 38-year period. This image was made with data from the Special Sensor Microwave Imager (SSM/I) on board the F13 satellite of the U.S. Defense Meteorological Satellite Program (DMSP).

Recent studies using satellite data have shown that the summer melting periods in many places in Greenland are now longer than average based on the preceding 38 years (Figure 1). Temperatures at nearby ground-based weather stations were correspondingly high. The average temperature between June and August 2008 was as much as 3 °C above average, with new record temperatures reported at many ground-based weather stations.

Comparable studies in Antarctica show melting is primarily confined to coastal areas and typically occurs over a 50-day period (Figure 2). However, long-term records show that areas of persistent ice melt have slowly increased in recent years, especially on

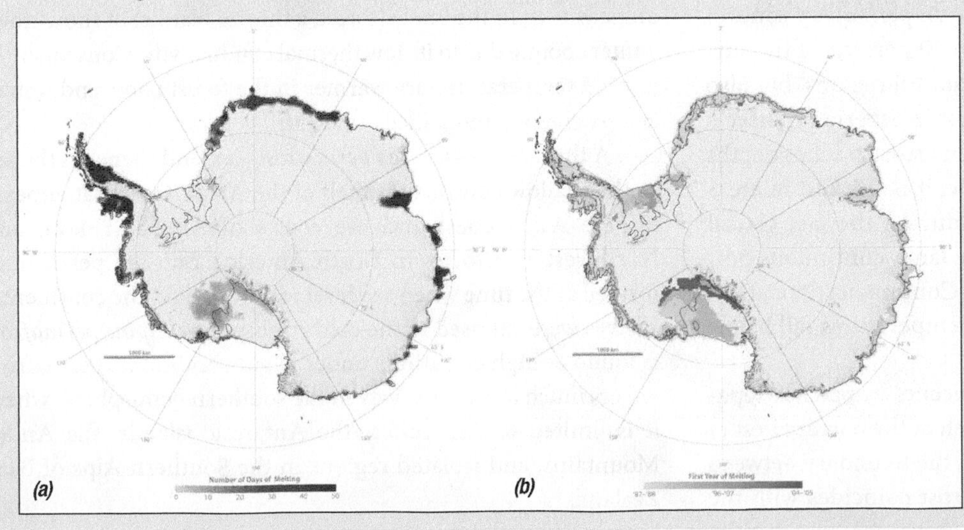

Figure 2 Ice melt data for Antarctica (a) Satellite imagery shows the number of Antarctic melting days for the 2004–2005 season. Areas where melting occurred for a greater number of days are indicated by the darker purple tones. (b) Observed dates of persistent melting detected from 1987 to 2006. Progressively later dates of persistent melting are shown in darker shades of green and blue.

the Ross Ice Shelf. Recent analysis of satellite and weather station data indicate that Antarctica has warmed on average at a rate of about 0.12 °C per decade since 1957. Warming has not been uniform across the continent. The greatest increase in temperature has occurred in western Antarctica, where some places have experienced a rate of warming approaching 0.1 °C per year (Figure 3). Some of the change can be linked to major calving events on the ice shelves so that subsequent temperatures were based on open water conditions rather than ice. However, across much of central and eastern Antarctica, general cooling has occurred; this may be caused by the ozone hole over the continent that facilitates chilling of the middle and upper atmosphere.

Recently, attention has focused on the West Antarctic Ice Sheet. Much of this vast expanse of Antarctic ice is "grounded" on a bedrock base that is well below sea level. Attached to the grounded ice sheet are the Ross, Ronne, and Filchner ice shelves, under which is sea water. They are also grounded at some points where the shelves enclose islands and overlay higher parts of the undersea topography. At present, the grounded ice shelves hold back the flow of the main part of the ice sheet. This represents an unstable situation, since melting or deterioration of the ice shelves would allow the main part of the ice sheet to advance and thin rapidly.

There is evidence to suggest that there has been at least one collapse of the West Antarctic Ice Sheet during a previous interglacial period. Sediments extracted from below the ice sheet show the presence of marine organisms that indicate freely circulating ocean water at some time within the last 2 million years. Moreover, temperatures at the time of the collapse are believed to have been not much greater than those today.

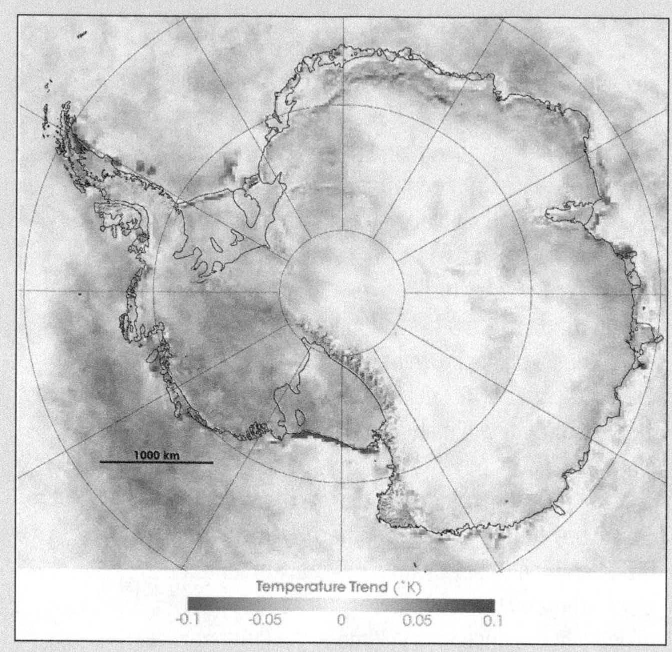

Figure 3 Temperature trends in Antarctica for 1981–2007 The data represent AVHRR thermal infrared observations of surface temperatures in the upper millimetre of the land, sea ice, or sea surface. Blue indicates cooling trends and red indicates warming trends.

especially on a slope. In this way, slight depressions accumulate water that cannot escape by sinking downward.

Four permafrost zones are recognized in the northern hemisphere (Figure 18.23). The *continuous permafrost* zone, in which permafrost underlies more than 90 percent of the surface, coincides largely with the arctic tundra regions, but also includes a large part of the boreal forest in Siberia. Transects through Alaska and Asia show that permafrost reaches depths of roughly 1,000 m near the North Pole; it is thickest in areas that were not covered by ice sheets during the last glacial advance. Where ice sheets covered the land, continuous permafrost reaches depths of 300 to 450 m. Continuous permafrost is found only where mean annual air temperatures fall below −6 to −8 °C.

Discontinuous permafrost, which occurs in patches separated by unfrozen ground, occupies much of the boreal forest of North America and Eurasia. In general, the boundary between continuous and discontinuous permafrost coincides with the tree line that separates boreal forest from tundra. In the tundra zone, a thin cover of blowing snow leaves much of the ground bare or nearly bare, allowing soil heat to escape during the long arctic winter. In the boreal forest, a thicker blanket of snow slows winter cooling due to its low thermal conductivity. Consequently, ground temperatures are warmer in the forest zone, and unfrozen patches are more likely to occur.

A third zone—*sub-sea permafrost*—extends beneath the sea in the shallow continental shelf of the Arctic Ocean. It appears off the Asian coast and the coasts of Alaska, Yukon, and Northwest Territories in North America. Sub-sea permafrost formed at the time when sea level was lower and the continental shelves were exposed to the cold atmosphere. *Alpine permafrost* is found at high elevations under frigid conditions. Permafrost occurs much less extensively in the southern hemisphere, where it is limited to Antarctica, the Antarctic islands, the Andes Mountains, and isolated regions in the Southern Alps of New Zealand.

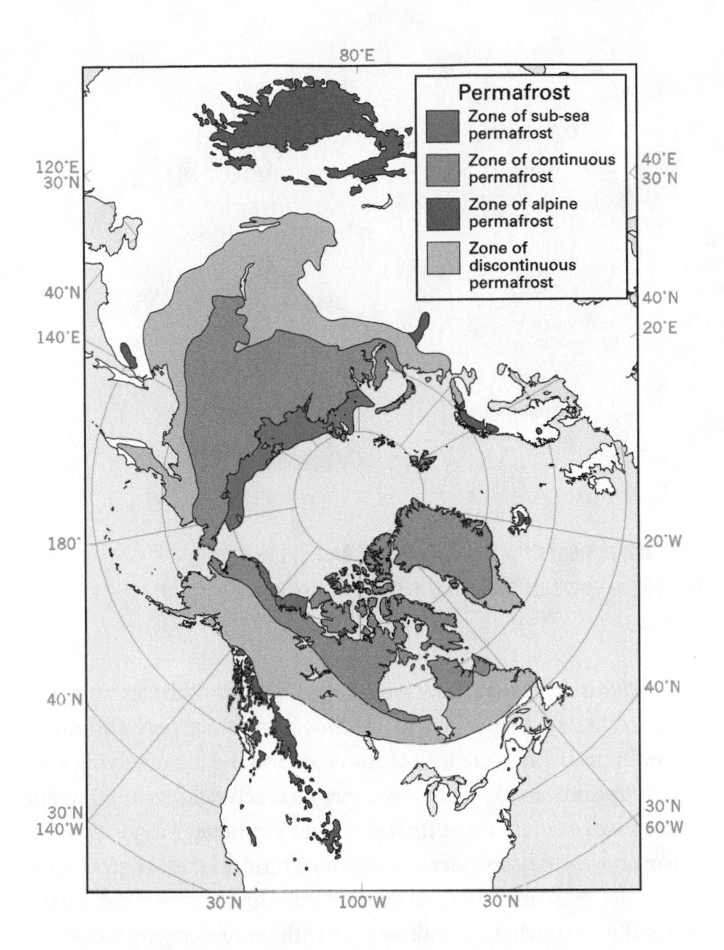

18.23 DISTRIBUTION OF PERMAFROST IN THE NORTHERN HEMISPHERE The general distribution of permafrost shows continuous permafrost in the northernmost regions, with bands of widespread discontinuous permafrost to the south. Note the effect of elevation, which extends permafrost southward in mountainous regions as alpine permafrost. Sub-sea permafrost formed along arctic coastlines where land was exposed by the lowered sea level during the Ice Age.

Whereas snow keeps the ground warmer, the latent heat properties of water (see Chapter 6) tend to keep the ground colder. In the warm season, when ice in the active layer melts, large amounts of heat are taken up during thawing. This heat comes from the atmosphere and is not available to raise ground temperatures. Evaporation of water from the surface of the soil, which occurs as long as the atmosphere is not saturated and the soil surface is moist, will also use heat that could otherwise warm the ground below. In addition, ice conducts heat about four times more rapidly than liquid water. Thus, in the winter, heat escapes more readily from a surface layer of ground saturated with ice than it enters in summer through water-saturated soil. In this way, water in the soil acts like a one-way thermal filter to let heat out of the ground in the winter and block the entry of heat in the summer. The net effect of water's latent heat and thermal properties is that wet environments have colder ground temperatures than dry environments and are thus more likely to have permafrost below.

Even in the areas of deepest continuous permafrost there are isolated pockets of ground, referred to as taliks, where temperatures never fall below the freezing point. Some taliks occur within permafrost as lenses of persistent unfrozen water. Others are large features underlying lakes and rivers where water temperature remains above 0 °C. The depth of the talik under a lake depends primarily on the size of the lake. Under a small lake, the talik may be bowl-shaped and extend downward some tens of metres. Under a large lake, the unfrozen zone may extend below the base of the adjacent permafrost.

FORMS OF GROUND ICE

When the active layer thaws, gravity causes water to accumulate at the bottom of the layer. As air temperature drops at the end of the warm season, freezing moves downward from the ground surface, trapping water at the top of the permafrost. In many cases, freezing also proceeds from the bottom up as the heat is slowly conducted to the cold permafrost below. This results in the formation of lenses of nearly pure ice at the base of the active layer.

Ice lenses are more or less horizontal layers of ice that form as the active layer freezes again at the end of the warm season. Freezing occurs as the *freezing front*—the location at which liquid water is changing to ice—moves downward into the soil and reaches the *capillary fringe*. Here, water is drawn upward from the saturated zone at the base of the active layer through tiny connected pores and pathways between mineral particles. As capillary water approaches the freezing front, it loses heat, becomes supercooled, and quickly freezes to the growing ice mass. Particles of sediment are pushed aside, allowing an ice lens to form. As the freezing front descends into the active layer, the ice lenses usually form bands, separated by soil layers. This *segregated ice* is the most common form of ground ice in seasonally frozen ground and is primarily responsible for *frost heave* of soils.

Another common type of ground ice is the **ice wedge** formed where ice accumulates in deep cracks in the sediment. The cracks develop when permafrost contracts during the extreme winter cold. Surface meltwater enters the crack during the spring, moves downward into the permafrost, and freezes there (Figure 18.24). As the active layer thaws, the crack is covered with surface material, but the new ice is preserved in the permafrost. In the following winter, the crack may reopen, repeating the cycle. Further cracking and filling with ice over successive seasons causes the ice wedge to thicken. Ice wedges can grow up to 3 m wide and 5 m deep. The extra volume created by an ice wedge displaces the ground upward to form low ridges on either side, while a trough forms between the ridges, above the wedge. Because the cracking relieves strain across a large area, the cracks

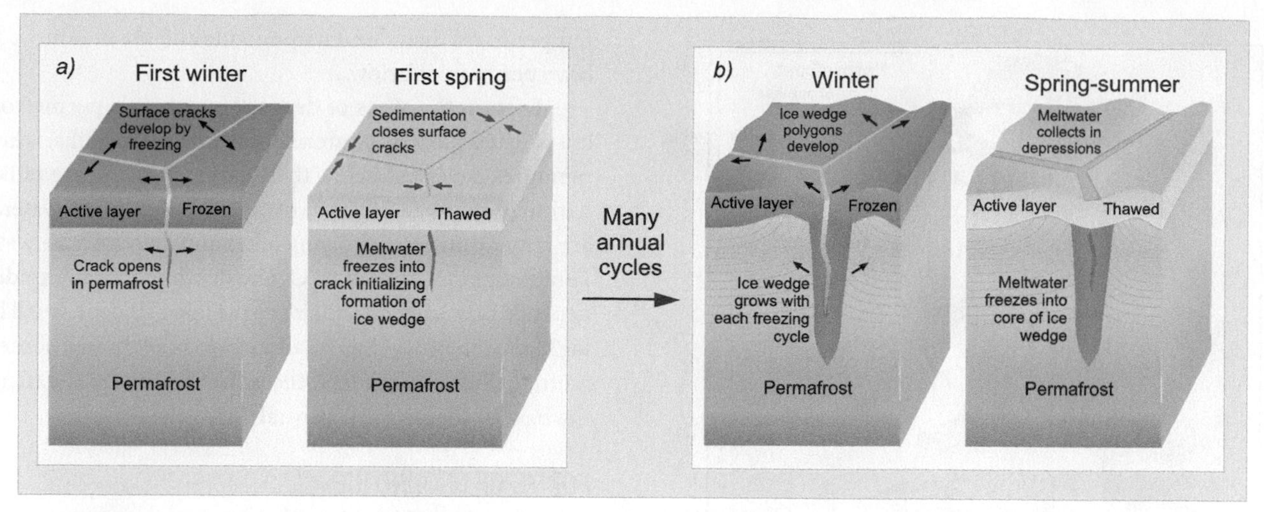

18.24 FORMATION OF AN ICE WEDGE (a) In the beginning stage, an open crack appears during the winter, and in the spring it fills with meltwater that immediately freezes and expands. (b) The ice wedge grows as the cycle is repeated over numerous seasons with water often filling the intervening depressions.

and wedges are often interconnected in a network of surface troughs, called *ice-wedge polygons* (Figure 18.25). Ice wedge polygons are generally 10 to 15 m wide, and are mainly associated with regions of continuous permafrost.

18.25 ICE-WEDGE POLYGONS These polygons in the Alaskan north slope, near the border of Alaska and Yukon, were formed by the growth of ice wedges.

Pingos are conspicuous ice-formed features of the arctic tundra (Figure 18.26). These conical mounds have an ice core and slowly grow in height as more ice accumulates, forcing the overlying sediment upward. In extreme cases, pingos reach heights of 50 m and have bases that are more than 600 m in diameter. Pingos typically form above sandy sediments on drained tundra lakes, as groundwater under pressure is pushed upward into the freezing front. Pingos typically form where a shallow lake with an underlying wet sandy talik is subsequently drained (Figure 18.27). Without its insulating cover of water, the former lake bottom begins to freeze and permafrost penetrates downward. As the water in the talik slowly converts

18.26 IBYUK PINGO Located on the Arctic coast near Tuktoyaktuk, Northwest Territories, this classic pingo began to form about 1,200 years ago in a drained lake bed. Since that time, coastline retreat has flooded the former lake basin, and it is now connected to Kugmallit Bay of the Beaufort Sea.

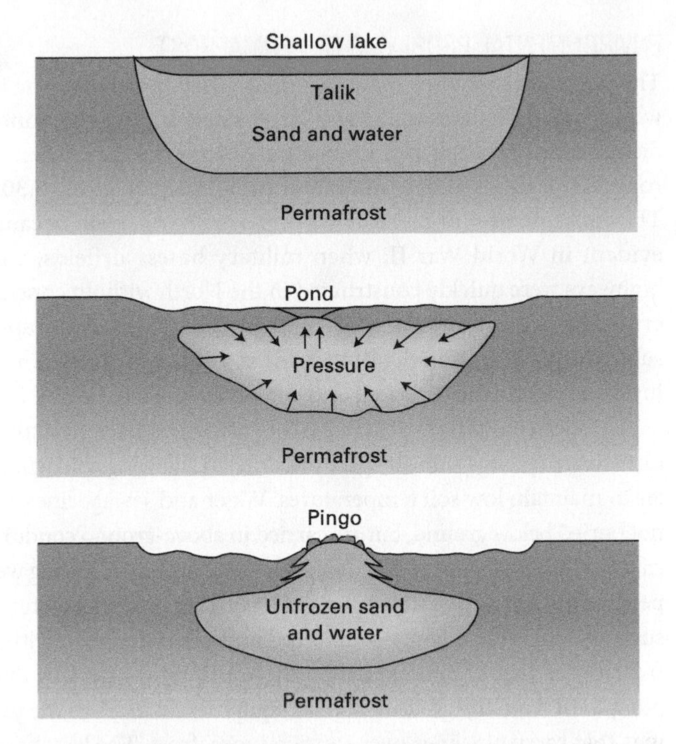

18.27 PINGO FORMATION The draining of a shallow lake underlain by wet, sandy sediments is the first step in the formation of a pingo. As the freezing of the lake bottom moves downward, expansion of ice places pressure on the unfrozen talik. The pressure causes water to well upward at a weak point in the permafrost layer, where it freezes into a pingo. (After A. L. Washburn.)

to ice, it expands, placing pressure on the remaining water. The pressure pushes water upward at a weak point in the upper permafrost to form an ice dome that grows into a pingo as permafrost slowly engulfs the talik.

THERMOKARST LAKES

Many areas of ice-rich permafrost are marked with shallow lakes formed by thawing of permafrost. These features are called *thermokarst lakes*. Topographic relief in thermokarst areas develops by melting of ground ice and subsequent settling of the ground. The majority of thermokarst lakes are less than 10 m deep and rarely more than 2 km wide. Thermokarst lakes typically begin to form after a chance event causes localized thawing, usually because the insulating vegetation cover has been disturbed. This allows the active layer to deepen and the ice-rich upper portion of the permafrost melts, reducing the volume of the previously frozen layer. A small water-filled depression forms as the ground settles. As thawing and subsidence proceed, the lake becomes deep enough to maintain a bottom temperature of 0 °C or higher year round, and a talik begins to form. The growth of thermokarst lakes eventually ceases when the supply of ice-rich permafrost is exhausted.

RETROGRESSIVE THAW SLUMPS

A unique form of mass wasting in permafrost terrain is the *retrogressive thaw slump*, which may develop where erosion exposes the ice-rich upper layer of permafrost or massive icy beds. These slumps typically occur along eroding river banks and lake and ocean shorelines (see Figure 17A.7). Melting of the ground ice and growth of the slump can be rapid during the thaw season. The headwalls of thaw slumps migrate at rates of up to 16 m per year, and can reach lengths of over 650 m from the point of initiation to the headwall. As such, they are the most rapidly developing geomorphic features in the North. In most cases, sediment ultimately buries the melting ice front, stabilizing the slump. However, coastal thaw slumps can be reinvigorated by wave action during storms, by re-exposing ice-rich permafrost. Heavy rains can similarly wash sediment from the ice face and renew rapid retrogression.

PATTERNED GROUND AND SOLIFLUCTION

The growth of ice lenses usually causes stones to move upward or sideways within the soil profile through a process of *cryoturbation*. In arctic environments, cryoturbation can produce regular surface forms, such as circles, polygons, and nets of stones, or fields of low mounds of soil (Figure 18.28). These features are collectively known as **patterned ground**. Circles, polygons, and nets are common features of permafrost terrain on flat or gentle slopes, but often become elongated into stripes on steeper slopes.

A common type of patterned ground in permafrost regions is the *mud hummock*, a low, rounded dome of soil found where fine-grained sediments overlie permafrost. Hummocks form as

18.28 STONE CIRCLES Sorted circles of gravel abound on this low, wet plain, underlain by permafrost in Broggerhalvoya, Spitsbergen, latitude 78° N. The circles in the foreground are 3 to 4 m across; the gravel ridges are 20 to 30 cm high.

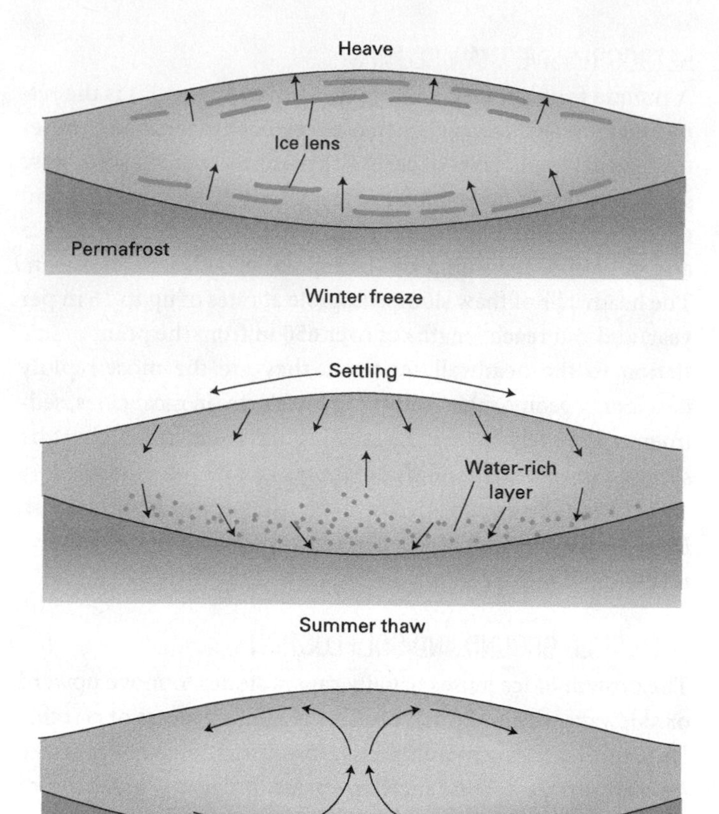

18.29 FORMATION OF MUD HUMMOCKS In the winter, the formation of ice lenses toward the top and bottom of the hummock causes an upward heaving motion. With the summer thaw, outward settling occurs at the top, and a water-rich layer of mud allows sediment movement inward and upward at the bottom of the hummock. (After J. R. Mackay.)

ice lenses develop in both the upper and lower parts of the active layer in winter, causing the top of the hummock to heave upward and outward (Figure 18.29). As the summer thaw sets in, the heaved soil settles back, leaving a small mound, and later, as the ice lenses at the permafrost table melt, a water-rich layer of soft mud forms. The pressure of overlying sediments moves the mud toward the bottom of the hummock floor, where it is pushed upward in the centre of the mound. The resulting circulation moves material inward along the bottom of the active layer, upward in the centre, and outward at the top.

A special type of earth flow characteristic of cold regions is **solifluction**. This occurs in late summer, when the ice-rich layer at the bottom of the saturated active layer melts to form mud. The active layer above then moves downslope as a single mass, typically at a rate of a few millimetres per year. Movement is facilitated by lubrication from the water-rich layer of sediment at the top of the permafrost. The process creates *solifluction terraces* and *solifluction lobes* that give tundra slopes a stepped appearance.

ENVIRONMENTAL PROBLEMS OF PERMAFROST

The permafrost environment is in many ways a delicate one in which small changes can have large impacts. Even a minor disturbance of the ground surface can change the thermal environment, leading to degradation of permafrost (Figure 18.30). The consequences of disturbance of permafrost terrain became evident in World War II, when military bases, airfields, and highways were quickly constructed in the North without considering the protection of the permafrost. Thawing led to rapid subsidence of the ground, tilting roads and runways, and swamping buildings as they cracked and settled irregularly.

Modern building practices in the North now place structures on insulated pilings with air space underneath, allowing cold winter air to maintain low soil temperatures. Water and sewage lines are not buried below ground, but are carried in above-ground conduits called utilidors (Figure 18.31). Roadways and runways crossing wet permafrost are built on thick pads of insulating gravel. Structures such as dams are carefully engineered and often designed with a frozen core that is maintained by thermosiphons—devices that pump heat from the ground. Similar equipment is used on the pilings that carry pipelines over ice-rich permafrost. The lessons of superimposing technology on a highly sensitive natural environment were learned the hard way—by encountering undesirable and costly effects that were not anticipated.

Climate change is expected to bring warmer temperatures and increased precipitation to much of the polar regions (see Chapter 10). Substantial warming has already occurred, and predictions are that by the end of this century, winter temperatures will have increased by 3 to 10 °C, with annual precipitation increasing by as much as 25 percent.

It seems certain that warming temperatures and increasing precipitation, especially in the form of snow, will deepen the active

18.30 DEGRADATION OF PERMAFROST Disturbance of vegetation and peat by vehicles reduces soil insulation and can quickly produce thermokarst subsidence.

18.31 ULTIDORS Utilidors are used for water, electricity, communication cables, and sewage lines in areas where burial is not feasible because of permafrost.

layer over broad areas of continuous permafrost. This is likely to be accompanied by extensive development of thermokarst terrain, where the ice-rich upper layers of permafrost melt and subside. Increasing mass movement is expected to occur on hill slopes, disrupting roads, pipelines, and other infrastructure. Sea-level rise along the shores of the Arctic Ocean will add to the number of retrogressive thaw slumps. The flow of rivers to the Arctic Ocean will decrease due to higher summer evaporation. Subsistence hunting by native peoples will become more difficult as populations of mammals, fish, and birds fluctuate and change their distribution. The winter overland and over-ice travel season will become shorter.

The border between continuous and discontinuous permafrost will also move northward in response to heavier snowfalls that will warm the ground under newly established forest. Discontinuous permafrost will decrease at the southern boundary, and the present zone of isolated permafrost will migrate to higher latitudes.

SEA ICE AND ICEBERGS

Free-floating ice on the sea surface is of two general types: *sea ice* and *icebergs*. Sea ice is formed by direct freezing of ocean water, which due to salinity, occurs at about −1.8 °C. Sea ice that has frozen along coasts is referred to as *fast ice*. *Pack ice* consists of extensive areas of freely floating sea ice that is detached from land. Wind and currents break up pack ice into smaller patches called *ice floes* (Figure 18.32). The narrow strips of open water between these floes are known as *leads*. Where winds forcibly bring ice floes together, the ice margins buckle and turn upward into pressure ridges. Sea ice conditions are reported daily by Environment Canada's Ice Service. Ice analysis charts provide an estimate of actual ice conditions based on data derived from polar orbiting NOAA satellites and RADARSAT imagery. Different ice conditions are described using the Canadian-designed International Standard Ice Code.

Icebergs are bodies of land ice that have broken away from glaciers and ice shelves. Because they are composed of glacial ice, icebergs are differentiated from sea ice. Being slightly less dense than sea water, icebergs float low in the water so that about 80 percent of their bulk is submerged. A major difference between sea ice and icebergs is thickness. Sea ice does not exceed 5 m, whereas icebergs may be hundreds of metres thick. Most of the icebergs that drift along Canada's coastlines originate from Baffin Island and Greenland, and are eventually carried by the Labrador Current as far south as Newfoundland. The southward drift rate of icebergs increases from about 0.15 kph in more northerly parts of Baffin Bay and Davis Strait, to about 0.6 kph (7.6 nautical miles per day) along the coast of Labrador. At this rate, it takes an average of two to three years to drift the 3,300 km from the west coast of Greenland to the Grand Banks off Newfoundland. The latter part of the route is often referred to as "Iceberg Alley." The size and number of icebergs decreases as they travel southward. This is mainly due to air and water temperatures and the strength and duration of the prevailing winds that affect wave action as well as sublimation rates. The rate of loss increases once icebergs move beyond the zone of ice pack because of warmer water temperatures and increased wave action. The southern limit of drift is generally marked by the northern edge of the warm North Atlantic Drift or Gulf Stream.

Sea ice forms and melts with the polar seasons, and plays an important role in regulating climate. Relatively warm ocean water is insulated from the cold polar atmosphere, except where cracks in the ice allow exchange of heat and water vapour. This affects regional cloud cover and precipitation. In the Arctic, some sea ice persists year after year. In late winter, Arctic sea ice typically covers about 13.5 million km^2 based on the average for 1979 to 2000, but decreases to about 7 million km^2 by the end of summer (see Figure 4.25). Over the past 30 years, sea ice has decreased significantly in the Arctic. The present long-term decline in end-of-summer sea ice is now estimated at

18.32 ICE FLOES A Canadian Coast Guard icebreaker in warmer weather.

approximately 9 percent per decade. The reduction in ice area is particularly noticeable off the coasts of Alaska and Siberia. Sea ice thickness likewise has declined by about 1.3 m over the past 40 to 50 years. Antarctic sea ice develops in the southern hemisphere during winter, and at its maximum extent covers about 18 million km^2 of the Southern Ocean; by summer's end it is reduced to about 4 million km^2. Unlike in the Arctic, the area of sea ice around Antarctica has remained fairly stable.

A Look Ahead

The chapters in Part 4 have reviewed landform-making processes that operate on the surface of the continents. Human influence on landforms is felt most strongly on surfaces affected by fluvial processes due to changes caused by agriculture and urbanization. Landforms shaped by wind and by waves and currents are also highly sensitive to changes induced by human activity. Only continental ice sheets maintain their integrity and are largely undisturbed to date. However, they are increasingly sensitive to climate changes induced by industrial activity. Part 5 discusses soils and vegetation, both of which have been greatly affected by human activity, mainly because of the ever-pressing demand for sufficient food and other agricultural products to sustain the growing human population.

CHAPTER SUMMARY

- Glaciers form when snow accumulates to a great depth, creating a mass of ice that is plastic in its lower layers and flows outward or downhill from a centre in response to gravity. As they move, glaciers can deeply erode bedrock by abrasion and plucking. The eroded fragments, incorporated into the flowing ice, leave depositional landforms when the ice melts.
- Alpine glaciers develop in cirques in high mountain locations. They flow down valleys on steep slopes, picking up rock debris and depositing it in lateral and terminal moraines. Through erosion, glaciers carve distinctive U-shaped glacial troughs in mountainous regions. Glacial troughs will become fjords if later submerged by a rising sea level.
- Ice sheets are extensive accumulations of ice that cover vast areas. They are present today in Greenland and Antarctica. The Antarctic Ice Sheet includes *ice shelves*—great plates of floating glacial ice.
- An ice age includes alternating periods of glaciation, *deglaciation*, and interglaciation. During the past 2 to 3 million years, the Earth has experienced the *Late-Cenozoic Ice Age*. During this ice age, continental ice sheets expanded and melted as many as 30 times. The most recent glaciation was the *Wisconsinan Glaciation*, in which ice sheets covered much of North America. Contemporaneous expansion of ice occurred in Europe, as well as parts of northern Asia and southern South America.
- Moving ice sheets create many types of landforms. Bedrock is grooved and scratched. Where rocks are weak, long valleys can be excavated to depths of hundreds of metres. The melting of glacial ice deposits glacial drift, which may be stratified by subsequent water transport or deposited directly as till.
- *Moraines* accumulate at ice edges. *Outwash plains* are built up by meltwater streams. Tunnels within the ice leave stream-bed deposits, or *eskers*. Till may spread out smoothly and thickly under an ice sheet, leaving a *till plain*. This may be studded with elongated till mounds, called *drumlins*. Many small lakes have developed on the undulating till plains, especially in parts of the Canadian Prairies.
- Meltwater streams built *glacial deltas* into lakes formed at the ice margin and cover lake bottoms with clay and silt. These features remain after the lakes have drained. Level *glaciolacustrine plains* are common features throughout the Canadian Prairies.
- Several factors have been proposed to explain the cause of present ice age glaciations and interglaciations. These factors include ongoing change in the global position of continents, an increase in volcanic activity, and a reduction in the sun's energy output. Individual cycles of glaciation seem strongly related to cyclic changes in Earth's orbit and axial tilt.
- The tundra environment is dominated by the *periglacial system* of distinctive landforms and processes related to freezing and thawing of water in the *active layer* of permafrost terrain. The *active layer* overlies permafrost and thaws each year during the warm season.
- *Taliks* are pockets of unfrozen ground surrounded by permafrost.
- *Thermokarst lakes* typically begin development when the surface cover is disrupted, triggering the melting of ice-rich permafrost. The ground settles to form a water-filled depression that expands by melting permafrost at its edges.
- *Retrogressive thaw slumps* occur where erosion exposes ground ice. *Patterned ground* includes circles, polygons, and nets that cover large areas of tundra and are the result of *cryoturbation*.
- *Solifluction* (soil flowage) occurs in late summer, when the bottom of the active layer thaws, releasing water that lubricates downhill movement of the active layer.
- *Icebergs* form when glacial ice flowing into an ocean breaks into huge masses and floats freely. *Sea ice*, which is much thinner and more continuous, is formed by direct freezing of ocean water.

KEY TERMS

ablation
alpine glacier
astronomical hypothesis
chattermark
cirque
crevasse
drumlin
eccentricity
equilibrium line
esker
fjord
glacial drift
glacial trough
glaciation
ground ice
ice lenses
ice sheet
ice wedge
interglaciation
interstadial
Late-Cenozoic Ice Age (Ice Age)
lateral moraine
obliquity
patterned ground
permafrost
precession
pressure ridge
solifluction
stratified drift
striation
terminal moraine
tidewater glacier
till
varves
zone of ablation
zone of accumulation

REVIEW QUESTIONS

1. How does a glacier form? What factors are important? Why does a glacier move?
2. Distinguish between alpine glaciers, ice sheets, and ice shelves.
3. What is a glacial trough and how does it form? What is its basic shape?
4. Where are ice sheets found today? How thick are they?
5. What areas were covered with ice sheets by the last glaciation? How was sea level affected?
6. What are moraines? How do they form? What types of moraines are there?
7. Identify the landforms and deposits associated with deposition underneath a moving ice sheet.
8. What is the Milankovitch curve? What does it show about warm and cold periods during the last 500,000 years?
9. What does the term *periglacial* mean?
10. Define and describe permafrost and some of its features, including ground ice, active layer, and permafrost table.
11. Identify the zones of permafrost and describe their general location, both globally and within Canada.
12. Identify the factors that affect the temperature of permafrost. Describe how and why each factor creates warmer or cooler ground temperatures.
13. What is a *talik*? How does the size of a talik under a lake depend on the width of the lake?
14. How and when do ice lenses form in the active layer? Where are they found within the active layer and why?
15. What is a *pingo*? How does it form?
16. Surface disturbance in permafrost terrain can trigger the formation of a thermokarst lake. Explain how this process occurs. What limits the growth of a thermokarst lake?
17. What does *solifluction* mean? How and when does it occur, and how does solifluction differ from other mass wasting processes?
18. Contrast sea ice and icebergs, including the processes by which they form.
19. What process seems to be offsetting the melting rate of ice sheets as the climate warms?
20. Why is the West Antarctic Ice Sheet considered unstable? What is the role of ice streams in maintaining the present size of the ice sheet?

VISUALIZATION EXERCISES

1. What are some typical features of an alpine glacier? Sketch a section along the length of an alpine glacier and label it in terms the processes associated with its role as both a matter and energy system.
2. What are some typical features associated with continental glaciation? Draw a landscape that shows these and label it.

ESSAY QUESTIONS

1. Imagine that you are planning a car trip to the Canadian Rockies from Winnipeg. What glacial landforms might you expect to find in the mountains? How do they differ from those you would encounter as you travel there?
2. At some time during the latter part of the Pliocene Epoch, the Earth entered an ice age. Describe the nature of this ice age and the cycles that occurred within it. What are the proposed explanations for the cause of an ice age and its cycles? What cycles have been observed since the last ice sheets retreated?

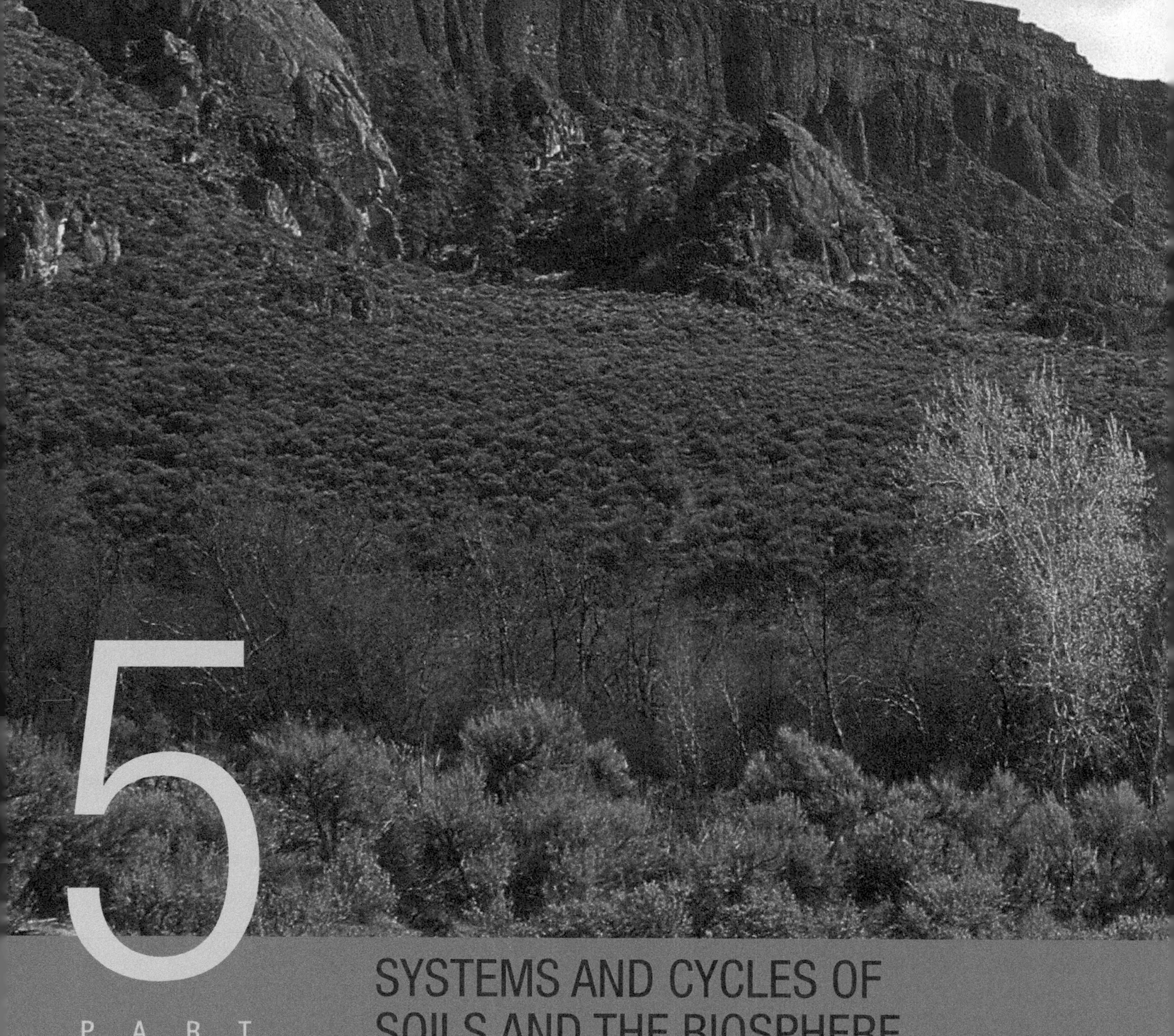

PART 5
SYSTEMS AND CYCLES OF SOILS AND THE BIOSPHERE

Sagebrush dominates the landscape in this dry intermontane environment, with occasional stands of aspen and ponderosa pine where soil conditions are more favourable.

Part 5 focuses directly on the life layer, where biological and physical processes interact. Soils are formed by the physical processes that weather and transport rock material, as well as the biological processes that result in the decay of organic matter. These processes are strongly influenced by temperature and the abundance of water at the surface; hence, they are functions of climate. Climate also describes the availability of sunlight, water, and heat, which affect photosynthesis and thus determine plant growth. Moreover, individual landforms present distinct habitats for the development of local soil types, as well as local biotic communities. Soils and the biosphere provide a logical conclusion to a study of physical geography because these topics relate to many of the systems, cycles, and processes introduced in earlier chapters. ● Chapter 19 provides an overview of the key characteristics of soils, the important factors in soil formation, the classification of soils, and the global distribution of soil types. ● Chapter 20 describes ecosystem processes, focusing on energy and the cycling of major nutrient elements within plant communities and the environment. ● Chapter 21 discusses the biogeographic processes that govern the distribution of plants in space and time. ● Finally, Chapter 22 examines the nature of biomes and their global distribution patterns, especially as they relate to climate and soils.

CHAPTER 19 SOIL SYSTEMS

Horsethief Butte State Park, Washington.

Although it is commonly assumed that soils occur everywhere, large expanses of continents possess a surface layer that cannot be called soil. For example, dunes of moving sand, exposed bedrock, and surfaces of recently solidified lava near active volcanoes have no soil layer. This chapter describes the processes that form soils and give them their distinctive characteristics, together with soil classification, and global distribution of soil types. The many different types of soils reflect the high variability of factors and processes that influence soil development over time.

THE NATURE OF SOIL

Soil, as the term is used in soil science (or *pedology)*, is the uppermost layer of the land surface that plants use and depend on for nutrients, water, and physical support. In the Canadian system of soil classification, soil is defined as the naturally occurring, unconsolidated mineral or organic material (at least 10 cm thick) that occurs at the Earth's surface and is capable of supporting plant growth. In its broadest sense, the term "naturally occurring" includes materials disturbed by human activities, such as cultivation and logging, but not materials such as gravel dumps and mine spoils. The Canadian definition includes soils submerged to a depth of 60 cm or less either at low tide in coastal areas or during the driest part of the year in inland areas. Dumped materials, such as earth fills and gravels, are defined as *non-soils*.

In addition to *mineral matter*, soils also contain varying amounts of *organic matter*. Organic matter is of biological origin and may be living or dead. Living matter in the soil consists of plant roots and also many kinds of organisms, including microorganisms. Dead and partially decomposed organic matter is an important constituent of soils, and in most cases accounts for about 10 percent of the dry weight. The finest organic debris forms *humus* and is important for soil fertility. Most humus rests on the soil surface, but some is carried deeper into the upper soil layers by percolating water.

As well as solid materials, soils also contain gaseous and liquid components. The *soil atmosphere* consists of oxygen and other gases that diffuse from the air, together with gases, such as CO_2 and methane, that are derived from respiration and decomposition processes. *Soil water* contains dissolved substances, including plant nutrients, in solution. The solid, liquid, and gaseous matter in soil is continuously interacting through chemical and physical processes, which contributes to its dynamic nature.

FACTORS OF SOIL FORMATION

It is generally agreed that soils are a function of five main factors: parent material, topography, climate, biological factors, and time (Figure 19.1). For most soils, the bulk of the material present is derived from disintegration of the underlying rock through a combination of physical and chemical weathering processes. Over time, weathering weakens, disintegrates, and breaks bedrock apart, forming a layer of *regolith*, or residual mineral matter. Alternatively, transported materials, such as glacial tills, river alluvium, and wind-transported loess may be deposited in sufficient thickness to act as parent material. The type of substrate from which the soil is derived constitutes the *parent material*, and has an important influence on the soil's physical and chemical properties. For example, soils formed from granites are often rich in clay. They may be water retentive and quite fertile because of the various minerals released from the weathered feldspars (see Chapter 11). Sandstone, on the other hand, typically forms sandy soils that drain rapidly and are comparatively low in nutrients. In most soils, the inorganic material present consists of fine mineral particles and typically accounts for some 40 to 60 percent of the soil volume.

The effect of *topography* is linked to slope angle and aspect. Gentle slopes are considered ideal for soil development, as drainage is good and the erosion rate can be compensated by the addition of weathered parent material. Surface erosion by overland flow and runoff is rapid where slopes are steep and little water infiltrates. As a result, soils generally are thicker on gentle slopes, but relatively thin on steep slopes where material is more rapidly removed by erosion. Poor drainage in low-lying areas also tends to slow soil development and may lead to *anaerobic* conditions and unusual soil chemistry. In areas of undulating terrain, it is common to find soils becoming progressively wetter downslope. The effect is characteristic of areas of hummocky terrain, such as occurs on the moraine and till deposits of the Canadian Prairies.

The relative balance between infiltration and runoff is a major factor in the formation of a *soil continuum* or *soil catena*, in which soil characteristics change along a topographic gradient (Figure 19.2). In low-lying areas, accumulated drainage water may saturate the soil. Under extreme conditions, this can lead to oxygen deficits in the soil, which can affect chemical and biological processes. Bog soils and mucks formed under such conditions exhibit distinct bluish, greyish, or olive-coloured lenses of sticky clay and an accompanying smell of hydrogen sulphide (H_2S) when disturbed.

Slope aspect affects the soil temperature and water regime. Slopes facing away from the sun are sheltered from direct insolation and tend to have cooler, moister soils. Slopes facing toward the sun are exposed to more intense solar radiation, raising soil temperature and increasing evapotranspiration. In the northern hemisphere, insolation is more direct on south-facing slopes.

Climate, measured by precipitation and temperature, is also an important factor in soil formation. Precipitation controls the downward movement of nutrients and other chemical compounds in soils and provides a medium for many chemical

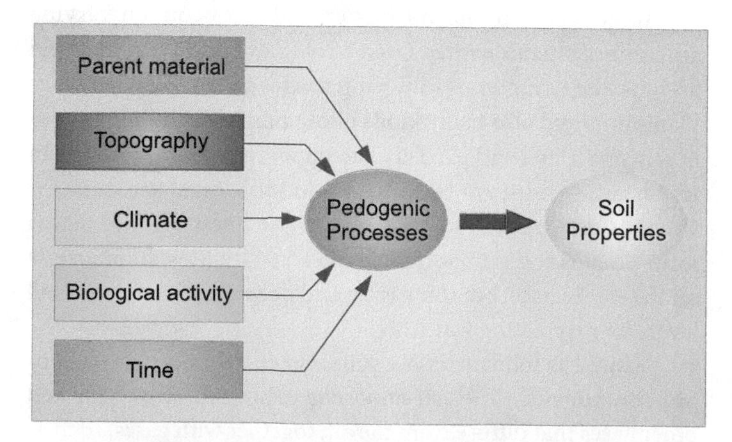

19.1 FACTORS IN SOIL FORMATION Interactions between five principal factors acting simultaneously and at varying rates influence soil formation at global, regional, and local scales.

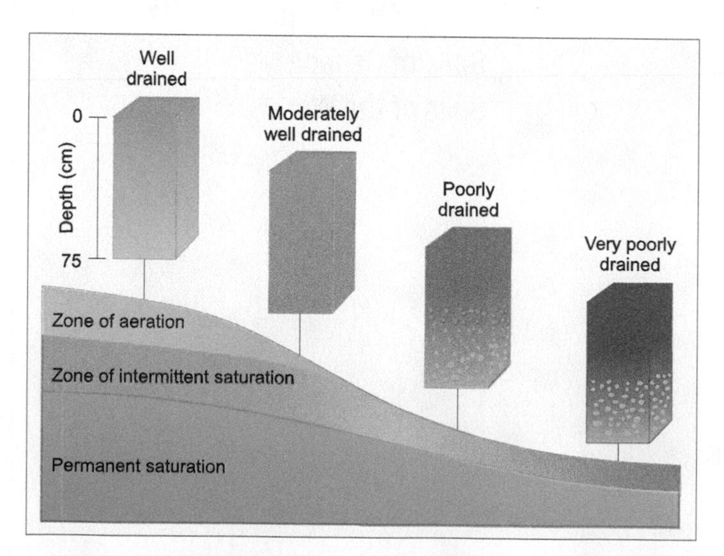

19.2 SOIL CATENA On undulating terrain, soil properties change in relation to topographic position, mainly because of drainage conditions and the influence they have on plant growth and other soil-forming processes. Mottlng, a pattern of light and dark patches or streaks of colour, can develop in some soils under anaerobic conditions.

reactions, such as hydrolysis and solution. Soil temperature influences the intensity of chemical processes affecting soil minerals, and also influences biologic activity. Below 10 °C, chemical activity is slowed, and below 0 °C, it mostly ceases.

Temperature controls the rate of decay of organic matter that falls to the soil from the plant cover or is provided to the soil by the death of roots. When conditions are warm, decay organisms work efficiently, readily consuming organic matter. Thus, organic matter is generally low in soils of the tropical and equatorial zones (Figure 19.3). Under cooler conditions, decay proceeds more slowly and organic matter is more abundant in the soil. Fallen leaves and stems tend to accumulate to form a thick surface layer, which is slowly decomposed and carried deeper into the soil. Vegetation growth is slow or absent in desert areas, so organic matter will be low, regardless of temperature conditions.

Biological activity arising from the presence of living plants and animals, as well as their non-living organic products, are essential for good soil development. Organic matter is present in various stages of decomposition from freshly fallen leaves and twigs, to completely decayed material in the form of humus. Humus is a stable component of the soil and is important for soil fertility because of its capacity to store the nutrients needed for plant growth. Humus helps to bind mineral particles together in ways that allow water and air to penetrate the soil freely, thereby providing a healthy environment for plant roots. The process of decomposition and humification add to the pool of soil organic matter.

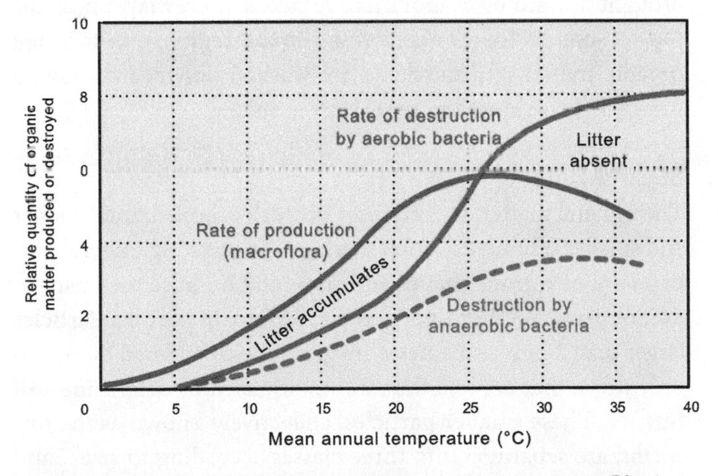

19.3 PRODUCTION AND DECOMPOSITION RATES OF PLANT MATERIALS Plant production increases with temperature to a maximum at about 25 to 30 °C and then declines due to high rates of plant respiration, which accelerate consumption of products formed in photosynthesis. Decomposition typically increases with temperature. The net effect is that litter tends to accumulate on the soil surface in colder climates, but not in warmer climates.

The types of organisms living in the soil range from bacteria to burrowing mammals. The role organisms play in soil formation is extremely important, especially when conditions are warm and moist enough to support large populations. Microscopic organisms, such as bacteria, are not only important for decomposition, but they also affect soil chemistry through processes such as nitrogen cycling. Earthworms continually rework the soil not only by burrowing, but also by passing soil through their intestinal tracts. They ingest large amounts of decaying leaf matter, carrying it down from the surface and incorporating it into the mineral horizons. Many forms of insect larvae perform a similar function. Larger animals, such as gophers, rabbits, and badgers, are also noted for their excavations and soil disturbance.

Many agricultural soils have been cultivated for centuries. As a result, both the structure and composition of these soils have undergone great changes. Agricultural soils are often recognized as distinct soil classes and are given the same status in soil classifications as natural soils.

The characteristics and properties of soils develop only after a long period of time, usually many centuries. The length of time that has elapsed is reflected by the properties of the soil, with younger soils more closely resembling their parent materials. No specific time can be defined for soil development; it varies according to rainfall, temperature, slope, and other factors that act on the parent material. For this reason, a *young soil* may be defined as one that is poorly developed. A *mature soil* is one that is in equilibrium with the forces and processes acting upon it. Because soils are considered to be a function of their environment, an old soil can never exist, as environmental factors and processes are continually acting and changing it. Thus, a soil is considered to be a *dynamic system.*

SOIL PROPERTIES

SOIL COLOUR

The most obvious feature of a soil is its colour. Soils that are black or dark brown in colour are invariably rich in humus; for this reason, the surface layers tend to be darker than the underlying subsoils. The colours of the subsoil are mainly determined by the presence and state of iron oxides. Red or yellow colours are indicative of good drainage and aeration, while greenish-blue colours can indicate very moist, anaerobic conditions. Munsell soil colour charts are used to standardize the assessment of soil colour. The charts are based on three components—*hue* (a specific colour), *value* (lightness and darkness), and *chroma* (colour intensity)—that are arranged in sets of colour chips (Figure 19.4). A sample of soil is visually matched to a colour chip and assigned the corresponding Munsell notation.

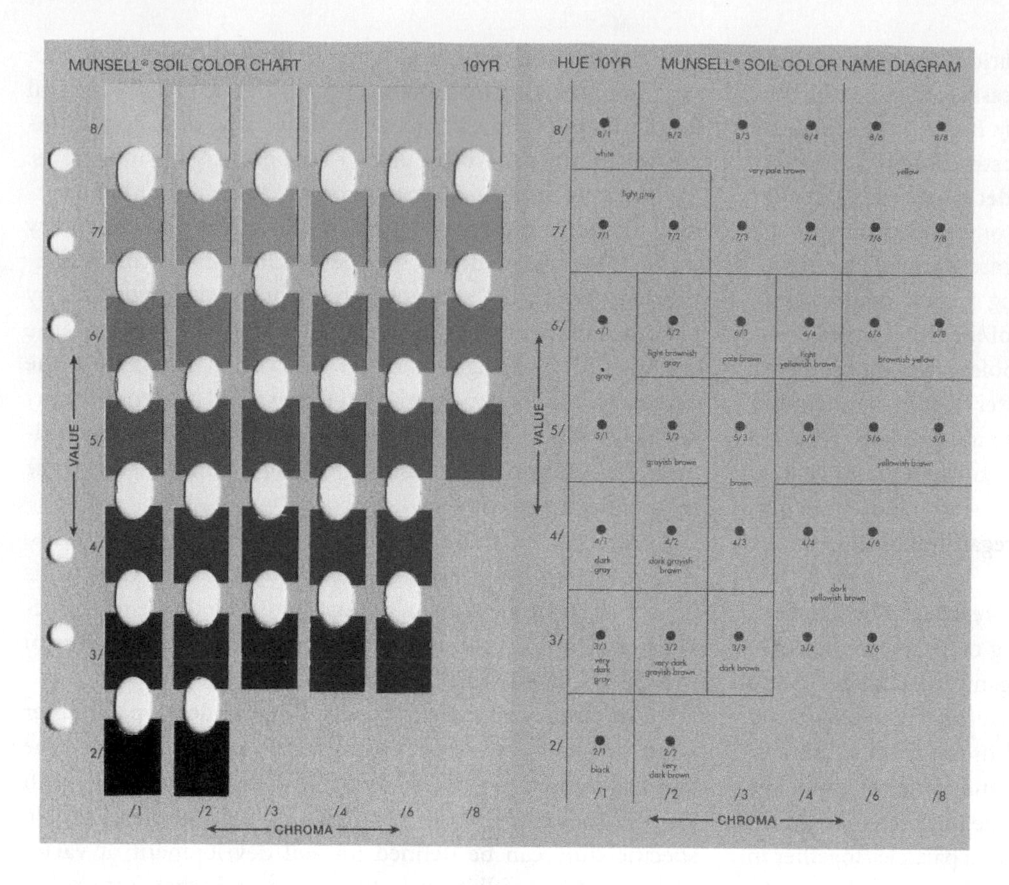

19.4 A SAMPLE PAGE FROM THE *MUNSELL SOIL COLOR BOOK* Three components are used to assess soil colour: hue refers to a specific colour and is represented by the pages in the book, in this case 10YR; value, or brightness, of that colour decreases from top to bottom on the page; and chroma, or intensity, changes across the page. Thus, a soil that matches the colour chip in the fourth row of the sixth column is assigned the Munsell notation 10 YR 5/8, and is described as yellowish brown.

In some areas, soil colour may be inherited from the parent mineral material, but more generally, soil colour is generated by soil-forming processes. For example, a white surface layer in soils of dry climates often indicates the presence of mineral salts brought upward by evaporation. A pale, ash-grey layer near the top of soils in the boreal forest climate region results when organic matter and nutrients are washed downward, leaving mostly light-coloured mineral matter behind.

SOIL TEXTURE

The mineral matter derived from bedrock is broken into smaller and smaller fragments by weathering. The resulting mineral particles are of various shapes and sizes, and because they usually do not pack together tightly, the soils remain porous. Particles larger than 2 mm in diameter are generally considered inert, and only those that are less than 2 mm are used to determine **soil texture**. These smaller particles, collectively known as the *fine earths*, are separated into three classes according to size. Sand particles range from 2 to 0.05 mm in diameter, silt particles from 0.05 to 0.002 mm in diameter, and clay particles, which are less than 0.002 mm in diameter (Figure 19.5). Soil particles smaller than 0.001 mm diameter are called **colloids**. The finer clays are therefore part of the soil colloid complement, as is finely divided and highly decomposed organic matter. The nature and properties of soil colloids is discussed later in the chapter.

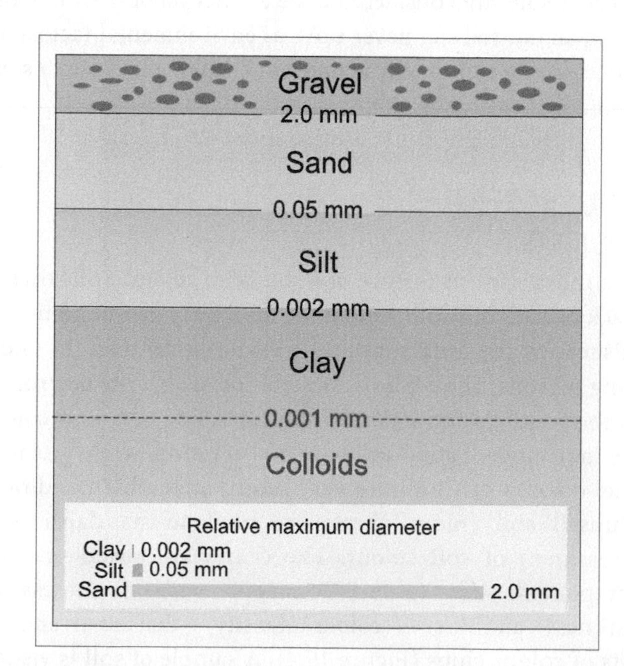

19.5 MINERAL PARTICLE SIZES Size grades, like sand, silt, and clay, refer to mineral particles within a specific size range.

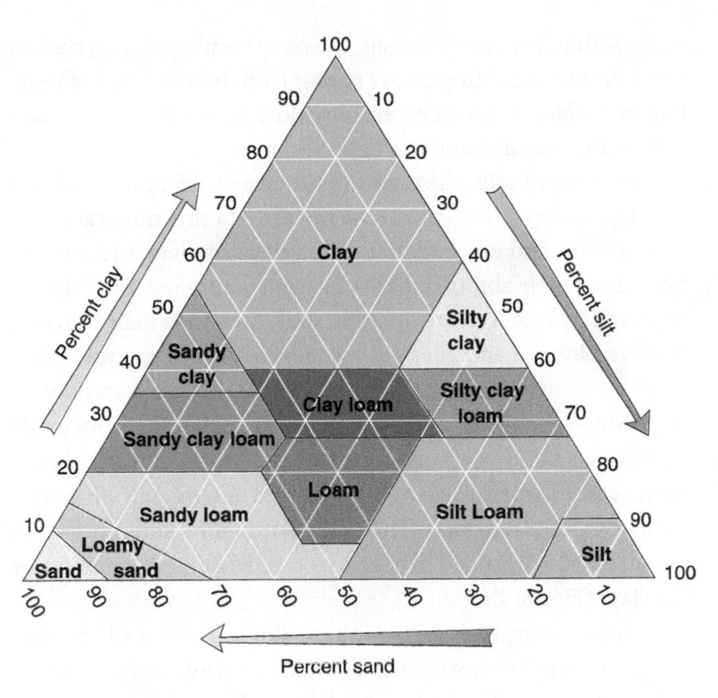

19.6 SOIL TEXTURE Once the proportions of sand, silt, and clay are known, the textural classes can be read from a graph.

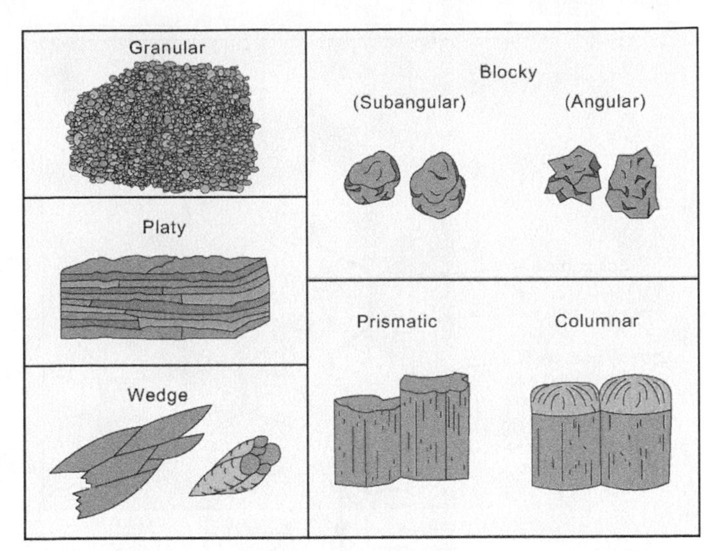

19.7 TYPES OF SOIL STRUCTURE Soil structure refers to how soil particles are combined to form crumbs or larger units.

Soil texture is based on the relative proportions of sand, silt, and clay present; each grade class is named according to the dominant particle sizes (Figure 19.6). For example, a silty soil is composed mainly of fine sand and silt; sandy soils contain mostly coarse sand. A *loam* soil contains from 28 to 50 percent silt, 7 to 27 percent clay, and less than 52 percent sand. Various classes of loams are distinguished depending on which size fractions are dominant. The amounts of gravel and other particles larger than 2 mm in diameter are used to qualify the textural classes; for example, a loam soil with 15 to 35 percent gravel would be described as a gravelly loam, or as an extremely gravelly loam if more than 60 percent gravel is present. With practice, the relative proportions of each size class can be determined by hand—sand feels gritty, silt is floury or slippery, and clay is sticky—but texture usually is determined in a laboratory using methods based on the settling rate of the different-sized particles.

Texture largely determines soil aeration and the ability of the soil to retain water. Coarse-textured (sandy) soils have many small passages between the mineral grains that quickly conduct gases and water to deeper layers through mass flow. If the soil is a fine-textured clay, these micropores are much smaller, and air and water penetrate more slowly.

SOIL STRUCTURE

Whereas soil texture refers to the sizes of the individual particles that make up the soil, *soil structure* refers to the way in which soil grains are grouped together into larger masses, called *peds*. These aggregates form as a result of cohesion between clay particles and between clays and larger particles. Organic materials and soil colloids also help bind particles together. Small peds, roughly shaped like spheres, give the soil a granular or crumb-like structure (Figure 19.7). Larger peds provide an angular, blocky structure; other soils may have a platy or columnar structure.

Peds form as a result of various processes, including seasonal climatic changes that cause wetting and drying, and freezing and thawing. Soils with a well-developed granular or blocky structure are easy to cultivate. Soils with high clay content can lack peds and so are considered to be structureless; the particles stick together in massive lumps. Conversely, sandy soils may also be structureless, but in this case it is because the sand grains do not readily adhere to each other. The ease with which soil aggregates can be broken down is termed *soil consistence*.

Good soil structure, particularly in fine-textured soils, increases soil porosity and facilitates movement of air and water. The spaces between the peds form a series of interconnected macropores that are important for root penetration, soil drainage, and aeration (Figure 19.8). A measure of pore space created by the textural and structural characteristic of the soil can be obtained by measuring the bulk density of the soil. *Bulk density* represents the mass of dry soil per unit volume (g cm^{-3}). This is distinguished from soil particle density that represents the mass of dry soil compacted to remove the voids, which eliminates the effects of texture and structure. Because the compacted soil occupies a lesser volume, particle density is higher than bulk density. Basically, particle density is determined by boiling a known mass

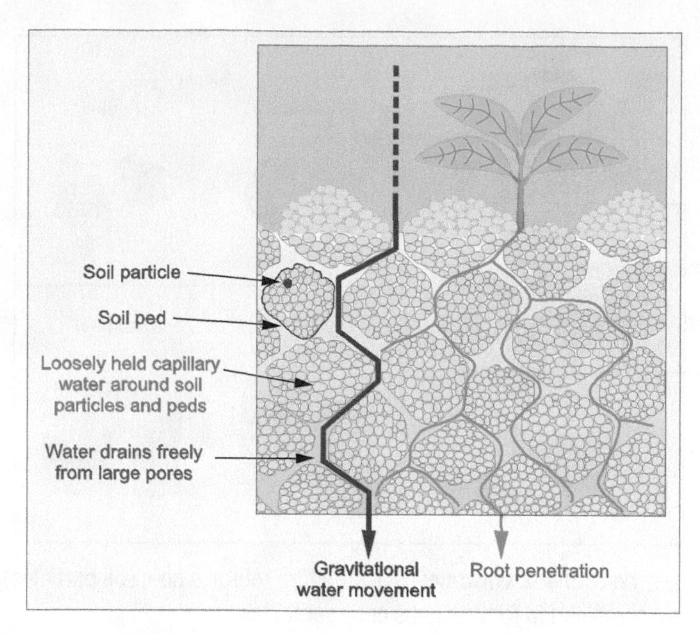

19.8 EFFECTS OF SOIL STRUCTURE Good soil structure affects porosity and allows water, air, and roots to permeate the soil.

of soil in water to obtain an air free sample. Soil porosity (%) is then obtained as 1 − (bulk density/particle density) × 100.

SOIL MINERALS

Most soils contain both primary and secondary minerals. **Primary minerals** are compounds present in the parent rock. **Secondary minerals** are formed through chemical alteration and decomposition of primary minerals by weathering. Primary minerals include quartz and silicate minerals, such as feldspars and micas, which contain varying proportions of aluminum, calcium, sodium, iron, and magnesium. Although a large fraction of the solid matter of most kinds of soils consists of primary minerals, they play no important role in sustaining plant or animal life. When primary minerals are exposed to air and water at or near the Earth's surface, their chemical composition slowly changes. This is part of the *mineral alteration* process that results chemical weathering (see Chapter 11). Mineral weathering occurs most rapidly in an acidic soil environment, and is facilitated by the release of CO_2 by respiration and decomposition which then reacts with water to form carbonic acid.

The removal of the least resistant minerals increases the abundance of more resistant minerals, and so the mineralogy of a soil slowly changes. Different weathering stages are recognized according to the primary and secondary mineral composition of the finest particles. Soils that have undergone minimal weathering, perhaps because of lack of moisture, will still contain less resistant minerals, such as calcite. Moderate weathering produces soils in which primary and secondary minerals are abundant, as is the case for most of the soils in temperate regions. Intensely weathered soils, such as those in the humid tropics, are almost entirely composed of resistant secondary minerals, especially oxides of iron (haematite) and aluminum (gibbsite) and kaolinite clay.

In terms of soils, the most important secondary minerals are the *clay minerals*. They form the majority of fine mineral particles in soils, and are essential to soil development and fertility because of their ability to hold plant nutrients. Several different types of clay minerals are found in soils, each with their own distinctive physical and chemical properties. However, most clays contain only two basic components arranged in different ways to produce two general structures: the silica tetrahedron, made up of one silicon atom and four oxygen atoms, and the aluminum octahedron, consisting of aluminum and oxygen atoms and hydroxyl (-OH) groups. The silica tetrahedra and aluminum octahedra are arranged in sheets to give what are termed 1:1 or 2:1 clays (Figure 19.9).

In a 1:1 clay, such as kaolinite, a single sheet of silica tetrahedra is joined to a single aluminum octahedral sheet. These paired sheets readily join with adjacent sheets through atomic bonding between the oxygen in the tetrahedral layer with the hydrogen of the hydroxyl group in the octahedral layer. This results in large, strongly bonded clay particles with a platy structure. Water and mineral ions can only attach to the external surfaces of these larger plates.

The 2:1 clay minerals, such as vermiculite and montmorillonite, have a different structure with spaces between adjacent

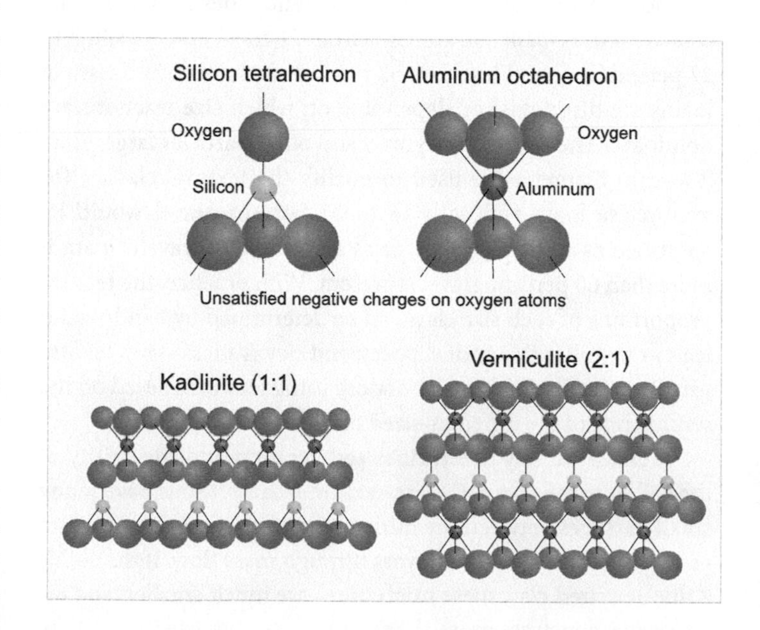

19.9 GENERAL STRUCTURE OF CLAY MINERALS Silica tetrahedra and aluminum octahedra are arranged in sheets to give clays with 1:1 structure such as kaolinite, or 2:1 structure such as vermiculite.

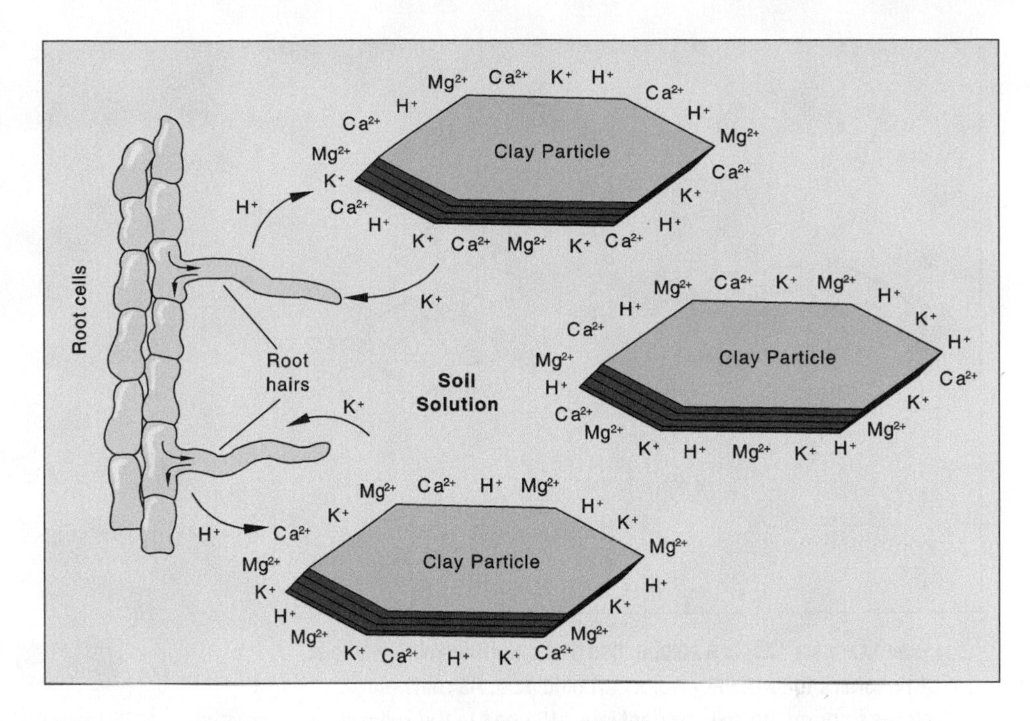

19.10 NUTRIENT EXCHANGE Plant nutrients can move from the adsorption sites on colloids into the soil water solution where they are available for root uptake. Colloidal particles have negative surface charges that attract and hold positively charged ions.

layers. Bonding is dependent on the oxygen atoms at the apices of the silica tetrahedra, replacing the -OH groups in the octahedral layer so that oxygen atoms are shared between the layers. Vacancies in the octahedral and tetrahedral layers lead to unsatisfied negative charges on the face of the clay plates, which can attract and hold positively charged mineral cations and water. As a result, 2:1 clays tend to shrink and crack when they dry out and swell when wet.

SOIL COLLOIDS

Soil colloids consist of particles smaller than 0.001 mm. Inorganic colloids, which mainly include fine particles of clay and hydrous oxides, typically make up the majority of colloids in a soil. When well mixed in water, these very small particles remain suspended indefinitely, giving the water a murky appearance. Organic colloids include soil humus, an amorphous substance derived from organic material that is resistant to decay.

The nature of the clay minerals in a soil determines its *base status*. The 2:1 clay minerals that can hold abundant cations give soils a *high base status* and generally will be highly fertile. Because 1:1 clay minerals hold a smaller supply of bases, these soils have a *low base status* and low fertility. Organic colloids are more chemically reactive than inorganic colloids, and typically have a greater influence on the concentration of plant nutrients held in a soil. Humus has a high capacity to hold bases, so its presence is usually associated with potentially high soil fertility.

Soil colloids attract nutrients that are dissolved in the soil water. Colloid surfaces tend to be negatively charged because of their molecular structure, and thus attract and hold positively charged cations derived from the dissociation of chemical compounds. One important group of cations is the **bases**, which include plant nutrients such as calcium (Ca^{2+}), magnesium (Mg^{2+}), and potassium (K^{+}). Although colloids retain these ions in the soil, they also release them to plants (Figure 19.10).

SOIL ACIDITY AND ALKALINITY

In humid regions, the charges on the colloids are mainly dominated by Ca^{2+}, H^{+}, and often Al^{3+}, resulting in acidic soils. As the soil becomes more acid, H^{+} and Al^{3+} become predominant. The cations Mg^{2+}, K^{+}, and Na^{+} are usually found in lesser amounts, while NH_4^{+} may be present in considerable quantities if the soil has been recently fertilized with ammonium fertilizers. In semi-arid and arid regions, Ca^{2+} usually dominates the cations, but Mg^{2+} and Na^{+} are often found in large quantities. H^{+} and Al^{3+} are usually present only in small concentrations.

Generally, cations are held tightly enough on adsorption sites to restrict their loss through leaching. However, any cation present in the soil solution can be exchanged with other cations attached to the colloids. Also, any cation in solution can potentially be removed from the soil by percolating water. The removal of plant nutrients from a soil through leaching is primarily associated with the abundance of hydrogen ions in the soil solution. Hydrogen (H^{+}), unlike the bases, is not considered a plant nutrient. Once displaced from the colloids, the bases can be gradually washed out of the soil (Figure 19.11).

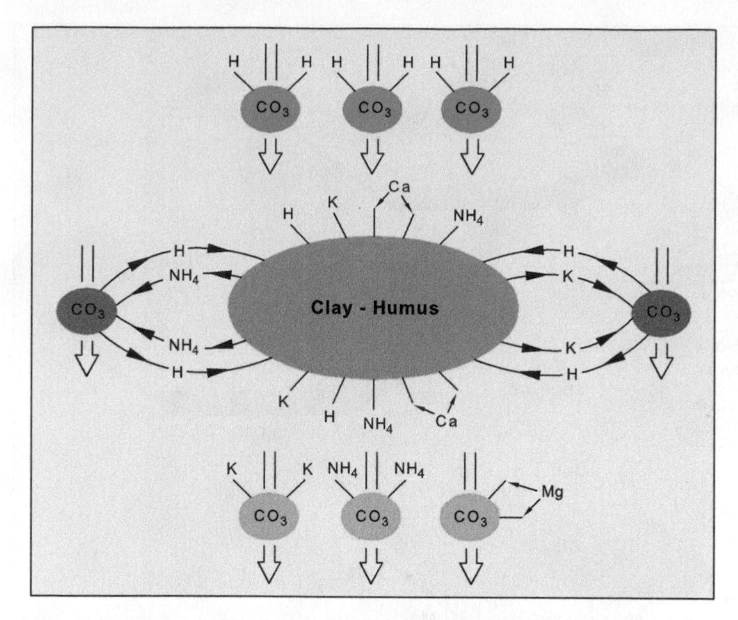

19.11 **SOIL LEACHING** CO_2 is a soluble gas that combines with water in the atmosphere to form very weak carbonic acid. As rainwater percolates through the soil, nutrient ions attached to the soil colloids are replaced by hydrogen ions. In this example, potassium, ammonium, and magnesium are removed as carbonates. Over time, the exchange of nutrients for hydrogen increases soil acidity and reduces soil fertility.

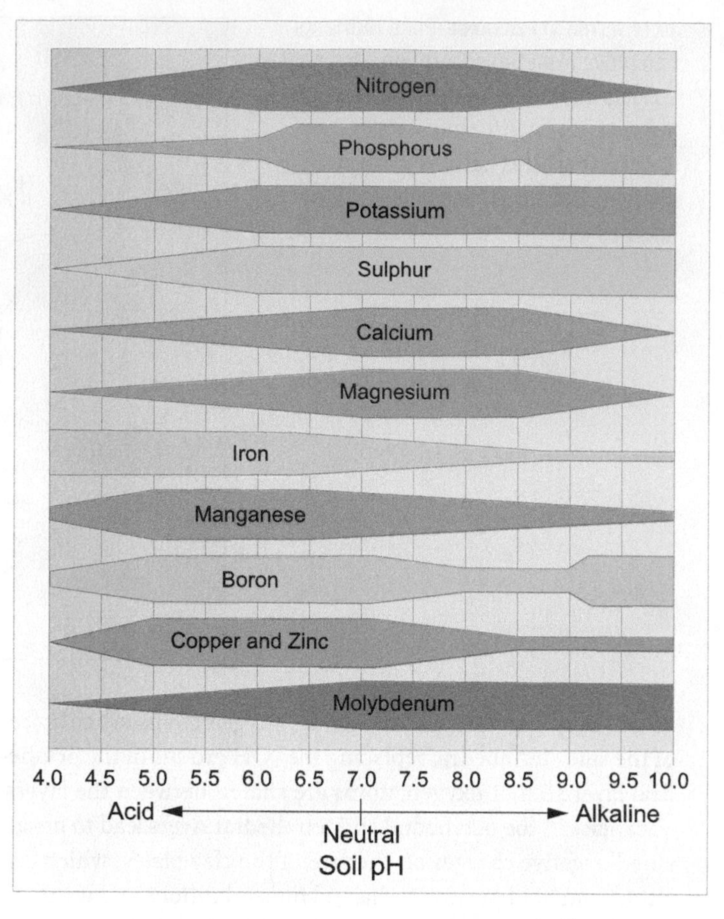

19.12 **EFFECT OF SOIL ACIDITY ON PLANT NUTRIENT AVAILABILITY** For most nutrients, availability is greatest between pH 6 to pH 8. (The wider the bar the greater the nutrient availability.)

The presence of H^+, and to a lesser degree Al^{3+}, in the soils of humid regions makes the soil solution acidic. This can affect soil fertility and plant growth by altering the availability of the essential plant nutrients. The displacement of nutrient bases from the soil colloids in turn increases soil acidity. The acidity or alkalinity of a soil is designated by its pH, with a value of 7 representing a neutral state. Values lower than pH 7 are in the acid range, values higher than pH 7 are in the alkaline range.

Soil pH is important from an agricultural perspective because it not only affects the availability of plant nutrients, but also influences nutrient toxicity and microorganism activity. Availability of most nutrients is greatest in the general range pH 6.5 to pH 7.5, although for some, such as iron, availability increases in more acidic soils; for others, such as molybdenum, availability increases under more alkaline conditions (Figure 19.12). High soil acidity is typical of cold, humid climates where organic acids are added to the soil from partially decomposed plant matter. Soils of high alkalinity are associated with semi-arid and arid regions and have an abundance of Ca^{2+} together with lesser amounts of Mg^{2+}, and in more extreme areas, Na^+.

In addition to carbon, hydrogen, and oxygen, which are derived from the atmosphere or from water, 13 mineral soil elements are essential for healthy plant growth (Table 19.1). The *macronutrients*, or primary nutrients, N, P, and K are needed in comparatively large quantities and are applied at high rates when added to soils as fertilizers. The concentrations of these three elements are listed on fertilizer bags as 12-8-8 or a similar string of numbers that represent the percentage of N, P, and K, respectively, in the product. The *secondary nutrients*, Ca, Mg, and S, are required in lesser amounts, while *micronutrients*, or *trace elements*, such as Zn and Cu, are needed only in very small quantities to bring about a response in plant performance.

SOIL MOISTURE

Besides providing nutrients for plant growth, the soil layer serves as a reservoir for the moisture that plants require. The soil receives water from rain and melting snow. Some of the water can run off the soil surface and flow into streams and

TABLE 19.1 ESSENTIAL PLANT NUTRIENTS: THEIR FUNCTIONS AND GENERAL SYMPTOMS CAUSED BY DEFICIENCY OR EXCESS

NUTRIENT CATEGORY	ELEMENT	FUNCTION IN PLANT	SYMPTOMS OF DEFICIENCY	SYMPTOMS OF EXCESS
Primary nutrients	Nitrogen (N)	Proteins, amino acids	Leaves are light green to yellow, especially older leaves; stunted plant growth and poor fruit development.	Dark green foliage that may be susceptible to lodging, drought, disease, and insect invasion. Fruit and seed crops may fail to yield.
	Phosphorus (P)	Nucleic acids, ATP	Leaves may develop purple colouration; stunted plant growth and delay in plant development.	Excess phosphorus may cause micronutrient deficiencies, especially iron or zinc.
	Potassium (K)	Catalyst, ion transport	Older leaves turn yellow initially around margins and die; irregular fruit development.	Excess potassium may cause deficiencies in magnesium and possibly calcium.
Secondary nutrients	Calcium (Ca)	Cell wall component	Reduced growth or death of growing tips; poor fruit development and appearance.	Excess calcium may cause deficiency in either magnesium or potassium.
	Magnesium (Mg)	Part of chlorophyll	Initial yellowing of older leaves between leaf veins, spreading to younger leaves; poor fruit development and production.	High concentration tolerated in plant; however, imbalance with calcium and potassium may reduce growth.
	Sulphur (S)	Amino acids	Initial yellowing of young leaves spreading to whole plant; similar symptoms to nitrogen deficiency but occurs on new growth.	Excess of sulphur may cause premature dropping of leaves.
Micronutrients	Iron (Fe)	Chlorophyll synthesis	Initial distinct yellow or white areas between veins of young leaves, leading to spots of dead leaf tissue.	Possible bronzing of leaves with tiny brown spots.
	Copper (Cu)	Component of enzymes and protein synthesis	Yellowing of leaves, reduced leaf growth, and wilting.	Leaves are bluish-green colour.
	Manganese (Mn)	Activates enzymes	Interveinal yellowing or mottling of young leaves.	Older leaves have brown spots surrounded by a chlorotic circle or zone.
	Zinc (Zn)	Activates enzymes	Interveinal yellowing on young leaves; reduced leaf size.	Excess zinc may cause iron deficiency in some plants.
	Boron (B)	Cell wall component	Death of growing points and deformation of leaves with areas of discolouration.	Leaf tips become yellow, followed by necrosis. Leaves get a scorched appearance and later fall off.
	Molybdenum (Mo)	Involved in N fixation	Leaves turn yellow.	Leaves turn orange.

Source: Based on Table 2, General Symptoms of Plant Nutrient Deficiency and Excess, http://www.cartage.org.lb/en/themes/Sciences/BotanicalSciences/PlantHormones/EssentialPlant/EssentialPlant.htm.

rivers, eventually reaching the sea. Water that percolates into the soil may return to the atmosphere as water vapour, either by evaporation or transpiration from plant leaves. In addition, some water can flow completely through the soil layer to recharge supplies of groundwater at depths below the reach of plant roots.

Precipitation that infiltrates and moistens the soil results in *soil water recharge*. Once the maximum capacity of a soil to hold water is reached, any additional water present in the pore spaces drains into the underlying substrate. When a soil has been saturated by water and then drains freely under gravity until no more water moves downward, the soil is said to be at *field capacity*; for most soils this is achieved within two or three days. Field capacity depends largely on the soil's texture (Figure 19.13). Fine-textured clay soils hold more water than coarse-textured sandy soils, because fine particles have a much larger surface area per unit of volume.

Permanent wilting point is an agricultural term that approximates the water storage level below which plants can experience moisture stress (Figure 19.13). The wilting point also depends on soil texture, because fine particles hold water more tightly, making it difficult for plants to extract. Thus, plants can wilt in fine-textured soils even though more soil water is present than in coarse-textured soils. The difference between the field capacity of a soil and its wilting point is the maximum available water capacity; this is greatest in loamy soils.

Free water or *gravitational water* will drain from a soil until the soil water potential reaches −30 kPa. *Soil water potential* (ψ) is a measure of the energy status of soil water and reflects how hard plants must work to extract moisture from the soil. Because it is essentially the amount of suction required to remove the water, soil water potential is recorded as negative pressure in kilopascals. Most soil water is held as *matric water* in the microscopic pores and attached to the surface of the solid soil particles (Figure 19.14).

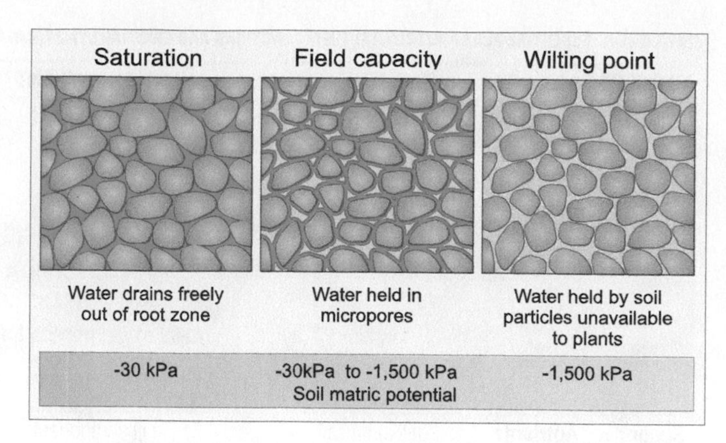

19.14 SOIL WATER AVAILABILITY The ability of plants to extract water from a soil decreases as soil moisture levels decline because the water is held more tenaciously by the soil particles as they dry. Soil water potential, which measures the suction required to extract the water, increases as soil passes from saturation to the permanent wilting point.

Gravitational water is not considered available to plants because it remains in the soil for only a short time. Once this surplus water has drained away and ψ reaches −30 kPa , the soil moisture level is at field capacity. This water clings to the soil particles through *capillary tension* and remains there until it evaporates or is absorbed by plant roots. As the soil dries, the energy needed to extract water increases until a point is reached where it can no longer be removed by plants. This is the permanent wilting point, which for most plants is reached when ψ is −1,500 kPa. Thus, the zone of available water ranges from −30 to −1,500 kPa. Some water remains in the soil at the permanent wilting point; in extreme drought conditions, this may continue to be lost through evaporation until the soil reaches an air dry state at −3,100 kPa.

Compared to sandy soils, clays and silts contain more moisture at the permanent wilting point because of the greater surface area created by their constituent small particles and fine network of pores (Figure 19.15). Although this water is less

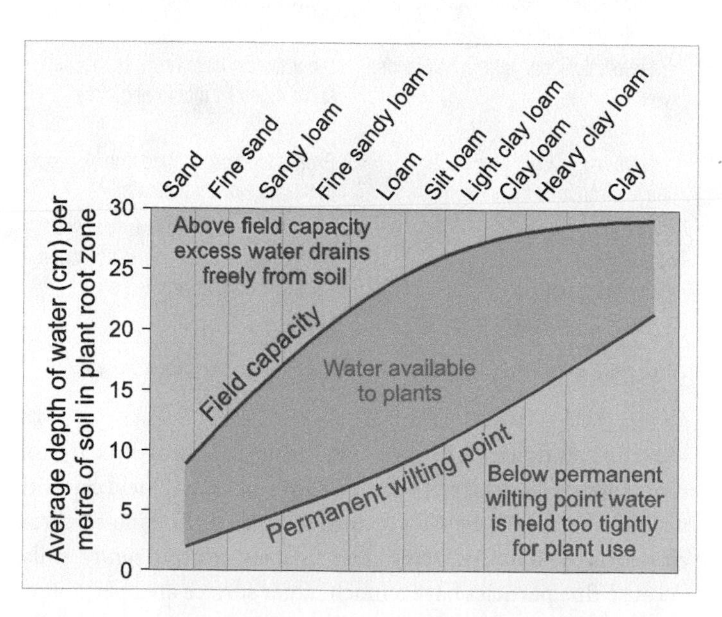

19.13 FIELD CAPACITY AND WILTING POINT ACCORDING TO SOIL TEXTURE Fine-textured soils hold more water. They also hold water more tightly, so plants wilt more quickly. The amount of water held between the field capacity and the permanent wilting point is termed the available water content and is a measure of the amount of water in the soil that is "potentially" available to plants.

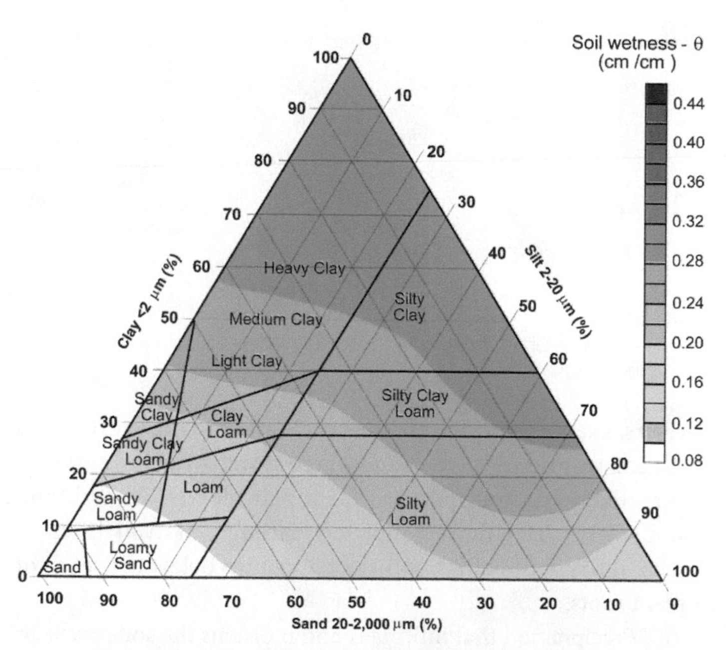

19.15 SOIL MOISTURE LEVELS IN SOILS AT PERMANENT WILTING POINT At the permanent wilting point, fine-textured soils hold the greatest amounts of water, whereas coarse sandy soils are essentially completely dry.

available to plants at low matric potentials, one benefit is that retention of water in finer textured soils can reduce the loss of nutrients by leaching. Conversely, in rapidly draining, coarse-textured soils there is greater potential for nutrient loss and reduced fertility.

THE SOIL WATER BALANCE

Water in the soil is a critical resource needed for plant growth. The amount of water available at any given time is determined by the *soil water balance*, which includes the gain, loss, and storage of soil water (Figure 19.16). Water held in storage in the soil water zone is increased by recharge during precipitation, but decreased by use through evapotranspiration. Surplus water is disposed of by downward percolation to the groundwater zone or by overland flow.

The rate at which water vapour returns to the atmosphere from the ground and its plant cover is called *actual evapotranspiration (Ea)*. *Potential evapotranspiration (Ep)* represents the water vapour loss under ideal conditions. Ideal conditions are represented by a complete, uniform cover of vegetation consisting of fresh green leaves and no bare ground, and an adequate water supply to maintain the soil's storage capacity at all times. This second condition can be fulfilled naturally by abundant and frequent precipitation, or artificially by irrigation. These terms can be simplified as follows:

- Actual evapotranspiration *(Ea)* is **water use.**
- Potential evapotranspiration *(Ep)* is **water need.**

Need refers to the quantity of soil water required to maximize plant growth under prevailing conditions of solar radiation, air temperature, wind speed, and the available supply of nutrients. Of these, the most important factor in determining water need is temperature. The difference between water use and water need is the *soil water shortage*, or *deficit*. This is the quantity of water that irrigation must supply to achieve maximum crop growth within an agricultural system.

A SIMPLE SOIL WATER BUDGET

The soil *water budget* consists of the amounts of water needed to satisfy each process in the soil water balance (Figure 19.17). These various amounts are calculated by adding and subtracting the mean monthly values for a given observing station. All terms of the soil water budget are stated in centimetres of water depth. The terms needed to complete the budget are the following:

- Precipitation, *P*
- Water need, *Ep*
- Water use, *Ea*
- Storage withdrawal, $-G$
- Storage recharge, $+G$
- Soil water shortage, *D*
- Water surplus, *R*

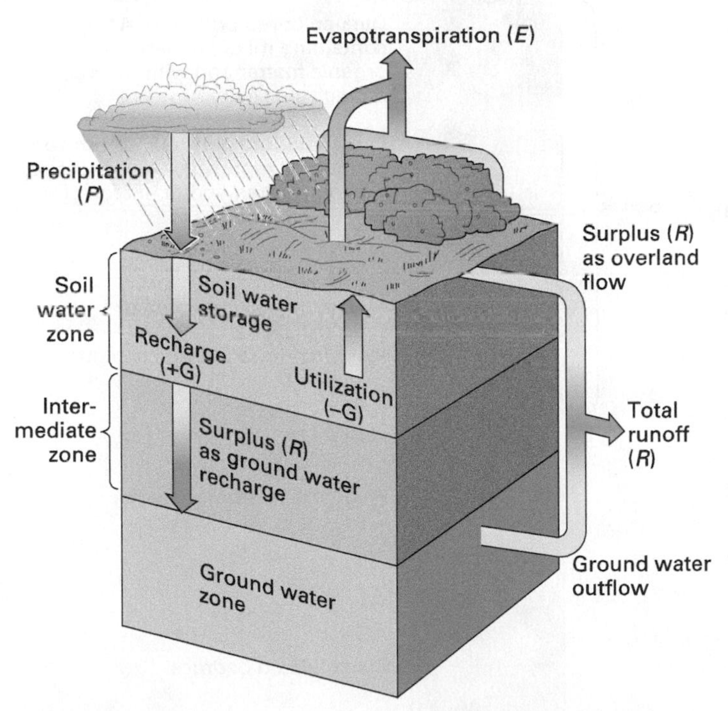

19.16 SOIL WATER BALANCE The amount of water held in storage in a soil is determined by differences in the rates at which water is gained through precipitation and lost through evapotranspiration and percolation. (© A. N. Strahler.)

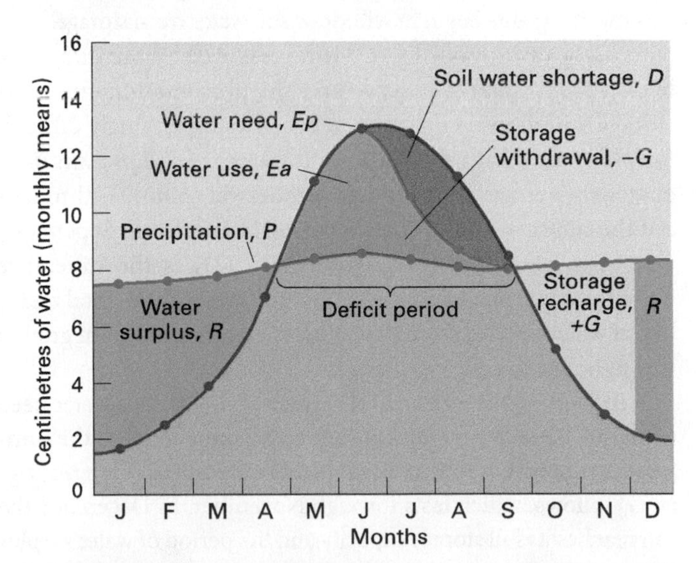

19.17 A SIMPLIFIED SOIL WATER BUDGET This soil water budget is typical of a midlatitude moist climate.

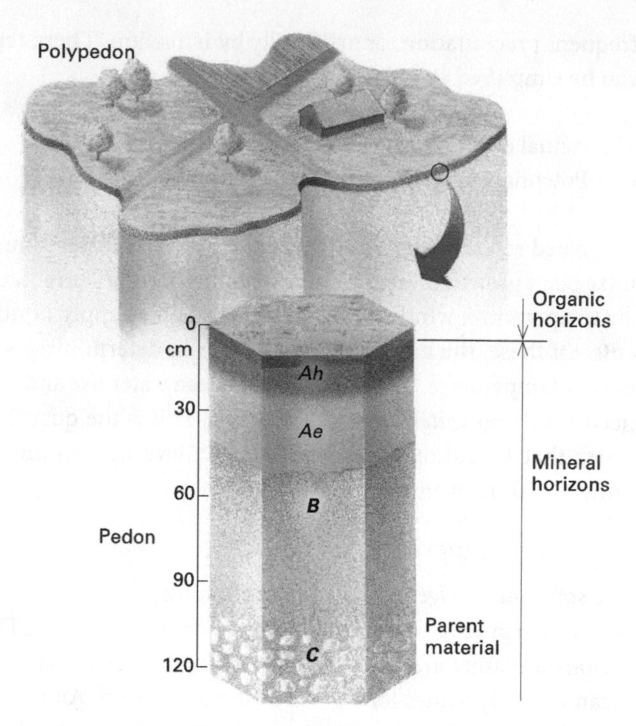

19.18 SOIL HORIZONS A column of soil will normally show a series of horizons, which are horizontal layers with different properties.

The points on the graph represent average monthly values of precipitation and water need. In this example, precipitation (*P*) is much the same in all months. In contrast, water need (*Ep*) shows a strong seasonal cycle, with low demands in winter and a high summer peak.

At the start of the year, precipitation greatly exceeds water use and a large water surplus (*R*) exists. Runoff disposes of this surplus. By May, water use exceeds precipitation, creating a water deficit. In this month, plants begin to withdraw soil water from storage.

Storage withdrawal (−*G*) is represented by the difference between the water-use curve and the precipitation curve. As storage withdrawal continues, it becomes increasingly difficult for plants to obtain soil water, and water need (*Ep*) eventually exceeds water use (*Ea*). Storage withdrawal continues throughout the summer, and the deficit period lasts through September. The area labelled soil water shortage (*D*) is the difference between water need and water use. It represents the total quantity of water needed from irrigation to ensure maximum growth throughout the deficit period.

In October, precipitation (*P*) once again exceeds water need (*Ep*), but the soil must first absorb an amount equal to the summer storage withdrawal; consequently, a period of *storage recharge* (+*G*) follows, which lasts through November. In December, the soil reaches its full storage capacity and the period of water surplus (*R*) continues through the winter. The *Working It Out • Calculating a Simple Soil Water Budget* feature at the end of this chapter explains the procedures used to calculate a water budget.

SOIL DEVELOPMENT

SOIL HORIZONS

Most soils possess **soil horizons**—distinctive layers that differ in physical and chemical composition, organic content, or structure (Figure 19.18). Soil horizons usually develop through selective removal or accumulation of ions, colloids, and chemical compounds by water moving through the soil. A **soil profile** is the full set of horizons exposed in a soil pit that is excavated down to the parent material. A soil column exhibiting these same properties is referred to as a pedon; it too extends from the surface to a lower limit in regolith or bedrock. A pedon represents the smallest distinctive division of the soil of a given area. It exhibits all the features needed to properly classify and describe the soil at that location. An area of soil of the same type, which is made up of many pedons, is called a *polypedon*.

The main types of horizons, which might occur in a typical boreal forest soil, are shown in Figure 19.19. Organic horizons

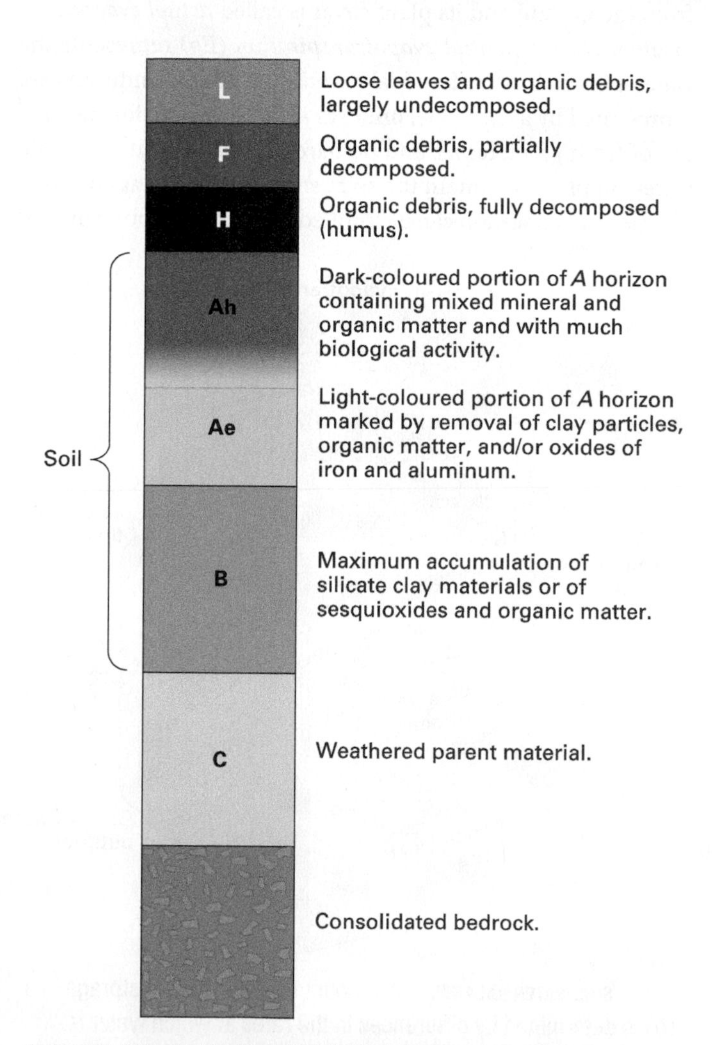

19.19 SOIL HORIZONS CHARACTERISTIC OF A BOREAL FOREST CLIMATE A sequence of horizons that might appear in a forest soil developed in a cool, moist climate.

overlie the mineral horizons and are formed from accumulations of material derived from plants and animals. Several types of organic layers are distinguished, depending on the state of decay. The Canadian System of Soil Classification recognizes three organic horizons that contain decayed matter ranging from slightly (*L or litter horizon*), to moderately (*F or fermentation horizon*), to very decomposed (*H or humus horizon*). In addition, horizons in peat soils, such as those in bogs and swamps, are designated O (*organic horizon*). Table 19.2 provides a listing and description of the major horizons recognized by the Canadian

TABLE 19.2 HORIZONS AND SUBHORIZONS OF THE CANADIAN SYSTEM OF SOIL CLASSIFICATION

	ORGANIC HORIZONS		
O	Organic horizon developed mainly from mosses, rushes, and woody materials (e.g., peat).	**F**	Same as L, except that the original structures are difficult to recognize.
L	Organic horizon characterized by an accumulation of organic matter derived mainly from leaves, twigs, and woody materials in which the organic structures are easily discernible.	**H**	Organic horizon characterized by decomposed organic matter in which the original structures are not discernible.
	MINERAL HORIZONS		
A	Mineral horizon found at or near the surface in the zone of leaching or eluviation of materials in solution or suspension, or of maximum *in situ* accumulation of organic matter, or both.	**C**	Mineral horizon comparatively unaffected by the pedogenic processes operating in A and B horizons. Gleying processes and the accumulation of calcium, magnesium, and more soluble salts can occur in this horizon.
B	Mineral horizon characterized by enrichment in organic matter, sesquioxides, or clay; by the development of soil structure; or by a change of colour denoting hydrolysis, reduction, or oxidation.		
	SUBHORIZONS (LOWERCASE SUFFIXES)		
b	Buried soil horizon.	**n**	Horizon in which the ratio of exchangeable calcium (Ca) to exchangeable sodium (Na) is 10 or less, as well as the following distinctive morphological characteristics: prismatic or columnar structure, dark coating on ped surfaces, and hard to very hard consistency when dry.
c	Irreversibly cemented pedogenic horizon, also known as a hardpan.	**p**	Horizon disturbed by human activities, such as cultivation, logging, and habitation.
ca	Horizon of secondary carbonate enrichment, in which the concentration of lime exceeds that in the unenriched parent material.	**s**	Horizon of salts, including gypsum, which may be detected as crystals, veins, or surface crusts of salt crystals.
cc	Horizon containing irreversibly cemented pedogenic concretions.	**sa**	Horizon with secondary enrichment of salts more soluble than calcium and magnesium carbonates; the concentration of salts exceeds that in the unenriched parent material.
e	Horizon characterized by the eluviation of clay, iron, aluminum, or organic matter, alone or in combination.	**ss**	Horizon containing *slickensides*—shear surfaces that form when one soil mass moves over another.
f	Horizon enriched with amorphous material, principally aluminum and iron combined with organic matter; reddish near upper boundary, becoming more yellow at depth.	**t**	Illuvial horizon enriched with silicate clay.
g	Horizon characterized by grey colours or prominent mottling, or both, indicating permanent or intense chemical reduction.	**u**	Horizon that is markedly disrupted by physical or faunal processes other than cryoturbation.
h	Horizon enriched with organic matter.	**v**	Horizon affected by disruption and mixing caused by shrinking and swelling of the soil mass.
j	Used as a modifier of suffixes *e, f, g, n,* and *t* to denote an expression of, but failure to meet, the specified limits of the suffix it modifies.	**x**	Loamy subsurface horizon of high bulk density and very low organic matter content; when dry, it is hard and seems to be cemented; also known as a fragipan.
k	Denotes the presence of carbonate as indicated by visible effervescence when dilute hydrogen chloride (HCl) is added.	**y**	Horizon affected by cryoturbation, as manifested by disrupted or broken horizons, incorporation of materials from other horizons, and mechanical sorting.
m	Horizon slightly altered by hydrolysis, oxidation, or solution, or all three to give a change in colour or structure, or both.	**z**	A frozen layer.

Source: Table compiled from Chapter 2 of *Canadian System of Soil Classification, 3rd ed.*, Ottawa: NRC Research Press, 1998, http://sis.agr.gc.ca/cansis/taxa/cssc3/chpt02.html#soilhorizons.

System of Soil Classification, including organic and mineral horizons. Also described are subhorizons, identified with lowercase letters; for example, the *Ah* and *Ae* subhorizons are used in the stylized boreal forest soil profile.

The mineral horizons below the organic layers at the surface are distinguished as the *A*, *B*, and *C* horizons. Plant roots readily penetrate the *A* and *B* horizons of most soils and influence soil development within them. In its scientific usage, soil refers only to the *A* and *B* horizons. The *A* horizon is the uppermost mineral horizon. In the stylized profile the upper part *(Ah)* is rich in organic matter, consisting of numerous plant roots and humus that have washed down from the overlying organic horizons. In the lower part of the *A* horizon *(Ae)*, clay particles and oxides of aluminum and iron, as well as plant nutrients, are removed by percolating water and leave behind grains of sand or coarse silt.

The *B* horizon receives the clay particles, aluminum and iron sequioxides, as well as any organic matter that has washed down from the *A* horizon. The filling of natural spaces with clays and sesquioxides makes the *B* horizon dense and hard. Beneath the *B* horizon is the *C* horizon, which consists of the parent mineral matter of the soil—the weathered regolith. Below the regolith lies unaltered bedrock or accumulated sediments.

SOIL-FORMING PROCESSES

Soils develop in response to four types of soil-forming processes (Table 19.3). *Soil enrichment* processes add material to the soil. For example, inorganic enrichment occurs when sediment is brought from higher to lower areas by overland flow. Stream flooding also deposits fine mineral particles on low-lying soil surfaces. Wind is another source of fine material that can accumulate on the soil surface. Organic enrichment occurs when humus, accumulating in organic surface layers, is carried downward to the *A* horizon below.

The second class of soil-forming processes consists of those that transport material out of the soil column. Such *removal* occurs when surface erosion carries sediment away from the soil's uppermost layer. Another important process is *leaching*, in which percolating water dissolves soil materials and moves them below the soil profile or into the groundwater.

In moist climates, surplus soil water moves downward to the groundwater zone. This water movement leaches calcium carbonate ($CaCO_3$) from the entire soil profile in a process called *decalcification*. Soils that have lost most of their calcium are also usually acidic and comparatively low in bases. The addition of lime or pulverized limestone not only corrects the acidic condition, but also restores calcium, which is a plant nutrient.

The third class of soil-forming processes involves *translocation*, in which materials are moved within the soil body, usually from one horizon to another. Two translocation processes that operate simultaneously are eluviation and illuviation. **Eluviation** consists of the downward transport of fine particles, particularly clays and colloids, from the uppermost part of the soil. Eluviation leaves behind grains of sand or coarse silt, forming the *Ae* horizon. **Illuviation** is the accumulation of materials that are brought downward, normally from the *Ae* horizon to the *B* horizon. The materials that accumulate may be clay particles, humus, or iron and aluminum sesquioxides.

The translocation of $CaCO_3$ is another important process. In many areas, the parent material of the soil contains a substantial proportion of $CaCO_3$ derived from the disintegration of limestone, a common variety of bedrock. Carbonic acid, which forms when CO_2 dissolves in rainwater or soil water, readily reacts with $CaCO_3$, which then goes into the solution as Ca^{2+} and CO_3^- ions. In dry climates, such as in the prairie grasslands of Saskatchewan and Alberta, $CaCO_3$ dissolves in the upper layers of the soil during periods of rain or snow melt when soil water recharge is taking place. The dissolved carbonates are carried down to the *B* horizon, where water

TABLE 19.3 SOIL-FORMING PROCESSES

Enrichment	Addition of material to the soil, for example, by deposition of mineral matter by water or wind action.
Removal	Removal of material from the soil, for example, by erosion of uppermost layers or by leaching of dissolved matter to lower layers or to groundwater.
Decalcification	Leaching of calcium carbonate from the soil to the groundwater below by large amounts of infiltrating precipitation in moist climates.
Translocation	Movement of materials upward or downward within the soil body.
Eluviation	Downward transport of fine materials from the upper part of the soil.
Illuviation	Accumulation of fine materials in a lower part of the soil.
Calcification	Accumulation of calcium carbonate by dissolution in upper layers and precipitation in the *B* horizon.
Salinization	Upward wicking of salt-laden groundwater toward the soil surface with evaporation to produce a layer of salt accumulation.
Transformation	Transformation of material in the soil body; for example, conversion of primary to secondary minerals.
Humification	Decomposition of organic matter to produce humus.

percolation reaches its limit. Here, the carbonate is precipitated through the process of *calcification*. The deposits of $CaCO_3$ take the form of white or pale-coloured grains, nodules, or plates in the *B* or *C* horizons where they form distinctive *calcic (ca)* subhorizons.

Another form of translocation associated with arid climates is *salinization*. Where groundwater lies close to the surface in a poorly drained area, evaporation draws up a continual flow of moisture by capillary tension. When evaporation occurs, any dissolved salts precipitate and accumulate as a distinctive *salic (sa) subhorizon*. Most of the salts are compounds of sodium, which, when present in large amounts, is associated with highly alkaline conditions and is toxic to many kinds of plants. When salinization occurs in irrigated lands in a desert climate, the soil can be ruined for further agricultural use.

A further soil-forming process involves the *transformation* of material within the soil body. One example is the conversion of minerals from primary to secondary types. Another is the decomposition of organic matter to produce humus, a process termed *humification*. In warm, moist climates, humification can reduce most of the organic matter to CO_2 and water, leaving virtually no humus in the soil.

THE GLOBAL SCOPE OF SOILS

Soils are the product of many processes acting at different rates. For this reason, soils exist as a continuum in which properties and characteristics are infinitely variable. Soils are therefore grouped into discrete classes according to specified parameters in much the same way as climates are arranged together. Soil classification systems group soils according to their inherent properties. One such global classification system, called the World Reference Base for Soil Resources (WRB), was developed by the United Nations Food and Agriculture Organization (FAO) with the support of the UN Environment Programme (UNEP) and the International Society of Soil Science.

The WRB uses soil morphology, including soil structural types, assemblages of soil constituents, and diagnostic soil horizons that are observable in the field. The highest level of the classification recognizes 32 Reference Soil Groups (RSGs) differentiated according to the primary pedogenic processes associated with their formation (Table 19.4). In most cases the RSGs are identified on the basis of some natural condition. For example, Cryosols and Leptosols are recognized because they have severe limitations for rooting. In the case of Cryosols, this is due to permafrost; for Leptosols, it is because they consist of shallow soils over bedrock, or are extremely gravelly or stony. The importance of human activity as a soil-forming factor is also recognized; for example, human influence is used to distinguish the Anthrosol and Technosol RSGs.

TABLE 19.4 KEY TO THE WRB REFERENCE SOIL GROUPS

1. Soils with thick organic layers	Histosols
2. Soils with strong human influence	
Soils with long and intensive agricultural use	Anthrosols
Soils containing many artefacts	Technosols
3. Soils with limited rooting due to shallow permafrost or stoniness	
Ice-affected soils	Cryosols
Shallow or extremely gravelly soils	Leptosols
4. Soils influenced by water	
Alternating wet-dry conditions, rich in swelling clays	Vertisols
Flood plains, tidal marshes	Fluvisols
Alkaline soils	Solonetz
Salt enrichment upon evaporation	Solonchaks
Groundwater affected soils	Gleysols
5. Soils set by Fe/Al chemistry	
Allophanes or Al-humus complexes	Andosols
Cheluviation and chilluviation	Podzols
Accumulation of Fe under hydromorphic conditions	Plinthosols
Low-activity clay, P fixation, strongly structured	Nitisols
Dominance of kaolinite and sesquioxides	Ferralsols
6. Soils with stagnating water	
Abrupt textural discontinuity	Planosols
Structural or moderate textural discontinuity	Stagnosols
7. Accumulation of organic matter, high base status	
Typically mollic	Chernozems
Transition to drier climate	Kastanozems
Transition to more humid climate	Phaeozems
8. Accumulation of less soluble salts or non-saline substances	
Gypsum	Gypsisols
Silica	Durisols
Calcium carbonate	Calcisols
9. Soils with a clay-enriched subsoil	
Albeluvic tonguing	Albeluvisols
Low base status, high-activity clay	Alisols
Low base status, low-activity clay	Acrisols
High base status, high-activity clay	Luvisols
High base status, low-activity clay	Lixisols
10. Relatively young soils or soils with little or no profile development	
With an acidic dark topsoil	Umbrisols
Sandy soils	Arenosols
Moderately developed soils	Cambisols
Soils with no significant profile development	Regosols

Source: Food and Agriculture Organization of the United Nations, IUSS Working Group WRB, 2006. *World reference base for soil resources 2006*. World Soil Resources Reports No. 103, FAO, Rome, Table 1, p. 6, ftp://ftp.fao.org/agl/agll/docs/wsrr103e.pdf.

Anthrosols are soils that have been profoundly modified through human activities such as the addition of organic materials or household wastes, irrigation, or cultivation. Anthrosols are found wherever people have practised agriculture for a long time, and normally their influence is restricted to the surface horizons. Different subtypes are noted according to specific agricultural practices. For example, *irragric horizons* develop in areas that have been under irrigation for centuries, and *hortic horizons* are found where soils have been fertilized with household wastes and manure.

The Technosols have only recently been added to the WRB. These are soils whose properties and pedogenesis are strongly influenced by human-made materials. Their *technical* origin is based on a significant amount of materials that have either been extracted from the Earth or manufactured by humans, such as landfill waste and refinery slag. Technosols are generally referred to as *urban* or *mine* soils. This group also includes soils constructed from human-made materials and underlain with geomembranes; reclamation sites in which materials are overlain with a layer of "natural" soil would be classed as Technosols. Many derelict brownfield sites contaminated by previous industrial activities would be included in this category.

The WRB was developed over several decades by successive international congresses of soil scientists; the latest edition was published in 2006. Its goal was to provide a framework through which existing soil classification systems developed by individual nations could be correlated and ongoing soil classification efforts could be harmonized. Its final objective was to identify a set of major soil groupings at the global level, as well as the criteria for defining and identifying them. The WRB serves not only as a framework for soil scientists from many countries and regions to use in their work, but also as a common ground for the inventory of land and natural resources.

Many countries have developed unique classification systems—Canada and the United States are two examples. The Canadian System of Soil Classification pays special attention to the soils of cold regions and permafrost terrains. The American system, known as the Comprehensive Soil Classification System (CSCS), is more general and can be applied globally.

Soil classification systems typically use the **soil order** as the highest-level grouping. In the Canadian system, the next subdivision is the *great group*, while the American system uses the term *suborder*. Soil orders and great groups are often distinguished by the presence of a *diagnostic horizon*. Each diagnostic horizon has some unique combination of physical properties (such as colour, structure, and texture) or chemical properties (for example, an abundance of calcium). Two basic types of diagnostic horizons are used. A diagnostic horizon that forms at the surface is called an *epipedon*; one that forms by processes occurring at various depths in the soil is simply referred to as a subsurface horizon.

SOILS OF CANADA

Although classification systems such as the WRB or the American CSCS cover the global range of soils, Canada has evolved a unique soil classification system that is especially suited to its own soils. Because Canada is located entirely north of lat. 40° N, it has no tropical and equatorial soil types. Moreover, nearly all of Canada experienced glaciation during the last ice age. Ice sheets and glaciers largely removed pre-existing soils, replacing them with unsorted glacial debris, transported sands and gravels, and lake-bottom deposits. As a result, the Canadian system emphasizes young soils of cold regions in more detail than do other systems.

Like the WRB and the CSCS, the Canadian System of Soil Classification (CSSC) uses a system of classes based on the properties of the soils themselves, rather than interpretations of various uses of the soils. Thus, the classes are based on generalized properties of real, not idealized, soils. Although the Canadian system's classes are defined according to actual soil properties that can be observed and measured, the system retains a genetic bias. Thus, it favours properties or combinations of properties that reflect processes of soil formation when distinguishing between higher divisions in the classification. Thus, the soils grouped under a single soil order are considered the product of a similar set of dominant soil-forming processes resulting from broadly similar climatic conditions. The basis of the CSSC is the recognition of horizons and subhorizons that have developed through distinctive processes (see Table 19.2). The nature and order of horizons and subhorizons found in the soil profile are primary determinants of the soil order and the great group class assigned to a soil. The Canadian System of Soil Classification includes 10 soil orders: Chernozemic, Brunisolic, Luvisolic, Podzolic, Cryosolic, Gleysolic, Organic, Solonetzic, Regosolic, and Vertisolic (Table 19.5). Figure 19.20 shows their distribution in Canada, together with representative soil profiles.

Chernozemic Order

Soils of the *Chernozemic order* are the predominant soils of the agricultural regions in the Canadian Prairies. They have thick, dark *A* horizons, rich in organic matter. These are designated as *Ah* horizons and originate from the decomposition of the grasses that grow abundantly in these regions. The *B* horizon is well aerated and is characteristically yellowish-brown in colour due to the presence of iron oxides, such as goethite (FeOOH) and haematite (Fe_2O_3). Typically, the *C* horizon contains varying amounts of calcium, which is deposited through evaporation

TABLE 19.5 SUMMARY OF THE SOIL ORDERS IN THE CANADIAN SYSTEM OF SOIL CLASSIFICATION

ORDER	DIAGNOSTIC HORIZON	COMMENTS
Chernozemic	*Ah*, *Ap*, *Ahe*	A grassland soil whose diagnostic horizon is formed by high levels of organic matter additions from the roots of grasses.
Brunisolic	*Bm*	A forest soil whose properties are not strongly enough developed to meet the criteria for the Luvisolic or Podzolic orders.
Luvisolic	*Bt*	A forest soil found in areas with parent materials derived from sedimentary rocks. Dominant process is eluviation of clay from the *Ae* horizon and its deposition in the *Bt* horizon.
Podzolic	*Bf* or *Bh*	A forest soil normally associated with coniferous vegetation on igneous rock-derived parent materials. High acidity in the *A* horizon results in formation of a bleached *Ae* horizon and deposition of iron and aluminum in the *B* horizon.
Cryosolic	*By*, *Cy*, *Cz*	A soil of arctic and tundra regions; characterized by presence of permafrost.
Organic	*O* horizon	Organic soils are associated with the accumulation of organic materials (peat) in water-saturated conditions. They are most commonly associated with boreal forest soils.
Gleysolic	*Bg*, *Cg*	Found throughout Canada wherever temporary or permanent water saturation cause formation of gleyed features in the profile.
Solonetzic	*Bn* or *Bnt*	A grassland soil with high sodium levels in the *B* horizon; usually associated with a clay-rich *B* horizon and often with saline *C* horizon material.
Regosolic	No *B* horizon	Found throughout Canada wherever pedogenic conditions prevent the formation of *B* horizons (unstable slopes, sand dunes, flood plains, etc.).
Vertisolic	*Bss*, or *Css* and *Bv*	Associated with high clay glacio-lacustrine landscapes; characterized by shrinking and swelling of clays.

Source: Soils of Canada, http://www.soilsofcanada.ca/orders/index.php.

during periods of drought. The presence of a *Cca* horizon is a noted characteristic of these soils. **Chernozemic soils** form in areas that have cold winters, hot summers, and low precipitation that quickly evaporates in the summer heat. These soils range in colour from black to various shades of brown and grey, depending on how much organic matter has been incorporated into the profile. The deepest and darkest soils have developed under tall-grass prairie in more easterly regions of the Prairies, especially in Manitoba where precipitation is a little higher. Toward the west, the soils become shallower and paler in colour. The amount of calcium in the profile increases and is found progressively closer to the surface as drought conditions become more intense and prolonged. Chernozems are rich in nutrients, have excellent structure, and have good water-holding capacity, and so are ideal for agriculture.

Brunisolic Order

Brunisolic soils typically lack the degree of horizon development found in many of the other soil orders. They are generally associated with coniferous or deciduous forests and occur in a wide range of climates throughout Canada, but especially in cooler, drier regions where mean annual temperatures are near 0 °C and precipitation is less than 700 mm. They are characteristically associated with sandy glacio-fluvial deposits and glacial tills. The Brunisols are regarded as a transitional soil type better developed than Regosols, and which may eventually acquire the properties of mature forest soils of the Podzolic and Luvisolic orders. A distinctive character of these soils is the presence of a *Bm* horizon that is brownish in colour and may contain small accumulations of aluminum and iron or clay; it has undergone minimal pedogenic alteration. Brunisols develop under a range of drainage conditions from good to imperfect drainage. Although suitable for agriculture, Brunisols tend to be somewhat acidic, with pH values often as low as 5.5.

Luvisolic Order

Luvisolic soils are characteristic of forested regions, underlain by loamy tills derived from glacially eroded sedimentary parent materials. These soils are rich in calcium and magnesium or fine lacustrine sediments. They are well- to imperfectly drained soils. The largest area of these soils occurs in the central to northern interior plains under deciduous, mixed, and coniferous forest. Luvisolic soils have a surface humus layer overlying a leached, light-greyish eluvial *A* horizon that is low in clay and iron-bearing minerals. Beneath this *Ae* horizon is an illuvial *Bt* horizon that is enriched with clay and often exhibits high levels of available nutrient ions, such as calcium, magnesium, and potassium. The genesis of Luvisolic soils is

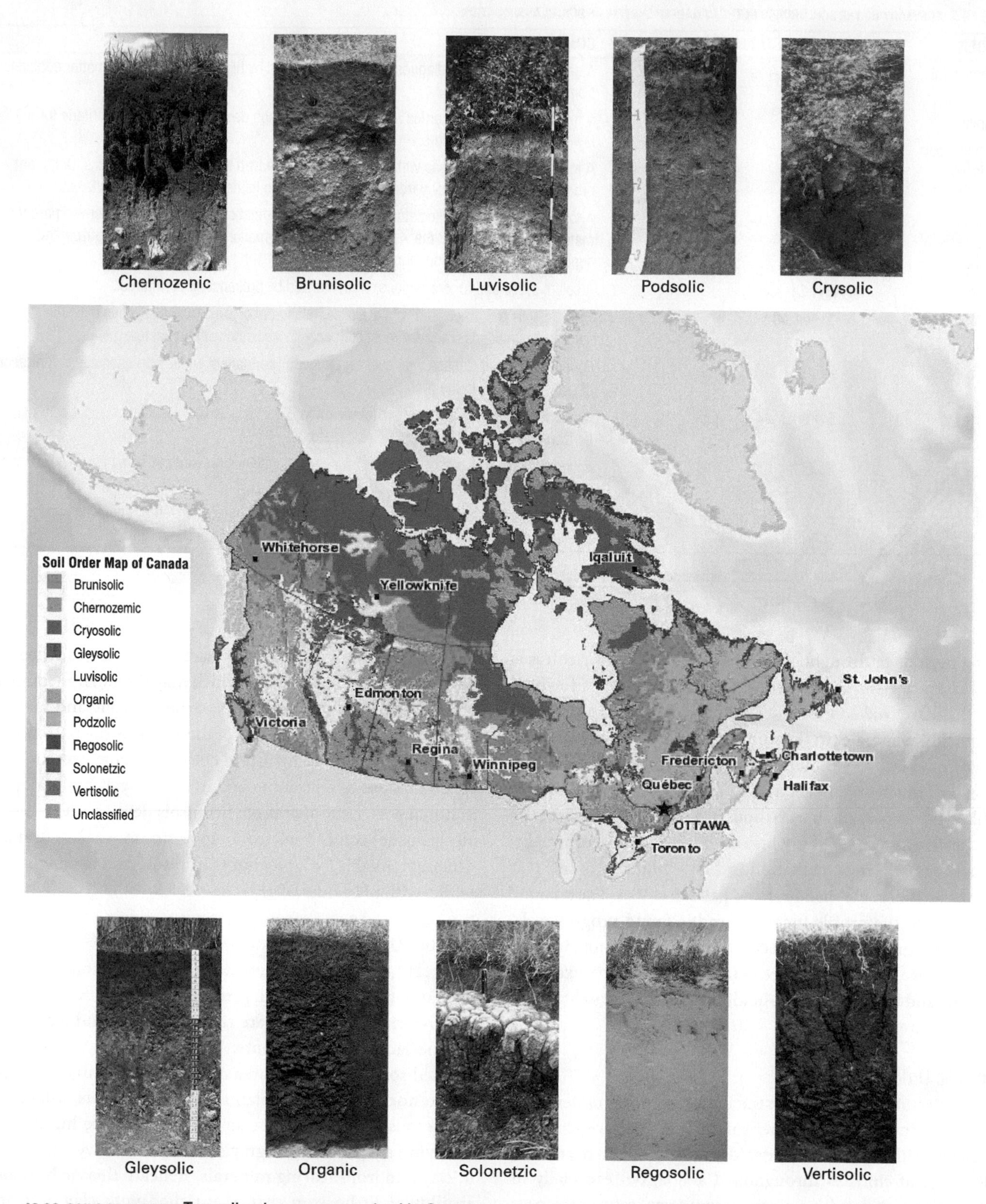

19.20 SOILS OF CANADA Ten soil orders are recognized in Canada.

thought to involve the suspension of clay in the soil solution near the soil surface. The clay is then carried down the profile through a process known as *lessivage* and deposited at a depth where downward motion of the soil solution ceases or becomes very slow. The transfer of clay from the surface to the *B* horizon begins after it has been removed by decalcification. The presence of $CaCO_3$ causes clay minerals to aggregate into larger particles that are not readily eluviated.

Podzolic Order

Podzolic soils typically form under coniferous forests in cool, moist regions where abundant precipitation causes pronounced leaching. Podzols are especially well-developed where parent materials, such as granite or sandstone, are rich in quartz. They are also found on sandy glacio-fluvial deposits. The cool, moist climates do not favour rapid breakdown of organic litter by soil organisms. Consequently, the soil surface is typically covered by coniferous needles in various stages of decomposition, which form distinctive *L*, *F*, and *H* layers of pronounced acidity, 5 to 10 cm thick. Organic acids, such as tannins dissolved from the litter, increase the leaching potential of percolating water.

The constant downward movement of water carries nutrients from the upper layers, and acidity characteristically is below pH 5.5. Podzols are usually distinguished by the presence of a marked eluviated horizon from which iron and aluminum sesquioxides, clays, and organic material have been removed. The resulting *Ae* horizon is characteristically pale ash-grey in colour and sandy in texture. Redeposition in the *B* horizon results in a dark-coloured *Bh* horizon that is enriched with organic matter. Beneath this is a rust-coloured *Bf* horizon, in which the iron and aluminum sesquioxides accumulate. If abundant, these minerals can form a hard, impermeable layer that restricts drainage of water through the soil. Where present, such hardpans are designated as *Bc* horizons.

Podzols are formed under conditions of severe leaching, which can leave the upper horizons virtually depleted of all soil constituents except quartz grains. Some of these constituents are re-deposited in the lower part of the soil profile, but most of the plant nutrients can be carried away in drainage waters. Consequently, podzols are not ideal for agriculture, although productivity can be improved with additions of lime and other fertilizers.

Cryosolic Order

Cryosolic soils occupy much of the northern third of Canada, where permafrost remains close to the surface of both mineral and organic deposits. Cryosols predominate in arctic tundra north of the tree line, but are also common in the open subarctic forests and extend into the boreal forest, especially in some organic materials. Cryosols are occasionally found in alpine areas of mountainous regions.

The formation of Cryosols is predominantly influenced by severe climatic conditions. Short, cool summers and long winters favour the accumulation of organic matter without decomposition because of limited microbial activity. A unique characteristic of Cryosols is the presence of a perennially frozen layer (*Cz*), typically 40 to 80 cm below the surface. This invariably leads to intense seasonal freeze–thaw activity. Above the *permafrost layer*, the upper portion of the soil thaws in the summer to form the *active layer*. The thickness of the active layer is controlled by factors such as soil texture, soil moisture regime, and thickness of the insulating organic surface layer.

At the end of the summer, freezing in the active layer occurs upward from the frost table and downward from the soil surface. Soil material trapped between these freezing fronts comes under *cryostatic pressure*, resulting in contorted and displaced soil horizons. Coarser rock fragments are also heaved and sorted, which further disturbs the soil profile and leads to the formation of patterned ground (see Chapter 18).

Weak leaching and little translocation of materials occur in permafrost soils because the frozen subsoil impedes drainage and cryoturbation mixes the soil materials in the active layer. In winter, the soils are frozen and firm, but during summer, the surface horizons become soft and pliable and can become waterlogged, leading to gleying in the *C* horizon (*Cg*). The pH of Cryosols largely depends on the composition of the parent material, and this is accentuated by cryoturbation, which continually mixes materials within the entire soil profile. The nutrient content of Cryosols is low, with most of the plant nutrients released slowly from organic matter.

Cryosols are sensitive to disturbance, especially when the insulating organic layer is removed, leading to increased thawing of the underlying permafrost and the potential for severe thermokarst development. Cryosols are an important store of organic carbon under present climate conditions. However, increased decomposition rates due to global warming could potentially release carbon dioxide and methane to the atmosphere and accelerate climate change.

Gleysolic Order

Gleysolic soils have features indicating periodic or prolonged saturation, and are usually associated with either a high groundwater table for part of the year, or temporary saturation above a relatively impermeable layer. They are particularly abundant in the low-lying river basins, but commonly occur in patches among other soils in the landscape.

Gley soils typically have a thick organic horizon (*Ah*) at the surface due to slower rates of decomposition imposed by anaerobic conditions in the saturated profile. The underlying *B* horizon is greyish in colour. When accompanied by streaks and patches of greenish-blue or orange-brown, the mottled horizons are designated *Bg*. Mottling develops when waterlogging restricts oxygen diffusion into the soil from the atmosphere. Under these conditions, anaerobic microorganisms extract oxygen from chemical compounds. For example, iron sesquioxide (Fe_2O_3), when reduced to ferrous oxide, imparts a greenish-blue colour to the soil (Figure 19.21). An orange-brown mottled colour is usually indicative of localized re-oxidation of ferrous salts and is often associated with root channels or cracks that develop during dry spells.

Organic Order

The Organic order includes most of the soils commonly known as peat, muck, or bog soils, representing accumulations of partially or completely decomposed plant materials formed under anaerobic conditions. **Organic soils** are often associated with poorly drained depressions and so are saturated with water for prolonged periods. However, they can also form under cool, wet climatic conditions, which in combination with high acidity and nutrient deficiency, restrict microbiological decomposition.

Typical organic soils have greater than 30 percent organic matter content and a surface organic horizons of 40 to 60 cm, although layers as thin as 10 cm sometimes develop directly over bedrock. Different types of Organic soils are recognized according to the degree of decomposition of the organic material. For example, fibrisols are composed mainly of organic matter that has not decomposed and are distinguished by an *Of* horizon; in mesisols, an *Om* horizon is present in which decomposition is

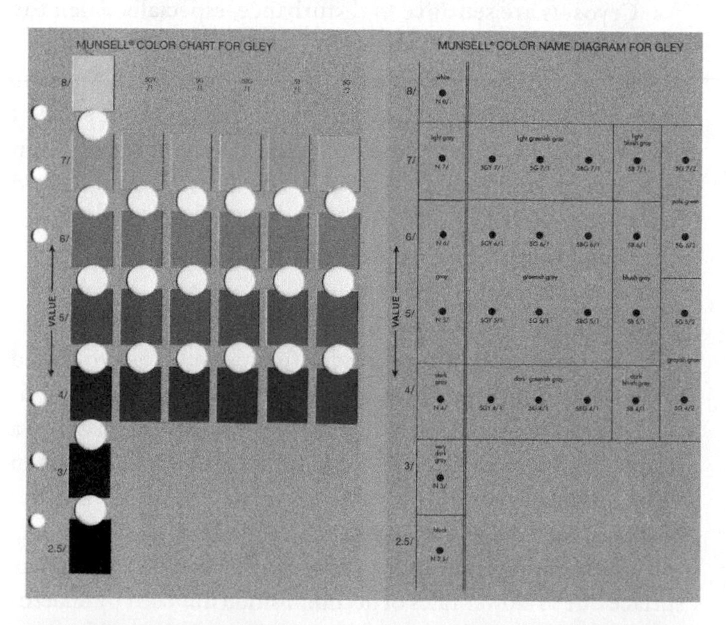

19.21 MUNSELL COLOUR CHART FOR GLEY Green, blue, and grey patches and streaks in a soil are characteristic of the gley conditions associated with waterlogged soil.

partially complete; and in humisols, most of the organic material has been broken down into a soft amorphous mass to form an *Oh* horizon.

Organic soils possess high water-holding capacity and low load-bearing strength. They are usually acidic, with a pH lower than 5, unless supplied with nutrient-rich drainage water that has percolated through adjacent mineral soils. Organic soils supplied with nutrients in this way support *fens*. Fens are biologically diverse communities of sedges, reeds, and rushes, and water-tolerant shrubs and trees such as willows and alder. Fen soils will commonly intergrade with marshes and gleyed mineral soils. Organic soils that are supplied only by precipitation develop into highly acidic *bogs*, often dominated with Sphagnum mosses.

Solonetzic Order

Soils of the *Solonetzic order* occur on saline parent materials in some areas of the semi-arid to subhumid prairies, in association with Chernozemic soils and, to a lesser extent, with Luvisolic and Gleysolic soils. Most **Solonetzic soils** are associated with a vegetation cover of grasses and forbs. They are primarily associated with glacial tills derived from sedimentary rocks. Solonetzic soils are distinguished by their high sodium content. They are formally identified by the presence of a *Bn* horizon in which the ratio of calcium to sodium is 1:10 or less. The high sodium content arises either directly from sodium-rich parent materials, such as marine shales, or because it has been brought into the soil by upward moving groundwater and subsequently deposited through evaporation.

The presence of sodium increases the mobility of clay minerals, which are carried downward by percolating water. In this way, the surface horizons are depleted of clay that is then deposited in pores in the *B* horizon, and soil drainage is progressively reduced. Solonetzic soils are characterized by indurated, almost impermeable *hardpans* that may vary from 5 to 30 cm or more below the surface. Hardpan severely restricts root and water penetration into the subsoil. Although these *subsurface* horizons are typically hard when dry, they swell to a sticky, plastic mass of low permeability when wet. This shrinking and swelling of the clay produces a distinctive prismatic or columnar structure in the soil, described as a *Bnt* horizon (Figure 19.22).

The productivity of Solonetzic soils is low, due to the limited depth of topsoil and hardness of the *Bnt* horizon. Moisture is usually the major limitation, but alkalinity is also a problem. Solonetzic soils tend to be low in organic matter, and reserves of nutrients are often small. Deep plowing and, more recently, subsoiling, have been developed to improve these soils. In deep plowing, calcium in the form of lime is mixed into the *Bnt* horizon to replace the sodium and prevent the hardpan from re-forming. Subsoiling accomplishes little mixing of the soil horizons, but may improve crop growth by shattering the hardpan.

19.22 **SOLONETZIC SOIL** High sodium content gives rise to the distinctive white-capped, columnar structure seen in this dry, prairie soil.

Regosolic Order

Regosolic soils (Figure 19.23) develop on unconsolidated parent materials, such as dune sands and alluvium. They have weakly developed profiles in which *B* horizons are absent. Typically, only a thin humic *Ah* horizon and a surface litter layer overlies the unaltered parent substrate, resulting in *AC* profiles. This lack of development may be due to the youthfulness of the parent material, as is the case for recently deposited alluvium, or instability, as occurs in colluvium on slopes subject to mass wasting. Drainage in Regosolic soils is quite variable and can be very rapid in materials such as dune sand, but slow in finer alluviums. Regosols occur in a wide range of vegetation and climates, but generally are of limited extent in Canada.

Vertisolic Order

Soils of the *Vertisolic order* occur in parent materials rich in montmorillonite clays that expand greatly when wet and then shrink excessively when dry. Shrinking and swelling is strong enough that horizons characteristic of other soil orders have either been prevented from forming or have been severely disrupted. This is because **Vertisolic soils** develop deep cracks as they dry, and material from the upper part of the soil profile falls into them. Thus, Vertisolic soils are self-mixing and lack distinctive horizons. Typically they have deep *Ah* horizons that merge with an indistinct *Bv* horizon, which is disrupted by shrinking and swelling of the soil mass; this process is called *argillipedoturbation*.

The Vertisolic order was only recently recognized in Canada. These soils are mostly found in the cool, semi-arid to subhumid grasslands of the interior plains of western Canada. Vertisolic soils typically develop on glacio-lacustrine clays, and are especially characteristic of flat, ancient glacial lake beds such as those that occur near Regina and Winnipeg. Minor areas of Vertisolic soils occur in valleys in the western mountains, in parts of the southern boreal forest, and in the cool temperate regions of central Canada.

19.23 **REGOSOL** Regosols occur extensively in arid and semi-arid regions where they are associated with coarse-textured, unconsolidated parent materials. Apart from a thin organic layer at the surface, they lack well-developed horizons.

Vertisols typically form from non-acidic parent materials in climates that are seasonally humid. In terms of colour and organic matter content, Vertisols are usually dark and may have a chernozemic-like *A* horizon at the surface, or they may have mottling in the *B* horizon that is a characteristic feature of soils of the Gleysolic order. However, they are distinguished from both of these by the presence in the lower *B* horizon of slickensides (*Bss*)—polished shear surfaces that form when one soil mass moves over another—and a vertic (*v*) horizon that has been strongly mixed by shrinking and swelling.

TABLE 19.6 SOIL ORDERS

GROUP I Soils with well-developed horizons or fully weathered minerals, resulting from long-continued adjustment to prevailing soil temperature and soil water conditions.			
Oxisols	Very old, highly weathered soils of low latitudes, with a subsurface horizon of mineral oxide accumulation and very low base status.	**Alfisols**	Soils of humid and subhumid climates with a subsurface horizon of clay accumulation and high base status. Alfisols range from equatorial to subarctic latitude zones.
Ultisols	Soils of equatorial, tropical, and subtropical latitude zones, with a subsurface horizon of clay accumulation and low base status.	**Spodosols**	Soils of cold, moist climates, with a well-developed *B* horizon of illuviation and low base status.
Vertisol	Soils of subtropical and tropical zones with high clay content and high base status. Vertisols develop deep, wide cracks when dry, and the soil blocks formed by cracking move with respect to each other.	**Mollisols**	Soils of semi-arid and subhumid midlatitude grasslands, with a dark, humus-rich epipedon and very high base status.
		Aridisols	Soils of dry climates, low in organic matter, and often having subsurface horizons of carbonate mineral or soluble salt accumulation.
GROUP II Soils with a large proportion of organic matter.			
Histosols Soils with a thick upper layer very rich in organic matter.			
GROUP III Soils with poorly developed or no horizons that are capable of further mineral alteration.			
Inceptisols Soils with weakly developed horizons, having minerals capable of further alteration by weathering processes.			
Entisols Soils lacking horizons, usually because their parent material has accumulated only recently.			
Andisols Soils with weakly developed horizons, having a high proportion of glassy volcanic parent material produced by erupting volcanoes.			

SOILS OF THE WORLD

The soils of the world present a much broader range of soil types and conditions than those found in Canada. The following survey of global soils uses the U.S. Comprehensive Soil Classification System (CSCS).

At the highest level, the CSCS system recognizes three groups of soil orders (Table 19.6). The largest group includes seven orders with well-developed horizons or fully weathered minerals. A second group includes a single soil order that is rich in organic matter. The last group includes three soil orders with poorly developed or no horizons. Although each order has several suborders, in this summary only the suborders associated with the Alfisols are discussed.

The world soils map (Figure 19.24) shows the general distribution of the soil orders and representative profiles of each. Note that the Alfisols have been subdivided into four important suborders that correspond well to four basic climate zones. Entisols, Inceptisols, Histosols, and Andisols are not shown as these orders, for the most part, only occur locally. In highland regions, the soil patterns are too complex to show at a global scale.

Three soil orders dominate the low latitudes: Oxisols, Ultisols, and Vertisols. Soils of these orders have developed over long time spans in a warm environment. Soil water is periodically abundant during a wet season or is available throughout the year because of high rainfall in all months.

Oxisols

Oxisols develop on stable land areas in equatorial, tropical, and subtropical regions with large water surpluses. They are principally associated with rainforests and occur throughout the wet equatorial climate zone in Africa, South America, and Asia.

Oxisols usually lack distinct horizons, except for darkened surface layers. Soil minerals are weathered to an extreme degree and are dominated by stable aluminum (Al_2O_3) and iron (Fe_2O_3) sesquioxides. Red, yellow, and yellowish-brown colours are normal. The base status of Oxisols is very low; nearly all of the nutrients required by plants have been removed from the soil profile, either through storage in the vegetation or leaching by percolating water. The soil is quite easily broken apart and allows easy penetration by rainwater and plant roots.

Ultisols

Ultisols are similar to the Oxisols in appearance and environment of origin. Ultisols are reddish to yellowish in colour. They have a subsurface horizon of illuviated clay, which is not found in the oxisols. Although forest is the characteristic native vegetation, the base status of the Ultisols is low like that of the Oxisols. Hence, the shallow rooting trees and shrubs must take up and recycle nutrients quickly.

Ultisols are widespread throughout Southeast Asia and the East Indies. Other important areas include South America,

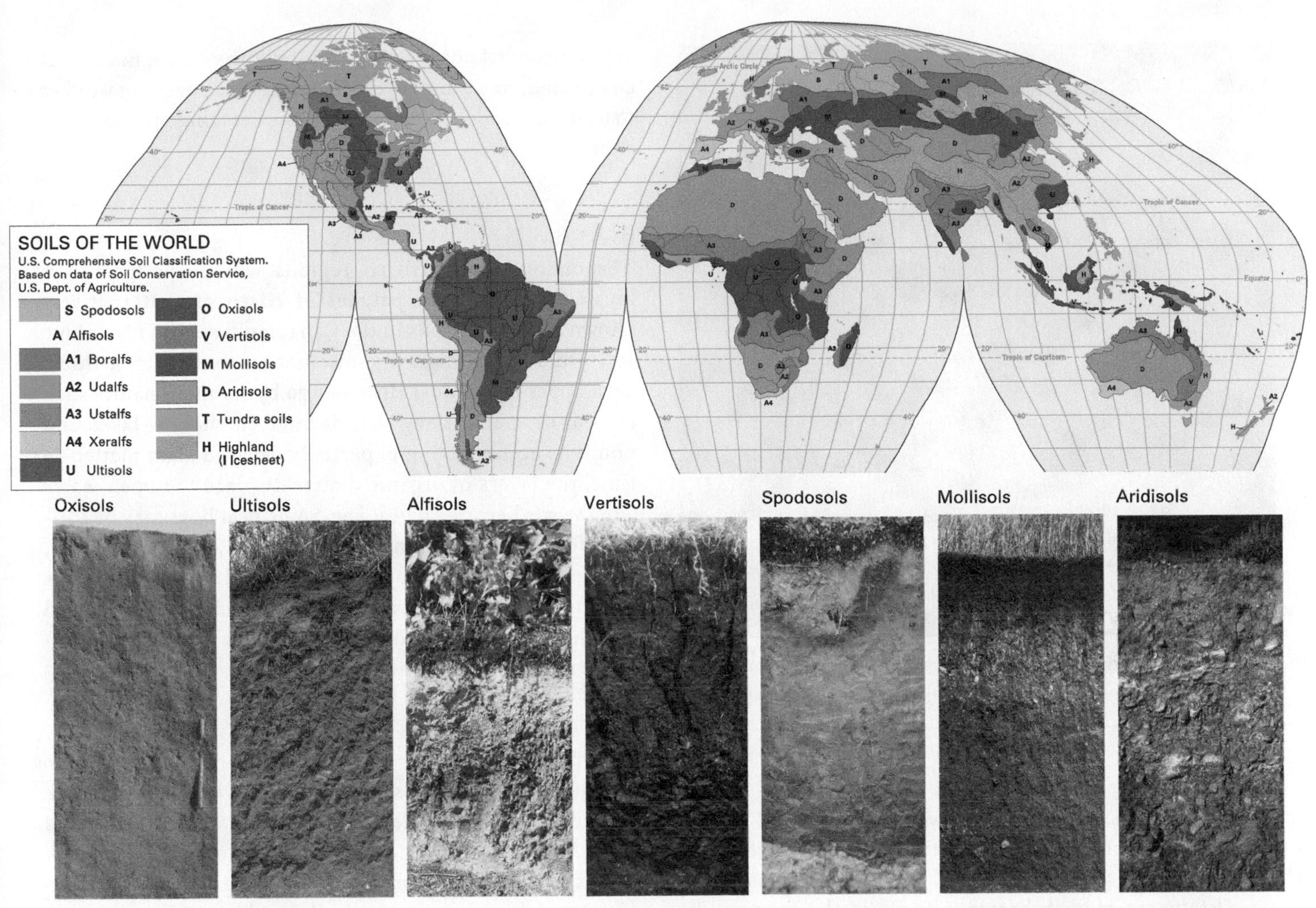

19.24 SOILS OF THE WORLD The map shows the global distribution of the soil orders used in the CSCS classification. The representative profiles are for the Group I soils of the CSCS.

Africa, and northeastern Australia. Ultisols also occur in the southeastern United States, where they correspond with the area of moist subtropical climate. In lower latitudes, Ultisols are identified with the wet–dry tropical climate and the monsoon and trade-wind coastal climate. Note that all these climates have a dry season, even though it may be short.

Vertisols

Vertisols are typically black in colour and have a high clay content. The predominant clay mineral is montmorillonite, which shrinks and swells with seasonal changes in soil water content. Wide, deep vertical cracks develop in the soil during the dry season (Figure 19.25). When moistened and softened by rain, some fragments of surface soil drop into the cracks before they close, so that the soil is constantly being mixed. Vertisols are equivalent to Vertisolic soils in the Canadian system.

Vertisols typically form under grass and savanna vegetation in subtropical and tropical climates with a pronounced dry season. Because Vertisols require a particular clay mineral as a parent material, the major areas of occurrence are scattered and show no distinctive pattern on the world map. An important region of Vertisols is the Deccan Plateau of western India, where basalt supplies silicate minerals that weather into montmorillonite clay.

Alfisols

Alfisols are characterized by a pale-coloured eluviated *A* horizon that is low in bases, clay minerals, and sesquioxides. These materials are concentrated by illuviation in the *B* horizon where the clay holds bases, such as calcium and magnesium. The base status of the Alfisols, therefore, is generally quite high.

Alfisols are distributed throughout the world, ranging from latitudes as high as 60° N in North America and Eurasia to the equatorial zone in South America and Africa. Because the Alfisols occur in many climate types, four important suborders are noted.

Boralfs are Alfisols of the cold (boreal) forest lands of North America and Eurasia. They have a grey surface horizon and a brownish subsoil. *Udalfs* are brownish Alfisols of the midlatitude zone. They are closely associated with the moist

19.25 **A VERTISOL IN AUSTRALIA** The clay minerals that are abundant in Vertisols shrink when they dry out, producing deep cracks in the soil surface.

continental climates in North America, Europe, and eastern Asia. These suborders are equivalent to Luvisolic soils in Canada.

Ustalfs are reddish-brown Alfisols of the warmer climates. They range from the subtropical zone to the equator, and are associated with the wet–dry tropical climate in Southeast Asia, Africa, Australia, and South America. *Xeralfs* are Alfisols of the Mediterranean climate, with its mild, moist winter and dry summer. The Xeralfs typically are brownish or reddish in colour.

Spodosols

Spodosols are formed in the cold boreal climate beneath coniferous forest. They are equivalent to the Podzolic soils of the Canadian system. Spodosols are distinguished by their acidic, reddish-orange *B* horizons. This horizon is called the *spodic* horizon and is composed of organo-aluminum and iron compounds brought downward by eluviation. Intensive leaching produces a conspicuous, ash-grey sandy horizon in the upper part of the soil profile, with a dark organic layer at the soil surface.

Spodosols are closely associated with regions recently covered by the great ice sheets of the Late-Cenozoic Era. These soils are therefore relatively young. Typically, the parent material is coarse sand, consisting largely of quartz, which does not weather to form clay minerals. Because of heavy leaching, spodosols are strongly acidic.

Histosols

Throughout the northern regions where Spodosols are located are countless patches of **Histosols**, alternatively known as Organic soils in the Canadian system. This unique soil order has a high organic matter content in a thick, dark upper layer. Most Histosols go by common names such as peat or muck. They have formed in shallow lakes and ponds by accumulation of partially decayed plant matter. In time, the layers of organic matter displace the open water and the site becomes a *peat bog*. Some peat bogs may be suitable for the cultivation of cranberries (cranberry bogs). Sphagnum peat from bogs is dried and baled for sale as mulch for use on suburban lawns and shrubbery beds. Dried peat has also been used for centuries in Europe as a low-grade fuel.

Entisols

Entisols are mineral soils that lack distinct horizons. They are soils in the sense that they support plants, and may be found in any climate under various types of vegetation. Entisols lack distinct horizons for two reasons. It may be the result of a parent material, such as quartz sand, in which horizons do not readily form. Alternatively, there may have been insufficient time for horizons to form in transported materials such as sand drifts and recent deposits of alluvium, or on actively eroding slopes. In the Canadian system, these soils are known as Regosolic soils.

Inceptisols

Inceptisols are soils with horizons that are weakly developed, usually because the soil is relatively young. Inceptisols occur quite extensively in the same regions as Ultisols and Oxisols. Especially important are the Inceptisols of the river flood plains and delta plains in Southeast Asia, which support dense populations of rice farmers. Here, annual river floods cover low-lying plains and deposit layers of fine silt. This sediment is rich in primary minerals that yield base nutrients as they weather chemically over time. Inceptisols are also found in glaciated areas, where young sediment is present. Equivalent soils in Canada are the Brunisolic soils.

Andisols

Andisols are soils in which more than half of the parent mineral matter is volcanic ash. They are, for the most part, fertile soils and, in moist climates, they support a dense natural vegetation cover. These soils are usually very dark in colour because of the high proportion of carbon contributed by decomposition of plant matter. Andisols form over a wide range of latitudes and climates, and are found in localized patches associated with individual volcanoes.

Mollisols

Mollisols are soils associated with the semi-arid and subhumid midlatitude grasslands. Mollisols have a thick, dark brown to black surface horizon with a loose, granular structure. Calcium is abundant in the *A* and *B* horizons, which together with other plant nutrients gives these soils a high base status. Mollisols are some of the most naturally fertile soils and now produce most of the world's commercial grain crop.

Mollisols are the dominant soils in the Great Plains region of North America and occur extensively in the Pampas of Argentina and Uruguay. In Eurasia, a great belt of Mollisols stretches across the steppes of Russia. The Chernozemic soils of the Prairie Provinces are the Canadian equivalent of the Mollisols.

Aridisols

Aridisols, the soils of desert regions, are dry for long time periods. Because the climate supports only very sparse vegetation, humus is lacking and the soil colour ranges from pale grey to pale red. Soil horizons are weakly developed, but there may be important subsurface horizons of accumulated calcium carbonate and other soluble salts, especially sodium. The salts give the Aridisols a high degree of alkalinity. The closest equivalent in the Canadian soil classification would be soils of the Solonetzic order.

Locally, where water supplies permit irrigation, Aridisols can be highly productive for a wide variety of crops. Irrigation in areas such as the Imperial Valley of the United States, the Nile Valley of Egypt, and the Indus Valley of Pakistan have greatly increased the agricultural value of Aridisols, but not without problems of salt buildup and waterlogging.

Tundra Soils

Tundra soils (Figure 19.26) are formed mainly of primary minerals, ranging in size from silt to clay, that have been broken down by frost and glacial action. Layers of peat are often present between mineral layers. Beneath the tundra soil is perennially frozen ground (permafrost). Because the annual summer thaw affects only a shallow surface layer, soil water cannot easily drain away. Thus, the soil is saturated with water over large areas. Repeated freezing and thawing of the active layer disrupts plant roots, so that only small, shallow-rooted plants can maintain a hold. In the Comprehensive Soil Classification System, tundra soils are classed as wet subtypes of Inceptisols, but in the Canadian system, they are recognized at the highest level as the Cryosolic soil order.

19.26 TUNDRA SOIL Tundra soils, classed as Cryosol in the Canadian system, are mineral soils that develop in regions of permafrost. They characteristically exhibit poor horizonation due to frost heaving, cracking, and mixing during the annual freeze-thaw cycle of the active layer.

A Look Ahead

Soil is the complex outermost layer that covers most of the Earth's continents. It is influenced by many factors, including the parent materials from which it is derived, the vegetation that it harbours, and the water regime of precipitation and evapotranspiration that it experiences. In many environments, soil-forming processes operate very slowly, and because they respond to ambient conditions, soils reflect a complex history of environmental change. By inducing soil erosion, human activities can rapidly strip the uppermost soil horizons, leaving less productive layers at the surface. Soil horizons and properties can also vary markedly over short distances. This not only affects crop production, but is an important factor controlling natural vegetation patterns and ecosystem processes. This is discussed in the three chapters that follow.

CHAPTER SUMMARY

- The soil layer is a complex mixture of solid, liquid, and gaseous components. It is derived from parent material, or regolith, that is produced from rock by *weathering*. The major factors influencing soil and soil development are parent material, topography, climate, vegetation, and time.
- The type of substrate from which the soil is derived constitutes the *parent material*. In most soils, the inorganic material present accounts for some 40 to 60 percent of the soil volume.
- The effect of *topography* is linked to slope angle and aspect that affect infiltration, erosion rates, and soil temperature.
- Precipitation controls the downward movement of nutrients and other chemical compounds in soils and provides a medium for many chemical reactions. Soil temperature affects the intensity of chemical processes affecting soil minerals, and also influences biologic activity.
- The characteristics and properties of soils usually develop only after a long period of *time*, usually many centuries.
- *Biological activity* arising from the presence of living plants and animals, as well as their non-living organic products, is essential for good soil development. Through *humification*, organic matter is broken down by bacterial decay. Animals, such as earthworms, can be very important in soil formation.
- Soil texture refers to the proportions of *sand*, *silt*, and *clay* that are present.
- Colloids are the finest particles in soils and are important because they help retain nutrients, or bases, that plants use. Soils show a wide range of *pH values*, from acid to alkaline. Soils with granular or blocky structures are most easily cultivated.
- In soils, *primary minerals* are chemically altered to form secondary minerals, which include oxides and *clay minerals*. The nature of the clay minerals determines the soil's base status. If *base status* is high, the soil retains nutrients. If low, the soil can lack fertility. When a soil is fully moistened by heavy rainfall or snow melt and allowed to drain, it reaches its *storage capacity*. Evaporation from the surface and transpiration from plants draws down the soil water store until precipitation occurs again to recharge it.
- The soil water balance describes the gain, loss, and storage of soil water. It depends on water need *(potential evapotranspiration)*, water use *(actual evapotranspiration)*, and precipitation. Monitoring these values on a monthly basis provides a *soil water budget* for the year.
- Most soils possess distinctive layers called horizons. These layers are developed by processes of *enrichment*, *removal*, *translocation*, and *transformation*. In downward translocation, materials such as humus, clay particles, and mineral oxides are removed by eluviation from an upper horizon and accumulate by illuviation in a lower one. In *salinization*, salts are moved upward by evaporating water to form a *salic horizon*.
- Soil classification systems typically group soils together according to inherent properties and formation process rather than by usage. The World Reference Base for Soil Resources (WRB) was developed by the United Nations Food and Agriculture Organization (FAO) as a comprehensive global classification system.
- Canada's soil classification system emphasizes soils of cold regions and permafrost terrains, while the American system is more global in scope. These systems use the presence and nature of *diagnostic horizons* to distinguish *soil orders* and *suborders* or *great groups*. The Canadian system recognizes 10 soil orders, while the American has 11 orders.
- Soils of Canada's *Chernozemic order* have a dark, rich, loose *A* horizon with carbonate accumulation in the *B* horizon. These prairie soils develop on the semi-arid grasslands of the Prairie Provinces.
- Soils of the *Brunisolic order* lack strong horizon development. They underlie northern vegetation covers ranging from boreal forest to heath and tundra.
- Soils of the *Luvisolic order* have a light-coloured *A* horizon from which clay has eluviated and a *B* horizon where clay has accumulated. They are found under forest, south of the permafrost zone.
- Soils of the *Podzolic order* typically have an organic layer overlying a pale grey *A* horizon, with aluminum and iron oxides accumulating in the *B* horizon. They are found under coniferous forest in cool to cold, moist environments.
- *Cryosolic* soils occupy permafrost terrains. Horizon development is typically disrupted by freeze–thaw mixing and soil flowage.
- Soils of the *Gleysolic order* are typically saturated with water during some or most of the annual cycle. They often show mottling in the *B* horizon, brought about by chemical-reducing conditions during saturation.
- *Organic* soils are composed largely of organic materials that decay very slowly in cool, wet environments.
- Soils of the *Solonetzic order* have *B* horizons that are very hard when dry and sticky and plastic when wet. They develop in semi-arid saline environments in interior plains.
- *Regosolic* soils have weakly developed horizons. Often they are fresh deposits of sediment; for example, recent alluvium and windblown sands.

- Soils of the *Vertisolic order* develop on clays that expand and shrink on wetting and drying, leading to mechanical mixing of the soil. They lack distinct horizons. In Canada, they are found at some locations in the interior plains under grassland.
- The soils of the world can be classified using the U.S. Comprehensive Soil Classification System.
- *Oxisols* are old, highly weathered soils of low latitudes. They have a horizon of mineral oxide accumulation and a low base status. *Ultisols* are also found in low latitudes. They have a horizon of clay accumulation and are also of low base status. *Vertisols* are rich in a type of clay mineral that expands and contracts with moistening and drying, and have a high base status.
- *Alfisols* have a horizon of clay accumulation like Ultisols, but they are of high base status. They are found in moist climates from equatorial to subarctic zones.
- *Spodosols*, found in cold, moist climates, exhibit a horizon of illuviation and low base status. *Mollisols* have a thick upper layer rich in humus. They are soils of midlatitude grasslands. *Aridisols* are soils of arid regions, marked by horizons of accumulation of carbonate minerals or salts.
- *Histosols* have a thick upper layer formed almost entirely of organic matter.
- Three soil orders have poorly developed or no horizons: *Entisols*, *Inceptisols*, and *Andisols*. Entisols have no horizons and often consist of fresh parent material. The horizons of Inceptisols are only weakly developed. Andisols are weakly developed soils occurring on young volcanic deposits.

KEY TERMS

Alfisols	Cryosolic soils	Inceptisols	primary minerals	soil profile	Vertisols
Andisols	eluviation	Luvisolic soils	Regosolic soils	soil texture	water need
Aridisols	Entisols	Mollisols	secondary minerals	Solonetzic soils	water use
bases	Gleysolic soils	Organic soils	soil	Spodosols	
Brunisolic soils	Histosols	Oxisols	soil horizons	Ultisols	
Chernozemic soils	illuviation	Podzolic soils	soil order	Vertisolic soils	

REVIEW QUESTIONS

1. What important factors condition the nature and development of soil?
2. Soil colour, texture, and structure are used to describe soils and soil horizons. Define each of these three terms, showing how they are applied.
3. Explain the concepts of acidity and alkalinity as they apply to soils.
4. Identify two important classes of secondary minerals in soils and provide examples of each class.
5. How does the ability of soils to hold water vary, and how does this ability relate to soil texture?
6. Define water need (potential evapotranspiration) and water use (actual evapotranspiration). How are they used in the soil water balance?
7. Identify the following terms as used in the soil water budget: storage, withdrawal, storage recharge, soil water shortage, water surplus.
8. What is a soil horizon? How are soil horizons named? Provide two examples.
9. Identify four classes of soil-forming processes and describe each.
10. What are translocation processes? Identify and describe four translocation processes.
11. What is a diagnostic horizon? Explain how soils are identified by diagnostic horizons, and provide several examples.
12. How many soil orders are there in the Canadian system? Name them all.
13. Chernozemic and Brunisolic soils are both found in interior Canada. Compare these two soil types.
14. Describe the process by which Luvisolic soils attain their distinctive properties.
15. What are the distinctive features of Podzolic soils? What environments are associated with Podzolic soils?
16. Where are Cryosolic soils located? What are their key features?
17. Gleysolic and Organic soils are both soils of wet environments. How do they differ?
18. What is unique about soils of the Solonetzic order? Where do they occur?
19. What factors contribute to the weak development of horizons in Regosolic soils?

20. What is unique about soils of the Vertisolic order? Where do they occur?
21. How many soil orders are there in the Comprehensive Soil Classification System (CSCS)? Name them all.
22. Name three CSCS soil orders that are especially associated with low latitudes. For each order, provide at least one distinguishing characteristic and explain it.
23. Compare Alfisols and Spodosols. What features do they share? What features differentiate them? Where are they found?
24. Where are Mollisols found? How do the properties of Mollisols relate to climate and vegetation cover?
25. Desert and tundra are extreme environments. Which soil order is characteristic of each environment? Briefly describe desert and tundra soils.
26. Identify five Canadian soil types that have good equivalents in the CSCS global system and name these equivalents.

VISUALIZATION EXERCISES

1. Sketch the profile of a cool, moist forest soil, labelling *L, F, H, Ah, Ae,* and *C* horizons. Indicate the movement of materials from the zone of eluviation to the zone of illuviation.
2. Examine the Canadian soils map (Figure 19.20) and identify three soil types that are found near your location. Develop a short list of characteristics that would help you tell them apart.

ESSAY QUESTIONS

1. Document the important role of clay particles and clay mineral colloids in soils. What is meant by the term *clay*? What are colloids? What are their properties? How does the type of clay mineral influence soil fertility? How does the amount of clay influence the water-holding capacity of the soil? What is the role of clay minerals in horizon development?
2. Using maps showing the global distributions of soils and climates, compare the pattern of soils on a north-south transect along the 20° E meridian with the corresponding climate patterns. What conclusions can you draw about the relationship between soils and climate? Be specific.

WORKING IT OUT • CALCULATING A SIMPLE SOIL WATER BUDGET

The soil water balance is a system in which an input of precipitation flows through soil pathways to be output as evapotranspiration or runoff. The amount of water returned to the atmosphere through evapotranspiration (*Ea*) is determined largely by temperature and the amount of vegetation cover. The amount of runoff will depend, in the long term, on how water use (*Ea*) compares with precipitation (*P*). If precipitation is greater than water use, then runoff will occur, both as overland flow and as groundwater outflow.

Water storage (*S*) in the soil layer is an important characteristic of the system. This storage provides a reserve of water for plants to draw upon, so that water use can exceed precipitation for part of the year. Later, when precipitation exceeds water use, the reserve can be recharged, and only after the reserve is replenished can runoff occur again in the annual cycle.

In this example, the monthly soil water budget is calculated for a station in the marine west-coast climate, which exhibits wet, cool winters and warm, drier summers. For each month, the basic data required are precipitation (*P*), water use (*Ea*), and water need (*Ep*), from which the other quantities are derived. In any month, precipitation can follow only three pathways: to the atmosphere as water use (*Ea*), to streams and rivers as runoff (*R*), or to storage

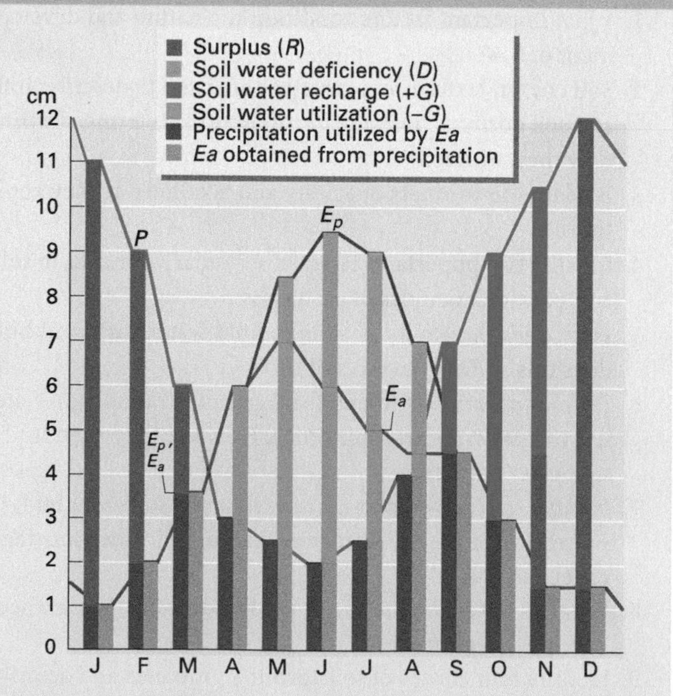

Figure 1 A model soil water budget Bars are scaled to the values in the table.

(+G) within the soil layer. In January, precipitation is 11 cm, and water use is 1 cm, leaving 10 cm for storage or runoff. At this time of the year, storage is full so the entire amount goes to runoff.

The flow of runoff persists until April, when water use (6 cm) exceeds precipitation (3 cm) by 3 cm. This amount is withdrawn from storage, giving the value of −3 for −G. Storage withdrawals continue through August, yielding a total withdrawal of 14.5 cm.

In September, precipitation (7 cm) again exceeds water use (4.5 cm), leaving a surplus (2.5 cm) for +G. This surplus begins to recharge the soil water storage through October and into November, when it is $10.5 - 1.5 = 9$ cm. However, only 6 cm is required to finally balance the total summer withdrawal of 14.5 cm, leaving $9 - 6 = 3$ cm for runoff (R). In December, the surplus $12.0 - 1.5 = 10.5$ cm goes entirely to runoff.

Soil water shortage (D) is the difference between water need (Ep) and water use (Ea). The two quantities are equal until May, when water use drops below water need and water storage withdrawals begin to occur. This condition persists until September, when sufficient precipitation allows water use to equal water need.

Note that, for practical applications in agriculture, the depth of storage water in the soil that is available to field crops is normally limited to 30 cm.

SIMPLIFIED EXAMPLE OF A SOIL WATER BUDGET (CM)

Equation	P =	Ea	−G	+G	+R	Ep	(Ep − Ea) = D
January	11.0 =	1.0			+10.0	1.0	0.0
February	9.0 =	2.0			+ 7.0	2.0	0.0
March	6.0 =	3.5			+ 2.5	3.5	0.0
April	3.0 =	6.0	− 3.0			6.0	0.0
May	2.5 =	7.0	− 4.5			8.5	1.5
June	2.0 =	6.0	− 4.0			9.5	3.5
July	2.5 =	5.0	− 2.5			9.0	4.0
August	4.0 =	4.5	− 0.5			7.0	2.5
September	7.0 =	4.5		+ 2.5		4.5	0.0
October	9.0 =	3.0		+ 6.0		3.0	0.0
November	10.5 =	1.5		+ 6.0	+ 3.0	1.5	0.0
December	12.0 =	1.5			+10.5	1.5	0.0
Totals	78.5 =	45.5	−14.5	+14.5	+33.0	57.0	11.5
	78.5 =	78.5					

QUESTION / PROBLEM

1. Below are monthly values and annual totals for precipitation, water need (Ep), and water use (Ea) at Rockcliffe, Ontario, in the moist continental climate. Prepare and plot a soil water budget for this station similar to Figure 1 at the beginning of this feature. This will involve determining soil water shortage (D), soil water use (−G), soil water recharge (+G), and water surplus (R) for each month, following the method described in feature box.

Find answers at the back of the book.

SOIL WATER BUDGET FOR ROCKCLIFFE, ONTARIO

Month	P=	Ea	−G	+G	+R	Ep	D
Jan	7.4	0.0				0.0	
Feb	5.5	0.0				0.0	
Mar	7.0	0.0				0.0	
Apr	6.9	2.6				2.6	
May	6.3	8.0				8.0	
Jun	8.9	11.8				11.8	
Jul	8.6	13.7				13.7	
Aug	6.5	6.8				11.4	
Sep	8.2	7.4				7.4	
Oct	7.4	3.3				3.3	
Nov	7.6	1.0				1.0	
Dec	6.6	0.0				0.0	
Total	86.9	54.6	−10.0	+10.0	32.3	59.2	4.6

CHAPTER 20

SYSTEMS AND CYCLES OF THE BIOSPHERE

A papaya tree in a tropical microcosm.

This chapter begins the section on **biogeography**, a branch of geography that focuses on the distribution of Earth's plants and animals. It identifies and describes the processes that influence plant and animal distribution patterns at varying scales of space and time. Biogeography encompasses two major themes. *Ecological biogeography* is concerned with how the environment affects the distribution patterns of organisms. *Historical biogeography* focuses on the origins of present spatial patterns of organisms; it studies the processes of evolution, migration, and extinction of species. This chapter examines how ecosystems cycle energy and matter. Chapter 21 discusses how organisms live and interact with their environment from the perspectives of ecological and historical biogeography. The final chapter of the book describes the major biomes of the world.

ENERGY FLOW IN ECOSYSTEMS

THE FOOD WEB

Energy transformations in an ecosystem occur through a series of *trophic levels*, or feeding levels, that are collectively referred to as a **food chain** or **food web** (Figure 20.1).

The plants and algae in a food web are the **primary producers** and make up the first trophic level. These organisms use light energy to convert CO_2 and water into carbohydrates, and eventually into other biochemical molecules needed to support life. The primary producers support the **consumers**—organisms that ingest other organisms as their food source. The *primary consumers* or herbivores are the lowest level of consumers. Next are the *secondary consumers* or carnivores, which feed on the primary consumers. More than one level of carnivore is usually present in an ecosystem, as larger predators normally feed on smaller carnivores. Some animals are **omnivores** and can feed on both plant and animal materials. Omnivores tend to be more selective in their plant food requirements than herbivores, and will normally feed on fruits and grains to supplement their intake of animal tissue. The **decomposers** use *detritus*, or decaying organic matter, derived from all feeding levels. This group includes insect larvae and worms, but is mainly represented by fungi, microorganisms, and bacteria.

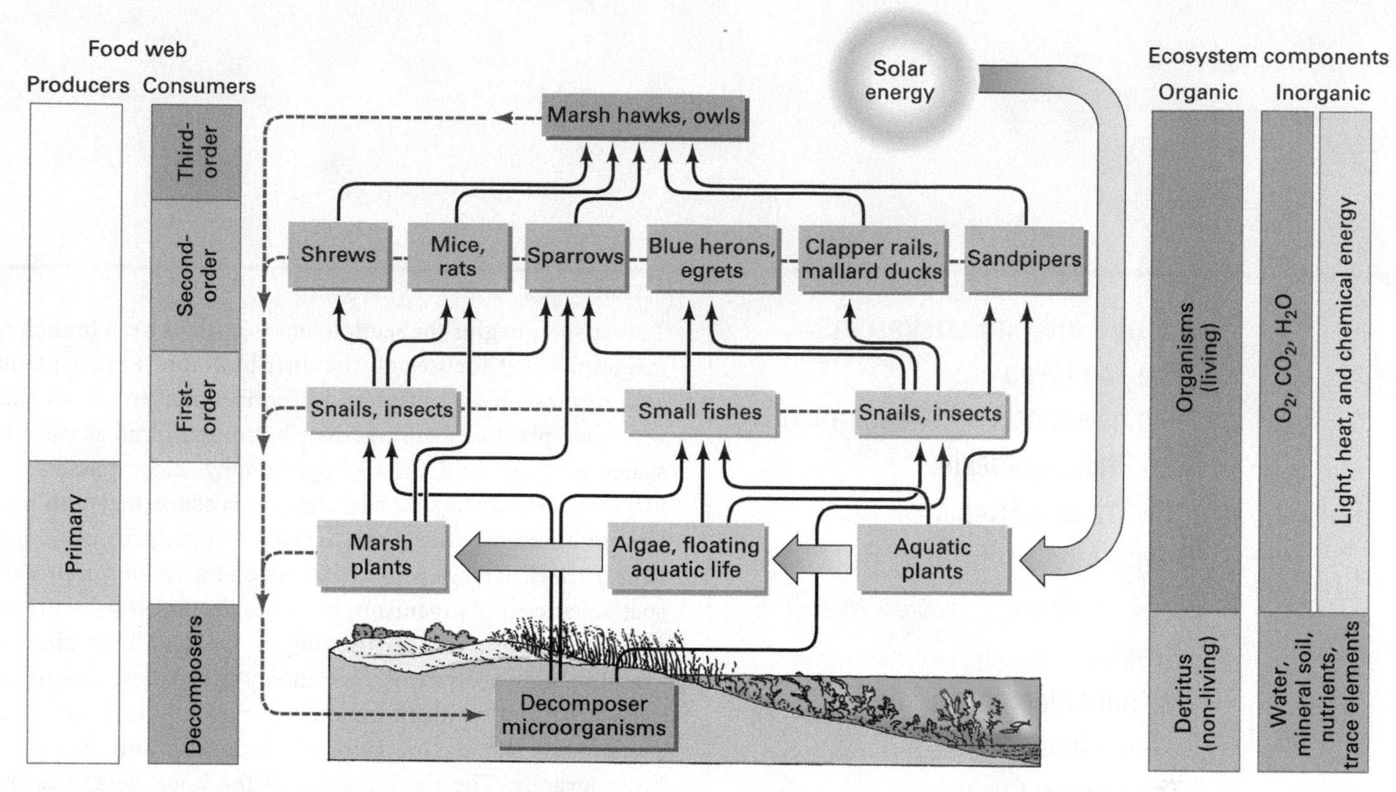

20.1 ENERGY FLOW DIAGRAM OF A SALT-MARSH ECOSYSTEM IN WINTER The arrows show how energy flows from the sun to producers, consumers, and decomposers. (Food chain after R. L. Smith, *Ecology and Field Biology*, New York: Harper and Row.)

The food web is an energy flow system, tracing the path of solar energy through the ecosystem. The primary producers absorb solar energy and store it in the chemical products of photosynthesis. As consumers eat and digest these organisms, chemical energy is released. This chemical energy is used to power new biochemical reactions, which convert it to other forms of chemical energy that is stored in the tissues of the consumers.

At each level of energy flow in the food web, energy is expended in respiration, and ultimately lost as waste heat. This means that both the numbers of organisms and their total biomass generally decrease at each successive level in a food web. Typically, between 10 and 50 percent of the energy stored in organic matter at one level can be passed to the next level, which effectively limits the system to four levels of consumers.

For each species, the number of individuals present in an ecosystem ultimately depends on the level of resources available to support that species' population. If these resources provide a steady supply of energy, the population size will stabilize. Normally, the population will expand exponentially until the resources become scarce. As this happens, an increasing proportion of the population is unable to sustain itself, so population growth slows and eventually stops. This type of population growth, in which the rate of increase declines and eventually reaches zero, is called *logistic growth* (Figure 20.2). At this point, the habitat's *carrying capacity* has been reached. Carrying capacity is the number of individuals a particular habitat can sustain without significant negative impact on the population or its environment. Logistic population growth is characteristic of most vertebrates and perennial plant species. If resources are unlimited or if there is no situation that adversely affects a population, then the population can grow at an exponential rate. This is the case for many insect species that can produce several generations during the summer, before the adults are killed by the onset of winter.

PHOTOSYNTHESIS AND RESPIRATION

Photosynthesis is the production of carbohydrate—a general term for a class of organic compounds consisting of the elements carbon (C), hydrogen (H), and oxygen (O) with the general molecular formula CH_2O. Common sugar or sucrose ($C_{12}H_{22}O_{11}$) is one example. Photosynthesis requires a series of complex biochemical reactions using water and CO_2 as well as light energy. Gaseous molecular oxygen (O_2) is a byproduct of

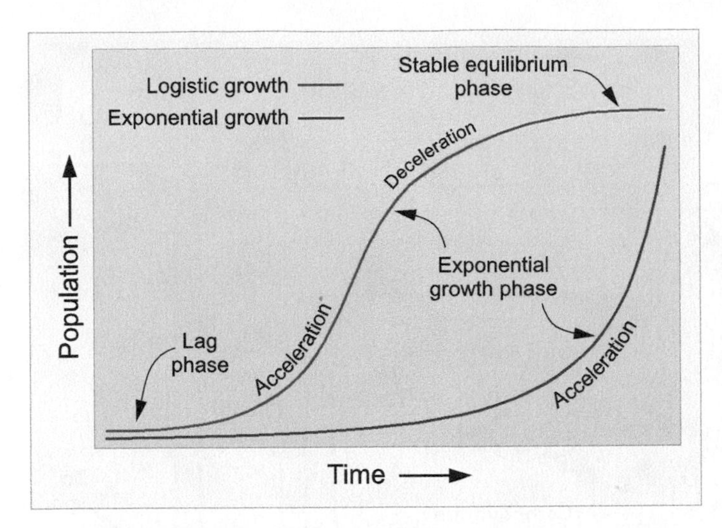

20.2 POPULATION GROWTH CURVES Logistic growth has three distinct phases. Population growth is initially slow during the lag phase, then accelerates during the exponential growth phase, before slowing during the deceleration phase. The population stabilizes at the carrying capacity with birth and death rates about equal. Exponential growth begins slowly, resulting in an extended lag phase before the population increases at an accelerating, exponential rate.

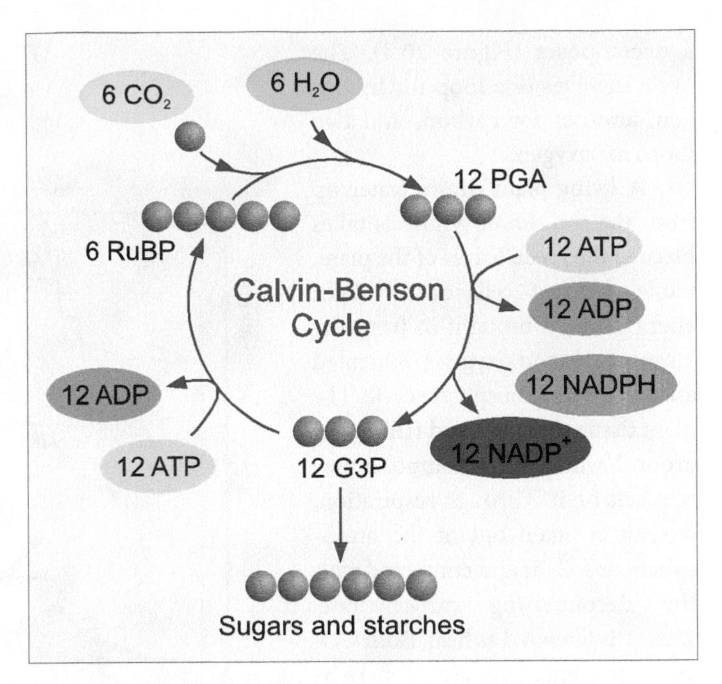

20.3 THE C_3 PHOTOSYNTHESIS PATHWAY OR CALVIN-BENSON CYCLE In this simplified representation, CO_2 diffuses into a leaf and combines with the 5-carbon sugar ribulose-1,5-biphosphate (RuBP) to produce the 3-carbon compound 3-phosphoglyceric acid (PGA). Energy derived from ATP (adenosine triphosphate) through the oxidation of foodstuffs, in conjunction with the co-enzyme NADPH (nicotinamide adenine dinucleotide phosphate), is used to remove the phosphate group from the PGA to produce the 3-carbon sugar glyceraladehyde-3-phosphate (G3P). Some G3P is used to regenerate RuBp, with the remainder used for synthesis of glucose, sucrose, starch, and other carbohydrates.

photosynthesis. A simplified chemical reaction for photosynthesis can be written as:

$$H_2O + CO_2 + \text{light energy} \longrightarrow CH_2O + O_2$$

Three different photosynthesis processes have been identified. About 95 percent of plants use the **C_3 pathway**, so designated because the first stable product manufactured from CO_2 is the 3-carbon compound *3 phosphoglyceric* acid (PGA) with chemical formula $C_3H_7O_7P$. Through a series of complex biochemical processes, the C_3 pathway, also known as the Calvin cycle or Calvin-Benson cycle, ultimately produces starch and sugars such as glucose and sucrose (Figure 20.3) About 1 percent of plants use the **C_4 photosynthetic pathway**. In C_4 plants, CO_2 is converted into oxaloacetic acid, a 4-carbon compound with the chemical formula $C_4H_4O_5$. This is used to form intermediate products before final processing via the standard C_3 pathway. Plants that use the **CAM (Crassulacean acid metabolism) pathway** initially manufacture oxaloacetic acid in the same way as C_4 plants. However, CAM plants take in CO_2 at night and store it for later processing via the C_3 pathway during daylight hours.

Each photosynthesis pathway has adaptive value. C_3 plants are more efficient than C_4 and CAM plants under cool and moist conditions and under normal light intensities typical of mid- and high latitude regions. C_4 plants have an adaptive advantage under high light intensity and high temperatures that are characteristic of low latitudes. They also tend to use water more efficiently. Many tropical grasses, as well as sugar cane and maize, are C_4 species. CAM plants include succulents, such as cacti and agaves. They open their stomata at night when it is cooler. Their transpiration rates are low, and they use water sparingly. This gives cacti and other CAM plants an advantage in desert environments.

Respiration is the opposite of photosynthesis in that carbohydrate is broken down and combined with oxygen to yield CO_2 and water. The reaction can be simplified as follows:

$$CH_2O + O_2 \longrightarrow CO_2 + H_2O + \text{chemical energy}$$

Photosynthesis and respiration act in a continuous cycle, which in its basic form can be represented by a primary producer and a decomposer (Figure 20.4). The cycle involves one loop for hydrogen, another for carbon, and two loops for oxygen.

A living plant draws water up from the soil. Photosynthesis takes place in the green leaves of the plant, while the leaf cells absorb light energy. CO_2 is brought in from the atmosphere, and oxygen is liberated and begins its atmospheric cycle. The plant tissue then dies and falls to the ground, where the decomposer system acts on it. Through respiration, oxygen is taken out of the atmosphere or soil air and combined with the decomposing carbohydrate. Energy is liberated as heat. Both CO_2 and water enter the atmosphere as gases during decomposition.

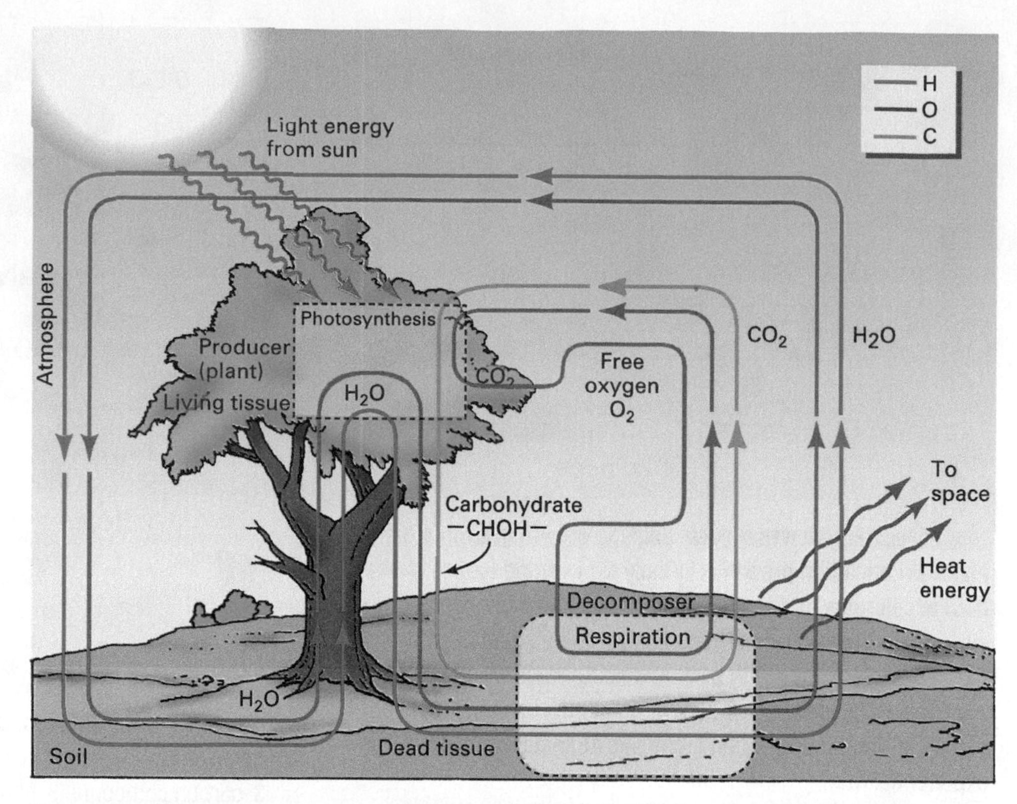

20.4 **PHOTOSYNTHESIS AND RESPIRATION CYCLES** A simplified flow diagram of the essential components of photosynthesis and respiration through the biosphere.

An important concept emerges from this flow diagram. Energy passes through the system. It comes from the sun and eventually returns to space. However, the material components—hydrogen, oxygen, and carbon—are recycled within the total system. Many other material components, including plant nutrients, are recycled in a similar way. Because the Earth as a planet functions as a closed system, the material components never leave the total system. However, they can be stored in different ways and in various compounds, making them unavailable for use by organisms for prolonged periods; coal and oil are examples.

NET PHOTOSYNTHESIS

Because photosynthesis and respiration both occur simultaneously in a plant, the amount of new carbohydrate placed in storage is less than the total carbohydrate being synthesized. A distinction, therefore, is made between gross photosynthesis and net photosynthesis. **Gross photosynthesis** is the total amount of carbohydrate produced by photosynthesis. **Net photosynthesis** is the amount of carbohydrate remaining after respiration has broken down enough carbohydrate to meet the plant's own needs. Thus:

$$\text{Net photosynthesis} = \text{Gross photosynthesis} - \text{Respiration}$$

Respiration accounts for most of the energy trapped by plants; the net result is an overall photosynthetic efficiency in the order of 3 to 6 percent of total available solar energy.

The rate of photosynthesis is strongly dependent on air temperature and the available light energy (Figure 20.5). At first, the photosynthesis rate rises rapidly as both temperature and light intensity increase. The rate then slows and reaches a maximum value. The light intensity needed for maximum net photosynthesis is only 10 to 30 percent of full summer sunlight for most green plants. Additional light energy is simply ineffective and can reduce photosynthesis through photoinhibition. High light intensity also heats the leaves, which increases the rate of respiration; this offsets gross production and decreases net photosynthesis.

Duration of daylight is closely related to the rate at which products of photosynthesis accumulate in plant tissues. The seasonal contrast in day length varies with latitude (see Chapter 2). In the tropics, the daylight period remains close to 12 hours throughout the year, whereas in subarctic regions continuous daylight may persist for several weeks. At high latitudes, photosynthesis can take place during most of the 24-hour day, which compensates for the shortness of the growing season.

NET PRIMARY PRODUCTION

Accumulated net production by photosynthesis is measured in terms of **biomass**, and is usually expressed as the dry weight of organic matter per unit of surface area within the ecosystem; for

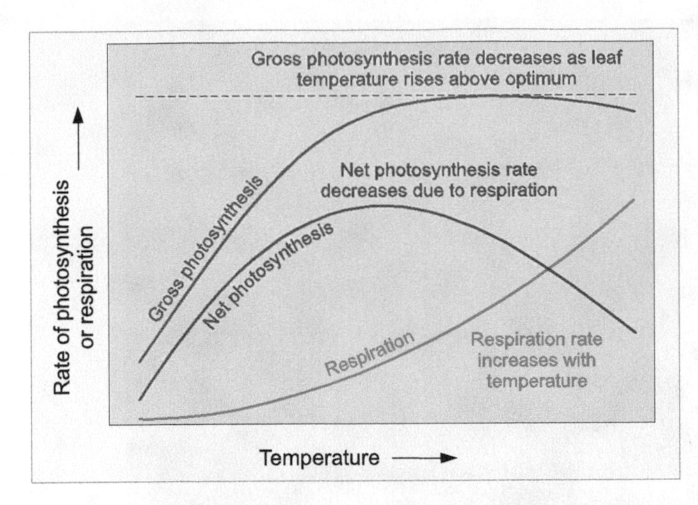

20.5 PHOTOSYNTHESIS AND ENERGY FLOW Photosynthesis varies with light intensity and temperature, but respiration is temperature-dependent. As a result, gross photosynthesis tends to reach a maximum plateau before declining under strong irradiance, whereas net photosynthesis declines sharply at higher temperatures.

example, as kilograms per square metre (kg m^2) or tonnes per hectare (t ha^{-1}; 1 ha $=10^4$ m^2). Forests have the greatest biomass because of the large amount of wood that the trees accumulate over time. Although biomass is an important indicator of photosynthetic activity, it can be misleading, especially in ecosystems where it is rapidly used by consumers and decomposers. From the viewpoint of ecosystem productivity, what is important is the annual yield of useful energy produced by the ecosystem, or the **net primary production (NPP)**. The highest NPPs are found in tropical forest regions such as Brazil, central Africa, and Indonesia, where values typically range from 1,000 to 5,000 gC m^{-2} a^{-1} with average values of 1,200 to 1,500 gC m^{-2} a^{-1} (Figure 20.6). In general, NPP in terrestrial ecosystems is inversely related to latitude and is negligible in polar regions where photosynthesis is primarily limited by temperature. Areas of low rainfall, such as the Sahara Desert, central Asia, and the southwestern United States also have low NPPs, generally lower than 100 gC m^{-2} a^{-1}. At a more restricted scale, freshwater swamps and marshes can be extremely productive, in some cases with NPPs as high as 4,000 gC m^{-2} a^{-1}.

Productivity of the oceans is generally low, accounting for about 30 percent of global primary production even though they cover about 70 percent of Earth's surface (Figure 20.7). The deep water oceanic zone, which comprises about 90 percent of the world ocean area, is the least productive of the marine ecosystems, with annual production less than 200 gC m^{-2} a^{-1}. Continental shelf areas and zones of upwelling, such as occur off the west coast of South America, are much more productive. Here production typically ranges from 600 to 800 gC m^{-2} a^{-1}, but these areas are comparatively small. Upwelling of cold water from ocean depths brings nutrients to the surface, which increases the growth of microscopic *phytoplankton*. These, in turn, serve as food sources for the animals in the marine food chain. Production in temperate oceans such as the northern Pacific and northern Atlantic Oceans is enhanced by seasonal overturns associated with warming and cooling of the

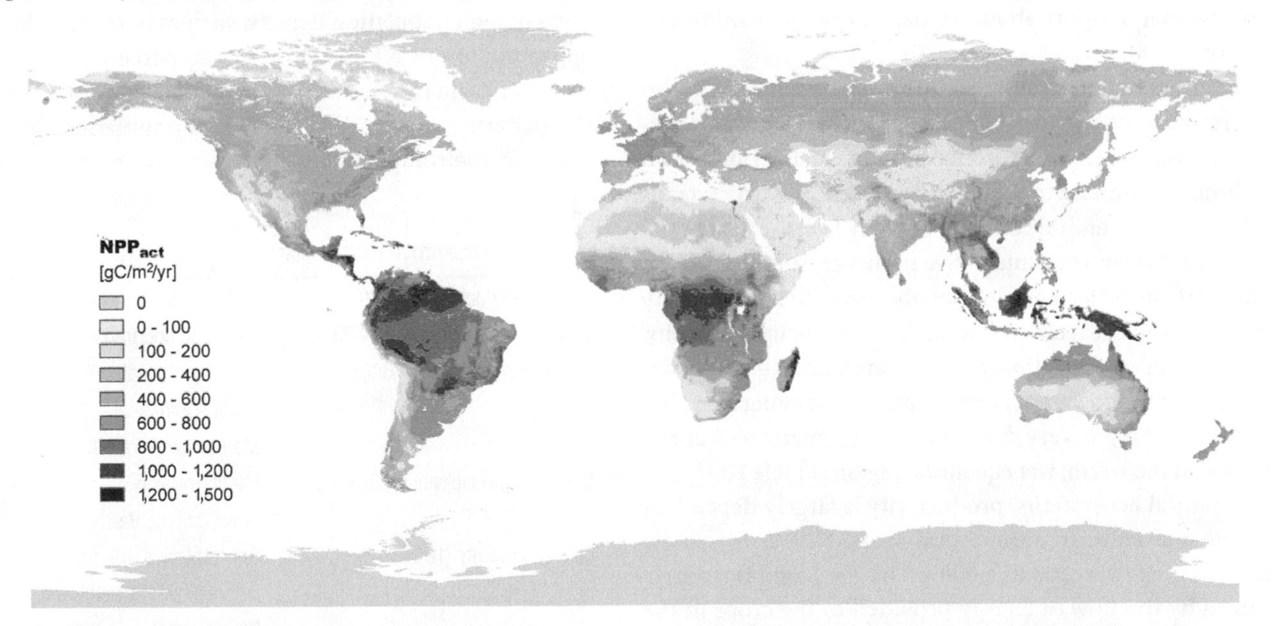

20.6 ANNUAL NET PRIMARY PRODUCTION IN TERRESTRIAL ECOSYSTEMS (gC m^{-2} a^{-1}) Net primary production is highest in tropical forests, but declines in arid regions where plant growth is limited by moisture and at high latitudes where the growing season is shortened by low temperatures.

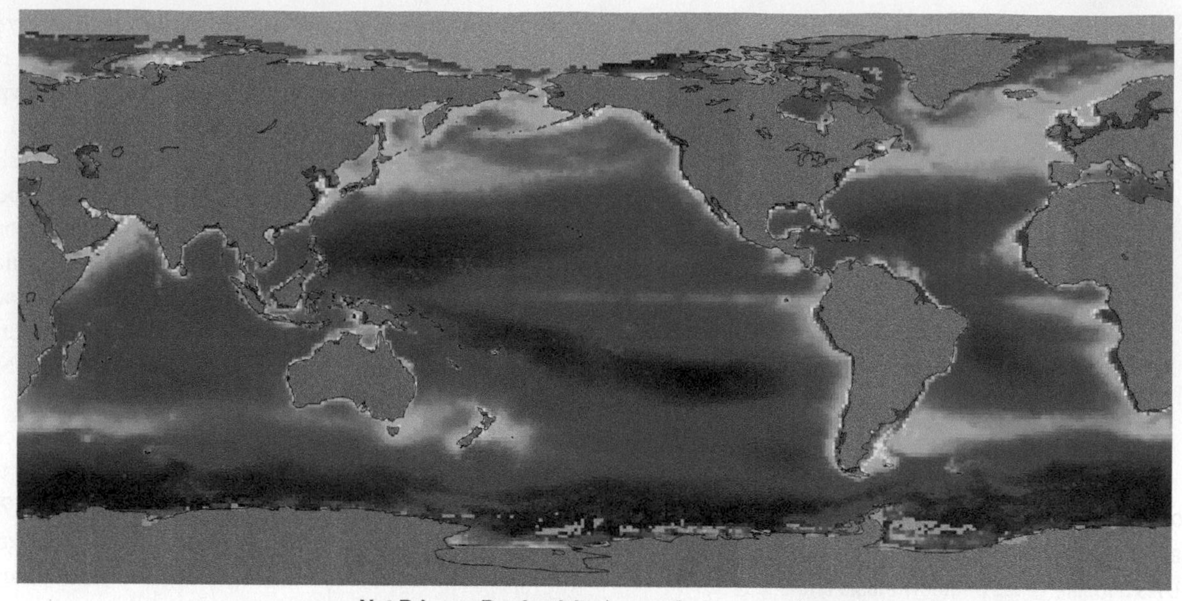

20.7 AVERAGE ANNUAL NET PRIMARY PRODUCTIVITY OF THE OCEANS (1997–2002) Production in oceans is mainly limited by nutrient availability. The warm surface water in tropical oceans prevents circulation of nutients from the depths. Production is enhanced in temperate oceans by the seasonal overturn that brings nutrients to the illuminated surface layer. Regions of high productivity are limited to narrow coastal zones of upwelling, such as along the west coast of South America and southwest Africa, and shallow continental shelves. Data acquired by the Sea-viewing Wide Field-of-view Sensor (SeaWiFS). Light grey areas indicate missing data.

epilimnion (see Chapter 5). Zones of upwelling and continental shelves together support about 99 percent of the world's fish production.

In addition to temperature and light, photosynthesis is also affected by water availability. Annual NPP increases with precipitation. The response is most pronounced in drier regions, from desert through semi-arid to subhumid climates; production tends to level off in more humid climates, possibly because a large soil water surplus has some counteractive influence, such as removal of plant nutrients by leaching, or possibly because light intensity is reduced by cloud cover. Combining the effects of light intensity, temperature, and precipitation, approximate values of productivity can be assigned to various climate regions, ranging from less than 100 $gC\ m^{-2}\ a^{-1}$ in very dry or very cold climates to over 800 $gC\ m^{-2}\ a^{-1}$ in the warm, wet equatorial regions (Table 20.1).

For natural ecosystems, productivity is largely dependent on climate and soils. In agricultural areas, NPP is extremely variable, ranging from 100 to 4,000 $gC\ m^{-2}\ a^{-1}$, and is strongly influenced by the flow of energy provided to the crops in the form of agricultural chemicals, irrigation water, and machinery. Much if not all of this energy is derived from the burning of fossil fuels, and so represents a conversion of fossil fuel energy to human foodstuffs that is not always efficient. In addition, the demand for food is not evenly distributed across the world. In areas of high population density, such as parts of India, human appropriation of NPP is close to 100 percent, whereas in the Amazon Basin it is less than 10 percent (Figure 20.8). However, the pattern is not entirely linked to population density. For example, the comparatively high demand in areas such as the

TABLE 20.1 PRODUCTIVITY AND CLIMATE

Highest productivity (> 800)	Wet equatorial
Very high productivity (600–800)	Monsoon and trade-wind coastal
High productivity (400–600)	Wet-dry tropical
	Moist subtropical
	Marine west-coast
Moderate productivity (200–400)	Mediterranean
	Moist continental
Low productivity (100–200)	Dry tropical, semi-arid
	Dry midlatitude, semi-arid
	Boreal forest
Very low productivity (<100)	Dry tropical, arid
	Dry midlatitude, arid
	Tundra

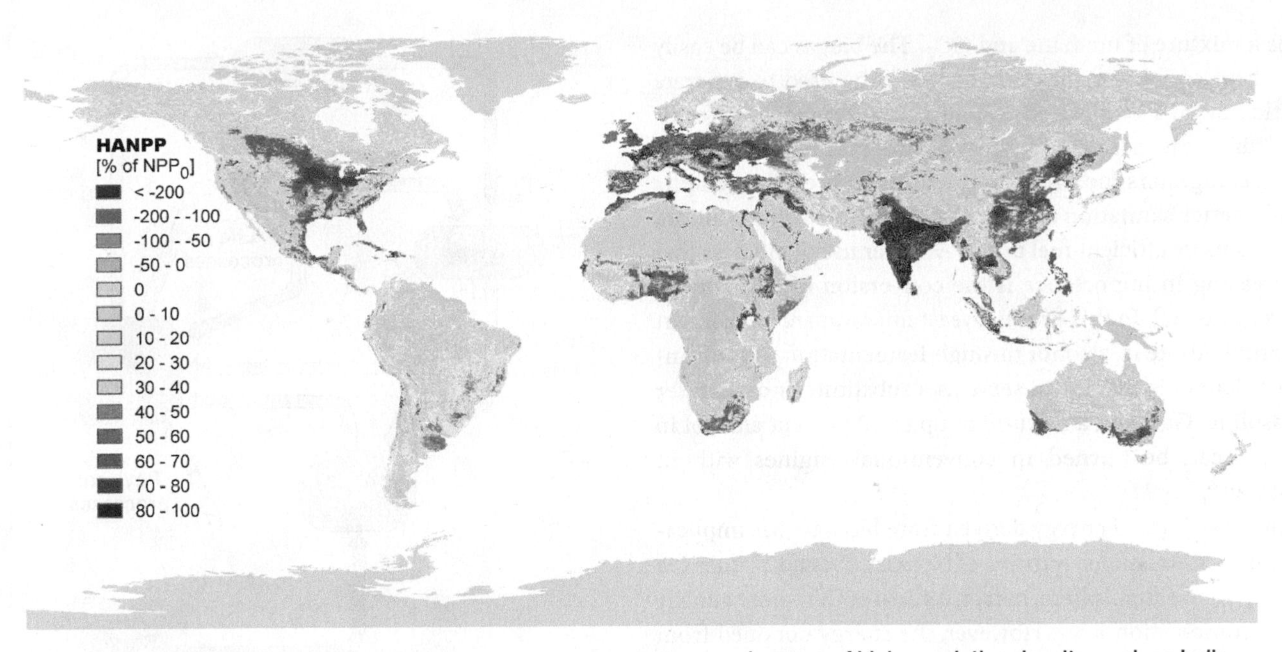

20.8 HUMAN APPROPRIATION OF NET PRIMARY PRODUCTION AS A PERCENTAGE OF NPP In areas of high population density, such as India, much of the energy is used regionally. In some areas, such as the Canadian Prairies, much of the production is exported to meet demands from elsewhere.

Prairie provinces is due to commercial agriculture rather than local consumption. Transport of food and fibre to meet demands elsewhere further reduces the efficiency of the global agricultural system.

Remote sensing is now widely used as a tool for mapping primary production on a global scale. Based on remote sensing data, terrestrial primary production has increased since about 1980, while primary production in the oceans has decreased. These trends are most likely linked to global climate changes, such as warmer temperatures at higher latitudes, reductions in cloud cover in equatorial regions, and decreasing winds over oceans. *Appendix 20.1 • Monitoring Global Productivity from Space* at the end of this chapter provides more information on this topic.

BIOMASS AS AN ENERGY SOURCE

Net primary production represents a source of renewable energy derived from the sun that can be exploited to fill human energy needs. The use of biomass as an energy source involves releasing solar energy that has been fixed in plant tissues through photosynthesis. This process can take place in a number of ways—the simplest is direct burning of plant matter as fuel. Other approaches involve the generation of intermediate fuels from plant matter; for example, charcoal from wood, ethanol from grains, and methane from anaerobic digestion of organic wastes. Biomass energy conversion is not particularly energy efficient. However, the abundance of terrestrial biomass is so great that, with proper development, biomass use could provide the energy equivalent of several million barrels of oil per day.

One important use of biomass energy is the burning of firewood for cooking in developing nations. The annual growth of wood in the forests of developing countries totals about half the world's energy production. However, fuelwood use exceeds production in many areas, creating local shortages and severe strains on some forest ecosystems. The forest-desert transition areas of thorntree, savanna, and desert scrub in central Africa south of the Sahara Desert are examples where heavy consumption has devastated local ecosystems. Even in closed stoves used for cooking, wood burning has an efficiency of only about 10 to 15 percent. However, the conversion of wood to charcoal or gas can boost efficiencies to 70 or 80 percent with appropriate technology. In this process, called *pyrolysis*, controlled partial burning in an oxygen-deficient environment reduces carbohydrate to free carbon (charcoal) and yields flammable gases, such as CO_2 and hydrogen. Charcoal is more energy efficient than wood, burns more cleanly, and is easier to transport. In addition, charcoal can be made from waste fibres and agricultural residues that would normally be discarded. Thus, charcoal is an efficient fuel that can help extend the firewood supply in areas where wood is in high demand.

A second method of extracting energy from biomass uses anaerobic digestion to produce *biogas*. In this process, animal and human wastes are fed into a closed digesting chamber, where anaerobic bacteria break down the waste to produce a gas that is a mixture of methane and CO_2. The biogas can be easily burned for cooking or heating, or it may be used to generate electric power. The digested residue is a sweet-smelling fertilizer. China now maintains a vigorous program of construction of biogas digesters for small family units to use. The benefits include better sanitation and reduced air and water pollution, as well as more efficient fuel usage. Another use of biomass that is increasing in importance is the conversion of agricultural wastes to alcohol. In this process, yeast microorganisms convert the carbohydrate to alcohol through fermentation. An advantage of alcohol is that it can serve as a substitute and extender for gasoline. Gasohol, a mixture of up to 10 percent alcohol in gasoline, can be burned in conventional engines without adjustment.

Increased use of energy derived from biomass has implications for CO_2 emissions. Burning of biomass does not reduce the CO_2 flow to the atmosphere; in fact, it releases CO_2 more quickly than decomposition does. However, the energy obtained from biomass burning will, in all likelihood, be a substitute for some quantity of fossil fuel, so the expected effect is a reduction in the quantity of CO_2 released to the atmosphere.

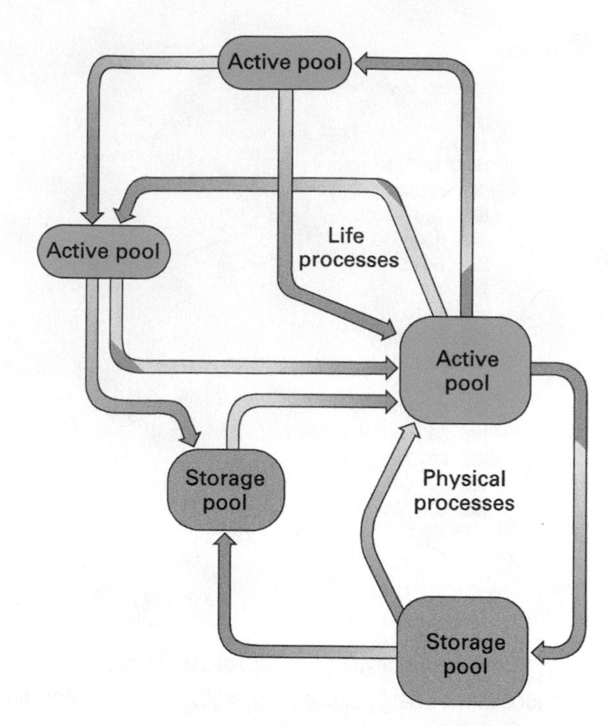

20.9 GENERAL FEATURES OF A BIOGEOCHEMICAL CYCLE A system of flow pathways links active and storage pools where matter is concentrated for varying lengths of time.

BIOGEOCHEMICAL CYCLES IN THE BIOSPHERE

Solar energy flows through ecosystems, passing from one part of the food chain to the next, until ultimately, it is lost to the biosphere by radiation to space. Matter also moves through ecosystems, but because of gravity, matter cannot be lost to space. As molecules are formed and reformed by chemical and biochemical reactions within ecosystems, the atoms that compose them are neither changed nor lost. Thus matter is conserved, and the constituent atoms and molecules can be used and reused, or cycled, within ecosystems.

Matter moves through ecosystems under the influence of both physical and biological processes. Each substance follows a specific **biogeochemical cycle** (sometimes referred to as a *material cycle* or *nutrient cycle*) that consists of various pools interconnected by flow pathways (Figure 20.9). A pool refers to any area or location of material concentration. There are two types of pools—**active pools**, where materials are in forms and places easily accessible to life processes, and **storage pools**, where materials are more or less inaccessible to living systems. A system of material flow pathways connects the various active and storage pools within the cycle. Life processes usually control pathways between active pools, while physical processes typically control pathways between storage pools.

Biogeochemical cycles exists in two forms: gaseous cycles and sedimentary cycles. In a *gaseous cycle*, the element or compound can be converted directly into a gas. The gas diffuses throughout the atmosphere and so can arrive over land or sea in a relatively short time, to be reused by the biosphere. The primary constituents of living matter—carbon, hydrogen, oxygen, and nitrogen—all move through gaseous cycles. In a *sedimentary cycle*, weathering releases the compound or element from rock. It then follows the movement of running water, either in solution or as sediment, to the sea where, eventually, it is converted into rock. The cycle is completed when the rock is uplifted and exposed to weathering.

The magnitudes of the total storage and total active pools can be very different. In some cases, the active pools are much smaller than storage pools, and as a result materials move more rapidly between active pools than between storage pools, and more rapidly than they move into and out of storage. For example, in the carbon cycle, photosynthesis and respiration will cycle all the CO_2 in the atmosphere (active pool) through plants in about 10 years. But it may be many millions of years before the carbonate sediments (storage pool) now forming as rock will be uplifted and decomposed to release CO_2 to the atmosphere.

NUTRIENT ELEMENTS IN THE BIOSPHERE

Fifteen elements are commonly present in living matter. The three principal components of a carbohydrate—hydrogen, carbon, and oxygen—account for 99.5 percent of all living matter. In addition to these **macronutrients** are secondary nutrients and micronutrients including nitrogen, calcium, potassium, magnesium, sulphur, and phosphorus (see Chapter 19).

THE CARBON CYCLE

The movements of carbon through the life layer are of great importance because all life is composed of carbon compounds of one form or another. Of the total carbon available, most lies in storage pools as carbonate sediments below the Earth's surface. Only about 0.2 percent is readily available to organisms as CO_2 or as decaying biomass in active pools.

In the gaseous portion of the cycle, carbon moves mainly as CO_2 in the form of a free gas in the atmosphere or as a dissolved gas in fresh and salt water (Figure 20.10). In the sedimentary portion of the cycle, carbon is in the form of carbohydrate molecules in organic matter, hydrocarbon compounds in rock (petroleum, coal), and mineral carbonate compounds, such as calcium carbonate ($CaCO_3$). The world supply of atmospheric CO_2 is about 2 percent of the carbon in active pools. This atmospheric pool is supplied by plant and animal respiration in the oceans and on the lands. Under natural conditions, some new carbon enters the atmosphere each year from volcanoes by out-gassing in the form of CO_2 and carbon monoxide (CO). Industry injects substantial amounts of carbon into the atmosphere through combustion of fossil fuels.

CO_2 leaves the atmospheric pool to enter the oceans, where phytoplankton uses it in photosynthesis. These organisms are primary producers in the ocean ecosystem and are consumed by marine animals in the food web. Phytoplankton also build skeletal structures of $CaCO_3$. This mineral matter settles to the ocean floor to accumulate as sedimentary strata, such as chalk and limestone, and forms an enormous storage pool not available to organisms until released later by rock weathering. Organic compounds synthesized by phytoplankton also settle to the ocean floor and eventually transform into the hydrocarbon compounds making up petroleum and natural gas reserves. On land, plant matter accumulating over geologic time forms layers of peat that ultimately transform into coal. Fossil fuels include petroleum, natural gas, and coal, and represent huge storage pools of carbon.

Human activity is currently having a significant effect on the carbon cycle. The burning of fossil fuels is releasing CO_2 to the atmosphere at a rate far beyond that of any natural process. *Eye on the Environment • Human Impact on the Carbon Cycle* documents how human activity has influenced the major flows within the carbon cycle.

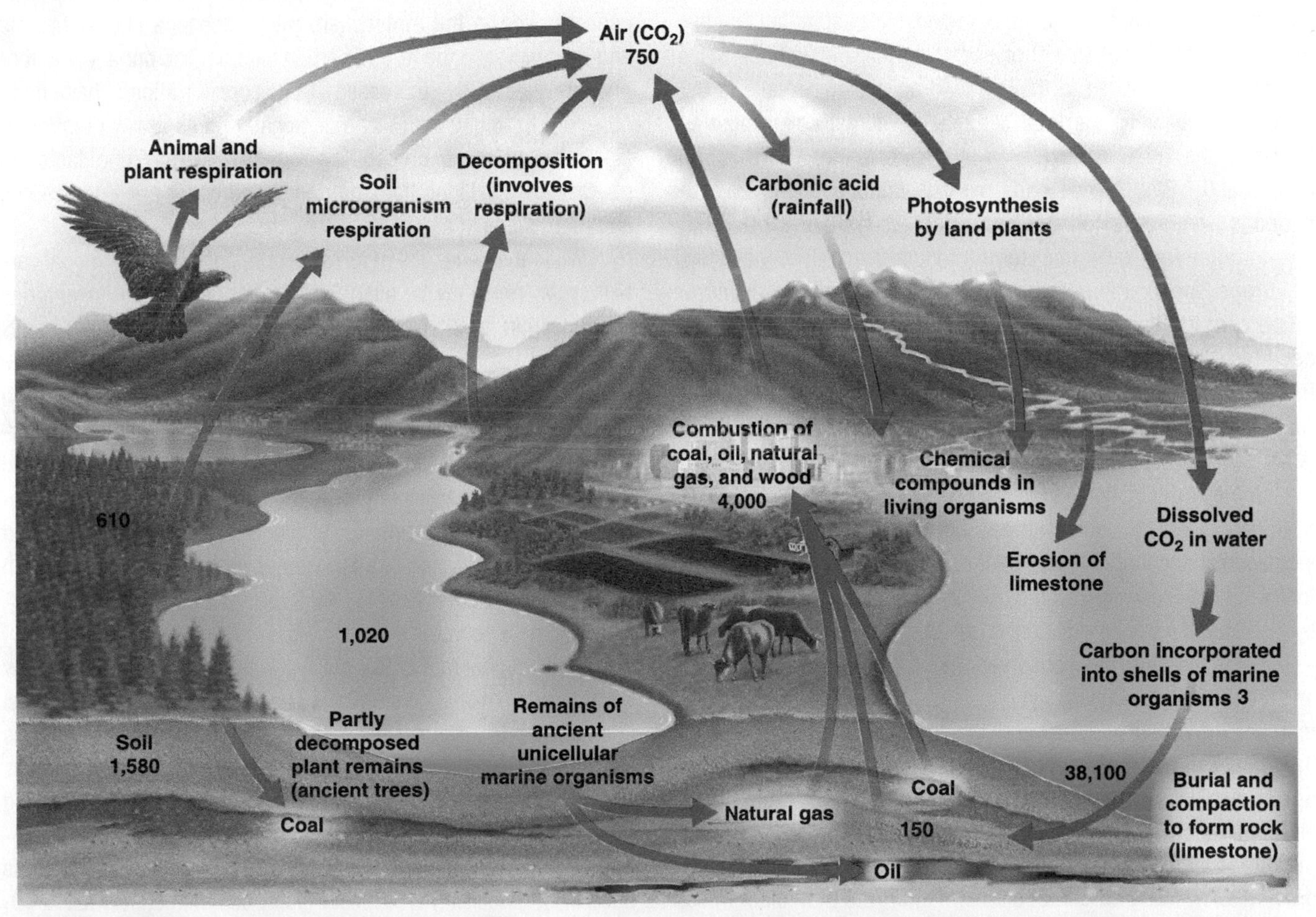

20.10 **THE CARBON CYCLE** The numbers indicate how much carbon is stored in various reservoirs, in billions of tonnes. (From Berg, Hager, Goodman, Baydack, *Visualizing the Environment*, Figure 5.9, p. 140. John Wiley & Sons Canada, Ltd., 2010.)

EYE ON THE ENVIRONMENT

HUMAN IMPACT ON THE CARBON CYCLE

Carbon cycles continuously among the land surface, atmosphere, and ocean in many complex pathways. However, human activity now strongly influences these flows. The most important human impact on the carbon cycle is the burning of fossil fuels. Another important impact arises from changing the Earth's land covers, such as clearing forests or abandoning agricultural areas, for example.

About half of the output of carbon by fossil fuel burning is taken up by the atmosphere. Of the remaining amount, about 65 percent is absorbed by the oceans and 35 percent flows into the biosphere. Thus, ecosystems are a *sink* for CO_2, accepting about 0.7 gigatonnes per year (Gt a^{-1}) of carbon. Ecosystems cycle carbon in photosynthesis, respiration, decomposition, and uncontrolled combustion through burning.

If the value of 0.7 Gt a^{-1} in carbon uptake is correct, the amount of terrestrial biomass must be increasing at that rate. However, forests are diminishing in area as they are logged or converted to farmland or grazing land. This conversion is primarily occurring in tropical and equatorial regions; it is estimated that this will release about 1.6 Gt a^{-1} of carbon to the atmosphere. Since this release is included in the net land ecosystem uptake of 0.7 gigatonnes per year, the Earth's remaining forest cover must be taking up the 1.6 Gt a^{-1} released by deforestation, as well as an additional 0.7 Gt a^{-1}. Thus, mid- and high-latitude forests are estimated to be increasing in area or biomass at a rate of 2.3 Gt a^{-1}.

Independent evidence seems to confirm this conclusion. In Europe, for example, forest statistics show an approximate 25 percent increase in the volume of living trees in the past few decades. This increase has been sustained despite damage to forests by air pollution, especially in eastern Europe. In North America, forest areas are increasing in many regions, and there is an increasing trend in reforestation following timber removal.

Some of the increase in global biomass may also be the result of enhancement of photosynthesis by warmer temperatures and increased atmospheric CO_2 concentrations. Another possible reason for increased ecosystem productivity is nitrogen fertilization of soils by washout of nitrogen-pollutant gases in the atmosphere.

Some foresters have observed that harvesting mature forests and replacing them with young, fast-growing timber should increase the rate of withdrawal of CO_2 from the atmosphere. Since the lumber of the mature forests goes into semi-permanent storage in dwellings and structures where it is protected from decay and oxidation to CO_2, it represents a withdrawal of carbon from the atmosphere. The young forests that replace the mature ones grow quickly, fixing carbon at a much faster rate than older trees, in which annual growth has slowed.

However, some reports indicate that the conversion of old-growth forests to young, fast-growing forests will not significantly decrease atmospheric CO_2. While 42 percent of the harvested timber goes into comparatively long-term storage (greater than five years) in building structures, much of the remainder is directly discarded on the logging site, where it is burned or rapidly decomposed. In addition, some biomass becomes waste in factory processing of the lumber, where sawdust and scrap are burned as fuel. Similarly, the manufacture of paper also results in short-term conversion of a large proportion of the harvested trees to CO_2. Thus, harvesting of old-growth forests, as now practised, appears to contribute substantially to atmospheric CO_2.

Some environmentalists have advocated increased tree planting as a way of enhancing CO_2 fixation. To take up the quantity of carbon now being released by fossil fuel burning would require some 7 million km^2 of new closed-crown broad-leaved deciduous forest—an area about the size of Australia. Higher CO_2 concentration in the atmosphere might enhance photosynthesis and thus increase the rate of carbon fixation. The enhancement of photosynthesis by increased CO_2 concentrations has been observed for many plants and demonstrated as a way of increasing yields of some crops. However, CO_2 is only one factor in photosynthesis. Light, heat, nutrients, and water are also needed, and restrictions in any of these would affect productivity.

While the dynamics of forests are important in the global carbon cycle, soils may be even more important. Recent inventories estimate that about four times as much carbon resides in soils than in above-ground biomass. The largest reservoir of soil carbon is in the boreal forest. In fact, there is about as much carbon in boreal forest soils as in all above-ground vegetation. This soil carbon has accumulated over thousands of years under cold conditions that have slowed its decay. However, there is now great concern that global warming, which is acting more strongly at high latitudes, will increase the rate of decay of this vast carbon pool, and that boreal forests, which are presently a sink for CO_2, will become a source.

Reducing the rate of CO_2 buildup in the atmosphere is a matter of great international concern and has been the subject of several international treaties designed to limit emissions of greenhouse gases. The latest was signed by about 170 countries at Kyoto, Japan, in December 1997 (see Figure 1). Since that time, nations—including Canada—have been struggling with the Kyoto Protocol's implementation. While a lot of good progress has been made, more work is necessary, specifically an effective global commitment to reduce CO_2 releases and control global warming.

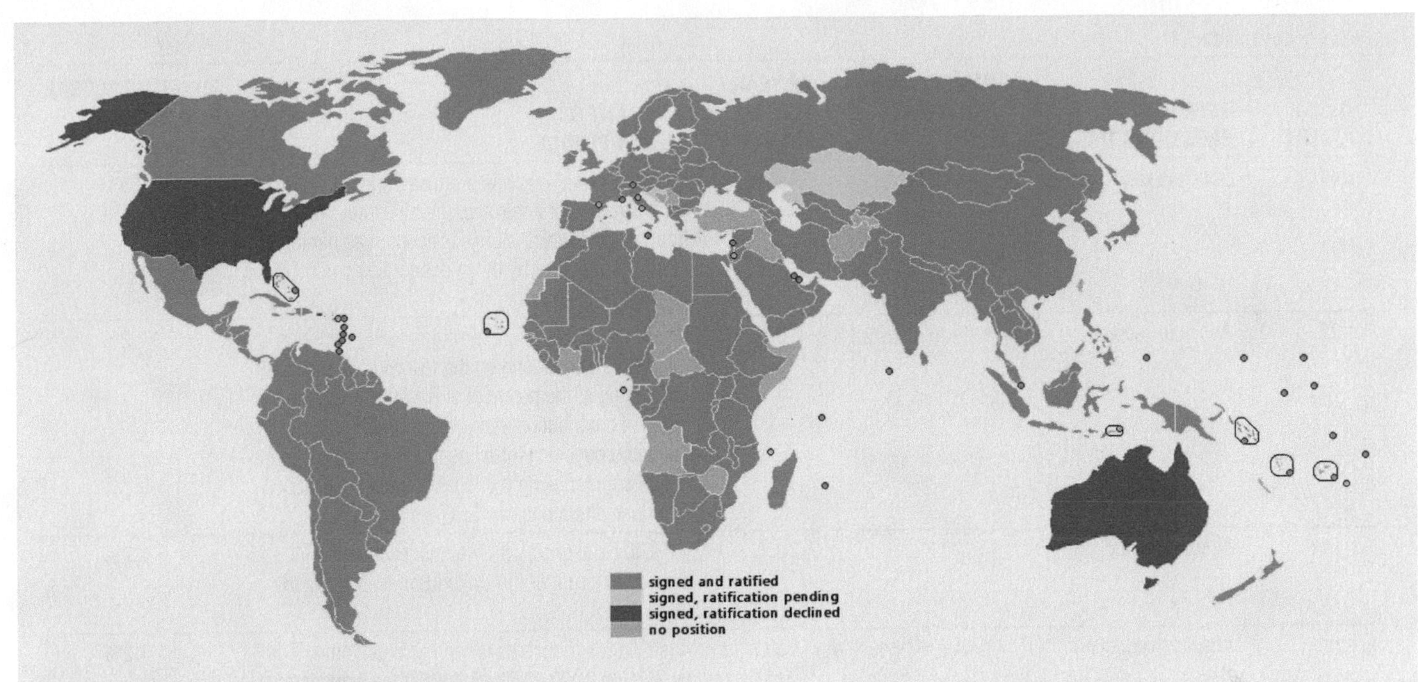

Figure 1 Signatories of the 1997 Kyoto Protocol

The first commitment period of the Kyoto Protocol expires at the end of 2012 and no follow-up treaty is yet in place. Table 1 shows the initial targets set by some of the countries that signed the agreement. Subsequent climate change conferences have failed to reach any agreement on CO_2 emission targets. Although the Kyoto Protocol remains in effect, several countries have amended their CO_2 emission targets (see Table 1). At the December 2009 United Nations Climate Change Conference held in Copenhagen, a proposal was made to limit temperature increases to 1.5 °C and cut CO_2 emissions by 80 percent by 2050. After much discussion, this was subsequently replaced by the Copenhagen Accord tabled by the United States, China, India, Brazil, and South Africa. The Copenhagen Accord recognized that climate change remains a concern and that actions are needed to keep future temperature increases to below 2 °C. The document does not contain any legally binding commitments for reducing CO_2 emissions, and was opposed by many countries.

TABLE 1 LIST OF SELECTED COUNTRIES' COPENHAGEN COMMITMENTS

COUNTRY	TYPE OF EMISSIONS TARGET	QUANTITATIVE TARGET FOR 2020	BASE YEAR / NATURE OF TARGET	SUMMARY OF TARGET PLEDGE	SHARE OF GLOBAL EMISSIONS IN 2005
United States	Absolute reduction	−17%	2005	In the range of 17%, in conformity with anticipated U.S. energy and climate legislation, recognizing that the final target will be reported to the Secretariat in light of enacted legislation.	14.3%
European Union (EU)	Absolute reduction	−20% to −30%	1990	As part of a global and comprehensive agreement for the period beyond 2012, the EU reiterates its conditional offer to move to a 30% reduction by 2020 compared to 1990 levels, provided that other developed countries commit themselves to comparable emission reductions and that developing countries contribute adequately according to their responsibilities and respective capabilities.	10.6%

TABLE 1 CONTINUED

COUNTRY	TYPE OF EMISSIONS TARGET	QUANTITATIVE TARGET FOR 2020	BASE YEAR / NATURE OF TARGET	SUMMARY OF TARGET PLEDGE	SHARE OF GLOBAL EMISSIONS IN 2005
Japan	Absolute reduction	−25%	1990	25% reduction, which is premised on the establishment of a fair and effective international framework in which all major economies participate and on agreement by those economies on ambitious targets.	2.8%
Russia	Absolute reduction	−15% to −25%	1990	The range of the greenhouse gas (GHG) emission reductions will depend on the following: appropriate accounting of the potential of Russia's forestry in terms of contribution in meeting the obligations of the anthropogenic emissions reduction; all major emitters undertaking the legally binding obligations to reduce anthropogenic GHG emissions.	4.2%
Canada	Absolute reduction	−17%	2005	17%, to be aligned with the final economy-wide emissions target of the United States in enacted legislation.	1.7%
Australia	Absolute reduction	−5% to −25%	2000	Australia will reduce its greenhouse gas emissions by 25% on 2000 levels by 2020 if the world agrees to an ambitious global deal capable of stabilizing levels of greenhouse gases in the atmosphere at 450 ppm CO_2-eq or lower. Australia will unconditionally reduce emissions by 5% below 2000 levels by 2020, and by up to 15% by 2020 if there is a global agreement that falls short of securing atmospheric stabilization at 450 ppm CO_2-eq and under which major developing economies commit to substantially restrain emissions and advanced economies take on commitments comparable to Australia's.	1.2%
China	Intensity reduction	−40% to −45%	Emissions intensity change 2005−2020	China will endeavour to lower its CO_2 emissions per unit of gross domestic product (GDP) by 40−45% by 2020 compared to the 2005 level, increase the share of non-fossil fuels in primary energy consumption to around 15% by 2020, and increase forest coverage by 40 million ha and forest stock volume by 1.3 billion m^3 by 2020 from the 2005 levels.	15.1%
India	Intensity reduction	−20% to −25%	Emissions intensity change 2005−2020	India will endeavour to reduce the emissions intensity of its GDP by 20−25% by 2020 in comparison to the 2005 level.	3.9%
Indonesia	Reduction below business as usual (BAU)	−26%	Reduction below BAU at 2020	26% reduction relative to BAU unilaterally, up to 41% reduction with international assistance. (In the quantitative analysis, only the 26% target is considered, as reductions above that appear to be credited toward credit-buying countries.)	4.3%
Brazil	Reduction below BAU	−36% to −39%	Reduction below BAU at 2020	Anticipation that reductions in deforestation and other sectors of the economy will lead to reductions of 36.1−38.9% relative to projected emissions at 2020.	6.0%

Mexico	Reduction below BAU	−39%	Reduction below BAU at 2020	Mexico aims at reducing its GHG emissions up to 30% with respect to the business-as-usual scenario by 2020, provided the provision of adequate financial and technological support from developed countries as part of a global agreement.	1.4%
South Korea	Reduction below BAU	−30%	Reduction below BAU at 2020	To reduce national greenhouse gas emissions by 30% from business-as-usual emissions at 2020.	1.2%
South Africa	Reduction below BAU	−34%	Reduction below BAU at 2020	A 34% deviation below the BAU emissions growth trajectory by 2020.	0.9%

Source: Frank Jotzo, *Comparing the Copenhagen emissions targets*. CCEP working paper 1.10, October 2010, Table 1, p. 26. http://ccep.anu.edu.au/data/2010/pdf/wpaper/CCEP-1-10.pdf.

THE OXYGEN CYCLE

The largest active pool of the *oxygen cycle* is found in the atmosphere, but a small active pool is also present in the oceans. The complete picture of the cycling of oxygen includes its movements and storage when combined with carbon as CO_2 and as organic and inorganic compounds as discussed in the carbon cycle.

Oxygen enters the active pool through release in photosynthesis, both in the oceans and on land. Each year, a small amount of new oxygen comes from volcanoes through out-gassing, principally as CO_2 and water. Loss through organic respiration and mineral oxidation balances the input to the atmospheric pool. Adding to the withdrawal from the atmospheric oxygen pool are domestic and industrial activities through the combustion of wood and fossil fuels. Forest fires and grass fires are another means of oxygen consumption. Some oxygen from the small, active pool dissolved in the oceans is continuously placed in storage as mineral carbonates in ocean-floor sediments.

Human activity reduces the amount of oxygen in the air by burning fossil fuels. Clearing and draining land also uses atmospheric oxygen by speeding up the oxidation of soil minerals and soil organic matter. The oxygen pool is further affected when forests are cleared for agriculture and by paving and covering previously productive surfaces, as this reduces photosynthesis. Fortunately, the oxygen pool is so large that human impact or potential impact is small at this time.

THE NITROGEN CYCLE

Nitrogen moves through the biosphere in the gaseous *nitrogen cycle* in which the atmosphere acts as a vast storage pool (Figure 20.11). Nitrogen in the atmosphere, in the form N_2, is an inert gas, and most plants or animals cannot assimilate it directly. The process by which nitrogen is converted into nitrogen compounds, such as ammonia (NH_3) and nitrates (NO_3^-), is called *nitrogen fixation*. In these forms, nitrogen is then available for various biochemical processes. Only certain microorganisms possess the ability to use nitrogen directly. Some of these are species of free-living soil bacteria; other nitrogen-fixing microorganisms live in symbiotic association on the roots of higher plants. Symbiotic bacteria of the genus *Rhizobium* are associated with about 200 species of trees and shrubs, as well as almost all members of the legume family, which includes important agricultural species such as clover, alfalfa, soybeans, peas, beans, and peanuts. The blue-green algae are an important group as they can fix nitrogen, both in soil and in the oceans. Nitrogen fixation by all biological processes totals approximately 175 million metric tonnes annually.

Nitrogen fixation also occurs as a result of lightning discharge, which contributes about 5 percent of the total nitrogen fixed by natural and commercial processes. Lightning can dissociate nitrogen molecules, enabling the atoms to combine with oxygen in the air to form nitric oxide (NO) and nitrogen dioxide (NO_2), which dissolves in rain and is carried to the Earth as nitrates. Atmospheric nitrogen fixation can be respresented as:

lightning

$$N_2 + O_2 \longrightarrow 2\,NO$$

The nitrous oxide then combines with oxygen to form nitrogen dioxide:

$$2\,NO + O_2 \longrightarrow 2\,NO_2$$

Nitrogen dioxide readily dissolves in water to produce nitric (HNO_3) and nitrous (HNO_2) acids:

$$2\,NO_2 + H_2O \longrightarrow HNO_3 + HNO_2$$

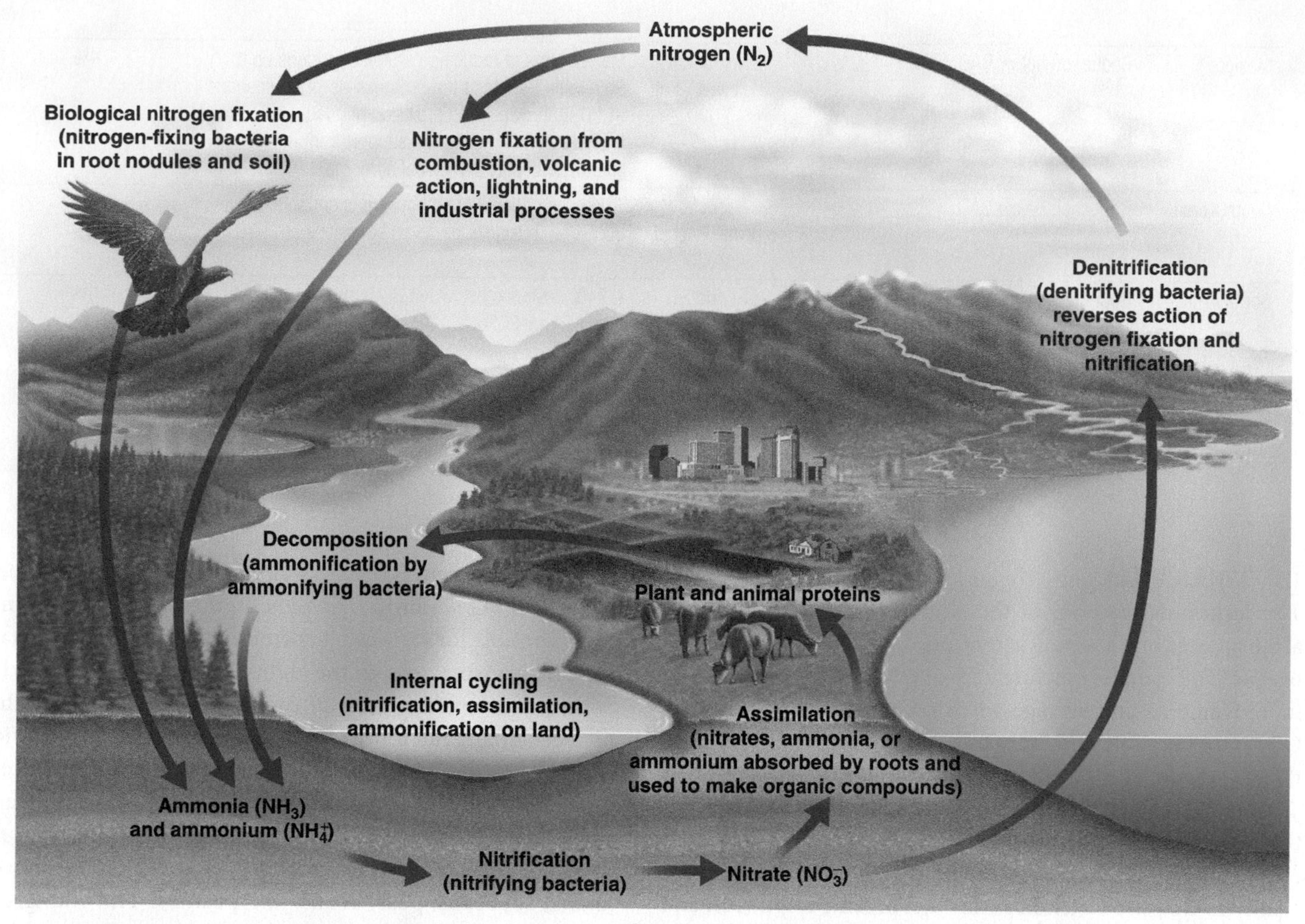

20.11 THE NITROGEN CYCLE (From Berg, Hager, Goodman, Baydack, *Visualizing the Environment*, Figure 5.11, p. 143. John Wiley & Sons Canada, Ltd., 2010.)

These acids release the hydrogen forming nitrate (NO_3^-) and nitrite (NO_2^-) ions, both of which are available to plants and microorganisms:

$$HNO_3 \longrightarrow H^+ + NO_3^-$$
$$HNO_2 \longrightarrow H^+ + NO_2^-$$

Industrial conversion of nitrogen to ammonia is carried out through the Haber-Bosch Process. This combines atmospheric nitrogen with hydrogen mainly derived from methane; the reaction uses an iron-based catalyst, very high pressure (~200 atmospheres), and high temperature (~450 °C). This process converts about 50 million metric tonnes of nitrogen annually, and accounts for about 25 percent of nitrogen fixation by all processes. The ammonia produced in this way may be applied directly under pressure as liquid fertilizer; once in the soil, the ammonia (NH_3) combines with hydrogen from soil water to form the ammonium cation (NH_4^+). Alternatively, ammonia produced by the Haber-Bosch Process may be further processed into ammonium nitrate (NH_4NO_3).

Ammonium is held in the soil through attachment to colloidal clay particles and humus and is subsequently converted to nitrate through the process of *nitrification*. Nitrification is a two-stage process in which ammonium is first oxidized to nitrite (NO_2^-) and then to nitrate (NO_3^-). Both processes are carried out by autotrophic bacteria; the first stage involves species of the genus *Nitrosomonas*, and the second stage species of the genus *Nitrobacter*.

Nitrogen in the form of NO_3^- is taken up by plants and passes through the food web in the tissues and wastes of grazing animals and carnivores. Most of the nitrogen in plants and animals is bound up in proteins. When plants and animals die, their remains are decomposed and the nitrogen is returned to the environment through the process of *ammonification*. Decomposer bacteria include species of *Bacillus, Clostridium, Proteus, Pseudomonas*, and *Streptomyces*. Through ammonification, organic nitrogen is converted into NH_3, which then dissolves in water to form NH_4^+. In this way, NH_4^+ is returned to the soil where it is once again available to the nitrifying bacteria and converted into NO_3^-.

Although most of the NO_3^- formed through nitrification is taken up by plants, some is converted to nitrogen through the process of *denitrification*. This is primarily an anaerobic process and is especially important in waterlogged soils and ocean sediments. *Denitrification* is carried out by many species of anaerobic bacteria, especially of the genus *Pseudomonas*, *Clostridium*, and *Thiobacillus*. The soil microorganisms can get the oxygen they need through reduction of NO_2^- and NO_3^-. In this way, NO_2^- and NO_3^- are converted to gaseous nitrogen (N_2) and nitrous oxide (N_2O), which return to the atmosphere, By returning nitrogen to the atmosphere, denitrification completes the organic portion of the nitrogen cycle.

Currently, nitrogen fixation far exceeds denitrification, and usable nitrogen is accumulating in the life layer. Nearly all of this excess can be attributed to human activities, including the manufacture of nitrogen fertilizers and oxidation of nitrogen in the combustion of fossil fuels. Widespread cultivation of legumes has also greatly increased worldwide nitrogen fixation. At present rates, nitrogen fixation attributable to human activity is approximately equal to all natural biological fixation.

DEAD ZONES

Much of the nitrogen fixed by human activities is carried from the soil into rivers and lakes where it can cause major water pollution problems by stimulating the growth of algae and phytoplankton to create algal blooms. Ultimately, this nitrogen reaches the oceans. The long-term impact of large amounts of nitrogen on the Earth's marine ecosystems remains uncertain, although nitrogen is implicated as the cause of the **dead zones** that have been reported in many coastal regions (Figure 20.12).

Dead zones are characterized by **hypoxia**, a condition in which oxygen is almost entirely depleted. Hypoxia develops because of

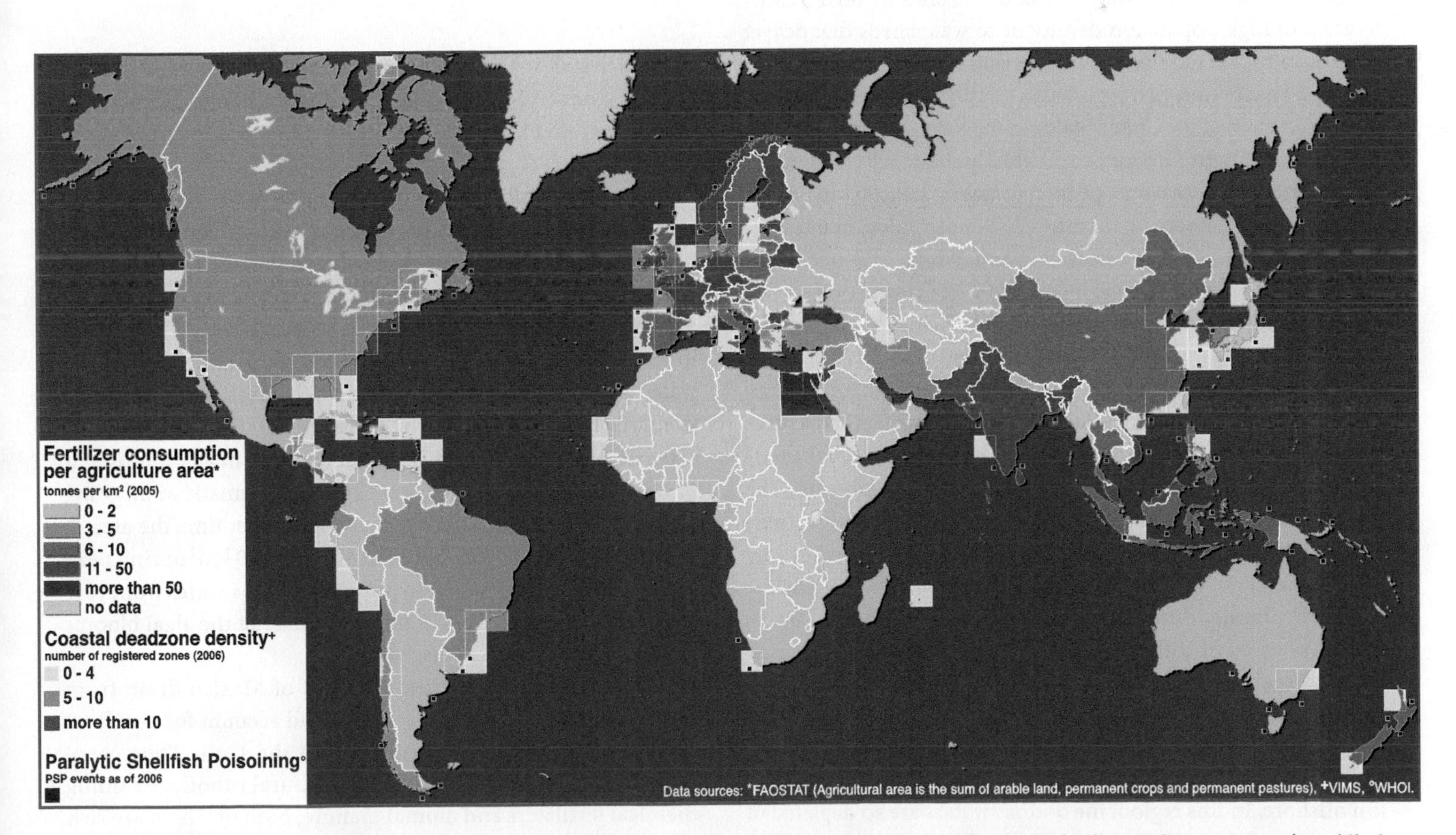

20.12 THE EARTH'S MARINE DEAD ZONES Dead zones are areas where concentrations of dissolved oxygen in the bottom waters are so reduced that the organisms that normally inhabit the region are unable to survive. Dead zones are associated with areas of high phytoplankton production caused by agricultural fertilizers carried to the sea by rivers.

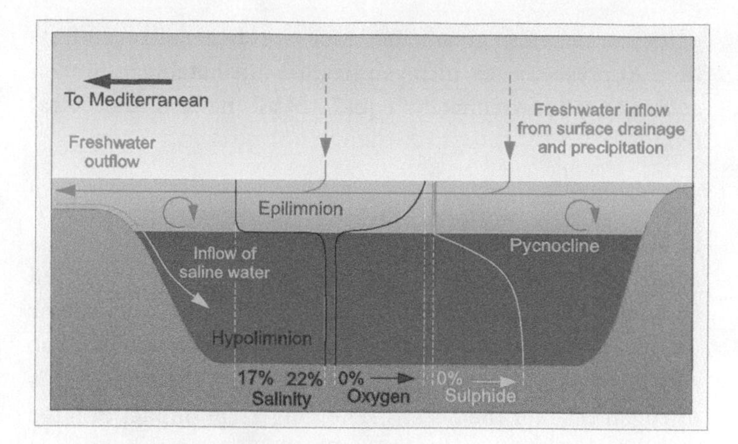

20.13 THE BLACK SEA DEAD ZONE The Black Sea below 150 m is considered the world's largest dead zone. In addition to nutrient influx, it is caused by the limited opportunity for water exchange due to basin configuration and the permanent *pycnocline* that prevents circulation of the water column and allows freshwater to drain directly from the system.

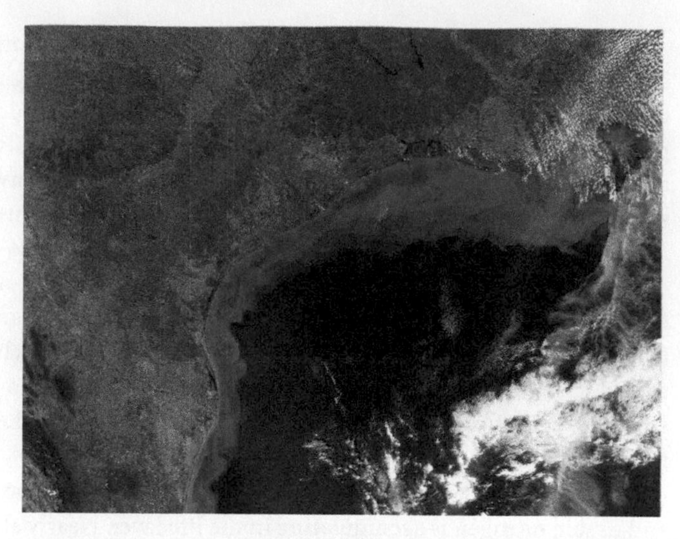

20.14 GULF OF MEXICO DEAD ZONE Brightly coloured waters in the Gulf of Mexico indicate the presence of sediment, detritus, and blooms of marine phytoplankton, which reduce dissolved oxygen in the water to crtically low concentrations. The Mississippi River Delta is at the right edge of the image. The image was acquired by the Moderate Resolution Imaging Spectroradiometer (MODIS) in September 2002.

high biological demand from decomposer bacteria that flourish on the abundant but short-lived algal blooms, which develop as a result of excess nutrients in the water. Thus, dead zones are linked either to areas of high population density or to watersheds that deliver large quantities of fertilizers and other nutrients to the oceans.

Dead zones were first reported in the 1970s in Chesapeake Bay on the east coast of the United States, in the Kattegat near Denmark, and in the northern Adriatic Sea. Currently about 400 locations are of concern, with a total area of approximately 250,000 km^2. Many dead areas are quite small; for example Saanich Inlet, an intermittently hypoxic zone off Vancouver Island, has an area of 65 km^2. Conversely, the hypoxic zone in the Baltic Sea covers over 60,000 km^2. The largest dead zone in the world is the deep waters of the Black Sea; it covers an area of 436,000 km^2 and descends to 2,245 m at its deepest point. The Black Sea dead zone arises because exchange of water with the Mediterranean Sea is limited by two factors. First, the only outlet of the Black Sea is through the narrow, shallow stretch of water known as the Bosphorus. The Bosphorus is 700 m wide at its narrowest point, and in mid-channel its depth ranges from 36 to 125 m. Second, mixing only takes place in the upper 150 m due to the permanent pycnocline caused by density differences related to salinity (Figure 20.13). Fresh water from inflowing rivers, being less dense than salt water, remains at the surface and drains directly through the Bosphorus.

The largest dead zone in North America is found off the northern coast of the Gulf of Mexico and extends as far as 125 km offshore. In this region, the bottom waters are so depleted of oxygen that fish and other forms of sea life are killed or forced to move away. Hypoxia develops when the oxygen concentration falls below 2 mg l^{-1}. The condition is linked to phytoplankton blooms that develop in response to increased nutrients in the water (Figure 20.14). The rate of production of organic matter is directly related to the nitrogen supply rate from the Mississippi River watershed.

The algae sink to the seabed when they die, along with the remains of the zooplankton that consumed them and other nitrogenous wastes. Dissolved oxygen in the deep water is progressively used by decomposer organisms to break down the copious supply of detritus from the surface layer. As is the case in the Black Sea, re-aeration is limited by stratification of the water column. In the Gulf of Mexico, the pycnocline develops primarily because of the influx of freshwater discharge from the Mississippi River. Hypoxia has been reported in all months, but is most severe in summer (Figure 20.15). Systematic monitoring of this dead zone began in 1985, and since that time the area of hypoxia has ranged from between 40 to 22,000 km^2 during peak months. The severity of the hypoxic conditions varies from year to year due to differences in the magnitude of the algal blooms and fluctuations in river discharge.

The rivers that flow into the Gulf of Mexico drain from about 40 percent of all U.S. land area and account for nearly 90 percent of the freshwater runoff into the Gulf. They carry excess nutrients generated from agricultural runoff containing chemical fertilizers and animal manure, both of which are rich in nitrogen. In drought years, the dead zone is relatively small due to the reduced influx of nitrates and nitrites transported to the Gulf of Mexico by the Mississippi River (Figure 20.16).

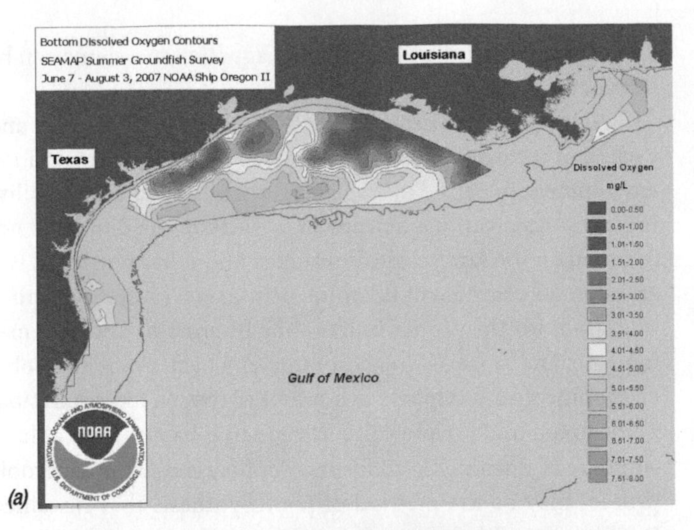

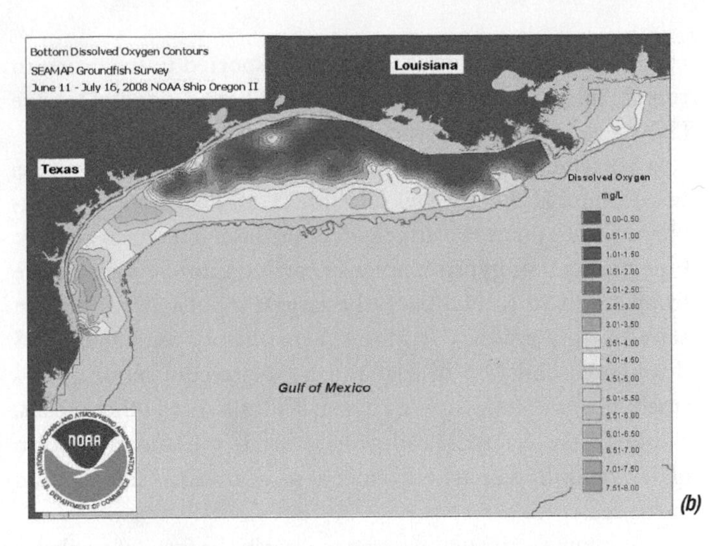

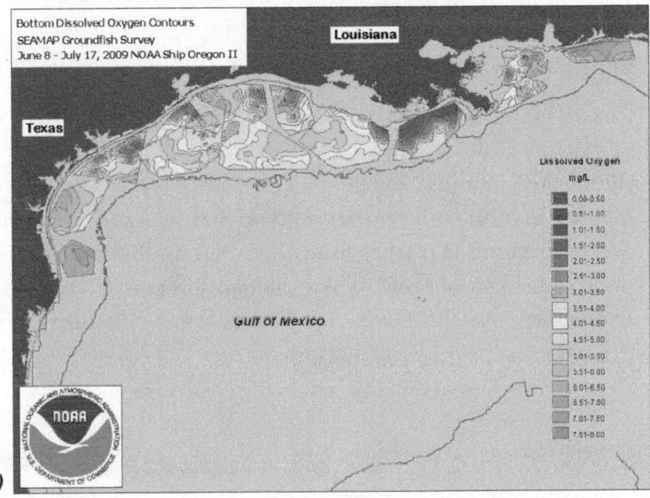

20.15 ZONE OF HYPOXIA IN THE GULF OF MEXICO Mid-summer dissolved oxygen concentrations in the bottom waters of the northern Gulf of Mexico for 2007–2009.

This type of ocean pollution is predicted to become more acute because industrial fixation of nitrogen in fertilizer manufacture is currently doubling about every six years. In the case of the Gulf of Mexico, the interagency Gulf of Mexico/Mississippi River Watershed Nutrient Task Force has a goal to reduce or make significant progress toward reducing the area of the dead zone to 6,000 km^2 or less by 2015. This represents about a 60 percent reduction compared to the average area of 17,200 km^2 recorded for the period 2000–2008.

THE SULPHUR CYCLE

Most of the Earth's sulphur is tied up in rocks and ocean sediments. A relatively small amount is also present in the atmosphere (Figure 20.17). Sulphur originates from igneous rocks, such as pyrite (FeS_2), and is also found in gypsum ($CaSO_4 \cdot 2H_2O$) and other sedimentary deposits. Long-term storage of sulphur occurs in both organic and inorganic forms, from which it is released by weathering and decomposition. Sulphur in mineral form can be mobilized through oxidation of sulphides to sulphate (SO_4^{2-}), which

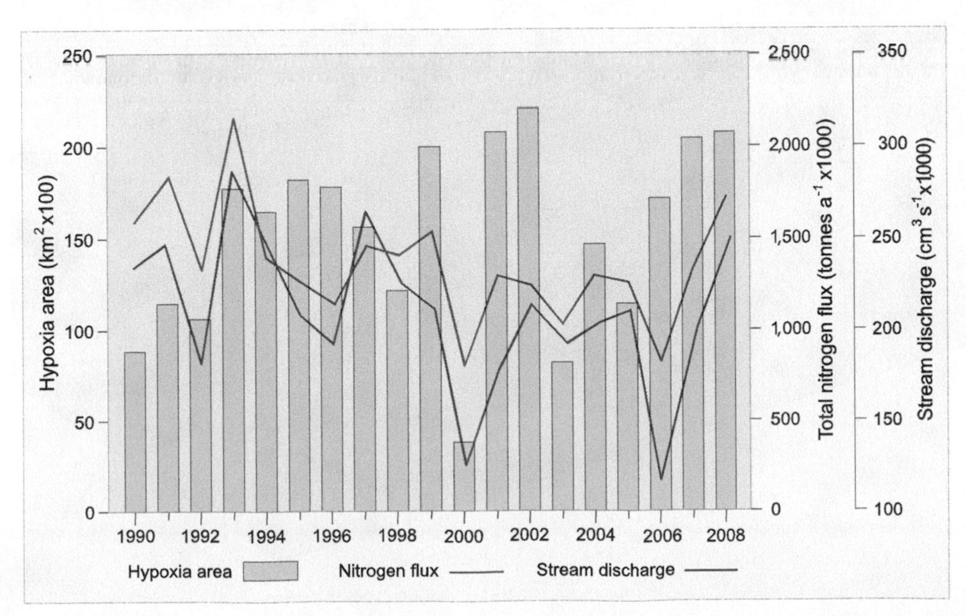

20.16 GULF OF MEXICO DEAD ZONE AND RIVER DISCHARGE Annual variation in the area of the hypoxic zone (where dissolved oxygen concentration is < 2.0 mg l^{-1}) closely follows the level of algal blooms and the combined flux of dissolved nitrogen in the form of nitrites and nitrates delivered to the Gulf of Mexico by river drainage.

may then go into solution and be transported to the ocean in runoff. Sulphur can also enter the atmosphere as sulphur dioxide (SO_2) and hydrogen sulphide (H_2S).

Volcanic activity releases sulphur gases to the atmosphere. In addition, H_2S, dimethyl sulphide (DMS), and carbonyl sulphide (COS) enter the atmosphere through biological activity. Fine particles of gypsum are also carried into the atmosphere from desert soils. However, the largest input is from human activity; this enters the atmosphere mainly as SO_2 or H_2S through combustion of coal and oil, petroleum refining, and smelting of sulphide ores of copper, lead, zinc, and other metals. Most of these terrestrial emissions return to the land in the form of dry deposition or as acid rain, but some transfer from the land to the sea.

Sulphur occurs in the ocean mainly in the form of sulphates that are derived from rivers and from atmospheric redeposition of sulphate-containing aerosols evaporated from the ocean surface. The principal gas originating from the oceans is DMS, produced by decomposing phytoplankton cells. DMS represents the largest natural source of gaseous sulphur in the atmosphere. It combines with water droplets to form acidic aerosols that act as cloud condensation nuclei. This has been suggested as a mechanism that might offset global warming (see Chapter 6).

A small component of the sulphur in river discharge comes from the natural weathering of pyrite and gypsum. However, most of the sulphur transported by rivers comes from human activity associated with mining, erosion, and air pollution. Acid mine drainage is a particular problem and arises where mining operations expose rocks containing ferrous sulphide (FeS_2). Reaction with air and water produces sulphuric acid (H_2SO_4) and other acidifying substances. This, in turn, can increase solubility of heavy metal contaminants, such as cobalt and arsenic. The effect of acid mine drainage on aquatic ecosystems can be devastating.

An important distinction between cycling of sulphur and cycling of nitrogen and carbon is that sulphur is present in the environment in a fixed or available form as sulphates, which living organisms can use immediately. Although sulphur is not required on the same scale as nitrogen and other nutrients, it is an essential component of amino acids, such as cystine, and is important for the proper functioning of proteins and enzymes in plants and animals. In addition, many bacteria in anaerobic environments use sulphate as a source of oxygen for respiration. Some green and purple sulphur-reducing bacteria are able to combine H_2S with CO_2 to form carbohydrates. This anaerobic process may be the method of photosynthesis that sustained Earth's earliest life forms.

SEDIMENTARY CYCLES

The carbon, oxygen, nitrogen, and sulphur cycles are referred to as gaseous cycles because in each case, the element is present in significant quantities as a gas in the atmosphere. Many other elements move in sedimentary cycles; that is, from the land to ocean in running water, returning after millions of years in uplifted terrestrial rocks. These elements are not present in the atmosphere, except in small quantities as blowing dust or condensation nuclei in precipitation.

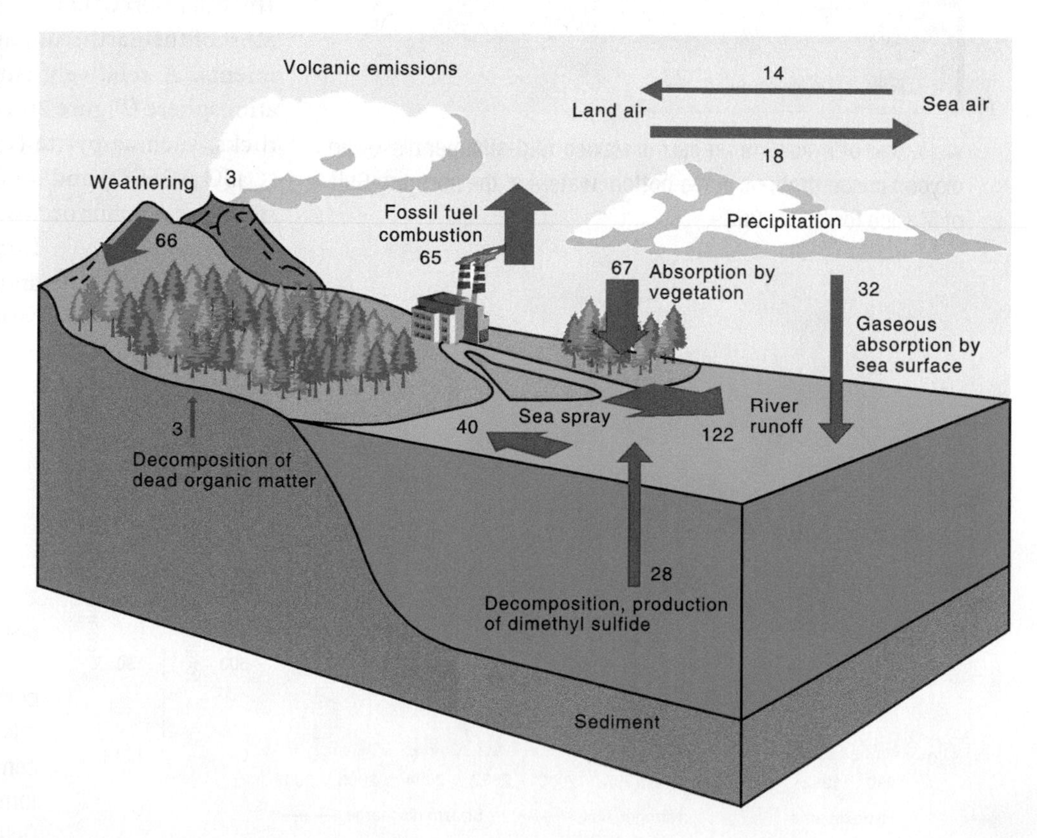

20.17 THE SULPHUR CYCLE The numbers represent transfer rates of sulphur, expressed as 10^9 kg of sulphur per year.

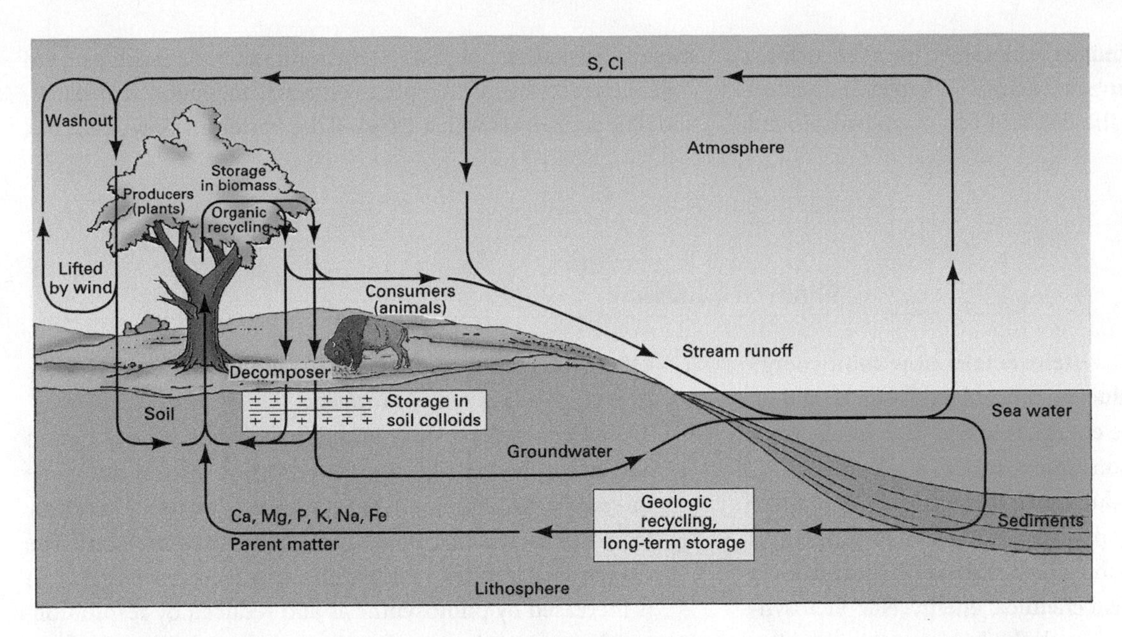

20.18 SEDIMENTARY CYCLES Sedimentary cycles move materials in and out of the biosphere and within the inorganic realms of the lithosphere and hydrosphere. The atmosphere plays only a minor role in sedimentary cycles.

Mineral nutrients, derived mainly from weathering of soil minerals and decomposition of organic residues, move in sedimentary cycles (Figure 20.18). Some are stored in nutrient pools within the soil, from which they may, to varying degrees, be extracted by plants or lost by leaching. Nutrients that are held as ions on the surfaces of soil colloids are readily available to plants; however, in other forms, they may be relatively insoluble and slowly become available over a long period of time. Although nutrients in solution can be taken up immediately by plants, they are also susceptible to leaching and removal by percolating water. In addition, some nutrients are lost in particulate form through overland flow and erosion. Other particles are lifted into the atmosphere by winds, sometimes falling back to the Earth great distances from their source.

The mineral elements required by plants are generally held as cations in a short-term storage pool and rapidly replenish nutrient ions as they are removed from the soil solution. The rate at which soil minerals are released from storage depends on their solubility. Minerals in the form of nitrates, carbonates, sulphates, and chlorides are readily taken up in the soil solution. Other forms may require a complex series of reactions before they are released from the mineral particles. For example, phosphates are strongly bound to clays and are released quite slowly. Soil organic matter also releases nutrients slowly as it decomposes. Most readily available nutrients are held as positively charged cations on the negatively charged surfaces of clay particles and organic matter (see Chapter 19). Cations are continually exchanged with other soluble ions until their supply is depleted.

Cycling of many plant nutrients closely follows the carbon cycle. When nutrient-rich plant residues return to the soil, this pool of carbon compounds is acted upon by decomposer organisms. As organic matter breaks down to simpler compounds, plant nutrients are released in available forms to growing plants. Rapid cycling of nutrients in this way is important in ecosystems, such as the tropical rain forest, where heavy rainfall can quickly leach essential cations from the soil.

Considerable element recycling occurs among the producers, consumers, and decomposers in an ecosystem. However, the elements used in the biosphere are continually escaping to the sea as ions dissolved in stream runoff and groundwater flow. Here, nutrient elements are held in enormous storage pools making them unavailable to organisms. As well as sea water, these storage pools include sediments on the sea floor and enormous accumulations of sedimentary rock beneath both land and oceans. These stored elements are eventually released to the soil through weathering.

A Look Ahead

The organizing principles of energy flow and matter cycling help to clarify some of the processes acting in the biosphere. Just as solar energy is the driving force for the circulation of the global atmosphere and ocean, it also provides the power source for photosynthesis, upon which all the world's organisms, including humans, ultimately depend for sustenance.

Organisms also rely on the smooth functioning of material cycles. Without the tiny fraction of the Earth's atmosphere comprising CO_2, no terrestrial photosynthesis would occur. If human activity enhances this component through CO_2 release, it is sure to impact the productivity of the biosphere directly, as well as indirectly through climate change. Similarly, humans influence the nitrogen and oxygen cycles without fully knowing the consequences.

The final two chapters examine the biosphere from the perspective of the spatial patterns of plants and animals. Chapter 21 discusses the processes that determine the distributions of individuals and species, including organism–environment relationships and dynamic processes such as species dispersal, migration, and extinction. The book closes with a survey of the world's major biome types.

CHAPTER SUMMARY

- The food web of an ecosystem details how food energy flows from primary producers, through consumers, and on to decomposers. Because energy is lost at each level, only a relatively few top-level consumers are normally present.
- Photosynthesis is the production of carbohydrate from water CO_2, and light energy by primary producers. Respiration is the opposite process, in which carbohydrate is broken down into CO_2 and water to yield chemical energy. Net photosynthesis is the amount of carbohydrate remaining after respiration has reduced gross photosynthesis. Net photosynthesis increases with increasing light and temperature, up to a point, then declines as optimum conditions are exceeded.
- Forests and estuaries are ecosystems with high rates of net primary production, while grasslands and agricultural lands are generally lower. Oceans are most productive in coastal and upwelling zones near continents. Among climate types, those with abundant rainfall and warm temperatures are most productive.
- Biomass is a form of solar-powered energy. Charcoal, biogas, and alcohol are biomass products that can be used as fuels.
- There are two types of biogeochemical cycles; *gaseous*, in which the element has an important gaseous phase and moves within the atmosphere, and *sedimentary*, when no important gaseous phase is involved. Biogeochemical cycles consist of *active pools* and *storage pools* linked by flow paths. Of most concern are the biogeochemical cycles of carbon, hydrogen, oxygen, and nitrogen.
- The *carbon cycle* includes an active pool of biospheric carbon and atmospheric CO_2, with a large storage pool of carbonate in sediments. Human activities have provided a pathway from storage to active pools by the burning of fossil fuel. The *oxygen cycle* features an active pool of atmospheric O_2, which is increased by photosynthesis and reduced by respiration, combustion, and mineral oxidation.
- The *nitrogen cycle* also has an important gas phase, but nitrogen is largely held in the form of N_2, which most organisms cannot use directly. *Nitrogen fixation* occurs when bacteria or blue-green algae convert N_2 to more useful forms, often in symbiosis with higher plants. Human activity has doubled the rate of nitrogen fixation, largely through fertilizer manufacture. Human activity has also greatly influenced the *sulphur cycle*, mainly through the deleterious effects of acid precipitation and acid mine drainage. Dead zones or zones of hypoxia, where dissolved oxygen for organisms is limited, are also linked to excessive nitrogen runoff to oceans.
- Sedimentary cycles involve nutrients, such as potassium, calcium, and magnesium, that do not have an important gas phase. These elements are held in active pools in living and decaying organisms and in soils. Storage pools include sea water, sediments, and sedimentary rocks.

KEY TERMS

active pool
biogeochemical cycle
biogeography
biomass
C_3 pathway
C_4 photosynthetic pathway
CAM (Crassulacean acid metabolism) pathway
consumers
dead zone
decomposers
food chain
food web
gross photosynthesis
hypoxia
macronutrients
net photosynthesis
net primary production (NPP)
omnivores
photosynthesis
primary producers
respiration
storage pool

REVIEW QUESTIONS

1. What is a food web or food chain? What are its essential components? How does energy flow through the food web of an ecosystem?
2. How is net primary production related to biomass? Identify some types of terrestrial ecosystems that have a high rate of net primary production and some with a low rate.
3. What is a biogeochemical cycle? What are its essential features? Identify and compare two types of biogeochemical cycles.
4. List three macronutrients, two secondary, and two micronutrients and identify those associated with gaseous and sedimentary cycles.
5. What are the essential features and flow pathways of the carbon cycle? How have human activities affected the carbon cycle?
6. What are the essential features and flow pathways of the nitrogen cycle? What role do bacteria play? How has human activity modified the nitrogen cycle?
7. What is a dead zone? How does it form, and how might it change over time?
8. What are the essential features and flow pathways of macronutrients in sedimentary cycles?
9. How is change in land use affecting the global carbon balance? Will replacing old-growth forests with younger faster-growing forests help remove CO_2 from the atmosphere?
10. Will increasing levels of atmospheric CO_2 have an effect on the rate of carbon fixation by ecosystems? How will increases in temperature affect the release of boreal soil carbon?

VISUALIZATION EXERCISE

1. Draw a diagram of the general features of a biogeochemical cycle in which life processes and physical processes link storage pools and active pools.

ESSAY QUESTIONS

1. Select one of the cycles described in the text (carbon, oxygen, nitrogen, sulphur, or sedimentary). Identify and describe the processes involved at each of the major pathways in the cycle.
2. Suppose atmospheric CO_2 concentration doubles. What will be the effect on the carbon cycle? How will flows change? Which pools will increase? Decrease?

APPENDIX 20.1 MONITORING GLOBAL PRODUCTIVITY FROM SPACE

As human activity drives CO_2 concentrations in the atmosphere to ever higher levels, it becomes increasingly important to understand and model the biosphere's photosynthetic activity. Analysis of the global carbon budget shows that the biosphere must be a *carbon sink*. In other words, for carbon flows to balance, global photosynthesis must exceed global respiration by a significant amount. This means that fixed carbon is accumulating in the biosphere, reducing the amount of CO_2 buildup in the atmosphere. However, this conclusion is not certain without some sort of direct measurement. Until recently, there was no way to measure the Earth's photosynthetic activity, but new techniques using remote sensing, meteorological observations, and models of biological productivity now make it possible.

The factors that control primary production at any point on Earth include the following:

- *The amount of photosynthetic material present.* For terrestrial plants, this is measured by the surface area of leaves above a square metre of ground. For oceanic phytoplankton, it is the chlorophyll concentration within a cubic metre at the ocean surface.
- *Light.* The amount of light absorbed through photosynthesis depends on two things: (1) the amount of illumination from the sun and sky, determined by such factors as season, latitude, and cloud cover; and (2) the amount of photosynthetic material (chlorophyll) present to absorb the available light energy.
- *Temperature.* The biochemical process of fixing carbon is sensitive to temperature. Warmer temperatures favour photosynthesis, but if temperatures are too high, photosynthesis shuts down.
- *Water.* Terrestrial plants transport water from their roots to their leaves in the process of transpiration. As the water evaporates from leaf pores, it cools the leaves and allows them to maintain high levels of photosynthesis, even under intense sunlight. Whereas water is freely available to aquatic phytoplankton at all times, terrestrial plants can experience water shortages.
- *Nutrients.* Nutrients can play an indirect role by reducing the health of organisms or limiting their development. In many cases, the productivity of phytoplankton is limited by a scarcity of nutrients, especially iron.

Production on Land

The "greenness" of the surface detected by remote sensing can be related to the amount of photosynthetic material. Land plants strongly absorb red light and strongly reflect near infrared light. Thus, locations that appear darker in a red band image and brighter in a near-infrared band image will have more leaf area.

Figure 20A.1a shows an image of leaf-area index for an area of North America that includes the United States and a portion of southeastern Canada. Leaf-area index (LAI) is the ratio of the area of the upper surface of a leaf to the area of ground below. Thus, a leaf-area index of two indicates that each square metre of ground surface is covered by two square metres of leaf area. Typical leaf-area indices can range from zero in barren deserts to six or seven in dense forest. The map of leaf-area index is derived from MODIS images, acquired May 1 to 10, 2003. At this time of year, the southeast and west-coast regions have considerably more functional leaf cover than other parts of the continent.

Figure 20A.1b shows the fraction of photosynthetically active radiation (FPAR) absorbed by the leaf canopy. As shown on the scale, this fraction ranges from 0 to 100 percent. It expresses the proportion of usable solar energy that is actually absorbed by the leaf canopy. By comparing this value with the solar energy falling on each point during each day, it is possible to estimate the energy available for photosynthesis.

The amount of photosynthesis that actually occurs also depends on both temperature and water availability, as noted above. Another factor is the efficiency of the type of vegetation cover; given the same environmental conditions and the same leaf area, some vegetation types are more productive than others. Gross primary production (GPP) can be calculated by combining temperature and rainfall from meteorological data sources with efficiency based on a map of vegetation cover type (Figure 20A.1c). The map shows the same general patterns as the others, with maximum GPP in the southeast and west-coast regions, but the fine detail shows the influence of rainfall, cloudiness, temperature, and type of vegetation.

The last step is to estimate respiration—the rate at which carbon is released to the atmosphere in plant metabolism. This will depend on many of the same factors as photosynthesis, including leaf area, temperature, and moisture. The result is net primary production (NPP), which is shown globally in Figure 20A.2.

(a)

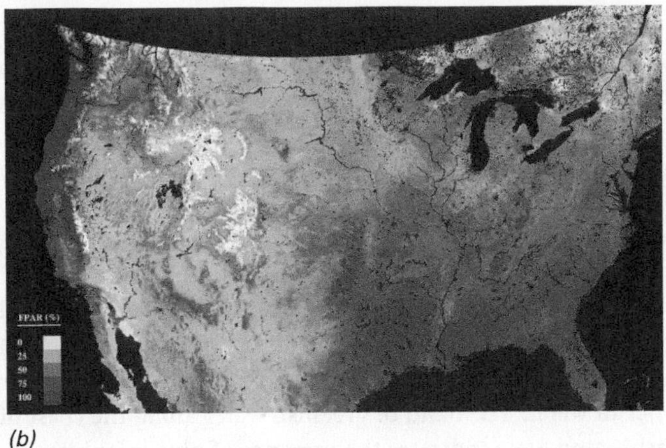

(b)

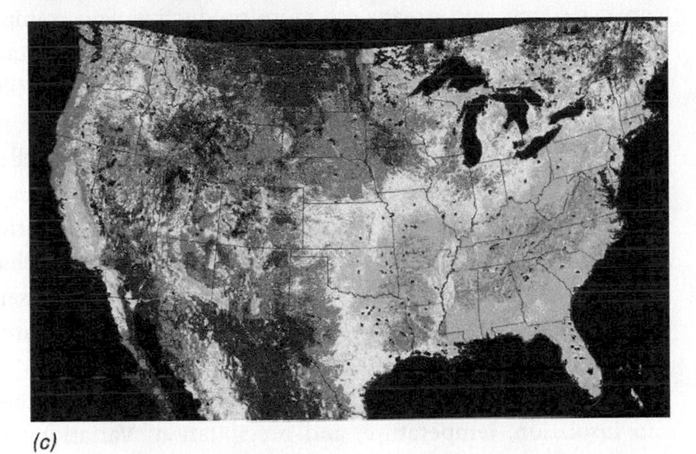

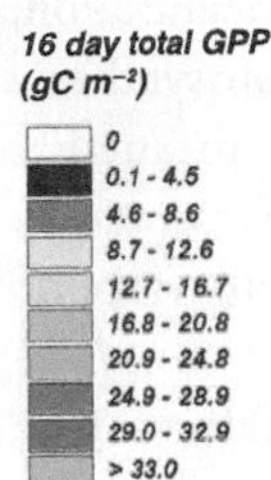

(c)

20A.1 MAPPING PRIMARY PRODUCTION IN NORTH AMERICA (a) Leaf-area index from MODIS, May 1 to 10, 2003. (b) Fraction of photosynthetically active radiation (FPAR) absorbed by vegetation, also derived from MODIS, for the same period. (c) Gross primary production, from MODIS and ancillary data, for the same period. (R. Nemani and S. W. Running, University of Montana, NTSG/NASA.)

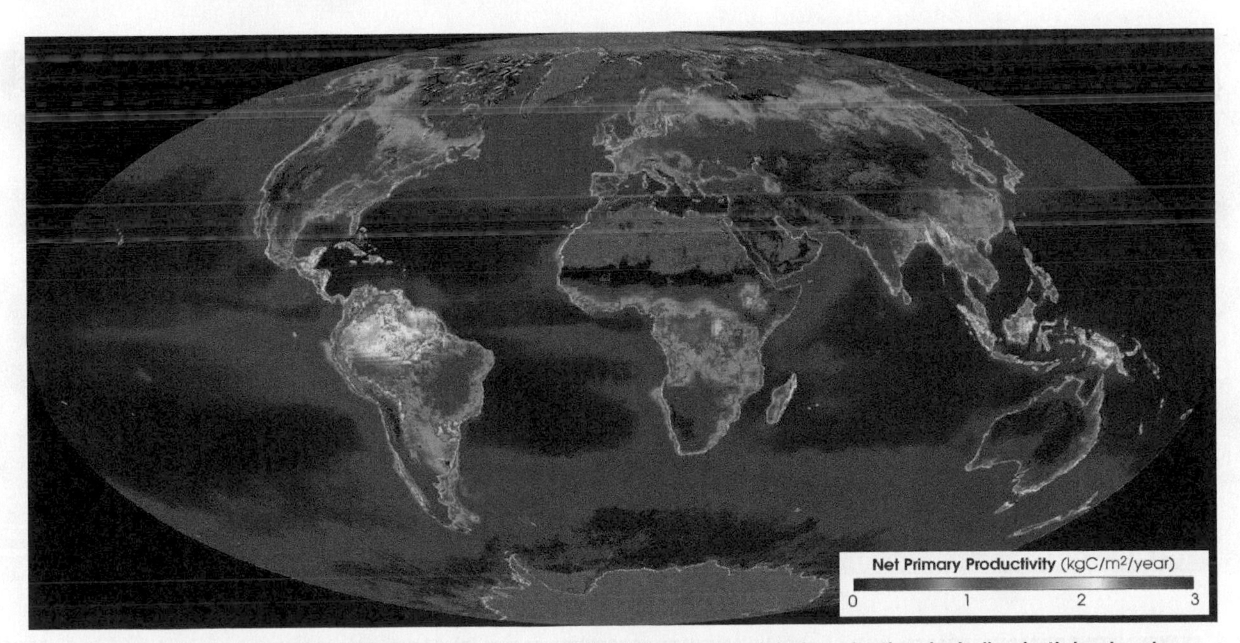

20A.2 GLOBAL NET PRIMARY PRODUCTION FROM MODIS FOR 2002 This image maps global production, including both land and oceans, as viewed by the MODIS instruments on NASA's Terra and Aqua satellite platforms. (MODIS Science Team/NASA.)

Primary Production of Oceans

For oceans, the amount of photosynthetic material present at a location is measured in terms of chlorophyll concentration expressed as milligrams of chlorophyll per cubic metre (mg m^{-3}). As with terrestrial plants, this quantity is determined through colour, detected by satellite instruments using special narrow spectral bands.

Figure 20A.3a shows a colour image of the waters around southern Vancouver Island and the Puget Sound area acquired by the SeaWiFS (Sea-viewing Wide Field-of-view Sensor) instrument, which is designed specifically to image the colour of ocean water. The band of greenish water along the coast of Washington State indicates the presence of phytoplankton. By carefully analyzing the spectral information acquired by SeaWiFS, it is possible to estimate chlorophyll concentration across the image (Figure 20A.3b). Chlorophyll concentration, like leaf-area index, is not sufficient to model photosynthesis accurately. The presence of chlorophyll doesn't mean that photosynthesis is proceeding at a maximum rate. For example, lack of nutrients such as iron may slow the photosynthetic process.

The efficiency of photosynthesis by phytoplankton can be assessed from MODIS satellite data. This satellite instrument has spectral bands that measure light emitted at particular wavelengths by chlorophyll fluorescence. Chlorophyll fluorescence occurs primarily when absorbed sunlight is not being used to fix carbon. In other words, when chlorophyll fluorescence is strong, photosynthesis is weak. Therefore, the strength of the fluorescence leaving ocean waters provides a measure of the photosynthetic efficiency of the chlorophyll that is present. With this missing link in the chain of information, mapping primary production of the ocean becomes possible.

Global Primary Production

Net primary production is generally higher on land than in oceans (see Figure 20A.2). However, there is a lot more ocean surface than land surface. Taking this into account, terrestrial and oceanic NPP are about equal. Large central areas of the oceans are unproductive (dark purple). This is largely due to lack of essential nutrients in surface waters. Near land and along coasts where nutrient-rich deep water rises from below, nutrients are in greater supply and oceans are more productive. For example, upwelling is particularly important along the west coast of South America from Peru southward and shows dark and light blue tones.

On land, the highest values of NPP are shown by the yellow and red tones of the Amazon basin and equatorial Asia. Lesser, but still high, values are evident in eastern North America, central Africa, eastern Asia, and Scandinavia. Low values characterize arid and semi-arid regions, such as the interior deserts of Australia and semi-arid western North America (grey areas show regions for which data are not available).

NPP varies through the year in response to seasonal changes in insolation, temperature, and precipitation. Variations are especially pronounced over land at higher latitudes. For example, NPP in the boreal forest of the northern hemisphere is high

(a)

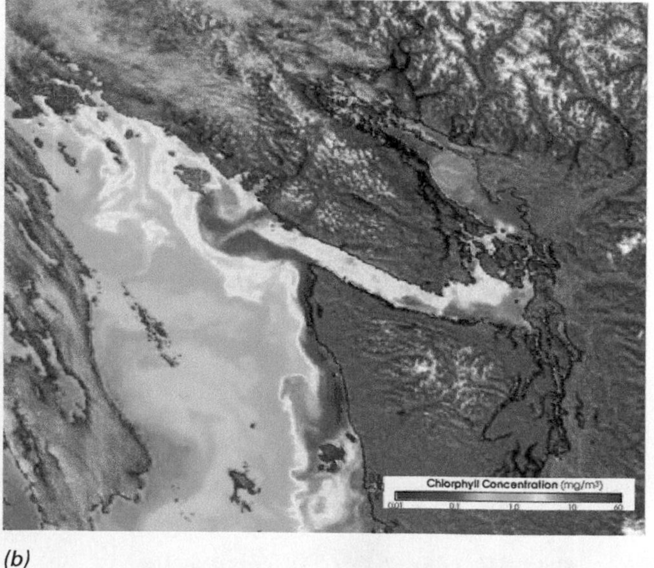

(b)

20A.3 SEAWIFS VIEWS OF PACIFIC COASTAL WATERS AROUND SOUTHERN VANCOUVER ISLAND (a) A true-colour image acquired on July 9, 2003. (b) Chlorophyll concentration on the same day. (SeaWiFS Project, NASA/GSFC/ORBIMAGE.)

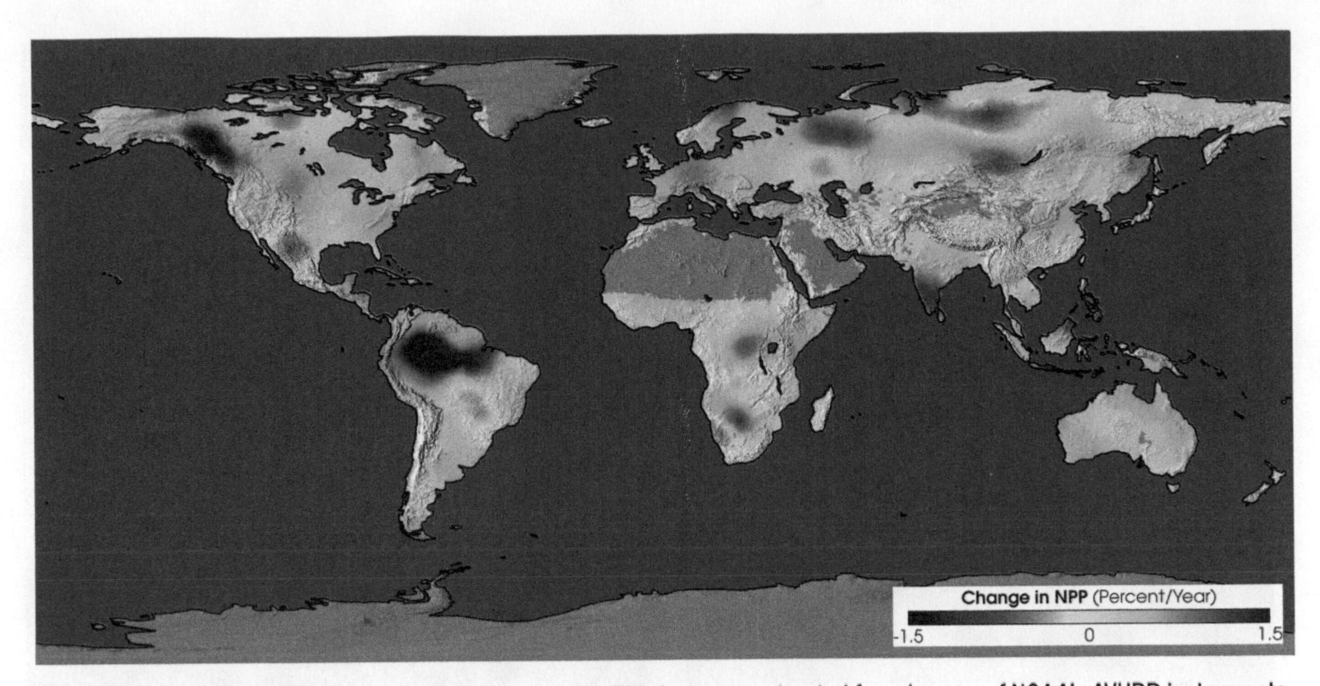

20A.4 CHANGE IN NET PRIMARY PRODUCTION FOR LAND, 1982–1999 This image, constructed from images of NOAA's AVHRR instruments acquired from 1982 to 1999, shows the change in net primary production as percent per year. (R. Nemani and S. W. Running, University of Montana, NTSG/NASA.)

in summer when long days and warm temperatures favour photosynthesis and essentially drops to zero during the long cold winter. Production remains high in the Amazon and equatorial regions. In December, net primary production is high in South America and sub-Saharan Africa, as well as along northern and eastern coasts of Australia. Meanwhile, most of northern North America and Eurasia are dormant. Production in the tropical and subtropical oceans is similar throughout the year, but at mid- and higher latitudes the oceans are considerably more productive during the summer months in each hemisphere.

Recent Changes in Global Productivity

Analysis of satellite images acquired by NOAA's Advanced Very High Resolution Radiometer (AVHRR) between 1982 and 1999 shows that terrestrial NPP increased by about 6 percent during that period (Figure 20A.4). The greatest increases occurred in northwestern North America, equatorial and subtropical South America, and India. Large increases were also noted in Africa and northern Russia. By comparing these results with meteorological data, it was concluded that the increase at low latitudes was due to reduced cloud cover, which allowed more light for photosynthesis. At high latitudes, the probable causes were increased temperature and, to some extent, increased water availability.

Over a similar time period, large changes in ocean phytoplankton concentration also took place (Figure 20A.5). The colour scale is logarithmic, with deeper tones of blue and red showing much larger changes than occurred in areas of lighter tones. The most striking feature observed was the decline in phytoplankton in northern oceans (blue), which amounted to about 30 percent in the North Pacific and 14 percent in the North Atlantic. In the equatorial zones, increases of up to 50 percent (red) were observed at some locations, but the increases were not large enough to account for the high latitude decreases. Thus, it was concluded that global concentrations of phytoplankton had decreased overall.

One explanation for this decline is that warmer sea-surface temperatures are increasing the duration and strength of the thermocline at high latitudes. This inhibits the mixing of nutrient-rich deep water with nutrient-poor surface water, and thus keeps the phytoplankton population in a nutrient-limited condition. Another possibility is that wind speeds are decreasing, which will also reduce mixing. Both of these changes have already been observed. Summer sea-surface temperatures in northern regions increased by 0.4 °C between the early 1980s and 2000, and average wind stress on the sea surface in spring has decreased by about 8 percent. However, it is not certain that these changes are

20A.5 CHANGE IN OCEAN CHLOROPHYLL CONCENTRATION FROM 1979–1986 TO 1997–2000 This image shows how summer ocean chlorophyll concentrations have changed over the past two decades. Data are for July to September, as acquired by NOAA's Coastal Zone Color Scanner (1979–1986) and SeaWiFS instrument (1997–2000). (NASA/NOAA.)

the result of global climate change or originate from a multi-year ocean cycle yet to be discovered.

Implications for the Global Carbon Budget

What are the implications of these changes in global productivity for the global carbon budget? First, the balance in global productivity between land and ocean seems to be shifting toward land. Although this may increase the amount of terrestrial biomass and decrease the rate of CO_2 buildup, it also means that ocean productivity is probably declining and altering oceanic ecosystems.

Second, soil respiration must also be considered. Soil respiration is analogous to the decay of organic matter in soils, which releases CO_2 to the atmosphere. Soil respiration is temperature dependent, and small increases in temperature can stimulate large increases in respiration. Considering that boreal forest soils are rich in organic matter and that global temperatures are increasing most rapidly at high latitudes, increased release of CO_2 by soil respiration could exceed increased biomass production and result in higher levels of atmospheric CO_2.

Third, it is not known if the increase in terrestrial productivity will continue. For example, with more sunlight and higher temperatures, moisture may become limited in equatorial and tropical forests, causing NPP to plateau or even decrease. Moreover, the large changes in temperature forecast for high-latitude regions will ultimately lead to reduced production as boreal forests come under increasing stress, and trees at their southern limits of distribution begin to die.

All these uncertainties emphasize the importance of the ability to map and monitor global production using remote sensing. While it may not yet be possible to predict exactly how the carbon cycle will behave in the future, at least current trends can be monitored. This will allow models of the carbon cycle to be refined over time, so that the full impact of human activity on the global carbon cycle can be assessed.

QUESTIONS

1. What factors control primary production at any point on the Earth's surface? Identify each factor and relate it to the process of photosynthesis.
2. What is leaf-area index? How and why can it be sensed remotely? How does it relate to the fraction of photosynthetically active radiation absorbed by the plants on the surface?
3. What factor is most important in determining ocean primary production? How is it mapped using remote sensing?
4. What are the main patterns of primary production over the oceans? Over land?
5. What changes in net primary production over land have occurred over the past two decades? How are they related to the controlling factors identified in Question 1?
6. What changes in net primary production over oceans have occurred over the past two decades? What are the possible causes of these changes?
7. What are the implications of changes in production for the carbon cycle? How will remote sensing be useful in assessing the impact of human activity on the global carbon cycle?

CHAPTER 21

BIOGEOGRAPHIC PROCESSES

- Ecological Biogeography
  - Water Need
  - Temperature
  - Other Climatic Factors
  - Bioclimatic Frontiers
  - Geomorphic and Edaphic (Soil) Factors
  - Disturbance
  - Interactions among Species
- Ecological Succession
- Historical Biogeography
  - Evolution
  - Speciation
  - Extinction
  - Dispersal
  - Distribution Patterns
  - Biogeographic Realms
- Biodiversity

Wetlands at Chicko River, British Columbia.

The preceding chapter focused on ecosystems from the perspective of energy capture and global cycling of carbon, nitrogen, and other nutrients. But ecosystems are composed of individual organisms that interact with their environment in different ways. Each organism has a range of environmental conditions that limits its survival, as well as a set of characteristic adaptations that it exploits to obtain the energy it needs to live.

This chapter begins with a discussion of ecological biogeography and examines how relationships between organisms and their abiotic environment determine where they are found. Ecological biogeography explains the spatial pattern of organisms on local, regional, and global scales, and on a time scale of a few generations.

Historical biogeography examines the distribution of organisms on continental or global scales over longer time periods. This branch of biogeography describes processes such as evolution, dispersal, and extinction of species through time. It is closely linked to geologic processes and events such as the breakup of landmasses through plate tectonics and environmental disruption associated with the Late-Cenozoic Ice Age. Together, ecological and historical biogeography provide a comprehensive framework for understanding and appreciating the diversity of the biosphere, and how this is threatened by human activities and environmental change.

ECOLOGICAL BIOGEOGRAPHY

The relationship between organisms and their physical environments is strongly influenced by landforms and soils. For example, upland soils are often stony and well drained, while the soils in valleys may be finer, richer in organic matter, and wetter. Such variations provide a range of **habitats** that are more or less suited to the needs and preferences of organisms or groups of organisms.

The concept of the **ecological niche** is related to habitat and expresses the functional role an organism plays as well as the physical space it inhabits. An organism's ecological niche describes how and where it obtains its energy and how it influences other species and the environment around it. The niche is a dynamic concept insofar as it recognizes an organism's tolerance and response to changes in environmental factors such as moisture, temperature, and soil chemistry. Although many different species can occupy the same habitat, no two species can occupy identical ecological niches. Thus, the theoretical limit for a species, as determined by its physiological requirements for all environmental factors, establishes its **fundamental niche**. The reduced environmental space that it actually occupies because of competition with other species is called its **realized niche** (Figure 21.1).

Every habitat is home to a distinct group of species, each occupying specific, but interrelated, ecological niches. A group of species that live in a particular habitat and interact with each other form a *community*. Because similar habitats in a region often contain similar assemblages of species, communities can be grouped together as *associations*. Such associations are defined by dominant species, as in the hemlock–white pine–northern hardwoods association found in the mixed-wood forests of Ontario and Quebec.

Plant associations are themselves grouped together as part of a hierarchical vegetation classification system (Table 21.1). Higher levels in the classification emphasize dominant life forms and general compositional similarities at the global scale. Mid-levels tend to reflect differences in vegetation form and composition at the continental scale, with differences at regional and local scales used to establish the lower levels.

In general, moisture and temperature are the most important environmental factors that determine where a species is

21.1 NICHE PARTITIONING IN WARBLERS Each species uses a different aspect of the tree's habitat to reduce competition with other species. This allows multiple species to co-exist in the same location.

TABLE 21.1 REVISED U.S. NATIONAL VEGETATION CLASSIFICATION STANDARD (2008) This natural vegetation hierarchy encompasses compositional similarity of life form and species at global, regional, and local scales.

Physiognomic levels	Formation class	Broad combination of dominant general life forms and structure that correspond to global moisture and temperature regimes
	Formation subclass	A combination of dominant general life forms reflecting global macroclimatic factors as affected by latitude or continental position
	Formation	A combination of general life forms reflecting global macroclimatic factors such as elevation, seasonality, and soil moisture conditions
	Division	A combination of general diagnostic life forms reflecting biogeographic differences at the global scale
	Group	A combination of broadly similar species occurring at sub-continental scales in response to regional differences in habitat
Floristic levels	Alliance	A characteristic group of species reflecting general habitat differences at the local scale
	Association	Species groupings reflecting specific differences in local habitats

Source: Adapted from NatureServe, "Contours of the Revised U.S. National Vegetation Classification Standard." Table 2, Bulletin of the Ecological Society of America, January 2009, http://www.natureserve.org/publications/pubs/Faber-Langendoen_Tart_Crawford_Revised%20NVC_ESA%20Bulletin%20January%202009.pdf.

found. Under conditions of extreme temperature or dryness, some organisms may persist as spores or cysts, but few if any organisms have limits that are not exceeded at least somewhere on the Earth at some time. At the global scale, temperature and moisture patterns are fundamental characteristics of climate. For this reason, there is a strong relationship between broad scale climate and vegetation patterns.

WATER NEED

The availability of water to terrestrial organisms at a particular point in time or space is determined by the balance between precipitation, evaporation, runoff, and infiltration. This balance is, in turn, affected by organisms, especially the plant cover. Through transpiration, plants return much of the soil water to the atmosphere. This process is important in tropical rain forests, where transpiration directly contributes as much as 50 percent of the regional rainfall. By obstructing overland flow and increasing soil porosity, plants reduce runoff and increase infiltration. Burrowing animals can also enhance infiltration. Although these local water flows may be important within individual habitats, their effects are generally small compared with those of the physical processes that control the major features of the water cycle. Thus, the overall dynamics of the atmosphere and oceans in the form of global climate still determine the basic pattern of water availability from one place to another.

Both plants and animals show a variety of adaptations that enable them to cope with a scarcity or abundance of water. Plants that are adapted to drought conditions are called **xerophytes** (Figure 21.2). While most xerophytes are associated with desert regions with scarce rainfall, some are adapted to habitats that dry quickly following rapid drainage of precipitation; for example, sand dunes, beaches, and rock surfaces. Many species adapt to drought by closing the stomatal cells in the leaf epidermis to limit transpiration. In some xerophytes, a thick layer of wax-like material on leaves and stems reduces water loss. The wax helps to seal water vapour inside the plant tissues. Others adapt by greatly reducing their leaf area or by bearing no leaves at all. In cacti, for example, the leaves are reduced to spines.

Adaptations of plants to water-scarce environments also include improved abilities to obtain and store water. Roots may extend several metres to reach soil moisture far below the surface. If the roots reach the groundwater zone, they will have a steady supply of water. Other desert plants develop a widespread, but shallow, root system. This enables them to absorb water from brief storms that saturate only the uppermost soil layer. Leaves of desert plants, such as aloes and agaves and stems of cacti, are greatly thickened by spongy tissue in which water can be stored. Plants with this adaptation are called *succulents*. This spongy tissue also plays a role in CAM photosynthesis, where plants use it to store the intermediate products formed by gas exchange during hours of darkness.

Another adaptation to extreme dryness is a very short life cycle. Many small desert plants are *ephemeral annuals*. Following germination, they leaf out, bear flowers, and produce seeds in a few weeks immediately following a heavy rain shower. Ephemeral annuals complete their life cycle when soil moisture is available, and survive the dry period as seeds that require no moisture.

In the wet–dry tropical climate, which has a yearly cycle with a pronounced dry season, trees and shrubs will often respond by dropping their leaves at the end of the moist season, becoming dormant during the dry season (see Figure 9.10). When water is available again, these *drought-deciduous* plants leaf out and grow quickly. In the Mediterranean climate, which experiences hot, dry summers, the plants typically retain their

(a)

(b)

(c)

21.2 PLANT ADAPTATIONS TO DROUGHT (a) Leaf succulents such as *Aloe ferox* can store water in their swollen leaves. (b) The creosote bush (*Larrea tridentata*) has small leaves that are naturally varnished to conserve water. (c) Leaves on the ocotillo (*Fouqueria splendens*) develop quickly after a rainstorm, but soon wither and fall from the stem as the soil dries out. In this way, the ocotillo can produce several crops of leaves each year.

tough, leathery leaves all year. Plants that retain their leaves through a dry or cold season have the advantage of being able to resume photosynthesis immediately when growing conditions become favourable, whereas deciduous plants have to grow a new set of leaves each growing season. This is particularly valuable for the evergreen conifers at high latitudes where growth is limited to a short growing season because of low temperatures.

Plants that grow in lakes, marshes, and bogs, are referred to as **hydrophytes** (Figure 21.3). These plants are adapted to cope with excessive moisture that invariably means they can tolerate low concentrations of soil oxygen. Oxygen diffuses very slowly into saturated soils; this results in anaerobic conditions and unusual soil chemistry. In particular, iron becomes very soluble and, potentially, could become toxic to plants. Plants that grow in shallow water environments, such as cattails (*Typha latifolia*), generally have abundant air space tissue that allows them to take in air through the leaves and quickly conduct it down the stems to the roots, where it passes into the mud immediately surrounding the roots. Through this process of *radial oxygen loss*, the iron is oxidized to insoluble forms and precipitated externally to the roots. This permits the preferential uptake of other essential nutrients. Iron toxicity can also be minimized by reducing the need for essential elements. This also reduces the uptake of iron, as all cations are absorbed from the soil solution by way of the transpiration stream. Species such as Labrador tea (*Ledum groenlandicum*) use this strategy and have limited transpiration through a reduction in leaf size. Floating leaves are another common adaptation of aquatic plants. The waxy, water-repellent leaves of water lilies (*Nymphaea odorata*) and water hyacinths (*Eichhornia crassipes*) keep their upper surfaces dry and maintain an optimum interface for rapid gas exchange between the plants and the atmosphere. Finally, the seeds of many hydrophytes float so they can be dispersed by lake or ocean currents and by stream flow.

To cope with water shortages, some invertebrate animals have evolved methods that are similar to those of ephemeral annual plants; they avoid the dry period by becoming dormant. When rain falls, they emerge to take advantage of the moisture and the newly developed, short-lived vegetation. Similarly, many bird species regulate their behaviour to nest only when the rains occur, as this is the time of most abundant food for their offspring. Tiny brine shrimp may wait many years in dormancy until normally dry lake beds fill with water, an event that occurs perhaps three or four times a century. The shrimp then emerge and complete their life cycles before the lake evaporates.

(a)

(b)

(c)

21.3 PLANTS OF WET HABITATS (a) Tall cattails (*Typha latifolia*) are able to conduct oxygen to their submersed roots. (b) The small leaves of Labrador tea (*Ledum groenlandicum*) have characteristics similar to those of desert shrubs; in both environments, the leaves have adapted to limit transpiration. (c) The floating leaves of water lilies (*Nymphaea odorata*) allow them to be fully exposed to air and sunlight.

Mammals are by nature poorly adapted to desert environments, but many survive through a variety of mechanisms that enable them to avoid water loss. Just as plants reduce transpiration to conserve water, many desert mammals do not sweat through skin glands. Instead, they rely on other methods of cooling, such as avoiding the sun and becoming active only at night. In this respect, they join most of the rest of the desert fauna, spending their days in cool burrows in the soil and their nights foraging for food. However, larger animals are less able to avoid the rigours of the desert environment and other strategies are needed. The principal adaptation of the camel is that its blood doesn't thicken through dehydration. Unlike humans, for example, the blood of the camel still circulates freely and prevents the animal's core temperature from becoming dangerously high. Kangaroo rats (*Dipodomys spp.*) can metabolize water from a diet of dry seeds, so they are essentially immune to prolonged drought.

Most animal groups include numerous species that have adapted to wet environments. Some, like amphibians, are able to exploit both terrestrial and aquatic habitats. However, problems can arise for some species when land areas are periodically flooded. Worms, for example, are often forced out of the soil during periods of saturation, and rodents in savanna grasslands can be drowned in their burrows during the wet season. The beaver (*Castor canadensis*), meanwhile, works industriously building dams to flood the land.

TEMPERATURE

The temperature of the air and soil acts directly on organisms by influencing the rates at which physiological processes take place in plant and animal tissues. For example, each plant species has an optimum temperature associated with photosynthesis, flowering, and seed germination. There are also limiting lower and upper temperatures for these functions and for the total survival of the plant itself. In general, the colder the climate, the fewer the species that are capable of surviving. For this reason, plant and animal species diversity is comparatively low in severe arctic and alpine environments. A plant's tolerance to cold is closely tied to the physical disruption that accompanies the growth of ice crystals inside cells. Cold-tolerant plant species can expel excess water from cells so that it freezes in the intercellular spaces where it does no damage.

The effects of temperature variations on animals are moderated by their physiology and their ability to seek shelter. Most animals lack a physiological mechanism for internal temperature regulation. These animals, including reptiles, invertebrates, and fish, are *cold-blooded animals*—their body temperatures passively follow the environment. With a few exceptions (notably fish), these animals are active only during the warmer parts of the year. They survive the cold winter weather by becoming dormant. Some vertebrates enter a dormant state called *hibernation*, in which metabolic processes virtually stop and body temperature closely parallels that of the surroundings. Most hibernators seek out burrows, nests, or other environments where winter temperatures are not extreme nor fluctuate rapidly. Burrows are particularly suited to hibernation because the annual range of soil temperatures gets smaller with depth below the surface.

Other animals maintain a relatively constant body temperature by internal metabolism. This group includes birds and mammals. These *warm-blooded animals* possess a variety of adaptations to regulate their body temperature. Fur, hair, and feathers act as insulation by trapping air next to the surface of the skin, thereby reducing heat loss. A thick layer of fat also provides excellent insulation. Other adaptations are needed for cooling; for example, sweating or panting use the high latent heat of vaporization of water to dissipate body heat. Heat loss is also facilitated by exposing blood-circulating tissues to the cooler surroundings. A seal's flippers, a bird's feet, and the long ears of a jackrabbit serve this function.

OTHER CLIMATIC FACTORS

At the global scale, *light* available for plant growth varies by latitude. Duration of daylight in summer increases rapidly with latitude and reaches its maximum within the Arctic and Antarctic circles, where the sun can remain above the horizon for 24 hours (see Figure 2.18). Although frost greatly shortens the growing season for plants at high latitudes, the prolonged daylight accelerates the rate of plant growth during the brief summer. Plants in the midlatitudes use changes in the duration of nighttime during the year as a cue for budding, flowering, and fruiting through the light-sensitive pigment mediated process of *photoperiodism*. Leaf shedding in deciduous species is controlled in the same way.

Three major groups of flowering plants are recognized according to their response to seasonal changes in the length of the dark period. *Long-day* plants flower only after exposure to periods of illumination longer than 12 hours, although 14 to 16 hours are usually required; hibiscus, larkspur, and delphinium are examples. Long-day plants grow naturally at mid- and high latitudes where the summer nights are short. *Short-day* plants, such as tulips, asters, poinsettias, and chrysanthemums flower when exposed to short periods of daylight less than 12 hours in duration; they are mainly represented by the spring- and fall-flowering species of midlatitudes. *Day-neutral* plants are not regulated by the photoperiod. Many tropical species are of this type. By controlling the flowering process, photoperiodism exerts a strong control on plant reproduction and consequently has an important influence on global distribution patterns.

The diurnal and seasonal cycles of illumination also influence animal behaviour. Activity of most animals is closely linked to the circadian rhythm of day and night. Birds, for example, are generally active during the day, whereas small mammals often forage at night. Changing day length over the course of a year induces seasonal activity in animals. In midlatitudes, as autumn days grow shorter, squirrels and other rodents hoard food for the coming winter season, and birds begin their migrations. In the spring, the lengthening photoperiod will trigger such activities as mating and reproduction.

Wind is an important environmental factor that can affect vegetation structure in highly exposed areas. Close to the timberline in high mountains and along the northern limits of tree growth in the arctic zone, wind has deformed trees in such a way that the branches project only from the lee side of the trunk (Figure 21.4). The wind causes excessive drying, damaging the exposed side of the plant. As a result, the altitudinal tree limit on mountainsides varies with the degree of exposure to strong prevailing winds, and extends higher on lee slopes and in sheltered pockets.

21.4 **THE EFFECT OF WIND ON TREES** The one-sided branching pattern on this alpine fir (*Abies lasciocarpa*) is caused by abrasive action of wind-blown ice crystals. The branches develop only on the side of the trunk that is protected from the prevailing winter winds.

BIOCLIMATIC FRONTIERS

Climatic factors, such as moisture and temperature, often limit the distribution of plant and animal species. Biogeographers recognize that there is a critical level of climatic stress beyond which a species cannot survive. A geographic boundary can thus mark the limits of the potential distribution of a species. Such a boundary is referred to as a **bioclimatic frontier**. Although the frontier is usually marked by a variety of climatic elements, it is sometimes possible to single out one climatic element that approximately coincides with it.

The distribution of ponderosa pine *(Pinus ponderosa)* in western North America is an example (Figure 21.5). In this mountainous region, annual rainfall varies sharply with elevation. The 500-mm total annual precipitation isohyet encloses most of the upland areas with ponderosa pine. The parallel arrangement of the isohyet with the forest boundary, rather than actual degree of coincidence, is significant. The sugar maple *(Acer saccharum)* is a more complex example (Figure 21.6). Here, the boundaries on the north, west, and south coincide roughly with selected values of annual precipitation, mean annual minimum temperature, and mean annual snowfall.

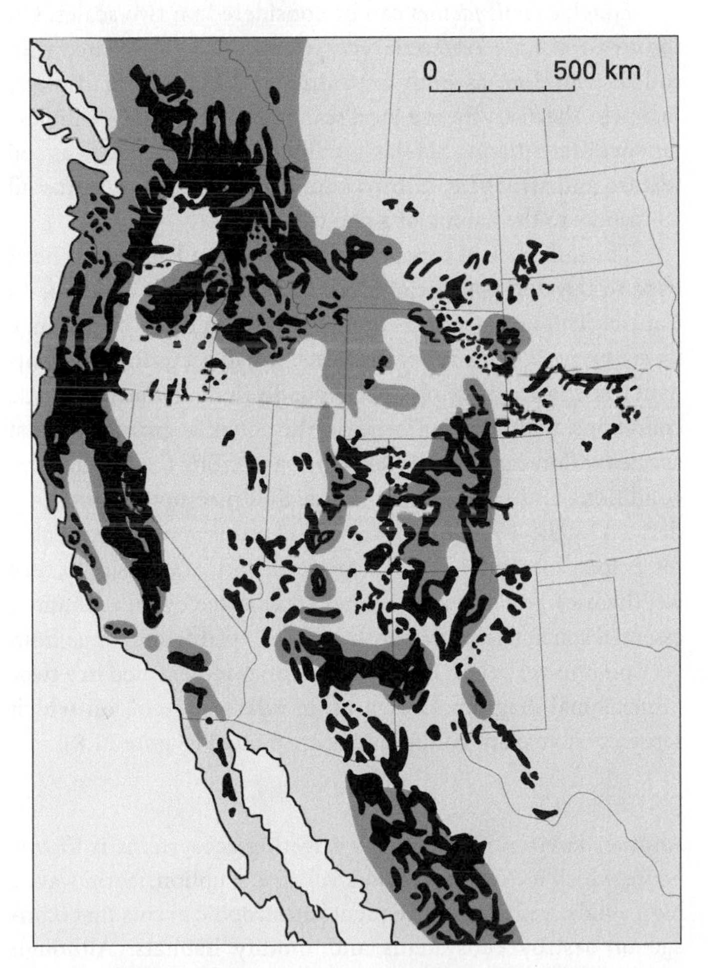

21.5 DISTRIBUTION OF PONDEROSA PINE IN WESTERN NORTH AMERICA Areas of ponderosa pine (*Pinus ponderosa*) are shown in black. The edge of the area shaded in green marks the 500-mm rainfall isohyet.

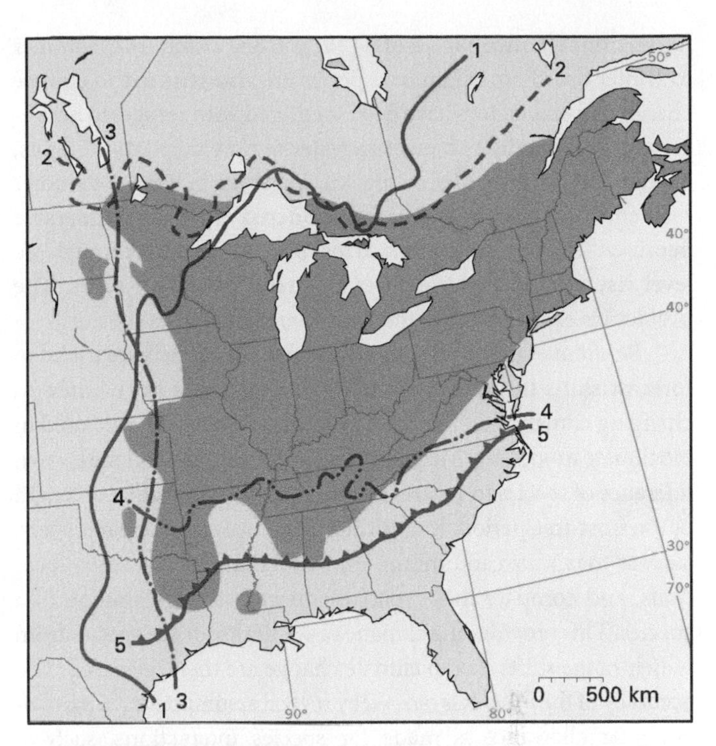

21.6 BIOCLIMATIC LIMITS OF SUGAR MAPLE The shaded area shows the distribution of sugar maple (*Acer saccharum*) in eastern North America. Line 1 represents annual precipitation of 750 mm. Line 2 represents a mean annual minimum temperature of −40 °C. Line 3 represents the eastern limit of the boundary between arid and humid climates. Line 4 represents a mean annual snowfall of 25 cm. Line 5 represents a mean annual minimum temperature of −10 °C.

Although bioclimatic limits must exist for all species, no plant or animal is necessarily found throughout the entire region designated by its bioclimatic frontier. Many other factors, such as diseases or predation, can limit the distribution of a species. Alternatively, a species (especially a plant species) may migrate slowly and may still be radiating outward from the location where it originated. In addition, some species may be dependent on another species and are therefore limited by the latter's distribution.

Identifying bioclimatic frontiers has particular relevance in a world that is experiencing climatic change. For example, the eggs of several species of marine turtles incubate successfully at nest temperatures between 25 and 32 °C. Cooler sand temperatures produce male hatchlings, with females coming from warmer areas. In recent years it has been observed that some beaches in Malaysia now produce mostly female turtles. In northern Australia, sand temperatures in summer on some beaches now exceed the upper lethal limit for incubation of 34 °C, causing a shift from summer to winter breeding. Migratory species are also sensitive to climate change. Several factors have been identified with respect to migratory birds, including changes in water regime through droughts, changes in the range of prey species, and shifts in habitat. Summer residents of Arctic regions are considered especially vulnerable because this area is susceptible to loss of tundra habitat through sea level rise and encroachment of different vegetation types. The problem is exacerbated by the limited area of land to the north.

Bioclimate models or species distribution models are useful to forecast shifts in geographic range of organisms in response to changing climatic parameters. Two general approaches are used in bioclimate modelling. Mechanistic models are based on the known tolerance of species to various factors, such as temperature or length of the frost-free period. Empirical models use climatic parameters, such as maximum and minimum temperatures from many locations, and compare these spatial patterns with distribution of a species. This provides the climatic range limits for the species from which range shifts due to climate change are then predicted. The accuracy of the models is limited by several assumptions. In particular, little allowance is made for species interactions, such as competition for resources. Similarly, no restrictions are usually imposed on the dispersal capacity of the species, nor is there any consideration of changes in the genetic characteristics of the species that might affect its tolerance limits.

GEOMORPHIC AND EDAPHIC (SOIL) FACTORS

Geomorphic factors influencing ecosystems include slope steepness, slope aspect, and relief. Slope steepness influences the rate at which precipitation drains from the surface. On steep slopes, surface runoff is rapid, and soil water recharge by infiltration decreases. On gentle slopes, much of the precipitation can penetrate the soil and be retained. More rapid erosion on steep slopes can result in thinner soils compared with those found on gentler slopes (see Chapter 19). Slope aspect has a direct influence on plants by increasing or decreasing the exposure to sunlight and prevailing winds. Slopes facing the sun have a warmer, drier environment than slopes that are shaded for much of the day. In midlatitudes, these slope-aspect contrasts can be so strong that they produce quite different biotic communities on north-facing and south-facing slopes (Figure 21.7).

21.7 SLOPE ORIENTATION AND HABITAT In this photo taken in the Santa Monica Mountains in southern California, dry south-facing slopes on the right support a community of low, xerophytic shrubs, while moister north-facing slopes on the left support an open woodland cover dominated by species of *Ceanothus*.

Geomorphic factors are partly responsible for the dryness or wetness of the habitat within a region that has the same general climate. On divides, peaks, and ridge crests, the soil tends to dry out because of rapid drainage and because the surfaces are more exposed to sunlight and wind. Conversely, valley floors are wetter because surface runoff over the ground and into streams causes water to collect there. In humid climates, the groundwater table in the valley floors may lie close to or at the ground surface to produce marshes, swamps, ponds, and bogs.

Edaphic (soil) factors can be considered on two scales. On the broadest scale, terrestrial ecosystems can be associated with soil distribution, as both are influenced by general climatic regimes. Alternatively, at a local scale, edaphic factors are important in differentiating habitat conditions. Properties such as soil texture and structure, humus content, and nutrient status all contribute to the habitat of a plant.

The influence of topography and soils on habitat has been used to develop provincial forest ecosystem classifications in Canada. This approach was first developed in British Columbia, where *biogeoclimatic zones* were established according to dominant tree species growing under broadly defined macroclimatic conditions. In the original scheme, the potential growth of forest species within each zone is described according to soil moisture conditions and soil nutrient status. Soil moisture regimes are defined by the average amount of soil water annually available for evapotranspiration, and range from very dry (xeric) to very wet (hydric). Soil nutrient regimes are assessed by the amount of essential soil nutrients that are available to plants and range from very poor to very rich. This information is represented in a two-dimensional diagram known as an **edatopic grid**, on which representative plant communities are plotted (Figure 21.8).

DISTURBANCE

Another environmental factor affecting ecosystems is *disturbance*, which includes fire, flood, volcanic eruption, storm waves, high winds, and other infrequent catastrophic events that damage or destroy ecosystems and modify habitats. Although disturbance can greatly alter the nature of an ecosystem, it is part of a natural cycle of regeneration that provides opportunities for short-lived or specialized species to grow and reproduce.

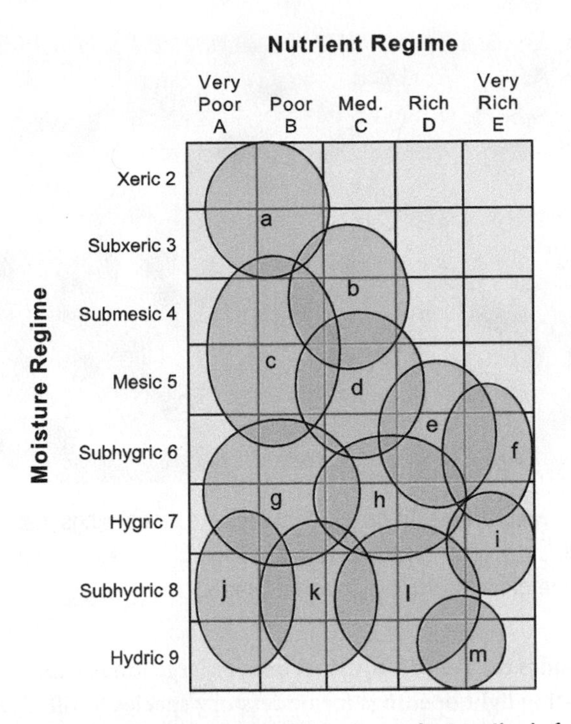

21.8 EDATOPIC GRID The edatopic grid is a convenient method of placing either sites or species (in this example, simply labelled a–m) into a two-dimensional ecological framework, which has direct implications for plant growth and diversity.

21.9 FOREST FIRES IN CANADA 1980–2009 The distribution of fires >200 ha shows that few areas of the boreal forest escape fire. In drier regions, such as northern Alberta and Saskatchewan, the average interval between successive fires is about 50 years, but in wetter coastal districts, the interval generally ranges from 100 to 200 years.

Fire is particularly important in the boreal forests of North America and has occurred throughout most of the region at some point in time (Figure 21.9). The frequency and intensity of fires in this region is primarily related to regional weather patterns which, over time, influence the nature and the diversity of the forest (see Chapter 10). Although economic losses from fire are considerable, the ecological impact is generally beneficial. Fire cleans out the understory and consumes dead and decaying organic matter, while leaving some of the overstory trees untouched. In converting biomass to ash, fire mobilizes nutrients, such as potassium and phosphorus, making them available for a new generation of plants. However, other elements, such as nitrogen and sulphur, may be lost through volatilization and removed as gases.

On the forest floor, mineral soil is exposed and fertilized with new ash following a fire, providing a productive environment for dormant seeds. Sunlight is also abundant, with shrubs and forbs no longer shading the soil. Among tree species, pines are typically well adapted to germinate under these conditions. For example, the jack pine of eastern Canada and the lodgepole pine of the intermontane west have cones that remain tightly closed until the heat of a fire opens them, releasing the seeds. Pines are a fire-resistant species and are directly dependent on fire disturbance to maintain their geographic range and importance in the ecosystem (Figure 21.10). In the montane and boreal forests of North America, there

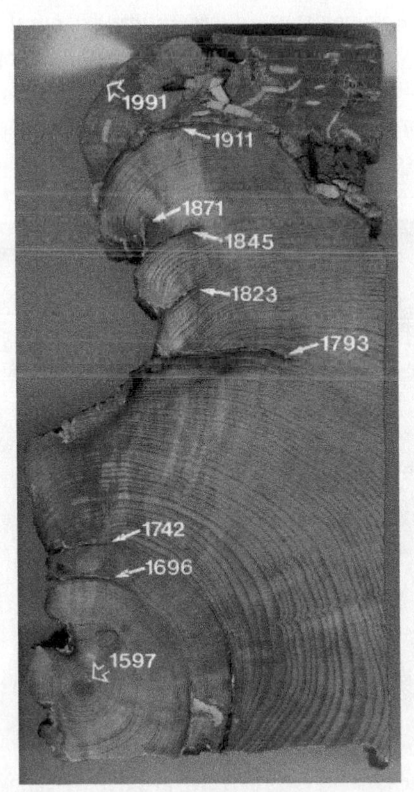

21.10 FIRE-SCARRED PONDEROSA PINE (*PINUS PONDEROSA*) This specimen of fire-resistant Ponderosa pine from Laramie, Wyoming, endured seven major fire events during its 400-year life.

are many patches, large and small, of jack pine and lodgepole pine of different ages that document a long history of fire. These burned stands provide specialized habitats for particular plants, insects, birds, and mammals.

Fires are also important to the preservation of grasslands. Grasses have extensive root systems with buds located at or just below the surface, making them quite fire-resistant. However, woody plants are not so resistant and are usually killed by grass fires. Most grasslands are dependent to some degree on fire for their maintenance. In Mediterranean climates, shrubby vegetation, such as the chaparral of California, is also adapted to regular burning during the dry summer period. However, increased runoff and soil erosion is not uncommon on steeper slopes when the rains return in the autumn and winter.

Fires can be remotely sensed in several ways. Thermal imagers detect active fires as bright spots because they emit more heat energy than normal surfaces. Similarly, the extent of a fire can be mapped using mid-infrared bands that respond to the change in surface energy balance of recently burned areas. Smoke plumes can also identify the location of fires, but are hard to distinguish from clouds in some images (Figure 21.11).

Fire is one of several forms of disturbance that can disrupt plants and animals. Strong winds and ice storms can bring down large areas of forest (Figure 21.12). High winds can generate destructive waves that adversely impact coral reefs. In other coastal environments, storm waves may overtop offshore bars and deposit large quantities of sediment in lagoons and marshes, while pronounced scouring can be detrimental to resident organisms on sandy beaches. At a fine spatial scale, disturbance includes the fall of individual trees or large limbs within forests, creating light openings for understory species to fill. Even animal burrows and wallows can provide unique habitats of disturbance.

21.11 LOCATIONS OF THE FIRES DETECTED BY SATELLITE Smoke from multiple large wildfires in Quebec is drifting to the south in this MODIS image from July 6, 2002. Active fire detections are indicated with red dots. The bright white area at the right of the image is clouds.

21.12 WIND DAMAGE Trees felled by wind create openings that allow understory species and seedlings to flourish. Wind is an important regeneration mechanism in some forests.

In addition to natural agents, the modern world is increasingly subject to human disturbance. The earliest societies mostly affected the environment through hunting. In some cases, this led to the extinction of species. Particularly susceptible were large flightless birds and other megafauna, such as the woolly mammoth. Humans have been able to dominate every environment on the Earth and have wrought inexorable change. Often this change has been direct and deliberate, such as through introduction of new species—often ones that have been bred specifically for foodstuffs. But equally often, humans have introduced species with devastating results, for example rabbits in Australia. In many cases, humans have inadvertently introduced species, such as zebra mussels, to regions beyond their natural range (see Appendix 15.1). The zebra mussel is an *invasive species*, one that takes over the habitat of other species, causing native species populations to decline or even disappear from their natural environment. Invasive species tend to be highly competitive and highly adaptive, and lack natural predators in their new habitat. Invasive species are always considered to be harmful to an ecosystem, whereas *introduced species*, perhaps naively, are not considered a serious threat in their new environment.

Direct and indirect human-induced environmental change is immense. Earlier chapters have covered some of these topics, such as air pollution and global climate change. The list can be extended to include soil erosion, water quality, desertification, loss of wetlands, logging, plantation agriculture, agrochemicals, and so on, all of which can and do impact the biosphere through *habitat loss* and degradation. Habitat is lost or degraded when natural or human-caused activities alter the environment in a way that reduces the number of species that can live there. This process represents the main threat to the world's endangered plant and animal species, as in most cases they have very specific habitat requirements and limited geographic ranges. The pervasive impact of human disturbance is difficult and, in many cases, impossible to rectify on a time scale that is relevant to society. It is too late for many species, regardless of conservation efforts that are being mounted worldwide.

INTERACTIONS AMONG SPECIES

Species interactions can be an important determinant of distribution patterns of plants and animals. Two species that are part of the same ecosystem can interact with one another in three ways: interaction may be negative to one or both species; the effect on both species may be neutral, not affecting either one; or interaction may be positive, benefiting at least one of the species.

Competition, a negative interaction, occurs whenever two species require a common resource that is in short supply. The concept is well illustrated by the distribution of two species of barnacles on rocky shorelines (Figure 21.13). In mixed communities, the smaller species, *Chthamalus stellatus*, lives on the higher shore that is exposed to the atmosphere for long periods during low tide. The larger species, *Semibalanus balanoides*, which is less tolerant of desiccation, occupies the lower shore. When only one species is present, their distribution on the shore changes. In the absence of *Semibalanus*, *Chthamalus* extends its range into the lower zone, but is pushed back when *Semibalanus* is reintroduced. However, if *Chthamalus* is absent from the upper shore, it is not replaced by *Semibalanus*. Thus, in mixed communities, competition restricts *Chthamalus* to a small portion (realized niche) of its possible fundamental niche. In situations where neither species has full use of the resource, both populations suffer, with lower growth rates than would normally occur if only one of the species were present. Therefore, competition is an unstable situation that may lead to the elimination of one of the species.

Predation and parasitism are also negative interactions between species. *Predation* occurs when one species feeds on another. The benefits are obviously positive to the predator species, which obtains energy for survival, and negative to the prey species. *Parasitism* occurs when one species gains nutrition from another, typically when the parasite organism invades or attaches to the body of the host in some way (Figure 21.14). Although predation and parasitism are usually regarded as negative processes that benefit one species at the expense of the other, these interactions may really be beneficial in the long term to the prey or host populations. Predation helps to maintain prey populations at levels that are sustainable in a particular environment. In addition to maintaining equilibrium population levels, predation and parasitism differentially remove less-fit individuals and help preserve species vigour.

A third type of interaction is *herbivory*. Some plant species have adapted well to grazing and can maintain themselves in the face of increased grazing pressure; others may be quite sensitive to this process. Plants are classified into three groups according to way they respond to grazing. *Decreasers* are species that decrease in abundance under grazing, *increasers* increase in abundance, and *invaders* enter an area when grazing is so severe

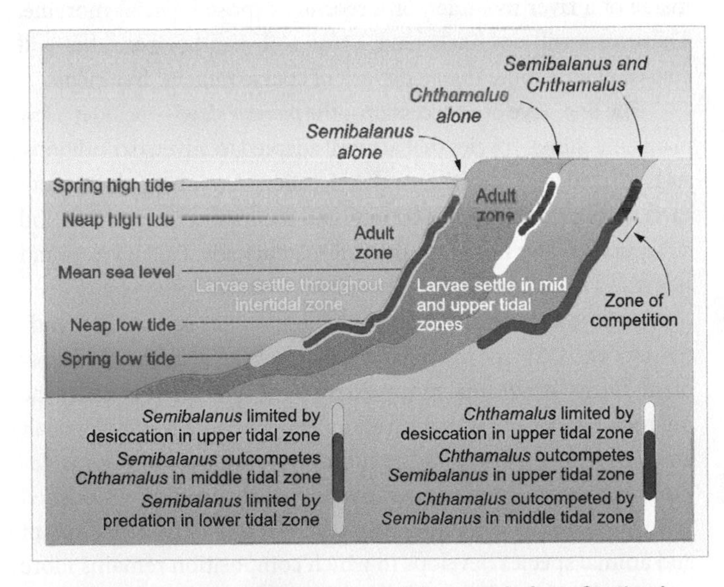

21.13 COMPETITION ON A ROCKY SHORE In this example from Scotland, the distribution of two species of barnacles changes in response to competition. When both species are present, *Chthamalus stellatus* only occupies the upper shore, but it will expand its range into the lower shore when *Semibalanus balanoides* is absent.

21.14 PARASITISM Parasitic dwarf mistletoe (*Arceuthobium americanum*) causes the clumped growth deformations known as witches' brooms on many conifer species.

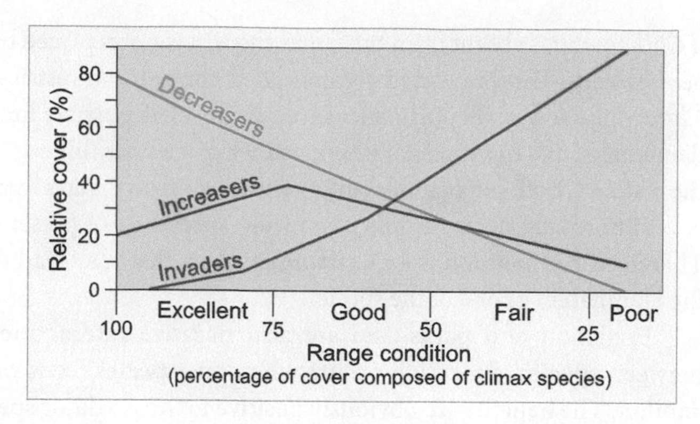

21.15 RANGELAND CONDITION The relative contribution of increaser, decreaser, and invader species in grassland communities is used as an indicator of range quality.

that the increasers are handicapped. When overgrazing occurs, these differing sensitivities can produce significant changes in the structure and composition of pasturelands, and their relative proportions are used to assess range quality (Figure 21.15).

Allelopathy is also a type of negative interaction; in this case, chemical toxins produced by one plant species inhibit the growth of others. Several shrub species in the California chaparral are allelopathic and exude toxins from their foliage. The allelopathic toxins accumulate in the soil until periodic fires break them down.

The term **symbiosis** includes three types of positive interactions between species: *commensalism, protoco-operation,* and *mutualism*. In commensalism, one of the species benefits and the other is unaffected. Examples of commensals include epiphytic plants, such as orchids or Spanish moss, which live on the branches of larger plants. These epiphytes depend on their hosts for physical support only. In the animal kingdom, small commensal crabs or fish seek shelter in the burrows of sea worms. Similarly, the remora is a commensal fish that attaches itself to a shark.

When the relationship benefits both parties but is not essential to their existence, it is called protoco-operation. The attachment of a stinging coelenterate, such as a jellyfish or sea anemone, to a crab is an example of protoco-operation. The crab gains camouflage and an additional measure of defence, while the coelenterate eats bits of stray food that the crab misses. A similar arrangement is seen between species of acacia and ants (Figure 21.16). The ants defend the plants against herbivorous animals and also chew away any encroaching plants, clearing areas as much as 4 m in radius. In return, the plant provides proteins and fats from specialized Beltian bodies and nectar. The ants hollow out thorns for their nests.

Where protoco-operation has progressed to the point that one or both species cannot survive alone, the result is mutualism. The association of the nitrogen-fixing bacterium *Rhizobium* with the root tissue of legumes is an example. The bacteria convert nitrogen gas to a form directly usable by the plant. The association is mutualistic because *Rhizobium* cannot survive alone.

21.16 ANT GUARDS The bullhorn acacia (*Acacia sp.*) is one of several acacia species that is occupied by colonies of stinging ants (*Pseudomyrmex ferruginea*).

ECOLOGICAL SUCCESSION

The Earth is a dynamic planet in which landscapes change continually. The phenomenon of change in plant and animal communities through time is referred to as **ecological succession**. If succession begins on newly deposited sediment or bare rock, it is called *primary succession*. If succession occurs on a previously vegetated area that has been recently disturbed by fire, flood, windstorm, or humans, it is referred to as *secondary succession*.

Sites on which primary succession occurs can originate in several ways, such as a sand dune or beach, the surface of a new lava flow or freshly deposited volcanic ash, the deposits of silt on the inside of a river meander, or a recently exposed glacial moraine. These sites will not likely have a true soil with horizons, but will consist of little more than a deposit of coarse mineral fragments.

The first stage of a succession—the *pioneer stage*—includes a few plant and animal species that are well adapted to adverse conditions, such as rapid water drainage and desiccation. As pioneer plants grow, their roots penetrate the substrate, and their subsequent death and decay contributes humus to the rudimentary soil. Fallen leaves and stems add an organic layer to the ground surface.

In time, conditions are favourable for other species to invade the area and displace the pioneers. The new arrivals may be larger plant forms providing more extensive cover. In this case, the microclimate near the ground alters, experiencing less extreme air and soil temperatures, higher humidity, and less intense insolation. Additional species now invade and thrive in the modified environment. Succession continues until a community of plant and animal species develops in which composition remains more or less stable. In traditional literature this is referred to as a **climax community**, but it is considered a theoretical condition because of frequent and recurring disturbance.

The colonization of a coastal sand dune provides an example of primary succession. Growing foredunes bordering the ocean

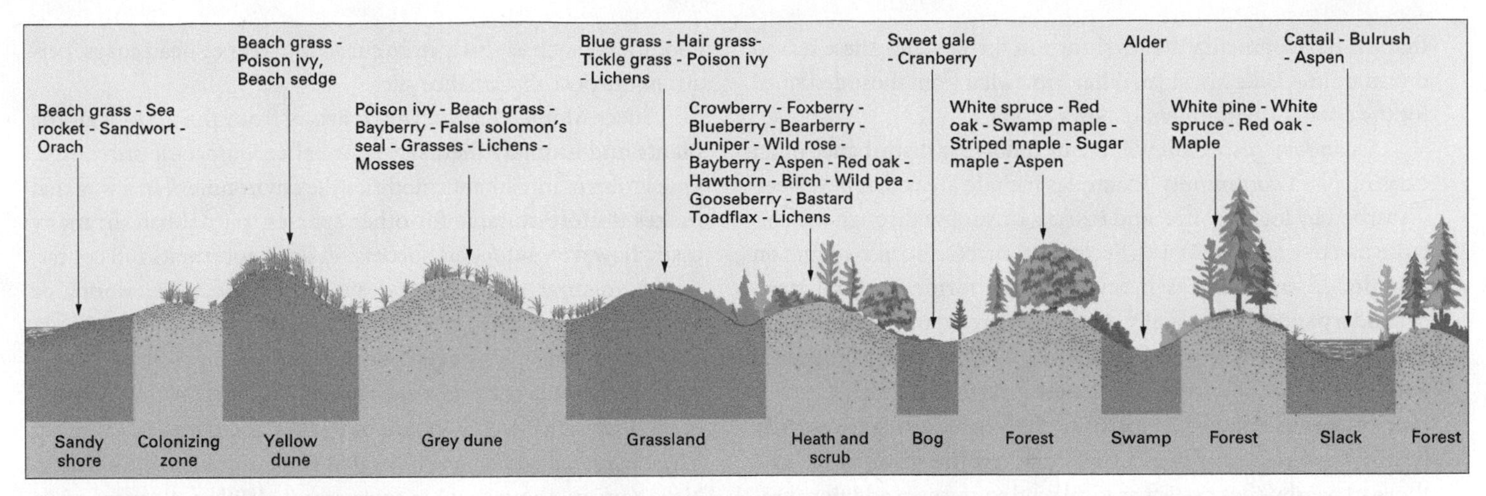

21.17 DUNE SUCCESSION IN NOVA SCOTIA Beach grass is a pioneer on beach dunes and helps stabilize the dune against wind erosion.

present a sterile habitat. The dune sand usually lacks nitrogen, phosphorus, and other nutrients, and its water-holding ability is very low. Under intense solar radiation, the dune surface becomes hot and dry. At night, radiation cooling in the absence of moisture produces low surface temperatures. One of the first pioneers of this extreme environment is beach grass (*Ammophila breviligulata*), which reproduces by sending out rhizomes (creeping underground stems). In this way, the plant slowly spreads over the dune. Unlike less specialized plants, beach grass does not die when moving sand buries it; instead, it sends up new shoots.

After colonization, the beach grass shoots begin to suppress sand movement, and the dune becomes more stable. With increasing stabilization and better water-holding capacity, other plants can establish. On beach and dune ridges in Nova Scotia, for example, the species that follow beach grass include beach pea, poison ivy, and sedge. Where active deposition of sand has ceased, the beach grass is gradually replaced by other species, such as scouring rush, bayberry, and lichens. Over time, heath and scrub take over; bearberry, blueberry, and juniper are common. Trees, such as aspen and red oak, may also be present. Eventually a forest of white pine, white spruce, oak, and maple develops (Figure 21.17).

Although this example has stressed the changes in plant cover, animal species also change as succession proceeds (Table 21.2). Note

TABLE 21.2 INVERTEBRATE SUCCESSION ON THE LAKE MICHIGAN DUNES

	SUCCESSIONAL STAGES				
INVERTEBRATE	BEACH GRASS–COTTONWOOD	JACK PINE FOREST	BLACK OAK DRY FOREST	OAK AND OAK–HICKORY MOIST FOREST	BEECH–MAPLE FOREST CLIMAX
White tiger beetle	X				
Sand spider	X				
Long-horn grasshopper	X	X			
Burrowing spider	X	X			
Bronze tiger beetle		X			
Migratory locust		X			
Ant lion			X		
Flat bug			X		
Wireworms			X	X	X
Snail			X	X	X
Green tiger beetle				X	X
Camel cricket				X	X
Sow bugs				X	X
Earthworms				X	X
Wood roaches				X	X
Grouse locust					X

Source: V. E. Shelford, as presented in E. P. Odum, *Fundamentals of Ecology*, Philadelphia: W. B Saunders Co., p. 259.

that the developmental stages shown in the table for these inland dunes around Lake Michigan differ somewhat from those described for the coastal environment of Nova Scotia.

Secondary succession will occur following disturbance of an existing plant community. Examples include areas that have been disturbed by logging, fire, and insects, or simply through the collapse of a tree (Figure 21.18). Secondary succession also occurs on abandoned farmland as it reverts to its former natural state. Secondary succession usually proceeds more rapidly than primary succession. One reason is that there is already a supply of seeds and roots in the soil from which new growth can quickly germinate or sprout. Likewise, not all of the vegetation is necessarily destroyed. Patches of plants that survived the disturbance and those in nearby sites can act as local seed sources. In addition, previous organisms have substantially modified the fertility and structure of the soil, compared with freshly deposited substrates like dune sand. This makes it more amenable for growth and colonization, especially if some of the other micro-environmental conditions, such as shade from surviving trees or dead snags, persist in the post-disturbance site.

Successional change, which arises from the actions of the plants and animals themselves, is called *autogenic succession*. One group of inhabitants modifies the environment in a way that makes it more suitable for other species to establish. In many cases, however, autogenic succession does not run its full course. Environmental disturbances, such as wind, fire, flood, or renewed clearing for agriculture, may divert the course of succession temporarily or even permanently.

Introduction of a new species can also greatly alter existing ecosystems and successional pathways. The parasitic chestnut blight (*Cryphonectria parasitica*) that was introduced from Asia to New York in about 1900 is an example. Within 40 years of its introduction, the blight had decimated the population of the American chestnut (*Castanea dentata*) that previously grew on the slopes of the Appalachian Mountains from New England to Georgia (Figure 21.19). This tree species, which may have accounted for as much as 25 percent of the mature trees in eastern forests, is now found only as small blighted stems sprouting from old root systems. Projects have been in place for several decades

(a)

(b)

21.18 SECONDARY SUCCESSION IN MIXED WOOD FOREST IN SASKATCHEWAN (a) Small-scale disturbance, such as the collapse of a tree, creates a gap in the overstory that light-demanding species, which are normally excluded from a mature forest stand, may initially occupy. (b) Fire killed most of the above-ground vegetation at this site, but new growth has developed from seeds and roots that survived in the soil.

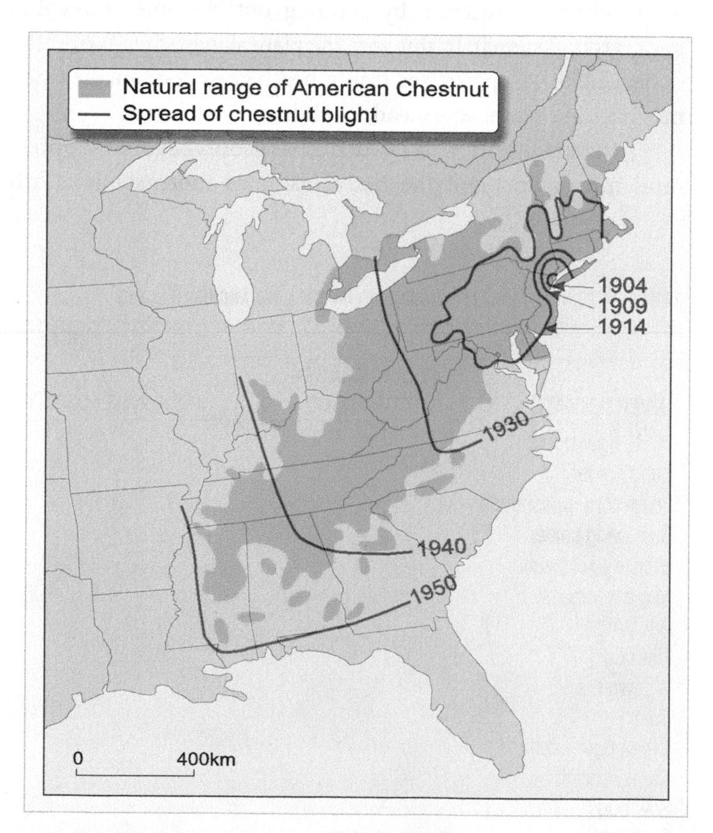

21.19 THE SPREAD OF CHESTNUT BLIGHT IN THE EASTERN UNITED STATES Prior to the introduction of the chestnut blight fungus about 1900, the American chestnut was widely distributed in the Appalachian Mountains. By 1950, it was all but eliminated.

to collect fruit and to carry out artificial pollination of blight-resistant individuals in an attempt to reestablish the species.

The example of the American chestnut shows that, while succession is a reasonable model to explain many of the changes seen in ecosystems, it may be more realistic to view the pattern of ecosystems on the landscape as a reflection of a spatial dynamic equilibrium. In this sense, a balance is established between autogenic forces of self-induced change and external forces of disturbance that reverse or redirect vegetation change temporarily or permanently. The biotic landscape can therefore be regarded as a mosaic of distinctive plant and animal communities with different biological potentials and different histories.

This view of the landscape assumes that all successional species are available to colonize new space or establish in existing communities. In this case, the nature of the biotic communities is determined by varying environmental and ecological factors that act within a new space created by physical and human processes. But not all species are available to colonize new spaces, particularly at continental and global scales, and especially over long time spans. On these broader spatial and temporal scales, the processes of migration, dispersal, evolution, and extinction are more important in determining patterns of distribution. These processes are within the realm of historical biogeography.

HISTORICAL BIOGEOGRAPHY

Historical biogeography examines four key processes that influence the distribution of species: evolution, speciation, extinction, and dispersal.

EVOLUTION

An astonishing number of organisms exist on the Earth—about 40,000 species of microorganisms, 350,000 species of plants, and 2.2 million species of animals, including some 800,000 insect species, have been identified and described. However, many organisms remain unclassified. Estimates suggest that species of plants will ultimately number about 540,000, and at present perhaps only a third of all insects have been classified. This great diversity has been achieved through *evolution*, as a result of the interaction between organisms and the environment. The process of evolution, as enunciated by Charles Darwin, is based on the premise that all life possesses *variation;* that is, differences arise between parent and offspring. The environment acts on variation in organisms in a way that ultimately favours individuals that are best suited to their environment. Darwin called this survival and reproduction of the fittest **natural selection**. Over time, this brings about the formation of new species that differ from their ancestral forms.

A *species* is defined as all individuals capable of interbreeding to produce fertile offspring. A *genus* is a collection of closely related species that share a similar genetic evolutionary history. Each species has a scientific name, which combines a generic name and a specific name. Thus, white spruce, a common conifer throughout the boreal forest of North America, is *Picea glauca*. The related black spruce is *Picea mariana*.

Although the true test of a species is the ability of all of its individuals to reproduce successfully, this criterion is not always easily applied. Instead, a species is usually defined by its *morphology* or outward form and appearance of its individuals. The *phenotype* of an individual is the morphological expression of its genetic information, or *genotype*, and includes all the physical aspects of its structure that are readily seen. Species, then, are usually defined by a characteristic phenotype and are identified by criteria, such as leaf shape (Figure 21.20).

SPECIATION

Speciation refers to the process by which species are differentiated and maintained. One way this can occur is through **geographic isolation**, which causes populations to become isolated from one another. For example, plate tectonics may uplift a mountain range that subsequently separates a population into two subpopulations isolated by a climatic or physical barrier. Chance long-distance dispersal can also establish a new population far from the original population. These are examples of *allopatric speciation*, in which populations are geographically isolated. With time, the populations gradually diverge and eventually lose the ability to interbreed.

In contrast, *sympatric speciation* occurs within a larger population. This may result from mutations that allow some members of a species to exploit the environment in slightly different ways. When subject to natural selection, these mutants will begin to evolve into different subpopulations, each adapted to its own environmental niche. Eventually, the subpopulations may become separate species. This type of evolution, in which there is the opportunity for the formation of many new species adjusted to different habitats, is called *adaptive radiation*. A good example is the 300 or more species of cichlid fishes that have evolved in the lakes of the African Rift Valley.

Another mechanism of sympatric speciation that can occur in plants is **polyploidy**. Normal organisms have two sets of genes and chromosomes; that is, they are *diploid*. Through accidents in the reproduction process, two closely related species can cross in such a way that the offspring have both sets of genes from both parents. These *tetraploids* are fertile but cannot reproduce with the populations from which they arose, and so are instantly isolated as a new species. According to some estimates, 70 to 80 percent of higher plant species have developed in this way.

EXTINCTION

Over geologic time, the fate of all species is *extinction*; it is estimated that about 98 percent of all the species that once existed on Earth are now extinct. It has long been assumed that extinction is

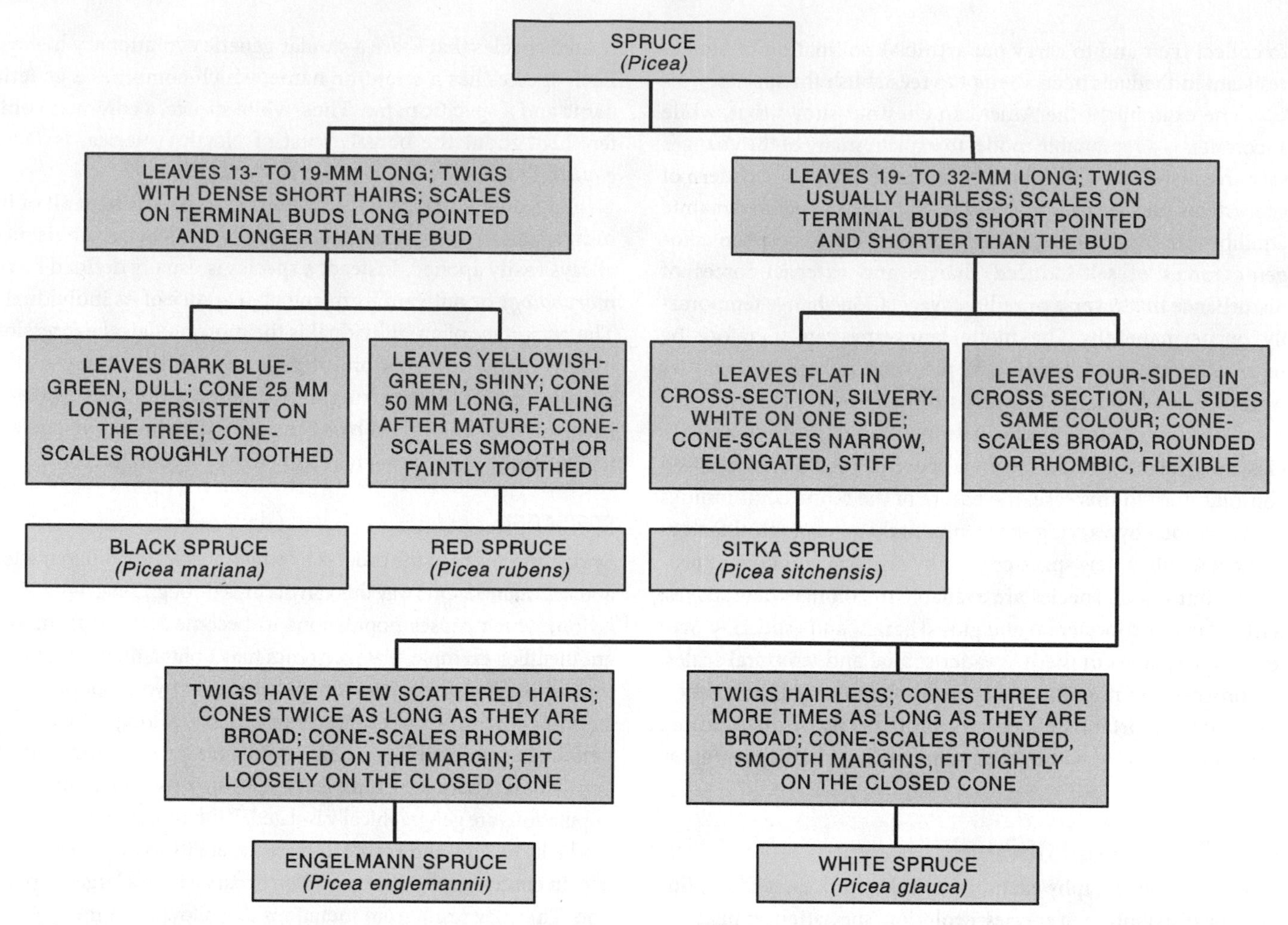

21.20 KEY TO THE IDENTIFICATION OF CANADIAN SPRUCES Five species of native spruces are recognized according to the size and shape of their leaves and cones and related characteristics.

a continuous process in that species are always exposed to some risk; however, there is increasing evidence that many extinctions are actually short-lived events of extreme stress. At least five major extinctions are documented in the geologic record. The most severe extinction occurred toward the end of the Permian and is dated at 251 Ma, marking the boundary between the Permian and Triassic periods (see Chapter 12). This so-called P-Tr extinction is credited with the loss of over 95 percent of all marine species and 70 percent of terrestrial invertebrates. Glaciation is suggested as a cause of the P-Tr extinction, but changes in the depth and distribution of oceans and the extent of continental shelves due to the formation of Pangaea is also a possibility. Volcanic activity, notably the formation of flood basalts in present-day Siberia, could also have contributed. Plant families are largely unaffected by mass extinctions, although the P-Tr event appears to have caused significant changes in their distributions.

Earlier extinction episodes are known from the Cambrian (542–510 Ma), which led to the disappearance of trilobites and ancient reef-building organisms known as the archaeocyathids. The Cambrian extinctions were possibly caused by an early glaciation, perhaps accompanied by oxygen depletion in the seas as they cooled and became shallower. Extensive diversification of marine organisms occured in the Ordovician (510–438 Ma), and is marked by the appearance of corals; gastropods, similar to the present-day snails; bivalve molluscs, like the modern-day clam; and cephalopods, of which octopus, squid, and the nautilus are modern representatives. However, many of these groups were lost during the Ordovician extinction dated at 440–450 Ma, and ranked second only to the P-Tr event. Also important was the Devonian extintion (375 Ma) that affected many coral organisms and the early fishes. Glaciation is again implicated, but some evidence suggests a meteorite impact.

The only mass extinction in the fossil record since the Permian occurred at the end of the Cretaceous (144–65 Ma) and resulted in the loss of 85 percent of known species. This K-T extinction (65 Ma) is associated with the disappearance of the dinosaurs and marine reptiles, as well as many species of lower plants. Microscopic marine plankton were severely affected, as were marine molluscs and fish. However, most mammals and birds appear to have been less affected by the K-T extinction, possibly because their ability to regulate body temperature made them less sensitive to a brief, but intense, period of cold. Seed plants and insects that spend part of their life cycle in a dormant state also seem to have survived.

The high concentration of iridium in sediments deposited at the Cretaceous–Tertiary boundary has led to speculation that the principal cause of the K-T extinction was a meteorite that impacted Earth in the vicinity of the Yucatan Peninsula. Another theory is linked to the massive volcanic eruptions associated with the formation of the Deccan Traps lava flows which laid down basalts over an area exceeding 100,000 km^2 and up to 150 m thick.

The recent impact of human activity on global ecosystems has been referred to in some reports as the Holocene mass extinction. In this regard, the most severely affected areas are the tropical rain forests, where some studies estimate that several hundred species are lost with every 200 km^2 of forest that is cleared for agriculture. When conditions change more quickly than populations can evolve new adaptations, population size decreases and becomes increasingly vulnerable to chance occurrences, such as a fire, a rare climatic event, or an outbreak of disease. Ultimately, the population succumbs, and the species becomes extinct. Some human-induced extinctions are very rapid. A classic example is that of the passenger pigeon, a dominant bird of eastern North America in the late nineteenth century. Flying in huge flocks and feeding on seeds and fruits, these birds were easily captured in nets and shipped to markets for food. By 1890, they were virtually gone. The last known passenger pigeon died in the Cincinnati Zoo in 1914.

DISPERSAL

Nearly all types of organisms have some **dispersal** mechanism; that is, a capacity to move from a location of origin to new sites. Often dispersal is confined to one stage in the life cycle, as in the dispersal of higher plants through seeds. Even in animals that are inherently mobile, there is often a developmental stage when movement from one site to the next is more likely to occur.

Dispersal does not normally change the geographic range of a species. Seeds fall near their source, and animals seek out nearby habitats to which they are adjusted. However, when new areas become available, dispersal moves colonists into the available habitat. For example, dispersal is a fundamental aspect of succession. But species also disperse by *diffusion*, the slow extension of range from year to year. Rare, long-distance dispersal events can also occur, and these are significant in establishing global distribution patterns.

Some species have propagation modes that are especially well adapted to long-distance dispersal. Mangrove species, which line coastal estuaries in equatorial and tropical regions, have seeds that ocean currents carry thousands of kilometres to populate far distant shores. Another example of a plant well adapted to oceanic dispersal is the coconut palm. Its large seed, housed in a floating husk, has made it a universal occupant of tropical beaches. Among the animals, birds, bats, and insects are frequent long-distance travellers. Generally, non-flying mammals, freshwater fishes, and amphibians are less likely to make long migrations, with rats and tortoises the exceptions.

The cattle egret demonstrates both long-distance dispersal and diffusion. This small heron crossed the Atlantic from Africa, arriving in northeastern South America in the late 1880s. One hundred years later, it had become one of the most abundant herons in the Americas (Figure 21.21).

21.21 **DIFFUSION OF THE CATTLE EGRET** After long-distance dispersal to northeastern South America, the cattle egret spread to Central and North America, as well as to coastal regions of western South America. (Map from Figure 6.2, p. 173 of *Biogeography*, Fourth Edition, by Mark V. Lomolino, Brett R. Riddle, Robert J. Whittaker, and James H. Brown (2010). © Sinauer Associates, Inc. Used with permission.)

Another example of diffusion is the northward colonization of trees following the retreat of the ice sheets at the end of the Late-Cenozoic Ice Age (Figure 21.22). The oaks reached their present northern limit in Canada about 8,000 years ago, the beech achieved a similar range about 4,000 years ago, and the chestnut spread to the southern shores of Lake Ontario only about 2,000 years ago.

Dispersal often means surmounting *barriers;* that is, regions a species is unable to occupy even for a short period. Oceans have proven to be effective barriers to long-distance dispersal for most species. But other barriers are not so obvious. For example, the basin and range country of Utah, Nevada, and California presents an expanse of desert with small patches of forest. While birds and bats may have no difficulty moving from one patch to the next, a small mammal may never be able to cross the desert. In this case, the barrier is one in which the physiological limits of the species are exceeded. But there may be ecological barriers, as well, such as a zone of intense predation or a region occupied by a vigorous competitor.

Just as there are barriers to dispersal, there also are corridors that facilitate dispersal. Central America forms a present-day *land bridge* connecting North and South America. It has been in place for about 3.5 million years. Other corridors of great importance to present-day species distribution patterns have existed in the recent past. For example, the Bering Strait region between Alaska and eastern Siberia was dry land during the early Cenozoic Era and during the Ice Age, when sea level dropped by more than 100 m. Many plant and animal species of Asia crossed this bridge and then spread southward into the Americas. One notable migrant species of the last continental glaciation was the aboriginal human. There is substantial evidence to support the hypothesis that the skilled hunters who crossed the Bering land bridge were responsible for the extinction of many of the large animals that disappeared from the Americas about 10,000 years ago.

The process of dispersal has been intensively studied by ecologists and biogeographers, most notably MacArthur and Wilson in the theory of *island biogeography*. They proposed that the number of species on an island is related to the rate at which new species arrive and the rate at which previously established species go extinct. The number of new arrivals was predicted to decrease over time, as the pool of candidate migrant species is finite. Conversely, the rate of extinction increases over time as available niches are exhausted and resources become progressively limited through population pressure and competition.

Ultimately, a dynamic equilibrium is established that is appropriate for the size and diversity of habitat on the island. The *Working It Out • Island Biogeography: The Species–Area Curve* feature at the end of this chapter provides more information on this important concept. The theory of island biogeography forms the basis of mathematical models that couple dispersal and extinction, and that have been used to explain the number of species that might be expected within a region of

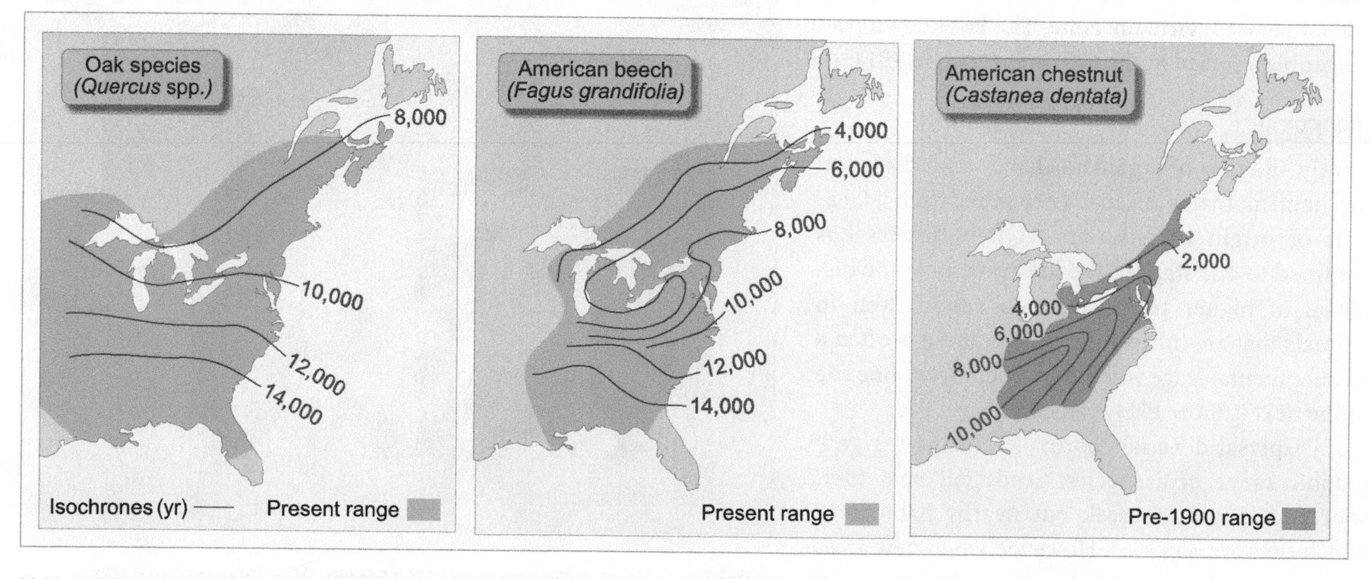

21.22 POST-GLACIAL MIGRATION OF DECIDUOUS TREES IN EASTERN NORTH AMERICA The oak (*Quercus spp.*) migrated quickly northward and reached its present limit about 8,000 years ago. Beech (*Fagus grandifolia*) was well established in southern Canada about 4,000 years ago, and chestnut (*Castanea dentata*) migrated more slowly, reaching its northern limit only about 2,000 years ago. Apart from a few small individuals, the chestnut has since disappeared from eastern North America because of chestnut blight.

21.23 **GINGKO TREE** The ginkgo tree is an eastern Chinese endemic that has survived millions of years with little evolutionary change. It has been introduced in many parts of the world. The fan-shaped leaves show an uncommon arrangement of parallel veins radiating from the leaf stem.

a given size. The theory has also been widely applied in studies of habit fragmentation and the determination of the minimal area needed to support viable breeding populations of birds and other species.

DISTRIBUTION PATTERNS

The processes of evolution, speciation, extinction, and dispersal have, over time, produced the many spatial patterns of species distribution seen on the Earth. One of the simplest patterns is that of the **endemic** species, which is restricted to one area and nowhere else. An endemic distribution can arise in one of two ways: as the result of a contraction of a broader range, or as the origin location of a species that has not dispersed widely. Some endemic species are ancient relics of species that have otherwise gone extinct. An example is the ginkgo tree (Figure 21.23), which was widespread throughout the Mesozoic Era but until recently was restricted to a small region in eastern China. It is now widely planted as an urban street tree, known for its hardiness.

In contrast to endemics are *cosmopolitan* species, such as dandelions, which are distributed globally in many habitats. Most cosmopolitan species have efficient propagating forms that are readily dispersed by wind or ocean currents and wide ecological tolerance.

Another interesting pattern is **disjunction**, in which one or more closely related species are found in widely separated regions. An example is the distribution of the tinamous and flightless ratite birds, which include the ostrich, emu, cassowary, and kiwi (Figure 21.24). This disjunct pattern is thought to result from an ancestral species that was widespread across the ancient continent of Gondwana. As plate tectonics split Gondwana into North and South America, Africa, Australia, and New Zealand, isolation and evolution differentiated the ancestral lineage into the diverse array of related species that now inhabit these continents.

BIOGEOGRAPHIC REALMS

As spatial distributions of species are examined on a global scale, certain common patterns emerge. From this, it is possible to delineate **biogeographic realms**, characterized by a distinct assemblage of plants and animals (Figure 21.25). The largest realm is the

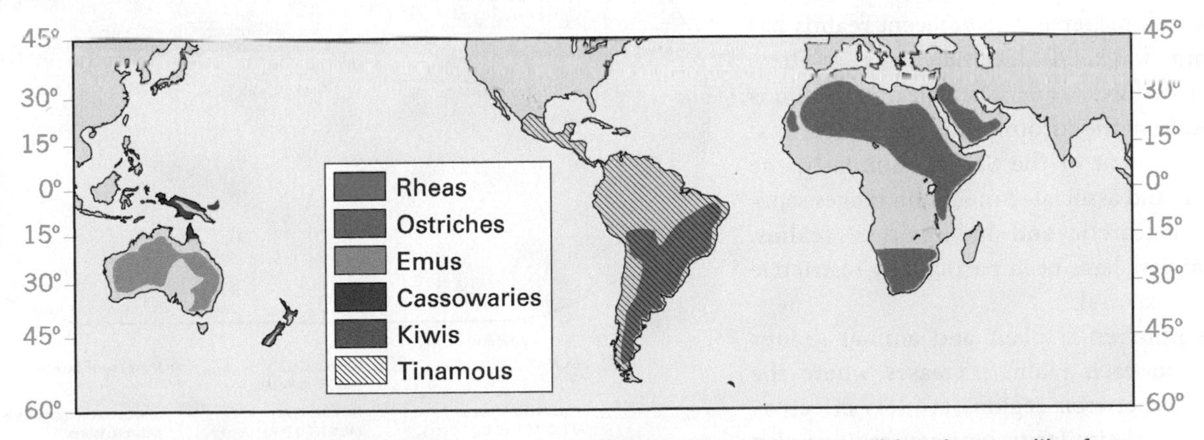

21.24 **DISJUNCT DISTRIBUTION** The distribution pattern of ratite birds and tinamous shows disjunctions resulting from isolation by continental movements. (From Figure 10.26, p. 396 of *Biogeography*, Fourth Edition, by Mark V. Lomolino, Brett R. Riddle, Robert J. Whittaker, and James H. Brown (2010). © Sinauer Associates, Inc. Used with permission.)

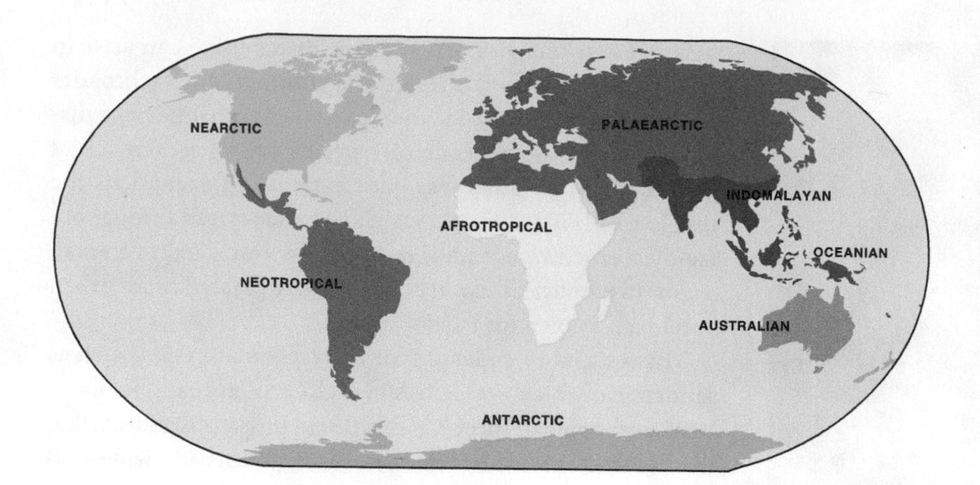

21.25 BIOGEOGRAPHIC REALMS Each biogeographic realm is characterized by a distinctive suite of plant and animal species that has arisen because of past and present barriers to dispersal.

Palearctic. It covers 54.1 million km^2 and includes most of Eurasia and North Africa. The Nearctic realm, which includes most of North America, covers 22.9 million km^2, and is similar in size to the Afrotropical realm (22.1 million km^2), which extends from the southern margin of the Sahara Desert to the southern tip of Africa. South America and the Caribbean are included in the Neotropical realm (19.0 million km^2). The remaining realms range from 7.7 million km^2 for Australasia to 0.3 million km^2 for the Antarctic.

The similarity in plant and animal distribution patterns suggests they have experienced related histories of evolution and environmental affinity. However, these patterns are not identical; therefore, distinct floristic and faunal regions are also recognized (Figure 21.26). The boundaries of each region are related to significant barriers across which dispersion to adjacent realms has been limited. The Australasian realm is effectively isolated by oceans. The Palearctic realm is separated from the Afrotropical realm by the hostile environment of the Sahara. Similarly, the Himalayas and associated mountain ranges separate the Palearctic and Indomalayan realms. These barriers have been particularly restrictive to animal dispersal.

The number of plant and animal groups common to each realm increases where the boundaries between regions are less effective. The degree of similarity between regions also reflects how recently the barriers were formed or broken. For example, India drifted into Asia about 45 million years ago, and North and South America became linked through Central America

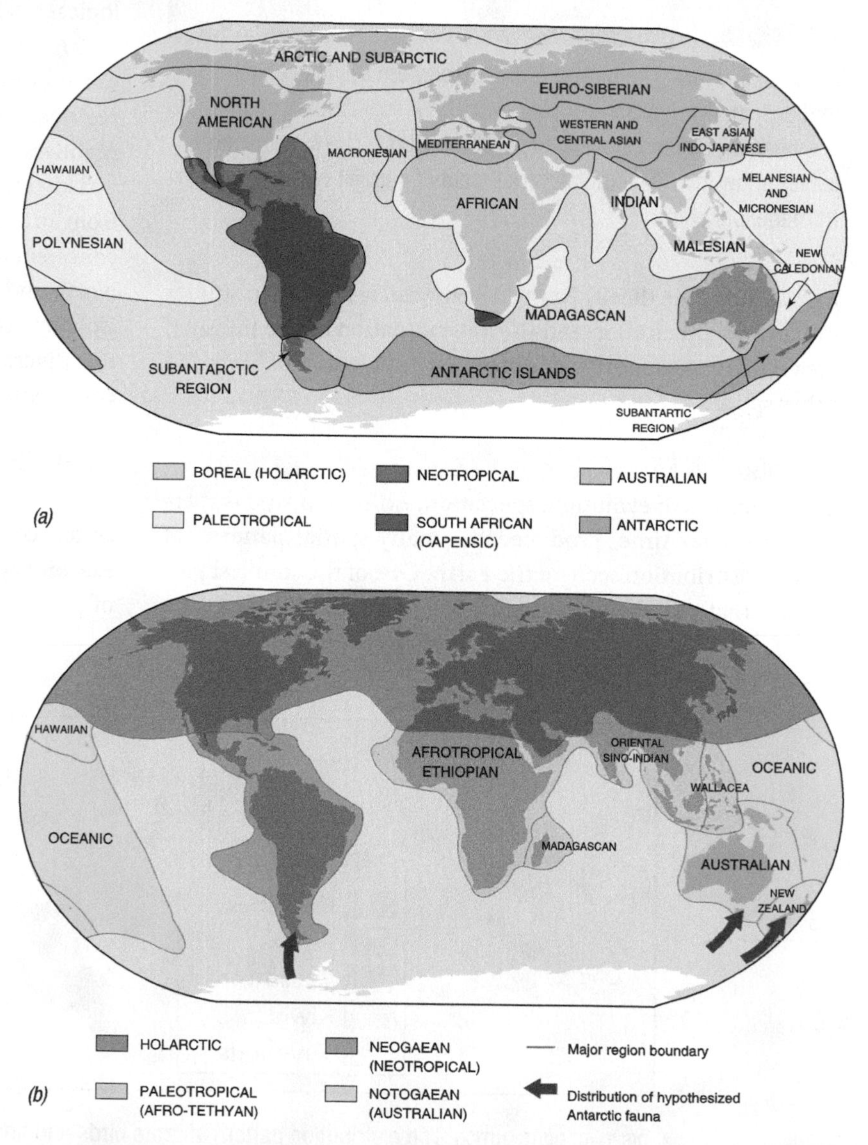

21.26 FLORISTIC AND FAUNAL REGIONS OF THE WORLD (a) Floristic regions are distinguished mainly on the basis of distinctiveness and degree of endemism in families and genera of flowering plants. (b) Faunal regions are distinguished by their distinctive animal assemblages, particularly their mammals.

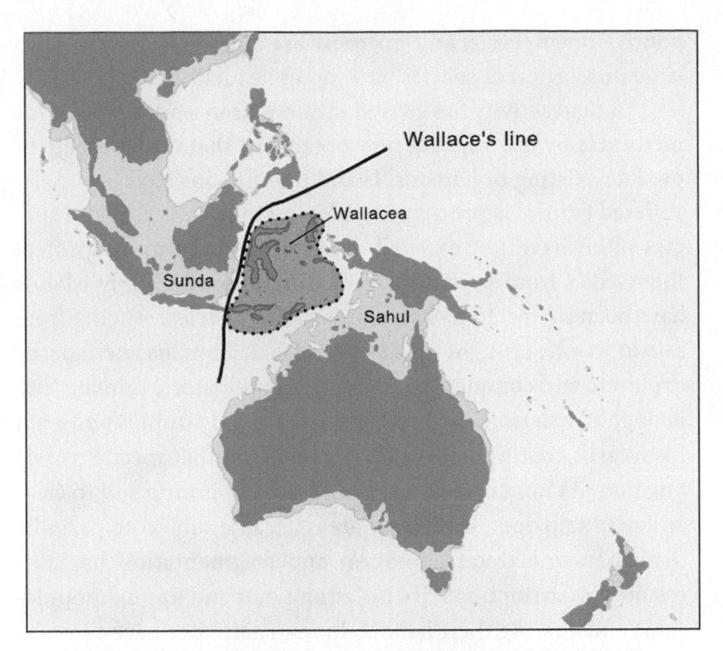

21.27 **WALLACE'S LINE** Wallace's Line separates the Oriental and Australasian faunal regions. The yellow areas are the Sunda and Sahul shelves, which were exposed when sea level dropped in the Late-Cenozoic Ice Age. The purple area, known as Wallacea, consists of a group of islands that are separated from these continental shelves by deep water.

about 3.5 million years ago. More recently, North America and Asia were joined by the Bering land bridge, a 1,500-km wide tract of steppe grassland, as recently as 12,000 to 15,000 years ago.

The boundary that has been the most contentious is the one separating Southeast Asia from Australia. Plant and animal groups on the various islands of the East Indies are transitional and have greater or lesser affinity with the adjacent continents. Original work by Wallace in 1876 noted a clear distinction in the bird families inhabiting two of the islands in the region—Bali and Lombok—that were separated by a distance of only 35 km. On Bali, the birds were related to those of Java, Sumatra, and mainland Malaysia. On Lombok, they more closely resembled those of New Guinea and Australia. Based on this, the Oriental and Australasian faunal regions were separated by Wallace's Line, drawn between Bali and Lombok and extended northward between Borneo and Sulawesi (Figure 21.27). The faunal and floral assemblages of the Oriental and Australian regions are clearly distinct at their centres, but species merge at their common boundary. This blurring can be traced to Late-Cenozoic sea level changes, which alternately exposed and submerged extensive areas of land in this transition zone.

BIODIVERSITY

Biodiversity—the variety of biological life on the Earth—depends on patterns of environmental variation and the processes of evolution, dispersal, and extinction through geologic time. Currently, the Earth's biodiversity is rapidly decreasing as a result of human activity. Humans now use some 20 to 40 percent of global primary productivity, as well as exploiting 70 percent of the marine fisheries to provide food. This demand for environmental resources has doubled the natural rate of nitrogen fixation, used more than half of the Earth's supply of surface water, and transformed more than 40 percent of the land surface.

In addition, recent human activity has ushered in a wave of extinctions unlike any that has been seen for millions of years. In

TABLE 21.3 COSEWIC STATUS DESIGNATIONS AS OF NOVEMBER 2010

TAXON	EXTINCT	EXTIRPATED	ENDANGERED	THREATENED	SPECIAL CONCERN	TOTALS
Mammals	2	3	20	17	27	69
Birds	3	2	29	24	20	78
Reptiles	0	4	17	11	9	41
Amphibians	0	1	9	4	7	21
Fishes	7	3	46	33	49	138
Arthropods	0	3	25	6	5	39
Molluscs	1	2	18	3	6	30
Vascular Plants	0	3	94	49	39	185
Mosses	1	1	8	3	4	17
Lichens	0	0	4	3	6	13
Totals	14	22	270	153	172	631

Source: Data from COSEWIC, http://www.cosepac.gc.ca/rpts/Full_List_Species.htm.

the last 40 years, biologists have documented the disappearance of several hundred land animal species, including 58 mammals, 115 birds, 100 reptiles, and 64 amphibians. Aquatic species have also been severely affected, with 40 species or subspecies of freshwater fish lost in North America alone in the last few decades. Botanists estimate that more than 600 plant species have become extinct in the past four centuries.

In 2010, the Committee on the Status of Endangered Wildlife in Canada (COSEWIC) identified 14 species as now extinct in Canada and 631 species or populations as extirpated, endangered, threatened, or of special concern (Table 21.3). The list of extinct species includes animals that were lost many years ago, such as the great auk, last seen in 1844, the Labrador duck (1875), sea mink (1894), and passenger pigeon (1914). More recent extinctions include two species of stickleback last seen in British Columbia in 1999. The majority of the species now designated to be at risk are in the southern parts of Canada, particularly southwestern and interior British Columbia, the southern Prairies, and southern Ontario.

The numerous extinctions that have been documented globally may represent only a fraction of the species that have been lost. Many species have not yet been discovered and so may become extinct without ever being identified. Some of these species could be important to a better understanding of evolutionary history. In the past decade, three new families of flowering plants and two new phyla of animals were discovered. Among animal groups, insects, spiders, and other invertebrates are the most poorly known. Fungi and microbes are other groups in which a large proportion of species have yet to be identified.

Human activity has caused extinctions in several ways. One method is by introducing new organisms that out-compete or predate existing organisms. Island populations have especially suffered from this process. Developing in isolation, island species often have not evolved defence mechanisms to protect themselves from predators, including humans. Many islands have been subjected to waves of invading species, ranging from rats to weeds, brought first by prehistoric humans and later by explorers and conquerors. Hunting by prehistoric humans was enough to exterminate many species, not only from islands, but from large continental regions as well. Another mechanism is burning. As humans learned to use fire in hunting and to clear and maintain open land, large areas became subject to periodic disturbance. Habitat alteration and fragmentation has also resulted in extinctions. By isolating plant and animal populations and altering their habitat, human activities have caused a reduction in numbers to a point where the species eventually is unable to tolerate further environmental stress.

Biodiversity is not uniform over the Earth's surface. In general, tropical and equatorial regions have more species and more variation in species composition between different habitats (Figure 21.28). In isolated areas, such as islands or mountaintops, species diversity tends to increase with the size of the isolated area and decrease with the degree of isolation from surrounding sources of colonists. Endemic species

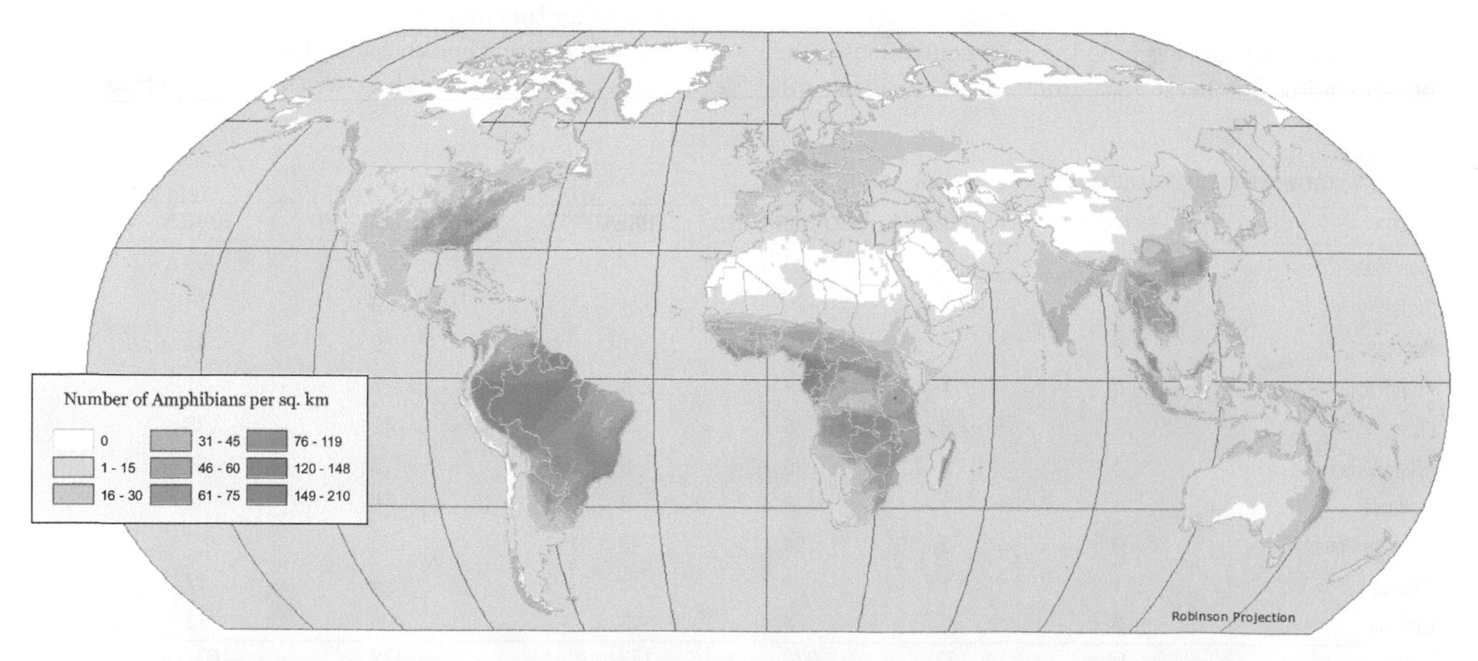

21.28 GLOBAL DISTRIBUTION OF AMPHIBIAN SPECIES Like most plant and animal groups, the diversity of amphibian species is highest in the low latitudes where environmental conditions have favoured evolution and adaptation.

contribute to much of this diversity. Geographic areas in which biodiversity is especially high are referred to as **hotspots**, a concept originally conceived by Myers in 1988. Hence, an important strategy for preservation global biodiversity is to identify hotspots and take conservation measures to protect them.

A hotspot is defined as a region in which at least 0.5 percent or 1,500 species of vascular plants are endemic, and that has lost at least 70 percent of its primary vegetation through human impact. Based on these criteria, 25 hotspots are currently recognized and nine other regions are under review (Figure 21.29). These are located principally in the tropics, in coastal areas, and in the Mediterranean climate regions of the world. Hotspots contain nearly 50 percent of all vascular plant species, but account for only 2 percent of the Earth's surface. However, these regions have collectively lost almost 90 percent of their original primary vegetation, with direct consequences on plant species diversity and the animals that rely on them.

In many hotspots, the numbers of endemic plant and vertebrate species are strongly correlated. More than half of all threatened plants and terrestrial vertebrates are endemic to biodiversity hotspots. Hotspots contain about 40 percent of terrestrial vertebrate species, many of which are threatened by habitat loss. Conservation of these important areas is therefore essential to minimize the threat of mass extinctions. However, hotspots are defined only on the basis of vascular plants; other taxa, including vertebrates, are not considered. Some regard this exclusion to be a serious deficiency of the concept. Other concerns have noted that hotspots do not include important areas of lesser diversity, neither do they allow for changing land use patterns that may seriously impact species diversity, although they are practised in relatively small areas. Hotspots were conceived as a tool to facilitate conservation strategies. Consequently, some evaluation of cost is needed before any plans can be implemented.

Nature, operating over millennia, has provided an incredibly rich array of organisms that interact with one another in a seamless web of life. Humans are an integral part of that web. When human activity causes the extinction of a species, a link in the web is broken. Ultimately, the web will become so impoverished that it could threaten the future of the human species and many other life forms. There is no way of knowing which organisms future humans will rely on. Thus, it seems prudent to increase efforts to conserve them.

One approach to promoting a balanced relationship between humans and the biosphere has been the development of **biosphere reserves**, designed to conserve examples of characteristic ecosystems. Biosphere reserves combine both conservation and sustainable use of natural resources. As of September 2010, there are 562 biosphere reserves in 109 countries. Currently there are 15 biosphere reserves in Canada (Figure 21.30); the first was established in 1978 at Mont Saint-Hilaire, and the most recent in 2007 at the Bay of Fundy.

Each biosphere reserve consists of a core area that is protected by law; for example, a nature reserve or national park (Figure 21.31). The core area is designed to protect sensitive

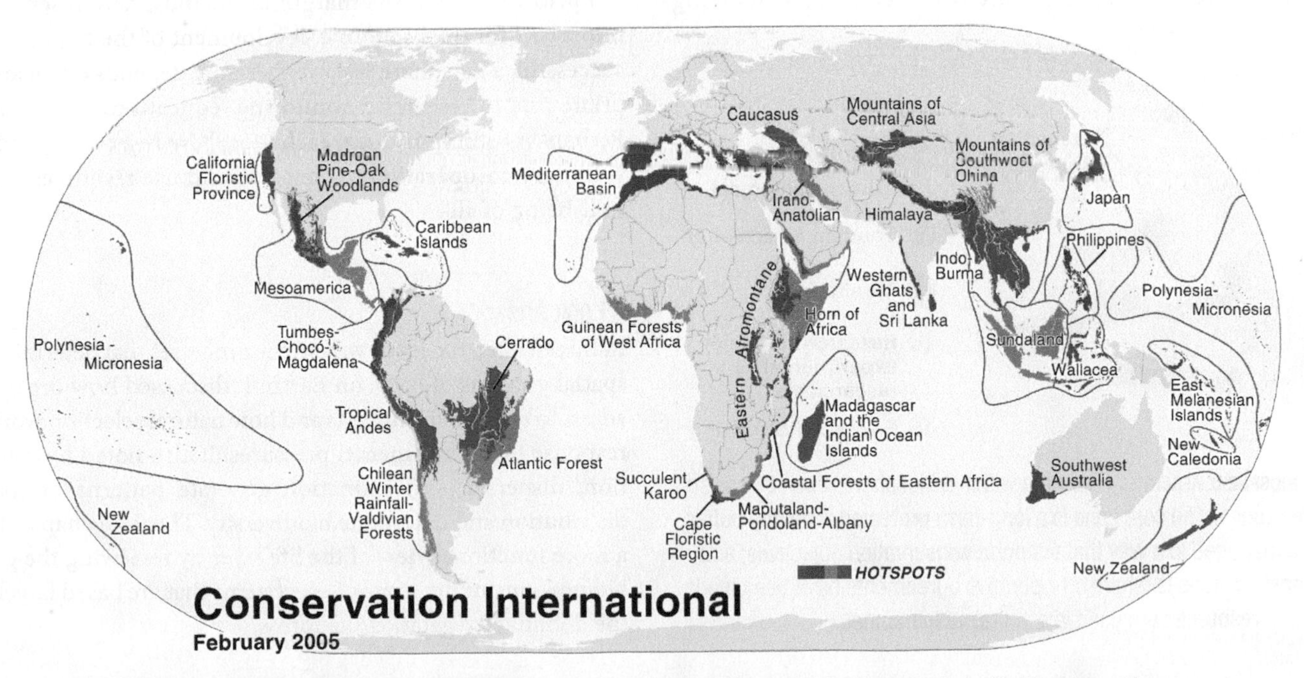

21.29 BIODIVERSITY HOTSPOTS These areas exhibit unusually rich biodiversity, but are also threatened with destruction.

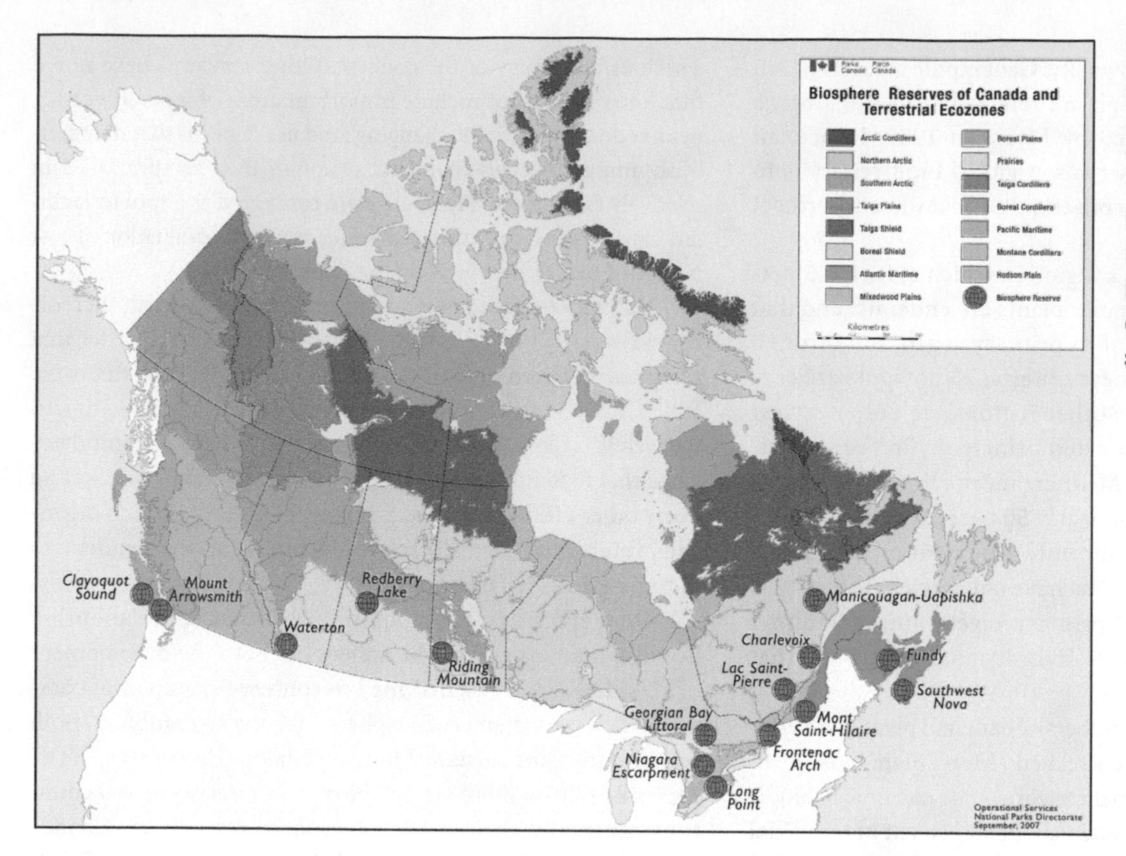

21.30 CANADIAN BIOSPHERE RESERVES Fifteen biosphere reserves have been established, covering a total area of 102,240 km^2 in eight provinces. Each is managed in a way that reduces biodiversity loss, improves livelihoods, and enhances environmental sustainability.

and valuable species. It may include original ecosystems, as well as those that are valuable to humans, such as grazing land. Core areas may also contain characteristic cultural features. Land and water use may occur in the core area, provided it is consistent with the aims of nature conservation. Surrounding a core area is a buffer zone in which activities and resource use are consistent with protection of the core. Restrictions in buffer zones are based on local voluntary agreements. Beyond the buffer zone is a transition area where sustainable development is a priority. This is the margin of the biosphere reserve; it is important for the economic development of the region. To be successful, a biosphere reserve therefore depends on an appropriate mix of research, monitoring, education, and training. Perhaps most importantly, each biosphere reserve depends on voluntary co-operation to conserve and use resources for the well-being of all.

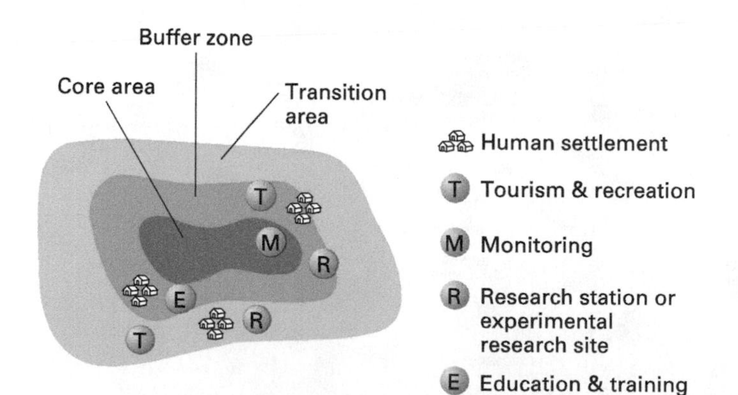

21.31 BIOSPHERE RESERVES The area within a biosphere reserve consists of three zones. The core zone is a long-term protected area; the buffer zone is managed in a way that supports conservation objectives; and the transition zone is where co-operative agreements have been made to ensure resources are used in a sustainable manner.

A Look Ahead

In this chapter, the focus was on the processes that determine the spatial patterns of biota on Earth. It discussed how organisms adjust to their environments and how natural selection works in response to environmental pressures. It also noted how evolution, dispersal, and extinction generate patterns of species distribution and determine biodiversity. The final chapter takes a more functional view of the life layer by reviewing the global biomes—major divisions of ecosystems that are based largely on the dominant vegetation life forms.

CHAPTER SUMMARY

- *Ecological biogeography* examines how relationships between organisms and the environment help determine when and where organisms are found. *Historical biogeography* examines how, where, and when species have evolved and how they are distributed over longer times and broader scales.
- A *community* of organisms occupies a particular environment, or habitat. *Associations* are often defined by characteristic species or vegetation life forms. Environmental factors influencing the distribution patterns of organisms include moisture, temperature, light, and wind.
- Organisms require water to live, so are limited by the availability of water. *Xerophytes* are adapted to dry habitats. They reduce water loss by having waxy leaves, spines instead of leaves, or no leaves at all. Some develop extensive root systems. *Hydrophytes* are adapted to wet environments such as bogs and swamps. *Xeric* animals include vertebrates that are nocturnal and have various adaptations to conserve water. *Invertebrates* such as brine shrimp can adjust their life cycle to prolonged drought.
- Temperature triggers and controls stages of plant growth, as well as limits growth at temperature extremes. Survival below freezing requires special adaptations; only a small proportion of plants are frost tolerant. *Cold-blooded animals* have body temperatures that follow the environment; however, they can moderate these temperatures by seeking warm or cool places. Mammals and birds are warm-blooded animals that maintain constant internal temperatures through a variety of adaptation mechanisms.
- The light available to a plant depends on its position in the structure of the community. Duration and intensity of light vary with latitude and season and serve as a cue to initiate growth stages in many plants. The day–night cycle regulates a lot of animal behaviour, as does the *photoperiod* (seasonal change in the duration of daylight). Wind deforms plant growth by desiccating buds and young growth on the windward side of the plant.
- A *bioclimatic frontier* marks the potential distribution boundary of a species.
- *Geomorphic factors* of slope steepness and slope orientation affect both the moisture and temperature environment of the habitat, and serve to differentiate the microclimate of each community.
- Soil, or *edaphic*, factors, such as soil texture, structure, acidity, and salinity, can also limit the distribution patterns of organisms or affect community composition.
- *Disturbance* includes catastrophic events that damage or destroy ecosystems. Fire is a common type of disturbance that influences forests, grasslands, and shrublands. Floods, high winds, and storm waves are others. Many ecosystems include specialized species that are well adapted to disturbance.
- Species interact in a number of ways, including *competition*, *predation* and *parasitism*, and *herbivory*. In *allelopathy*, plant species literally poison the soil environment against competing species. Positive (beneficial) interaction between species is called *symbiosis*.
- Ecological succession comes about as ecosystems change through time. *Primary succession* occurs on new soil substrate, while *secondary succession* occurs in disturbed habitats. Although succession is a natural tendency for ecosystems to change with time, it is countered by natural disturbances and limited by local environmental conditions.
- *Historical biogeography* focuses on the influence of evolution, speciation, extinction, and dispersal on the distribution patterns of species.
- A species is best defined as a population of organisms that is capable of interbreeding successfully, but is usually defined by a typical *morphology* or *phenotype*.
- *Speciation* is the process by which species are differentiated and maintained. It includes mutation and natural selection. *Geographic isolation* involves the isolation of subpopulations of a species, allowing genetic divergence and speciation to occur. In *sympatric speciation*, adaptive pressures force a breeding population to separate into different subpopulations that may become species. Sympatric speciation of plants includes polyploidy.
- *Extinction* occurs when populations become very small and thus are vulnerable to chance occurrences of fire, disease, or climate anomalies. Rare but extreme events can cause mass extinctions. An example is the asteroid impact that the Earth suffered about 65 million years ago.
- Species expand their ranges by *dispersal*. Plants generally disperse through seeds, while animals often disperse through their own movements and migrations. Long-distance dispersal, though rare, is important in establishing biogeographic patterns. *Barriers*, often climatic or topographic, inhibit dispersal and induce geographic isolation.
- *Endemic* species are found in one region or location and nowhere else. They arise by either a contraction of the range of a species or a recent speciation event. *Cosmopolitan* species

are widely dispersed and nearly universal. Disjunction occurs when one or more closely related species appear in widely separated regions.

- *Biogeographic realms* capture patterns of species occurrence that arise from common histories and similar environmental preferences. The boundaries between realms are often quite marked where significant change in habitat occurs. For this reason, oceans are effective barriers to terrestrial organisms. Less distinct boundaries occur where adjacent land areas were joined comparatively recently.
- *Biodiversity* is rapidly decreasing as human activity progressively affects the Earth. Extinction rates for many groups of plants and animals are as high or higher today than they have been at any time in the past. Humans act to disperse predators, parasites, and competitors, disrupting long-established evolutionary adjustments of species with their environments. Hunting and burning have exterminated many species. Habitat alteration and fragmentation also lead to extinctions.
- Preservation of global biodiversity includes a strategy of protecting *hotspots* where diversity is greatest. *Biosphere reserves* provide a functional compromise between conservation and sustainable land use.

KEY TERMS

bioclimatic frontier	competition	edatopic grid	hotspot	speciation
biodiversity	disjunction	endemic	hydrophyte	symbiosis
biogeographic realm	dispersal	fundamental niche	natural selection	xerophyte
biosphere reserve	ecological niche	geographic isolation	polyploidy	
climax community	ecological succession	habitat	realized niche	

REVIEW QUESTIONS

1. What is a habitat? What are some of the characteristics that differentiate habitats? Compare *habitat* with *niche*.
2. Contrast the terms *community* and *association*.
3. Although water is a necessity for terrestrial life, many organisms have adapted to dry environments. Describe some of the adaptations that plants and animals have evolved to cope with desert environments.
4. How does the annual variation in temperature influence plant growth, development, and distribution? How do animals cope with variation in temperature?
5. How does the ecological factor of light affect plants and animals?
6. What does the term *bioclimatic frontier* mean? Provide an example.
7. How do primary succession and secondary succession differ?
8. Distinguish between allopatric and sympatric speciation. Provide examples of each.
9. Describe the effects of barriers and corridors in the dispersal process.
10. How are biogeographic realms differentiated?
11. What is biodiversity? How has human activity affected biodiversity?

VISUALIZATION EXERCISE

1. Compare parts (a) and (b) of Figure 21.26. Which boundaries are similar, and which are different? Speculate on possible reasons for the similarities and differences.

ESSAY QUESTION

1. Imagine yourself as a biogeographer discovering a new group of islands. Select a global location for your island group, including climate and geologic substrate, and indicate its proximity to nearby continents or land masses. What types of organisms would you expect to find within your island group and why?

WORKING IT OUT • ISLAND BIOGEOGRAPHY: THE SPECIES–AREA CURVE

The basic principle of island biogeography is simply that the larger the island, the more species it contains. That is, the number of species systematically increases with island area. This finding is true whether studying lists of all plant and animal species, or just those species within a particular genus, family, or other group of related organisms.

The table shows the areas of some islands of the Lesser and Greater Antilles in the Caribbean, as well as the number of species of amphibians and reptiles found on each island.

Island	Area km^2	Species*
Redonda	1.3	3
Saba	13	5
Montserrat	102	9
Puerto Rico	9,104	40
Jamaica	10,991	39
Hispaniola	76,480	84
Cuba	110,860	76

*Data of P.J. Darlington, 1967.

These data are plotted in Figure 1. In Figure 1a, as the area of the island increases, the species count at first increases steeply. But for larger islands, the number of species increases more gradually. This *species–area curve* follows a power function; the number of species is related to the area raised to a power. Expressed as an equation:

$$S = cA^z$$

where S is the number of species, A is the area of the island, and c is a constant. The variable z, the power to which A is raised, lies between 0 and 1 to provide a curve of this shape.

When a power function is plotted on logarithmic axes, it follows a straight line, as shown in Figure 1b. For these data, the resulting best-fit equation for the power function is:

$$S = 2.47\ A^{0.30}$$

The second example, shown in Figure 2, plots area against species counts of land birds and freshwater birds on 23 islands of the Sunda group, the Republic of the Philippines, and New Guinea. These islands are located in a broad arc off the southeast coast of

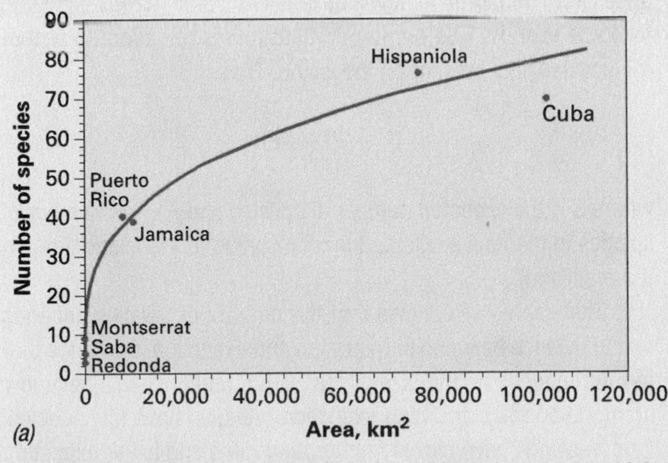

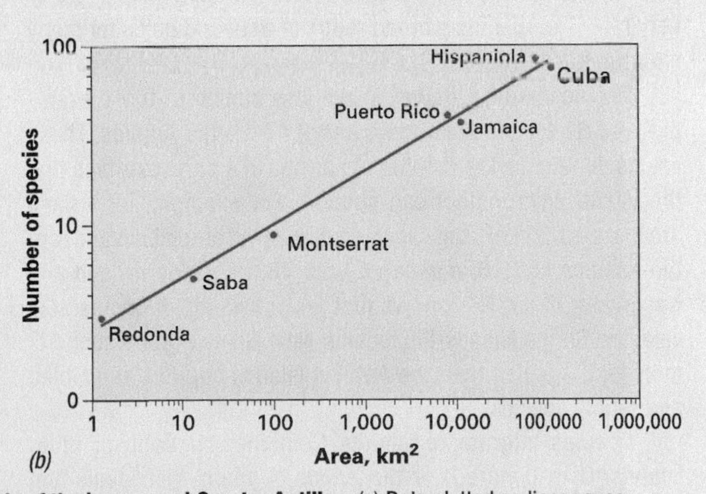

Figure 1 Species counts of reptiles and amphibians on selected islands of the Lesser and Greater Antilles (a) Data plotted on linear axes. (b) Data plotted on logarithmic axes.

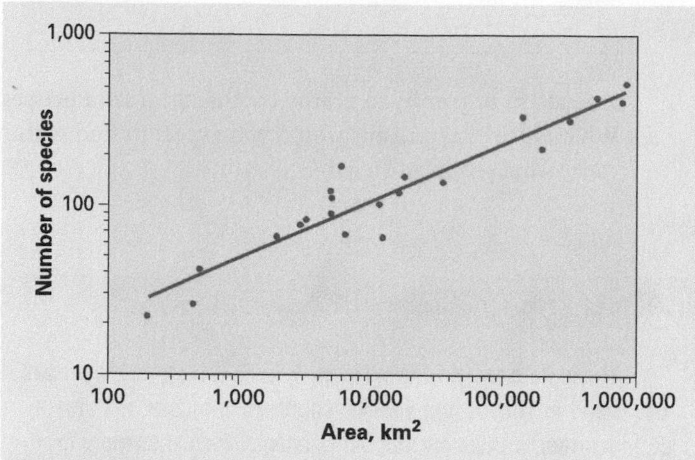

Figure 2 Species–area plot for land and freshwater bird species Data are shown for 23 islands of the Sunda group, the Republic of the Philippines, and New Guinea. (Data are from sources as cited in R. H. MacArthur and E. O. Wilson, Evolution, vol. 17, 1963, p. 374.)

Asia. As is the case for the Antilles, there is again a strong straight-line relationship between species and area. For these data, the best-fit values of c and z are:

$$S = 4.61\,A^{0.34}$$

Note that the values of z are somewhat similar in these two examples. A number of studies of species–area curves for different islands and different species groups show that values of z are generally in the range of 0.20 to 0.35. Mathematical ecologists have determined that z, theoretically, is expected to be about 0.26 if species follow a log-normal rule of abundance. This rule states that, in any given region, only a few species are relatively common, while most species are moderately rare or very rare. The rule fits a broad range of data and holds for many situations.

The values of c, however, are less similar in these examples—4.61 for the Asian data and 2.47 for the Antilles. These values depend on the number of species of a particular type that the island environment can support. For example, for a fixed area, say 1,000 km^2, the expected number of amphibian and reptile species is 19.6, compared with 48.3 species of land and freshwater birds. This shows that, given islands of comparable size, the Sunda Islands-Philippines-New Guinea group supports more bird species than the Antilles islands support amphibian and reptile species.

Species migrate to islands from the mainland or other islands. Over hundreds or thousands of years, individuals find their way to islands by chance—a typhoon, for example, may carry a bird to an island hundreds of kilometres from its normal habitat. Or a flood may sweep animals out to sea on floating debris that comes to rest on a far island shore. In this way, islands receive new immigrants.

The immigration rate depends on the area of the island—a large island is more likely to receive more immigrants simply because it has more area to intercept accidental travellers. Also, a large island is more likely to have a greater diversity of habitats, so more immigrants can successfully establish themselves once they arrive. So if the rate of immigration (I) is defined as the number of new species per year that an unpopulated island could support, that rate will be low for small islands and high for large islands.

At the same time, species on islands suffer extinction, which reduces the total number of species present. The extinction rate also depends on the island's area. On a small island, the population of a species will tend to be small, and a catastrophic event, like a typhoon or volcanic eruption, is more likely to wipe out that population. On a larger island, there is a better chance for survival of enough individuals to keep the species in existence there. The extinction rate (E) is the number of species per year that are lost (assuming that all possible species are present); it will be high for small islands and low for large islands. These situations are shown in the table below.

	Small Island	**Large Island**
Immigration rate (I)	Low	High
Extinction rate (E)	High	Low
$I/(I+E)$	Small	Large

Combining the immigration and extinction rates gives the *turnover rate* ($I + E$). This rate expresses the number of species that either immigrate or become extinct in a given year. The expression $I/(I+E)$ is then the proportion of the turnover rate that arises from immigration; it will be small for small islands and large for large islands. This proportion determines the average size of the species pool present on the island. Thus:

$$S = P[I/(I+E)]$$

where S is the expected number of species and P is the number of species in the pool available for colonization of the island (i.e., on the mainland).

This expression shows that the number of species depends on the balance between immigration and extinction, a balance that in turn depends on the size of the island. Small islands have low immigration rates and high extinction rates, so have few species. Larger islands have higher immigration rates and lower extinction rates, so have more species.

QUESTIONS / PROBLEMS

1. A zoogeographer assembles the following data for area and species counts of termites and butterflies on a set of (fictitious) islands several hundred kilometres off the coast of Brazil.

Island	Area (km^2)	Termites	Butterflies
Alhambra	156	11	23
Bonarote	845	16	41
Carlo	1,746	26	47
Delore	3,550	31	64
Edmundo	14,323	42	101
Fonseca	71,420	68	151

Analysis shows that the species–area curve for termites best fits the expression $S = 2.07A^{0.31}$, while the curve for butterflies best fits the expression $S = 4.56A^{0.32}$. Plot the data in two ways: on arithmetic (linear) axes and on logarithmic axes. Show data for both termites and butterflies on each graph. Sketch smooth curves to fit the data plotted on the arithmetic graph and straight lines for the data plotted using the logarithmic axes.

2. Use the expressions for the two species–area curves to find the predicted number of species of termites and butterflies for Carlo Island. Compare these with the observed values in the table. Are the observed values higher or lower than predicted? Identify some possible reasons why this difference might occur.

Find answers at the end of the book.

CHAPTER 22 THE EARTH'S TERRESTRIAL BIOMES

The arctic tree line near the headwaters of the Thelon River, Northwest Territories.

The previous two chapters focused on ecology and biogeography, examining the principles and processes that determine the distribution of plants and animals on the Earth's surface. This concluding chapter describes the broad global distribution patterns and characteristics of the Earth's major biomes. It emphasizes vegetation because it is a visible and obvious part of the landscape. Also, the largest division of ecosystems—the biome—is defined primarily by the characteristics of the vegetation cover. Vegetation tends to be related to climate, so the occurrence of major biomes coincides, in many cases, with broad climate types.

NATURAL VEGETATION

Over the last few thousand years, human societies have come to dominate much of the land area and, in many regions, have changed the natural vegetation—sometimes drastically, other times more subtly. **Natural vegetation** is a plant cover that develops with little or no human interference. It is subject to natural forces of modification and destruction, such as storms or fires. Natural vegetation still occurs in some remote areas, such as the arctic and alpine tundra or parts of the wet equatorial rain forests.

In contrast, much of the land surface in midlatitudes has been totally affected by human activities, through intensive farming or urbanization. Some areas of natural vegetation appear to be untouched, but are actually dominated by human activity in a subtle way. For example, most national parks were protected from fire for many decades. In the past, when lightning started a forest fire, it was extinguished as quickly as possible. However, periodic burning is part of the natural cycle in many regions, and, in recent years, suppressing wildfires has been stopped in some parks, allowing them to develop in response to more natural processes.

Humans have influenced vegetation by moving plant species from their original habitats to foreign lands and different environments. The eucalyptus tree is a striking example. From Australia, species of eucalyptus have been transplanted to various places, including California, North Africa, and India. Sometimes exported plants thrive like weeds, forcing out natural species and becoming a major nuisance. For example, brome grass (*Bromus inermis*), originally imported to North America to improve forage quality, has spread aggressively in many prairie habitats, reducing the diversity of native species.

Nevertheless, all plants have limited tolerance to the environmental conditions imposed by factors such as temperature, soil water availability, and soil nutrients. Consequently, the structure and appearance of the plant cover conforms to basic environmental controls, and each vegetation type, whether forest, grassland, or desert, is associated with a characteristic geographical region. At the global scale, vegetation patterns are closely releated to specific climates and soils.

STRUCTURE AND LIFE FORM OF PLANTS

Botanists recognize and classify plants by genus and species. However, plant geographers are often less concerned with individual species and more concerned with the vegetation as a whole. In describing the vegetation, plant geographers refer to the **life form** of the plants, which emphasizes their physical structure, size, and shape (Table 22.1). This method of classifying plants can provide information on the relationship between the species and their environments. Thus, vegetation in widely separated regions is often remarkably similar in appearance, even though there are few if any common species. For example, the trees that grow in the boreal forests of Canada look very similar to those growing in Scandinavia and Russia, but in fact no tree species occur naturally in both regions. They are, however, similarly adapted to cope with the rigorous climate and thin soils that characterize northern Canada and northern Eurasia.

Trees and shrubs are erect, *perennial* woody plant forms that endure from year to year. Most have lifespans of many years. *Trees* typically have a single upright main trunk, often with few branches on the lower part but many in the upper part to form a crown. *Shrubs* usually have several stems branching from a base near the soil surface so their foliage is close to the ground. *Lianas* are also woody plants, but they take the form of vines supported on trees and shrubs. Lianas include not only the heavy, rope-like vines of the wet equatorial and tropical rain forests, but also more delicate vines found in some mid-latitude forests.

Herbs make up a major class of plant life forms. They lack woody stems so are usually smaller plants. Some are *annuals*, living only for a single season; others are *perennials* and live for multiple seasons. Some herbs are broadleaf forbs, and others are narrow-leaf, such as grasses. Herbs are usually the predominant life form in the ground layer. As well as forbs and grasses, ferns, mosses, and horsetails are included in this class.

TABLE 22.1 A GENERALIZED LIFE FORM CLASSIFICATION OF PLANTS

1. ***Vascular plants***
a. *Trees*: woody plants, generally with a single main trunk b. *Shrubs*: woody plants, generally possessing several major stems branching out from the ground or very close to it; most shrubs are smaller trees c. *Lianas*: woody plants with a climbing habit causing them to seek support from trees or other plants d. *Herbs*: plants with little or no woody development and typically small in size Sub-types include: i. *Forbs* – broadleaf herbs, e.g., dandelion ii. *Graminoids* – grasses characterized by long, thin leaves with lower part wrapping around stem iii. *Ferns* – reproduce by spores and do not flower iv. *Club mosses* – generally small plants with sprawling horizontal stems resembling miniature conifers but reproducing through spores, not seeds v. *Horsetails* – jointed stems with small scaly leaves that grow in whorls around stem joints; spore-producing cones borne at tips of the stems
2. ***Non-vascular plants***
a. Byrophytes: small plants living close to the ground or on tree trunks; carry out photosynthesis and reproduce by spores Subdivided into: i. *True mosses* (e.g., sphagnum) – structures that resemble leaves, stems, and roots, but with no vascular tissue ii. *Liverworts* – resemble leaves on the ground, with no roots or stems; photosynthesis goes on at the top surface of the plant, with manufactured foodstuff stored directly beneath iii. *Hornworts* – irregular, greasy-looking blue-green "leaves" from which long, slender spore-bearing stalks develop
3. ***Epiphytes***
Epiphytes include vascular and non-vascular plants that live on other plants without soil contact for at least part of their lives; note: epiphytes are not parasitic
4. ***Lichens***
Lichens are dual organisms composed of fungi living in close association with green algae or cyanobacteria

Source: Adapted from National Vegetation Classification Standard, Version 2, February 2008. http://www.fgdc.gov/standards/projects/FGDC-standards-projects/vegetation/NVCS_V2_FINAL_2008-02.pdf/.

The layering and spacing of the various life forms determines the structure of the vegetation cover. In forests, the trees grow close together. Crowns are in contact, so the foliage shades the ground. Many forests in moist climates have at least three life form layers (Figure 22.1). Tree crowns form the uppermost layer, shrubs form an intermediate layer, and herbs form a lower layer. There is often a fourth, lowest layer that consists of mosses and other small plants. In **woodland**, the trees are widely spaced so that their crowns are separated by open areas that usually support grasses, low herbs, or a shrub layer. *Lichens* are another life form that can be abundant in the ground layer. In some alpine and arctic environments, lichens grow in profusion and dominate the vegetation (Figure 22.2).

22.1 **VERTICAL STRUCTURE OF A FOREST** Most forests consist of one or more layers, including an understory of shrubs and a ground layer of herbs and grasses. Dead and decaying plants are also an important component of this forest ecosystem in northern Saskatchewan.

22.2 **LICHEN** Despite its name, the reindeer moss (*Cladonia* sp.) shown here is actually a common variety of lichen.

TERRESTRIAL ECOSYSTEMS—THE BIOMES

From the viewpoint of human use, ecosystems are natural resource systems. Food, fibre, fuel, and structural material are products of ecosystems and are manufactured by organisms using energy derived from the sun. **Terrestrial ecosystems** are directly influenced by climate and interact with the soil. At the global scale, the largest recognizable subdivision within terrestrial ecosystems is the **biome**. Although the biome includes the total assemblage of plant and animal life interacting within the life layer, green plants dominate the biome physically because of their enormous biomass compared with that of other organisms.

There are five principal biomes. The **forest biome** is dominated by trees, which form a closed or nearly closed canopy. Forests require an abundance of soil water, so they are found in moist climates. Temperatures must also be suitable, requiring at least a warm season, if conditions are not warm year round.

The **savanna biome** is transitional between forest and grassland. It supports an open cover of trees with grasses and herbs beneath. The **grassland biome** develops in regions with moderate shortages of soil water. Temperatures must provide adequate warmth during the growing season. The semi-arid regions of the dry tropical, dry subtropical, and dry midlatitude climates are associated with the grassland biome.

The **desert biome** includes organisms that have adapted to the moderate to severe water shortages that occur for most, if not all, of the year. The characteristic plants are xerophytes, species that can survive with limited water. Temperatures in deserts range from very hot, as in the Sahara Desert, to extremely cold, as is the case in parts of the Gobi Desert in central Asia during winter. The **tundra biome** is limited by cold temperatures. Only small plants that grow quickly when temperatures rise above freezing during the short growing season can survive.

Biogeographers subdivide biomes into smaller vegetation units, called *formation classes*, based on plant life forms. For example, different types of forests, such as evergreen needleleaf, broadleaf deciduous, and broadleaf evergreen, are easily distinguished within the forest biome. Similarly, there are at least three kinds of temperate grasslands, including tall-grass prairie, mixed prairie, and steppes. Deserts, too, span a wide range in terms of plant abundance and representative life forms, which, in addition to cacti, include drought-tolerant trees, shrubs, and herbs. The formation classes are major, widespread units that are clearly associated with specific climate types; their general distribution is shown in Figure 22.3. With remote sensing, it is possible to map global land cover more accurately. *Geographer's Tools • Mapping Global Land Cover by Satellite* describes how data from NASA's MODIS instrument is used for this purpose.

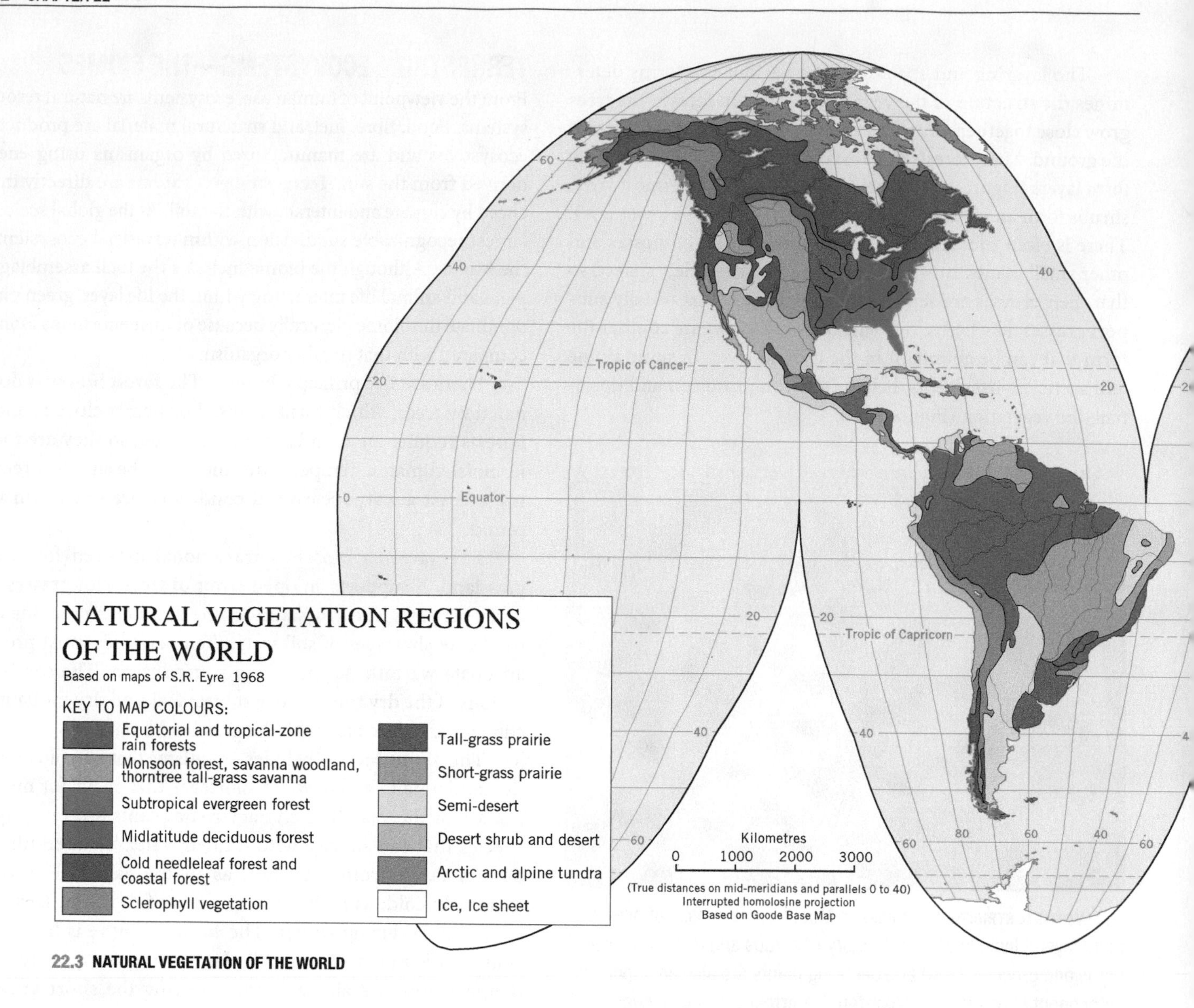

22.3 NATURAL VEGETATION OF THE WORLD

FOREST BIOME

Within the forest biome, there are six major formations: low-latitude rain forest, monsoon forest, subtropical evergreen forest, midlatitude deciduous forest, needleleaf forest, and sclerophyll forest.

Low-Latitude Rain Forest

Low-latitude rain forest, found in the equatorial and tropical latitude zones, consists of tall, closely set trees. The crowns of the trees are generally arranged in two or three layers (Figure 22.4). The highest layer consists of scattered "emergent" crowns that protrude from the closed canopy below, often rising to 40 m. Below the layer of emergents is a second, continuous layer, which is 15 to 30 m high. A third, lower layer consists of small, slender trees 5 to 15 m high, with narrow crowns. Together the trees form a continuous canopy of foliage and provide shade to the understory (Figure 22.5a). Some of the tall emergent species develop wide buttress root systems for support (Figure 22.5b). The trees are characteristically smooth-barked and have no branches on the lower two-thirds of their trunks. Tree leaves are large and evergreen—thus, the equatorial rain forest is often described as a *broadleaf evergreen forest*.

Woody lianas supported by the trunks and branches of trees are common in low-latitude rain forests. Some are slender, like ropes, while others reach diameters of 20 cm. They climb high into the trees to the upper canopy where light is available, and develop numerous branches of their own.

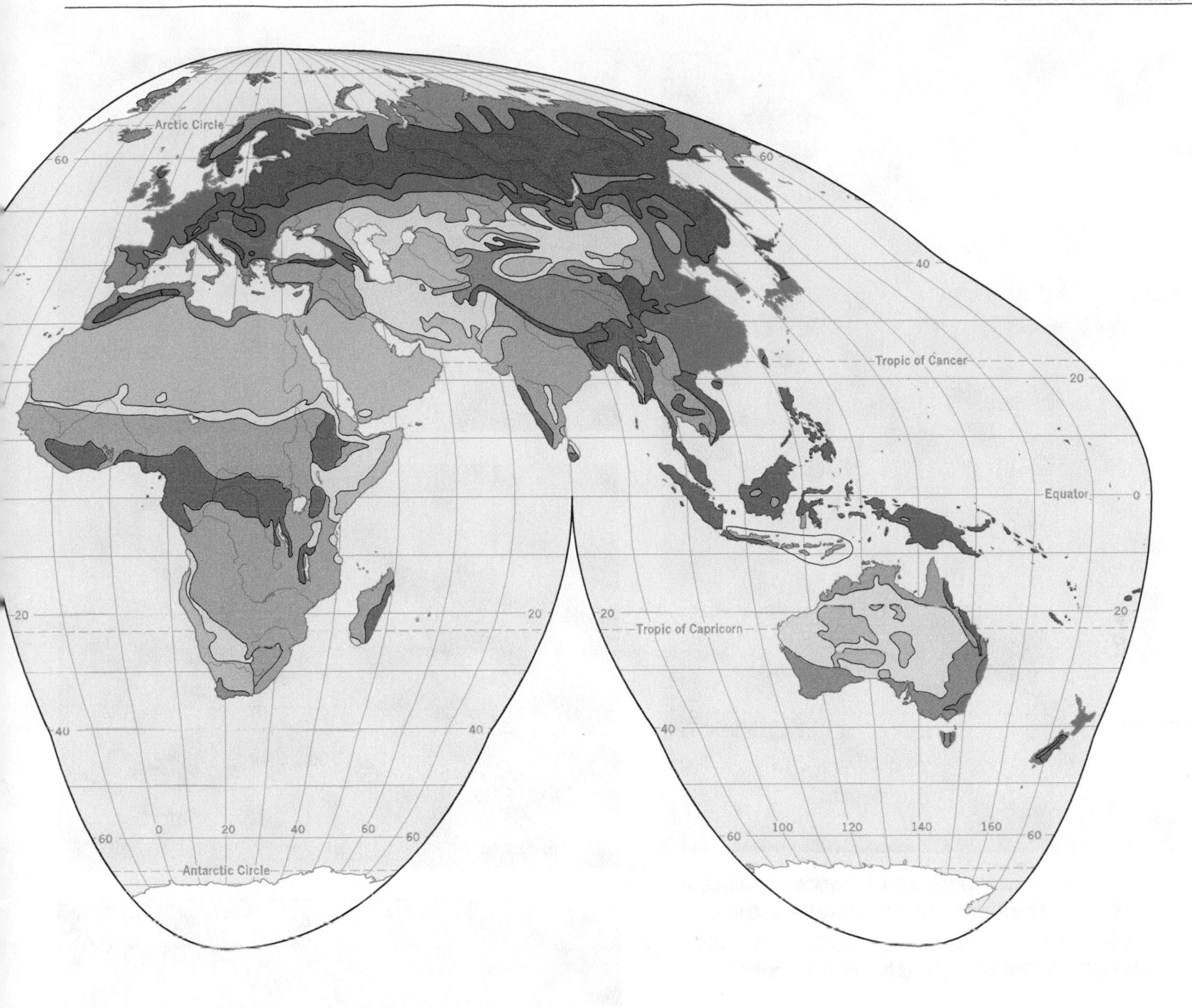

Epiphytes also grow here. These plants attach themselves to the trunks, branches, or foliage of trees and lianas, which they use solely as a means of physical support. Epiphytes include many different types of plants, especially ferns, orchids, mosses, and lichens (Figure 22.5c).

An important characteristic of the low-latitude rain forest is the large number of species of trees that coexist. In some regions, as many as 3,000 species may be found in a few square kilometres, although individuals of a given species are often widely separated. It is widely accepted that many species of plants and animals in this diverse ecosystem are still undocumented.

Equatorial and tropical rain forests are not impenetrable jungles. Rather, the floor of the low-latitude rain forest is usually so shaded that branches and foliage are comparatively sparse nearer the ground. This opens up the forest, making it easier to travel within its interior. The ground surface is generally covered only by a thin litter layer of leaves. Dead plant matter rapidly decomposes because the warm temperatures and abundant moisture promote breakdown by bacteria and termites. Roots quickly absorb the nutrients released by decay. As a result, the organic matter content and fertility of the soil is low. Most lowland rain forest soils are classed as Oxisols and are characteristically fine textured and reddish or orange in colour (see Chapter 19). Soil drainage under the natural forest is generally good. However, once the tree cover is removed, problems can arise as the soils are prone to developing impervious layers or hard pans that can also make them difficult to cultivate.

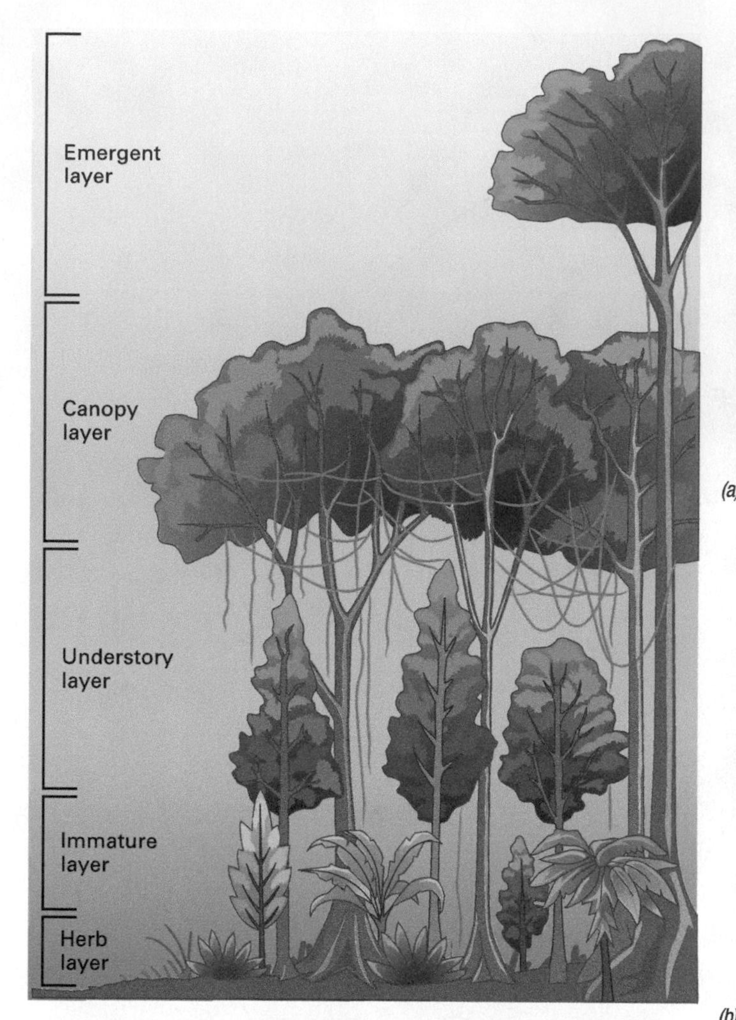

22.4 **MULTI-LAYERED STRUCTURE OF A RAIN FOREST** In a typical low latitude rain forest the tallest trees of the A-stratum emerge above the continuous canopy of the B-stratum, with smaller trees forming the discontinuous C- stratum beneath, together with layers of tree ferns and herbs.

(a)

(b)

(c)

22.5 **TROPICAL RAIN FOREST** (a) This general view of a tropical forest shows the dense canopy and straight, branchless lower trunks of the tallest trees. (b) Buttress roots provide support at the base of a large tree in Daintree National Park, Queensland, Australia. (c) Epiphytes adorn a rain forest tree in Iguazu National Park, Argentina.

Large herbivores are uncommon in the low-latitude rain forest; those that do live in this environment include the African okapi and the tapir of South America and Asia. Herbivores are mainly represented by the primates—monkeys and apes that move noisily among the branches. Tree sloths spend their life hanging upside down as they browse the forest canopy. Many species of birds, including toucans and parrots, as well as bats, depend on the wide selection of fruits and seeds produced by the forest plants. Many of these species, like the insects that pollinate the flowers, have evolved very specialized traits to take advantage of the myriad of niches afforded by the high plant diversity. There are few large predators. Notable are the leopards of African and Asian forests and jaguars and ocelots in South America.

Low-latitude rain forest develops in a climate that is continuously warm, frost-free, and has abundant precipitation in all months of the year (or, at most, has only one or two drier months). These

GEOGRAPHER'S TOOLS

MAPPING GLOBAL LAND COVER BY SATELLITE

One of the striking things about the Earth's surface when viewed from space is its colour. Deserts appear in shades of brown, dotted with white salty playas. Equatorial forests are green, dissected by branching lines of dark rivers. Shrublands are marked with a greenish tinge. Some regions show substantial change throughout the year. In the midlatitude zone, deciduous forests go from intense green in the summer to brown as leaves drop from the trees. In the tropical zones, grasslands and savannas change from brown to green as the rainy season comes and goes. Thus, the colour of the land surface can provide information about the type of plant cover present at a location.

Ever since they received the first satellite images of the Earth, scientists have used colour—that is, the spectral reflectance of the surface—as an indicator of land cover type. They have produced land cover maps using individual Landsat images for more than 30 years. Detailed maps can be generated from the spectral signatures of known plant types, and these are subsequently used to classify the vegetation in an area.

Most satellite-derived vegetation maps now use the normalized difference vegetation index (NDVI). This Index is based on the absorption and reflectance characteristics of plant tissues for different wavelengths of light. The pigment chlorophyll strongly absorbs visible light (from 0.4 to 0.7 μm), whereas leaf surfaces and other plant tissues strongly reflect near-infrared (NIR) light (from 0.7 to 1.1 μm). The difference in the amount of energy absorbed and reflected in these wavelengths is used to calculate NDVI.

$$NDVI = \frac{NIR - Red}{NIR + Red}$$

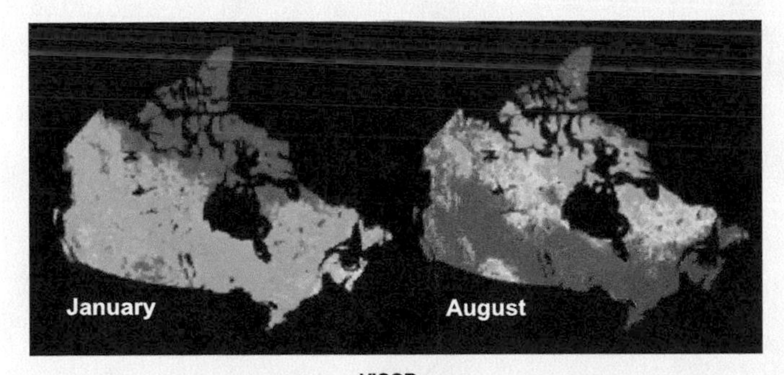

Figure 1 January and August normalized difference vegetation indices (NDVI) for Canada The NDVI provides a measure of the density and vigour (photosynthetic activity) of the vegetation cover. It uses data derived from NOAA's AVHRR sensor.

NDVI provides an assessment of plant density and vigour and is extremely useful in comparing seasonal changes in vegetation cover. From Figure 1, it is evident that vegetation in Canada is mostly dormant in the winter months, but grows actively in the summer.

Global mapping of land cover requires instruments that can observe the surface on a daily or near-daily basis, thus maximizing the opportunity to image the surface in areas that are frequently covered in cloud. These instruments have a coarser ground resolution than Landsat, with pixels typically representing areas about $1{,}000 \times 1{,}000$ m in size; however, that is not a deficiency when working on a global scale. Figure 2 shows a map of global land cover produced from MODIS images acquired mainly in 2001. The legend recognizes 17 types of land cover, including forests, shrublands, savannas, grasslands, and wetlands.

Evergreen broadleaf forest dominates the equatorial belt, stretching from South America, through Central Africa, to south Asia. Adjoining the equatorial forest belt are regions of savannas and grasslands, which have seasonal wet-dry climates. The vast desert region running from the Sahara to the Gobi is barren or sparsely vegetated. It is flanked by grasslands on the west, north, and east. Broadleaf deciduous forests are prominent in eastern North America, western Europe, and eastern Asia. Evergreen needleleaf forests span the boreal zone from Alaska and northwest Canada to Siberia. Croplands are found throughout most regions of human habitation, except for dry desert regions and cold boreal zones.

The map was constructed using both spectral and temporal information. The graph in Figure 3 shows how these information sources are used. It depicts reflectance values in red and near-infrared (NIR) spectral bands for three land cover types, as observed in the southeastern United States: evergreen needleleaf forest, cropland/natural vegetation mosaic, and deciduous broadleaf forest. The reflectance values are shown as they change over the course of about a year and a half, from October 31, 2000, to April 2002.

The top three curves, which are near-infrared values, show the patterns of the three types most clearly. During the winter, values are generally low, with deciduous broadleaf forest, evergreen needleleaf forest, and cropland in increasing order of reflectance. As spring begins, cropland reflectance rises before deciduous broadleaf forest, but deciduous broadleaf forest reaches a higher peak. Evergreen needleleaf forest also shows a spring green-up, but it is later than the others and peaks at a lower value. Reflectance gradually drops during the fall, and the three types reach about the same reflectance levels as in the prior year. The three lower curves, which display the red band, also have distinctive trends, but they are less obvious on this graph.

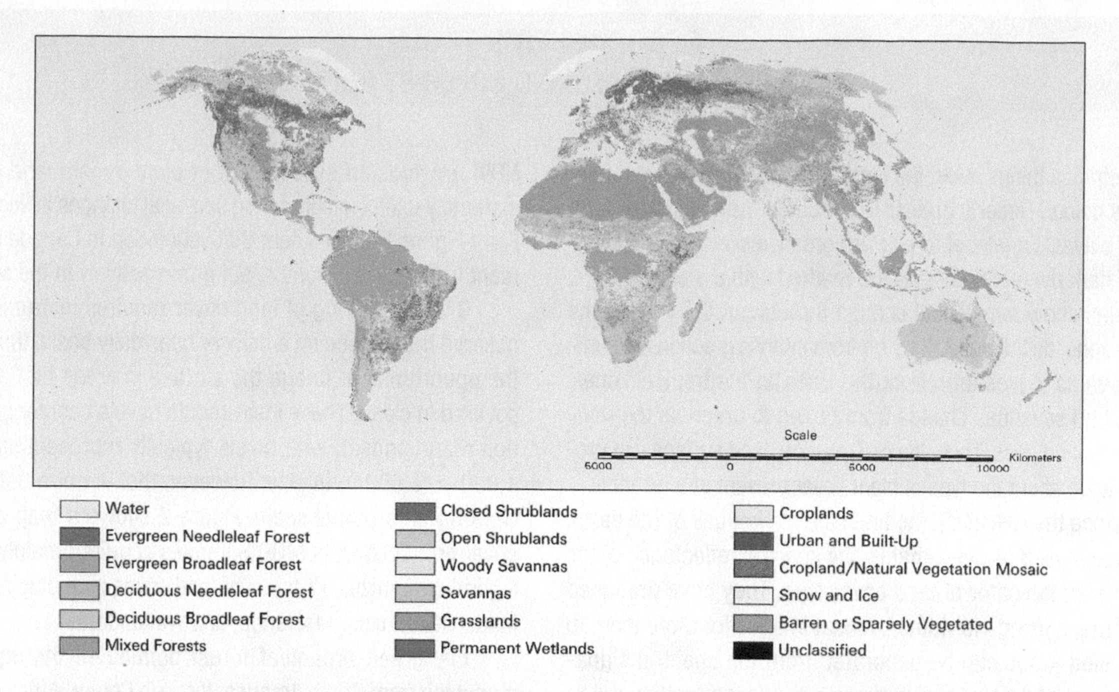

Figure 2 Global land cover from MODIS This map of global land cover types was constructed from MODIS data acquired mainly during 2001. The map has a spatial resolution of 1 km^2; that is, each square kilometre of the Earth's land surface is independently assigned a land cover type label. (A. H. Strahler, Boston University/NASA.)

The process by which each global pixel is given a label is referred to as *classification.* In short, a computer program is presented with many examples of each type of land cover. It then "learns" the examples and uses them to classify pixels, depending on their spectral and temporal pattern. The MODIS global land cover map was prepared with more than 1,500 examples of the 17 land cover types. It is estimated to be about 75 to 80 percent accurate.

Land cover mapping is a common application of remote sensing. Given the ability of space-borne instruments to image the Earth consistently and repeatedly, classification of remotely sensed data is a natural way of extending our knowledge from the specific to the general to provide valuable new geographic information.

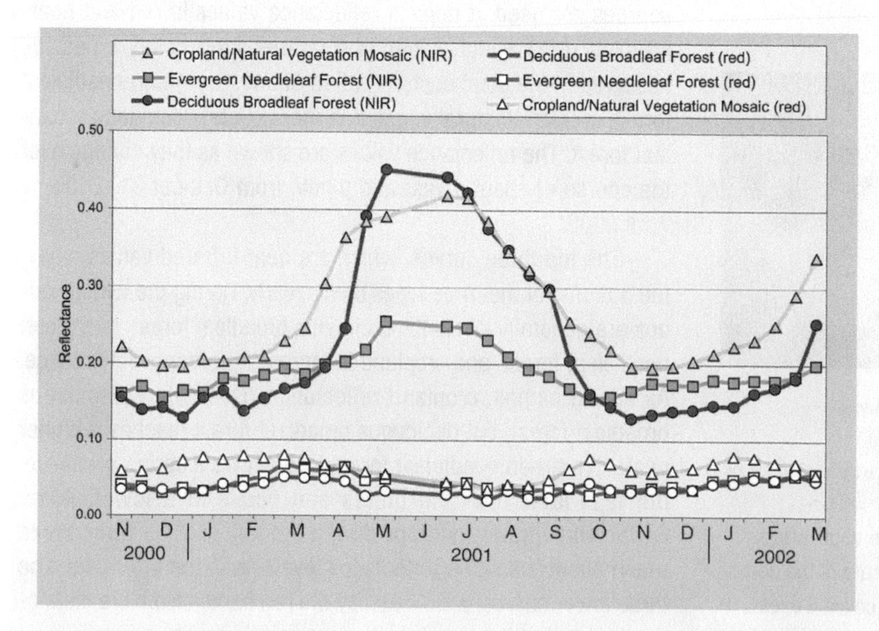

Figure 3 Spectral and temporal reflectance patterns The graph shows how the spectral reflectance in red and near-infrared bands of three land cover types in the southeastern United States varies during the period from October 2000 to April 2002. (A. H. Strahler, Boston University/NASA.)

conditions occur in the wet equatorial climate and the monsoon and trade-wind coastal climate (see Chapter 9). Without a cold season or dry season, plant growth continues throughout the year. In this uniform environment, some species grow new leaves and shed old ones continuously. Other plants shed their leaves according to their own physiological schedule, responding to the slight changes in daylight temperature or moisture that occur over the course of the year.

The *equatorial rain forest* lies astride the equator, and occurs extensively in South America where it occupies much of the Amazon lowland. In Africa, this forest type is found in the Congo lowland and in a coastal zone extending westward from Nigeria to Guinea. The island of Sumatra marks the western boundary of the Asian lowland rain forest region that stretches eastward to the islands of the Pacific.

The *tropical-zone rain forest* is similar in appearance to the low-latitude equatorial forest. It represents a poleward extension of the rain forest biome through the tropical zone (lat. 10° to 25° N and S) in coastal regions that receive abundant moisture from monsoon rainfall, or where the trade-winds bring moist air from the oceans. The tropical-zone rain forest thrives in the monsoon and trade-wind coastal climates that experience a short dry season. In these regions, the dry season is not intense enough to deplete soil water (see Figure 9.5).

In the northern hemisphere, tropical-zone rain forests are found along trade-wind coasts in the Philippines and the eastern coasts of Central America and the West Indies. These are rugged highland areas where an orographic effect increases rainfall in the belt of the trade winds. The eastern mountains of Puerto Rico provide a good example of a tropical-zone rain forest. In southeast Asia, tropical-zone rain forests occur in Vietnam and Laos, southeastern China, and on the western coasts of India and Myanmar. In the southern hemisphere, belts of tropical-zone rain forest extend down the coast of Madagascar and in the northeastern coastal region of Australia. The corresponding belt of coastal rain forest in eastern Brazil has largely been cleared for agriculture.

22.6 CLOUD FORESTS ON MOUNT KINABALU, BORNEO These cloud forests grow at elevations above 2,200 m in a cool, humid tropical environment.

Within the regions of low-latitude rain forest are many highland regions where the climate is cooler and the orographic effect again increases rainfall. Rain forest extends up the rising mountain slopes, but at elevations between 1,000 and 2,000 m, its structure gradually changes and it eventually becomes *montane forest*. The canopy of montane forest is more open and tree heights are lower than in the rain forest. Tree ferns and bamboos are numerous, and epiphytes are particularly abundant. As elevation increases, mist and fog become persistent, giving high-elevation montane forest the name *cloud forest* (Figure 22.6).

The low-latitude rain forest is under increasing human pressure; it is being logged and cleared for grazing and cropland. *Eye on the Environment • Exploitation of the Low-Latitude Rain Forest Ecosystem* describes this process in more detail.

EYE ON THE ENVIRONMENT

EXPLOITATION OF THE LOW-LATITUDE RAIN FOREST ECOSYSTEM

Many of the world's equatorial and tropical regions are home to the rain forest ecosystem. This ecosystem is the most diverse on the Earth; it possesses more species of plants and animals than any other. Large tracts of rain forest still exist in South America, south Asia, and some parts of Africa. Ecologists regard this ecosystem as a genetic reservoir of many species of plants and animals. But as human populations expand and the quest for agricultural land continues, low-latitude rain forests are being threatened with clearing, logging, cultivation of cash crops, and grazing.

In the past, native peoples farmed low-latitude rain forests using the *slash-and-burn* method—cutting down all the vegetation in a small area, then burning it (Figure 1). In a rain forest ecosystem, most of the nutrients are held within living plants rather than in the soil. Burning the vegetation on the site releases the trapped nutrients, returning a portion of them to the soil. Here, the nutrients are available to growing crops. The supply of nutrients derived from the original vegetation cover is small, however, and the harvesting of crops rapidly depletes the nutrients. After a few seasons of cultivation, a new field is cleared, and the old field is abandoned. Rain forest plants re-establish their hold on the abandoned area, and eventually, the rain forest returns to its former state. In this way, primitive slash-and-burn agriculture can maintain the rain forest ecosystem.

Modern intensive agriculture, on the other hand, uses large areas of land and is not compatible with the rain forest ecosystem. When large areas are abandoned, seed sources are so far away that the original forest species cannot take hold. Instead, secondary species dominate, often accompanied by species from other vegetation types. These species are good invaders, and once they enter an area, they tend to stay. The dominance of these secondary species is permanent, at least on the human time scale. Thus, we can regard the rain forest ecosystem as a resource that, once cleared, may never return. The loss of low-latitude rain forest will result in the disappearance of thousands of species of organisms—a loss of millions of years of evolution, together with the destruction of the most complex ecosystem on Earth.

In Amazonia, the transformation of large areas of rain forest into agricultural land has included the use of heavy machinery to carve out major highways, such as the Trans-Amazon Highway in Brazil. Once the initial route was established, innumerable secondary roads and trails were developed (Figure 2). Large fields for cattle pasture or commercial crops have been created by cutting, bulldozing, clearing, and burning the vegetation. In some regions, the great broadleaf rain forest trees are being removed for commercial lumber.

Figure 1 Slash-and-burn clearing This rain forest in Maranhão, Brazil, has been felled and burned in preparation for cultivation.

Recent computer simulations indicate that if the Amazon rain forest is entirely removed and replaced with pasture, surface and soil temperatures will increase by 1 to 3 °C. Precipitation in the region will decline by 26 percent and evaporation by 30 percent. The deforestation will change weather and wind patterns so that less water vapour enters the Amazon Basin from outside sources, making the basin even drier. In areas already experiencing a marked dry season, the dry period will become longer. Although these models contain simplifications and are subject to error, the results support the pessimistic forecast that once large-scale deforestation has occurred, it may be impossible to artificially restore a rain forest comparable to the original.

According to a report by the UN Food and Agriculture Organization (FAO), about 0.6 percent of the world's rain forest is lost annually by conversion to other uses. More rain forest land, about 2.2 million ha, is lost annually in Asia than in Latin America and the Caribbean, where 1.9 million ha are converted every year. Africa's loss of rain forest is estimated at 470,000 ha per year. Among individual countries, forest conversion in Brazil and Indonesia accounts for nearly half of this loss. Deforestation in low-latitude moist deciduous forests, dry deciduous forests, and hill and montane forests in these regions is also very serious.

Although deforestation rates are rapid in some regions, many nations are working to reduce the loss of rain forest environment. However, rain forests can provide agricultural land, minerals, and timber, so the pressure to allow deforestation continues.

Figure 2 Landsat photos of Rondonia, Brazil These photos show the amount of deforestation that has occurred as highways, secondary roads, and trails have been developed through the rain forest over the course of just 17 years.

Monsoon Forest

The *monsoon forest* of the tropical latitude zone differs from the tropical rain forest because it is deciduous. Most of the monsoon forest trees shed their leaves during the dry season as a result of moisture stress. The monsoon forest is usually open, with patches of shrubs and grasses interspersed among the trees (Figure 22.7). Because trees are some distance apart, light easily penetrates to the understory, so the ground layer is better developed than in the rain forest. Tree heights are also lower. Typically, as many as 30 to 40 tree species are present in a small tract of monsoon forest, which is less than half of what is encountered in most rain forests. The trees can be massive, generally with thick, rough bark and branches found lower down their trunks than on lowland rain forest species.

The monsoon forest develops in the wet-dry tropical climate in which a long rainy season alternates with a dry, rather cool season. These conditions are most strongly developed in Southeast Asia, but are not limited to that area (see Chapter 9). The typical regions for monsoon forest are in Myanmar, Thailand, and Cambodia. Teak (*Tectona grandis*) was once abundant in monsoon forests throughout the region, but because it is very resistant to termite attack and rot, it has been used extensively for construction and shipbuilding. Natural stands have been severely depleted and today teak is mostly found on plantations. In addition to Southeast Asia, large areas of monsoon forest also occur in south-central Africa and in Central and South America, bordering the equatorial and tropical rain forests.

Subtropical Evergreen Forest

The *subtropical evergreen forest* is found in regions of moist subtropical climate, where winters are mild and there is ample rainfall throughout the year. There are two forms of this forest, broadleaf and needleleaf. The *subtropical broadleaf evergreen forest* has fewer and shorter species of trees than the low-latitude broadleaf evergreen rain forest. In addition, the leaves of subtropical broadleaf evergreen species tend to be smaller and more leathery, and the canopy less dense. This forest often has a well-developed lower layer of vegetation that may include tree ferns, small palms, bamboos, shrubs, and herbaceous plants. Lianas and epiphytes are abundant.

22.7 MONSOON WOODLAND This woodland is in the Khao Yai National Park in Thailand. The photo was taken in the rainy season, with trees in full leaf.

In the northern hemisphere, the subtropical evergreen forest consists of broadeaf trees such as evergreen oaks and trees of the laurel and magnolia families. Because laurel is often the dominant or co-dominant genus, these ecosystems are commonly known as "laurel forests." They are associated with the moist subtropical climate in the southeastern United States, southern China, and southern Japan (see Chapter 10). However, these regions are under intense cultivation because of their favourable climate. The land has been cleared of natural vegetation for centuries, and little natural forest remains.

The *subtropical needleleaf evergreen forest* occurs only in the southeastern United States. Here it is called the southern pine forest, since it is dominated by species of pine (*Pinus* spp.). It is found on the wide belt of sandy soils that borders the Atlantic and Gulf coasts. These soils, which are classed as Ultisols in the CSCS system, are typically of low nutrient status, so the native trees are shallow rooting like those in the lowland rain forest and are similarly dependent on rapid nutrient cycling (see Chapter 19). Because the soils are coarse textured, water drains away quickly, leaving them quite dry. During infrequent drought years, these forests may burn. Since pines are well adapted to droughts and fires, they form a stable vegetation cover over large parts of the region. Timber companies have taken advantage of this natural adaption of fast-growing pines, creating many plantations that yield valuable lumber and pulp (Figure 22.8).

Subtropical evergreen forests do not grow extensively in the southern hemisphere; remnant stands are still present in coastal regions of Brazil, and similar structural and life form types grow in southeastern Australia and parts of New Zealand. Although the forests of New Zealand are almost entirely temperate in nature, many of the plants have close affinities with species and genera in tropical regions (Figure 22.9).

Midlatitude Deciduous Forest

The *midlatitude deciduous forest* is the native forest type of western Europe, eastern Asia, and eastern North America. It is dominated by tall, broadleaf trees that provide a continuous and dense canopy in summer but shed their leaves completely in the winter (Figure 22.10). Small trees and shrubs provide a distinctive understory in some areas. The trees resume growth in the spring in response to warmer temperatures and longer days. Many herbaceous species flower early, while the forest floor remains bright and sunny. Ferns and other shade-loving plants thrive beneath the leafy canopy when

22.8 **PINE PLANTATION** This plantation of longleaf pine grows on the sandy soil of the southeastern coastal plain near Waycross, Georgia.

22.9 **SUBTROPICAL EVERGREEN FORESTS OF NEW ZEALAND** The oceanic climate of New Zealand favours the growth of broad-leaf evergreen trees, particularly species of southern beech (*Nothofagus* spp.) often in association with conifer species including kauri (*Agathis australis*), matai (*Podocarpus spicatus*), and rimu (*Dacrydium cupressinum*) with smaller hardwoods and tree ferns in the understory.

it is fully developed in midsummer. The end of the growing season is marked by vivid autumn colours, and soon after the leaves fall to the ground and the trees enter their winter leafless state.

Midlatitude deciduous forests are found almost entirely in the northern hemisphere. Throughout much of its range in eastern North America and eastern Asia, this forest type is associated with the moist continental climate. Precipitation is adequate in all months, normally with a summer maximum due to incursions of moist mT (see Chapter 10). There is a strong annual temperature cycle, with warm summers alternating with cool winters as cP and cA air masses push into the region. The length of the growing season is about 120 days in northerly regions, increasing southward to more than 250 days.

In western Europe, the midlatitude deciduous forest is associated with the marine west-coast climate where moist air from the Atlantic Ocean is brought onshore by the prevailing southwesterly winds accompanied by the frequent passage of low-pressure systems (see Chapter 10). Oak trees have been planted extensively and are the dominant species in many areas, although birch, beech, and ash are also common. Elm was particularly widespread in southern Britain, but its distribution has been drastically reduced by Dutch elm disease.

Because of intense land-use pressures, the deciduous forests of eastern Asia are mostly found in hilly areas that are not easily cultivated. The forests are composed of oak, elm, maple, lime, and ash. Numerous flowering shrubs are found in undisturbed areas. A small area of deciduous forest is found in the drier parts of Patagonia near the tip of South America. This forest is composed almost entirely of southern beeches (*Nothofagus* spp.).

In North America, deciduous forests extend from the Great Lakes region south to the Gulf of Mexico. Many deciduous forest trees are nutrient-demanding and grow on soils of medium to high base saturation. In the United States, the principal soils in the most diverse deciduous forest regions are classed as Alfisols, which characteristically have a thin surface horizon darkened by humus underlain by a paler, bleached *Ae* horizon and an illuvial clay layer. Similar soils in Canada are classed as Luvisols. Considerable quantities of nutrients are returned to the soil and contribute to the rich humus layer when the leaves are shed each autumn, with much of the material being fully decomposed by the following growing season.

The greatest diversity is in the southern Appalachians, and includes beech, tulip-tree, basswood, sugar maple, buckeye, and oaks, all of which can grow to heights of 35 to 40 m (Figure 22.11a). The number of tree species decreases westward until, in

22.10 TEMPERATE DECIDUOUS FOREST The leafy summer foliage falls to the ground in autumn and the trees go through winter in a leafless state.

the driest areas bordering the prairie grasslands, only oaks and hickories are common. Beech and sugar maple are the dominant species in the deciduous forests of southern Ontario and Quebec (Figure 22.11b). Other species include elm, basswood, ash, and walnut. The northern limit of the deciduous forests occurs in a mixed-wood forest region that runs from northern Minnesota to the Maritime provinces (Figure 22.11c). Climate and soils are less favourable for deciduous species in this transitional region, and conifers increase in abundance. Here, pure deciduous stands, composed mainly of trembling aspen and birch, are generally limited to sites that have been disturbed by fire.

The deciduous forest includes a variety of animal life, much of it stratified according to layers in the canopy that typically include the upper canopy, lower canopy, understory, shrub layer, and ground layer. Because the ground layer presents a more uniform environment in terms of humidity and temperature, it contains the largest concentration of organisms and the greatest diversity of animal species. Many small mammals burrow in the soil for shelter or to feed on soil invertebrates. Among this burrowing group are ground squirrels and shrews, as well as some larger animals—foxes, woodchucks, and rabbits. Most of the larger mammals feed on ground and shrub layer vegetation. Large herbivores that graze in the deciduous forests include the red deer of Eurasia and the white-tailed deer of North America. However some, such as the brown bear, are omnivorous and prey on the small animal as well. Voles, mice, and squirrels are common species of smaller herbivores. Predators include bears, lynx, and wolves, as well as owls.

Even though birds possess the ability to move through the layers at will, many restrict themselves to one or two layers. For example, the wood peewee is found in the lower canopy, and the red-eyed vireo forage in the understory. Above them, in the upper canopy, are species such as the scarlet tanagers, and below are the ground dwellers, such as grouse and warblers. Flying insects often show similar patterns of stratification.

(a)

(b)

(c)

22.11 DECIDUOUS FORESTS (a) The most diverse forest, and often the biggest trees, occur in the southern Appalachian region of the United States. (b) Beech-maple forests occur extensively in Ontario and Quebec. (c) A zone of mixed-wood forest marks the transition between the deciduous and the boreal forest to the north.

Needleleaf Forest

Needleleaf forest refers to a forest composed mainly of conifer trees with comparatively short branches and small, narrow, needle-like leaves. Most are evergreen, retaining their needles for several years before shedding them. When the needleleaf forest is dense, it casts a continuous and deep shade; lower layers of vegetation are sparse, except perhaps for a thick carpet of mosses. Species diversity is generally low, with large tracts of the needleleaf forest overstorey consisting almost entirely of only one or two tree species.

Boreal forest is the cold-climate needleleaf forest of high latitudes. It occurs in two great continental belts, one in North America and one in Eurasia. These belts correspond closely to the boreal forest climate region (see Chapter 10). The boreal climate is dominated by cold, dry, polar air masses. Summers are cool and short, although this is offset by long hours of daylight. At higher latitudes where there is continuous daylight, the diurnal temperature range is generally 10 to 15 °C due to the change in the angle of the incoming solar beam over the course of the day. The frost-free season is generally between 50 and 90 days.

Winters are long and cold, with temperatures often falling below −40 °C. Annual precipitation varies across this broad zonal region, and is normally in the range of 300 to 600 mm, much of it falling as snow. The soils throughout the boreal forest region are relatively young, having formed since the retreat of the Late-Cenozoic ice sheets. In southern regions they are predominantly Brunisols and Podsols, which characteristically develop on medium- to coarse-textured substrates under cool, humid conditions that promote leaching (see Chapter 19). The plant litter that accumulates on the forest floor is low in nutrients and has a high content of resins, waxes, and lignin. Humification is slow, and a thick layer of coarse, acidic humus typically develops. Discontinuous or sporadic permafrost occurs throughout the southern boreal forest and more continuously at higher latitudes, favouring the development of Cryosols (see Chapter 18).

The structure of the boreal forest changes as growing conditions become increasingly severe at higher latitudes. The *forest–tundra transition* extends beyond the Arctic tree line, where it is represented by small patches of conifers that survive in sites protected from the rigorous climate (Figure 22.12a). Further south lies the open *lichen woodland* in which trees are several metres apart, and a thick carpet of feathery lichen, commonly known as *reindeer moss*, covers the intervening ground (Figure 22.12b, also see Figure 22.2). The dense stands of closely spaced trees that form a continuous cover in the southernmost section of the boreal forest make up what is called *close forest* (Figure 22.12c).

The Eurasian boreal forests, or *taiga*, extend from Scandinavia across Siberia to the Pacific Ocean. Scots pine is the dominant species in the west, but eastward it is replaced successively by Norway spruce, Siberian spruce, and Siberian larch.

(a)

(b)

(c)

22.12 **BOREAL FOREST** (a) At the northernmost edge of the boreal forest, trees are increasingly restricted to sheltered sites. (b) Open lichen woodland consists of scattered trees, mostly black spruce, with reindeer moss forming the predominant cover in the adjacent openings. (c) In the close-forest structural type, little light penetrates the canopy and the understory is predominantly hardy shrubs and mosses.

East of the Ural Mountains, Dahurian larch increases in abundance. Larches shed their needles in winter and thus are deciduous needleleaf trees. Broadleaf deciduous trees, such as aspen and birch, are usually restricted to burned areas, and in time are replaced by conifers.

The boreal forests of North America form a continuous belt across the continent from Alaska to Newfoundland but, like the taiga, they are composed of relatively few species. White spruce and black spruce are widely distributed, but other tree species have more limited ranges. In Alaska and the Yukon, the forests are composed mainly of white spruce, growing in pure stands or mixed with birch and aspen. Black spruce and tamarack are common in wetter settings. In central Canada, drier, sandy soils support stands of jack pine, and balsam fir appears in response to moister climatic conditions. The diversity of the boreal forest increases to the east of the Great Lakes. Although white spruce and balsam fir remain the dominant species over much of eastern Canada, other conifers, such as red pine, eastern white pine, eastern hemlock, and eastern white cedar, are common. Black spruce regains its importance in the rugged, windswept areas of Newfoundland and Labrador, where the character of the forest resembles the open lichen woodlands of the subarctic.

Needleleaf evergreen forest extends into lower latitudes along mountain ranges and high plateaus. For example, in western North America, montane forests are found on the Rocky Mountains and the Sierra Nevada. There are four relatively distinct regions of montane needleleaf forest, based on species distributions. White spruce is the dominant species in the north, but the presence of lodgepole pine and subalpine fir distinguishes the montane forest from adjacent boreal forest. Engelmann spruce largely replaces white spruce at about lat. 53° N. Further to the south, Engelmann spruce and alpine fir persist at higher elevations (Figure 22.13a), but below about 2,500 m, they are replaced by Douglas fir and ponderosa pine. Giant sequoias and ancient bristlecone pines are distinctive montane species in the south (Figure 22.13b, c).

Mammals of the boreal needleleaf forest in North America include deer, moose, elk, black bear, marten, mink, wolf, and wolverine. Numerous bird species, such as jays, ravens, chickadees, nuthatches, and warblers are also present. The caribou, lemming, and snowshoe rabbit inhabit both needleleaf forest and the adjacent tundra biome. The boreal forest often experiences large fluctuations in animal species populations, a result of the low diversity and highly variable environment.

The *coastal forest* extends in a narrow zone from southern Alaska to northern California, and closely corresponds with the west-coast marine climate (see Chapter 10). Here, in a band of heavy orographic precipitation, mild temperatures, and high humidity, are some of the densest of forests on Earth. Although the soils are heavily leached acidic Podsols, they support the world's largest trees, which commonly grow 50 to 80 m in height, with massive trunks 2 to 3 m in diameter. Hemlock and Sitka spruce are the dominant species in Alaska and northern British Columbia. The coastal forests of southern British Columbia

(a)

(b)

(c)

22.13 **MONTANE NEEDLELEAF FOREST** (a) Engelmann spruce and alpine fir are common in high-elevation sites in the Canadian Rockies. (b) Giant sequoias occur on the mid-elevation slopes of the Sierra Nevada. (c) Bristlecone pine grow to a considerable age in the harsh climate near the tree line; the oldest known specimen, aged 4,800 years, grows in the White Mountains of California.

22.14 **GIANTS OF THE COASTAL FOREST** Huge trees, including Douglas fir, cedar, and redwood thrive in the cool, moist environment of western North America.

support hemlock, Douglas fir, and western red cedar (Figure 22.14), with Douglas fir increasing in importance in the forests of Washington and Oregon. The southern limit of the coastal forest is marked by stands of redwoods in California, which depend on coastal fogs to supplement precipitation during the dry summer months.

Sclerophyll Forest

The native vegetation of the Mediterranean climate is adapted to survival through the long summer drought (see Chapter 9). Shrubs and trees that can survive these annual droughts are characteristically equipped with small, tough leaves that resist water loss through transpiration; this is the characteristic leaf form of sclerophyll vegetation. Sclerophyllous leaves are also considered to provide an adaptive advantage in areas of high light intensity and nitrogen deficient soils, which are characteristic of the Xeralfs of this region.

The *sclerophyll forest* consists of trees that are generally thick barked, with low branches that give a gnarled and twisted appearance. The formation class includes *sclerophyll woodland*, a more open form of forest in which only 25 to 60 percent of the ground is covered by trees. Also included are extensive areas of *scrub*, a plant formation type consisting of relatively close-growing shrubs. The trees and shrubs are evergreen, retaining their thickened leaves despite a severe annual drought.

Sclerophyll vegetation includes forest, woodland, and scrub types. Being closely associated with the Mediterranean climate, sclerophyll forest is limited to west coasts between lat. 30° and 45° N and S; in Europe, it forms a narrow, coastal belt around the Mediterranean Sea. The European formation includes species such as cork oak and olive. Over the centuries, human activity has reduced the sclerophyll forest to woodland, or has destroyed it entirely. Today, large areas of this former forest consist of dense scrub.

California is the only other location in the northern hemisphere where this type of vegetation is found. Here, the sclerophyll forest or woodland is typically dominated by live oak and white oak (Figure 22.15a). Grassland occupies the open ground between the scattered oaks. Much of the remaining vegetation is sclerophyll scrub known as *chaparral* (Figure 22.15b), which varies in composition with elevation and exposure. Chaparral may contain wild lilac, manzanita, mountain mahogany, poison oak, and evergreen live oak, but in most places chamise (*Adenostoma fasciculatum*) is the dominant species.

In central Chile and the Cape region of South Africa, sclerophyll vegetation has a similar appearance, but the dominant species are quite different (Figure 22.15c). Important areas of sclerophyll forest, woodland, and scrub are also found in southeast, south central, and southwest Australia, consisting mainly of species of eucalyptus and acacia (Figure 22.15d).

SAVANNA BIOME

The savanna biome is usually associated with the tropical wet-dry climate of Africa and South America (see Chapter 9). This climate has a distinct dry season during the low-sun period when at least one month receives less than 60 mm of rain. In most of the savanna regions, mean annual temperature never falls below 24 °C, with maximum temperatures normally occurring at the end of the wet season under cloudless skies. Precipitation varies from about 250 mm in drier areas to 1,500 mm or more, but water shortages develop during the dry season throughout this biome (see Figure 9.7). Regional differences in precipitation are reflected in the diversity of the savanna biome, which includes vegetation formation classes ranging from woodland in the moister areas to grassland.

Fire is a frequent occurrence in the savanna during the dry season. Periodic burning of the savanna grasses helps to protect the grassland from the invasion of forest. Fire does not kill the underground parts of grass plants, but it limits the tree cover to fire-resistant species. Grazing by animals kills many young trees, which also helps to maintain the grasslands.

Savannas occur extensively on old plateaus that are variously interrupted by escarpments and further dissected by rivers. Many areas are underlain by layers of laterite or gravelly veneers. Prolonged weathering has produced nutrient-poor Oxisols in wetter regions that are subject to drought due to the pronounced seasonality of the rainfall. Alfisols are more common in the drier

22.15 **SCLEROPHYLL VEGETATION** (a) Trees in the sclerophyll woodland in California are mainly species of oak and often grow into rather twisted forms. (b) Shrub-dominated communities, such as the chamise (*Adenostoma fasciculatum*) chaparral of California and (c) Ceanothus chaparral, occur extensively in all Mediterranean climate regions. (d) Species of Eucalyptus compose forest, woodland, and scrub formations in the Mediterranean regions of Australia.

savannas. The Oxisols are generally fine-textured soils with much of the clay in the form of kaolinite, with a 1:1 lattice structure that affords few sites for exchangeable cations. High concentrations of aluminum produced by clay decomposition can also affect the nutrient balance of the soils; phosphorus deficiency is a common problem as this element becomes immobilized as iron-aluminum phosphate. The Alfisols are characteristically rather light-textured, sandy soils, derived from crystalline parent materials with a high quartz content. This, together with the predominance of kaolinite clays, result in soils that are naturally low in plant nutrients.

The savanna biome is most widely distributed in Africa, where it covers about 65 percent of the continent. It forms a broad belt around the tropical forests and changes progressively from woodland to semi-arid forms in response to rainfall regimes. In *savanna woodland*, despite a relatively high annual precipitation, the trees are spaced rather widely apart because there is not enough soil moisture during the dry season to support a full tree cover. The open spacing between trees facilitates the development of a dense lower layer mainly consisting of grasses and herbs, giving savanna woodland an open park-like appearance. In Africa, the trees are of medium height. Tree crowns are flattened or umbrella-shaped, and the trunks have thick, rough bark. Some species of trees are xerophytic; their small leaves and thorns are a distinctive characteristic of species that have adapted to drought. Others are broadleaf deciduous species that shed their leaves in the dry season. In this respect, the savanna woodland resembles the monsoon forest.

The savanna woodland grades into a belt of *thorntree-tall-grass savanna*, a formation class that provides a transition to the desert biome. The thorntrees are predominantly species of acacia. These leguminous, pod-bearing plants enhance soil fertility through the activity of nitrogen-fixing microbes associated with their roots. In the thorntree tall-grass savanna, trees are more widely scattered, and the open grassland is more extensive than in the savanna woodland (Figure 22.16). The thorntree tall-grass savanna is closely identified with the semi-arid subtype of the dry tropical and subtropical climates. In the semi-arid climate, soil water storage is adequate to meet the needs of plants only during the brief rainy season. The onset of the rains is quickly followed by the greening of the trees and grasses. For this reason, vegetation of the savanna biome is described as *rain-green*, a term that also applies to the monsoon forest.

The African savanna is widely known for the diversity of its large grazing mammals, which include numerous species of antelopes, wildebeest, zebra, and giraffe. Studies have shown that each species has a particular preference for different parts of the grasses—leaf blade, leaf sheath, and stem. They also feed on these grasses at different times of the year during their regular seasonal migrations. Grazing stimulates the grasses to continue to grow, and so the ecosystem is more productive when grazed than when left alone. With these grazers comes a large variety of predators, such as lions, leopards, cheetahs, hyenas, and jackals. Elephants are the largest animals of the savanna and adjacent woodland regions. However, as in the tropical forests, termites and other soil-dwelling organisms are the most abundant animal groups in the savanna ecosystem (Figure 22.17).

22.16 TROPICAL SAVANNAS The tropical grasslands that form the extensive undulating plains of East Africa are home to large herds of grazing animals. Trees, mainly acacias with their distinctive spreading canopy, are often present and give the ecosystem an open park-like structure.

GRASSLAND BIOME

The grassland biome includes two major formation classes—tall-grass prairie and short-grass prairie or steppe; between these lies a broad transition known as mixed prairie. The grasslands are concentrated mainly in North America and Eurasia. In the southern hemisphere they are represented by the Pampas of Argentina and Uruguay and the *veldt* region of South Africa. Grasslands are found in midlatitude and subtropical zones with pronounced winter and summer seasons. The global distribution corresponds

22.17 TERMITE MOUNDS Termite mounds are a distinctive feature in the savanna landscape, although not all of them are inhabited. The internal structure can be very complicated, with numerous tunnels and shafts connecting various parts of the mound with what is usually an even larger underground chamber. Termites are an important part in the savanna ecosystem because of their role as decomposer organisms.

well with the semi-arid subtype of the dry continental climate and parts of the moist continental climate (see Chapter 10). Spring rains nourish the grasses, which grow rapidly until early summer. By midsummer, the grasses are usually dormant.

Tall-grass prairie tend to occur in the moister parts of the continental climate region, where soil water can be in short supply briefly during the summer months; it is gradually replaced by mixed prairie where summer moisture is more limited. The simplified climate diagrams for Neepawa, Manitoba (Figure 10.17a), and Lethbridge, Alberta (Figure 10.14b), illustrate this change in moisture between the tall-grass prairie and mixed-prairie regions in Canada. Mixed prairie is in turn replaced by steppe grasslands in areas where conditions are even drier and droughts are a frequent hazard.

Tall-grass prairie consists mainly of grasses that can exceed 2 m in height at the end of the growing season when they come into flower and produce seed. Many *forbs*, or broadleaf herbs, are also present; their showy flowers develop either in the spring or fall (Figure 22.18a). Trees and shrubs are rare in tall-grass prairie, being generally restricted to narrow patches of woodland in stream valleys. The grasses are deeply rooted and form a thick and continuous turf. When European settlers first arrived in North America, the tall-grass prairies extended in a belt from the Texas Gulf coast northward to southern Manitoba. A broad peninsula of tall-grass prairie extended eastward into Illinois, where conditions are somewhat moister. Since this time, these prairies have been converted almost entirely to agricultural land. Another major area of tall-grass prairie is the Pampa region of South America, which occupies parts of Uruguay and eastern Argentina. The Pampa region falls into the moist subtropical climate, with mild winters and abundant precipitation (see Chapter 10).

Between tall-grass prairie and steppe is a band of *mixed prairie*. In North America, mixed prairie occupies much of the central Great Plains and stretches from Saskatchewan to Texas. It is the most extensive grassland type in Canada and extends across southeastern Alberta, southern Saskatchewan, and eastern Manitoba. Mixed prairie is composed of medium-height and short-grass species, and the relative dominance of each life form reflects variations in growing conditions. On well-drained soils throughout the Canadian Prairies, the dominant species include wheat grasses (*Agropyron* spp.) and spear grasses (*Stipa* spp.). In drier areas, drought-resistant species, such as blue grama (*Bouteloua gracilis*), increase in abundance, while June grass (*Koeleria cristata*) is characteristically associated with the heavy clays of glacial lake beds.

Herbaceous plants, such as prairie crocuses and lilies, add diversity to the grasslands, but shrubs and trees are generally restricted to moister valleys and depressions. Groves of aspen increase toward the north and ultimately form a zone of

(a)

(b)

(c)

22.18 **TEMPERATE GRASSLANDS** (a) The moist tall-grass prairie contains a rich diversity of herbs, such as the purple-flowered blazing star (*Liatris pycnostachya*). In drier mixed prairie, herbs such as the yellow prairie cone flower (*Ratibida columnifera*) add occasional colour. (b) Sloughs, small water bodies associated with the rolling topography created by the continental ice sheets, provide excellent waterfowl habitat in the northern mixed prairie. (c) Shorter grasses of the steppe grasslands often grow in association with small shrubs, such as the silver sage (*Artemisia cana*).

woodland adjacent to the boreal mixed-wood ecosystem. In addition, the northern glacial plains are marked with water-filled depressions (sloughs) that form valuable waterfowl habitat (Figure 22.18b). Various aquatic plants grow in the sloughs, which are often bordered by small willows and aspen.

Steppe, also called *short-grass prairie*, is a vegetation type dominated by sparse clumps or bunches of grasses. The dominant species in short-grass prairie are drought resistant grasses that typically grow to a maximum height of about 30 cm (Figure 22.18c). Shrubby plants, such as sagebrush, are also common; some deep-rooting forbs are also found in these dry grasslands, but they are less conspicuous than those in the tall-grass and mixed prairies. Steppe grasslands grade into semi-desert where the moisture deficit is more severe.

Before the exploitation of the North American grasslands for cattle ranching, large grazing mammals were abundant. These included pronghorn antelope, elk, and bison, which ranged widely from tall-grass prairie to steppe. Now nearly extinct, bison once numbered 60 million. By 1889, however, the herds had been reduced to 800. Several small herds of bison remain in the Prairies, and their reintroduction into the Grasslands National Park of Saskatchewan in 2006 adds an important element to the regional ecosystem. Throughout much of the prairie, rodents and rabbits join cattle as the major grazers in the grasslands.

The grassland ecosystem supports some rather unique adaptations to life. A common adaptive mechanism is jumping or leaping, assuring an unimpeded view of the surroundings. Jackrabbits and jumping mice are examples. The pronghorn antelope combines the leaping trait with great speed, which allows it to avoid predators and fire. Burrowing is also another common habit; the soil provides the only shelter in the exposed grasslands. Examples are burrowing rodents, including prairie dogs, gophers, and field mice. Rabbits use old burrows for nesting or shelter. Invertebrates also seek shelter in the soil where moisture and temperature are more moderate. The burrowing owl, an endangered species in Canada, has adapted to the prairie environment in a similar way. These little birds feed voraciously on mice and insects, and a similar has been promoted as an asset to landowners, who are being encouraged to maintain the species on their lands.

DESERT BIOME

The desert biome accounts for about one-third of the Earth's land surface. This vast region includes several formation classes that are transitional between grassland and savanna biomes and the arid desert. Desert climates are characterized by the scarce and variable nature of their precipitation. In hot regions, *effective precipitation* can be further reduced by high evaporation rates, and some classifications recognize extremely arid, arid, or semi-arid deserts on this basis. In extreme deserts, precipitation is so low that only the hardiest plants and animals can survive, but even in semi-arid regions, moisture surplus is small and available for only a brief part of the year (see Figure 9.15). Annual temperature regimes vary markedly between different arid regions. In the dry tropical climate, characteristic of regions like the Sahara Desert, maximum temperatures commonly exceed 45 °C (see Chapter 9). Conversely, in the dry midlatiude climate, which occurs in areas such as the Gobi Desert, mean monthly temperatures can fall below −20 °C (see Chapter 10).

Distinctive weathering and erosional processes induced by arid conditions have resulted in a certain uniformity in desert landscapes throughout the world. Despite severe moisture deficits, water is an important geomorphological agent. Rain splash and sheet wash cause pronounced erosion during infrequent rainstorms because there is little vegetation to protect the surface. Dry stream channels fill quickly and scour the land, but flow is short-lived, and the heavy sediment loads are deposited on the channel bed or as alluvial fans at the base of intricately dissected highlands. These poorly sorted gravels, sands, and silts contrast with the rubbly talus that accumulates at the base of escarpments and mesas and the sand sheets and dunes molded by the wind. Most desert soils are poorly developed Aridisols and Entisols. With insufficient moisture to leach out soluble salts, concentrations of calcium and magnesium carbonates can develop hard layers, known as *caliche*, in the profile. Where groundwater is close to the surface, sodium salts can form evaporitic crusts.

Semi-desert is a transitional formation class that is distributed over a wide range of latitude from the tropical zone to midlatitudes. Semi-desert consists of sparse xerophytic shrubs. One example is the sagebrush vegetation of the middle and southern Rocky Mountain region and Colorado Plateau (Figure 22.19a). Recently, as a result of overgrazing and trampling by livestock, semi-desert shrub vegetation has expanded widely into areas of the western United States that were formerly steppe grasslands.

The *thorntree semi-desert* of the tropical zone consists of xerophytic trees and shrubs that have adapted to a climate with a long, hot dry season and a brief, but intense, rainy season. These conditions are found in the semi-arid and arid subtypes of the dry tropical and dry subtropical climates. The thorny trees and shrubs are mainly deciduous plants that shed their leaves in the dry season. The shrubs typically grow close together to form dense thickets. Cacti are present in some locations.

(a)

(b)

(c)

(d)

22.19 **VEGETATION OF THE ARID LANDS** (a) Sagebrush (*Artemisia tridentata*) dominates the landscape in many semi-desert regions in North America. (b) The giant saguaro cactus (*Cereus giganteus*) is a distinctive species in parts of the American southwest. (c) Welwitschia (*Welwitschia mirabilis*) is an ancient and rare plant of Namibia. (d) *Lithops* spp. resemble the pebbles and stones among which they grow in the deserts of southwest Africa.

Dry desert is a formation class of sparsely distributed xerophytic plants that have adapted to the arid conditions in several different ways. The most common perennial plants are drought-enduring small, hard-leaved or spiny shrubs, such as the creosote bush of the American southwest. Succulent drought-enduring plants, including cacti and aloes, are also well represented in the deserts of North and South America. For example, the tree-like saguaro cactus is a prominent species in the Mojave and Sonoran deserts of the southwestern United States (Figure 22.19b). In addition, many species of small, drought-avoiding annual plants may appear sporadically, after a rare, but heavy, downpour. In some deserts, there is almost no plant cover at all because of shifting dune sands or sterile salt flats. The ancient and unique Welwitschia persists under extreme conditions in the Namibian Desert (Figure 22.19c); the small, inconspicuous flowering stones (*Lithops* spp.) are also distinctive plants in this region (Figure 22.19d).

Desert animals, like the plants, are well adapted to the inhospitable conditions, restricting water loss and regulating body temperature through various physiological and behavioural mechanisms. Insects are abundant, as are insect-eating bats and birds. Venomous arthropods, such as the scorpion, and reptiles, including lizards and snakes, are also common. Important desert herbivores include kangaroo rats, jackrabbits, and the mule deer. Most small desert mammals are nocturnal and escape the daytime heat by resting in burrows. Larger mammals, such as kangaroos and camels, have few opportunities to avoid the rigorous environment. The camel is, perhaps, the best adapted of the large mammals; its legendary survival skill is largely due to the way it can dissipate body heat by maintaining good blood circulation, even when dehydrated.

TUNDRA BIOME

In the *Arctic tundra*, plants grow during the long days of the brief summer and then endure the long (or continuous) darkness of winter. In the summer months, air temperature rises above freezing, and a shallow layer of ground ice thaws at the surface. Across much of the tundra climate region, the mean temperature of the warmest month is generally about 10 °C, but winters are long and cold with temperatures commonly falling to between −30 and −40 °C (see Chapter 10). The summer warmth is not sufficient to thaw the permafrost below a general depth of about 1 m, and the frozen subsoil restricts meltwater drainage (see Chapter 18). This creates a saturated environment over wide areas for at least part of the growing season.

Because plant remains decay slowly in the cold meltwater, layers of organic matter can build up in the moist ground. Frost in the soil fractures and breaks roots, thus restricting the size of tundra plants. In winter, wind-driven snow and extreme cold also injure plant parts that project above the snow.

Arctic tundra grows over a wide range of latitudes. Its extreme northern limit is at 85.5° N. The decrease in summer warmth northward is marked by progressive changes in the vegetation. In North America, the most southerly division is known as the *low arctic*, and is characterized by a continuous plant cover in all but the driest or most exposed sites. Northward, this division merges with the polar semi-desert of the *middle arctic*. Beyond this is the *high arctic*—a polar "desert" where vascular plants are restricted to a few favourable sites.

The rolling terrain of the low arctic tundra supports a thin cover of short grasses and sedges (Figure 22.20a). Small shrubs, such as crowberry and Labrador tea, may grow to heights of 15 to 30 cm, and mosses and lichens form a discontinuous ground

22.20 **ARCTIC TUNDRA** (a) Short grasses and sedges dominate much of the tundra landscape. The patch of tall grass seen here has developed in response to higher soil nitrogen levels around an animal burrow. (b) The dwarf shrub arctic heather (*Cassiope tetragona*) survives only where it is protected by snow during the severe winter. (c) The high arctic landscape affords little opportunity for plant growth, although purple saxifrage (*Saxifraga oppositifolia*) is well adapted to these conditions. (d) In extreme locations, only lichens survive.

cover. Taller dwarf birches and willows form scrub communities in sheltered sites. In drier upland sites, mosses and lichens are more conspicuous, and patches of hardy, drought-resistant evergreen shrubs, such as arctic heather, are present (Figure 22.20b). The distribution of the shrubs is dependent on the depth and duration of the winter snowpack. In the most exposed sites, the shrubs are replaced by low-growing cushion plants, such as moss campion and purple saxifrage. Cotton grass is common in moist locations.

Similar plants grow in the middle arctic, but their distribution and productivity are more limited and the vegetation cover is much more open. In the high arctic, the vegetation cover is sparse (Figure 22.20c). Small mats of Arctic willow take advantage of the warmer temperatures near the ground surface, and patches of heath occasionally grow on warmer slopes. Lichens are common in exposed areas (Figure 22.20d).

Tundra also occurs in the southern hemisphere. In the south polar region, tundra is limited to ice-free areas in the Antarctic peninsula and rocky coastlines. Only two species of flowering plants are native to Antarctica, although higher plant diversity occurs in the sub-antarctic islands, where the characteristic cover is predominantly tussock grasses.

Alpine tundra is found at high elevations, above the limit of tree growth and below the vegetation-free zone of bare rock and perpetual snow (Figure 22.21a). In North America, alpine tundra occurs discontinuously from Alaska to Mexico and less extensively in the mountains of eastern Canada. The elevation of the tree line is related to temperature conditions during the growing season. For example, the tree line is some 500 to 600 m higher in the Rockies than in the Pacific Coast Ranges because summer temperatures in the mountains are higher inland than on the coast. The density of the montane forest cover decreases as it approaches the tree line, and the trees become arranged in clumps known as tree islands (Figure 22.21b). Still higher, near the limit of tree growth, harsh winter conditions cause the trees to develop misshapen flagged krummholz forms, mostly because blowing ice crystals abrade the branches on the upwind side of the trunk (see Figure 21.4).

(a)

(b)

22.21 ALPINE TUNDRA (a) Above the tree line, mostly sedges, grasses, and perennial herbs replace woody plants. (b) At higher elevations in the Rockies, the closed coniferous forests are replaced by tree islands mainly consisting of Engelmann spruce and subalpine fir.

Many alpine tundra species also grow in the Arctic, but some species are endemic to specific mountain areas. This is especially the case in Europe and Asia where mountain ranges, such as the Alps, Pyrenees, and Carpathians, are separated geographically. Vegetation patterns in alpine tundra are closely related to the complex mountain topography and develop in response to snow depth, timing of snowmelt, exposure, and drainage. At high elevations, solar radiation can be intense, and many of the plants must cope with periods of summer drought in combination with drying winds. The plants are mostly perennial herbs and dwarf shrubs, with grasses, sedges, and mosses found in wetter locations. Vegetative reproduction becomes increasingly important as growing conditions deteriorate, and many species spread by rhizomes and runners. However, flowering is generally conspicuous in the alpine tundra, more so than in the Arctic. Some species are pollinated by the wind, but most are pollinated by insects that spend their days in the alpine meadows and then descend into the warmer valleys at night. These excursions are facilitated by mountain and valley breezes that characterize diurnal air movements in the mountains (see Chapter 7).

The alpine flora of tropical mountains is very distinctive. The tree line occurs at 3,500 to 4,000 m at low latitudes, with the upper limit of the alpine zone at about 4,800 m. The tropical alpine zone experiences marked diurnal temperature fluctuations, with strong daytime heating and rapid heat loss at night. Tropical alpine vegetation consists mainly of tussock grasses and small-leaf shrubs, but the most distinctive species are the giant rosette plants that grow on the high volcanoes in East Africa and in the Andes Mountains (Figure 22.22).

As is often the case in particularly dynamic environments, species diversity in the tundra is low, but the number of individuals is high. Among the animals, vast herds of caribou (in North America) and reindeer (in Eurasia) roam the tundra, lightly grazing on the lichens and plants. A smaller number of muskox are also primary consumers of the tundra vegetation. Wolves and wolverines, as well as arctic foxes and polar bears, are predators, although some of these animals may feed directly on plants as well. Among the smaller mammals, snowshoe rabbits and lemmings are important herbivores. Invertebrates are scarce in the tundra, except for a small number of insect species. Black flies, deerflies, mosquitoes, and "no-see-ums" (tiny biting midges) are all abundant and can make July on the tundra most uncomfortable. Reptiles and amphibians are rare. The boggy tundra, however, presents an ideal summer environment for many migratory birds such as waterfowl, sandpipers, and plovers.

The food web of the tundra ecosystem is simple and direct. The important producer is reindeer moss, the lichen *Cladonia*

22.22 GIANT ROSETTE PLANTS OF TROPICAL ALPINE TUNDRA The giant groundsel (*Senecio keniodendron*), shown here near Mount Kenya, can reach a height of 5 m with a 1-m long spike of yellow flowers.

rangifera. In addition to caribou and reindeer, lemmings, ptarmigan (arctic grouse), and snowshoe rabbits are lichen grazers. During the summer, the abundant insects provide food for the migratory waterfowl populations.

CLIMATIC GRADIENTS AND VEGETATION TYPES

Climate is an important control of vegetation, especially at the global scale represented by the major formation classes. Several general classification schemes have been devised on this basis. In most cases, global vegetation patterns are examined in terms of temperature and precipitation gradients, with the dominant growth forms, such as trees, shrubs, and grasses, shown in relation to these changing climatic parameters (Figure 22.23). As vegetation adjusts to climate changes with latitude or longitude, corresponding transformations also occur in soils.

The manner in which vegetation changes with latitude and longitude is illustrated by the three transects shown in Figure 22.24. Note, for these transects the effects of highland regions on climate and vegetation are not considered. The upper transect runs through Africa from the equator to the Tropic of Cancer. Across this region, climate ranges through all four low-latitude climates: wet equatorial, monsoon and trade-wind coastal, wet–dry tropical, and dry tropical. Vegetation changes accordingly from equatorial rain forest, savanna woodland, and savanna grassland, to tropical scrub and tropical desert.

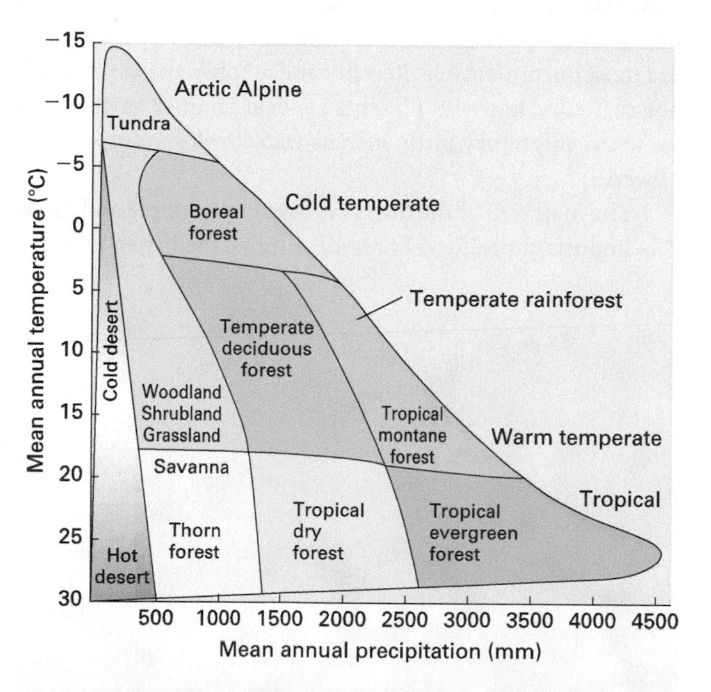

22.23 CLASSIFICATION OF BIOME TYPES ACCORDING TO AVERAGE TEMPERATURE AND ANNUAL PRECIPITATION Average temperature and annual precipitation provide only a very generalized assessment of bioclimatic conditions, since temperatures during extreme seasons and seasonal distribution of precipitation can also be important. Within the red-dashed line, expected climatic patterns can be strongly influenced by other factors, such as fire and grazing. Note that the temperature axis is reversed. (After Wittaker, 1970.)

The middle transect extends from the Tropic of Cancer to the Arctic Circle through Africa and Eurasia. Climates include many of the mid- and high-latitude types: dry subtropical, Mediterranean, moist continental, boreal forest, and tundra. The vegetation cover ranges from tropical desert, through subtropical steppe, to sclerophyll forest in the Mediterranean region. Further north is the midlatitude deciduous forest that transforms into boreal needleleaf forest, subarctic woodland, and finally tundra.

The lower transect runs across the United States, from Nevada to Ohio. This transect begins in the west with the dry midlatitude climate, and progresses to moist continental as precipitation gradually increases eastward. The vegetation changes from midlatitude desert and steppe to short-grass prairie, tallgrass prairie, and midlatitude deciduous forest.

The changes in climate and vegetation along these transects are largely gradational rather than abrupt. Yet, the global maps of both vegetation and climate show distinct boundaries from one region to the next. Which is correct? The changes shown on the transects are more accurate. Maps must necessarily have boundaries to communicate information. However, climate and vegetation have no specific boundaries. Instead, they are classified into recognized classes for convenience in assessing their spatial patterns. When studying any map of natural features, keep in mind that boundaries are always approximate and gradational. For vegetation, these transitional zones are called *ecotones*.

ALTITUDE ZONES OF VEGETATION

Generally, temperature decreases and precipitation increases with elevation, and this is reflected in a systematic change in the vegetation cover. The changes that occur with altitude are similar to those related to latitude, and many of the vegetation forms seen in the major biomes recur in the bands of vegetation associated with mountain areas. The concept of altitudinal zonation originated from studies carried out in Arizona, where, over a short distance, elevation ranges from about 700 m above sea level at the bottom of the Grand Canyon to over 3,800 m at the summit of the San Francisco Peaks district, about 100 km away (Figure 22.25). Annual precipitation ranges from 120 to 250 mm in the desert scrub vegetation type, to 800 to 900 mm in the Engelmann spruce forest that grows up to the tree line. Because corresponding trends were noted between plants and animals, the concept of **life zones** was proposed. The life zones that occur along geographically restricted elevational gradients are therefore analogous to the biomes that have developed in response to environmental gradients determined by latitude and longitude.

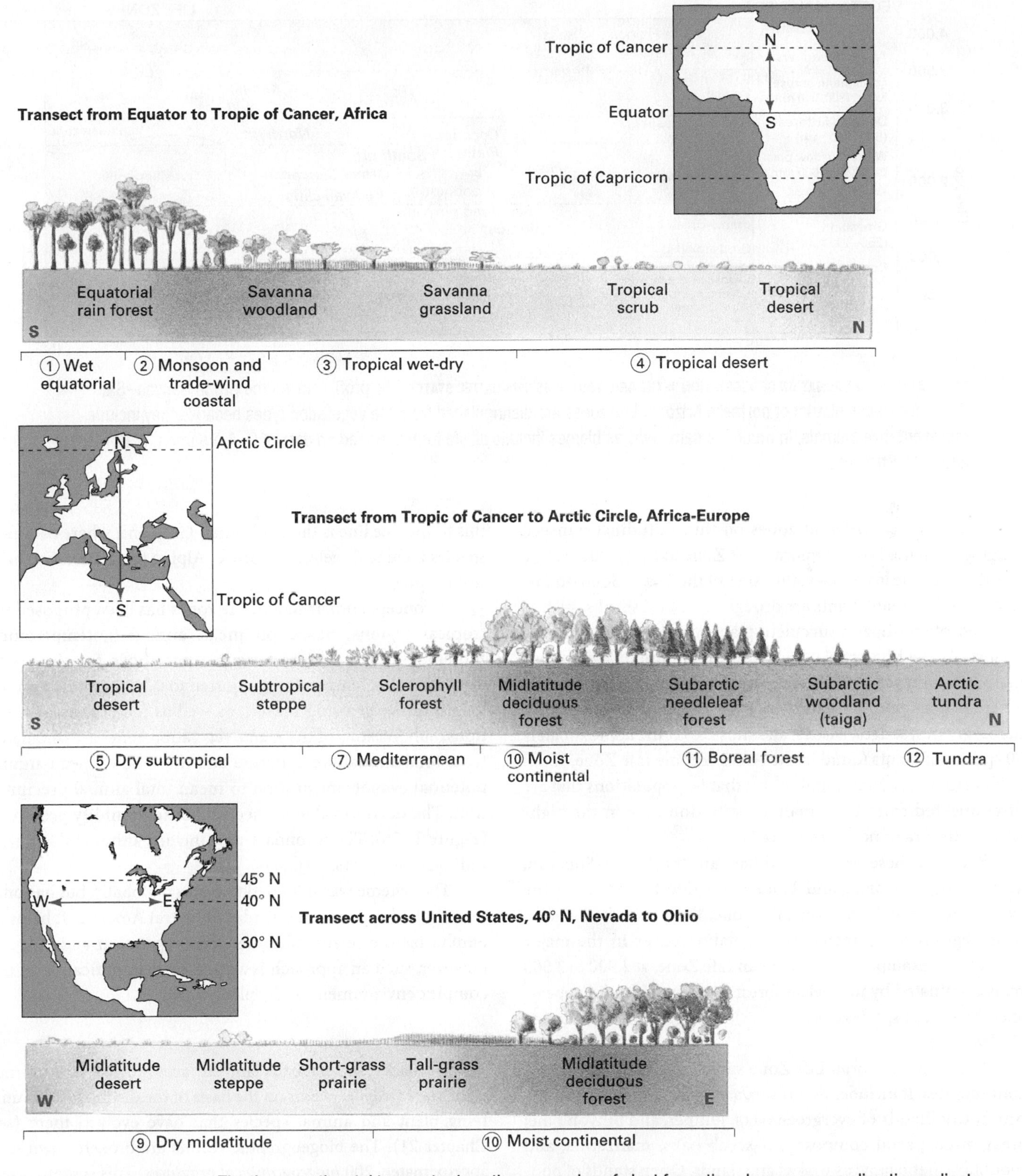

22.24 VEGETATION TRANSECTS The three continental transects show the succession of plant formation classes across climatic gradients.

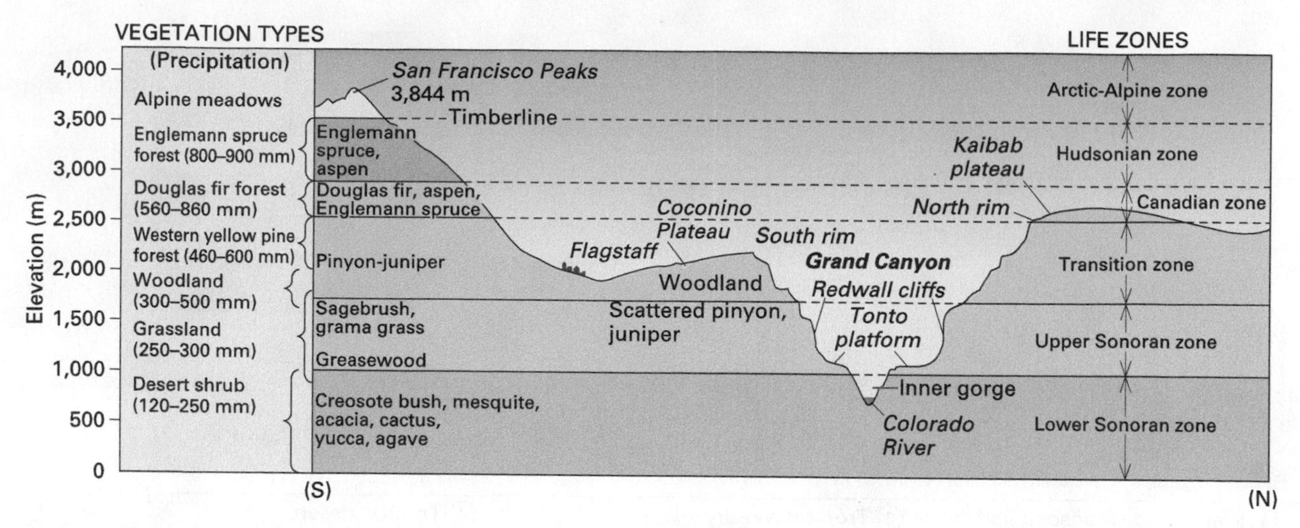

22.25 ALTITUDINAL ZONATION OF VEGETATION IN THE ARID SOUTHWESTERN UNITED STATES The profile shows the Grand Canyon–San Francisco Peaks district of northern Arizona. Life zones are distinguished from the vegetation types because they include representative animals, in much the same way as biomes include all life forms. (Based on data of G. A. Pearson, C. H. Merriam, and A. N. Strahler.)

There are six different zones on this altitudinal transect, ranging from the Lower Sonoran Life Zone to the Arctic-Alpine Life Zone. In the lowest elevation sites of the Lower Sonoran Life Zone, the dominant plants are drought-resistant shrubs, such as the creosote bush, and succulent plants typical of the Sonoran Desert. At the highest elevations on the mountain peaks, two main habitat types are found: rocky outcrops and boulder fields where lichens predominate, and areas of alpine meadow with herbs, grasses, sedges, rushes, and mosses. Of the approximately 50 species of plants found in the Arctic-Alpine Life Zone, about half occur as disjunct populations; that is, populations that are disconnected from their main distribution area in the high-latitude tundra of northern Canada.

Between these extreme habitats are the Upper Sonoran, Transition, Canadian, and Hudsonian Life Zones. With the exception of the Transition Life Zone, they take their names from regions of characteristic vegetation cover in the major biomes. For example, the Hudsonian Life Zone, at 2,900 to 3,500 m, is dominated by needleleaf forest that resembles (in appearance, but not in species composition) the boreal forest bordering Hudson Bay.

The Upper Sonoran Life Zone supports a diversity of plant communities. It includes desert scrub, mainly composed of sagebrush, woodlands of evergreen oaks, juniper, and pinyon pine; areas of chaparral composed of scrub oaks, manzanita, and mountain mahogany; as well as grasslands. Open stands of ponderosa pine forest are representative of the Transition Life Zone. In the Canadian Life Zone, the composition of the forests changes to Douglas fir and white fir, interspersed with deciduous trees, such as Gambel oak and trembling aspen. Rising above this to the tree line is the Hudsonian Life Zone, where common species include Engelmann spruce, Alpine fir, and ancient bristlecone pines.

A concept similar to altitude zones has been proposed for tropical regions, based on mean annual biotemperature. Biotemperature refers to temperatures above freezing, with below freezing temperatures adjusted to 0 °C; it is based on the length of the growing season, as well as temperature conditions. Subdivisions of the major vegetation zones are calculated from total annual precipitation and the ratio of mean annual potential evapotranspiration to mean total annual precipitation. The derived value is used to define humidity provinces (Figure 22.26). The secondary axes include altitudinal and latitudinal descriptors.

The scheme was intended to be used globally, but has only seen limited application outside of Central America. It helps to emphasize the relationship between vegetation and climate; however, such an approach is a gross oversimplification of the complex environment of the plant world.

ECOZONES

At the broadest scale, Earth's land surfaces are now divided into eight *biogeographic realms* on the basis of the distinctive and unifying plant and animal species that have evolved there (see Chapter 21). The biogeographic realms are broken down into approximately 200 *biogeographical provinces*. This system, originally developed by Udvardy in 1975 as part of UNESCO's Man and the Biosphere Programme, is now used by the United Nations Environment Programme World Conservation Monitoring Centre (UNEP-WCMC) for establishing World Heritage Sites.

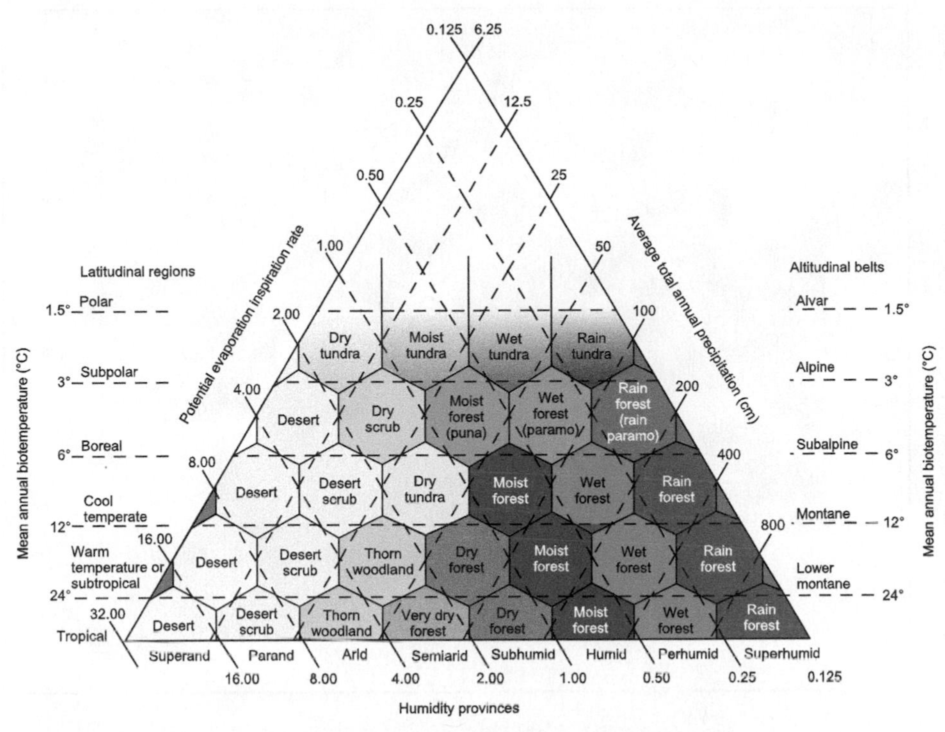

22.26 THE HOLDRIDGE SYSTEM OF LIFE ZONES Thirty community types are defined according to growing season temperatures and available moisture. In this system, a potential evapotranspiration (PE) ratio of 1.00 indicates that potential moisture demand equals precipitation; the climate becomes more humid as PE ratio drops below 1.00 and more arid as PE ratio increases above 1.00.

Under the World Heritage Convention, designation as a World Heritage Site is made on the basis of:

- natural features consisting of physical and biological formations or groups of such formations, which are of outstanding universal value from the aesthetic or scientific point of view;
- geological and physiographical formations and precisely delineated areas that constitute the habitat of threatened species of animals and plants of outstanding universal value from the point of view of science or conservation; and
- natural sites or precisely delineated natural areas of outstanding universal value from the point of view of science, conservation, or natural beauty.

For a site to be included on the World Heritage List, the World Heritage Committee must find that it has "outstanding universal value" (OUV). The draft revised Convention Operational Guidelines define OUV as:

> *"[A piece of] cultural and/or natural significance which is so exceptional as to transcend national boundaries and to be of common importance for present and future generations of all humanity. As such, the permanent protection of this heritage is of the highest importance to the international community as a whole."*
>
> (IUCN Evaluation of World Heritage Nominations, 2009, p. 3)

The term *ecozone* has recently come into use, which similarly is differentiated on the basis of distinctive plants and animals in association with a general land form type associated with specific climate conditions. The National Ecological Framework for Canada (Ecological Stratification Working Group 1996) distinguishes 15 terrestrial and five marine ecozones. Ecozones are part of a hierarchical classification with progressively finer details incorporated at the *ecoregion* and *ecodistrict* levels; over 200 ecoregions are listed in Canada. A summary of Canada's ecozones, showing how each of the four great spheres in physical geography—lithosphere, atmosphere, hydrosphere, and biosphere—are reflected in the landscape, is presented below and provides a fitting conclusion to this book.

Terrestrial Ecozones of Canada

An ecozone is an area of the Earth's surface that has characteristic landforms and climate and is distinguished from other areas by its unique flora and fauna. This concept acknowledges that all species contribute to a complex dynamic system that is in equilibrium with its physical surroundings. Ecozones represent the highest level in the hierarchical ecoregion approach to landscape classification. The ecozone level assesses ecological diversity at the subcontinental level by describing major ecosystems. Ecozones are large and very generalized.

Ecozones are subdivided into ecoregions on the basis of distinctive communities associated with particular climatic conditions or landforms. Ecoregions in turn are subdivided into ecodistricts that reflect differences in the physical landscape, such as soil, relief, and water availability, which can be related to locally discrete assemblages of plant and animals. The following section describes the 15 terrestrial ecozones defined by the National Ecological Framework for Canada (Ecological Stratification Working Group 1996). The distribution of the Canadian ecozones is shown in Figure 22.27; the descriptions summarize information presented on the Parks Canada Terrestrial Ecozones of Canada website (www.pc.gc.ca).

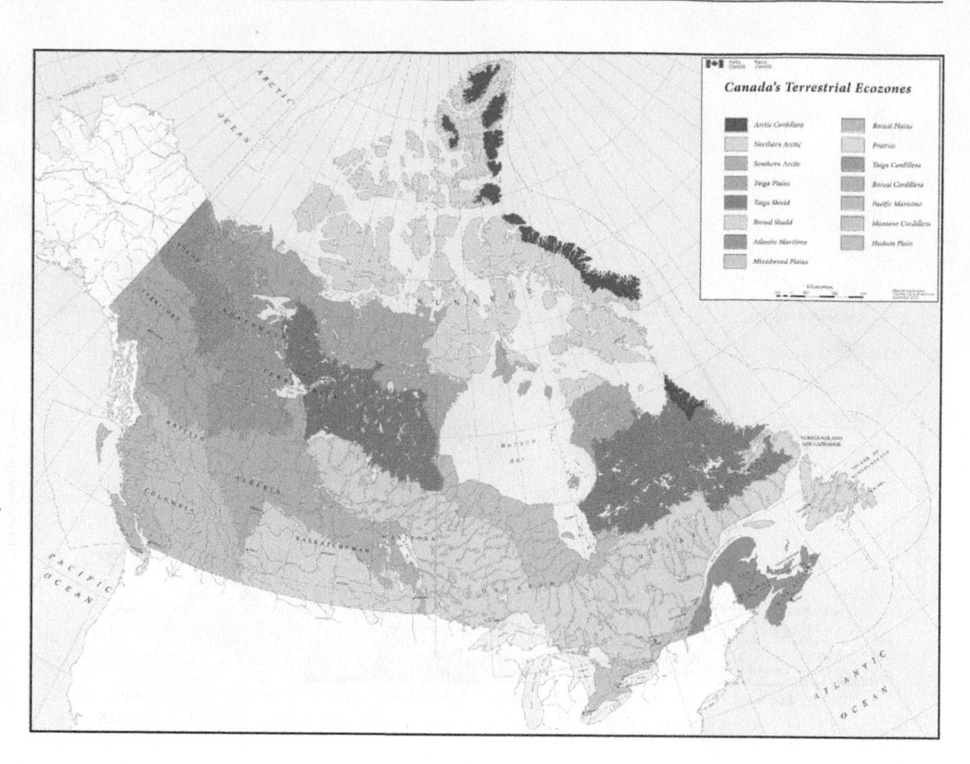

22.27 **TERRESTRIAL ECOZONES OF CANADA**

1. **Arctic Cordillera Ecozone**
The climate is very cold and dry. Mean daily January temperatures range from −25.5 °C in the south to −35 °C in the north, and mean daily July temperatures are about 5 °C. Precipitation generally varies from 200 to 300 mm, with higher totals on exposed eastern slopes and at lower latitudes. Vegetation is sparse and over much of the region is largely absent due to the permanent ice and snow. Apart from polar bears in coastal areas, there are few terrestrial mammals. Birds species such as the northern fulmar and snow bunting are summer residents.

2. **Northern Arctic Ecozone**
This region consists of lowland plains covered with glacial moraine, high plateaux, and rocky hills. Mean daily January temperatures range from −30 °C to −35 °C, with daily July temperatures between 5 °C and 10 °C. Annual precipitation ranges from 100 to 200 mm. Herb and lichen form the dominant vegetation cover and provide food for a range of herbivores, including the barren-ground caribou, muskox, arctic hare, and lemming. These in turn are preyed upon by wolf and arctic fox. Birds are represented by the loon, goose, ptarmigan, and snowy owl.
3. **Southern Arctic Ecozone**
The landscape throughout this region is predominantly one of strongly rolling lowland plains, mantled by glacial moraines. The climate is typically arctic with short, cool summers and long, cold winters. Mean daily July temperatures are about 10 °C, with the mean daily January temperature around −30 °C. Precipitation varies from 200 mm in northern areas to 400 mm in the south. The dominant vegetation is shrub tundra consisting of arctic heather, dwarf birch, and willows with grasses, sedges, and lichens. This region is an important calving ground for caribou; other mammals include moose, muskox, wolf, arctic fox, grizzly bear, and arctic hare. The area is a major breeding and nesting ground for birds, such as loon, snow goose, gyrfalcon, ptarmigan, snowy owl, and snow bunting.
4. **Atlantic Maritime Ecozone**
Topographically this ecozone area ranges from the interior Appalachian upland to the Atlantic coastal plain. It experiences a cool, moist maritime climate, with mean annual precipitation ranging from 1,000 mm inland to 1,425 mm along the coast. Mean daily January temperatures range between −2.5 °C in coastal regions and −10 °C inland; the mean daily July temperature is about 18 °C. The dominant vegetation is mixed forest, consisting of coniferous and deciduous species, such as red spruce, balsam fir, white pine, yellow birch, and sugar maple. Representative mammals include the white-tailed deer, moose, black bear, and raccoon. The blue jay, eastern bluebird, and rose-breasted grosbeak are distinctive bird species.
5. **Boreal Cordillera Ecozone**
This is a physiographically diverse ecozone including high mountain peaks, extensive plateaus, and intermontane plains. Winters are long and cold, with mean January temperatures ranging from −15 to −27 °C. Summers are short and warm, with the mean daily July temperature typically around 12 to 15 °C. Precipitation is about 400 mm per year in the intermontane plateau areas, but increases to between 1,000 and 1,500 mm in the mountains. The vegetation is typical of the boreal forest, with white spruce, black spruce, alpine fir, lodgepole pine, trembling aspen, balsam poplar, and white birch present in varying proportions, depending on elevation. Characteristic mammals include the woodland caribou, moose, mountain goat, black bear, grizzly bear, and lynx. Ptarmigan and spruce grouse are distinctive birds.
6. **Boreal Plains Ecozone**
This a heavily glaciated region of subdued relief. It experiences a moist climate, with annual precipitation totalling 400 to 500 mm. Winters are cold, with mean daily January

temperatures ranging from −17.5 to −22.5 °C. Summers are moderately warm, with mean daily July temperatures ranging from 12.5 to 17.5 °C. White spruce, black spruce, and jack pine are the dominant conifer species, while broadleaf trees, particularly white birch, trembling aspen, and balsam poplar are also common. Characteristic mammals include the woodland caribou, mule deer, and coyote. Numerous birds are represented, including the great horned owl, blue jay, and brown-headed cowbird.

7. **Boreal Shield Ecozone**
   In this extensive, heavily glaciated region of the Canadian Shield, the topography provides a diverse landscape of outcropping bedrock, glacial moraines, and sandy outwash plains, interspersed with lakes and poorly drained muskeg. Precipitation ranges from 400 mm in westerly areas and increases to 1,000 mm eastward in response to moist airflows from Hudson Bay and the Atlantic Ocean. Mean daily January temperatures are between −10 and −20 °C, with daily July temperatures averaging 15 to 18 °C. White spruce and balsam fir are the dominant conifer species on moist, well-drained soils, black spruce and tamarack occur in wetter areas, with jack pine often forming almost pure stands on drier, sandy soils. Broadleaf trees, particularly white birch, trembling aspen, and balsam poplar, are characterisitically associated with disturbed sites, with deciduous forests of sugar maple and beech increasingly dominant in more southerly regions. Representative mammals include the woodland caribou, white-tailed deer, moose, black bear, lynx, bobcat, and raccoon. The boreal owl, great horned owl, blue jay, and pileated woodpecker are found throughout this forested ecozone.
8. **Hudson Plains Ecozone**
   This low plain, which rises to an elevation of about 180 m along its southern border, is mantled with glacial and marine sediments. Drainage is poor over much of the region, and the low-lying areas of the landscape are characterized by marshes, peat bogs, and numerous small lakes. Broad, gently sloping tidal mud flats occur along the coast of Hudson Bay. The climate of this low relief ecozone is strongly influenced by the cool, moist air masses from Hudson Bay. Mean daily temperatures in January are about −19 °C, rising to 12 to 16 °C in July. Annual precipitation varies from 400 to 700 mm. Woodland caribou, moose, black bear, and polar bear are found in this ecozone. Representative bird species include the snow goose, Canada goose, and peregrine falcon.
9. **Prairie Ecozone**
   This is a region of broadly rolling topography, mainly associated with unsorted glacial tills. It experiences a moist, continental climate. Mean daily January temperatures are around −22.5 to −25 °C, with mean daily July temperatures ranging from 15 to 17.5 °C. Annual precipitation varies from about 350 mm in the west to 550 mm in the east; summer droughts are not uncommon. Characteristic mammals of the Prairie Ecozone include the mule deer, white-tailed deer, pronghorn antelope, coyote, and skunk. Bird species include the ferruginous hawk, prairie chicken, and sage grouse, while the burrowing owl has been reintroduced to some areas.
10. **Mixed-wood Plains Ecozone**
    This region has a low relief associated with glacial tills and lacustrine deposits. Extending east from Lake Huron through the St. Lawrence Valley, climatic conditions in this ecozone reflect the proximity of the Great Lakes. Summers are warm and humid, with mean July temperatures ranging from 18 to 22 °C. Winters are cool and damp, with mean daily January temperatures of −3 to −12 °C. The area receives between 720 mm to 1,000 mm of precipitation annually. The forest in the northern region is mainly a mixture of coniferous and deciduous trees, with species including white pine, red pine, eastern hemlock, oak, maple, and birch. Characteristic mammals include the white-tailed deer, raccoon, skunk, eastern chipmunk, and the grey and black squirrel. Representative birds are the great blue heron, red-shouldered hawk, whippoorwill, red-headed woodpecker, blue jay, eastern bluebird, and oriole.
11. **Montane Cordillera Ecozone**
    Physiographically, this region consist of rugged mountain ranges separated by intermontane valleys in the south, which broaden out northward to form more extensive plains. The diverse topography is reflected in a variety of climatic conditions, but generally it experiences moderately long, cold winters and warm summers. Annual precipitation is about 800 mm over the northern section, 1,200 mm along the British Columbia–Alberta border, and decreasing to 500 mm in the southern interior of British Columbia. The mean daily January temperature ranges from −7.5 to −17.5 °C, with the mean daily July temperature ranging from 13 to 18 °C. The vegetation is equally variable, ranging from alpine lichen and shrub associations, subalpine forests of Engelmann spruce and alpine fir, with lodgepole pine characteristically found on the lower slopes. Ponderosa pine and grasses give the drier valleys a distinctive parkland appearance. Various mammals are found in these diverse habitats, including bighorn sheep and hoary marmot in higher sites, elk, grizzly bear, and black bear at mid-elevations, and deer, coyote, skunk, and badger favouring more open conditions in the valleys. Typical bird species in the forests include owls, blue grouse, Stellar's jay, and woodpeckers.
12. **Pacific Maritime Ecozone**
    This coastal region consists of glaciated mountains cut through by numerous fjords and glacial valleys, and

bordered by a narrow coastal plain. Because of its maritime location, the climate in this ecozone is characteristically mild and wet. Average July temperatures vary from 12 to 18 °C. Winters are comparatively warm, with average January temperatures of 4 to 6 °C, although conditions are quite variable because of the range of elevation. Topography also affects annual precipitation patterns, with as little as 600 mm of precipitation in southern coastal regions, increasing to 3,000 mm in the mountains. The western coastal forest is composed mostly of western red cedar, western hemlock, Douglas fir, mountain hemlock, amabilis fir, Sitka spruce, and alder. Characteristic mammals include the white-tailed deer, black bear, grizzly bear, and mountain lion. Bird species are distributed according to major habitat conditions; numerous seabirds and shorebirds are found in coastal areas, including the black oystercatcher, black puffin cormorant, and sandpiper. The marshes are occupied by waterfowl, such as the heron, Canada goose, trumpeter swan, and mallard. The forest birds include quail, grouse, and woodpecker.

13. **Taiga Cordillera Ecozone**
This is predominantly a region of steep, mountainous topography dissected by narrow valleys, but also includes the Arctic coast of the Yukon. Away from the coast, the climate is generally dry and cold. Annual precipitation averages about 300 mm, with mean daily temperature ranging from −25 to −30 °C in January, rising to 12 to 15 °C in July. In the north, the dominant vegetation includes tundra species such as arctic heather, bearberry, and saxifrage; in the south, this is replaced by forests of spruce, lodgepole pine, alpine fir, birch, and aspen, with alpine tundra at higher elevations. Representative mammals include caribou, moose, mountain goat, black bear, grizzly bear, and lynx. Owls, eagles, hawks, and gyrfalcon are found throughout the region, along with various songbirds, waterfowl, grouse, and ptarmigan.

14. **Taiga Plains Ecozone**
Low rolling terrain in this heavily glaciated region, together with permafrost, restrict drainage, resulting in extensive areas of muskeg and numerous lakes. The region experiences short warm summers with mean July temperatures decreasing with latitude from about 14 °C in the south to 10 °C in the north. Winters are long and cold, with mean daily January temperatures ranging from −22.5 to −35 °C. Annual precipitation varies from about 400 mm in the south to about 200 mm in the north. Dwarf birch, Labrador tea, willow, bearberry, cranberry, sphagnum moss, and sedges are associated with the arctic tundra environment. Upland areas and southerly districts, which are typically better drained and warmer, support mixed-wood forests of white spruce, black spruce, tamarack, white birch, trembling aspen, balsam poplar, and lodgepole pine. Characteristic mammals include moose, woodland caribou, bison, wolf, black bear, marten, and lynx. Some common birds include the bald eagle, peregrine falcon and osprey, numerous waterfowl, such as loons, geese, swans, and ducks, and forest species like the grey jay, raven, and chickadee.

15. **Taiga Shield Ecozone**
Flat to gently rolling terrain has developed on heavily glaciated shield rocks, which are either exposed in outcrops or thinly covered by soils. Much of the area is waterlogged, so muskeg and lakes dominate the landscape. The climate is subarctic continental, but regional differences occur because this ecozone is separated by Hudson Bay. Thus, mean daily January temperatures range from −17.5 °C in the east to −27.5 °C in the west, with corresponding mean daily July temperatures of 7.5 to 17.5 °C. Precipitation is generally less than 200 mm in the west, increasing to 1,000 mm along the coast of Labrador. The northern boundary of this ecozone coincides approximately with the poleward limit of tree growth. Here, open lichen woodland merges into arctic tundra. Southward stands of mainly black spruce, tamarack, alder, willow, and tamarack are found in the fens and bogs, with open, mixed-wood associations of white spruce, balsam fir, and trembling aspen in drier sites. Characteristic mammals include caribou, moose, wolf, snowshoe hare, arctic fox, black bear, grizzly bear, and lynx. The bald eagle is a representative bird of the Taiga Shield with ospreys, loons, ducks, and gulls in coastal areas. Grouse and ptarmigan are found in the forests, with warblers and many other songbirds resident in summer.

A Look Back

This chapter concludes our synopsis of physical geography as the science of global environments. The overview began with an examination of weather and climate systems, powered largely by solar energy. This was followed by a description of the systems of the solid Earth, and the dramatic forces associated with it. The primary land forms created by tectonic forces are sculpted more finely into a diversity of landscapes by geomorphic agents that derive their energy from gravity and solar power. Lastly, we covered systems of soils and ecosystems, grouped together by their dependence on climate and substrate, and powered largely by solar energy.

As a discipline, physical geography stresses the interrelationships among its many component sciences and focuses on the environment in an integrated way. Originally, physical geography emphasized natural processes, but today few of these are completely immune from the impacts of human activity. As the human population continues to grow, its influence on the Earth will become more and more profound. This book provides information on natural systems as the benchmark against which human impact can be evaluated.

CHAPTER SUMMARY

- Natural vegetation is a plant cover that develops with little or no human interference. Although much vegetation appears to be in a natural state, humans influence the vegetation cover through fire suppression and the introduction of new species.
- The life form of a plant refers to its physical structure, size, and shape. Life forms include *trees*, *shrubs*, *lianas*, *herbs*, and *lichens*.
- The largest unit of terrestrial ecosystems is the biome: forest, grassland, savanna, desert, and tundra. The forest biome includes a number of important forest formation classes. The *low-latitude rain forest* exhibits a dense canopy and open floor with a large number of species. *Subtropical evergreen forest* occurs in *broadleaf* and *needleleaf* forms in the moist subtropical climate. The *monsoon forest* is largely deciduous, with most species shedding their leaves after the wet season.
- *Midlatitude deciduous forest* is associated with the moist continental climate. Its species shed their leaves before the cold season. The *needleleaf forest* consists largely of evergreen conifers. It includes the *coastal forest* of the Pacific Northwest, the *boreal forest* of high latitudes, and needleleaf mountain forests. The *sclerophyll forest* is composed of trees with small, hard, leathery leaves, and is found in the Mediterranean climate region.
- The savanna biome consists of widely spaced trees with an understory, often of grasses. Dry-season fire is frequent in the savanna biome, limiting the number of trees and encouraging the growth of grasses.
- The grassland biome of midlatitude regions includes the *tall-grass prairie* in moister environments and *short-grass prairie*, or steppe, in semi-arid areas.
- Vegetation of the desert biome ranges from thorny shrubs and small trees to dry desert vegetation composed of drought-adapted species.
- Tundra biome vegetation consists predominantly of low herbs that are adapted to the severe drying cold experienced at high latitudes.
- Since climate changes with altitude, vegetation typically occurs in altitudinal life zones. Climate also changes gradually with latitude, so biomes normally merge through ecozones rather than across abrupt boundaries.

KEY TERMS

biome	grassland biome	natural vegetation	terrestrial ecosystem
desert biome	life form	savanna biome	tundra biome
forest biome	life zones	steppe	woodland

REVIEW QUESTIONS

1. What is natural vegetation? How do humans influence vegetation?
2. Plant geographers describe vegetation by its overall structure and by the life forms of individual plants. Define and differentiate the following terms: forest, woodland, tree, shrub, herb, liana, perennial, deciduous, evergreen, broadleaf, and needleleaf.
3. What are the five main biome types that ecologists and biogeographers recognize? Describe each briefly.
4. Low-latitude rain forests occupy a large region of the Earth's land surface. What are the characteristics of these forests? Include forest structure, types of plants, diversity, and climate.
5. The monsoon forest and midlatitude deciduous forest are both deciduous, but for different reasons. Compare the characteristics of these two formation classes and their climates.
6. Subtropical broadleaf evergreen forest and tall-grass prairie are two vegetation formation classes that have been greatly altered by human activities. How was this done and what impact has it had on these ecosystems?
7. Distinguish between the various types of needleleaf forest. What characteristics do they share? How are they different? How do their climates compare?
8. What general characteristics do woodland and scrub types associated with the Mediterranean climate exhibit? How have these vegetation types adapted to the Mediterranean environment?
9. How do traditional agricultural practices in the low-latitude rain forest compare to present-day practices? What are the implications for the rain forest environment?
10. What are the effects of large-scale clear-cutting on the rain forest environment?

11. Describe the formation classes of the savanna biome. Where is this biome found and in what climate types? What role does fire play in the savanna biome?
12. Compare the two formation classes of the grassland biome. How do their climates differ?
13. What is meant by the term "mixed prairie"? Where is it found? What are its characteristics?
14. Describe the vegetation types of the desert biome.
15. What are the features of arctic and alpine tundra? How does the cold tundra climate influence the vegetation cover? How does alpine tundra differ from arctic tundra?
16. How does elevation influence vegetation? Provide an example of how vegetation zones are related to elevation. Compare altitudinal zonation with vegetation patterns related to latitude.

VISUALIZATION EXERCISE

1. Forests often contain plants of many different life forms. Sketch a cross-section of a forest, including typical life forms, and label it.

ESSAY QUESTION

1. Figure 22.24 shows some vegetation transects along with associated climates. Construct and describe a similar transect starting at Calgary, then passing through Winnipeg, Sudbury, Montreal, and Halifax. Discuss how this vegetation pattern is related to climate.

ANSWERS TO WORKING IT OUT PROBLEMS

Chapter 2: Distances from Latitude and Longitude

1. If the two cities are separated by 44° latitude and each degree of latitude is equal to 111 km, then:

$$\text{Distance} = 44° \text{ lat} \times \frac{111\text{km}}{1° \text{lat}} = 4{,}884 \text{ km}$$

2. The two cities are on the 52° parallel. Therefore the number of kilometres equal to one degree longitude is:

$$\cos(52)° \times 111 \text{ km} = 0.62 \times 111 \text{ km} = 68.3 \text{ km}$$

Therefore, the distance between the two cities is:

$$\text{Distance} = (106° - 1°) \times 68.3 \text{ km} = 7{,}176 \text{ km}$$

3. To find the area of the map, convert 1° latitude and 1° longitude to kilometres. At the equator, both 1° latitude and 1° longitude are equal to 111 km. The size of the area at the equator is therefore 111 km × 111 km = 12,321 km^2. For the area near Churchill, 1° latitude is still 111 km; however, the number of kilometres equal to 1° longitude must be calculated:

$$\cos(60°) \times 111 \text{ km} = 0.5 \times 111 \text{ km} = 55.5 \text{ km}$$

The area is therefore 6,161 km^2.

Chapter 3: Radiation Laws

1. To calculate the flow rate of energy coming from the sun, use the Stefan-Blotzmann Law, which states that the flow rate is a function of the surface temperature of a blackbody. The sun's surface temperature is 5,950 K.

$$M = \sigma T^4 = (5.67 \times 10^{-8} \text{ W m}^{-2} \text{ K}^{-4}) \times (5{,}050 \text{ K})^4$$
$$M = \sigma T^4 = (5.67 \times 10^{-8} \text{ W m}^{-2} \text{ K}^{-4})\, 1.25 \times 10^{15} \text{ K}^4$$
$$M = 7.11 \times 10^7 \text{ W m}^{-2}$$

The wavelength of greatest radiance is given by Wien's Law:

$$\lambda_{max} = b / T = \frac{(2{,}892\ \mu\text{mK})}{(5{,}950 \text{ K})} = 0.49\ \mu\text{m}$$

2. Again, use the Stefan-Boltzmann Law here, using the surface temperature of the Earth in kelvin, not Celsius: 15.4 °C + 273.15 = 288.55 K.

$$M = \sigma T^4 = (5.67 \times 10^{-8} \text{ W m}^{-2} \text{ K}^{-4}) \times (288.55)^4$$
$$M = \sigma T^4 = (5.67 \times 10^{-8} \text{ W m}^{-2} \text{ K}^{-4})\, 6.9 \times 10^9 \text{ K}^4$$
$$M = 392 \text{ W m}^{-2}$$

The wavelength of maximum irradiance for the Earth's surface:

$$\lambda_{max} = b/T = \frac{(2{,}892\ \mu\text{mK})}{(288.15 \text{ K})} = 10\ \mu\text{m}$$

3. The ratio of the flow rate of energy emitted by the sun's surface to that of Earth's:

$$\frac{(M_{sun})}{(M_{Earth})} = \frac{(7.11 \times 10^7 \text{ W m}^{-2})}{(392 \text{ W m}^{-2})} = 1.8 \times 10^5$$

4. Since the radiation flow to Venus is 1.92 times greater than the radiation flow to Earth, the solar constant for Venus will be 1.92 × 1.37 kW m^{-2} = 2.63 kW m^{-2}. If the radius of Venus is 6,050 km, the area the planet presents to the sun will be that of a disk with area πr^2 = 3.14 × (6,050 km)^2 = 3.14 × 3.65 × 10^7 km^2 = 1.15 × 10^8 km^2 = 1.15 × 10^{14} m^2. The total energy flow intercepted is equal to Venus's solar constant times the area it presents to the sun, or = 2.63 kW m^{-2} ×1.15 × 10^{14} m^2 = 3.03 × 10^{14} kW. For outflows, 65 percent of the incoming solar radiation is directly reflected back, providing a short-wave radiation flow rate to space of 0.65 × 3.03 × 10^{14} kW = 1.97 × 10^{14} kW. The remaining portion, 35 percent, or 0.35 × 3.03 × 10^{14} kW = 1.06 × 10^{14} kW, flows outward to space as long-wave radiation emitted by the planet and its atmosphere.

Chapter 5: Pressure and Density in the Oceans and Atmosphere

1. Since P (kg m^{-2}) = 1,000 D (m), the pressure at the bottom of the diving pool will be 5,000 kg m^{-2}. Adding atmospheric pressure (1,000 kg m^{-2}) gives 6,000 kg m^{-2}. The fraction due to water is then 5/6, while the fraction due to the atmosphere is 1/6.
For the deep-sea diver, the pressure will be:

$$P \text{ (kg m}^{-2}) = 1{,}000\, D \text{ (m)} + 1{,}000 \text{ (for atmosphere)}$$
$$= (1{,}000 \times 100) + 1{,}000 = 101{,}000 \text{ kg m}^{-2}$$

The percentage contribution from the ocean water is 100,000/101,000 = 99% and for the atmosphere 1,000/101,000 = 1%.

2. For Yellowhead Pass, z = 1,110 m = 1.11 km. The pressure at the pass is the calculated as:

$$P_z = 101.4 \text{ kPa} \times [1 - (0.0226 \times 1.11 \text{ km})]^{5.26}$$
$$P_z = 101.4 \text{ kPa} \times [1 - 0.0251)]^{5.26}$$
$$P_z = 101.4 \text{ kPa} \times 0.975^{5.26}$$
$$P_z = 101.4 \text{ kPa} \times 0.875$$
$$P_z = 88.73 \text{ kPa}$$

For the summit of Mount Robson, Z = 3,954 m = 3.954 km. The pressure for the summit is then:

$$P_z = 101.4 \text{ kPa} \times [1 - (0.0226 \times 3.954 \text{ km})]^{5.26}$$
$$P_z = 62.0 \text{ kPa}$$

3. The equation for atmospheric pressure must be rearranged to find the altitude at which a pressure of 97.9 kPa would be found under normal conditions. Before the passage of the hurricane, the normal sea level pressure condition was 101.6 kPa.

$$P_z = 101.6\,\text{kPa} \times [1-(0.0226 \times z)]^{5.26}$$

$$\frac{P_z}{101.6\,\text{kPa}} = [1-(0.0226 \times z)]^{5.26}$$

$$\left[\frac{P_z}{101.6\,\text{kPa}}\right]^{\frac{1}{5.26}} = [1-(0.0226 \times z)]$$

$$0.0226 \times z = 1 - \left[\frac{P_z}{101.6\,\text{kPa}}\right]^{\frac{1}{5.26}}$$

$$z = \frac{1 - \left[\frac{P_z}{101.6\,\text{kPa}}\right]^{\frac{1}{5.26}}}{0.0226}$$

Now substitute the pressure value of 97.9 kPa for P_z:

$$z(km) = \frac{1 - \left(\frac{97.9\,\text{kPa}}{101.6\,\text{kPa}}\right)^{\frac{1}{5.26}}}{0.0226}$$

$$= \frac{1-(0.964)^{\frac{1}{5.26}}}{0.0226}$$

$$\frac{1-0.993}{0.0226} = 0.311\,\text{km} = 311\,\text{m}$$

Alternative Calculation:

$$P_z = 101.6\,\text{kPa} \times [1-(0.0226 \times z)]^{5.26}$$
$$97.9\,\text{kPa} = 101.6\,\text{kPa} \times [1-(0.0226 \times z)]^{5.26}$$
$$0.964 = [1-(0.0226 \times z)]$$
$$(0.964)^{\frac{1}{5.26}} = [1-(0.0226 \times z)]$$
$$0.993 = 1-(0.0226 \times z)$$
$$-0.007 = -0.0226 \times z$$
$$0.311\,\text{km} = z$$

Therefore, under normal conditions before the passage of the hurricane, a pressure of 97.9 kPa would be found at an altitude of 311 m.

Chapter 6: The Lifting Condensation Level

1.

$$H = 1{,}000 \times \frac{25-18}{8.2} = 1{,}000 \times \frac{7}{8.2} = 854\,\text{m}$$

$$T = 25 - \left(854 \times \frac{10}{1{,}000}\right) = 25 - 8.5 = 16.5\,°\text{C}$$

2. Equation (1) states:

$$T_0 - H \times R_{DRY} = T_{DEW} - H \times R_{DEW}$$

Placing terms with T on the left and terms with H on the right,

$$T_0 - T_{DEW} = H \times R_{DRY} - H \times R_{DEW}$$

Factoring,

$$T_0 - T_{DEW} = H\,(R_{DRY} - R_{DEW})$$

Solving for H, we have:

$$H = \frac{T_0 - T_{DEW}}{R_{DRY} - R_{DEW}} = \frac{T_0 - T_{DEW}}{\frac{10}{1{,}000} - \frac{1.8}{1{,}000}} = 1{,}000\frac{T_0 - T_{DEW}}{8.2}$$

which is Equation (2).

Substituting 30 °C for T_0 and 18 °C for T_{DEW}, then:

$$H = 1{,}000\frac{(30-18)}{8.2}$$

$$= 1{,}463\,\text{m}$$

Chapter 6: Energy and Latent Heat

Following the example, we can easily find that 4.19 × (100 − 15) + 2,260 = 4.19 × 85 + 2,260 = 2,616 kJ/kg are required. Thus, we have:

$$490\,\text{km}^3 \times \left[\frac{10^3\,\text{m}}{1\,\text{km}}\right]^3 \times \frac{10^3\,\text{kg}}{1\,\text{m}^3} \times \frac{2{,}616\,\text{kJ}}{1\,\text{kg}} = 1.28 \times 10^{18}\,\text{kJ}.$$

Chapter 8: Averaging in Time Cycles

1. Average precipitation for 2001–2005 in Dryden, Ontario, in mm:

Jan	29.8	**May**	102.0	**Sept**	108.6
Feb	13.6	**June**	119.8	**Oct**	59.2
Mar	26.3	**July**	77.1	**Nov**	29.4
April	46.7	**Aug**	102.4	**Dec**	31.5

2. During the autumn and winter months, the values for the 2001–2005 average and the long-term average are similar, especially in October, December, and January. However, in the spring, the 2001–2005 data show higher precipitation, especially in May. The 2001–2005 data also show large variability through the summer months, with a pronounced minimum in July. The long-term mean shows less variability in June, July, and August, and in fact decreases steadily from June through August. Both data sets show a maximum in June and a minimum in February.

Chapter 9: Cycles of Rainfall in the Low Latitudes

1. The mean is 164 mm, and the mean deviation is 43 mm.
2. The relative variability is 0.26, which makes San Juan about the same as Padang, but it is less variable than Abbassia and Mumbai.

Chapter 10: Standard Deviation and Coefficient of Variation

1. For Gander, Newfoundland:

Statistic	January	July
Sample Std. Dev. (mm)	42.47	28.12
Mean (mm)	108.9	88.8
Coeff. of variation	0.39	0.32

Precipitation in January has both the largest standard deviation and the largest coefficient of variation.

Chapter 12: Radioactive Decay and Radiometric Dating

1. The half-life is 1.28 b.y., $k = 0.693/1.28 = 0.542$; so:

$$P(t=1) = e^{-0.542\times1} = e^{-0.542} = 0.582 = 58.2\%$$
$$P(t=3) = e^{-0.543\times3} = e^{-1.626} = 0.197 = 19.7\%$$

2. Here, the half-life is 14.1 b.y., $k = 0.693/14.1 = 0.0491$, and:

$$P(t=5) = e^{-0.0491\times5} = e^{-0.245} = 0.782 = 78.2\%$$
$$P(t=10) = e^{-0.0491\times10} = e^{-0.491} = 0.612 = 61.2\%$$
$$P(t=15) = e^{-0.0491\times15} = e^{-0.737} = 0.479 = 47.9\%$$

3. For ^{14}C, $k = 0.693/5730 = 1.21\times10^{-4}$. Then:

$$P(t) = e^{-1.21\times10^{-4}t}.$$

In this case, $P(t)$ is known and equal to 0.1, while t is unknown. Solving for t, start with

$$P(t) = e^{-kt}$$

Take the log to the base e of both sides

$$\ln(P(t)) = -kt$$

Then solve for t:

$$t = -\frac{\ln(P(t))}{k}$$

Substituting values for $P(t)$ and k in this formula yields:

$$t = \frac{\ln(P(t))}{k} = \frac{\ln(0.1)}{1.21\times10^{-4}} = \frac{-2.30}{1.21\times10^{-4}}$$
$$= 1.90\times10^{-4} = 19{,}000 \text{ yrs}$$

4. $$t = \frac{1}{k}\ln\left[\frac{D}{M}+1\right] = \frac{1}{0.155}\ln[0.448+1]$$
$$= \frac{0.370}{0.155} = 2.39 \text{ b.y.}$$

5. For this decay sequence, $H = 7.04\times10^{8}$ yr $= 0.704$ b.y., and $k = 0.693/H = 0.693/0.704 = 0.985$ for time in b.y. Thus:

$$t = \frac{1}{0.985}\ln[9.56+1] = \frac{2.36}{0.985} = 2.40 \text{ b.y.}$$

The results of the two analyses are therefore quite consistent.

Chapter 14: The Power of Gravity

1. To apply the formula, first convert the velocity of the rock mass from kph to m s^{-1}:

$$\frac{150\text{ km}}{\text{hr}}\times\frac{1\text{ hr}}{60\text{ min}}\times\frac{1\text{ min}}{60\text{ s}}\times\frac{10^{3}\text{ km}}{\text{km}} = 41.7\text{ m s}^{-1}$$

Then,

$$E = \frac{1}{2}mv^{2} = \frac{1}{2}\times7.56\times10^{10}\text{ kg}\times\left(\frac{41.7\text{ m}}{8}\right)^{2}$$
$$= 6.57\times10^{13}\text{ J}$$

The total energy released is 2.82×10^{14} J, so the ratio of kinetic to total energy is:

$$\frac{6.57\times10^{13}\text{ J}}{2.82\times10^{14}\text{ J}} = 0.223 = 22.3\%$$

2. The free-fall velocity will be:

$$v = \sqrt{2gd} = \sqrt{2\times\frac{9.8\text{ m}}{s^{2}}\times500\text{ m}} = 99.0\text{ m s}^{-1}$$

The ratio of the velocity of the Madison Slide to the free-fall velocity is then:

$$\frac{41.7\text{ m s}^{-1}}{99.0\text{ m s}^{-1}} = .421 = 42.1\%$$

Thus, the slide moves less than half as fast as a free-falling body. The slide moves more slowly because it encounters friction in moving.

Chapter 16: River Discharge and Suspended Sediment

1. Applying the suspended load-discharge relationship for Oldman River and setting Q to 50 $m^{3}\,s^{-1}$:

$$S = 0.015Q^{2.30} = 0.015\times(50\text{ m}^{3}\text{ s}^{-1})2.30 = 0.015\times8{,}084$$
$$= 121 \text{ tons per day}$$

If $Q = 500\text{ m}^{3}\text{ s}^{-1}$ then:

$$S = 0.015Q^{2.30} = 0.015\times(500\text{ m}^{3}\text{ s}^{-1})^{2.30} = 0.015\times1{,}612{,}988$$
$$= 24{,}195 \text{ tons per day}$$

If there is an original discharge, X, the suspended load, S_1 would be:

$$S_1 = 0.015Q^{2.30} = 0.015\times(X\text{ m}^{3}\text{ s}^{-1})^{2.30}$$

If the discharge is doubled to 2X, calculate the effect on the suspended load quantity without knowing the actual value of X by substituting 2X for Q:

$$S_2 = 0.015Q^{2.30} = 0.015\times(2X\text{ m}^{3}\text{ s}^{-1})^{2.30}$$

Separate out the factor of 2:

$$S_2 = 0.015\times(X)^{2.30}\times(2)^{2.30}$$

Simplify:

$$S_2 = S_1\times(2)^{2.30} = S_1\times4.92$$

Therefore, if the discharge is doubled, then the suspended sediment load increases by a factor 4.92.

Chapter 19: Calculating a Simple Soil Water Budget

1. Soil water budget for Rockcliffe, Ontario

Month	*P*	*Ea*	−*G*	+*G*	R	*Ep*	D
Jan	7.4	0.0			+7.4	0.0	0.0
Feb	5.5	0.0			+5.5	0.0	0.0
Mar	7.0	0.0			+7.0	0.0	0.0
Apr	6.9	2.6			+4.3	2.6	0.0
May	6.3	8.0	−1.7			8.0	0.0
Jun	8.9	11.8	−2.9			11.8	0.0
Jul	8.6	13.7	−5.1			13.7	0.0
Aug	6.5	6.8	−0.3			11.4	4.6
Sep	8.2	7.4		+0.8		7.4	0.0
Oct	7.4	3.3		+4.1		3.3	0.0
Nov	7.6	1.0		+5.1	+1.5	1.0	0.0
Dec	6.6	0.0			+6.6	0.0	0.0
Total	86.9	54.6	−10.0	+10.0	32.3	59.2	4.6

Chapter 21: Island Biogeography: The Species–Area Curve

1.

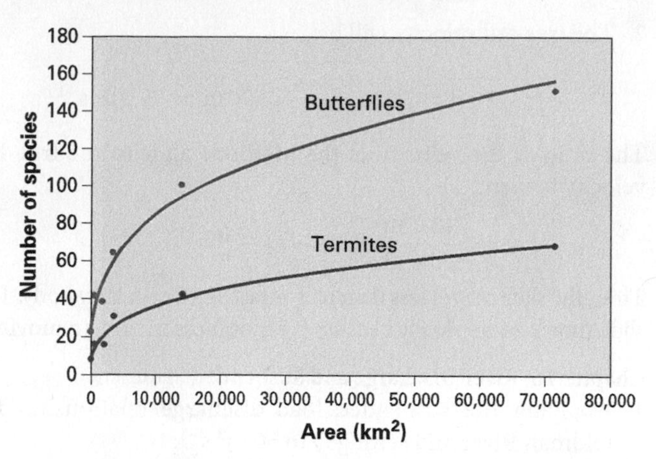

2.

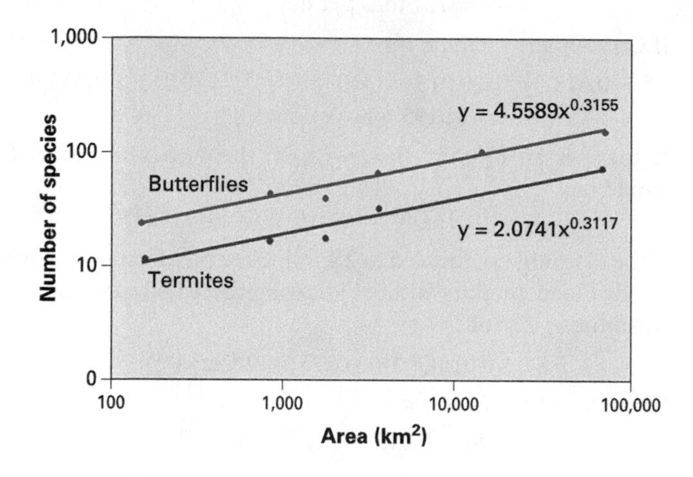

$$\text{Termites: } S = 2.07\, A^{0.31}$$
$$= 2.07(1{,}746)^{0.31}$$
$$= 2.07(10.12)$$
$$= 2.09$$

The two predicted values are somewhat higher than those actually observed. Perhaps Carlo Island has a less diverse environment than the other islands, given their sizes, and so has fewer termite and butterfly species.

$$\text{Butterflies: } S = 4.56\, A^{0.32}$$
$$= 4.56(1{,}746)^{0.32}$$
$$= 4.56(10.90)$$
$$= 49.7$$

Or it may be that the island has not been fully explored and all species present might not have been found. Or perhaps a recent catastrophic event wiped out some species and they have not yet returned by migration.

GLOSSARY

This glossary contains definitions of terms shown in the text in boldface.

A

ablation A wastage of glacial ice by both melting and evaporation.

abrasion Erosion of bedrock of a stream channel by impact of particles carried in a stream and by rolling of larger rock fragments over the stream bed; abrasion is also an activity of glacial ice, waves, and wind.

absolute instability A condition that develops when the rising air remains warmer and less dense than the surrounding air.

absolute stability A condition that develops when the temperature of the surrounding air is warmer than the air parcel.

absorption Conversion of electromagnetic energy into heat energy within a gas or liquid through which the radiation is passing or at the surface of a solid struck by the radiation.

active pool Type of pool in a biogeochemical cycle in which materials are in forms and places easily accessible to life processes. (See also *storage pool.*)

adiabatic process Change of temperature within a gas because of compression or expansion, without gain or loss of heat from the outside.

advection Horizontal movement of air or water from one place to another.

aeolian The processes collectively referring to the geomorphic work of wind in terrestrial environments. This includes erosion, transportation, and deposition of Earth materials.

aerosol Tiny particle present in the atmosphere, so small and light that the slightest movements of air keeps it aloft.

aggradation Rising of stream channel by continued deposition of bed load.

air mass Extensive body of air within which upward gradients of temperature and moisture are fairly uniform over a large area.

air pollutant An unwanted substance injected into the atmosphere from the Earth's surface by either natural or human activities; includes aerosols, gases, and particulates.

Air Quality Health Index On a scale of 1 to 10, a measurement of the risk associated from ozone, particulate matter, and nitrogen dioxide in the atmosphere.

air temperature Temperature of air, normally observed by a thermometer under standard conditions of shelter and height above the ground.

albedo Percentage of downwelling solar radiation reflected upward from a surface.

Alfisols Soil order in the Comprehensive Soil Classification System of the United States consisting of soils of humid and subhumid climates, with high base status and an argillic horizon.

alluvial fan Gently sloping, conical accumulation of coarse alluvium deposited by a braided stream undergoing aggradation below the point of emergence of the channel from a narrow gorge or canyon.

alluvial river Stream of low gradient flowing upon thick deposits of alluvium and experiencing overbank flooding of the adjacent flood plain in most years.

alluvial terrace Bench-like landform carved in alluvium by a stream during degradation.

alluvium Any stream-laid sediment deposit found in a stream channel and in low parts of a stream valley subject to flooding.

alpine glacier Long, narrow, mountain glacier on a steep downgrade, occupying the floor of a trough-like valley.

andesite Extrusive igneous rock consisting primarily of plagioclase feldspar; it is equivalent in composition to the plutonic rock diorite.

Andisols A soil order in the Comprehensive Soil Classification System of the United States that includes soils formed on volcanic ash; often enriched by organic matter, yielding a dark soil colour.

Annual Greenhouse Gas Index (AGGI) A measure of the total radiative forcing due to greenhouse gases relative to the reference baseline of 1.00, based on 1990 greenhouse gas concentrations.

Antarctic Circle Parallel of latitude at 66½° S.

antecedent conditions The state of a system prior to some event; in geomorphology, it generally refers to soil moisture levels prior to a rainstorm.

anticlines Upfold of strata or other layered rock in an arch-like structure; a class of folds. (See also *syncline.*)

anticyclone Centre of high atmospheric pressure.

Arctic Circle Parallel of latitude at 66½° N.

Aridisols Soil order in the Comprehensive Soil Classification System of the United States consisting of soils of dry climates, with or without argillic (clay) horizons, and with accumulations of carbonates or soluble salts.

artesian well Drilled well in which water rises under hydraulic pressure above the level of the surrounding water table and may reach the surface.

asthenosphere Soft layer of the upper mantle, beneath the rigid lithosphere.

astronomical hypothesis Explanation for glaciations and interglaciations making use of cyclic variations in the solar energy received at the Earth's surface.

atmosphere Envelope of gases surrounding the Earth, held by gravity.

atoll Circular or closed-loop coral reef enclosing an open lagoon with no island inside.

B

barchan dune Crescentic sand dune with a sharp crest and a steep lee slip face, with crescent points (horns) pointing downwind.

barometer Instrument for measurement of atmospheric pressure.

basalt A hard, dense extrusive igneous rock consisting primarily of mafic minerals; occurs as lava.

bases Certain positively charged ions in the soil that are also plant nutrients; the most important are calcium, magnesium, potassium, and sodium.

beach Accumulation of sand, gravel, or cobbles in the zone of breaking waves.

bedrock Solid rock in place with respect to the surrounding and underlying rock and relatively unchanged by weathering processes.

Bergeron (Bergeron-Findeisen) process The formation of precipitation by ice crystal growth in the cold clouds of mid- and upper latitudes.

bioclimatic frontier Geographic boundary corresponding with a critical limiting level of climate stress beyond which a species cannot survive.

biodiversity The variety of biological life on Earth or within a region.

biogeochemical cycle Total system of pathways by which a particular type of matter (a given element, compound, or ion, for example) moves through the Earth's ecosystem or biosphere; also called a material cycle or nutrient cycle.

biogeographic realm Region in which the same or closely related plants and animals tend to be found together.

biogeography The study of the distributions of organisms at varying spatial and temporal scales, as well as the processes that produce these distribution patterns.

biomass Dry weight of living organic matter in an ecosystem within a designated surface area; units are kilograms of organic matter per square metre.

biome Largest recognizable subdivision of terrestrial ecosystems, including the total assemblage of plant and animal life interacting within the life layer.

biosphere All living organisms of the Earth and the environments with which they interact.

biosphere reserve An area designed to conserve examples of characteristic ecosystems.

blackbody Ideal object or surface that is a perfect radiator and absorber of energy; absorbs all radiation it intercepts and emits radiation perfectly according to physical theory.

blowout Shallow depression produced by continued deflation.

boreal forest climate Cold climate of the subarctic zone in the northern hemisphere with long, extremely severe winters and several consecutive months of zero potential evapotranspiration (water need).

Brunisolic soils A class of moderately developed forest soils in the Canadian soil classification system with brownish B horizon.

butte Prominent, steep-sided hill or peak, often representing the final remnant of a resistant layer in a region of flat-lying strata.

C

C_3 pathway Also known as the Calvin cycle, it is the principal process of photosynthesis in which carbon dioxide is initially converted into a 3-carbon compound before final transformation to carbohydrate.

C_4 photosynthetic pathway One of the three processes of photosynthesis. In this process, carbon dioxide is initially converted into oxaloacetic acid, which is a 4-carbon compound.

caldera Large, steep-sided circular depression resulting from the explosion and subsidence of a stratovolcano.

CAM pathway One of the three processes of photosynthesis. In this process, like the C4 pathway, the plant initially manufactures oxaloacetic acid. What makes this process different is that at night, carbon dioxide is taken in and stored for later processing during daylight hours.

carbonation A chemical weathering process in which carbonic acid reacts with minerals in rocks (especially limestone), transforming them into carbonates.

carbon cycle Biogeochemical cycle in which carbon moves through the biosphere; includes both gaseous pathways and sedimentary pathways.

cartography The science and art of making maps.

channel Long, narrow trough-like depression occupied and shaped by a stream moving to progressively lower levels.

chattermark A crescentic gouge on a bedrock surface caused by the impact of large rocks that are rolled along under a glacier.

chemical energy Energy stored within a molecule and capable of being transformed into other forms of energy.

chemical weathering Chemical change in rock-forming minerals through exposure to atmospheric conditions in the presence of water; mainly involves oxidation, hydrolysis, carbonic acid action, or direct solution.

chemically precipitated sediment Sediment consisting of mineral matter precipitated from a water solution in which the minerals have been transported in the dissolved state as ions.

Chernozemic soils A class of grassland soils in the Canadian soil classification system with a thick A horizon rich in organic matter.

chlorofluorocarbons (CFCs) Synthetic chemical compounds containing chlorine, fluorine, and carbon atoms that have been widely used as coolant fluids in refrigeration systems.

cirque Bowl-shaped depression carved in rock by glacial processes and holding the firn of the upper end of an alpine glacier.

clastic sediment Sediment consisting of particles broken away physically from a parent rock source.

clay mineral Class of minerals with plastic properties when moist, produced by alteration of silicate minerals.

climate change The variation in the Earth's global climate or in regional climates over time.

climate Generalized statement of the prevailing weather conditions at a given place, based on statistics of a long period of record and including mean values, departures from those means, and the probabilities associated with those departures.

climatology The science that describes and explains the variability in space and time of the heat and moisture states of the Earth's surface, especially its land surfaces.

climax community Stable community of plants and animals reached at the end point of ecological succession.

climograph A graph on which two or more climatic variables, such as monthly mean temperature and monthly mean precipitation, are plotted for each month of the year.

closed flow system Flow system that is completely self-contained within a boundary through which no matter or energy is exchanged with the external environment. (See also *open flow system*.)

cloud Dense concentrations of suspended water or ice particles in the diameter range of 20 to 50 μm.

coastal foredune A narrow belt of dunes in the form of irregularly shaped hills and depressions, typically found on the landward margin of sand beaches. They normally bear a cover of beach grass and other species of plants capable of survival in the severe environment.

coastal geography The study of the geomorphic processes that shape shores and coastlines and their application to coastal development and marine resource use.

coastline Zone in which coastal processes operate or have a strong influence.

cold front Moving weather front along which a cold air mass pushes underneath a warm air mass, causing the latter to be lifted.

collision-wake capture During precipitation, the process by which larger water droplets drag smaller droplets down behind them, which arises because of the range of droplet sizes naturally present in clouds.

colloids Particles of extremely small size, capable of remaining indefinitely in suspension in water; may be mineral or organic in nature.

colluvium Deposit of sediment or rock particles accumulating from overland flow at the base of a slope and originating from higher slopes where sheet erosion is in progress. (See also *alluvium*.)

competition Form of interaction among plant or animal species in which both draw resources from the same pool.

condensation nuclei A tiny particle of solid matter (aerosol) in the atmosphere on which water vapour condenses to form a tiny water droplet.

conditional instability Arises when the environmental temperature lapse rate (ETLR) is between the dry adiabatic lapse rate (DALR) and the saturated adiabatic lapse rate (SALR); associated with moist air.

conduction Transmission of sensible heat through matter by transfer of energy from one atom or molecule to the next in the direction of decreasing temperature.

constant pressure surface A surface in the atmosphere across which pressure is the same at all points.

consumers Animals in the food chain that live on organic matter formed by primary producers or by other consumers.

continental crust Crust of the continents, of felsic composition in the upper part; thicker and less dense than oceanic crust.

continental scale Scale of observation that recognizes continents and other large Earth surface features, such as mountain ranges and ocean currents.

continental shields Ancient crustal rock masses of the continents; largely igneous rock and metamorphic rock, and mostly of Precambrian age.

continental suture Long, narrow zone of crustal deformation, including underthrusting and intense folding, produced by a continental collision (e.g., Himalayan Range, European Alps).

convection A process by which a fluid is heated by a warm surface, expands, and rises, creating an upward flow. This flow moves heat away from the surface.

convectional precipitation A form of precipitation induced when warm, moist air is heated at the ground surface, rises, cools, and condenses to form water droplets, raindrops, and eventually rainfall.

convergence The horizontal inflow of air into a region of low pressure, which subsequently tends to produce vertical uplift.

convergence precipitation Occurs when moist air flows come together and then rise; also associated with hurricanes.

Coordinated Universal Time (UTC) The legal time standard recognized by all nations and administered by the Bureau International de l'Heure, located near Paris. It is a high-precision atomic time standard that is periodically adjusted by leap seconds to compensate for discrepancies in the Earth's rotation.

coral reef Rock-like accumulation of carbonates secreted by corals and algae in shallow water along a marine shoreline.

core Spherical central mass of the Earth composed largely of iron and consisting of an outer liquid zone and an interior solid zone.

Coriolis effect Effect of the Earth's rotation tending to turn the direction of motion of any object or fluid toward the right in the northern hemisphere and to the left in the southern hemisphere.

counter-radiation Long-wave radiation from the atmosphere directed downward to Earth's surface.

creep Very slow movement of soil or rock material over a period of several years.

crevasse A vertical crack that develops in a glacier as a result of friction between valley walls, tension forces of extension on convex slopes, or compression forces on concave slopes.

crust Outermost solid shell or layer of the Earth, composed largely of silicate minerals.

Cryosolic soils A class of soils in the Canadian soil classification system associated with strong frost action and underlying permafrost.

curvature effect Limits the growth of small, pure water droplets to conditions of supersaturation. It is related to the size of the droplet and arises because the surface of a small droplet has greater curvature than the surface of a large droplet.

cycle of rock transformation Total cycle of changes in which rock of any one of the three major rock classes—igneous rock, sedimentary rock, metamorphic rock—is transformed into rock of one of the other classes.

cyclone Centre of low atmospheric pressure. (See also *tropical cyclone, wave cyclone.*)

cyclonic precipitation A form of precipitation that occurs as warm moist air is lifted by air motion occurring in a cyclone.

cyclonic storm Intense weather disturbance within a moving cyclone generating strong winds, cloudiness, and precipitation.

D

Dalton's law of partial pressure The pressure exerted by a mixture of gases is equivalent to the sum of the pressures exerted by the individual gases.

Daylight Saving Time Time system under which clocks are advanced by one hour with respect to the time of the prevailing standard meridian.

dead zone A region on the seabed in which the concentration of dissolved oxygen is so low that most lifeforms cannot survive.

declination Latitude at which the sun is directly overhead; varies from −23½° (23½° S lat.) to +23½° (23½° N lat.)

decomposers Organisms that feed on dead organisms from all levels of the food chain; most are microorganisms and bacteria that feed on decaying organic matter.

deflation Lifting and transport in turbulent suspension by wind of loose particles of soil or regolith from dry ground surfaces.

delta Sediment deposit built by a stream entering a body of standing water and formed from the transported stream load.

dendrochronology A method of scientific dating based on the analysis of tree-ring growth patterns.

deposition The change of state of a substance from a gas (water vapour) to a solid (ice); in the science of meteorology, the term sublimation is used to describe both this process and the change of state from solid to vapour.

depression Low-pressure centres, particularly those that bring cloudy conditions and precipitation in midlatitudes.

depression storage The retention of water in natural depressions in the surface of the land.

desert biome Biome of the dry climates normally consisting of widely dispersed plants that may be shrubs, grasses, or perennial herbs, but few if any trees.

desert pavement Surface layer of closely spaced pebbles and coarse sand from which finer particles have been removed.

dewpoint temperature Temperature of an air mass at which the air holds its full capacity of water vapour.

diffuse radiation Solar radiation that has been scattered (deflected or reflected) by minute dust particles or cloud particles in the atmosphere.

dip Acute angle between an inclined natural rock plane or surface and an imaginary horizontal plane of reference; always measured perpendicular to the strike. Also a verb, meaning to incline toward.

discharge Volume of flow moving through a given cross-section of a stream in a given unit of time; commonly given in cubic metres per second ($m^3 s^{-1}$).

disjunction Geographic distribution pattern of species in which one or more closely related species are found in widely separated regions.

dispersal The capacity of a species to move from a location of birth or origin to new sites.

Dobson unit (DU) The standard way to express ozone amounts in the atmosphere.

drainage basin Total land surface occupied by a drainage system, bounded by a drainage divide or watershed.

drainage system A branched network of stream channels and adjacent land slopes, bounded by a drainage divide and converging to a single channel at the outlet.

drumlin Hill of glacial till, oval or elliptical in basal outline and with a smoothly rounded summit, formed by plastering of till beneath moving, debris-laden glacial ice.

dry adiabatic lapse rate (DALR) Rate at which rising air is cooled by expansion when no condensation is occurring; 10 °C per 1,000 m. (See also *saturated adiabatic lapse rate.*)

dry deposition The process by which atmospheric gases and particles are transferred to the surface as a result of random turbulent air motions.

dry midlatitude climate Dry climate at latitudes of 35 to 50° N with a strongly developed annual temperature cycle and limited precipitation, often due to the blocking effect of mountain ranges that restrict the passage of maritime air masses.

dry subtropical climate Dry climate of the subtropical zone associated with cT air of the subtropical high pressure zones; transitional between the dry tropical climate and the dry midlatitude climate.

dry tropical climate Climate of the tropical zone with large total annual potential evapotranspiration (water need).

dust storm Heavy concentration of dust in a turbulent air mass.

E

earth flow Moderately rapid downhill flow of masses of water-saturated soil, regolith, or weak shale, typically forming a step-like terrace at the top and a bulging toe at the base.

Earth rotation The counter-clockwise turning of the Earth on its axis, an imaginary straight line passing through the centre of the planet and joining the North and South poles.

earthquake A trembling or shaking of the ground produced by the passage of seismic waves.

easterly wave A weak, slowly moving trough of low pressure within the belt of tropical easterlies; causes a weather disturbance with rain showers.

eccentricity The amount by which the orbit of an astronomical body, such as Earth, deviates from a circle.

ecological niche A concept that includes the role an organism occupies and the function it performs in an ecosystem.

ecological succession Sequential change over time of distinctive plant and animal communities occurring within a given area of newly formed land or land cleared of plant cover by burning, clear cutting, or other agents.

economic geography The study of the location, distribution, and spatial organization of economic activities across Earth.

edatopic grid A method of placing sites or species into a two-dimensional ecological space, which has direct implications for forest growth, composition, and management.

electromagnetic radiation Wavelike form of energy radiated by any substance possessing heat; travels through space at the speed of light.

electromagnetic spectrum The total wavelength range of electromagnetic energy.

eluviation Soil-forming process consisting of the downward transport of fine particles, particularly the soil colloids (both mineral and organic), carrying them out of an upper soil horizon.

emergence Exposure of submarine landforms by a lowering of sea level or a rise of the crust, or both.

endemic A species found only in one, usually geographically restricted, region or location.

Entisols Soil order in the Comprehensive Soil Classification System of the United States consisting of mineral soils lacking well-developed soil horizons.

environmental pollution Unchecked human activity, such as fuel consumption, fertilizer runoff, toxic industrial production waste, and acid mine drainage, which degrades environmental quality.

environmental temperature lapse rate Rate of temperature decrease upward through the troposphere; standard value is 6.4 °C km^{-1}.

eon A major division in the geologic time scale, specifically the Phanerozoic (from present to 542 million years), Proterozoic (beginning 2.5 billion years BP), Archean (beginning 3.8 billion years BP), and Hadean (extending back to the formation of the Earth, 4.5 billion years BP).

epicentre The point on Earth's surface directly above the focus of an earthquake.

equator Parallel of latitude occupying a position midway between the Earth's poles of rotation; the largest of the parallels—the only one that is a great circle—is designated as latitude 0°.

equilibrium In flow systems, a state in which flow rates continually adjust to maintain a state of balance over a specified period.

equilibrium line Marks the boundary between the accumulation zone and the ablation zone.

equilibrium tide The hypothetical tide that would be produced by the lunar and solar tidal forces in the absence of ocean basin constraints and dynamics.

equinox Instant in time when the subsolar point falls on the Earth's equator and the circle of illumination passes through both poles. Vernal equinox occurs on March 20 or 21; autumnal equinox occurs on September 22 or 23.

esker Narrow, often sinuous embankment of coarse gravel and boulders deposited in the bed of a meltwater stream enclosed in a tunnel within stagnant ice of an ice sheet.

evaporation Process in which water in liquid state or solid state passes into the vapour state.

evaporites Class of chemically precipitated sediment and sedimentary rock composed of soluble salts deposited from saltwater bodies.

evapotranspiration Combined water loss to the atmosphere by evaporation from the soil and transpiration from plants.

exfoliation Process of removal of overlying rock load from bedrock by processes of denudation, accompanied by expansion, and often leading to the development of sheeting structure.

extreme events Catastrophic events, such as floods, fires, hurricanes, and earthquakes, that have great and long-lasting impacts on human and natural systems.

extrusive igneous rock Rock produced by the solidification of lava or ejected fragments of igneous rock (tephra).

eye The circular region of relatively light winds and fair weather found at the centre of a severe tropical cyclone.

F

fallout Return of atmospheric particulates to Earth's surface under the influence of gravity.

feedback In flow systems, a linkage between flow paths such that the flow in one pathway acts either to reduce or increase the flow in another pathway.

felsic Quartz and feldspars treated as a mineral group of light colour and relatively low density. (See also *mafic minerals.*)

fetch Distance over water that wind blows, creating wind waves.

fjord A deep, steep-walled, U-shaped valley formed by erosion by a glacier and submerged with sea water.

flood Stream flow at a stage so high that it cannot be accommodated within the stream channel and must spread over the banks to inundate the adjacent flood plain.

flood plain Belt of low, flat ground, present on one or both sides of a stream channel, subject to inundation by a flood and underlain by alluvium.

flow systems A physical system in which matter, energy, or both move through time from one location to another.

fluvial landforms Landforms shaped by running water.

fluvial processes Geomorphic processes in which running water is the dominant fluid agent, acting as overland flow and stream flow.

fog Cloud layer in contact with or very close to the land or sea surface.

food chain Organization of an ecosystem into steps or levels through which energy flows as the organisms at each level consume energy stored in the tissues of organisms of the next lower level.

food web (See *food chain.*)

forest biome Biome that includes all regions of forest over the lands of the Earth.

fossil fuels Naturally occurring hydrocarbon compounds that represent the altered remains of organic materials enclosed in rock; examples are coal, petroleum (crude oil), and natural gas.

front Surface of contact between two unlike air masses. (See also *cold front, occluded front, polar front, warm front.*)

frost point When the dewpoint temperature is below freezing.

fundamental niche The theoretical limit for a species, as determined by its physiological requirements for all environmental factors.

G

genetic classification system A classification system (of climate, soils, etc.) based on the causes of the differences in observed properties or characteristics.

geographic cycle A theory that views landscapes as stages in a cycle, beginning with rapid uplift and followed with erosion by streams.

geographic grid Complete network of parallels and meridians on the surface of the globe, used to fix the locations of surface features.

geographic information systems (GIS) A system for acquiring, processing, storing, querying, creating, and displaying spatial data; normally computer-based.

geographic isolation Speciation process in which a single breeding population is split into parts by an emerging geographic barrier, such as an uplifting mountain range or a changing climate.

geography The study of the evolving character and organization of Earth's surface.

geography of soils The study of the distribution of soil types and properties and the processes of soil formation.

geoid Refers to the shape that the Earth would assume if it were entirely covered with water and responding to the forces acting upon it.

geologic column A complete sequence of all sedimentary strata with the oldest rocks of the of the Cambrian Period at its base and the youngest Quaternary rocks at the top.

geomorphology Science of Earth surface processes and landforms, including their history and processes of origin.

glacial drift General term for all varieties and forms of rock debris deposited in close association with ice sheets of the Pleistocene Epoch.

glacial trough Deep, steep-sided rock trench of U-shaped cross-section formed by alpine glacier erosion.

glaciation (1) General term for the total process of glacier growth and landform modification by glaciers. (2) Single episode or time period in which ice sheets formed, spread, and disappeared.

Gleysolic soils A class of soils in the Canadian soil classification system characterized by indicators of periodic or prolonged water saturation.

global energy balance A full accounting of all the energy flows among the sun, the atmosphere, Earth's surface, and space.

Global Positioning System (GPS) A satellite-based system constantly sending radio signals to Earth with information that allows a GPS receiver to calculate its position on the Earth's surface.

global scale Scale at which we are concerned with the Earth as a whole; for example in considering Earth–sun relationships.

global warming potential A measure of how much a given mass of greenhouse gas is estimated to contribute to global warming.

gneiss Variety of metamorphic rock showing banding and commonly rich in quartz and feldspar.

Goode projection An equal-area map projection, often used to display areal thematic information, such as climate or soil type.

graded stream Stream (or stream channel) with stream gradient adjusted so it achieves a balanced state in which average bed load transport matches average bed load input; an average condition over periods of many years' duration.

granite Intrusive igneous rock consisting largely of quartz, potash feldspar, and plagioclase feldspar, with minor amounts of biotite and hornblende; a felsic igneous rock.

grassland biome Biome consisting largely or entirely of herbs, which may include grasses, grass-like plants, and forbs.

great circle Circle formed by passing a plane through the exact centre of a sphere; the largest circle that can be drawn on the surface of a sphere.

greenhouse effect Accumulation of heat in the lower atmosphere through the absorption of long-wave radiation from the Earth's surface.

greenhouse gases Atmospheric gases such as CO_2 and chlorofluorocarbons (CFCs) that absorb outgoing long-wave radiation, contributing to the greenhouse effect.

gross photosynthesis Total amount of carbohydrate produced by photosynthesis by a given organism or group of organisms in a given unit of time.

ground ice Frozen water within the pores of soils and regolith or as free bodies or lenses of solid ice.

ground inversion An air layer with its base at the ground surface and in which temperature increases with height.

groundwater Subsurface water occupying the saturated zone and moving under the force of gravity.

gyre Large circular ocean current systems centred under the oceanic subtropical high-pressure cells.

H

habitat Subdivision of the environment according to the needs and preferences of organisms or groups of organisms.

Hadley cell Atmospheric circulation cell in low latitudes involving rising air over the equatorial trough and sinking air over the subtropical high-pressure belts.

hairpin dune A long, narrow dune with parallel sides.

hammada A desert surface from which wind has removed most of the regolith, leaving only bedrock surfaces scattered with large rocks.

hazards assessment A field of study blending physical and human geography focusing on the perception of risk of natural hazards and on developing public policy to mitigate that risk.

heat energy A form of energy in which the atoms and molecules within a solid (or liquid, or gas) are in rapid motion.

heat island Persistent region of higher air temperatures centred over a city.

helical flow The spiralling or corkscrew-like path described by water as it flows around a meander.

highland climates Cool to cold, usually moist, climates that occupy mountains and high plateaus.

Histosols Soil order in the Comprehensive Soil Classification System of the United States consisting of soils with a thick upper layer of organic matter.

hotspot Geographic region of high biodiversity.

human geography The part of systematic geography that deals with social, economic, and behavioural processes that differentiate places.

hydrograph Graphic presentation of the variation in stream discharge with elapsed time, based on stream gauge data at a given station on a stream.

hydrologic cycle Global processes of movement, exchange, and storage of Earth's free water in gaseous, liquid, and solid states.

hydrophyte A plant that grows in lakes, marshes, and bogs that is adapted to cope with excessive moisture.

hydrosphere Total water realm of the Earth, including the oceans, surface waters of the lands, groundwater, and water held in the atmosphere.

hypoxia A condition in which oxygen is almost entirely depleted; characteristically associated with ocean dead zones.

I

ice lenses More-or-less horizontal layer of segregated ice formed by capillary movement of soil water toward a freezing front.

ice nuclei Particles that act as the nuclei for the formation of ice crystals in the atmosphere.

ice sheet Large, thick mass of glacial ice moving outward in all directions from a central region of accumulation.

ice sheet climate Severely cold climate, found on the Greenland and Antarctic ice sheets, with potential evapotranspiration (water need) effectively zero throughout the year.

ice wedge Vertical, wall-like body of ground ice, often tapering downward, occupying a shrinkage crack in silt of permafrost areas.

ideal gas law A theoretical expression in which the density of a gas is directly related to changes in temperature and pressure.

igneous rocks Rock solidified from a high-temperature molten state; rock formed by cooling of magma. (See also *extrusive igneous rock, felsic igneous rock, intrusive igneous rock, mafic igneous rock, ultramafic igneous rock.*)

illuviation Accumulation in a lower soil horizon (typically, the B horizon) of materials brought down from a higher horizon; a soil-forming process.

Inceptisols Soil order in the Comprehensive Soil Classification System of the United States consisting of soils having weakly developed soil horizons and containing weatherable minerals.

individual scale The finest level of scale that focuses on individual landscape features, such as a coastal sand dune.

infiltration Absorption and downward movement of precipitation into the soil and regolith.

infrared Electromagnetic energy in the wavelength range of 0.7 to 1 mm.

inputs Flow of matter or energy into a system.

inselbergs An isolated hill, knob, or small mountain that rises abruptly from a gently sloping or virtually level surrounding plain.

insolation Interception of solar energy (shortwave radiation) by an exposed surface.

interglaciation Within an ice age, a time interval of mild global climate in which continental ice sheets were largely absent or were limited to the Greenland and Antarctic ice sheets; the interval between two glaciations. (See also *glaciation.*)

International Date Line The 180° meridian of longitude, together with deviations east and west of that meridian, forming the time boundary between adjacent standard time zones that are 12 hours fast and 12 hours slow with respect to Greenwich standard time.

interstadial A relatively warm period during a glacial epoch marked by the retreat of the ice sheets.

intertropical convergence zone (ITCZ) Zone of convergence of tropical easterlies (trade winds) along the axis of the equatorial trough.

intrusive igneous rock Igneous rock body produced by solidification of magma beneath the surface, surrounded by pre-existing rock.

island arcs Curved lines of volcanic islands associated with active subduction zones along the boundaries of lithospheric plates.

isohyet Line on a map drawn through all points having the same numerical value of precipitation.

isoline A general term for a line on a map connecting points of equal value—a contour line joining points of equal elevation is a specific type of isoline.

isotherm Line on a map drawn through all points having the same air temperature.

J

jet streams High-speed airflow in narrow bands within the upper-air westerlies and along certain other global latitude zones at high levels in the troposphere.

K

karst A landscape or topography dominated by surface features of limestone solution and underlain by a limestone cavern system.

kilopascal The standard unit, for meteorological purposes, used to measure pressure.

kinetic energy Form of energy represented by matter (mass) in motion.

kopje Rocky outcrops of old granite which, because of erosion and weathering, have broken up into a rough and jumbled surface.

L

lake Body of standing water that is enclosed on all sides by land.

landform Configurations of the land surface taking distinctive forms and produced by natural processes (e.g., hill, valley, plateau).

landslide Rapid sliding of large masses of bedrock on steep mountain slopes or from high cliffs.

lapse rate Rate at which temperature decreases with increasing altitude. (See also *environmental temperature lapse rate, dry adiabatic lapse rate, wet adiabatic lapse rate.*)

Late-Cenozoic Ice Age The series of glaciations, deglaciations, and interglaciations experienced during the Late-Cenozoic Era.

latent heat Heat absorbed and held in storage in a gas or liquid during the processes of evaporation, melting, or sublimation; distinguished from sensible heat.

latent heat transfer Flow of latent heat that results when a substance (usually water) absorbs heat to change from a liquid or solid to a gas and then later releases that heat to new surroundings by condensation or deposition.

lateral moraine Moraine forming an embankment between the ice of an alpine glacier and the adjacent valley wall.

latitude Arc of a meridian between the equator and a given point on the globe.

lava Magma emerging on Earth's solid surface, exposed to air or water.

life form Characteristic physical structure, size, and shape of a plant or of an assemblage of plants.

life layer Shallow surface zone containing the biosphere; a zone of interaction between atmosphere and land surface, and between atmosphere and ocean surface.

life zones Series of vegetation zones describing vegetation types that are encountered with increasing elevation, especially in the southwestern U.S.

lifting condensation level The altitude at which the air will become saturated because of adiabatic cooling (caused by expansion).

limestone Non-clastic sedimentary rock in which calcite is the predominant mineral, and with varying minor amounts of other minerals and clay.

lithosphere Strong, brittle outermost rock layer of the Earth, lying above the asthenosphere.

lithospheric plates Segment of lithosphere moving as a unit, in contact with adjacent lithospheric plates along plate boundaries.

littoral drift Transport of sediment parallel with the shoreline by the combined action of beach drift and longshore current transport.

local scale Scale of observation of the Earth in which local processes and phenomena are observed.

loess Accumulation of yellowish to buff-coloured, fine-grained sediment, largely of silt grade, upon upland surfaces after transport in the air in turbulent suspension (i.e., carried in a dust storm).

longitude Arc of a parallel between the prime meridian and a given point on the globe.

longitudinal dune Class of sand dunes in which the dune ridges are oriented parallel with the prevailing wind.

long-wave radiation Electromagnetic energy emitted by the Earth, largely in the range from 3 to 50 μm.

Luvisolic soils A class of forest soils in the Canadian soil classification system in which the B horizon accumulates clay.

M

macronutrients Essential chemical elements needed by all life in large quantities for it to function normally. The three principal elements are hydrogen, carbon, and oxygen.

mafic Minerals, largely silicate minerals, rich in magnesium and iron, dark in colour, and of relatively great density.

magma Mobile, high-temperature molten state of rock, usually of silicate mineral composition and with dissolved gases.

mantle Rock layer or shell of the Earth beneath the crust and surrounding the core, composed of ultramafic igneous rock of silicate mineral composition.

map A paper representation of space showing point, line, or area data.

map projection Any orderly system of parallels and meridians drawn on a flat surface to represent the Earth's curved surface.

marine west-coast climate Cool moist climate of west coasts in the midlatitude zone, usually with a substantial annual water surplus and a distinct winter precipitation maximum.

mass wasting Spontaneous downhill movement of soil, regolith, and bedrock under the influence of gravity, rather than by the action of fluid agents.

material cycle A closed matter flow system, in which matter flows endlessly, powered by energy inputs. (See also *biogeochemical cycle.*)

mathematical modelling Using variables and equations to represent real processes and systems.

meander Sinuous bends of a graded stream flowing in the alluvial deposit of a flood plain.

mechanical energy Energy of motion or position; includes kinetic energy and potential energy.

Mediterranean climate Climate type of the subtropical zone, characterized by the alternation of a very dry summer and a mild, rainy winter.

Mercator projection Conformal map projection with horizontal parallels and vertical meridians and with map scale rapidly increasing with increase in latitude.

meridian North–south line on the surface of the global oblate ellipsoid, connecting the North Pole and South Pole.

mesa Plateau of comparatively small extent bounded by cliffs and occurring in a region of flat-lying strata.

mesocyclone A large rotating atmospheric circulation that generally produces heavy rain, large hail, and lightning; may lead to tornado activity.

metamorphic rocks Rock altered in physical structure and/or chemical (mineral) composition by action of heat, pressure, shearing stress, or infusion of elements, all taking place at substantial depth beneath the surface.

metamorphism Recrystallization of pre-existing rock through intense pressure and/or heat generated during tectonic activity.

meteorology Science of the atmosphere, particularly the physics of short-term fluctuations that occur in the lower atmosphere.

microburst Brief onset of intense winds close to the ground beneath the downdraft zone of a thunderstorm cell.

Mie scattering Scattering of light by atmospheric particles larger than the wavelength of light; mainly produces whitish glare. Does not occur in clean, particle-free air.

mineral Naturally occurring inorganic substance, usually having a definite chemical composition and a characteristic atomic structure. (See also *felsic, mafic.*)

mineral alteration Chemical change of minerals to more stable compounds upon exposure to atmospheric conditions; same as chemical weathering.

moist continental climate Moist climate of the midlatitude zone with strongly defined winter and summer seasons, adequate precipitation throughout the year, and a substantial annual water surplus.

moist subtropical climate Moist climate of the subtropical zone, characterized by a moderate to large annual water surplus and a strongly seasonal cycle of potential evapotranspiration (water need).

Mollisols Soil order in the Comprehensive Soil Classification System of the United States consisting of soils with a mollic horizon and high base status.

moment magnitude scale A scale that measures the total energy released by an earthquake. It is based on the movement that occurs on a fault, not on how much the ground shakes during an earthquake.

monadnocks Prominent, isolated mountain or large hill rising conspicuously above a surrounding peneplain and composed of a rock more resistant than that underlying the peneplain; a landform of denudation in moist climates.

monsoon System of low-level winds blowing into a continent in summer and out of it in winter, controlled by atmospheric pressure systems developed seasonally over the continent.

monsoon and trade-wind coastal climate Moist climate of low latitudes showing a strong rainfall peak in the season of high sun and a short period of reduced rainfall.

mountain arc Curving section of an alpine chain occurring on a converging boundary between two crustal plates.

mudflow A form of mass wasting consisting of the downslope flowage of a mixture of water and mineral fragments (soil, regolith, disintegrated bedrock), usually following a natural drainage line or stream channel.

N

natural levee Belt of higher ground paralleling a meandering alluvial river on both sides of the stream channel and built up by deposition of fine sediment during periods of overbank flooding.

natural selection Evolution of organisms by environmental conditions and processes.

natural vegetation Stable, mature plant cover characteristic of a given area of land surface, largely free from the influences and impacts of human activities.

neap tide When the tidal range between high and low water is at its lowest.

net photosynthesis Carbohydrate production remaining in an organism after respiration has broken down sufficient carbohydrate to power the metabolism of the organism.

net primary production (NPP) Rate at which carbohydrate is accumulated in the tissues of plants within a given ecosystem; units are kilograms of dry organic matter per year per square metre.

net radiation Difference in intensity between all incoming energy (positive quantity) and all outgoing energy (negative quantity) carried by both short-wave radiation and long-wave radiation.

neutral stability Occurs when the environmental temperature lapse rate equals the dry adiabatic lapse rate or the saturated adiabatic lapse rate.

normal fault Variety of fault in which the fault plane inclines (dips) toward the downthrown block and a major component of the motion is vertical.

normal temperature lapse rate Measures the drop in temperature in stationary air, averaged for the entire Earth over a long period.

O

obliquity The tilt of the axis of an astronomical body, such as Earth.

occluded front Weather front along which a moving cold front has overtaken a warm front, forcing the warm air mass aloft.

oceanic crust Crust of basaltic composition beneath the ocean floors, capping oceanic lithosphere. (See also *continental crust.*)

oceanography The study of the Earth's oceans and seas.

omnivores Animals that feed on plant materials as well as animals.

open flow system System of interconnected flow paths of energy and/or matter with a boundary through which energy and/or matter can enter and leave the system.

organic sediment Sediment consisting of the organic remains of plants or animals.

Organic soils A class of soils in the Canadian soil classification system that is composed largely of organic materials.

orographic precipitation Precipitation induced by the forced rise of moist air over a topographic barrier.

outputs The flow of matter or energy out of a system.

overland flow Motion of a thin surface layer of water over a sloping ground surface at times when the infiltration rate is exceeded by the precipitation rate; a form of runoff.

Oxisols Soil order in the Comprehensive Soil Classification System of the United States consisting of very old, highly weathered soils of low latitudes, with an oxic horizon and low base status.

ozone A form of oxygen with a molecule consisting of three atoms of oxygen, O_3.

ozone layer layer in the stratosphere, mostly in the altitude range of 20 to 35 km, in which a concentration of ozone is produced by the action of solar ultraviolet radiation.

P

P waves Produced by earthquakes, they are the highest velocity of all seismic waves. They can travel through both solid and liquid rock material, as well as ocean water.

Pangaea Hypothetical parent continent, enduring until near the close of the Mesozoic Era, consisting of the continental shields of Laurasia and Gondwana joined into a single unit.

parabolic dune Isolated low sand dunes of parabolic outline, with points directed into the prevailing wind.

parallel East–west circle on the Earth's surface, lying in a plane parallel with the equator and at right angles to the axis of rotation.

particulates Solid and liquid particles capable of being suspended for long periods in the atmosphere.

pathways In an energy flow system, a mechanism by which matter or energy flows from one part of the system to another.

patterned ground General term for a ground surface that bears polygonal or ring-like features, including stone circles, nets, polygons, steps, and stripes; includes ice wedge polygons; typically produced by frost action in cold climates.

peneplain Land surface of low elevation and slight relief produced in the late stages of denudation of a landmass.

permafrost Soil, regolith, and bedrock at a temperature below 0 °C, found in cold climates of arctic, subarctic, and alpine regions.

photochemical reaction A chemical reaction that is induced by light.

photosynthesis Production of carbohydrate by the union of water with carbon dioxide while absorbing light energy.

physical geography The part of systematic geography dealing with the natural processes occurring at Earth's surface that provide the physical setting for human activities; includes the broad fields of meteorology, climatology, geomorphology, coastal geography, geography of soils, and biogeography.

physical weathering Breakup of massive rock (bedrock) into small particles through the action of physical forces acting at or near the Earth's surface. (See also *weathering*.)

place In geography, a location on the Earth's surface, typically a settlement or small region with unique characteristics.

plane of the ecliptic Imaginary plane in which the Earth's orbit lies.

plate tectonics Theory of tectonic activity dealing with lithospheric plates and their activity.

plateau Upland surface, more or less flat and horizontal, upheld by resistant beds of sedimentary rock or lava flows and bounded by a steep cliff.

playa Flat land surface underlain by fine sediment or evaporite minerals deposited from shallow lake waters in a dry climate in the floor of a closed topographic depression.

pluton Any body of intrusive igneous rock that has solidified below the surface, enclosed in preexisting rock.

podzolic soils A class of forest and heath soils in the Canadian soil classification system in which an amorphous material of humified organic matter with Al and Fe accumulates.

polar front Line of contact between cold polar air masses and warm tropical air masses, often associated with the upper-air westerly jet stream.

polar projection Map projection centred on Earth's North Pole or South Pole.

polyploidy Mechanism of speciation in which entire chromosome sets of organisms are doubled, tripled, quadrupled, etc.

potential energy Energy of position; produced by gravitational attraction of the Earth's mass for a smaller mass on or near the Earth's surface.

power source Flow of energy into a flow system that causes matter to move.

precession A change in the direction of the Earth's axis of rotation in space.

precipitation Particles of liquid water or ice that fall from the atmosphere and may reach the ground. (See also *orographic precipitation, convectional precipitation, cyclonic precipitation*.)

pressure gradient Change of atmospheric pressure measured along a line at right angles to the isobars.

pressure ridge The undulating surface of a glacier caused when ice is supplied at a faster rate than it can be removed.

primary minerals In pedology (soil science), the original, unaltered silicate minerals of igneous rocks and metamorphic rocks.

primary producers Organisms that use light energy to convert carbon dioxide and water to carbohydrates through the process of photosynthesis.

progradation Shoreward building of a beach, bar, or sandspit by addition of coarse sediment carried by littoral drift or brought from deeper water offshore.

Q

quartz Very common mineral composed of silicon dioxide.

quartzite Metamorphic rock consisting largely of the mineral quartz.

R

radiant energy Generally the form of energy that is emitted from the sun; Earth's principal source of energy.

radiation Energy that travels in the form of waves, such as the flow of short-wave energy from the sun to Earth through space.

radiation balance Condition of balance between incoming energy of solar short-wave radiation and outgoing long-wave radiation emitted by Earth into space.

radiative forcing The relative effectiveness of greenhouse gases at restricting the flow of terrestrial long-wave radiation to space, as determined by the ability of the gas to absorb energy and its concentration in the atmosphere.

Rayleigh scattering Scattering of light by atmospheric gas molecules less than one-tenth the wavelength of the incoming solar beam; strongly wavelength-dependent, causes the blue colour of the sky.

realized niche The reduced environmental space that a species actually occupies because of competition with other species.

realms The major components of the planet comprising the atmosphere, lithosphere, hydrosphere, and biosphere.

reflection Outward scattering of radiation toward space by the atmosphere and/or Earth's surface.

regional geography That branch of geography concerned with how the Earth's surface is differentiated into unique places.

regional scale The scale of observation at which subcontinental regions are discernable.

regolith Layer of mineral particles overlying the bedrock; may be derived by weathering of underlying bedrock or be transported from other locations by fluid agents.

Regosolic soils A class of soils in the Canadian soil classification system that exhibits weakly developed horizons.

relative humidity Ratio of water vapour present in the air to the maximum quantity possible for saturated air at the same temperature.

remote sensing Measurement of some property of an object or surface by means other than direct contact; usually refers to the gathering of scientific information about the Earth's surface from great heights and over broad areas, using instruments mounted on aircraft or orbiting space vehicles.

respiration The oxidation of organic compounds by organisms that powers bodily functions.

retrogradation Cutting back (retreat) of a shoreline, beach, marine cliff, or marine scarp by wave action.

reverse fault A type of fault in which one fault block rides up over the other on a steep fault plane.

revolution Motion of a planet in its orbit around the sun, or of a planetary satellite around a planet (e.g., the moon about the Earth).

rhyolite Extrusive igneous rock of granite composition; occurs as lava or tephra.

Richter magnitude scale Scale of magnitude numbers describing the quantity of energy released by an earthquake.

ridge-and-valley landscape Assemblage of landforms developed by denudation of a system of open folds of strata and consisting of long, narrow ridges and valleys arranged in parallel or zigzag patterns.

rock Natural aggregate of minerals in the solid state; usually hard and consisting of one, two, or more mineral varieties.

Rossby waves Horizontal undulations in the flow path of the upper-air westerlies; also known as upper-air waves.

runoff Flow of water from continents to oceans by way of stream flow and groundwater flow; a term in the water balance of the hydrologic cycle. In a more restricted sense, refers to surface flow by overland flow and channel flow.

S

S waves Produced by earthquakes, they move slower than P waves, and shear through rock sideways, at right angles to the direction they are travelling. They cannot propagate in liquid materials.

saltation Leaping, impacting, and rebounding of sand grains transported over a sand or pebble surface by wind.

sand dune Hill or ridge of loose, well-sorted sand shaped by wind and usually capable of downwind motion.

sandstone Variety of sedimentary rock consisting largely of mineral particles of sand grade size.

saturated (wet) adiabatic lapse rate (SALR) Rate at which rising air is cooled by expansion when condensation occurs; varies with moisture content of the air, but is usually around 6 °C per 1,000 m. (See also *dry adiabatic lapse rate.*)

saturation vapour pressure (SVP) The maximum amount of water vapour that can be present at a given temperature.

savanna biome Biome that consists of a combination of trees and grassland in various proportions.

scattering Turning aside of radiation by an atmospheric molecule or particle so that the direction of the scattered ray is changed.

schist Metamorphic rock with layered structure in which mica flakes are typically found oriented parallel with the layers.

secondary minerals In soil science, minerals that are stable in the surface environment, derived by mineral alteration of the primary minerals.

sediment Finely divided mineral matter and organic matter derived directly or indirectly from pre-existing rock and from life processes. (See also *chemically precipitated sediment, organic sediment.*)

sedimentary dome Up-arched strata forming a circular structure with domed summit and flanks with moderate to steep outward dip.

sedimentary rocks Rock formed from accumulation of sediment.

seismic moment A quantity used to measure the size of an earthquake; calculated by measuring the total length of fault rupture and then factoring in the depth of rupture, total slip along the rupture, and the strength of the faulted rocks.

semidiurnal lunar tide A tide that has two high and two low tides each tidal day.

sensible heat Heat measurable by a thermometer; an indication of the intensity of kinetic energy of molecular motion within a substance.

sensible heat transfer Flow of heat from one substance to another by direct contact.

shadow zone An area of the Earth from which waves do not emerge or cannot be recorded.

shale Sedimentary rock of mud or clay composition, showing layered structure that can be easily split apart.

sheet erosion Type of accelerated soil erosion in which thin layers of soil are removed without formation of rills or gullies.

shield volcano Low, often large, dome-like accumulation of basalt lava flows emerging from long radial fissures on flanks.

shoreline Shifting line of contact between water and land.

short-wave radiation Electromagnetic energy in the range from 0.2 to 3 μm, including most of the energy spectrum of solar radiation.

small circle Circle formed by a plane passing through a sphere without passing through its exact centre.

smog Mixture of aerosols and chemical pollutants in the lower atmosphere, usually found over urban areas.

soil Natural terrestrial surface layer containing living matter and supporting or capable of supporting plants.

soil creep Extremely slow downhill movement of soil and regolith as a result of continued agitation and disturbance of the particles by such activities as frost action, temperature changes, or wetting and drying of the soil.

soil horizons Distinctive layer of the soil, more or less horizontal, set apart from other soil zones or layers by differences in physical and chemical composition, organic content, and structure, or a combination of those properties, produced by soil-forming processes.

soil order A soil class at the highest category in the classification of soils.

soil profile Display of soil horizons on the face of a freshly cut vertical exposure through the soil.

soil texture Descriptive property of the mineral portion of the soil based on varying proportions of sand, silt, and clay.

solar constant Previous name for total solar irradiance.

solar noon Instant at which the subsolar point crosses the meridian of longitude of a given point on the Earth; instant at which the sun's shadow points exactly due north or due south at a given location.

solar path diagram A circular graph that shows the solar zenith angle and day length at different times of the year; usually used to show the shading effect of topographic features and buildings.

solar power tower A type of solar furnace using a tower to receive focused sunlight.

solifluction Tundra (arctic) variety of earth flow in which sediments of the active layer move in a mass slowly downhill over a water-rich plastic layer occurring at the top of permafrost; produces solifluction terraces and solifluction lobes.

Solonetzic soils A class of soils in the Canadian soil classification system with a B horizon of sticky clay that dries to a very hard condition.

solute effect Allows water droplets to develop when air is below saturation; occurs because cloud droplets form around condensation nuclei, and are rarely composed of pure water.

solution A weathering process in which rock materials are slowly dissolved and removed by water.

speciation The process by which species are differentiated and maintained.

specific heat capacity The heat required to raise the temperature of a unit mass of substance by one degree; usually 1 gram of substance by 1 degree Celsius.

specific humidity (q) Mass of water vapour contained in a unit mass of air.

spodosols Soil order in the Comprehensive Soil Classification System of the United States consisting of soils with a spodic horizon, an albic horizon, with low base status, and lacking in carbonate materials.

spring tide A tide that has the greatest range between high and low water.

stable air Air mass in which the environmental temperature lapse rate is less than the dry adiabatic lapse rate, inhibiting convectional uplift and mixing.

standard time system Time system based on the local time of a standard meridian and applied to belts of longitude extending 7½° (more or less) on either side of that meridian.

statistics A branch of mathematical sciences that deals with the analysis of numerical data.

Stefan-Boltzmann Law Describes the relationship between the flow of energy from an object and that object's surface temperature.

stemflow Flow of water intercepted by vegetation along branches and then to the ground via stems and trunks.

steppe Semi-arid grassland occurring largely in dry continental interiors.

storage pool Type of pool in a biogeochemical cycle in which materials are largely inaccessible to life. (See also *active pool.*)

stored energy (See *potential energy.*)

storm surge Rapid rise of coastal water level accompanying the onshore arrival of a tropical cyclone.

strata Layers of sediment or sedimentary rock in which individual beds are separated from one another along bedding planes.

stratified drift Glacial drift made up of sorted and layered clay, silt, sand, or gravel deposited from meltwater in stream channels, or in marginal lakes close to the ice front.

stratosphere Layer of atmosphere lying directly above the troposphere.

stratovolcano Volcano constructed of multiple layers of lava and tephra (volcanic ash).

stream Long, narrow body of flowing water occupying a stream channel and moving to lower levels under the force of gravity. (See also *graded stream*.)

stream deposition Accumulation of transported particles on a stream bed, upon the adjacent flood plain, or in a body of standing water.

stream erosion Progressive removal of mineral particles from the floor or sides of a stream channel by tractive force of the moving water, by abrasion, or by corrosion.

stream load Matter carried by a stream in dissolved form (as ions), as solids in turbulent suspension, and as bed load.

stream transportation Down-valley movement of eroded particles in a stream channel in solution, in turbulent suspension, or as bed load.

striation Linear scratches and grooves formed by rock fragments at the base of a glacier being scraped across the underlying bedrock.

strike Compass direction of the line of intersection of an inclined rock plane and a horizontal plane of reference. (See also dip.)

structure The pattern of the pathways and their interconnections within a flow system.

subduction Descent of the downbent edge of a lithospheric plate into the asthenosphere so it passes beneath the edge of the adjoining plate.

sublimation Process of change of ice (solid state) to water vapour (gaseous state); in meteorology, sublimation also refers to the change of state from water vapour (liquid) to ice (solid), which is referred to as deposition in this text.

submergence Inundation or partial drowning of a former land surface by a rise of sea level or a sinking of the crust or both.

subsidence inversion A temperature inversion caused when air aloft warms through compression as it sinks to a lower elevation.

subtropical high-pressure belts Belts of persistent high atmospheric pressure trending east–west and centred approximately on lat. 30° N and S.

summer solstice Solstice occurring on June 21 or 22, when the subsolar point is located at 23½° N; in this sense, used with respect to the northern hemisphere.

supercell A severe thunderstorm with a strong rotating updraft, often favouring the development of tornadoes.

surface retention Water from a precipitation event that is held on the surface of plants, on the irregular surfaces of rocks and stones, or in small depressions in the soil surface.

surface wave A wave that transmits energy from an earthquake's epicentre along the Earth's surface. Along with S waves, these produce the strongest vibrations and are the source of the most damage.

suspension Stream transportation in which particles of sediment are held in the body of the stream by turbulent eddies. (Also applies to wind transportation.)

symbiosis Form of positive interaction between species that is beneficial to one of the species and does not harm the other.

syncline Downfold of strata (or other layered rock) in a trough-like structure; a class of folds. (See also *anticline*.)

synoptic weather type classification The classification of weather conditions according to the general characteristics of the prevailing air masses originating from specific source regions.

systematic geography The study of the physical, economic, and social processes that differentiate the Earth's surface into places.

systems approach The study of the interconnections among natural processes by focusing on how, where, and when matter and energy flow in natural systems.

T

tectonic activity Process of bending (folding) and breaking (faulting) of crustal mountains, concentrated on or near active lithospheric plate boundaries.

temperature inversion Upward reversal of the normal environmental temperature lapse rate, so that the air temperature increases upward.

terminal moraine Moraine deposited as an embankment at the terminus of an alpine glacier or at the leading edge of an ice sheet.

terracette A narrow step-like feature formed on a hillside through the imperceptibly slow downslope movement of soil particles.

terrestrial ecosystem Ecosystems of land plants and animals found on upland surfaces of the continents.

thalweg The deepest part of a stream channel.

thermal conductivity The ability of a substance to conduct heat through its mass in the direction of the decreasing temperature gradient.

thermal wind The change in vertical wind shear caused by the horizontal pressure gradient associated with the relative thicknesses of cold air and warm air generally across a latitudinal gradient from polar to tropical latitudes.

thermally direct In atmospheric circulation, an increase in density that results from a loss of energy.

threshold velocity The velocity required to entrain a particle of a given diameter; increases with the square root of the particle size.

throughfall Water that drips from a vegetation canopy.

thunderstorm Intense, local convectional storm associated with a cumulonimbus cloud and yielding heavy precipitation, also with lightning and thunder, and sometimes hail.

tide-generating force The cyclically varying resultant effect of lunar and solar gravitational forces and their directions.

tidewater glacier Glaciers that end in the ocean.

till Heterogeneous mixture of rock fragments ranging in size from clay to boulders, deposited beneath moving glacial ice or directly from the melting in place of stagnant glacial ice.

time cycle In flow systems, a regular alternation of flow rates with time.

time zone A zone or belt of given east–west (longitudinal) extent within which standard time is applied according to a uniform system.

tornado Small, very intense wind vortex with extremely low air pressure at its centre, generally formed beneath a dense cumulonimbus cloud in proximity to a cold front.

total solar irradiance Intensity of solar radiation falling upon a unit area of surface held at right angles to the sun's rays at a point outside the Earth's atmosphere; equal to an energy flow of about 1370 W/m^2.

traction Transport of sediment by wind or water in which the sediment remains in contact with the ground or bed of the stream, moving by rolling or sliding.

transpiration Evaporative loss of water to the atmosphere from leaf pores of plants.

transverse dune Field of wavelike sand dunes with crests running at right angles to the direction of the prevailing wind.

Tropic of Cancer Parallel of latitude at 23½° N.

Tropic of Capricorn Parallel of latitude at 23½° S.

tropical cyclone Intense travelling cyclone of tropical and subtropical latitudes, accompanied by high winds and heavy rainfall (in North America more commonly referred to as a hurricane).

troposphere Lowermost layer of the atmosphere in which air temperature falls steadily with increasing altitude.

tsunami Train of sea waves set off by an earthquake (or other seafloor disturbance) travelling over the ocean surface.

tundra biome Biome of the cold regions of arctic tundra and alpine tundra, consisting of grasses, sedges, flowering herbs, dwarf shrubs, mosses, and lichens.

tundra climate Cold climate of the arctic zone with eight or more consecutive months of zero potential evapotranspiration (water need).

U

Ultisols Soil order in the Comprehensive Soil Classification System of the United States consisting of soils of warm soil temperatures with an argillic (clay) horizon and low base status.

ultraviolet (UV) radiation Electromagnetic energy in the wavelength range of 0.2 to 0.4 μm.

unstable air Air with substantial content of water vapour, capable of breaking into spontaneous convectional activity leading to the development of heavy showers and thunderstorms.

V

vapour pressure Refers to the contribution that water vapour makes to the pressure exerted by the atmosphere.

varves Annual layer of sediment on the bottom of a lake or the ocean, marked by a change in colour or texture of the sediment.

Vertisolic soils A class of soils in the Canadian Soil Classification System with a high clay content, developing deep, wide cracks when dry and showing evidence of movement between aggregates; similar to vertisols.

Vertisols Soil order in the Comprehensive Soil Classification System of the United States consisting of soils of the subtropical zone and the tropical zone with high clay content, developing deep, wide cracks when dry, and showing evidence of movement between aggregates.

visible light Electromagnetic energy in the wavelength range of 0.4 to 0.7 μm.

volcano Conical, circular landform built by accumulation of lava flows and tephra. (See also *stratovolcano, shield volcano.*)

volumetric heat capacity The ability of a substance to store heat energy in response to a given change in temperature.

W

Walker circulation The large-scale east–west circulation cell over the tropical Pacific Ocean that strengthens and weakens in response to the southern oscillation index and El Niño events.

warm front Moving weather front along which a warm air mass slides up over a cold air mass, leading to production of stratiform clouds and precipitation.

washout Removal of atmospheric particulates by precipitation.

water need Ideal or hypothetical rate of evapotranspiration estimated to occur from a complete canopy of green foliage of growing plants continuously supplied with all the soil water they can use; a real condition reached in those situations where precipitation is sufficiently great or irrigation water is supplied in sufficient amounts.

water resources A field of study that couples information about the location, distribution, and movement of water with use and quality of water for human use.

water table Upper boundary surface of the saturated zone; the upper limit of the groundwater body.

water use Actual rate of evapotranspiration at a given time and place.

watershed A land area that drains water into a river system or other body of water.

wave cyclone A travelling cyclone of the midlatitudes involving the interaction of cold and warm air masses along sharply defined fronts.

wave-cut notch Rock recess at the base of a marine cliff where wave impact is concentrated.

weather system Recurring pattern of atmospheric circulation associated with characteristic weather, such as might occur in a cyclone or anticyclone.

weathering Total of all processes acting at or near the Earth's surface to cause physical disruption and chemical decomposition of rock. (See also *chemical weathering, physical weathering.*)

wet equatorial climate Moist climate of the equatorial zone with a large annual water surplus, and with uniformly warm temperatures throughout the year.

wet-dry tropical climate Climate of the tropical zone characterized by a very wet season alternating with a very dry season.

Wien's Law States that there is an inverse relationship between the wavelength of the peak of the emission of a blackbody and its temperature.

wind Air motion, predominantly horizontal relative to the Earth's surface.

wind erosivity The ability of wind to erode.

winter solstice Solstice occurring on December 21 or 22, when the subsolar point is at 23½° S; in this sense used with respect to the northern hemisphere.

woodland Plant formation class, transitional between forest biome and savanna biome, consisting of widely spaced trees with canopy coverage from 25 to 60 percent.

X

xerophytes Plants adapted to a dry environment.

Z

zone of ablation The area in which annual loss of snow through melting, evaporation, iceberg calving, and sublimation exceeds annual gain of snow and ice on the surface.

zone of accumulation The area above the firn line, where snowfall accumulates and exceeds the losses from ablation.

PHOTO AND ILLUSTRATION CREDITS

Note: EOTE = Eye on the Environment boxes; GT = Geographer's Tools boxes

Material copyright A.N. Strahler reprinted with permission of John Wiley & Sons, Inc.

Chapter 1 Part Opener: © Oleksandr Buzko/Alamy. Chapter Opener: © Ryan Lindsay/iStockphoto.com. Fig. 1.2: Robin Karpan/Parkland. Fig. 1.3: Victor Last. Fig. 1.4a: Reproduced with the permission of Natural Resources Canada 2011, courtesy of the Atlas of Canada. Fig. 1.4b: Petr Cizek, Cizek Environmental Services, Gillies Bay, British Columbia. Fig. 1.4c: NASA image by Jeff Schmaltz, MODIS Rapid Response Team, Goddard Space Flight Center. Fig. 1.5: Frank Slide Interpretive Centre, Alberta Culture and Community Spirit. Fig. 1.6: CP/Larry MacDougal. Fig. 1.7: Eric Gay/ASSOCIATED PRESS. Fig.1.8: K.E. Archibold. Fig. 1.9: © Alan Crawford/iStockphoto.com. Fig 1.10: NASA. Fig 1.13: © Darren Pinkoski/iStockphoto.com. Fig. 1.14: © Bart Sadowski/iStockphoto.com. Fig. 1.15: © David Morgan/iStockphoto.com.

Chapter 2 Chapter Opener: D. Archibold. 2.4: Reproduced with the permission of Air Canada. Fig. 2.5: Garmin.com. Fig. 2.6a & b: D. Archibold. Fig 2.11: National Research Council Canada: http://time5.nrc.ca/maps/. Fig. 2.12: Paul Eggert/Wikipedia. Fig. 2.14: © MoonConnection.com. Fig. 2A.1: INCA/Global Chemical Weather Forecast. Fig 2A.3a & b: © Department of Natural Resources Canada. All rights reserved. Fig. 2A.5a & b: D. Archibold. Fig 2A.7: Adapted from: Alberta Agriculture and Food, 2005.

Chapter 3 Part Opener: © Christopher Purcell/Alamy. Chapter Opener: © Radius Images/Alamy. EOTE Fig. 1: NASA. Fig. 3.2: NASA. Fig. 3.5: Ted Kinsman/Photo Researchers, Inc. Fig. 3.6: © Environment Canada, 2011. Fig. 3.7: RADARSAT-2 Data and Products © MacDONALD, DETTWILER AND ASSOCIATES LTD (2008) – All Rights Reserved. RADARSAT is an official mark of the Canadian Space Agency. Fig. 3.12: Victor Last. GT Fig. 1: NASA. Fig 3.14: NASA Earth Observatory. Fig 3A.2: Courtesy Bow River Irrigation District, Land Division. Fig. 3A.3: NASA.

Chapter 4 Chapter Opener: CORBIS/Lowell Georgia. Fig. 4.4: Mark Boulton/Photo Researchers, Inc. Fig. 4.5: © Environment Canada, 2011. Fig. 4.10: Camilo Pérez Arrau (2008) www.urbanheatislands.com. Fig. 4.12: Courtesy Cooperative Institute for Research in the Environment. Fig. 4.13: V.B. Mendes, University of Lisbon. Fig. 4.14: Developed by Vitali Fioletov, Experimental Studies Section of Envronment Canada. © Environment Canada, 2011. Fig. 4.16: D. Archibold. Fig. 4.17 (map): D. Hartmann and M. Michelsen, University of Washington. Fig. 4.19: Courtesy of European Centre for Medium-Range Weather Forecasts (ECMWF). Fig. 4.25a, b, c, & d: Fetterer, F., K. Knowles, W. Meier, and M. Savoie. 2002, updated 2009. Sea Ice Index. Boulder, Colorado USA: National Snow and Ice Data Center. Digital media. Fig. 4.27: These plots and data are courtesy of the Global Monitoring Division of NOAA's Earth System Research Laboratory, Boulder, Colorado, USA. Fig. 4.28: Berg, Hager, Goodman, Baydack, *Visualizing the Environment,* Figure 10.38. John Wiley & Sons Canada, Ltd. 2010. Fig. 4.29: Climate Change 2007: Synthesis Report. Contribution of Working Groups I, II and III to the Fourth Assessment Report of the Intergovernmental Panel on Climate Change, Figure 2.4. IPCC, Geneva, Switzerland. Fig. 4.30: Fröhlich, C., "Evidence of a long-term trend in total solar irradiance," Astronomy and Astrophysics 501, L27-L30, 2009, Fig. 1. Reproduced with permission © ESO, and courtesy of Claus Fröhlich, Davos. Fig. 4.31a: Photodisc/Getty Images. Fig. 4.32: National Climatic Data Center, NOAA Satellite and Information Service, NESDIS. Fig 4.33: Dr. Henri D. Grissino-Mayer. Fig. 4A.1–4A.4: Climate Change 2007: Synthesis Report. Contribution of Working Groups I, II and III to the Fourth Assessment Report of the Intergovernmental Panel on Climate Change. IPCC, Geneva, Switzerland. (Specific figure numbers from Climate Change 2007): Fig. 4A.1: Figure 3.1. Fig. 4A.2: Figure 3.2. Fig. 4A.3: Figure 1.1. Fig 4A.4: Figure 3.3.

Chapter 5 Chapter Opener: © Aurora Photos/Alamy. Fig. 5.1a: © Artur Synenko/iStockphoto.com. Fig 5.1b: © sciencephotos/Alamy. Fig. 5.1c: Photo courtesy of Vaisala. Fig. 5.3: © Emil Pozar/Alamy. Fig. 5.5: Risien, C.M., and D.B. Chelton, 2008: A Global Climatology of Surface Wind and Wind Stress Fields from Eight Years of QuikSCAT Scatterometer Data. J. Phys. Oceanogr., 38, 2379-2413. Available online at http://cioss.coas.oregonstate.edu/scow. Fig. 5.11a & b: Martin Jakobsson/Photo Researchers, Inc. Fig. 5.20: North Pacific Sea Planes, Prince Rupert, BC. Fig. 5.21: NASA. EOTE Fig. 1: Canadian Wind Energy Association. EOTE Fig. 2: © Source: Environment Canada, Canadian Wind Energy Atlas, 2003. Fig. 5.29: © Environment Canada, 2011. Fig. 5.30: Aqua-MODIS nighttime sea surface temperature image courtesy of the Ocean Biology Processing Group at NASA's Goddard Space Flight Center. Fig. 5.33: © Weldon Schloneger/iStockphoto.com. Fig. 5.34: Courtesy of Otis B. Brown, Robert Evans, and M. Carle, University of Miami, Rosenstiel School of Marine and Atmospheric Science, Florida, and NOAA/Satellite Data Services Division. Fig. 5.35: NASA. Fig. 5.36: NASA/JPL.

Chapter 6 Chapter Opener: Wolfgang Kaehler/CORBIS. Fig. 6.3: Global Precipitation Climatology Project. Fig. 6.5: © mauritius images GmbH/Alamy. Fig. 6.7a: Courtesy of ProSource Scientific. Fig 6.7b: K.E. Archibold. Fig. 6.7c: Courtesy of Wilh. LAMBRECHT GmbH. Fig. 6.9: © Subic/iStockphoto.com. Fig. 6.12a: © Dave White/iStockphoto.com. Fig. 6.12b: Marc Moritsch/National Geographic/Getty Images. Fig. 6.12c: © Dean Turner/iStockphoto.com. Fig. 6.12d: Arnar Birgisson/Flickr/Getty Images. GT Fig. 1: NASA-GSFC, data from NOAA GOES. GT Fig. 2a & b: NOAA National Environmental Satellite, Data and Information

Service (NESDIS) Regional and Mesoscale Meteorology Branch at the Colorado State University Cooperative Institute for Research in the Atmosphere. Fig. 6.13: © arturbo/iStockphoto.com. Fig. 6.14: © Emmanuel Rondeau/Alamy. Fig. 6.16: "Mean Number of Days/ Freezing Rain, Canada" from The Climates of Canada, 1990, pp. 60-61. Reproduced with the permission of the Minister of Public Works and Government Services Canada, 2007. Fig 6.17: Public Safety and Emergency Prepardness Canada, Keeping Canadians Safe. Average number of days with hail, http://www.publicsafety.gc.ca/images/hail_map_e.gif. Fig. 6.18: Doug Sokell/Visuals Unlimited, Inc. Fig. 6.19a: © Per Karlsson, BKWine 2/Alamy. Fig. 6.19b: © Copyright Design Analysis Associates, a division of YSI Inc. Fig. 6.19c: Sveta Stuefer, University of Alaska Fairbanks. Fig. 6.19d: The University of Waterloo Weather Station. Fig 6.20: Reproduced with the permission of Natural Resources Canada 2011, courtesy of the Atlas of Canada. Fig. 6.21: "Mean Number of Days/Blowing Snow, Canada" from The Climates of Canada, 1990, pp. 60-61. Reproduced with the permission of the Minister of Public Works and Government Services Canada, 2007.

Chapter 7 Chapter Opener: ©Stuart Monk/Shutterstock. Fig. 7.11a: O.W. Archibold. Fig. 7.11b: © PeterAustin/iStockphoto.com. Fig. 7.11c: O.W. Archibold. Fig. 7.11d: Feature Pics. Fig. 7.11e: © rotofrank/iStockphoto.com. Fig. 7.11f: © Blackout Concepts/Alamy. Fig. 7.11g: O.W. Archibold. Fig. 7.11h: O.W. Archibold. Fig. 7.11i: O.W. Archibold. Fig. 7.15: GOES-12 satellite, NASA, NOAA. Fig. 7.16: Provided courtesy of University of Wisconsin-Madison Space Science and Engineering Center. Fig. 7.17: Wikimedia Commons. Fig. 7.19a & b: NASA. Fig. 7.21: USGS Coastal and Marine Geology Program. EOTE Fig. 1: ROBYN BECK/AFP/Getty Images. EOTE Fig. 2: Justin R. Glenn, Southeast Regional Climate Center, Columbia, SC. Fig. 7.24: NOAA National Climatic Data Center. Fig. 7.25: NOAA National Severe Storms Laboratory. Fig. 7.26: © Clint Spencer/iStockphoto.com.

Chapter 8 Chapter Opener: © Peter Haigh/Alamy. Fig 8.4: Simplified and modified from Plate 3, World Climatology, Volume 1, The Time Atlas, Editor John Bartholomew, The Times Publishing Company, Ltd. London, 1958. Fig. 8.6: Courtesy U.S. National Arboretum, USDA-ARS, Washington, DC 20002; USDA Miscellaneous Publication No. 1475 (Issued January 1990).

Chapter 9 Chapter Opener: © Terry Hewitt/Alamy. Fig. 9.8. Calvin Jones. EOTE Fig. 2: © frans lemmens/Alamy. Fig. 9.9b: © FLORIN IORGANDA/X02105/Reuters/Corbis. Fig. 9.10a & b: O.W. Archibold. Fig. 9.11: O.W. Archibold. Fig. 9.12a & b: Earth Observatory, NASA. Fig. 9.13a & b: Visible Earth, NASA. Fig. 9.16a: S. Hanusch/Shutterstock. Fig. 9.16b: ©Michael Fogden/DRK PHOTO/drkphoto.com. Fig. 9.16c: © Danita Delimont/Alamy. Fig. 9.17a: NASA image created by Jesse Allen, using Landsat data provided by the United States Geological Survey. Fig. 9.19: © Graeme Shannon/iStockphoto.com.

Chapter 10 Chapter Opener: © mlwinphoto/iStockphoto.com. Fig. 10.4: © Environment Canada, 2011. Fig. 10.6: SIMON FRASER/ SCIENCE PHOTO LIBRARY. Fig. 10.9a: © Gary Crabbe/Alamy. Fig. 10.9b: Mark L. Kaufman. Fig. 10.10a: © Mark Salter/Alamy. Fig. 10.10b: © Jennifer Hart/Alamy. Fig. 10.10c: © Thomas Shjarback/Alamy. Fig. 10.12: Darryl Dyck/The Canadian Press. Fig. 10.13: David Muench Photography. Fig. 10.15: NASA Earth Observatory. EOTE Fig. 2: Agriculture and Agri-Food Canada. Fig. Fig. 10.18: Image provided by the SeaWiFS Project, NASA/Goddard Space Flight Center, and ORBIMAGE. 10.19: str/Robert Galbraith/The Canadian Press. Fig. 10.23a & b: Canadian Forest Service (Natural Resources Canada), reproduced with permission. Fig. 10A.1–10A.13: Climate Change 2007: The Physical Science Basis. Working Group I Contribution to the Fourth Assessment Report of the Intergovernmental Panel on Climate Change. Cambridge University Press. (Specific figure numbers from Climate Change 2007): Fig. 10A.1: Figure 10.9 (Temperature). Fig. 10A.2: Figure 10.11 (b). Fig. 10A.3: Figure 10.19 (b), (d) and (f). Fig. 10A.4: Figure 11.12 (top three maps). Fig. 10A.5: Figure 10.10 (a) and (b). Fig. 10A.6: Figure 10.12 (a). Fig. 10A.7: Figure 10.18 (b) and (d). Fig. 10A.8: Figure 10.12 (b) (c) and (d). Fig. 10A.9: Figure 11.12 (middle three maps). Fig. 10A.10: Figure 10.9 (SL Pressure). Fig. 10A.11: Figure 10.14. Fig. 10A.12: Figure 11.13. Fig. 10A.13: Figure 10.32.

Chapter 11 Part Opener: © Simon Wilkinson/Alamy. Chapter Opener: Russ Heinl/All Canada Photos. Fig. 11.1: U.S. Geological Survey. Fig. 11.3a: Photo by Mary Sutherland, James Madison University Mineral Museum. Fig. 11.3b: © Mike Danton/Alamy. Fig. 11.4a: Scientifica/Visuals Unlimited, Inc. Fig. 11.4b: Mark A. Schneider/Photo Researchers, Inc. Fig. 11.4c: © Geology.com. Fig. 11.5: Trevor Clifford Photography/Photo Researchers, Inc. Fig. 11.7: Chris Joseph/All Canada Photos. Fig. 11.8: A.G.E. Foto Stock/First Light. Fig. 11.9: © LOOK Die Bildagentur der Fotografen GmbH/Alamy. Fig. 11.12 a, b & d: © Geology.com. Fig. 11.12c: Joel Arem/Photo Researchers, Inc. Fig. 11.13a & b: © Geology.com. Fig. 11.15: Dr. Christian Scheibner. Fig. 11.16: Gregory G. Dimijian, M.D./Photo Researchers, Inc. Fig. 11.18: Michael Szoenyi/ Photo Researchers, Inc. Fig. 11.20: © moodboard/Alamy. Fig. 11.21: © Eye Ubiquitous/Alamy. Fig. 11.22: © Kevin Schafer/Alamy. Fig. 11.23: © Tom Till Photography. Fig. 11.24: © Worldwide Picture Library/ Alamy. Fig. 11.25: © nobleIMAGES/Alamy. Fig. 11.27: Canadian Centre for Energy Information www.centreforenergy.com. Fig. 11.28: John Kaprielian/Photo Researchers, Inc. Fig. 11.29: Leo Batten/FLPA/Photo Researchers, Inc. GT Fig. 2a, b, & c: NASA.

Chapter 12 Chapter Opener: © Rainer Albiez/istockphoto.com. Fig. 12.2: U.S Geological Survey. Fig. 12.3a: © Josemaria Toscano/iStockphoto.com. Fig. 12.4: Reproduced with the permission of Natural Resources Canada 2011, courtesy of the Geological Survey of Canada. Fig. 12.6: LES INC./First Light. Fig. 12.11: Harvey Lloyd/Taxi/Getty Images. Fig. 12.13: © Peter Arnold, Inc./Alamy. Fig. 12.15a & b: © Russ Heinl/All Canada Photos/Corbis. Fig. 12.19: Space Shuttle radar topography image by NASA. Fig. 12.20: Photo courtesy of Alexandria L. Guth. Fig. 12.23: Provided by the SeaWiFS Project, NASA/Goddard Space Flight Center, and ORBIMAGE.

Chapter 13 Chapter Opener: © Tryphosa Ho/Alamy. Fig. 13.2: The Canadian Press. Fig. 13.4: USGS Photograph taken March 21, 2001, by Ken McGee. Fig. 13.5: Dr. Richard Roscoe/Visuals Unlimited, Inc. Fig. 13.6: WALTER MEAYERS EDWARDS/National Geographic Stock. Fig. 13.7: epa/Corbis. Fig. 13.8: © Greg Vaughn/Alamy. Fig. 13.13: © Gary Fiegehen/All Canada Photos. Fig. 13.14: JOE SCHERSCHEL/National Geographic Stock. Fig. 13.15: NORBERT ROSING/National Geographic Stock. Fig. 13.16: National Oceanic and Atmospheric Administration (NOAA). EOTE Fig 1: © gprentice/iStockphoto.com. Fig. 13.18: Dana Stephenson/Getty Images. GT Fig. 1 & Fig. 2: NASA. GT Fig. 3: U.S. Geological Survey (USGS). Fig. 13.19: Phil Coale/ASSOCIATED PRESS. Fig. 13.23a & b: U.S. Geological Survey. Fig. 13.24: Reproduced with the permission of Natural Resources Canada 2011, courtesy of Earthquakes Canada. Fig. 13.25: Reproduced with the permission of Natural

Resources Canada 2011, courtesy of Earthquakes Canada. Fig. 13.26: National Oceanic and Atmospheric Administration (NOAA). Fig. 13.27: NOAA National Data Buoy Center.

Chapter 14 Part Opener: D. Archibold. Chapter Opener: K.E. Archibold. Fig. 14.2: D. Archibold. GT Fig. 1: USGS/LANDSAT. GT Fig. 2: NASA/JPL. GT Fig. 3: USGS/NASA. GT Fig. 4: NASA. Fig. 14.4: © Steven Vidler/Eurasia Press/Corbis. Fig. 14.5: NASA/GSFC/MITI/ERSDAC/JAROS, and U.S./Japan ASTER Science Team. Fig. 14.7: Photo by L. J. Maher, Jr. Fig. 14.8: Photo by Dinah Kruze. Fig. 14.9: © luoman/iStockphoto.com. Fig. 14.10: Francois Gohier/Photo Researchers, Inc. Fig. 14.11: © ilona bila/iStockphoto.com. Fig. 14.12: © Charles & Josette Lenars/CORBIS. Fig. 14.13: Planetary and Space Science Centre, University of New Brunswick. Fig. 14.15a: Chris R. Sharp/Photo Researchers, Inc. Fig. 14.15b: Photo by David Albeck, http://www.davidalbeck.com/photos. Fig. 14.16: Susan Wells Rollinson. Fig. 14.17: © Robert Shantz/Alamy. Fig. 14.18: Gregory G. Dimijian/Photo Researchers, Inc. Fig. 14.19: K.E. Archibold. Fig. 14.20a: © zbindere/iStockphoto.com. Fig. 14.20b: © Dave Porter/Alamy. Fig. 14.21a: Rob Macdonald. Fig 14.21b: Photo by islandhikes.com. Fig. 14.22a: © Chris Mattison/Alamy. Fig. 14.22b: © Frans Lanting/Corbis. Fig. 14.24: © Laurence Parent. All rights reserved. Fig. 14.25: © David Wall/Alamy. Fig. 14.27: © Peter Mukherjee/iStockphoto.com. Fig. 14.30a: NOAA National Geophysical Data Center/B. Bradley, University of Colorado. Fig. 14.30b: Ralph Lee Hopkins/Photo Researchers, Inc. Fig. 14.31: D. Archibold. Fig. 14.32a: Courtesy of Kansas Geological Survey (www.kgs.ku.edu). Fig. 14.32b: Tom Myers/Photo Researchers, Inc. Fig. 14.33: Phil Carpenter, The Gazette (Montreal) © 2010. Fig. 14.34: USGS Photograph taken on March 21, 1982, by Tom Casadevall, USGS/Cascades Volcano Observatory. Fig. 14.35 Robert Leighty.

Chapter 15 Chapter Opener: K.E. Archibold. Fig. 15.9a & b: D. Archibold. Fig. 15.11: National Resources Canada. Fig. 15.18b: Reproduced with the permission of Natural Resources Canada 2011, courtesy of the Geological Survey of Canada (Photo 2011-012 by Greg Brooks). Fig. 15.18c: Steven Fick, Ken Francis/Canadian Geographic; source (Hood extents) Manitoba Centre for Remote Sensing. Fig. 15.18d: THE CANADIAN PRESS/Joe Bryksa. EOTE Fig. 2: U.S. Environmental Protection Agency. Fig. 15.21: Robert Gilbert. Fig. 15A.1 & 15A.2: Environment Canada and the U.S. Environmental Protection Agency. 2009. State of the Great Lakes 2009 Highlights. ISBN 978-1-100-12213-7, EPA 950-K-09-001, Cat No. En161-3/2009E.

Chapter 16 Chapter Opener: Wolfgang Weber, Canadian Arctic Gallery. Fig. 16.2: Neil Preedy. Fig. 16.3: National Resources Canada. Fig. 16.4: Bernard Bauer. Fig 16.5: Nir Alon/Alamy. Fig 16.7a: John Quinton, Lancaster University. Fig. 16.7b: Roger De La Harpe/Animals Animals/maXximages.com. Fig. 16.8 Charlotte Swanson/istockphoto.com. Fig 16.9: O.W. Archibold. Fig. 16.11: Atlas of Canada/Natural Resources Canada. Fig. 16.13: Keren Su/China Span/Getty Images. Fig.16.14: www.ec.gc.ca/eau-water, Environment Canada. Reproduced with the permission of the Minister of Public Works and Government Services, 2011. Fig. 16.15a: © Cornel Stefan Achirei/iStockphoto.com. Fig. 16.15b: © Third Eye Images/Alamy. Fig. 16.15c: © canadabrian/Alamy. Fig. 16.16: J.A. Kraulis/All Canada Photos. Fig.16.19: O.W. Archibold. Fig. 16.21: Win McNamee/Getty Images. Fig. 16.22: Copyright © Michael Collier. Fig. 16.24: Photo by Andrew Stacey. Fig. 16.27: Keith Douglas/All Canada Photos. Fig. 16.28: Time & Life Pictures/Getty Images. Fig. 16.30: Larry Dale Gordon/Getty Images. Fig. 16.31: Fenykepez/Istock. Fig. 16.32: Mark Zanzig/www.zanzig.com. Fig. 16.33: Dr. Manaan Kar Ray/World of Stock. Fig. 16.34: Donna Ikenberry/maXximages.com. Fig. 16.36: T.A. Wiewandt/DRK Photo. Fig. 16.38: © imagebroker/Alamy.

Chapter 17 Chapter Opener: Copyright © Adrian Heisey. Fig. 17.3: © Mlenny Photography/istockphoto.com. Fig. 17.4: Tom Nebbia/Corbis Images. Fig. 17.5: Satellite image courtesy of GeoEye. Fig 17.6: Photograph by Patrick A. Hesp. Fig 17.7b: ©Johnathan Bascom. Fig. 17.8: © George Steinmetz/Corbis. Fig. 17.9: © Atlantide Phototravel/Corbis. Fig. 17.10: D. Archibold. Fig 17.12: © Minden Pictures/SuperStock. Fig. 17.14: ©Michael Fenton. Image courtesy of AGI Earth Science World Image Bank. Fig. 17.15: Photo by D.A. Rahm. © EPIC The Easterbrook Photo/Image Center. Fig. 17.16: ©Brian J. McMorrow. Fig. 17.17: ©George Steinmetz. Fig. 17.18: J.A. Kraulis/Masterfile. Fig. 17.20: Jim Richardson/National Geographic/Getty Images. Fig. 17.21: NASA/JPL. Fig. 17.23: ©2011 Alex S. MacLean/Landslides Aerial Photography. Fig. 17.26: © Les Gibbon/Alamy. Fig. 17.27: ©Andrew Stacey. Fig. 17.29: © OceanwideImages.com. Fig. 17.30: Photo by Jinx McCombs. Fig. 17.31: © Dr. Frank M. Hanna/Visuals Unlimited. Fig. 17.32: Tony Hopewell/Taxi/Getty Images. Fig. 17.37: CLS/Legos (http://www.aviso.oceanobs.com). Fig. 17.38a: John Stumbles/Wikipedia. Fig. 17.38b: © Tom Till/Alamy. Fig. 17.39a: Charles Bowman/Robert Harding World Imagery/Getty Images. Fig. 17.39b: Roy Toft/Getty Iimages. Fig. 17.41a: © NASA/Corbis. Fig. 17.41b: Image courtesy of USGS National Center for EROS and NASA Landsat Project Science Office. Fig. 17.41c: Phillippe Giraud/Sygma/Corbis. Fig. 17.42: © F1online digitale Bildagentur GmbH/Alamy. Fig. 17A.1a: © CSIRO http://www.cmar.csiro.au/sealevel/index.html. Fig. 17A.1b: CLS/Cnes/Legos http://www.aviso.oceanobs.com/en/news/ocean-indicators/mean-sea-level. Fig. 17A.2: National Snow and Ice Data Center. Fig. 17A.3: Stephen Crowley/New York Times Pictures 05465092. Fig. 17A.4: Stephen Rose – Rainbow Getty Images. Fig. 17A.5: Erik Sampers/Getty Images. Fig. 17A.6: NOAA Coral Reef Watch. Fig. 17A.7: © Steven J. Kazlowski/Alamy.

Chapter 18 Chapter Opener: Dean Conger/Corbis. Fig. 18.1a: © Ron Niebrugge/www.wildnatureimages.com. Fig. 18.1b: Courtesy of the Geological Survey of Canada. Photo by Dr. William Shilts. Fig. 18.1c: © Tibor Bognar/Alamy. Fig. 18.1d: Icefield Helicopter Tours, Canada. Fig. 18.3: Lindley Hanson, Department of Geological Sciences, Salem State University. Fig. 18.5a: USGS/photo by Bruce Molnia. Fig. 18.5b: Icefield Helicopter Tours, Canada. Fig. 18.6a: H.J.A. Berendsen. Fig. 18.6b: John T. Andrews. Fig. 18.7a: © Michael T. Sedam/CORBIS. Fig. 18.7b: James Emmerson/RobertHarding/GettyImage. Fig. 18.7c: © Nick Hanna/Alamy. Fig. 18.7d: Lakes Gliding Club, www.lakesgc.co.uk. Fig. 18.8: Arnold Bloom, Global Climate Change: Convergence of Disciplines, (Sunderland, MA: Sinauer Associates, 2010) p. 21, Fig. 2.11. Reproduced with permission. Fig. 18.9: Yvon Maurice. Fig. 18.11: J.B. Krygier. Fig. 18.12: Juerg Alean, Eglisau, Switzerland, Glaciers online. Fig. 18.13: Reproduced with the permission of Natural Resources Canada 2011, courtesy of the Geological Survey of Canada (Photo 2001-079 by Lynda Dredge). Fig. 18.14: NASA/The Visible Earth. Fig. 18.17a: Thomas & Pat Leeson Photo Researchers, Inc. Fig. 18.17b: Georg Gerster/Photo Researchers, Inc. Fig. 18.17c: © George D. Lepp/Corbis. Fig. 18.18: Dave Reede/All Canada Photos. EOTE Fig. 1, Fig. 2, & Fig. 3: NASA/Robert Simmon. Fig. 18.25: Stephen J. Krasemann/DRK Photo. Fig. 18.26: Steve McCutcheon, McCutcheon Collection; Anchorage Museum, B1990.014.4.pingoes.35761. Fig. 18.28: Bernard Hallet, Periglacial Laboratory, Quaternary Research Center. Fig. 18.30: © Hugh

Rose/AccentAlaska.com. Fig. 18.31: © John Sylvester/Alamy. Fig. 18.32: Canada Foundation for Innovation.

Chapter 19 Part Opener: Photo by Alan L. Bauer/www.alanbauer.com. Chapter Opener: Photo by Alan L. Bauer/www.alanbauer.com. Fig. 19.4: Munsell is a registered trademark of X-Rite, Incorporated. Fig. 19.6: © Queen's Printer for Ontario, 2006. Reproduced with permission. Fig. 19.15: Dr. John Triantafilis, School of Biological, Earth and Environmental Sciences, The University of New South Wales. http://www.terragis.bees.unsw.edu.au. Fig. 19.20a (map): Minister and Department of Agriculture and Agri-Food Canada (AAFC). © Government of Canada (2010). Figure 19.20 (specific soil photos, letters start clockwise from top left): Fig. 19.20b, d, e, f, g, & i: © Agriculture and Agri-Food Canada, Canadian Soil Information Service. Fig. 19.20c, h, j, & k: USDA Natural Resources Conservation Service. Fig. 19.21: Munsell is a registered trademark of X-Rite, Incorporated. Fig. 19.22: © Agriculture and Agri-Food Canada, Canadian Soil Information Service. Fig. 19.23: USDA Natural Resources Conservation Service. Fig. 19.24 (all photos): USDA Natural Resources Conservation Service. Fig. 19.25: © Johan Furusjö/Alamy. Fig. 19.26: Photo by Skip Walker.

Chapter 20 Chapter Opener: Bruce G. Marcot. Used by permission. Fig. 20.6: Reprinted from: Haberl, Helmut, et al., 2007. "Quantifying and mapping the human appropriation of net primary production in earth's terrestrial ecosystems," Proceedings of the National Academy of Sciences of the USA 104, 12942-12947. Copyright (2007) National Academy of Sciences, USA. Reprinted with permission of the National Academy of Sciences, USA, and the authors. Fig. 20.7: Robert Simmon, NASA GSFC Earth Observatory, based on data provided by Watson Gregg, NASA GSFC. Fig. 20.8: Reprinted from: Haberl, Helmut, et al., 2007. "Quantifying and mapping the human appropriation of net primary production in earth's terrestrial ecosystems," Proceedings of the National Academy of Sciences of the USA 104, 12942-12947. Copyright (2007) National Academy of Sciences, USA. Reprinted with permission of the National Academy of Sciences, USA, and the authors. Fig. 20.10: Berg, Hager, Goodman, Baydack, *Visualizing the Environment*, Figure 5.9. John Wiley & Sons Canada, Ltd. 2010. Fig. 20.11: Berg, Hager, Goodman, Baydack, *Visualizing the Environment*, Figure 5.11. John Wiley & Sons Canada, Ltd. 2010. Fig. 20.12: UNEP/DEWA/GRID-Europe. Fig. 20.14: Jacques Descloitres, MODIS Rapid Response Team, NASA/GSFC. Fig. 20.15: NOAA NESDIS National Coastal Data Development Center. Fig. 20A.1: R. Nemani and S. W. Running, University of Montana, NT SG/NASA. Fig. 20A.2: MODIS/NASA. Fig. 20A.3: NASA/GSFC/ORBIMAGE. Fig 20A.4: R. Nemani and S. W. Running, University of Montana, NT SG/NASA. Fig. 20A.5: NASA/NOAA.

Chapter 21 Chapter Opener: Joel W. Rogers/Corbis. Fig. 21.1: Berg, Hager, Goodman, Baydack, *Visualizing the Environment*, p. 148. John Wiley & Sons Canada, Ltd. 2010. Fig. 21.2a: © Anthony Bannister; Gallo Images/CORBIS. Fig. 21.2b: ©2007, Doug Von Gausig, Critical Eye Photography. Fig. 21.2c: © Jo Ann Snover/iStockphoto.com. Fig. 21.3a: © Matt Matthews/iStockphoto.com. Fig. 21.3b: © Erin Paul Donovan/scenicNH.com Photography. Fig. 21.3c: © Mike Bousquet/iStockphoto.com. Fig. 21.4: O.W. Archibold. Fig. 21.7: Noel Elhardt/Wikimedia Commons. Fig. 21.9: Canadian Forest Service (Natural Resources Canada), reproduced with permission. Fig. 21.10: Photo by Peter M. Brown, Rocky Mountain Tree-Ring Research. Fig. 21.11: Image courtesy Jacques Descloitres, MODIS Land Rapid Response Team at NASA GSFC. Fig. 21.12: © Christoph Achenbach/iStockphoto.com. Fig. 21.14: Daniel Mosquin. Fig. 21.16: © Dan L. Perlman/EcoLibrary.org. Fig. 21.18a & b: O.W. Archibold. Fig. 21.21b: © Frank Leung/iStockphoto.com. Fig. 21.23: © 2008 by Kevin C. Nixon. Fig. 21.28: © 2008 The Trustees of Columbia University in the City of New York. Http://sedac.ciesin.columbia.edu/species. Fig. 21.29: © Conservation International Foundation, www.conservation.org. Fig. 21.30: © Parks Canada Agency.

Chapter 22 Chapter Opener: © Galen Rowell/Corbis. Fig. 22.1: D. Archibold. Fig. 22.2: Steve McCutcheon, McCutcheon Collection; Anchorage Museum, B90.14.4.21717. Fig. 22.5a: © Paul A. Souders/Corbis. Fig. 22.5b: © Ferrero/Labat/Auscape International Pty. Ltd. Fig. 22.5c: Terry Whittaker/Photo Researchers, Inc. GT Fig. 1: NOAA. GT Fig. 2 & Fig. 3: A.H. Strahler/NASA. Fig. 22.6: Wikimedia Commons. EOTE Fig. 1: © Jacques Jangoux/Stone/Getty Images. EOTE Fig. 2: NASA. Fig. 22.7: © H Lansdown/Alamy. Fig. 22.8: © Kenneth Murray/Photo Researchers. Fig. 22.9: Stomac/Wikimedia Commons. Fig. 22.10a: © Kevin Schafer/CORBIS. Fig. 22.10b: © Chris Taylor; Cordaiy Photo. Fig. 22.11a: Stephen Alvarez/National Geographic/Getty Images. Fig. 22.11b: © Raymond Gehman/CORBIS. Fig. 22.11c: © Gunter Marx Photography/CORBIS. Fig. 22.12a: Ron Sanford/Photo Researchers/First Light. Fig. 22.12b: © Arcticphoto/Alamy. Fig. 22.12c: © Bill Brooks/Alamy. Fig. 22.13a: D. Archibold. Fig. 22.13b: © Soren Breiting/Alamy. Fig. 22.13c: © David Samuel Robbins/CORBIS. Fig. 22.14: Fletcher & Baylis/Photo Researchers. Fig. 22.15a: © David Muench/Corbis. Fig. 22.15b: Ken Lucas/Visuals Unlimited, Inc. Fig. 22.15c: Mark W. Skinner @ USDA-NRCS PLANTS Database. Fig. 22.15d: © Bill Bachman/Alamy. Fig. 22.16: Arthur Morris/Corbis. Fig. 22.17: © Nico Smit/iStockphoto.com. Fig. 22.18a: © Annie Griffiths Belt/Corbis. Fig. 22.18b: John W. Bova/Photo Researchers, Inc. Fig. 22.18c: John E Marriott/All Canada Photos. Fig. 22.19a: © William Manning/Corbis. Fig. 22.19b: © George D. Lepp/Corbis. Fig. 22.19c: © Peter Johnson/Corbis. Fig. 22.19d: © Emilio Ereza/Alamy. Fig. 22.20a: O.W. Archibold. Fig. 22.20b: © Scott T. Smith/Corbis. Fig. 22.20c: © David Whitaker/Alamy. Fig. 22.20d: © David Muench/Corbis. Fig. 22.21a: © Michael T. Sedam/Corbis. Fig. 22.21b: O.W. Archibold. Fig. 22.22: M. Phillip Kahl/Photo Researchers Inc. Fig. 22.27: © Parks Canada Agency.

INDEX

A

B

D

E

F

H

I

J

K

L

N

O

P

Q

R

S

T

U

X

Y

Z